Carbon Dioxide Capture for Storage in Deep Geologic Formations – Results from the CO_2 Capture Project

Volume 4

CCS Technology Development and Demonstration Results (2009-2014)

Edited by

Karl F. Gerdes

cplpress

2015

Published in the UK by:

CPL Scientific Publishing Services Ltd trading as CPL Press (CPL)
Liberty House, Greenham Business Park, Thatcham, Berks RG19 6HS
Tel/Fax: +44 1635 817359 Email: press@cplpress.com http://www.cplpress.com

for:

CO_2 Capture Project (CCP) Phase 3 members: BP, Chevron, Eni, Petrobras, Shell and Suncor. For further information on the work of the CO_2 Capture Project please visit http://www.co2captureproject.org

The Climate Change and Emissions Management (CCEMC) Corporation of Edmonton Alberta co-funded work in Chapter 15.

First edition 2015

British Library Cataloguing in Publication Data
A catalogue record is available from the British Library

ISBN: 978-1-872691-68-8

Copies of this publication can be obtained via the CPL Press Online Bookshop: http://www.cplbookshop.com

Volumes 1 & 2 of this series were published by Elsevier BV ISBN: 0-08-044570-5 (2 volume set)
Volume 3 of this series was published by CPL Press ISBN: 978-1-872691-49-7

First printed in the UK by CPI Antony Rowe

PREFACE

Nigel Jenvey[1] - Chair, Vincent Kwong[2] - Vice Chair
CO_2 Capture Project Executive Board
[1]BP Corporation North America Inc., Houston, USA
[2]Chevron Energy Technology Company, USA

It has been over a decade since the CO_2 Capture Project (CCP) was established to help CO_2 capture and storage (CCS) technology become a viable option for use by the oil and gas industry to potentially manage future CO_2 emissions. Since then, the CCP has seen its members and partners collaborate to share expertise, facilities and assets to close gaps in knowledge and verify relevant technologies from an owner's perspective.

It has always been clear that for CCS to realise its potential there has to be a high level of knowledge transfer between the different players. As a group founded on the principles of co-operation, we have been committed to sharing knowledge beyond our member companies. In this volume you will find detailed reports from our four Teams – Capture, Storage, Policy, and Communications – covering the third phase of our activity (CCP3 2009–2014).

Since Phase 3 began in 2009, the outlook for CCS in general has remained uncertain. However, some important milestones were achieved, notably CCS being recognized by the United Nations Clean Development Mechanism and sitting alongside renewable technologies. Also, more recently we saw a landmark accomplishment with US Department of Energy supported projects having safely captured 10 million metric tons of carbon dioxide. However, at the same time globally we saw a slowdown in implementation, with some major integrated demonstration projects stalling, or being cancelled. The IEA, in its technology roadmap in 2013, acknowledged that CCS had been developing at a much slower rate than anticipated. But crucially, it emphasised that CCS has to be an integral part of any low-cost mitigation scenario alongside energy efficiency, nuclear and renewables.

During this time of uncertainty our approach continued to be practical. We focused on finding solutions to establish if CCS can become a viable option for use by our industry – for oil refining, for heavy oil production and for gas power generation. We continued to work closely with government, academic bodies and innovative technology suppliers. And we are proud to have together delivered a number of important demonstrations and field trials to make a major contribution to our industry's understanding of the application of CCS.

Capture Overview

The Capture Team focused on reducing the cost of CO_2 capture from refining, in-situ extraction of bitumen and natural gas power generation sources, with some of these capture technologies also applicable to natural gas processing/LNG facilities. Following the screening of over 200 capture technologies in Phase 1 (2000–2003), and further development and assessment in Phase 2, the first capture demonstration was held in 2012 – an oxy-firing test at a pilot-scale Fluid Catalytic Cracking (FCC) unit. Held at Petrobras' research facilities in Brazil, the test indicated the technical viability of retrofitting an FCC unit to enable CO_2 capture from the catalyst regenerator.

The Team was also involved with an oxy-firing combustion technology pilot to reduce CO_2 emissions from once-through steam generator (OTSG) boilers, the primary source of CO_2 emissions in the in-situ production of heavy oil. The project was undertaken in collaboration with Cenovus Energy, the Climate Change Emissions Management Corporation, Devon Canada, Praxair, Statoil and MEG Energy. The demonstration was hosted by Cenovus at their Christina Lake facility, and consisted of retrofit of an existing OTSG and operation in both air and oxy-firing modes. The demo was successfully completed in April, 2015. This test completed the CCP3 Capture Demonstration program planned for Phase 3.

The Team also carried out testing and development of novel technologies including Chemical Looping Combustion, Membrane Water Gas Shift and a number of solvent and enzyme-based technologies. Importantly, the Team delivered an economic model with greater clarity – comparing baseline costs for oxy-firing, post-combustion and pre-combustion across the main oil & gas capture scenarios.

Highlights of the Capture Team's work through the three phases of CCP are shown in Table 1.

Table 1. Capture Team Highlights 2000–2014.

CCP1: 2000–2004 **Screening proof of concept**	**CCP2: 2004–2009** **Intensive development**	**CCP3: 2009–2014** **Demonstration**
• Screened 200 capture technologies. • Identified pre-combustion, post-combustion and oxy-firing technologies to deliver significant capture cost reduction.	• Further developed most promising capture technologies from CCP1, with two potentially ready for field demonstration. • Identified novel capture technologies.	• Field demonstrated FCC and OTSG oxy-firing capture technologies. • Continued development of novel technologies including – Chemical Looping Combustion and Membrane Water Gas Shift. • Developed economic baseline modelling for all scenarios.

Storage, Monitoring & Verification Overview

The aim of the Storage Team was to close specific knowledge gaps around the storage of CO_2 and to trial monitoring technologies. Providing stakeholders with assurance that secure CO_2 storage can be achieved has been vital for the entire industry. The Storage Team, with its heritage in the upstream oil and gas industry, was in a unique position to contribute to this understanding. The Team drew on the decades of expertise of its members in subsurface exploration, production and site decommissioning. Highlights of work by the Storage Team are shown in Table 2.

Table 2. Storage Team Highlights 2000–2014.

CCP1: 2000–2004 **Screening proof of concept**	**CCP2: 2004–2009** **Intensive development**	**CCP3: 2009–2014** **Demonstration**
• Pioneered risk-based approach for geological site selection, operation and closure. • Developed new storage monitoring tools.	• Strengthened the science of storage with a focus on well integrity, through systematic R&D. • Published a definitive book on storage.	• Storage R&D and trialling of field monitoring technologies. • Developed Certification Framework. • Completed modelling of storage contingencies.

In terms of building better understanding of CO_2 behaviour in the subsurface, one study demonstrated the effects of impurities in the CO_2 stream on injection, migration and pressure evolution. Other important studies covered areas such as relative permeability, capillary entry pressure and geomechanical hysteresis. For monitoring, an important investigation was carried out into the effectiveness of a number of technologies, including satellite, modular borehole, 3D/4D VSP and electromagnetic based systems.

In Phase 3, a contingencies program was established to increase confidence around storage integrity by improving the industry's ability to manage unexpected migration of CO_2 from storage sites.

Policy & Incentives Overview

The Team focused on providing the technical and economic insights needed by stakeholders to inform the development of legal and policy frameworks. This work has been shared with the industry at conferences, including the UN climate change conferences, and was made available online in the CCP's publication section. The Team has provided several updates for its regulatory study: *Challenges and key lessons learned from real-world development of projects*. This study has provided a practical and focused update on CCS projects that have undergone or progressed significantly through the regulatory process, and found that pathways for the regulatory approval of a CCS project do exist. The scope of work carried out by the Policy & Incentives Team during the three phases of CCP is characterized by the highlights shown in Table 3.

Table 3. Policy & Incentives Team Highlights 2000–2014.

CCP1: 2000–2004 **Screening proof of concept**	**CCP2: 2004–2009** **Intensive development**	**CCP3: 2009–2014** **Demonstration**
• Surveyed existing policies, regulations and incentives affecting CCS projects. • Identified gaps in the regulatory and policy framework that inhibit CCS deployment. • Established outreach effort to inform the policy debate.	• Conducted study to assess issues of financing a CO_2 pipeline infrastructure. • Surveyed issues and concerns of stakeholders, including general public, policy makers, NGOs, investors and others.	• Updated survey of regulatory issues facing CCS. • Documented lessons learned from real CCS projects from a variety of jurisdictions. • Updated stakeholder concerns and added granularity. • Studied types and effectiveness of local benefit sharing as a means to gain support for CCS projects.

Communications Overview

The communications program was more ambitious in scope in Phase 3 than ever before. Whilst it continued to share the knowledge and insights from project teams with the CCS and oil and gas industries, it also played a greater role by helping to translate the science of CCS and make it accessible to non-technical audiences. The highlight was the creation of the CCS Browser (www.ccsbrowser.com) – the industry's first, multi-platform, digital tool dedicated to explaining CCS to a wide range of stakeholders. Highlights of the Communication Team's work through the three phases of CCP are shown in Table 4.

Table 4. Communications Team Highlights 2000–2014.

CCP1: 2000–2004 **Screening proof of concept**	**CCP2: 2004–2009** **Intensive development**	**CCP3: 2009–2014** **Demonstration**
• Launched an official website with a vast resource and registrant database. • Participated in inter-national conferences. • Published a book of the CCP Phase 1 results. • Developed a Phase 1 results brochure.	• Published a comprehensive technical guide to CO_2 geological storage. • Developed a meter-long In Depth leaflet that provides a spatial perspective of CO_2 storage. • Published a book of the CCP Phase 2 results. • Participated in inter-national conferences.	• Developed regular project overview factsheets. • Participant at key UNFCCC COP, GHGT and CCUS conferences. • Launched an educational website that explains CCS to the broader audience. • Created an interactive, digital version of the In Depth leaflet. • Published a book of the CCP Phase 3 results.

Concluding Remarks

The CCP team continues to lead the oil and gas industry response for CCS technology research, development and demonstration. During CCP Phase 3, we saw improvements in post-combustion capture technologies for retrofitting large-scale, gas-fired power generating plants and also in solutions that promise to dramatically reduce capture costs of industrial operations further. We have also demonstrated that CO_2 storage can be performed safely and securely, and options to increase the scale of deployment are now being considered. We are very proud of these accomplishments that have been achieved in collaboration with our partners, and look forward to advancing this further in CCP Phase 4.

ACKNOWLEDGEMENTS

Mark Crombie, BP International Ltd
Program Manager, CO_2 Capture Project Phase 3

As a partnership of major energy companies, the CCP provides a unique, collaborative forum for those companies to develop practical CCS knowledge and solutions specifically relevant to the oil and gas industry. Since it was founded in 2000, the CCP's expert Technical Teams, comprised of engineers, scientists and geologists from member companies, have undertaken almost 200 projects to increase understanding of the science, economics and practical engineering applications of CCS. The Teams cover four areas of activity (1) Storage, Monitoring and Verification, (2) Capture (underpinned by CCP-derived economic modelling methodology), (3) Policy and Incentives and (4) Communications.

To meet its objectives, CCP's work is being conducted in phases, with the partnership having now completed its third phase. During Phase 1, eight corporate partners engaged with governments to develop the initial program. Phase 2 followed in 2004 with eight corporate members to the full program, two additional partners and four participating governments. Phase 3 began in late 2009 with seven corporate members: BP, Chevron, ConocoPhillips, eni, Petrobras, Shell, Suncor, and Associate member: EPRI (ConocoPhillips and EPRI later withdrew from the program). In this phase, the CCP has worked closely with a wide range of national and regional bodies, industry partners, academic organizations and global research institutes. It has been recognised by the Carbon Sequestration Leadership Forum (CSLF) for its contribution to the advancement of CCS.

The Phase 3 achievements have involved more than 100 technical experts seconded from the member organizations to CCP working teams. These teams have been led by Ivano Miracca – Capture Team Lead (eni), Scott Imbus Storage Team Lead (Chevron), Arthur Lee – Policy & Incentives Team Lead (Chevron) and myself as Communications Team Lead. The teams are governed by the Executive Board with representatives from each corporate member: Nigel Jenvey (CCP Chairman, BP), Rodolfo Dino (Petrobras), Stephen Kaufman (Suncor), Vincent Kwong (Chevron), John MacArthur (Shell) and Mario Vito Marchionna (eni). Additionally, the teams are guided by an independent Technical Advisory Board (TAB) made up of CCS industry experts, led by Vello Kuuskraa. The TAB participants include: Olav Bolland, Michael Celia, Pierpaolo Garibaldi, Christopher Higman, Larry Myer and Dale Simbeck. The Advisory Board conducts independent peer reviews on CCP activities and respective programs to ensure efforts remain true to the partnership's aims and future direction.

Sincere thanks are extended to all of CCP's external collaborators for their support through generous government contributions, expert research and development, and technical project associations. Phase 3 will be remembered for the successful delivery of major demonstration trials and significant R&D projects. Without this global network working together, many of the achievements would not have been possible.

Special thanks to all reviewers of the technical contributions found in this book as their voluntary efforts have helped the book provide a good representation of CCP's work during Phase 3. Finally, this volume has been edited by Karl F Gerdes and on behalf of the CCP, thank you for managing and coordinating all the documentation of the work by so many researchers and technology providers.

Please see the section "Contributors to CCP3" for a listing of the individuals from the CCP Member Companies who planned and guided this work, a listing of all technology providers contributing to the success of CCP3, and a listing of the individuals who provided peer review of the technical chapters. Individual chapter authors are listed in the Author Index.

CONTENTS

Field Trialing **611**

Contingencies **791**

Carbon Dioxide Capture for Storage in Deep Geological Formations, Volume 4
Karl F Gerdes (Editor)

INTRODUCTION

Mark Crombie[1] and Karl F Gerdes[2]
[1]Program Manager, CO_2 Capture Project, BP International Limited, Sunbury on Thames, UK
[2]Editor, Karl F Gerdes Consulting, Davis, California, USA

BACKGROUND

The CO_2 Capture Project (CCP) is an international partnership of major energy companies, working alongside specialists from industry, technology providers and academia, to advance technologies and improve operational approaches to help make Carbon Capture and Storage (CCS) a viable option for CO_2 mitigation in the oil and gas industry.

The CCP has now completed its third phase of activity and a fourth phase is set to commence in 2015. Phase 1 of the project, CCP1 (2000-2004), involved screening almost 300 capture and storage technologies to identify those that could deliver significant capture cost reduction and increased assurance of long term storage. Other achievements included pioneering a risk-based approach for geological site selection, operation and closure, and identifying storage monitoring tools for further development [1, 2].

Phase 2 of the project, CCP2 (2004-2009), further developed the capture technologies identified as promising, with two technologies made ready for demonstration. During this phase, the CCP also identified promising novel capture technologies, developed the concept of a Certification Framework for storage site viability assessment, and achieved a landmark CO_2-exposed well study elucidating the type and extent of alteration, preservation and factors influencing alteration of the well integrity [3].

Phase 3 of the project, CCP3 (2009-2014), has focused on implementing selected capture and monitoring technologies at the demonstration stage. This phase has seen significant progress resulting in an array of demonstrations, field trials and studies. A few of the key milestones from CCP3 are outlined below:

- Demonstrated oxy-firing capture technology for Fluid Catalytic Cracking catalyst regeneration at a refinery research facility in Brazil.
- Demonstrated oxy-firing capture technology for a once-through steam generator (OTSG) at the Cenovus Christina Lake facility in Canada.
- Economic baselines for CO_2 capture using state-of-the-art post-combustion solvent technology were defined for oil refinery, heavy oil and natural gas power scenarios, to be used as a basis of comparison for advances in technology.
- Modular borehole monitoring system successfully deployed at Citronelle site.
- Contingencies modelling study for CO_2 storage completed, with significant conclusions drawn regarding detection, characterization and intervention approaches for unexpected CO_2 leakage from a storage reservoir.
- Documented application of the Certification Framework, providing a consistent means of storage site leakage risk assessment.

- Deployed InSAR satellite technology to detect surface deformation from CO_2 storage at Decatur, USA.
- Completed a study of the effects of impurities in injected CO_2 on injection, migration and pressure evolution.

Full industry members in CCP3 were BP, Chevron, eni, Petrobras, Shell and Suncor, with previous member ConocoPhillips and associate member EPRI having served until the end of December 2011. Government agencies provided funding for selected projects during the phase. Examples include: the INNOCUOUS chemical looping combustion project (Chapter 16) part funded by the EU; the Partnership for CO_2 Capture (Chapter 22), which received funding from the US Department of Energy; and the oxy-firing OTSG project (Chapters 15) with funding from the Climate Change and Emissions Management (CCEMC) Corporation which receives funds from the Province of Alberta.

Phase 4 of the project, CCP4 (2015-2019), will continue to develop pioneering CCS technology and knowledge for potential application in the oil and gas industry. The CCP4 program aims to continue work on CO_2 capture solutions and understanding of their application in the context of scenarios identified from previous CCP phases (refinery, heavy oil and natural gas power generation), together with a new scenario – CO_2 separation from natural gas production. The storage program will continue to demonstrate safe and secure geological containment through field-based monitoring and developing robust intervention protocols.

ABOUT THIS BOOK

This book is a collection of peer-reviewed scientific chapters that describe the technological advances made during CCP3. This is the fourth in a series of volumes, with the first two published in 2005, documenting progress during CCP1 [1, 2], and the third volume published in 2009 covering CCP2 [3].

The CCP Program Manager and Editor have ensured that all scientific papers presented in this volume are of the highest standard and all the relevant technical papers have been peer reviewed by at least one referee.

All chapters of this book are available to download if you are registered on the CCP website at www.co2captureproject.org.

Introductory Material

This introductory chapter provides background information on the CCP3, including its achievements and organizational structure. Following this introduction is a chapter by the chair of CCP's independent Technical Advisory Board, Vello Kuuskraa, which offers the views and recommendations of the Advisory Board.

CO_2 Capture

Thereafter, a section follows on **CO_2 Capture**, comprising 23 chapters. Chapter 1 is an executive summary of the CCP3 capture program and results. Chapter 2 is an introduction to the capture program and describes the rationale for the various scenarios chosen by CCP as a framework for developing and assessing capture technologies. This is followed by two technical chapters which describe the methodology for the technical studies (Chapter 3) and for the economic assessments (Chapter 4) of the capture technologies selected for study under CCP3. This is a critical aspect of the Capture program – that a consistent methodology was used for all the techno-economic assessments. Thereafter are three groupings of chapters dedicated to the CO_2 capture application scenarios of CCP3:

- **Refining Scenario** (Chapters 5 – 12). Chapters 5-7 describe the extensive work done by CCP3 to assess options for and execute a pilot demonstration of CO_2 capture from a Fluid Catalytic Cracking (FCC) unit, which is often the major emissions source in a refinery. Chapters 8-11 cover the assessments, development and testing of several approaches to mitigation of CO_2 emissions from the numerous and, usually, widely-dispersed fired heaters and boilers in a refinery. Chapter 12 documents assessment of capture from a modern Steam Methane Reformer (SMR) for hydrogen production, a large point source of CO_2 emissions.
- **Heavy Oil Scenario** (Chapters 13-20). Chapters 13-15 document technology assessment, design, and retrofit of oxy-firing capture technology applied to the large steam generators that are used for thermal heavy oil production. Chapters 16-19 describe technology development efforts and progress on natural gas fuelled Chemical Looping Combustion, which is a promising technology for application to steam generation and other heating applications, that inherently captures CO_2 released by the fuel combustion. Chapter 20 contains the economic assessments completed for the many options considered for CO_2 capture in the Heavy Oil scenario.
- **NGCC Scenario** (Chapters 21-22). These chapters document economic assessments and development work aimed at mitigating CO_2 emissions from Natural Gas-fired Combined Cycle (NGCC) power generation. A key part of this work was a techno-economic analysis of the current state of the art post-combustion capture technology using solvent scrubbing.

The Capture section is completed by Chapter 23, which offers conclusions and recommendations for further work.

Storage, Monitoring and Verification

The next section of the book describes CCP3 work on **Storage, Monitoring and Verification** (SMV). Chapter 24 provides an executive summary of the SMV work by CCP3. This is followed by a series of chapters grouped by themes.

- **Subsurface Processes** Four chapters (25-28) summarize work which was aimed at assessing current assumptions and providing new insights into physico-chemical processes that impact CO_2 injectivity, migration-trapping and long-term containment of CO_2 in storage systems.
- **Monitoring and Verification** Chapters 29-31 document CCP3's work on M&V, a core process for the long term management and confirmation of the safe geologic storage of CO_2. Since there are a wide variety of M&V deployments being undertaken globally in pilot storage projects, this work by CCP3 sought to identify common challenges and demonstrate practical processes and design elements necessary for planning monitoring processes. This section provides context to the various monitoring activities in the Field Trialling theme.
- **Optimization** The Certification Framework (CF) is a leakage risk assessment framework developed around the concept of effective trapping. CCP has supported development of the CF since early in CCP2. Chapter 32 documents a number of case studies illustrating the effectiveness and value of the CF. Chapter 33 looks beyond the work related to storage assurance. This chapter summarizes existing knowledge of

issues surrounding the use of CO_2 for enhanced unconventional gas and oil recovery, and outlines gaps with recommended approaches to addressing them.

- **Field Trialling** This 8 chapter section (34-41) documents the considerable success CCP3 achieved in accessing, operating and deriving useful (and often remarkable) findings from field trialling of M&V technology. This was accomplished by leveraging third party sites, which have characterization, reservoir modelling and other M&V data available for comparison.
- **Contingencies** This seven chapter section (42-48) documents work for CCP3 that is the first comprehensive study to address detection, characterization and intervention approaches for unexpected CO_2 leakage from a storage reservoir. The program includes modelling of storage projects to assess the detectability and characterization of leakage, passive (stop injection) and active (e.g., hydraulic controls, sealant injection) controls and plans to test intervention at the bench-field scale.

The final chapter of the SMV section, Chapter 49, offers a summary of key findings from the work of CCP3, technology gaps remaining and recommendations for addressing them.

Policy and Incentives / Communications

The book concludes with three chapters focusing on the work completed by CCP's Policy and Incentives (P&I) and Communications teams.

Chapter 50 documents three key studies that helped to inform CCP and the broader CCS community of the challenges for real-world CCS projects which are going through permitting processes, in the face of regulatory schemes still under development, analysing stakeholders who can impact CCS project development, and developing mechanisms for sharing benefits from CCS projects.

Chapter 51 documents work for CCP3 to develop a process for identifying, framing and, in some cases, resolving concerns about potential regulatory hurdles to the demonstration and deployment of technologies for the carbon capture utilization and storage (CCUS) through individual projects. The focus is on the CCUS regulatory framework being created by the United States Environmental Protection Agency (EPA) under its existing statutory authorities.

Chapter 52 describes the efforts which CCP undertook for its most ambitious communications program during CCP3. The effort included continuing to share results and insights arising from the CCP work program with members and the CCS industry. However, it also involved creating digital tools to help make the science of CCS accessible to a wider, non-technical, audience. The creation of the CCS Browser (www.ccsbrowser.com) – the industry's first, multi-platform, digital tool dedicated to explaining CCS was at the centre of this.

CCP3 STRUCTURE

CCP3 had two levels of membership:

- Full members took part in all CCP3 activities, its working teams and had representation on the Executive Board
- Associate members had the right to attend SMV and Policy meetings, annual meetings of the Executive and Technical Advisory Boards, NGO outreach meetings and public/government meetings. Associate members also had the opportunity to participate in planned technology demonstrations.

Structure and Management Process

The CCP3 structure is illustrated below:

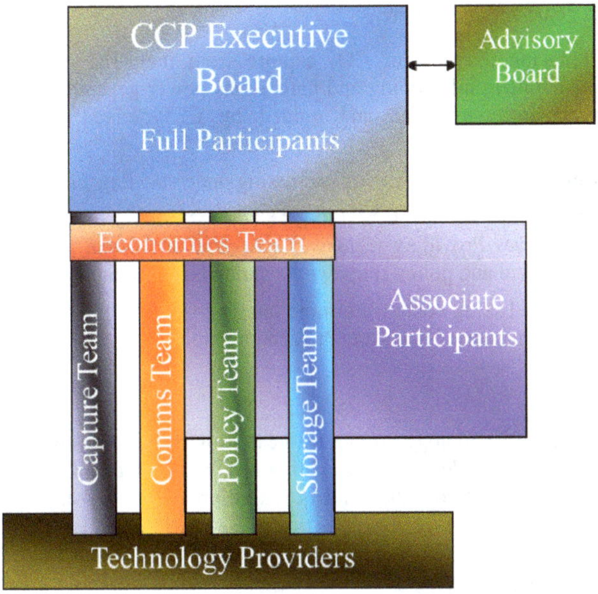

Figure 1. CCP3 Structure.

Executive Board

Each full member organisation was represented on the CCP Executive Board, which exercised overall control and governance of the program, and prioritisation of specific projects to be undertaken and funded. The Board typically met quarterly.

Technical Advisory Board

The CCP was supported by a Technical Advisory Board (TAB) responsible for conducting independent peer reviews on the activities of the CCP teams and their respective programs. The resulting TAB recommendations helped to ensure the program remains true to its aims and shape its future direction. The TAB was comprised of seven independent, international technical experts from industry and academia.

Program Management

The Program Manager provided direction to the teams for the delivery of the program goals within the overall budget and timeframe set by the Executive Board.

Teams

The details of CCP were planned and carried out by four teams – Capture; Storage, Monitoring & Verification; Policy & Incentives; and Communications. The teams were comprised of geologists and other subsurface specialists, engineers, policy and regulatory specialists and communications experts drawn from each of the member organisations.

Each of the technical teams is led by a designated team leader. The technical teams were assigned the following broad goals within their areas:

- **Capture** Reducing the cost of CO_2 capture from industrial operations for in-situ extraction of bitumen, oil refining and natural gas power generation. The work of the Capture Team was supported and guided by CCP-derived economic modelling methodology.
- **Storage, Monitoring & Verification** Increasing understanding and developing methods for safely storing and monitoring CO_2 in the subsurface.
- **Policy & Incentives** Providing technical, economic and social insights to inform the development of legal and policy frameworks and to help public understanding.
- **Communications** Taking content from the ongoing work of other teams and delivering it to government, industry, NGOs and the general public.

Technology providers

CCP3 engaged a variety of technology providers to carry out the specifics of the technical program. The technology providers include academic institutions, research companies, CCP3 participants, national laboratories and expert consulting firms. These providers worked together with the CCP teams to deliver the program results.

CCS TECHNOLOGIES IN CCP3

As noted in the Background section of this chapter, the technology program of CCP3 was built on results from the previous phases of the project. The guiding principles used by CCP throughout for selecting technologies on which to work have been:

- Likelihood to achieve target cost reduction relative to existing (baseline) technology.
- Health, Safety and Environmental requirements – safe for people and the environment, and reliable – must be met.
- Relevant to the CCP Member companies' operations.
- Likelihood to meet the time schedule set for CCP3.
- Acceptable to government and public stakeholders.

The CCP3 Capture Technology Portfolio

The portfolio of CO_2 capture technology consisted of six projects involving lab, pilot or demonstration unit testing: three of which were oxy-firing, one for solvent-base post-combustion (variety of solvents tested), one for pre-combustion, and one for gas-fuelled chemical looping combustion. In addition, significant techno-economic studies were completed to define baseline economics for state of the art post-combustion solvent capture applied to heaters and boilers, FCC catalyst regeneration, steam methane reforming, and natural gas combined cycle power generation. Additional techno-economic studies included state of the art, large-scale H_2 fuel production with CO_2 capture, and technology-specific economic assessments for each of the developmental technologies.

The CCP3 SMV Technology Portfolio

The Storage, Monitoring, and Verification portfolio consisted of 20 technical projects organized around five themes: 1) Subsurface Processes, 2) Monitoring and Verification, 3) Optimization, 4) Field Trialling, and 5) Contingencies. Chapter 24 summarizes the results.

The CCP3-SMV program made considerable progress in supporting the technical case for CO_2 storage assurance. Accomplishments across a range of CO_2 storage technical topics include: 1) challenges were raised to fundamental interpretation of lab experiments in terms of what they can tell us about subsurface processes, 2) field trialling of several new M&V technologies, 3) feasibility of conducting bench-field scale experiments at underground labs and 4) launch of a comprehensive Contingencies program to guide detection, characterization and mitigation of unexpected fluid leakage.

REFERENCES

1. Thomas, D.C. (editor), (2005) *Carbon Dioxide Capture for Storage in Deep Geologic Formations – Results from the CO_2 Capture Project*, Vol 1 *Capture and Separation of Carbon Dioxide From Combustion Sources*; ISBN 0-08-044570-5, Elsevier Publishing, UK.
2. Benson, S.M. (editor, (2005) *Carbon Dioxide Capture for Storage in Deep Geologic Formations – Results from the CO_2 Capture Project*, Vol 2 *Geologic Storage of Carbon Dioxide with Monitoring and Verification*; ISBN 0-08-044570-5, Elsevier Publishing, UK.
3. Eide, L.I. (editor), (2009) *Carbon Dioxide Capture for Storage in Deep Geologic Formations – Results from the CO_2 Capture Project*, Vol 3 *Advances in CO_2 Capture and Storage Technology - Results (2004-2009)* ; ISBN 0-08-044570-5, CPL Press, UK.

Carbon Dioxide Capture for Storage in Deep Geological Formations, Volume 4
Karl F Gerdes (Editor)

CCP3 TECHNICAL ADVISORY BOARD REPORT: APPRAISING THE PERFORMANCE AND ACCOMPLISHMENTS OF THE CO_2 CAPTURE PROJECT - PHASE III

Vello A. Kuuskraa
Advanced Resources International, Inc.
4501 Fairfax Drive, Suite 910, Arlington, VA 22203 USA

ABSTRACT: Since its inception in 2000, the CO_2 Capture Project (CCP) has pursued two main themes: (1) identify and further develop technologies that would lead to lower costs for capture of CO_2; and (2) improve the performance and public acceptance of CO_2 storage. In addition to overall management of the project by the CCP Executive Board (containing representatives from each of the participating companies), the CCP established a Technical Advisory Board (TAB) as part of its commitment to quality project management. This chapter transmits the CCP TABs appraisal of the performance and accomplishments of Phase III of the CO_2 Capture Project (CCP3).

KEYWORDS: CO_2 capture; CO_2 storage; CCS; refinery FCC unit; natural gas combined cycle (NGCC); pre-combustion CO_2 capture; post-combustion CO_2 capture; oxy-firing

BRIEF HISTORY OF THE CO_2 CAPTURE PROJECT

During Phase I of the CO_2 Capture Project (CCP), which encompassed four years (2000 to mid-2004), the CCP conducted in depth reviews of nearly 200 novel CO_2 capture technologies. Based on these reviews, CCP1 identified a handful of "preferred" CO_2 capture technologies for further appraisal and development. In parallel, CCP1 also sponsored a series of studies that helped identify the high priority barriers facing CO_2 storage.

During Phase II of the CO_2 Capture Project (mid-2004 to mid-2009), the CCP supported more intensive research and investigation of this handful of "preferred" CO_2 capture technologies and pursued high priority topics for improving the performance and reliability of CO_2 storage, monitoring and verification.

The purpose of Phase III of the CO_2 Capture Project (mid-2009 to end of 2014) has been to further advance the science, technology and performance of CO_2 capture and storage, addressing CO_2 emission sources of most relevance to the oil and gas industry. This Chapter transmits the TAB's appraisal of the performance and accomplishments of Phase III of the CCP.

The current CCP3 TAB members are: Vello Kuuskraa, Chairman, Dale Simbeck, Chris Higman, Olav Bolland, Pierpaolo Garibaldi, Larry Myer and Michael Celia.

PURPOSE OF THE CO_2 CAPTURE PROJECT

The scientific evidence overwhelmingly shows that the continued release of carbon dioxide (CO_2) from use of oil and gas plus other fossil fuels contributes significantly to global warming. Yet, the use of oil and gas provides (and is projected to continue to provide) efficient, economically viable energy to the world's economy. As such, advanced technologies that help reduce the emissions of CO_2 from oil and gas extraction and conversion, such as CO_2 capture, utilization and storage

(CCUS), provide an important path forward for the industry. Pursuing advances in CCUS technologies, establishing their reliability, and reducing their costs are at the heart of the CO_2 Capture Project.

THE CCP3 PROGRAM FOR CO_2 CAPTURE

CCP3 Objectives for CO_2 Capture

Phase III of the CCP's research and demonstration program for CO_2 capture had five main objectives:

1. Demonstrate the use of oxy-firing for capturing CO_2 from a refinery fluidized catalytic cracking (FCC) unit.
2. Examine alternative options for capturing CO_2 from refinery heaters and boilers.
3. Demonstrate the use of oxy-firing for capturing CO_2 from natural gas-fired steam generation as part of heavy oil recovery.
4. Identify and test low cost CO_2 capture options for Natural Gas Combined Cycle (NGCC) power plants.
5. Pursue lower cost options for CO_2 capture from refinery hydrogen plants.

A comprehensive technical and economic analysis of the costs of CO_2 capture accompanied each of these five topics.

TAB's Appraisal of the CCP3 Program for CO_2 Capture

Overall Comments and Observations

The TAB finds that the CO_2 capture work by CCP3 is of high quality and continues to add valuable information to the knowledge base on CCS. In addition, the TAB finds that the CCP3 CO_2 Capture Team has completed significant work that sets the foundation for further advances to be pursued as part of Phase IV of the CCP (CCP4). Most notably, CCP3 has conducted pilot scale testing of alternative CO_2 capture options for refineries (notably the FCC unit) and for steam generation (as part of heavy oil recovery). The CCP3 CO_2 Capture Team has also significantly advanced the understanding of the cost-efficiency of alternative solvents and adsorbents for use by post-combustion capture of CO_2 from natural gas-fired power plants.

Demonstration of Oxy-Firing at a Refinery Fluidized Catalytic Cracking (FCC) Unit

The TAB finds that the demonstration of oxy-firing in a refinery FCC unit established the technical viability and operating efficiency of this CO_2 capture option. The pilot test was successfully conducted in the 33 barrels per day FCC test unit operated by Petrobras in Paraná, Brazil. This pilot test represents a major step forward for developing a reliable technology for mitigating CO_2 emissions from oil refineries.

The TAB believes that the analysis and testing of oxy-fuel combustion CO_2 capture for the oil refinery fluid catalytic cracking (FCC) unit should continue. Retrofitting existing FCC units also provides benefits that help mitigate the costs of CO_2 capture at an oil refinery. These benefits include higher throughput capacity, higher light distillates yield, and enhanced ability to effectively process heavier feedstocks to higher value products. At higher oil prices and higher distillate-to-residue prices, the costs and economics of oxy-fuel CO_2 capture at FCC units appear to be better than post-combustion CO_2 capture. For retrofit of existing FCC units where space is limited, oxy-fuel combustion has the advantage of off-site oxygen production.

Alternative Options for Capturing CO_2 from Refinery Heaters and Boilers

The development of cost effective options for capturing CO_2 from refinery heaters and boilers (given their widely distributed locations) remains a challenge, even with the rigorous work performed on this topic by CCP3. Large scale, closely located refinery heater and boiler systems may lend themselves to post-combustion capture of CO_2. However, reduction of CO_2 emissions from widely distributed refinery heater and boiler systems may require the use of more costly H_2, rather than natural gas or refinery off-gas, as the fuel. More fundamental design changes to the placement of refinery heaters and boilers may be required to achieve lower cost CO_2 capture from these sources.

Demonstration of Oxy-Fired OTSG (Once Through Steam Generation)

CCP3 has set forth a sound research protocol and field implementation plan for testing the use of oxy-firing for capturing CO_2 from natural gas-fired steam generation as part of heavy oil recovery. The TAB looks forward to the results from the field trials, being conducted during 2015.

Identify Lower Cost CO_2 Capture Options for Capture of CO_2 from Natural Gas Combined Cycle (NGCC) Power Plants

The CCP work during Phase 2, in collaboration with the CACHET project, established that post-combustion rather than pre-combustion is the preferred technology option for CO_2 capture from natural gas-fired power plants. CCP3 sponsored studies showed lower reductions in power generation efficiency for post-combustion CO_2 capture of 8 to 10% loss of efficiency compared to 14 to 16% loss of efficiency using pre-combustion CO_2 capture. The CCP2/CACHET effort also noted that use of post-combustion would avoid the need to develop new turbine components to handle higher hydrogen concentrations and higher temperatures required by pre-combustion CO_2 capture from NGCC power plants.

Based on this finding, the follow-on work by CCP3 concentrated on three research pathways for CO_2 capture from NGCC power plants.

1. Identify, evaluate and advance the development of emerging post-combustion technologies that would deliver lower cost CO_2 capture.
2. Conduct techno-economic feasibility analyses to establish a cost benchmark for currently "best available" post-combustion CO_2 capture from NGCC power plants.
3. Conduct additional pilot scale tests of selected, promising post-combustion solvent technologies.

The TAB finds that CCP3 has achieved a significant accomplishment in identifying alternative, potentially lower cost post-combustion solvent systems and in establishing today's benchmark costs of CO_2 capture from NGCC power. Future work on solvent testing should take place in facilities suitable to handling dilute, methane-based flue gas including consideration of other locations, such as at Mongstad/Norway, the RWE facility in Germany, or the NCCC in Alabama.

Pursuing Lower Cost Options for Capture of CO_2 from Hydrogen Plants

The growing need for oil refinery hydrogen (H_2) makes H_2 generation an important pre-combustion option for CO_2 capture. This is especially true for certain H_2 plant designs that generate a pure CO_2 vent. The incremental costs for adding CCS to these H_2 plant designs is low, involving primarily the cost of CO_2 compression and pipeline transport to a geologic storage site. Unfortunately, the bulk of today's SMR-based hydrogen plants emit a dilute concentration of flue gas, requiring more costly options for capturing CO_2.

CCP3 has completed an effective analysis of oxygen (O_2) blown natural gas autothermal reforming (ATR) for H_2 pre-combustion capture of CO_2. Nevertheless there are other H_2 process alternatives with pure CO_2 vents. These included several commercially proven heat exchange reformer (HER) designs where the exothermic O_2-blown ATR supplies the heat for the endothermic steam methane reformer (SMR). These H_2 process designs may have economic and efficiency advantages at large scale. There is also a high pressure O_2-HER H_2 design that integrates a gas turbine for the O_2 generation or air separation unit's (ASU) power needs and promises overall cost and efficiency improvements. Thus, additional research on H_2 generation pre-combustion CO_2 capture would be useful to pursue during CCP4.

THE CCP3 PROGRAMS FOR CO_2 STORAGE

CCP3 Objectives for CO_2 Storage

Phase III of the CCP's research and demonstration program for CO_2 storage had three main objectives:

1. Develop more rigorous understanding of the fundamental science and processes involved with safe, secure storage of CO_2.
2. Advance the state of the art of CO_2 storage monitoring and verification.
3. Develop improved methods of CO_2 leakage detection and intervention, including developing technology for sealing natural fracture conduits.

TAB Appraisal of the CCP3 Program for CO_2 Storage

Overall Comments and Observations

The TAB is favorably impressed with CCP3's work on CO_2 storage. A number of the projects and research topics addressed by CCP3 (some of which were continuation of work started in CCP2) have come to their logical conclusion and are documented in this volume for broad dissemination. Still, a series of CCP3's more recent storage research topics – improved sub-surface monitoring technology, assessment and remediation of CO_2 leakage, and the integrated fracture sealing experiment – provide a solid foundation for continuation as part of CCP4, as further discussed below.

Advancing the Understanding of Fundamental Storage Mechanisms.

The TAB applauds the CCP3 for undertaking additional experiments to establish the residual saturation of CO_2 (pore space CO_2 trapping), rather than distributing the results from the initial work. Still, further information on residual CO_2 saturations is needed for alternative rock systems and reservoir settings.

Advanced Sub-Surface Monitoring Technology.

The development and field validation of subsurface monitoring technology, such as the Modular Borehole Monitoring (MBM) package, remains an important research goal. While considerable work on this topic is being conducted by the U.S. DOE Regional Partnerships, the CCP4 should continue its work with these partnerships to help advance the state of the art, particularly for defining a thorough but practical suite of monitoring technologies for large-scale CO_2 injection operations.

CO_2 Leakage Detection and Intervention.

The TAB finds the information from the CO_2 leakage detection and intervention modeling by Stanford University to be of high quality and of high value. The TAB's view is that for CO_2 storage to be broadly accepted as a safe and reliable technology, the storage practitioners need to have scientifically sound methods for early detection and early remediation of any CO_2 leakage. Importantly, the detection of leakage needs to be in the deep subsurface and not after the CO_2 has reached potable water or the surface.

The TAB is a strong proponent of the use of extensive pressure monitoring and a carefully designed intervention plan as part of a rapid-response strategy for all large scale CO_2 storage efforts. In addition, the TAB notes that achieving the full potential of time lapse seismic, with its extensive processing and interpretation challenges, requires considerable reductions in time of delivery to be accepted as a rapid-response monitoring technology.

The TAB recommends that the next phase of the CO_2 leakage detection and intervention program involve basin-scale modeling and large-scale CO_2 injection (6 million tons per year) at several locations in the basin, accompanied by a comprehensive suite of monitoring and measurement protocols.

Mont Terri Fracture Sealing Experiment.

The TAB finds the planning and test design for the fracture sealing experiment at Mont Terri to be comprehensive and rigorous. This includes the modeling efforts used to help design the project, the incorporation of previous experiences on rock fracturing conducted at the test facility, and the selection of candidate sealing materials for the test. Given the costs of the next phase of the Mont Terri fracture sealing work and its potential for providing high value outcomes, the TAB strongly recommends pursuing a partnership with the U.S. DOE on the upcoming experiment. A productive next step would be discussing this option directly with the U.S. DOE/NETL Carbon Sequestration Program.

GUIDANCE AND RECOMMENDATIONS FOR CCP4

The first three phases of the CCP provided its member companies an in-depth understanding of CO_2 capture and storage sciences and technologies of highest relevance to the oil and gas industry. Given this impressive progression of effort and success in building the scientific and technical foundations for CO_2 capture and storage, the TAB recommends that CCP4 take the next logical step – preparing its member companies for large-scale implementation of CCS by year 2020.

Future CCP4 Capture Program.

The TAB recommends that CCP4 undertake a comprehensive CO_2 capture effort at a "representative", fully integrated refinery, including its FCC unit, its heaters and boilers, its hydrogen plant and its dedicated natural gas-fired combined cycle (NGCC) power plant. Of particular value would be understanding the synergies resulting from an integrated refinery CO_2 capture system and its impact on refinery product costs.

- The first phase (years 2015-2017) would involve process modeling, systems integration and cost assessment.
- The second phase (years 2018-2020) would involve implementation of the integrated retrofit CO_2 capture system at an actual refinery.

The TAB also recommends continuing CCP's efforts on CO_2 capture from steam generation (OTSG) for heavy oil recovery. After bringing the current field pilot to completion, the TAB recommends taking a step back to identify options for significantly reducing CO_2 capture investment costs from OTSG before proceeding with a follow-on field demonstration.

While the integrated commercial-scale refinery CO_2 capture project and pursuit of lower cost CO_2 capture from steam generation at heavy oil fields would be the highlights of the CO_2 capture portion of CCP4, the TAB recommends continued research support and assessment of promising, emerging CO_2 capture options.

Future CCP4 Storage Program

In parallel with preparation for commercial-scale CO_2 capture, the TAB recommends undertaking a comprehensive, modeling-based analysis of commercial-scale CO_2 storage involving the injection of large volumes of CO_2 at a series of CO_2 injection sites in a "representative" basin. The large-scale injection analysis would incorporate basin-scale monitoring, installation of early warning systems of CO_2 leakage, and implementation of effective intervention strategies.

The TAB recommends continuing with the next phases of the Mont Terri Fracture Sealing Experiment and seeking additional sponsors (e.g., DOE's CO_2 Carbon Sequestration Program, others). It also recommends continuing its research support for geomechanics and advanced monitoring systems relevant to CO_2 storage.

Finally, the TAB recommends that the CCP4 Storage Team work closely with the CCP4 Policies and Regulatory Team to enable the CO_2 that is stored as part of CO_2 enhanced oil recovery to become accepted by regulators as "bona fide" CO_2 storage.

Carbon Dioxide Capture for Storage in Deep Geological Formations, Volume 4
Karl F. Gerdes (Editor)

Section 1

CO_2 CAPTURE

Capture Overview

Carbon Dioxide Capture for Storage in Deep Geological Formations, Volume 4
Karl F. Gerdes (Editor)

Chapter 1

CCP3 – CO_2 CAPTURE EXECUTIVE SUMMARY

Ivano Miracca[1]and Jonathan Forsyth[2]
[1]Saipem S.p.A. (Eni) – Via Martiri di Cefalonia, 67 – I-20097 San Donato Milanese, Italy
[2]BP International Limited, ICBT Chertsey Road, Sunbury-on-Thames, Middlesex, TW16 7LN, United Kingdom

ABSTRACT: The CO_2 Capture Project (CCP) is reporting the results from a five year programme of CO_2 capture technology development which has included field-based technology demonstration, as well as development of promising new technologies. These activities have been complemented by technical and economic assessments of new and existing technologies applied to capturing CO_2 arising from CCP member companies' industrial operations.

KEYWORDS: CO_2 Capture Project; CCP; CCP3; CO_2 Capture and Storage; CCS; CO_2 Capture Technologies

INTRODUCTION

The CO_2 Capture Project (CCP) is a collaborative partnership of world-leading energy companies. The initiative undertakes research and develops technologies to help make CO_2 capture and geological storage (CCS) a practical reality for reducing global CO_2 emissions and tackling climate change – one of the greatest international challenges of our time.

CCP has progressed through three sequential phases. Phase 1 began in 2000 and was followed by Phase 2 which completed in 2009 [1,2,3]. Phase 3 is now complete and the findings are reported in this 4th volume of results from CCP. During the life-time of the CCP initiative it has been evident that countries and states differ in how they perceive the risk of climate change, and what measures they are prepared to take in response. As a result of these differences, the pace and stringency of policy and regulatory development has been variable, but nevertheless there has been a consistent direction which has served to sustain the need for the CCP effort.

CCP3 CO_2 CAPTURE OBJECTIVES

CCP Phase 3, or CCP3 as it is referred to in this book, included both capture and geological storage activities in the programme of work. The intent for the CO_2 capture programme was to confirm the technical and economic performance of CO_2 capture technologies applied to the specific applications of interest to the member companies and to identify and advance the development of promising new technologies. The CCP3 CO_2 capture objectives covered the breadth of the technology readiness scale, involving large-scale demonstration, pilot plant, and laboratory testing with supporting modelling and detailed application studies. The capture objectives may be summarised as follows:

1. Deliver successful demonstrations at representative industrial scale of key technologies of interest.
2. Support a shortlist of new technologies to advance their development towards readiness for field-based pilot.

3. Scan the landscape for emerging new technologies and understand their potential.
4. Evaluate the application of state of the art technology for specific applications.

CO_2 CAPTURE DEMONSTRATION PROJECTS

Technology demonstration projects were a key part of the CCP3 CO_2 capture work programme. Two such projects were carefully selected to advance CO_2 capture technologies for industrial applications of particular relevance to CCP member company operations: Refinery Fluid Catalytic Cracking (FCC) and steam generation for heavy oil production.

FCC Oxy-fuel Demonstration at Petrobras Shale Industrial Business Unit

This technical demonstration of oxy-firing the catalyst regenerator of a 33 BPSD fluid catalytic cracking (FCC) pilot unit proved that oxy-firing is a feasible method of CO_2 capture in this application. The test was hosted and carried out by Petrobras at the SIX FCC pilot plant in São Mateus do Sul, Paraná state, Brazil. Steady and controllable operation of the pilot unit with the regenerator in oxy-firing mode was successfully demonstrated. The findings from extended testing showed that the CO_2-rich gas recycle system must be carefully designed to avoid corrosion issues. A side benefit of oxy-firing the regenerator is the potential to debottleneck the FCC by improving the feed conversion rate. For the future, detailed study of the flue gas recycle system is recommended, including different types of compressors and scale up to full commercial capacity.

OTSG Oxy-fuel Demonstration at Cenovus Oil Sands Production Operations

The second CCP3 demonstration-scale project was successful oxy-firing operation of a once-through steam generator (OTSG) for CO_2 capture in an oil sands production operation. The testing was hosted by Cenovus Energy at Christina Lake, northern Alberta, Canada, and involved the retrofit of a 50 MMBTU/h OTSG for oxy-firing. An important aspect of learning from this project related to cleanliness and the design and construction of oxygen systems. The demonstration showed that the OTSG could be operated safely in oxy-firing mode and that the transitions between oxy-firing and air-firing could be carried out smoothly and safely. This result confirmed that existing commercial OTSGs can be retrofitted for CO_2 capture with relatively little modification and can provide operational flexibility by smoothly transitioning from oxy-firing to air-firing whilst maintaining constant steam output.

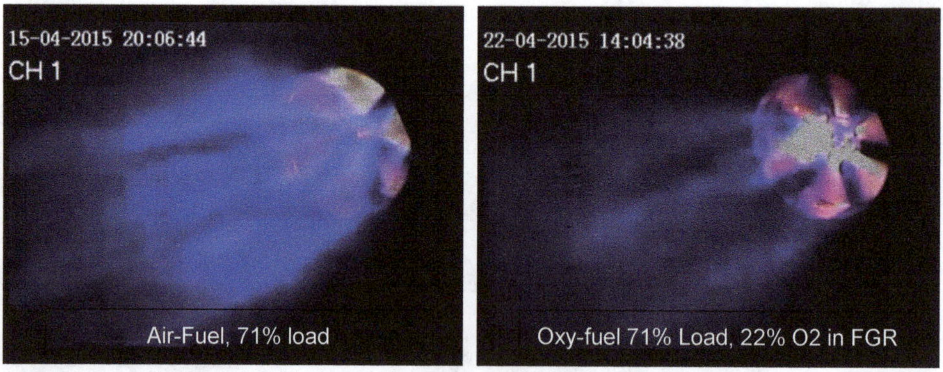

Figure 1. OTSG burner at 71% load. (Left) Air-fired with small flue gas recycle flow. (Right) Oxy-fired with 2.5% J-burner (a small, pure oxy-fuel burner retrofit to the main burner). Source: Cenovus Energy Inc.

FURTHER DEVELOPMENT OF PROMISING NEW TECHNOLOGIES

Furthering the development of promising technology has been an enduring theme of the CO_2 capture work programme. Chemical Looping Combustion (CLC) is a promising technology for CO_2 capture, which involves the use of a regenerable solid material as an oxygen carrier to produce conditions for combustion in the absence of atmospheric nitrogen. CCP has supported the development of this technology over a number of years. Two other emerging technologies were also developed through CCP3: Oxy-firing of burners for refinery process heaters and also continued development of hydrogen permeable membranes for application in pre-combustion CO_2 capture.

Chemical Looping Combustion – New Carrier Development and Performance Testing

Much previous CLC development has focussed on nickel-based oxygen carrier material. Such material has been shown to be an excellent oxygen carrier technically, but the material presents challenges in terms of safety and environmental impact. In CCP3, efforts were directed to find and develop suitable high-performance alternative materials. Copper-based material was synthesised and tested and shown to have good levels of activity, but unfortunately not good durability. Carrier materials containing Calcium and Manganese with different promoters were found to have good activity, good resistance to attrition and acceptable cost. The next steps would be to scale up production of these carriers for use in a future pilot unit.

Oxy-firing of Burners for Process Heaters

To ascertain the feasibility of oxy-firing in typical refinery service heaters, burner tests were conducted for oxy-firing with flue gas recirculation. Computer simulations predicted upright, well-structured flames for most burner/fuel combinations, which the burner tests confirmed. Flue gas recirculation achieved stable flames without modifications to burners. The test results showed that preventing air-ingress into the system is important to reduce the need for CO_2 purification.

Hydrogen Permeable Membranes

High temperature hydrogen-permeable membranes can be used to separate hydrogen from reformer syngas. This operation produces both a decarbonised fuel suitable for distribution to refinery fired equipment, such as process heaters, and also a CO_2-rich stream for storage. CCP3 supported development and testing of Palladium alloy membrane material mounted on stainless steel support tubes. A multi-tube membrane module was produced and tested to provide data for evaluation of a commercial-scale membrane separator. The fabrication of the membrane material and testing was successful; however, the costs for the large capacity membrane separator were found to be unattractive compared to alternative solvent separation technology.

SCANNING AND UNDERSTANDING THE POTENTIAL OF NEW TECHNOLOGIES

The CO_2 capture technology landscape has changed during the period of CCP3, with established providers continuing to develop their technologies and also new-entrants emerging. Innovative new developments have been identified and evaluated by CCP for their potential impact.

Post-Combustion Capture Solvent Testing

CCP3 collaborated in post-combustion solvent testing at pilot scale with the PCO2C Partnership, a program co-funded by the US Department of Energy and run by the Energy and Environment Research Center (EERC) of the University of North Dakota. It proved to be a technically difficult challenge to achieve genuinely representative test condition; however, some promising candidate solvents were identified from the data produced, which CCP believes have the potential to offer performance improvements. In a parallel effort, CCP conducted desktop feasibility studies, which sought to identify novel post-combustion solvent technologies that could materially reduce CO_2 capture costs. Based on these comparative assessments, which made use of the pilot scale test data, a solvent from ION Engineering was identified as holding promise to reduce costs relative to the state of the art solvent.

Laboratory testing of post-combustion solvent systems using enzymes as rate promoters was also supported in the CCP3 program. Practical guidance for further development and evaluation of the potential of this interesting approach to reduce cost was delivered by CCP3.

Post-Combustion Capture Adsorbents and Advanced Technology Oxy-fuel

Solid adsorbents represent an interesting alternative to solvents for post-combustion capture. CCP supported testing and evaluation of this approach but did not, from the limited results achieved, succeed in confirming the full potential of this approach. CCP3 also continued effort from earlier phases of CCP to evaluate advanced technologies, such as oxy-fuel/CO_2-rich turbine power cycles, and found that there was potential for performance improvement; however, a significant development programme to produce the high-pressure combustion, heat exchange and rotating equipment to practice the technology would be needed.

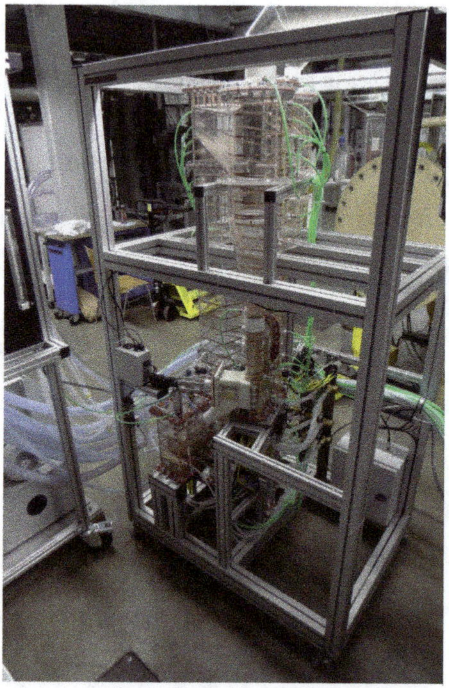

Figure 1. CLC – Cold Model test equipment at Technical University of Vienna.

Chemical Looping Combustion – Scale-up

To complement the CLC oxygen carrier development, the CCP3 developed the outline design of a CLC once-through steam generator (CLC-OTSG) for the Canadian oil sands environment, to model performance, and to generate a cost estimate. Tasks included conceptual design of OTSG with CLC technology, process and reaction engineering design, boiler engineering and equipment design, balance of plant engineering and equipment design and system performance modelling and cost estimation. Quantified results showed CLC to be more efficient and to achieve a much higher capture rate than alternative methods; however CLC was found to be expensive in terms of capex. Priority future work should look to scale-up high-capacity and durable oxygen carrier particles.

EVALUATION OF STATE OF THE ART TECHNOLOGY

CCP studied CO_2 capture from various processes, which were representative of the main emissions sources from member company operations. Ultimately, five application scenarios were considered:

1. Refinery Fluidised Catalytic Cracker (FCC)
2. Refinery Fired Heaters

3. Oil Sands Once Through Steam Generators (OTSGs)
4. Gas-Fired Power Generation – Natural Gas-Fired Combined Cycle (NGCC)
5. Refinery Hydrogen Plant – Steam Methane Reformer (SMR)

All cases taken together form a comprehensive picture of CO_2 capture from relevant industrial processes, including both new-build and retrofit of the CO_2 capture plant, by established and alternative concepts. The key results from these studies are summarized in Table 1. The CO_2 capture cost and CO_2 avoided cost are expressed on a first year rather than levelized cost basis:

Table 1. Summary of key CCP3 technical and economic study results.

Application Scenario and Case Description	Fuel	CO_2 captured		CO_2 avoided	CO_2 capture cost	CO_2 avoided cost
		t/h	%	%	$/t	$/t
Refinery – US Gulf Coast						
FCC – Post-Combustion	Carbon	55.5	85.5	65.5	94.2	122.9
FCC Oxy-fuel Retrofit (99.5% O_2)	Carbon	64.8	100.0	83.5	108.3	129.7
Fired Heater Post-Combustion	Fuel gas	26.6	85.0	65.0	118.6	156.5
Fired Heaters Pre-Combustion	Fuel gas	284.0	90.0	76.0	111.1	160.1
Refinery SMR with Post-Combustion	Nat. gas	36.1	85.5	65.5	95.9	123.3
Oil Sands Steam Generation – Fort McMurray						
OTSGs Post-Combustion	Nat. gas	67.4	90.0	76.0	170.7	237.9
OTSGs CLC	Nat.gas	63.3	100.0	86.0	195.7	236.4
Gas-Fired Power Generation – US Gulf Coast						
NGCC – Post-Combustion	Nat. Gas	126.1	85.5	73.7	97.9	113.6

The results showed that generally post-combustion capture was the most economic method in each application scenario; however a large plot space would be required to site the absorber adjacent to the CO_2 source in each case. In practical applications, where the CO_2 capture equipment may be required to be retrofitted into already congested plant areas, sufficient space may not be available and pre-combustion capture may be the preferred method. The various post-combustion cases all used a single train configuration (all CO_2 sources fed to a single absorber/regenerator and CO_2 compressor) and there was a marked economy of scale effect, as specific cost reduced with increasing capacity. All oil sands steam generation cases were based on a Northern Alberta Canada location and this leads to high costs for all CO_2 capture methods. For this application scenario, chemical looping showed promising results, particularly in terms of efficiency and capture rate, but complexity and capital cost are challenging. Mainly due to the very large scale relative to the other application scenarios, capture from natural gas fired power generation showed the lowest specific costs.

MAIN CONCLUSIONS AND NEXT STEPS

Phase 3 of the CCP partnership has successfully delivered a portfolio of CO_2 capture activity, including significant demonstration projects, which has advanced the technology to readiness for first commercial-scale application. The FCC oxy-fuel demonstration project showed the potential for not just effective CO_2 capture but also efficiency and through-put increases. The oxy-firing demonstration for OTSGs showed that such fired equipment can successfully be retrofit for CO_2 capture using this approach. The economic assessment work for OTSG in the northern Canadian environment shows how challenging it will be to practice economic CO_2 capture in a high capital cost scenario. Alongside the demonstration projects, CCP has also supported significant progress in the development of three new technologies and each of these has succeeded in achieving milestones toward readiness for field-based pilot testing. In particular chemical looping combustion technology is emerging as a credible alternative to more established methods. The technology development projects have been complemented by an active evaluation and appraisal effort directed at identifying promising emerging technology. More than 10 such technologies were evaluated by laboratory and pilot plant testing and simulation studies, with many showing interesting attributes for further development. There has also been a sustained effort to maintain and improve understanding of the technical performance and cost of state-of-the-art and high-readiness level technologies as they would be applied to capture CO_2 from CCP member company operations. It is notable that the state-of-the-art benchmark has not stood still over the past 5 years and developmental technologies must be compared with this moving target. In all, more than 20 separate cases have been evaluated. The consolidated results of the CCP3 CO_2 capture work programme have helped to provide CCP members with deep insight into the performance of state-of-the-art and emerging and new technologies, which will help them to prioritise and guide technology development activity and support evaluation of CO_2 capture in their own operations.

The recommended next steps for the development of each technology have been considered and documented in this book. The cost of CO_2 capture remains high and a barrier to the uptake of CCS across CCP member company operations and beyond. However, the progress that has been achieved in this third phase of the CCP in developing technology and insights in its use has underscored the value of collaboration, both between the CCP member companies and more widely with technology providers and other stakeholders. The commitment of the CCP member companies to reducing the cost and risk of CO_2 capture implementation remains as strong as ever.

REFERENCES

1. Thomas, D.C. (editor), (2005) *Carbon Dioxide Capture for Storage in Deep Geologic Formations – Results from the CO_2 Capture Project*, Vol 1 *Capture and Separation of Carbon Dioxide From Combustion Sources*; ISBN 0-08-044570-5, Elsevier Publihsing, UK.
2. Benson, S.M. (editor, (2005) *Carbon Dioxide Capture for Storage in Deep Geologic Formations – Results from the CO_2 Capture Project*, Vol 2 *Geologic Storage of Carbon Dioxide with Monitoring and Verification*; ISBN 0-08-044570-5, Elsevier Publishing, UK.
3. Eide, L.I. (editor), (2009) *Carbon Dioxide Capture for Storage in Deep Geologic Formations – Results from the CO_2 Capture Project*, Vol 3 *Advances in CO_2 Capture and Storage Technology - Results (2004-2009)* ; ISBN 978-1-872691-49-7, CPL Press, UK.

Carbon Dioxide Capture for Storage in Deep Geological Formations, Volume 4
Karl F Gerdes (Eds.)

Chapter 2

INTRODUCTION TO THE CCP3 CAPTURE TECHNOLOGY PORTFOLIO

Ivano Miracca
Saipem S.p.A. (Eni), Viale de Gasperi 16, I-20097 San Donato Milanese, Italy

ABSTRACT: This chapter describes the portfolio of CO_2 capture technologies studied by the CCP3 (2009-2014) and provides a summary of program objectives, project descriptions and key results achieved by a number of industry and academic research partners. The CCP Capture Program is organised around three industrial applications scenarios – Oil Refineries, Heavy Oil production, and Natural Gas Combined Cycle power generation – for which the team developed a number of technology options.

INTRODUCTION

Carbon dioxide (CO_2) capture technologies have been in commercial use for many years in the oil and gas industry. The original challenge for the CCP at its inception in 2000 was to identify the next generation technologies that were relevant to the members' varied operations and which could be developed at scale for the purpose of geological storage. The high cost of capture continues to be a concern, so a key objective of CCP3 was to demonstrate significant cost savings (up to 60%) for selected new technologies relative to existing state-of the-art post-combustion and gasification technologies, while also delivering performance improvements.

The aim of this chapter is to:

- Offer an overview of the context in which the CCP3 capture activity took place (following CCP1 and CCP2).
- Provide a summary of the application scenarios for which the CCP selected technologies have been evaluated.
- Summarise the objectives, activities and results of the most significant individual projects.

The Capture segment of this book is divided into four sections – an overview, three technology scenarios (Oil Refineries, Heavy Oil, and Natural Gas Combined Cycle) – followed by a concluding chapter which summarizes progress from the program and recommendations for future work.

THE PHASES AND SCENARIOS OF THE CCP CAPTURE PROGRAM

The CCP has now completed three phases of work since its inception in 2000. The highlights and milestones of the Capture program are shown below:

CCP1: 2000-2004 – Screening/Proof of Concept [1]

- 200 capture technologies were screened.
- A small number of pre-combustion, post-combustion and oxy-firing technologies were identified with the potential to deliver significant reductions in the cost of CO_2 capture.

*CCP2: 2004-2009 – **Intensive Development*** [2]

- Further development of the most promising capture technologies from CCP1 was undertaken, with two projects identified that could be made ready for field demonstration during CCP3.
- Novel capture technologies with promise were identified.

*CCP3: 2009-2014 – **Demonstration***

- Field demonstrations on a Fluid Catalytic Cracker (FCC) and Once Through Stem Generator (OTSG), which were both converted to oxy-firing were planned, and the FCC demonstration was undertaken in 2012, while the OTSG demonstration was completed during 1H 2015.
- An assessment of the development of novel technologies was completed.
- Economic baseline modelling was completed for the three scenarios.

CCP3 SCENARIOS

Three industrial application scenarios were selected for the work program in CCP3, for which a range of oxy-firing, pre-combustion and post-combustion CO_2 capture options were assessed, including a number of novel capture technologies. The scenarios defined were:

- Oil refineries.
- Heavy oil/oil sands production.
- Natural Gas Combined Cycle power generation.

An economic evaluation has been completed for each technology in each of the three scenarios, allowing comparison against a state-of-the-art baseline cost also developed by the CCP.

The work carried out by the CCP Capture Team was conducted with the support of a number of industry and academic partners. It has evolved into a significant body of knowledge which should prove of real use in the future development of cost-effective capture solutions for the oil and gas industry.

CCP3 PROGRAM SUMMARY

CO_2 Capture from Oil Refineries

Refineries produce some 6% of total emissions of CO_2 from stationary sources worldwide. Power stations account for more than 80% of total CO_2 emissions from stationary sources. Refinery emissions amount to 0.8 billion tons per year. The emissions from a single refinery depend on several factors, but are in the range to a moderate-sized, 600 MW coal power station.

A typical refinery has a wide range of CO_2 emission sources, often widely dispersed, including fired heaters, boilers and process units, with multiple stacks emitting flue gas to the atmosphere. In this multi-source environment, two major CO_2 emitters have been identified:

- The catalyst regenerator of the Fluid Catalytic Cracking (FCC) unit.
- The hydrogen production unit (usually a steam reformer).

The combination of these two process units (if present) may account for up to 50% of CO_2 emitted from a refinery. Therefore, they are the primary targets for reducing refinery greenhouse gas (GHG) emissions. Remaining sources of CO_2 in a refinery are represented by boilers and fired heaters (multiple units, grouped in to process clusters and discharging through a number of stacks).

In this scenario the CCP3 selected the FCC unit and a typical process unit using fired heaters as the subjects of two R&D projects to identify the most promising solutions for CO_2 capture.

Demonstration: FCC (Oxy-Firing)

The FCC process catalytically converts heavy oil fractions to lighter products such as Liquefied Petroleum Gas (LPG) and gasoline using a fluidized bed reactor containing small catalyst particles. During the reaction step, coke is formed as a by-product and deposits on the surface of the catalyst, which then becomes deactivated. The coke is burned off in the air-blown, fluidized bed catalyst regenerator to restore catalyst activity. Oxy-firing is the only alternative to post-combustion for CO_2 capture from the FCC regenerator. In the oxy-combustion process, the air normally used for combustion of the coke and bed fluidization is replaced by pure oxygen, which is diluted with recycled CO_2 to maintain the thermal balance and catalyst fluidization in the regenerator.

Figure 1. The FCC multipurpose test unit (courtesy of Petrobras).

The CCP has previously conducted a technical-economic evaluation of oxy-firing and post-combustion amine absorption for CO_2 capture from the FCC regenerator [3]. Both processes were able to achieve the required CO_2 recovery level and purity specification for storage.

The CCP and Petrobras set up an oxy-firing field demonstration at Petrobras' SIX testing facility in Brazil, testing different operational conditions and product feedstocks to demonstrate that stable operation may be maintained when operating the regenerator in oxy-combustion mode and to obtain reliable data for potential scale-up of this technology.

The pilot FCC unit (Figure 1) can simulate a commercial scale unit by replicating the operational parameters, including the energy balance. The retrofit of the pilot unit involved the design, construction and installation of an oxygen supply system (OSS) and a CO_2 recycle system (CRS).

The test program, performed in 2011-2012, demonstrated steady operation in oxy-firing mode, achieving CO_2 concentrations up to 95% vol. in the flue gas. CO_2 has a much higher specific heat than nitrogen (N_2) and, as a consequence, more heat may be absorbed in the regenerator. This enables operation of the FCC unit with a higher feed throughput (up to 10% by volume) or with lower cost feedstocks (chapter 6). This potentially adds an economic benefit for the oxy-fired option to somewhat offset the cost of capture, as shown in the economic evaluation (chapter 7).

Among the lessons learned during the demonstration of oxy-firing is the need for particular care in the design of the flue gas recycle system to avoid corrosion in the recycle compressor. While the root cause for corrosion experienced during the tests is fully understood, an overall optimization of the recycle system would be advisable before proceeding to a large scale demonstration.

Pilot Testing: Burners for Process Heaters (Oxy-Firing)

While specific burners for oxy-combustion are under development by several vendors, in the case of a retrofit, the use of the existing burner system with some modifications, (if required), would be preferable. The cost for the retrofit might be reduced, and the option of operating in either air or oxygen combustion modes might be possible. CCP therefore, commissioned the John Zink Company to conduct oxy-fired testing on two of their conventional process heater burners – an SFG staged gas low NOx burner and a COOLstar® Ultra-Low NOx burner.

Figure 2. Furnace for oxy-combustion burner tests (courtesy of John Zink Co).

John Zink carried out single burner testing of both burners at their test facility in Tulsa, Oklahoma (Figure 2). Both burners were tested with natural gas and simulated Refinery Fuel Gas (RFG), which contains hydrogen. The two burners performed satisfactorily under oxy-firing conditions with no performance issues. Transition between air and oxygen firing was successfully performed several times. The level of air ingress into the process heater was higher than expected. As a result of this air ingress, achieving levels of N_2 in the flue gas of less than 10% vol. is likely to be a major challenge, adversely affecting downstream CO_2 purification costs. Oxy-firing remains a more viable option for systems with boilers, which operate at a positive pressure in the firebox. More details about this project may be found in Chapter 9.

Novel Technology: Membrane Water Gas Shift (Pre-Combustion)

Pre-combustion capture (conversion of hydrocarbon fuel to H_2 for use as fuel with CO_2 capture) has been identified by CCP as an interesting option for the refinery. For most existing refineries, there is little open plot space, and to implement post-combustion or oxy-firing, space for the capture unit would be required close to the many existing refinery units. In contrast, for pre-combustion, the capture unit can be remotely located, and the hydrogen fuel can be distributed to the refinery units. Refineries are already producing hydrogen (H_2) for use in hydro-treatment processes, so the technology for pre-combustion is already familiar.

During previous phases of CCP, several technologies have been supported for development to be applied in pre-combustion schemes (See Chapters 9-12 in [2]). Based on this, CCP3 supported further development of the Membrane-enhanced Water Gas Shift (MWGS) concept as a way to lower the cost of conventional pre-combustion capture. For the current state-of-the-art technology for large-volume H_2 production, which would utilize an oxygen-blown ATR for reforming, followed by WGS, the CO_2 would be captured downstream of the WGS unit by solvent absorption. To implement the MWGS concept, H_2 permeable membranes are used to move the shift reaction closer to completion with little residual CO. The H_2 would be permeated through the membrane, leaving a stream of highly concentrated CO_2 at relatively high pressure (20-30 bars). This scheme is not as attractive for power generation, since the H_2 fuel is at low pressure.

CCP commissioned the Pall Corporation to perform intensive testing of their palladium alloy membranes, first in the form of single 12 inch long tubes and later in modules composed of twelve of these tubes. These multi-tube modules could be the basic building block for scale-up to a larger pilot. Test results confirmed the technical feasibility of this scheme, with the membranes maintaining their performance over 1000 hours of testing and with no significant degradation recorded. A detailed description of the results of this project is reported in Chapter 10.

CO_2 Capture from Oil Sands/Heavy Oil Production

The in-situ extraction of bitumen from oil sands requires large quantities of steam for the Steam Assisted Gravity Drainage (SAGD) technique. A pair of horizontal wells is drilled into the reservoir, one a few meters above the other. Steam is continuously injected into the upper wellbore to heat the bitumen and reduce its viscosity, causing the heated bitumen to drain into the lower wellbore, where it is pumped out.

Once-through steam generators (OTSG) are the primary technology used in oil sands operation for producing this steam. These boilers use natural and/or associated gas, and are expected to be a major source of growth of GHG emissions in Canada, since upwards of 85% of the country's oil sands are suited to in-situ methods of extraction. Development of CO_2 capture technology suited for use with OTSGs could significantly reduce the GHG emissions from these operations, both for new build and by retrofit units.

CCP3 has worked both on capture from OTSG and on the development of novel technologies, such as Chemical Looping Combustion that could be deployed to generate steam for SAGD production.

Demonstration: OTSG (Oxy-Firing)

CCP and partners, Cenovus Energy, Devon Canada, MEG Energy, Praxair, and Statoil, executed a successful demonstration of oxy-fuel combustion technology to reduce CO_2 emissions from an OTSG. The project carried out a feasibility study to assess the cost of a retrofit for a typical commercial scale SAGD operation (4 OTSG boilers, each rated at 250 million BTU/hr). The boiler chosen for the demonstration is rated at 50 million BTU/hr and is operated by Cenovus at their Christina Lake facility (Figure 3).

Figure 3. OTSG boiler for demo run in Christina Lake (courtesy of Cenovus Energy Inc.).

Following the feasibility study the project moved to the demonstration phase, which has been partly funded by the Government of Alberta through the Climate Change and Emission Management Corporation. An oxygen storage and supply system and flue gas recycle system was designed and constructed as part of the retrofit. The test program of 15 days used the original burner designed for conventional air firing, with only slight modifications made for oxy-firing. The test program was carried out during 1H 2015. Results of feasibility study and a description of the preparation for the demonstration are reported in Chapter 14 and Chapter 15.

Novel Technology: Chemical Looping Combustion (Oxy-Firing)

Chemical Looping Combustion (CLC) is a variation on oxy-firing combustion technology, where a solid particle oxygen carrier is used to transfer oxygen from the combustion air to the fuel, thus avoiding direct contact between air and fuel. The oxygen carriers are metal oxide particles, and chemical looping is implemented by circulating the solids between two reactors for the purpose of transferring oxygen from an "air reactor" to a "fuel reactor." In the fuel reactor, the fuel reacts with the metal oxide generating CO_2 and Water (H_2O), and in the air reactor the reduced metal oxide is re-oxidized by oxygen contained in the air. CCP has focused on use of natural gas as the fuel.

During the first two phases of the CCP, CCP supported the formation of a Partnership that was co-funded by the European Community and moved the technology from the almost pure conceptual level (2000) to the development of a very active Ni-based oxygen carrier and operation of a 10 kW unit with continuous solid circulation (2003) and finally to operation of a 120 kW unit (2007) [1,2].

A CLC boiler system uses two vessels and, therefore, requires a larger plot space than does a conventional boiler. This could be strong constraint for installation in an existing refinery, in which CO_2 capture will mostly probably be a matter of retrofitting existing equipment. In contrast, the extraction of heavy oil is expanding, and new-built facilities will play a major role. In this case, plot space constraints are less likely. CLC could therefore become a practical solution for next generation capture technology in this application.

Development of novel oxygen carriers that are cheaper than Ni-based carriers was the main task of CCP3, and was undertaken through parallel projects (Chapters 17-18). For the Heavy Oil scenario, designs for the next scale demonstration unit (10 MW) and for commercial units of the size currently used for existing SAGD extraction operations, were completed to enable a thorough evaluation of the economic benefits of the technology.

CO_2 Capture from Natural Gas Power Generation

CO_2 capture is difficult for NGCC power generation, since the low concentration of CO_2 in the flue gas (4% vol.) and high excess oxygen (10+% vol.) penalizes post-combustion solvent capture due to larger equipment size and possibly higher energy consumption compared to coal-based power generation. The CCP2 carried out a thorough investigation of state-of-the-art and emerging pre-combustion technology in search of more cost effective alternatives to post-combustion. This effort with several technology providers, working within the EU-funded project CACHET, concluded that for natural gas fired power generation, pre-combustion cannot challenge post-combustion due to lower efficiency and high capital cost [2]. Evaluations in CCP3 have confirmed that commercial state-of-the-art post-combustion technology already tested at a relatively large scale (though still one order of magnitude smaller than needed for a 400 MW power station) would cause a drop in efficiency of approximately 12% (from 57% to about 50%). The CCP3 consequently focused efforts on improving post-combustion technology, through testing of novel solvents, including non-aqueous solvents and enzyme-accelerated solvents.

Novel Technology: Testing of Novel Solvents at EERC (Post-Combustion)

As part of this process the CCP joined the PCO2C Partnership, a program co-funded by the US Department of Energy and run by the Energy and Environment Research Center (EERC) of the University of North Dakota. The main focus of PCO2C is pilot scale testing of novel post-combustion technologies, including novel solvents characterized by low energy consumption and/or enhanced kinetics, multiphase systems, solid sorbents, high efficiency packings and novel contactors applying process intensification principles (Figure 4) [4]. These results were used in combination with independent work by CCP on other projects, to evaluate specific technologies. An advanced solvent from ION Engineering has been identified as a promising technology. The CCP3 also undertook an economic evaluation and comparison against a current state-of-the-art technology to assess the potential of emerging solvent technologies. (See Chapters 21 and 22)

Figure 4. The pilot test unit for novel solvents at EERC (Courtesy of EERC).

Technical Studies and Economic Evaluations

An assessment of the performance and economics of state-of-the-art post-combustion, oxy-firing and pre-combustion capture technologies, applied to refining operations, steam production for heavy oil extraction and natural gas power generation, has been carried out. Amec Foster Wheeler (AFW) was commissioned to conduct technical evaluation studies, while an in-house economic model was developed, which included capture, transportation, and storage costs. The objective of these studies was to provide material for a CO_2 capture handbook to be used to support CO_2 capture technology selection in the oil and gas industry.

For each scenario the studies by AFW included a post-combustion baseline, using a state-of-the art commercial technology from MHI, and the most relevant alternative and novel technologies. All of the technologies were compared based on identical plant of the same size and application, and using the same assumptions. The methodology of technical studies is reported in Chapter 3.

The results were further expanded by the CCP using an in-house methodology, to develop consistent economic assessments of CO_2 capture and avoidance costs. The methodology for economic evaluations is reported in Chapter 4.

The combined technical and economic results for each scenario are reported in the dedicated chapters for each scenario (Chapters 7, 11, 20 and 22).

REFERENCES

1. Thomas, D.C. (editor), (2005) *Carbon Dioxide Capture for Storage in Deep Geologic Formations – Results from the CO_2 Capture Project, Vol 1 Capture and Separation of Carbon Dioxide From Combustion Sources*; ISBN 0-08-044570-5, Elsevier Publishing, UK.
2. Eide, L.I. (editor), (2009) *Carbon Dioxide Capture for Storage in Deep Geologic Formations – Results from the CO_2 Capture Project,* Vol 3 *Advances in CO_2 Capture and Storage Technology - Results (2004-2009)* ; ISBN 0-08-044570-5, CPL Press, UK.
3. De Mello, L.F., Moure, G.T., Pravia, O.R.C., Gearhart, L., and Milios, P.B., "Oxy-Combustion for CO_2 Capture from Fluid Catalytic Crakers," Chapter 3 in Eide, L.I. (editor), (2009) *Carbon Dioxide Capture for Storage in Deep Geologic Formations – Results from the CO_2 Capture Project,* Vol 3 *Advances in CO_2 Capture and Storage Technology - Results (2004-2009)* ; ISBN 0-08-044570-5, CPL Press, UK.
4. Brandon Pavlish et al., Partnership for CO_2 Capture – Phase II, Final Report, 2013-EERC-05-03 (May 2013).

Carbon Dioxide Capture for Storage in Deep Geological Formations, Volume 4
Karl F. Gerdes (Editor)

Chapter 3

TECHNICAL STUDIES METHODOLOGY

Mark Crombie[1], Richard Beavis[1], Jonathan Forsyth[1], Tony Tarrant[2], Tim Bullen[2] and Tim Abbott[2]
[1]BP International Limited, ICBT Chertsey Road, Sunbury-on-Thames, Middlesex, TW16 7LN, United Kingdom
[2]Amec Foster Wheeler, Shinfield Park, Reading, Berkshire, RG2 9FW, United Kingdom

ABSTRACT: The members of the CO_2 Capture Project (CCP) have collaborated to produce a series of technical studies of CO_2 capture from various processes which are representative of the main emissions sources from member company operations. The studies involved typical energy-intensive oil production, refining and power generation processes with CO_2 capture by post-combustion, oxy-fuel and pre-combustion technologies. The methodology that was followed is explained along with the main assumptions that were adopted and the key results and conclusions that were reached.

KEYWORDS: CO_2 Capture Project; CCP; CO_2 capture; post-combustion; oxy-fuel; pre-combustion

INTRODUCTION

The CO_2 Capture Project (CCP) is a partnership of six of the world's leading energy companies (BP, Shell, Eni, Petrobras, Chevron and Suncor), undertaking research and developing technologies to help make CO_2 capture and geological storage (CCS) a practical reality for reducing global CO_2 emissions. In the last few years CCP has had an increased focus on how to capture CO_2 using the best technology available now or in the relatively near term. To support this focus a series of technical studies was performed to evaluate the technical and economic performance of CO_2 capture applied to typical industrial processes. This paper describes the scope of these studies and the objectives that were set by CCP. The methodology that was followed is explained along with the main assumptions that were adopted and the key results and conclusions that were reached.

OBJECTIVES OF THE TECHNICAL STUDIES

The purpose of performing the technical studies was to:

1. Create a technical definition for the reference cases where it was needed.
2. Provide a quantified performance assessment for the state-of-the-art CO_2 capture technology and to establish the CO_2 capture base line for each application scenario.
3. Provide a quantified performance assessment for alternative CO_2 capture methods and emerging novel technology approaches.
4. Develop insight into the performance of state-of-the-art and emerging technologies to help CCP to prioritise and guide technology development activity.
5. Deliver a package of CO_2 capture technology application knowledge to CCP members to support evaluation of CO_2 capture from their operations.

SCOPE OF THE TECHNICAL STUDIES

CCP studied CO_2 capture from various processes which were representative of the main emissions sources from member company operations. Ultimately five application scenarios were considered:

1. Refinery Fluidised Catalytic Cracker (FCC)
2. Refinery Fired Heaters
3. Oil Sands Once Through Steam Generators (OTSGs)
4. Gas-Fired Power Generation – Natural Gas-Fired Combined Cycle (NGCC)
5. Refinery Hydrogen Plant – Steam Methane Reformer (SMR)

The CCP studies were conducted in four sequential phases: Phase 1 considered post-combustion CO_2 capture applied to each of the application scenarios; Phase 2 considered mainly oxy-fuel applications; Phase 3 mainly pre-combustion CO_2 capture and Phase 4 considered mainly new emerging technologies. This series of technical studies was distinct from the application studies undertaken in CCP in earlier years which have been previously reported in CCP results books in 2005: Carbon Dioxide Capture for Storage in Deep Geological Formations, Volume 1 and Volume 2 [1,2], and in 2009: Carbon Dioxide Capture for Storage in Deep Geological Formations, Volume 3 [3].

For each application scenario, a series of technical and quantified performance and implementation evaluations, and cost estimates were produced (each referred to as a "Case"). The technical configuration of the application scenario without CO_2 capture was referred to as the "Reference case". Post-combustion CO_2 capture by aqueous amine scrubbing from an established technology licensor was applied to the reference case to create a benchmark for CO_2 capture performance referred to as the "Base case". Alternative CO_2 capture methods or more advanced concepts formed the basis of further cases which were compared to the base case for each application scenario.

All cases taken together build into a comprehensive picture of CO_2 capture from relevant industrial processes including both new build and retrofit of the CO_2 capture plant by established and alternative advanced concepts.

Refinery Fluidised Catalytic Cracker (FCC) Cases

- ***Phase 1, Case 3 – Refinery FCC Base Case – Post-Combustion CO_2 Capture***
  Exhaust gas from a 60,000 BPSD FCC is routed to post-combustion CO_2 capture. Capture is by a single train of CO_2 absorption, stripping, dehydration and compression.
- ***Phase 2, Case 2 – Refinery FCC Oxy-fuel Retrofit***
  The 60,000 BPSD FCC in Phase 1 Case 3 is retrofitted with oxy-fuel CO_2 capture. 97% purity oxygen and recycled CO_2 is supplied to the FCC regenerator, a stream of CO_2 is dehydrated, compressed and purified in a cryogenic processing unit.
- ***Phase 3, Case 4 – Refinery FCC Oxy-fuel Retrofit with Alternative O_2 Purity and CO_2 Purification***
  The 60,000 BPSD FCC with retrofitted oxy-fuel CO_2 capture in Phase 2 Case 2 is re-evaluated with increased oxygen purity (99.5%) and thermal catalytic CO_2 purification.

Refinery Fired Heaters Cases

- ***Phase 1, Case 1 – Refinery Fired Heater Base Case - Post-Combustion CO_2 Capture***
  Exhaust gas from 3x100 MMBTU/h refinery heaters fired on refinery fuel gas is collected into a combined stream for post-combustion CO_2 capture. Capture is by a single train of CO_2 absorption, stripping, compression and dehydration.

- ***Phase 1, Case 2 – Refinery Fired Heater Upsize Base Case - Post-Combustion CO_2 Capture***
  Exhaust gas from 4x150 MMBTU/h refinery heaters fired on refinery fuel gas is collected into a combined stream for post-combustion CO_2 capture. Capture is by a single train of CO_2 absorption, stripping, compression and dehydration.
- ***Phase 2, Case 1 – Refinery Fired Heater Oxy-fuel Retrofit***
  The 4x150 MMBTU/h refinery heaters in Phase 1 Case 2 are retrofitted with oxy-fuel CO_2 capture. 97% purity oxygen and recycled CO_2 is supplied to the heater furnace, a stream of CO_2 is dehydrated, compressed and purified in a cryogenic processing unit.
- ***Phase 2, Case 3 – Refinery Fired Heaters Pre-Combustion CO_2 Capture Retrofit with aMDEA***
  Refinery fuel gas and natural gas is converted into a hydrogen rich fuel (5,000 MMBTU/h) for refinery heaters and boilers. Note that all the pre-combustion cases are at the scale of refinery-wide fuel decarbonisation. The overall process scheme for this case involves autothermal reforming of refinery fuel gas with supplementary natural gas. The process comprises refinery fuel gas compression, sulphur removal, autothermal reforming, water gas shift, acid gas removal by amine absorption, CO_2 compression and dehydration and hydrogen product expansion power recovery
- ***Phase 3, Case 2 – Refinery Fired Heaters Pre-Combustion CO_2 Capture Retrofit with Pressure Swing Adsorption (PSA) and Thermal Oxidiser***
  The 5,000 MMBTU/h hydrogen-rich fuel plant in Phase 2 Case 3 is re-evaluated with hydrogen purification/acid gas removal by Pressure Swing Adsorption (PSA) and processing of the PSA tail gas stream in a thermal oxidiser with recovery of waste heat.
- ***Phase 3, Case 2A – Refinery Fired Heaters Pre-Combustion CO_2 Capture Retrofit with PSA and Cryogenic Purification Unit***
  The 5,000 MMBTU/h hydrogen-rich fuel plant in Phase 2 Case 3 is re-evaluated with hydrogen purification/acid gas removal by processing of the PSA tail gas stream in a cryogenic processing unit.
- ***Phase 4, Case 2.1 – Refinery Fired Heaters Pre-Combustion CO_2 Capture Retrofit with H_2 Permeable Membrane with Cryogenic CO_2 Purification Unit***
  The 5,000 MMBTU/h hydrogen-rich fuel plant in Phase 2 Case 3 is re-evaluated with hydrogen purification/acid gas removal by high temperature, hydrogen permeable metal membrane and processing of the retentate gas stream in a cryogenic purification unit.
- ***Phase 4, Case 2.2 – Refinery Fired Heaters Pre-Combustion CO_2 Capture Retrofit with H_2 Permeable Membrane with Cryogenic CO_2 Purification Unit and PSA***
  The 5,000 MMBTU/h hydrogen-rich fuel plant in Phase 2 Case 3 is re-evaluated with hydrogen purification/acid gas removal by high temperature, hydrogen permeable metal membrane and processing of the retentate gas stream in a cryogenic purification unit and PSA unit.
- ***Phase 4, Case 2.3 – Refinery Fired Heaters Pre-Combustion CO_2 Capture Retrofit with Cryogenic CO_2 Separation Unit with aMDEA Scrubbing***
  The 5,000 MMBTU/h hydrogen-rich fuel plant in Phase 2 Case 3 is re-evaluated with acid gas removal by cryogenic CO_2 separation unit followed by aMDEA amine absorption.

Refinery Hydrogen Plant – Steam Methane Reformer Cases

- ***Phase 1, Case 4 – Refinery SMR with Post-Combustion CO_2 Capture***
  Exhaust gas from a modern steam methane reformer producing 50,000 Nm^3/h of hydrogen is routed to CO_2 capture. This SMR uses PSA to purify H_2 and routes the tail gas to the furnace to utilize the heating value. Capture is by a single train of CO_2 absorption, stripping, compression and dehydration.

Oil Sands Once Through Steam Generators (OTSGs) Cases

- ***Phase 4, Case 3.1 – Oil Sands OTSGs Uncontrolled Reference Case***
  4x250MMBTU/h OTSGs for steam assisted gravity drainage in the Canadian oil sands with no CO_2 capture.
- ***Phase 1, Case 6 – Oil Sands OTSGs Base Case - Post-Combustion CO_2 Capture***
  4x250MMBTU/h OTSGs for steam assisted gravity drainage in the Canadian oil sands with post-combustion CO_2 capture. Capture is by a single train of CO_2 absorption, stripping, dehydration and compression.
- ***Phase 4, Case 3.3 – Oil Sands OTSGs Base Case - Higher Rate Post-Combustion CO_2 Capture***
  4x250MMBTU/h OTSGs with post-combustion CO_2 capture in Phase 1 Case 6 is re-evaluated with reduced flue gas by-pass and capture from auxiliary boiler exhaust to give higher capture rate. Capture is by a single train of CO_2 absorption, stripping, dehydration and compression.
- ***Phase 2, Case 4 – Oil Sands OTSGs Oxy-fuel New Build***
  The 4x250MMBTU/h OTSGs in Phase 4 Case 3.1 are evaluated as new build with oxy-fuel CO_2 capture. 97% purity oxygen and recycled CO_2 is supplied to the OTSG furnaces, a stream of CO_2 is dehydrated, compressed and purified in a cryogenic processing unit.
- ***Phase 2, Case 5 – Oil Sands OTSGs Oxy-fuel Retrofit***
  The 4x250MMBTU/h OTSGs in Phase 4 Case 3.1 are evaluated as retrofitted with oxy-fuel CO_2 capture. 97% purity oxygen and recycled CO_2 is supplied to the OTSG furnaces, a stream of CO_2 is dehydrated, compressed and purified in a cryogenic processing unit.
- ***Phase 4, Case 3.2 – Oil Sands OTSGs CO_2 Capture by Chemical Looping Combustion (CLC)***
  The 4x250MMBTU/h OTSGs with CO_2 capture by CLC. Circulating oxygen-carrying particles are alternately oxidised in an air reactor and then reduced in a fuel reactor to produce a CO_2-rich gas and oxygen-depleted air exhaust. The CO_2-rich gas stream is treated to increase CO_2 purity before dehydration and compression.
- ***Phase 3, Case 3 – Oil Sands OTSGs Pre-Combustion CO_2 Capture with aMDEA***
  Natural gas is converted into a hydrogen rich fuel for 4x250MMBTU/h oil sands OTSGs. The overall process scheme for this case involves autothermal reforming of natural gas, comprising natural gas compression, sulphur removal, autothermal reforming, water gas shift, acid gas removal by amine absorption, CO_2 compression and dehydration and hydrogen product expansion power recovery.

Gas-Fired Power Generation – Natural Gas-Fired Combined Cycle Cases

- ***Phase 1, Case 5A – NGCC Uncontrolled Reference Case***
  A single natural gas fired GE Frame 9FA gas turbine generates power for export and the hot exhaust gas is fed to a heat recovery steam generator (HRSG). The steam produced from the HRSG is fed to a single steam turbine which also generates power for export. Following heat recovery in the HRSG, the flue gas is emitted to atmosphere via a downstream stack.
- ***Phase 1, Case 5 – NGCC Base Case - Post-Combustion CO_2 Capture***
  The single train NGCC from Phase 1, Case 5A is integrated with a CO_2 capture unit. Flue gas from the HRSG is routed to the CO_2 capture unit in a nearby location of suitable size. It is assumed that 5% of the total flue gas flow bypasses the CO_2 capture plant via the stack. The CO_2 capture unit removes 90% of the CO_2 in the flue gas it receives, with the cleaned flue gas returning to the stack. Capture is by a single train of CO_2 absorption, stripping, dehydration and compression.
- ***Phase 4, Case 4.1 – NGCC Post-Combustion CO_2 Capture by Enzyme-Promoted Absorption***
  This case re-evaluates Phase 1 Case 5 post-combustion CO_2 capture applied to the natural gas combined cycle plant using an enzyme promoted absorption process. Capture is by a single train of CO_2 absorption, stripping, dehydration and compression.

- ***Phase 4, Case 4.2 – NGCC Post-Combustion CO_2 Capture by Novel Solvent Absorption***
  This case re-evaluates Phase 1 Case 5 post-combustion CO_2 capture applied to the natural gas combined cycle plant using a novel solvent absorption process. Capture is by a single train of CO_2 absorption, stripping, dehydration and compression.

Each case is described in more detail in the chapters of this volume as follows:

- Refinery FCC cases – Chapter 5
- Refinery fired heaters cases – Chapter 8
- Refinery hydrogen plan SMR cases – Chapter 12
- Oil sands OTSG cases – Chapter 13
- Gas-fired power generation NGCC cases – Chapter 21

METHODOLOGY USED TO DELIVER THE TECHNICAL STUDIES

A technical basis was created for each application scenario, and this was applied consistently for all of the cases through the sequential phases of the study to ensure that the different cases studied for each application scenario were comparable. Whilst not specific to any asset, the CCP members made sure that the application scenarios were representative of real operations.

An engineering contractor, Amec Foster Wheeler (AFW), was brought in to create process descriptions, designs, technical performance models, cost estimates and execution plans for each case. AFW sought and obtained input from technology providers and equipment suppliers as follows:

- Mitsubishi Heavy Industries (MHI) provided state-of-the-art, solvent-based post-combustion CO_2 capture input
- Praxair provided oxy-fuel input
- Air Liquide provided Air Separation Unit (ASU) input
- UOP provided Pressure Swing Adsorption (PSA) input
- Pall provided input on use of H_2-permeable metal membranes for H_2 purification and CO_2 capture
- Bertsch Energy and Vienna University of Technology provided input on Chemical Looping Combustion (CLC) technology
- MAN Turbo provided input on CO_2 compressors
- ABB Randall Gas Technologies provided input on the FCC base case.

Deliverables

The deliverables for each case studied included:

1. Outline basis of design – scale, capture level, utility availability and conditions, boundary conditions, meteorological conditions
2. Process description
3. Process flow diagrams
4. Heat and mass balance
5. Utility requirements
6. Chemical requirements
7. Emissions and effluent summary
8. Sized equipment list (with cost breakdown),
9. Typical plot plan
10. Costs - Capital and operating costs (based upon CCP assumptions) +/- 30% accuracy

11. Identification of specific application issues (e.g. limited plot size)
12. Highlight issues associated with retrofit
13. Scale: a) Limitations of single trains, b) Capital cost curve for each technology and scenario (capex as a function of selected input variables/specifications), c) Fully built up cost curves (Cost of CO_2 derived from CCP assumed costs)
14. Project Management: a) Main practical implementation issues (Constructability, logistics), b) Health, Safety and Environment (HSE) considerations and safety issues associated with technology choice,
    c) Permitting and environmental issues, d) Project execution timescale (typical schedule), e) Identification of long lead items.

Basis of Design

A basis of design was developed in order to ensure that clearly defined, consistent basis and input data were applied to each case and group of cases. The basis of design provided the flue gas characteristics which provided the initial reference for the design of the post-combustion cases. The only exceptions were new build NGCC cases, for which the fuel gas and air compositions supplied by CCP were used by AFW to perform an outline design of the combined cycle plant and calculate the flue gas characteristics. For the oxy-fuel and pre-combustion cases CCP provided design data to enable AFW to calculate the heat and material balances of the capture plant.

Technical Execution Model

AFW's execution model for undertaking the CCP technical studies was based on their experience in the complete lifecycle of engineering projects from concept through Front End Engineering Design (FEED) to engineering, procurement and construction/commissioning phases. This level of project engagement creates a recycle loop that maintains up-to-date project technical, cost and schedule data from all phases through to construction. This was particularly important when importing knowledge into the studies.

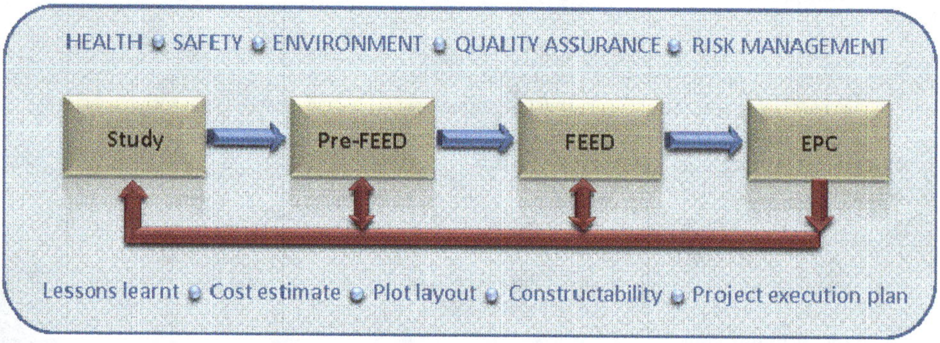

This capability gave assurance that technical content, costs and schedules generated were credible and benchmarked against current engineering and construction market information available in-house to AFW.

Process Definition

Through the sequential phases of the technical studies the basis of design was maintained and developed as needed to cover any new requirement of the technology being studied.

Following formal definition of the design basis for each phase, the development of process descriptions, process flow diagrams, process simulations, and heat and material balance calculations proceeded.

The utility and chemical requirements, emissions and effluents, variable operating costs and sized equipment list were calculated based on the heat and material balances. The capital cost and layout were then developed based on the sized equipment list, also allowing the fixed operating costs to be determined. This information was then used to develop a capital cost curve.

Special considerations were identified and discussed in the study reports with respect to retrofit and scale up and down, including currently understood limitations on maximum train size for each stage of the capture process. Practical issues such as constructability and logistics as well as permitting and HSE considerations were outlined for each case. An approximation of the timescale involved in project implementation was evaluated along with associated identification of long lead items.

Capital Cost Estimate

Through current and recent projects undertaken, AFW had access to the current market conditions for global equipment procurement and labour costs, including those specific to US Gulf Coast and Fort McMurray locations, and these were reflected within their in-house cost database. All cost estimates included AFW's best assessment of these markets for the stated estimate validity.

AFW prepared capital cost estimates for each case based on the technical definition developed for each option to an estimate accuracy of +/- 30%. The cost estimate included assessment of equipment, materials, labour, engineering, procurement, installation / construction and pre-commissioning of the new facility on an instantaneous basis. Excluded from the AFW estimates are forward escalation, Outside Battery Limits (OSBL) costs, contingencies, owners costs, land lease and taxes / duties, which were all estimated by CCP. The capital cost estimates were complemented by an operating cost estimate that was based on AFW's in-house experience for each of the cases. The operating cost estimate included feed-stocks, utilities, catalysts and chemicals, direct labour and maintenance. The operating cost estimate excluded business rates, property costs, land costs, insurance and leases and waste disposal.

The capital cost estimates were largely based on AFW "Indexed" Aspen Capital Cost Estimator ("ACCE") computer programme. The ACCE programme includes "pre-defined" P&ID models, and was used to generate the base equipment & bulk material costs and direct labour man-hours. The prime inputs to the cost estimate were the process definition, sized equipment lists, site plan and overall execution strategy. The AFW benchmarked ACCE output was checked against in-house costs and statistical data from a variety of sources. AFW has an effective feedback and benchmarking system which ensures that the cost database reflects current information on all aspects of the cost estimate. The cost estimates were fully benchmarked against other similar CCS studies and projects which AFW had undertaken.

All capital cost estimates were based on a stick-build construction methodology rather than modular pre-assembled units.

Plot Plan

AFW developed an equipment level layout and plot area definition for each of the cases based on the sized equipment lists and process flow diagrams generated.

The work was based on desktop study and did not include any site visits. The following activities were undertaken for each case:

- Layout basis and methodology
- Layout narrative
- Definition of plot area requirements
- Equipment layout requirements.

For the retrofit cases, CCP provided details of the available plot areas and constraints.

Construction

Construction input was also provided to assist with the development of the equipment layouts. A section of the study report outlined key opportunities and issues related to constructability and construction logistics.

Overall Implementation Plan

AFW developed high level overall implementation plans. These included an overall schedule and associated narrative with consideration of the relevant option specifics including FEED, engineering, procurement and construction elements.

Procurement

AFW undertook an assessment of long lead delivery equipment and used this assessment as input to the overall execution plans which they prepared. The assessment was made from the equipment details contained within the sized equipment lists, and AFW's database of equipment delivery times for their current projects and a subjective assessment of future market movements. AFW did not make any specific equipment enquiries to the marketplace.

AFW considered the health, safety and environmental aspects of each option. Key risk areas, including HSE issues were highlighted in the final reports.

BASIS OF THE TECHNICAL STUDIES

The key assumptions were agreed upon between CCP members and consistently applied to all of the cases in the study. These assumptions included the capacity of the reference case, CO_2 capture rate and CO_2 specification, fuels specifications, utilities and the general site conditions and constraints.

This section summarises a set of key technical data and assumptions serving as the basis for the technical studies:

- Cost basis
- Scenario ambient conditions
- Fuel gas supply, quality, composition, and specifications
- CO_2 delivery pressure and purity
- Utility information

Cost Basis

The cost basis for the refinery retrofit cases was at a generic US Gulf Coast brown field site location.

The cost basis for the oil sands steam generation case was at a generic Fort McMurray, Alberta Canada brown field site location.

For the oil sands steam generation cases, capital cost estimate sensitivities were also provided on a generic US Gulf Coast location. Each of these estimates used the same technical basis as the corresponding Fort McMurray scenario. The site was assumed to be a green-field location at Fort McMurray, Alberta Canada. A clear, level, obstruction-free site (both under and above ground), without the need for any required special civil works was assumed.

The cost basis for the Natural Gas Fired Power Generation case was at a generic US Gulf Coast brown field site location.

All of the cases were estimated using an end of 2009 basis.

Climatic Data

		US Gulf Coast	Fort McMurray, Canada
Temperatures Dry Bulb (°C)	Average (design)	20*	1*
	Minimum	-10	-30
	Maximum	40	30
Barometric Pressure (mmHg)	Estimated Mean	760*	727*
Humidity (%)	Estimated Mean	60*	57*
	Minimum	10	20
	Maximum	100	100
Elevation (m)	Altitude	10*	369*

The climatic conditions marked (*) were considered reference conditions for plant performance evaluation.

Air Composition

The following air composition was used for all scenarios:

	Composition (mole fraction)
CO_2	0.00033
N_2	0.77288
O_2	0.20733
Ar	0.00924
H_2O	0.01022
Total	1.00000

Natural Gas Feeds

The following natural gas feed composition and battery limits conditions were used for the make-up natural gas composition for the refinery heater scenarios and for the primary natural gas conditions for the natural gas fired power generation cases:

	Composition (vol%)
CH_4	79.76
C_2H_6	9.68
C_3H_8	4.45
$i\text{-}C_4H_{10}$	0.73
$n\text{-}C_4H_{10}$	1.23
$i\text{-}C_5H_{12}$	0.21
$n\text{-}C_5H_{12}$	0.20
C_6H_{14}	0.21
CO_2	2.92
N_2	0.61
H_2S	5 ppmv

Pressure: 31 bara
Temperature: 16°C
Molecular Weight: 20.74 (kg/kmol);
LHV: 42.06 MJ/Nm^3 (39.88 MJ/Sm^3 =11.08 kWh/Sm^3, = 45,452 kJ/kg)

The following natural gas feed composition and battery limits conditions were used for the OTSG cases:

	Composition (vol %)
CH_4	98.87
C_2H_6	0.04
C_3H_8	0.01
N_2	0.69
CO_2	0.36
C_4H_{10}	0.01
C_6H_{14}	0.01

Pressure: 4 bara;
Temperature: ambient;
HHV: 1000.2 BTU/SCF

Refinery Fuel Gas Feed

The following refinery fuel gas feed composition and battery limits conditions were used:

	Composition (vol %)
CH_4	69.7
C_2H_6	0.9
C_2H_4	0.1
H_2	29.1
N_2	1.23

Pressure: 2.5 bara;
Temperature: 20°C;

CO_2 Capture and Quality

In all cases CO_2 was purified and compressed for end delivery to either pure storage or EOR customers. The table below lists key characteristics for the delivered CO_2:

Minimum CO_2 (mol.%)	97
Pressure (barg)	150
Temperature (°C)	40
Moisture Content (ppmv. max)	50
SOx (ppm)	<75.0[1]
NOx	No spec
Inerts (mol%)	<4.0
Oxygen (ppm)	<75.0[2]

[1]Praxair report on OTSG oxy-firing: SOx <50ppm using distillation for CO_2 purification.
[2]Praxair report on OTSG oxy-firing: Oxygen <50ppm using distillation for CO_2 purification.

Carbon Capture Rate was typically designed to achieve a CO_2 capture rate of 90%, defined as:

$$CO_2 \text{ capture rate } (\%) = \frac{100 \text{ x Moles of carbon contained in the } CO_2 \text{ product}}{\text{Moles of carbon in flue gas feed to } CO_2 \text{ processing unit}}$$

The refinery fired heaters retrofit cases were typically designed to achieve a CO_2 capture rate of 90% defined as:

$$CO_2 \text{ capture rate } (\%) = \frac{100 \text{ x Moles of carbon contained in the } CO_2 \text{ product}}{\text{Moles of carbon in the feed (RFG + NG)}}$$

Cooling Systems

Cooling water was available at the following conditions:

		US Gulf Coast	Fort McMurray, Canada
Sea water	Supply Temperature (°C)	11	Air Cooled
	Return Temperature (°C)	22	
	Pressure (barg)	3.5	
Cooling water	Supply Temperature (°C)	32	
	Return Temperature (°C)	43	
	Pressure (barg)	3.5	

A pressure drop of 1.5 bar was assumed through heat exchangers and piping.

Steam Systems

Typical steam pressure levels (superheated) in NGCC plants:

- LP: 5 bara, 290°C
- MP: 30 bara, 320°C
- HP: 124 bara, 550°C

Typical steam pressure levels in a refinery:

- LP: 4.5 bara, (20°C superheat)
- MP: 11 bara, (20°C superheat)
- HP: 42 bara, (20°C superheat)

Where significant quantities of steam were generated as a result of the CO_2 capture process for a particular case, a steam turbine system was included within the scope to generate power and minimise the quantity of steam crossing the scenario battery limits.

Demineralised Water Supply

Demineralised water was supplied at 15.6°C and 7.9 bara.

Compressed Air

Both instrument and plant air was distributed at a pressure of 9 bara at consumers battery limit (at a water dew point of -40°C).

Inert Gas

Nitrogen was assumed available be supplied at up to 14 bara and up to 99.5% purity.

Fresh and Potable Water

Fresh water was distributed for general use at hose stations as well as for steam system boiler feed water make-up (after treatment). Fresh water source was dedicated for the power plant.

Potable water was distributed from the public distribution network for sanitary use at the plant, including any emergency showers.

Effluent Treatment

Any oily water effluents were collected. Oil was separated through gravity and flotation before further treatment for pH-control and particulate removal.

Power Grid

Open access to an external power grid (300 kV, 60 Hz) with sufficient capacity for exporting any plant output power was assumed. All power required by the CO_2 capture processes was imported from the grid except for the NGCC cases.

Transport Criteria for Fort McMurray, Alberta Canada Scenarios

Shipping envelope guidelines for OSTGs as provided by CCP:

Location		Dimensions (ft)	Weight (tonnes)
Ex Edmonton	The largest size and weight that can be achieved for loads destined to Fort MM (without specific transport engineering feasibility review)	L: 120 W: 32 H: 32	400
	A more routine envelope	L: 100 W: 24 H: 24	150
Saskatoon	Constraint is the bridges that must be crossed to get back into AB	L: 100 W: 20 H: 20	45 (winter) 65 (rest of year)
Great Lakes	Via Port of Duluth then over the road	L: 100 H: 15 H: 14	90
Gulf Coast		L: 80 W: 20 H: 14	65 to 75
Rail Shipments	From either Coast and including Gulf	L: 80 W: 12 H: 12	80

KEY PERFORMANCE AND COST RESULTS OF THE TECHNICAL STUDIES

The technical studies were provided by AFW to CCP members as four stand-alone reports covering each sequential phase of the work. The key results are summarised in the following tables:

CCP3 Results Summary		Application Scenario 1 - Refinery FCC			Application Scenario 2 - Refinery Fired Heaters		
Phase		1	2	3	1	1	2
Case		3	2	4	1	2	1
Title		Refinery FCC Base Case Post-Combustion CO_2 Capture	Refinery FCC Oxyfuel Retrofit	Refinery FCC Oxyfuel Retrofit with Alternative O_2 Purity and CO_2 Purification	Refinery Fired Heater Base Case Post-Combustion CO_2 Capture	Refinery Fired Heater Upsize Base Case Post-Combustion CO_2 Capture	Refinery Fired Heater Oxyfuel Retrofit
Location		US Gulf Coast	US Gulf Coast	US Gulf Coast	US Gulf Coast	US Gulf Coast	US Gulf Coast
Capacity		60,000 BPSD	60,000 BPSD	60,000 BPSD	3 x 100 MMBTU/h process heating	4 x 150 MMBTU/h process heating	4 x 150 MMBTU/h process heating
Fuel type:							
FCC catalyst		✗	✓	✓	✗	✗	✗
Refinery fuel gas		✓	✗	✗	✓	✓	✓
Natural gas		✗	✗	✗	✗	✗	✗
Fuel consumption (LHV)	GJ/h	166.1	3.5	2.8	40.6	81.2	3.1
CO_2 capture technique:							
Post-combustion		✓	✗	✗	✓	✓	✗
Oxyfuel		✗	✓	✓	✗	✗	✓
Pre-combustion		✗	✗	✗	✗	✗	✗
CO_2 captured	t/h	55.5	58.3	64.8	13.3	26.6	31.5
Total installed capital cost	M$	67.9	145.8	128.6	32.3	45.7	133.5
Specific TIC	M$/t/h	1.2	2.5	2.0	2.4	1.7	4.2
Total annual operating cost	M$/y	21.6	30.2	30.8	7.3	11.8	23.6

CCP3 Results Summary		Application Scenario 2 - Refinery Fired Heaters					
Phase		2	3	3	4	4	4
Case		3	2	2A	2.1	2.2	2.3
Title		Refinery Fired Heaters Pre-Combustion CO_2 Capture Retrofit with aMDEA	Refinery Fired Heaters Pre-Combustion CO_2 Capture Retrofit with PSA and thermal oxidiser	Refinery Fired Heaters Pre-Combustion CO_2 Capture Retrofit with PSA and cryogenic purification unit	Refinery Fired Heaters Pre-Combustion CO_2 Capture Retrofit with H_2 Permeable Membrane with Cryogenic CO_2 Purification Unit	Refinery Fired Heaters Pre-Combustion CO_2 Capture Retrofit with H_2 Permeable Membrane with Cryogenic CO_2 Purification Unit and PSA	Refinery Fired Heaters Pre-Combustion CO_2 Capture Retrofit with Cryogenic CO_2 Separation Unit with aMDEA Scrubbing
Location		US Gulf Coast	US Gulf Coast	US Gulf Coast	US Gulf Coast	US Gulf Coast	US Gulf Coast
Capacity		5,000 MMBTU/h	5,000 MMBTU/h	5,000 MMBTU/h	5,000 MMBTU/h	5,000 MMBTU/h	5,000 MMBTU/h
Fuel type:							
FCC catalyst		×	×	×	×	×	×
Refinery fuel gas		✓	✓	✓	✓	✓	✓
Natural gas		✓	✓	✓	✓	✓	✓
Fuel consumption (LHV)	GJ/h	6261.0	7116.0	6379.0	6410.0	6378.0	6320.0
CO_2 capture technique:							
Post-combustion		×	×	×	×	×	×
Oxyfuel		×	×	×	×	×	×
Pre-combustion		✓	✓	✓	✓	✓	✓
CO_2 captured	t/h	284.0	330.5	291.1	300.5	304.6	293.9
Total installed capital cost	M$	528.6	795.9	646.4	2818.6	2869.0	500.2
Specific TIC	M$/t/h	1.9	2.4	2.2	9.4	9.4	1.7
Total annual operating cost	M$/y	398.5	379.6	375.0	1497.0	1512.3	394.5

CCP3 Results Summary		Application Scenario 3 - Oil Sands Steam Generation Scenario					
Phase		4	1	4	2	2	4
Case		3.1	6	3.3	4	5	3.2
Title		Oil Sands OTSGs Uncontrolled Reference Case	Oil Sands OTSGs Base Case Post-Combustion CO_2 Capture	Oil Sands OTSGs Base Case Higher Rate Post-Combustion CO_2 Capture	Oil Sands OTSGs Oxyfuel New Build	Oil Sands OTSGs Oxyfuel Retrofit	Oil Sands OTSGs CO_2 Capture by CLC
Location		Fort McMurray	Fort McMurray	Fort McMurray	Fort McMurray	Fort McMurray	Fort McMurray
Capacity		4 x 250 MMBTU/h boilers	4 x 250 MMBTU/h boilers	4 x 250 MMBTU/h boilers	4 x 250 MMBTU/h boilers	4 x 250 MMBTU/h boilers	4 x 250 MMBTU/h boilers
Fuel type:							
FCC catalyst		×	×	×	×	×	×
Refinery fuel gas		×	×	×	×	×	×
Natural gas		✓	✓	✓	✓	✓	✓
Fuel consumption (LHV)	GJ/h	1156.0	218.2	191.0	5.8	5.8	1152.0
CO_2 capture technique:							
Post-combustion		×	✓	✓	×	×	×
Oxyfuel		×	×	×	✓	✓	✓
Pre-combustion		×	×	×	×	×	×
CO_2 captured	t/h	0.0	58.4	67.4	56.2	56.2	63.3
Total installed capital cost	M$	133.8	189.1	204.5	300.0	335.1	353.9
Specific TIC	M$/t/h	-	3.2	3.0	5.3	6.0	5.6
Total annual operating cost	M$/y	52.3	27.4	27.6	36.7	38.5	75.9

CCP3 Results Summary		Application Scenario 3 Oil Sands Steam Generation	Application Scenario 4 - Gas Fired Power Generation				Application Scenario 5 Refinery Steam Methane Reformer
Phase		3	1	1	4	4	1
Case		3	5A	5	4.1	4.2	4
Title		Oil Sands OTSGs Pre-Combustion CO_2 Capture with aMDEA	NGCC Uncontrolled Reference Case	NGCC Base Case Post-Combustion CO_2 Capture	NGCC Post-Combustion CO_2 Capture by Enzyme Promoted Absorption	NGCC Post-Combustion CO_2 Capture by Novel Solvent Absorption	Refinery SMR with Post-Combustion CO_2 Capture
Location		Fort McMurray	US Gulf Coast	US Gulf Coast	US Gulf Coast	US Gulf Coast	US Gulf Coast
Capacity		1,091 MMBTU/h Hydrogen-rich fuel	1 x GE Frame 9FA + ST	1 x GE Frame 9FA + ST	1 x GE Frame 9FA + ST	1 x GE Frame 9FA + ST	50,000 Nm^3/h Hydrogen
Fuel type:							
FCC catalyst		×	×	×	×	×	×
Refinery fuel gas		×	×	×	×	×	×
Natural gas		✓	✓	✓	✓	✓	✓
Fuel consumption (LHV)	GJ/h	1386.0	2488.4	2488.4	Not disclosed		120.9
CO_2 capture technique:							
Post-combustion		×	×	✓	✓	✓	✓
Oxyfuel		×	×	×	×	×	×
Pre-combustion		✓	×	×	×	×	×
CO_2 captured	t/h	68.6	0.0	126.1	Not disclosed		36.1
Total installed capital cost	M$	346.8	354.3	133.3	Not disclosed		45.7
Specific TIC	M$/t/h	5.1	0.0	1.1	Not disclosed		1.3
Total annual operating cost	M$/y	73.8	151.1	14.4	Not disclosed		14.3

Notes:

1. For most cases the basis of design includes an auxiliary steam boiler plant in the scope of the capture plant to provide steam for solvent regeneration and stripping and for regeneration of CO_2 dehydration equipment. The exceptions to this general rule are made for the natural gas combined cycle cases and the pre-combustion CO_2 capture cases where steam to the CO_2 capture process is available and provided from a waste heat recovery boiler on the main process unit.
2. For most cases the figure in the table for fuel consumption is the fuel burned in the auxiliary boiler. The exceptions to this general rule are made for the natural gas combined cycle cases and the pre-combustion CO_2 capture cases and the un-controlled cases (no CO_2 capture) where the figure in the table gives the fuel consumption of the main process unit.
3. The figure in the table for the total annual operating cost includes the value for the fuel either consumed in the auxiliary boiler or main process unit.

Some observation based upon these key results include:

Application Scenario 1: Refinery FCC

Oxy-fuel delivers a higher rate of capture than post-combustion capture; however oxy-fuel appears to be significantly more expensive in terms of both capex and opex than post-combustion CO_2 capture.

Application Scenario 2: Refinery Fired Heaters

Two capacities of post-combustion capture (both reflecting localized clusters of heaters) were studied, which revealed strong economies of scale. As capture capacity doubled, specific capex dropped by 30%. Oxy-fuel delivers a higher rate of capture than post-combustion, however oxy-fuel appears significantly more expensive in terms of both capex and opex than post-combustion CO_2 capture. A variety of larger-scale pre-combustion capture configurations were studied, which were at the scale of complete refinery fuel system decarbonization. The pre-combustion case using conventional amine combined with cryogenic purification for CO_2 separation from H_2 had similar specific capex as the larger scale post-combustion cases. The cases using high temperature, metal membranes were very expensive.

Application Scenario 3: Oil Sands Steam Generation Scenario

All cases showed high specific capex due to the location factors and relatively small scale. The higher percentage capture post-combustion case showed advantage over the lower capture rate case. Oxy-fuel and pre-combustion capture appeared expensive compared to post-combustion capture.

Application Scenario 4: Gas Fired Power Generation

The largest capacity post-combustion capture system studied by CCP (nearly five times bigger than the largest refinery heaters case) showed the lowest specific capex of all the cases studied.

Application Scenario 5: Refinery Steam Methane Reformer

Relative to other cases studied the specific capex for CO_2 capture from the SMR was among the lowest found. This post-combustion application benefits from a higher-CO_2 content flue gas due to the configuration of the SMR. The tail gas from the PSA H_2 purification step is routed to the SMR furnace to capture the fuel value of residual CO and CH_4. This stream also contains all the CO_2 generated by the reforming and water gas shift operations.

Further Analysis and Reporting of the Results from the Technical Studies

The results from the technical studies were used in the CCP economic model to create fully built-up and escalated capex and opex values. CO_2 captured and avoided were estimated, including specific CO_2 intensity for imported power and steam. The fully built-up, adjusted and aligned results are presented and discussed in a series of separate chapters each dedicated to one of the application scenarios.

CONCLUSIONS FROM THE TECHNICAL STUDIES

The technical studies achieved their defined objectives:

1. A technical definition for the reference case was created and developed for each of the five application scenarios where needed.
2. A quantified performance assessment for the state-of-the-art CO_2 capture technology - CO_2 capture base line for each application scenario was delivered.
3. Quantified performance assessments were completed for alternative CO_2 capture methods and emerging novel technology approaches, including oxy-fuel, pre-combustion CO_2 capture, chemical looping combustion, gas-permeable membranes and novel post-combustion solvent systems.
4. The results have provided insight into the performance of state-of-the-art and emerging technologies, allowing CCP to prioritise and guide technology development activity.
5. The CO_2 capture technology application knowledge developed by this work supports CCP members in evaluating options for CO_2 capture from their operations.

For technologies where there was a good level of definition, the study proved useful. However, for some newer technologies at lower readiness level, it was found that the definition available was insufficient to justify detailed assessments.

Rigorous application of consistent basis of design made comparisons possible, and revealed interesting new insights into the relative effectiveness and competitiveness of different technologies.

This type of scenario-based evaluation was powerful and effective and yielded valuable insights which helped to inform CCP members to guide the CCP program of work and set targets for further research and development.

REFERENCES

1. David C. Thomas (ed.), *Carbon Dioxide Capture for Storage in Deep Geological Formations – Results from the CO_2 Capture Project , Volume 1: Capture and Separation of CO_2 from Combustion Sources*, (2005) Elsevier Publishing.
2. Sally M. Benson (ed.), *Carbon Dioxide Capture for Storage in Deep Geological Formations – Results from the CO_2 Capture Project, Volume 2: Geologic Storage of CO_2 with Monitoring and Verification*, (2005) Elsevier Publishing.
3. Lars Ingolf Eide (ed.), *Carbon Dioxide Capture for Storage in Deep Geological Formations – Results from the CO_2 Capture Project, Volume 3, Advances in CO_2 Capture and Storage Technology Results (2004-2009)*, (2009) CPL Press.

Carbon Dioxide Capture for Storage in Deep Geological Formations, Volume 4
Karl F. Gerdes (Editor)

Chapter 4

FINANCIAL STUDY METHODOLOGY

David Butler[1]
[1]Calgary, Alberta, Canada

ABSTRACT: This chapter summarizes the methodology used to complete the economic analysis of the carbon capture cases considered in this phase of CCP. Efforts were made to establish the full incremental cost of completing each carbon capture project. In addition to the direct and indirect costs provided by the engineers, costs for OSBL, contingencies, owners costs and financing during construction were included. All costs were derived on a First-of-a-Kind basis. A first year unitized cost methodology similar to that used by the National Energy Technology Laboratory was used. Unitized costs were derived by taking O&M, fuel, CO_2 transportation and storage costs, and amortized capital recovery costs in the first year and dividing them by the output of the facility in a given year. This output could be tonnes of steam or MWh for instance. These unitized costs were used to calculate the cost of CO_2 captured and the avoided cost of CO_2. Since data for the reference cases was not available, a methodology for calculating the avoided costs based on the incremental cost of adopting carbon capture was adopted.

KEYWORDS: economics; avoided cost; capture cost; costs

ECONOMIC ANALYSIS CONVENTIONS AND ASSUMPTIONS

Basis for Economic Results

For this phase of work, cost estimates were used to assess numerous CO_2 capture project cases. These cost estimates included all of the expected costs to capture, transport and store CO_2 for each of the cases considered. The incremental costs for capital, operating and maintenance (O&M), fuel and utilities required to capture CO_2 plus costs for transportation and storage of CO_2 were estimated for each case. Amec Foster Wheeler (AFW) provided the direct and indirect capital costs, and cost estimates for O&M for the capture plants. They also estimated incremental fuel, steam and power requirements for the capture plants. For most cases AFW provided just these incremental capital and O&M costs. For some cases AFW provided costs for a technology with and without carbon capture. The incremental costs were then derived by taking the difference in costs between these two cases.

AFW provided the values required to estimate the cost to capture CO_2. The CCP Economics Team estimated the costs for CO_2 transportation and storage. No costs were included for regulatory requirements to mitigate GHG emissions or for the sale of GHG credits. Taxes have not been included in the cost estimates. Costs for all projects were based on a US Gulf Coast location. However, some of the steam production cases were assumed to be located in Fort McMurray, Alberta, Canada. The NGCC cases were assumed to be located in Europe, assuming a European price for natural gas and considering that capital cost of power plants in Europe and USA are similar. All costs are reported on a First-of-a-Kind (FOAK) basis. No attempt was made to apply learning curve assumptions to estimate Nth-of-a-Kind (NOAK) costs.

The in-service date for all CO_2 capture projects is Q1 2014. The capital and O&M cost estimates provided by AFW were provided on a Q4 2009 basis. These costs were subsequently escalated to Q1 2014 using IHS CERA indices [1]. The data in these indices were interpolated and the slopes were used to estimate escalation rates over the period of interest. These escalation rates were applied to extrapolate costs from Q4 2009 to Q1 2012. For the period Q2 2012 to Q1 2014 construction is under way, and this cost is therefore no longer escalated at the same rate. Half the escalation rate used in the previous period was used during the construction period based on the assumption that capital would be spent evenly during this period. The overall DCCI (Downstream Capital Cost Index) index was used for all capture capital costs. The escalation value used for these costs over this period was 18.2%. The UOCI (Upstream Operation Cost Index) index was used to estimate an escalation rate applied to operating and maintenance costs for the period Q4 2009 to Q1 2014. The escalation value used for these costs over this period was 19.4%. The UCCI (Upstream Capital Cost Index) index was used to estimate an escalation rate applied to CO_2 transportation and storage costs for the period Q4 2009 to Q1 2014. The escalation value used for these costs over this period was 19.8%.

All results and costs are expressed in $US unless otherwise stated.

Capital Costs

AFW estimated the direct and indirect bare erect capital costs plus costs for EPC contractor services for all the cases. The anticipated accuracy of the capital cost estimates was reported as +/- 30%. AFW's cost estimate excluded several other types of costs, such as owner's costs and OSBL costs normally included in a Total As-Spent Cost (TASC) estimate. For this reason, adjustments were made to the AFW cost estimates to include additional costs to derive Total As-Spent Costs. The adjustments used on the AFW costs are described below.

The National Energy Technology Laboratory (NETL) recently published two reports [2,3] describing four definitions of capital costs. Figure 1 shows a schematic describing these four types of capital cost. One of these reports [3] also describes the methodology NETL uses to estimate the first year and levelized costs for a project.

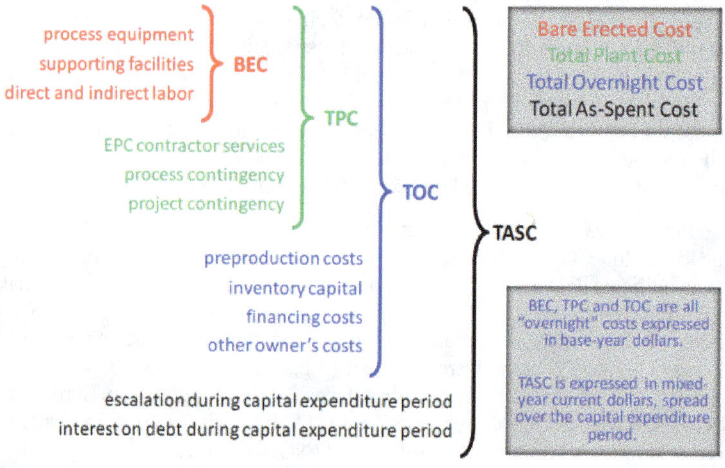

Figure 1. Capital Cost Estimates.

Since the AFW reports included the Bare Erect Costs (BEC) plus EPC contractor services, the remaining costs components required to estimate TASC were estimated using a version of this NETL methodology. The methodology described below is based on the approach used in several NETL reports.

NETL also recently published a study [4] reviewing the costs of retrofitting an existing NGCC unit with carbon capture. In this study NETL applied several factors and cost references to go from Bare Erected Costs to Total As-Spent Costs. Those adjustments are also described below.

Total Plant Costs (TPC): OSBL and escalation costs were added to the BEC provided by AFW. OSBL was assumed to be 30% of the direct and indirect costs provided by AFW except for air separation unit costs. To arrive at Total Plant Costs, escalation, process and project contingencies were added to the BEC. NETL appears to have applied process contingencies against those parts of the project which have not been constructed routinely in the past. The CCP used a process contingency of 10% for technologies already demonstrated but not implemented at the industrial scale. A 20% process contingency was used for technologies which have not been demonstrated. Specifically, process contingencies were not applied to capital cost components which are widely used in industry. Process contingencies are assumed to be 10 or 20% of BEC, EPC fees, OSBL costs and escalation. Project contingencies are assumed to be 20% of BEC, EPC fees, OSBL costs, escalation and process contingency [5].

Total Overnight Costs (TOC): The following table describes the owner's costs included in the TOC. It was assumed that 5 acres will be set aside for the carbon capture plant. NETL included other costs in owner's costs which are not capital costs. Those costs are not described below. NETL simplified the first item in the table below to 6 months of fixed O&M and the second to fourth items to 1 month of variable O&M.

Table 1. NETL Capital Cost Adjustments.

Owner's Cost	Comprising
Pre-production Costs	• 6 months O&M, and administrative & support labour • 1 month maintenance materials • 1 month non-fuel consumables • 1 month of waste disposal costs • 25% of one month's fuel costs @100% Capacity Factor • 2% of TPC
Inventory Capital	• 60 day supply of consumables @100% CF • 0.5% of TPC (spare parts)
Land	• $3,000/acre (100 acres for greenfield NGCC)
Financing Costs	• 2.7% of TPC
Other Owner's Costs	• 15% of TPC

Total As-Spent Costs (TASC): NETL applies a single Allowance for Funds During Construction (AFUDC) factor of 1.078 on the TOC to account for the cost of financing during construction. This factor is meant for high risk projects. Another NETL report [5] indicates that this factor is appropriate for capital expenditure periods of 3 years. This appears appropriate given the capital spending profiles provided by AFW.

Unitized Costs

Capital and operating costs were translated into various unitized costs such as the cost of power in \$/MWh or the cost of capturing CO_2 in \$/t. Unitized costs were derived to include all of the costs which must be recovered in a given year to allow a project to break even on an NPV basis over its whole project life.

Unitized costs can be derived by completing present value calculations on cash flows projected out into the future. A simpler factored approach to calculating unitized costs based on the costs and revenues in the first year of operation was used instead for this study. This approach is similar to that employed by NETL and other organizations when reporting unitized costs. Unitized costs were derived to include operating costs for the first year of operation plus a portion of the upfront capital cost.

A capital recovery factor was employed to determine how much capital is required to be recovered in each year. The amount of capital recovered in each year is expected to increase by inflation so that over the project life all the capital will be recovered. The per unit capital costs recovered in a given year were based on a capital recovery factor multiplied by the TASC. The formula for the capital recovery factor is:

$$i \times (1 + i)^n \,/\, ((1 + i)^n - 1)$$

where i is the discount rate and n is the number of years the project operates.

This standard formula leads to a levelized cost annuity. Therefore the capital recovery factor was multiplied by the levelization factor to derive a first year capital recovery factor. The levelization factor is:

$$[((1 + i)^n - 1)/\, i \times (1 + i)^n] \,/\, [(1 - (1 + ir)^n \times (1 + i)^{-n})/(i - ir)]$$

where i is the discount rate, ir is the inflation rate and n is the number of years the project operates. A discount rate of 10% and an inflation rate of 2.0% were used in these calculations. The discount rate acts like a weighted average cost of capital (WACC) in the calculations. Since taxes are not included the discount rate, the discount rate would be a WACC based on the interest rate without the tax effect and using a pre-tax ROE.

The premise behind a unitized cost is that it indicates how much additional revenue must be recovered on a per unit basis to set the NPV of a project equal to zero. The first year cost basis assumes that the recovery of the cost in question will be based on a set of unitized revenues which escalate by inflation through the life of the project. Levelized costing assumes that costs will be recovered on a constant unit revenue basis through the life of the project. One can think of the first year basis cost as real levelized cost. For example, for the non-CCS NGCC case in Europe, the first year cost of power is expected to be \$110.6/MWh. That is, for the NGCC project to obtain a NPV = 0, all the power in the first year must be sold at \$110.6/MWh, and all power sold in the second year must be sold at a price which is greater than \$110.6/MWh by the assumed inflation rate, and so on. The levelized cost for this case is calculated to be \$129.3/MWh. All power produced through the life of the project must be sold for this price for the project to have a NPV = 0.

Therefore when unit costs are reported, they are reported on a first year basis and it is assumed that the value for the next year and all subsequent years will need to increase by inflation to recover all costs incurred. Roughly a two year construction period was assumed for all cases. Unitize costs were determined on a first year basis at Q1 2014. Unless otherwise stated all results are reported in US dollars. Table 2 provides the base financial assumptions used in the calculation of the unitized costs in the model.

Table 2. Base Financial Assumptions.

Assumptions	Units	Values
Exchange Rate	USD/EUR	1.35
Exchange Rate	USD/CND	0.98
Escalation Rate	pct	2.0%
Discount factor, pre-tax	pct	10.0%
Time horizon	years	25
Levelization Factor		0.856
1st Year Capital Recovery Factor		9.43%

Other Assumptions

For some cases an estimate of the GHG emissions associated with the incremental fuel consumed is required. Table 3 below shows the GHG emission intensities used for the various fuels.

Table 3. GHG Emissions for Fuel.

GHG Emission Intensity of Fuel	t CO_2/GJ HHV
Nat Gas - USGC	.0537
Nat Gas - Alberta	.0499
Refinery Fuel Gas (RFG)	.0439

A conversion between LHV and HHV fuel definitions is required in the model. A value of 1.12 is multiplied by the LHV values to derive the HHV value for RFG, while a value of 1.1 is multiplied by the LHV values to derive the HHV value for natural gas – reflecting the differences in composition.

Fuel and Utilities Costs

For the NGCC cases, it was assumed that all power and steam required for the capture process would be supplied by the NGCC unit itself. For all the other cases, power required by the capture process was provided over the fence from the power grid, and steam required to provide heat to the scrubbing process was supplied by a new boiler. Fuel costs are related to incremental fuel required by the boilers to produce steam. Incremental natural gas may also be required for the cases producing hydrogen used as a low carbon fuel. Refinery fuel gas was assumed to be used as fuel for all but the NGCC, ATR and steam generation cases where natural gas was assumed to be the fuel. Table 4 describes the fuel and power price assumptions used in the model.

Table 4. Fuel and Power Assumptions.

Fuel	Units	Cost
Fuel gas price - US	USD/GJ	4.50
Fuel gas price - CND	USD/GJ	4.50
Fuel gas price - Europe	USD/GJ	10.90
Electricity price - US	USD/MWh	70.00
Electricity price - CND	USD/MWh	60.50

The prices ending in "US" were based on US Gulf costs forecasts. The prices ending in "CND" were based on forecasts for Alberta. All gas prices are on a HHV basis.

The price for power in the model changes as the price of natural gas changes, but only if the GHG emission intensity for a scenario drops below 0.5 t CO_2/MWh in a market with significant natural gas power generation.

For the USGC the price of power is:

$$33 + 5.5 \times Gas\ Price$$

For Alberta the price of power is:

$$44 + 5.5 \times Gas\ Price$$

Operating and Maintenance Costs

O&M costs were provided by AFW at Q4 2009 and escalated to Q1 2014. The O&M costs were comprehensive and in most cases were not revised. The capacity factor for all cases was assumed to be 90%.

Transportation and Storage Costs

The CO_2 pipeline costs were based on a 200 km pipeline carrying 5 Mt/yr of CO_2. WorleyParsons [6] published a report for the Australian government on the estimated sizing of CO_2 pipelines given distances and mass flow rates. Based on this data a 16 inch diameter pipe was chosen.

Two sources for the cost to transport CO_2 were used in the model: i) Cost estimates used by the National Energy Technology Laboratory (NETL) and, ii) Recent cost information reported for natural gas pipeline in the Oil and Gas Journal [7].

NETL Capital Cost Estimates

NETL [8] has established an approach to estimate the components of the capital cost of a CO_2 pipeline. These equations were originally developed by the University of California [9]. The original equations were modified by NETL to include escalation to bring the costs to June-2007 year dollars.

$$Materials = 64{,}632 + 1.85 \times L \times (330.5 \times D^2 + 686.7 \times D + 26{,}960)$$

$$Labour = 341{,}627 + 1.85 \times L \times (343.2 \times D^2 + 2{,}074 \times D + 170{,}013)$$

$$Misc = 150{,}166 + 1.58 \times L \times (8{,}417 \times D + 7{,}234)$$

$$Right\ of\ Way = 48{,}037 + 1.2 \times L \times (577 \times D + 29{,}788)$$

$$CO_2\ Surge\ Tank = 1{,}150{,}636$$

$$Pipeline\ Control\ System = 110{,}632$$

Where D is diameter in inches and L is the length of pipe in miles. Table 5 shows rough estimates for the costs of pipelines for various terrains. It was decided that on-shore costs for transportation and storage would be multiplied by 2 to derive the off-shore transport and storage costs.

Table 5. Terrain Factors.

Location	Kinder-Morgan ($/inch/mile) [9]	Cost Multiplier[10]
Flat, Dry	$50,000	1.0
Mountainous	$85,000	2.5
Marsh, Wetland	$100,000	
River	$300,000	
High Population	$100,000	
Offshore (150'-200'depth)	$700,000	
Desert		1.3
Forest		3.0
Offshore (< 500m depth)		1.6
Offshore (>500m depth)		2.7

Oil and Gas Journal Information

The November 1, 2010 *Oil and Gas Journal* [7] published costs for natural gas pipelines over the past decade. The data suggests that the cost of pipelines in 2010 have doubled compared to the costs in 2008 and 2009 and increased by a factor of 4 since 2006. Since various terrains are included in these values, the cost estimate for this study will be for a blend of on-shore terrains. The NETL cost data was multiplied by 1.5 to escalate it from 2007 to 2009. For the diameter of pipe used in this study both sets of data give similar cost results. This cost data was also used to estimate costs for a collector system to feed the back bone pipeline. O&M was estimated to be 5% of capital per year.

The capital costs were multiplied by a first year capital recovery factor and added to the O&M costs in a given year. This sum was divided by the mass flow of CO_2 per year in the pipeline to determine an estimate for the pipe toll. This toll was estimated to be $6.75/tonne on a first year basis (USGC location).

The Wabamun Area CO_2 Sequestration Project (WASP) [11] carried out a rigorous estimate of CO_2 storage costs for Alberta. This report estimated the costs of storage to be between $C2.65 and $3.38/tonne. To be conservative, a value of $2.94/tonne (USGC) was assumed for the economic model used in the present study.

CO_2 Footprint

The capture cost for a project was based on the incremental costs to capture CO_2 including the costs for steam and power required by the capture plant. For the NGCC cases the effect of the derate of the unit was included in the denominator of the unit cost calculations. For all cases except the NGCC cases, power required by the capture unit was supplied by the grid and steam was provided by a new boiler. The amount of CO_2 emitted to produce the steam and power required by the capture plant was estimated by AFW. The CO_2 intensity for power supplied from the grid was assumed to be 0.6 t/MWh for all USGC cases. The CO_2 intensity used for new power supplied by the grid in Alberta was assumed to be 0.65t/MWh, in line with near term forecasts for that area [12].

Capture Cost Calculations

The literature provides a standard equation for deriving the cost of CO_2 capture based on unitized costs of production with and without capture.

The standard CO_2 capture cost formula is:

$$(COE_{capture} - COE_{ref}) \times Output_{capture}/CO_{2\ captured}$$

where COE is the unitized cost of output. For a power plant COE is the cost of electricity and the output would be MWh in a given year. This approach was used to calculate the cost of capture for the NGCC cases.

However, for most cases the CCP did not develop information on the cost or mass balance of CO_2 of the reference case without CO_2 capture. For these cases only the incremental cost of capture was estimated, along with the corresponding mass balance of CO_2 for the case with CO_2 capture. Therefore, since no information on the reference case was available, the standard capture cost equation had to be modified. If the assumption is made that the output for the reference and capture case is the same, then the above formula can be restated as shown below. This is a valid assumption given that the basis of design specified no change in output for most of the cases.

$$(COE_{capture} \times Output - COE_{ref} \times Output)/CO_{2\ captured}$$

COE is the yearly cost for a case divided by the Output. If this definition is inserted into COE in the above equation we are left with the following:

$$\left(\frac{Yearly\ Cost_{capture}}{Output} \times Output - \frac{Yearly\ Cost_{ref}}{Output} \times Output\right)/CO_{2\ captured}$$

Since the yearly Output cancels out, what remains is difference between the yearly costs of the two cases in the numerator. This is the incremental cost of capture. Therefore the above equation simplifies to:

$$Incremental\ Cost\ of\ Capture/CO_{2\ captured}$$

The incremental cost of capture is based on the first year capital recovery factor multiplied by the incremental Total As-Spent Capital cost plus the incremental fuel, O&M, CO_2 transportation and storage costs. Each of these components can be divided by the mass of CO_2 captured in a given year to illustrate how each component contributes to the total CO_2 capture cost. The cost of CO_2 capture has been estimated based on this approach for all but the NGCC cases.

Avoided Cost Calculations

The standard calculation of avoided cost provided in the literature is:

$$(COE_{capture} - COE_{ref})/(Intensity_{ref} - Intensity_{capture})$$

Intensity is defined as the amount of CO_2 emitted divided by the appropriate unit of output. COE is the unitized cost of producing a given unit of output.

For those cases studied where AFW estimated the costs and the mass balance for CO_2 for both the CO_2 capture case and the reference case without capture, the above relationship can be used to calculate avoided cost. However, for most cases only incremental costs to derive the cost of capture CO_2 were provided along with the CO_2 mass balance for just the case with CCS since estimates regarding the costs and CO_2 balance for the reference case were not requested of AFW. For this reason, the standard avoided cost formula had to be modified to address this issue.

More explicitly the standard CO_2 avoided cost formula is:

$$\left(\frac{Yearly\ Cost_{capture}}{Output_{capture}} - \frac{Yearly\ Cost_{ref}}{Output_{ref}}\right) / \left(\frac{CO_2\ Emitted_{ref}}{Output_{ref}} - \frac{CO_2\ Emitted_{capture}}{Output_{capture}}\right)$$

The formula above explicitly takes into account the fact that the output for the capture case may be less than the reference case. If the assumption is made that the output for both the reference and capture case is the same, then the formula above simplifies to the following equation:

$$\left(\frac{Yearly\ Cost_{capture} - Yearly\ Cost_{ref}}{Output}\right) / \left(\frac{CO_2\ Emitted_{ref} - CO_2\ Emitted_{cap}}{Output}\right)$$

This is a valid assumption given that the basis of design specified no change in output for most of the cases. This is simplifies to:

$$\frac{Yearly\ Cost_{capture} - Yearly\ Cost_{ref}}{CO_2\ Emitted_{ref} - CO_2\ Emitted_{cap}}$$

The numerator is the incremental cost of CO_2 capture and the denominator is the mass of CO_2 avoided. Therefore the avoided cost of CO_2 becomes:

$$\frac{Yearly\ Incremental\ Cost\ of\ CO_2\ Capture}{Mass\ of\ CO_2\ Avoided}$$

This formula has been used to calculate the avoided cost of CO_2 for all cases where information on a reference case is not available. The mass of CO_2 avoided is the mass of CO_2 generated in the non-capture reference case less CO_2 emissions in the capture case less CO_2 which is effectively produced to capture this CO_2. The following is a definition of the mass of CO_2 avoided. More specifically the mass of CO_2 avoided is:

- CO_2 in reference non-CCS case (CO_2 generated)
- Less CCS emission
- Less CO_2 associated with power used to capture CO_2
- Plus CO_2 associated with waste heat boiler operating on extra heat generated by the capture process
- Plus CO_2 emissions associated with reduced fuel usage
- Less uncaptured CO_2 associated with steam use

However, for many cases little information was available regarding the nature of the reference case without CO_2 capture. There are other equally valid ways to calculate the avoided mass of CO_2. The mass of CO_2 avoided can also be expressed as the mass of CO_2 captured less CO_2 which was effectively produced to capture this CO_2. More specifically the mass of CO_2 avoided is:

- CO_2 captured
- Less CO_2 associated with power used to capture CO_2
- Plus CO_2 associated with waste heat boiler operating on extra heat generated by the capture process
- Plus CO_2 emissions associated with reduced fuel usage
- Less uncaptured CO_2 associated with steam use
- Less CO_2 emissions associated with extra fuel usage

This definition of the mass of CO_2 avoided was used for all cases were information on the reference case was not available. In this definition the mass of CO_2 captured usually includes CO_2 associated with the extra fuel used. This additional CO_2 must be deducted from the mass of CO_2 captured in this version of the avoided mass calculation, since it is not avoided. All the approaches described above should yield the same avoided cost of CO_2 capture.

Other Results

One important way to assess the cost impact of carbon capture is to estimate its effect on the production of the key output(s). For example, the increased cost of one of the key outputs because of the additional costs associated with CO_2 Capture can be compared to the cost of this output without CO_2 capture. This is commonly done for power projects with and without CO_2 capture. For this study, this sort of analysis can be completed for the OTSG cases, since the cost to capture CO_2 can be translated into an increase in the cost of steam production. This simple comparison is difficult to complete for the other scenarios, such as refinery process heaters and fluid catalytic cracking units, where a single natural output is not easily defined. When employing oxygen-firing for CO_2 capture for fluid catalytic cracking, it is possible to change the product slate and perhaps increase output, so the base operation is changed in multiple ways. Given the data available, only a cursory estimate of the potential benefit of using oxygen-firing for FCC can be approximated.

Another important metric is the carbon capture cost expressed for a project on a yearly basis. Economic results are shown in the economic sections for individual process technologies chapters.

REFERENCES

1. IHS Indexes, http://www.ihs.com/info/cera/ihsindexes/index.aspx
2. J. Klara, Current and Future IGCC Technologies, A Pathway Study Focused on Carbon Capture Advanced Power Systems R&D Using Bituminous Coal – Volume 2, DOE/NETL – 2009/1389, 10/ 2010, 2-4 – 2-5.
3. Cost Estimation Methodology for NETL Assessments of Power Plant Performance, DOE/NETL-2011/1455, 12/2010.
4. E. Grol, Cost and Performance of Retrofitting Existing NGCC Units for Carbon Capture, DOE/NETL-401/080610, 10/2010, 35-37.
5. J. Black, Cost and Performance Baseline for Fossil Energy Plants, DOE/NETL-2007/1281, Volume 1: Bituminous Coal and Natural Gas to Electricity Final Report (Original Issue Date, May 2007), Revision 2, 11/2010.
6. DRET CCS Task Force Support, Pipeline and Injection Pumping Study, WorleyParsons, 8/2009.
7. C. Smith, Transportation Special Report, Oil & Gas Journal, vol 108.41, 11/2010, 102 – 123.
8. Estimating Carbon Dioxide Transport and Storage Costs, DOE/NETL-2010/1447, 03/2010.
9. N. Parker, Using Natural Gas Transmission Pipeline Costs to Estimate Hydrogen Pipeline Costs, University of California, UCD-ITS-RR-04-35, 2004.
10. CO_2 Pipeline Infrastructure: An analysis of global challenges and opportunities, Element Energy, IEA GHG, 04/2010.
11. R. Nygaard, R Lavoie, Project Cost Estimate, Wabamun Area CO_2 Sequestration Project (WASP), 12/2009.
12. Technical Guidance for Offset Protocol Developers, Environment Alberta, ISBN: 978-0-7785-8810-8 (On-line), 2011.

Carbon Dioxide Capture for Storage in Deep Geological Formations, Volume 4
Karl F. Gerdes (Editor)

Section 1

CO_2 CAPTURE

Oil Refineries

Carbon Dioxide Capture for Storage in Deep Geological Formations, Volume 4
Karl F. Gerdes (Editor)

Chapter 5

CO_2 CAPTURE FROM A FLUID CATALYTIC CRACKING UNIT

Leonardo F. de Mello[1], Rodrigo Gobbo[2], Gustavo T. Moure[1], Ivano Miracca[3]
[1] Petrobras, Av. Horácio Macedo, 950 – Cidade Universitária, Rio de Janeiro 21941-915, Brazil
[2] Petrobras, Rodovia do Xisto BR 476, c.p. 28 – São Mateus do Sul, Paraná 83900-000, Brazil
[3]Saipem S.p.A. (Eni) – Via Martiri di Cefalonia, 67 – I-20097 San Donato Milanese, Italy

ABSTRACT: Fluid Catalytic Cracking (FCC) is a key operation in planning CO_2 mitigation from oil refineries, since the FCC is often the largest single point source in a refinery. The work carried out during CCP3, which consisted of technical studies, to compare post-combustion vs. oxy-firing as capture options from the FCC catalyst regenerator, and large-scale pilot plant testing of oxy-firing the FCC catalyst regenerator, is breifly summarized.

KEYWORDS: FCC; CCS; CO_2 Capture, post-combustion; oxy-combustion; oxy-firing

INTRODUCTION

The fluid catalytic cracking (FCC) process is one of the main sources of greenhouse gas emissions in the oil refining industry as a sngle source, often accounting for 20-30% of total CO_2 emissions from a typical refinery. Capturing CO_2 from FCC flue-gas is therefore an important step in mitigating CO_2 emission from a refinery as a whole.

The conventional FCC process catalytically converts heavy oil fractions to lighter products such as liquid petroleum gas (LPG) and gasoline. A medium size FCC unit has a feed rate of 60,000 barrels per day (bpd), and emits to the atmosphere about 0.5 million tons per year of CO_2. The feed capacity of the largest units is well above 100,000 bpd.

A simplified process scheme of an FCC unit is shown in Figure 1.

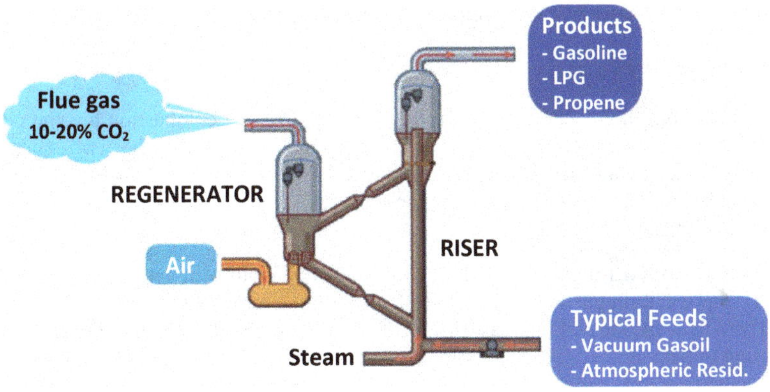

Figure 1. Conceptual scheme of FCC.

During the reaction step, byproduct coke is also formed and deposited on the surface of the catalyst, which is then deactivated. To maintain catalyst activity, the coke is burned off in the regenerator an air-blown, fluidized bed. The flue-gas from the regenerator contains CO_2 at typical concentrations of 10 – 20 vol % vol. The FCC process uses a fluidized circulating bed type mechanism, with the catalyst continuously flowing from the regenerator vessel to the reactor and back. The FCC unit is thermally balanced, which means that the heat generated by burning the coke in the regenerator provides the heat necessary for feed vaporization and for the endothermic cracking reactions in the reactor. Since the catalyst serves to carry the heat released in the regenerator to the reaction side, the amount of catalyst that circulates is dictated by this thermal balance. The process is typically operated in the pressure range of 1.5-2.5 barg, with reaction temperature in the 500-550°C range, and with the regeneration temperature close to 700°C.

Based on the characteristics of the FCC process, two possible ways to capture CO_2 from FCC units are post-combustion technologies, such as CO_2 absorption, and oxy-combustion. In a previous study [1] carried out during CCP2, the oxy-combustion concept using flue gas recycle to moderate combustion temperature was shown to be a cost-effective alternative for CO_2 capture from FCC units and also small, bench-scale test results showed that this technology was ready to be taken to a larger scale.

Pre-combustion is not applicable to this scenario, because the fuel is the coke deposited on the surface of the catalyst and therefore cannot be decarbonized before combustion.

THE FOCUS OF CCP3

Based on the results of the study carried out in CCP2, two lines of work were followed in parallel during CCP3:

- A demonstration run was organized and carried out to establish the technical viability of operating an FCC unit in the oxy-combustion mode with flue gas recycle, to test the start-up and shut down procedures, and also to measure the impact of oxy-combustion operation on the main process variables and product yields. Petrobras retrofitted its large pilot scale FCC unit for operation at oxy-combustion conditions and tests were performed during 2012 (details and results in Chapter 6).
- As part of the engineering studies carried out by Amec Foster Wheeler, optimized process schemes to capture CO_2 from an FCC unit by state-of-the-art post-combustion technology and by oxy-firing were developed. Technical issues were identified and CO_2 avoidance cost calculated for each option (Details and results in Chapter 7).

SUMMARY OF CONCLUSIONS

For CO_2 capture from the regenerator of a refinery FCC unit, oxy-combustion has the potential for being competitive with post-combustion. CO_2 avoidance costs for the two techniques are comparable, but oxy-firing provides more flexibility to the operation of the FCC unit, and also requires a smaller plot area close to the unit than does post-combustion, which may be an important factor inside a refinery.

The field demonstration run has confirmed the technical viability of retrofitting an FCC unit to enable CO_2 capture through oxy-firing. The demonstration proved that the FCC unit can work steadily in oxy-firing mode and that oxy-firing can enable a higher throughput or allow switching to the processing of heavier feeds while keeping the same product yield.

Commercial demonstration of post-combustion capture might in principle be possible in the short term, considering the maturity of the technology.

For oxy-firing, more activities at the pilot plant level might be required before proceeding with the scale-up to a commercial demonstration, notably:

- Testing of different types of compressors for recycle of flue gas and optimized recycle schemes to verify that the corrosion problems observed during testing have been actually solved.
- Testing of a CO_2 purification section based on catalytic combustion, because this is novel application for this technology.

REFERENCES

1. L. de Mello, G.T. Moure, O.R.C. Pravia, L. Gearhart, P.B. Milios – "Oxy-combustion for CO_2 capture Fluid Catalytic Crackers" in "Carbon Dioxide Capture for Storage in Deep Geologic Formations – Results from the CO_2 Capture Project" – Volume 3 – pp. 31-42 (Ed. Lars Ingolf Eide), CPL Press [2009].

Carbon Dioxide Capture for Storage in Deep Geological Formations, Volume 4
Karl F. Gerdes (Editor)

Chapter 6

FCC OXY-FUEL DEMONSTRATION AT PETROBRAS SHALE INDUSTRIAL BUSINESS UNIT

Leonardo F. de Mello[1], Rodrigo Gobbo[2], Gustavo T. Moure[1], Ivano Miracca[3]
[1] Petrobras, Av. Horácio Macedo, 950 – Cidade Universitária, Rio de Janeiro 21941-915, Brazil
[2] Petrobras, Rodovia do Xisto BR 476, c.p. 28 – São Mateus do Sul, Paraná 83900-000, Brazil
[3]Saipem S.p.A. (Eni) – Via Martiri di Cefalonia, 67 – I-20097 San Donato Milanese, Italy

ABSTRACT: Fluid Catalytic Cracking is a major CO_2 source in most refineries and therefore, is a key unit in planning CO_2 mitigation from oil refineries. A major accomplishment during CCP3 was confirming the technical viability of retrofitting an FCC unit to enable CO_2 capture through oxy-firing the FCC regenerator, with flue gas recycle for temperature moderation. A series of tests were executed by Petrobras over a range of process conditions and feeds on a large-scale, fully integrated FCC pilot unit.

KEYWORDS: FCC, CCS, CO_2 Capture, post-combustion, oxy-combustion, oxy-firing

INTRODUCTION

The CCP's first large-scale capture demonstration project took place at a Petrobras research facility in Paraná, Brazil. Testing of oxy-fired capture technology on a pilot-scale FCC unit was conducted during 2011 and early 2012. Full assessment of the results was completed in early 2013, with some key presentations of those results made at the ERTC-17 (European Refinery Technology Conference) and GHGT-11 [2] (Greenhouse Gas Technologies) conferences.

The main goals of the project were the following:

- Test start-up and shut-down procedures
- Maintain stable operation of the FCC unit in oxy-firing mode
- Test different operational conditions and process configurations
- Obtain reliable data for scale-up

UNIT DESCRIPTION AND RETROFIT

The large scale pilot FCC unit (U-144) was built in 1993 and has been continuously updated since then. The unit is located at the Shale Industrialization Business Unit (SIX), in São Mateus do Sul (Paraná State, Brazil). SIX is also part of Petrobras Research Complex and operates as an advanced research center within the framework of Petrobras research organization.

U-144 can process up to 200 kg/h (32 bbl/d) of vacuum gasoil (VGO) or 100 kg/h of atmospheric residue (ATR) and operates with a catalyst inventory of 300 kg. In terms of feed flow rate, the unit is about 200 times larger than a typical FCC pilot unit, but at least 500 times smaller than a commercial unit. Nevertheless, the size of the unit allows close approximation to a commercial unit, including the use of feed nozzles, cyclone systems, structured packing in the stripper and pseudo-adiabatic operation through the use of electrical resistance heaters on each vessel to counterbalance thermal losses. The main dimensions of the unit are reported in Table 1. The process flow diagram for the unit is shown in Figure 1. The reaction products from the riser are sent to a set of condensers where the liquid product is collected, weighted and analyzed. Gaseous products are quantified and sent to flare.

Table 1. Main dimensions of the pilot FCC unit.

	Diameter (mm)	Height (m)
RISER	50	18.2
REGENERATOR		
Lower part (Dense phase)	445	2.1
Upper part (Dilute phase)	650	2.0
STRIPPER	260	2.4

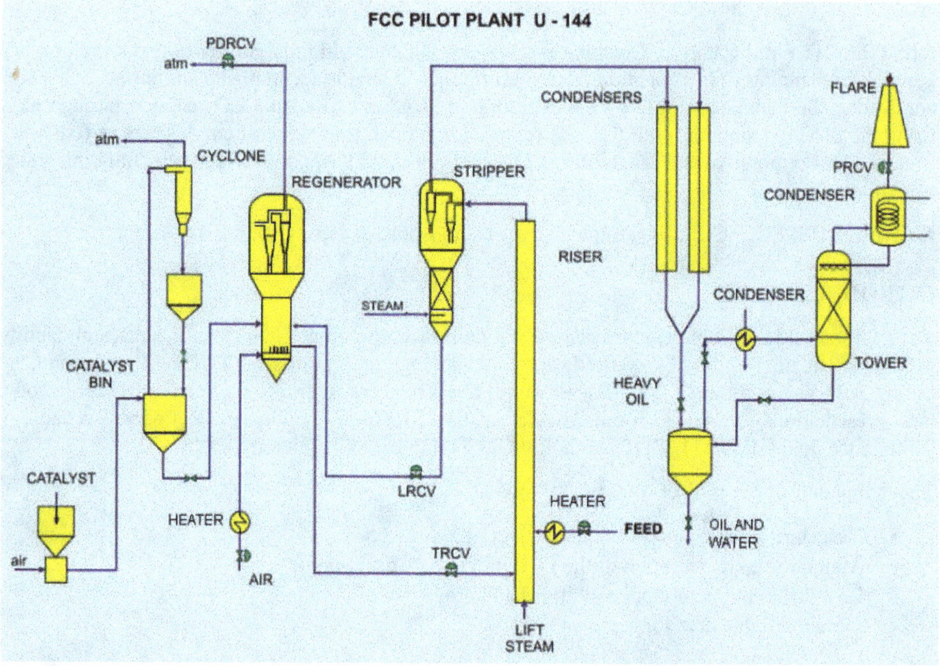

Figure 1. Process Flow Diagram of the FCC Pilot Unit.

The unit has been retrofitted to allow operation in oxy-fuel mode with flue gas recycle in the catalyst regenerator. Two sections were designed, constructed, installed and integrated in the existing unit:

- Oxygen Supply System (OSS), including liquid oxygen tank, vaporizer system, flow and pressure control skid, injector and piping.
- CO_2 Recycle System (CRS), including catalyst fines and sulfur oxides removal unit, recycle compressor, storage tank, gas analyzers and piping.

Due to the relatively brief duration of the tests, high purity (> 99.5 vol%) oxygen was supplied from a cryogenic tank at 10 barg. Two pairs of vaporizers work alternately to vaporize oxygen from the tank. The general arrangement of the OSS is represented in Figure 2.

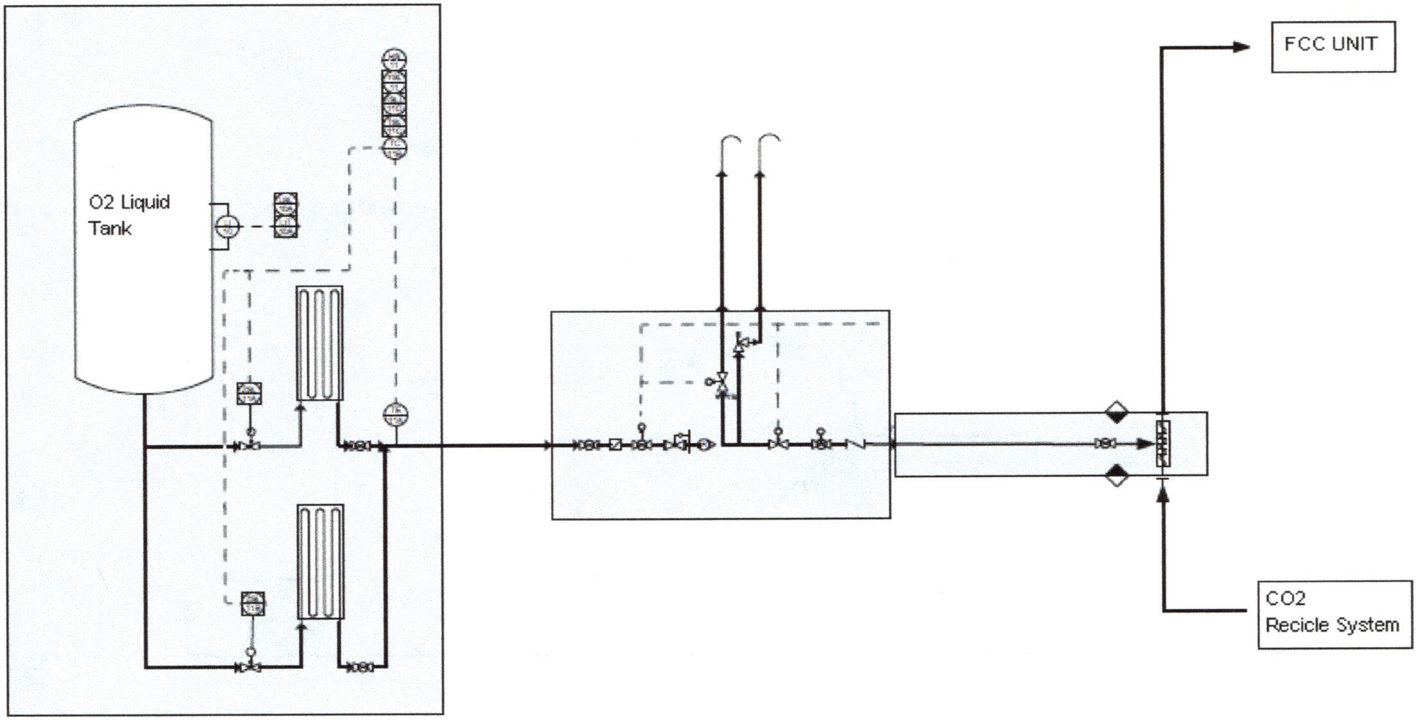

Figure 2. Oxygen Supply System general arrangement.

The CO_2 recycle system was designed to receive the flue gas from the regenerator, treat it and recycle it back to the regenerator. The system is mainly composed of an alkaline washing tower followed by cold water washing to remove SOx and catalyst fines, a flue gas compressor to allow gas recycle, a pressurized buffer tank to prevent pressure oscillations and a chiller to supply cold water to the wash tower. The entire system is skid mounted. The process flow diagram is shown in Figure 3.

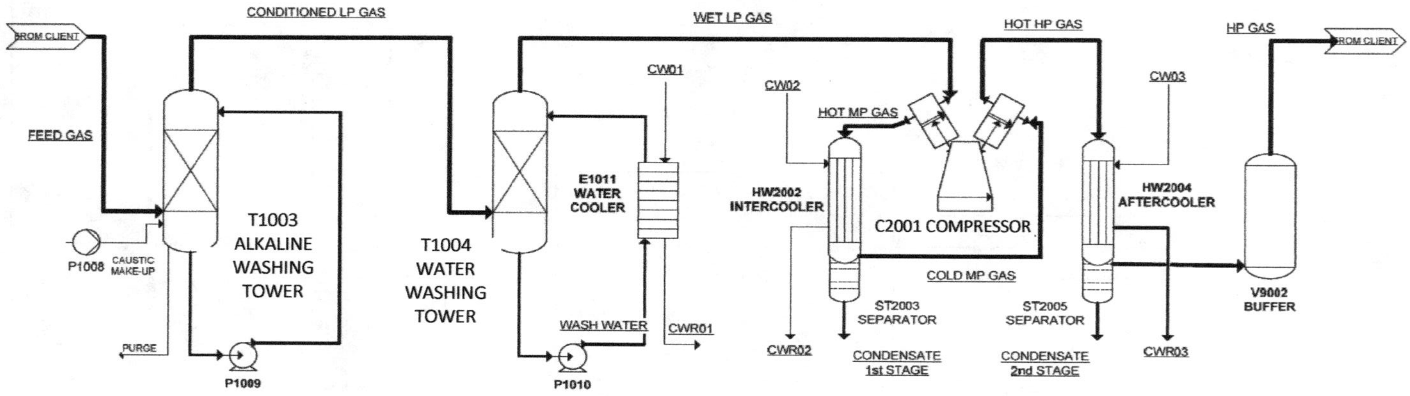

Figure 3. CO_2 Recycle System: Process Flow Diagram.

An overall lay-out of the pilot unit after retrofitting to oxy-firing mode is shown is Figure 4.

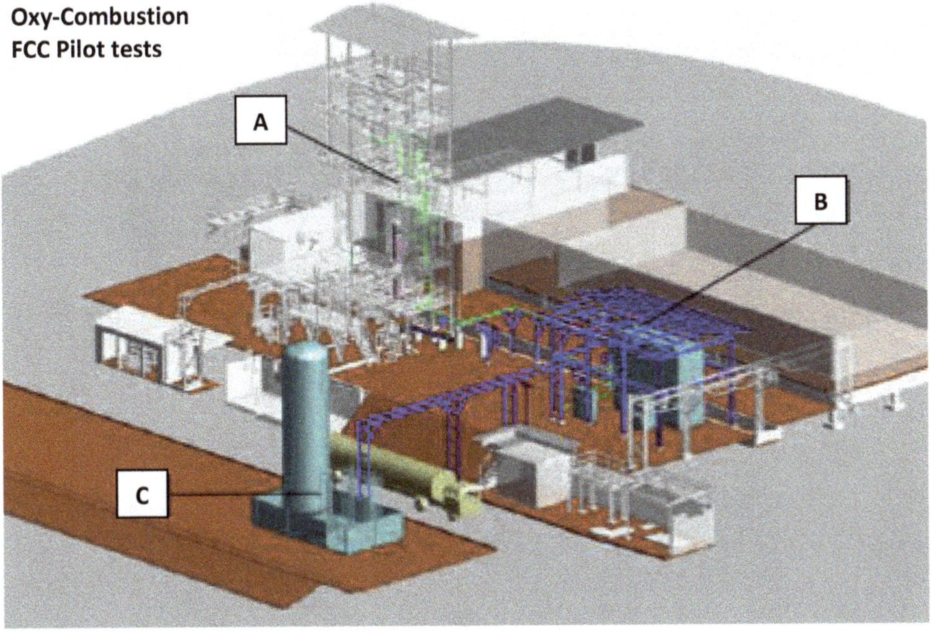

Figure 4. Overall lay-out of the unit after retrofit: (A) FCC unit, B(CRS), C(OSS).

COMMISSIONING, START-UP AND TROUBLESHOOTING

Commissioning commenced in March 2011. As usual in these cases, several minor problems were evidenced and corrected.

The first start-up in normal operating mode followed by switch to oxy-firing took place on March 30, 2011. From an operational point of view, switching from air to oxygen represents a critical step, due to the different thermal and physical properties of CO_2, which is the main component in the recycled flue gas compared to nitrogen:

- CO_2 has higher specific heat than nitrogen. This may result in a temperature decrease in the regenerator during the switch phase.
- CO_2 has a higher density than nitrogen. This may cause higher entrainment of the catalyst, potentially overloading the cyclones.

The FCC unit operates in dynamic thermal balance between the reactor and regenerator, because the heat released by coke combustion represents the primary source of energy for the endothermic reaction of hydrocarbon cracking. The switch may therefore lead to unit upset if not properly done.

The sequence of a typical switch can be seen in the charts of Figure 5.

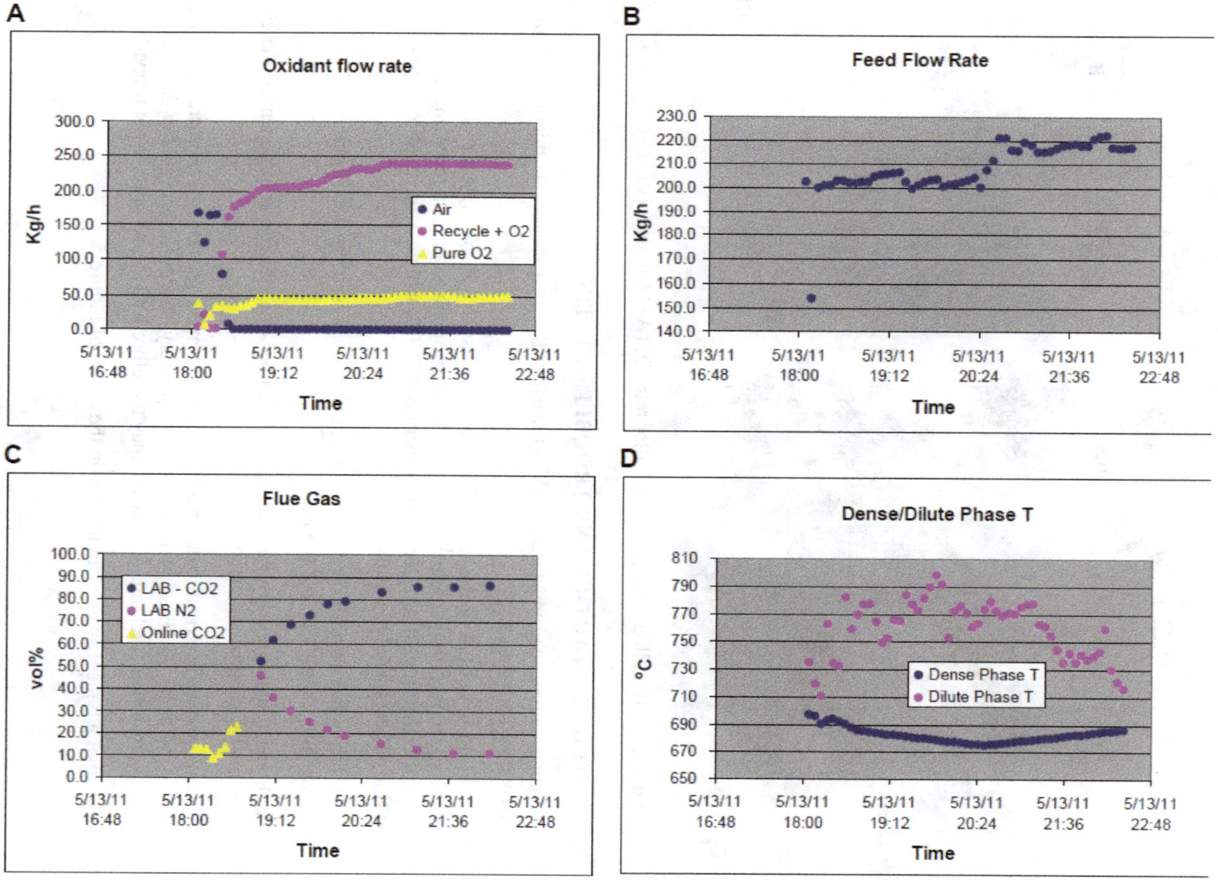
A
Oxidant flow rate
Kg/h
300.0
250.0
200.0
150.0
100.0
50.0
0.0
Air
Recycle + O2
Pure O2
5/13/11 16:48
5/13/11 18:00
5/13/11 19:12
5/13/11 20:24
5/13/11 21:36
5/13/11 22:48
Time
B
Feed Flow Rate
Kg/h
230.0
220.0
210.0
200.0
190.0
180.0
170.0
160.0
150.0
140.0
5/13/11 16:48
5/13/11 18:00
5/13/11 19:12
5/13/11 20:24
5/13/11 21:36
5/13/11 22:48
Time
C
Flue Gas
vol%
100.0
90.0
80.0
70.0
60.0
50.0
40.0
30.0
20.0
10.0
0.0
LAB - CO2
LAB N2
Online CO2
5/13/11 16:48
5/13/11 18:00
5/13/11 19:12
5/13/11 20:24
5/13/11 21:36
5/13/11 22:48
Time
D
Dense/Dilute Phase T
°C
810
790
770
750
730
710
690
670
650
Dense Phase T
Dilute Phase T
5/13/11 16:48
5/13/11 18:00
5/13/11 19:12
5/13/11 20:24
5/13/11 21:36
5/13/11 22:48
Time

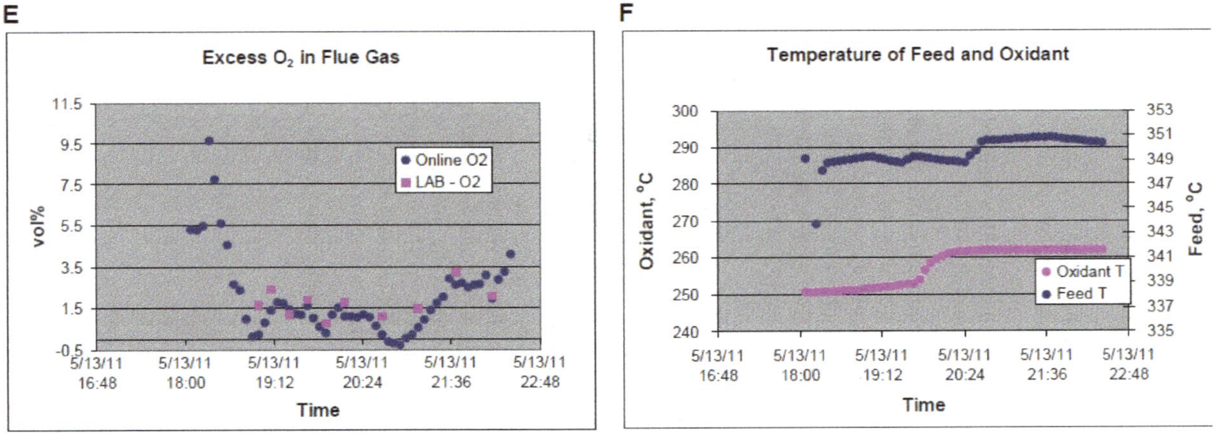

Figure 5. Charts representing the switch sequence from air-firing to oxy-firing.

Chart A shows that the switch from full air to full oxygen and recycle was very quick - taking about 20 minutes. CO_2 concentration in the flue gas increased steadily to almost 90% vol. within a couple of hours, while nitrogen concentration decreased to about 10% in the same time interval (chart C). The higher specific heat of CO_2 led to a decrease in the dense phase temperature of the regenerator by about 20°C (chart D). Since temperature in the reactor is controlled by the catalyst circulation rate from the regenerator, the system automatically increased the circulation rate to try to keep the set temperature. This caused a strong increase in the amount of coke for combustion with consequent higher consumption of oxygen (chart E). At the same time the temperature in the upper part of the regenerator (dilute phase) increased by about 80°C (chart D). This increase in temperature is related to "after-burning", caused by the low temperature of the dense phase. Increasing the flow-rate of oxygen solved the problem, making enough oxygen available for coke burning in the dense phase and restoring an acceptable level of excess oxygen. As already anticipated by simulations performed by ConocoPhillips, FCC operation in oxy-firing mode is not feasible at the "same volumetric flow rate" to the regenerator, as oxygen feed must be slightly increased, so that this condition may be defined as "same *inert* volumetric flow rate".

Overall, the experience gained from these first trials made in the pre-operation/start-up phase resulted in the following lessons learnt:

- The switch from air to oxygen is relatively fast.
- Excess oxygen in the flue gas must be closely monitored by constantly adjusting (increasing) the oxygen flow rate during the switch.
- The switch may increase the "after-burn". The use of a combustion promoter during this phase is therefore strongly advised.
- The starting temperature of the dense phase is critical for stable operation. This temperature should be kept as high as possible when the switch procedure is started to give more time for operators to adjust all of the parameters.
- The targeted condition should not be of "same volumetric flow rate" of CO_2 + O_2 as air, but "same inerts volumetric flow rate" for CO_2 compared to N_2.

The main upsets during operation were caused by severe corrosion inside the recycle compressor due to the acidic nature of the flue gas and the conditions found in that specific equipment. The compressor is an electrically driven oil-free two-stage intercooled reciprocating model. At the exit of the second stage, the compressor supplies flue gas at pressure of about 12 barg, much higher than actually needed. The body of the compressor is made of cast iron, with some valves made of carbon steel. Pistons, jackets and the other valves are made of stainless steel or aluminum.

The corrosion problems inside the recycle compressor were first detected after the second start-up test on April 7th, 2011. Upon inspection, a considerable amount of moisture and solid deposits, mainly in the pistons and valves of both compression stages, was found (pictures in Figure 6). Analysis of the solid deposits revealed that they were mainly composed of iron, confirming that the solids were mainly the result of corrosion. The high percentage of sulfur in the solids suggested that condensation of sulfuric acid inside the compressor could be the root cause of the corrosion.

It is known that part of the SO_2 formed in the regenerator may be oxidized to SO_3 (it is assumed that 2 to 5% of SO_2 is converted to SO_3). Therefore, the gas entering the recycle system contains both species of SOx (predominantly SO_2). The efficiency of the alkaline washing tower in removing SO_2 was proven to be of 96%, so that a small amount of SO_2 (3 ppm) remained in the recycled flue gas. However, the efficiency in removing SO_3 could not be measured. The dew-point temperature for the flue gas mixture entering the compressor was estimated at about 90°C, confirming that condensation inside the compressor could not be avoided (a compressor inlet temperature of > 90°C would have resulted in a discharge temperature considerably above the design temperature).

In addition, recent results published by Air Products [1] highlighted the role of the following reactions:

$$NO + \frac{1}{2} O_2 \leftrightarrow NO_2$$

$$NO_2 + SO_2 \leftrightarrow NO + SO_3$$

$$SO_3 + H_2O \leftrightarrow H_2SO_4$$

These reactions are thermodynamically favored at low temperature, and reaction rate increases with the third power of pressure. The overdesign of the compressor in terms of exit pressure contributed to a fast production of SO_3, in addition to the SO_3 already present in the flue gas, and subsequent formation and condensation of sulfuric acid followed by corrosion. This also explains why the corrosion was limited to the compressor.

Figure 6. Corrosion inside the recycle compressor.

Upon understanding of the reasons for corrosion, it was realized that preventing acid formation with the existing equipment and process scheme was not possible within the constraints of time and budget. The only alternative would be changing the compressor parts to more resistant ones. The following modifications were made:

- Design and installation of a new knock-out vessel upstream the compressor inlet with a demister and solid filter.
- Upgrade alkaline solution of the washing tower to a more efficient one.
- Steam trace the gas upstream of the first and second stage compressor inlet and to increase its temperature from 12°C to 35-40°C.
- Install a purge line with dry air to clean both stages of compression from the acid moisture at each shut-down.
- Replace the original valves with new ones in which seals are made of thermoplastic material.
- Upgrade the metallurgy of some parts of the compressor.
- Apply aluminum alloy coating to the cast iron parts.
- Upgrade metallurgy of suction and discharge valves.

After performing all of these changes, the efficiency of the recycle system improved significantly and operation became much more stable. Although some oscillation in pressure was still observed, the compressor efficiency did not drop over time. Carrying out the planned test program was therefore made possible.

The knowledge gained by troubleshooting the recycle compressor would be applied to the design of the flue gas recycle system for a large demonstration or commercial application, drastically reducing the risks of incurring this type of problems.

OXY-COMBUSTION TEST PLAN

The test program was designed to cover various pre-established conditions that were considered to be the most relevant and representative to measure the impact of oxy-combustion on the FCC unit operation. A total of 15 conditions were tested with two different feeds: a typical Vacuum Gasoil (VGO) and an Atmospheric Residue (ATR), as summarized in Table 2 and described below. Most of the conditions presented in Table 2 were previously simulated by a FCC simulator designed by ConocoPhillips, using feed and catalyst properties supplied by Petrobras. The simulations were performed to help predict what would happen with the pilot FCC unit in each mode of operation.

Table 2. Plan of testing.

Test	Feed	Mode of operation	Goal
1	VGO	Air	Base Case
2	VGO	Oxy	Same Heat Removal
3	VGO	Oxy	Same Inerts Flow Rate
4	VGO	Oxy	Same Solids
5	VGO	Oxy	Maximum Throughput
6	VGO	Air	Catalyst Balance
7	VGO	Oxy	Catalyst Balance - Same Inerts Flow Rate
8	VGO	Air	Catalyst Deactivation
9	VGO	Air + O_2 enrich.	Catalyst Deactivation
10	VGO	Oxy	Catalyst Deactivation
11	ATR	Air	Base Case
12	ATR	Oxy	Same Heat Removal
13	ATR	Oxy	Same Inerts Flow Rate
14	ATR	Oxy	Same Solids
15	ATR	Oxy	Maximum Throughput

Base Cases (Air fired Regeneration) - VGO and ATR feeds

The purpose of these tests was to get performance data from the FCC unit to be used as reference for comparison with the oxy-combustion cases.

Same Heat Removal (Oxy-fired Regeneration) – VGO and ATR feeds

For this test, the goal was to remove the same amount of heat from the regenerator during oxy-fired operation as in operation with air. Since CO_2 has a higher heat capacity than N_2, a lower volumetric flow rate of recycled CO_2 is needed when compared to the volumetric flow rate of N_2 in air. Figure 7 illustrates this condition. Since the heat balance of the FCC unit is not affected in this condition, the operating conditions should remain the same, with minimum impact on product yields. However, since the oxygen demand remains the same (the amount of oxygen is that necessary to burn the coke), the lower volumetric flow rate of CO_2 during oxy-combustion implies that the oxidant gas mixture will have a higher partial pressure of oxygen compared to air and this could in principle affect the catalyst deactivation rate by a mechanism involving steam and highly oxidized species of vanadium (a metal contained in the feed which is a poison for catalyst).

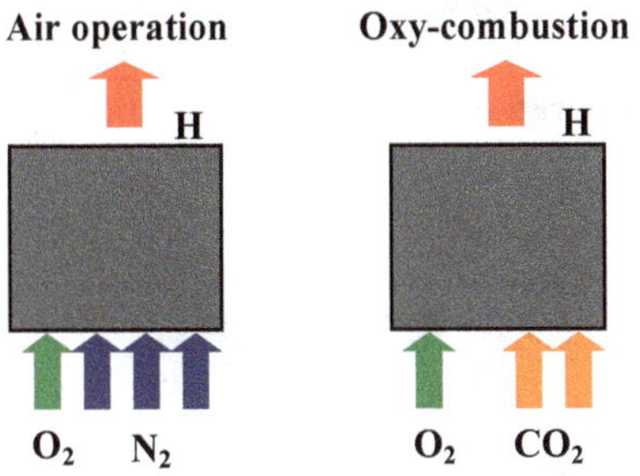

Figure 7. Oxy-combustion operation in the same heat balance condition.
H = heat removed from the regenerator by the flue gas.

Same Inerts Flow Rate (Oxy-fired Regeneration) – VGO and ATR feeds

For this test, the volumetric flow rate of recycled CO_2 is set to be exactly the same as that of N_2 during air operation. Due to the larger heat capacity of CO_2, the amount of heat removed from the regenerator (H_1) is greater compared to the operation with air. Figure 8 illustrates this condition. In this type of operation, the thermal balance of the unit is affected by the extra heat removed from the regenerator. In order to reestablish thermal equilibrium, catalyst circulation is increased to account for the heat demand from the reaction side. In this operating condition, it is expected that the catalyst-to-oil ratio (CTO), which is the amount of catalyst per mass of feed, will increase, and this may represent a gain in total feed conversion rate. However, due to the higher specific gravity of CO_2, the total amount of solid entrainment is also expected to increase. This may affect the loss of catalyst fines through the regenerator.

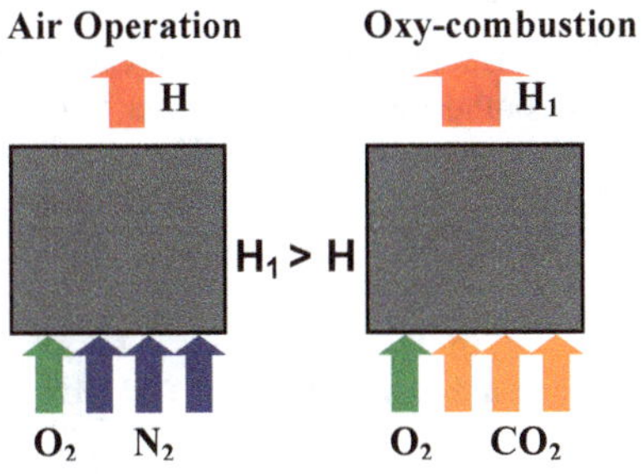

Figure 8. Oxy-combustion operation in the same inerts volumetric flow rate condition.
H and H_1 = heat removed from the regenerator by the flue gas.

Same Solid Entrainment (Oxy-fired Regeneration) – VGO and ATR feeds

Condition for this test were set to have the same solids entrainment rate for oxy as for air operation in the regenerator. Since CO_2 has a higher specific gravity than N_2, the volumetric flow rate of recycled CO_2 is less than that of N_2 during air operation, however it is still higher than for the same heat balance condition. This is therefore an intermediate oxy-combustion mode of operation when compared to the two previous ones, and both the oxygen partial pressure and the heat removed from the regenerator (H_2) have intermediate values. This condition was suggested by ConocoPhillips and anticipated as a possible solution for the solids balance (entrainment) in oxy-combustion mode. Figure 9 illustrates this condition. To estimate the solid entrainment rate and the operating point for this condition, the PSRI (Particulate Solid Research, Inc.) correlations were used.

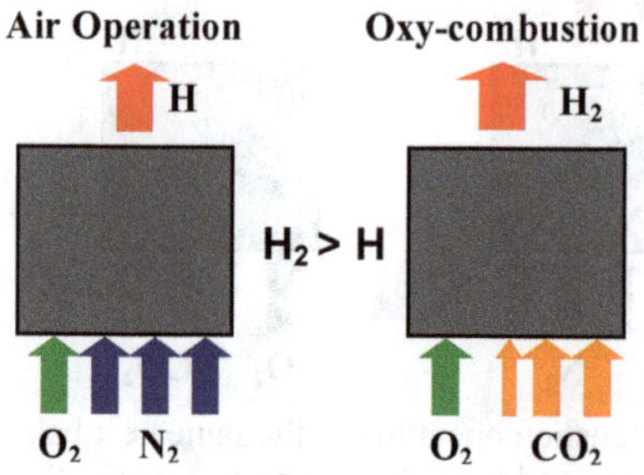

Figure 9. Oxy-combustion operation in the same solids entrainment condition.
H and H_2 = heat removed from the regenerator by the flue gas.

Maximum Throughput (Oxy-fired Regeneration) – VGO and ATR feeds

The goal for this test was to define the maximum possible increase in feed while operating in the same inerts volumetric flow rate condition, taking advantage of the higher amount of heat removed from the regenerator. As described previously, when operating in the same inerts flow rate condition, the lower regenerator temperature will increase the catalyst to oil ratio and consequentially the feed conversion. The idea here is to trade the gain in feed conversion for an additional amount of feed processing, with little or no change in product yields when compared to the base case air operation.

Catalyst Balance (Air and Oxy-fired Regeneration) – VGO feed

This test condition was designed to measure the differences in catalyst loss through the regenerator cyclones when operating in the oxy-combustion mode. In order to do that, long term runs (5 to 10 days) in both air (base case) and oxy modes are needed. The same inerts flow rate condition was chosen because this is the most severe condition regarding catalyst entrainment. The catalyst inventory was calculated by collecting and measuring loss through the third stage cyclones and by monitoring regenerator level differences.

Catalyst Deactivation (Air and Oxy-fired Regeneration) – VGO feed

This test condition was designed to evaluate whether there would be a higher catalyst deactivation rate while operating in oxy mode due to the higher oxygen partial pressure. Similar to the previous case, in order to measure catalyst activity decay, long term runs (5 to 10 days) in both air (base case) and oxy modes were needed. The same heat balance condition was chosen as for the previous test. Two additional tests were designed using air plus oxygen enrichment with the oxygen partial pressure in the same level as that obtained during the oxy mode to try to isolate the effect of oxygen partial pressure in catalyst deactivation.

OXY-COMBUSTION TEST RESULTS

The entire test program outlined above, designed to test the impacts of this technology on the FCC process, was carried out. The main findings are summarized below:

- The CO_2 concentration in the flue gas reached values close to 95% (dry basis), which was one of the targets of testing the oxy-combustion technology. A comparison of typical compositions between oxy-firing and normal operation is shown in Table 3.

Table 3. Comparison between flue gas compositions.

Flue Gas Composition (dry)	Oxy-firing	Normal operation
CO_2 (vol. %)	93.0	15.0
Oxygen (vol. %)	2.8	3.7
Nitrogen (vol. %)	2.7	79.7
CO (vol. %)	0.3	0.1

Given that the O_2 purity was 99.5%, the N_2 contamination was most likely caused by pure nitrogen used to fluidize the catalyst in the standpipe of the pilot unit. Since steam is used for this purpose in industrial units, the CO_2 content in the flue gas is expected to be even higher than achieved in these tests. Therefore, it is likely that oxygen is the only major impurity to be removed before transportation/storage when applying oxy-firing in a commercial unit. A specific purification technique, based on catalytic combustion of excess oxygen could therefore be applied. Incidentally this would also take care of the carbon monoxide.

- It is possible to run the FCC unit in the oxy-fuel mode at the same heat balance condition as air operation with no impact on the thermal balance and with little changes in product yields (Figure 10). However the CO_2 recycle rate used in this condition resulted in higher O_2 partial pressure, in the range of 27-30%vol. These results were in agreement with the ConocoPhillips simulations.

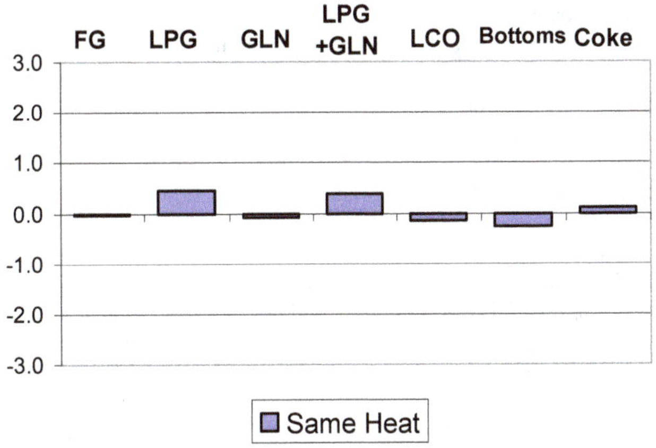

Figure 10. Same Heat Balance – Comparison oxy-firing vs. air firing - absolute yield change.

- It is also possible to run the unit in the same inert volumetric flow rate condition, where the CO_2 recycle rate matches the N_2 flow rate during air operation. In this condition, the thermal balance of the unit is affected, resulting in a decrease in dense phase temperature, higher catalyst circulation and gain in feed conversion (Figure 11). In this case the distribution of the products improves considerably with an increase of about 2 absolute points for gasoline and LPG, and corresponding reduction in the heavy bottoms. The trends observed were all in agreement with the simulations, although the absolute values differed from those predicted.

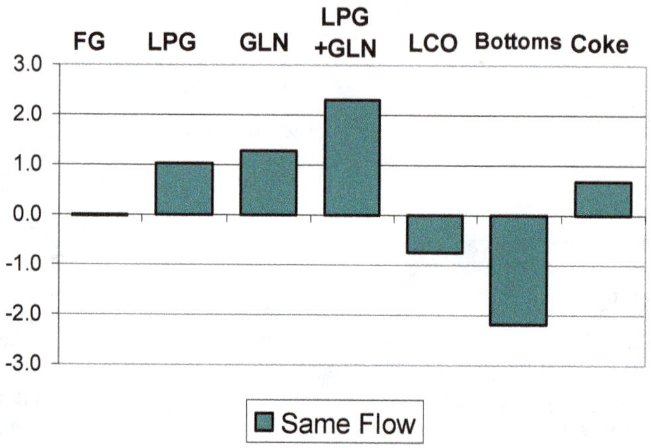

Figure 11. Same Inert Flow rate – Comparison oxy-firing vs. air firing - absolute yield change.

- The catalyst entrainment rates compared to air operation varied from lower values (for the same heat balance condition) to higher values (for the same inert flow rate condition). An intermediate condition, called the "same solid entrainment" was tested and confirmed that it is possible to operate the unit in oxy mode with the same amount of solids entrained as in air operation. This condition turned out to be the most difficult to achieve and the simulations were very helpful in anticipating the values of some operating variables. In this condition, the heat removed from the regenerator is still higher and the gains from an increased catalyst circulation may still be obtained. It is estimated that a gain in production capacity of about 3% compared to normal operation is achievable, while keeping the entrainment rate at the same level as with air operation. This case will be considered to evaluate possible cost mitigation of capture related to debottlenecking of the FCC unit.
- Achieving the design production capacity using a heavier, cheaper feedstock, could be an alternative option to exploit the increased flexibility given by oxy-firing the regenerator. This option was not tested during this project and a quantification of potential benefits would be very site specific.
- The long term deactivation tests showed that the increase in oxygen partial pressure when operating in the oxy mode did not appear to significantly increase catalyst deactivation, indicating that for industrial operation there will not be a need to increase catalyst make-up.
- The catalyst balance tests showed that the higher entrainment rates obtained while operating in oxy mode with a CO_2 recycle rate above that of the same solid entrainment condition will in fact result in a higher cyclone loading and higher catalyst loss (Figure 12). For an industrial unit this may represent a faster degradation of the cyclone's internals and the need for higher catalyst make-up. Calculations showed that the catalyst loss was increased by 80% compared to standard operation. Analysis of the fines collected in the 3rd cyclone stage also showed an increase in the mean size of the particles, confirming the higher level of catalyst loss. This would also lead to significant cyclone erosion and possible increased wear in other flue gas components. This would either add maintenance costs, require cyclone modifications or reduced run length between maintenance turnarounds.

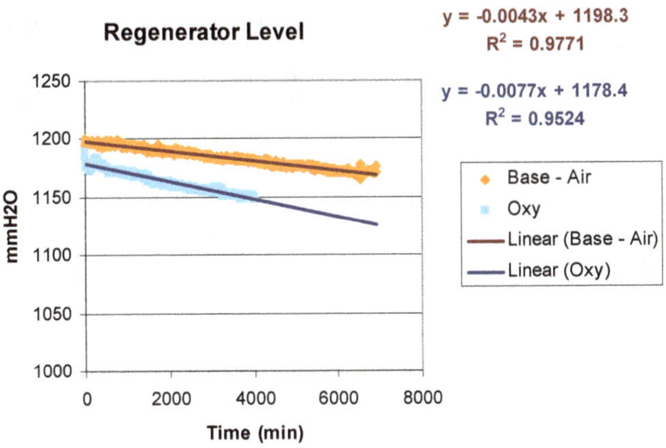

Figure 12. Regenerator level variation vs. time in operation.

- During oxy-fuel operation, the recycling of the flue gas will lead to a build-up in NO_x. Preliminary mass balance calculations had shown that the increase of NO_x in the flue gas could be as high as 6 times the base level. Figure 13 shows the variation of NO_x levels in flue gas during transition to oxy-firing in a typical run. The increase in NOx levels varied from 1.6 to 3.3 times the base level, which is considerably lower than predicted with the mass balance calculations. In any case, the overall NO_x emissions from the FCC shall decrease, as the NO_x would be captured (and possibly stored) along with the CO_2.

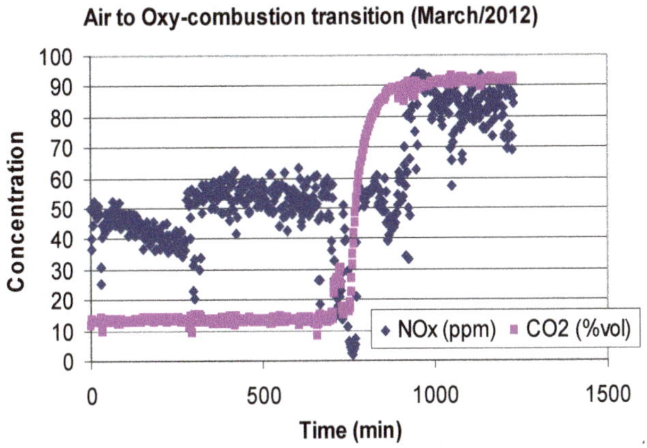

Figure 13. NOx concentration in flue gas vs. time.

FINAL CONSIDERATIONS

The field demonstration run has confirmed the technical viability of retrofitting an FCC unit to enable CO_2 capture through oxy-firing with CO_2 recycle. The demonstration proved that the FCC unit can work steadily in oxy-firing mode and that oxy-firing can enable a higher throughput or allow switching to the processing of heavier feeds.

Commercial demonstration of post-combustion might in principle be possible in the short term, considering the maturity of the technology.

For oxy-firing, more activities at the pilot plant level might be required before proceeding with the scale-up to a commercial demonstration, notably:

- Testing of different types of compressors for recycle of flue gas and optimized recycle schemes to verify that the corrosion problems have been actually solved.
- Testing of a purification section based on catalytic combustion, because this is novel application for this technology.

ACKNOWLEDGEMENT

The CCP acknowledges the outstanding contribution of ConocoPhillips during the preparation of the FCC Oxy-fuel demonstration. Special thanks to the team composed by Julio Rincon, Kening Gong and Paul Meier.

REFERENCES

1. V. White, K. Fogash, F. Petrocelli; “Air Products Oxyfuel CO_2 Compression and Purification Developments”, 2nd Oxyfuel Combustion Conference (OCC2), Yeppoon, Queensland, Australia, [2011].
2. L. de Mello, R. Pimenta, G. Moure, O. Pravia, L. Gearhart, P. Milios, T. Melien; “A technical and economical evaluation of CO_2 capture from FCC units; GHGT11; *Energy Procedia* 1 (2009) 117-124.

Carbon Dioxide Capture for Storage in Deep Geological Formations, Volume 4
Karl F. Gerdes (Editor)

Chapter 7

CO_2 CAPTURE FROM A FLUID CATALYTIC CRACKING UNIT: TECHNICAL/ECONOMIC EVALUATION

Ivano Miracca[1], David Butler[2]
[1]Saipem S.p.A. (Eni) – Via Martiri di Cefalonia, 67 – I20097 San Donato Milanese (Italy)
[2]Calgary, Alberta, Canada

ABSTRACT: The Fluid Catalytic Cracking (FCC) is a key unit in planning CO_2 mitigation from oil refineries, since it is often the largest single emission source in a refinery. This chapter summarizes the technical and economic evaluation of CO_2 capture from the effluent of the fluid catalytic cracking (FCC) catalyst regenerator using oxy-firing and state-of-the-art post-combustion solvent scrubbing.

KEYWORDS: fluid catalytic cracking (FCC); CCS; CO_2 capture; post-combustion; oxy-combustion; oxy-firing

THE FCC SCENARIO IN CCP3

The basis selected for the CCP3 scenario is a medium size FCC unit, with a feed rate of 60,000 barrels per day (bpsd). FCC units may be much larger than this - with a single line record capacity of 180,000 bpsd (Reliance, India). Several units are in the 100,000-130,000 bpsd range. The unit studied in CCP2 was 120,000 bpd [1]. In CCP3 the size was reduced to consider a more common size. The main features of the CCP3 FCC unit concerning carbon capture are reported in Table 1.

Table 1. Main features of the CCP3 FCC unit.

Capacity (bpd)	60,000
Flue Gas Flow (kg/hr)	272,000
Flue Gas Pressure (barg)	0.0
Flue Gas Temperature (°C)	200
Flue Gas Composition (mol%)	
CO_2	16.29
SO_X	220 ppm
NO_X	200 ppm
CO	100 ppm
N_2	74.38 (balance)
Ar	0.89
O_2	1.04
H_2O	7.38

The FCC unit selected for CCP3 emits to the atmosphere about 510,000 tons of CO_2 per year, based on an on-stream factor of 90% at max capacity. The captured CO_2 is compressed to 150 barg for transportation.

For this scenario Amec Foster Wheeler (AFW) studied three configurations:

- Post-combustion
- Oxy-firing with 97% oxygen purity
- Oxy-firing with 99.5% oxygen purity

Pre-combustion is not applicable to this operation (the "fuel" is coke on the catalyst).

POST-COMBUSTION CAPTURE STUDY

Post-combustion represents the "conservative" way to deal with CO_2 capture in this scenario. Modifications to the existing plant are located downstream from the FCC unit and do not have any direct interactions with the process itself. A schematic representation of the application of post-combustion to the regenerator of an FCC unit is shown in Figure 1.

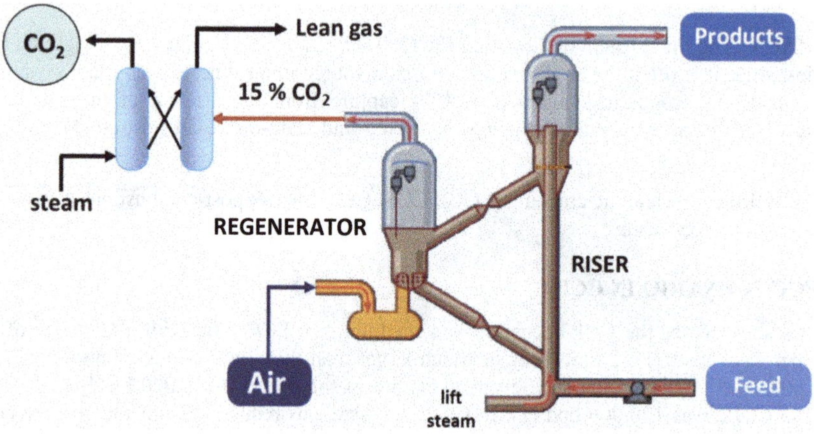

Figure 1. Post-combustion capture in an FCC unit.

The flue gas is fed to an amine scrubbing system in which CO_2 is chemically absorbed by an aqueous amine solution. Almost pure (> 99.5% vol.) CO_2 is desorbed in the reboiled stripper unit. It is assumed that 5% of the total flue gas flow bypasses the CO_2 capture plant via the stack to assure safe operation by keeping the stack warm and swept of air. Flue gas is delivered to the CO_2 capture unit by a flue gas blower. In the event of a failure or trip of the flue gas blower, the flue gas is emitted directly to the atmosphere through the existing stack with no impact on operation. A continuous flow of flue gas to the stack is always maintained to avoid fresh air being drawn into the flue gas duct via the stack. The CO_2 capture unit removes 90% of the CO_2 in the flue gas it receives, with the clean flue gas returning to the stack. A single train of CO_2 absorption, stripping, compression and dehydration was considered the most appropriate for the total volume of gas to be treated in the case. A stand-alone steam boiler package (not shown) supplies the steam required for the stripper reboiler and solvent reclaimer as well as for regeneration of the dehydration beds. The auxiliary steam boiler is fed by Refinery Fuel Gas (RFG). A combination of sea and fresh cooling water is used and electrical power is supplied over the fence to the CO_2 capture area.

A commercial state-of-the-art post combustion capture technology using a proprietary solvent has been used for this study (see chapter 3). The provider (Mitsubishi Heavy Industries) was directly involved and estimated sizing, capital cost, utility and chemical consumption for the capture system. AFW supplied the balance of plant components and costs. The key information related to CO_2 captured, CO_2 avoided and related power and utility consumption are reported in Table 2.

Table 2. Main results of post-combustion study.

CO_2 In Flue gas	kg/h	64938
CO_2 to Capture Plant	kg/h	61691
CO_2 Captured	kg/h	55522
CO_2 to Water Streams	kg/h	6
CO_2 Emitted from Stack	kg/h	9416
CO_2 Capture across Capture Plant	%	90%
CO_2 Capture from Total Flue Gas	%	85%
Total Power Usage	MW	8.01
Total Steam Usage	kg/h	64871
Total Cooling Water Usage	te/h	5094
Additional CO_2 Generated for Utilities		
CO_2 due to Power use	kg/h	4807
CO_2 due to Steam use	kg/h	8167
Overall CO_2 Avoided	%	65.5%
Overall CO_2 Avoided	kg/h	42548
Total Steam Usage per unit CO_2 Avoided	kg/kg	1.52
Total Steam Usage per unit CO_2 Captured	kg/kg	1.17

A plot space of 80x53 meters is needed close to the FCC unit to accommodate the capture unit. An additional plot space of 43x43 meters is needed for CO_2 dehydration, compression and for the auxiliary boiler needed to generate the stripping steam.

OXY-FUEL CAPTURE STUDIES

The oxy-firing technique involves use of pure oxygen in a combustion operation to produce a flue gas that is mainly CO_2 and H_2O. For the FCC unit, oxygen is fed to the regenerator rather than air to burn the coke deposited on the catalyst surface. Since pure oxygen combustion would result in extremely high temperature, partial recycle of the flue gas is necessary to maintain the regenerator temperature within normal operating limits. A schematic representation of the application of oxy-firing to the catalyst regenerator of an FCC unit is shown in Figure 2. Air is fed to a cryogenic separation unit (ASU), producing oxygen with a specified purity. Oxygen is then mixed with part of the dehydrated flue gas and sent to the regenerator. The flue gas at the regenerator exit contains about 85% CO_2 by volume (dry). Depending on the presence of other impurities (nitrogen, oxygen) further purification (not shown in the figure) may be needed before the CO_2 may be considered ready for transportation and injection. In the oxy-firing option, no partial bypass to the stack is required, since the flue gas is compressed. This gives oxy-firing an inherent boost on CO_2 recovery compared to post-combustion.

When setting the basis for a study applying oxy-fuel capture to an FCC unit, there are several options that may affect the cost of capture. The two main ones are:

- ***Oxygen purity:*** cryogenic air separation may be designed to supply oxygen of a requested level of purity, commonly ranging from 90% to 99.9% by volume. According to information supplied Praxair, one of the major vendors of ASU's, overall costs of ASU are quite flat in the 90-97% purity range, increase by about 10% going to 99.5% purity from 97% purity and ramp very steeply for higher purity (See Chapter 19).
- ***CO_2 purification system:*** cryogenic purification is typically included in purification schemes for oxy-fuel capture. The case of the FCC is somewhat unique, because oxygen is by far the main impurity and might alternatively be removed by catalytic combustion.

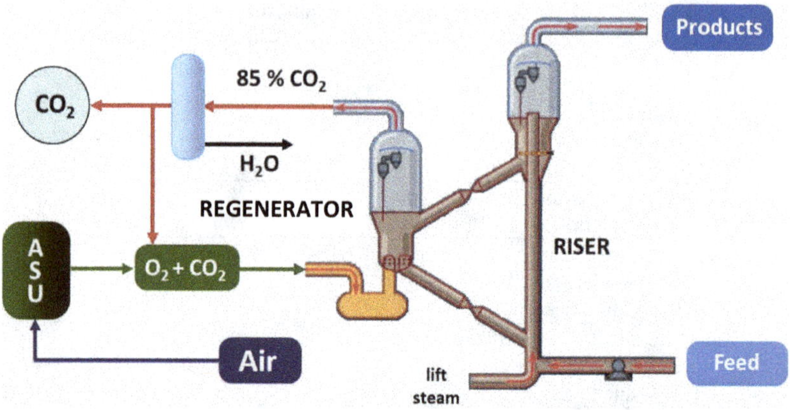

Figure 2. Oxy-fuel capture in an FCC unit.

Two cases have therefore been studied:

- Case 1 with 97% oxygen purity, cryogenic purification of CO_2 and 90% CO_2 capture rate.
- Case 2 with 99.5% oxygen purity and purification via catalytic combustion with greater than 99% CO_2 capture rate.

The > 99% capture rate for Case 2 was selected because the combination of higher purity oxygen and catalytic combustion of residual O_2 has negligible additional cost compared to Case 1.

The costs for the ASU were supplied directly by a vendor. AFW calculated the costs for the balance of plant, including flue gas dehydration and recycle, CO_2 purification and compression.

A block flow diagram for Case 1 is shown in Figure 3. Case 1 is comprised of a waste heat steam generator, SO_2 scrubber (with limestone), a direct contact quench cooler, air separation unit, flue gas recycle compressor, and CO_2 compressor, purification and dehydration unit. The recycle back to the FCC regenerator is 82% of the treated and cooled flue gas. This recycle maintains the same volumetric flow rate through the FCC regenerator as in normal air operation ("same inert flow" mode. See Chapter 6). The recycle gas is compressed from 1.14 bar (16.5psia) to 3.95 bar (57 psia) in a 2-stage compressor with a cooling water intercooler before mixing with the oxygen for combustion in the FCC. The net flue gas stream is compressed (4 stages), then sent to dehydration, followed by cryogenic purification. The cryogenic purification condenses carbon dioxide in two stages using two sequential plate-fin heat exchangers. The condensed CO_2 is revaporized, providing cooling in the purification process, with the vapor returned to the final CO_2 compressor stages. Oxygen is removed in a stripper column and the oxygen-depleted CO_2 is then pumped to the specified 151 bar (2190 psia) delivery pressure. A non-condensable stream is reheated to ambient temperature and vented. The resulting overall capture rate is 90%. In accordance with the approach described above regarding steam self-sufficiency of the carbon capture unit, a natural gas fired package boiler provides MP steam to the ASU (to drove the air compressor) and carbon dioxide dehydration mol sieve dryers. CO_2 is not captured from the flue gas from this boiler.

The same block flow diagram of Figure 3 may be considered representative for Case 2. Case 2 is comprised of a waste heat steam generator, SO_2 scrubber (with limestone), a direct contact quench cooler, air separation unit, flue gas recycle compressor, and CO_2 compressor, catalytic purification and dehydration unit. The cryogenic air separation unit (ASU) with integrated oxygen compressor produces 99.5% purity gaseous oxygen at approximately 3.75 bara (54 psia) which is sent to the FCC regenerator. The recycle back to the FCC regenerator is 78.4% of the treated and cooled flue gas. This recycle maintains the volumetric flow rate through the FCC regenerator close to normal operating mode ("same inert flow" mode, as described in Chapter 6).

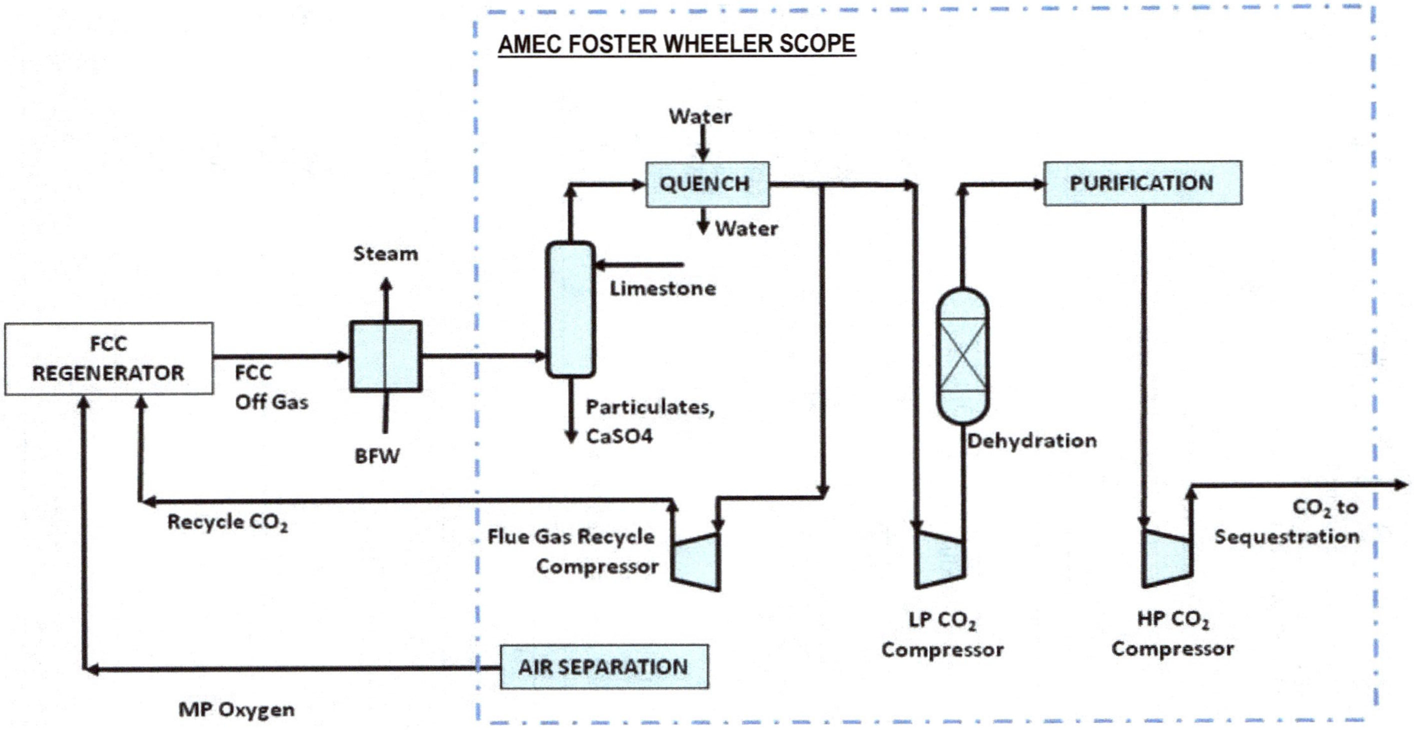

Figure 3. Block flow diagram of oxy-fuel capture in FCC – Case 1.

Carbon dioxide purification is accomplished in a catalytic reactor. The oxygen content in the gas is reacted with hydrogen over a catalyst, to meet the oxygen requirement for transportation and injection, conservatively assumed to be 75 ppm. The information related to CO_2 captured, avoided and related power and utility consumption for both cases are reported in Table 3. In Case 2 the capture rate of CO_2 is 99.98% for the CO_2 fed to the capture step, with only minor losses of CO_2 to water during condensation.

Table 3. Main results for the FCC oxy-fuel cases.

Description		Case 1 97% Oxygen Cryogenic sep.	Case 2 99.5% Oxygen Catalytic comb.
Capacity		60,000 BPSD	60,000 BPSD
Oxygen Consumption	te/day	1417	1376
	lb/hr	130164	126389
Fuel Consumption – Auxiliary Boiler	GJ/h LHV	3.5	2.8
	MMBTU/hr	3.3	2.6
Flue Gas CO_2 Concentration	mol %	89.5	88.2
Process CO_2 Balance			
CO_2 In Flue gas	kg/h	64776	64768
	lb/hr	142806	142781
CO_2 to Capture Plant	kg/h	64776	64768
	lb/hr	142806	142781
CO_2 Captured	kg/h	58335	64755
	lb/hr	128607	142752
CO_2 to Water Streams	kg/h	8	13
	lb/hr	18	29
CO_2 Emitted from Capture Plant	kg/h	6433	0
	lb/hr	14182	0
CO_2 Capture across Capture Plant	%	90.0	99.98%
Total Power Usage	MW	34.2	32.5
Total Steam Usage	kg/h	0	1275
	lb/hr	0	2811
Total Cooling Water Usage	te/h	4309	4452
	lb/hr	9499708	9814968
Additional CO_2 Generated for Utilities			
CO_2 due to Power use	kg/h	22062	18198
	lb/hr	48638	40120
CO_2 due to auxiliary boiler	kg/h	171	62
Total CO_2 Balance			
Total CO_2 Generated	kg/h	87009	83028
	lb/hr	191822	183035
Total CO_2 Captured	kg/h	58335	64755
	lb/hr	128607	142752
Total CO_2 Emitted	kg/h	28674	18273
	lb/hr	63215	40283
Overall CO_2 Avoided	%	71.6%	85.4%

Comparing the two cases, Case 2 has lower power consumption in spite of the higher purity oxygen used. The details of power consumption for the two cases are reported in Table 4. For Case 1, the energy requirement of the cryogenic purification and the higher recycle rate exceeds the increased power consumption for Case 2 in the ASU to achieve the higher purity oxygen. Note that the incremental power consumption for both oxy-fuel cases is about 10.5 MW lower than that shown in Table 4, since the air blower for conventional operation is not used.

For both cases, a plot space of about 50x50 meters is needed close to the FCC unit for the flue gas quenching and recycle system. Another plot of the same size is needed for CO_2 purification and compression (possibly less in the case of catalytic combustion), while the ASU needs a 65x50 meter plot, possibly in a separate area away from the FCC unit.

Table 4. Breakdown of power consumption for the oxy-fuel cases.

Power consumption (MW)	Case 1	Case 2
Air Separation	11.4	15.9
Oxygen compression	2.9	2.7
Flue Gas Recycle comp.	9.8	7.5
CO_2 purification & compression	9.1	6.4
TOTAL	**34.2**	**32.5**

TECHNICAL COMPARISON BETWEEN POST-COMBUSTION AND OXY-FUEL

Electric power is by far the main energy consumption for the oxy-fuel scheme, while post-combustion mainly relies on fuel for steam. The cost of different energy sources in different locations and times may therefore play an important role in selecting the lowest cost capture technique. A sensitivity analysis of economics may help in understanding the relative cost of energy which may favor one technique.

The optimal capture rate for post-combustion solvent scrubbing is not higher than 85-90%, and the overall rate will be lower due to the requirement for 5% of the flue gas to by-pass the capture unit for safety reasons. The optimal capture rate for oxy-fuel operation may be higher than 99% of the CO_2 in the flue gas fed to the capture step. This may give an advantage to oxy-fuel, depending on the overall level of mitigation required.

The overall plot space required by oxy-fuel is larger than for post-combustion, but the plot space close to the unit, which may be most critical, is roughly half for oxy-fuel compared to post-combustion.

Finally, post-combustion may be considered as a relatively mature technology (already widely commercialized and applied, though at a smaller scale) and does not impact the FCC process directly, while application of oxy-fuel to the FCC unit has a direct impact on the FCC process and needs to be proven and properly scaled-up before commercial application. For this reason, the CCP carried out a demonstration of FCC oxy-firing in a large pilot unit, described in Chapter 6.

SUMMARY OF ECONOMC EVALUATIONS

Table 5 summarizes the key inputs and outputs of each case developed by AFW and used to derive the economic results of CO_2 capture technologies employed on a Fluidised Catalytic Cracker (FCC). All cases are assumed to have a feedstock input rate of 60,000 bpd and are located on the USGC.

The descriptions of the cases in Table 5 are:

Post-combustion:

- **Post C.-Amine:** FCC retrofit – Amine Scrubbing using state-of-the-art commercial technology

Oxy-firing:

- **Oxy C. (97%):** FCC retrofit – Oxy-firing; using 97% pure O_2
- **Oxy C. (99.5%):** FCC retrofit – Oxy-firing; using 99.5% pure O_2

Table 5 shows the incremental fuel and power used for capture, the CO_2 directly generated by the industrial process and the CO_2 captured and emitted. This is followed by the values used to calculate the mass of CO_2 avoided. Of note, the oxy-fuel cases show an avoided CO_2 benefit associated with a fuel savings. The hot gases sent to the waste heat boiler (WHB) will contain more CO_2 when the FCC is fired on oxygen compared to when it is fired on air. This higher concentration of CO_2 will make the WHB more efficient - producing an estimated 13.28 t/hr of additional steam (about 25% more). The production of this additional steam will mean that less fuel will need to be consumed elsewhere in the refinery, and CO_2 emissions will be avoided. This accounts for the negative values for Steam for WHB shown in Table 5 below. Subtracting a negative value increases avoided CO_2 emissions. The post-combustion cases produce additional CO_2 because of steam required by the capture process. The Oxy C. (99.5%) case produces almost no CO_2 emissions at the plant site.

Table 5. Key Metrics of CO_2 Capture Technologies Applied to FCC.

Characteristic	**Units**	**Post C. -Amine**	**Oxy C. (97%)**	**Oxy C. (99.5%)**
Fuel Type		RFG	RFG	RFG
Throughput	bpd	60,000	60,000	60,000
Incremental Fuel - HHV	GJ/hr	185.5	-42.6	-43.4
Incremental Power	MW	8.0	23.9	21.9
CO_2 Intensity on Throughput wo/Capture	kg/kb	26.0	25.9	25.9
CO_2 Intensity on Throughput w/Capture	kg/kb	9.0	7.4	4.3
CO_2 Generated	Mt/yr	0.51	0.51	0.51
CO_2 Captured	Mt/yr	0.44	0.46	0.51
CO_2 Emitted by Plant	Mt/yr	0.07	0.05	0.00
Reference Emissions - No CCS	Mt/yr	0.51	0.51	0.51
Less: CCS Emissions	Mt/yr	0.07	0.05	0.00
Less: Capture Power Use	Mt/yr	0.04	0.11	0.10
Less: Steam for WHB	Mt/yr	-	-0.02	-0.02
Less: Capture Steam Use	Mt/yr	0.06	0.00	0.00
Avoided Emissions	Mt/yr	0.34	0.37	0.43

Table 6 shows the key financial results for each of the cases. The "% Avoided" is defined as the mass of CO_2 avoided divided by the mass of CO_2 generated in the base reference case without CCS. The Levelized Capture Cost is based on levelized incremental costs rather than on first year costs. The Additional Yearly Cost row shows the cost associated with capturing CO_2 each year. It includes an amortized annual cost for capital recovery.

Table 6. Financial Results.

Characteristic	Units	Post C. -Amine	Oxy C. (97%)	Oxy C. (99.5%)
Incremental Fuel	MUSD/yr	6.6	-1.5	-1.5
Incremental Capital	MUSD	174.4	312.9	275.6
Incremental O&M	MUSD/yr	18.2	30.4	30.9
% Capture	%	85.5%	90.1%	100.0%
% Avoided	%	65.5%	71.6%	83.5%
Capture Cost	USD/tCO_2	94.2	126.9	108.3
Avoided Cost	USD/tCO_2	122.9	159.7	129.7
Levelized Capture Cost	USD/tCO_2	110.0	148.3	126.6
Levelized Avoided Cost	USD/tCO_2	143.6	186.6	151.6
Additional Yearly Cost	MUSD/yr	41.2	58.4	55.3
Incremental Cost per bbl	USD/bbl	2.1	3.0	2.8

The Post C.-Amine case has the lowest capture cost and additional yearly cost. The avoided costs for the Post C.-Amine and Oxy C. (99.5%) are similar. It appears that there is an economic benefit associated with using higher purity oxygen in the oxy-fuel configuration. FCC units include a compressor to feed air to the regenerator. This compressor consumes 10.5 MW for the FCC studied. This compressor is no longer required for the oxy-fuel cases, because oxygen is supplied at the required pressure by the air separation unit, and that power requirement is accounted for. Therefore, the 10.5 MW of power for the FCC air blower was deducted from the power requirements for the oxy-fuel cases. Since these are retrofit cases, no cost associated with an un-needed air blower have been deducted from the capital costs. The air blowers are assumed to be retired. However, on a greenfield basis, the cost of this air blower can be avoided.

Figure 4 shows the cost segments which make up the total cost to capture CO_2 for the cases considered. Capital costs tend to dominate the costs for most of the cases. The oxy-fuel case has a large CapEx associated with the air separation unit, which also increases the power cost and associated CO_2 footprint. The Post C.-Amine case shows a significant requirement for natural gas fuel which is used to produce steam required for the CO_2 capture process. This case also benefits from substantially lower capital and power costs compared to the other cases.

For the oxy-fuel cases there is a significant fuel saving since a significant amount of waste heat is used to produce extra steam. This accounts for the negative fuel values in Figure 4. The fuel savings is a relatively small effect, however, and does not change the competitiveness of the oxy-fuel option relative to the post combustion case. Transportation and storage are expected to represent a small contribution to the cost.

Figure 5 shows the component costs making up the avoided cost of CO_2 for the cases considered.

The field demonstration project has shown that oxy-firing may increase the flexibility of the operation of an FCC unit. This may be achieved either by increasing the feed flow rate (up to about 3% without any negative effects) or decreasing the quality of the feedstock. In each case, an additional operating margin would be achieved, somewhat mitigating CO_2 capture costs. For instance, considering a 3% increase in production capacity, with an operating margin of $8 per additional barrel processed (mean world value for the year 2012, according to the BP Statistical Review of World Energy 2013), an additional annual profit of about $5 million would be achieved, which is about 35% of O&M costs related to capture.

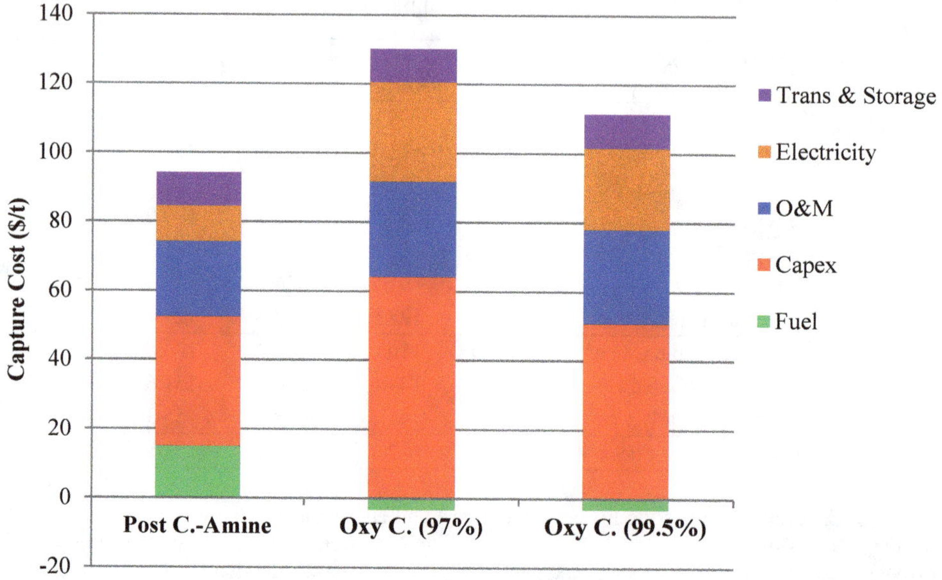

Figure 4. Capture Cost Components.

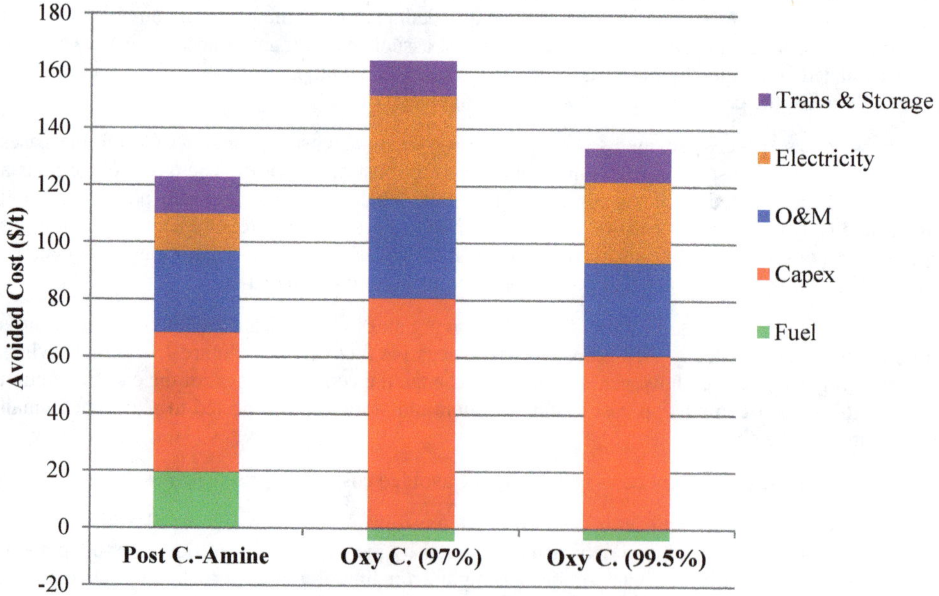

Figure 5. Avoided Cost Components.

Figure 6 shows the impact of this extra capacity on the avoided cost of CO_2 for the three cases evaluated.

Capital costs for air separation units were provided by a third party on a turn-key basis. The CCP assumed that these costs include Outside Battery Limits (OSBL). Since the extent of OSBL inclusion is not clear, Figure 7 shows the impact of adding OSBL to the air separation unit capital costs.

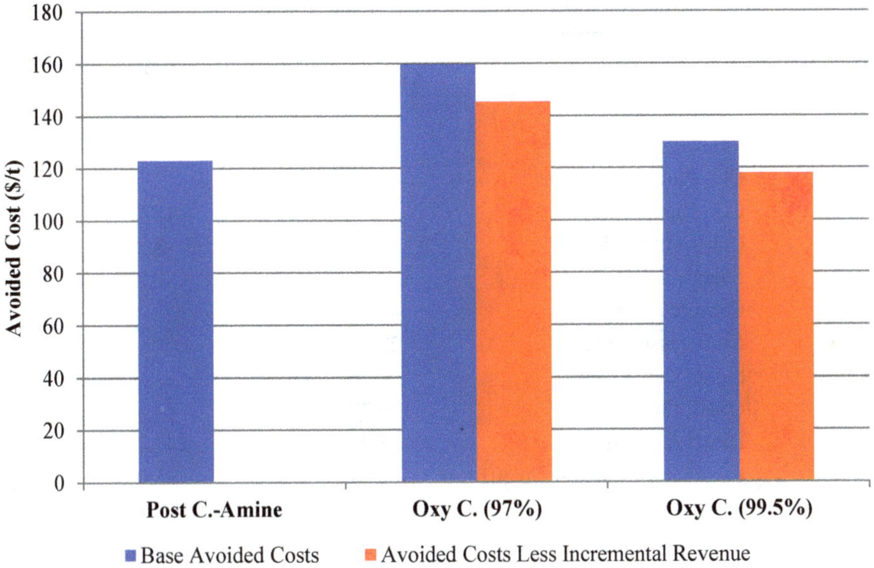

Figure 6. Impact of Extra Throughput Associated with Oxy-Combustion.

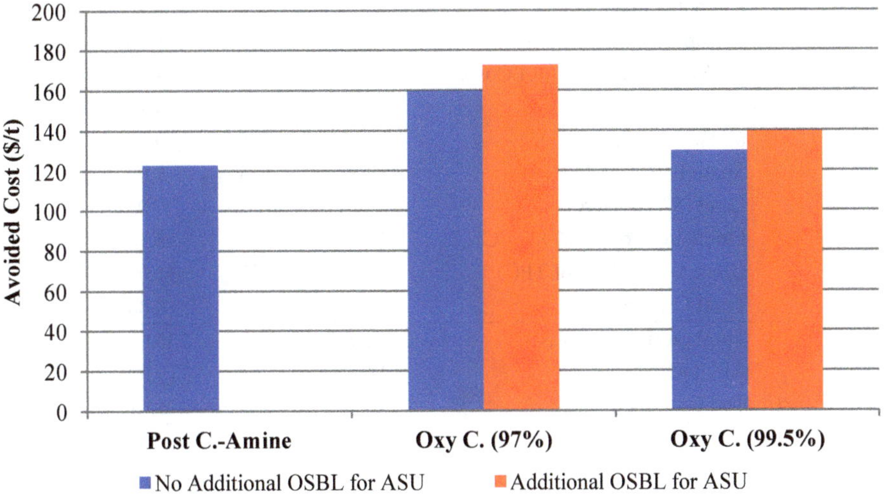

Figure 7. Impact of Including OSBL on Air Separation Capital Costs.

Table 7 shows the components that contribute to the total incremental capital cost to capture CO_2. The top part of the table shows the direct costs for each case. This is followed by indirect OSBL costs. The base capital costs have been adjusted to include escalation from Q1 2009 to the assumed in-service date of Q1 2014. This is followed by estimates for contingencies. To this, estimates for Owners Costs are added. Finally AFUDC during construction are included. The bottom row shows the total as spent costs expected just before the plant is commissioned.

Table 7. Capital Cost Components.

Cost	**Post C. -Amine**	**Oxy C. (97%)**	**Oxy C. (99.5%)**
Oxygen Supply		89.70	80.33
Flue Gas Treat		27.78	21.42
CO_2 Purification		5.97	2.59
CO_2 Compression		14.39	15.70
CO_2 Dehydration		7.60	8.23
Steam Gen		0.36	0.37
Capture Plant	40.74		
Compression Plant	22.69		
Steam Gen & BOP	4.46		
OSBL (30%)	20.37	16.84	14.49
Direct & Indirect	88.26	162.64	143.13
Escalation from 2009 to 2014	16.07	29.60	26.05
Process Contingency	4.82	3.99	3.14
Project Contingency (20%)	21.83	39.25	34.46
Total Plant Costs	130.98	235.48	206.79
Owners Costs			
6 Months Fixed O&M	2.92	5.79	5.25
1 Month Variable O&M	0.68	1.19	1.28
25% of 1 Month of Fuel Costs	0.14	-	-
2% of TPC	2.62	4.71	4.14
60 Days of Cons.	0.62	0.19	0.56
Spare Parts (.5% TPC)	0.65	1.18	1.03
Land ($3,000/acre)	0.02	0.02	0.02
Financing Costs (2.7% TPC)	3.54	6.36	5.58
Other Costs (15% of TPC)	19.65	35.33	31.02
Total Owners Costs	30.82	54.77	48.88
Total Overnight Costs	161.80	290.25	255.67
AFUDC	12.62	22.64	19.94
Total As-Spent Cost	174.42	312.89	275.61

Table 8 shows the operating and maintenance costs for each case. The O&M costs have been adjusted for escalation or capacity factor and are reported in millions of $2014.

Table 8. O&M Costs.

Cost	Post C.-Amine	Oxy C. (97%)	Oxy C. (99.5%)
Fixed Costs			
Direct Labour	1.37	2.21	2.21
G&A	0.42	0.67	0.67
Maintenance	2.44	5.22	4.61
Insurance and P Taxes	1.62	3.49	3.07
Total Fixed Costs	5.85	11.58	10.55
Variable Costs			
Capture & Compression	8.11	14.33	15.42
Transportation	2.95	3.10	3.44
Storage	1.28	1.35	1.50
Total Variable Cost	12.35	18.80	20.36
Total O&M	18.21	30.38	30.91

Figure 8 shows how the cost of capturing CO_2 varies as the cost of natural gas changes. The Post C.-Amine case shows an increase in capture cost as the cost of natural gas increase. This is because Post C.-Amine case uses a significant amount of fuel to produce steam required to capture CO_2. The relative slopes are related to the fuel usage. The decrease in capture cost for the oxy-fuel cases as natural gas prices rises is due to the benefit of reduced fuel consumption associated with extra steam production. Therefore as the gas price increases, the economic benefit associated with using less fuel increases.

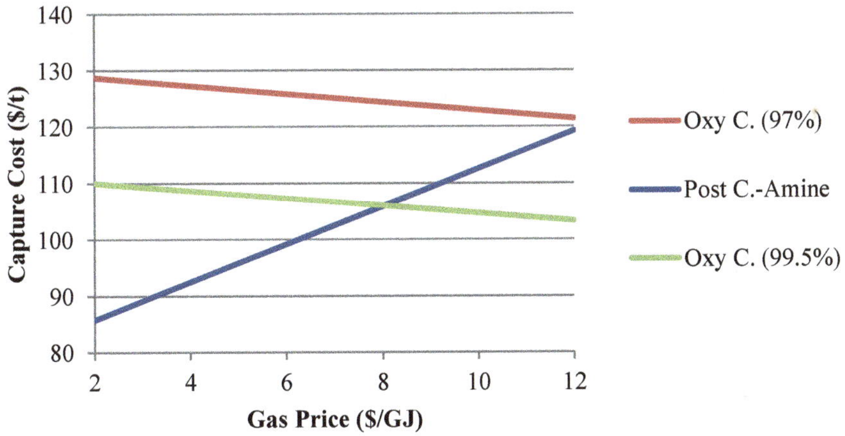

Figure 8. Capture Cost Vs Gas Price for Selected Technologies.

Figure 9 shows how the avoided costs change as gas prices change. As with the graph above, the oxy-fuel case shows a reduced avoided cost as the price of gas increases.

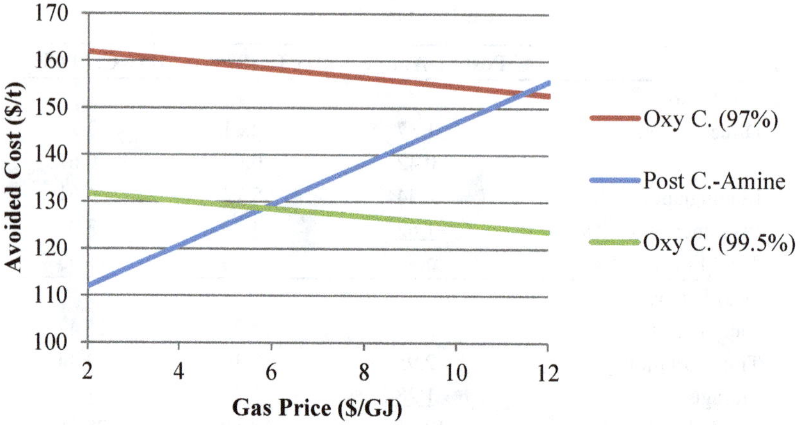

Figure 9. Avoided Cost vs Gas Price for Selected Technologies.

Figure 10 shows how the avoided cost changes with changes in GHG emission intensity of the externally purchased power. As discussed in Chapter 4, if the GHG emission intensity dips below 0.5 t/MWh, the power price becomes a function of the gas price. That accounts for the inflexion in the graph. This adjustment is meant to show that lower GHG emission intensities are likely a reflection of heavy reliance on natural gas and therefore gas price should have a significant effect on power prices. As the GHG emission intensity increases this indicates a higher reliance on coal. Therefore the price of power in the market may change as a result. The avoided cost for the oxy-fuel case is likely to be significantly impacted by changes in GHG emission intensity of the power employed, since power is a significant component of the avoided cost.

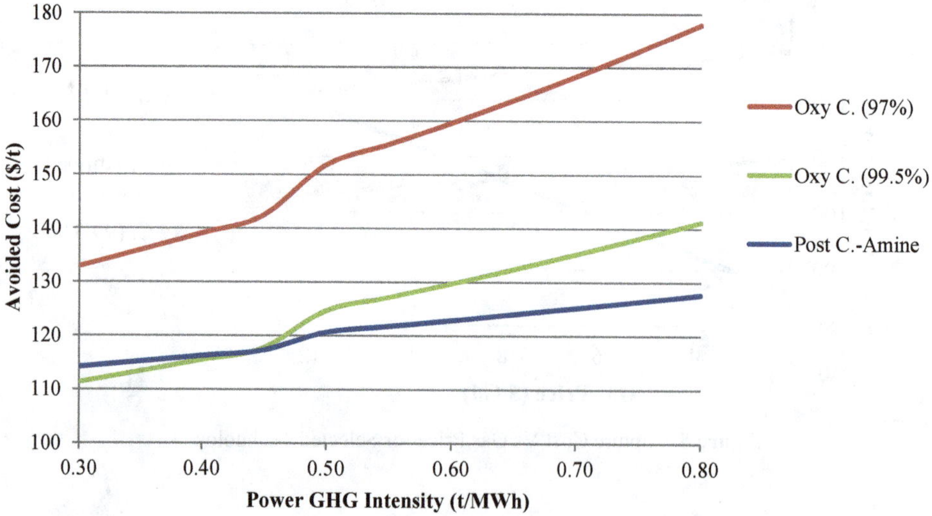

Figure 10. Avoided Cost vs Power GHG Emission Intensity for Selected Technologies.

For the USGC the price of power is 33+5.55 x Gas Price. As the cost of natural gas increase, there is the expectation that the cost of power will increase and the generation mix in a given area may be influenced more by new coal plants than otherwise would be expected. The point of this graph is to indicate which technologies are likely to be sensitive to changes in gas price, which influences power cost and generation mix.

Figure 11 shows how the avoided costs for a project change as the price of power changes. Power is a significant cost component for the oxy-fuel case. Therefore, the avoided costs for these cases are very sensitive to changes in the price of power. The Post C.-Amine case is less sensitive to change in power prices since this case uses less power to capture a ton of CO_2 than the oxy-fuel cases.

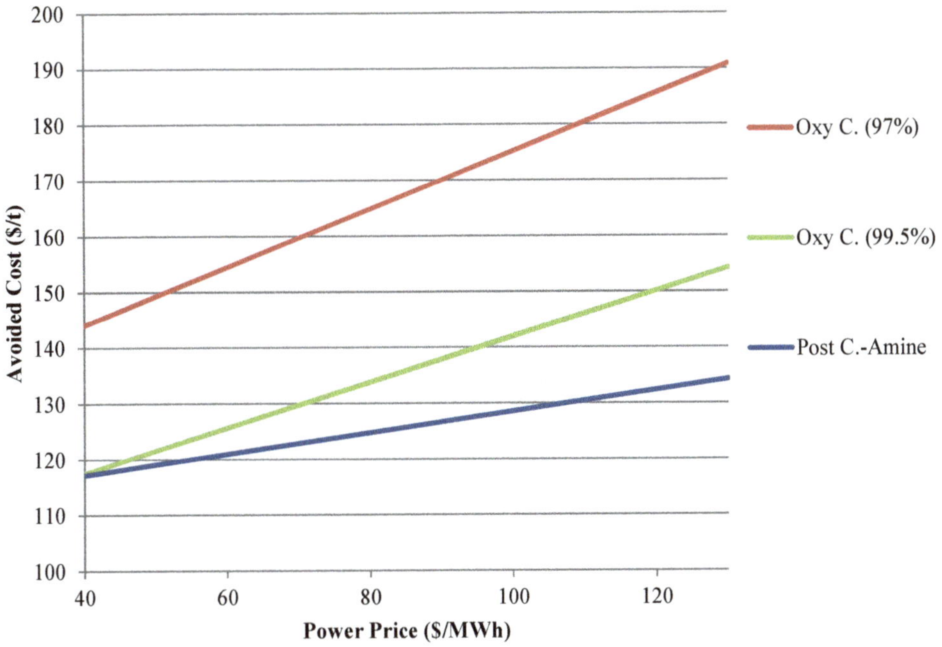

Figure 11. Avoided Cost vs Power Price for Selected Technologies.

CONCLUSIONS

For CO_2 capture from the regenerator of a refinery FCC unit, oxy-combustion has the potential for being competitive with post-combustion. CO_2 avoidance costs for the two techniques are comparable, but oxy-firing provides more flexibility to the operation of the FCC unit, and also requires a smaller plot area close to the unit, which may be an important factor inside a refinery.

The use of high purity oxygen (99.5%) increases the cost of air separation, but requires less purification of the captured CO_2 before compression and transportation, resulting in a lower cost of CO_2 avoided.

Oxy-combustion is favoured by high cost of fuel and low carbon footprint of imported energy. For a cost of natural gas roughly above $6 /GJ, oxy-combustion becomes the lowest cost option. The same is true if a consistent fraction of renewables is included in the imported power mix.

REFERENCES

1. L. de Mello, G. Moure, O. Pravia, L. Gearhart, P. Milios; "Oxy-combustion for CO_2 Capture from Fluid Catalytic Crackers (FCC)"; Chapter 3 in *Carbon Dioxide Capture for Storage in Deep Geologic Formations – Results from the CO_2 Capture Project,* Vol. 3; L.I. Eide (editor); 2009.

Carbon Dioxide Capture for Storage in Deep Geological Formations, Volume 4
Karl F. Gerdes (Editor)

Chapter 8

RETROFIT OPPORTUNITIES FOR CARBON DIOXIDE CAPTURE FROM REFINERY HEATERS AND BOILERS

Betty K. Pun[1]
[1]Chevron Energy Technology Company, 100 Chevron Way, Richmond, CA 94801

ABSTRACT: In a typical oil refinery, fuel combustion in process heaters and boilers accounts for more than 60% of the CO_2 emissions. These sources tend to be distributed throughout the refinery, with resulting implications on retrofitting for CO_2 capture.

Several CO_2 capture options have been evaluated for retrofitting individual heaters and boilers, including post- and pre-combustion capture and oxy-combustion:

- John Zink demonstrated the feasibility of retrofitting process burners to oxy-combustion (Details in Chapter 9 of this volume).
- Amec Foster Wheeler (AFW) conducted techno-economic analyses of six configurations of pre-combustion CO_2 capture, two of which involved the use of novel palladium alloy membranes to separate H_2 (Details in Chapter 11)
- CCP3 supported Pall Corporation in the laboratory development of palladium alloy membranes for H_2 separation from syngas (Details in Chapter 10).

The John Zink study established the feasibility of retrofitting a general service refinery heater for oxy-firing using flue gas recirculation to reduce film and tube metal temperatures to acceptable levels. The modified operation resulted in an increase in heater efficiency. John Zink tested a low NO_x burner and an ultra-low NO_x burner and operated both successfully under oxy-firing conditions (with flue gas dilution of fuel/oxygen mix) without modifications to the burners. For retrofitting a few heaters and boilers, the techno-economic results of the oxy-firing cases were at a disadvantage against the state-of-the-science post-combustion, solvent-based approaches. A major cost element for oxy-firing is the air separation unit for both capital and operating costs.

For pre-combustion capture, major equipment to produce H_2 fuel and capture CO_2 may be placed away from the refinery processes, while H_2 fuel is distributed through the existing fuel gas infrastructure. This approach has potential advantages over CO_2 capture retrofits of individual sources in space-constrained refineries. Centralized pre-combustion approaches benefit from the economy of scale (5,000 MMBtu/hr fuel demand for a large refinery) in comparison to retrofit of individual combustion units. Within the H_2 fuel production system, amine absorption and cryogenic separation offer competitive CO_2 capture and avoided costs. Two pre-combustion systems using pressure swing absorption (PSA) were assessed for CO_2 removal, and were found to be higher cost than the solvent-based options, even though the economics of the PSA approach are enhanced by power recovery. The option of making high purity H_2 for feedstock use in a refinery, in addition to producing H_2-rich fuel, using a PSA scheme may be attractive in some circumstances

Pall Corporation measured data to establish the membrane design to separate H_2 from syngas and tested the long-term stability of the membrane modules. Several configurations of pre-combustion capture using the membranes were assessed. However, Pd membrane-based approaches are not economically competitive at this time.

KEYWORDS: refinery; retrofit; heaters; boilers, post-combustion; pre-combustion, oxy-fuel combustion

INTRODUCTION

In a typical refinery, there are many point sources of CO_2, including fluidized catalytic crackers, hydrogen plants, and numerous combustion sources. As a group, heaters and boilers account for more than 60% of the total CO_2 emissions from a typical refinery [1]. The refinery heaters and boilers scenario was created to evaluate opportunities to retrofit combustion sources in a refinery setting. The location being considered is the Gulf Coast of the U.S.

Two approaches were considered: capturing CO_2 from individual heaters and boilers (nominal size 100-150 MMBtu/h each), or converting the refinery fuel system to a de-carbonized hydrogen-based fuel, with carbon capture at the centralized fuel production facility. Several CO_2 capture options were evaluated for retrofitting individual heaters and boilers.

Post-combustion CO_2 capture was considered at two size scales:

- A small unit to capture CO_2 from three 100-MMBtu/hr boilers.
- A nominal case with four boilers with each having an as-fired capacity of 150 MMBtu/hr (commensurate in size with cases studied in CCP2 [2]: 2 million ton/y).

The **oxy-combustion case** examined retrofitting four 150-MMBtu/hr production steam boilers. A single air separation unit supplies oxygen to all fired units. In all post-combustion and oxy-combustion cases, flue gas is collected from individual heaters and boilers into a single flue gas duct to facilitate CO_2 capture.

Key parameters for the refinery heaters and boilers are summarized in Table 1.

Table 1. Key parameters for the existing refinery heaters and boilers

Refinery heaters and boilers retrofit	Nominal
Fired capacity (mmbtu/hr)	4 x 150
Total flue gas flow (kg/hr)	259,606
Flue gas flow to CO_2 capture plant	246,626
Pressure (bara)	1.013
Temperature (°C)	150
Flue gas composition	
CO_2 (mole%)	7.5
SO_x (ppm)	4
NO_x (ppm)	30
CO (ppm)	0
N_2 (mole%)	70.38
Ar (mole%)	0.84
O_2 (mole%)	2.46
H_2O (mole%)	18.81

In **pre-combustion configurations**, a centralized hydrogen plant reforms refinery fuel gas (RFG) plus supplemental natural gas (NG) to produce a mostly decarbonized H_2 fuel at a capacity of about 5000 MMBtu/hr. The H_2 fuel is distributed using the existing fuel utility infrastructure to various heaters and boilers within the refinery. CO_2 is separated from the hydrogen product and captured at the reformer.

For all cases, the CO_2 product must meet the following specifications for export:

- Minimum CO_2: 97 mol%
- Pressure: 150 barg
- Temperature: 40°C
- Maximum moisture: 50 ppm
- Maximum inerts: 4 mole %
- Maximum oxygen: 75 ppm (A specification of 3% was used for the state-of-the-art amine post-combustion technology, but it is expected to meet 75 ppm specification without additional treatment).

LIST OF STUDIES AND PROJECTS FOR THE REFINERY HEATERS AND BOILERS SCENARIOS

Amec Foster Wheeler Technical Studies

- Post-combustion capture using Best Available Technology (amine CO_2 capture): small size case base case (Phase 1)
- Post-combustion capture using Best Available Technology: nominal case (Phase 1)
- Oxy-firing case (Phase 2; including oxy-fired heater modelling)
- Pre-combustion: CO_2 separation by amine (Phase 2)
- Pre-combustion: H_2 purification by PSA; thermal oxidizer/waste heat recovery for CO_2 (Phase 3)
- Pre-combustion: H_2 purification by PSA; cryogenic purification for CO_2 (Phase 3)
- Pre-combustion: H_2 permeable membrane; cryogenic purification for CO_2 (Phase 4)
- Pre-combustion: H_2 permeable membrane; cryogenic purification for CO_2; PSA for additional H_2 recovery (Phase 4)
- Pre-combustion: Cryogenic CO_2 separation and additional CO_2 removal by amine (Phase 4)

These studies, together with economic modelling performed by CCP3's economic team, are discussed in Chapter 11.

John Zink: Oxy-firing of Process Heaters Feasibility Study

To test the feasibility of retrofitting oxy-firing, CCP3 sponsored a multi-task project. OnQuest, under subcontract with John Zink, conducted heater modelling to determine oxy-fuel firing conditions. Using the modelling results as guidance, John Zink carried out pilot scale testing (June 11-25, 2012) of a single burner in a test furnace that had been modified for flue gas recirculation. Computational Fluid Dynamics (CFD) modelling was performed for single- and multiple-burner configurations. The following reports document this work:

- CFD Report: COOLstar™ Oxy Firing Simulations in Test Furnace 9 (June 2013)
- CFD Report: COOLstar™ Oxy Firing Simulations: Multi-burner Cabin Heater (June 2013)
- CFD Report: COOLstar™ Oxy Firing Simulations: Multi-burner Vertical Cylindrical Heater (June 2013)
- CFD Report: PSFG Oxy Firing Simulations in Test Furnace 9 (June 2013)
- CFD Report: PSFG Oxy Firing Simulations: Multi-burner Cabin Heater (June 2013)
- CFD Report: PSFG Oxy Firing Simulations: Multi-burner Vertical Cylindrical Heater (June 2013)

- Oxy-firing of COOLSTAR®-13 and PSFG-14 burners: A performance demonstration (January 2013)
- Oxy-firing of Process Heaters Feasibility Study: Executive Summary (June 2013)

A summary of the project is provided in Chapter 9.

Pall: Development of Palladium Alloy Membrane for CO_2 Capture and H_2 Recovery

In a separate project, CCP3 provided support to Pall Corporation to develop palladium (Pd) alloy membranes for hydrogen (H_2) recovery from syngas as part of a pre-combustion approach. A two-stage water gas shift/H_2 recovery configuration was the basis of the development program. Pall conducted a series of tasks, including:

- Production of Pd alloy tubes.
- Modelling and testing of membrane performance with and without sweep gas.
- Manufacture of modules.
- Testing of modules with and without sweep gas.
- Determination of the membrane area for each stage.
- Design of module housing with and without sweep gas.

Task reports were generated at the conclusion of each task. Module and housing cost information was supplied by Pall to AFW to support the techno-economic evaluation of pre-combustion H_2 membrane cases. A summary of the Pall project results has been compiled in Chapter 10.

MAIN CONCLUSIONS AND LESSONS LEARNED

For the nominal 4x150 MMBtu/h case, the benchmark post-combustion costs for CO_2 capture and avoided are \$120/tonne CO_2 and \$157/tonne, respectively. In the previous phase of CCP, oxy-fuel combustion was shown to be an attractive technology for retrofitting heaters and boilers. In the CCP3, the technical feasibility of oxy-fuel combustion with flue gas recycling was demonstrated for heaters in typical refinery service. However, the cost of the air separation unit remains significant, accounting for more than half of the total installed cost; and the associated power consumption contributes to both operating cost and carbon footprint. These factors render oxy-fuel combustion not competitive at this time when compared to the state-of-the-art post-combustion carbon capture based on amine absorption. The conclusion does not change even when the increased efficiency of heaters when firing with O_2 is taken into account.

In addition to CO_2 collection ducts, which are also required for post-combustion capture from individual heaters and boilers, oxy-fuel combustion requires oxygen piping, flue gas recycle retrofit, and sealing of the furnace against air ingress. There will also be additional safety considerations and controls associated with the use of 97% oxygen. Therefore, logistically, oxy-fuel combustion is more difficult to implement when compared to other options.

Many refineries are space constrained. Post-combustion and oxy-fuel retrofits require extra space for ducting and equipment (e.g., absorber/regenerator, air separation unit), which may not be readily available close to combustion sources. In addition, the distributed nature of the combustion sources makes it difficult to implement either of those solutions if a high degree of CO_2 abatement is necessary. In this case, pre-combustion CO_2 capture (i.e., conversion of the refinery to use decarbonized fuel) has inherent advantages. Refinery fuel plus supplementary natural gas is re-formed into syngas, which is converted to H_2 and CO_2 through the water shift reaction. CO_2 is captured and H_2 is distributed throughout the refinery as fuel, using the existing infrastructure. Because of centralized production of fuel for the entire refinery, pre-combustion approaches are large (5,000 MMBtu/h) in scale and can take advantage of equipment (such as the autothermal reformer) that would not be economically feasible in smaller scales.

CCP3 sponsored development of Pd membranes for H_2 separation from syngas. The integration of H_2 separation with the water-gas shift reaction facilitates H_2 production by altering the equilibrium. Pall designed and tested a Pd-alloy membrane for this purpose. For this work, a staged reaction and separation sequence was the basis, rather than the more-difficult integration of membrane in the reactor. However, the cost of the membrane modules is too high at this point to allow for their deployment at scale. Further membrane improvement and/or process intensification will be needed for Pd-alloy membrane-based pre-combustion configurations to be competitive with other alternatives.

The most promising pre-combustion approaches are also the simplest technologies as far as CO_2 capture is concerned. A combination cryogenic-methyldiethanolamine (MDEA) absorption approach and MDEA solvent absorption alone offer capture and avoided costs of \$103-\$111/tonne CO_2 and \$148-\$160/tonne, respectively. The small difference in cost between these approaches is well within the uncertainty of the current cost estimates (30%). These costs are also commensurate with the post-combustion approach for small scale applications.

Two pre-combustion configurations involving PSA have been studied. Because H_2 is generated at high pressure as a product of the PSA, power recovery is possible. Thermal oxidation or cryogenic purification were assessed for purification to pipeline quality CO_2. In the case of thermal oxidation, even though it has the highest O_2 demand, the energy recovered after thermal oxidation contributes to a large amount of avoided CO_2. It is important to note that the PSA units in this scheme are optimized to recover the fuel value of the H_2 fuel, and the resulting H_2 stream purity, approximately 95-96%, is lower than the purity typically achieved by a modern H_2 plant. For future work, it may be of interest to investigate designs that could provide optionality to produce reactant grade hydrogen (98+%), in addition to H_2-rich fuel.

ACKNOWLEDGEMENTS

The author wishes to thank internal and external reviewers for their helpful comments.

REFERENCES

1. Available and emerging technologies for reducing greenhouse gas emissions from the petroleum refining industry; U.S. EPA, Office of Air and Radiation, Research Triangle Park, NC, October 2010 (http://www.epa.gov/nsr/ghgdocs/refineries.pdf)
2. Eide, L.I. (editor), Carbon Dioxide Capture for Storage in Deep Geologic Formations – Results from the CO_2 Capture Project, Volume Three: Advances in CO_2 Capture and Storage Technology Results (2004-2009); CPL Press (2009).

Carbon Dioxide Capture for Storage in Deep Geological Formations, Volume 4
Karl F. Gerdes (Editor)

Chapter 9

OXY-FIRING OF PROCESS HEATERS – A FEASIBILITY STUDY

Jamal Jamaluddin[1], Jamie A. Erazo, Jr.[2], Charles E. Baukal[2]
[1]Shell Global Solutions (US), Inc., Houston, TX, U.S.A.
[2]John Zink Co. LLC., Tulsa, OK, U.S.A.

ABSTRACT: The CO_2 Capture Project (CCP) commissioned John Zink Company to assess the feasibility of retrofitting process heaters for oxy-firing. Modeling studies established that a flue gas recycle rate of 72% was optimal, and that the overall heater efficiency would increase by several percentage points. Successful single burner tests were conducted with a low NO_x PSFG burner and an ultra-low NO_x COOLstar™ burner, with no modification to the burner design. Computational fluid dynamic models were constructed to simulate oxy-firing of multiple burners in typical heater geometries. In most cases, the flame patterns were predicted to be upright and well structured. Based on the test data and the simulations, we conclude that oxy-firing can be applied to process heaters using conventional burners to achieve temperature and heat transfer profiles similar to air firing. Preventing air ingress is an important consideration for carbon dioxide capture.

KEYWORDS: oxy-firing; refinery; heaters; burners; computational fluid dynamics

INTRODUCTION

Oxy-firing refers to the concept of combustion with oxygen (O_2) instead of air [1]. While O_2 separation is a pre-requisite, the benefit of oxy-firing is that the flue gas consists mainly of water and CO_2, and is nearly nitrogen free. A CO_2-rich stream suitable for sequestration can thus be obtained by simply cooling the flue gases and condensing the water out. Some purification may be necessary to increase the CO_2 concentration level [2] or to reduce O_2 concentration, to meet pipeline specifications (See Chapter 11).

In the first phase of the CO_2 Capture Project (2000-2003), it was determined that oxy-firing of refinery heaters and boilers showed significant potential for lower cost of avoided CO_2 when compared to post-combustion capture [3,4]. Although the capex requirements were higher for oxy-firing (primarily due to the addition of an air separation plant), the significantly lower energy requirements resulted in lower overall avoided CO_2 costs. The feasibility of using an alternative method of generating O_2, based on high temperature, ceramic ion transport membranes [5], was also investigated. Consequently, the CCP pursued oxy-firing tests in this, the third phase of the project (2009-2013). While previous studies had investigated the feasibility of using oxy-firing for CCS (CO_2 capture and sequestration), this technology had never been tested in a simulated process heater.

The objectives of CCP's oxy-fired process heater program were to:

- Assess the feasibility of utilizing conventional process heater burners for oxy-firing.
- Confirm the feasibility of oxy-firing in process heaters by conducting single burner testing with flue gas recycle.
- Construct computational fluid dynamics (CFD) models to simulate oxy-firing in typical multi-burner heater geometries.

SCOPE OF WORK

The CO_2 Capture Project commissioned the John Zink Company to conduct the oxy-firing feasibility study. The program consisted of the following tasks.

1. Heater performance modelling to define the operating conditions, particularly the flue gas recycle requirements.
2. Single burner testing to test two conventional process heater burners, a PSFG staged fuel low NOx burner and a COOLstar™ Ultra-Low NOx burner, under air and oxy-firing conditions, using two different fuels.
3. Computational fluid dynamics (CFD) modelling to simulate the single burner tests, and subsequently to predict heater performance under multi-burner process heater operating scenarios.

RESULTS

Heater modelling

John Zink engaged OnQuest (a heater vendor) to provide process heater modeling using the FRNC-5 software [6]. This first round of heater modeling was conducted to identify feasible operating conditions in a typical process heater. The heater modeling exercise focused on the following:

- Overall heater efficiency
- Maximum film temperature and tube metal temperature limitations
- Radiant/ convection heat absorption ratio
- Flue gas recycle requirements

Several oxy-firing conditions were evaluated, and a base case with ambient air was used for comparison purposes. As a starting point, the first set of simulations varied the level of excess oxidant without any flue gas recirculation. The findings indicated that very high levels of excess oxidant would be required to meet the allowable film and tube metal temperature design limits.

The subsequent round of simulations maintained the range of excess oxidant between 0.5% and 2%, while the amount of flue gas recirculation was varied until the constraint on film temperature and tube metal temperature were satisfied. From these simulations, two feasible operating conditions were identified. Both conditions demonstrated that the process design constraints could be met at 0.5 - 2% (vol. wet) oxygen concentration in the flue gases and a high flue gas recirculation rate of 72%.

Compared to the base ambient air case, these oxy-firing conditions provided a ~14% improvement in heater efficiency, and met the maximum film and tube metal temperature constraints with small changes to the radiant/convection section duty split. Additionally, these operating conditions required the minimal use of oxygen.

Single burner tests

Approach

The burner tests were conducted at the John Zink test facility in Tulsa, Oklahoma. Two John Zink burners were tested: the PSFG is a low NO_x, diffusion flame burner incorporating staged fuel injection for NO_x reduction; while the COOLstar™ burner is an ultra-low NO_x diffusion flame burner that uses internal flue gas recirculation (FGR) and staged fuel injection to reduce NO_x.

Each burner was tested in a rectangular test furnace with the following internal dimensions: 13 ft (4 m) wide, 7 ft (2.1 m) deep and 31 ft (9.4 m) tall. The test furnace had been modified (as discussed below) to accommodate both air and oxy-firing. A picture of the test furnace is shown in Figure 1. The furnace was cooled by single-pass water tubes in the radiant section, and insulated to provide a nominal, mid-furnace flue-gas temperature of 1600 deg F (870 deg C). Unlike conventional process heaters, the test furnace did not have a convection section. An external boiler was therefore installed to cool the flue gas prior to entering the induced draft fan used for flue gas recirculation, since the fan components had carbon steel metallurgy. The oxygen was stored in a large liquid vessel equipped with a vaporizer.

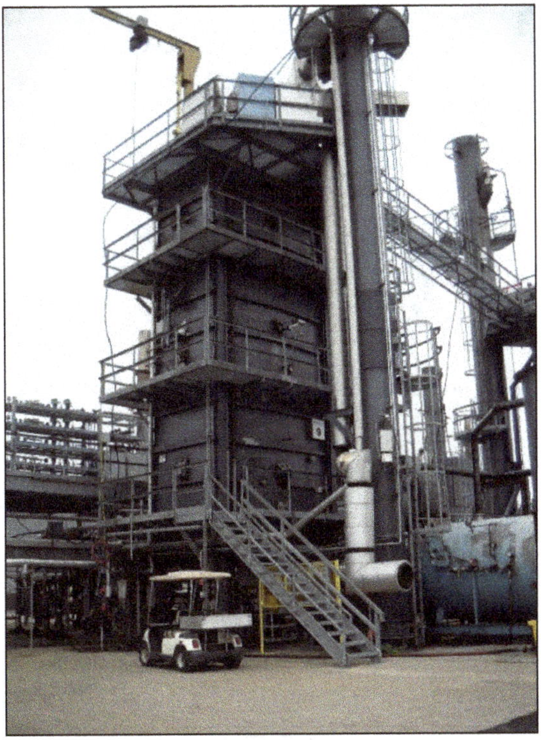

Figure 1. Test furnace.

Figure 2 shows a simplified schematic diagram of the oxy-firing setup. For natural draft operation, the flue gas damper was closed and the air inlet valve was opened so that combustion air could be drawn in and the flue gases could leave through the stack. During oxy-firing tests, the air inlet was closed and the flue gas damper was opened to control the amount of flue gas recirculation to the burner. Oxygen was injected into the ductwork downstream of the flue gas damper and upstream of the burner plenum. The flow of oxygen was measured and controlled through a flow skid.

Three conditions were tested:

- Ambient air-fired operation for a performance baseline
- Oxy-fire with 5.2% (vol. dry) oxygen in the flue gas (Condition A)
- Oxy-fire with 1.3% (vol. dry) oxygen in the flue gas (Condition B)

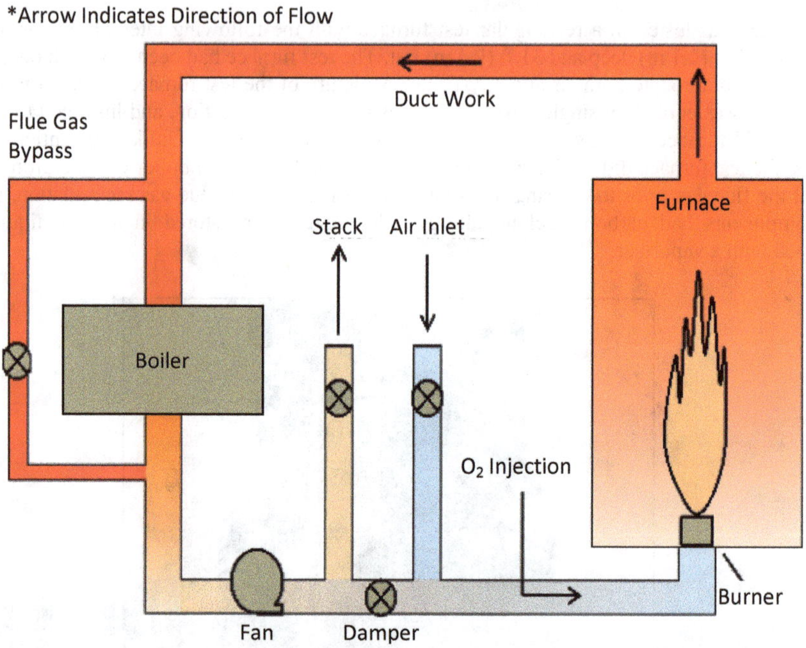

Figure 2. Simplified schematic diagram of oxy-fire setup.

Two fuel gases were tested: Tulsa natural gas (TNG) and a simulated refinery gas (RFG) mixture. The test protocol followed API-535 guidelines [7]. In addition to visual observation of flame appearance at all operating conditions, flame length was verified via CO probing and the incident heat flux profile [8] was measured with a calibrated flux-probe. Emissions measurements of CO, NO_x, wet and dry O_2 were made at the exit of the test furnace. Measurements of oxygen on a wet basis were also made before and after the oxygen injector. In addition, dry measurements of O_2 and CO_2 were made just upstream of the burner inlet to determine the nitrogen content in the oxidant stream. Furnace flue gas temperature was monitored using a velocity thermocouple.

Table 1. Operating conditions for the burner tests

Parameters (units)	Natural Draft	Oxy-fire
Heat release (MMBtu/h)		
Maximum	7.68	6.45
Normal	6.4	5.37
Minimum - PSFG	2.56	2.56
Minimum - COOLstar^{TM}	1.92	1.92
Oxygen concentration (%,v) in oxidant	20.9	22.2/20.6
Recirculated flue gas (%)	N/A	~72
Oxygen concentration in flue (%)	~3	5.2/1.3
Oxidant temperature (deg F)	ambient	~500

Results

The John Zink staged gas low NO_x burner and COOLstar™ ultra-low NO_x burner performed satisfactorily during oxy-fire operation, with no burner modification. The burners were demonstrated from maximum to minimum heat release with no performance issues. The transition from ambient air to oxy-fire operation was successfully demonstrated several times with each burner. Satisfactory turndown was obtained with both burners and both fuels, with 72% flue gas recycle.

The flue gas re-circulation (FGR) rate and the oxygen concentration in the oxidant stream were the two important operating parameters. Changes in the FGR fraction can have a significant impact on burner performance. Larger FGR flow rates can push the combustion process to the flammability limit, especially if operating under low oxygen conditions. A less "reactive" fuel, e.g., natural gas, may need a higher percent of oxygen in the oxidant during oxyfiring. This was confirmed by testing, in which TNG required 24% O_2, while the simulated RFG required 20.6% - 22.2%.

A significant reduction in NO_x emissions was measured under oxy-firing conditions, as shown in Table 2.

Table 2. NO_x emissions (ppm) summary
(corrected to 3% O_2 vol. dry and 1600 deg F bridgewall temperature).

Oxidant - Fuel	**PSFG**	COOLstar™
Air - TNG	24.9	9
Oxy -TNG	14.5	8.7
Air - RFG	41.4	12.7
Oxy A - RFG	14	6.1
Oxy B - RFG	10.6	6

On average, about an hour was needed to purge the initial nitrogen out of the system. A boiler was used to simulate the convection section of a heater and control the FGR temperature; however, it had a high pressure drop. During oxy-firing, the arch draft was -2 to -5 inches of H_2O compared to an arch draft of -0.1 in H_2O in a typical refinery heater. The negative pressure resulted in higher air in-leakage and elevated N_2 levels. Typical N_2 levels were 15-20% (vol. wet). Even under certain low draft conditions, the measured N_2 concentrations were as high as 11% (vol. wet) in the recirculated flue gas. Therefore, proper sealing would be essential to minimize air in-leakage in order to make oxy-firing effective in process heaters.

Flame appearance was documented with digital photography. For both burners, few differences were noticed between the air and oxy-firing modes of operation. Figure 3 shows a comparison of the COOLstar™ burner under ambient air and oxy-firing (A) conditions. Some sections of the tile appear to be glowing more brightly under oxy-fire conditions than with ambient air operation. Under Oxy-fire A conditions, the concentration of oxygen in the oxidant stream is higher than that of air which results in more intense local combustion of the fuel gas.

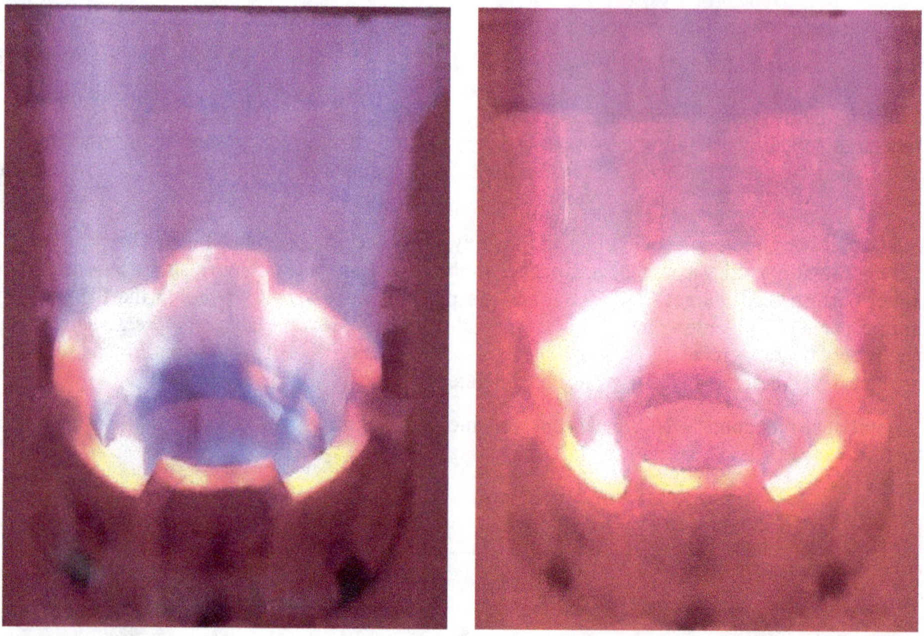

Figure 3. Appearance of flames of the COOLstar™ burner under natural draft (left) and oxy-firing (Condition A, right) operations with RFG.

Computational Fluid Dynamics Modelling

CFD Simulations of the Test Furnace

CFD simulations of the test furnace were performed using ANSYS FLUENT v13.0. Figure 4 depicts the PSFG and COOLstar™ burners, as modelled. The conditions simulated were (1) air operation (two fuels), (2) Oxy A operation with RFG, (3) oxy-firing with 24% O_2 concentration with TNG.

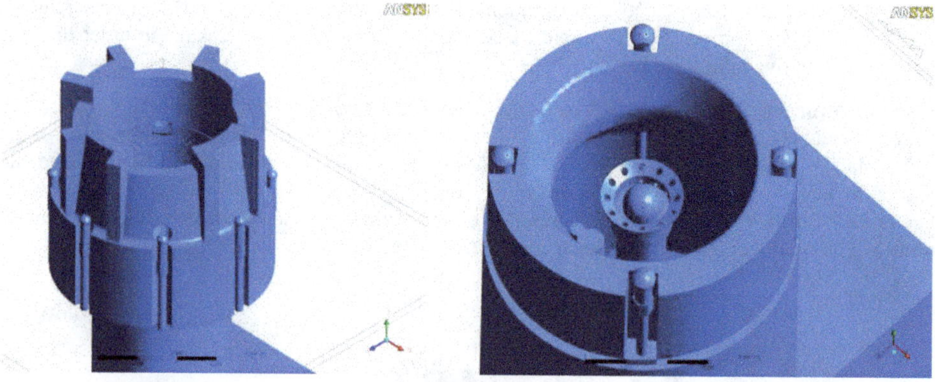

Figure 4. John Zing PSFG and COOLstar™ burners, as modelled in CFD simulations.

The default model within FLUENT used to determine flue gas emissivity and absorption coefficients was acceptable for combustion simulation using air as the oxidant source. However, for the oxy-fuel conditions, the default model was not able to provide accurate predictions for the emissivity and absorption coefficients for the CO_2 and H_2O-rich combustion products. A user-defined function was developed and used under oxy-firing conditions to provide more realistic predictions of the incident heat flux profile in the furnace.

Modeled bulk flue gas flow patterns and temperature distribution in the test furnace agreed with the observations and measurements. The contour of 2000 ppmv CO was used to represent flame length in simulations. The CFD simulation under-predicted the COOLstar™ burner flame length by an average of 10%, while the PSFG burner flame length was over-predicted by an average of 30%. The discrepancy is partially attributed to experimental uncertainty from the CO probe. The simulated heat flux profiles and the normalized profiles measured from the single burner tests are shown in Figure 5 for both burner types, firing RFG. The agreement between the two results is good and within the experimental uncertainty of the measurements from the test furnace. In most cases, peak flux location and general profile predictions agree well with experimental measurements.

CFD Simulations of Multi-burner Process Heaters

Two heater configurations were utilized in the CFD simulations of multi-burner process heaters: a vertical cylindrical (VC) heater with eight burners, and a cabin style heater with six burners. These typical heater geometrics are representative of refinery application heaters. Sketches of these heaters are shown in Figure 6a and 6b.

In total, 16 simulations were conducted: 2 burners (PSFG and COOLstar™) x 2 geometries (VC and cabin) x 2 fuels (TNG and RFG) x 2 conditions (air and oxy). The simulations predicted that burners would perform well under all of these scenarios.

In the case of the COOLstar™ burner in the VC geometry, CFD predicted flame merging for both ambient and oxy-firing conditions (see Figure 7). The reason for the flame merge is due to an imbalance of flue gas recirculation in the heater. Strong flue gas flow along the circumference of the heater disrupts fuel/oxidant mixing and pushes the flames towards the centre of the heater, thus elongating the flames. Better balance can be achieved by optimizing the burner circle diameter to reduce or eliminate flame interaction. The rest of the simulations show flame patterns that are upright and well-structured (Figures 8, 9 and 10) due to symmetric flue gas recirculation patterns.

In general, there was good agreement between the bulk flue gas temperature in the CFD model and the FRNC-5 heater simulation performed by OnQuest for all cases. One example is shown in Table 3 for the cabin heater with COOLstar™ burners.

Table 3. Bulk flue gas temperature and radiant efficiency comparison between FRNC heater simulation and CFD for cabin heater and COOLstar™ burner.

Parameter	Firing condition	FRNC-5 Simulation	CFD - TNG	CFD - RFG
Bridgewall temperature (deg C)	air	796	791	794
	oxy	762	773	779
Radiant efficiency (%)	air	55	52.7	50
	oxy	67.7	61.9	64.5

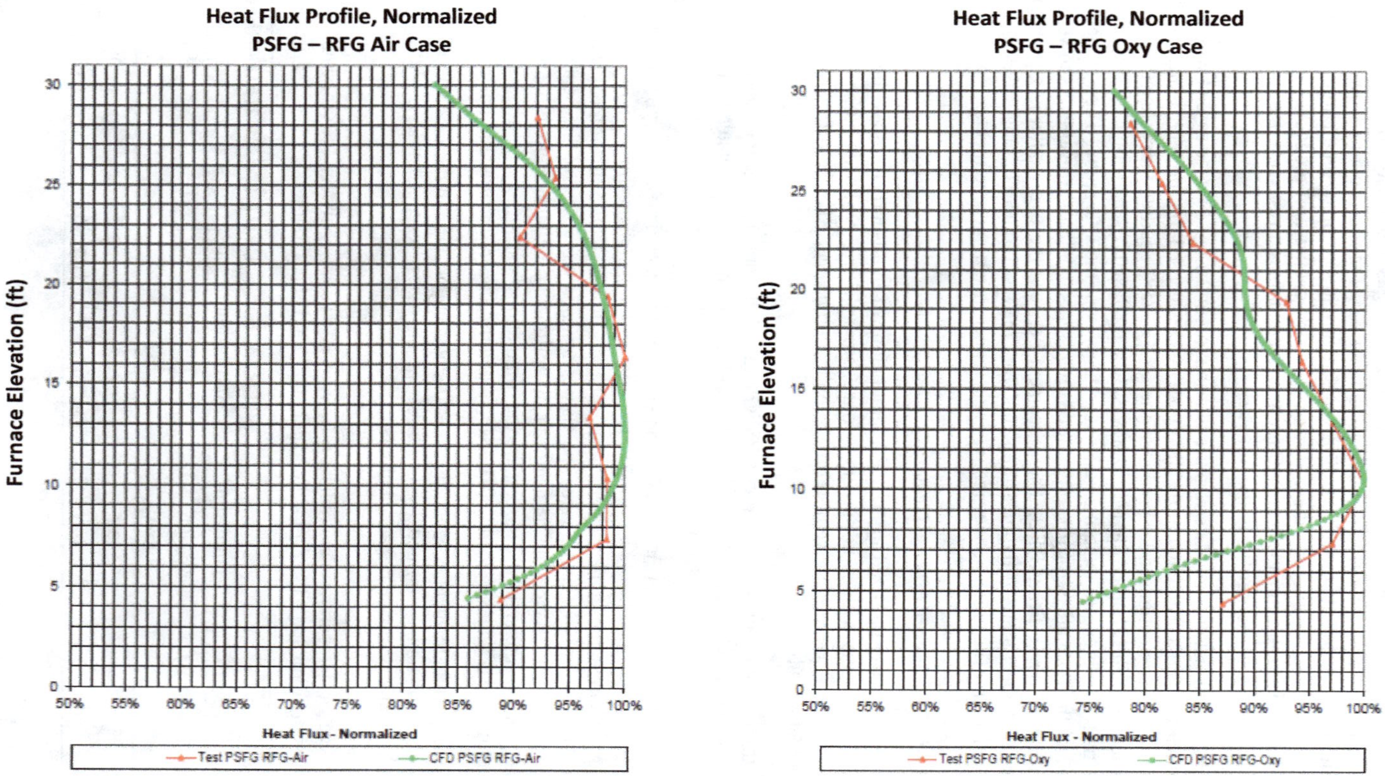

Figure 5a. Measured and predicted heat flux profiles for John Zing PSFG firing RFG - Test (Red), CFD (Green).

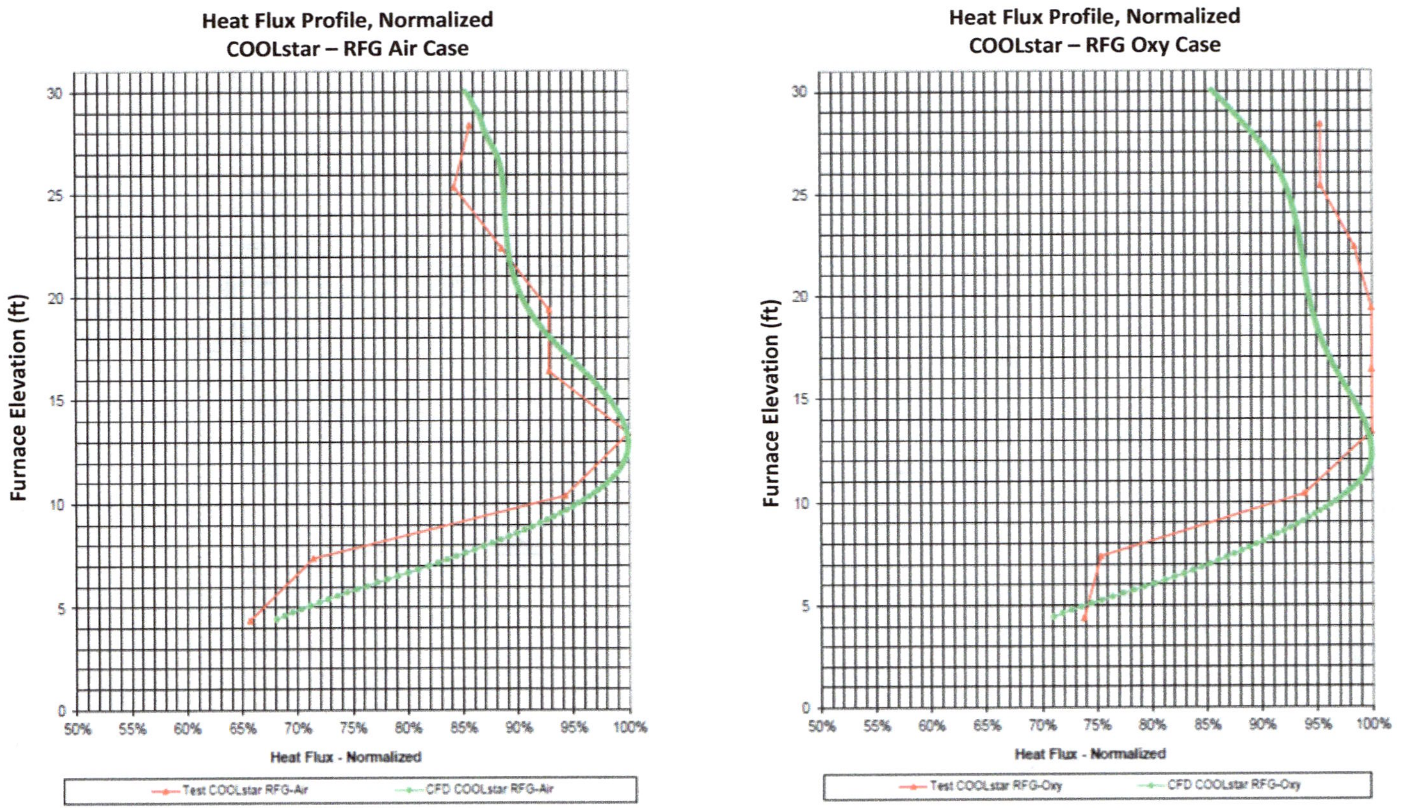

Figure 5b. Measured and predicted heat flux profiles for COOLstar^{TM} firing RFG - Test (Red), CFD (Green).

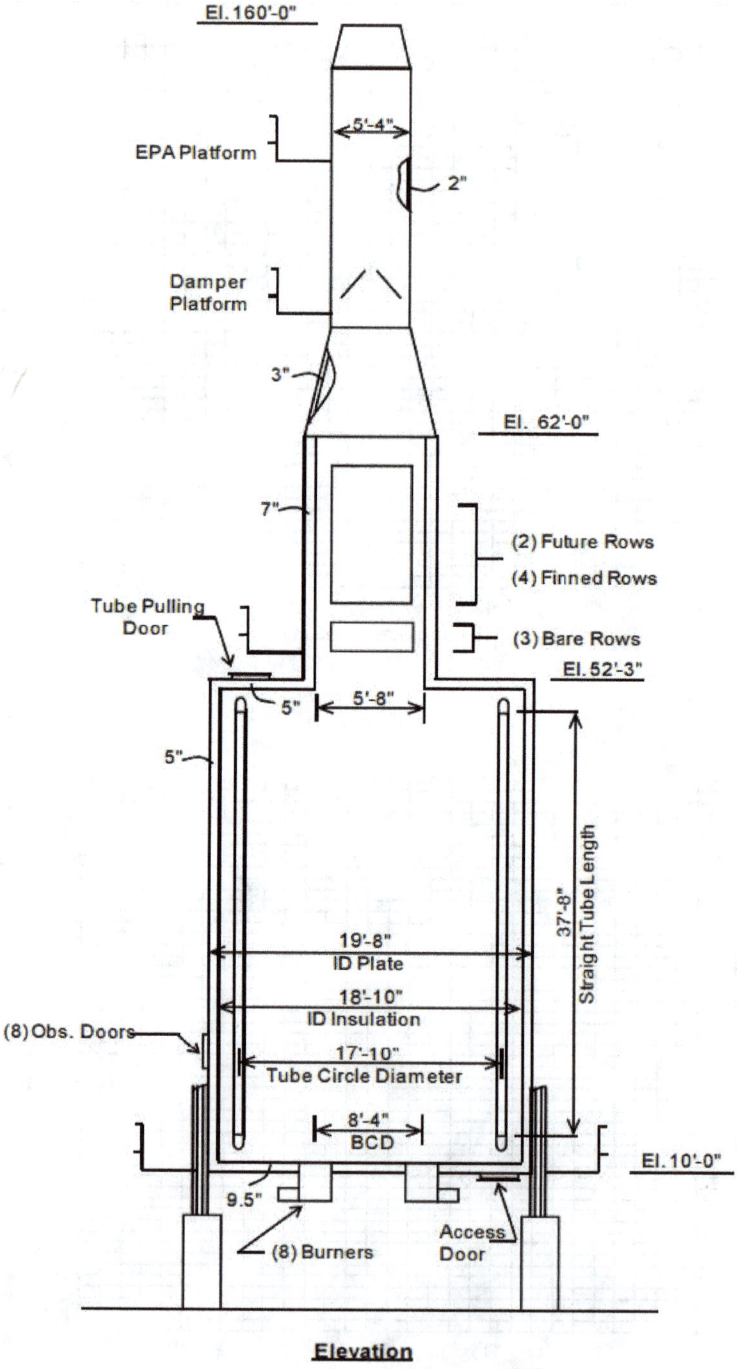

Figure 6a. Sketch of the VC heater, one of the two heaters modelled.

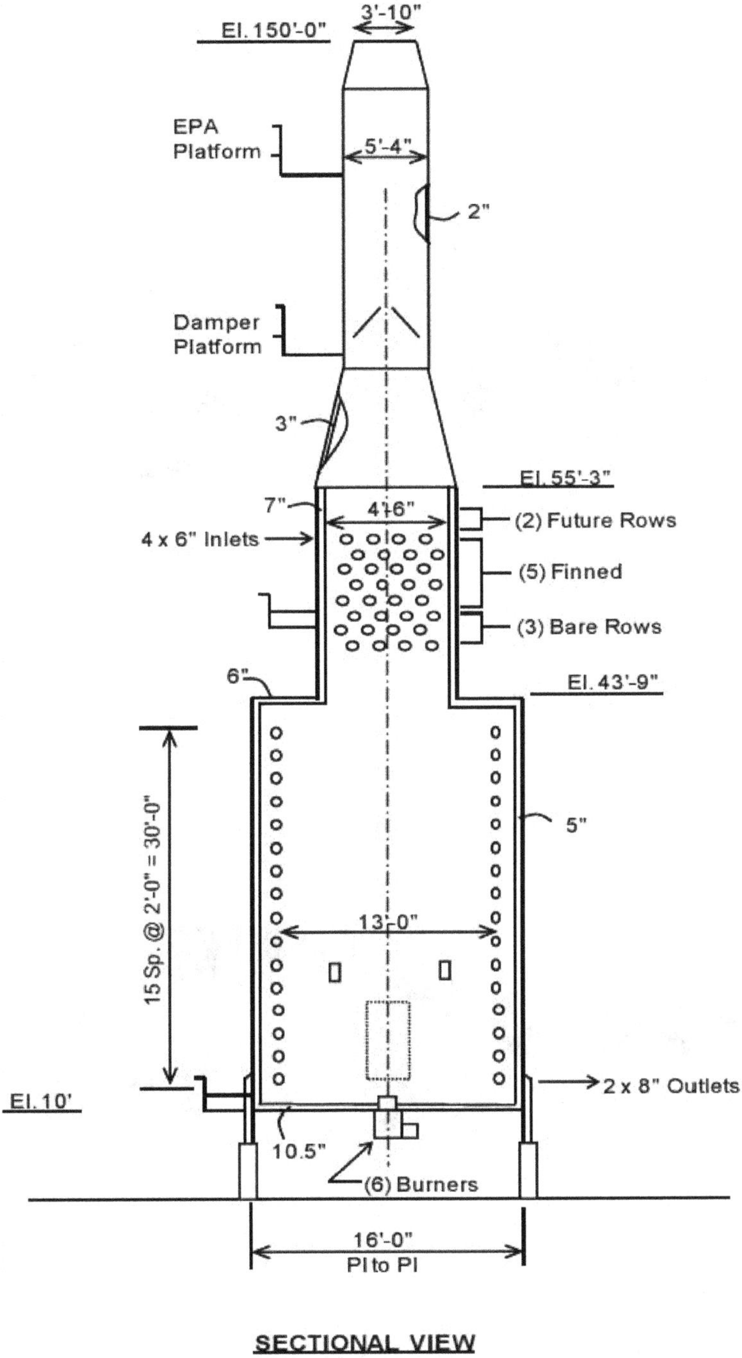

Figure 6b. Sketch of the cabin style horizontal box heater, one of the two heaters modelled.

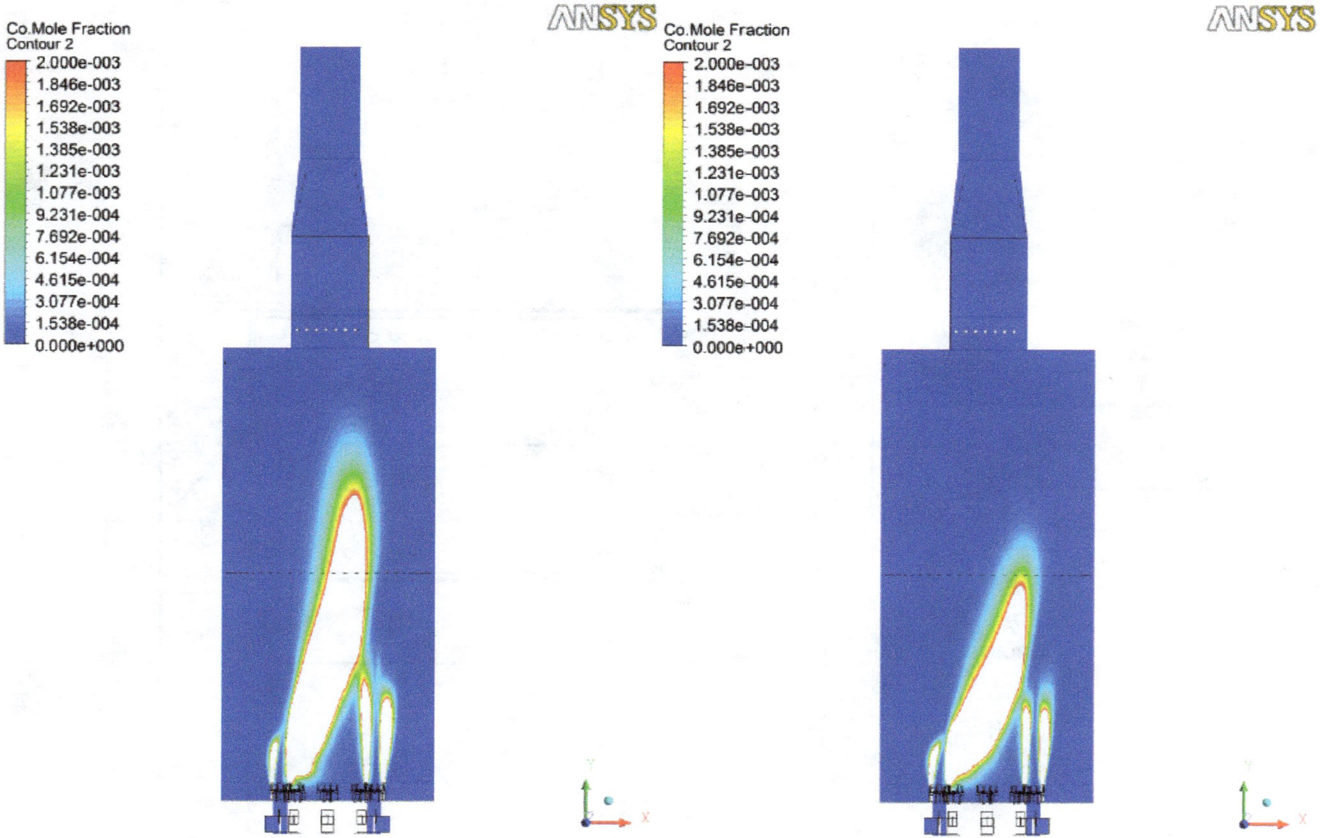

Figure 7a. Predicted flame shapes for VC heater employing COOLstar^{TM} burners - TNG Firing.

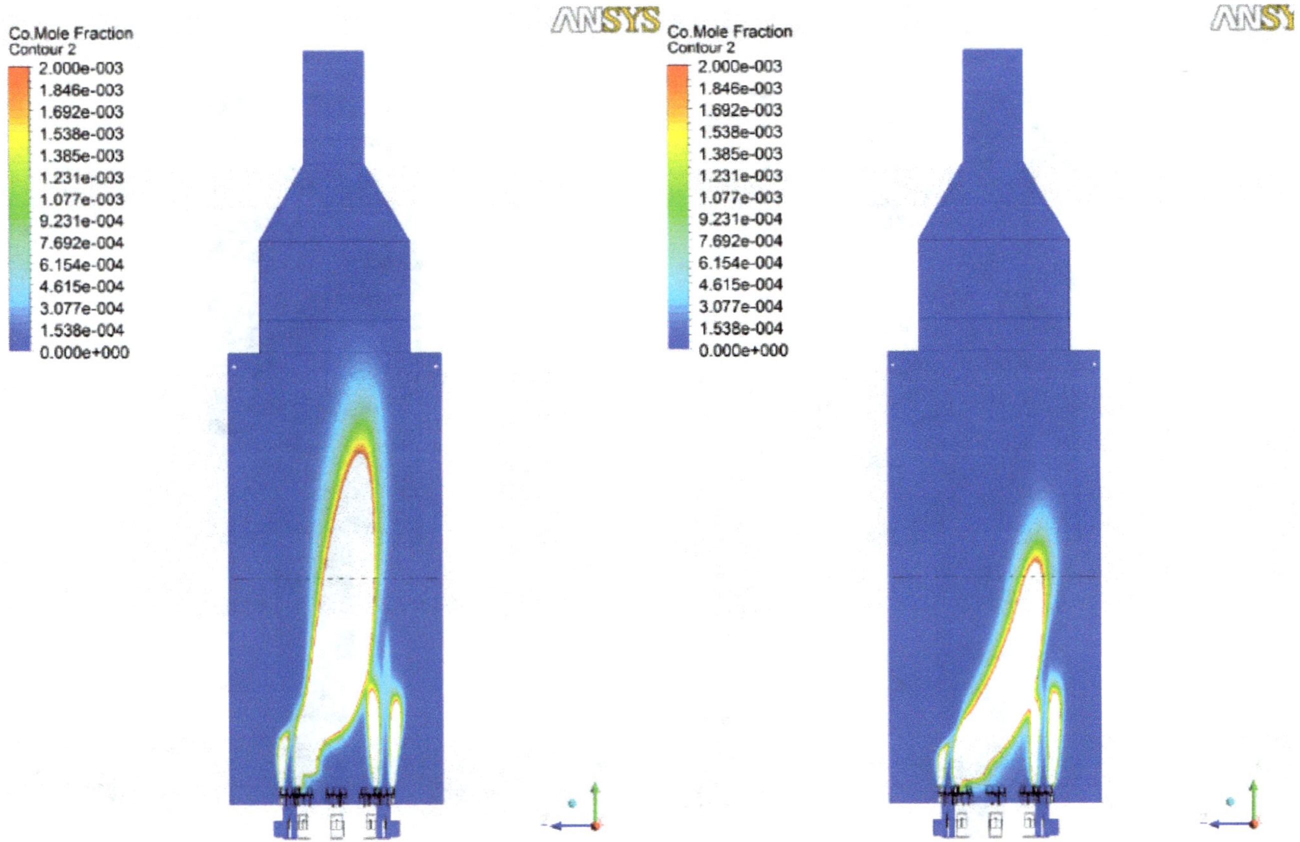

Figure 7b. Predicted flame shapes for VC heater employing COOLstar^{TM} burners - RFG Firing.

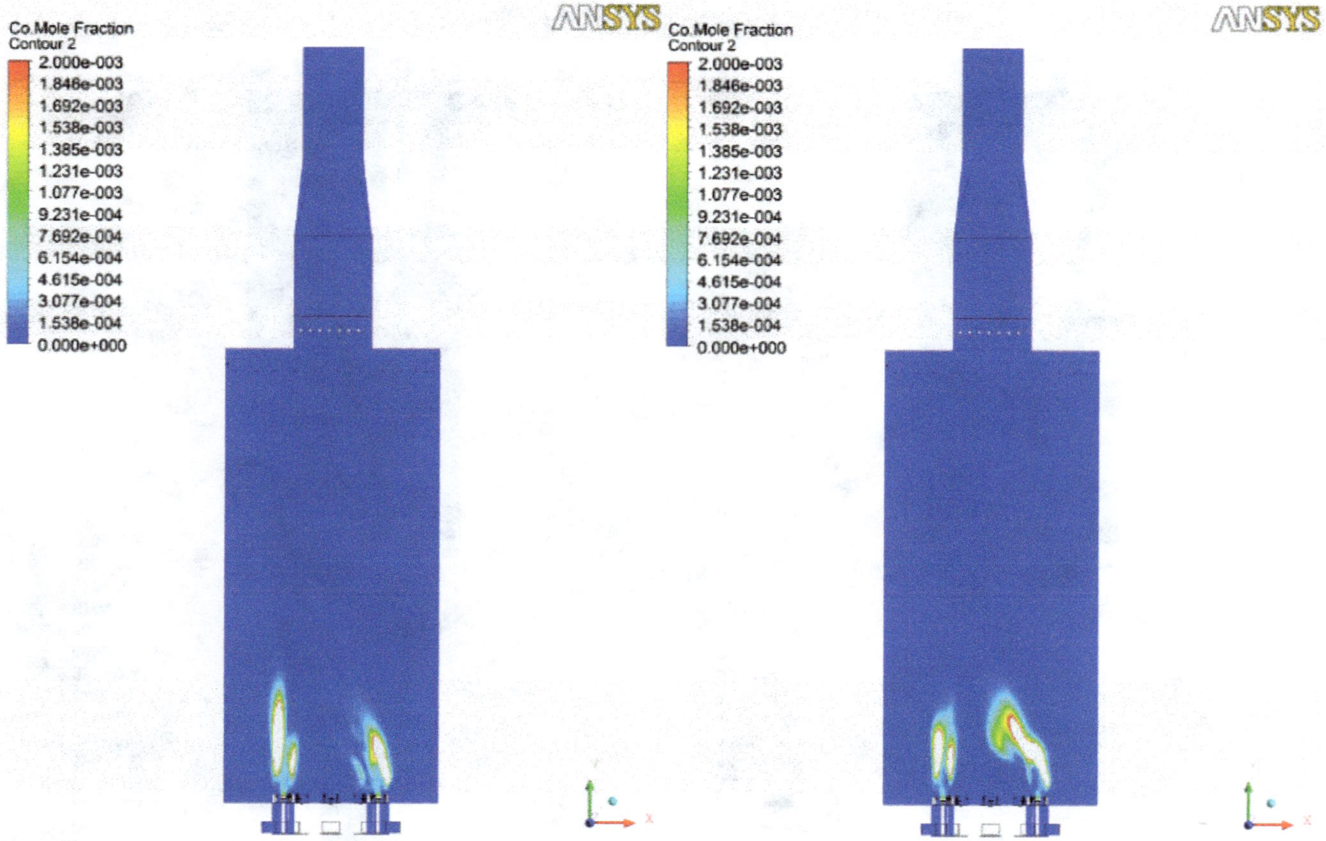

Figure 8a. Predicted flame shapes for VC heater employing PSFG burners - TNG Firing.

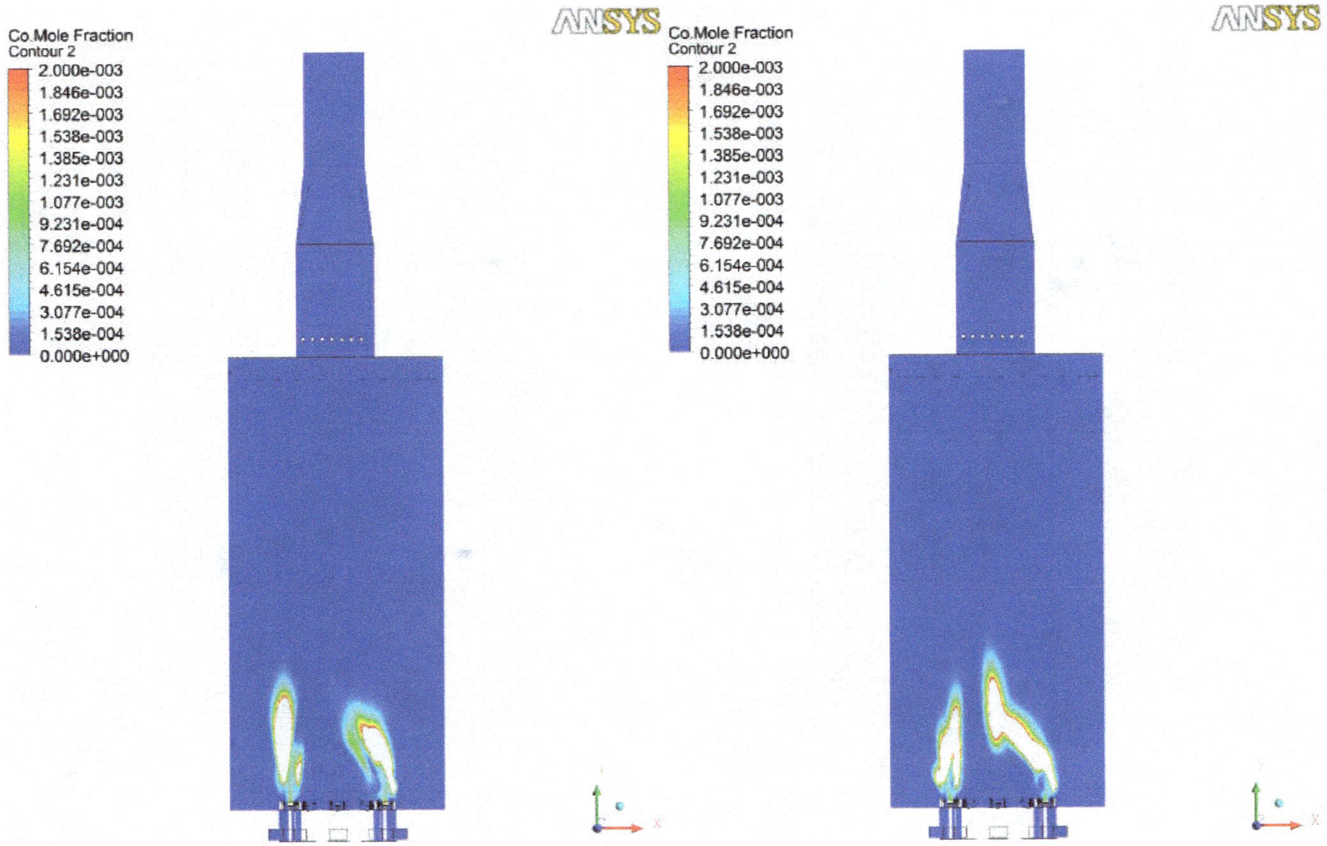

Figure 8b. Predicted flame shapes for VC heater employing PSFG burners - RFG Firing.

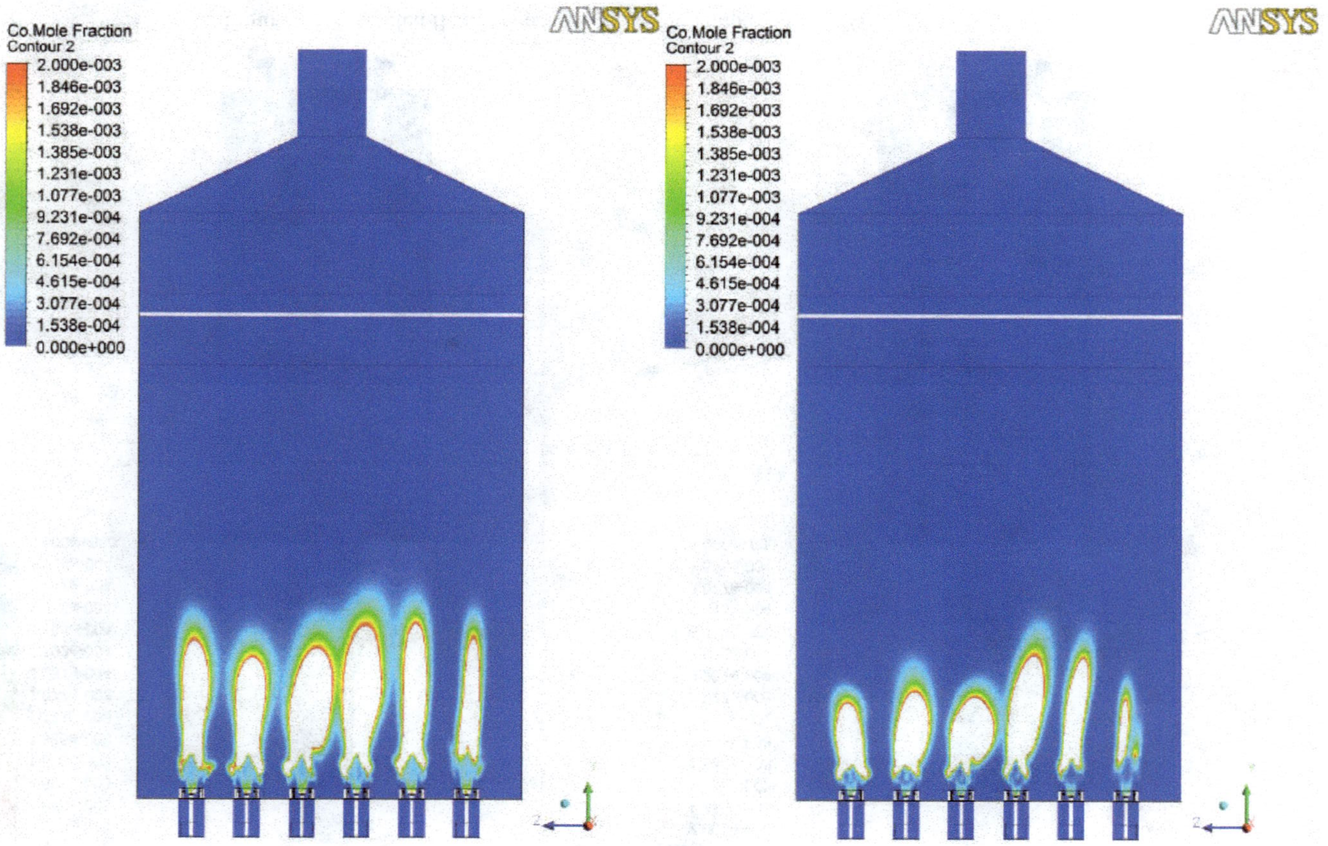

Figure 9a. Predicted flame shapes for Cabin heater employing PSFG burners - TNG Firing.

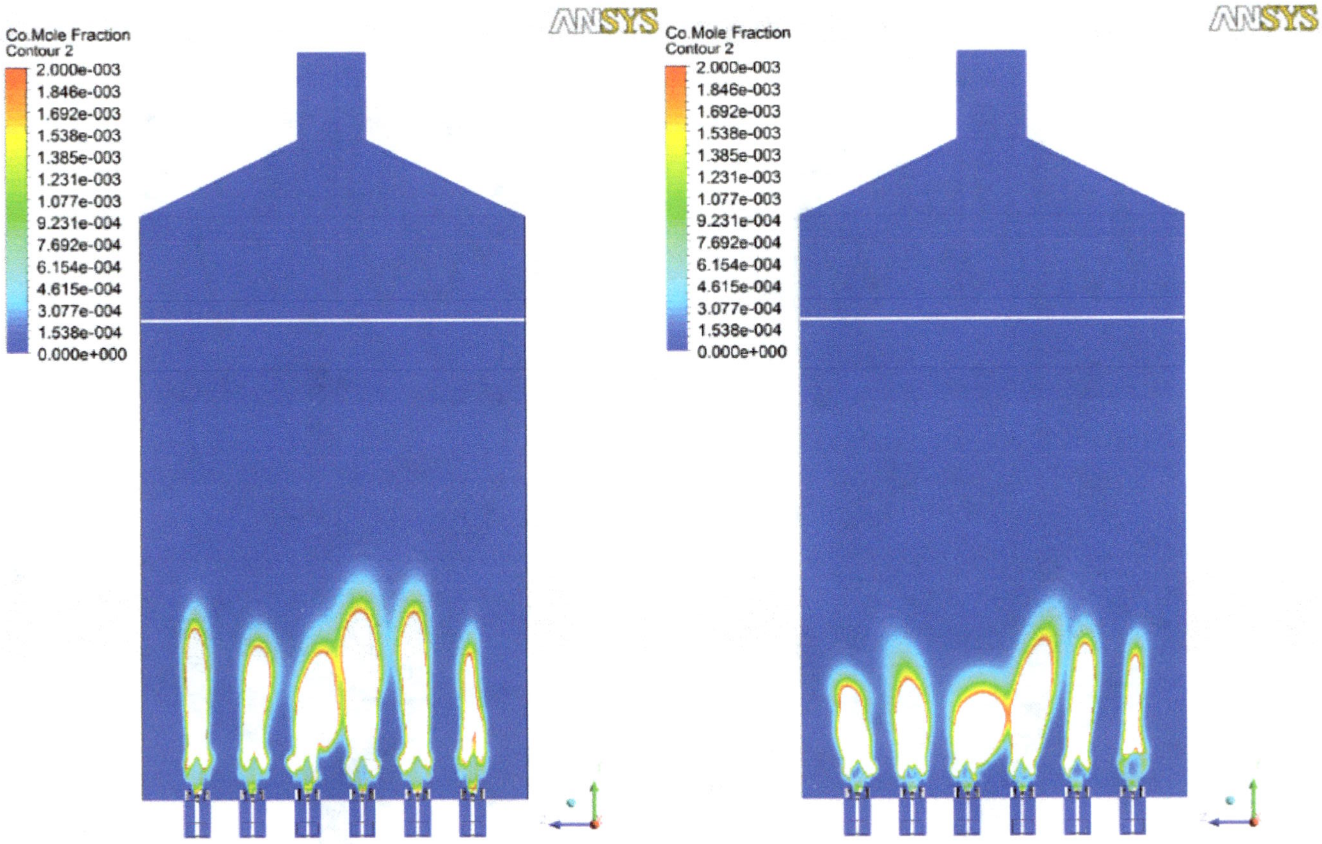

Figure 9b. Predicted flame shapes for Cabin heater employing PSFG burners – RFG Firing.

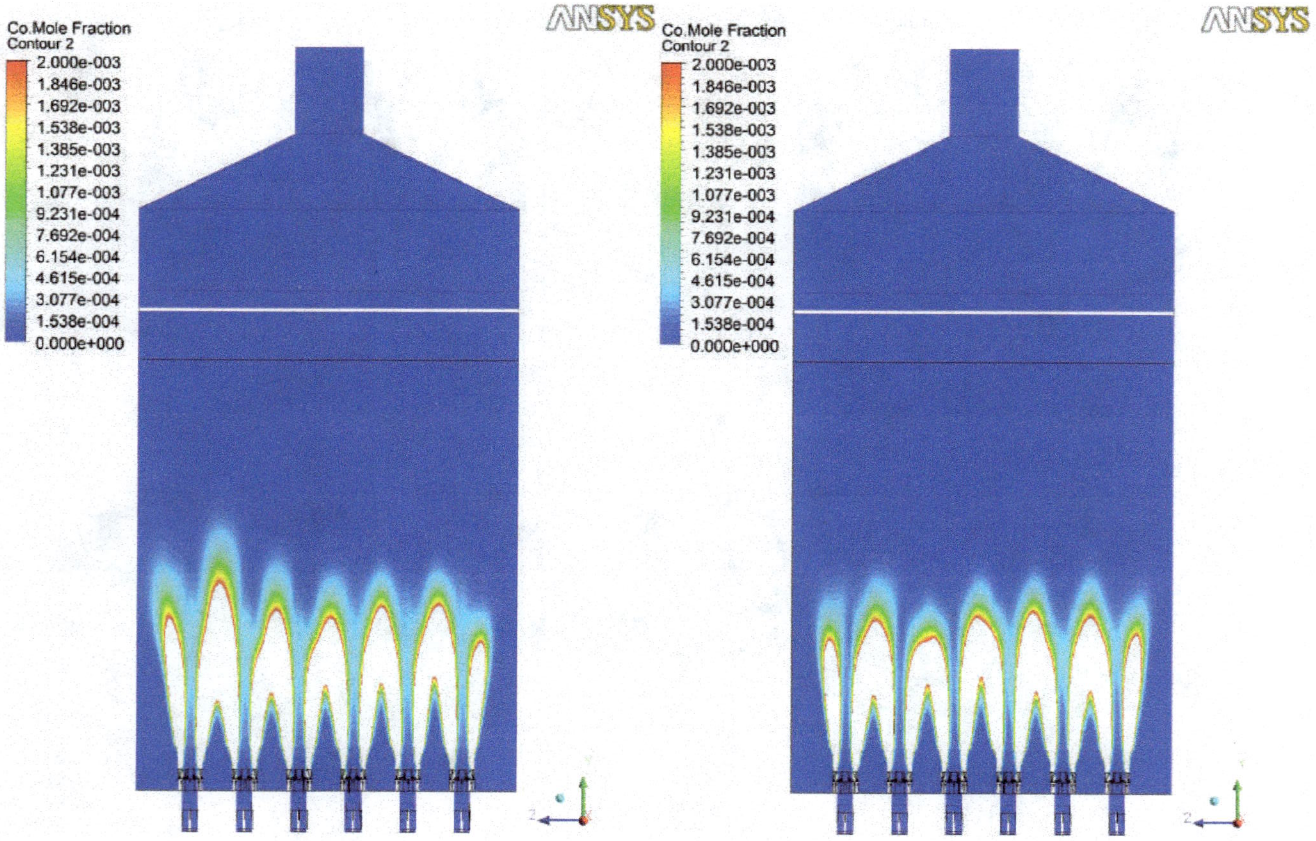

Figure 10a. Predicted flame shapes for Cabin heater employing COOLstar™ burners – TNG Firing.

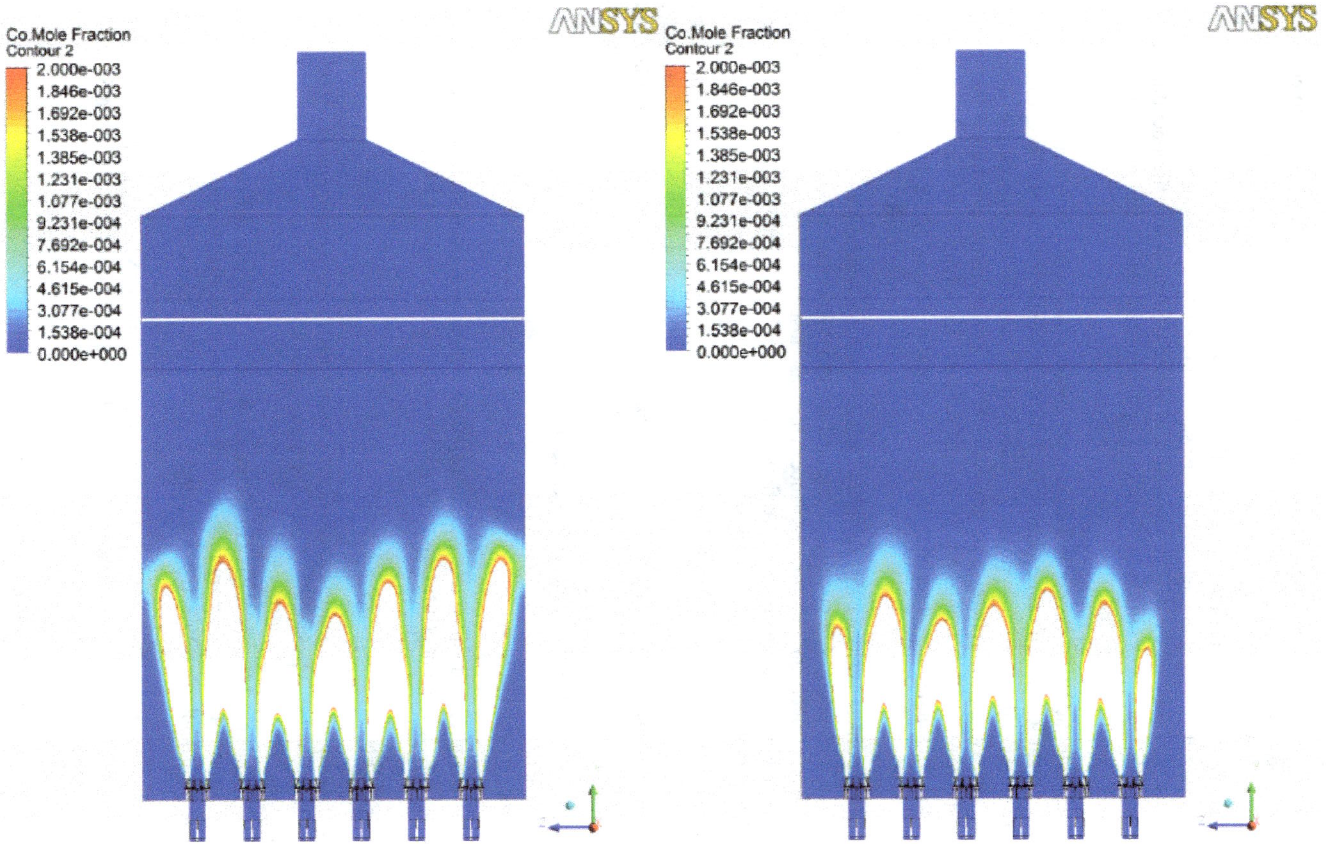

Figure 10b. Predicted flame shapes for Cabin heater employing COOLstar^{TM} burners – RFG firing.

CONCLUSIONS

A John Zink staged gas low NOx burner and a COOLstar™ Ultra-Low NOx burner were tested under oxy-fire conditions at the John Zink test facility. Flame appearance, flame stability, stack emissions, flame length and incident heat flux profile measurements were conducted for both burners under natural draft, ambient air conditions and forced draft, oxy-firing conditions. Suitable oxy-firing conditions were defined in a heater simulation study by OnQuest. Film and tube metal temperature were used as the design constraints to identify potential oxy-fire operating conditions. Parameters such as heater efficiency, heat duty split between radiant and convection section and heater draft were also evaluated.

Air leakage into the heater system is a significant concern as it may jeopardize the attainment of a CO_2-rich stream for ease of sequestration. High nitrogen concentrations will result in more costly downstream CO_2 purification systems. Unlike boilers, process heaters operate under negative pressure relative to atmosphere. The test furnace had some unique conditions that resulted in additional pressure drop compared to typical heaters in refinery service. In a commercial application, consideration will need to be given for proper sealing of the heater to minimize air leakage.

The process burners operated satisfactorily under oxy-firing conditions. Important performance variables such as flue gas re-circulation rate and oxygen concentration in the system were varied without adverse effects on heater performance. This was most evident during the transition between ambient air and oxy-fire operation where flow-rate, oxygen concentration and flue gas temperature were all changing.

A modified algorithm was used in CFD simulations to predict the radiative properties of the flue gas rich in CO_2 and water. CFD simulations were verified against single-burner test data and subsequently applied to modeling multi-burner heater performance. Temperature and heat transfer profiles similar to air-fired operation can be achieved when oxy-firing with flue gas recirculation. Some optimization of the process heater geometry (e.g., burner circle diameter in VC heaters) may reduce flame interaction, and improve heater performance.

ACKNOWLEDGEMENTS

Betty Pun compiled and edited this chapter based on reports, presentations, and papers by authors and collaborators, including Cliff Lowe of Nick Brancaccio of Chevron [9,10,11].

REFERENCES

1. C.E. Baukal (ed.), Oxygen-Enhanced Combustion, CRC Press, Boca Raton, FL, 1998.
2. S. Ferguson and M. Stockle, Carbon capture options for refiners, Petroleum Technology Quarterly, 17(3), 77-87, 2012.
3. M.B. Wilkinson, J.C. Boden, T. Gilmartin, C. Ward, D.A. Cross, R.J. Allam, and N.W. Ivens, CO_2 Capture from Oil Refinery Process Heaters Through Oxyfuel Combustion, Greenhouse Gas Control Technologies, Vol. 1, J. Gale and Y. Kaya (eds.), Elsevier, Oxford, UK, 2003, pp. 69-74.
4. M.B. Wilkinson, M. Simmonds, R.J. Allam, and V. White, Oxyfuel Conversion of Heaters and Boilers for CO_2 Capture, presented at the Second National Conference on Carbon Sequestration, May 5-8, 2003, Washington, DC.
5. R.J. Allam, C.J. McDonald, V. White, V. Stein and M. Simmonds, Oxyfuel Conversion of Refinery Process Equipment Utilising Flue Gas Recycle for CO_2 Capture, Greenhouse Gas Control Technologies, Vol. 1, E.S. Rubin, D.W. Keith and G.F. Gilboy (eds.), Elsevier, Oxford, UK, 2005, pp. 221-229.

6. FRNC-5 General Purpose Heater Simulation Program, available from PFR Engineering Systems, Inc.; Sante Fe Springs, CA 90670 U.S.A.
7. API RP-535: Burners for Fired Heaters in General Refinery Services, 3rd Edition, issued May 2014, American Petroleum Institute, Washington, DC.
8. C.E. Baukal, Heat Transfer in Industrial Combustion, CRC Press, Boca Raton, FL, 2000.
9. Jamaluddin, A.S. (Jamal), Lowe, C., Brancaccio, N., Erazo, J. and Baukal, C., "Technology Assessment of Oxy-Firing of Process Burners", Paper presented at the GHGT11 Meeting, Kyoto, Japan, November, 2012.
10. C. Lowe, N. Brancaccio, J. Jamaluddin, J. A. Erazo, Jr.; C.E. Baukal, Jr.; Technology assessment of oxy-firing of process heater burners, Energy Procedia, 37, 7793–7801, 2013.
11. Jamaluddin, A.S. (Jamal), Erazo, J. and Baukal, C., "Oxy-firing of Fired Process Heaters: CFD Analyses and Comparison with Data", Paper presented at the 3rd International Oxy-fuel Combustion Conference, Ponferrada, Spain, September 2013.

Carbon Dioxide Capture for Storage in Deep Geological Formations, Volume 4
Karl F. Gerdes (Editor)

Chapter 10

DEVELOPMENT OF PALLADIUM ALLOY MEMBRANE FOR H_2 RECOVERY FROM AUTOTHERMAL REFORMERS

Yeny Hudiono[1], Mike Whitlock[1] and Raja Jadhav[2]
[1]Pall Corporation – Finger Lake Center of Excellence, 3669 State Route 281, Cortland, NY 13045
[2]Chevron Energy Technology Company, 100 Chevron Way, Richmond, CA 94802

ABSTRACT: Pre-combustion is a promising CO_2 capture technology in which the CO_2 is captured from the fuel before it is burned. There are several options for separating CO_2 from the H_2 fuel, ranging from solvents to adsorbents to membranes. One option is to integrate high temperature H_2 membranes with the water gas shift operation. This approach has the potential to improve the efficiency of H_2 production while recovering CO_2 at high pressure, which reduces compression requirements to transport and store the CO_2.

This CCP project carried out a techno-economic evaluation of Palladium (Pd) alloy-based membranes. The basis for this assessment was performance of Pd alloy membranes for H_2 separation from a simulated syngas. Both single tube and multi-tube modules were fabricated and tested by Pall Corporation as part of this work. An empirical model was developed and calibrated to match the performance of the membranes and was used to estimate the membrane surface area required for large-scale H_2 separation. Finally, a detailed design and cost estimate of the membrane separator was completed. The estimated cost of CO_2 capture using the membrane separation process is reported in Chapter 11.

KEYWORDS: pre-combustion capture; palladium (Pd) alloy membrane; water gas shift (WGS); autothermal reformer (ATR)

INTRODUCTION

Heater and boilers (H&Bs) are a significant source of CO_2 in a refinery, and they are usually scattered throughout a refinery. Therefore, post-combustion and oxy-combustion technologies, which require a large amount of piping and processing equipment in the vicinity of the CO_2 source, are not generally practical in an existing, space-constrained refinery. Oxy-combustion may also require significant boiler modifications. For pre-combustion capture, major equipment to produce H_2 fuel and capture CO_2 may be placed away from the refinery processes, while H_2 fuel is distributed through the existing fuel gas infrastructure. A centralized pre-combustion scheme will benefit from economy of scale compared to retrofit of individual combustion units.

Hydrogen in a refinery is generally supplied by a steam methane reformer (SMR), which is currently limited in single-train H_2 production capacity to 200–250 MMSCFD of H_2. For a typical refinery, however, the equivalent fuel demanded by H&Bs requires a much higher volume of H_2. Autothermal reformers, therefore, provide a suitable alternative to SMRs to produce large volumes of H_2, up to 400 MMSCFD in a single train. Additionally, most of the CO_2 produced by an ATR-based H_2 plant is at high pressure in the reformer syngas. It is likely a lower cost option to capture a high fraction of the CO_2 from an ATR compared to an SMR in which about 40-50% of the total CO_2 produced is at atmospheric pressure and low concentration in the reformer furnace flue gas.

Several technologies can be employed for capturing CO_2 from ATRs. Examples include amine absorption, pressure swing adsorption (PSA), cryogenic purification, membrane separation, or a combination of these. Of these, CO_2 capture by absorption in an amine such an activated MDEA (or aMDEA) solution is commercial technology. To reduce the cost of CO_2 capture, alternatives to the conventional amine-based capture technology are sought. Separation of H_2 using Pd alloy-based membranes provides several advantages over the amine-based technology – (a) the CO_2 is produced at higher pressures, thus reducing the cost of CO_2 compression to the pipeline delivery pressures, (b) the dense Pd membranes produce H_2 at very high purity, typically in excess of 99.9 mol%, and (c) the H_2 is delivered at the low pressures needed at the burner head in the H&Bs.

Process Configuration

A simplified block flow diagram of a process in which Pd-based membranes are used to separate H_2 and produce pure CO_2 stream is shown in Figure 1.

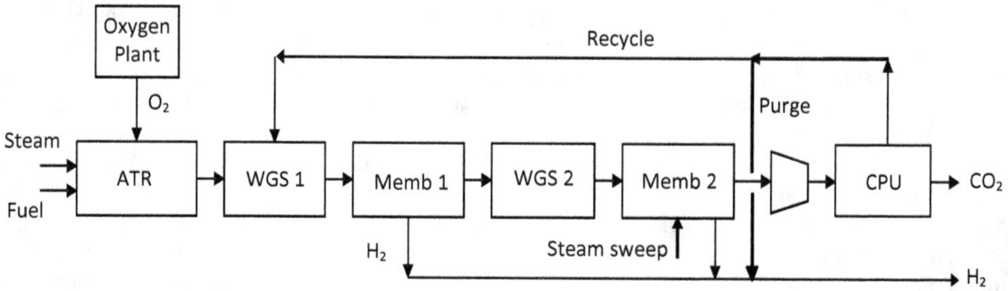

Figure 1. Simplified process block flow diagram.

As shown in the figure, the natural gas feed, which may be combined with refinery fuel gas (RFG), is converted using oxygen and steam into syngas primarily consisting of CO, CO_2, H_2 and unreacted CH_4 in the ATR reactor. After the syngas stream is cooled by generating high pressure steam in a waste heat boiler (WHB), not shown here, the syngas is sent to water gas shift (WGS) reactors and membrane separators. Two membrane stages are used and each is preceded by a high-temperature shift reactor to convert CO into H_2 and thus increase the overall H_2 recovery in the membranes. The performance of Pd membranes is inhibited by the presence of high concentrations of CO, so the shift reactor is upstream of the membrane stage to lower CO while increasing the concentration of H_2 and the driving force for H_2 permeation. A sweep of steam is used in the second stage to maximize the performance of this membrane unit. Both membranes operate in a syngas environment at 400°C. In principle, the membrane and shift reaction could be combined in a single unit by embedding the shift catalyst in the membrane stage in a configuration known as Membrane Water Gas Shift, or MWGS reactor. However, the combination of reaction and separation imposes several design and maintenance challenges for a commercial-scale reactor with minimal benefits over the series configuration shown in Figure 1.

The two membrane separators together remove about 90% of the overall H_2 produced. The retentate from the second membrane unit contains the remaining hydrogen, CO_2, CO and methane, which is processed in a cryogenic purification unit (CPU) to recover a high purity CO_2 product stream. The reject from the CPU, which contains primarily H_2 and CO_2, is recycled to the inlet of the first WGS reactor to improve the overall recovery of CO_2 and H_2. A purge is taken out of the recycle stream to avoid buildup of inert in the process. The hydrogen permeate product, after condensing the sweep steam (not shown in the diagram), is mixed with the purge stream and distributed to the H&Bs as fuel. The purge stream contains combustible components CO and CH_4 in addition to H_2. This satisfies part of the fuel demand and reduces the net amount of H_2 fuel produced in the ATR, while still satisfying the specified overall 90% CO_2 capture rate for the H&Bs. The CO_2 product, after compression and drying, is sent to a suitable storage location.

The overall goal of this project was to test the performance of Pd alloy-based membranes in separating H_2 from simulated syngas and use the performance data along with a model to design and estimate the cost of a commercial-scale membrane system for producing 5,000 MMBtu/h (LHV) of low-carbon H_2-rich fuel in an ATR process, while producing a purified CO_2 stream at high pressure suitable for sequestration.

EXPERIMENTAL

The main goal of the experimental program was to evaluate the H_2 separation performance of the membrane, as characterized by the permeability and the recovery of H_2. To accomplish this, the performance of a 12-inch single-tube Pd alloy membrane, with an effective length of 10 inches, was tested with two different syngas compositions, which corresponded to operating an ATR at steam-to-carbon ratios (S/C) of 1.8 and 1. These two syngas compositions with different pressures were selected to evaluate membrane performance in widely different ATR operating conditions. However, in this chapter, the focus is on results for a S/C ratio of 1.8, as this condition is thought to be more representative of the current commercial ATR offering.

The summary of the experimental conditions for the 12" single tube Pd alloy membrane is shown in Table 1. The feed to Stage 2 of the membrane was estimated based on the expected H_2 recovery in Stage 1, which was modelled using an in-house simulation (described later), and the simulated performance of the 2nd stage of the WGS reactor. Due to the low H_2 driving force in the second stage, steam sweep was used to reduce the partial pressure of H_2 in the permeate stream. The effect of the sweep is investigated by running Stage 2 with and without the sweep.

Table 1. Summary of experimental conditions for single tube measurement.

	Steam to Carbon = 1.8			Steam to Carbon = 1		
	Stage 1	Stage 2	Stage 2 w/sweep	Stage 1	Stage 2	Stage 2 w/sweep
Feed composition (mol fraction)						
H_2O	0.2421	0.3472	0.3472	0.173	0.2161	0.2161
H_2	0.5515	0.3104	0.3104	0.595	0.3539	0.3539
CO	0.0475	0.0245	0.0245	0.084	0.0537	0.0537
CO_2	0.1464	0.2971	0.2971	0.137	0.3542	0.3542
Total feed flow (SLPM)	11.9	7.8	7.8	14.8	9.8	9.8
Expected hydrogen recovery	83%	65%	67%	88%	78%	80%
Feed pressure (psig)	294.2	278.2	278.2	405.8	405.8	405.8
H_2 permeate pressure (psig)	21.75	21.75	19.3	21.75	21.75	19.3
Total permeate pressure (psig)	21.75	21.75	29	21.75	21.75	29
Sweep flow rate (SLPM)	-	-	0.8	-	-	1.4

Note that in Table 1, Stage 1 and Stage 2 refer to Memb 1 and Memb 2 in Figure 1, respectively.

Experimental Procedure

Permeation experiments were conducted using a constant pressure permeation apparatus (Figure 2). The as-made Pd alloy membrane was extended using ¼" stainless steel tubing and was installed into the sample module (Figure 3). The membrane was exposed to the feed conditions at 400°C, and the pressure was adjusted to reach the selected experimental condition. The permeation flux was monitored for approximately 2 hours while the composition of the permeate stream was measured using gas chromatography (GC).

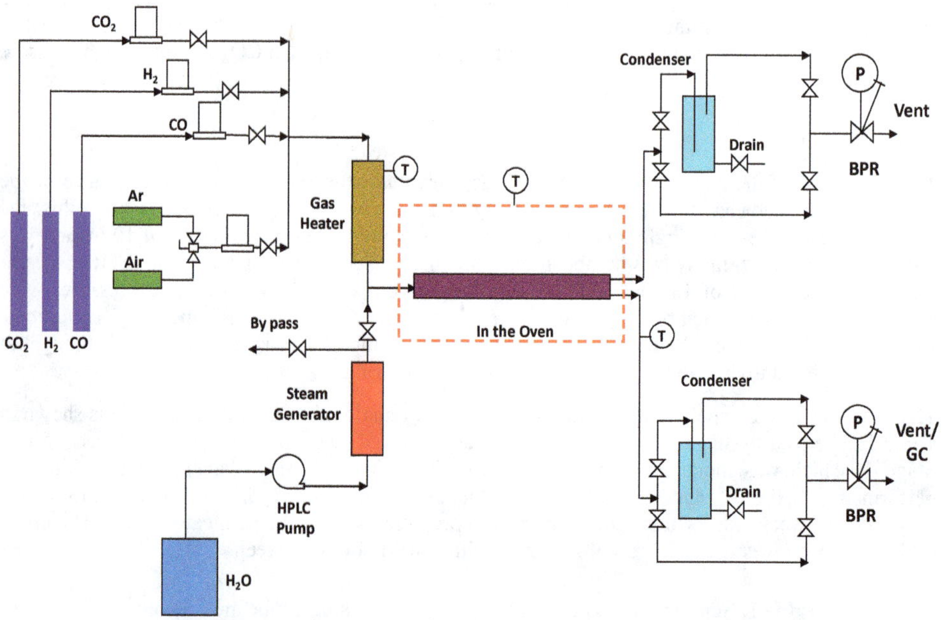

Figure 2. Schematic of the constant pressure permeation apparatus.

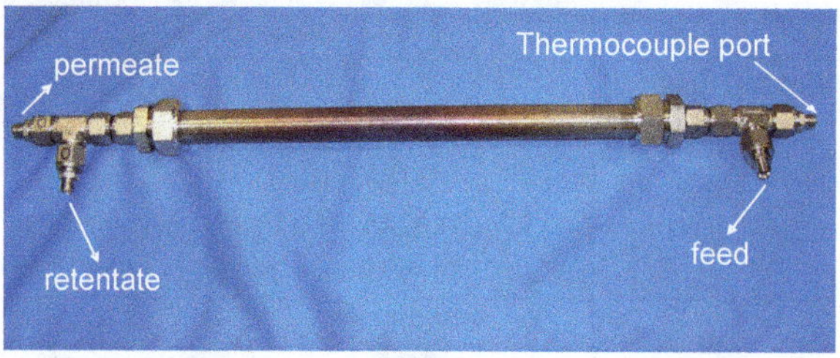

Figure 3. A sample membrane module containing a 12-inch Pd alloy single tube membrane.

Experimental Results

The membranes exhibited good integrity during testing, which was checked by the argon leak rate, and confirmed by high hydrogen purity (99.5%) before and after the tests.

Figure 4 shows the behavior of the H_2 flow rate and recovery as a function of time for S/C = 1.8, Stage 1 condition.

The experimental and simulated performances of the 12-inch Pd alloy membranes for different experiment conditions are shown in Table 2.

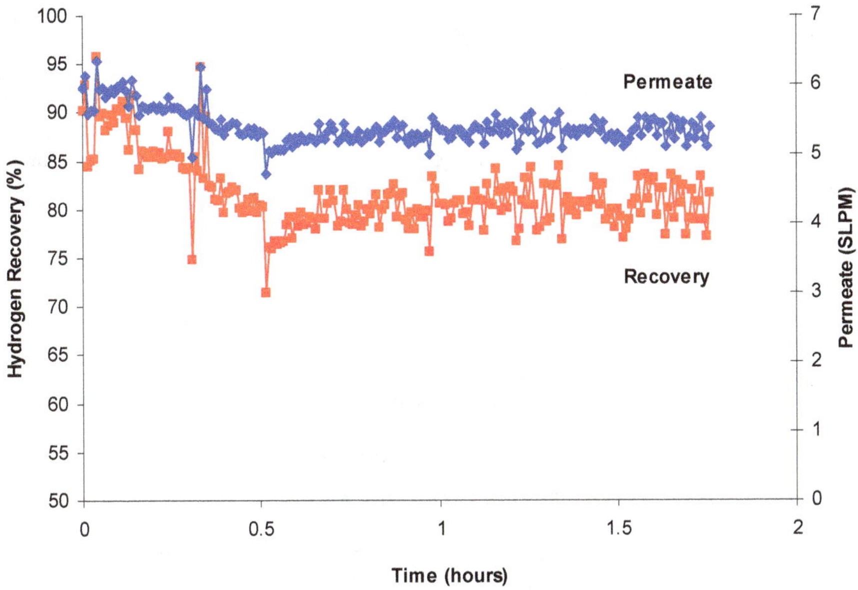

Figure 4. H_2 flow rate and recovery as a function of time.
Time zero equals the time when the feed condition was introduced to the membrane.

Table 2. H_2 permeation performance of 12-inch-single-tube Pd alloy membrane with various feed conditions.

Experimental Conditions	H_2 Purity (%)	H_2 permeance ($\times 10^{-4}$ mol/m^2-s-Pa$^{1/2}$)	H_2 Recovery (%) Experimental	Simulation
S/C = 1.8, stage 1	99.95	7.33	81	83
S/C = 1.8, stage 2, with sweep	99.18	4.20	75	67
S/C = 1.8, stage 2, without sweep	99.15	4.27	72	65
S/C = 1, stage 1	99.90	7.66	84	88
S/C = 1, stage 2, with sweep	99.57	3.97	77	80
S/C = 1, stage 2, without sweep	99.57	4.01	75	78

As shown in Table 2, the H_2 permeance through Pd alloy membrane varies with the experimental condition. This is due to the effect of mass transport resistance on the apparent membrane performance. In order to calculate the intrinsic H_2 permeance through the Pd alloy membrane, one needs to consider the true H_2 concentration on the Pd surface. Since the H_2 concentration on the surface cannot be measured from the experiment, it was calculated using Equation 1. The hydrogen permeation through the Pd alloy membrane depends on the mass transport resistance on the feed surface of the membrane, as described by the Stefan-Maxwell equation (left side of Equation 1), and the solution diffusion transport of hydrogen through the membrane, as described by the modified Sieverts' Law (right side of Equation 1).

$$k_x \ln\left(\frac{1-x_{H_{2,w}}}{1-x_{H_{2,f}}}\right) = P\left(p^n_{H_{2,w}} - p^n_{H_{2,p}}\right) \qquad \text{Equation 1}$$

In Equation 1, P is the membrane permeability and n is a constant. The values of P and n were estimated by fitting to the pure hydrogen permeation data and are 1.69 SLPM/psia^n (1.36×10^{-10} m^3-m/m^2-s-Pa^n) and 0.5212, respectively. The $x_{H_{2,w}}$ and $x_{H_{2,f}}$ are the mole fractions of H_2 on the Pd membrane surface and in the feed stream, respectively. The $x_{H_{2,w}}$ is used to account for concentration polarization close to the membrane surface (boundary layer). k_x is the mass transfer coefficient in the boundary layer, which depends on the module geometry and is estimated by fitting Equation 1 to the experimental data. Thus, the mass transfer coefficients calculated from single tube testing and from module testing would be different.

The comparison of measured H_2 recoveries and simulated values is shown in Table 2. Overall, the agreement between the experiment and the model results is very good except for the S/C = 1.8 Stage 2, for which the experimental recovery values are somewhat higher than the values predicted by the simulation. This is thought to be indirectly due to the sensitivity of membrane performance to CO. The diffusivity coefficients of the non-permeating gases in the boundary layer were obtained by fitting previous in-house permeation data from the tests with high CO (8–20 mol %) feed. Thus, the simulation better predicts the membrane performance for feed conditions with CO levels between 4.75 and 8.4 mol % than for the low-CO feed from S/C =1.8 stage 2 (CO = 2.5 mol %).

As detailed later in the chapter, the large scale membrane separator consisted of a cascade arrangement of small scale membrane modules, each containing 12-tube Pd membranes. Therefore, to estimate the membrane area for the large scale membrane system, the performance of a 12-tube Pd membrane module was measured for the syngas compositions corresponding to S/C = 1.8 and S/C =1 at Stage 1 of the membrane separators. The performance data for the 12-tube Pd membrane module was used to adjust the fitting parameters (mass transfer coefficient) used in the model. The competitive adsorption of CO on the Pd membrane was also separately accounted for in the updated model. The hydrogen flux through the Pd membrane, accounting for inhibition by CO, is described as below:

$$J_{H_2}^{inhibited} = (1 - IC)J_{H_2}^{clean} \qquad \text{Equation 2}$$

$$J_{H_2}^{clean} = P\left(p_{H_{2,w}}^n - p_{H_{2,p}}^n\right) \qquad \text{Equation 3}$$

IC is the effect of CO inhibition, which is described by the Langmuir isotherm as:

$$IC = \frac{\alpha b P_{CO,b}}{1+bP_{CO,b}} \qquad \text{Equation 4}$$

where, α and b are the fitting parameters, which represent the intrinsic properties of CO adsorption on the membrane. By fitting the experimental data, parameters α and b were estimated to equal 0.5426 and 4.11×10^{-5}/Pa, respectively. $P_{CO,b}$ is the partial pressure of CO in the feed stream.

Experimental results, not shown here, indicate that the membrane permeabilities for both feed conditions are comparable even though the feed H_2 partial pressure for the stream with S/C ratio of 1 is much higher than for S/C of 1.8 (see Table 1). Since syngas for S/C=1 contains a much higher concentration of CO, this confirms that the presence of high CO inhibits the H_2 flow rate through the Pd membrane.

A comparison between the experimental and simulated recovery of H_2 using a module containing 12 tubes of 12-inch single tube Pd membranes (12-tube membrane module) is shown in Table 3. The results show that the old model with old fitting parameters predicts the H_2 permeate and recovery at low feed flow rate reasonably well, with about 15% error. However, at feed flow rate of 400 SLPM, the model over-estimates the H_2 permeate flow rate and recovery by nearly 34%. This is because the fitting parameter used in the old model was fitted using the previous experiment data with low feed flow rate. By adjusting the Sieverts' parameter, mass transfer coefficient and accounting for CO adsorption in the model, the new model can predict the experiment data within + /- 15% accuracy.

Table 3. The 12-tube membrane module performance for S/C = 1.8

Feed Rate (SLPM)	H_2 Purity (%)	% H_2 Recovery (expt)	% H_2 Recovery (old model)	% error	% H_2 Recovery (updated model)	% error
400	99.97	31.23	47	33.6	31	-0.8
100	99.90	67.35	79	14.7	79	14.7

A constant pressure permeation apparatus was also used to test membrane durability under condition corresponding to the S/C=1. This condition was chosen since it included the CO at high pressure, which could negatively impact the H_2 transport through the membrane by reducing the membrane permeability as a result of competitive adsorption on the membrane surface. The short term durability test showed that the membrane survived the S/C = 1 condition for 3 days, with a H_2 permeate purity of 99.8%.

Long-term Durability Tests

The ability of Pd alloy membranes to maintain performance over a long duration is critical for their commercial application. To test durability, the H_2 purity and permeate flow rate of the membrane were measured over 1,000 hours using a Pall hydrogen separation module that consisted of 3 tubes. This smaller capacity module was used to conserve the amount of fuel and water used during the experiment. For these long-duration experiments, the syngas stream was produced using a methanol reformer.

The test consisted of an 800-hour segment, where the membrane was exposed to a feed stream corresponding to the inlet of the Stage 1 of the membrane for an ATR feed steam to carbon (S/C) ratio of 1.8, followed by a 200-hours segment, where the feed stream corresponded to the S/C ratio of 1. See Table 1 for the gas compositions. The retentate was analyzed by a gas chromatograph and the permeate was analyzed using an infrared absorption gas analyzer, which was able to detect low gas concentration. The pressure on the feed side was kept at 200 psig, and the permeate side was maintained at 1.5 psig.

S/C = 1.8 syngas test

For the S/C=1.8 syngas condition, the total feed flow rate into the membrane module was 95 SLPM and the H_2 output (permeate) flow rate was 16 SLPM, the H_2 recovery was 32%. The H_2 purity was equal to 99.97%. After the system had run for about 155 hours, the feed flow rate was reduced to 67 SLPM, which increased the H_2 recovery, and increased the gas residence time on the membranes, and thus it sets the membranes in the more severe conditions. The H_2 purity remained at 99.97%.

The H_2 recovery and permeate flow were stable throughout the course of the test. Towards the end of the test, the H_2 permeance was slightly reduced by 6%. This can potentially be due to the formation of carbon (coke) on the membrane over time. The H_2 flux was restored by flushing the membranes with air.

S/C = 1 syngas test

The membrane was next operated for 200 h using the conditions corresponding to S/C = 1 at stage 1. The H_2 purity and permeate flow rate for the 200 hours test were 99.86% and 18 SLPM, respectively.

The H_2 permeate flow decreased over the 200 hours test. This was due to the presence of CO on the Pd membrane surface that dissociated and adsorbed on the membrane surface and thus reduced the H_2 adsorption sites. The permeate purity was initially reduced as the feed condition was changed; however, it increased with time and leveled out at about 99.8%. The slight reduction of permeate purity may be caused by the increasing amount of CO in the feed stream.

The long term test results show that the Pall Pd membrane can potentially be used for H_2 purification from syngas corresponding to S/C = 1.8. However, the conditions corresponding to S/C = 1 may be too harsh and the membranes may lose productivity due to the high CO content in the feed stream.

DESIGN AND COST OF A COMMERCIAL SCALE MEMBRANE SEPARATOR

The main goal of this study was to estimate the cost of a membrane separator to produce 5,000 MMBtu/h (LHV) of H_2 fuel from a commercial scale ATR plant. To achieve this, the design of the membrane separator was built up from smaller multi-tube modules as described below.

Arrangement of the Membrane Tubes in a Commercial Unit

A Pd alloy membrane tube 1 cm in diameter and 40.5 cm in active length is the basic building block for the separation of H_2. Multiple membrane tubes are housed in a self-contained module that comprises the smallest field-serviceable unit of membrane area, with each module containing 14 tubes. In this module, similar to the 12-tube module shown in Figure 5, the feed gas is introduced to the outside surface of the membrane tubes. A hydrogen rich stream permeates through each tubular membrane element and exits the module. The hydrogen depleted retentate stream exits the module on the upstream end of the membrane array. The diameter of the module is 7.6 cm and the length is 65.5 cm. The estimated membrane area is about 0.18 m^2 per module.

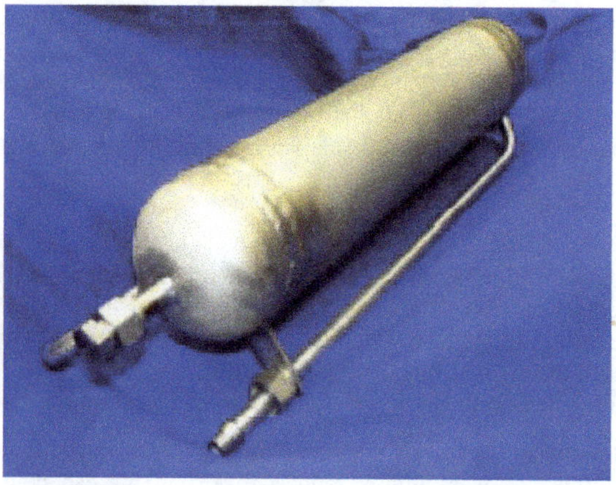

Figure 5. Pall's small scale membrane separation module (12 tube, 25.4 cm active membrane length).

To achieve separation of H_2 from the syngas of a large scale ATR, the small scale membrane modules are arranged in a combination of parallel and series configuration as shown in Figure 6, to form a membrane stage (e.g., membrane stage 1 as shown in Figure 1). The Pall membrane modules are packaged in parallel in a large scale vessel. Each box shown in Figure 6 can also be viewed as representing a single vessel consisting of 14-tube modules. In order to efficiently fabricate, install, and service the modules, a nominal 122 cm diameter for the vessel was chosen. Each 122 cm vessel can accommodate approximately 220 membrane modules per tube sheet for a total of 440 modules per vessel, which results in nominally 70 m^2 of membrane area. As shown in Figure 6, the retentate flow from each vessel was combined and equally redistributed to form a feed to the next "layer" in series.

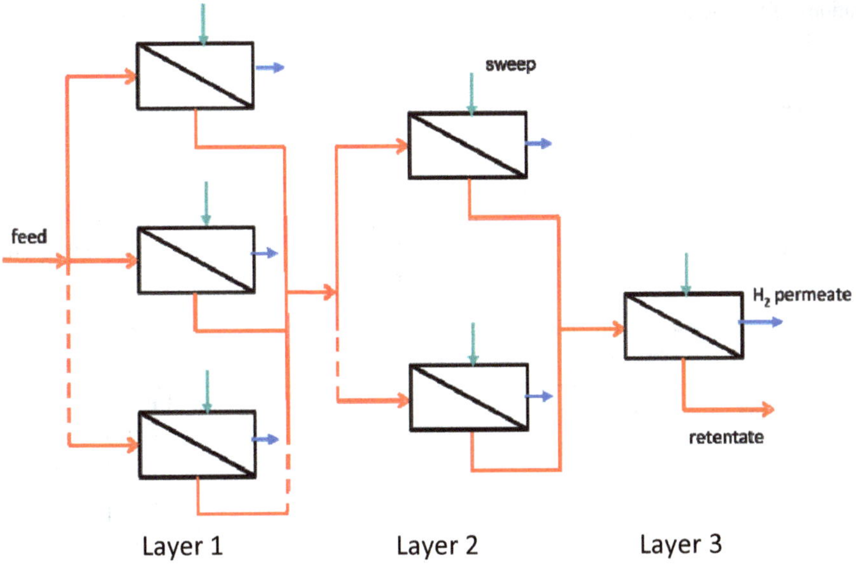

Figure 6. The arrangement of Pall's small scale H_2 purification units to achieve the required membrane area for large scale H_2 production. The schematic shows the incorporation of sweep gas, which is used only for the Stage 2 membranes.

Estimation of the Total Membrane Area

In order to estimate the total number of membrane modules and vessels required in a commercial scale unit, it is first necessary to estimate the total area of the membrane surface that is needed to produce 5,000 MMBTU/h (LHV) of H_2 fuel from a commercial scale ATR plant. The empirical model, which was developed and tested for the 12-tube Pd membrane module, was used for the estimation of the membrane area. As described earlier, the model included mass transfer resistance in the boundary layer and accounted for the reduction of H_2 permeation across the membrane due to the presence of carbon monoxide (CO). The model predicted the experimental data within +/- 15% as reported earlier.

The membrane area was calculated based on the 14 tube Pd membrane module described earlier. In the simulation model, the feed flow rate was equally distributed to several membrane modules to form the first layer such that the feed flow rate per module was less than 650 SLPM. The feed flow rate was specified such that it was not too high for the gas to flow across the tubing/port configuration, and the pressure drop along the module did not exceed 10 psi. Also, the required amount of sweep steam per module for Stage 2 was designed to be less than ~246 SLPM.

To achieve the target capacity of 5,000 MMBtu/h of H_2, the membrane arrangement shown in Figure 6 was simulated in combination with the overall reforming and separation process shown in Figure 1. The feed flow rate, the number of modules in parallel per layer and the number of layers were adjusted such that the hydrogen recovery per membrane stage (stage 1 or 2) reached 70%, which was estimated to be the optimum H_2 recovery rate for this process. An iterative solution procedure was required since the membranes impact the performance of the shift reactors and the CPU unit by changing the feed composition to these units, and there is a recycle from the CPU to the inlet of the first shift reactor. The overall process simulation was set up in Aspen Hysys, whereas, the membrane simulation was set up in MS Excel. This necessitated independent simulations of the membrane and the overall process and several iterations to achieve the target H_2 production.

Based on the simulations, the detailed material balance across the layers within the two membrane stages is shown in Tables 4-7 for the syngas composition corresponding to S/C=1.8. The required membrane areas for Stages 1 and 2 are estimated to be 13,764 and 12,903 m^2, respectively. For Stage 2, the steam sweep rate was set to maintain a ratio of permeate pressure to partial pressure of hydrogen in the permeate at 2, which was considered to be an optimized value taking into account the balance between the membrane area (i.e., the capex) and the steam flow rate (i.e., the opex). The feed pressures for Stage 1 and 2 were 290 and 280 psig, respectively. The overall hydrogen recoveries for Stages 1 and 2 were 70.7% and 70.2%, respectively.

Table 4. Overall mass balance across the membrane contactor for syngas composition corresponding to S/C=1.8

Stage	Layer	# Vessels	% Hydrogen Recovery	Total Flow (× 10^7 SLPM)		
				Feed	Permeate	Retentate
1	1	68	37.4	1.64	0.315	1.32
	2	55	33.7	1.32	0.178	1.15
	3	48	29.3	1.15	0.102	1.04
2	1	60	45.8	1.04	0.278	0.905
	2	52	32.4	0.905	0.107	0.852
	3	49	18.7	0.852	0.0417	0.831

Table 5. Feed composition to the two membrane stages and different layers within the stages.

Stage	Layer	Total Flow (× 10^7 SLPM)	Feed Composition (mol fraction)						
			H_2	CO	CO_2	H_2O	CH_4	N_2	Ar
1	1	1.64	0.514	0.057	0.172	0.199	0.025	0.018	0.015
	2	1.32	0.398	0.070	0.213	0.246	0.031	0.022	0.018
	3	1.15	0.304	0.081	0.246	0.284	0.035	0.025	0.021
2	1	1.04	0.291	0.035	0.325	0.259	0.040	0.028	0.023
	2	0.905	0.181	0.040	0.374	0.298	0.046	0.032	0.026
	3	0.852	0.130	0.042	0.398	0.317	0.049	0.034	0.028

Table 6. Retentate composition from the two membrane stages and different layers within the stages.

Stage	Layer	Total Flow (× 10^7 SLPM)	Retentate Composition (mol fraction)						
			H_2	CO	CO_2	H_2O	CH_4	N_2	Ar
1	1	1.32	0.398	0.071	0.213	0.246	0.031	0.022	0.019
	2	1.15	0.305	0.082	0.246	0.285	0.036	0.026	0.021
	3	1.04	0.237	0.090	0.270	0.312	0.039	0.028	0.024
2	1	9.05	0.182	0.040	0.374	0.298	0.046	0.032	0.027
	2	0.852	0.131	0.043	0.398	0.317	0.049	0.034	0.028
	3	0.831	0.109	0.044	0.408	0.325	0.050	0.035	0.029

Table 7. Permeate composition from the two membrane stages and different layers within the stages.

Stage	Layer	Total Flow ($\times 10^6$ SLPM)	Permeance ($\times 10^{-4}$ mol/m^2/s/Pa$^{0.5}$)	Pressure (psig)	Composition (mol fr.) Steam	H_2
1	1	3.15	7.61	21.75	0	1
	2	1.78	6.68	21.75	0	1
	3	1.02	5.79	21.75	0	1
2	1	2.78	7.58	26.1	0.5	0.5
	2	1.07	6.71	26.1	0.5	0.5
	3	0.417	6.18	26.1	0.5	0.5

As explained earlier and shown in Figure 1, after water removal, the permeate H_2 stream is mixed with the purge stream and sent to the H&Bs as fuel to achieve an overall target CO_2 capture rate of 90% for the H&Bs. The molar composition of the fuel stream sent to the H&Bs is: H_2 (91.6%), CH_4 (0.54%), CO (0.47%), CO_2 (1.33%), H_2O (5.37%), N_2 (0.38%), and Ar (0.31%). The molar flow rate of the fuel stream is 23,194 kmol/h with the LHV of 5,000 MMBTU/h.

Table 8 provides a summary of results for the large-scale membrane separator.

Table 8. Results summary for the large-scale membrane separator.

Feed	**Stage 1**	**Stage 2**
Feed flow rate ($\times 10^7$ SLPM)	4.1	2.8
Feed composition (mol fraction)		
H_2	0.514	0.291
CO	0.057	0.035
CO_2	0.172	0.325
H_2O	0.199	0.259
CH_4	0.025	0.04
N_2	0.018	0.028
Ar	0.015	0.023
Feed pressure (psig)	289.8	279.7
Partial pressure of H_2 in the feed (psig)	149.0	81.4
Permeate	**Stage 1**	**Stage 2**
Total H_2 permeate flow rate ($\times 10^7$ SLPM)	0.60	0.43
Permeate pressure (psig)	21.75	26.1
Partial pressure of H_2 in the permeate (psig)	21.75	13.05
Membrane	**Stage 1**	**Stage 2**
Operating temperature (°C)	400	400
Membrane area (m^2)	13764	12903
Membrane cost (million $)	947	916

Membrane Separator Costs

As shown in Figure 6, each stage of the membrane separator consists of a parallel and series arrangement of multiple membrane vessels. As described earlier, each vessel consists of 440 membrane modules, with each module consisting of 14 Pd alloy coated membrane tubes, each with 40 cm active length.

In order to efficiently fabricate, install, and service the modules, a nominal 122 cm diameter vessel designed per the ASME Sect. VIII Div. 1 Boiler & Pressure Vessel Code was chosen. While the diameter was somewhat arbitrarily selected, the complications of the operations above this size increase rapidly as vessels become larger. It also provides a small enough building block such that stages and sub-stages (or layers) could be sized without greatly exceeding the required membrane area. A horizontal vessel arrangement with membrane modules mounted to two opposing tube sheets maximizes the utilization of space, while providing service access at each end and a reduction in the amount of piping. As described earlier, each 122 cm vessel can accommodate approximately 220 membrane modules per tube sheet for a total of 440 modules per vessel which results in nominally 70 m^2 of membrane. The vessel is further designed to be universally applied through all stages, sub-stages, and layers within each sub-stage. The membrane modules within the vessels are also a universal design, such that the required amount of maintenance spares can be kept at an absolute minimum.

Vessel Cost

A detailed mechanical design of the membrane vessel that houses 440 membrane modules was carried out. The diameter of each of the vessels was 122 cm, and the length was 305 cm. Cost quotations for 122 cm ASME Sect. VIII Div. 1 vessels were solicited for purchase in quantities of 10 vessels at a time. Cost for a vessel including tubesheet assemblies (minus modules) and 122 cm flange gaskets is $305,000 each, without steam sweep (for Stage 1) and $306,130 each, with steam sweep (for Stage 2).

Module Cost

A module is defined as an individual multi-tube membrane assembly that is the smallest field serviceable unit. This module incorporates the Pall hydrogen separation membranes and associated hardware that is assembled in quantity into the vessel tubesheets. While as many common components and design features are incorporated as possible, there are separate module designs for use with and without steam sweep. Costing for each module design was based on quantities required for (10) full 122 cm separation vessels, or 440 x 10 = 4,400 modules to be fabricated at a time. The module costs are $11,894 each, without steam sweep and $12,240 each, with steam sweep.

Assembly Cost

Since each vessel consists of 440 membrane modules, the total cost for a fully assembled 122 cm separation vessel including tubesheet assembly is $5,538,360 each, without steam sweep and $5,691,730 each, with steam sweep.

Since there are a total of 171 and 161 vessels in stages 1 and 2, respectively, the total assembly cost is $947 million for stage 1, and $916 million for stage 2, which makes the overall membrane cost $1.86 billion. The economic evaluation for estimating the cost of CO_2 capture needs to account for membrane replacement as an operating expense. With an assumed lifetime of 2-years, the annual opex to replace the membranes is $930 million.

Although the membrane cost estimated here is much higher than other conventional CO_2/H_2 separation processes (such as aMDEA, PSA, etc.), it should be noted that the pre-combustion CO_2 capture using palladium membrane is a relatively new process. Thus, the capital cost presented in this report is based on the relatively limited test data. Performance and fabrication can be improved over time. For example, the 2-year Pd membrane lifetime is a conservative estimation which is based on the limited testing conditions. Additionally, the manufacturing cost of the Pd membrane will be reduced as the technology becomes mature.

The carbon capture technology using Pd membrane also has the following characteristics:

1. The CO_2 emission rate is very low. As confirmed by both short and long-term experiments, the hydrogen purity is about 99.6%, which means that only hydrogen is recovered and used as the fuel source.
2. The recovered high purity CO_2 is already at high pressure, which will require less compression for transportation and subsequent storage.

SUMMARY

In this project, the performance of Pd alloy-based membranes was evaluated for different syngas compositions with the overall objective of estimating the cost for separating H_2 from a large-scale ATR process. Long-term testing confirmed the performance of Pd alloy-based membrane for a S/C ratio of 1.8. A simulation model was developed and calibrated to match the performance of the single and multi-tube membranes. This model was used to estimate the membrane surface area for producing 5,000 MMBTU/h of H_2 fuel (LHV) in an ATR. Finally, a detailed design and cost estimate of the membrane separator was completed. The overall membrane separator cost was estimated to be about $1.86 billion.

ACKNOWLEDGEMENTS

The authors would like to acknowledge the following individuals at Pall Corporation – Keith Rekczis for supplying the Pd alloy membranes and assisting with the Pd alloy membrane test, Matthew Keeling for assisting with the Pd alloy membrane test, Geoffrey Bradshaw for assisting with the design of the large scale vessel, and Scott D. Hopkins for overseeing the entire project. The test for the small scale module was completed at TDA Research Inc. assisted by Michael Bonnema. Amec Foster Wheeler, carried out the mass and energy balance of the overall reforming process. The authors would also like to acknowledge Professor Jerry Y.S. Lin of Arizona State University for reviewing and providing useful comments on the manuscript.

REFERENCE

1. Caravella, A., Scura, F., Barbieri, G., Drioli, E., Inhibition by CO and Polarization in Pd-Based Membranes: A Novel Permeation Reduction Coefficient, J. Phys. Chem. B., 2010, 114 (38), p. 12264-12276

Carbon Dioxide Capture for Storage in Deep Geological Formations, Volume 4
Karl F. Gerdes (Editor)

Chapter 11

TECHNOECONOMIC EVALUATION OF CARBON DIOXIDE CAPTURE FROM REFINERY HEATERS AND BOILERS

Betty K. Pun[1], David Butler[2]
[1]Chevron Energy Technology Company, 100 Chevron Way, Richmond, CA 94801
[2]Calgary, Alberta, Canada

ABSTRACT: Post-combustion, oxy-fuel combustion, and pre-combustion options to retrofit refinery heaters and boilers are evaluated for a typical U.S. Gulf Coast refinery. Amec Foster Wheeler (AFW) was engaged to design each process, and provide a process flow diagram, technical data on the mass balance, energy and utility requirement, operating costs, equipment and installed capital costs, and a preliminary layout. Using these data, CCP conducted economic analyses to evaluate all processes on common bases of \$/tonne CO_2 captured and \$/tonne CO_2 avoided.

Post-combustion and oxy-fuel combustion capture were assessed for the retrofit of individual heaters and boilers. Post-combustion capture costs per tonne of CO_2 were found to be 23% lower from four 150-MMBtu/hr (as fired) heaters than from three 100-MMBtu heaters, indicating significant economies of scale. Oxy-fuel combustion is associated with capture and avoided costs that are 1.8 and 2.3 times, respectively, those associated with post-combustion capture at the same scale. Both the capital and operating costs of air separation need to improve for oxy-fuel combustion to become a competitive solution for CO_2 capture.

In pre-combustion capture, major equipment to produce H_2 fuel and capture CO_2 may be placed away from the refinery processes, while H_2 fuel is distributed via the existing fuel gas infrastructure. This layout has potential advantages over retrofitting post-combustion or oxy-firing to individual heaters and boilers for refineries. Since the pre-combustion approach was based on a refinery-wide application, these cases benefit from economy of scale, and avoided costs per tonne are competitive with post-combustion. Several options for separating CO_2 from H_2 were assessed. The simpler approaches using amine absorption and cryogenic separation provided the lowest capture cost. Separation approaches based on palladium membranes are not economically competitive at this time due to high membrane costs.

KEYWORDS: refinery; CO_2 capture retrofit; heaters; boilers; post-combustion; pre-combustion; oxy-fuel combustion

INTRODUCTION

The life-cycle CO_2 emissions from fossil fuels include not only CO_2 resulting from the combustion of the fuel product, but also emissions associated with the production and refining of crude oil into fuel. End user emissions typically dominate the "Well to wheel" emissions [1]. The refining step accounts for only about 8-10% of the well-to-wheel CO_2 emissions associated with fossil fuel. However, oil refining represents a major industrial source after power, and on par with cement, and iron/steel production [2]. In a typical refinery, combustion sources are numerous and spatially distributed. Combined, heaters and boilers account for a significant fraction of the refinery's CO_2

inventory. Depending on the CO_2 reduction target of the facility, different strategies may be pursued to capture CO_2 from these combustion sources.

The benchmark technology for CO_2 capture from refinery heaters and boilers chosen for this study is post-combustion solvent absorption using a state-of-the-art amine currently commercially available from Mitsubishi Heavy Industries. This technology is applicable for retrofitting existing heaters and boiler stacks. Retrofitting each individual heater or boiler with a stand-alone capture system is considered infeasible within a refinery setting, both due to space constraints and to the relatively small scale of each source. Therefore, the nominal scenario is to capture flue gas from a group of three of four heaters (100-150 MMBtu/hr each) that are located in close proximity to each other and to the capture plant. Oxy-fuel combustion with flue gas recycle is assessed under a similar scenario as post-combustion, because oxy-fuel combustion was shown to be an attractive technology for heaters and boilers retrofit in a previous phase of CCP [3].

Most refineries are space constrained. Post-combustion and oxy-fuel retrofits require space for extra ducting (e.g., CO_2 collection, flue gas recycle, O_2 delivery), and large equipment (e.g., absorber/regenerator and air separation unit), which may not be readily available next to combustion sources. Therefore, if a high degree of capture is required, a comprehensive solution may be to decarbonize the fuel gas in the refinery (pre-combustion capture). The production of low carbon fuel takes place at a centralized location, where the resulting CO_2 can be captured. The low carbon, hydrogen-based fuel is then distributed throughout the refinery in the existing fuel gas utility system. Here, a large refinery (5000 MMBTU/hr fired duty) is considered, so a direct comparison to costs of the small-scale post- and oxy-fuel applications is not appropriate.

AFW was engaged to design each capture process based on a common basis. The methodology of the technical studies is summarized in Chapter 3 of this volume. AFW provided process flow diagrams and technical data on the mass balance, stream properties, energy and utility requirements, operating costs, equipment and installed capital costs, and a preliminary layout. In a few selected cases, AFW worked with technology providers to develop the design, enabling the use of proprietary technologies in these analyses. Using the technical data, CCP conducted economic analyses to evaluate all processes on common bases of \$/tonne CO_2 captured and \$/tonne CO_2 avoided. Some of the key questions addressed in this study are:

- What are the carbon capture and avoided costs associated with the benchmark amine absorption system?
- What is effect of scale on the costs of post-combustion capture?
- Is oxy-fuel combustion an attractive alternative?
- Are pre-combustion configurations technically and economically competitive in a refinery setting?
- What are the best options for carbon capture in this scenario at various scales?

Other considerations, as well as critical development needs, are discussed for each technology in the following sections.

TECHNICAL EVALUATIONS

Post-Combustion

Post-combustion capture is expected to be the first CO_2 capture technology deployed at scale for the oil and gas industry. While many consider MEA to be the default technology, several state-of-the-art amine solvents are commercially available - Mitsubishi Heavy Industries KS1^{TM}, Shell CANSOLV, and Fluor's Econamine FG Plus^{SM} - which offer significant advantages over MEA and are commercially ready for large-scale deployment. For CCP3, MHI's KS1^{TM} solvent was selected as the benchmark. This technology is based on the common absorber/regenerator type configuration that is used in many current amine-based CO_2 technologies (Figure 1). The specific solvent composition, equipment, and operating conditions are proprietary information for this technology, and AFW incorporated "black box" designs from the vendor into the technical analyses.

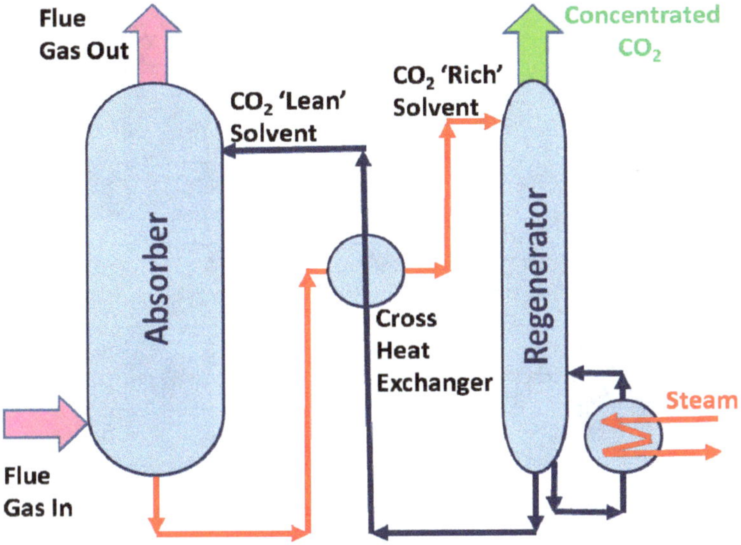

Figure 1. Schematic of generic post-combustion CO_2 capture.

Flue gas from 4 x 150 MMBtu/hr (nominal case) or 3 x 100 MMBtu/h (small size case) refinery fired heaters is collected into a single combined stream for CO_2 capture, as shown in Figure 2. Each heater is fired on refinery fuel gas (RFG), which contains methane (69 mole %) and hydrogen (29%) as the two main components. The heaters are assumed to be located in close proximity to one another (e.g., within 20 m). Plot space is assumed to be available in the vicinity on which to locate the CO_2 capture, compression, and dehydration equipment. The flue gas contains 7.5% CO_2 and 2.5% O_2, which is less excess O_2 than typical of gas-fired turbines. Five percent of the flue gas bypasses the capture plant via the stack to ensure safe operation. A single train of CO_2 capture, compression, and dehydration is considered to be appropriate for the total volume of gas treated in both cases. The design CO_2 removal rate is 90%, and the clean flue gas is returned to stack. A stand-alone package boiler supplies the steam required for the stripper reboiler, solvent reclaimer, and regeneration of the dehydration beds. Emissions from this boiler are not captured. A combination of sea water and fresh water is used for cooling and electricity is supplied over the fence for CO_2 capture, compression, and dehydration.

The techno-economic analysis was carried out by a collaborative effort of AFW and MHI. The technology provider supplied the equipment list and sizing, capital cost, utility, and chemical consumption for the capture system. AFW modelled the balance of plant components and costs, including the auxiliary boiler to supply steam for the regeneration of amine. In applying post-combustion technology, no modification is made to the heaters. Hence, the CO_2 in flue gas emitted from the status-quo heaters provides a suitable reference case. Power requirements are dominated by the compression of the CO_2 product. These additional steam and power requirements are associated with CO_2 emissions, which must be accounted for in the calculation of avoided CO_2 emissions. The technical performance results are summarized in Table 1.

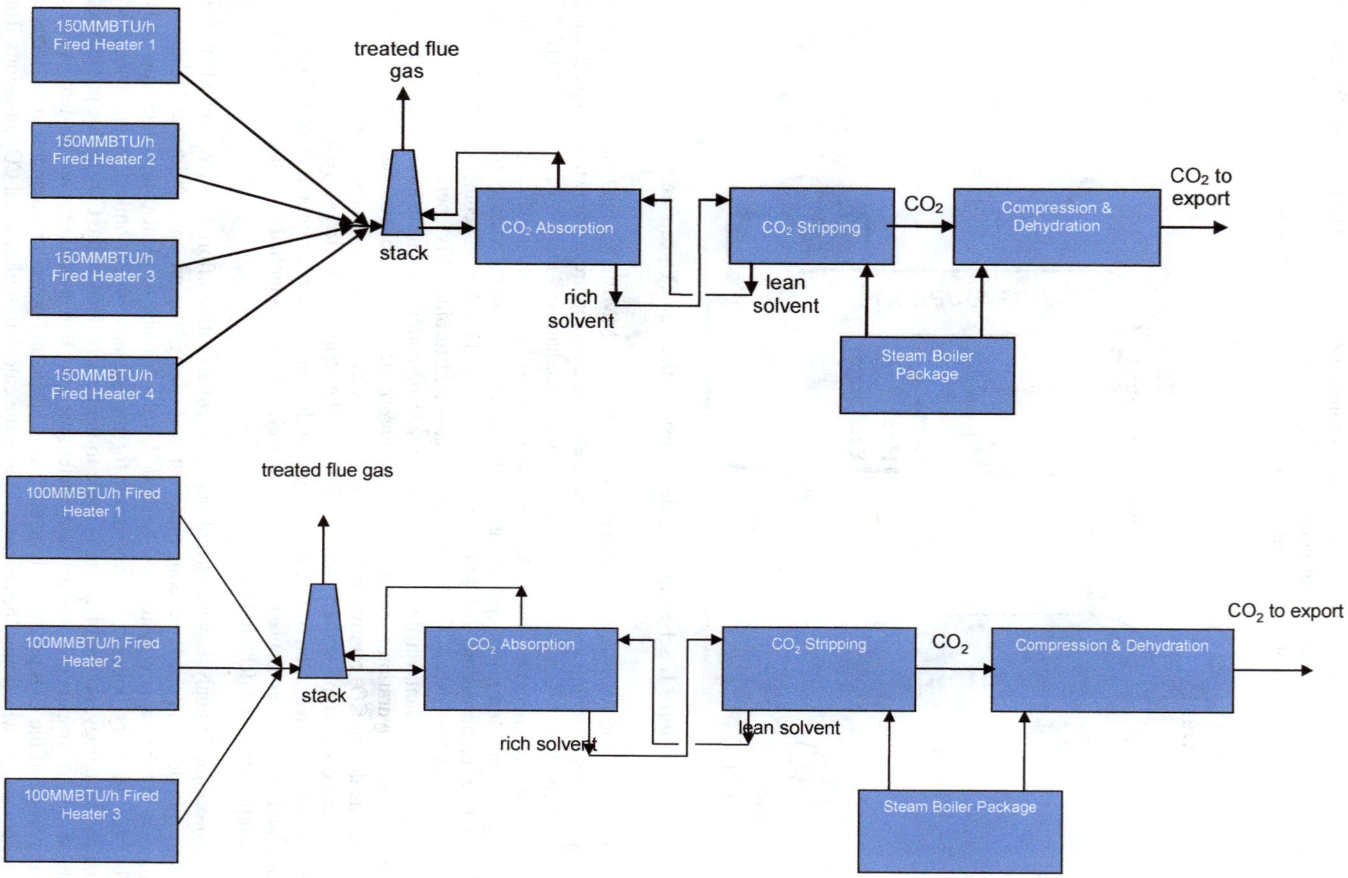

Figure 2. Post-combustion CO_2 capture in the refinery fired heater cases 4 x 150 MMBtu/h (top) and 3 x 100 MMBtu/h (bottom).

Table 1. Refinery heaters and boilers: performance results of post-combustion cases.

Fired heater **Duty (MMBtu/h, as fired)**	**Unit**	**Nominals** **4 x 150**	**Small size** **3 x 100**
Flue gas CO_2 concentration	%	7.5	7.5
Process CO_2 balance			
CO_2 In Flue gas	kg/h	31124	15562
CO_2 to Capture Plant	kg/h	29568	14784
CO_2 Captured	kg/h	26607	13303
CO_2 to Water Streams	kg/h	1	1
CO_2 Emitted from Capture Plant	kg/h	2960	1480
CO_2 Emitted from Stack	kg/h	4517	2259
CO_2 Capture across Capture Plant	%	90	90
CO_2 Capture from Total Flue Gas	%	85	85
Total Power Usage	MW	4.07	2.07
Total Steam Usage	kg/h	31667	15834
Total Cooling Water Usage	te/h	3781	1890
Additional CO_2 Generated for Utilities			
CO_2 due to Power use (1)	kg/h	2444	1244
CO_2 due to Steam use	kg/h	3994	1996
Total CO_2 Balance			
Total CO_2 Generated	kg/h	37563	18802
Total CO_2 Captured	kg/h	26607	13303
Total CO_2 Emitted	kg/h	10956	5499
CO_2 avoided (2)	kg/h	20169	10063
	%	64.8	64.7
Specific CO_2 emissions	gCO_2/MMBtu	18259	18330

(1) Assumed carbon footprint of 600 kg/MWh for power in the U.S. Gulf Coast
(2) CO_2 in flue gas (status quo) – total CO_2 emitted

The plot space requirements are shown in Table 2, where the physical dimensions of key equipment are also compiled. Six stages of compression are required to compress CO_2 from 0.5 to 150 barg. Reciprocating compressors are used for the small size case but centrifugal compressors are required for the nominal case. While the plot space requirement is fairly modest, at about 5000 m^2 (1.2 acres) for the nominal case, it will likely not be feasible to duplicate this system at distributed locations for various groups of heaters in a typical refinery.

Table 2. Space requirements for post-combustion equipment

Fired heater **Duty (MMBtu/h, as fired)**	**Unit**	**Nominal** **4x150**	**Small size** **3x100**
Absorber/regenerator	m^2	68 x 47	62 x 44
Compression/dehydration/steam generation	m^2	42 x 44	42 x 40
CO_2 absorber (SS and CS)			
Internal diameter	m	5.1	3.6
Tangential height	m	39.6	37.7
CO_2 regenerator (SS)			
Internal diameter	m	3.0	2.2
Tangential height	m	31.0	31.0

Note: SS = Stainless Steel; CS = Carbon Steel

Oxy-fuel Combustion

In the oxy-fuel combustion case, a central air separation unit (ASU) is installed to supply 97% purity O_2 to 4 x 150 MMBtu/h heaters and boilers (absorbed duty or steam output basis). Existing burners are assumed to be compatible with oxy-firing, provided flue gas is recirculated and pre-mixed to meet the O_2 concentration specifications for the existing burners. Detailed burner requirements and retrofit options were not evaluated by AFW. The feasibility of using existing burners for oxy-fuel firing with the use of flue gas recirculation (FGR) was verified in the CCP study by John Zink [Chapter 9]. Flue gas is collected from several heaters; a portion of which is recycled, while the remainder is sent to the CO_2 capture unit. A schematic is shown in Figure 3.

AFW conducted modelling to establish requirements for oxy-fuel combustion as follows:

- Oxygen content of combustion medium: 19.9 mol% wet basis (72% flue gas recirculation)
- Oxygen content exit burners: 2.5 mol% wet basis
- Air in leakage: 28.0 lb / MM Btu process duty (50.4 kg/Gcal)
- Flue gas temperature outlet convection section (process coil): 450 °C (842 °F)

Oxy-fired flue gas has a higher temperature than that in the unmodified heaters and boilers; therefore, a new medium pressure (MP) steam generator is added to the common flue gas duct for heat recovery. Accounting for the recovered heat would increase the equivalent thermal efficiency for oxy-firing from 91.7% to 94.9% (e.g., in the form of oxygen pre-heating). In comparison, air firing with pre-heat achieves an efficiency of 90.3%. Table 3 compares the characteristics of flue gas from unmodified heaters and boilers vs. those fired with oxygen.

After dehydration using molecular sieve adsorption, cryogenic purification is applied to recover 90% of the CO_2. A reboiled stripping column is used to reduce O_2 to 75 ppm in the CO_2 before compression to export specification. To provide steam for the CO_2 capture plant, a refinery gas fired package boiler is provided to generate the small requirement for MP steam required for regeneration of the molecular sieve dryers in the ASU and CO_2 dehydration units. CO_2 in the flue gas from this small boiler is not captured.

Table 3. Comparison of post- and oxy-fuel combustion flue gas characteristics and O_2 requirements.

Fired heater	Units	Nominal Post-Combustion	Oxy-fuel Combustion
Capacity	MMBtu/h	4 x 150 (fuel value)	4 x 150 (steam output)
Total flue gas flow	kg/hr	259,606	291,774
Flue gas flow to CO_2 capture plant	kg/hr	246,626	
Pressure	bara	1.013	0.99
Temperature	°C	150	250
Flue gas composition (wet)			
CO_2	mole %	7.5	25.8
SO_x	ppm	4	
NO_x	ppm	30	
CO	ppm	0	
N_2	mole %	70.38	7.84
Ar	mole %	0.84	1.74
O_2	mole %	2.46	2.5
H_2O	mole %	18.81	62.01
Oxygen flow rate	kg/h		58,484
O_2 Pressure	bara		1.2
O_2 concentration	%		97

Note: Slightly different basis used for post-combustion (fuel heating value) vs. oxy-firing (absorbed duty or steam value). Absorbed duty for post-combustion case should be 4 x 150 x 90.3%, accounting for burner efficiency.

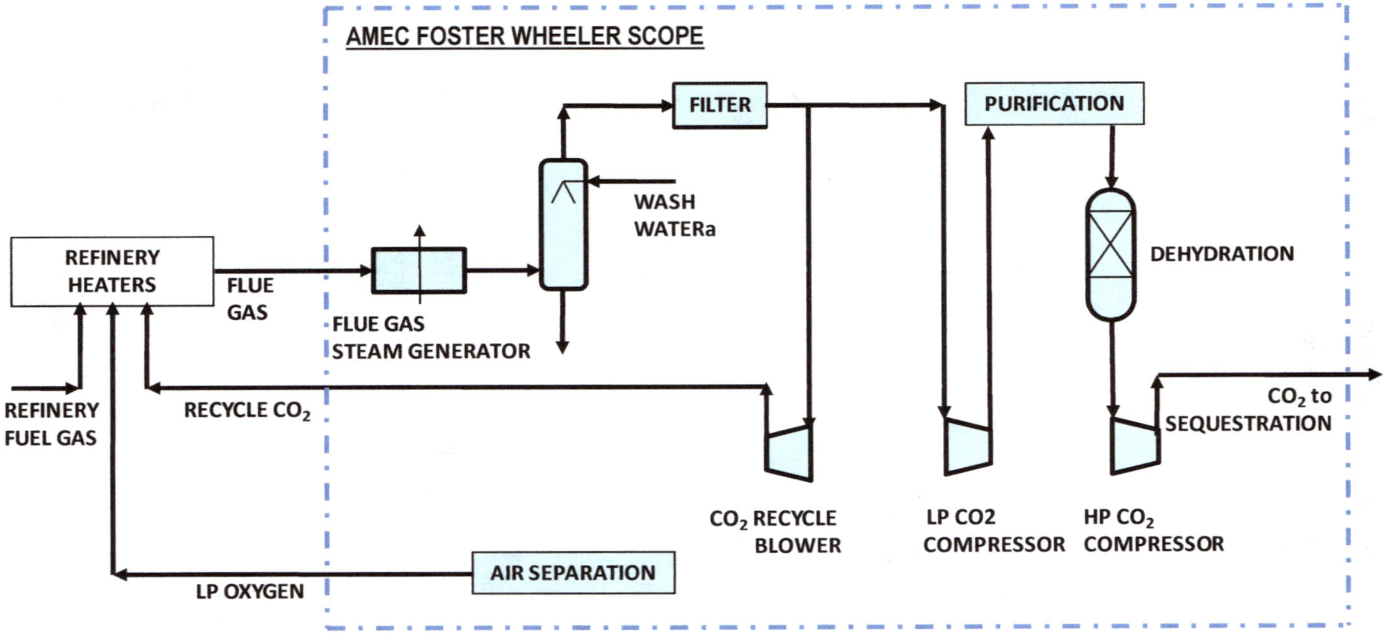

Figure 3. Oxy-fuel combustion for refinery heaters and boilers.

The key technical results for the oxy-fuel combustion capture case are shown in Table 4.

Table 4. Refinery heaters and boilers: performance results of the oxy-fuel combustion case.

Fired heater	**Unit**	**Nominal 4 x 150 MMBtu/hr absorbed duty**
Flue gas CO_2 concentration (wet)	%	25.9
Process CO_2 balance		
CO_2 In flue gas	kg/h	34969 (1)
CO_2 to capture plant	kg/h	34969
CO_2 captured	kg/h	31478
CO_2 to water streams	kg/h	24
CO_2 emitted from capture plant	kg/h	3467
CO_2 emitted from stack	kg/h	n/a
CO_2 capture across capture plant	%	90
CO_2 capture from total flue gas	%	90
Total power usage	MW	23.5
Total steam usage	kg/h	0
Total cooling water usage	te/h	4549
Additional CO_2 generated for utilities		
CO_2 due to power use (2)	kg/h	14121
CO_2 due to steam use	kg/h	0
Total CO_2 balance		
Total CO_2 generated	kg/h	49241
Total CO_2 captured	kg/h	31476
Total CO_2 emitted	kg/h	17763
CO_2 avoided (3)	kg/h	18854
	%	53.9
Specific CO_2 emissions	gCO_2/MMBtu	29605

(1) The in flue gas CO_2 content in oxy- case 34969 is higher than post-combustion case due to different basis, as explained in footnote of Table 3.
(2) Assumed carbon footprint of 600 kg/MWh for power in the U.S. Gulf Coast
(3) Reference case used in economic analysis based on fuel consumption of underlying process without CCS. AFW modelling was not carried out for an exact reference case. In the heater modelling study, oxy-firing with pre-heating was found to be 4.3% more efficient than conventional boiler, thus a ~4.5% fuel savings was applied in the avoided CO_2 calculation.

The plot space requirement for oxy-fuel combustion and CO_2 capture from 4 x 150 MMBtu/hr case is as follows: 65 m x 50 m for the ASU, 75 m x 86 m for the flue gas treatment and recycle compressor, and downstream units for CO_2 purification, dehydration, and compression. It is possible to locate the ASU away from the CO_2 sources; however, additional O_2 piping may present a logistical challenge for the refinery. A single space 212 m x 86 m would be preferable.

The major equipment consists of a turn-key, electrically driven ASU, the cryogenic purification unit, 3 beds of molecular sieve for dehydration, an O_2 stripping column (1.5 m ID, 65 m tangent-to-tangent [T/T] height), and 7 stages of centrifugal compressors. The ASU and CO_2 compressors are key users of electrical power.

Pre-combustion

The scale considered for pre-combustion capture is different from the post- and oxy-fuel combustion cases. In this case, hydrogen (H_2) is distributed as fuel (in lieu of RFG) to the entire refinery, not just to three or four process heaters and boilers. Therefore, the pre-combustion approaches analysed here benefit from economies of scale likely not available to post- and oxy-combustion approaches in this application. The rationale is that additional very large piping and headers are required by post-combustion and oxy-combustion retrofits and these may not be feasible at many refineries that are space constrained, considering the distributed nature of the combustion sources. Major equipment for pre-combustion configurations can be located away from the main processing units, and H_2 fuel can be distributed by the existing RFG infrastructure. Table 5 summarizes the design basis for a 5000 MMBtu/hr supply of H_2 fuel.

Table 5. Key design considerations of the pre-combustion cases.

Output capacity (fuel value)	MMBtu/h	5000
Fuel		Refinery fuel gas (5000 MMBtu/hr); balance natural gas
H_2-rich fuel gas product		
Flow rate	kg/h	73846
Pressure	bara	2.50
Temperature	°C	41.9
Example composition (1)	mol%	
H_2		95.05
CO		0.99
CO_2		0.75
N_2		0.52
Ar		0.53
Methane		0.96
O_2 product from ASU		
Flowrate	kg/h	136326
Pressure	bara	24
Temperature	°C	122.1
O_2 composition	%	97

(1) There is no specification of the H_2 fuel based on composition; a range of H_2 contents may be acceptable for fuel application as long as heating value specification is met.

Some energy loss is incurred in the conversion of refinery fuel gas (RFG) to hydrogen-based fuel. Therefore, supplemental natural gas (NG), compared to the uncontrolled case, is added to the RFG to produce 5000 MMBtu/hr of decarbonized fuel. Compressed RFG, together with supplemental NG, is converted to syngas in an auto-thermal reformer (ATR), which is used for this large-scale application. For smaller refineries, other syngas technologies may be appropriate. A cryogenic ASU supplies O_2, which is compressed prior to the ATR. A steam-to-carbon ratio of 1.8 is employed. The ATR feed is at 550°C and 24 bar (O_2) and 28 bar (fuel), and the outlet is at 950°C and 23 bar. These conditions are representative of current commercial practice. From the hot ATR effluent, waste heat is recovered to pre-heat the feed stream and to generate superheated steam for electricity production. Next, the syngas undergoes water-gas shift (WGS) reaction to maximize the production of H_2 and CO_2. In the high temperature shift reactor, ~65% of the CO content is converted to CO_2 over a Fe/Cr-based catalyst at 445°C. Subsequently, in the low temperature WGS reactor (Cu/Zn catalyst), an overall 90% conversion of the CO is achieved, with an exit temperature of 245 °C. Extensive heat exchange and recovery are used. Except in the H_2 permeable membrane configurations, the H_2 product is expanded from 18.5 - 18.8 bar to 2.5 bar for power recovery.

For pre-combustion capture, CO_2 is separated from H_2 in syngas at high pressure, which provides additional partial pressure driving force for the separation compared to post-combustion approaches. Several technology options are considered for H_2 purification or CO_2 removal in the six configurations summarized in Table 6.

Table 6. Pre-combustion technology options.

Short-hand	H_2 Purification	CO_2 Purification
Pre-C. - MDEA	None	Amine solvent
Pre-C. - PSA/Ox	PSA	Thermal oxidizer with waste heat recovery
Pre-C. - PSA/Cryo	PSA	Cryogenic
Pre-C. - Mem/Cryo	H_2 membrane	Cryogenic
Pre-C. - Mem/Cryo/PSA	H_2 membrane/PSA	Cryogenic
Pre-C. - Cryo/MDEA	None	Cryogenic, amine solvent

Schematics of the various cases are shown in Figure 4. A single train of auto-thermal reformer (ATR), water-gas shift (WGS), and CO_2 and/or H_2 separation equipment is installed to provide H_2 fuel to refinery heaters and boilers. No modification is assumed for existing refinery boilers and heaters to use H_2 fuel instead of RFG [4]. CO_2 is compressed and dried to specification given previously. Technical performance of the pre-combustion cases is summarized in Table 7.

Table 7. Refinery heaters and boilers: performance results of pre-combustion cases.

Pre-C	Unit	MDEA	PSA/ Ox	PSA/ Cryo	Mem/ Cryo	Mem/ Cryo/PSA	Cryo/ MDEA
O_2 consumption	te/day	3272	5014	3360	3362	3342	3323
Fuel consumption (LHV)	GJ/h	6261	7117	6380	6410	6378	6320
Process CO_2 balance							
CO_2 in flue gas	kg/h	n/a	n/a	n/a	n/a	n/a	n/a
CO_2 to capture plant (1)	kg/h	315198	366945	323450	326345	324447	319796
CO_2 captured	kg/h	283951	330517	291143	300534	304610	293937
CO_2 to water streams	kg/h	0	61	3	0	0	0
CO_2 emitted from capture plant	kg/h	31247	35575	31617	25811	19837	25859
CO_2 emitted from stack	kg/h	671	792	687	749	532	n/a
CO_2 capture across capture plant	%	90.1	90.1	90.0	92.1	93.9	91.9
CO_2 capture from total flue gas	%	90.1	90.1	90.0	92.1	93.9	91.9
Total power usage	MW	50.0	-25.6	39.9	83.8	108.1	46
Total steam usage	kg/h	0	0	0	0	0	0
Total cooling water usage	te/h	20558	16040	14025	20424	21866	15199
Additional CO_2 from utilities							
CO_2 due to power use (2)	kg/h	29994	-15346	23938	50305	64862	27609
CO_2 due to steam use	kg/h	0	0	0	0	0	0
Total CO_2 balance							
Total CO_2 generated	kg/h	345192	351599	347388	376651	389310	347405
Total CO_2 captured	kg/h	283951	330517	291143	300534	304610	293937
Total CO_2 emitted	kg/h	61241	21082	56245	76117	84700	53468
CO_2 avoided (3)	kg/h	198126	238285	203122	183250	174667	205899
	%	76.4	91.9	78.3	70.7	67.3	79.4
Specific CO_2 emissions	gCO_2/MMBtu	12248	4216	11249	15223	16940	10694

(1) Equivalent CO_2, including carbon content from hydrocarbons and CO
(2) Assumed carbon footprint of 600 kg/MWh for power in the U.S. Gulf Coast
(3) CO_2 avoided = CO_2 from burning 5000 MMBtu/h RFG (e.g., in typical refinery heaters and boilers) – CO_2 emitted

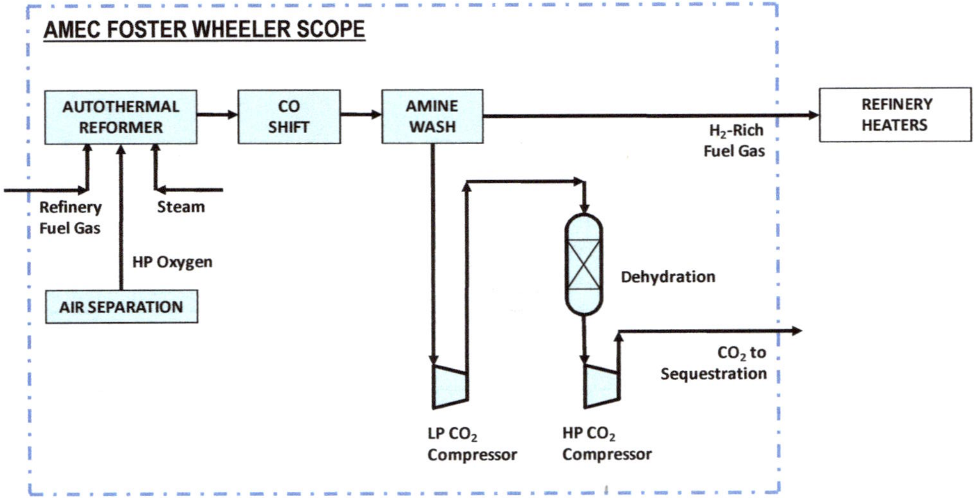

Figure 4a. Oxy-fuel combustion for refinery heaters and boilers - Pre-C. – MDEA.

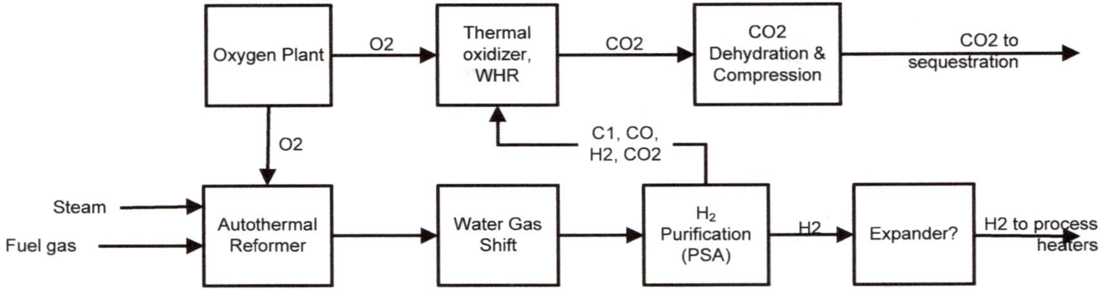

Note: catalytic purification occurs part way through the CO_2 compression train

Figure 4b. Oxy-fuel combustion for refinery heaters and boilers - Pre-C. - PSA/Ox.

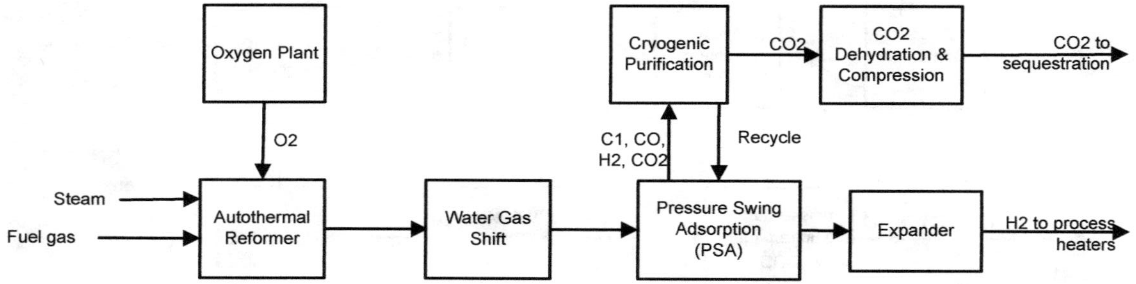

Figure 4c. Oxy-fuel combustion for refinery heaters and boilers - Pre-C. - PSA/Cryo.

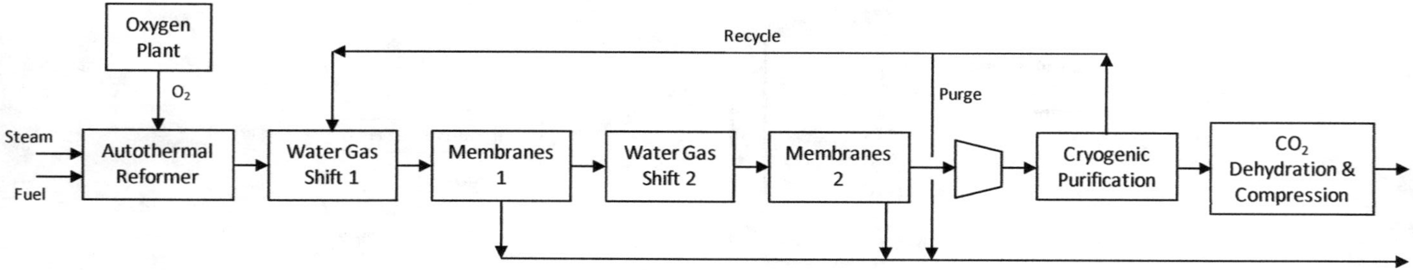

Figure 4d. Oxy-fuel combustion for refinery heaters and boilers - Pre-C. - Mem/Cryo.

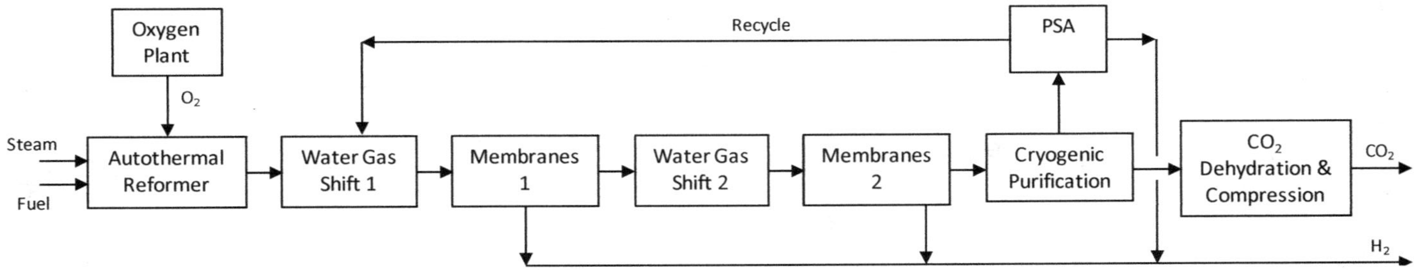

Figure 4e. Oxy-fuel combustion for refinery heaters and boilers - Pre-C. - Mem/Cryo/PSA.

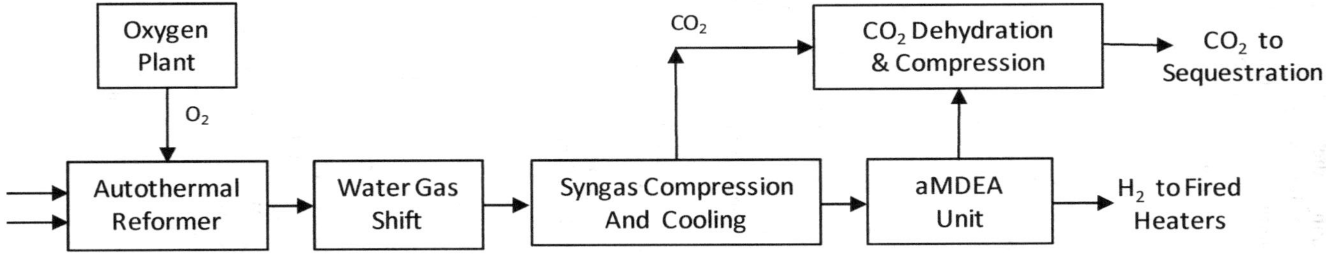

Figure 4f. Oxy-fuel combustion for refinery heaters and boilers - Pre-C. - Cryo/MDEA.

All pre-combustion configurations require an ASU, compressors for fuel gas (3 stages) and O_2 (5 stages, reciprocating), the ATR, WGS reactors, and associated heaters, coolers, boilers, and heat exchangers. The O_2 requirement for the ATR is about 3000 tonne per day for all cases, except for the PSA/Ox combination (~5000 tpd). The ASU and O_2 compressors require significant power input, which is assumed to be supplied over the fence. The ATR reactor vessel is 6.1 m ID, 17.1 m tangent-to-tangent height, and the WGS reactors are smaller: 5.8 m ID, 5.9 m tangent-to-tangent height for the high temperature reactor, and 6.8 m ID, 6.6 m tangent-to-tangent height for the low temperature reactor. Land requirements differ for different configurations. Both the 3000 tonne-per-day ASU unit and the 5000 tonne-per-day unit require an area of 60 m x 115 m.

Pre-C.-MDEA

Syngas exiting the WGS reactor is cooled via a combination of heat recovery, heat integration, and simple cooling before entering the acid gas removal unit (AGRU) at 60 °C and 19 bara. The AGRU inlet stream contains 23.2 mol% CO_2, 0.76 mol% CO, 0.74 mol% CH_4, and 73.3 mol% H_2 (for comparison, see Table 1 for flue gas characteristics entering CO_2 capture from post- and oxy-fuel combustion). Due to higher partial pressure of CO_2 (both system pressure and CO_2 concentration are higher), and absence of O_2, a methydiethanolamine (MDEA)-based solvent can be used for this application, instead of the advanced amine used in the post-combustion application for CO_2 removal. The equipment used is similar to the absorber/regenerator system in the post-combustion case, except a flash and a stripper column are used in the regeneration step. The treated gas contains 95.05 mol% H_2, 1.35 mol% CO_2, and CO and CH_4 just below 1 mol% each. The system was designed to achieve 90% overall CO_2 capture. AFW noted that this type of acid gas removal system is capable of achieving 0.1 mole% CO_2 slip with only a small increase in reboiler duty for an overall capture rate of 94%.

Process heat is recovered as superheated steam, which is used to drive an electric generator producing about 24 MW of electrical power. Energy is also recovered by expanding the H_2 gas.

The major equipment for CO_2 capture is the absorber, flash column, and stripping column. The absorber is 8.6 m ID, 58 m in height. The CO_2 stripper is 5.7 m ID and 24 m in height. The flash column is 9.6 m ID and 28 m in height. The compression of CO_2 is accomplished in 6 stages of centrifugal compressors.

The CO_2 separation, compression and dehydration area is assumed to be located adjacent to the ATR and WGS reactors. Including the ASU (60 m x 115 m), the total space requirement is approximately 151 m x 188 m, with some stacking of equipment such coolers and heat exchangers.

Pre-C.-PSA/Ox

In this case, the ASU supplies O_2 to not only the ATR, but also to a thermal oxidizer. The total O_2 requirement is 5000 tonnes per day, split 75% to the ATR and 25% to the thermal oxidizer.

Syngas exiting the WGS reactor is cooled with a combination heat recovery, heat integration, and simple cooling steps before entering pressure swing adsorption unit at 40 °C and 19.4 bara. The PSA inlet stream contains 23.8 mol% CO_2 and 73.5 mol% H_2. In the PSA unit, 90% of the hydrogen is recovered in the product gas and 90% of the carbon is captured in the tail gas. The H_2 product is 95.4 mol% pure, containing 2.4 mol% CO_2, 0.81 mol% CO, and 0.47 mol% CH_4. The CO_2-rich tail gas contains 72.0 mol% CO_2, but also contains 23.9 mol% H_2, 1.35 mol% CH_4, and 0.81 mol% CO.

The CO_2-rich tail gas from the PSA unit is heated from about 30 °C to 240 °C before entering the thermal oxidizer. O_2 from the ASU is used in the process to oxidize CO to CO_2 and H_2 to water. In theory, sub-stoichiometric operation could minimize O_2 consumption, and eliminate a catalytic O_2 purification step downstream, while still meeting the specification for CO_2. However, oxidizer performance needs to be verified for this service before such a process scheme is considered. Steam is generated using the hot effluent from both the oxidizer and the ATR, and used to drive a steam turbine for power. In conjunction with power recovery by the hydrogen expander, this configuration results in a net production of approximately 25.6 MW of electricity, despite a significant electrical load due to the ASU. CO_2 is compressed in 6 stages, passing through a catalytic purification unit after the third stage to reduce the O_2 content by reaction with H_2, to meet the 75 ppm oxygen requirement.

Key equipment is supplied in two packages: the PSA unit and catalytic purification unit. The total space requirement is 151 m x 320 m. With three parallel trains, the space requirement for the PSA package is quite significant (~60 m x 100 m).

Pre-C.-PSA/Cryo

In this scheme, cryogenic purification is applied to the tail gas of the PSA unit. The H_2 product is 95.8 mol% pure, containing 1.8 mol% CO_2, 0.94 mol% CO, and 0.58 mol% CH_4. As in the above PSA case, power recovery is possible using a hydrogen expander. The CO_2-rich tail gas (73.7 mol% CO_2, 21.9 mol% H_2, 1.2 mol% CO) from the PSA is compressed in 4 stages, dried, further compressed to 64 bara and sent to the cryogenic unit, where the stream is cooled to -35 °C to condense CO_2. Ninety percent of the CO_2 is recovered as vapour (96% purity, with 1.9 mol% H_2), while the lighter components are recycled to the PSA unit in a stream that contains 64.7 mol% H_2 and 25.5 mol% CO_2.

Because of the volume of recycled gas entering the first stages of compression with the CO_2-rich PSA tail gas, 8 stages of centrifugal compression is required in this process (as opposed to 6 in the pre-combustion-absorption and pre-combustion-PSA-catalytic oxidation schemes).

Plot space requirements are similar to the previous case (151 m x 320 m).

Pre-C.-Mem/Cryo

This case assesses a developmental high-temperature hydrogen permeable membrane technology (see Chapter 10). A key advantage of using palladium (Pd) alloy membranes is that a carbon-free hydrogen fuel is produced. A membrane separator is placed after each stage of water gas shift (WGS) reaction (see Figure 4d). By removing hydrogen between WGS stages, the equilibrium of the WGS reaction is shifted towards the production of H_2 and CO_2 in the second WGS reactor and slippage of carbon is minimized by recovering more pure H_2 after the second WGS stage. This configuration does not reduce catalyst volume as much as use of an integrated membrane reactor might accomplish. However, the engineering and fabrication issues are greatly reduced by use of staged membrane separators. In addition, at the current state of development, the membranes are sensitive to the high CO levels found in the raw ATR effluent, making H_2 separation ahead of the first WGS reactor, or integration of the steps into a membrane reactor problematic.

In the first WGS reactor, ~57% of the CO content of the reformed gas is converted to CO_2 (as opposed to ~65% conversion in the non-membrane cases). The partially shifted gas is cooled to 400 °C before entering the first stage of hydrogen membrane. Approximately 70% of the H_2 is removed, with the hydrogen permeate at 2.5 bara. The retentate stream, containing 23.6 mol% H_2, 8.9 mol% CO, 31.3 mol% H_2O, and 27.1 mol% CO_2 is further cooled to 350°C prior to entering the second stage of WGS, where an overall conversion of 83% of the CO from the ATR outlet is achieved (note that the overall conversion is lower than the 90% value used in designs without H_2 membranes). The outlet gas from the second WGS stage is fed into a second stage of hydrogen membranes, where an additional 70% of the remaining hydrogen is removed. In the second stage, steam is used to sweep the permeate, which leaves the membrane module at 2.8 bara. The hydrogen streams from the two membrane stages are combined, cooled for water removal, and then recompressed to 2.5 bar for use in the refinery.

The CO_2-rich retentate is cooled to remove water and then dehydrated prior to the cryogenic purification section. Part of the condensate is purified and used for boiler feed water (BFW) to generate steam used for stripping gas in the second membrane unit. The dry CO_2 is cooled to -49.5 °C at 29 bara (compared to 64 bara and -35 °C in the PSA/Cryo case for ~89% recovery of CO_2) to condense ~75% of the CO_2. The vapour (30.2 mol% H_2 content) is recycled back to the first WGS reactor, except for a 10% purge to prevent the build-up of CO and N_2. (A simple change to this flow sheet to redirect the purge can result in a high purity H_2 stream, if desired).

The new equipment in this configuration consists of the two membrane separation steps, one after each WGS stage. Each membrane step has 3 stages in series to achieve the recovery target, and sufficient parallel modules in the front stages to handle the throughput (see Chapter 10)). Because a part of the tail gas is recycled into the shift reactor, the WGS reactor sizes are larger relative to the cases without recycle (high temperature reactor is 7.1 m ID and 10.1 m T/T height; low temperature reactor is 6.5 m ID and 7.9 m T/T height).

Total plot space requirement is 151m x 385m. A significant portion of space is devoted to the two membrane separation steps, each of which has a footprint of 20m x 116m. There are 195 vessels for the first step and 182 vessels for the second step, arranged in three levels, with access for maintenance. The sheer number of vessels presents a significant logistics challenges in design and maintenance. Careful consideration will be required to ensure proper mixing of the retentate from parallel and series elements of each stage to avoid performance issues of the membrane unit. In addition, maintenance will be a continuous operation, assuming membrane replacement is required every two years and takes two days for each vessel. Membrane durability or mechanical designs to reduce the number of membrane vessels (increase membrane area per vessel) are key improvements needed to ensure operability.

Pre-C.- Mem/Cryo/PSA

Similar to Pre-C.-Mem/Cryo case, this case also utilizes Pd-alloy hydrogen permeable membrane technology. The key difference is the addition of a PSA unit to increase the H_2 recovery, and indirectly, CO_2 recovery. As shown in Table 7, for the production of 5000 MMBtu/hr of H_2 fuel, the methane fuel usage is lower for this case than for the H_2 membrane case without PSA, and the CO_2 capture percentage is higher.

The WGS reactors and Pd membrane units operate under largely similar conditions as in the Pre-C.-Mem/Cryo case. There are small differences in flow rate and composition, which result in small differences in reactor sizes. Feed conditions for CO_2 recovery in the cryogenic purification unit are 49.4 bara and -45.5 °C, in which 80% of the CO_2 is recovered.

The vapour (24.5 mol% H_2) from the cryogenic purification enters the PSA unit to produce additional H_2 product (83% of the H_2 entering the PSA recovered) at 48 bara. The PSA tail gas, containing 6.2 mol% H_2, leaves the unit at 1.2 bar. It is compressed in 4 stages of centrifugal compressors and then recycled to the first WGS reactor. The PSA unit and associated compressors for recycle gas are key additional process equipment in this case.

Plot requirements for the two Pd membrane cases are similar, but the PSA unit fills up the empty space in the previous case. Similar operability concerns arise due to a large number of membrane vessels.

Pre-C.- Cryo/MDEA

The Pre-C.-Cryo/MDEA case is similar to the Pre-C.-MDEA case, with an addition of a cryogenic purification step upstream of the MDEA scrubber. Conditions at the ATR, and WGS units are similar; with small differences in flow rates due to H_2 lost as impurities in CO_2 product.

Syngas exiting the WGS reactors is cooled in a series of boilers, heat exchangers, and coolers and water vapour is removed by condensation and dehydration prior to entering the cryogenic unit. The shifted gas is first compressed to 80 bar (2 stages) before it is cooled to -49 °C to separate out approximately 50% of the CO_2.

The vapour stream exiting the cryogenic unit is expanded to 19 bara prior to the MDEA absorption unit. A flash column and a stripper column are used to regenerate the rich amine, achieving approximately 96% removal of CO_2 in the absorber unit, as in Pre-C./MDEA case (~84% capture of carbon, including methane, CO).

Because of the cryogenic pre-treatment, the inlet stream into the absorption unit in this case is reduced by approximately 13% compared to Pre-C - MDEA. The CO_2 concentration is also reduced from 23% to 13%. A two-section design is used for the absorber and stripper. The absorber is 4.2/6.2 m ID, 41.7 m height. The CO_2 stripper is 7.1/4.1 m ID and 17.3 m in height. The flash column is 6.9 m ID and 20.1 m in height. Total plot space requirement is 151 m x 360 m.

ECONOMIC EVALUATIONS

The methodology employed in the economic evaluation is discussed in Chapter 4. Table 8 below summarizes the key data from AFW used to derive the economic results of CO_2 capture technologies employed on heaters and boilers. The nomenclature used in the top row of Table 8 is: Post-C for post-combustion, nominal scale; Post-C. (ss) for post-combustion, small scale (i.e., 3 x 100 MMBtu/h); Oxy-C for oxy-fuel combustion. The nomenclature for pre-combustion cases is given in Table 6.

The top part of Table 8 shows fuel consumption, incremental fuel and power used for capture. It also shows the CO_2 directly generated by the industrial process, the CO_2 captured and emitted. This is followed by the values used to calculate the mass of CO_2 avoided. Of note, the oxy-fuel cases show an avoided benefit associated with a fuel savings. This fuel savings is based on the assumption that using oxygen instead of air will require less fuel to produce the same amount of steam. Only the post-combustion cases have additional CO_2 associated with the production of steam required by the capture process.

Table 9 shows the key financial results for each of the cases. Percent avoided is defined as the mass of CO_2 avoided divided by the mass of CO_2 generated in the base reference case without CCS. The Levelized Capture Cost is based on levelized incremental costs rather than on first year costs. Additional Yearly Cost is the cost of capturing CO_2 for the first year of operation and includes the fuel, power, O&M, CO_2 transportation and storage costs and amortized capital cost recovery for CO_2 capture equipment incurred for a year.

The Post-C. and Oxy-C. cases capture about one tenth (one twentieth for the small scale Post-C case) the quantity of CO_2 compared to the Pre-C. cases. The pre-combustion cases were meant to assess the benefits from economies of scale associated with the opportunity to supply large amounts of fuel to a large facility. For retrofitting a small numbers of heaters and boilers, the oxy-fuel option requires significantly higher investments than post-combustion amine. Given that the air separation unit is characterized by good economy of scale, the Oxy-C option might be less disadvantaged than Post-C if a larger size application can be considered within the constraints of space and extra piping.

Table 8. Key metrics of CO_2 capture technologies applied to refinery heaters and boilers.

Characteristic	Units	Post-C. (ss)-Amine	Post-C.-Amine	Oxy-C.	Pre-C.-MDEA	Pre-C.-PSA/Ox	Pre-C.-PSA/Cryo	Pre-C.-Mem/Cryo	Pre-C.-Mem/Cryo/PSA	Pre-C. -Cryo/MDEA
Fuel consumption - HHV	GJ/h	353.5	707.1	797.5	7,027.0	7,948.6	7,125.3	7,160.0	7,124.2	7,059.4
Incremental fuel – HHV	GJ/h	45.3	90.7	3.5	1,081.8	2,057.4	1,234.3	1,248.8	1,232.1	1,167.3
% Increase in fuel	%	11.5%	11.5%	0.4%	13.8%	23.2%	15.5%	15.6%	15.5%	14.8%
Incremental power	MW	2.1	4.1	23.5	50.0	-25.6	39.9	83.8	108.1	46.0
CO_2 intensity on fuel wo/capt	kg/GJ	44.0	44.0	43.8	43.4	43.8	43.8	43.7	43.8	43.8
CO_2 intensity on fuel w/capt	kg/GJ	15.6	15.5	20.2	10.3	3.6	9.5	12.9	14.4	9.1
CO_2 generated	Mt/y	0.12	0.25	0.28	2.49	2.89	2.55	2.57	2.56	2.52
CO_2 captured	Mt/y	0.10	0.21	0.25	2.24	2.61	2.30	2.37	2.40	2.32
CO_2 emitted by plant	Mt/y	0.02	0.04	0.03	0.25	0.29	0.25	0.20	0.16	0.20
Base emissions - no CCS	Mt/y	0.12	0.25	0.28	2.04	2.04	2.04	2.04	2.04	2.04
Less: CCS emissions	Mt/y	0.02	0.04	0.03	0.25	0.29	0.25	0.20	0.16	0.20
Add: reduced fuel usage	Mt/y	-	-	0.01	-	-	-	-	-	-
Less: capture power use	Mt/y	0.01	0.02	0.11	0.24	-0.12	0.19	0.40	0.51	0.22
Less: capture steam use	Mt/y	0.02	0.03	0.00	-	-	-	-	-	-
CO_2 avoided	Mt/y	0.08	0.16	0.15	1.55	1.87	1.59	1.44	1.37	1.62

Table 9. Financial results.

Characteristic	Units	Post-C.(ss)-Amine	Post-C.-Amine	Oxy-C.	Pre-C.-MDEA	Pre-C.-PSA/ Ox	Pre-C.-PSA/ Cryo	Pre-C.-Mem/ Cryo	Pre-C.-Mem/Cryo /PSA	Pre-C. -Cryo/MDEA
Incremental fuel	MUSD/y	1.6	3.2	-1.2	38.4	73.0	43.8	44.3	43.7	41.4
Incremental capital	MUSD	83.2	117.5	286.5	1,227.2	1,902.3	1,535.1	8,001.1	8,117.4	1,149.0
Incremental O&M	MUSD/y	6.8	10.6	27.0	94.7	125.7	145.6	1,281.1	1,298.3	89.8
% capture	%	85%	85%	90%	90%	90%	90%	92%	94%	92%
% avoided	%	65%	65%	54%	76%	92%	78%	71%	67%	79%
Capture cost	USD/tCO_2	154.8	118.6	213.2	111.1	145.1	145.5	877.7	877.5	103.3
Avoided cost	USD/tCO_2	204.6	156.5	355.9	160.1	202.1	209.7	1,447.6	1,539.5	148.3
Levelized capture cost	USD/tCO_2	180.9	138.7	249.1	129.9	169.5	170.1	1,025.7	1,025.4	120.8
Levelized avoided cost	USD/tCO_2	239.1	182.9	415.9	187.1	236.2	245.1	1,691.6	1,799.0	173.3
Additional yearly cost	MUSD/y	16.2	24.9	52.9	248.8	378.0	334.1	2,079.7	2,107.3	239.5

The six pre-combustion options for generating cleaner-burning hydrogen fuels for the entire refinery are associated with a wide range of capture and avoided costs per tonne. The Pre-C.-Cryo/MDEA case has the lowest capture cost at \$103.4/t CO_2. On the other hand, the two membrane approaches show the significant cost challenges associated with the capital cost of palladium membranes, which are estimated to be \$2.2B based on Pall's membrane design (see Table 10 for comparative costs). In addition to the extremely high capital costs, the replacement costs for these membranes, roughly every two years, also drive up the O&M costs. No credit has been assumed for reclaiming PD from spent membranes. The pre-combustion CO_2 capture using palladium membrane is an emerging technology. The capital cost presented in this report is based on relatively limited performance data and fabrication experience, which may improve over time. For example, the 2-year Pd membrane lifetime is a conservative estimate based on the limited testing conditions. Additionally, the manufacturing cost of the Pd membrane may be reduced as the technology becomes mature.

For the membrane cases, the balance of plant costs are higher than for the other H_2 separation schemes due to the inefficiencies of the separation achieved by the present generation of technology. Therefore, even if the cost of the membrane modules was \$0, the per-tonne capture costs would still be higher than the Pre-C.-MDEA case, due to added capital cost and electricity cost. The difference in avoided costs is smaller, but the present membrane options show no advantage over the Pre-C.-MDEA case. While the process scheme proposed by Pall/AFW could be further optimized, process optimization will not bring the capture and avoided costs to a competitive level unless membrane performance is improved and costs are reduced by at least an order of magnitude.

Figure 5 (top) shows the breakdown of the CO_2 capture cost for the cases considered. The equivalent information for avoided costs is presented in Figure 5 (bottom). Capital costs tend to dominate the costs for most of the cases. The oxy-fuel case has large Capex associated with the air separation unit, which also increases the power usage and associated CO_2 footprint. All of the cases except the oxy-fuel cases show a significant requirement for natural gas fuel. For the amine cases this fuel is used to produce steam required by the CO_2 capture process. For the ATR cases, this fuel is used to supplement RFG to produce high hydrogen content fuel.

The oxy-fuel cases burn fuel more efficiently and generate waste heat. Based on limited data, the use of oxygen rather than air results in a fuel savings of about 4.5% for the Oxy-C case, even accounting for the auxiliary boiler to capture CO_2. This accounts for the negative fuel values for Oxy-C in Figure 5. The fuel savings is a relatively small effect, however, and does not change the competitiveness of the oxy-fuel option relative to the post-combustion base case. In a sensitivity analysis, a 10% fuel savings was assumed for the Oxy-C. case. This reduces the CO_2 intensity on fuel with capture from 20.2 kg/GJ to 17.4 kg/GJ, and increases the amount of CO_2 avoided from 0.15 Mt/yr to 0.17 Mt/y. The resulting effect on the capture cost per tonne CO_2 is small (\$213.2 to \$206.9) but more important on the avoided cost (from \$355.9 to \$308.8 per tonne CO_2).

Overall, transportation and storage are expected to represent a small contribution to the total cost of CO_2 capture and storage.

Three cases stand out as having relatively low CO_2 capture costs. The Post-C.-Amine case has a fairly low capture cost. The Pre-C.-MDEA case has a lower capture cost than all of the other pre-combustion cases except the Pre-C.-Cryo/MDEA.

Table 10 shows the components that contribute to the total incremental capital cost to capture CO_2. The top part of the table shows the direct costs for each case. This is followed by indirect, outside of boundary (OSBL) costs. The base capital costs have been adjusted to include escalation from Q1 2009 to the expected in-service date of Q1 2014. This is followed by estimates for contingencies. To this, estimates for Owners Costs are added. Finally, allowances for funds used during construction (AFUDC) are included. The bottom row shows the total as spent costs expected just before the plant is commissioned.

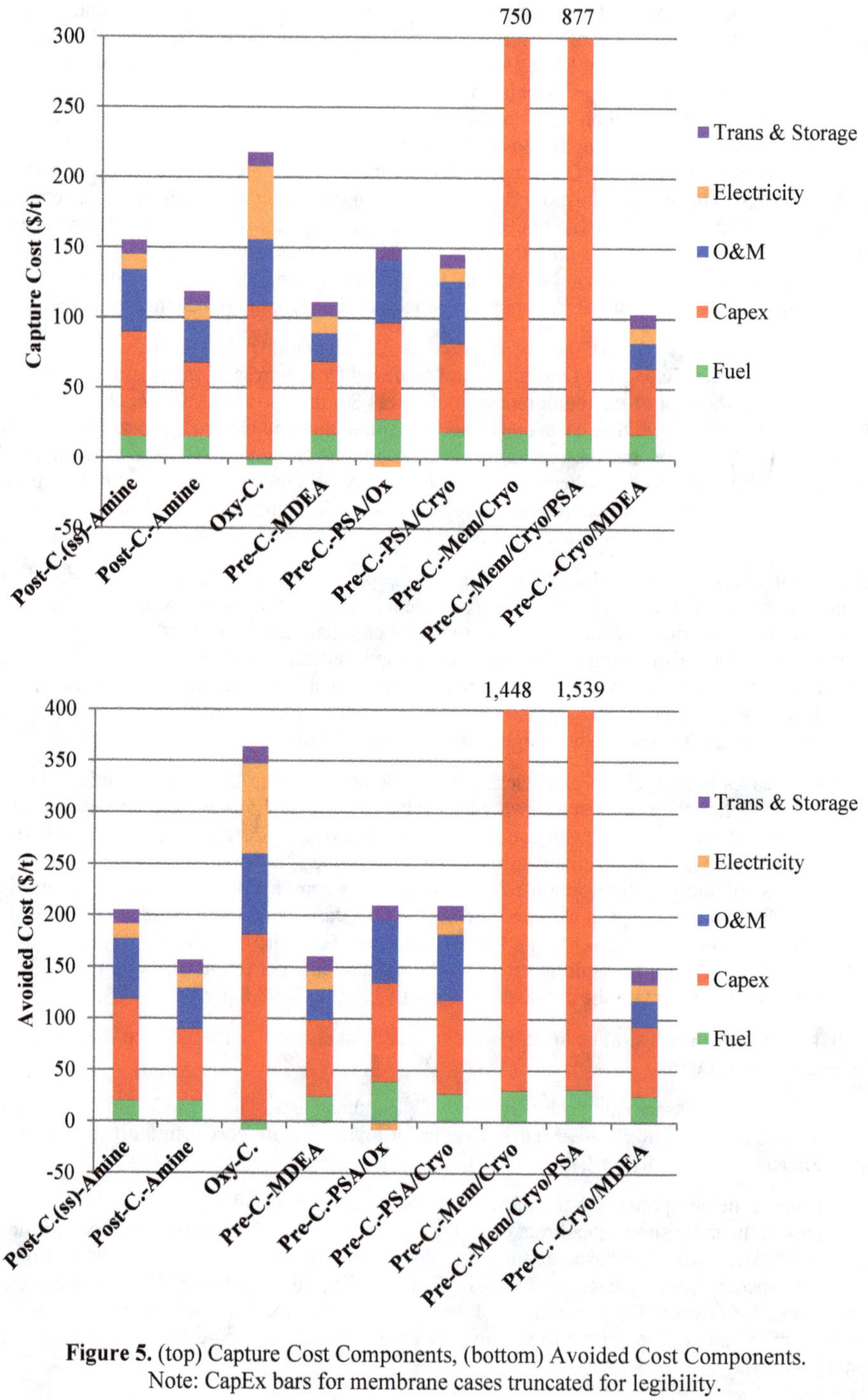

Figure 5. (top) Capture Cost Components, (bottom) Avoided Cost Components. Note: CapEx bars for membrane cases truncated for legibility.

Table 10. Capital cost components.

Cost	Post-C.(ss)-Amine	Post-C.-Amine	Oxy-C.	Pre-C.-MDEA	Pre-C.-PSA/Ox	Pre-C.-PSA/Cryo	Pre-C.-Mem/Cryo	Pre-C.-Mem/Cryo/PSA	Pre-C. -Cryo/MDEA
Capture plant	19.78	27.63							
Compression plant	10.59	15.29							
Steam gen & BOP	1.97	2.78							
Oxygen supply			78.46	153.71	200.20	160.10	153.17	153.54	158.53
ATR. WGS and membranes			-	-	-	-	2,629.19	2,610.22	-
Flue gas treat			7.51	-	-	-	-	-	-
ATR and WGS			-	136.16	155.95	147.44	-	-	147.01
CO_2 purification			11.51	-	2.85	-	-	-	60.76
CO_2 removal			-	106.50	-	-	-	-	-
CO_2 compression			14.86	53.20	87.28	120.50	15.27	14.02	28.88
CO_2 dehydration			5.42	24.21	30.20	32.41	-	-	17.33
H_2 fuel expansion			-	35.16	36.99	36.55	-	-	24.41
Steam gen			0.35	19.64	119.76	76.94	16.34	16.41	21.94
PSA			-	-	71.10	71.10	-	70.45	-
Thermal oxidizer			-	-	91.55	-	-	-	-
Cryo purification			-	-	-	1.41	4.67	4.39	41.33
Retrofit items			15.44	-	-	-	-	-	-
OSBL (30%)	9.69	13.71	16.52	112.47	178.71	145.92	799.63	814.64	102.50
Direct & indirect	42.03	59.41	150.07	641.05	974.59	792.37	3,618.27	3,683.67	602.69
Escalation from 2009 to 2014	7.64	10.81	27.31	116.67	177.38	144.20	658.52	670.42	109.69
Process contingency	2.34	3.27	2.25	14.60	41.38	23.05	622.09	617.64	10.66
Project contingency (20%)	10.39	14.70	35.93	154.47	238.67	191.91	979.77	994.34	144.61
Total plant costs	62.36	88.19	215.56	926.79	1,432.02	1,151.53	5,878.65	5,966.06	867.65
Owners costs									
6 months fixed O&M	1.86	2.26	5.42	18.27	26.27	21.79	86.77	88.28	17.43
1 month variable O&M	0.17	0.34	1.15	3.04	3.99	6.65	90.38	91.54	2.71
25% of 1 month of fuel costs	0.03	0.07	-0.02	0.80	1.52	0.91	0.92	0.91	0.86
2% of TPC	1.25	1.76	4.31	19.31	29.83	23.99	122.47	124.29	18.08
60 days of cons.	0.15	0.30	0.13	1.49	10.34	9.62	173.06	173.14	1.18
Spare parts (.5% TPC)	0.31	0.44	1.08	4.63	7.16	5.76	29.39	29.83	4.34
Land ($3,000/acre)	0.02	0.02	0.02	0.02	0.02	0.02	0.02	0.02	0.02
Financing costs (2.7% TPC)	1.68	2.38	5.82	25.02	38.67	31.09	158.72	161.08	23.43
Other costs (15% of TPC)	9.35	13.23	32.33	139.02	214.81	172.72	881.79	894.91	130.15
Total owners costs	14.83	20.79	50.24	211.61	332.60	272.55	1,543.53	1,563.99	198.19
Total overnight costs	77.19	108.98	265.80	1,138.40	1,764.62	1,424.08	7,422.18	7,530.05	1,065.83
AFUDC and escalation	6.02	8.50	20.73	88.80	137.64	111.07	578.93	587.34	83.14
Total as-spent cost	83.21	117.48	286.53	1,227.19	1,902.27	1,535.16	8,001.10	8,117.40	1,148.97

Table 11. O&M costs.

Cost	Post-C.(ss)-Amine	Post-C.-Amine	Oxy-C.	Pre-C.-MDEA	Pre-C.-PSA/ Ox	Pre-C.-PSA/ Cryo	Pre-C.-Mem/ Cryo	Pre-C.-Mem/ Cryo/ PSA	Pre-C. -Cryo/ MDEA
Fixed costs									
Direct labour	1.37	1.37	2.21	3.82	3.82	3.82	4.06	4.06	3.82
G&A	0.42	0.42	0.67	1.15	1.15	1.15	1.22	1.22	1.15
Maintenance	1.16	1.64	4.79	18.94	28.51	23.15	100.96	102.77	17.92
Insurance and taxes	0.78	1.09	3.19	12.62	19.01	15.44	67.31	68.51	11.94
Total fixed costs	3.73	4.51	10.85	36.52	52.49	43.56	173.55	176.56	34.83
Variable costs									
Power	1.14	2.25	12.99	27.59	-14.13	22.02	46.25	59.66	25.39
Sea water	0.02	0.05	0.46	0.55	1.49	0.74	0.30	0.29	0.49
Cooling water	0.49	0.99	0.32	4.77	4.64	4.01	5.27	5.72	3.34
Absorbent	0.02	0.04	0.02	0.20	0.19	0.19	-	-	0.16
Solvent	0.18	0.34	-	-	-	-	-	-	-
Chemicals, disposal	0.20	0.38	-	3.43	55.72	52.81	1,032.78	1,032.81	3.08
Transportation	0.71	1.42	1.67	1.23	1.43	1.26	1.30	1.32	1.27
Storage	0.31	0.62	0.73	20.44	23.79	20.96	21.64	21.93	21.16
Total variable cost	3.07	6.08	16.20	58.21	73.15	101.99	1,107.53	1,121.72	54.90
Total O&M	6.80	10.60	27.05	94.74	125.63	145.55	1,281.08	1,298.28	89.73
Fuel costs	1.61	3.22	-1.15	38.38	72.99	43.79	44.31	43.71	41.41

Capital costs for air separation units were provided by a third party vendor on a turn-key basis and, as a default, no OSBL was added for this unit in the Capital Cost calculations. Recognising the potential bias in favour of all cases involving ASU, Figure 6 shows the impact of adding OSBL to the air separation unit capital costs to selected oxy-fuel and pre-combustion cases. The impact on the OSBL accounting is modest for the oxy-fuel case, because the ASU accounts for a large fraction of the equipment cost. The effects on pre-combustion cases are fairly negligible.

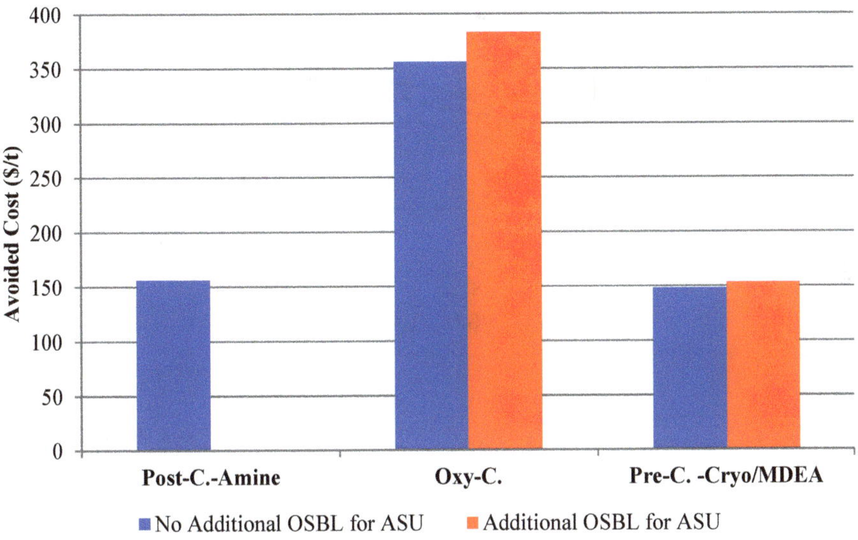

Figure 6. Impact of Including OSBL on Air Separation Capital Costs.

Table 11 shows the operating and maintenance costs for each case. The O&M costs have been adjusted for escalation or capacity factor and are reported in 2014 US$. The periodic replacement of palladium membranes is factored into the Catalyst and Chemical expense for those cases.

Figure 7 (top) shows how the cost of capturing CO_2 varies as the cost of natural gas changes. All cases except the oxy-fuel case show an increase in capture cost as the cost of natural gas increases. These cases use a significant amount of fuel to produce steam required to capture CO_2 or as supplemental fuel. The relative slopes are related to the fuel usage. The decrease in capture cost for the oxy-fuel case as natural gas prices rises is due to the benefit of reduced fuel consumption relative to the reference burner. Therefore, as the gas price increases, the economic benefit associated with using less fuel increases. Figure 7 (bottom) shows how the avoided costs change as gas prices change. As with the capture cost, the oxy-fuel case shows a reduced avoided cost as the price of gas increases. However, even as the cost gap narrows at high fuel costs, oxy-fuel remains less attractive than the best Post-C and pre-C options.

Figure 8 shows how the avoided cost changes with changes in GHG emission intensity for externally supplied power. As the GHG intensity of power increases, avoided carbon emissions decrease; therefore, the avoided cost per tonne of CO_2 increases. The inflexion in the graph is a result of the assumed relationship between power intensity and NG price. A low GHG emission intensity is indicative of natural gas being a larger share of the energy production mix. Therefore, when GHG emission intensity is below 0.5 t/MWh in a sensitivity scenario, the gas price affects the power price (for the USGC, the price of power is 33+5.55 x Gas Price).

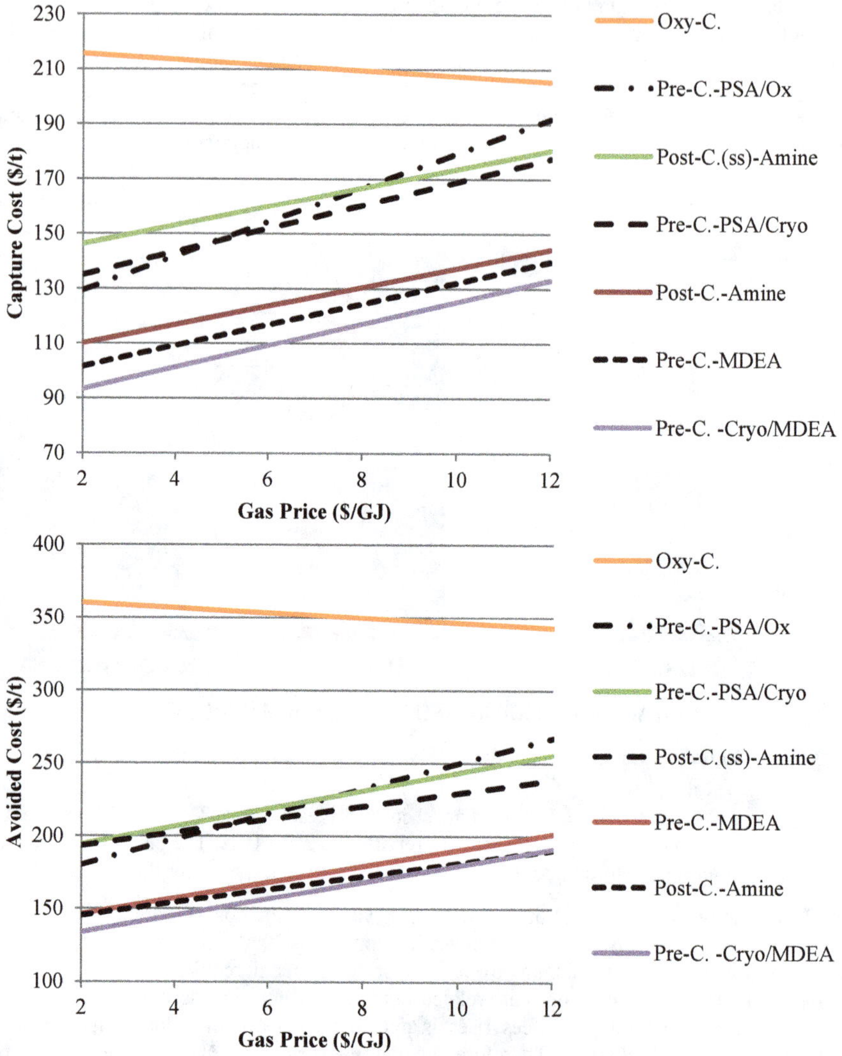

Figure 7. (top) Capture cost vs. gas price for selected technologies; (bottom) avoided cost vs. gas price for selected technologies.

As the cost of NG increases, the expectation is that cost of power will increase, but the generation mix may start shifting towards coal. High GHG emission intensity indicates a higher reliance on coal, and the power price becomes decoupled from the gas price. Only the avoided cost for the oxy-fuel case is likely to be significantly impacted by changes in GHG emission intensity of the power employed, since this is the only case with power as a significant component of the avoided cost.

Figure 9 shows how the avoided costs change as the price of power changes. For most cases, power is not a significant cost component of the avoided cost. Therefore, changes in power price have little impact on avoided costs. However, power is a significant cost component for the oxy-fuel case. Therefore, the avoided cost for this case is very sensitive to changes in the price of power.

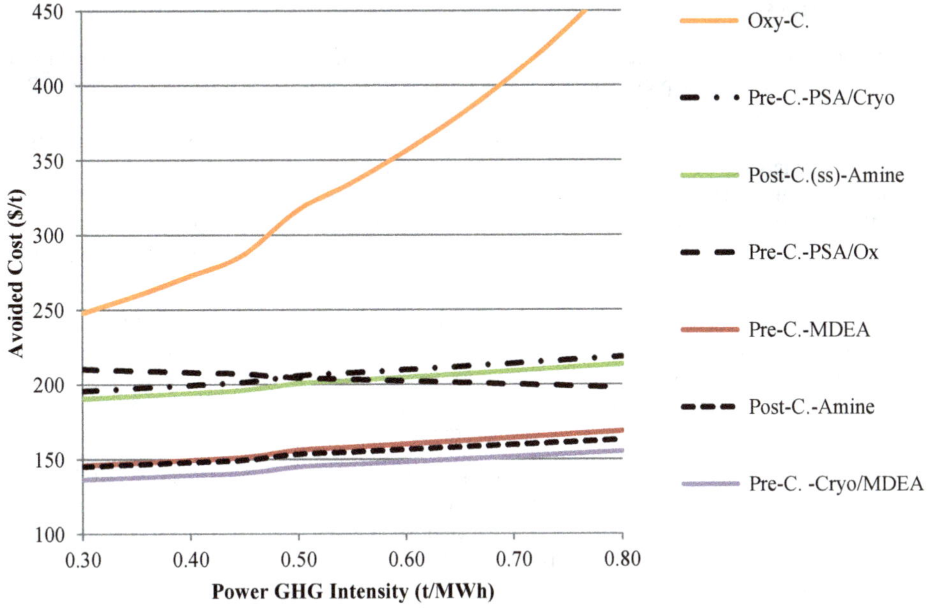

Figure 8. Avoided cost vs. power GHG emission intensity for selected technologies.

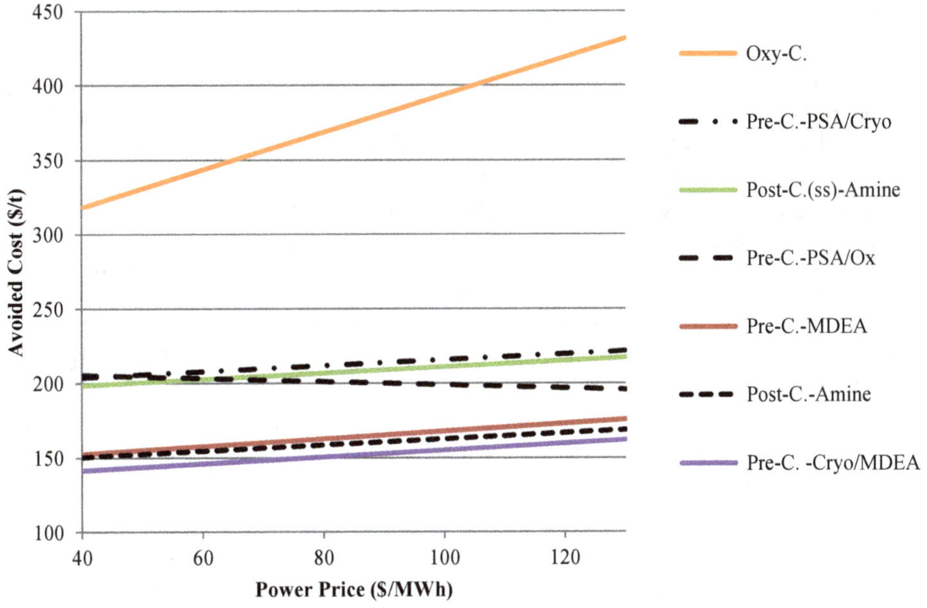

Figure 9. Avoided cost vs power price for selected technologies.

CONCLUSIONS

In a refinery, there are many heaters and boilers distributed in various locations. Post-combustion, oxy-fuel combustion, and pre-combustion approaches were considered as retrofit options. Post-combustion and oxy-fuel combustion can be applied at relatively small scale – 4 x 150 MMBtu/hr is considered the nominal size in this work. For these approaches, the new piping requirements for collecting flue gas for treatment (and distributing O_2 in the case of oxy-fuel combustion) will be a key factor limiting the sources that can be retrofitted. For pre-combustion, the entire refinery is converted to use H_2 as fuel, and a hydrogen fuel plant is connected to the fuel system. The difference in scale must be considered as a caveat to the comparison of the retrofit approaches investigated here.

Comparing the post-combustion and oxy-fuel combustion approaches, both applied to the 4 x 150 MMBtu/h scenario, the flue gas compositions are very different. In post-combustion, CO_2 is a minor component at 7.5%. In oxy-fuel combustion, after the removal of water, CO_2 becomes the major component at 68%. This difference drives the choice of absorption for post-combustion vs. cryogenic purification for oxy-fuel as the key separation step.

In this study, the capture plant removes 90% of the CO_2 from flue gas that is treated but 5% of the post-combustion flue gas is allowed to by-pass the capture plant to the stack. Thus, more CO_2 is captured by the oxy-fuel approach than the post-combustion cases. The oxy-fuel process has a larger electricity requirement, primarily to drive the ASU. The additional CO_2 generated for power is 14121 kg/h for oxy-fuel combustion but only 2444 kg/h for post-combustion. Despite the higher capture rate in oxy-combustion, the net amount of CO_2 emitted in the oxy-fuel case is substantially larger than the CO_2 emitted in the post-combustion case. When compared to the status quo of 4 x 150 MMBtu/hr heaters/boilers, the amount of CO_2 avoided is much less (54%) in the oxy-fuel case compared to the post-combustion case (65%).

Although oxy-firing increases the efficiency of the heater compared to conventional air firing, additional energy is required to drive the ASU. Therefore, the amount of CO_2 avoided is less than for the post-combustion retrofit. Considering the additional equipment, oxygen piping, and associated land requirement for the reduced CO_2 avoidance benefit, oxy-firing seems to have no advantage over post-combustion for refinery heaters and boilers retrofit.

The pre-combustion case with the lowest CO_2 emissions and highest CO_2 avoided % is the PSA/Catalytic oxidation configuration. In this case, fuel consumption is high but there is net electricity generation. Waste heat is recovered from the ATR (as in all pre-combustion configurations), the hydrogen expander (in all non-membrane cases), as well as from the steam turbine used to recover waste heat from the oxidation unit. There is a net power generation despite high power consumption of the ASU to generate extra O_2 beyond that needed for the ATR for both the thermal and catalytic oxidation units.

The palladium hydrogen permeation membrane technology is early in development compared to the other H_2/CO_2 separation technologies considered here for the pre-combustion approach. With the support of CCP3, Pall Corporation designed and fabricated membrane modules and conducted long-term tests of membranes under pre-C relevant conditions. Unfortunately, the two H_2 membrane cases conceived based on the laboratory data are associated with the highest CO_2 emissions and lowest CO_2 capture %. The membrane cases were not cost-effective compared to the baseline pre-combustion configuration (with amine capture) in AFW's techno-economic studies. The primary driver of the capital cost was the cost of the membranes and vessels. Power consumption is high and land requirement is significant. Cost and operability improvements are expected as the technology matures.

The H_2 fuel product contains impurities in the form of CO, CH_4, and CO_2. These carbon-containing species will be emitted as CO_2 when the fuel is burned. Based on this study, the membrane/cryogenic/PSA process and MDEA process perform the best for removing carbon from the H_2 product stream, while the PSA-based processes leave the most carbon in the H_2 fuel. PSA is a standard technology for high purity H_2 in today's refineries. In pre-combustion applications for fuel, some compromise on the recovery is made in lieu of H_2 purity. It may be beneficial to explore potential designs to generate optionality to co-produce high-purity H_2 to help de-bottleneck a hydrogen constrained refinery.

The pre-combustion approaches have a range of CO_2 performance from 4.2 kg CO_2/mmbtu to 16.9 kg CO_2/mmbtu. CO_2 avoided numbers range from 70.7 to 91.9% compared to the status quo pre-capture scenario of burning 5000 MMBtu/hr RFG in heaters and boilers. The PSA/Ox configuration consumes the most fuel, but generates electricity with heat recovery, resulting in an advantageous performance. However, simpler CO_2 removal options, such as cryogenic/MDEA and MDEA, are also attractive because the avoided CO_2 benefits are less sensitive to accounting of power benefits. The membrane processes represent the worst performers in terms of CO_2 avoidance.

Another way to compare the CO_2 performances of post-combustion, oxy-fuel combustion, and pre-combustion approaches, despite the difference in application scale, may be to look at the amount of CO_2 emitted per MMBtu fuel burned in heaters and boilers. With no capture, 51.9 kg CO_2 is emitted per MMBtu of fuel burned. As discussed, oxy-combustion does not provide any advantages over post-combustion solvent absorption, which has a CO_2 emission of 18.3 kg/MMBtu. Pre-combustion approaches emit less CO_2 per MMBtu than post-combustion absorption (4.2 for PSA/Ox to 16.9 kg/MMBtu for Mem/Cyro/PSA). The post-combustion approach may be suitable for retrofitting a few heaters and boilers, when only modest mitigation is required for a refinery. Similar or better CO_2 avoidance benefits can be obtained using H_2 as fuel for retrofitting the entire refinery. The best technical approach may well depend on the distribution of the heaters and boilers and nearby space availability; but economics will be the overriding consideration for this case.

Capturing CO_2 from refinery heaters and boilers costs \$118.6 to \$154.8 per tonne CO_2 in the post-combustion cases. The lower cost per tonne of CO_2 captured shows the effects of economies of scale when doubling the size of the capture plant. The avoided cost of CO_2 is as low as \$156.5/tonne for the post-combustion amine capture case involving 4 x 150 MMBtu/h heaters and boilers.

At the scale studied here, oxy-fuel combustion requires more than double the capital for post-combustion technology. In addition, electricity is an important contributor to the overall high capture and avoided costs for oxy-fuel. The fuel saving characteristics of the oxy-fuel approach is unique among the technologies considered. However, even at high fuel prices, the oxy-fuel is still disadvantaged over other post-combustion in terms of capture and avoided costs. Oxy-fuel technology, as well as ATR-based pre-combustion technologies, will benefit from advances in O_2 separation that lower both the CapEx and OpEx.

The pre-combustion cases considered here are much larger scale (5000 MMBtu/hr) compared to those considered for retrofitting individual heaters and boilers (nominal 600 MMBtu/h). Of the four price-competitive pre-combustion process options (MDEA, PSA/Ox, PSA/Cryo, Cryo/MDEA), the capture costs range from \$103.3 to 145.5 per tonne and the avoided costs range from \$148.3 to \$209.7 per tonne. The membrane-based configurations will require significant development to reduce the capture and avoided costs.

How a refinery might approach retrofitting heaters and boilers depends not only on the target CO_2 reduction but also on other factors, such as the availability of space, especially around boiler and heater units where additional duct work will be necessary for some configurations. For retrofitting a few heaters and boilers that may be in close proximity, post-combustion amine absorption is an obvious option. Oxy-fuel combustion was considered, and shown to be technically feasible with controlled flue gas recirculation. However, this technology offers no cost advantage over post-combustion capture for this application. The capital and operating costs of the air separation unit will need to be reduced to improve the economics of oxy-firing options.

For a refinery-wide solution, pre-combustion configurations offer the opportunity to switch fuel to clean-burning hydrogen. Each refinery is different in size, product/crude slate, and in demand for H_2, fuel, and power. Several pre-combustion configurations were investigated for a relatively large-scale refinery of 5000 MMBTU/h. At first glance, the MDEA and cryogenic/MDEA CO_2 removal configurations seem most attractive. However, the PSA options may offer an additional opportunity to co-produce high-purity process H_2, along with the decarbonized fuel. For many H_2-short refineries, this may be an attractive proposition.

ACKNOWLEDGEMENTS

AFW's project team was led by Tony Tarrant, Tim Bullen and Tim Abbott. The multi-phase project was managed by Richard Beavis and Jonathan Forsyth on behalf of the CCP capture team. The authors would like to acknowledge the CCP capture team, led by Ivano Miracca, for helpful discussions and review of draft documents.

REFERENCES

1. Eriksson, M. and S. Ahlgren (2013). LCAs of petro and diesel, a literature review, Report 2013:058 ISSN 1654-9406, The Swedish Knowledge Centre for Renewable Transportation Fuels.
2. Intergovernmental Panel on Climate Change (IPCC) (2005) Special Report on Carbon Dioxide Capture and Storage (SRCCS) https://www.ipcc.ch/report/srccs/
3. Thomas, D.C. (Editor), Carbon Dioxide Capture for Storage in Deep Geologic Formations – Results from the CO_2 Capture Project, Volume One: Capture and Separation of Carbon Dioxide from Combustion Sources; Elsevier (2005).
4. Lowe, C., N. Brancaccio, D. Batten, C. Leung, and D. Weibel; Technology Assessment of Hydrogen Firing of Process Heaters, Poster 147; 10th International Conference on Greenhouse Gas Control Technologies 2010 (GHGT-10), Elsevier (2011).

Carbon Dioxide Capture for Storage in Deep Geological Formations, Volume 4
Karl F. Gerdes (Editor)

Chapter 12

TECHNO-ECONOMIC ASSESSMENT OF DEEP CO_2 CAPTURE FROM HYDROGEN PLANTS

Mahesh Iyer[1], Richard Beavis[2], Tony Tarrant[3] and Tim Bullen[3] and David Butler[4]
[1]Shell International Exploration & Production Inc.,
200 N. Dairy Ashford Rd., Houston, TX 77079, USA
[2]BP International Ltd, ICBT Chertsey Road,
Sunbury-on-Thames, Middlesex, TW16 7LN, UK
[3]Amec Foster Wheeler, Shinfield Park, Reading, Berkshire, RG2 9FW, UK
[4]Calgary, Alberta, Canada

ABSTRACT: Hydrogen production is energy intensive and the Hydrogen Manufacturing Unit (HMU) can be a substantial single point source of CO_2 emission in a refinery complex. Steam Methane Reforming (SMR) is the most pervasive technology used for refinery hydrogen production. CO_2 emissions from an SMR system can be captured from the process (pre-combustion capture), as well as from the furnace flue gas (post-combustion capture). Pre-combustion capture will result in capturing only about 50% of the total CO_2 from the HMU. The state-of-the-art SMR uses a PSA system to produce high purity H_2 from shifted syngas. The PSA tail gas contains significant heating value, so this stream is used as fuel in the firebox of the SMR furnace. Hence a post-combustion scheme on the SMR furnace flue gas can capture both the process-generated and combustion-generated CO_2 to achieve deep capture. MHI-KS1 solvent technology was employed on the furnace flue gas for this study.

An SMR with a capacity of 50,000 Nm^3/h (45.6 MSCFD) hydrogen production, which is representative of typical refinery scale, was the basis for this study. The CO_2 capture rate is 0.28 Mt/year with 66.5% CO_2 avoided. CO_2 was not captured from the capture unit boiler. The economic analysis was based on a US Gulf Coast location. The CO_2 avoided cost was $123/tonne with an assumed natural gas price of $4.50/GJ. CAPEX plays a critical role in the avoided cost, accounting for nearly 40% of the total. O&M costs and fuel costs make up the remainder.

KEYWORDS: refinery; hydrogen plant; CO_2 capture; post-combustion; pre-combustion; steam methane reforming; techno-economics; avoided cost

INTRODUCTION

Hydrogen production is energy intensive, and the overall process including the reforming and water gas shift reactions produces a large amount of CO_2. Hydrogen is used in numerous processes, such as hydro-treating, hydrocracking and hydro-desulfurization in refineries; or ammonia synthesis in chemicals plants; or syngas composition adjustments for either methanol production or Fischer-Tropsch liquid production processes. Hydrogen is typically produced and used on-site. Therefore, Hydrogen Manufacturing Units (HMUs) are found in many petrochemical, gas-to-liquids and refinery complexes. While CCP's objective was to understand the cost of CO_2 abatement on HMUs from a refinery standpoint, the results could be applied, to other applications using a similar type of HMU.

A typical refinery has many distributed sources of CO_2 emission, and the HMU is often one of the largest single point sources, which can account for 5-20% of the refinery CO_2 emissions [1]. Hence it is attractive to study capture options for HMUs. Hydrogen from fossil fuels can be typically produced by several processes such as steam methane reforming, partial oxidation or auto-thermal reforming. Of these, the Steam Methane Reformer (SMR) is the industrial workhorse for producing hydrogen in refineries as well as petrochemical complexes. Hence, this study has focused on CO_2 capture from SMR units.

PROCESS SCHEME

An SMR is primarily used to convert methane and other lower hydrocarbons (C_2-C_4) to hydrogen-rich syngas. A simplified process block diagram for the current state-of-the-art SMR systems in operations is shown in Figure 1. Natural gas or refinery fuel gas is first treated to remove sulphur-containing compounds in a desulfurization system using zinc oxide guard beds. The process gas is then mixed with steam and fed to a pre-reformer where the lower hydrocarbons (C_2-C_4) and some small amount of lower olefins are converted to methane. In addition, some steam methane reforming occurs in the pre-reformer to produce syngas. The process gas is then reheated and fed to the reformer, which consists of reactor tubes packed with nickel-based reforming catalysts on alumina support. The tubes are placed in the radiant box of the reformer furnace, which provides energy for the endothermic reforming reaction. The hot syngas is then cooled and conditioned and is sent to the water-gas shift units for further conversion of CO to hydrogen and CO_2. There is extensive heat integration in the SMR system where different levels of steam can be generated from the hot flue gas from the furnace and hot syngas from the reformer tubes. The SMR tubes are typically operated at moderate to high pressures (15-30 bar) and at high temperatures around 900°C or above to sustain typical methane conversions of 70-80% [2]. However, the system pressures and operating temperatures can be modified as required in accordance with thermodynamic equilibrium of the system.

The shifted syngas, which contains predominantly H_2, CO_2, H_2O, some CO and unconverted CH_4, is then cooled and the water is knocked out. The process stream is then sent to a Pressure Swing Adsorption (PSA) unit for H_2 purification. Very pure H_2 with purities around 99.99% and above is generated from the PSA unit at high pressure, with very little pressure drop from the feed pressure. The low pressure reject stream from the PSA, referred to as "Tail gas," contains some H_2 and nearly all of the CO, CO_2 and unconverted CH_4. This stream is used as fuel in the reformer furnace, supplemented with the natural gas.

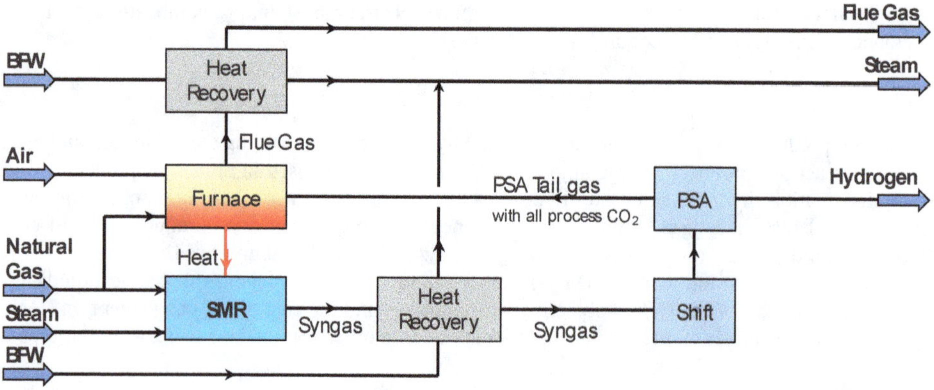

Figure 1. Simplified process block diagram for the current state-of-the-art SMR scheme.

CO_2 capture location

In the SMR process there are two sources of CO_2 generation:

(a) CO_2 produced due to the reforming and shift steps in the main process stream.
(b) CO_2 produced due to fuel combustion in the reformer furnace in the flue gas stream.

The CO_2 produced in the main process stream is under pressure (15-30 bar) with the CO_2 partial pressures ranging from about 2-5 bar depending on the mode of operation. Capturing CO_2 from this stream, upstream of the PSA, is easier than capture from flue gas, where the CO_2 is at much lower partial pressure and oxygen from excess air is present. The pre-combustion approach is used in two SMR CCS demonstration projects: the Air Products' Port Arthur US and Shell Edmonton Canadian "Quest" projects. However, for SMRs the process-generated CO_2 comprises only 50-60% of the total CO_2 generated, with the balance generated in the furnace by combustion [2]. With a 90% pre-combustion CO_2 capture rate, only 45-55% of the total CO_2 produced would be captured. The CO_2 avoidance rate would then be approximately in the 35-40% range. In addition, removing CO_2 from the process gas will affect the PSA tail gas composition and in some cases can affect the furnace combustion characteristics, which will need some additional investigation, especially for retrofit cases.

On the other hand, post-combustion capture from the SMR flue gas potentially tackles all the CO_2 produced in the SMR system, since the PSA tail gas is routed to the furnace. Typically the CO_2 concentration in this flue gas, which includes CO_2 from the PSA, can range from 18-22 vol%. Hence, CCP decided to study the post-combustion capture configuration as shown in Figure 2, with a target to achieve 80-90% of total capture. A benefit of this approach is that the hydrogen process system availability should not be impacted by the carbon capture plant, which is an end-of-pipe capture system. Integrating capture into syngas flow of the SMR would require additional scope to maintain availability.

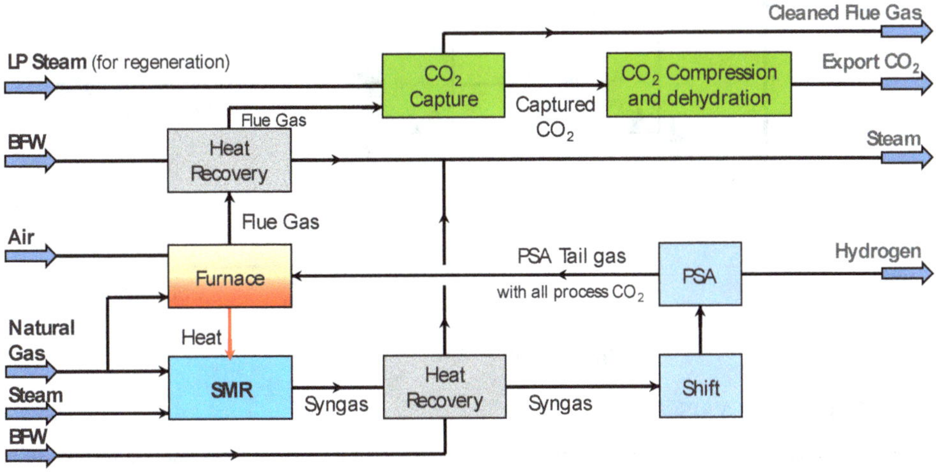

Figure 2. Simplified process block diagram for SMR with post-combustion capture.

SMR PROCESS WITH CAPTURE: TECHNICAL PERFORMANCE RESULTS

CCP defined the scope and managed the technical study, which was delivered by Amec Foster Wheeler. The main deliverables were the process design, simulations to derive mass and heat balances and capital cost estimates. The details of the study methodology are described in Chapter 3. For this study, a typical refinery-scale SMR with 50,000 Nm^3/h (45.6 MSCFD) hydrogen production was chosen as the basis. Flue gas from the SMR furnace is routed to the CO_2 capture unit in a nearby plot space of suitable size (Figure 3). The design was such that 5% of the total flue gas flow bypasses the CO_2 capture plant via the stack to accommodate bypass of the capture system, if needed. The CO_2 capture unit removes 90% of the CO_2 in the flue gas it receives, and the scrubbed flue gas is routed to the stack.

MHI provided performance and cost information for their proprietary KS1^{TM} solvent system package, which was selected by CCP as being representative of the current state-of-the-art post-combustion technology. A single train of CO_2 absorption, stripping, compression and dehydration was considered the most appropriate for the total volume of gas to be treated in the case. A stand-alone steam boiler package supplies the steam required for the stripper reboiler and solvent reclaimer, as well as for regeneration of the CO_2 dehydration beds. The CO_2 emissions from the boiler are not captured. This choice in the basis has a strong impact on CO_2 avoided. In some refinery/SMR applications, excess waste heat steam may be available for use in the capture plant, at the expense of lost power generation from low pressure (LP) steam balancing turbines if the system uses a cogen unit. A combination of sea and fresh water cooling is used and electrical power is assumed to be supplied over the fence to the CO_2 capture area.

The figure below shows a pictorial representation of the high level CO_2 balance from Table 1.

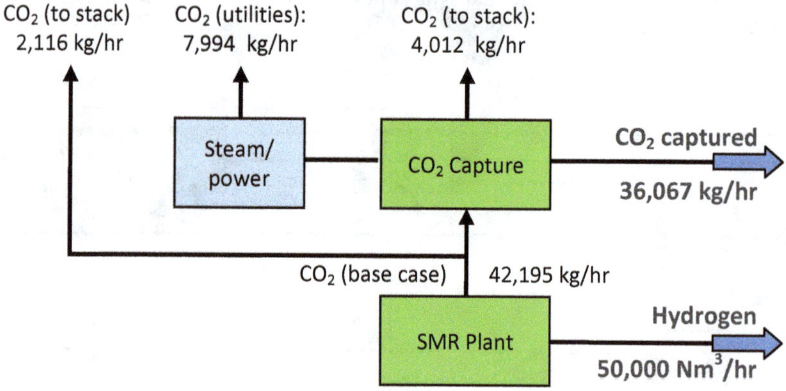

Figure 3. Schematic of CO_2 flows.

ECONOMIC ANALYSIS

Table 2 summarizes the key technical performance data used to derive the economic results of the post-combustion CO_2 capture technology employed on an SMR with a capacity of 50,000 Nm^3/h hydrogen production. The location used for this study was US Gulf Coast. The details of the economic evaluation methodology are discussed in Chapter 4. Specifically, Table 2 summarizes the fuel consumption, incremental fuel and power used for capture. It also shows the CO_2 directly generated by the SMR process (unabated), the CO_2 captured and emitted. This is followed by the values used to calculate the mass of CO_2 avoided.

Table 1. Refinery SMR Case Performance Results.

Description: H_2 SMR Case for Refinery Capacity: 50,000 Nm^3/hr hydrogen production		
Fuel Type Used		RFG
Fuel Consumption	t/h	2.33
Fuel Consumption	MW	33.58
Fuel Consumption(3)	GJ/h	120.90
Heat Rate (LHV)	GJ/MWh	n/a
Efficiency (LHV)	% points	n/a
Flue Gas CO_2 Concentration	mol %	19.60
Process CO_2 Balance		
CO_2 In Flue gas	kg/h	42195
CO_2 to Capture Plant	kg/h	40085
CO_2 Captured	kg/h	36067
CO_2 to Water Streams	kg/h	6
CO_2 Emitted from Capture Plant	kg/h	4012
CO_2 Emitted from Stack	kg/h	6128
CO_2 Capture across Capture Plant	%	90%
CO_2 Capture from Total Flue Gas	%	85%
Power for Capture Plant	MW	0.70
Power for Compression & Dehydration	MW	3.94
Power for Steam Package	MW	0.05
Power for PI Balance of Plant	MW	0.00
Total Power Usage	MW	4.69
Steam for Capture Plant	kg/h	41000
Steam for Compression & Dehydration	kg/h	91
Total Steam Usage	kg/h	41091
Cooling Water for Capture Plant	t/h	3000
Cooling Water for Compression & Dehydration	t/h	516
Cooling Water for Power Island	t/h	0
Total Cooling Water Usage	t/h	3516
Additional CO_2 Generated for Utilities		
Assumed grid power carbon footprint	kg/MWh	600
CO_2 due to Power use(1)	kg/h	2813
CO_2 due to Steam use(2)	kg/h	5181
Total CO_2 Balance		
Total CO_2 Generated (including utilities)	kg/h	50188
Total CO_2 Captured	kg/h	36067
Total CO_2 Emitted	kg/h	14121
Overall CO_2 Captured	%	71.9%
Overall CO_2 Avoided	%	66.5%
Overall CO_2 Avoided	kg/h	28073
Total Steam Usage per unit CO_2 Avoided	kg/kg	1.46
Total Steam Usage per unit CO_2 Captured	kg/kg	1.14
Specific CO_2 Emissions	g CO_2/Nm^3H_2	282
Specific CO_2 Emissions Avoided	g CO_2/Nm^3H_2	561

1. Carbon footprint for power based upon typical grid-average.
2. Carbon footprint due to steam use calculated from outline simulation of the auxiliary boiler package.
3. Fuel consumption is given on an LHV basis.

Table 2. Key Metrics of post-combustion CO_2 Capture applied to SMR.

Characteristic	Units	Post C.- Amine (MHI)
Fuel Type		RFG
H_2 Production	Nm3/hr	50,000
Incremental Fuel - LHV	GJ/hr	120.9
Incremental Power	MW	4.7
CO_2 Intensity on H_2 Production (wo/Capture)	kg/Nm^3	0.84
CO_2 Intensity on H_2 Production (w/Capture)	kg/Nm^3	0.28
CO_2 Generated	Mt/yr	0.33
Less: CO_2 Captured	Mt/yr	0.28
CO_2 Emitted by Plant	Mt/yr	0.05
Reference Emissions - No CCS	Mt/yr	0.33
Less: CCS Emissions	Mt/yr	0.05
Less: Capture Power Use	Mt/yr	0.02
Less: Capture Steam Use	Mt/yr	0.04
Avoided Emissions	Mt/yr	0.22

Table 3 below shows the key financial results. % avoided is defined as the mass of CO_2 avoided divided by the mass of CO_2 generated in the base reference case without CCS. The Levelized Capture Cost is based on levelized incremental costs rather than on first year costs. The Additional Yearly Costs show the cost associated with capturing CO_2 each year. It includes an amortized annual cost for capital recovery.

Table 3. Key Financial Results.

Characteristic	Units	Post C.- Amine
Incremental Fuel	MUSD/yr	4.8
Incremental Capital	MUSD	117.1
Incremental O&M	MUSD/yr	11.4
% Capture	%	85.5%
% Avoided	%	66.5%
Capture Cost	USD/tCO_2	95.9
Avoided Cost	USD/tCO_2	123.3
Levelized Capture Cost	USD/tCO_2	112.1
Levelized Avoided Cost	USD/tCO_2	144.0
Additional Yearly Cost	MUSD/yr	27.3
Incremental Cost of H_2	USD/Nm^3	0.07
Incremental Cost of H_2	USD/t	769

Figure 4 shows the cost breakdown to capture a tonne of CO_2. Capital costs tend to dominate the capture costs. Transportation and storage are expected to represent a small contribution to the cost.

Figure 5 shows the cost breakdown for the avoided cost of CO_2. Thus from Figures 4 and 5 it is clear that CAPEX plays an important role in the capture and avoided cost, accounting for almost 40% of the total. O&M costs (~22%) and then fuel (16%) costs are also important in their contribution to the avoided costs.

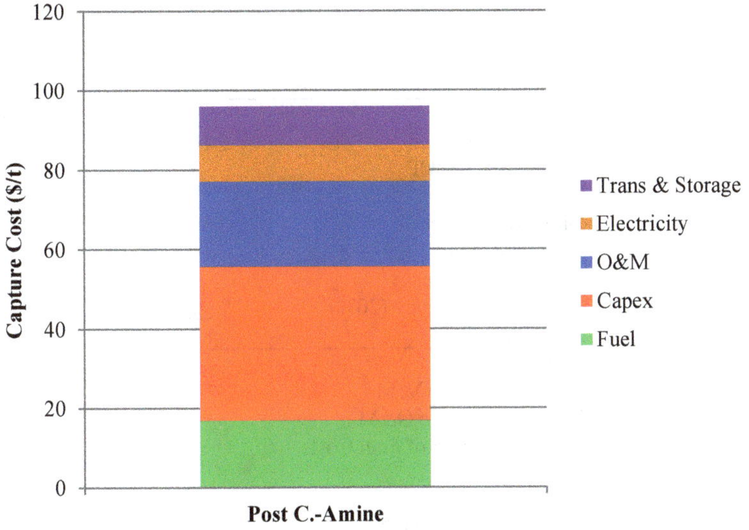

Figure 4. Capture Cost Components.

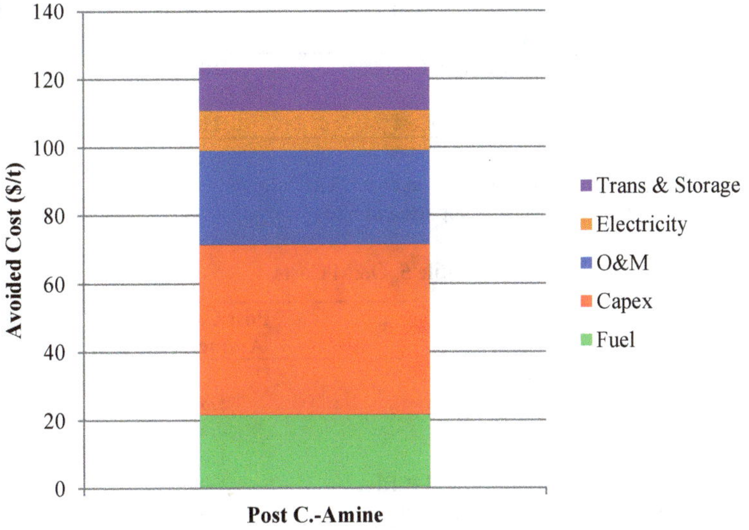

Figure 5. Avoided Cost Components.

Table 4 shows the components that contribute to the total incremental capital cost to capture CO_2. The top part of the table shows the direct costs for each case. This is followed by indirect Outside Battery Limits (OSBL) costs. The base capital costs have been adjusted to include escalation from Q1 2009 (the original estimate basis date) to the assumed in-service date of Q1 2014. This is followed by estimates for contingencies. To this, estimates for Owners Costs are added. Finally AFUDC during construction are included. The bottom row shows the total as spent costs expected just before the plant is commissioned.

Table 4. Capital Cost Components.

Cost	Post C.-Amine
Capture Plant	25.68
Compression Plant	16.95
Steam Gen & BOP	3.09
OSBL (30%)	13.71
Direct & Indirect	59.43
Escalation from 2009 to 2014	10.81
Process Contingency	3.04
Project Contingency (20%)	14.65
Total Plant Costs	87.91
Owners Costs	
6 Months Fixed O&M	2.26
1 Month Variable O&M	0.35
25% of 1 Month of Fuel Costs	0.10
2% of TPC	1.76
60 Days of Cons.	0.27
Spare Parts (.5% TPC)	0.44
Land ($3,000/acre)	0.02
Financing Costs (2.7% TPC)	2.37
Other Costs (15% of TPC)	13.19
Total Owners Costs	20.74
Total Overnight Costs	108.65
AFUDC	8.47
Total As-Spent Cost	117.13

Table 5 shows the operating and maintenance costs. The following O&M costs have been adjusted for escalation or capacity factor and are reported in $2014.

Table 5. O&M Costs.

Cost	Post C.-Amine
Fixed Costs	
Direct Labour	1.37
G&A	0.42
Maintenance	1.64
Insurance and P Taxes	1.09
Total Fixed Costs	4.51
Variable Costs	
Capture & Compression	4.18
Transportation	1.92
Storage	0.83
Total Variable Cost	6.93
Total O&M	11.44

Figure 6 shows how the cost of capturing CO_2 varies as the cost of natural gas changes. There is an increase in capture cost as the cost of natural gas increases, since there is significant fuel used to produce steam required to capture CO_2. The slope of the line is related to the fuel usage.

Figure 7 shows how the avoided costs change as gas prices change.

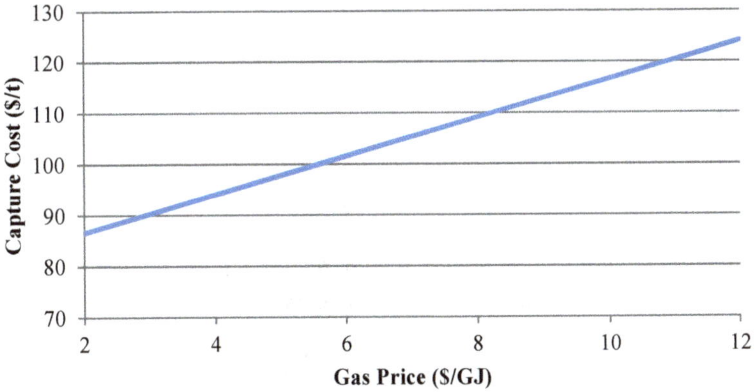

Figure 6. Capture Cost vs Gas Price for Selected Technologies.

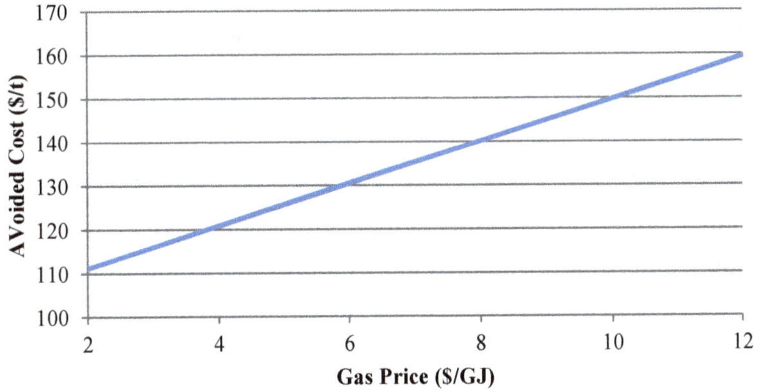

Figure 7. Avoided Cost vs Gas Price for Selected Technologies.

Figure 8 shows how the avoided cost changes with changes in the CO_2 emission intensity of the electric power from the grid. As discussed previously, if the CO_2 emission intensity drops below 0.5 t/MWh the power price is assumed to be a function of the gas price. This adjustment is meant to show that lower CO_2 emission intensities for grid power are likely a reflection of strong reliance on natural gas in the generating mix and therefore in such circumstances gas price would be expected to have a significant effect on power prices. As the CO_2 emission intensity of the grid power increases this indicates a likely higher reliance on coal. Therefore the price of power in the market may be less reliant on gas price.

Figure 9 shows how the avoided costs change as the price of power changes.

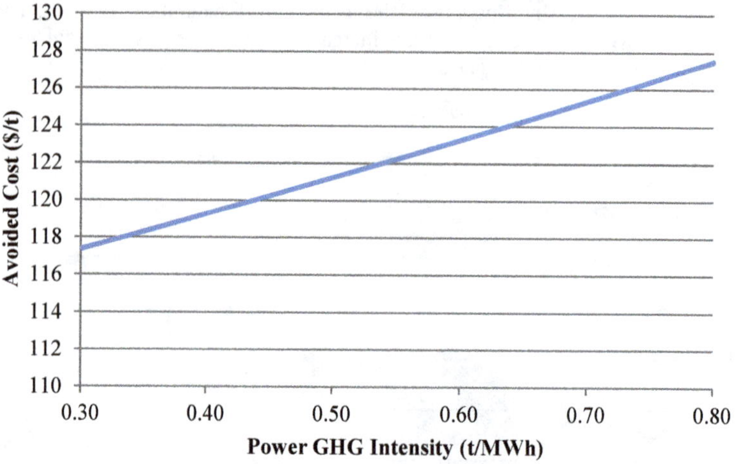

Figure 8. Avoided Cost vs Power GHG Emission Intensity for Selected Technologies.

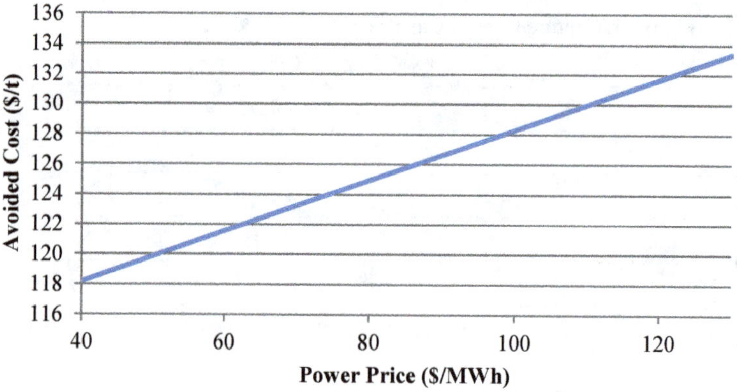

Figure 9. Avoided Cost vs Power Price.

CONCLUSIONS

This study shows that post-combustion capture from SMR furnace flue gas can achieve a high degree of capture. The CO_2 avoidance in this case was a modest 66.5% due to the choice to not capture emissions from the capture process boiler. A different basis would result in a higher capture fraction and some increase in capture CAPEX, but would likely move the high avoided cost closer to the capture cost. Post-combustion achieves a higher extent of capture from an SMR than only pre-combustion capture. Due to the small scale (0.3 Mt/y capture rates), capture costs do not achieve the economy of scale of typical large post-combustion capture systems. Hence the CO_2 avoided cost for these units are rather high. Economic sensitivity analyses show that the CAPEX contribution to the avoided cost plays an important role. O&M costs and fuel costs play a medium role in the cost. The capture cost is moderately sensitive to gas prices and not strongly affected by power price or power CO_2 intensity.

Finally, if a high CO_2 capture rate from the HMU is not needed, pre-combustion capture only may deliver lower CO_2 avoided cost. However, the final CO_2 avoided will be less than 50% of the overall HMU carbon footprint.

RECOMMENDATIONS FOR FUTURE WORK – THE VALUE OF INTEGRATON

PSA separation technology is used in many refineries to generate high purity, reactant grade hydrogen for various uses. In present CCP analysis, pre-combustion approaches were assessed for refinery heaters and boilers, as discussed in Chapter 8. Those studies focused on producing low carbon fuel. Therefore, the PSA units in those cases were optimized for recovery of heating value in the fuel gas and not for hydrogen purity. There are opportunities to combine production of decarbonized fuel for heaters and boilers, with generation of high purity hydrogen. The cases in Chapter 8 illustrated that the volume of hydrogen produced for use as decarbonized fuel can be easily an order of magnitude higher than the demand for pure hydrogen used for various refinery hydroprocessing requirements as discussed in this chapter. Hence, a "polishing PSA" step can be added to treat a slip stream of the "hydrogen fuel" stream to produce a 99.9%+ purity hydrogen stream to debottleneck an existing hydrogen plant or replace an older hydrogen plant. Figure 10 below illustrates how the refinery heaters and boilers Pre-C.-PSA/Cryo case might be modified to provide higher purity hydrogen optionality. The tail gas of the polishing PSA may be recycled to the fuel gas PSA or to the cryogenic purification step, depending on where the concentration is most compatible. If only a small slip stream is processed, the tail gas can be simply mixed back into the fuel gas, with some slight changes to the specification of the fuel PSA.

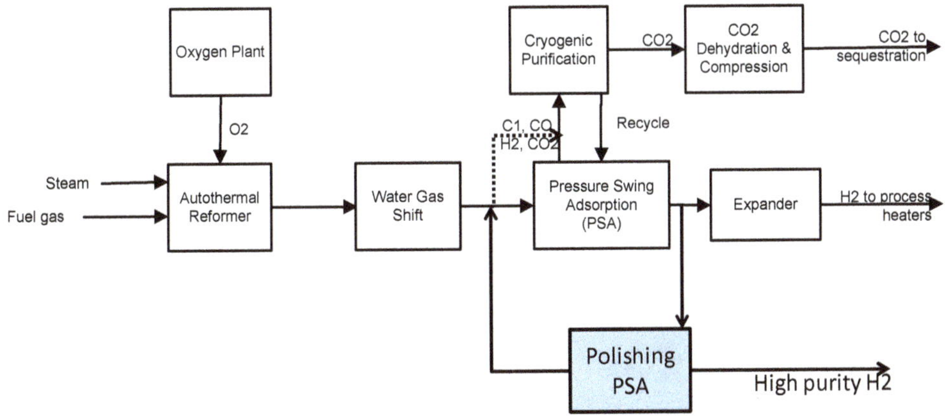

Figure 10. Process scheme to integrate fuel H_2 production for heaters/boilers and process H_2 replacing SMR.

Other integration/optimisation opportunities include:

- Combined capture of SMR furnace and "CO_2 capture" steam boiler flue gasses.
- Capture of SMR, steam boiler, process and utilities flue gases in dedicated absorbers with combined regeneration and compression.
- Use of excess steam in existing steam systems, potentially coupled with a "primary load shedding" position with a modest impact on overall capture system availability.
- Use of a higher capture ratio (95%) in the capture plant.

REFERENCES

1. Straelen J. van, Geuzebroek F., Goodchild N., Protopapas G., and Mahony L.; "CO_2 capture for refineries, a practical approach;" *Energy Procedia* 1 179-185 (2009).
2. Anderson N. U. and Olsson H.; "The hydrogen generation game" *Hydrocarbon Engineering*, 49-52 July (2011).

Carbon Dioxide Capture for Storage in Deep Geological Formations, Volume 4
Karl F. Gerdes (Editor)

Section 1

CO_2 CAPTURE

Heavy Oil

Carbon Dioxide Capture for Storage in Deep Geological Formations, Volume 4
Karl F, Gerdes (Editor)

Chapter 13

TECHNOLOGIES FOR CO_2 CAPTURE FROM STEAM GENERATORS IN THE CANADIAN OIL SANDS

Dan Burt[1], Iftikhar Huq[1] and Jonathan Forsyth[2]
[1]Suncor Energy, 150-6th Avenue SW, Calgary, AB, Canada
[2]BP International Limited, ICBT Chertsey Road, Sunbury-on-Thames, Middlesex, TW16 7LN, United Kingdom

ABSTRACT: Thermal recovery of Canadian oil sands can produce large CO_2 emissions. These emissions are mainly from combustion operations, such as steam generation, and typically are in the form of flue gas at low pressure and low CO_2 concentration. CO_2 capture is technically feasible but the cost is high. CCP3 work reported here includes the development and demonstration of oxy-firing for CO_2 capture from once-through steam generators, and further development of natural gas-fired chemical looping combustion. The remoteness, ambient conditions and engineering/construction resource limitations affecting all capital projects in the Canadian oil sands mean that none of the capture technologies is economically feasible for deployment at this time. Future RD&D should focus on technology that has the potential for large economic performance improvement compared to the state-of-the-art.

KEYWORDS: heavy oil; oil sands; Steam Assisted Gravity Drainage (SAGD); post-combustion capture; oxy-firing; Once Through Steam Generator (OTSG)

INTRODUCTION

The province of Alberta in Canada has proven oil sands reserves of around 168 billion barrels. This represents the third-largest proven crude oil reserves in the world, after Saudi Arabia and Venezuela. Oil sands are either loose sand or partially consolidated sandstone containing a naturally occurring mixture of sand, clay, and water, saturated with a dense and extremely viscous form of petroleum technically referred to as bitumen. In Alberta the oil sands underlie 142,200 square kilometres of land in the Athabasca, Cold Lake and Peace River areas (Figure 1). About 3% of the oil sands area in Alberta can be currently mined, whereas 97% of the oil sands are too deep to be mined and must be extracted using other methods [2].

Figure 1. Location of Canadian Oil Sands [1].

Thermal Recovery

The deeper unmineable oil sands reserves, mentioned above, are typically produced using thermal methods that involve injecting steam into the reservoir to reduce the viscosity of the bitumen and allow it to flow through production wells to the surface for processing. The produced fluid typically comprises a mixture of bitumen, condensed steam, ground water, water vapour and hydrocarbon gas. Processing facilities on the surface separate the bitumen, water and gas. The bitumen may be mixed with a light hydrocarbon diluent, or alternatively, some upgrading may be performed to make it suitable for pipeline transportation to a refinery. The water is treated and recycled for steam production, whilst the produced gas is conserved, mixed with imported natural gas and used as fuel in the steam generation process.

Steam Generators

The steam generators typically employed in oil sands operations are designed to produce high pressure steam from feed-water containing a high proportion of dissolved solids and need to perform reliably in the field conditions of Alberta in ambient conditions ranging from -40 C to +35 C. Most operators use a steam generator design which has once-through forced water circulation (because of dissolved solids) with water tubes arranged in a series of coils around a gas-fired furnace. The oil sands operations are typically sited in remote locations and present particular challenges in terms of logistics. The Once-Through Steam Generator (OTSG) is executed as a modular concept which is mostly shop-fabricated and shipped to site for final assembly (Figure 2).

Figure 2. Source: Firebag SAGD Facility, Suncor Energy Inc., October 19, 2011.

CO_2 Emissions

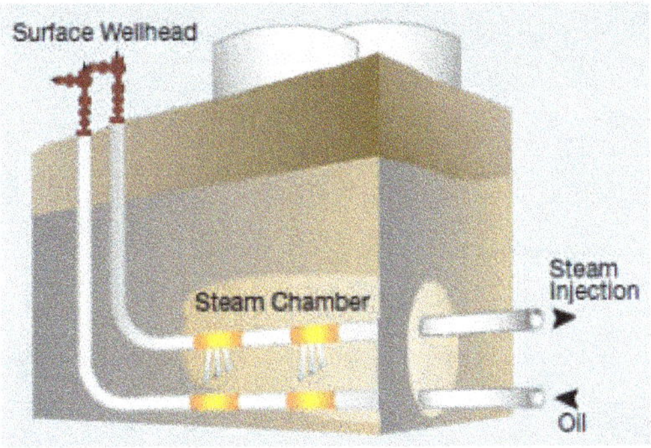

Figure 3. SAGD Well Pair and Steam Chamber. [1]

A typical nominal OTSG rating is 250 million BTU per hour (lower heating value basis) producing steam at 110 bar(a) and 78% steam quality. Such a unit would emit around 64 tonnes of CO_2 per hour in the stack exhaust at a concentration around 9%. Typical oil sands production operations use multiple OTSGs to support a thermal recovery technique known as Steam Assisted Gravity Drainage (SAGD) in which the steam is injected continuously into the reservoir to create and maintain a well-defined underground steam chamber in the unconsolidated bitumen-bearing sands. Such operations may emit of the order of 0.5 to 2 million tonnes of CO_2 per annum at typical bitumen production facility rates, and the number of these operations is increasing as more production capacity is developed. Recent published figures show total greenhouse gas emissions from all mined and steam-extracted oil sands production operations to be around 55 million tonnes of CO_2 (equivalent) per annum with CO_2 emissions from the combustion of fuel for SAGD making up a growing share of the total.

THE OIL SANDS APPLICATION SCENARIO IN CCP3

CCP3 studied CO_2 capture from various processes that were representative of the main emissions sources from member company operations. One of the application scenarios developed was based upon Canadian oil sands production using OTSGs for SAGD. The Canadian oil sands application scenario considered a production block located in the Fort McMurray area of Northern Alberta, comprising four natural gas-fired OTSGs, each rated 250 million BTU per hour (fired duty). This application scenario is the basis for CCP3 participation in a number of activities, the most significant of these is a field-based oxy-firing CO_2 capture technology demonstration project. CCP3 also supported the further development of gas-fired chemical looping combustion technology for this application. These efforts were complemented by a set of CO_2 capture techno-economic studies, using state-of-the-art and emerging technologies, which provided insight into performance and cost.

Oxy-firing CO_2 Capture Demonstration Project

Oxy-firing is a well-established concept for CO_2 capture. Fuel is burned with high-purity oxygen, rather than with air, to minimize dilution of CO_2 with nitrogen. The resulting exhaust gas consists predominantly of a mixture of CO_2 and water vapour, from which the water can easily be separated by cooling and condensation. The oxy-firing concept can be applied to gas-fired OTSGs by supplying high-purity oxygen from an air separation unit to the burner. In common with most other oxy-firing applications, a proportion of the CO_2-rich exhaust gas is recirculated to the combustion chamber to provide dilution and reduce the flame temperature to a level similar to burning the fuel in air. This recirculation results in flow and heat transfer in the oxy-firing case similar to that produced when air is used as the oxidant. This similarity in operating conditions means that it is reasonable for existing OTSGs to be retrofitted for oxy-firing service without extensive changes. Consequently, both oxy-firing and conventional air-combustion capability can be retained as alternative operating modes. An attractive co-benefit of this system is the relatively small foot-print of the added equipment required to be placed at the OTSG location. Oxygen must be piped-in, recirculation ducting must be added with a suitable fan and control/isolation dampers, and CO_2-rich exhaust gas must be piped-out to create a functional system (Figure 4). This creates a smaller footprint adjacent to the OTSG than would be required for post-combustion CO_2 capture that, in turn, would require large vessels/columns close to the OTSGs.

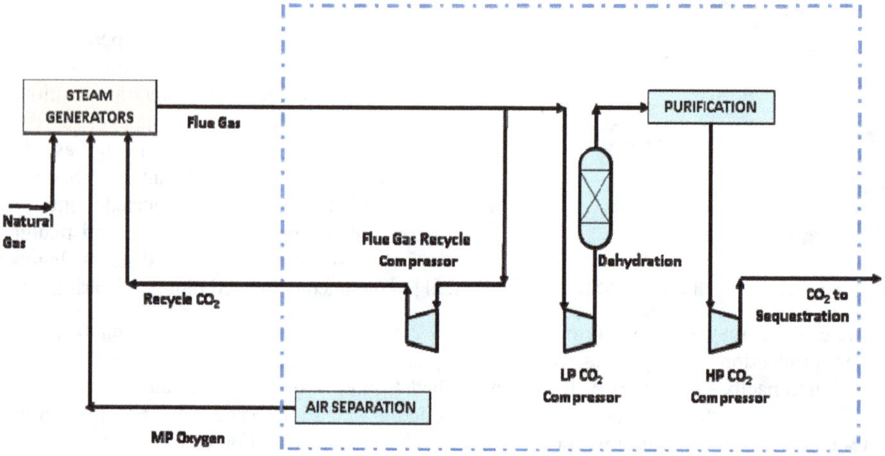

Figure 4. Oxy-Fired Steam Generators for Heavy Oil / Oil Sands New Build Case.

To advance the technology readiness level of oxy-firing OTSGs, CCP3 supported burner design and testing and the integrated design of an oxy-firing OTSG which is discussed in greater detail in Chapter 14 of this volume. This work then lead to the design and construction of a field-based oxy-firing demonstration utilizing a retrofitted operational 50 million BTU per hour OTSG (Figure 5). The burner and oxygen systems were provided by Praxair, and the demonstration was hosted at the Cenovus oil sand production site located at Christina Lake, Alberta. This significant demonstration project is described in detail in Chapter 15 of this volume.

Figure 5. Cenovus Energy Inc., Oxy-fired 50 million BTU per hour Test Boiler.

Chemical Looping Combustion Technology Development

Chemical Looping Combustion (CLC), a variation on oxy-firing, is a promising technology for CO_2 capture that involves the use of a regenerable solid material as an oxygen carrier, to produce conditions for combustion in the absence of atmospheric nitrogen. CLC can be applied to replace gas-fired OTSGs by using gas as the fuel in the CLC process and recovering the heat released to produce steam in a series of heat exchangers (Figure 6).

Previous work in earlier stages of the CCP has shown very promising results for CLC as a CO_2 capture technology. The process has been found to be highly efficient and capable of achieving high rates of CO_2 capture, however the process is very different from conventional combustion and retrofit of CLC to existing OTSGs would not be feasible. CCP3 supported research and development work aimed at identifying and developing suitable oxygen carrier particles for CLC, created a design for a CLC OTSG, modeled performance and estimated the cost of a commercial-scale unit. This work is described in more detail in Chapters 16 to 19 of this volume.

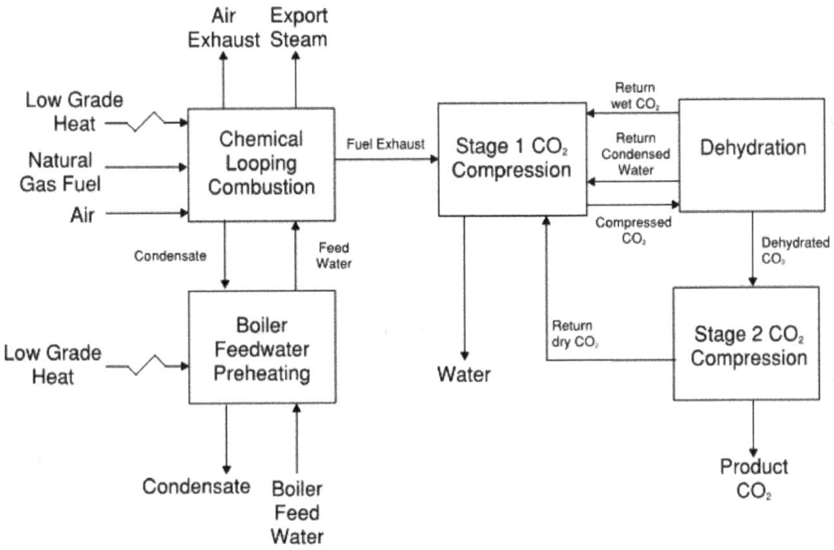

Figure 6. Chemical Looping Combustion Base Case.

Technical and Economic Studies

Current state-of-the-art post-combustion capture uses a suitably selective solvent to absorb CO_2 from a flue gas or other exhaust stream. The CO_2-laden solvent is then removed from contact with the exhaust gas and heated to release the CO_2. The CO_2 is compressed and sent to utilization or storage facilities. The solvent is then cooled and recirculated to the absorption stage of the process. This is a commercially available process available from several technology providers, which uses a suitable aqueous amine solution as the solvent. CCP3 obtained performance and cost data from a leading technology licensor, Mitsubishi Heavy Industries (MHI), and developed a series of comparative engineering studies to investigate the relative performance and cost of this technology applied to the gas-fired OTSG, relative to other capture technologies, including oxy-firing and CLC. While care was taken to achieve as high a degree of accuracy as practical, the rankings and relative values are more important and insightful than the absolute value of the capture and avoidance costs, which depend on the project-specific assumptions chosen for these studies. In order for the absolute value of these study results to be compared to the results of other studies, these assumptions need to be understood.

Pre-combustion CO_2 capture involves chemical conversion of the fuel by reforming or partial oxidation followed by water gas shift reaction to produce a hydrogen-rich fuel and CO_2. This is felt to have better potential for higher efficiency CO_2 capture than lower-pressure post-combustion processes, especially for multiple dispersed sources, if the fuel conversion step can be designed to have lower energy intensity and cost. The CO_2 can be separated at a central location and the hydrogen-rich fuel can be distributed to various combustion units. This technique has the advantage that a large-capacity decarbonisation plant can benefit from economies of scale, and the modification to the combustion units can be minimal with no specific equipment for CO_2 capture needing to be located at the combustion unit. CCP3 studied application of pre-combustion CO_2 capture to OTSGs by evaluating the performance and cost of a large gas-fired autothermal reformer unit producing hydrogen-rich fuel for multiple OTSGs. All of the technical and economic studies performed by CCP3 for the Heavy Oil Scenario are described in more detail in Chapter 20 of this volume.

DISCUSSION OF RESULTS

Oxy-firing Demonstration

Earlier work by CCP identified oxy-firing as a promising approach for CO_2 capture from OTSG operations. To better define the performance and economics, a field demonstration was determined to be the next logical step. This demonstration was carried out in two phases. In Phase I, detailed analyses were completed for both a commercial and test scale OTSG operating under air and oxy-fuel modes. Results of computational fluid dynamic models combined with rigorous heat and mass balance models indicate that oxy-fuel combustion technology can be applied to OTSGs while maintaining key operating parameters (temperature, heat flux) that are similar to operation with air by using flue gas recirculation. Therefore, through proper control integration OTSGs can be retro-fittable and capable of air or oxy-fuel operation.

The successful Phase II Oxy-Fuel Combustion Field Test provided the data confirming the Phase I result that OTSGs can be retrofit for oxy-firing. There have been several key learnings and takeaways that can be applied to oxy-firing projects in the future, particularly those involving OTSG retrofits. Safety in design is critical, because oxygen is explosive in nature and a substance not normally handled in large volumes in field operations in Alberta. Having adequate site support and buy-in from all levels will ensure that the local Operations staff is comfortable taking on a project involving oxygen. Ensuring proper integration between all systems is key, and having a single system control the combustion process is a possible improvement that could be applied to a future project. Cleanliness of the oxygen system is also of concern. Additional learnings from this project revolve around proper attention to the damper controls for air-firing the boiler, transitioning the boiler from air firing to oxy-firing, and collecting data from the boiler during oxy-firing operation.

Chemical Looping Combustion

The main achievements of CCP3 for the CLC technology are:

- Development at the pilot scale of non-Ni based carriers with positive features for further scale-up.
- Design of the next scale 10 MW demonstration unit.
- Design and economic evaluation of a cluster of CLC boilers (4 x 80 MW) for application in the production of steam for oil sands extraction by the SAGD technique.

Overall study results indicate that the CLC CO_2 capture and avoidance costs are competitive with post-combustion capture in the Heavy Oil Scenario in the context of Northern Alberta SAGD applications. Since CLC is a relatively immature technology compared to PCC, there is significant potential for further cost reductions as CLC progresses along its technology learning curve. However, since Northern Alberta is a high-cost environment and CLC is capital intensive, more work must be done to demonstrate clear benefits for selecting the CLC technology over the more proven post-combustion CO_2 capture technologies in this application.

Technical and Economic Studies

CO_2 capture from gas-fired OTSGs was shown to be technically feasible using pre-combustion, post-combustion and oxy-firing technologies. Capture rates of at least 90% were found to be possible. However the economic feasibility of CO_2 capture in the oil sands application is particularly challenged. Northern Alberta is a high-cost location due to the remoteness, challenging physical environment and large number of concurrent new project developments. This is reflected in CO_2 capture costs which are particularly high – ranging between \$200 and \$300 /tonne under the parameters of this comparative study.

Including the carbon footprint of utilities used by the carbon capture system had a significant impact on the overall CO_2 avoidance level. This is particularly significant for oxy-firing, where the assumptions made for the carbon intensity of the electric power can have a strong effect on the CO_2 avoided cost, because the large power usage by the air separation unit. In general, the capital cost is clearly a function of the total CO_2 captured, but this is far from being the only factor making a significant impact. The cost per tonne of capturing CO_2 is shown to decrease with increasing CO_2 capture flow rate - demonstrating a strong economy of scale effect. This economy of scale effect does have a practical limit, depending on the specific technology in question, beyond which different equipment or process flow schemes and parallel trains may be required. The total capital cost of the CO_2 capture plant also varies with the concentration of CO_2 in the flue gas.

The operating costs show that there is significant fuel required for additional steam generation needed by the capture equipment itself, as well as the cost of electrical power compared to other O&M cost components. Variations in local fuel and electrical power prices will therefore significantly affect the total operating costs at each project site. CO_2 concentration in the flue gas has minimal impact on operating costs and a lower impact on capital costs than might be expected.

SUMMARY OF CONCLUSIONS

Thermal recovery techniques used in the production of oil sands are major emitters of CO_2. A large proportion of the CO_2 emissions from such operations arise from the combustion of natural gas in Once Through Steam Generators (OTSGs). The exhaust gases from these units are at low pressure (more or less ambient) and the CO_2 concentration is low (less than 10%) with the remainder of the exhaust gases comprising mainly atmospheric nitrogen. The low pressure and dilute nature of the CO_2 means that, though technically feasible to capture about 90% of CO_2 emitted on-site, post-combustion CO_2 capture technology options applied to OTSGs inherently involve large equipment and have large energy requirements. The large size of the capture equipment results in high capital cost and the large energy requirements push up operating cost. The ambient conditions, remote location and engineering/construction resource limitations found in the Canadian oil sands region combine to further push CO_2 capture costs to some of the highest evaluated by CCP3. To address the challenge of high CO_2 capture cost, CCP3 progressed development of oxy-firing technology for application to OTSGs. The field demonstration showed that there are no technical limitations for retrofitting an OTSG for oxy-firing and no operational issues were encountered during oxy-firing. However, in the CCP3 study, the economic performance of oxy-firing did not show any benefit compared to post-combustion CO_2 capture. The capital cost of the cryogenic air separation unit and the assumed CO_2 footprint of the electric power required to run it were significant factors in the evaluation. To address the cost challenge posed by the cryogenic air separation unit, CCP3 also continued with the development of Chemical Looping Combustion (CLC) technology. The results showed progress towards the development of a viable industrial process which could be applied to

OTSGs with high levels of CO_2 capture and overall process efficiency compared to post-combustion capture alternatives. However, the CLC OTSG is a complex process unit, and capital costs were found to be somewhat higher than those for post-combustion capture technologies. Hence, for future CLC development, focus needs to be on reduction of complexity and cost.

In summary, CCP3 found CO_2 capture from the Canadian oil sands technically feasible, but with particularly high capital costs. The driver for the high capital cost was not specific to CO_2 capture but inherent to all capital projects in the Canadian oil sands. To become economically viable for deployment in this application scenario, CO_2 capture costs need to be reduced significantly. Hence, future RD&D effort should focus on technologies with the potential for large economic performance improvement compared to the state-of-the-art.

REFERENCES

1. Canadian Association of Petroleum Producers, (June, 2013), "About Canada's Oil Sands", http://www.capp.ca/library/publications/crudeOilAndOilSands/pages/pubInfo.aspx?DocId=228182
2. Alberta Government, (Nov. 2014), "Oil Sands Greenhouse Gases" Fact Sheet, http://oilsands.alberta.ca/FactSheets/Greenhouse_Gas_factsheetNov_2014.pdf

Carbon Dioxide Capture for Storage in Deep Geological Formations, Volume 4
Karl F. Gerdes (Editor)

Chapter 14

OXY-FUEL / CO_2 CAPTURE TECHNOLOGY FOR ONCE-THROUGH STEAM GENERATORS

Stefan Laux[1], Minish Shah[1], Kenneth Burgers[1],
Todd Pugsley[2], Dan Burt[2], Iftikhar Huq[2] and Mark Bohm[2]
[1]Praxair Inc., 175 East Park Drive, Tonawanda, NY, USA
[2]Suncor Energy, 150-6th Avenue SW, Calgary, AB, Canada

ABSTRACT: In Alberta, Once-Through Steam Generators (OTSGs) are expected to be one of the largest sources of growth in GHG emissions for in-situ bitumen recovery as operations are expanded. The CO_2 Capture Project (CCP), oil industry partners, and Praxair developed a multiphase program to evaluate oxy-fired combustion as applied to OTSGs with the objective of developing a reliable technical solution for capturing CO_2 at the lowest cost. Phase I results from assessment studies of commercial scale oxy-firing of OTSGs and design of a retrofit for a pilot scale demonstration of oxy-firing (Phase II) and CO_2 purification (Phase III) are documented. Phase I studies confirm that CO_2 capture by oxy-firing of OTSG is technically feasible and potentially a cost-competitive alternative to post-combustion capture. With the Phase II demo run now completed, the next steps are to evaluate the data and incorporate the learnings into the design of the Phase III test, and then execute Phase III if Phase II results generate sufficient interest.

KEYWORDS: oxygen firing; burner design; air separation

INTRODUCTION

The potential advancement of environmental policy to reduce Greenhouse Gas (GHG) emissions has prompted the oil industry to evaluate CO_2 capture options at their Canadian in-situ bitumen recovery operations. In Alberta, Once-Through Steam Generators (OTSGs) are expected to be one of the largest sources of growth in GHG emissions as operations are expanded. Under the leadership of the CO_2 Capture Project (CCP), leading oil producers and Praxair developed a program to evaluate oxy-fuel combustion as applied to OTSGs with the objective of developing a reliable technical solution for capturing CO_2 at the lowest cost.

This chapter summarizes Phase I of a multiphase project to demonstrate that oxy-fuel combustion is a safe, reliable and cost-effective technology for CO_2 capture from OTSGs. There are three main tasks in Phase I [1]:

- Task 1: Commercial boiler design basis and cost estimate – the cost and performance were evaluated for retrofitting commercial-scale OTSGs with oxy-fired combustion for CO_2 capture. The results were presented at conferences [2, 3].
- Task 2: Pilot Boiler Design Basis – the technical feasibility was assessed and a detailed design developed in preparation for Phase II, the pilot scale demonstration of oxy-fired combustion by retrofitting on an existing OTSG test boiler.

- Task 3: Pilot oxy-fired combustion with purification and compression – developed the engineering design, cost estimates and test plan for the demonstration-scale CO_2 processing unit (CPU).

Phase II is a short-duration oxy-fired combustion demonstration operation using a 50 MMBTU/hr pilot OTSG at Cenovus' Christina Lake facility, described in Chapter 15 of this book. It was completed in Q2 of 2015. The goal of Phase III is to demonstrate long-term continuous operation of a CO_2 compression and purification system connected to a single 50 MMBtu/hr test OTSG retrofitted for oxy-fuel operation.

The main objective of the work described in this chapter was to develop an accurate design basis and realistic cost estimates for both pilot and commercial scale OTSGs in heavy oil extraction service. The three Phase I Tasks were successfully completed in 2010 and showed potential for oxy-firing OTSGs to be competitive with traditional post-combustion capture technology.

Oxy-fuel technology uses high purity oxygen produced by an air separation unit (ASU), instead of air, for combustion in a modified boiler, thus producing a flue gas stream concentrated in CO_2 and water vapour. This flue gas stream requires minimal clean-up in a carbon dioxide processing unit (CPU) to produce CO_2 suitable for sequestration. The three main system components - the ASU, boiler, and CPU - must be carefully optimized in order to develop a cost effective, safe, reliable, integrated system. Figure 1 illustrates the difference between air and oxy-fuel combustion.

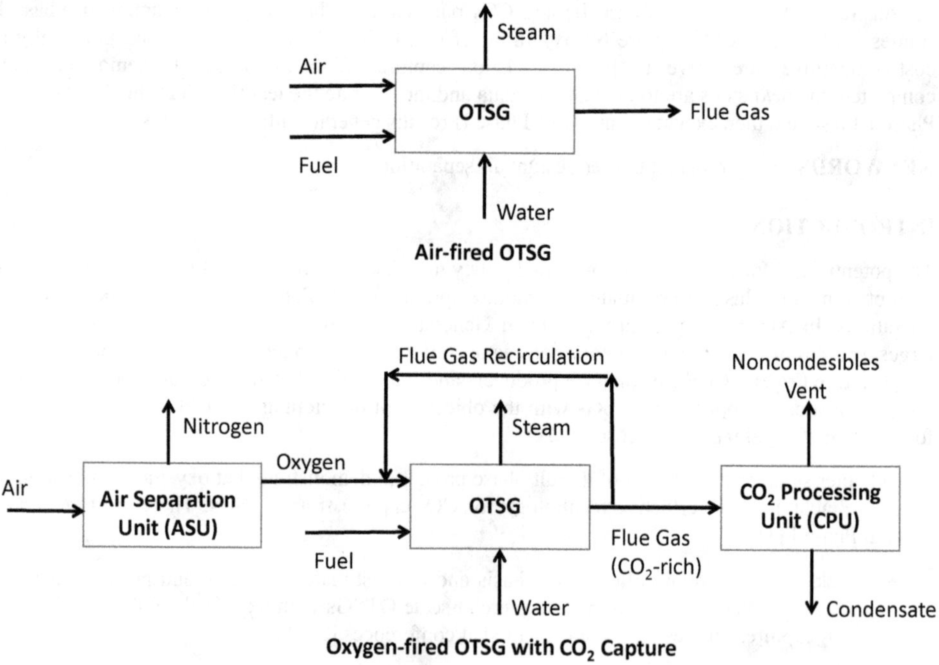

Figure 1. Schematics of Air-fired OTSG and Oxy-fired OTSG with CO_2 Capture.

More detail for the three phase program to fully evaluate the applicability of oxy-fuel combustion to SAGD OTSGs is shown in Table 1.

Table 1. Program Phases.

<table>
<tr><td colspan="7">Phase I (Process Design/Cost)</td></tr>
<tr><td rowspan="3">Design Basis</td><td>Commercial Boiler</td><td>CFD Modeling</td><td>Process Optimization</td><td colspan="2">Process Design Oxy-fuel Combustion/CO_2 Capture & Compression</td><td>Capital/Operating Estimates</td></tr>
<tr><td rowspan="2">Test Boiler</td><td>CFD Modeling</td><td colspan="4">Develop Process Design / Costs / Plan for Oxy-fuel Demonstration (Phase II)</td></tr>
<tr><td colspan="5">Develop Process Design / Costs for CO_2 Capture & Compression (Phase III)</td></tr>
<tr><td colspan="7">Phase II (Oxy-fuel Combustion Test)</td></tr>
<tr><td>Test Boiler</td><td>Final Process / Equipment Design & Costing</td><td>Control System Design & Integration</td><td>Safety Assessment</td><td>Procurement / Fabrication</td><td>Installation & Commissioning</td><td>Start-up / Testing</td></tr>
<tr><td colspan="7">Phase III (CO_2 Capture Demonstration)</td></tr>
<tr><td>Test Boiler (with CO_2 capture)</td><td>Process / Equipment Design & Costing</td><td>Control System Design & Integration</td><td>Safety Assessment</td><td>Procurement / Fabrication</td><td>Installation & Commissioning</td><td>Start-up / Testing</td></tr>
</table>

Phase I of this program was initiated in October 2009. The main participants included: the CCP (BP, Chevron, ConocoPhillips, ENI SPA, Petrobras, Shell and Suncor), Devon, Cenovus Energy and Statoil. Phase II began in late 2010/early 2011 with funding from the Climate Change and Emissions Management (CCEMC) Corporation and the stated participants. The Phase II operations were completed in April, 2015. Phase III is planned to commence after Phase II, depending on results obtained from that phase. Additional participants may be added.

PHASE 1 TASK 1: STUDY OVERVIEW OF COMMERCIAL SYSTEM

Oxy-fuel Technology Overview

An existing set of four OTSGs, capable of producing a total of 12,000 tonnes/day of steam, was selected as the basis for study of a Commercial System. The specific facility chosen for study is Suncor's Firebag operation near Fort McMurray, Alberta. The cost and performance were evaluated for retrofitting the OTSGs with oxy-fuel combustion for CO_2 capture. Schematics showing the major equipment and system requirements for OTSGs using (a) air-fired combustion and (b) oxy-fuel combustion are shown in Figure 1.

Air-fired OTSG

In air-fired operation, air and fuel enter the boiler burner(s). The fuel is burned to produce wet steam inside heat exchange tubes located within the boilers. The combustion products (flue gas) exit the boilers and are vented to atmosphere. The flue gas contains mostly nitrogen, and carbon dioxide, water vapor, argon, and oxygen plus trace components. Due to the high N_2 content, commercially available technology for CO_2 purification of this gas is limited to solvent based absorption systems, such as amines. However, amines are very energy and capital intensive. Since oxy-fuel technology generates flue gas with very high concentration of CO_2, which is relatively easy to purify for sequestration, the technology holds promise for lower costs.

Oxy-fired OTSG with CO_2 Capture

In oxy-fuel operation, a cryogenic air separation unit (ASU) separates air into an oxygen product stream for combustion and a waste nitrogen stream which is vented to atmosphere. The boiler produces flue gas, consisting largely of water vapor and CO_2, which is split into two streams. The majority of the flue gas is recirculated back through the boiler to maintain the boiler mass flow rates near those occurring in air-fired operation, and the balance is sent to the CO_2 Processing Unit (CPU). Most of the water vapour in the flue gas is condensed in the CPU and exits as slightly acidic condensate. In the CPU, impure CO_2 is compressed and processed by partial condensation or distillation to separate inert gases and to produce CO_2 at a purity of either >95% or >99.5%, respectively. The inert gases are vented to atmosphere. The purified CO_2 product is further compressed to the pressure required for pipeline transport and underground sequestration.

Although air separation, oxy-fuel combustion, and CO_2 processing have been practiced for many years, the specific processes to be implemented represent significant state-of-the-art advances. The ASU will use an advanced process optimized across the entire system to produce oxygen with the best purity and pressure for low energy consumption and overall lowest CO_2 capture costs. The oxy-fuel combustion process will incorporate advanced burner technology. The CPU will incorporate an auto-refrigeration process that eliminates the need for large external refrigeration. The entire process will be integrated and optimized to minimize overall costs, while maximizing safety and reliability.

Detailed process descriptions of the Air Separation Unit, Boiler Modifications, and CO_2 Processing Unit are given below.

Design Basis and Assumptions

The ASU and CPU will be retrofitted to an existing SAGD installation comprised of four OTSGs, producing a total of 12,000 tonnes/day of steam. Figure 2 depicts the conceptual layout of major components necessary to retrofit the East side boilers at Firebag with Praxair's oxy-fuel technology for CO_2 capture. As shown in Figure 2, the ASU and CPU will be located ~ 500 meters from the center of the boilers. They will be connected to the boilers via new pipelines. Equipment used in the ASU and CPU will meet all applicable standards, codes, and safety requirements such as ASME B31.3 used by Praxair in commercial designs of air separation units and CO_2 purification systems. Equipment located within the SAGD facility will meet all applicable standards, codes, and safety requirements for a typical SAGD facility in Alberta, Canada.

Turndown Capabilities

The OTSGs typically run year round at full load supplying steam to the SAGD operations. Occasionally, one of the boilers may be taken off line for maintenance. In such instance, the ASU and CPU will be required to operate at 75% of full capacity. For the purpose of this study, both the air separation unit (ASU) and CO_2 processing unit (CPU) were designed to be single train plants. The single train plants can typically achieve 75% of the design capacity while maintaining high efficiency, which matches well with the requirement of one boiler taken off-line. If further system turndown requirements are anticipated, the ASU and CPU designs can be modified as long as they are identified before the designs are finalized. The turndown to 75% of design capacity is inherent in a single train design. It is possible to achieve a turndown as low as 40 – 50% of design capacity with alternate configurations.

Transition from Oxy to Air Mode

In practice, the retrofitted system will include control system modifications that will automatically transition the boilers from oxy-firing to air-firing as required. A one hour supply of back-up oxygen was included in the ASU scope to allow for transition from oxy-firing to air operation, in the event of an ASU shutdown.

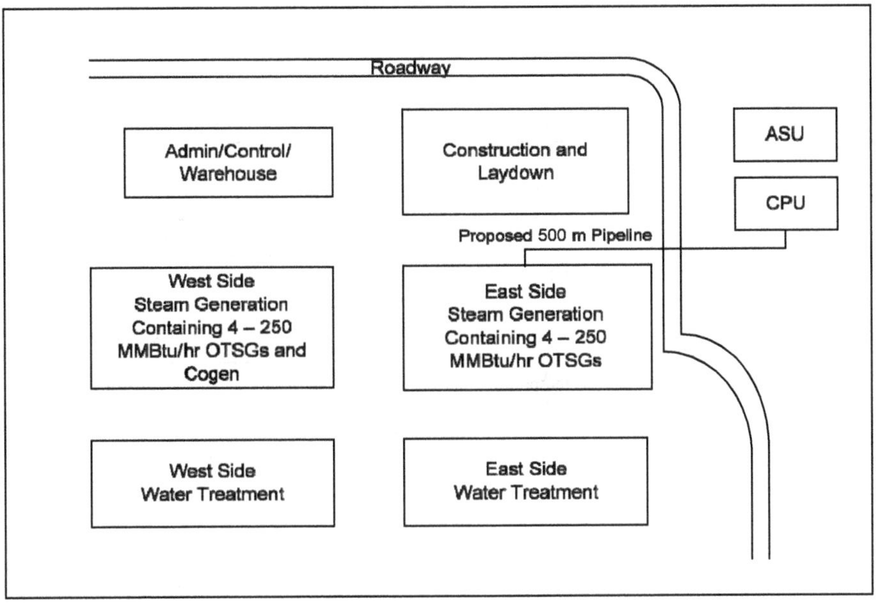

Figure 2. Conceptual Layout of Suncor's Firebag Facility with ASU and CPU.

CO_2 Purity

Praxair offers two types of CPU processes and both options were evaluated. The first type uses a partial condensation process in the CPU cold box to produce CO_2 at a purity of >95%. The second type uses a distillation process to produce CO_2 at a purity of >99.5%. Specifications for the CO_2 product from the CPU are shown in Table 2.

Table 2. CO_2 Product Specifications.

CPU Process	Partial Condensation	Distillation
CO_2, mol%	>95%	>99.5%
O_2, ppmv	TBD	<50
Total SOx, ppmv	<50	<50
Total NOx, ppmv	<50	<50
H_2O, ppmv	<10	<10
Outlet Conditions		
Pressure, psia	2190	2190
Temperature, F	110	110

Site Ambient Conditions

Site ambient conditions are shown in Table 3 and Table 4 and are representative of Northern Alberta, Canada. Equipment will be designed for the hot day conditions described in Table 4. System performance is evaluated for average day conditions.

Table 3. Ambient Conditions – Location and Elevation.

Location - Northern Alberta, Canada	
Seismic Zone	0
Elevation (ft)	1211
Barometric Pressure (psia)	14.1

Table 4. Ambient Conditions – Temperature and Relative Humidity Data.

Design Cases	Hot Day 10%	Avg Day 50%	Cold Day 10%
Dry Bulb Temperature (°F)*	73	36	-7
Wet Bulb Temperature (°F)*	63	33	-8
Relative Humidity (%)	63	75	N/A

* To accommodate the low temperature extremes the system was designed to utilize a water/glycol mix for all cooling streams. Additionally the flue gas is heated before entering an insulated transfer pipe to avoid water condensation and freezing.

Site Utilities

Electrical

Electric power is available at 13.8 KV. No provision has been included to transform from grid voltage to 13.8 KV. Lower voltage needs will be supplied by transformers.

Fuel

In addition to the fuel required to operate the OTSGs, some fuel is required in the ASU and CPU processes to regenerate feed stream contaminant removal systems. This study utilized pipeline natural gas (NG) as fuel, with specifications shown in Table 5. In actual operation, NG may be blended with produced gas, which may contain higher levels of sulphur. To meet the CO_2 purity specification defined earlier, the limit on sulphur content expressed as mercaptans in the combined feed was estimated.

Table 5. Fuel Specification.

Methane	98.87	vol%
Ethane	0.04	vol%
Propane	0.01	vol%
Nitrogen	0.69	vol%
Carbon Dioxide	0.36	vol%
Butane	0.01	vol%
Hexane	0.01	vol%
Assumed Mercaptans	5	ppmv
HHV	1000.2	Btu/ SCF

Note: Fuel specification was obtained from TIW Western.
Standard Cubic Feet is defined as gas at 14.7 psia, 60F, 0% RH.

Cooling System Definition

The cooling system will be a recirculating glycol/water system designed to avoid freezing in the cold climate. Heat is removed from the cooling system using forced air type heat exchangers. Minimal make up water is required.

Instrument Air

Instrument air for the ASU and CPU will be supplied with the ASU and CPU systems.

PHASE I TASK 2: OXY-FUEL COMBUSTION TEST

Test Outline and Objective

This section describes the findings and results from the work in preparation for Phase II, the demonstration of oxy-fired combustion on an OTSG test boiler. It includes the description of the test boiler, the design and operation of the oxy-fuel combustion systems as well as the results of a numerical simulation of the test boiler. An outline of the combustion tests, scope and costs of the proposed modifications conclude this section.

The objective of Phase II was to safely demonstrate oxy-fuel combustion on an existing boiler and to quantify changes in operation and performance. Process data was be collected for analysis to compare air combustion and oxy-fuel combustion.

The tests were done at the 50 MMBtu/hr OTSG located at Cenovus' Christina Lake facility in Alberta, Canada. Due to the short duration of the test (approximately 3 weeks) oxygen was supplied to the test site as liquid transported by truck. See Chapter 15 in this volume for further details.

Oxy-fuel Combustion Process

OTSGs consist of a radiant section, or fire box, that utilizes a gas fired burner to vaporize the feed water to steam of approximately 80% quality (20% of the feed water leaves the boiler as liquid droplets). The flue gas leaves the fire box and enters the convective section (economizer) where the feed water is pre-heated. The relative heat transfer area in the two sections is set by the process conditions (flue gas composition, properties, flow rate and temperature). For an existing conventional OTSG, the design was set for air-firing. Operating the OTSG under conditions different from the original design, such as in oxy-firing mode, may result in variations from the design heat transfer, including changes in tube wall temperature and shifting of absorbed heat duty from one section to another. This must be carefully assessed, as equipment and material limitations may restrict retrofitting and operating in oxy-firing mode.

When pure oxygen is used for combustion, the nitrogen in the air is eliminated, resulting in a significantly lower flue gas flow, which lowers the heat transfer in the economizer and increases the heat transferred in the radiant section. The use of pure oxygen results in very high flame temperatures and high radiant heat transfer, in the absence of dilution with flue gas recirculation. High radiant heat fluxes in this situation are a concern for boiler operation due to the potential for high tube wall temperatures, boiler tube dry out and subsequent tube damage.

To balance the boiler heat transfer and to approximate the heat transfer pattern of the original air-fuel combustion system, part of the flue gas is recycled from downstream of the economizer back to the firebox. The external flue gas recirculation system (FGR) includes dampers that can isolate the system when the boiler is operated with air. When the boiler is operated with oxygen, the existing forced draft (FD) OTSG fan is used as a FGR fan. In this case oxygen is mixed with the recycled flue gas to replace the air fed to the burner. The existing burner is used for both air and oxy-fuel combustion. For the demonstration test, the oxygen for oxy-fuel operation is stored in a cryogenic tank in liquid form, evaporated to a gas and used in the process. Figure 3 shows the modifications necessary to enable the oxy-fuel test in the boiler. Commercial installations would likely use an onsite air separation unit to produce oxygen. The modifications shown below are applicable to a commercial retrofit, as well.

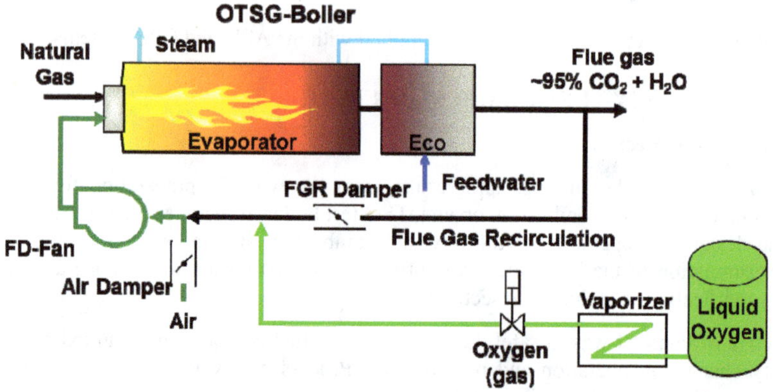

Figure 3. Oxy-fuel Combustion Test System.

Due to the high water content in the flue gas from oxygen combustion and the subsequently higher water dewpoint than is typical, it is desirable to start up the oxy-fuel boiler as a conventional air fired boiler to avoid condensation of water. Once the boiler is at operating temperature, the air damper is closed and the FGR damper is opened in a stepwise fashion. At the same time oxygen is admitted to the boiler to assure complete combustion of the fuel. The transition is discussed in detail in Chapter 15.

Test Boiler Description

To demonstrate the oxy-fuel technology performance on a SAGD boiler, a 50 MMBtu/hr natural gas fired Once-Through Steam Generator (OTSG) was modified for oxy-fuel tests. The test boiler is located at Cenovus Energy's Christina Lake SAGD operation in Alberta. See Chapter 15 in this volume for further details.

Boiler Design

The ITS/Thermotics OTSG consists of a cylindrical horizontal furnace section and a vertical economizer with a short stack (left in Figure 4). A single burner is installed in a windbox supplied with pre-heated combustion air by a forced draft (FD) fan. The horizontal furnace/evaporator has 56 tubes arranged in a single pass serpentine coil on the circumference. The convection section is fabricated from both bare and finned type tubes. Flue gas from the radiant section enters the convection section flowing across the bare tubes first and then across the finned tubes, and ultimately exits through the stack above the economizer. The radiant tubes lining the chamber walls are spaced approximately two tube diameters apart (center to center) and one diameter away from the refractory lined wall. The radiant and convective coils are made of 3" SCH80 SA-106-GR.B seamless pipe. For the purpose of the study only natural gas is considered as fuel. The boiler is equipped with a Coen QLN gas burner.

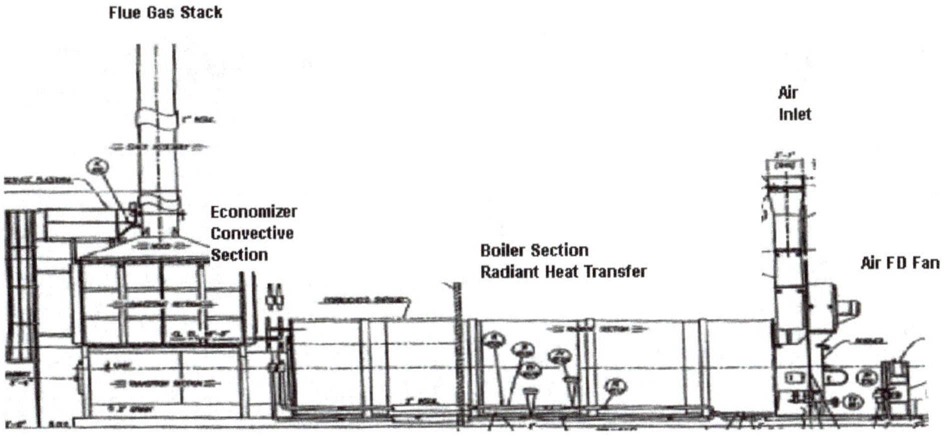

Figure 4. Test Boiler General Arrangement.

Air Burner

The test boiler uses a Coen QLN 3.2 ultra low-NOx burner. This burner consists of a refractory quarl with a 32" inner diameter and a metal burner insert that is installed into a round windbox opening so that the raised part of the quarl is flush with the refractory front wall of the boiler. This pre-mixed burner design uses multiple fuel injection locations to reduce the flame temperatures and minimize thermal NOx generation (Figure 5).

Figure 5. Coen QLN® Low NOx Burner.

Approximately half of the natural gas is injected though six burner nozzles penetrating the outer ring of the refractory quarl. The balance of the fuel is injected through inner gas spuds into the combustion air flow that enters the furnace through rectangular slots oriented similar to spokes in the burner quarl. Each of the six injectors has four nozzles. Six nozzles in the burner center (core spuds, not shown in Figure 5) inject only about 5% of the total gas flow. A liquid fuel gun on the burner axis is included with the design for fuel flexibility however it is not being utilized and therefore is not considered in this study. The pressure difference between the windbox and the furnace is used to drive the flow of air through the slots as well as past the outer natural gas nozzles. The gas flow can be biased between the outer, inner and core spuds to optimize flame shape and NOx emissions.

Operational Data

The operational data for the test boiler is summarized in the Data Sheet in Table 7.

Table 7. Test Boiler General Data.

Type/Model		Christina Lake, Alberta, operated by Cenovus		
Boiler Location		Energy ITS Engineered System, Inc.		
Manufacturer		ITS/Thermotics Steam Flooded		
Sweet Natural Gas				
Fuel	Methane	98.6	vol %	may also fire on mixed gas
	Ethane	0.1	vol %	
	Propane	0.05	vol %	
	C_4H_{10}	0.05	vol %	
	Nitrogen	0.9	vol %	
	CO_2	0.3	vol %	
	HHV	982.3	Btu / 1000ft^3	@ standard conditions
	Temperature	40	°F	
	Specific Gravity	0.563		
Boiler	Fuel Flow	58,562	scfh	
	Steam flow	67,386	lb/hr	
	Feedwater pressure	1827	psig	minimum required at inlet
	Eco inlet pressure	1812	psig	
	Eco outlet pressure	1800	psig	
	Steam pressure	1737	psig	@ discharge to header
	No of parallel steam passes	1		
	steam quality boiler outlet	80	%	
	Feedwater temperature	325	°F	
	Economizer inlet temperature	325	°F	
	Economizer outlet temperature	580	°F	
	Steam temperature boiler outlet	617	°F	
	Combustion air temperature	122	°F	
	Flue gas temperature furnace exit	1725	°F	
	Flue gas temperature eco exit	375	°F	
	Excess air flue gas (eco outlet)	15	%	
Geometry	Furnace length	41.167	ft	
	Furnace diameter	10.167	ft	tube center to tube center
	Diameter flue opening	5.896	ft	
	burner opening	5.083	in	
	evaporator tube OD	3.5	in	
	evaporator tube ID	2.9	in	
	tube center spacing from wall	4.46	in	
	tube center spacing on circumference	6.84	in	
FD fan	Design flow rate	63575	lb/hr	max
	Design pressure increase	15	in wg	

PHASE I TASK 3: DEMONSTRATION-SCALE CO_2 PROCESSING UNIT

The main objective of this task is to develop the engineering design, cost estimates and test plan for the demonstration-scale CO_2 processing unit (CPU) that is planned for installation and operation in Phase III of the program. The goal of Phase III is to demonstrate operation of a CO_2 compression and purification system connected to a single 50 MMBtu/hr test OTSG retrofitted for oxy-fuel operation. The CO_2 purification system will be designed to recover up to 99% CO_2 at >99.5% (by vol.) purity. The Phase III demonstration represents an intermediate step necessary prior to commercialization of the technology on a multiple (four – 250 MMBtu/hr) OTSG system anticipated for large scale implementation.

During Phase III, process data will be collected at several operating conditions to determine the optimum conditions for achieving CO_2 purities of interest. The data will be analyzed and compared with simulated output to confirm computer model accuracy. The model will be used to predict the CPU performance and to provide scale up information required for final design of a typical commercial unit described and cost estimated in Phase I Task I.

The Phase III demonstration plans are based on using the 50 MMBtu/hr OTSG located at Cenovus Energy's Christina Lake facility which was retrofit for the Phase II oxy-firing demonstration.

Specific Phase III goals include:

- Design, fabrication, and installation of the demonstration system.
- System operation to validate expected performance.
- Long term operability demonstration for at least a year.
- Verification of process scale factors.

This section of the chapter includes a description of the integrated system, the estimated capital costs and utilities, a schedule for construction and operation of the integrated system and the proposed tests for one to two years of operation. The required system will be comprised of an air separation unit (ASU), a once- through-steam-generator (OTSG) modified for oxy-fuel combustion, and a carbon dioxide processing unit (CPU). An Air Separation Unit is included in this project phase because the duration of planned operations and the remote location make trucked in oxygen, used for Phase II, impractical.

The cost estimates prepared are for budgeting purposes only and are considered to be in the +/- 20% range for the overall system. The CPU cost estimate is in the +/- 25% range. The reduced accuracy of this estimate is the result of uncertainty related to the project time frame (the plant would not be built for several years), and lack of site specific details. Some detail work, such as a specific control system integration plan, and a site specific safety review have been deferred as well, and will become part of the future Phase III project scope.

Design Basis

Design Basis and Assumptions

The ASU and CPU will be single train plants retrofitted to an existing installation comprising one OTSG, producing a total of 600 tonne/day of steam. The retrofitted system will be capable of automatically reverting to air-firing under desired circumstances. For the purpose of this study it is assumed that the ASU and CPU will be located ~ 50 meters from the boiler. Actual equipment location with appropriate design and cost adjustments will be required when the project is authorized in order to develop a +/- 10% cost estimate. Equipment used in the ASU and CPU will meet all applicable standards, codes, and safety requirements used by Praxair in commercial designs of air separation units and CO_2 purification systems. For example this includes ASME B 31.3. Equipment located within the SAGD facility will meet all applicable standards, codes, and safety requirements for a typical SAGD facility located in Alberta, Canada.

Site Ambient Conditions

Table 8 lists the ambient conditions for the region.

Equipment design will be based on hot day. System performance calculated for average conditions.

Table 8. Ambient Conditions.

Location	Christina Lake, Alberta Canada		
Seismic Zone	0		
Approximate Elevation (ft)	1800		
Barometric Pressure (psia)	13.8		
Design Cases	Hot Day	Avg Day	Cold Day
Dry Bulb Temperature (F)	73	36	-7
Wet Bulb Temperature (F)	64	33	-8
Relative Humidity %	63	75	N/A

Site Utilities

- Electrical: Electrical power will be supplied from the grid.
- Fuel: see Table 5 for fuel specifications
- Cooling System Definition: Cooling Water System will be a closed-loop system with heat rejection to atmosphere via fans.
- Instrument Air: Instrument air for the ASU and CPU will be supplied with the ASU and CPU systems.

Process and System Description

Integrated System

The overall configuration of the demonstration system will be similar to the commercial scale system described in Task 1. The main system components include an air separation unit (ASU) to provide the combustion oxygen, an OTSG capable of operating in both air- and oxy-firing modes and a CO_2 Processing Unit (CPU) for purifying flue gas to produce > 99.5% purity CO_2 (Figure 6).

Because of the relative scale of the demonstration CPU compared to the commercial CPU (50 vs. 1,000 MMBtu/hr), it was not practical to utilize exactly the same equipment types in both systems. For example, the demonstration CPU compressors will be reciprocating machines, while centrifugal types are suitable for the commercial scale. The emphasis in this demonstration will be on maintaining operating process conditions similar to those of the commercial system. This is not expected to introduce significant variations from the commercial design.

Oxygen produced from the onsite ASU is delivered at pressure to the OTSG where it is mixed with recycled flue gas and combusted with natural gas to produce steam and to pre-heat boiler feed water.

The flue gas produced consists mainly of carbon dioxide and water because nitrogen has been removed in the ASU. At the boiler exit the flue gas is split into two streams. One stream is recycled back to the boiler and the other is sent to the CPU for purification. The recycle stream is required to maintain the heat flux and temperature profiles in the boiler similar to those in the air-fired OTSG. The other portion of flue gas is delivered to the CPU via a short pipeline with the aid of a blower. Within the CPU it is compressed and purified to produce >99.5% CO_2. Because the objective of this project phase is only to demonstrate the operation of the CPU, no CO_2 compression has been included. The purified CO_2 stream that is produced in the CPU is vented to a safe location. Table 9 provides a stream summary for the integrated system.

In actual operation, the boiler will be started up using air and switched to oxy-fuel operation. The CO_2 Processing Unit (CPU) will be placed into service once the boiler operation has stabilized from the transition.

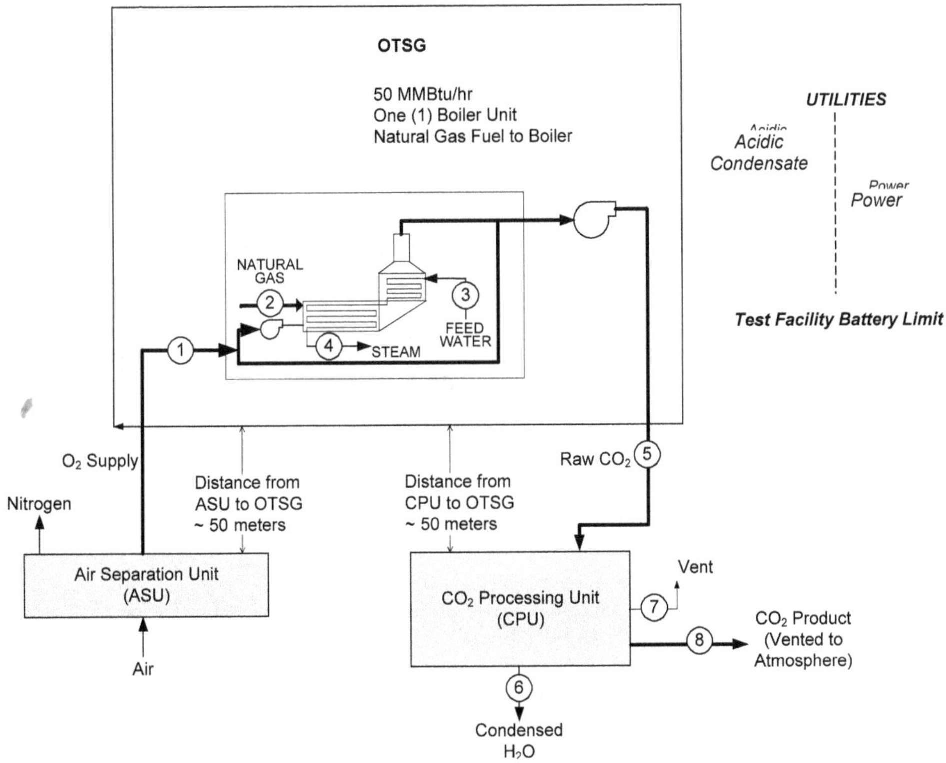

Figure 6. Demonstration System Process Schematic.

Table 9. Integrated System Stream Summary.

Stream	1	2	3	4	5	6	7	8
Description	Oxygen to Boiler	Natural Gas to Boiler	Boiler Feed Water	Steam from Boiler	Flue Gas from Boiler	Process Condensate	CPU Vent (Dry Basis)	CO_2 Product
Temperature, °F	50	36	371	605	405	110	50-400	40
Pressure, psia	15	20	2006	1605	13.8	15	15	15
Vapor Fraction	1	1	0	0.78	1	0	1	1
Mass Flow, lb/hr	10377	2488	73092	73092	12880	5469	781	6626
Molar Flow, lb mol/hr	322	153	4057	4057	476	303	22	151
Mole Fraction	1	2	3	4	5	6	7	8
CH_4		0.988794						
C_2H_6		0.000400						
C_3H_8		0.000100						
C_4H_{10}		0.000100						
C_6H_{14}		0.000100						
CH_4S		0.000005						
N_2	0.001433	0.006900			0.003331	0	0.073151	0
CO_2	0	0.003600			0.320416	0.000378	0.081357	0.999946
CO	0	0.000001			0.000050	0	0.001099	0
O_2	0.970000	0			0.018997	0	0.417360	0.000010
Ar	0.028567	0			0.019346	0	0.425150	0
SO_2	0	0			0.000002	0	0	0.000006
NO	0	0	1	1	0.000091	0	0.001883	0.000012
NO_2	0	0			0.000009	0.000002	0	0.000025
H_2O	0	0			0.637758	0.999620	0	0

Oxygen Supply System

Integrated, long-term oxy-fuel operation of the 50 MMBtu/hr boiler studied and tested under Phase II of this program will require the installation of an oxygen supply system capable of providing a continuous flow of 110 tonne/day O_2 at 45 psig. To accomplish this, an air separation unit will be installed at the site. The system proposed is a conventional cryogenic system as shown in Figure 7. The major equipment items required for the ASU will be similar to those needed in the ASU designed for a commercial-scale system: air compressor, air cooler, pre-purifier, booster compressor and cold box. The main difference will be in the cold box which will be designed to produce oxygen at 99.5% purity. Selection of O_2 at a higher purity than the optimum purity (~97%) will allow test flexibility with respect to purity of O_2 supplied to the OTSG. Test purities of 97% can be achieved by blending air with the higher purity oxygen. The use of air blending will provide the ability to study boiler and CPU performance at a range of oxygen purities, if desired.

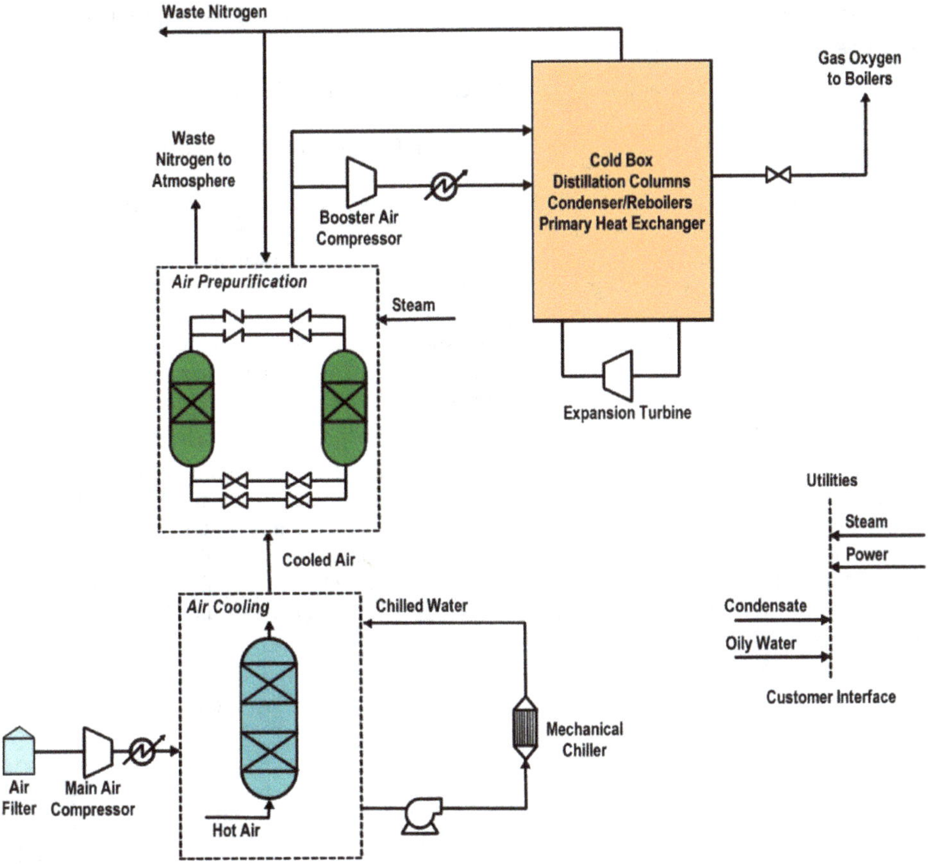

Figure 7. Air Separation Unit Block Flow Diagram.

Boiler Island

The oxy-fuel combustion system will consist of control systems for fuel gas and oxygen and a burner to combust fuel gas with oxygen and recirculated flue gas. Flexibility to operate on both air and O_2 will be included. The OTSG studied and tested under Phase II is proposed for this demonstration, since the required modifications for flue gas recirculation will already be in place (see Figure 3).

CO_2 Processing Unit Process Description

The demonstration scale CO_2 Processing Unit is a smaller version of the commercial CPU described in the Task 1 section of this chapter, with a design flow capacity ~ 5% of the commercial CPU. Because it is 20 times smaller than the commercial unit, the demonstration system does not utilize identical equipment.

The Task 3 Demonstration Scale CPU is different from the Task 1 Commercial Scale CPU in the following ways:

- Flue gas fed to the Task 3 CPU is collected from a single 50 MMBtu/hr OTSG, instead of four 250 MMBtu/hr boilers.
- The Task 3 CPU is assumed to be located 50 m from the OTSG, instead of 500 m as in the Task 1 CPU.
- The raw gas compressors for the Task 3 CPU are multi-stage reciprocating type compressors instead of the multi-stage centrifugal type used for the commercial system. Parts of the compression train in contact with raw CO_2 stream will be made from appropriate stainless steel grade resistant to corrosion by acidic species that could form during compression.
- The Task 3 CPU does not include compressors for the product CO_2 required for sequestration. Instead the product CO_2 will be vented to atmosphere at a safe location.
- Regeneration gas for the Task 3 CPU dryers is heated by an electric heater instead of a natural gas heater.

A process flow diagram for the demonstration CPU to treat flue gas from a single 50 MMBtu/hr oxy-fired OTSG is shown in Figure 8. Table 10 is a stream summary for the CPU, which uses a distillation process to produce CO_2 at a purity of >99.5%. The Demonstration Scale process flow diagram in Figure 8 is identical to the Commercial Scale process flow diagram except for the smaller flow capacity and absence of product compressors. CO_2 produced from the Demonstration Scale process will be vented to atmosphere. Appropriate conditioning of the CO_2 product will be done to prevent ice formation at the vent.

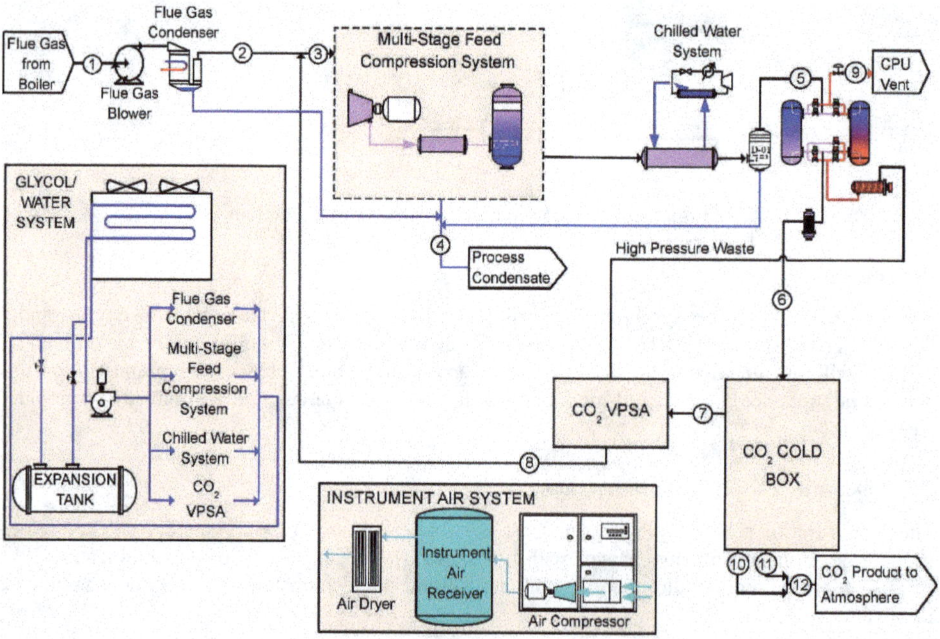

Figure 8. Demonstration CPU Process Flow Diagram.

It should be noted that since completion of Phase I, Praxair and others have shown that SOx and NOx species that are present in the CO_2-rich flue gas will partly be removed in the cooling and compression equipment of the CPU as acidic condensate. As a result, SOx and NOx concentration in the final purified CO_2 product would be much lower than those listed in Table 9 (stream 8) and Table 10 (stream 12), and process condensate (stream 6 in Table 9 and stream 4 in Table 10) would contain sulphuric and nitric acid species.. In addition, an activated carbon process developed by Praxair can remove any residual SOx and NOx from compressed raw CO_2 gas, which will result in product CO_2 essentially free of SOx and NOx species. With these changes, the CPU vent stream (stream 7 in Table 9 and stream 9 in Table 10) would contain negligible amount of NOx species.

Table 10. CPU Stream Summary.

Stream	1	2	3	4	5	6
Description	Flue Gas from Boiler	Flue Gas from FG Condenser	Raw Gas Compressor Suction	Process Condensate	Dryer Inlet	Cold Box Inlet
Temperature, F	405	135	133	110	50	50
Pressure, psia	13.8	15	15	15	359	358
Mass Flow, lb/hr	12880	7687	8473	5468	8197	8194
Molar Flow, lb mol/hr	476	188	206	303	191	191
Mole Fraction	1	2	3	4	5	6
N_2	0.003331	0.008446	0.005896	0	0.009348	0.009355
CO_2	0.320416	0.811999	0.873701	0.000378	0.881607	0.882289
CO	0.000050	0.000127	0.000089	0	0.000140	0.000141
O_2	0.018997	0.048168	0.038430	0	0.053341	0.053382
Ar	0.019346	0.049053	0.048863	0	0.054329	0.054371
SO_2	0.000002	0.000005	0.000007	0	0.000005	0.000005
NO	0.000091	0.000231	0.000295	0	0.000437	0.000438
NO_2	0.000009	0.000021	0.000022	0.000002	0.000020	0.000020
H_2O	0.637758	0.081950	0.032697	0.999620	0.000773	0

Stream	7	8	9	10	11	12
Description	CO_2 VPSA Inlet	Recycled CO_2 from CO_2 VPSA	CPU Vent (Dry Basis)	Low Pressure CO_2 Product	Medium Pressure CO_2 Product	CO_2 Product to Atmosphere
Temperature, F	24	110	50-400	24	40	40
Pressure, psia	257	20	15	910	222	15
Mass Flow, lb/hr	1568	787	781	3313	3313	6626
Molar Flow, lb mol/hr	40	18	22	75	75	151
N_2	0.044531	0.010841	0.073151	0	0	0
CO_2	0.439884	0.861920	0.081357	0.999946	0.999946	0.999946
CO	0.000669	0.000163	0.001099	0	0	0
O_2	0.254069	0.061853	0.417360	0.000010	0.000010	0.000010
Ar	0.258811	0.063007	0.425150	0	0	0
SO_2	0.000000	0.000000	0	0.000006	0.000006	0.000006
NO	0.002036	0.002216	0.001883	0.000012	0.000012	0.000012
NO_2	0	0	0	0.000025	0.000025	0.000025
H_2O	0	0	0	0	0	0

CONCLUSIONS AND RECOMMENDATIONS

Study analysis has been completed for both a commercial and test scale OTSG operating under air and oxy-fuel combustion mode. Results of computational fluid dynamic models combined with rigorous heat and mass balance models indicate that oxy-fuel combustion with flue gas recirculation can be applied to OTSGs while maintaining key operating parameters (temperature and heat flux) that are similar to operation with air. Therefore, through proper control integration, OTSGs will be retrofittable and capable of air or oxy-fuel operation.

The Phase II Oxy-fuel combustion test data will be analysed to confirm the Phase I results. Phase III, the CO_2 Purification Demonstration, is well in the future and will require significant capital and resource expenditures. Program requirements and cost estimates for Phase III will be reviewed now that Phase II is complete.

REFERENCES

1. Bohm, M. (2011), 'Application of Oxy-Fuel CO_2 Capture for In-Situ Bitumen Extraction from Canada's Oil Sands' presented at GHGT-10, 19-23 September 2010, Amsterdam, The Netherlands. Energy Procedia 4, pp 958–965
2. Shah, M M (2007), 'Oxy-Fuel Combustion for CO_2 Capture for New and Existing PC Boilers', paper presented at Electric Power 2007 Conference, Chicago, IL, USA, May 1-3, 2007, Electric Power.
3. Shah, M M (2005), 'Capturing CO_2 from Oxy-Fuel Combustion Flue Gas', paper presented at 1st Workshop of the International Oxy-Fuel Combustion Research Network, Cottbus, Germany Nov. 29-30, 2005, IEA GHG.

Carbon Dioxide Capture for Storage in Deep Geological Formations, Volume 4
Karl F. Gerdes (Editor)

Chapter 15

OTSG OXY-FUEL DEMONSTRATION PROJECT

Carlyn McGeean[1], Candice Paton[1], Jason Grimard[1], Todd Pugsley[2], Kelly Tian[3], Stefan Laux[3], Larry Bool[3] and Larry Cates[3]
[1]Cenovus Energy Inc., 500 Centre Street SE, Calgary, AB T2P 0M5, Canada
[2]Suncor Energy Inc., 150 6th Ave. SW, Calgary, AB T2P 3E3, Canada
[3]Praxair Inc., 175 East Park Drive, Tonawanda, NY, USA

ABSTRACT: A conventional 50 MMBTU/h once through steam generator (OTSG) was retrofitted and operated in oxy-fuel mode with flue gas recirculation at Cenovus' Christina Lake in-situ oil sands production facility. This successful demonstration showed that a conventional OTSG can be retrofitted for oxy-firing with minimal changes and retain the ability to operate in standard air-firing mode. No technical limitations were identified for such retrofitting, and oxy-firing can be considered as a viable technical alternative to post-combustion CO_2 capture. The project is a collaborative effort between several large multinational energy companies and the Climate Change and Emissions Management Corporation (CCEMC), which receives funds from the Province of Alberta that are generated through a carbon levy on large industrial emitters. Key learnings for this first-of-its-kind facility in an in-situ oil sands facility in Alberta have centred on the importance of ensuring process safety in an operating facility. The test encountered no operational issues during oxy-fuel firing. Fuel consumption was about 5% lower for oxy-fuel firing compared to air-firing at the same steam production rate and steam quality.

KEYWORDS: oxy-fuel; combustion; OTSG; carbon capture; SAGD; oxygen; flue gas recirculation

INTRODUCTION

The oil sands industry has identified carbon capture and storage (CCS) technology as a key opportunity for significantly reducing greenhouse gas (GHG) emissions, but current solutions are very expensive to implement on a broad scale. The CO_2 Capture Project (CCP) has assessed options for CO_2 mitigation and concluded that the current state-of-the art, post-combustion capture, is very costly for this application. Among CO_2 capture alternatives that might lower the cost, oxy-firing was identified by CCP as an option that is amenable to retrofitting existing once through steam generators (OTSGs) as documented in Chapters 13 and 14 in this volume. There are uncertainties regarding the technical issues of retrofitting existing OTSGs and operating them in oxy-firing mode. To address these issues, Cenovus Energy as Site Participant and Project Leader, Praxair as Site Participant and Project Leader, and Suncor Energy as Project Administrator and Project Manager together with their partners (the CO_2 Capture Project, Devon Canada Corporation, Statoil and MEG Energy) are leading a project to examine the feasibility of retrofitting an existing once through steam generator (OTSG) with oxy-fuel technology. The Climate Change and Emissions Management Corporation (CCEMC), which receives funds from the Province of Alberta provides funding to support the project.

Cenovus' Christina Lake facility, which is the host site for the OTSG oxy-fuel retrofit project, is located about 150 kilometres south of Fort McMurray, Alberta, and was started up in 2000. The oil at Christina Lake is located about 375 metres below the surface and requires specialized technology, steam-assisted gravity drainage (SAGD), to drill and pump it to the surface. This project is fifty percent owned by ConocoPhillips.

Production at Christina Lake will be grown in carefully planned phases so learnings are applied from one phase to the next. Christina Lake has eight phases planned so far. The first five phases are in operation. Plans for the next two phases have received regulatory approval, and an application for the eighth was submitted for regulatory review in 2013. With additional expansion phases and optimization work, total gross production capacity is believed to have the potential to reach 310,000 barrels per day.

Project Overview

The OTSG oxy-fuel retrofit project has completed its second phase. See Chapter 14 for a description of the overall project and results from the Phase I assessment studies of commercial scale oxy-firing of OTSGs and design of the retrofit for the demonstration.

Phase II of the project focused on testing oxy-fired combustion on a 50 MMBTU/hr OTSG unit at Cenovus' Christina Lake site. The scope included modifying the existing OTSG for oxy-fuel combustion (without capture and compression) and operating the OTSG for three weeks to demonstrate feasibility and provide essential data to design a full scale system. Since the planned period of operation for this project phase was of limited duration, the bulk oxygen for the test was supplied by trucked-in liquid oxygen from an existing air separation plant. Construction for Phase II of the project was completed in 2014, and commissioning was completed during the first quarter of 2015. The field test work was completed in April 2015.

The demonstration test had the following goals:

- Identify changes in heat transfer pattern and flame length between air-fuel and oxy-fuel combustion.
- Find the best combustion solution for oxy-fuel operation with respect to heat transfer, emissions, oxygen concentration and the flame stability of the existing burner.
- Determine the volumes of flue gas recirculation for the operation with oxy-fuel.
- Establish safe and reliable operating procedures, interlocks, and control methods.

OXY-FUEL TEST OVERVIEW

Boiler Modifications

To enable oxy-fuel combustion with the existing air combustion system, a flue gas recirculation system was added to the boiler (Figure 1). The flue gas recycle was routed from the economizer outlet via a new duct connecting to the windbox. Oxygen is mixed into the recycled flue gas prior to the existing air-fuel burner.

Two louver-type air dampers were installed at the combustion air inlet. One damper was designed as a low leakage damper to avoid air ingress into the system when the boiler was operated in full oxy-fuel operation, since it is very important to minimize the nitrogen content in the flue gas resulting from air ingress. High nitrogen in the captured flue gas increases the cost of CO_2 purification. The other air damper was designed as an opposed blade damper to allow good control of the air flow during transitions. The original air pre-heater had not been used for the existing air-fuel operation and was removed to allow space for the recirculation system.

The forced draft (FD) fan discharges into the windbox through a flow control damper. For the demonstration, the fan was replaced with a larger model to handle the larger volumetric flow rates produced during oxy-fired operation. When the boiler is operated using oxygen and flue gas recirculation, the actual volumetric flow rate to the burner is increased due to the higher temperature of the mixture (approximately 180°C) in comparison to the air at ambient temperature. The fan was equipped with a Variable Frequency Drive (VFD) motor to allow better adjustment to the flow. During commissioning, it was found that the flow control with the control damper after the fan was entirely adequate and the VFD was set to 60% for all tests.

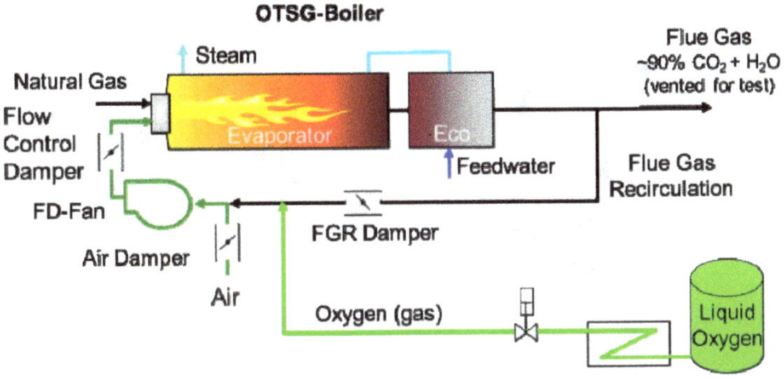

Figure 1. Main features of OTSG retrofit.

The recirculation duct was tied into the plenum above the economizer at the base of the stack. Similar to the air dampers, two FGR dampers were installed to control the flow in the duct. One damper was designed as a low leakage shut-off device and the other designed as a control device. The total oxidant mixture flow, whether combustion air or oxygen mixed with recirculated flue gas, was controlled by the damper downstream of the FD fan. For oxy-firing operation, the two dampers in the recirculation duct were opened and the air dampers closed. An oxygen mixing device was installed in the FGR duct to mix the oxygen into the flue gas (mostly CO_2 and water vapor). The oxygen concentration was measured and controlled to the target value by adjusting the amount of oxygen fed to the sparger via a flow control valve. The total oxidant mixture flow was controlled by the control damper downstream of the FD fan through the Cenovus DCS via a cross limiting network between fuel and oxidant flows.

Oxygen and moisture analyzers provided the oxygen concentration of the oxidant and the dewpoint temperature for control purposes. A stack gas analyser collected emission performance data. The existing burner fuel supply system was used essentially unchanged. However, a small burner to support flame stability (J-Burner) was installed in the center of the existing air burner and separately supplied with natural gas and oxygen (see Oxy-fuel Combustion System section below).

Oxygen Delivery

Oxygen was delivered by Praxair to the site via truck. The oxygen was transferred into on-site liquid storage tanks provided by Praxair for use as needed during the test. The installation was comprised of two large cryogenic storage trailers, a transfer pump, a high pressure buffer tank, and ambient-air vaporizers to provide gaseous oxygen (Figure 2). Oxygen piping conveys the gas through the plant into the boiler house.

The flow of oxygen into the system is controlled by an oxygen flow control skid designed by Praxair (Figure 3). A portion of the flue gas recirculation ductwork is shown in Figure 4. Please see

Chapter 14 for a description of the flue gas recirculation system necessary to operate the boiler with oxygen.

Figure 2. Oxygen storage and supply system (courtesy of Cenovus and Praxair).

Figure 3. Praxair flow control skid (courtesy of Cenovus and Praxair).

Figure 4. Flue gas recirculation duct (courtesy of Cenovus).

Oxygen Control

An oxygen control valve skid measured and regulated oxygen flow, and provided oxygen safety interlock and shutoff functions. This pre-fabricated skid was located in the boiler building close to the burner. The skid components were selected and cleaned for oxygen service and were designed for a maximum pressure of 300 psig (20.7 bar). Pressure at the inlet was maintained at approximately 150 psig (10.7 bar).

The oxygen flow was measured with an orifice flow meter incorporating pressure and temperature compensation. The skid included pressure switches and double block valves for safely shutting down oxygen flow. The automatic safety functions of the skid were controlled through hard wired interlocks and by the PLC that was interfaced with the boiler control DCS system to receive permissives and setpoints and communicate process data. In the event of power interruption, instrument air failure, or safety shutdown, all valves are designed to close automatically.

Downstream of the block valves the oxygen flow was split into the main flow to the sparger in the flue gas recycle duct and a much smaller flow to the stabilizing J-Burner. The main flow was controlled with an automatic flow control valve through the PLC. The oxygen flow to the stabilizing burner was manually controlled through a small valve. This flow was measured with a second orifice flow meter similar in design to the upstream flow meter mentioned above.

Oxy-fuel Combustion System

The oxygen mixing device (sparger) mounted in the flue gas recirculation duct after the FGR damper consisted of a 4" pipe with multiple holes drilled perpendicular to the FGR flow. Computational Fluid Dynamic (CFD) modelling was used to design the sparger to insure complete mixing of the oxygen into the flue gas, and included the effect of the straight duct immediately after the mixer. An oxygen concentration measurement after the sparger was used to control the oxygen concentration in the oxidant mixture.

A modelling study executed in Phase I predicted that the flame stability of the existing Coen burner may be affected by the FGR/oxidant mixture because the actual volumetric flow is higher and the combustion properties of the mixture is different than when combustion air is used [1]. The numerical study also provided guidance that a small, pure oxy-fuel burner in the center of the main burner is a successful strategy to overcome this problem. A small Praxair J-Burner was inserted into the center guide pipe of the air burner to provide flame stability. This patented solution is a simple method for providing flame stability by adding heat in the burner center [2]. The burner firing capacity was approximately 700kW (3 MMBtu/hr), about 6% of the total heat input.

Measurements

Process measurements from the dedicated PLC system, as well as the existing boiler control DCS system, were recorded at 15-second intervals throughout the trial. Data included:

- Boiler feedwater flow, pressure, and temperature
- Steam flow, temperature, pressure, and quality.
- Stack oxygen, CO, and moisture concentration.
- Fuel gas flow, temperature, and pressure.
- Combustion air flow and temperature.
- Flue gas flow, temperature, and pressure.
- Flue gas oxygen concentration, before and after sparger.
- Oxygen flow and pressure to sparger.

- Oxygen flow to "J" burner.
- Heat flux and boiler tube temperature.

The original plan was to capture stack flow rate and emission measurements (O_2, CO_2, CO, NOx, SO_2) at the top of the stack. However, the low flue gas flow during the oxy-fuel operation (only about 25% of the comparable flue gas flow during air combustion) resulted in stack velocities below the detection limit of the system. As a result, the flue gas volume during oxy-fuel tests was calculated from a boiler mass balance with the fuel and oxygen input and the excess oxygen. During oxy-firing operation, the flue gas has high moisture content (>50 vol %) and dew point (>80°C, or 180°F) which requires proper flue gas sample handling and line insulation to obtain reliable data. The sampling system diluted the flue gas with a known amount of nitrogen at the stack to avoid water condensation. A heated sample line conveyed the sample to a heated trailer for analysis. Diluting the sample to a very low dewpoint allowed measuring the wet gas directly in the analyzers without any drying, but requires accurate analyzers. Throughout the test, daily calibration and proper maintenance to the emission equipment was performed remotely to ensure that the systems were functioning properly.

Three Commercial heat flux probes were installed for the demonstration in the boiler evaporator wall at 19, 44 and 63% of the relative evaporator length measured from the burner. The probes use a Schmidt-Boelter sensor, which measures total heat flux (convection and radiation) at the probe location. Past experience has shown that such probes are sensitive to the environment and are not reliable for long-term operational measurements. Unfortunately, only the center probe at 44% of the length was functional at the time of the demonstration. The other two probes had failed earlier due to the long exposure to the operation after installation. Although the presence of all probes would have made possible a better characterization of the radiant flux distribution in the furnace, the average flux could be calculated from the boiler performance. Visual observation and the single mid-furnace flux reading was sufficient for detection of potential issues.

Control Integration

A key aspect of the project from a safety and operation perspective was the integration of the oxy-fuel combustion system into the existing boiler controls. The control system for the oxy-fuel demonstration has three components:

- Burner Management System (BMS) – this mandatory and code compliant safety component for any combustion system was upgraded for the demonstration to include more interlocks. The basic function is to monitor the flame and ensure safe boiler startup and operation.
- Distributed Control System (DCS) – the operational control of all equipment at the plant is integrated in the DCS. Additional communication with the oxy-fuel system controls was added to the existing DCS system for the boiler.
- Programmable Logic Controller (PLC) – A new PLC handled most of the additional components for the oxy-fuel demonstration and their controls. A visual Human Machine Interface (HMI) system was used to facilitate operating the PLC controlled components, with monitors located both in the remote control room and in a lab located near the boiler. The oxy-fuel PLC communicated with the DCS via the plant Ethernet network and through several direct wired I/O signals. Due to the temporary nature of the demonstration the separate PLC was used. In a commercial implementation, these functions would be permanently incorporated into the existing DCS.

Description of the Control System

The boiler operation is primarily controlled by the DCS system that monitors and controls the boiler startup and shutdown, and all process variables (i.e. steam quality, steam generation rate, ramp up/down, etc.), while the local PLC managed the oxy-fuel system operation.

The boiler firing rate (fuel gas flow rate) is automatically adjusted by the DCS according to the feedwater rate through the boiler, in order to indirectly control the steam quality at the boiler outlet. The combustion air flowrate is controlled by the pre-set air to fuel ratio, cross limited to the fuel gas flow rate, and trimmed to maintain the desired excess O_2 level in the exhaust stack. In order to provide a safety margin during transitions between air-fuel and oxy-fuel operation, the excess O_2 trim control was used to slightly increase the oxygen to fuel ratio to avoid fuel-rich conditions in the boiler. During transients, cross-limiting of the combustion air and fuel flow ensured that the above stoichiometric operation was maintained.

Safety critical shutdowns, such as interruptions to the fuel gas supply or a loss of flame due to flame instability, trigger a shutdown of the boiler by the BMS. The BMS receives input from a flame sensor that detects the presence of flame through UV radiation from the combustion process.

Prior to starting the burner and after a shutdown, code prescribed purge sequences were completed. These purge sequences and interlocks were coordinated through the BMS. The DCS controlled the operation of the forced draft fan, the downstream air flow control damper, and the upstream air isolation damper. The PLC controlled the operation of the flue gas dampers and the upstream air modulating damper.

When operating in oxy-firing mode, the PLC controlled the oxygen concentration in the recirculated flue gas. The oxygen concentration was measured in the flue gas duct downstream of the sparger and controlled through a proportional-integral-derivative controller (PID) loop.

HMI System

The HMI system provided for the test was programmed with several screens to facilitate operation and monitoring of the process:

- An Operation Screen provided a view of the “faceplate” for the key controllers – O_2 skid start/stop, Sparger O_2 flow control, Flue Gas flow control, Flue Gas isolation damper and Air Inlet damper. On this same screen, Process Variable Indicators displayed values for monitored parameters including J-Burner O_2 flow, O_2 flow set point, various measured O_2 concentrations, FGR moisture, and stack conditions.
- The Process Graphics screen provided a status overview with key measured variables superimposed on a graphical flow sheet of the oxy-fired system.
- An Interlocks screen provides the status of interlocks and their associated alarms.
- PID Faceplate screens are provided for each control loop in the system for tuning.
- Alarm Summary screen.
- A graphical Trending screen is available for any of the logged variables.

OPERATING RESULTS

Overview of Operation

The testing of the modified boiler began on 15 April 2015 with the focus on optimizing the transition from air to oxy-fuel. Several iterations were required to define the adjustment of the air and flue gas recirculation dampers to successfully achieve a smooth transition without loss of flame. Oxy-fired and air-fired data collection then proceeded. By the end of the test operation, the transition between operating modes became somewhat routine and required about 30 minutes.

Table 2 shows a summary of the test conditions for air and oxy-fired. The boiler steam flow load was 53 to 82% for the air-fuel tests and 54 to 78% for the oxy-fuel tests. The upper range of steam load for oxy-fuel was restricted due to several operational issues discussed below. Typical operation for this boiler is at approximately 80% of theoretical design steam flow. The operation with oxy-fuel requires less fuel flow at the same boiler feedwater flow due to the higher efficiency of oxy-fuel combustion. The maximum possible fuel flow for this boiler is 28 t/d which corresponds to 93% of theoretical design steam flow.

Table 2. Range of test conditions.

	Air-Fuel	Oxy-Fuel
Test Period	14 to 26 April 2015	17 to 23 April 2015
Feedwater Flow, t/d	390 - 601	395- 571
Boiler Load, % (see note)	53 - 82%	54 - 78%
Boiler Fuel Flow, t/d	14.1 - 24.6	13.6 -20.3
Steam Quality, %	68.2-69.1% & 77.6-78.7%	68.0-69.9% & 77.9-78.4%
Excess O_2, %	2.8 – 4.4%	3.8 - 5.9%

Note: Boiler load is expressed as % of design steam flow (733 t/d).

Testing was performed at approximate levels of 70% and 80% of steam quality for both air-fuel and oxy-fuel to allow easier comparison of the data. Although excess oxygen in the flue gas was about 1% higher during oxy-fuel tests compared to air-fuel operation, this was primarily due to the control system tuning for the equivalent air to fuel ratio during oxy-fuel operation. There is no question that the control system could be further optimized to match the air-fuel operation excess O_2 for a permanent installation.

Typical flue gas composition at high boiler load is shown in Table 3.

Table 3. Typical flue gas composition for high boiler load.

	Air-Fuel	Oxy-Fuel
Carbon Dioxide, CO_2	8.2%	27.5%
Water Vapor, H_2O	15.5%	52.0%
Nitrogen, N_2	72.9%	16.2%
Oxygen, O_2	3.4%	4.2%
Sulfur Dioxide, SO_2	0.0%	0.0%
Mol. Weight of Mixture	27.9	27.4
Dewpoint of Mixture	55°C	83°C

Theoretically, oxy-fuel eliminates all nitrogen in the flue gas. However, there was sufficient air damper leakage (equivalent to about 4-6% of oxidant flow to the boiler), coupled with the minor volume of cooling air purge for the camera and other instruments, that the resulting N_2 concentration at full load was approximately 16%. This concentration is affected by the internal boiler pressure and the quality of the air dampers. For a commercial system, it will be important to minimize air leakage. Even with this level of air ingress the N_2 in the flue gas was significantly reduced during oxy-firing. The CO_2 concentration increased from 8.2 to 27.5% when the boiler operation is switched to oxy-fuel.

Boiler Load and Steam Quality

The results for the boiler performance discussed here are expressed in terms of feedwater flow as the measure for boiler load. Figure 5 and Figure 6 show the fuel flow and steam quality for air-fuel and oxy-fuel versus the boiler load for steam quality levels of about 70 and 80%. Higher steam quality requires higher heat input for the same feedwater rate to evaporate the additional water.

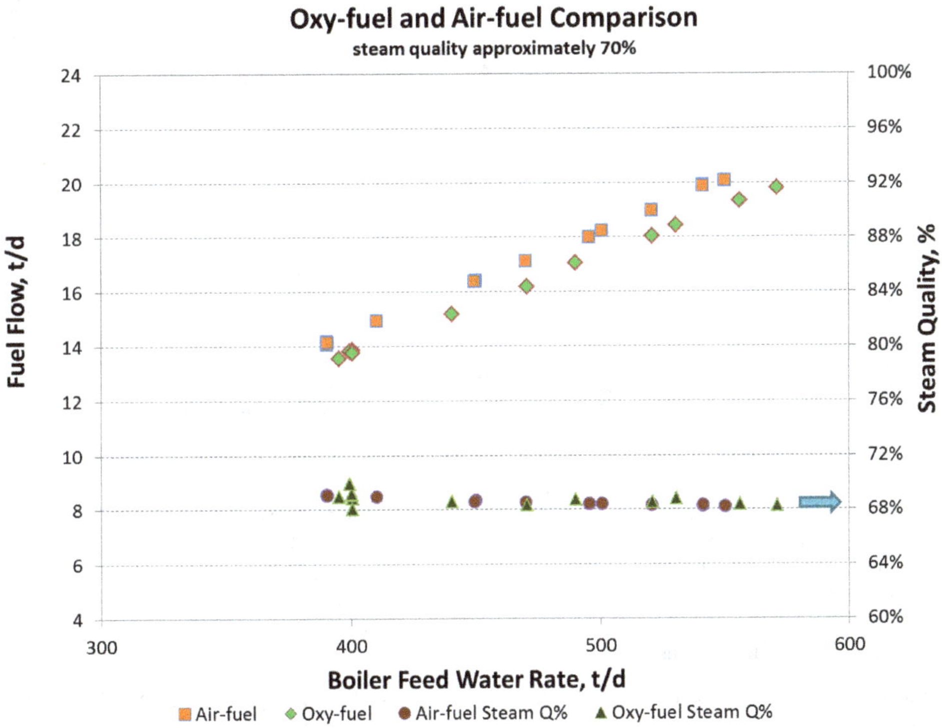

Figure 5. Fuel Flow at 70% Steam Quality vs Boiler Feedwater Flow.

There is no steam quality difference between air-fuel and oxy-fuel combustion. As both Figures show, the fuel flow for a given load is lower for oxy-fuel combustion. This is expected as oxy-fuel combustion is inherently more efficient than air fuel combustion. Air contains 78% nitrogen, which is entering the boiler at ambient temperature (~15°C at the time of testing) and leaves at the stack temperature (~200°C), and heating the N_2 causes an efficiency loss. The test boiler did not have an air pre-heater due to space constraints. The combustion with O_2 does not introduce N_2 to the process, which reduces the flue gas amount by approximately 75% and the resulting stack loss is significantly reduced. During the demonstration, the N_2 concentration in the stack was about 16%, due to air leakage through the air dampers and the use of cooling air for the instruments in the boiler.

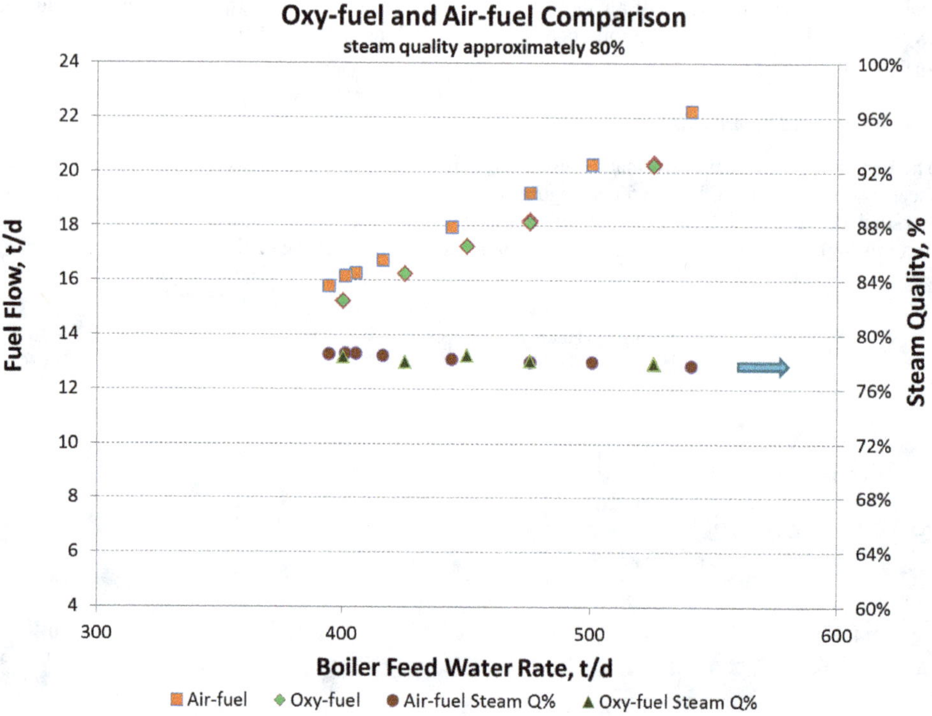

Figure 6. Fuel Flow at 80% Steam Quality vs Boiler Feedwater Flow.

The variation of excess oxygen in the flue gas is shown in Figure 7. With air-fuel combustion the average excess O_2 was varied between about 4% at low load and 3% at maximum test load. The variation shown in the figure is larger with oxy-fuel testing, mainly due to deliberate process variable testing and no excess O_2 trim correction was used for this short demonstration test. There is no technical reason that excess O_2 cannot be controlled to a tighter band and to a lower level comparable to air-fuel combustion.

In a commercial implementation of oxy-fuel combustion for carbon capture, it is very desirable to minimize excess O_2. Like N_2, O_2 is a non-condensable gas and would be separated from the CO_2 in the CO_2 Processing Unit (CPU) prior to storage. With the appropriate controls, oxy-fuel combustion will allow for stack excess O_2 to be maintained near 1%. Reduced excess O_2 results in both lower ASU and CPU power usage, since less total O_2 is required and fewer impurities will be sent to the CPU.

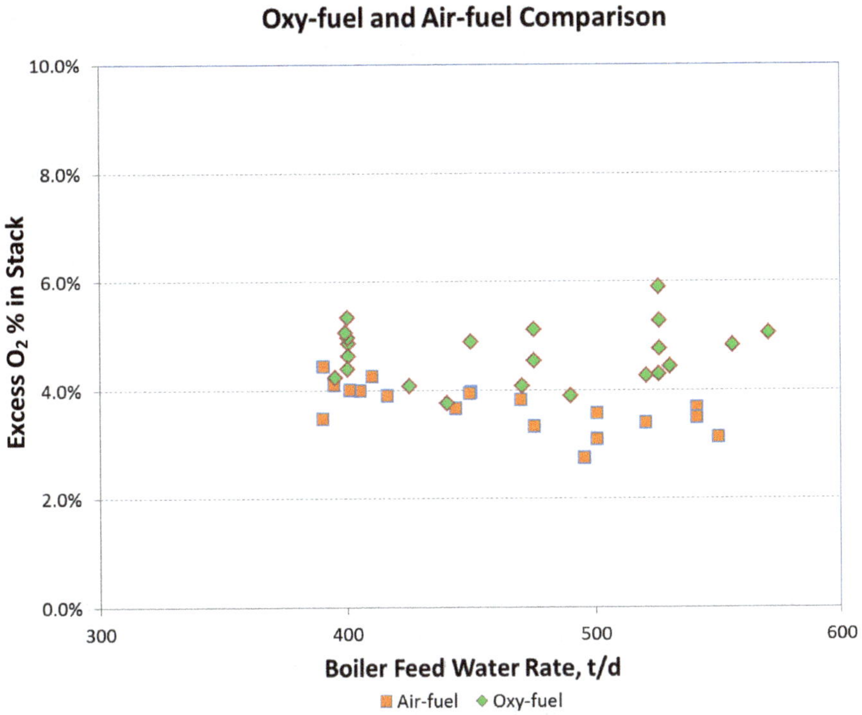

Figure 7. Stack excess O_2 versus boiler load.

Figure 8 shows the mass flow of flue gas through the evaporator and economizer and the concentration of O_2 in the flue gas recirculation versus the boiler load at 70% steam quality. Most of the experiments were conducted with an FGR O_2 concentration of 23%. This oxygen concentration produced stable combustion while keeping burner temperatures acceptably low.

The flue gas mass flow for oxy-fuel is generally lower than for air-fuel at comparable load and similar boiler performance, which means that the boiler heat transfer is similar under these conditions although less flue gas mass is flowing through the boiler. It should be noted that for oxy-fuel the mass flow of gas through the boiler is significantly higher than the stack mass flow as most of the gas flowing through the boiler is flue gas recirculation. The differences in heat transfer characteristics for oxy-fuel will be discussed in subsequent sections.

Figure 8 also shows the variation of the O_2 concentration in the flue gas recirculation from 23% down to 21.2% at low load. The operation at 21.2% at the lower boiler load was at the burner stability limit and any further reduction in oxygen concentrations would have resulted in a loss of the flame. For the same boiler load (and fuel rate) the amount of O_2 required for combustion is the same, so the lower O_2 concentration is achieved by increasing the FGR, which results in a higher mass flow through the boiler. The reduction of the O_2 concentration in the FGR by approximately 2% leads to an increase of the boiler flue gas mass flow to the level of the air combustion. The same increase in mass flow occurs when stack excess O_2 is increased by about 2%. Increased gas mass flow shifts heat transfer from the radiant section to the economizer.

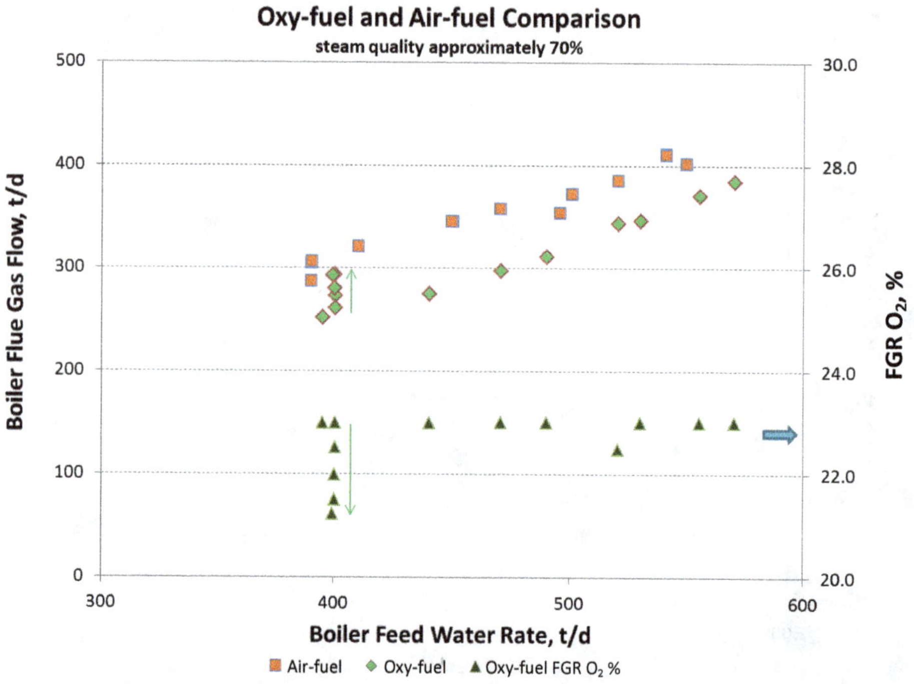

Figure 8. Stack excess O_2 vs boiler load.

Transitions between Air-fired and Oxy-fired Operation

Transitioning between air-fuel and oxy-fuel was accomplished by first slowly opening the FGR dampers to introduce a mixture of FGR and oxygen, followed by slowly reducing the air flow to the boiler with the air dampers until the two air dampers were completely closed. It proved essential to place the J-Burner in the center of the air burner into service for the transitions in and out of oxy-fuel, as well as during full oxy-fuel operation. The energy from the small pure oxy-fuel burner increased the local temperature at the flame root and provided stable ignition conditions for the fuel of the main burner. Transition back to air-fuel was accomplished in a similar fashion but in reversed steps.

Transitions were attempted at 70% and 50% boiler load. However, at 70% load, it was found that changing the damper positions induced pulsations in the flow that prevented a successful transition. With the recycle loop in the system, combustion disturbances are fed back to the burner inlet after the gas has travelled through the system. In the case of oxy-fuel combustion with flue gas recirculation, approximately 75% of the combustion gas is recycled. For example, if the burner starts to become unstable and the flame starts pulsing, the resulting pressure and flue gas composition fluctuations are fed back to the burner inlet where they further disturb the combustion. Eventually the flame is lost.

However, after several initial flame failures and boiler shutdowns, the transition procedure was optimized at 50% load, and transitioning became somewhat routine. The first successful transition took approximately two hours while at the end of the demonstration test a transition to oxy-fuel was completed in less than 30 minutes. The transition back to air-fuel was accomplished in a similar time frame.

Figure 9 shows several boiler parameters versus time for the transitions from oxy-fuel to air-fuel combustion and back. The O_2 concentration in the recycle flue gas was 22% during the transition and on oxy-fuel. The left vertical axis shows the mass flow of air, O_2 and fuel. The brown line at the top shows the "Total Air", which is the actual combustion air flow during air-fuel operation, and the equivalent air calculated based on the oxygen flow rate during oxy-fuel operation. The black line displays the fuel flow times 10 to fit the common scale. The O_2 flow is shown in green and the actual combustion air flow in orange. The right axis shows the excess O_2 (gray line) in the stack gas.

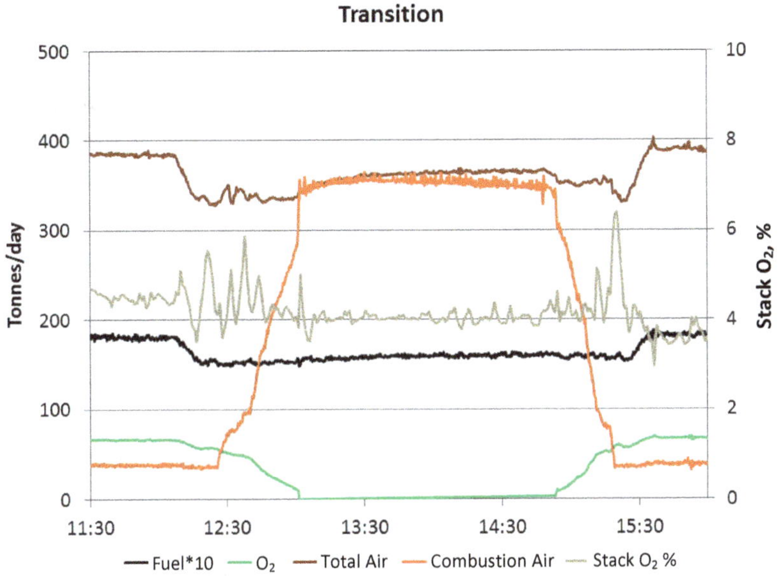

Figure 9. Transition from Oxy-Fuel to Air-Fuel and back.

The boiler is in stable oxy-fuel mode at the beginning of the time period. The air flow measurement (orange) is non-zero due to the measurement location and an offset of the flow transmitter at low or zero flows. However, the actual flow of air was reduced to small air ingress through the closed air dampers. To initiate the transition, the boiler fuel input was reduced to approximately 15 t/h (50%), and the flow of oxygen and FGR with O_2 followed suit. At 12:27h the air dampers were opened and the flow of air started to increase while the oxygen flow is reduced. At 13:02h the transition to air-fuel was completed and the air flow measurement matched the total air flow. The boiler was held at a constant low load for approximately 2 hours before the transition from air-fuel to oxy-fuel was started at 14:52h. Here the process is reversed and initiated by opening the FGR dampers and starting FGR flow. The control system starts to increase the oxygen flow as needed to maintain appropriate oxygen concentration in FGR. When the FGR dampers were open the combustion air damper started to close further reducing air flow. The control system increases FGR and oxygen flows to maintain required combustion stoichiometry. The transition was completed at 15:20h and load was increased.

Although the excess oxygen in the flue gas (grey) shows increased fluctuations during the transitions, this was due to control system tuning for the air-to-fuel ratio. The fluctuations were within an acceptable range. The main reason for the sensitivity is due to large air flow change with small damper changes when the air dampers are almost closed.

Further process trends for the transition are shown in Figure 10. The O_2 flow in green is used as a reference for oxy-fuel operation. Initially, the mixture of FGR and O_2 is at 162°C, and the humidity is about 40%. The on-line steam quality measurement shows values just below 80%. When the boiler is transitioning to air-fuel, the temperature and humidity of the oxidant flowing to the burner drops as the oxidant is gradually switched from flue gas/O_2 to air. As mentioned above, the air pre-heater was removed from the boiler for the demonstration. The humidity drops during the transition due to the low humidity of the increasing volume of ambient air mixing into the FGR.

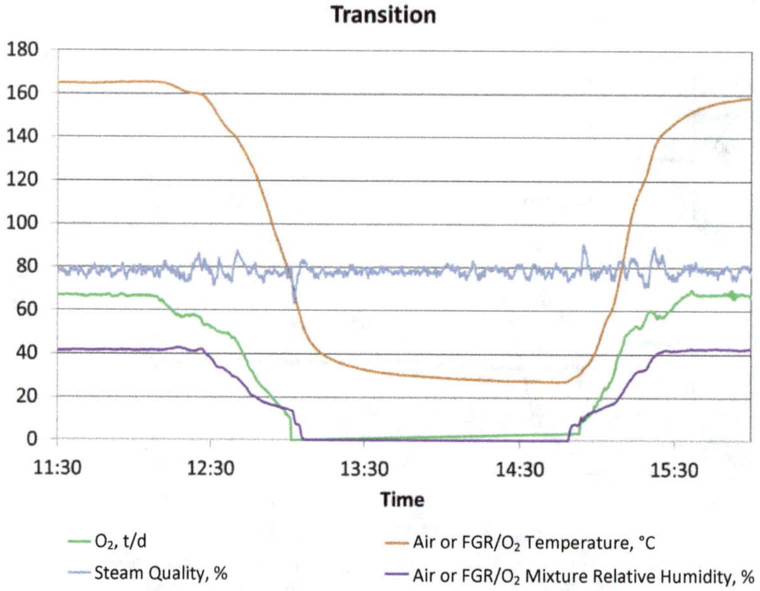

Figure 10. Other process data during transition.

The transition of the boiler from air-fuel to oxy-fuel reverses the trends. Although there are a few larger fluctuations of the steam quality during the transition periods, it remains essentially unchanged during switches from oxygen to air and back.

A video camera allowed constant flame monitoring and was invaluable for operation of the tests. The changes in flame shape, potential localized overheating and beginning of flame instability could be easily detected and corrective action taken. Figure 11 shows four flame images during the transition from air-fuel to oxy-fuel. The air-fuel flame shows bright flames near the pie-shaped burner refractory tiles. The fuel injected through the core fuel nozzles of the burner burns in the recirculation zone of the tiles that develops due to the air from the windbox that is coming from the slots between the tiles. The tiles glow orange and the increased temperature in this area provides appropriate ignition conditions. However, the bulk of the pre-mixed fuel is burning with a blue flame which fills nearly the entire image. The tips of this flame are brighter from small amounts of soot that burn out in this region. This soot combustion illuminates the boiler internally.

The image at the top right of Figure 11 shows the oxy-fuel J-Burner located in the center of the air burner in service. The heat input of this small support burner is only 3% of the total combustion heat input. The very luminous oxy-fuel flame illuminates the outer burner parts, and the ring of the outer fuel nozzles becomes visible which makes the burner appear to be larger. Due to the auto-gain setting of the camera the increased brightness reduces the visibility of the air-fuel flame. The brightness of the burner tiles appear less for the same reasons.

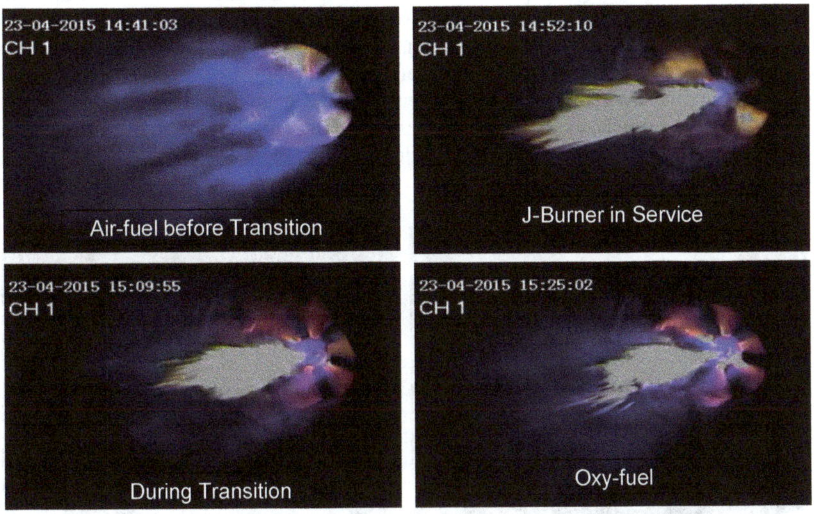

Figure 11. Flame images for transition from air-fuel to oxy-fuel.

During the transition (Figure 11, bottom left image) the flame above the tiles has nearly disappeared, and the J-Burner flame is shorter. The mixture of air, oxygen, flue gas recirculation and natural gas burns with a dark blue flame. The shorter J-Burner flame is a result of higher overall burner velocities due to the increased temperature of the incoming mixture with the flue gas recirculation. The forward momentum of the small J-Burner competes with the local recirculation near the burner axis which provides a backwards flow near the burner. At very high burner loads this resulted in a very short flame from the J-Burner.

The oxy-fuel flame after the transition is complete is shown in the bottom right image (Figure 11). The burner tiles glow orange with the exception of the immediate vicinity of the core gas nozzles which are very bright. The J-Burner flame has further shortened and is more slender. As a result the burner is less illuminated and the outer refractory ring is less visible. The pre-mixed fuel burns with the oxygen in the flue gas recirculation with a dark blue flame. The overall brightness in the boiler was decreased. However, the brightness or visible light emissions are not necessarily related to the local heat flux (see below).

Flame Characteristics

The comparison of the oxy-fuel flame to air combustion at higher load (Figure 12) shows that the air flame has a large bulb-shaped area of blue flame close to the burner. The tips of this part of the flame are sometimes more luminous due to combustion of small amounts of soot. The tiles are glowing orange indicating that there are high combustion temperatures present in the ignition zone. Adjacent to some tiles more luminous flame can be seen. In the video this luminous flame is not stationary and jumps from tile to tile. In the center of the burner the tip of the J-Burner is glowing orange due to the high temperatures in this area. The J-Burner was designed for these conditions and there was no damage noticed on the burner after the demonstration.

The oxy-fuel flame image on the right in Figure 12 shows a much darker main flame which appears to be more slender. This lower luminosity of the main flame resulted in less light inside the boiler, which made observation through inspection doors a bit more challenging. However, the lower light conditions should not be confused with the radiative heat flux that the boiler tubes receive. The heat flux is generally higher with oxy-fuel combustion, because the products of combustion, water vapor and CO_2, are radiating gases with much more effective heat transfer. The center of the oxy-fuel flame is illuminated by the flame of the J-Burner. The heat from this burner and the fuel injected close to the center through the core spuds results in a very bright burner center with high temperatures. The rest of the burner tiles are glowing orange in a similar fashion than with air combustion.

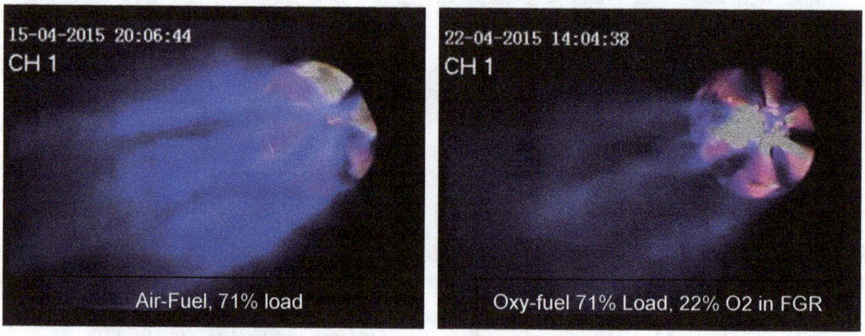

Figure 12. Air-fuel and Oxy-fuel flame images for high load.

Figure 13 compares flame images for oxy-fuel operation at different loads, with an O_2 concentration of 22% in the FGR blend. At low load, the J-Burner flame is noticeably longer. As discussed in the description of the transition above, the forward momentum of the small J-Burner is not strong enough to compete with the internal recirculation flows of the larger flame at higher load. This "pulls" the small oxy-fuel flame towards the burner. The burner tile appeared hotter at high loads and the temperatures at the center of the burner increased considerably. This was confirmed by inspection of the tile steel backing plates during the test period. Due to concern for the burner integrity, higher loads were not tested. With relatively simple burner design changes, this limitation can be overcome. The darker blue flame parts of the main flame are brighter at high load and the flame is longer and more slender.

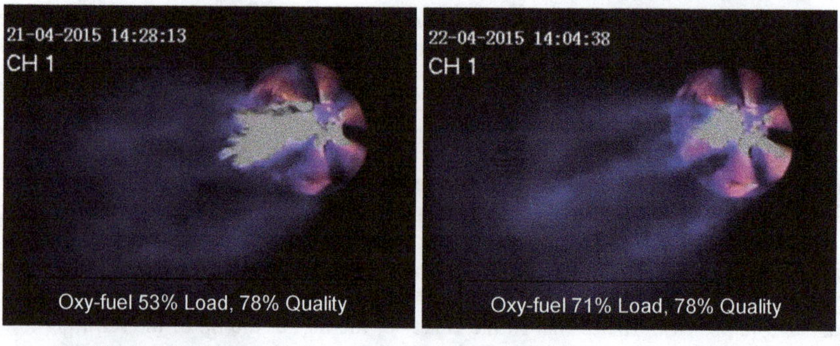

Figure 13. Oxy-fuel flame images for low and high load operation.

Flame Stability

Figure 14 shows a comparison of two flame images for oxy-firing with different concentration of O_2 in the flue gas recirculation at low boiler load (53%). The left image shows combustion with a low O_2 concentration, or higher FGR flow at the same O_2 flow. This image shows combustion at the stability limit. It was not possible to operate the burner below a concentration of 21% O_2 in the FGR. The video shows that the flame exhibits noticeable flickering. The blue section of the flame is not well shaped and "pumping" with larger dark parts. The burner tiles are not very bright. The burner produced significantly higher CO emissions when it was close to the stability limit, which is an indication that parts of the flame are getting too cold for complete fuel combustion. The flame extinguished several times while determining the lowest possible O_2 concentration in the FGR.

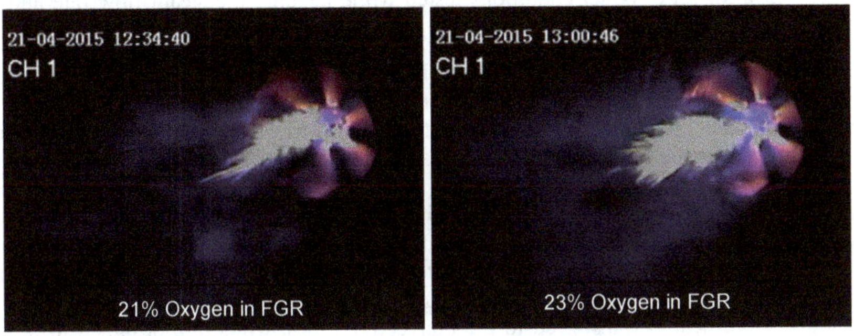

Figure 14. Oxy-fuel flame images for varying O_2 concentration.

In contrast, the image at 23% oxygen in the FGR shows that the tile is hotter and the flame is well shaped. There were no noticeable stability issues in the video and the blue part of the flame is well developed.

It is difficult to define a single parameter that affects flame stability. It is a function of burner velocity and local recirculation of hot combustion gas back to the flame root, where sufficiently high local temperatures at the ignition point near the tile are necessary to maintain the flame. These local temperatures are influenced by many factors such as firing rate, burner fuel distribution and O_2 concentration in the oxidant, as well as the heat capacity of the gas. Higher velocities through the burner negatively influence the ignition process, because local velocities at the ignition point must be below the flame speed, so that the ignition front can propagate back.

The design of the ultra-low NOx burners pre-mixes the majority of the fuel with the air (or the oxygen/FGR mixture) that is flowing through the burner. This results in very low combustion temperatures and therefore low NOx. However, the air flow is intentionally fast to avoid early combustion behind the burner tiles. The normal operation of the burner is close to the stability limit.

The mixture of oxygen and hot recycle flue gas entered the windbox at 160 to 180°C. In contrast, the combustion air temperature was much lower, due to the absence of pre-heating. The differences in mass flow and the combustion oxidant density difference resulted in gas velocities at the burner 33% to 40% higher for oxy-fuel versus air-fuel. This appreciable increase in velocity contributes to the burner exhibiting lower stability for oxy-fuel combustion under certain operational conditions. However, for most conditions the oxy-fuel combustion showed a stable flame, and the demonstration of oxy-fuel combustion with this burner was quite successful.

The increase of burner velocities was also experienced when large SAGD boilers in commercial operation were retrofitted with flue gas recirculation systems for NOx reduction. Mixing flue gas recirculation (approximately 15%) into the combustion air lowers local flame temperatures and reduces NOx, but it also increases the gas mass flow through the burner. The latest generation of SAGD boilers with FGR for NOx control has relatively larger burners compared to similar boilers without this control method.

Heat Flux and Boiler Performance

It is well known that oxy-fuel combustion can increase furnace and boiler efficiencies due to reduced sensible heat loss (no N_2 heating) and improved radiative heat transfer. This increase in efficiency allows the boiler operator to reduce the firing rate while maintaining the same steam output and quality. Data from the demonstration showed that the fuel consumption rates were reduced by ~ 5% for all oxy-fuel conditions tested while keeping the steam flow rate and quality constant.

The design of the OTSG makes the heat duty split between the radiant section and the convective section a critical performance parameter. The common design practice for OTSG is based on low radiative heat flux and incomplete evaporation of the boiler feed water in the radiant section to avoid solids precipitation in the boiler tubes, due to use of low quality boiler feed water. For oxy-fuel operation, it is important to keep the split similar to that of air-fuel operation as well as to keep the heat transfer in the radiant section below the design limit in order to avoid added maintenance and operational issues caused by radiant tube dry out. If the heat transfer to the radiant tubes is high enough to cause the coils to dry out, the feed-water solids that deposit act to inhibit inner tube wall heat transfer and results in higher tube temperatures, which can lead to tube failures. A constraint also occurs if more heat is transferred to the convective tubes, which are designed to heat liquid phase boiler feed water. Too much heating could cause steam evaporation in the economizer, which is not designed for that service.

Since N_2 from the air is eliminated with oxy-fuel combustion, the net flue gas flow is significantly lower than for air-fuel combustion. The emissivity of the oxy-fuel flue gas is significantly higher than that from air-fuel combustion, which results in higher heat transfer in the radiant section. Therefore, in order to better match the oxy-fuel radiant duty in the boiler with the design for air-fuel combustion, flue gas recirculation is used to increase gas mass flow through the boiler to moderate temperature and to transport more heat to the convective section.

Data for heat duty splits for oxy-fuel combustion and air-fuel combustion for various test conditions are plotted against boiler fuel load in Figure 15. The duty split was calculated by a heat and mass balance on the boiler feed water / steam side. A slight increase in heat transfer in the boiler evaporative (radiant) section can be seen with oxy-fuel operation compared to that of air-fuel, however, in general there is little difference. The higher heat transfer in the evaporator is consistent with the fact that the O_2 concentration in FGR actually used in the operation is higher than originally designed, which led to slightly less flue gas flow. The heat duty split between the radiant and convective section is load dependent and at lower firing rates more heat is transferred in the radiant section. This could be seen in the same plot by the slight trending down in the duty split as firing rate is increased.

The higher heat transfer in the radiant section during oxy-fuel combustion relative to air-fuel is shown in Figure 16 with heat flux measurements at 44% of the relative evaporator length versus firing rate. The standard OTSG design limits the maximum heat flux in the radiant section to 39,000 to 40,000 Btu/ft^2-hr, or 123 KW/m^2. The firing rates in these test operations were not close to 100% of design capacity. The data in Figure 16 shows that the heat flux for oxy-fuel is slightly higher than that of the base air-firing case. This is due largely to the increase in the gas emissivity from ~0.30 (air-fuel) to ~0.52 (oxy-fuel). Therefore, even if the gas temperature profile is similar between the two combustion environments, the heat flux will be larger with oxy-firing.

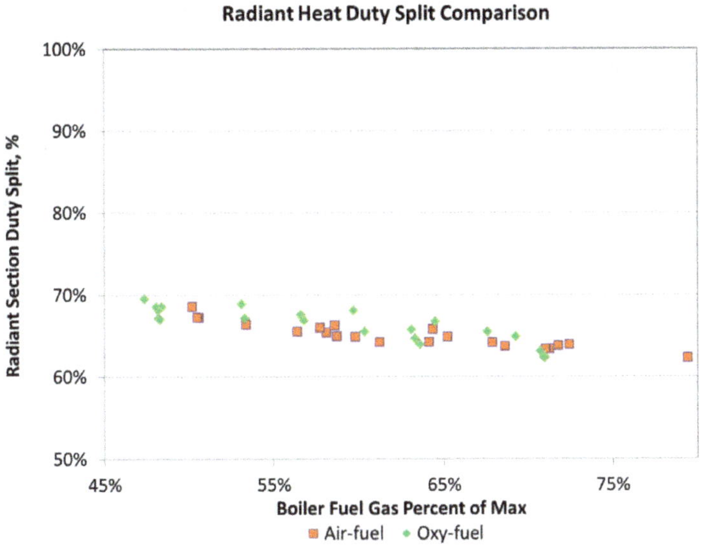

Figure 15. Heat duty split versus boiler load.

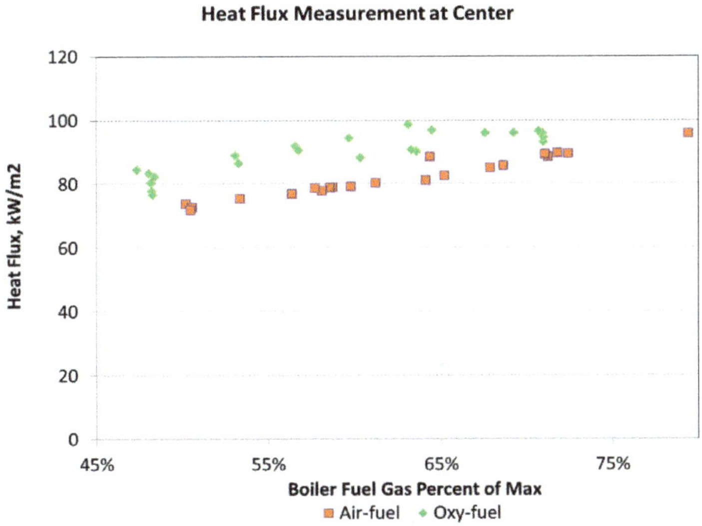

Figure 16. Total heat flux at Center Probe.

When the boiler tubes were inspected during operation at high oxy-fuel load, a more visible glow of the tube hangers in the evaporator center was noticed, compared to air-fired operation. This was likely a combination of the hangers being more noticeable due to the generally "darker" boiler under oxy-fuel conditions and due to the increase of the heat flux in the center by approximately 10-15%. Higher FGR flow would have reduced the radiant flux in this area, but would have also lowered the O_2 concentration in the FGR, which has negative flame stability impact.

The most direct way to control the heat absorption balance is to optimize the flue gas recirculation rate. Since the flow of O_2 is fixed at a given fuel consumption, the flue gas recirculation rate also defines the O_2 concentration in the oxidant mix fed to the burner. Our calculations show that the excess O_2 in the exhaust stack has strong correlation with the radiant/convective duty balance. This is illustrated in Figure 16, which shows the O_2 concentration in the FGR required to match the air-fuel combustion heat transfer as a function of excess O_2 in flue gas. During Phase I of the project, the boiler was simulated using CFD. The calculations showed that the O_2 concentration in the FGR has to be 19% to match the design heat transfer split between the radiant and convective section. The model, however, was based on the assumption that the flue gas has 1% excess O_2. However, as shown in Figure 17, at 1% excess O_2, the FGR must contain between 17-18% O_2 to balance the heat transfer in the two sections, which is consistent with the CFD model. During the oxy-fuel tests, the excess O_2 was at 4-5%, which means that the FGR O_2 concentration would have to be 21 to 23%, which is consistent with the experimental observations.

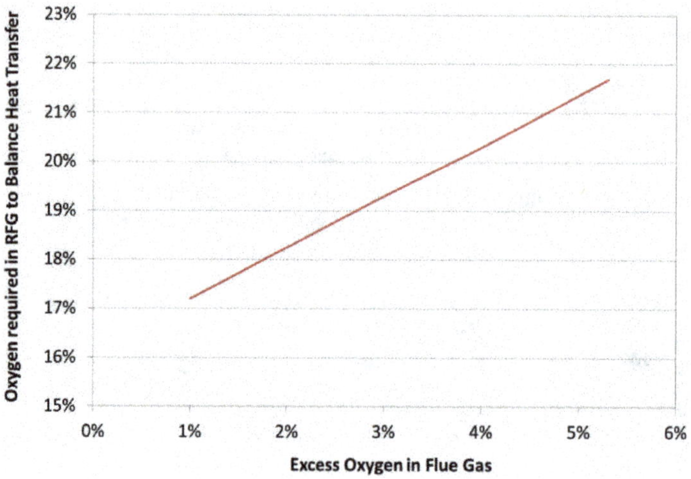

Figure 17. Oxygen required in FGR to balance heat transfer versus excess O_2 in flue gas.

Emissions

The NOx emissions for all tests regardless of parametric variations are shown in Figure 18. The volumetric concentration was converted to the mass-based unit commonly used in North America, lb NOx per MMBtu on a higher heating value basis. Reporting emissions based on mass, rather than based on concentration makes sense for oxy-fuel, because the volume of flue gas is much lower and there is no nitrogen that acts as a diluent as is the case for conventional air-fuel combustion.

The oxy-fuel NOx emissions are only about 15% of those measured for air combustion. The variation of the emissions under different operating conditions for both air-fuel and oxy-fuel operation is surprisingly small.

The CO emissions at the stack were below detection for both air-fuel and oxy-fuel operation. Significant CO emissions were only briefly measured when the burner was close to losing flame or immediately after flame was lost. Unburned fuel in the flue gas is expected when flame stability is compromised.

Figure 19 shows the NOx and CO_2 concentration and the stack temperature measured at the stack during the transition from oxy to air and back discussed previously. The stack temperature dropped slightly during the transition from oxy-fuel to air-fuel, but it is relatively constant during the entire timeframe. Since the fuel flow changes only a few percent between oxy-fuel and air-fuel operation, the CO_2 mass flow stayed essentially the same. However, as expected, the CO_2 concentration dropped during air-fuel combustion. The two anomalies in CO_2 concentration are due to the analyser going off line for automatic calibration.

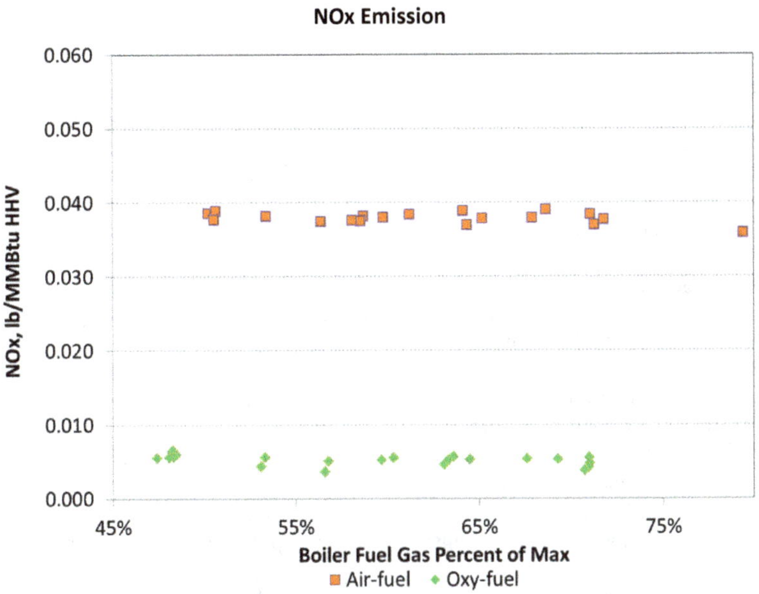

Figure 18. NOx emission for air-fuel and oxy-fuel operations versus boiler load.

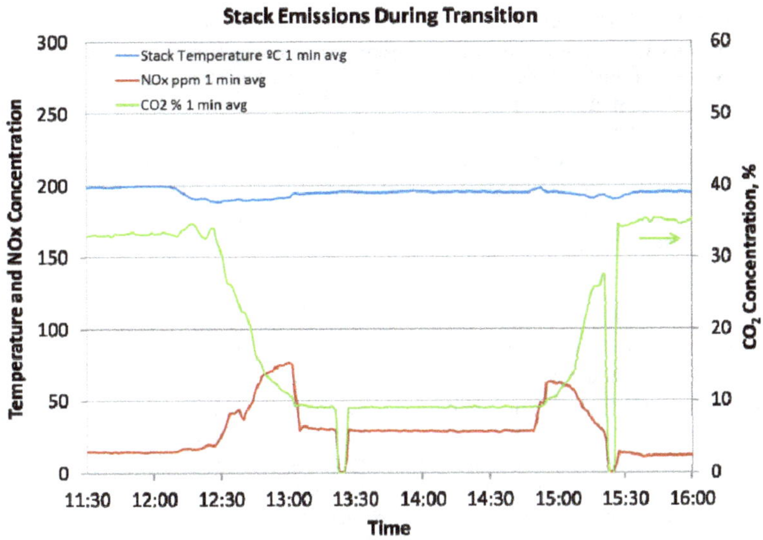

Figure 19. Stack emissions during operating mode transitions.

CONCLUSIONS

This section briefly summarizes the key results and limitations of the demonstration at Christina Lake. The demonstration test resulted in a number of recommendations for a full scale commercial implementation of oxy-fuel on a SAGD boiler that are documented here.

Successes

- The operation of an OTSG in SAGD service with oxy-fuel combustion and flue gas recirculation was safely demonstrated.
- The demonstration clearly showed that there are no technology limitations for retrofitting a SAGD OTSG boiler for oxy-fuel combustion and operating it continuously. Boiler performance was nearly identical for air-fuel and oxy-fuel operation.
- There were no operational issues with oxy-fuel combustion. The boiler achieved the same steam quality and steam flow compared to loads under air operation.
- On average, the fuel flow was reduced with oxy-fuel combustion by approximately 5% compared to air-fuel combustion at the same load and steam quality. This is an inherent advantage of oxy-fuel combustion.
- The use of a small, pure oxy-fuel pilot burner in the main burner center proved to be invaluable for the stabilization of the oxy-fuel flame. The heat from this burner resulted in improved combustion temperatures at the main burner flame root.
- After initial challenges, the transitions between air and oxy-firing were optimized very well so that the transition process was performed almost routinely.
- Transitions are best performed at low boiler load at moderate flows for air and oxygen. This minimizes the potential for pulsations to develop that could result in the loss of the flame.
- The retrofit of the boiler used common components that are available to the industry. The cost of the boiler retrofit is small compared to the cost of the entire carbon capture system.
- The mass-based oxy-fuel NOx emissions were on average only 15% of those measured with air-fuel combustion. No CO emissions were measureable with air or O_2 firing. However, in a future oxy-fuel application for carbon sequestration, the flue gas is fed to the CO_2 Processing Unit that will remove NOx, SOx and CO. In other words, the "classic" hazardous air pollutant emissions are virtually eliminated.
- A comprehensive safety process was executed following a hazard and operability study. The technical and operations team worked together to implement the oxygen supply and oxy-fuel process safely at a site that processes hydrocarbons. The demonstration resulted in an increased comfort level with oxygen storage and use at such a site.

Limitations

- The air pre-heater had to be removed for the demonstration to make room. This changed the temperatures at the windbox for air-fuel vs oxy-fuel operation.
- Pre-mix low NOx burners normally operate close to the stability limit to combust with low flame temperature for NOx reduction. The demonstration has shown that operation with O_2 and flue gas recirculation is not very stable and required higher than expected O_2 concentrations in the FGR (21.2% minimum). Due to material compatibility issues, the upper range of testing was 23.5% oxygen in FGR. This narrowed the range of oxygen concentration in the flue gas recirculation that could be tested.

- The gas volumes through the burner with oxy-fuel operation were higher than with air-fuel operation due to higher oxidant temperatures, which had the effect of making the burner undersized for this application. This made it impossible to reach full boiler load under oxy-fuel in this test. However, this is a demonstration test limitation and not a technology problem. It can be mitigated with relatively simple design changes to the burner.
- The main burner was not stable without operating the J-Burner serving as a pilot burner in the burner center. However, the relatively small pilot flame competes against the local flow of the main flame recirculation and is shortened substantially at higher loads. The resulting localized overheating of the main burner at high loads was an operational concern. A more integrated design of the pilot burner with the main burner needs to be developed for a full scale commercial burner.
- There was no limitation operating the boiler with flue gas recirculation, but it made robust control more challenging. FGR adds an internal process feedback loop for concentration and pressure fluctuations that negatively impacted burner stability.
- The tested boiler did not have a modulating stack damper to adjust the boiler pressure during operation (it was possible manually). The very low oxy-fuel flue stack flows resulted in a load limitation, since the duct pressure at the air inlet damper increased with higher load and can exceed the local ambient pressure. The load could not be further increased when FGR started leaking out of the air inlet damper due to a positive pressure inside the duct at that location, since leaking FGR into the boiler house is an asphyxiation hazard.
- Due to leakage mainly at the air intake damper, the air ingress was relatively high. Thus, the nitrogen concentration in the flue gas was higher than expected.
- High dew point in the FGR and stack flue gas poses a condensation challenge and dripping of water was noticed from inspection doors at the economizer. Since the test was completed in April, no safety issue was created by ice buildup.
- Two of three heat flux probes failed before the demonstration test which made the intended comparison of the heat flux profiles for air-fuel and oxy-fuel impossible. The data from the center probe indicated the expected increase in local heat flux at this location.
- The oxy-fuel combustion equipment and retrofit of the duct arrangement had to be integrated into an existing boiler which required making certain design compromises. As discussed in this chapter, this resulted in a few limitations in how the system could be operated, but the team learned quickly to understand the system response and safely work around these issues. This experience resulted in a few of the recommendations in the following section.

Scale-up Considerations

Although the demonstration boiler is smaller (50 MMBtu/hr) than typical full-scale SAGD boilers (250 to 300 MMBtu/hr), the findings can be applied to larger boilers. There are no technical scale-up barriers to implementing oxy-fuel combustion commercially for carbon capture in this application.

- Air ingress needs to be minimized to eliminate N_2 in the flue gas as much as possible to increase CO_2 concentration for carbon capture. This requires tight shutoff dampers for the air inlet (gate dampers). Condensation in the dead space of the air inlet behind the damper must be managed. Cleaned and dry CO_2 gas from the outlet of the CO_2 processing plant may be a suitable purge gas.

- There are no limitations to fully automate the air-fuel to oxy-fuel transition process in the future. The demonstration test did not have this goal. Dual combustion capability (air and oxygen) is simple to implement and provides maximum operational flexibility. Steam production would not be in jeopardy if any components of a future carbon capture system became unavailable. To automate the transition procedure and achieve a smooth transition, the flue gas recirculation and air control dampers require precise positioners. The control dampers should be opposed blade dampers with a low leakage and a nearly linear flow characteristic.
- Differential pressure based flue gas flow meters are not ideal during the low flow conditions of the transitions, especially if the transition process is automated. High turndown instrumentation such as thermal dispersion mass flow meters should be investigated.
- The burner design should be modified to manage higher velocities during oxy-fuel combustion with FGR. An alternative could be to introduce some of the recirculation gas through the boiler front wall outside of the burner at high flow conditions. In addition, the integration of a pilot burner or other means of stabilizing the main burner under oxy-fuel combustion should be investigated.
- The stack damper should be retrofitted with a modulating damper to easily control boiler internal pressure over the entire load range.
- Condensation management because of high oxy-fuel flue gas dew point is essential. The duct system needs to be designed with appropriate drain points, system cold spots must be avoided by adding insulation and boiler access door seals must be appropriately designed for the presence of liquid water.
- Liquid water corrosion or weak acid corrosion due to high CO_2 or residual SOx content of the flue gas was not characterized as part of the demonstration study. These issues can be managed very well through appropriate design measures such as "warm" casing, avoiding standing water, appropriate insulation and material choices. However, a separate design study is recommended before the oxy-fuel combustion technology is used on a more widespread basis on SAGD boilers. Studies and experience with coal-fired oxy-fuel demonstrations projects may provide good guidance.
- The oxygen piping and skid for the demonstration were stainless steel. This choice was made because of the presence of sour gas in the facility. However, copper and carbon steel piping are viable choices for oxygen systems if certain safe design practices are observed. These material choices may be more economic for a full scale installation and should be kept in mind.

ACKNOWLEDGEMENTS

The funding support provided to this project by Climate Change and Emissions Management Corporation (CCEMC), which receives funds from the Province of Alberta, the CO_2 Capture Project Phase 3 (CCP3), Cenovus Energy, Devon Energy Canada, Praxair, Statoil, and MEG Energy is gratefully acknowledged.

REFERENCES

1. Bohm, M. (2011), 'Application of Oxy-Fuel CO_2 Capture for In-Situ Bitumen Extraction from Canada's Oil Sands' presented at GHGT-10, 19-23 September 2010, Amsterdam, The Netherlands. Energy Procedia 4, pp 958–965.
2. Praxair, US Patent 9,091,430 B2 (2015), *Stabilizing combustion of oxygen and flue gas*, July 28, 2015.

Carbon Dioxide Capture for Storage in Deep Geological Formations, Volume 4
Karl F. Gerdes (Editor)

Chapter 16

CHEMICAL LOOPING COMBUSTION SUMMARY OF CCP3 ACHIEVEMENTS

Ivano Miracca[1], Anders Lyngfelt[2], Juan Adánez[3], Otmar Bertsch[4],
Karl Mayer[5], Frans Snijkers[6], Gerald Sprachmann[7], Gareth Williams[8]
[1]Eni S.p.A. – Via Martiri di Cefalonia, 67 – I-20097 San Donato Milanese, Italy
[2]Chalmers University of Technology, SE-41296 Gothenburg, Sweden
[3]Instituto de Carboquímica (ICB-CSIC), Miguel Luesma Castán 4, E-50018, Zaragoza, Spain
[4]Josef Bertsch Gesellschaft MBH & Co KG, Herrengasse 23, O-6700, Bludenz, Austria
[5]Vienna University of Technology, Getreidemarkt 9/166, O-1060, Wien, Austria
[6]Flemish Institute for Technological Research (VITO), Boeretang 200, B-2400, Mol, Belgium
[7]Shell Global Solutions International B.V., Grasweg 31, 1031 HW, Amsterdam, The Netherlands
[8]Johnson Matthey Technology Centre, Blount's Court, Sonning Common, Reading, RG4 9NH, UK

ABSTRACT: This chapter summarizes work supported by CCP3 to further develop Chemical Looping Combustion technology that utilizes natural gas as fuel. The specific application is production of steam for use in heavy oil extraction, while capturing the CO_2 generated by the operation. This work had several parallel activities, mainly focusing on the development of novel oxygen carriers and technology scale-up.

KEYWORDS: CO_2 capture; CCS; chemical looping combustion; oxygen carriers; steam assisted gravity drainage (SAGD); oxy-combustion; INNOCUOUS

INTRODUCTION

The Chemical Looping Combustion (CLC) concept is to utilize a solid carrier to extract oxygen from air and transport it to a reaction environment to oxidize a carbonaceous feedstock, which can be in solid, liquid or gaseous form. The carrier is a metal oxide or mixed metal oxide. This concept substitutes the circulating solid oxygen carrier for a cryogenic air separation unit in this variation on oxy-fuel combustion to achieve CO_2 capture. Chemical Looping Combustion is a flameless combustion technology where the chemically active solid material is circulated between two fluidized bed reactors (Figure 1).

In the "fuel reactor," a hydrocarbon fuel is oxidized by oxygen released from the oxygen carrier. If complete oxidation is achieved, CO_2 and H_2O are the products in the fuel reactor. In the "air reactor," the oxygen carrier is re-oxidized with air. The air reactor vents oxygen-depleted hot air. Somewhat counterintuitive is the fact that the fuel oxidation is slightly endothermic for most metal oxides that have been studied for CLC, and it is the re-oxidation of the metal by air that is highly exothermic. The net heat effect of the two reactors yields the normal heat of combustion. Such a CLC system typically operates at temperatures of 800-1000°C. In practice, the extent of fuel oxidation depends on the availability of oxygen (sufficient carrier circulation) and on the kinetics of the elementary gas-solid reactions (sufficient gas-solid contact time). As noted, energy is mainly released in the air reactor, and the solid oxygen carrier serves as a heat transfer agent to drive the endothermic fuel reactions. The air reactor is equipped with internal tube bundles, similar to a conventional boiler. Additional energy is recovered from the hot gaseous effluents of both reactors.

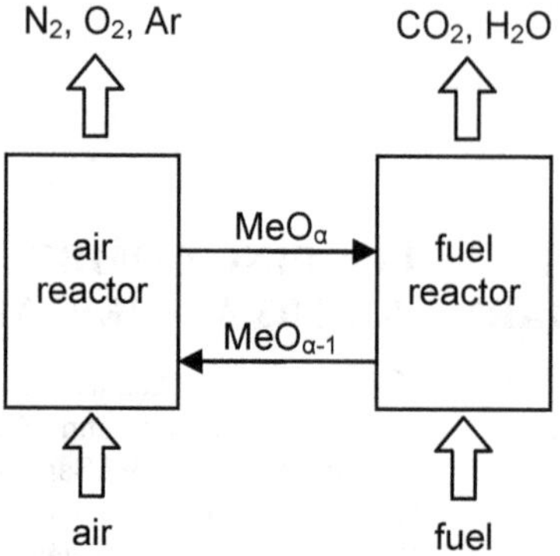

Figure 1. Conceptual scheme of Chemical Looping Combustion.

A key requirement for industrial feasibility of CLC is the availability of oxygen carriers suitable from both the technical and the economic point of view. During previous phases of the project, CCP supported formation of a Partnership, including Chalmers University of Technology, Consejo Superior de Investigaciones Científicas (CSIC), Technical University of Vienna (TUV), the Flemish Institute for Technological Research (VITO) and Alstom Boilers. This Partnership, in the frame of R&D projects co-funded by the European Union (EU), advanced the technology beyond the almost pure conceptual level (2000) with development of a very active nickel-based oxygen carrier and operation of an integrated 10 kW unit with continuous solid circulation at Chalmers (2003) [1]. Scale-up progressed further with operation of a 120 kW unit at TUV (2007) [2]. Evaluations by the CCP confirmed that this technology might result in considerable reduction in capture costs compared to oxy-firing or post-combustion capture.

A CLC boiler system, which requires two vessels, has a larger plot space requirement, and possibly higher overhead clearance, than does a conventional boiler. This is a serious constraint for application industrial facilities, such as oil refineries, in which CO_2 capture will mostly be a matter of retrofitting existing equipment in an often tightly constrained space. In contrast, the extraction of heavy oil by steam injection into the reservoir (e.g. Steam Assisted Gravity Drainage – SAGD), is an expanding application where new-build facilities will play a major role and plot space constraints are milder. CLC could therefore become the ideal solution for next generation capture technology in this field, replacing conventional Once Through Steam Generators (OTSGs) currently used in heavy oil operation in Canada and the USA.

THE FOCUS OF CCP3

Nickel materials have been demonstrated to be viable oxygen carriers, as they are very reactive and stable. The high performance of nickel-based oxygen carriers for hydrocarbon conversion is related to the catalytic activity of nickel for reforming reactions. The HSE risk assessments previously carried out within the EU-supported, CLC GAS POWER project concluded that use of nickel is not a "show stopper," although it was recommended to reduce potential particle emissions to a minimum.

Moreover, the CCP2 project showed that the relatively high cost of nickel should not be a major barrier, but it is clear that the successful development of nickel-free, or low nickel content, materials could provide important cost reductions. The cost reduction is not only due to lower material costs, but also for solids handling costs, where nickel materials would necessitate additional HES precautions.

Based on the above cost incentives, the development of novel carriers with low or no Nickel content has been the main focus for the research activity on CLC in CCP3. This effort was able to achieve financial leverage by working with the developers in the context on their ongoing work.

The CCP3 directly funded a project carried out by CSIC, VITO, TUV and Bertsch. Main tasks of this project were:

- Evaluation of a Cu-based carrier, developed by CSIC (made by impregnation) in a research project funded by the Spanish Government, in comparison to the Ni-based carrier, using the 120 kW pilot unit in Vienna.
- Preparation by VITO of a batch of spray-dried Cu-based particles.
- Design, construction and operation of a cold flow system by TUV to obtain useful data for design of larger scale and commercial CLC units.
- Preliminary design of a commercial CLC unit for application to the CCP heavy oil scenario, to be included in the CLC case prepared by Amec Foster Wheeler for the CCP economic evaluations.

In parallel, a consortium was formed by the same academic partners that worked with the CCP in the previous phases (see above), and the industrial partners Shell Global Solutions, Johnson Matthey and Bertsch Energy. This consortium carried out the EU-supported INNOCUOUS project (Innovative Oxygen Carriers Uplifting Chemical Looping Combustion). The project had a duration of 36 months (September 2010 – August 2013) with a budget of about 4 M€ (2.7 M€ funded by the European Union FP7 Program). Key objectives were:

- Develop new reactive oxygen carriers based on metals other than nickel.
- Develop non-nickel, mixed metal oxide materials with low fraction of added nickel catalyst.
- Optimization of scale-up ready particle manufacture.
- Testing of new particles under relevant conditions.
- Overall process integration and CLC next scale design.

While the activities and achievement of the project directly funded by the CCP are described in detail in chapters 17, 18 and 19, the remainder of this chapter is dedicated to the description of the main results of INNOCUOUS.

MAIN RESULTS OF INNOCUOUS

Development of novel oxygen carriers with low or no nickel

Ten oxygen carrier materials were tested in a 300W reactor for oxygen release performance and for fuel conversion with natural gas. The project identified a number of interesting non-Ni based oxygen carriers. Among these are two calcium manganates, denoted C14 and C28, which have been scaled up and tested successfully, both in Chalmers 10 kW unit and in Vienna's 120 kW pilot. Both carrier candidates also contain MgO, while C28 also includes titanium. Some of the main conclusions are:

- Both C14 and C28 show essentially complete fuel conversion – achieving very low concentrations of methane, carbon monoxide and hydrogen at the reactor outlet (see figure 2), in contrast to nickel oxide carriers, which are equilibrium-limited to 98-99% conversion.
- Both carriers showed low attrition rates in the 10 kW unit operation, comparable to previous results with Ni-based materials. An approximate life time of 12,000 hours in operation was estimated for the C14 carrier, and 9,000 hours for the C28 carrier.

- There is a clear deactivation in the oxygen transfer capability for the C14 particles after exposure to fuel at higher temperatures. This is not seen for the C28, indicating that titanium has positive implications with respect to stability of reaction.
- The rate of oxygen release was determined and was found to be somewhat greater for C28 compared to C14 and was in the range 0.0003-0.0006 kg O_2/kg of carrier.

These carriers may release oxygen directly to the surrounding atmosphere in the fuel reactor (Chemical Looping with Oxygen Uncoupling), though the direct fuel-carrier reactions are dominating the transfer mechanism at conditions used in this work. The release of oxygen was evident when operating in full conversion mode.

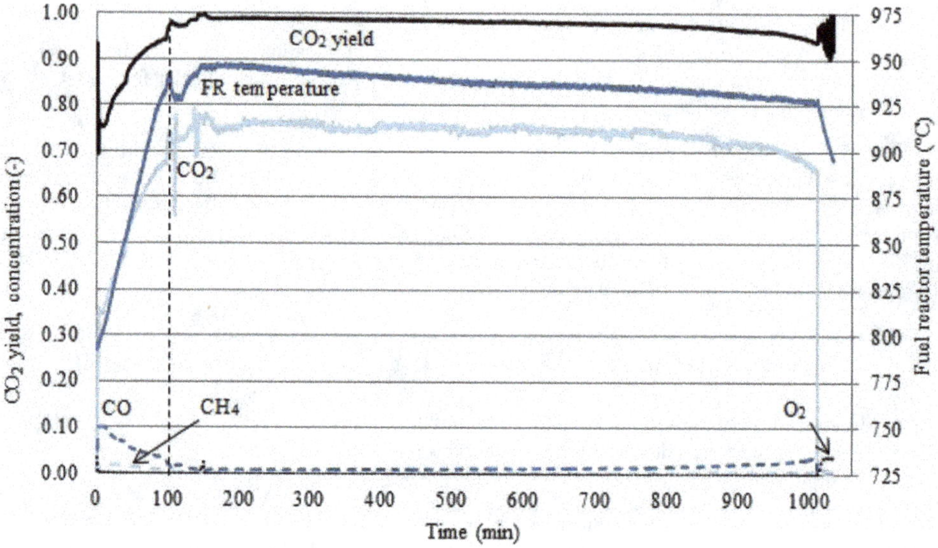

Figure 2. CO_2 yield, reaction temperature, mole concentrations of CO_2, CO, CH_4, in the fuel reactor effluent during a typical test on the 10 kW unit vs. time of operation.

These results are very positive, because they extend the range of potential carriers for large scale CLC application to materials other than Ni-based carriers. Results from laboratory experiments and operation in three pilots are given in [3-7].

In a parallel effort, a low fraction of NiO was added to different mixtures of low cost materials containing metal oxides and potentially active for CLC. The most promising materials were evaluated in a continuous 500 W CLC unit to identify oxygen carriers with reduced NiO content, while still exhibiting high performance. For most of the materials tested, however, the addition of Ni did not result in the expected improvement in performance.

Scale-up techniques for carrier preparation

The preparation route and associated manufacturing costs for the commercial (multi-tonnage) production of different CLC carriers were studied in detail by Johnson Matthey. The study included a Ni-based carrier developed within the CLC GAS POWER project, and novel Mn-, Fe-, and Cu-based carriers considered within INNOCUOUS. Both spray drying and impregnation routes were assessed. The analysis showed that, mainly due to the lower cost of raw materials, all of the novel carriers may be prepared commercially at about 15-30% of the cost of Ni-based carriers. Although not quantified, there is cost associated with safe containment and handling infrastructure for the Ni-based materials for the manufacturing plant when compared with the others; however, this is not judged to be a serious technical or economic barrier at present. This result confirms that the development of novel carriers may potentially lead to reduction of the CO_2 avoidance costs of the CLC technology compared to cost estimates based on Ni-carriers.

These data have been included as a sensitivity in the economic analysis of this technology (see chapter 20).

Intensive testing of novel carriers in the 120 kW unit

Four carriers were selected for an intensive test program in the 120 kW pilot unit at TUV.

Blend of Ilmenite and Ni oxide Particles

The first carrier was a physical mixture of ilmenite (a cheap and abundant natural material, containing iron and titanium oxides, known for its reactivity toward CO and H_2) and the baseline Ni carrier. The expectation was that addition of a small percentage of Ni/Al_2O_3 particles could result in complete conversion of CH_4 in the fuel reactor, resulting in a cost effective, yet high performance oxygen carrier. The experiments showed that CH_4 conversion was not as high as expected, and increased only gradually with further catalyst addition from 40% to 50%. We conclude that this type of mixture did not fulfil the expectations and is not a suitable oxygen carrier system for practical application.

Iron Oxide on Alumina

The second carrier tested is a Fe-based material impregnated on alumina. The use of Al_2O_3 as support material makes the Fe^{2+}/Fe^{3+} redox system accessible. Experiments have shown that the particles leaving the air reactor are always fully oxidized, indicating that the oxidation of the oxygen carrier is not limiting. The fuel conversion was, however, lower in the 120kW unit than reported in lab tests (60% max conversion achieved), meaning that a reactor with a larger inventory of catalyst or having longer gas-phase residence time would be needed for this carrier, compared to the baseline nickel-based carrier. Analysis of the particles after the experiments showed a significant loss of the active oxygen carrier material. Further development and optimization work for this carrier is being carried out in another EU-funded project (SUCCESS), started in September 2013.

Calcium Manganate-based Carrier C14

The C14 oxygen carrier material was tested in a series of experiments encompassing about 40 hours of CLC operation. With this oxygen carrier, stable operation was achieved. The fuel reactor temperature was varied in the range of 900 to 965°C. At 900°C the methane conversion was significantly lower (about 15 percentage points less) than for 950°C. At 965°C, a further increase of less than 5 percentage points in methane conversion was measured. Operation at 965°C fuel reactor temperature is at the limit of thermal stability for the pilot reactor design (stainless steel, no refractory lining). Since the higher temperature resulted in little difference in methane conversion compared to 950°C, the fuel reactor temperature for the remaining experiments was set to 950°C.

A series of pilot experiments with the C14 material was used to study system performance. One run used air feed diluted with nitrogen to lower the oxygen concentration, while maintaining the fluidizing gas flow to allow comparable solids circulation rates. Lowering the oxygen concentration in the air reactor led to lower fuel conversion in the fuel reactor. Increasing the solids circulation rate and higher oxygen concentrations in the air reactor increases the oxygen carrier to fuel ratio, which achieved full fuel conversion (Figure 3a). Also a higher specific inventory in the fuel reactor increases fuel conversion (Figure 3b). With a specific inventory of about 350kg/MW in the fuel reactor, full fuel conversion was possible. For operating conditions with incomplete fuel conversion, low concentrations of CO and H_2 were present in the fuel reactor off gas. Under operating conditions with full methane conversion, no CO was detected. Oxygen in the fuel reactor off gas was measureable for operating conditions with high fuel conversion (above 95%), but was at the lower end of the detection limit of the oxygen analyzer. For operating conditions resulting in full fuel conversion, excess O_2 was measured in the fuel reactor off gas stream at significant levels (up to 0.4 vol.%, wet basis).

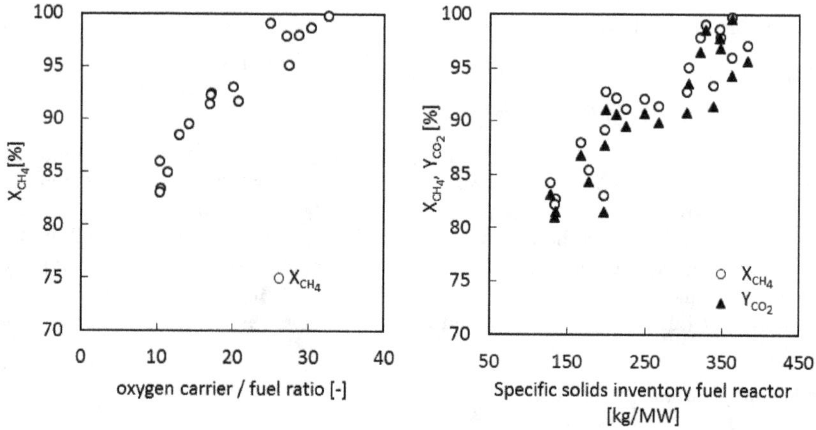

Figure 3. Carrier C14: (a) CH_4 conversion vs. oxygen carrier/fuel ratio, (b) CH_4 conversion and CO_2 mole % in the flue gas vs. specific solids inventory in the fuel reactor.

Based on the test data, carrier C14 has similar reactivity to Ni-based carriers with no equilibrium limitations in fuel conversion and has comparable attrition rates. This carrier is expected to be much cheaper than Ni-carriers (about 80% cost reduction), so that its use might result in cheaper operation.

Calcium Manganate-based Carrier C28

With the C28 oxygen carrier material, a similar testing program to that of C14 was carried out to enable easy comparison between the two quite similar oxygen carriers.

The temperature dependency for C28 fuel conversion was similar to the C14 material - high temperatures increase fuel conversion. The reduction of the oxygen content in the air reactor showed a decrease of the reactivity in the fuel reactor. The system behavior with C28 material was similar to that with the C14 oxygen carrier. Increasing the solids circulation rate, and as a consequence, the oxygen carrier-to-fuel ratio, led to full fuel conversion in the fuel reactor (Figure 4a). Compared to the C14 results, almost no partial oxidation occurred in the fuel reactor (no detectable CO and H_2). Even for operating conditions with low fuel conversion, partial oxidation was not detected. Similar to C14 behavior, O_2 in the fuel reactor off gas was measureable for operating conditions with high fuel conversion (above 95%) but was at the lower end of the detection limit of the oxygen analyzer. A significant increase in molecular O_2, like that observed with the C14 material, did not occur at complete fuel conversion with the C28 oxygen carrier. The specific inventory to reach full fuel conversion was also in the same range as for C14. About 320kg/MW of oxygen carrier in the fuel reactor are necessary to fully convert the fuel in the pilot unit (see Figure 4b).

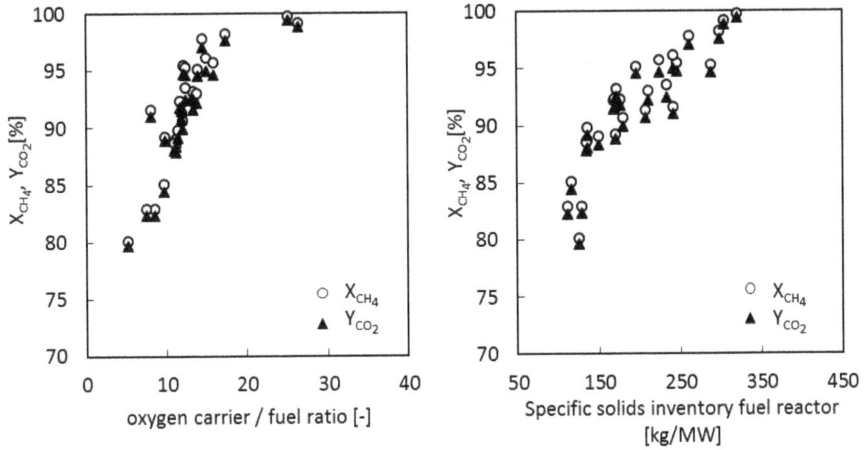

Figure 4. Carrier C28: (a) CH_4 conversion and CO_2 mole % vs. oxygen carrier/fuel ratio, (b) CH_4 conversion and CO_2 mole % in the flue gas vs. Specific solids inventory in the fuel reactor.

In summary, the C28 carrier is slightly more active than C14 (about 10% less inventory needed). The oxygen slip in the flue gas is lower for C28. Both C14 and C28 may be scaled-up as a suitable alternative to Ni-based carriers.

Overall integration and CLC Next Scale design

A CLC unit with heat production of 10 MW (thermal) is considered a suitable size to produce performance data reliable for scale-up. This compares to the required thermal duty of 80-100 MW for commercial application to produce steam for SAGD extraction of oil sands.

The sizing of the main equipment for a 10 MW demonstration unit is:

- Air Reactor
  - Cross section upper part: 1200 x 1200 mm
  - Diameter lower part: 900 mm
  - Height, approx.: 15000 mm
- Fuel Reactor
  - Cross section upper part: 900 x 900 mm
  - Diameter lower part: 550 mm
  - Height, approx.: 15000 mm

The budget price for delivery and erection of the CLC unit was estimated at about double the cost of a conventional boiler of the same size. This is in line with expectations, since a CLC boiler has two parallel vessels vs. a single fire box for conventional boilers.

Time for delivery and erection of this unit has been estimated at 24 months. Estimated operating costs are 15% higher than for conventional boilers, assuming the use of a Ni-based carrier and its full replacement after 1 year of operation. For a Ni-based carrier, this estimate may be considered as conservative, although intermittent operation during demo runs might negatively affect carrier consumption. Use of C14 or C28 carriers might result in lower operating costs.

A preliminary environmental assessment of available carrier materials was carried out by Shell. As expected, Ni-based carriers may potentially be hazardous to personnel due to the carcinogenic activity of nickel oxide. Cu-based carriers may be dangerous to the environment, due to the solubility of copper oxide in water. C14 and C28, as well as Fe-based carriers would only be the source of small or negligible hazards.

CONCLUSIONS

The main achievements of CCP3 towards further development of CLC technology are:

- Development at the pilot scale of non-Ni based carriers with positive features for further scale-up
    - Similar reactivity of Ni-carriers.
    - Lower production cost.
    - Low impact to humans and the environment.
    - Attrition rate comparable to Ni-carriers.
- Design of the next scale 10 MW demonstration unit.
- Design and economic evaluation of a cluster of CLC boilers (4 x 80 MW) for application in the production of steam for oil sands extraction by the SAGD technique.

Detailed results of the economic evaluations for application of the CLC technology to the extraction of oil sands are shown in Chapter 20. CLC is estimated to have the lowest CO_2 avoidance cost for this scenario. CLC could in principle be applied to oil refineries, but the large plot plan required, compared to conventional boilers, make it unlikely for replacement of existing units.

REFERENCES

1. P. Hurst, I. Miracca - "Chemical Looping Combustion (CLC) Oxyfuel Technology Summary", in *Carbon Dioxide Capture for Storage in Deep geologic Formations – Results from the CO_2 Capture Project – Volume* 1 – pp. 583-587 (Ed. David C. Thomas), Elsevier (2005).
2. J. Assink, C. Beal – "Chemical Looping Combustion (CLC) Technology Summary", in *Carbon Dioxide Capture for Storage in Deep Geologic Formations – Results from the CO_2 Capture Project – Volume 3* – pp. 67-75 (Ed. Lars Ingolf Eide), CPL Press (2009).
3. Källén, M., Rydén, M., Dueso, C., Mattisson, T., and Lyngfelt, A., $CaMn_{0.9}Mg_{0.1}O_{3-\delta}$ as Oxygen Carrier in a Gas-Fired 10 kWth Chemical-Looping Combustion Unit, *Industrial & Engineering Chemistry Research* 52 (2013) 6923-6932
4. Hallberg, P., Källén, M., Mattisson, T., Rydén, M., and Lyngfelt, A. (2014), Overview of operational experiences with calcium manganate oxygen carriers in chemical-looping combustion, in *3rd International Conference on Chemical Looping,* Göteborg, Sweden.
5. Hallberg, P., Källén, M., Jing, D., Snijkers, F., van Noyen, J., Rydén, M., and Lyngfelt, A., Experimental investigation of $CaMnO_{3-\delta}$ based oxygen carriers used in continuous Chemical-Looping Combustion, *International Journal of Chemical Engineering* Volume 2014, Article ID 412517, 9 pages, http://dx.doi.org/10.1155/2014/412517
6. Mayer, K., Penthor, S., Pröll, T., and Hofbauer, H. (2014), The different demands of oxygen carriers on the reactor system of a CLC plant – results of oxygen carrier testing in a 120kW pilot plant, in *3rd International Conference on Chemical Looping*. Göteborg, Sweden.
7. de Diego, L.F., Abad, A., Cabello, A., Gayán, P., García-Labiano, F., and Adánez, J., (2014), Reduction and Oxidation Kinetics of a $CaMn_{0.9}Mg_{0.1}O_{3-\delta}$ Oxygen Carrier for Chemical-Looping Combustion, *Industrial & Engineering Chemistry Research*., 53 (2014) 87–103.

Carbon Dioxide Capture for Storage in Deep Geological Formations, Volume 4
Karl F. Gerdes (Editor)

Chapter 17

SUITABILITY OF Cu-BASED MATERIALS FOR THE SCALE-UP OF CHEMICAL LOOPING COMBUSTION

J. Adánez, P. Gayán, A. Abad, F. García-Labiano and L.F. de Diego
Department of Energy and Environment, Instituto de Carboquímica (ICB-CSIC),
Miguel Luesma Castán 4, 50018 Zaragoza, Spain

ABSTRACT: Copper-based particles prepared by impregnation have been shown be suitable for combustion of gaseous fuels, e.g. natural gas or syngas, when they were tested in Chemical Looping Combustion (CLC) units at the scale 0.5 to 10 kW_{th}. In previous work, impregnated materials with high reactivity and avoidance of agglomeration were developed, however, low CuO load (<20 wt.%) was necessary to avoid agglomeration. The objective of this study was to assess the suitability of Cu-based materials as an oxygen carrier for the scale-up of the process. Two different approaches to oxygen carrier particle synthesis were pursued – spray drying and impregnation. Promising Cu-based oxygen carriers were prepared by the spray drying and achieved good reactivity and fluidization properties (avoidance of agglomeration), but exhibited high attrition rates. Particles prepared by impregnation were found to have suitable reactivity and attrition rates. The relatively modest Cu-loading (14 wt.% CuO on Al_2O_3) required adjustments to the solids inventory of both the fuel reactor and air reactor to achieve full fuel conversion. A validated theoretical model was developed to obtain the basic design a 10 MW_{th} CLC with this oxygen carrier.

KEYWORDS: CO_2 capture; chemical looping combustion; oxygen carrier; copper oxide; impregnation; spray-drying; modelling; design

INTRODUCTION

At the heart of the chemical looping combustion (CLC) process lies the oxygen carrier. A practical oxygen carrier must meet many criteria:

- Sufficient oxygen transfer capability.
- Favourable thermodynamics regarding the fuel conversion to CO_2 and H_2O in CLC.
- High reactivity for reduction and oxidation reactions, to reduce the solids inventory in the reactors, and stable reactivity during many successive redox cycles.
- Resistance to attrition to minimize losses of elutriated solids.
- Negligible carbon deposition that would release CO_2 in the air reactor thus reducing CO_2 capture efficiency.
- Good fluidization properties (no presence of agglomeration).
- Limited cost, including environment, health and safety cost.

The first two characteristics are intrinsically dependent on the redox system. Suitable properties are metal oxides based on Cu, Ni, Fe and Mn [1]. The cost and the environmental characteristics are also related to the type of metal oxide used. The quality of the other required characteristics must be experimentally determined for each specific material. Normally, the pure metal oxides do not fulfil the above characteristics and reaction rates quickly decrease in a few cycles. Supports have been shown to address the stability issue. A porous support provides a higher surface area for reaction, a binder for increasing the mechanical strength and attrition resistance, and also increases the ionic conductivity of solids. In this sense, the method used in the preparation of the materials strongly affects the properties of the oxygen carrier. The distribution of the metal oxide on the support and the possible interaction between them will affect the oxygen-carrier reactivity, as well as the strength and material stability during repeated redox cycles. Several preparation methods can be found in the literature [1]. In some methods, powders of metal oxide and support are mixed. Such methods include mechanical mixing and extrusion, freeze granulation, spray drying, or spin flash drying. In other methods, a solution of the active metal and support is the starting point in the preparation. In this case, the solid materials are generated by precipitation (co-precipitation, dissolution, sol-gel, solution combustion). Finally, there is the impregnation method, where a solution containing the active metal is deposited on a resistant and porous solid support.

An important aspect of the oxygen-carrier material is its suitability to be used in continuous CLC units during long periods of time. In this sense, the Chemical Looping Combustion (CLC) process has been successfully demonstrated for gaseous fuels with a wide range of oxygen carriers in multiple prototype plants, most of them with the configuration of two interconnected fluidized-bed reactors ranging from 0.3 to 140 kW_{th} [1]. Among candidate oxygen carriers, a Cu-based material prepared by the impregnation method has shown promise for scale-up production. The impregnation method is suited for production of large amounts of solid particles at low cost. The research group at Instituto de Carboquímica (ICB-CSIC) has optimized the impregnation method to produce highly reactive Cu-based oxygen carrier materials for use in the scale-up of the CLC process [2,3]. Materials prepared by incipient wetness impregnation on commercial porous γ-Al_2O_3 had suitable properties regarding reactivity, low attrition and avoidance of agglomeration during multiple redox cycles in a fluidized bed. Testing identified a maximum allowable CuO loading of 20 wt.% in the particles to avoid bed agglomeration. Moreover, the copper extracted from the fines produced during CLC operation may be re-used for carrier impregnation. Re-use of this material showed improved performance compared to original material [4].

A Cu-based material containing 14 wt.% CuO impregnated on γ-Al_2O_3, was tested in several CLC facilities. This oxygen carrier showed complete combustion of methane, syngas and light hydrocarbons during continuous operation in CLC units ranging of 0.5 and 10 kW_{th} [5-7], as well as an adequate resistance to attrition with proper operation, achieving an estimated lifetime of about 2500 hours. Moreover, this material did not suffer from deactivation when small amounts of H_2S were present in the fuel gas [8]. Attrition performance at high temperature, above 1173 K, was analysed using different supports [9,10]. The addition of a small amount of Ni to the support improves the particle lifetime. The CLC tests were characterized by low gas velocity in the fuel reactor, assuring bubbling fluidization conditions, solids inventories in the fuel reactor higher than 300 kg/MW_{th} and complete oxidation of the oxygen carrier in the air reactor.

A similar material prepared by CSIC, 'CuAlImp,' has been tested in the 120 kW_{th} CLC unit at Vienna University of Technology (TUV). This unit has a Dual Circulating Fluidized Bed (DCFB) system, which consists of two hydraulically connected circulating fluidized bed reactors. Complete combustion was not reached in any of the conditions tested. For example, maximum methane conversion values close to 80% were reached with solids inventories in the fuel reactor about 200 kg/MW_{th} of copper material [11]. The ultimate reason for the low methane conversion is unclear yet, but it was suggested that the low solids inventory in the fuel reactor is limiting the fuel conversion. A mathematical model can predict fuel conversion and can be used to evaluate relevant design parameters and operating conditions which would allow safe combustion of the fuel in a CLC unit.

An important aspect of the carrier preparation method is its suitability for scale up. Based on the different preparation methods described in the literature, oxygen carriers prepared using spray drying and impregnation techniques were selected for development of a scale-up ready process.

After testing Cu-based impregnated particles in a 10kW_{th} CLC unit, a lifetime of 2500 hours was estimated. Particles having an irregular shape are prone to suffer increased attrition relative to spherical particles. One option to improve the attrition performance and to increase the lifetime of CuO oxygen carriers is to use spray-drying as the preparation method. This method has the additional advantage that it is a one-step process compared to two steps for impregnation, which could reduce preparation costs. Spray dried particles might allow use of a higher CuO loading.

In order to evaluate the suitability of Cu-based impregnated or spray-dried materials as an oxygen carrier, two main tasks were carried out:

1. To develop an oxygen carrier prepared by the spray-drying method with higher CuO content than impregnated materials and suitable properties to be used as an oxygen carrier (i.e. high reactivity, avoidance of agglomeration and a long particle lifetime). In this work, several Cu-based materials prepared by spray-drying were evaluated. The effect of support material and the calcining conditions on its physical and chemical properties were analysed.
2. To optimize the design of a CLC unit and to define the operating conditions for the use of an impregnated Cu-based material for CH_4 combustion with CO_2 capture. To achieve this objective, a useful tool is a mathematical model of the main processes occurring in the fuel and air reactors. In this work, a detailed fluid dynamic and kinetic model was developed for both the air and fuel reactors to obtain a global simulation tool of the CLC process. The CLC model was validated against experimental results obtained in the 120 kW_{th} DCFB unit at TUV burning natural gas. The model provided insight into the ultimate reasons for the low performance of impregnated material in the DCFB unit. The scale-up of the process was then accomplished using the validated model to design a 10 MW_{th} CLC unit which achieves complete combustion of methane or natural gas using the impregnated Cu-based material as oxygen carrier.

SCREENING OF Cu-BASED OXYGEN CARRIERS PREPARED BY SPRAY DRYING

Preparation of materials

Impregnated Oxygen Carriers

The method of impregnation is suited for production of large amounts of solid particles at low cost. The research group at Instituto de Carboquímica (ICB-CSIC) has optimized the impregnation method to develop highly reactive Cu-based oxygen carriers [3] which do not show agglomeration tendencies. Among them, CuAlImp particles were considered a promising oxygen carrier for the scale-up of the CLC process, and 70 kg were prepared to be used in the DCFB unit at TUV [11].

Commercial γ-Al_2O_3 particles (Puralox NWa-155, Sasol Germany GmbH) with a size range of +100-300 μm, a density of 1300 kg/m^3 and a porosity of 55.4% were used as a support. A copper-based oxygen carrier was prepared by incipient wetness impregnation, with a CuO content of 14 wt.%. The impregnation was done by addition of a volume of copper nitrate solution (293 K, 5.4 M) corresponding to the total pore volume of the support particles (0.42 cm^3/g). The aqueous solution was slowly added to the alumina particles, with thorough stirring at room temperature. The particles were calcined in air for 30 min at 823 K using a muffle furnace, to decompose the copper nitrate into insoluble copper oxide. Finally, the oxygen carrier was calcined in air for 1 hour at 1123 K. Similar particles had been successfully tested in a 10 kW_{th} unit for more than 100 hours [5].

Table 1 shows the main properties of the fresh oxygen carrier. The presence of CuO and $CuAl_2O_4$ was confirmed by XRD analysis. The CuAlImp particles show a relatively high porosity and BET surface area. The crush strength is adequate for operating in a circulating fluidized bed. Also the oxygen transport capacity, which was calculated using the actual active CuO content for the redox process, is high enough for consideration as an oxygen carrier for CLC [12].

Table 1. Properties of the CuAlImp oxygen carrier.

Average particle size (μm)	235
Theoretical CuO content (%)	14
XRD phases	CuO, $CuAl_2O_4$, γ-Al_2O_3
R_{OC} (%) - Oxygen Transport Capacity	2.4
BET area (m^2/g)	94.2
Porosity (%)	49.6
Skeletal density (kg/m^3)	3175
Apparent density (kg/m^3)	1600
Crush strength (N)	2.0

Spray-Dried Oxygen Carriers

A number of samples with varying supports were prepared by spray-drying at VITO (Belgium) using mixtures of several materials (γ-Al_2O_3, $Al(OH)_3$, $MgAl_2O_4$, $CaAl_2O_4$, and kaolin). After the spray-drying process, a particle size distribution was measured. The distribution is dependent on the raw materials and processing steps, including suspension preparation, and the spray-drying specifics - mainly nozzle orifice diameter and orifice airflow. For use in chemical looping combustion, a specific size range of the spray-dried granules is selected, in this case between +100-300 μm. Preparation and calcination conditions were optimized to produce highly reactive particles with good crush strength (CS). Oxygen carriers with copper contents of 30 and 40 wt. % were made. Calcining conditions were varied from 4 to 12 h and 1223 to 1473 K. In total, 21 different samples were prepared, following an iterative process to develop a suitable copper-based spray dried oxygen carrier.

Table 2 shows the samples prepared by spray drying and the measured crush strength, oxygen transport capacity and rate index. For comparison reasons, the properties of the impregnated Cu-based material, CuAlImp are also included.

Table 2. Cu-based oxygen carriers prepared by spray-drying.

Sample	**CuO (wt.%)**	**Support (wt.%)**	**Calcining Conditions T (K)**	**t (h)**	**CS (N)**	***R_{OC}* (%)**	**Rate index (%/min) R.[(1)]**	**O.[(2)]**
CuAlImp	14	86 γAl_2O_3	1123	1	2.0	2.4	15.0	5.3
CuAlSD_1	30	70 γAl_2O_3	1223	6	Melt	-		
CuAlSD_2	30	35 γAl_2O_3- 35 $Al(OH)_3$	1223	6	0.1	-		
CuAlSD_3	30	35 γAl_2O_3- 35 $Al(OH)_3$	1273	6	0.3	-		
CuAlSD_4	30	35 γAl_2O_3- 35 $Al(OH)_3$	1323	6	0.2	-		
CuAlSD_5	30	35 γAl_2O_3- 35 $Al(OH)_3$	1373	6	0.2	-		
CuAlSD_6	30	35 γAl_2O_3- 35 $Al(OH)_3$	1423	6	0.3	-		
CuAlSD_7	30	21 γAl_2O_3 - 49 $CaAl_2O_4$	1373	6	Soft	-		
CuAlSD_8	30	21 γAl_2O_3 - 49 $CaAl_2O_4$	1423	6	Soft	-		
CuAlSD_9	30	21 γAl_2O_3 - 49 $CaAl_2O_4$	1473	6	Soft	-		

Table 2 continued.

Sample	CuO (wt.%)	Support (wt.%)	Calcining Conditions T (K)	t (h)	CS (N)	R_{OC} (%)	Rate index (%/min) R.[(1)]	O.[(2)]
CuAlSD_10	30	50 γAl_2O_3 - 20 kaolin	1223	6	0.6	-		
CuAlSD_11	30	50 γAl_2O_3 - 20 kaolin	1373	6	3.8	5.0	9.5	6.6
CuMgAlSD_1	30	70 $MgAl_2O_4$	1323	4	0.9	-		
CuMgAlSD_2	30	70 $MgAl_2O_4$	1323	12	1.4	-		
CuMgAlSD_3	30	70 $MgAl_2O_4$	1373	12	2.3	5.4	13.3	8.7
CuMgAlSD_4	30	70 $MgAl_2O_4$	1423	12	2.0	4.6	9.4	8.2
CuMgAlSD_5	40	60 $MgAl_2O_4$	1373	12	1.3	7.6	18.9	10.3
CuMgAlSD_6	30	70 $MgAl_2O_4$ milling time 15 min	1373	12	2.6	5.4	12.9	8.1
CuMgAlSD_7	30	70 $MgAl_2O_4$ milling time 90 min	1373	12	2.7	5.4	11.2	6.2
CuMgAlSD_8	30	60 $MgAl_2O_4$ - 10 kaolin	1373	12	2.8	4.7	4.9	5.9
CuMgAlSD_9	30	55 $MgAl_2O_4$ - 15 kaolin	1223	6	0.1	-		
CuMgAlSD_10	30	55 $MgAl_2O_4$ - 15 kaolin	1373	6	2.4	3.9	5.2	4.7

(1) Reduction: 15 vol.% CH_4 (2) Oxidation: 10 vol.% O_2

Experimental Facilities and Data Evaluation

The crush strength (CS) of the samples, determined with a Shimpo FGN-5X apparatus, was taken as the average value of 20 measurements of the force needed to fracture a particle.

The oxygen transport capacity and reactivity of the oxygen carriers were determined in a thermobalance (CI Electronics Ltd.) [2]. The operating conditions were 1073 K, and a gas composition for reduction of 15 vol.% H_2 in N_2 for oxygen transport capacity determination and 15 vol.% CH_4 with 20 vol.% H_2O in N_2 for reactivity tests. For oxidation air was used.

The oxygen transport capacity of the oxygen carrier, R_{OC}, was defined as the mass fraction that can be used in the oxygen transfer process:

$$R_{OC} = \frac{m_o - m_r}{m_o} \qquad \text{Equation 1}$$

where m_o and m_r are the mass of oxidized and reduced particles, respectively, determined by TGA experiments. The mass fraction of active metal oxide in the particles, x_{MeO}, can be calculated considering the theoretical oxygen transport capacity of the metal oxide, R_0, as:

$$x_{MeO} = \frac{R_{OC}}{R_0} \qquad \text{Equation 2}$$

The active fraction of metal oxide in particles can be lower than the theoretical fraction loaded in particles because of formation of low reactive compounds which were not reduced in TGA tests. The conversion of solids, X_s, was calculated for the reduction or oxidation period, respectively:

$$X_{s,r} = \frac{m_o - m}{R_{OC}\, m_o} \qquad \text{Equation 3}$$

$$X_{s,o} = \frac{m - m_r}{R_{OC}\, m_o} \qquad \text{Equation 4}$$

where m is the actual mass of the sample in TGA experiments.

In order to compare the reactivity of different materials, a rate index for the reduction and oxidation reactions was calculated as a normalized rate expressed in %/min [13].

$$\text{Rate index (\%/min)} = 100 \cdot 60 \cdot R_{OC} \left. \frac{dX_s}{dt} \right|_{X_s=0} \qquad \text{Equation 5}$$

The reaction rates with 15 vol.% CH_4 and 10 vol.% O_2 at 1073 K of the different Cu-based oxygen carriers with high enough mechanical strength were measured, and used to calculate the rate index value.

Candidate materials with high reactivity were futher characterized in a batch fluidized bed reactor to determine the gas product distribution during redox cycles using similar operating conditions to that existing in a CLC process. The fluidization behaviour of the oxygen carrier with respect to agglomeration and attrition phenomena was also analysed in the same equipment. The experimental set-up for testing the oxygen carriers consisted gas feed system, a fluidized bed (FB) reactor, a trap to recover the solids elutriated from the fluidized bed, and a gas analysis system. The reactor was 54 mm I.D. and 500 mm height [3]. The typical operating conditions used in these tests were 1073 K with an inlet gas velocity of 0.15 m/s. The feed gas composition for reduction was 25 vol.% CH_4 in N_2, and for oxidation was 10 vol.% O_2 in N_2. A 2-minute purge of N_2 was used after each reducing and oxidizing period to avoid mixing CH_4 and air.

Characterization Results of Spray Dried Materials

Evaluation of the Mechanical Strength of the Particles

A first batch, designated CuAlSD_1, was prepared with a composition of 30 wt.% CuO – 70 wt.% Al_2O_3,. The sample was sintered at 1223 K for 6 hours. Undesirable sintering behaviour was observed, with a large amount of fines formed. In order to improve the sintering behaviour of the CuO on alumina based oxygen carriers, $Al(OH)_3$ was substituted for some of the Al_2O_3. The batches CuAlSD_2 to CuAlSD_6 contained 30 wt.% CuO - 35 wt.% $Al(OH)_3$ - 35 wt.% Al_2O_3. The hydroxide will decompose during calcination and sintering treatment, and it was hoped this would improve the properties of the oxygen carrier. An initial sintering temperature of 1223 K was tested, but the crush strength was well below 1 N, the target value for oxygen carrier evaluation in fluidized bed reactors [14]. In order to improve the crush strength of the oxygen carrier, higher sintering temperatures were tried. However, increasing the sintering temperature resulted in particles with a crush strength of only 0.3 N, well below the target value.

Another possible route to obtain strong and active alumina-copper based oxygen carriers is the use of alumina with additives. In order to test this, samples CuAlSD_7 to CuAlSD_9 were prepared with a composition of 30 wt.% CuO – 49 wt.% $CaAl_2O_4$ – 21 wt.% γ-Al_2O_3. Calcium aluminate is a well-known compound in ceramic processing, and it is, for example, used as cement, with excellent mechanical properties. However, the crush strength values of the samples sintered at 1373, 1423 and 1473 K for 6h were too low for testing. Therefore, the route using γ-Al_2O_3 and $CaAl_2O_4$ was discarded.

An alternative additive frequently used in ceramic processing to improve the mechanical properties of a ceramic component is kaolin, a clay mineral with the chemical composition $Al_2Si_2O_5(OH)_4$. The kaolin phase undergoes structural transformations upon calcination and sintering into mullite and cristobalite phases under the right conditions, depending also on the exact chemical composition of the kaolinite. The mullite phase is well known for its mechanical properties and its thermal shock resistance. Samples CuAlSD_10 and CuAlSD_11 were prepared with 30 wt.% CuO – 50 wt.% Al_2O_3 – 20 wt.% $Al_2Si_2O_5(OH)_4$. Sample CuAlSD_10 was sintered at 1223 K for 6 h and resulted in a soft material. Sample, CuAlSD_11 was calcined at higher temperature, 1373 K for 6 h and the resulting material had the highest crush strength of the candidate materials (3.8 N).

Another approach to improve the mechanical properties of the oxygen carrier is to assess different support materials. Magnesium aluminate spinel as a support for copper based oxygen carriers has been previously evaluated for solid fuels [15]. Sintering temperature and time were scanned to optimize the crush strength of the particles in Samples CuMgAlSD_1 to CuMgAlSD_4 with 30 wt.% CuO -70 wt.% $MgAl_2O_4$. Sample CuMgAlSD_5, with 40 wt.% CuO and 60 wt.% $MgAl_2O_4$, was prepared to analyse the effect of the copper content on particle fluidization behaviour. A sintering temperature of 1323 K for 4 hours resulted in particles with crush strength approaching the 1 N target. Increasing the sintering time to 1373 K improved the crush strength considerably, with an average crush strength above 2 N. These particles were suitable for testing in fluidized bed reactors. Increasing the sintering temperature to 1423 K did not improve the strength further. The optimized sintering treatment was 12 hours isothermal at 1373 K.

For reactions involving gas and solid, a higher surface area means higher reactivity. In order to improve the activity of the particles before sintering, the slurry was milled for a longer period, 15 and 90 minutes, with the intent that smaller particle size of the raw materials should improve the crush strength of the final carrier particle. The CuMgAlSD_6 and CuMgAlSD_7 particles were sintered for 12 h at 1373 K. The crush strength of carrier particles made from a slurry that was milled for an extended period was only slightly improved, 2.7 N versus 2.6 N. This improvement was too small to justify further developments in this direction, so this route was abandoned.

The effect of adding kaolin to $MgAl_2O_4$ support was also assessed. Sample CuMgAlSD_8 was composed of 30 wt.% CuO – 60 wt.% $MgAl_2O_4$ – 10 wt.% $Al_2Si_2O_5(OH)_4$, and sintered for 12 h at 1373 K. The sample had a measured crush strength of 2.8 N, sufficient for evaluation in fluidized bed reactors. Different proportions of kaolin were evaluated for the effect on strength and activity of the sample. Both samples CuMgAlSD_9 and CuMgAlSD_10 were composed of 30 wt.% CuO – 55 wt.% $MgAl_2O_4$ – 15 wt.% $Al_2Si_2O_5(OH)_4$, but were sintered for 6 h at 1223 K and 1373 K, respectively. The sample sintered at 1223 K had a low crush strength of 0.1 N. The sample sintered at 1373 K had a crush strength of 2.4 N, which is above target and high enough for further testing.

Reactivity Tests

The oxygen transport capacity and reactivity of the samples with adequate crush strength were analysed in the TGA. These samples were mostly particles prepared with $MgAl_2O_4$ as the only or major inert compound. For samples which used Al_2O_3 as base support, only Sample CuAlSD_11, which included kaolin, had high enough crush strength to warrant further characterization. Oxygen transport capacity was characterized by reduction with H_2 using a TGA. Then, fuel reactivity was studied by reduction with CH_4 and oxidation with O_2. These reactivity results were compared to those for impregnated Cu-based CuAlImp.

Table 2 (above) shows the effective oxygen transport capacity of the different samples, based on TGA results. Based on high values for R_{OC}, the mass fraction that reacts, almost all the copper in the material is active for redox reactions. The reactivity of the spray dried carrier particles with methane and air is shown in Figure 1, together with the reactivity of the impregnated material, CuAlImp. The reactivity of the spray dried material was high, but lower than that of the impregnated material. The effect of calcination conditions can also be seen in this figure. Increased calcination temperature decreased the reactivity of the sample. Figure 1 also shows the effect of increasing the copper content while using the same calcination conditions. Similar results were found for samples CuMgAlSD_3 and CuMgAlSD_5. The addition of kaolin to the $MgAl_2O_4$ as support slightly increased CS from 2.3 to 2.8 N, (CuMgAlSD_8), but negatively affected reactivity. Reactivity of particles with alumina and kaolin as support, CuAlSD_11, showed improved reactivity relative to sample CuMgAlSD_10, but had lower reactivity than for most of the other materials prepared with $MgAl_2O_4$ as support.

In conclusion, samples with acceptable crush strength and high reactivity have been produced by spray drying using alumina or $MgAl_2O_4$ as support with a higher copper content than the impregnated material, CuAlImp. Among spray dried candidates, samples CuMgAlSD_3 and CuMgAlSD_5 were selected for further testing in a fluidized bed reactor. These materials required mild sintering conditions and easier preparation than other materials with $MgAl_2O_4$, as well as CuAlSD_11,which contained a mixture of CuO, Al_2O_3 and kaolin.

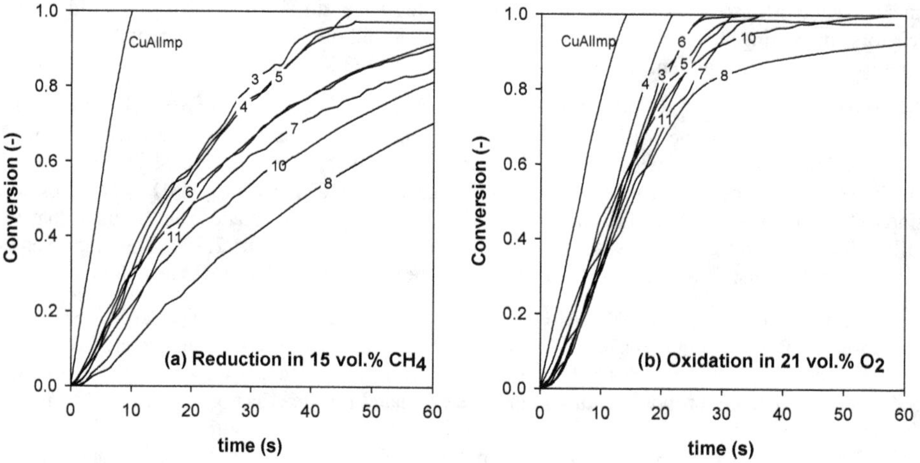

Figure 1. Conversion vs. time curves of different spray-dryied particles for (a) reduction with 15 vol.% CH_4 and (b) oxidation in air obtained in TGA. CuAlImp material was included for comparison purposes. T = 1073 K. Legend: 3 to 10 refer to number i for CuMgAlSD_i; 11 is CuAlSD_11 (see Table 2).

Fluidized Bed Tests

Selected samples from the reactivity characterization were further evaluated in the batch fluidized bed facility to analyse the product gas distribution during typical redox cycles with methane, the attrition rate, and possible agglomeration behaviour. Particle agglomeration must be avoided because it can lead to bed defluidization that causes solids circulation disturbances and gas channelling through the bed, which makes gas-particle contact less effective.

Bed agglomeration issues appeared for sample CuMgAlSD_5, the material with the highest copper content (40 wt.% on $MgAl_2O_4$), Previous work has shown that the agglomeration behavior of an oxygen carrier could depend on the copper content and preparation conditions [3]. Sample CuAlSD_11, with 30 wt. % of copper on Al_2O_3 and kaolin, also suffered from strong agglomeration. Product gas composition during continuous redox cycles of these samples could not be measured due to the appearance of agglomerates at the beginning of the reduction periods. Therefore, these samples were rejected for further investigation.

For sample CuMgAlSD_3, with a 30 wt.% of copper content on $MgAl_2O_4$, no defluidization problems were observed, despite the high Cu content, even during redox cycles with long reduction periods, when solids were highly reduced to base metal. Product gas composition obtained in a typical redox cycle showed complete combustion of CH_4 during the first minutes of the cycle, and there was no CH_4, CO or H_2 detected at the reactor outlet. Moreover, there no carbon deposition occurred, as reflected in the lack of CO or CO_2 detected during the oxidation step. The gas product distribution measured was stable during all the 80 redox cycles performed.

Fines elutriated during batch fluidized bed redox cycles were recovered with filters located downstream of the reactor. Fines are considered to be particles with a size below 40 μm. The attrition rate for sample CuMgAlSD_3 after 20 hours at hot conditions was calculated and plotted in Figure 2. In the early cycles, high attrition rates may be due to powder stuck to the spherical particles. The attrition rate stabilized to a mean value of 0.08 wt. % /hour, which corresponds to an average lifetime of 1200 h, which is less than that measured for impregnated material CuAlImp. The crush strength of the particles after redox cycles in the fluidized bed reactor showed significant decrease after 20 hours at 1073 K, from 2.0 to 0.4 N.

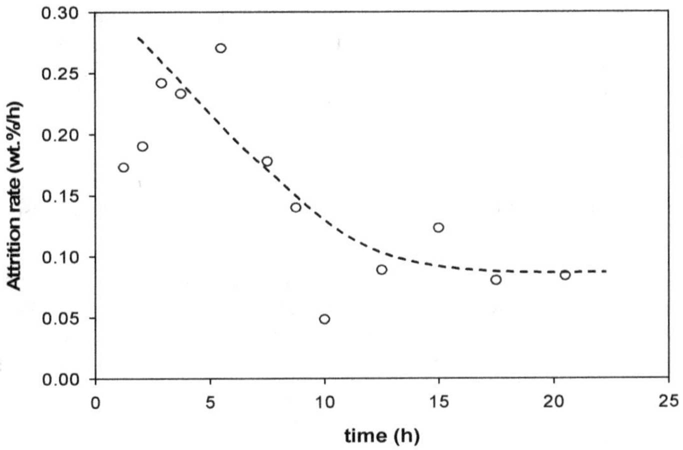

Figure 2. Attrition rate of sample **CuMgAlSD_3** measured in the batch fluidized bed facility during 80 redox cycles alternating methane and air. T=1073 K.

The state-of-the-art copper-based oxygen carrier (CuAlImp) is synthesized by impregnation on a commercial alumina support. This material has enough oxygen transport capacity and reactivity for full methane combustion and a particle lifetime of 2500 h [5-10]. To improve the attrition performance and to increase the copper content of Cu-based oxygen carriers, spray-drying was proposed as an alternate manufacturing method. During the screening carried out in this work, a spray-dried material was identified with 30 wt.% of copper oxide on $MgAl_2O_4$ and calcined at 1373 K for 12 hours (**CuMgAlSD_3**). This oxygen carrier had a high oxygen transport capacity and reactivity, although particle lifetime is half that measured using the CuAlImp oxygen carrier prepared by impregnation [3,5].

OPTIMIZING THE CLC DESIGN USING A CU-BASED OXYGEN CARRIER

To optimize the design of a CLC unit and to define the operating conditions for the use of a Cu-based oxygen carrier for CH_4 combustion with CO_2 capture, the impregnated CuAlImp oxygen carrier was selected due to its excellent reactivity and acceptable particle lifetime. A mathematical model was developed for both the air and fuel reactors, providing a global simulation for the CLC process. The model was validated against the experimental results from combustion of natural gas with the impregnated CuAlImp material in a pilot CLC unit.

There are a number of CLC reactor modelling studies in the literature, as reviewed in Adánez *et al.* [1]. Modelling of fluidized-bed reactors can be divided into three closely related fields:

- Fluid dynamics.
- Reaction scheme and kinetics.
- Heat balance.

Fluid dynamics, mass balances and heat balances in the reactor must be solved simultaneously because of interaction of reaction rates, temperature and gas properties. Validation of the models against experimental results obtained in continuously operated CLC systems is an important step before use them for design, optimization and scale-up purposes. However, few models have been validated against experimental results in continuously operated CLC units [1]. For impregnated Cu-based oxygen carriers, a theoretical model was validated against experimental results obtained during CH_4 combustion in a 10 kW_{th} CLC unit with an oxygen carrier containing 14 wt.% CuO on Al_2O_3. In this case, the fuel reactor was a fluidized bed operated in the bubbling regime. Modelling results predicted complete combustion of methane with a solids inventory of 130 kg/MW_{th} assuming a high degree of oxidation is achieved in the air reactor [16]. However, the solids inventory required in the fuel reactor significantly increases as the extent of oxidation of the oxygen carrier in the air reactor decreases [16], because the reactivity for fuel oxidation is reduced [12]. Therefore, both the fuel and air reactor performance must be addressed to optimize the operating and design conditions of a CLC unit.

In this work, a theoretical model for both air and fuel reactors was developed in Fortran®. The model is based on the reactor geometry of the 140 kW_{th} CLC unit at TUV [17]. The model included the main processes affecting the reaction of fuel gas or oxygen with the oxygen carrier, such as reactor fluid dynamics, reactivity of the oxygen carrier and mechanism of the reaction. The reaction mechanism and reactivity depends on the pair gas-oxygen carrier considered, whereas the fluid dynamics is linked to the design and operating conditions of the reactor [17]. The oxygen carrier modelled was the impregnated Cu-based oxygen carrier CuAlImp tested in the DCFB unit [11].

Theoretical Model of the CLC Unit

Fluid Dynamics

Both air and fuel reactors are circulating fluidized beds in the DCFB unit at TUV. Thus, the fluid dynamic model considers the gas and solids flows inside the reactor and the gas-solids mixing patterns in the different regions, including the macroscopic gas and solid distribution in the high-velocity regime of a fluidized bed reactor [18]. The reactor was divided into two vertical regions with respect to axial concentration and backmixing of solids: a dense region in the bottom bed with a high and roughly constant concentration of solids; and a dilute region, where there is a pronounced decrease in the concentration of solids as height increases. Gas distribution and mixing between the emulsion and bubbles in the dense region was taken into account. Thus, the gas flow in the dense region is split between the emulsion and bubble phases, with gas mixing between the two phases controlled by diffusion. The dilute region includes a cluster phase and a transport or dispersed phase. Both the cluster and transport phases were superimposed but with different mixing behaviour. The cluster phase has strong solids backmixing with solids in the dense region. The transport phase was characterized by a core/annulus flow structure. It is worth noting that there is not a proper dense region in the air reactor because of the high gas velocity and low solids inventory. Therefore, only the transport phase is considered for the air reactor. The global solids distribution in the reactors was calculated by fitting the total pressure drop in each reactor from the predicted solids concentration profile.

Mass Balances

Mass balances for the different reacting compounds and products were developed for each phase in the dense region and the dilute region. The reaction for CH_4 conversion in the fuel reactor is shown by Equation (6) and the corresponding reaction in the air reactor is shown by Equation (7).

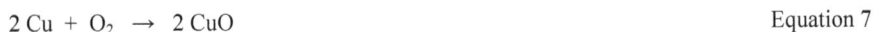

$$4\,CuO + CH_4 \rightarrow 4\,Cu + CO_2 + 2\,H_2O \qquad \text{Equation 6}$$

$$2\,Cu + O_2 \rightarrow 2\,CuO \qquad \text{Equation 7}$$

Oxygen carrier reduction and oxidation kinetics were determined by TGA and included in the model.

Integration of the Fuel Reactor with the Air Reactor

The fuel and air reactor models simulate steady state conditions. However, solids conversion in one reactor affects the performance of the other reactor. After fuel addition or a change on operation conditions, the CLC operation is transitional until steady state is reached in the global system. Considering the model initial state, particles were completely oxidized when fuel addition started. During the transition period, oxygen for combustion is taken from oxygen in air, and from oxygen initially present in the fully oxidized oxygen carrier. The oxygen carrier loses oxygen during the transition period. After fuel addition, the fuel and air reactor models were run consecutively until convergence was reached. The convergence metric was that oxygen transferred in the fuel reactor was equal to oxygen transferred in the air reactor, indicating steady state. This condition was satisfied for specific values of solids conversion both in the fuel and air reactors. At steady state, complete oxidation of the oxygen carrier could not be reached.

The main outputs of the model were: the fluid dynamics structure of the reactor (e.g. height of the dense bed and profiles of concentration and flow of solids in the freeboard); the axial profiles of gas composition and flows; the axial profiles of average conversion for the oxygen carrier; and the gas composition and solids flow in the reactor exit to the cyclone. From these outputs, the performance of the CLC system was assessed by calculating the CH_4 conversion in the fuel reactor (X_{CH4}) and the solids conversion at both the fuel reactor (X_{FR}) and air reactor (X_{AR}) outlets.

Kinetic Determination of Oxygen Carrier Reduction with CH_4 and Oxidation with O_2

Several samples of the oxygen carrier were tested in a TGA during reduction and oxidation cycles using several concentrations of CH_4 as reducing agent and oxygen as oxidant. Various conditions were tested, since the oxygen carriers will be exposed to different environments during their transit through the fluidized bed fuel reactor - ranging from 100 vol.% of fuel gas at the inlet to 0 vol.% at the outlet for complete conversion. Similarly, during particle oxidation, the O_2 concentration will vary from 21 vol.% at the inlet to an oxygen concentration as low as possible at the outlet. The reactors can operate at different temperatures. Therefore, the reactivity of the samples was analysed with the experimental conversion-time curves at various isothermal conditions (973-1173 K) or various gas concentration (5-60 vol.% for CH_4 or 5-21 vol.% for oxygen). To avoid carbon formation, 20 vol.% H_2O (N_2 to balance) was used during the reduction step with CH_4. Every experimental condition was evaluated starting with a new fresh sample, and three redox cycles were carried out. The reaction rate was found to be stable with cycle number.

Experimental conditions in the TGA were selected to avoid undesirable effects of diffusional processes. The effect of external film mass transfer and inter-particle diffusion on the observed reaction rate was minimized by using 50 mg of solids sample and a gas flow around the sample of 25 L/h (STP) [20]. Moreover, the particle size (100-300 μm) was small enough to consider uniform conversion and temperature inside the particle [21].

Effect of Gas Concentration

To determine the effect of gas concentration on kinetics of the reduction reaction, several experiments at 1073 K were carried out with different gas concentrations. Figure 3 shows the results for reduction with CH_4 and oxidation with O_2. There was a direct relationship between the fuel concentration and the reaction rate. As expected, an increase in the reacting gas concentration produces an increase in the reaction rate. In all cases, the reaction rate was very fast. In the reduction with CH_4 less than 20 seconds were needed to achieve a solid conversion higher than 80%, even for the lower concentration of 5 vol.%. However, more time was needed to reach the same conversion value during oxidation by O_2, because the particles exhibit a lower reactivity towards this reaction.

Effect of Reaction Temperature

The effect of temperature on the reaction rate of CuAlImp was investigated in the range 973-1173 K. Figure 4 shows the conversion vs. time curves for the reduction with CH_4, as well as the oxidation with O_2. In all cases, an increase of temperature produces an increase in the reduction rate, although this effect was less pronounced for oxidation.

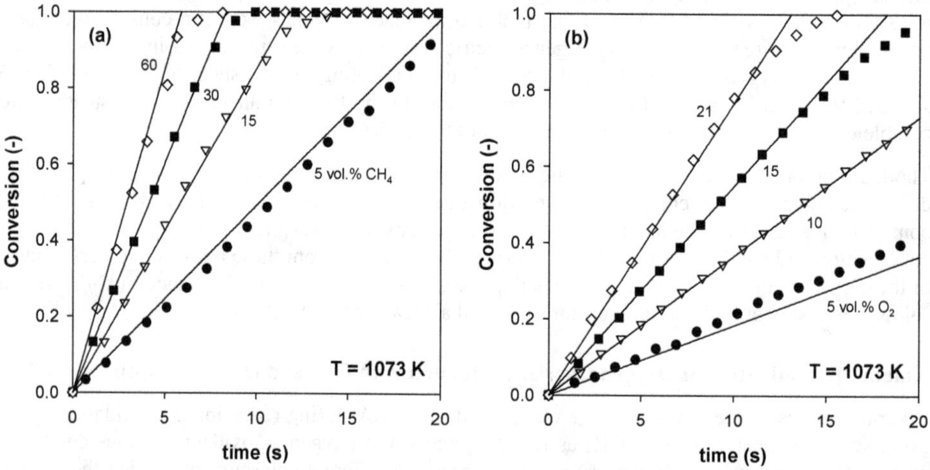

Figure 3. Effect of gas concentration on the (a) reduction reaction with CH_4; and (b) the oxidation reaction with O_2 with CuAlImp. T = 1073 K. Continuous line: model predictions.

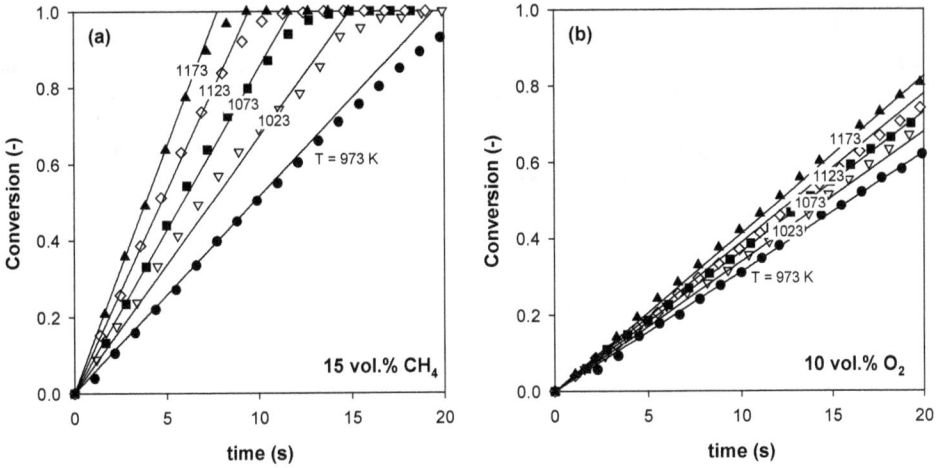

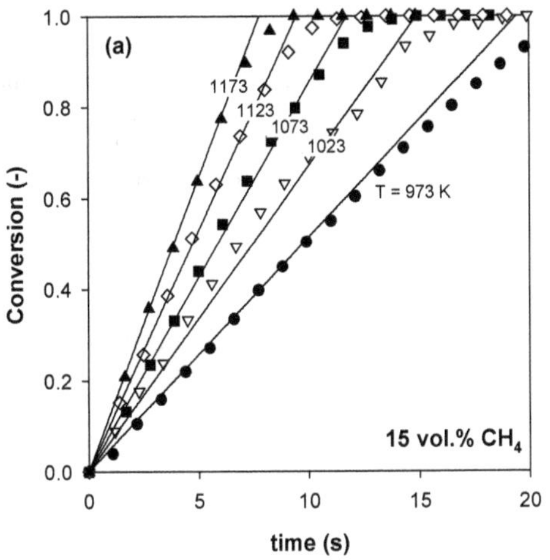

Figure 4. Effect of temperature on the (a) reduction reaction with 15 vol. % CH_4; and (b) the oxidation reaction with 10 vol.% O_2 with CuAlImp. Continuous line: model predictions.

Kinetic Model

Since it was shown in previous work using a similar Cu-based oxygen carrier, it was assumed here that the reduction of CuO proceeds towards Cu in one step [20]. The partial reduction to Cu_2O was not considered because the reduction of CuO to Cu_2O is not differentiated from reduction of Cu_2O to Cu. In addition, the presence of CuO and $CuAl_2O_4$ was confirmed by XRD analysis, and the reduction of CuO was indistinguishable from reduction of $CuAl_2O_4$. Thus, the chemical reaction was considered to be reduction or oxidation between Cu^{2+} and Cu^0, regardless of the chemical state of Cu.

Experimental conversion showed a linear dependence with time for both CuO reduction and Cu oxidation. Similar to previous work [16,20], for the kinetic model the particle is assumed to be composed of grains of CuO with a plate-like geometry on the porous surface of the support material. Every grain reacts following the shrinking core model, with chemical reaction control at the grain surface. The dependence of the solids conversion with time is given by:

$$X_s = \frac{t}{\tau} \qquad \text{Equation 8}$$

Parameter τ is the time for complete conversion of the CuO present in the particle. This parameter depends on the apparent kinetic constant, k_s , and the concentration of the reacting gas, C_g, as

$$\tau = \frac{L}{bV_M k_s \cdot C_g^{\ n}} \qquad \text{Equation 9}$$

The mean thickness of the layer, L, and molar volume, V_M, of CuO should be used in Equation 9 for kinetic determination of the reduction reaction, whereas these values for Cu should be used for the oxidation reaction (see Table 3).

Determination of Kinetic Parameters

The proposed kinetic model was used to extract the kinetic parameters from the data for the reduction reaction with CH_4, and oxidation with O_2. The time for complete conversion, τ, for all experiments was obtained from the slope of the plot of X_s versus t. The reaction order, n, with respect to each reacting gas (CH_4 or O_2) was obtained from values of τ at several gas concentrations, C_g. Equation 8 can be rearranged as follows:

$$\ln\left(\frac{L}{bV_M\tau}\right) = \ln(k_s) + n\ln(C_g) \qquad \text{Equation 10}$$

A plot of $\ln(L/bV_M\,\tau)$ vs. $\ln(C_g)$, has a slope equal to the reaction order, n, with $\ln(k_s)$ as the intercept. Figure 5(a) shows this plot for CH_4 and O_2. The calculated reaction order for each reaction is shown in Table 3.

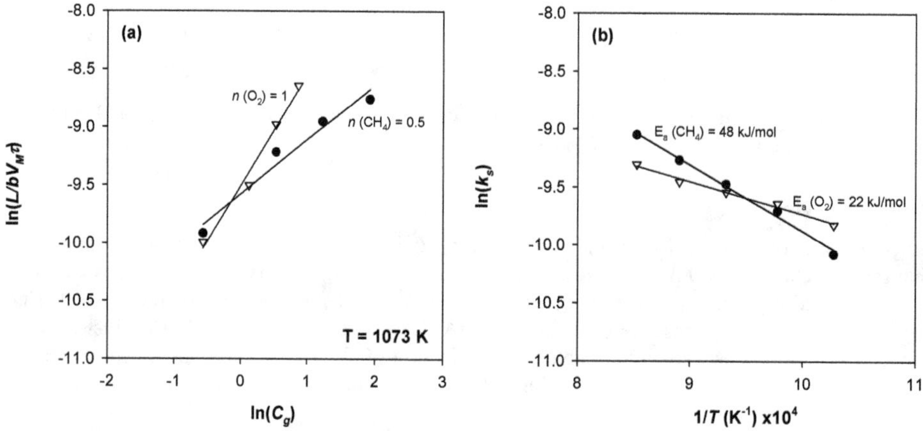

Figure 5. (a) Plot to obtain the reaction order with respect to CH_4 and O_2; and (b) Arrhenius plot for the kinetic constant k_s for the reaction of CH_4 and O_2 for CuAlImp.

From the experiments carried out at different temperatures, values for the time for complete conversion, τ, and the value of the chemical reaction kinetic constant, k_s, as a function of the temperature were obtained. The dependence on the temperature of the kinetic constant was assumed to be Arrhenius type, as follows:

$$k_s = k_{s,0}e^{-E_a/R_gT} \qquad \text{Equation 11}$$

Figure 5(b) shows the Arrhenius plot of $\ln(k_s)$ values as a function of $1/T$. The slope of the curves is E_a/R_g, whereas the intercept value is $\ln(k_{s,0})$. The values of the kinetic parameters for the reduction of the oxygen carrier with CH_4 and oxidation with O_2 are given in Table 3.

The conversion-time curves predicted by the reaction model with the estimated kinetic parameters are shown in Figures 3 and 4. The model with kinetic parameters determined in this work adequately predicts the conversion-time curves obtained in TGA for all temperatures and gas concentration analysed.

Table 3. Kinetic parameters for reaction of CuAlImp with CH_4 and O_2.

		units	CH_4	O_2
b	Stoichiometric coefficient	mol solid/mol gas	4	2
L	Layer thickness	m	$5.6\cdot10^{-8}$	$3.2\cdot10^{-8}$
V_M	Molar volume	m^3/mol	$12.4\cdot10^{-6}$	$7.1\cdot10^{-6}$
n	Order of the reaction	-	0.5	1
$k_{s,0}$	Pre-exponential factor	$mol^{1-n}\, m^{3n-2}\, s^{-1}$	$1.6\cdot10^{-2}$	$8.5\cdot10^{-4}$
E_a	Activation energy	kJ/mol	48	22

Model Validation

The CLC model was validated using experimental data from the 120 kW_{th} integrated CLC prototype at TUV (Figure 6). Dimensions for the DCFB unit at TUV are listed for both the fuel and air reactor [17], in Table 4. The CLC system combines two circulating fluidized bed reactors, with direct hydraulic communication via loop seals. Very high solids circulation flow rates are attainable with this CLC concept.

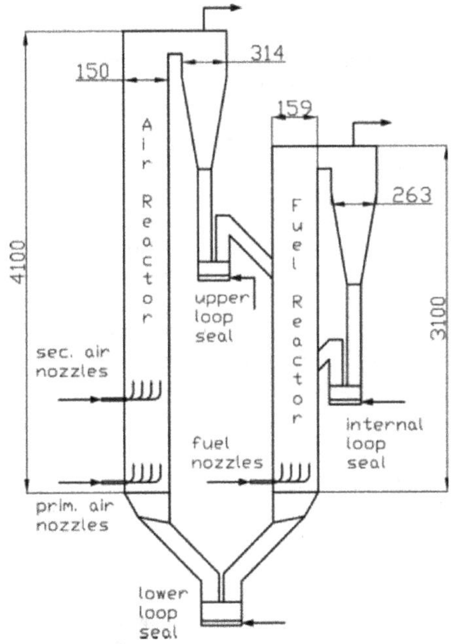

Figure 6. Schematic diagram of the DCFB unit at TUV [17].

This facility was operated with the Cu-based oxygen carrier, CuAlImp, which was prepared by the incipient impregnation method. The CuO loading was 14 wt.% supported on γ-Al_2O_3. A detailed description of the method of preparation can be found elsewhere [3]. The main physicochemical properties of the oxygen carrier particles are shown in Table 1. The fuel gas was mainly methane (98.2 vol. %). A series of experiments was done under varying operating conditions. Temperature, fuel flow rate, solids inventory and solids circulation flow rate were varied. Table 5 shows the main experimental conditions in the fuel and air reactors, as well as the CH_4 conversion measured with this oxygen carrier.

Table 4. Geometrical parameters of the fuel and air reactors in the DCFB unit [17].

	Fuel reactor	Air reactor
Height, H_r (m)	3.0	4.1
Diameter, d_{react} (m)	0.159	0.150
Height inlet from loop seals $^{(1)}$ (m)	0.35	0.0
Height inlet from loop seals $^{(2)}$ (m)	2.0	-

$^{(1)}$ Fuel reactor: steam from internal loop seal; Air reactor: steam from lower loop seal
$^{(2)}$ Fuel reactor: steam and solids from upper loop seal

Table 5. Main operating conditions for the tests carried out at TUV [11].

Temperature in Fuel Reactor (K)	1123
Fuel feed (kW_{th})	70
Air to fuel ratio (-)	1.5-1.8
Oxygen Carrier to fuel ratio (-)	2.7-5.1
Fuel Reactor solids inventory (kg/MW_{th})	130-270
Solid circulation (kg/h)	2930-5547
CH_4 conversion (%)	65-85

Figure 7(a) compares the methane conversion measured during the experimental campaign at TUV using the CuAlImp oxygen carrier with model predictions. Complete combustion was achieved for any of the conditions tested, although the methane conversion increased with the solids inventory. In general, there is good agreement between experimental results and model predictions for methane conversion at all conditions tested. The model enforces that solids conversion at the outlet of each reactor must correspond to equalize the oxygen transferred in both reactors. A comparison of experimental and predicted values for oxygen carrier conversion at the air reactor outlet can be used to assess confidence in the air reactor model (Figure 7(b)). Good agreement was found between experimental and predicted values. At steady state, methane conversion, X_{CH4}, is related to the fraction of oxygen taken in the air reactor, Ω_{AR}. Predicted values for both parameters showed deviations lower than 5% in all of cases with respect to the experimental values (Figure 7(c)). Therefore, the CLC model properly describes the behaviour of each reactor individually, as well as predicted with high accuracy, their interrelation in the DCFB system.

There are several possible explanations for incomplete combustion of methane in the CLC unit for these experiments. Insufficient solids inventory in the fuel reactor could be a factor [11]. However, higher methane conversion values might be expected, based on results from previous work [5,16]. For example, complete combustion was reached with a solids inventory of 350 kg/MW_{th} in the fuel reactor of a 10 kW_{th} CLC unit, although the fuel reactor was a bubbling fluidized bed. Furthermore, model predictions from a validated model for that system showed that a solids inventory as low as 130 kg/MW_{th} could be enough to allow complete methane conversion if a high degree of oxidation is achieved in the air reactor. In the present work, complete fuel conversion was not reached in the DCFB unit with CuAlImp material as oxygen carrier, although solids inventory in fuel and air reactor were as high as 270 and 170 kg/MW_{th}, respectively. These facts could suggest that a change in the fluidization regime and/or the solids residence time in the air reactor was affecting to the methane conversion.

Modelling showed that the required oxygen was not transferred in the fuel reactor. Modelling results show that the dilute zone is more efficient converting the fuel than the dense region [16,22,23]. So, low fuel conversion is due to a lack of solids inventory, mainly in the dilute region, due to the relatively low difference between gas velocity in the fuel reactor (0.4 m/s bottom and 1.2 m/s up) and the terminal velocity of particles (0.2-1.4 m/s for the 100-300 μm interval of particle size). This fact suggests that a smaller particle could be appropriate to increase the fuel conversion in the CLC unit.

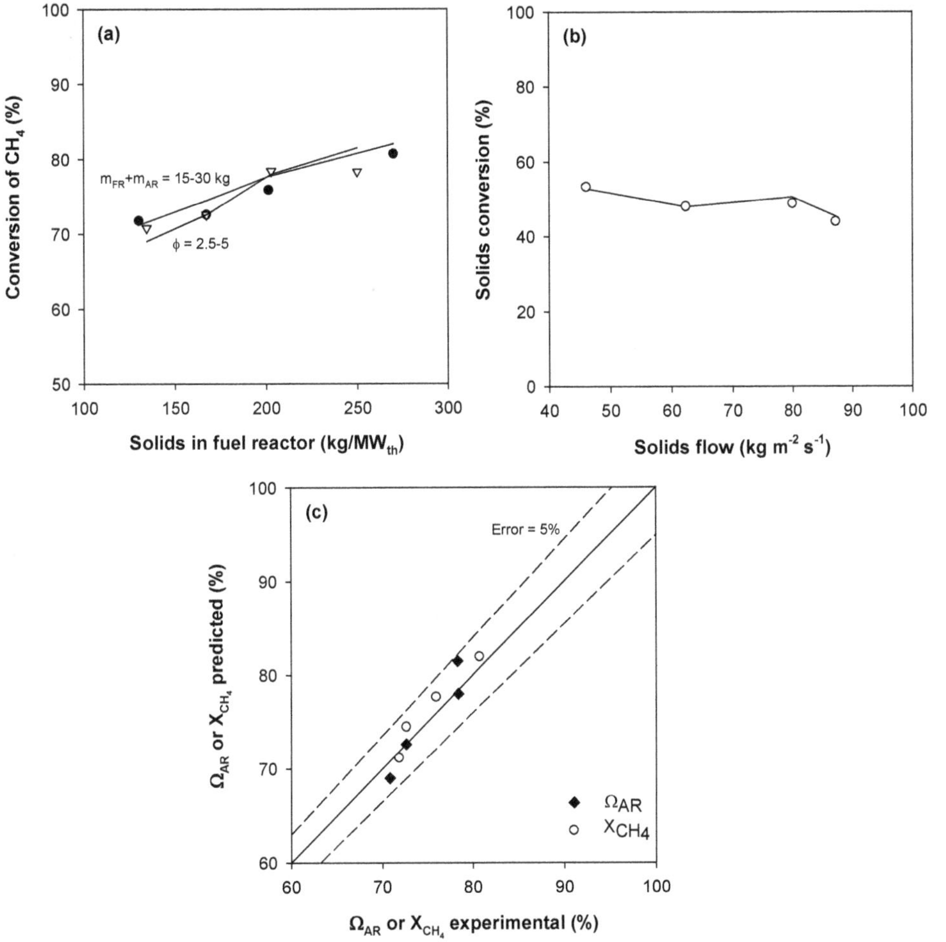

Figure 7. Comparison between experimental and predicted values in the DCFB unit with CuAlImp: (a) CH_4 conversion as a function of the solids inventory in the fuel reactor; (b) air reactor solids conversion as a function of the solids circulation rate; (c) comparison of methane conversion, X_{CH_4}, and fraction of oxygen taken in the air reactor, Ω_{AR}. Symbols: experimental; continuous line: model predictions

There was a deficiency of oxygen transferred from air to oxygen carrier in the air reactor, and as a result, the degree of particle oxidation was limited to less than 50% at the air reactor exit for these tests. It can be argued that an increase in the solids inventory of the air reactor would increase the quantity of oxygen taken from air by the oxygen carrier.

The mathematical model, after validation, can be used to optimize the operating conditions in the CLC unit. Methane conversion was calculated for varying solids inventory in the fuel reactor (i.e. the pressure drop), the solids circulation flux (i.e. the oxygen carrier to fuel ratio), the degree of oxidation for carrier particles, the fuel reactor temperature, the fuel power, and the oxygen carrier particle size. Results are shown in Figures 8(a)-(f).

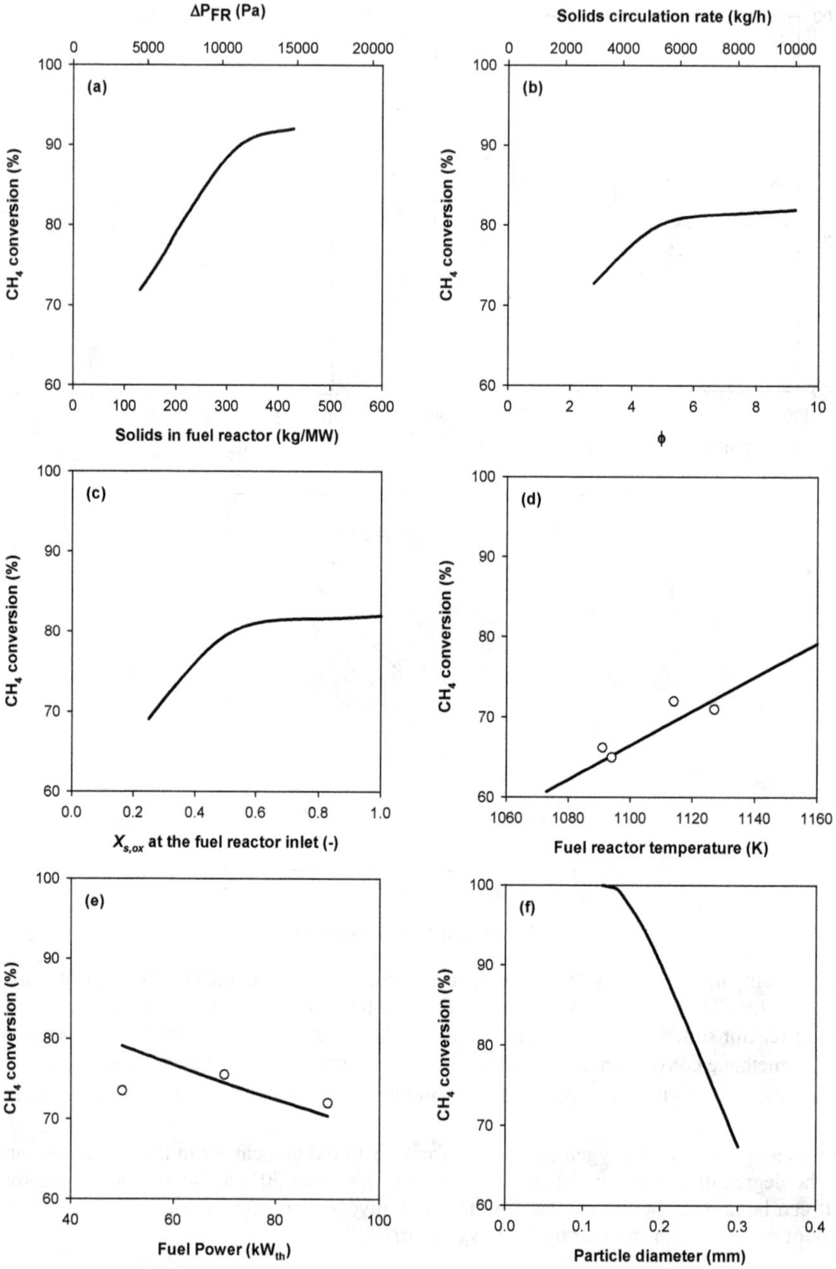

Figure 8. Methane conversion predicted by the model for CuAlImp as a function of (a) solids inventory in fuel reactor; (b) oxygen carrier to fuel ratio, ϕ; (c) inlet solids conversion to fuel reactor; (d) fuel reactor temperature; (e) fuel power; and (f) particle size. Experimental results: symbols.

It can be seen that full combustion of methane can only be reached in the DCFB unit when the particle size of the CuAlImp oxygen carrier was less than 140 μm. The fuel conversion increases when the particle sizes decreases due to the improved gas-solid contact in the dilute region compared to the dense bed.

Design of a 10 MW_{th} CLC Unit for an Impregnated Cu-based Oxygen Carrier

The scale-up of the process can be accomplished using the validated model as a tool for the proper design of a 10 MW_{th} CLC unit burning methane as fuel and using the selected Cu-based impregnated oxygen carrier, CuAlImp. From previous analysis, the fuel reactor temperature was selected to be 1123 K. Optimum values for particle size must be smaller than those tested in the DCFB unit. Taking into consideration fluidization properties of particles in circulating fluidized beds, a mean particle size of 150 μm was selected. The basic design parameters of a CLC unit to be specified are the cross-section area and gas velocity in reactors, solids circulation flow rate, pressure drops and solids inventory. These parameters are closely related, which is of help for modelling purposes with gaseous fuels. Thus, the gas velocity in the fuel reactor is fixed for a specified value of cross section area per MW_{th}, S_{MW}.

For an initial evaluation of the fuel reactor, the gas velocity was constrained between 0.6 and 3 m/s, corresponding to a cross section from 0.04 to 0.2 m^2/MW_{th}. Note that the gas velocity has to be high enough to assure the existence of a transport phase in the dilute region, which limits the maximum value of S_{MW}. The air excess ratio was fixed at $\lambda = 1.2$, which determines the air flow. The gas velocity in the air reactor was assumed to be between 5 and 15 m/s. The corresponding cross-section area of the air reactor was between 0.1 and 0.3 m^2/MW. Variations in the fluid dynamics caused by the changes in gas velocity were accounted for by the model.

Similar to the design of the DCFB system for the CLC unit at TUV, the solids entrained from the air reactor sets the solids circulation flowrate, which must be high enough to transport the required solids to the fuel reactor. Thus, the gas velocity in the air reactor was higher than the gas velocity in the fuel reactor in order to have a high enough solids circulation rate between reactors. Once the cross section area per MW_{th} was determined, the solids inventory in each reactor depends on the pressure drop. The pressure drop in the reactors was limited to 5 - 25 kPa to ensure enough solids and low pumping costs.

To evaluate the suitability of the CuAlImp material for a 10 MW_{th} CLC unit, we modelled the effect of changing the solids inventory in fuel reactor by modifying the pressure drop in the fuel reactor or the cross-section. Figure 9 shows that methane conversion increased when the pressure drop increased. Thus, the model predicted complete methane conversion when the pressure drop in the fuel reactor was 26 kPa, corresponding to 170 kg/MW_{th}. The solids inventory can also be increased by increasing the cross-section area of the fuel reactor. However, methane conversion was lower when the cross-section was increased to high values due to the low gas velocity, which minimized the amount of solids in the dilute region.

In addition to maintaining the solids inventory in each reactor, the CLC unit must be able to transport the required solid flux from the air reactor to the fuel reactor. Simulation analysis showed that complete combustion could be achieved with the CuAlImp oxygen carrier when a solids flux of 45 kg $m^{-2}s^{-1}$ was assumed, corresponding to oxygen carrier to fuel ratio, $\phi = 2.5$. This condition is well below the upper limit for the solids flux in the air reactor [11,17]. As a summary, Table 6 shows the main design and operating conditions for a 10 MW_{th} CLC unit which achieves complete combustion of methane using the Cu-based impregnated material as the oxygen carrier. The solids inventory needed in the air reactor was higher than the one needed in the fuel reactor. This fact shows the relevance of a proper design of the air reactor, accounting for kinetic differences between oxidation and reduction, to be able to transfer the required oxygen from air to oxygen carrier.

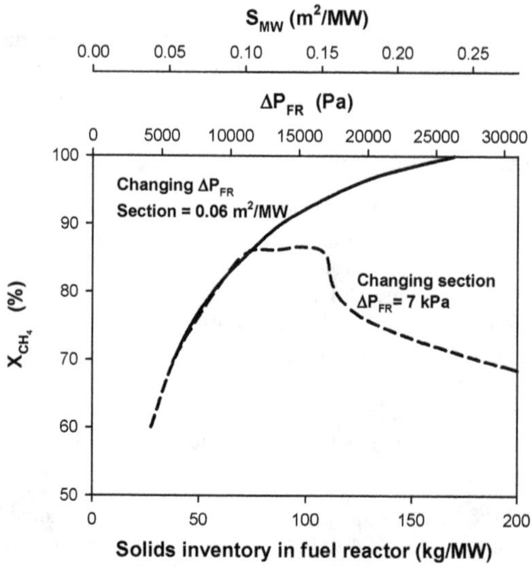

Figure 9. Methane conversion predicted by the model as a function of solids inventory in the fuel reactor by changing the pressure drop or the cross-section area. Oxygen carrier: CuAlImp.

Table 6. Main design and operating conditions of a 10MW_{th} CLC unit for CuAlImp.

	Fuel reactor	Air reactor
Solids flux (kg m^{-2} s^{-1})	-	47
Cross section, S_{MW} (m^2/MW_{th})	0.11	0.22
Pressure drop (kPa)	12	15
Solids inventory (kg/MW_{th})	140	350

CONCLUSIONS

An analysis of the performance of Cu-based particles prepared by impregnation or by spray-drying methods to be used as an oxygen carrier in chemical looping combustion (CLC) of gaseous fuels was done. Particles prepared by spray-drying were designed with the objective of increasing both the copper oxide fraction and the lifetime of particles compared to impregnated particles. An iterative process of preparation and performance testing was followed in order to develop a suitable copper-based spray dried oxygen carrier. The influence of the different materials (Al_2O_3, $Al(OH)_3$, $MgAl_2O_4$, $CaAl_2O_4$, and kaolin) and processing parameters such as, sintering temperature and time (1223 to 1473 K and 6 to 12 h) on the crush strength and reactivity of the oxygen carrier particles was assessed. Candidate particles with enough crush strength (>1 N) and high reactivity were characterized in a batch fluidized bed reactor. Samples prepared with 30 wt. % of CuO and 70 wt. % of $MgAl_2O_4$ and calcined for 12 h at 1373 K had sufficient reactivity and crush strength to be considered as a promising oxygen carrier. Despite the high copper content of this material, no agglomeration was detected during high temperature fluidization. The attrition rate measured after 20 h of operation at 1073 K stabilized at 0.08 wt.%/h, which corresponds to an average lifetime of 1200 hours, which is less than that showed for a Cu-based impregnated material. However, good results for reactivity and agglomeration encourage the possibility of further work to improve attrition. The use of other preparation methods or the identification of a proper support material can be key factors to obtain durable materials.

The Cu-based oxygen carrier, CuAlImp, prepared by impregnation, was selected for a basic assessment of scale-up of the process. A theoretical model of a CLC unit was used as a tool for the design of a 10 MW_{th} CLC unit for natural gas combustion using CuAlImp. A model describing fuel and air reactors of a CLC unit was developed, which considered the fluid dynamics of both reactors, as well as kinetics of reduction and oxidation of the oxygen carrier. The model was validated against experimental results obtained in the 120 kW_{th} CLC unit at Vienna University of Technology (TUV). For the conditions tested using CuAlImp at TUV, the simulation model was used to determine that insufficient solids inventory, in both fuel and air reactors, was responsible for low methane conversion observed. A sensitivity analysis identified the benefits of using a smaller particle size. The basic design of a 10 MW_{th} CLC unit was done using the validated model. The reactor cross sections are 0.11 m^2/MW_{th} for the fuel reactor and 0.22 m^2/MW_{th} for the air reactor. Oxygen carrier inventory required using the Cu-based material was 140 kg/MW_{th} and 350 kg/MW_{th} in the fuel reactor and the air reactor, respectively. This analysis showed that it is feasible to use the impregnated Cu-based oxygen carrier for methane combustion in a 10 MW_{th} CLC unit.

ACKNOWLEDGEMENTS

This work was supported by the European Commission, (INNOCUOUS Project, Contract 241401), and the CO_2 Capture Project (Contract CAP-011). The authors thank to Tobias Pröll, Karl Mayer and Stefan Penthor for sharing information from CLC tests in the DCFB facility. The authors also thank VITO (Belgium) for providing the solid material used in this work.

REFERENCES

1. J. Adánez, A. Abad, F. García-Labiano, P. Gayán, L.F. de Diego, Progress in Chemical-Looping Combustion and Reforming technologies, *Progress in Energy and Combustion Science* **38** (2012) 215.
2. L.F. de Diego, F. García-Labiano, J. Adánez, P. Gayán, A. Abad, B.M. Corbella, J.M. Palacios, Development of Cu-based oxygen carriers for chemical-looping combustion, *Fuel* **83** (2004) 1749.
3. L.F. de Diego, P. Gayán, F. García-Labiano, J. Celaya, A. Abad, J. Adánez, Impregnated CuO/Al_2O_3 Oxygen Carriers for Chemical-Looping Combustion: Avoiding Fluidized Bed Agglomeration, *Energy Fuels* **19** (2005) 1850.
4. F. García-Labiano, P. Gayán, J. Adánez, L.F. de Diego, C.R. Forero, Solid Waste Management of a Chemical-Looping Combustion Plant using Cu-Based Oxygen Carriers, *Env. Sci. Tech.* **41** (2007) 5882.
5. J. Adánez, P. Gayán, J. Celaya, L.F. de Diego, F. García-Labiano, A. Abad, Chemical Looping Combustion in a 10 kW_{th} Prototype Using a CuO/Al_2O_3 Oxygen Carrier: Effect of Operating Conditions on Methane Combustion, *Ind. Eng. Chem. Res.* **45** (2006) 6075.
6. C.R. Forero, P. Gayán, L.F. de Diego, A. Abad, F. García-Labiano, J. Adánez, Syngas combustion in a 500 W_{th} Chemical-Looping Combustion system using an impregnated Cu-based oxygen carrier, *Fuel Proc. Tech.* **90** (2009) 1471.
7. P. Gayán, C.R. Forero, L.F. de Diego, A. Abad, F. García-Labiano, J. Adánez, Effect of gas composition in Chemical-Looping Combustion with copper-based oxygen carriers: Fate of light hydrocarbons, *Int. J. Greenhouse Gas Control* **4** (2010) 13.
8. C.R. Forero, P. Gayán, F. García-Labiano, L.F. de Diego, A. Abad, J. Adánez, Effect of gas composition in Chemical-Looping Combustion with copper-based oxygen carriers: Fate of sulphur, *Int. J. Greenhouse Gas Control* **4** (2010) 762.
9. P. Gayán, C.R. Forero, A. Abad, L.F. de Diego, F. García-Labiano, J. Adánez, Effect of Support on the Behavior of Cu-Based Oxygen Carriers during Long-Term CLC Operation at Temperatures above 1073 K, *Energy Fuels* **25** (2011) 1316.

10. C.R. Forero, P. Gayán, F. García-Labiano, L.F. de Diego, A. Abad, J. Adánez, High temperature behaviour of a $CuO/\gamma Al_2O_3$ oxygen carrier for chemical-looping combustion, *Int. J. Greenhouse Gas Control* **5** (2011) 659.
11. F. Zerobin, Evaluation of a CuO/Al_2O_3 oxygen carrier for Chemical Looping Combustion, *Master Thesis presented at Vienna University of Technology* (2013).
12. A. Abad, J. Adánez, F. García-Labiano, L.F. de Diego, P. Gayán, J. Celaya, Mapping of the range of operational conditions for Cu-, Fe-, and Ni-based oxygen carriers in chemical-looping combustion, *Chem. Eng. Sci.* **62** (2007) 533.
13. T. Mendiara, R. Pérez, A. Abad, L.F. de Diego, F. García-Labiano, P. Gayán, J. Adánez, Low-Cost Fe-Based Oxygen Carrier Materials for the iG-CLC Process with Coal. 1, *Ind. Eng. Chem. Res.* **51** (2012) 16216.
14. M. Johansson, T. Mattisson, A. Lyngfelt, Investigation of Fe_2O_3 with $MgAl_2O_4$ for chemical-looping combustion, *Ind. Eng. Chem. Res.* **43** (2004) 6978.
15. I. Adánez-Rubio, P. Gayán , A. Abad, L.F. de Diego, F. García-Labiano, J. Adánez J., Evaluation of a Spray-Dried $CuO/MgAl_2O_4$ Oxygen Carrier for the Chemical Looping with Oxygen Uncoupling Process, *Energy Fuels* **26** (2012) 3069.
16. A. Abad, J. Adánez, F. García-Labiano, L.F. de Diego, P. Gayán, Modelling of the chemical-looping combustion of methane using a Cu-based oxygen-carrier, *Combustion Flame* **157** (2010) 602.
17. T. Pröll, P. Kolbitsch, J. Bolhár-Nordenkampf, H. Hofbauer, A novel dual circulating fluidized bed system for chemical looping processes, *AIChE J.* **55** (2009) 3255.
18. A. Abad, P. Gayán, L.F. de Diego, F. García-Labiano, J. Adánez, Fuel reactor modelling in chemical-looping combustion of coal: 1. model formulation, *Chem. Eng. Sci.* **87** (2013) 277.
19. D. Pallarés, F. Johnsson, Macroscopic modelling of fluid dynamics in large-scale circulating fluidized beds, *Prog. Energy Combust. Sci.* **32** (2006) 539.
20. F. García-Labiano, L.F. de Diego, J. Adánez, A. Abad, P. Gayán, Reduction and Oxidation Kinetics of a Copper-Based Oxygen Carrier Prepared by Impregnation for Chemical-Looping Combustion, *Ind. Eng. Chem. Res.* **43** (2004) 8168.
21. F. García-Labiano, L.F. de Diego, J. Adánez, A. Abad, P. Gayán, Temperature variations in the oxygen carrier particles during their reduction and oxidation in a chemical-looping combustion system, *Chem. Eng. Sci.* **60** (2005) 851.
22. A. Abad, P. Gayán, L.F. de Diego, F. García-Labiano, J. Adánez, K. Mayer, S. Penthor, Modelling a CLC process improved by CLOU and validation in a 120 kW unit, *Proc. 11th Int. Conf. Circulating Fluidized Bed (11CFB)*, Beijing, China (2014).
23. A. Abad, P. Gayán, F. García-Labiano, L.F. de Diego, J. Adánez, Relevance of Oxygen Carrier Characteristics on CLC Design for Gaseous Fuels, *Proc. 3rd Int. Conf. Chemical Looping*, Göteborg, Sweden (2014).

Carbon Dioxide Capture for Storage in Deep Geological Formations, Volume 4
Karl F. Gerdes (Editor)

Chapter 18

PERFORMANCE OF A COPPER BASED OXYGEN CARRIER FOR CHEMICAL LOOPING COMBUSTION OF GASEOUS FUELS AND FLUID DYNAMIC INVESTIGATIONS OF A NEXT SCALE REACTOR DESIGN

Stefan Penthor[1], Klemens Marx[1], Florian Zerobin[1], Michael Schinninger[1], Tobias Pröll[2] and Hermann Hofbauer[1]
[1]Vienna University of Technology, Getreidemarkt 9/166, 1060 Vienna, Austria
[2]University of Natural Resources and Life Sciences Vienna, Peter-Jordan-Straße 82, 1190 Vienna, Austria

ABSTRACT: Two important aspects of chemical looping process development have been covered in the present study: the investigation of the performance of a specific oxygen carrier in a 120 kW continuously operating pilot plant and the fluid dynamic investigation of a reactor concept for a next scale demonstration plant based on interconnected circulating fluidized beds. The oxygen carrier was prepared by impregnation of copper on an inert Al_2O_3 support and showed good performance in previous tests in pilot plants up to 10 kW fuel power. During the experiments in the larger unit, the copper particles showed good performance regarding conversion of CO and H_2 (almost full conversion), but only moderate conversion of CH_4 (up to 80%) was achieved. Three process parameters - fuel reactor temperature, solids circulation between air and fuel reactor, and solids inventory - have been identified as significant parameters for fuel conversion, i.e. increasing one of these parameters improves fuel conversion. The cold flow model studies of the next scale design, based on the design of the 120 kW pilot plant with an additional bed material cooler, underline the hydrodynamic feasibility of the design and verify a high degree of flexibility with respect to system control, part load operation and performance of the oxygen carrier. The solids inventory in the two reactors has a big effect on solids circulation between them. Gas staging, i.e. the use of primary and secondary air, can be used to control the solids circulation between the reactors.

KEYWORDS: chemical looping combustion; carbon capture and storage; fluidized bed systems; oxygen carrier; copper based oxygen carrier; cold flow modeling; reactor design

INTRODUCTION

Chemical looping combustion (CLC) is an innovative combustion technology with high potential to dramatically reduce carbon capture costs [1-3]. In CLC, the combustion process takes place in two different reactors, air and fuel reactor, so that air and fuel are never mixed (see Figure 1). A so-called oxygen carrier (OC), a metal oxide, is circulating between the two reactors providing the necessary oxygen for oxidation of the fuel. The oxygen carrier is oxidized in the air reactor by the combustion air, transported to the fuel reactor and is then reduced by the fuel. Thus, CLC yields two separate exhaust gas streams, where the one from the air reactor (AR) consists of oxygen-depleted air, and the one from the fuel reactor (FR) consists ideally only of H_2O and CO_2. By applying the CLC concept, inherent carbon capture can be realized with nearly no energy penalty.

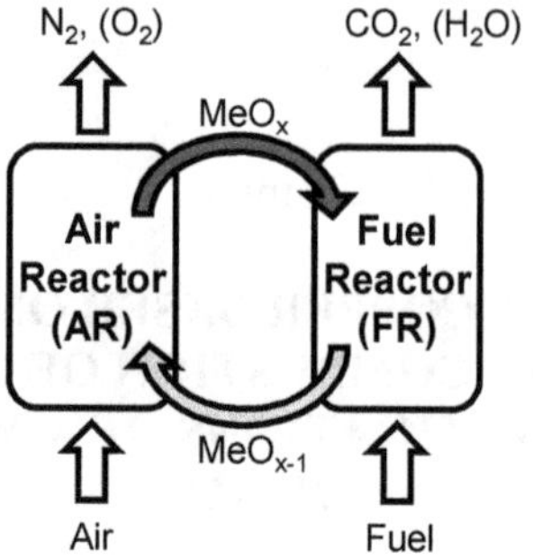

Figure 1. Chemical looping combustion of gaseous fuels.

Research and development in CLC has to be performed in two different areas in parallel: development of high performance oxygen carrier materials, and development and scale-up of reactor systems which are designed to meet the special characteristics and needs of the specific oxygen carrier. The CLC system must provide the necessary solids inventory and residence time in both reactors, as well as solids circulation rate between AR and FR [4].

THE DUAL CIRCULATING FLUIDIZED BED CONCEPT

Most reactor concepts for CLC are based on fluidized bed technology. AR and FR are designed as fluidized beds and the bed material is acting as OC circulating between the two reactors [2]. Other reactor designs are based on fixed beds and/or moving beds [5-7]. Advantages of fluidized beds over fixed beds are better gas-solid-contact, less thermal stress for particles and a lack of complicated plant equipment like high temperature valves [2]. Fluidized beds are also used in the dual circulating fluidized bed reactor concept (DCFB, see Figure 2). In this design, the AR is operated in the fast fluidization regime whereas the FR is operating in turbulent fluidization regime [8]. The two reactors are connected via steam fluidized loop seals (upper loop seal and lower loop seal) preventing gas leakage between them. The internal loop of the FR is closed via an internal loop seal.

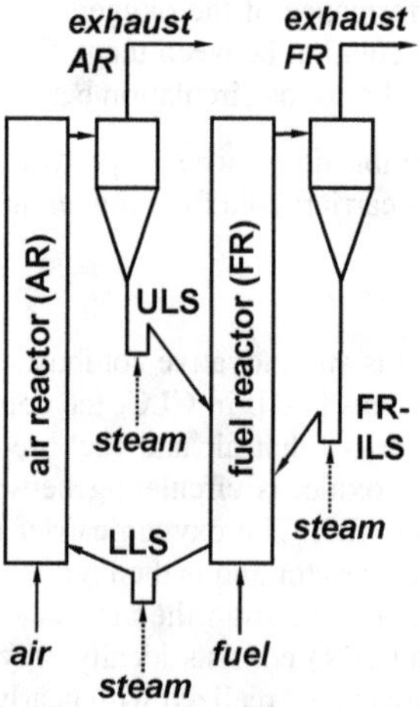

Figure 2. Dual circulating fluidized bed (DCFB) reactor system
(ULS = upper loop seal, LLS = lower loop seal, FR-ILS = fuel reactor internal loop seal).

The arrangement of upper and lower loop seal enables operation in a broad range since solids circulation between the two reactors is only governed by the solids entrainment of the AR, i.e. the amount of air introduced into the AR. The OC particles are oxidized in the AR, separated from the exhaust gas via a cyclone and transported to the FR via the upper loop seal. In the FR, which is fluidized by the gaseous fuel itself, particles are reduced and transported back to the AR via the lower loop seal.

EXPERIMENTAL

120kW Pilot Plant and the Copper-based Oxygen Carrier

The 120 kW pilot plant at Vienna University of technology is based on the DCFB concept and has been in operation since early 2008 (see Figure 3) [9]. It was originally designed for a nickel based OC and natural gas as fuel. In the AR, combustion air can be introduced at two different stages. Oxidized OC particles are separated from the AR exhaust gas stream via a high efficiency cyclone and transported to the FR via the steam fluidized upper loop seal. The FR is fluidized by the gaseous fuel itself and reduced particles are transported back to the AR via the steam fluidized lower loop seal. Particles entrained from the FR are re-introduced to the FR via the steam fluidized internal loop seal. Cooling jackets are arranged along the height of the AR in order to extract heat and control the temperature of the reactor system. The temperature profiles of the reactors and the temperature difference between the two reactors depends on solids circulation and conversion, as well as on the nature of the OC itself. OCs for which the reduction reaction in the FR is endothermic have different temperature profiles than OCs for which reduction is exothermic. Since the original design particles where Ni based, for which the reactions in the FR are endothermic, cooling jackets are attached to the AR.

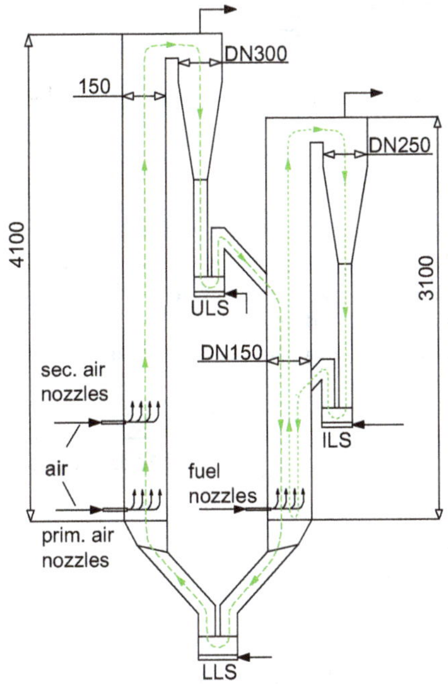

Figure 3. 120 kW CLC pilot unit for gaseous fuels
(ULS=upper loop seal, LLS=lower loop seal; ILS = internal loop seal; dimensions in mm).

Extensive analytical equipment provides a comprehensive overview of the system. Both exhaust gas streams are continuously monitored with respect to O_2, CO and CO_2 for the AR), and CH_4, O_2, CO, CO_2 and H_2 for the FR. A gas chromatograph is used to detect N_2, C_2H_4 and Ar in the FR exhaust gas. N_2 and Ar in the FR exhaust are markers for gas leakage from air reactor to fuel reactor via the upper loop seal. In order to determine the oxidation state of the particles, solids are sampled on a regular basis from the upper and lower loop seals. An error assessment has been performed by Kolbitsch *et al.* [10], where the effect of potential sources for error have been taken into account (e.g. accuracy of the scale), The typical error regarding the oxidation state is in the order of magnitude of 10^{-2} to 10^{-4}. The pilot plant has been operated in auto thermal CLC for several hundred hours using different OCs and fuels under a wide range of operating conditions [11-17].

For the experiments, a copper based oxygen carrier prepared by wet impregnation on a highly porous γ-Alumina support has been used (Cu14, see Table 1). By using this method and by limiting the copper loading on the support, agglomeration in the fluidized bed can be avoided. The preparation method has been described in detail by de Diego *et al.* [18, 19]. The OC had previously been tested extensively in pilot plants up to 10 kW fuel power and showed very good conversion performance [20-22]. Although the OC contains copper, the material is not able to release gaseous oxygen in the FR. Some OC materials can release free oxygen and that type of process is referred to as chemical looping combustion with oxygen uncoupling (CLOU). Due to interaction between the active CuO and the inert alumina support, the equilibrium partial pressure is lowered and spontaneous release of O_2 is not possible [18, 23]. The most important particle properties for the OC studied in this project are summarized in Table 1.

Table 1. Oxygen carrier particle properties.

Parameter	Unit	Value
Active Cu content	wt%	14.2
Inter support	-	γ-Al_2O_3
Oxygen transport capacity	kg/kg	0.0263
Porosity	%	50
Apparent particle density	kg/m^3	1500
Mean particle size	µm	235

The 120 kW pilot plant was originally designed for Ni-based particles produced by spray drying. Since reaction mechanisms of Ni and Cu based OCs are different [4, 15], the operating conditions like fuel power and operating temperature had to be adapted for the Cu based particles. Additionally, the impregnated Cu14 particles show different fluid dynamic characteristics, which had to be taken into account as well. The changes are summarized in Table 2.

Table 2. Comparison of design particles and Cu14 particles.

Parameter	Unit	Ni (design case)	Cu14
Fuel power	kW	120	70
FR temperature	°C	850-950	775-850
Total bed material	kg	65-75	35-45
Active bed material	kg	25-28	22-25

10 MW Reactor Concept and Scaled Cold Flow Model

The reactor design for a next scale CLC demonstration unit is based on the DCFB concept, used for the 120 kW pilot plant, with two circulating fluidized beds and steam fluidized loop seals for a nickel based OC. In order to hold the desired process temperature (800-1000 °C), it is necessary to withdraw heat from the system [9]. At typical CLC operation conditions (air/fuel ratio = 1.1-1.2, 90-100% fuel conversion), about 50% of the heat input has to be directly extracted from the reactor system [24]. In the present design, a combination of direct cooling of the AR via water walls and of heat removal from hot solids is used. An external fluidized bed heat exchanger, where bed material is extracted from the ULS and recirculated to the AR (see Figure 4) is used to extract heat from the circulating solids. The amount of heat extracted from the bed material cooler can be controlled via the amount of solids recirculated to the AR. The most important design parameters are summarized in Table 3.

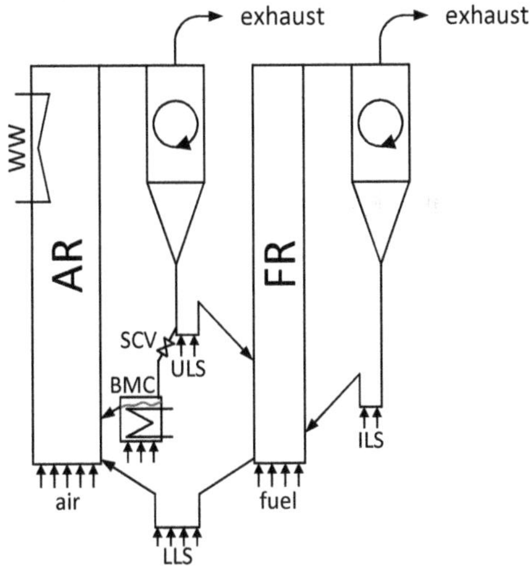

Figure 4. Proposed reactor system arrangement based on the dual circulating fluidized bed concept. Locations where fluidization is applied are indicated by arrows. (LLS = lower loop seal, ULS = upper loop seal, ILS = internal, BMC = bed material cooler; SCV = solids flow control valve)

Table 3. Most important design parameters for 10 MW concept.

Item	**Value**	**Unit**
Fuel power input	10	MW
Lower heating value of fuel	50	MJ/kg
Design air/fuel ratio	1.1	-
Design FR temperature	900	°C
Design AR temperature	960	°C
OC active material	NiO/Ni	-
Active NiO content	40	wt%
OC mean particle size	200	µm
OC particle apparent density	3425	kg/m^3
Estimated AR solids entrainment	74	kg/m^2s
AR and FR riser height	15	m
Estimated FR solids entrainment	26	kg/m^2s
Estimated solids inventories		
AR riser	2400	kg
FR riser	1560	kg
Total solids inventory	12.14	t

Cold flow models are a powerful tool used during process development and scale-up to investigate the characteristics and fluid dynamic properties of a reactor design. A scaled model of the original reactor design can be operated at ambient conditions if certain scaling criteria are applied. Fulfilling these scaling criteria guarantees similarity between the original hot design and the scaled cold flow model. Scaling criteria are usually based on dimensionless numbers and the most common ones for fluidized beds have been proposed and simplified by Glicksman [25, 26].

A schematic drawing of the cold flow model is depicted in Figure 5, including the air and fuel reactor risers, the cyclones, the upper, lower, and internal loop seals and the bed material cooler.

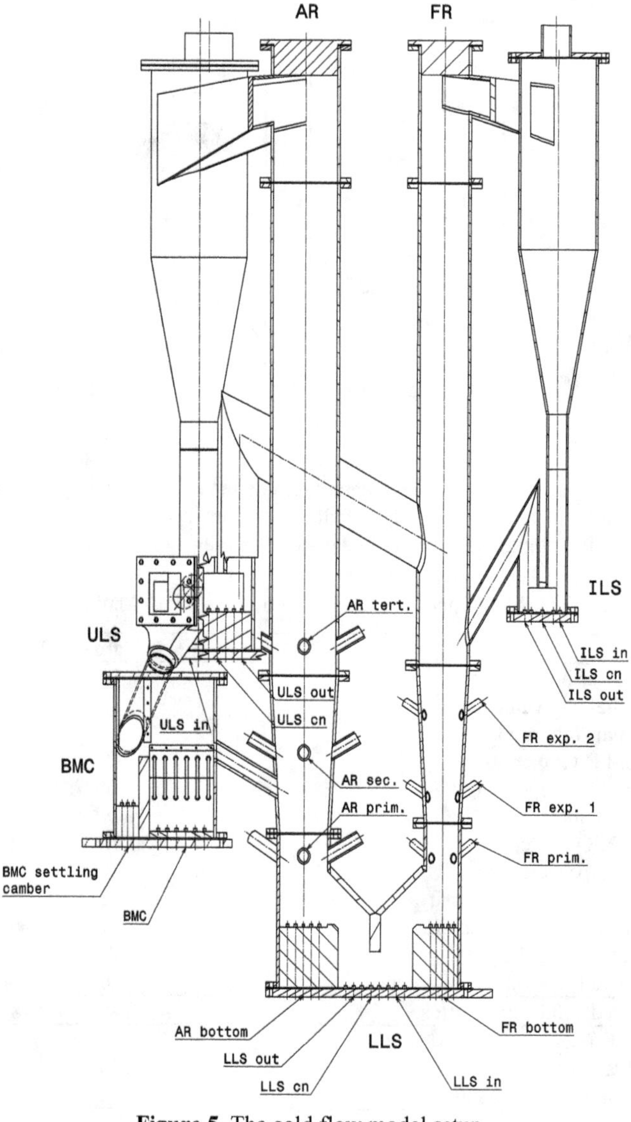

Figure 5. The cold flow model setup.

The dimensions were derived from hydrodynamic scaling using a scaling factor of 1:10 from the 10 MW chemical looping combustion demonstration unit, as discussed above. The model is built of transparent acrylic glass allowing visual observation of the fluid dynamic pattern inside. The air at the bottom of the reactors, at the bottom of the loop seals and at the bottom of the BMC is introduced via small nozzles. Air is supplied in stages to the air reactor to control the solids flow rate to the fuel reactor. To model the increase of the gas volume inside the fuel reactor of the hot unit, two additional gas injection ports are attached (FR exp 1 and FR exp 2). For an ideal CLC system, the gas volume increase in the fuel reactor can be as high as three times the inlet flow rate, which will have a significant influence on the hydrodynamics. The loop seal area is split into three parts: the inlet, center, and outlet sections. Air is provided separately to each section via separate flow meters. To control the solids flow to the bed material cooler, a solids flow control valve is attached to the upper loop seal.

Measurements for the cold flow model include volumetric flow rates of the inlet air, pressures, riser solids flow rates and solids elutriation rates from the cyclone. Additionally, at the outlet of the cyclones, ports are available for connection to a gas analyser. The air is delivered at 8 barg to the cold flow model air distributor system where variable area flow meters are used to quantify the air flow rates. Extensive pressure measurements are included in the cold flow model via 52 pressure measurement ports. The pressure ports are equally distributed along the height of the two risers with additional, closely-spaced ports where high pressure gradients are expected.

RESULTS AND DISCUSSION

Performance of the Cu14 Oxygen Carrier

In order to conveniently describe the process performance and compare different operating conditions, a key set of metrics is defined. Due to the high CH_4 concentration in the natural gas (98.5 vol%), the methane conversion X_{CH4} describing how much of the CH_4 entering the FR is converted to CO_2, CO and H_2 is a good indicator for process performance:

$$X_{CH_4} = 1 - \frac{\dot{n}_{FR,exh} \cdot y_{CH4,FR,exh}}{\dot{n}_{FR,feed} \cdot y_{CH4,FR,feed}} \qquad \text{Equation 1}$$

The global solids circulation rate G_S between AR and FR is the solids mass flow related to the riser cross section of the AR:

$$G_S = \frac{\dot{m}_{OC}}{A_{AR}} \qquad \text{Equation 2}$$

Compared to OCs based on Ni, Fe and Mn, the high temperature is limited when it comes to copper because of its lower melting point. In the present study, the system has been operated at temperatures up to 850 °C. Operation above 850 °C was avoided due to high temperature attrition [21]. The positive effect of increasing temperature on CH_4-conversion can be seen in Figure 6.

The specific active inventory, the active inventory of the system related to the fuel power, turned out to be critical for performance as well. The inventory of a fluidized bed also directly influences solids entrainment. In the case of the AR, increased inventory also means increased solids circulation. Figure 7 shows the effect of solids circulation between the reactors for different FR inventories. Both, solids circulation and inventory have a positive effect on fuel conversion.

Figure 8 shows the concentrations of CO, H_2 and CH_4 in the FR exhaust gas. Full conversion of CH_4 has never been achieved, but conversion of CO and H_2 was nearly complete above solid circulation rates of 50 kg/m^2s (CO) and 75 kg/m^2s (H_2), respectively. Here, two reasons have been identified for the higher conversion rates of H_2 and CO compared to CH_4: The reaction rate for conversion of the three species is the highest for H_2 followed by CO and then by CH_4 [27]. Further, Cu acts as promoter of the water gas shift reaction accelerating conversion of CO [2, 28].

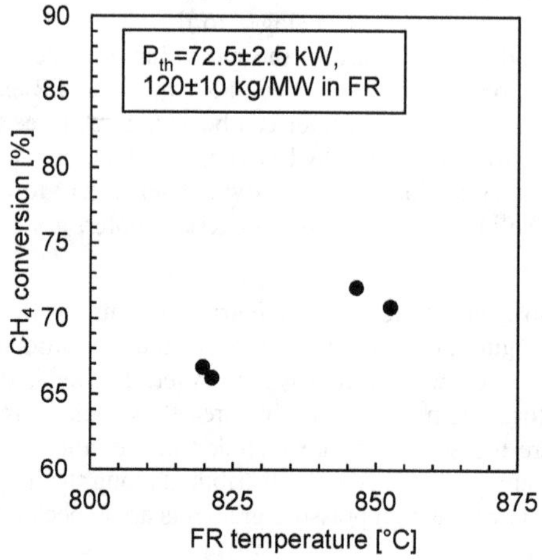

Figure 6. Influence of FR temperature on fuel conversion.

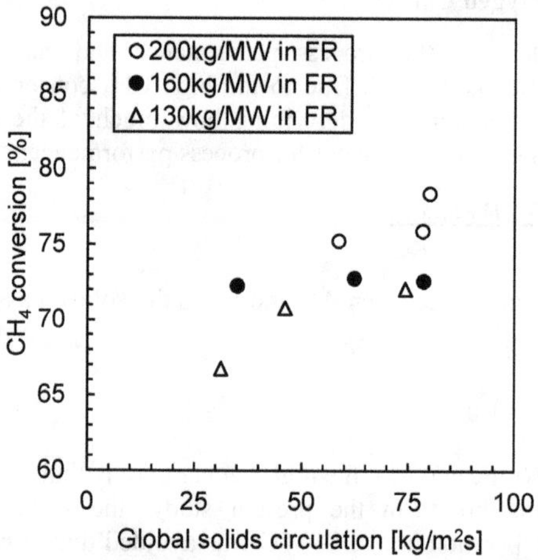

Figure 7. Influence of the global solids circulation rate on the methane conversion for three different specific solid inventories in the FR.

The Cu14 particles showed reasonable performance in the pilot plant used in this work, but did not perform as well as in smaller pilot plants with different designs. The present results show that increasing temperature, active inventory or global solids circulation improves fuel conversion. Comparing the results from the 120 kW pilot plant with 250 kg/MW maximum inventory with results from literature (up to 600 kg/MW), the low inventory can be identified as one reason for limited fuel conversion (see also Figure 7). Thus, in an optimized design for these copper-impregnated particles, the specific inventory in both reactors has to be increased.

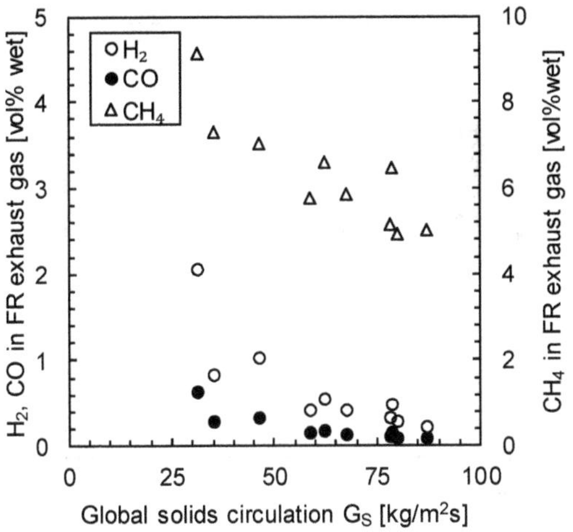

Figure 8. CH_4, H_2 and CO exhaust gas concentration for different solid circulation rates.

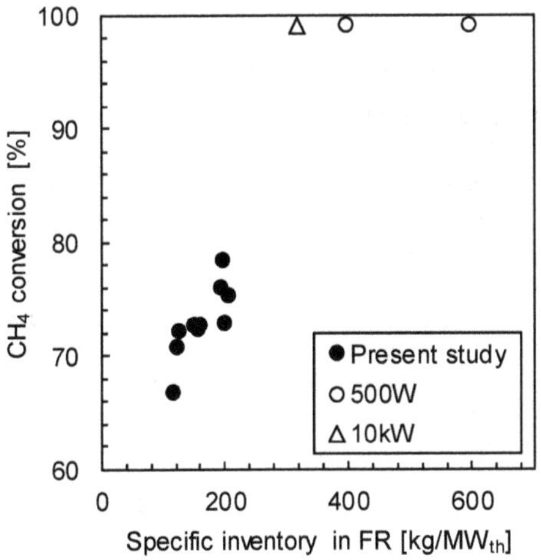

Figure 7. Comparison of process performance with data from literature.
500W data from [21] and [22], 10 kW data from [29].

OC particles were regularly sampled during the whole experimental campaign. Analysis of the particles showed that the active CuO content and thus, the oxygen transport capacity decreased during operation, but stabilized after about 30 hours at 9 wt% active CuO (see Figure 8). A further decrease was not observed. These observations are in accordance with previous results in the literature, although the initial loss was faster than in smaller pilot units. The initially higher loss of active copper is an indication that attrition comes from CuO material on the external surface of the particles and that the Al_2O_3 structure was not severely affected.

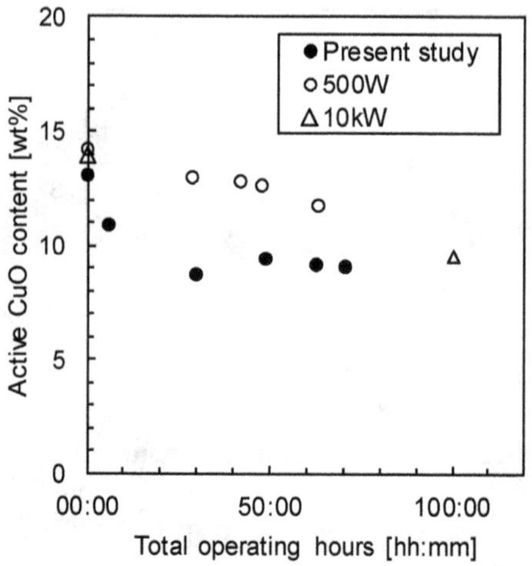

Figure 8. Change of active CuO content with increasing operating time. The operating hours include start-up and shut-down procedures. 500W data from [21] and [22], 10 kW data from [29].

Fluid Dynamic Investigation of a 10 MW Reactor Concept

The pressure profile in fluidized bed systems is useful for the identification of design problems and for an assessment of the solids distribution within the system. A typical pressure profile for operation of the risers under design conditions (listed in Table 3) is shown in Figure 9. High gradients were found at the bottom of the risers, corresponding to a high solids fractions, typical for bubbling fluidized beds. The solids distribution curves in the air reactor and fuel reactor at levels above the bubbling bottom zone are typical for circulating fluidized risers. For stable operation of loop seals, the pressure in the loop seal has to be higher than at the outlet to the riser. Additionally, the operational stability of dual circulating fluidized bed systems with respect to pressure fluctuations at the cyclone outlets is governed by the depth of the upper and lower loop seals. Deeper loop seals are able to compensate for such pressure fluctuations better than shallow seals and guarantee gas tightness between the two reactors.

A critical parameter of a CLC reactor system is the solids inventory of the reactors, since it is significant for fuel conversion and very important for the fluidization characteristics of the fluidized bed system. Inadequately selected inventories can lead to malfunction in circulating fluidized bed operation by the appearance of choking. Also, solids residence time and thus, reaction time, correlates with the solids inventory. Reactor systems for chemical looping applications, therefore, should be capable of handling varying solids inventories to meet the specific requirements of the oxygen carrier.

The effect of the solids inventory on the operation was investigated for total inventories of $m_{sol} = 35 \pm 5$ kg (Figure 10). As expected, the solids inventory had a large impact on the solids flow rates in the system. An increase of the inventory by about 15% was enough to double the solids flux in the air reactor and increase the fuel reactor solids flux by more than 50%. However, no operational difficulties were observed. If the global solids circulation rate needs to be limited, switching gas from lower sections in the riser to higher sections can help to meet the changed conditions and reduce the solids flux.

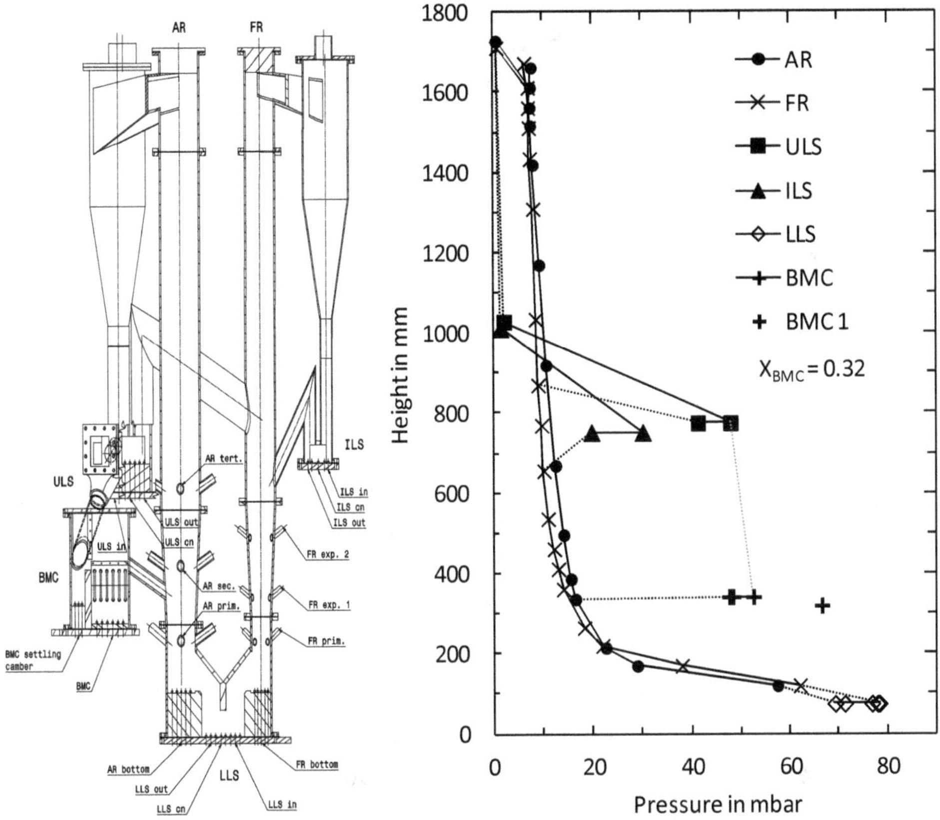

Figure 9. Overall pressure profile of the cold flow model under design operating conditions (hot unit P_{fuel} = *10 MW*, T_{FR} = *900°C* and $\lambda = 1.1$) and for a total solids inventory of 35 kg. Solids fluxes are $G_{sAR} = 42.3\ kg/m^2s$ and $G_{sFR} = 15.6\ kg/m^2s$. Total flow rates are $\dot{V}_{AR} = 168\ Nm^3/h$, $\dot{V}_{FR} = 48\ Nm^3/h$. Solids flow control vale position corresponding to a solids split-up ratio of $X_{BMC} = 0.32$.

The proposed reactor design and the cold flow model allow introduction of air to the AR at three different stages to have control over the solids entrainment rates and thus, global solids circulation. The effect of level of gas introduction to the air reactor on the global solids loop (Gs_{AR}) is illustrated in Figure 11 for a total solids inventory of m_{sol} = 35 kg. The gas flow rate to the primary and the bottom fluidization has the greatest effect on the solids flux. This occurs because increased gas flow to the lower section of the riser provokes an axial expansion of the lower dense section and carryover of solids to the upper section of the riser. These solids can be then transported more easily to the exit of the riser, which increases solids flux. It is important to note that although the effect of bottom fluidization in the air reactor seems to be less pronounced in Figure 11, it has to be kept in mind that these are relative values. In terms of absolute flow rate, the increase in gas flow to the bottom section is small compared to the pronounced effect on the solids entrainment. In addition, the tertiary fluidization has only a minor effect on the fluid dynamics. Therefore, to reduce the costs of such units, applying only bottom, primary and secondary gas injection nozzles seems to be meaningful.

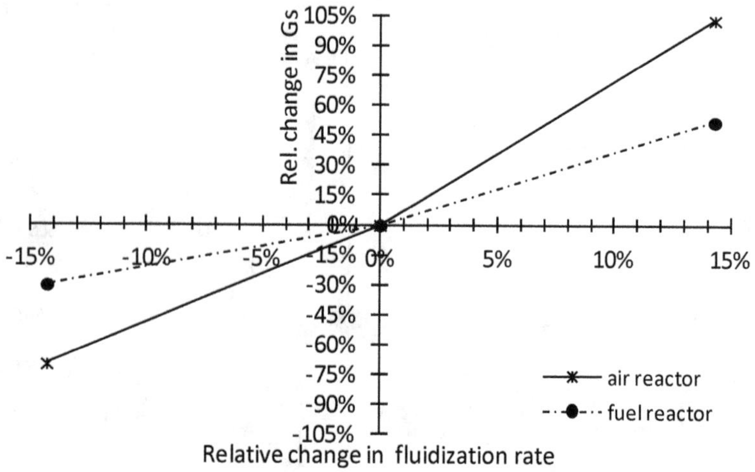

Figure 10. Effect of the solids inventory on the global (G_{sAR}) and internal (G_{sFR}) solids loop with the gas flow rates in the risers under design operating conditions.

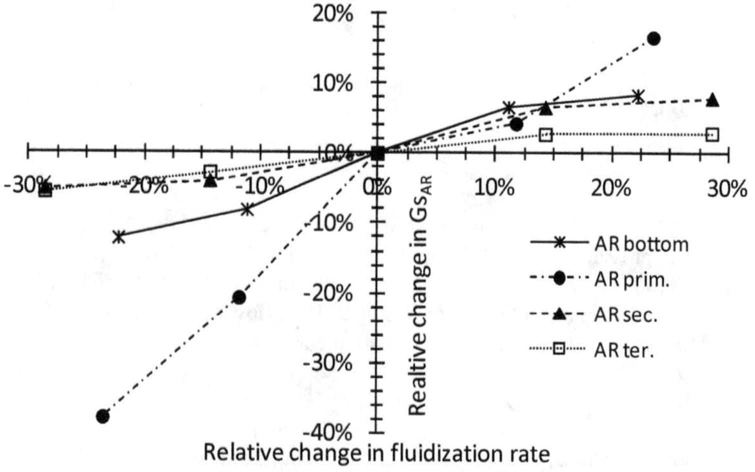

Figure 11. Effect of gas staging in the air reactor on the global solids loop (G_{sAR}) at a total solids inventory of m_{sol} = 35 kg.

Operation with recirculation of solids back to the AR via the BMC was stable and did not affect the overall fluid dynamics negatively. Figure 12 shows the influence of the solids flow control valve (SCV) on the solids split-up ratio of the BMC, i.e. the amount of solids going to the BMC related to all solids entrained from the AR. A linear correlation between SCV opening position and split-up ratio can be found between 0.6 mm to 2 mm corresponding to split-up ratios between 7 and 53%. For valve opening larger than 3.35 mm, all solids were directed to the BMC.

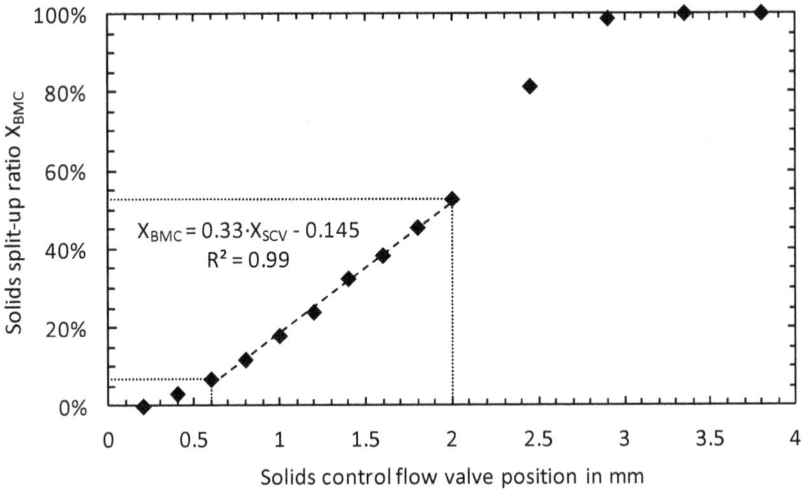

Figure 12. Effect of the SCV position on the solids split.up ratio of the BMC.

CONCLUSION AND OUTLOOK

In the present study, the performance of a copper based oxygen carrier for chemical looping combustion and the results of fluid dynamic investigations of a next scale reactor design are presented.

The experiments show that the temperature in the fuel reactor, the global solids circulation between the two reactors and the specific active inventory in the fuel reactor are significant for process performance. Though the particles showed good performance and were able to almost fully convert CO and H_2, full conversion of CH_4 was not achieved. This is in contrast to results with smaller pilot plants with different designs, where full fuel conversion was achieved using this OC. The lower specific active inventory in the fuel reactor compared to the other units was identified as the main reason for lower fuel conversion. Thus, the design of the reactor system in this 120 kW test unit, which was originally been done for a nickel based oxygen carrier, has to be adapted to allow higher solids inventories to reach full fuel conversion. Investigation of used particles shows that there is an initial loss of active Cu in the first 30 hours of operation, resulting in a stabilized active Cu content of 9 wt%. This is most probably caused by attrition of active material from the external surface of the particles.

The reactor design for a next scale 10 MW chemical looping combustion demonstration plant features a concept for efficient heat integration of the process using water walls in the air reactor and a fluidized bed material cooler. Fluid dynamic investigations using a scaled cold flow model show that the fluidized bed system works well with a broad operating range. The pressure profile of the reactors has the expected shape, and cyclones and loop seals work well. As expected, the solids inventory in the system had a strong effect, producing much higher solids flow rates. Although the system was not sensitive to much higher solids fluxes, shifting fluidization gas from the bottom areas to higher levels can help to counteract this.

ACKNOWLEDGEMENTS

This work was part of phase 3 of the CO_2 capture project (CCP). Financial support by CCP is gratefully acknowledged.

NOTATION

G_S	global solids circulation between air and fuel reactor	$kg \cdot m^2 \cdot s^{-1}$
$\dot{n}$	molar flow	$mol \cdot s^{-1}$
P_{th}	fuel power	W
T	temperature	K
$\dot{V}$	volume flow	$Nm^3 \cdot s^{-1}$
X	chemical conversion based on initial quantity	–
X_{BMC}	Solids split-up ratio between for bed material cooler	%
y	mole fraction in gas phase	$mol \cdot mol^{-1}$

Greek symbols

λ	global stoichiometric air to fuel ratio	–

Indices and abbreviations

AR	air reactor
BMC	bed material cooler
FR	fuel reactor
ILS	internal loop seal of the fuel reactor
LLS	lower loop seal
OC	oxygen carrier
SCV	solids flow control valve
ULS	upper loop seal

REFERENCES

1. Bolland O. Fundamental thermodynamic approach for analyzing gas separation energy requirement for CO_2-capture processes. Proceedings of the 8th International Conference on Greenhouse Gas Control Technologies (GHGT-8), Trondheim, Norway2006.
2. Adánez J, Abad A, García-Labiano F, Gayán P, de Diego LF. Progress in Chemical-Looping Combustion and Reforming technologies. Prog Energy Combust Sci. 2012;38:215–82.
3. Lyngfelt A. Oxygen Carriers for Chemical Looping Combustion - 4 000 h of Operational Experience. Oil & Gas Science and Technology – Revue d'IFP Energies nouvelles. 2011;66:161–72.
4. Pröll T, Schöny G, Mayer K, Hofbauer H. The impact of reaction mechanism on the design of continuously operating CLC systems - nickel based versus copper based oxygen carriers. Proceedings of the 2nd International Conference on Chemical Looping 2012.
5. Fernández JR, Abanades JC. Conceptual design of a Ni-based chemical looping combustion process using fixed-beds. Applied Energy. 2014;135:309-19.
6. Hamers HP, Gallucci F, Cobden PD, Kimball E, van Sint Annaland M. CLC in packed beds using syngas and CuO/Al2O3: Model description and experimental validation. Applied Energy. 2014;119:163-72.
7. Kimball E, Lambert A, Fossdal A, Leenman R, Comte E, van den Bos WAP, *et al.* Reactor choices for chemical looping combustion (CLC) — Dependencies on materials characteristics. Energy Procedia. 2013;37:567–74.
8. Pröll T, Kolbitsch P, Bolhàr-Nordenkampf J, Hofbauer H. A novel dual circulating fluidized bed system for chemical looping processes. AIChE J. 2009;55:3255–66.
9. Kolbitsch P, Pröll T, Bolhar-Nordenkampf J, Hofbauer H. Design of a Chemical Looping Combustor using a Dual Circulating Fluidized Bed (DCFB) Reactor System. Chemical Engineering & Technology. 2009;32:398-403.

10. Kolbitsch P, Pröll T, Bolhàr-Nordenkampf J, Hofbauer H. Characterization of Chemical Looping Pilot Plant Performance via Experimental Determination of Solids Conversion. Energy & Fuels. 2009;23:1450–5.
11. Bolhàr-Nordenkampf J, Pröll T, Kolbitsch P, Hofbauer H. Performance of a NiO-based oxygen carrier for chemical looping combustion and reforming in a 120 kW unit. Energy Procedia. 2009;1:19–25.
12. Díaz-Castro W-I, Mayer K, Pröll T, Hofbauer H. Effect of Sulfur on Chemical Looping Combustion of Natural Gas using a Nickel based Oxygen Carrier. Proceedings of the 21st international Conference on Fluidized Bed Combustion. Naples, Italy2012. p. 277–84.
13. Kolbitsch P, Bolhàr-Nordenkampf J, Pröll T, Hofbauer H. Comparison of Two Ni-Based Oxygen Carriers for Chemical Looping Combustion of Natural Gas in 140 kW Continuous Looping Operation. Industrial & Engineering Chemistry Research. 2009;48:5542–7.
14. Kolbitsch P, Bolhàr-Nordenkampf J, Pröll T, Hofbauer H. Operating experience with chemical looping combustion in a 120kW dual circulating fluidized bed (DCFB) unit. International Journal of Greenhouse Gas Control. 2010;4:180–5.
15. Mayer K, Penthor S, Pröll T, Hofbauer H. The different demands of oxygen carriers on the reactor system of a CLC plant - results of oxygen carrier testing in a 120 kW pilot plant. In: Linderholm C, Aronsson J, Källén M, Lyngfelt A, Mattisson T, editors. Proceedings of the 3rd International Conference on Chemical Looping 2014.
16. Pröll T, Kolbitsch P, Bolhàr-Nordenkampf J, Hofbauer H. Chemical Looping Pilot Plant Results Using a Nickel-Based Oxygen Carrier. Oil & Gas Science and Technology – Revue d'IFP Energies nouvelles. 2011;66:173–80.
17. Pröll T, Mayer K, Bolhàr-Nordenkampf J, Kolbitsch P, Mattisson T, Lyngfelt A, *et al.* Natural minerals as oxygen carriers for chemical looping combustion in a dual circulating fluidized bed system. Energy Procedia. 2009;1:27–34.
18. de Diego LF, García-Labiano F, Adánez J, Gayán P, Abad A, Corbella BM, *et al.* Development of Cu-based oxygen carriers for chemical-looping combustion. Fuel. 2004;83:1749–57.
19. de Diego LF, Gayán P, García-Labiano F, Celaya J, Abad A, Adánez J. Impregnated CuO/Al2O3 Oxygen Carriers for Chemical-Looping Combustion: Avoiding Fluidized Bed Agglomeration. Energy & Fuels. 2005;19:1850–6.
20. Forero CR, Gayán P, García-Labiano F, de Diego LF, Abad A, Adánez J. Effect of gas composition in Chemical-Looping Combustion with copper-based oxygen carriers: Fate of sulphur. International Journal of Greenhouse Gas Control. 2010;4:762–70.
21. Forero CR, Gayán P, García-Labiano F, de Diego LF, Abad A, Adánez J. High temperature behaviour of a CuO/γAl2O3 oxygen carrier for chemical-looping combustion. International Journal of Greenhouse Gas Control. 2011;5:659–67.
22. Gayán P, Forero CR, de Diego LF, Abad A, García-Labiano F, Adánez J. Effect of gas composition in Chemical-Looping Combustion with copper-based oxygen carriers: Fate of light hydrocarbons. International Journal of Greenhouse Gas Control. 2010;4:13–22.
23. Gayán P, Adánez-Rubio I, Abad A, de Diego LF, García-Labiano F, Adánez J. Development of Cu-based oxygen carriers for Chemical-Looping with Oxygen Uncoupling (CLOU) process. Fuel. 2012;96:226–38.
24. Marx K. Next scale Chemical Looping Combustion, PhD Thesis: Vienna University of Technology; 2013.
25. Glicksman LR. Scaling relationships for fluidized beds. Chem Eng Sci. 1984;39:1373–9.
26. Glicksman LR, Hyre M, Woloshun K. Simplified scaling relationships for fluidized beds. Powder Technol. 1993;77:177–99.

27. García-Labiano F, de Diego LF, Adánez J, Abad A, Gayán P. Reduction and Oxidation Kinetics of a Copper-Based Oxygen Carrier Prepared by Impregnation for Chemical-Looping Combustion. Industrial & Engineering Chemistry Research. 2004;43:8168–77.
28. Abad A, Adánez J, García-Labiano F, de Diego LF, Gayán P. Modeling of the chemical-looping combustion of methane using a Cu-based oxygen-carrier. Combust Flame. 2010;157:602-15.
29. Adánez J, Gayán P, Celaya J, Diego LFd, García-Labiano F, Abad A. Chemical Looping Combustion in a 10 kW th Prototype Using a CuO/Al2O3 Oxygen Carrier: Effect of Operating Conditions on Methane Combustion. Industrial & Engineering Chemistry Research. 2006;45:6075–80.

Carbon Dioxide Capture for Storage in Deep Geological Formations, Volume 4
Karl F. Gerdes (Editor)

Chapter 19

OUTLINE DESIGN AND COST ESTIMATION OF A CHEMICAL LOOPING COMBUSTION ONCE THROUGH STEAM GENERATOR BOILER PLANT

Jonathan Forsyth[1], Gerald Sprachmann[2], Tobias Pröll[3], Klemens Marx[3], Otmar Bertsch[4], Tony Tarrant[5] and Tim Abbott[5]
[1]BP International Ltd, ICBT Chertsey Road, Sunbury-on-Thames, Middlesex, TW16 7LN, UK
[2]Shell Global Solutions International BV, PO Box 38000, 1030 BN Amsterdam, The Netherlands
[3]Formerly Vienna University of Technology (Currently University of Natural Resources and Life Science, Peter-Jordan-Straße 82, 1190 Wien, Austria)
[4]Josef Bertsch Gesellschaft mbH & Co KG, Herrengasse 23, O-6700, Bludenz, Austria
[5] Amec Foster Wheeler, Shinfield Park, Reading, Berkshire, RG2 9FW, United Kingdom

ABSTRACT: A conceptual design and cost estimate of a chemical looping combustion, once-through steam generator are described. The design is based upon theoretical modelling, backed by practical test results, and heuristic design methods. A nickel-based oxygen carrier material is used in a dual circulating fluid bed configuration with heat exchangers for pre-heating feed streams and producing high pressure steam, whilst practically all of the carbon content of the fuel is captured as a high purity CO_2 stream suitable for geological storage.

KEYWORDS: Chemical Looping Combustion (CLC); Once Through Steam Generator (OTSG); Canadian oil sands

INTRODUCTION

This study represents an important practical investigation to determine the performance and cost of a once-through steam generator (OTSG) with CO_2 capture utilizing chemical looping combustion (CLC) technology. The study was produced by the effective collaboration of a number of organisations. BP and Shell worked with the other CCP members to define the overall conceptual scope and to manage the delivery of the study. The Vienna University of Technology undertook modelling and experimentation to produce the process and reaction engineering design. Bertsch Energy then developed the boiler engineering, equipment design and equipment cost estimate before Amec Foster Wheeler added balance of plant equipment design, overall system performance modelling and overall installed cost estimation.

CONCEPTUAL DESIGN OF OTSG WITH CLC TECHNOLOGY

Chemical looping combustion is a promising technology for CO_2 capture which involves the use of a re-generable solid material as an oxygen carrier to produce conditions for combustion in the absence of atmospheric nitrogen. CCP has been supporting development of CLC, with emphasis on use of gaseous fuel since CCP1 [1,2]. The results from the most recent project (described in Chapter 16), an EU co-funded effort under FP7, called INNOCUOUS [3], were utilized in this study of the use of CLC technology with a once-Through steam generator in the CCP's Canadian oil sands application scenario. This was an important study for CCP because the previous work had involved reaction modelling, materials R&D and pilot testing, and there was a need to extrapolate this information to a full-scale commercial application of the technology in a relevant operating environment.

The objective of this study was to develop the design of a practical CLC-OTSG appropriate for the Canadian oil sands environment and to use the design information to model performance and generate a cost estimate. The Canadian oil sands are one of the world's largest hydrocarbon reserves, and the production of this oil presents particular challenges from CO_2 emissions which are described in more detail in chapter 13.

The OTSG used in the Canadian oil sands is a particular type of boiler which has been developed to cope effectively with poor water quality. Produced fluid from the oil reservoir is a hot mixture of oil and water. After separation, the produced water is treated and then heated to produce steam. The steam is then injected into the reservoir to heat the oil in the formation and to reduce its viscosity and allow it to flow. As the water is cycled around, it picks up oil and other dissolved and suspended contaminants. Some contaminants are removed in treating units, but it is not economic to remove all of them. Hence, the boiler is required to be tolerant to these contaminants in the water. As well as the water quality constraint, the application scenario has logistical constraints arising from the remote location. The study was based upon a steam boiler plant nominal duty of: 4x250 MMBTU/h which is equivalent to 4x80 MW (fuel LHV) units.

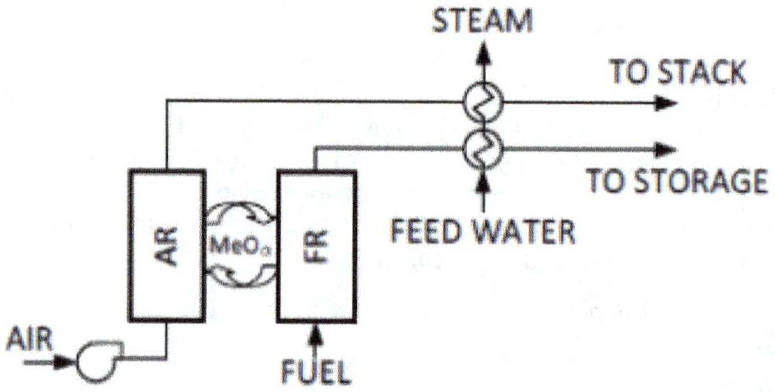

Figure 1. Conceptual arrangement of the CLC-OTSG steam boiler plant.

The CLC concept is based on the transfer of oxygen from combustion air to a fuel by means of a 'looping' oxygen carrier. The oxygen carrier is often a metal oxide. The CLC system is implemented using an "air reactor," where the oxygen carrier is oxidised and a 'fuel reactor,' where the oxygen carrier is reduced, and the fuel is oxidised by the carrier. The total amount of heat evolved from reactions in the air and fuel reactor is equivalent to stoichiometric combustion. The temperatures of the exit streams from the two reactors, the circulating solids, and any internal heat transfer surface govern the heat balance. Importantly, the exit gas stream from the fuel reactor contains mainly CO_2 and H_2O. Almost pure CO_2 is obtained when the water is condensed, facilitating a very high degree of CO_2 capture. The conceptual arrangement of the CLC-OTSG steam boiler plant is shown in Figure 1. The heat evolved in the air and fuel reactors is used to heat feed-water to produce steam whilst a high-purity CO_2 stream is captured from the exhaust of the fuel reactor. More description of the CLC concept, and the involvement of the CCP in its development is described in chapter 14 and also in Eide 2009 [2].

PROCESS AND REACTION ENGINEERING DESIGN

The Vienna University of Technology developed process models for the CLC-OTSG process. The models were validated against experimental testing reported in chapter 18. The simulation study performed by the University provided the basis for a preliminary design and sizing of the CLC-OTSG, along with performance calculations for the operation of the unit.

Modelling Methodology

Flow-Sheet Simulation Software

The SimTech Simulation Technology IPSEpro software is a flow sheet simulation tool for use in the simulation of power generation cycles and chemical processes. IPSEpro modules can be used for calculation of heat balances, prediction of performance at design or other conditions, verification and validation of measurements during acceptance tests, monitoring and optimization of plant performance on-line, and estimation of design costs. IPSEpro simulation results represent a steady-state operating point of a process. The Newton-Raphson algorithm is used to solve the multi-dimensional equation system. The equation-oriented approach enables rapid convergence with an average calculation time of just a few seconds.

The advanced energy technology library (AET-Lib) is a proprietary model library developed at the Vienna University of Technology for use with ISPEpro to model CLC processes. The organic species modelled in the AET-Lib are made up of six elements: C, H, O, N, S and Cl. Additionally, Ar is included, forming together with the organic molecules, the basis for all gaseous compounds modelled. Furthermore, 40 inorganic substances are modelled from the elements Na, K, Mg, Ca, Ti, Zr, Mn, Fe, Co, Ni, Cu, Zn, Al and Si. The material streams consisting of mixtures of pure molecules are defined by the mass or molar fractions of the different species. This allows the calculation of the mean molar mass, the mean standard enthalpy of formation, and methods for calculation of temperature and pressure dependent properties of the mixtures (enthalpy, entropy and density). The IAPWS-IF97 formulation [4] is used for calculating the properties of water and steam. The ideal gas data is formulated according to Burcat *et al.* [5], and the inorganic solids data are calculated from polynomials fitting the data tables reported in Barin and Platzki [6].

Thermodynamic Equilibrium Formulation

Thermodynamic equilibria were determined by calculating the thermodynamic equilibrium constant via minimization of the Gibbs free enthalpy.

$$\ln K_p\,(T) = -\frac{\Delta G_R^0(T)}{R \cdot T} \qquad \text{Equation E1}$$

The actual state of a substance combination with respect to the equilibrium of a specific chemical reaction:

$$\textstyle\sum_i v_i \cdot A_i = 0 \qquad \text{Equation E2}$$

was expressed by the logarithmic deviation from the equilibrium:

$$p\delta_{eq} = log_{10}\frac{\prod_i (p_i)^{v_i}}{\prod_i (p_i^*)^{v_i}} = log_{10}\frac{\prod_i (p_i)^{v_i}}{K_p(T)} \qquad \text{Equation E3}$$

This thermodynamic equilibrium description allowed consistent enforcement of equilibrium constraints in the simulation. For a known gas composition at a defined temperature and pressure, the value of $p\delta_{eq}$ was calculated from Equation E3. Values of $p\delta_{eq}$ below zero meant that the actual gas composition was on the left hand side of the reaction equation (where $v_i<0$) and values above zero implied an actual composition on the right hand side, i.e. the product side, of the reaction (where $v_i>0$). By setting the deviation from the equilibrium to zero the system of equations is forced to adjust the gas composition in a way that the Gibbs free enthalpy is minimal, which describes thermodynamic equilibrium.

Redox Systems and Reactions Modelled

Six different main redox systems are implemented in the AET-Lib based on Cu, Fe, Mn, Ni, Co and CaS as oxygen carriers. For this specific project the Ni/NiO system was used.

For an adequate description of the fuel reactor, the global reactions:

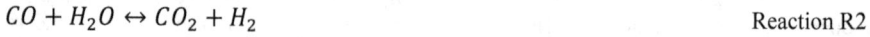

$$C_xH_y + xH_2O \leftrightarrow xCO + \left(x + \frac{y}{2}\right) H_2 \quad \text{Reaction R1}$$

$$CO + H_2O \leftrightarrow CO_2 + H_2 \quad \text{Reaction R2}$$

were divided into more elementary steps. The hydrocarbon conversion was modelled by defining conversion rates for CH_4, C_2H_4, C_2H_6 and C_3H_8. Under typical operating conditions in a CLC system, the equilibrium calculation yields practically complete conversion of hydrocarbons. Therefore, no equilibrium formulations were included involving these species. To describe the oxidation of CO and H_2 in the fuel reactor, reactions R2 and R3 were used:

$$CO + MeO_\alpha \leftrightarrow CO_2 + MeO_{\alpha-1} \quad \text{Reaction R3}$$

Only in the theoretical case where no carbon is present in the fuel reactor inlet gas, the direct oxidation of H_2 by the oxygen carrier is formulated instead of reactions R2 and R3.

$$H_2 + MeO_\alpha \leftrightarrow H_2O + MeO_{\alpha-1} \quad \text{Reaction R4}$$

In addition to hydrocarbon conversion the so-called CO_2 yield is defined:

$$Y_{CO_2} = \left.\frac{Y_{CO_2}}{Y_{CO}+Y_{CO_2}+Y_{CH_4}}\right|_{FR_{exit}} \quad \text{Equation E4}$$

The CO_2 yield is commonly used in the literature for characterization of the ability of a CLC system to completely convert hydrocarbons to CO_2 and water vapour.

In the air reactor the oxidation reaction:

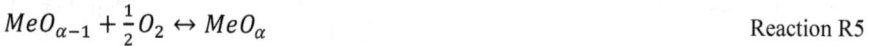

$$MeO_{\alpha-1} + \frac{1}{2}O_2 \leftrightarrow MeO_\alpha \quad \text{Reaction R5}$$

was formulated directly. A detailed description of the model can be found in Bolhár-Nordenkampf *et al* [16].

Site Conditions and Assumptions Used as the Basis for the Study

The site location assumed for the study was a brown-field oil sands production operation located near Fort McMurray, Alberta, Canada. The ambient conditions and assumptions are documented for the CCP oil sands application scenario reported in Chapter 3.

CLC-OTSG Conceptual Process Design

In CLC the combustion process is separated into two steps: first, separation of oxygen from combustion air and, second, oxidation of the fuel. The first takes place in the "Air Reactor" (AR), where a metal-based oxygen carrier is oxidized, and the second takes place in the "Fuel Reactor" (FR), where the oxidized oxygen carrier is reduced by the fuel (Figure 1). In this study the selected oxygen carrier was a physical mixture, in the ratio of 1:1 based on weight, of two nickel-based oxygen carriers. The nickel system is the most intensively tested oxygen carrier for gaseous fuels [7]. Detailed information on the oxygen carrier is given in Table 1.

Table 1. Oxygen carrier characterization.

Item	Symbol	NiO-VITO I	NiO-VITO II	1:1 mixture	Unit
Inert support material	-	Al_2O_3	Al_2O_3 + MgO	Al_2O_3, Al_2O_3 + MgO	-
Active Nickel content	-	40	41.3	40.6	wt%
Mean particle diameter	d_p	149	129	139	µm
Apparent density	ρ_p	3600	3250	3425	kg/m^3

The reactions in both the AR and FR are exothermic overall and thermal energy from the reactor exhaust gas streams is used to heat and boil water to produce high pressure steam for use in thermal oil recovery operations. Under typical CLC conditions, the heat produced in the reactor system is greater than the sensible heat removed from the reactors by the two exhaust gas streams. Thus, to maintain thermal neutrality in the reactor system, cooling of the air reactor riser using water walls and cooling of the circulating solids in a Bed Material Cooler (BMC) is required. The BMC is positioned in the pathway of the solids material as it moves from the AR to the FR to regulate the temperatures in the reactors. The amount of heat transferred from the circulating bed material is controlled in the process by routing only a proportion of the solids elutriated from the AR to the BMC. This could be achieved by using a dynamic cone valve which is common in circulating fluidised bed applications. Loop seals are used to allow the flow of solids material, while limiting the entrainment of gases with the solids material. Cyclones are included to separate and recover entrained solids from the reactor exhaust streams. The process flow diagram, shown in Figure 2a, with the simplified mass and energy balance, shown in Table 2, shows the AR, FR and BMC with an arrangement of heat exchangers, loop-seals, cyclones, fans/blowers and pumps in a conceptual CLC-OTSG process design. This design is not optimized in terms of output or energy efficiency but does embody the main features needed in a full commercial-scale CLC application to an industrial steam boiler plant.

The inputs to the CLC-OTSG are ambient air, natural gas fuel, boiler feed-water (190 °C) and feed-water (80 °C). The outputs from the CLC are AR exhaust, FR exhaust, high purity steam and steam condensate to disposal. Natural gas is provided at elevated pressure, and is throttled-down to a suitable pressure to feed into the FR. Air at ambient conditions is compressed by the air blower and fed into the AR riser with sufficient head to overcome the pressure drop across the air distributor, riser and cyclone.

Boiler feed-water (190 °C) is used to pre-heat air and natural gas in two heat exchangers before those streams reach the AR and FR, respectively (Figure 2b). The boiler feed-water is then fed to a network of heat exchangers which variously recover heat from the AR and FR exhaust streams, and to tubes in the walls of the AR and BMC. In this pathway the boiler feed-water is heated and boils to form steam. Impurities present in the boiler feed-water from the oil sands recovery process limit the proportion of vapour to liquid (dryness) of the steam produced to 78%. The dryness of the steam required for export to the oil sands recovery process is 98% so a water/steam separator vessel is included in the process design. Steam from the separator vessel is suitable for export. Water discharged from the separator vessel passes through a number of heat exchangers to recover heat to feed-water and air before it is discharged to disposal. The quantity of heat that can be transferred from the water discharged from the separator vessel is not sufficient to heat all of the boiler feed-water that is required from 80 °C to 190 °C and additional sources of heat are assumed. In practice these could be provided by heat exchange with produced fluids from the oil sands recovery process, for example.

Stream identifier		A1	F1	W1	W12	A7	F7	W15	W17
Title		Air	Fuel	BFW	Feedwater	AR Exhaust	FR Exhaust	Steam to Well	Water Disposal
Description		Ambient air charge for the AR	Natural gas charge for the FR	Boiler feedwater for the CLC-OTSG	Feedwater for loop-seal steam	Exhaust from the AR	Exhaust from the FR (raw CO2)	Steam export to the field	Condensate from main separator to disposal
Temperature	C	1.00	1.00	190.00	80.00	175.36	174.97	318.76	90.00
Pressure	bar(a)	0.97	4.00	138.30	2.20	0.97	0.97	111.00	111.00
Steam content	wt%	-	-	0.00	0.00	-	-	0.98	0.00
Normal volume flow	Nm^3/h	84851.00	8120.69	-	-	71075.90	26309.20	-	-
Operational volume flow	m^3/h	-	-	178.07	3.56	-	-	1946.38	33.14
Mass flow rate	kg/h	109493.00	5887.47	157544.00	3461.47	88444.40	30397.50	125392.00	32151.80
Sensible heat (ref 25 C)	kJ/kg	-24.15	-51.48	708.60	230.16	158.14	207.87	2574.47	280.63
LHV (ref 25 C)	kJ/kg	0.00	48917.40	-	-	0.00	75.19	-	-
HHV (ref 25 C)	kJ/kg	5.87	54281.00	-	-	56.48	1248.98	-	-
Total enthalpy (ref 25 C)	kJ/kg	-61.67	-4700.23	-15157.40	-15635.80	-158.74	-10833.50	-13291.05	-15585.30
Energy	kW	-734.46	79915.80	-75883.80	-2127.31	3885.23	2390.08	4593.03	-19308.70
Exergy	kW	493.54	82593.10	6937.19	19.97	1249.20	4001.00	36363.40	344.61
Composition:									
Ar	mol%	0.93	0.00	-	-	1.11	0.00	-	-
N2	mol%	77.78	0.69	-	-	92.86	0.21	-	-
O2	mol%	20.86	0.00	-	-	2.40	0.00	-	-
H2	mol%	0.00	0.00	-	-	0.00	0.48	-	-
H2O	mol%	0.39	0.00	-	-	3.58	68.60	-	-
CO	mol%	0.00	0.00	-	-	0.00	0.27	-	-
CO2	mol%	0.04	0.36	-	-	0.05	30.43	-	-
CH4	mol%	0.00	98.84	-	-	0.00	0.00	-	-
H2S	mol%	0.00	0.00	-	-	0.00	0.00	-	-
SO2	mol%	0.00	0.00	-	-	0.00	0.00	-	-

Table 2. Simplified mass and energy balance – 80 MW CLC OTSG with thermodynamic equilibrium in the fuel reactor.

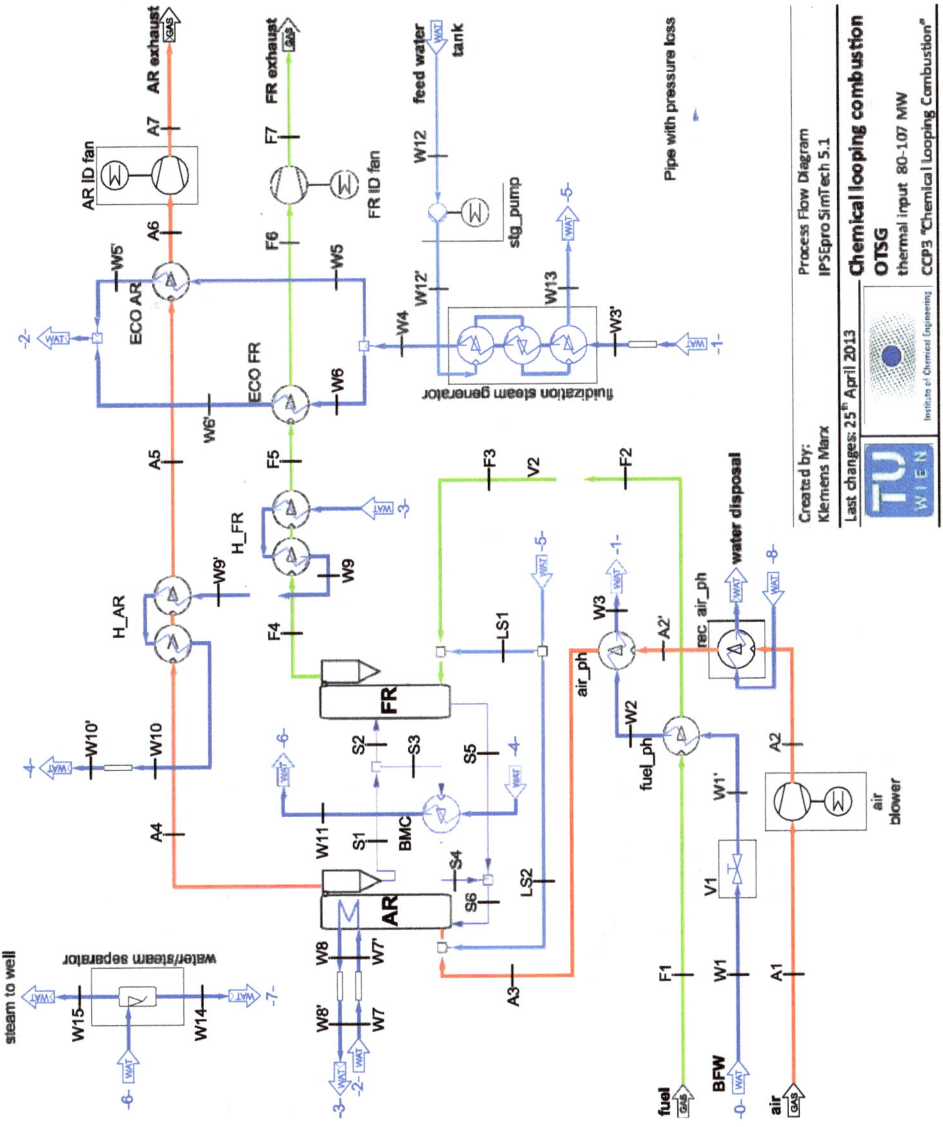

Figure 2a. CLC-OTSG process flow diagram.

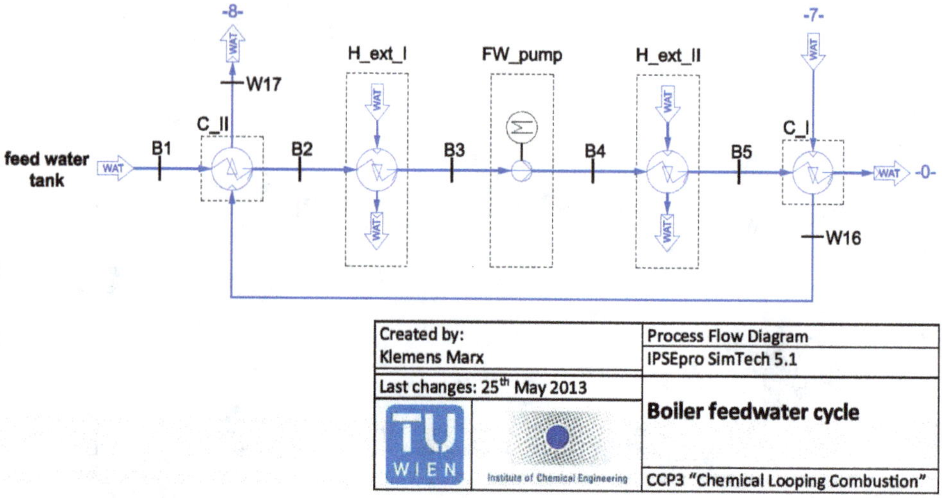

Figure 2b. Boiler feedwater heating flow diagram.

To compensate for the pressure drop in the heat recovery section of the fuel and air reactor, two induced draft fans are needed: the fuel reactor fan (FR ID fan) and the air reactor fan (AR ID fan).

The air reactor exhaust gas is oxygen depleted air and can be released to the surroundings without major concerns. The gas leaving the fuel reactor is a concentrated CO_2 stream ready for sequestration after water condensation.

CLC Reactor System Sizing

Apart from the oxygen carrier material, the reactor system design is a key issue in CLC applications. Typically, the extent of fuel gas conversion is dependent on the heterogeneous gas-solids reaction. Adequate gas-solid contact time is a key challenge for efficient CLC operation. Fluidized beds are well known for their excellent gas-solid mixing potential and, thus, are expected to be ideal for CLC. Different types of fluidized bed operating regimes are known, and the adoption of the right regime within the reactor system is crucial. Sufficiently high solids circulation from the air reactor to the fuel reactor is needed for oxygen and heat transport. Consequently at least one reactor (e.g. the air reactor) has to be operated in a regime that will transport the solids, which is either fast fluidization or pneumatic conveying in terms of fluidized bed regimes. The required solids circulation rate is mainly determined by the cyclic oxygen transport capacity of the carrier. Typical solids entrainment rates observed in high velocity turbulent or fast fluidized bed regimes are satisfactory for almost all of the oxygen carriers developed for CLC.

The reaction rate in the fuel reactor is an order of magnitude slower than in the air reactor. Therefore, the gas-solids contact is significantly more important in this reactor. Bubbling fluidized bed fuel reactors have risk of gas bypass through the bubble phase, which can be minimised by low fluidization numbers and deep beds. However, very deep bubbling beds suffer from operational constraints, as the bed pressure in the lower bed regions is far different than that in the upper bed, which runs the risk of de-fluidized dead zones at the bottom of the bed or appearance of the slugging fluidized bed regime. Furthermore, low fluidization numbers will result in a large fuel reactor cross-sectional area, high solids inventory and a large bed pressure drop. Additionally, most of the reactor volume will be lost because no relevant reaction can be expected in the practical particle-free freeboard of the bed. Turbulent fluidized bed or fast fluidized bed regimes allow gas-solids contact over the whole height of the reactor and permit operation with lower solids inventories, which is particularly important for increased plant capacity. Thus, finding the optimum solution between reduced gas residence time at an increased gas velocity and less intensive solids distribution over the reactor is a major challenge in the design of a chemical looping reactor unit. In addition to the fluid dynamics, the reactor system has to be simple in operation to avoid the occurrence of uncontrollable conditions.

In summary, the reactor system has to fulfil the following requirements:

- Optimal gas-solids contact in both reactor zones.
- High global solids circulation for sufficient oxygen transport.
- Low reactor footprint for scale-up.
- Simple operation.

The dual circulating fluidized bed (DCFB) reactor concept, specifically developed for chemical looping systems [8], is based on these requirements. It is composed of two interconnected circulating fluidized bed reactors. The design focuses on scale-up related issues such as low solids inventories and low reactor footprint, while maintaining the focus on high gas-solids contact. The schematic is shown in Figure 3.

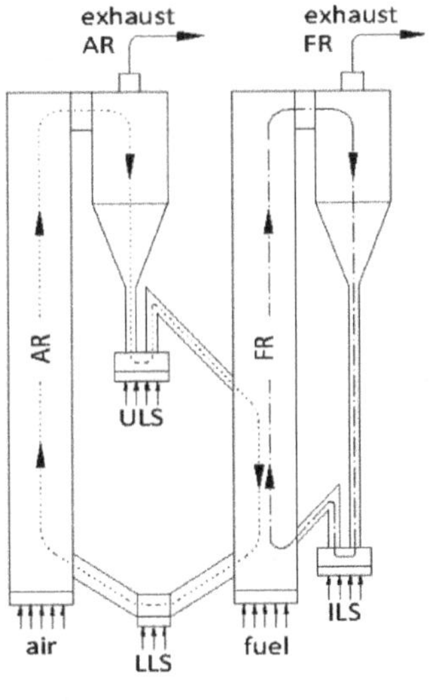

Figure 3. Principle setup of the dual circulating fluidized bed reactor system, with the global solids loop indicated by the dotted line and the internal solids loop indicated by the dash-dotted line. AR – air reactor, FR – fuel reactor, LLS – lower loop seal, ULS – upper loop seal, ILS – internal loop seal.

The DCFB concept involves two solids loops: the global loop and the internal loop. The solids entrained from the air reactor are separated by a cyclone and sent via the Upper Loop Seal (ULS) to the fuel reactor. The ULS, fluidized by steam, provides a proper sealing between the reactors to avoid gas leakage. The circulating solids are transported back to the air reactor via the Lower Loop Seal (LLS), closing the global solids loop. The fuel reactor includes an internal solids loop, which is closed by the fuel reactor cyclone, and the Internal Loop Seal (ILS). A key feature of the DCFB concept is the LLS. Solids inventories are inherently stabilised by the LLS which acts as a hydraulic link. Solids accumulation in any part of the system is avoided as long as proper fluidization is provided to the loop seal. One feature of the DCFB concept with the LLS is that the two solids loops are largely independent of each other, and as such the fuel reactor can be operated with respect to maximum fuel conversion.

Air Reactor (AR)

The AR is designed based on two requirements:

- Oxidation of the reduced oxygen carrier coming from the fuel reactor.
- Sufficient oxygen carrier entrainment for oxygen transport.

Therefore the AR is designed as a fast fluidized bed. The onset of fast fluidization is described as the regime where significant solids entrainment from a lower dense section of the fluidized bed is observed. This limit is the superficial velocity at which solids begin to be entrained significantly, U_{se}, setting an upper limit on conventional fluidized bed operation [9]. The Reynolds number at solids entrainment, Re_{se}, is defined as the Reynolds number at U_{se}, expressed as:

$$Re_{se} = 1.53 \cdot Ar^{0.50} \quad (2 < Ar < 4 \cdot 10^6) \qquad \text{Equation E5}$$

Ar is the Archimedes number, which is used in buoyancy/natural circulation calculations, and expresses the ratio of external forces (density differences between solid and fluid to internal viscous forces.

$$Ar = \frac{gL^3\rho_l(\rho-\rho_l)}{\mu^2} \qquad \text{Equation E6}$$

In the above, μ is fluid viscosity, ρ_l is density of fluid, and ρ is solid density. For proper operation the air reactor superficial gas velocity range is:

$$U_{AR} = 7 - 8\ \frac{m}{s} \qquad \text{Equation E7}$$

This value corresponds to approximately 1.2 times U_{se} for the chosen oxygen carrier. The required air reactor cross-section for the 80 MW capacity CLC-OTSG was calculated to be 12.29 m^2.

Fuel Reactor (FR)

The main design criteria for the FR is to maximise the gas solids contact while keeping the solids inventory low. Turbulent fluidized beds are characterised by bubble break-up instead of bubble growth as in bubbling fluidized beds. Small voids and particle clusters dart to and from the indistinct top surface of the bed. Thus, potential gas-slip through the bubble phase is minimised, while gas solids contact is expected to increase. The onset of the turbulent fluidization regime is defined by the critical velocity U_c where the standard deviation of the pressure fluctuation reaches a maximum. It is believed that this point reflects a dynamic balance between bubble coalescence and break-up. The turbulent regime is defined by the velocity range U_c to U_{se}, with U_c calculated from [9]:

$$Re_c = 1.24 \cdot Ar^{0.45} \quad (2 < Ar < 1 \cdot 10^8) \qquad \text{Equation E8}$$

The fuel reactor design superficial gas velocity is chosen to be in the middle of the turbulent regime. Thus, the fuel reactor superficial gas velocity range is:

$$U_{FR} = 5.25 - 5.75\ \frac{m}{s} \qquad \text{Equation E9}$$

This value corresponds to approximately 4 times U_c and is about 70% of U_{SE} at design operating conditions. The required FR cross-sectional area was calculated as 5.97 m^2 for the 80 MW rated CLC-OTSG.

The operating points at nominal load of AR and FR are shown in the flow regime map suggested by Grace [10] in Figure 4.

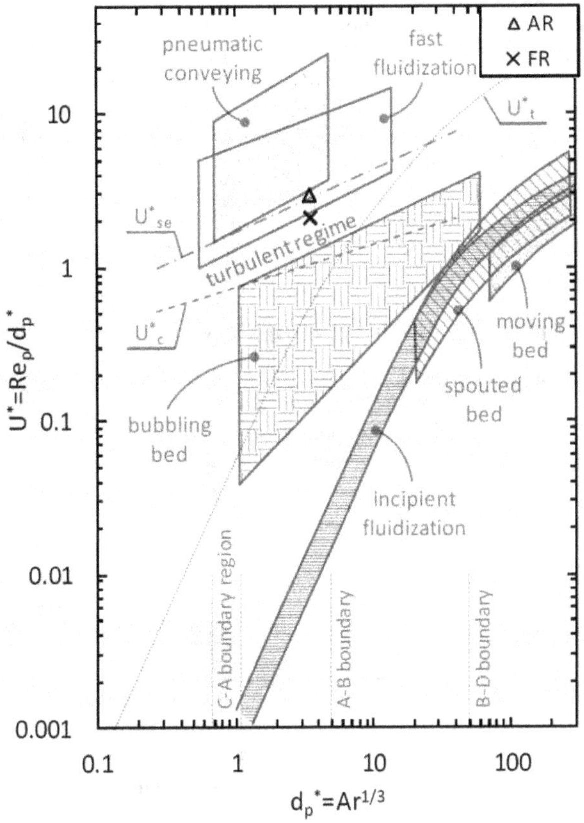

Figure 4. AR and FR fluidised bed operating state depicted in the grace [8] diagram.

Loop Seals

The loop seals are non-mechanical devices that allow the flow of solids from the entrance to the outlet. For proper operation, fluidization has to be applied to the loop seal to overcome a friction force, provoking the motion of the solids. Different types of loop seal are used in CFB applications, with loop seals of L-, and J-types being the most commonly used. A rectangular shaped J-type loop seal is applied to the design of the upper and internal loop seal, while a modified V-shaped version of a J-type loop seal is used in the lower loop seal section. The pressure balance and typical vertical and horizontal solids velocities are considered with the minimum free cross-section, and are determined by an average solids flux of:

$$G_S = 300\ \frac{kg}{m^2 s} \qquad \text{Equation E10}$$

Such solid flux values, and even higher, values have been measured in the 140 kW pilot at Vienna University of Technology [11] and are in the range proposed by Basu and Fraser [12].

Cyclones

Efficient gas-solids separation at the exit of an entrainment riser is important for the performance of circulating fluidized bed systems. Such separators should provide high separation efficiency while requiring small pressure drops. The separation of solids from a fluid can be based on either external, e.g. gravity or electrostatic, or internal forces, e.g. inertia or centrifugal. Cyclone separators, based on centrifugal forces, are typically used in circulating fluidized beds since they are simple and provide a high degree of separation. Although the shape of cyclone separators is simple, computation of the specific performance is difficult. Several dimensional ratios for cyclone separators can be found in the literature [13]. The design of Hugi [14] (see also [15]) focuses on cyclones with high solids entrance loads. Due to the high solids circulation rates, high solids entrance loads appear at the inlet of the cyclones in chemical looping applications. Therefore the design for this application is based on the cyclone separator geometry which showed the best performance in Hugi's tests.

BOILER ENGINEERING AND EQUIPMENT DESIGN

The process and reaction engineering design produced by the Vienna University of Technology was used by Bertsch Energy as the technical basis to develop a practical CLC-OTSG design. The deliverables from their work included an equipment list with descriptions, duties, sizes and materials of constructions and equipment costs. As a highly experienced designer and provider of complete industrial boiler systems Bertsch made allowances for all needed equipment items for the boiler island, which is summarised in Table 3.

Table 3. OTSG steam boiler plant for oil sands production – Equipment List.

Equipment list	Size & Material of Construction
Reactors	
Air Reactor (AR) – water-tube cooled wall vessel	3.8 x 3.8 x 25.0 m. Volume 360 m^3. MoC – 1.5415
Fuel Reactor (FR) – internally insulated vessel	3.5 x 3.5 x 25.0 m. Volume 240 m^3. MoC – 1.0481
Heat Exchangers	
Rec air pre-heater – finned tube	Area 3762 m^2
Air pre-heater – finned tube	Area 3050 m^2
Fuel pre-heater – plane tube	Area 14 m^2
BMC – plane tube	Area 170 m^2
AR1 – plane tube	Area 388 m^2
AR2 – plane tube	Area 430 m^2
AR3 – plane tube	Area 966 m^2
FR economiser – plane tube	Area 465 m^2
Loop seal fluidisation steam evaporator – plane tube	Area 15 m^2
Loop seal steam superheater – plane tube	Area 8 m^2
Water/water heat exchangers x 4	

Table 3. continued.

Equipment list	Size & Material of Construction
Blowers	
AR ID fan – radial fan with electric driver	Nominal capacity 79,000 Nm^3/h
FR ID fan – radial fan with electric driver	Nominal capacity 29,000 Nm^3/h
Air blower – radial fan with electric driver	Nominal capacity 94,000 Nm^3/h
Pumps	
Boiler feed pumps	Nominal capacity 168 m^3/h
Pumps for loop seal fluidization steam	Nominal capacity 3.7 m^3/h
Vessels	
Main steam separator vessel	1.3 x 1.8 m. MoC – 1.6368
Drains tank	1.2 x 1.5 m. MoC – 1.0425
Steam drum	1.2 x 3.0 m. MoC – 1.0481
Miscellaneous equipment	
Cyclone at the AR outlet - tangential separation cyclone	4.4 m diameter x 11.7 m. MoC - 1.0038
Cyclone at the FR outlet - tangential separation cyclone	4.0 m diameter x 11.7 m. MoC - 1.0038
Start-up burners - natural gas	2 x 10 MW
Stack - double skin	1.8 m diameter x 40 m. MoC - 1.0038/1.4301
Silo for oxygen carrier	30 m^3. MoC - 1.0345
AR loop seal - steel casing with air-nozzle bottom	1.4 x 1.9 x 1.9 m. MoC – 1.0038
FR loop seal - steel casing with air-nozzle bottom	1.1 x 1.7 x 1.9 m. MoC – 1.0038
Oxygen carrier flow control valve at the BMC entry (pneumatically actuated)	0.9 m diameter

BALANCE OF PLANT ENGINEERING AND EQUIPMENT DESIGN

Engineering contractor Amec Foster Wheeler has used the work produced by the Vienna University of Technology and Bertsch Energy to develop the overall system design of the CLC-OTSG-based steam boiler plant with CO_2 capture for the CCP oil sands steam generation application scenario (See Figure 5).

Four CLC-OTSGs, each with a nominal rating of 250 MMBTU/h (which is equivalent to 4x273 MMBTU/h actual fired fuel LHV), were incorporated into a plant design with complete boiler feed-water pre-heating and CO_2 drying/dehydration and compression to a pressure suitable for pipeline transportation and geological storage.

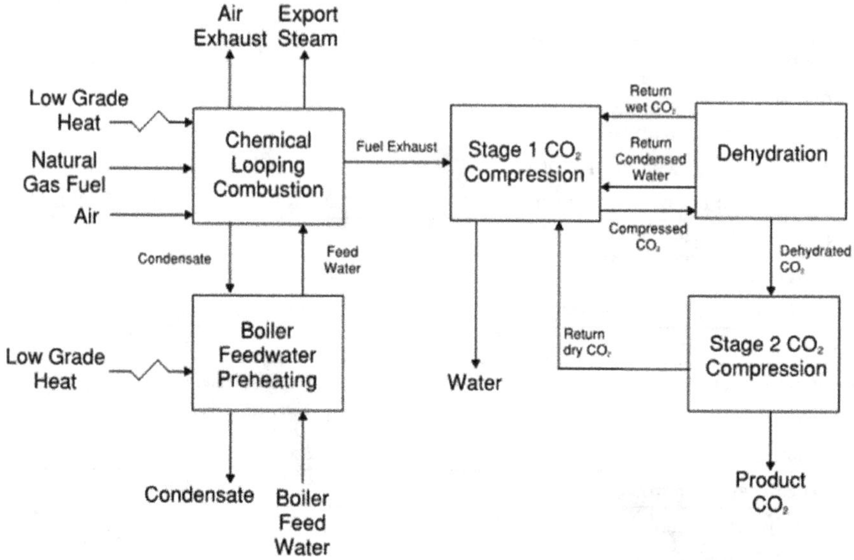

Figure 5. Block-flow diagram of the CLC-OTSG steam boiler plant with CO_2 capture.

CLC-OTSG process overview

The inputs to the CLC system are fuel, air, feed water, and boiler feed water. The fuel is natural gas, provided at elevated pressure. Air at ambient conditions is compressed for feed to the AR. Both the air and fuel streams are pre-heated before reaching their associated reactors. The air feed is pre-heated in a heat recovery step with the AR exhaust gas, and both streams pass through pre-heaters that use an outside battery limit source of low-grade heat, such as condensate or hot glycol (heat transfer fluid), to provide heat. Feed water is supplied at pressure, coming from the boiler feed water cycle. The output streams from the CLC are steam (to field for well injection), condensate (exported), the AR exhaust and FR exhaust. Both exhaust gases pass through a heat recovery steam generator before the air exhaust is released to atmosphere, and the fuel exhaust is sent forward to CO_2 compression and dehydration before being stored.

Fuel Gas Supply

Natural gas fuel is supplied from outside of the battery limit at 1°C and 3 barg. The flow rate is calculated to match the steam output of the plant design. The natural gas feeds the FR, where it reacts with oxygen extracted from the air supply stream by the circulating oxygen-carrier.

Fuel Gas Pre-treatment

The fuel gas undergoes heating via indirect heat exchange with a low-grade heat source such as condensate or hot glycol. The heated gas is then throttled to reduce the pressure, bringing it close to atmospheric conditions for feed to the FR.

Air Supply

Air at ambient conditions, 1°C and atmospheric pressure, is compressed for feed to the AR, where it is contacted in a fluidised bed with the circulating oxygen-carrier.

Air Pre-treatment

Similar to the fuel gas supply, the air stream passes through a heat exchanger after being compressed in an air blower to account for pressure loss through the AR system.

Reactor System

The reactor system comprises two parts, an AR and a FR, which are connected by a solids transfer system for oxygen-carrier material. The solids loop includes a bed material cooler. In both reactors, the oxygen-carrier is recovered from the exiting gas streams by use of a cyclone.

Air Reactor (AR)

The air reactor is the first stage of the CLC reaction, allowing the separation of oxygen from air using a metal-based oxygen-carrier, which is oxidized. The oxygen-carrier for this study is a physical mixture of two nickel based oxygen-carriers in a 1:1 weight ratio. The AR is a fluidised bed reactor, which has a high gas-solid mixing potential, and allows sufficent gas-solid contact time to enable greater reaction intensity. The reaction in the AR is generally faster than the reaction taking place in the FR by one order of magnitude. The AR incorporates a water wall to provide heat removal from the CLC system, in addition to that provided by the BMC, and provide heating for BFW.

Fuel Reactor (FR)

In the FR, the oxygen-carrier is reduced by reaction with the fuel. The oxygen-carrier is then cycled back to the AR. The resultant exhaust gas is a mixture of CO_2 and H_2O in high purity. Depending on reaction conditions, there may be low concentrations of CO and H_2. This exhaust gas is transferred to the CO_2 compression and dehydration stages after heat recovery.

Bed Material Cooler (BMC)

As the re-oxidized oxygen carrier recycles to the FR from the AR, some of the solids pass through the BMC. The BMC not only cools the bed material, keeping the CLC system under stable thermal conditions, but also generates the majority of steam for the system. The BMC allows for control of operating conditions inside the reactors by varying the amount of bed material passing through the BMC (in proportion to the amount bypassing the cooler) by use of a dynamic cone valve or similar design proven for application in circulating fluidized bed boiler plant.

Fuel Reactor Exhaust Gas and Heat Recovery

The exhaust gas from the FR is mainly a mixture of H_2O and CO_2. The stream will be sent to CO_2 compression and dehydration stages before being stored, but first it undergoes several stages of heat exchange. On exiting the reactor, the FR exhaust stream passes is cooled in a heat exchanger by a water stream coming from the AR water wall. From here the fuel exhaust is further cooled in a heat exchanger with boiler feed water. Finally the exhaust gas passes through an air cooler which reduces the stream temperature to approximately 40 °C. The heat recovery system imposes a pressure drop on the stream, so an induced draft fan is used to move the gas to the first CO_2 compression stage.

Air Reactor Exhaust Gas and Heat Recovery

Similar to the FR exhaust, the AR exhaust passes through several stages of heat exchange. The air exhaust stream is first used to further heat the water that has undergone heat transfer with the fuel exhaust, and a significant amount of the water is converted into steam here. The exhaust air further provides heat to incoming feed water, and more heat is extracted by boiler feed water, in parallel to the BFW heating by the FR exhaust. Finally, the air exhaust provides pre-heat to the air feed. By pre-heating the air supply, overall performance of the system is improved. Finally the air is released to atmosphere, assisted by an induced draft fan.

Boiler Feed Water Cycle

Feed water is supplied from the feed water tank containing a mix of treated condensate and make-up. The boiler feed water passes through several heat exchangers before being heated by cross-process exchange and being converted to steam. The water is converted to approximately 80% steam by mass (avoiding 100% vaporization to prevent build-up of solid formation). A separator splits the mixture, with steam going to the field, and the water is sent to the boiler feed water cycle where it passes through heat exchangers to heat feed water going to the process. To supplement the heating of the feed water in the boiler feed water cycle, two further exchangers are used, both using low-grade heat from outside the boiler plant battery limits to provide additional heating. This arrangement heats the feed water from 80 °C to 190 °C, which is common in OTSG systems.

CO_2 Compression and Dehydration Unit

A single train of CO_2 compression and dehydration is sufficient for the volume of exported CO_2 in this case study. Air cooling is used and electrical power is supplied over the fence to the CO_2 capture area. CO_2 from the FR exhaust must be compressed and dried before it can be exported from the facility. A six-stage, electric-motor driven, centrifugal compressor with inter-stage air cooling is provided to compress the CO_2 from atmospheric pressure to the delivery pressure of 150 barg. After four stages of compression, the CO_2 is at a pressure of approximately 39 barg, which mimimizes the water content of the gas prior to dehydration. The dehydration system consists of a set of three molecular sieve beds operating cyclically. At any point in the cycle, one of the three beds is drying the CO_2, one is being regenerated and the third is in transition: either being cooled down or heated up. A slipstream of the dried CO_2 is diverted and heated using steam for regeneration of the mol sieves. The water-laden regeneration stream is cooled, condensing much of the water, and is then routed back to the suction of the 4th compression stage. The dried CO_2 is routed through the last two stages of compression, metered and exported from the plant.

SYSTEM PERFORMANCE MODELLING AND COST ESTIMATION

The objective of this study was to take a full system-level approach to evaluate the performance and cost of the CLC-OTSG steam boiler plant with CO_2 capture. The fuel and electric power consumption, useful steam output and high-pressure purified CO_2 streams were quantified.

The technical performance modelling of the CLC-OTSG steam boiler plant with CO_2 capture was used as the basis for estimation of the capital equipment and operating costs of the plant. Bertsch Energy provided an equipment list, including descriptions, dimensions, materials of construction and an accompanying cost estimate on an ex-works basis for the CLC-OTSG. Amec Foster Wheeler then validated this cost information and used the information to create a total installed cost estimate for the complete plant, based on an Alberta Canada location and 2009 costs (Table 5). The results were used as the basis for a detailed analysis of the economics of CLC-OTSG steam boiler plant and comparison with post-combustion, pre-combustion and conventional oxy-fuel CO_2 capture techniques which are reported in Chapter 20.

Table 4. CLC-OTSG steam boiler plant for oil sands production with CO_2 capture – Performance Results.

Capacity – Nominal Basis	MMBTU/h	4x250
Fuel Type Used	Natural Gas	
Fuel Consumption	GJ/h LHV	1152
Flue Gas CO_2 Concentration	mol %	30.5
Process Carbon Balance		
CO_2 in Flue Gas	kg/h	63383
CO_2 to Capture Plant	kg/h	63383
CO_2 Captured	kg/h	63339
CO_2 to Water Streams	kg/h	43
CO_2 Emitted from Capture Plant	kg/h	n/a
CO_2 Emitted from Stack	kg/h	n/a
CO_2 Capture Across Capture Plant	%	99.9
Total Power Usage	MW	14.0
Total Steam Usage	kg/h	0
Total Cooling Water Usage	t/h	n/a
Additional CO_2 Generated for Utilities (CO_2 emission due to low-grade heat is assumed to be zero)		
Assumed Power CO_2 Footprint	kg/MWh	650
CO_2 due to Power Use	kg/h	9129
Total CO_2 Balance		
Total CO_2 Generated	kg/h	72512
Total CO_2 Captured	kg/h	63339
Total CO_2 Emitted	kg/h	9172
Overall CO_2 Captured	%	87.3
Export Steam Generation (Dry Basis)	t/h	500
Specific CO_2 Emissions	kg CO_2/t Steam	18.3

Table 5. CLC-OTSG steam boiler plant for oil sands production with CO_2 capture – Cost Results.

Capacity – Nominal Basis	MMBTU/h	4x250
Location	Fort McMurray	
Major Equipment CAPEX (Q4 2009 basis)		
200 CO_2 Compression & Dehydration	MUS$	14.64
700 CLC-OTSGs	MUS$	62.55
Unit Installed Cost		
200 CO_2 Compression & Dehydration	MUS$	77301
700 CLC-OTSGs	MUS$	276.91
Total Installed Capital Cost (TIC)	MUS$	353.9
TIC per Unit CO_2 Captured	MUS$/t/h	5.59
Operating & Maintenance Costs		
Electricity Power Price	US$/MWh	68.00
Natural Gas Price	US$/GJ	4.20
O&M fixed costs	MUS$/y	19.8
O&M variable costs	MUS$/y	56.1
Total Annual Operating Costs (inc fuel)	MUS$/y	75.9
Total Annual Operating Costs (excl fuel)	MUS$/y	33.5
Total Annual Fuel Costs	MUS$/y	42.4

CONCLUSION

The objective of this study was to develop the design of a practical CLC-OTSG steam boiler plant appropriate for the Canadian oil sands environment and to use the design information to model performance and generate a cost estimate. This objective was achieved successfully. CLC was evaluated as a feasible alternative to conventional OTSGs with CO_2 capture, which would need to be fitted either with post-combustion or conventional oxy-fuel capture systems.

Quantified results showed that CLC could provide a high level of efficiency and CO_2 capture. In fact virtually all CO_2 generated was captured, with only the CO_2 associated with the imported electric power consumption of the plant (mainly compressor, fans and pump drives) being emitted to atmosphere.

The results from this study were used further and extended to compare CLC-OTSG steam boiler plant with the alternative conventional CO_2 capture techniques, and to evaluate certain CLC-OTSG performance sensitivities. These further studies are reported in Chapter 20.

REFERENCES

1. Thomas, D.C. (ed.), (2005) Section 3B Chemical Looping Combustion (CLC) Oxyfuel Technology, in *Carbon Dioxide Capture for Storage in Deep Geological Formations – Results from the CO_2 Capture Project, Volume 1 Capture and Separation of Carbon Dioxide from Combustion Sources*. Elsevier.
2. Eide, L.I. (ed.), (2009) *Carbon Dioxide Capture for Storage in Deep Geological Formations – Results from the CO_2 Capture Project, Volume 3, Advances in CO_2 Capture and Storage Technology Results (2004-2009)*. CPL Press.

3. Lyngfelt, A., (2013) *Final Report Summary - INNOCUOUS (Innovative Oxygen Carriers Uplifting chemical looping combustion)*, Project No. 241401 http://cordis.europa.eu/result/rcn/140887_en.html
4. Wagner, W., Kruse, A.,(1998) *Properties of Water and Steam: The industrial standard IAPWS-IF97 for the thermodynamic properties and supplementary equations for other properties: tables based on these equations*. Springer-Verlag, Berlin, New York, ISBN 3540643397.
5. Burcat, A., McBride, B., (1997) ideal gas thermodynamic data for combustion and air-pollution use. Technion Israel Institute of Technology, Haifa, Israel.
6. Barin, I., Platzki, G., **1995**. Thermochemical Data of Pure Substances. 3rd ed. *VCH Publishers, Inc.*, New York, NY, USA, ISBN: 3527287450.
7. Linderholm, C., Mattisson, T., Lyngfelt, A., (2008). Long-term integrity testing of spray-dried particles in a 10-kW chemical-looping combustor using natural gas as fuel. *Fuel 88(11)*, pp. 2083-2096, doi: 10.1016/j.fuel.2008.12.018.
8. Pröll, T., Kolbitsch, P., Bolhàr-Nordenkampf, J., Hofbauer, H., (2009). A novel dual circulating fluidized bed system for chemical looping processes. *AIChE Journal 55(12)*, pp. 3255-3266, doi: 10.1002/aic.11934.
9. Bi, H.T., Grace, J.R., (1995). Flow regime diagrams for gas-solid fluidization and upward transport. *International Journal of Multiphase Flow 21(6)*, pp. 1229-1236, doi: 10.1016/0301-9322(95)00037-x.
10. Grace, J.R., **1986**. Contacting Modes and Behavior Classification of Gas - Solid and Other 2-Phase Suspensions. *Canadian Journal of Chemical Engineering 64(3)*, pp. 353-363.
11. Pröll, T.; Kolbitsch, P.; Bolhàr-Nordenkampf, J.; and Hofbauer, H., 2009, "Demonstration of CLC at Relevant Operating Conditions," Chapter 7 in *Carbon Dioxide Capture for Storage in Deep Geologic Formations – Results from the CO_2 Capture Project, Vol 3*; L.I.Eide (editor); ISBN 0-08-044570-5.
12. Basu, P., Fraser, S.A., (1991). *Circulating Fluidized Bed Boilers - Design and operations*. Butterworth-Heinemann, Boston, USA, ISBN: 0-7506-9226-X.
13. Hoffmann, A.C., Stein, L.E., (2002). *Gas cyclones and swirl tubes*. Springer, Berlin, Germany, ISBN: 3-540-43326-0.
14. Hugi, E., (1997). *Auslegung hochbeladener Zyklonabscheider für zirkulierende Gas/Feststoff-Wirbelschicht-Reaktorsysteme*. VDI-Verlag, Düsseldorf, Germany, written in German, ISBN: 3-18-350203-8.
15. Hugi, E., Reh, L., (1998). Design of Cyclones with High Solids Entrance Loads. *Chemical Engineering & Technology 21(9)*, pp. 716-719, doi: 10.1002/(Sici)1521-4125(199809)21:9<716::Aid-Ceat716>3.0.Co;2-Y.
16. Bolhár-Nordenkampf, J., Pröll, T., Kolbitsch, P., Hofbauer, H., 2009. Comprehensive Modeling Tool for Chemical Looping Based Processes. Chemical Engineering & Technology 32(3), pp. 410-417, doi: 10.1002/ceat.200800568.

Carbon Dioxide Capture for Storage in Deep Geological Formations, Volume 4
Karl F. Gerdes (Editor)

Chapter 20

TECHNOECONOMIC EVALUATION OF CARBON DIOXIDE CAPTURE FROM STEAM GENERATORS IN HEAVY OIL APPLICATIONS

David Butler[1], S Ferguson[2], G Skinner[2], R Conroy[2], J Hill[2],
Ivano Miracca[3], Dan Burt[4], Iftikhar Huq[4]
[1] Calgary, Alberta, Canada
[2]Amec Foster Wheeler, Shinfield Park, Reading, Berkshire, RG2 9FW, UK
[3]Eni S.p.A. – Via Martiri di Cefalonia, 67 – I-20097 San Donato Milanese, Italy
[4]Suncor Energy, 150-6th Avenue SW, Calgary, AB, Canada

ABSTRACT: This Chapter describes the economic analysis of the Heavy Oil Scenario cases investigated by the CO_2 Capture Project, Phase 3. Amec Foster Wheeler developed the heat and material balances and estimated capital costs for the various CO_2 capture cases studied, which are described in Chapter 13 of this volume. Based on this information, the avoided cost of CO_2 was calculated for the chosen technologies, based on Fort McMurray and US Gulf Coast economic conditions. For solvent-based, post-combustion capture, lower avoided costs are achieved when emissions from the capture unit auxiliary boiler are also captured (High Capture Rate Case). Chemical looping combustion (CLC) costs were comparable to post-combustion capture, and, as a relatively immature technology, it presents the best opportunity for future cost reductions as this technology progresses along its learning curve. Oxy-firing and pre-combustion capture had the highest capture costs of the technologies studied. All cases showed high specific capex due to the location factors and relatively small scale.

All of the cases have estimated CO_2 capture costs exceeding \$190/t except the Post-Combustion High Capture Rate Case. This case also has the lowest increase in steam cost. The pre-combustion case has the highest capture cost and highest incremental cost of steam due to high capital costs. The oxy-fuel cases have a cost of capture and incremental costs of steam about 35% greater than the Post-C.-Hi Cap case due to the high cost of supplying high purity oxygen with current air separation technology. The CLC cases have the second lowest capture cost and increased cost of steam, which can be expected to improve as the technology matures.

KEYWORDS: post-combustion capture; pre-combustion capture; oxy-firing; chemical looping; advantaged technology assessment; economic assessment; heavy oil

INTRODUCTION

This Chapter describes the economic analysis of the Heavy Oil Scenario cases investigated by CCP3. The bitumen-bearing oil sands of northern Alberta, Canada are one area where production of heavy and highly viscous petroleum reserves requires higher-than-average energy inputs in the form of steam. Heavy Oil production, consequently, has a higher greenhouse gas intensity to extract and refine than conventional sources of crude. The CO_2 Capture Project engaged Amec Foster Wheeler to develop heat and material balances and estimate capital costs for the various CO_2 capture technologies studied, which are described in Chapter 13 of this book. These results were used to calculate the avoided cost of CO_2 for chosen technologies based on the economic environments of both Fort McMurray and the US Gulf Coast. In the "no-capture" base case, a group of new build, natural gas fired, once through steam generators (OTSGs), consisting of 4 x 250 MMBTU/h boilers is required to supply a fixed quantity and quality of steam to be used in the processing of heavy oil or oil sands. Although cogeneration of steam and power is widespread in the Alberta oil sands, the cases studied here did not include cogen equipment and focused on OTSGs that generate steam only.

OTSG Economic Summary

The capture technology schemes described below were investigated to determine CO_2 capture and avoidance costs relative to the no-capture base case. Table 2 below summarizes the key data developed by Amec Foster Wheeler that was used to derive the economic results of CO_2 capture technologies employed on Once-Through Steam Generators (OTSGs). All cases are assumed to be located in Fort McMurray, Alberta, Canada except for the cases identified as being located in the US Gulf Coast (USGC) for economic comparison purposes.

Description of cases studied by AFW

Post-combustion

For the post-combustion cases, the flue gas from the steam generator is routed to conventional post-combustion amine CO_2 capture facilities located nearby on a plot of suitable size. It is assumed that 5% of the total flue gas flow bypasses the CO_2 capture plant via the stack for safety purposes. The CO_2 capture unit removes 90% of the CO_2 in the flue gas it receives, with the treated flue gas returning to the stack.

A single carbon capture process flow train of CO_2 absorption, stripping, compression and dehydration equipment was considered the most appropriate for the total volume of gas to be treated in the case. A stand-alone steam boiler package supplies the steam required for the stripper re-boiler and solvent reclaimer as well as for regeneration of the dehydration beds, as OTSG steam capacity is retained for production process steam requirements. Air cooling is used throughout and electrical power is supplied over the fence to the CO_2 capture area.

Oxy-firing

The oxy-fired cases consist of a group of new build, natural gas and air-or-oxygen fired once-through steam generators, consisting of 4 x 250 MMBTU/h (4 x 63.0 Gcal/h) fired duty boilers required to supply steam for used in the processing of heavy oil sands, plus an air separation plant to supply oxygen. A sensitivity case for retrofit of a similar set of existing OTSGs was also done. Flue gas recycle is used to moderate firebox temperature.

Key features of the configuration include:

- Oxy-firing of Boilers
- Flue Gas Cooling and Recycle
- Air Separation Unit
- Carbon dioxide compression and dehydration
- Carbon dioxide cryogenic purification
- Auxiliary steam generation

Pre-combustion

The overall process scheme for the pre-combustion case is production of a hydrogen-rich fuel gas by autothermal reforming of natural gas, integrated with CO_2 removal from the shifted syngas. The main components are natural gas compression, sulphur removal, autothermal reforming, CO shift unit, acid gas removal, CO_2 compression and dehydration and hydrogen product expansion power recovery. The individual heaters and boilers burn CO_2-free fuel, while the CO_2 emissions can be captured from the single large acid gas removal off-gas steam at the fuel processing facility. Key features of the proposed configuration include:

- Natural gas compression
- Sulphur removal
- Autothermal Reformer
- Air Separation Unit
- Shift Unit
- Acid Gas Removal Unit
- Carbon Dioxide Compression and Drying Units
- Hydrogen Product Expansion Turbine

Chemical Looping Combustion

Chemical looping combustion is an alternative oxy-fuel approach in which oxygen is supplied to the combustion process by using a solid oxygen carrier instead of conventional air separation. The baseline (see Case 3.2 in Chapter 3) adopted for Chemical Looping Combustion (CLC) in the heavy Oil scenario is based on a typical cluster of 4 boilers supplying a duty of 273 MMBTU/hr (80 MW thermal) each. The same amount of useful process steam is supplied here as for the 4x250 MMBTU/hr OTSG without capture. The differences between the nominal CLC and conventional OTSG fuel usage are due to variations in the assumed efficiency and also how the fired duty versus steam heat duty is represented. The CLC fired duty was adjusted so the CLC and OTSG cases have the same delivered steam volume and are, therefore, fully comparable cases. This case is designated as CLC-4X273. These 4 CLC units are connected to a single train of compression and dehydration for the CO_2 effluent. Air cooling is used throughout and electrical power is supplied over the fence to the capture area. Fuel and air are pre-heated with a glycol heat transfer fluid to the entry temperature of 55°C, as is normally done in SAGD operation with conventional boilers. Achievement of chemical equilibrium in the flue gas leaving the fuel reactor is the main assumption used in sizing the CLC system. In this case, the concentration of CO_2 in the CO_2 product gas after dehydration and compression is 97.55 vol%, with small amounts of hydrogen, nitrogen and CO, which is considered suitable for transportation and storage. A CO_2 capture rate of 99.9% is achieved but the CO_2 avoidance rate for this case is equal to 87% due to the CO_2 emissions released by the operation of the CLC equipment itself, and upstream emissions associated with the energy inputs (electric power). Solids material handling is a significant contributor to the unique energy requirements of CLC.

A sensitivity case to investigate economy of scale for CLC consists of a boiler cluster with 3 x 365 MMBtu/hr (107 MW_{th}) CLC units (Case 3.2a in Chapter 3). This CLC-3X365 case is based on the maximum size that could be transported via road into the Fort McMurray, Alberta location. The total cost estimated by a potential CLC manufacturer, Bertsch, for the three larger units is higher than the cost for the 4 smaller units, possibly indicating that a larger CLC unit becomes more difficult to construct and engineer. If this can be confirmed by more detailed future studies, an optimal scale for CLC units may be identified.

Another sensitivity investigates the impact if the base assumption of equilibrium reached in the CLC fuel reactor is not achieved (Case 3.2b in Chapter 3). Equilibrium has not been achieved to date in the 120 kW tests at the Technical University of Vienna. Technology developers believe that proper sizing of a large scale unit will result in a very close approach to equilibrium. The impact of incomplete combustion on the process scheme and economics was assessed using the best results actually achieved in the Vienna test unit. In this CLC-4X273 Pilot case, the raw gas from the CLC fuel reactor after dehydration is only 81% CO_2, with large amounts of CO, hydrogen and unburnt fuel. This case requires an additional cryogenic purification unit (CPU) to separate the unoxidized components from the CO_2 product stream, together with a thermal oxidizer to complete combustion of the waste gas before releasing it to the atmosphere. Alternatively, it may be beneficial to recycle the waste gas to the fuel reactor instead of incineration, which may allow additional capture of CO_2. The thermal efficiency of the process is reduced by 6% and the CO_2 avoidance rate falls to 69%.

A final sensitivity case CLC-4X273 Oxy considers the impact of air leakage into the CO_2 product due to sections of the plant handling exhaust gases from the CLC fuel reactor operating at a slight negative pressure (Case 3.2c in Chapter 3). The air leakage requires an additional oxy-catalytic bed to convert the trace levels of oxygen into CO_2 and H_2O, so that the oxygen specification can be met in the CO_2 product stream. Even though the calculation of air ingress for this case is a deliberate worst case estimate, the impact on capital cost of the additional catalyst bed is minor and operating costs would only be affected by the periodic replacement of Palladium-based catalyst with a very small overall impact on economics. The best case for air ingress would likely require an oxy-catalyst bed to meet export CO_2 specification. Thus the inclusion of an oxy-catalyst bed should not be viewed as optimal, at least at this stage of CLC demonstration or for first commercial application.

SUMMARY OF ALL CASE RESULTS

Table 1 provides the name and a brief description of each case analysed for this project.

Table 1. Case Study Nomenclature

Case Name	Description	Capacity
Pre-combustion		
Pre-C.	Pre-combustion capture with conventional(aMDEA) CO_2 solvent	Production of hydrogen-rich fuel gas with total LHV 1,091 MM BTU/hr
Pre-C. (USCG)	Pre-combustion capture with conventional(aMDEA) CO_2 solvent	Production of hydrogen-rich fuel gas with total LHV 1,091 MM BTU/hr
Post-combustion		
Post-C.	Natural gas fired	4 x 250 MMBTU/h
Post-C. (USCG)	Natural gas fired	4 x 250 MMBTU/h
Post-C. - Hi Cap.	Natural gas fired	4 x 250 MMBTU/h high CO_2 capture rate design, capture from both the OTSG and the auxiliary boiler
Post-C. - Hi Cap. (USCG)	Natural gas fired	4 x 250 MMBTU/h high CO_2 capture rate design, capture from both the OTSG and the auxiliary boiler
No CCS	Natural gas fired	4 x 250 MMBTU/h base case without CCS for high capture case – Reference case
No CCS (USCG)	Natural gas fired	4 x 250 MMBTU/h base case without CCS for high capture case – Reference case
Oxy-firing		
Oxy-C.- New	New build, natural gas oxy-fired	4 x 250 MMBTU/hr
Oxy-C.- Retro	Retrofit build, natural gas oxy-fired	4 x 250 MMBTU/hr
Chemical Looping		
CLC-4X273	4 CLC Steam Generators	Operating at 273 MMBTU/h – Base Case
CLC-4X273 (USGC)	4 CLC Steam Generators	Operating at 273 MMBTU/h
CLC-3X365	3 CLC Steam Generators	Operating at 365 MMBTU/h
CLC-4X273 Pilot	4 CLC Steam Generators	Operating at 273 MMBTU/h – Cryogenic Processing Unit for increased CO_2 purity on pilot plant data
CLC-4X273 Oxy	4 CLC Steam Generators	Operating at 273 MMBTU/h – Oxy-catalytic bed is used to convert residual O_2 in CO_2 to CO_2 and H_2O

Table 2 summarizes the key data provided by Amec Foster Wheeler used to carry out the economic assessment of CO_2 capture technologies employed on Once Through Steam Generators (OTSGs). Table 3 contains the financial metrics for the Alberta cases. All cases are assumed to be located in Fort McMurray, Alberta, Canada unless identified as being located in the US Gulf Coast (USGC). The nomenclature used in the top row of Table 2 is given in Table 1.

Table 2. Key metrics of CO_2 capture technologies applied to OTSG in Alberta.

Characteristic	Units	Post-C.	Oxy-C. - New	Oxy-C. - Retro	Pre-C.	CLC-4X273	CLC-3X365	CLC-4X273 Pilot	CLC-4X273 Oxy	Post-C. - Hi Cap.
Fuel Type		Nat Gas	Nat Gas	Nat Gas	Nat Gas	Nat Gas	Nat Gas	Nat Gas	Nat Gas	Nat Gas
Fuel Consumption - HHV	GJ/hr	1,269.3	1,163.7	1,163.7	1,275.1	1,270.7	1,275.1	1,270.7	1,270.7	1,275.1
Incremental Fuel - HHV	GJ/hr	240.6	6.4	6.4	253.7	-	-	-	-	210.7
% Increase in Fuel	%	19.0%	0.5%	0.5%	19.9%	0.0%	0.0%	0.0%	0.0%	16.5%
Incremental Power	MW	11.2	32.5	32.5	11.2	10.0	10.0	11.4	11.4	17.2
CO_2 Intensity on Fuel wo/Capture	kg/GJ	53.81	53.58	53.58	66.61	50.21	50.04	50.21	50.21	50.04
CO_2 Intensity on Fuel w/Capture	kg/GJ	21.74	21.69	21.69	14.59	7.20	7.17	15.54	7.91	14.13
CO_2 Generated	Mt/yr	0.54	0.49	0.49	0.60	0.50	0.50	0.50	0.50	0.59
CO_2 Captured	Mt/yr	0.46	0.44	0.44	0.54	0.50	0.50	0.42	0.50	0.53
CO_2 Emitted by Plant	Mt/yr	0.08	0.05	0.05	0.06	0.00	0.00	0.08	0.00	0.06
Reference Emissions - No CCS	Mt/yr	0.54	0.49	0.49	0.54	0.50	0.50	0.50	0.50	0.50
Less: CCS Emissions	Mt/yr	0.08	0.05	0.05	0.06	0.00	0.00	0.08	0.00	0.06
Add: Reduced Fuel Usage	Mt/yr	-	0.02	0.02	-	-	-	-	-	-
Less: Capture Power Use	Mt/yr	0.06	0.17	0.17	0.06	0.05	0.05	0.06	0.06	0.06
Less: CO_2 Related to Derate	Mt/yr	-	-	-	-	0.02	0.02	0.02	0.02	-
Less: Capture Steam Use	Mt/yr	0.08	0.01	0.01	-	-	-	-	-	-
Avoided Emissions	Mt/yr	0.32	0.29	0.29	0.42	0.43	0.43	0.34	0.42	0.38

Table 3. Financial results for Alberta cases.

Characteristic	Units	Post-C.	Oxy-C. - New	Oxy-C. - Retro	Pre-C.	CLC-4X273	CLC-3X365	CLC-4X273 Pilot	CLC-4X273 Oxy	Post-C. - Hi Cap.
Incremental Fuel	MUSD/yr	8.5	-1.6	-1.6	9.0	-0.2	-	-0.2	-0.2	7.5
Incremental Capital	MUSD	555.7	633.2	718.2	771.0	703.7	673.7	792.0	706.2	587.9
Incremental O&M	MUSD/yr	39.2	41.8	43.9	44.8	31.5	29.6	33.6	32.3	27.9
% Capture	%	85%	90%	90%	90%	100%	100%	85%	100%	90%
% Avoided	%	60%	55%	55%	64%	86%	86%	69%	84%	76%
Capture Cost	USD/t CO_2	217.6	225.7	248.6	233.7	195.7	186.0	255.2	197.7	170.7
Avoided Cost	USD/t CO_2	312.1	341.7	376.3	301.9	236.4	225.6	323.7	242.8	237.9
Levelized Capture Cost	USD/t CO_2	254.3	263.8	290.5	273.1	228.7	217.4	298.2	231.1	199.5
Levelized Avoided Cost	USD/t CO_2	364.8	399.3	439.7	352.8	276.2	263.6	378.2	283.7	278.0
Steam Production	te/hr	523.0	523.0	523.0	523.0	500.0	500.0	500.0	500.0	523.0
Increase in Steam Cost	$/te	24.3	24.2	26.7	30.7	24.8	23.6	27.4	25.0	22.0
Additional Yearly Cost	MUSD/yr	100.2	100.0	110.1	126.5	97.7	93.1	108.1	98.7	90.8

The top section of Table 2 (Alberta) and Table 4 (USGC) show fuel consumption, incremental fuel and power used for capture. It also shows the CO_2 directly generated by the industrial process, the CO_2 captured and emitted. This is followed by the values used to calculate the mass of CO_2 avoided. Of note, the oxy-fuel cases show an avoided benefit associated with fuel savings. This fuel saving is based on the assumption that using oxygen instead of air will require less fuel to produce the same amount of steam. The chemical looping combustion (CLC) cases produce 23 t/hr less steam than the reference OTSG cases without CCS. Additional CO_2 emissions associated with this 23 t/hr of steam produced in the reference OTSG case without CCS have been added, and is identified as CO_2 Related to Derate in Table 2, in order to keep the cases comparable.

Table 3 (Alberta) and Table 5 (USGC) show the key financial results for each of the cases. The "% avoided" is defined as the mass of CO_2 avoided divided by the mass of CO_2 generated in the base reference case without CCS. The Levelized Capture Cost is based on levelized incremental costs rather than on first year costs. The table shows the expected increase in the cost of steam related to CCS. The base cost of steam for the OTSG without CCS is $19.46/t for Alberta. For all the cases adding CCS will double the cost of steam.

The Additional Yearly Cost row shows the annual cost associated with capturing CO_2 each year. It includes an amortized annual cost for capital recovery.

All of the cases have capture costs exceeding $190/t except the Post-C.-Hi Cap case. This case also has the lowest increase in steam cost. The pre-combustion case Pre-C has the highest capture cost and highest incremental cost of steam. The oxy-fuel cases have a cost of capture and incremental costs of steam about 35% greater than the Post-C.-Hi Cap case. The CLC-3X107 case has the second lowest capture cost and increased cost of steam.

Table 4. Key metrics of CO_2 capture technologies applied to OTSG in USGC.

Characteristic	Units	Post-C. (USGC)	Pre-C. (USGC)	CLC-4X273 (USGC)	Post-C. - Hi Cap. (USGC)
Fuel Type		Nat Gas	Nat Gas	Nat Gas	Nat Gas
Fuel Consumption - HHV	GJ/hr	1,269.3	1,275.1	1,286.8	1,291.3
Incremental Fuel - HHV	GJ/hr	240.6	253.7	-	213.3
% Increase in Fuel	%	19.0%	19.9%	0.0%	
Incremental Power	MW	11.2	11.2	10.0	16.3
CO_2 Intensity on Fuel wo/Capture	kg/GJ	53.81	66.61	49.58	50.04
CO_2 Intensity on Fuel w/Capt	kg/GJ	21.30	14.04	6.56	13.97
CO_2 Generated	Mt/yr	0.54	0.60	0.50	0.59
CO_2 Captured	Mt/yr	0.46	0.54	0.50	0.53
CO_2 Emitted	Mt/yr	0.08	0.06	0.00	0.06
Reference Emissions - No CCS	Mt/yr	0.54	0.54	0.50	0.50
Less: CCS Emissions	Mt/yr	0.08	0.06	0.00	0.06
Add: Reduced Fuel Usage	Mt/yr	-	-	-	-
Less: Capture Power Use	Mt/yr	0.05	0.05	0.05	0.06
Less: CO_2 Related to Derate	Mt/yr	-	-	0.02	-
Less: Capture Steam Use	Mt/yr	0.08	-	-	-
Avoided Emissions	Mt/yr	0.33	0.42	0.43	0.38

Table 5. Financial results for USGC cases.

Characteristic	Units	Post-C. (USGC)	Pre-C. (USGC)	CLC-4X273 (USGC)	Post-C. - Hi Cap. (USGC)
Incremental Fuel	MUSD/yr	8.5	9.0	0.4	7.6
Incremental Capital	MUSD	268.8	450.0	344.1	426.3
Incremental O&M	MUSD/yr	25.8	35.5	24.4	28.0
% Capture	%	85%	90%	100%	90%
% Avoided	%	60%	65%	87%	76%
Capture Cost	USD/t CO_2	129.7	160.7	114.8	104.0
Avoided Cost	USD/t CO_2	183.5	205.4	139.2	144.9
Levelized Capture Cost	USD/t CO_2	151.5	187.7	134.1	121.5
Levelized Avoided Cost	USD/t CO_2	214.5	240.0	162.7	169.4
Steam Production	te/hr	523.0	523.0	500.0	523.0
Increase in Steam Cost	$/te	14.5	21.1	14.5	13.4
Additional Yearly Cost	MUSD/yr	59.7	86.9	57.3	55.3

Figure 1 (Alberta) and Figure 2 (USGC) show the breakdown of cost categories which make up the total cost to capture CO_2 for the cases considered. Capital costs tend to dominate the cost of capture for heavy oil applications. The oxy-fuel cases have a small fuel savings due to increased steam generation efficiency, which is shown as a negative value in the graphs.

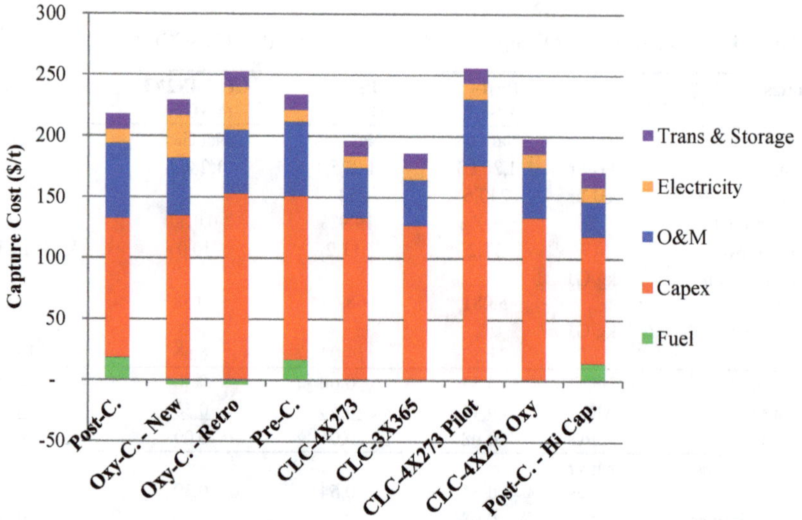

Figure 1. Capture cost components for Alberta cases.

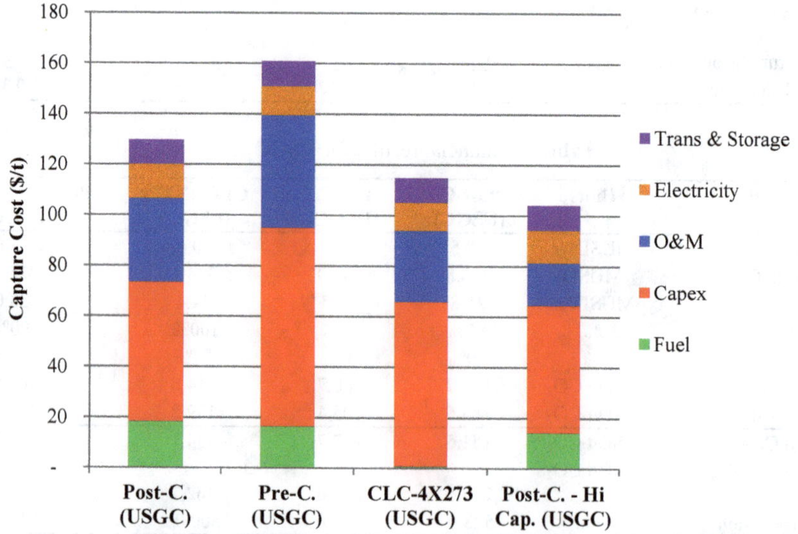

Figure 2. Capture cost components for USGC cases.

The up-front capital cost to build projects in Alberta is much higher than the USGC, making high capital cost technologies problematic in Alberta. Figure 1 suggests that post-combustion capture, the

Post-C – Hi Cap process has the lowest capture cost in Alberta. However, given the accuracy of the cost estimates, many of the CLC cases may ultimately have capture costs similar to the best post-combustion capture case, especially as the technology matures.

Figure 3 (Alberta) and Figure 4 (USGC) show the component costs which make up the avoided cost of CO_2 for the cases considered:

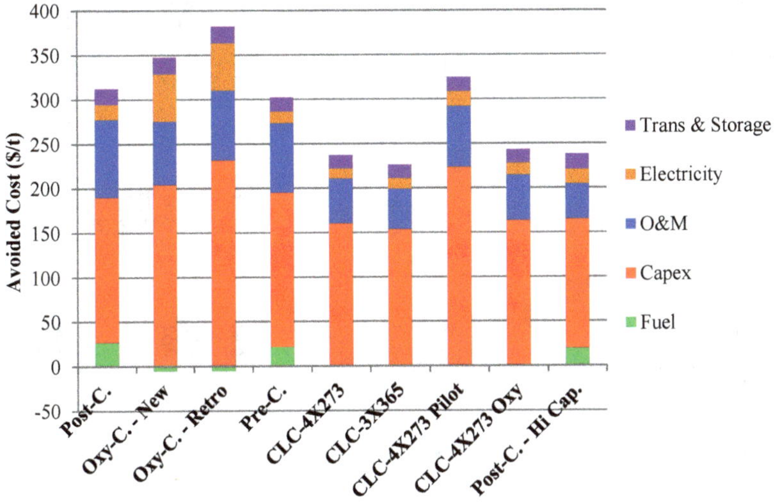

Figure 3. Avoided cost components for Alberta cases.

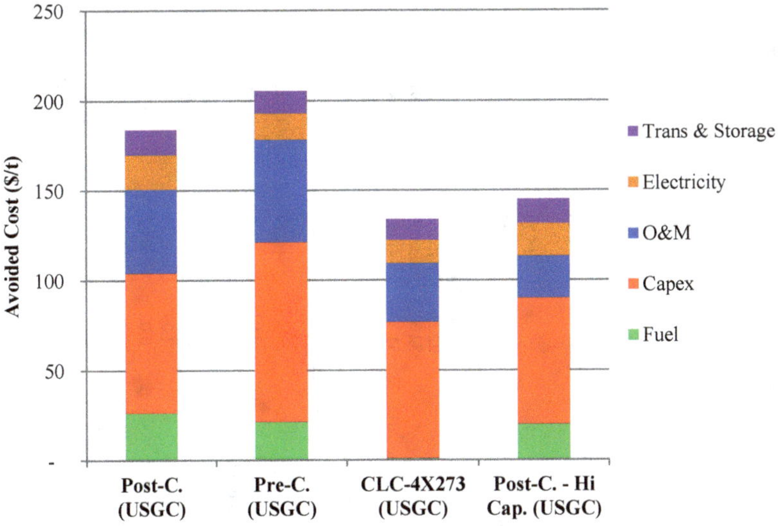

Figure 4. Avoided cost components for USGC cases.

Table 6. Incremental CCS capital cost components for Alberta cases.

Cost	Post-C.	Oxy-C. - New	Oxy-C. - Retro	Pre-C.	CLC-4X273	CLC-3X365	CLC-4X273 Pilot	CLC-4X273 Oxy	Post-C. - Hi Cap.
Capture Plant	145.64				77.01	77.78	117.58	78.68	129.29
Compression Plant	62.22								64.83
Steam Gen & BOP	9.31				276.91	265.61	273.32	276.37	38.39
Base OTSG					-105.76	-105.76	-105.76	-105.76	
Oxygen Supply		191.89	191.89	122.10					
Flue Gas Treat		11.68	11.68						
Ref. Precom. and Shift				72.17					
CO_2 Purification		25.58	25.58						
CO_2 Removal				62.87					
CO_2 Compression		46.70	46.70	28.02					
CO_2 Dehydration		14.84	14.84	13.76					
H_2 Fuel Expansion		7.72	7.72	24.22					
Steam Gen		1.55	1.55	16.03					
OSBL (30%)	65.14	32.42	42.95	66.52	74.44	71.29	85.54	74.77	69.75
Direct & Indirect	282.31	332.38	342.91	405.69	322.60	308.92	370.68	324.06	302.26
Escalation from 2009 to 2014	51.37	60.49	68.80	73.84	58.71	56.23	67.46	58.97	58.97
Process Contingency	13.90	6.23	6.23	5.73	76.26	36.78	73.03	87.63	76.60
Project Contingency (20%)	69.51	79.82	90.61	97.05	91.51	87.64	105.15	91.91	91.91
Total Plant Costs	417.03	478.93	543.66	582.31	536.10	513.53	603.31	537.98	447.08
Owners Costs									
6 Months Fixed O&M	13.07	10.39	11.44	12.84	7.46	7.16	8.60	7.46	6.98
1 Month Variable O&M	0.62	1.30	1.30	1.04	0.88	0.76	0.93	0.94	0.62
25% of 1 Month of Fuel Costs	0.18	-0.03	-0.03	0.19	-0.94	-0.94	-0.94	-0.94	0.16
2% of TPC	8.34	9.58	10.87	11.65	10.72	10.27	12.07	10.76	8.94
60 Days of Cons.	0.35	0.02	0.02	1.19	0.96	0.73	0.96	0.97	0.18
Spare Parts (.5% TPC)	2.09	2.39	2.72	2.91	2.68	2.57	3.02	2.69	2.24
Land ($3,000/acre)	0.02	0.02	0.02	0.02	-	-	-	-	0.02
Financing Costs (2.7% TPC)	11.26	12.93	14.68	15.72	14.47	13.87	16.29	14.53	12.07
Other Costs (15% of TPC)	62.56	71.84	81.55	87.35	80.42	77.03	90.50	80.70	67.06
Total Owners Costs	98.48	108.44	122.57	132.90	116.64	111.45	131.41	117.11	98.26
Total Overnight Costs	515.52	587.37	666.23	715.21	652.74	624.98	734.72	655.08	545.34
AFUDC	40.21	45.81	51.97	55.79	50.91	48.75	57.31	51.10	42.54
Total As-Spent Cost	555.73	633.18	718.20	770.99	703.66	673.73	792.03	706.18	587.88

The capital cost values in Table 6 through 9 are in millions of dollars. Table 6 (Alberta, above) and Table 7 (USGC) show the components which contribute to the total incremental capital cost to capture CO_2. The top part of the table shows the direct costs for each case. This is followed by indirect OSBL costs. The base capital costs have been adjusted to include escalation from Q1 2009 to the assumed in-service date of Q1 2014. This is followed by estimates for contingencies. To this, estimates for Owners Costs are added. Finally AFUDC during construction are included. The bottom row shows the total as spent costs expected just before the plant is commissioned. The CLC capital costs provided by Amec Foster Wheeler included the whole plant including the CLC steam generator and the CCS components. The incremental capital cost for the CLC cases is derived by deducting the costs for the OTSG reference case without CCS from the cost of the CLC cases with CCS.

Table 7. Incremental CCS capital cost components for USGC cases.

Cost	Post-C. (USGC)	Pre-C. (USGC)	CLC-4X273 (USGC)	Post-C. - Hi Cap. (USGC)
Capture Plant	71.76		37.35	65.38
Compression Plant	28.49			31.11
Steam Gen & BOP	4.37		140.54	16.08
Base OTSG			-58.23	
Oxygen Supply		66.71		
Flue Gas Treat				
Ref. Precom. and Shift		42.62		
CO_2 Purification		36.92		
CO_2 Removal		18.71		
CO_2 Compression		8.02		
CO_2 Dehydration		14.04		
H_2 Fuel Expansion		9.63		
Steam Gen				
OSBL (30%)	31.31	38.98	35.90	33.76
Direct & Indirect	135.93	235.62	155.56	146.29
Escalation from 2009 to 2014	24.69	42.87	28.31	26.62
Process Contingency	7.15	3.32	36.78	6.54
Project Contingency (20%)	33.50	56.35	44.13	35.89
Total Plant Costs	200.99	338.16	260.53	215.34
Owners Costs				
6 Months Fixed O&M	6.54	8.48	4.18	3.94
1 Month Variable O&M	0.69	1.11	0.94	0.66
25% of 1 Month of Fuel Costs	0.18	0.19	-	0.16
2% of TPC	4.02	6.76	5.21	4.31
60 Days of Cons.	0.35	1.19	0.96	0.18
Spare Parts (.5% TPC)	1.00	1.69	1.30	1.08
Land ($3,000/acre)	0.02	0.02	-	0.02
Financing Costs (2.7% TPC)	5.43	9.13	7.03	5.81
Other Costs (15% of TPC)	30.15	50.72	39.08	32.30
Total Owners Costs	48.38	79.28	58.70	48.45
Total Overnight Costs	249.36	417.45	319.24	263.79
AFUDC	19.45	32.56	24.90	20.58
Total As-Spent Cost	268.82	450.00	344.14	284.37

The incremental costs for the Post-C. - Hi Cap case can be added to the costs for the No CCS case (reference OTSG costs without CCS) to show the full costs for the OTSG case with CCS. Table 8 shows the costs components for the reference OTSG case without CCS. These costs were also used to help determine the incremental cost of the CLC cases compared to an OTSG plant without CCS.

Table 8. Capital cost components for the reference OTSG cases without CCS.

Cost	No CCS	No CCS (USGC)
Base OTSG	105.77	58.23
OSBL (30%)	31.73	17.47
Direct & Indirect	137.49	75.70
Escalation from 2009 to 2014	25.02	13.78
Process Contingency	-	-
Project Contingency (20%)	32.50	17.89
Total Plant Costs	195.02	107.37
Owners Costs		
6 Months Fixed O&M	4.36	2.45
1 Month Variable O&M	0.16	0.18
25% of 1 Month of Fuel Costs	0.94	-
2% of TPC	3.90	2.15
60 Days of Cons.	-	-
Spare Parts (.5% TPC)	0.98	0.54
Land ($3,000/acre)	0.02	0.02
Financing Costs (2.7% TPC)	5.27	2.90
Other Costs (15% of TPC)	29.25	16.11
Total Owners Costs	44.87	24.33
Total Overnight Costs	239.88	131.70
AFUDC	18.71	10.27
Total As-Spent Cost	258.60	141.98

Capital costs for air separation units were provided by a third party on a turn-key basis. It was not clear whether additional OSBL costs should be included in Table 6 and Table 7. Therefore Figure 5 shows the impact of adding OSBL to the air separation unit capital costs. Table 9 shows the total capital costs for the CLC cases with CCS. It also shows the capital costs for the OTSG cases with high capture. These values were derived by adding the OTSG costs without CCS to the incremental costs of completing CCS.

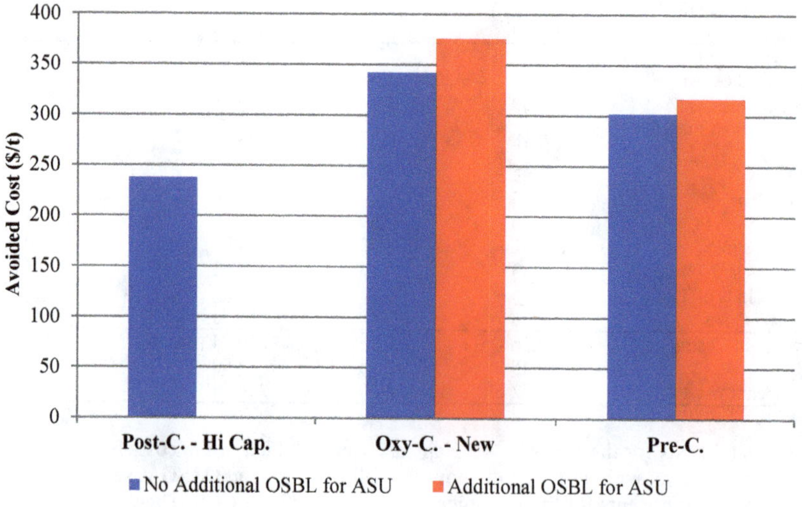

Figure 5. Impact of including OSBL on air separation capital costs.

Table 9. Total capital cost of CLC and high capture OTSG cases in Alberta and USGC.

Cost	CLC-4X273	CLC-3X365	CLC-4X273 Pilot	CLC-4X273 Oxy	Post-C. - Hi Cap.	CLC-4X273 (USGC)	Post-C. - Hi Cap. (USGC)
Capture Plant	77.01	77.78	117.58	78.68	129.29	37.35	65.38
Compression Plant	-	-	-	-	64.83	-	31.11
Steam Gen & BOP	276.91	265.61	273.32	276.37	38.39	140.54	16.08
Base OTSG	-	-	-	-	105.77	-	58.23
OSBL (30%)	106.17	103.02	117.27	106.50	101.48	53.37	51.23
Direct & Indirect	460.09	446.41	508.17	461.55	439.76	231.26	221.99
Escalation from 2009 to 2014	83.73	81.25	92.49	83.99	83.99	42.09	40.40
Process Contingency (20%)	76.26	36.78	73.03	87.63	76.60	36.78	6.54
Project Contingency (20%)	124.01	120.14	137.66	124.42	124.42	62.03	53.78
Total Plant Costs	731.12	708.55	798.32	732.99	642.10	367.90	322.71
Owners Costs	-	-	-	-	-	-	-
6 Months Fixed O&M	11.82	11.52	12.95	11.82	11.34	6.63	6.39
1 Month Variable O&M	1.03	0.92	1.09	1.10	0.78	1.12	0.84
25% of 1 Month of Fuel Costs	-	-	-	-	1.10	-	0.16
2% of TPC	14.62	14.17	15.97	14.66	12.84	7.36	6.45
60 Days of Cons.	0.96	0.73	0.96	0.97	0.18	0.96	0.18
Spare Parts (.5% TPC)	3.66	3.54	3.99	3.66	3.21	1.84	1.61
Land ($3,000/acre)	0.02	0.02	0.02	0.02	0.03	0.02	0.03
Financing Costs (2.7% TPC)	19.74	19.13	21.55	19.79	17.34	9.93	8.71
Other Costs (15% of TPC)	109.67	106.28	119.75	109.95	96.31	55.19	48.41
Total Owners Costs	161.51	156.32	176.28	161.97	143.13	83.04	72.79
Total Overnight Costs	892.63	864.87	974.60	894.97	785.23	450.94	395.50
AFUDC	69.63	67.46	76.02	69.81	61.25	35.17	30.85
Total As-Spent Cost	962.25	932.33	1,050.62	964.77	846.48	486.11	426.35

Table 10 and Table 11 show the operating and maintenance costs for each case. The following O&M costs are in millions of dollars per year and have been adjusted for escalation or capacity factor and are reported in $2014.

Table 10. Incremental CCS O&M costs for Alberta cases.

Cost	Post-C.	Oxy-C. - New	Oxy-C. - Retro	Pre-C.	CLC-4X273	CLC-3X365	CLC-4X273 Pilot	CLC-4X273 Oxy	Post-C. - Hi Cap.
Fixed Costs									
Direct Labour	2.75	2.21	2.21	3.82	1.37	1.37	1.37	1.37	1.37
G&A	0.84	0.67	0.67	1.15	0.42	0.42	0.42	0.42	0.42
Maintenance	13.54	10.75	12.00	12.42	7.89	7.51	9.22	7.93	7.33
Insurance and P Taxes	9.03	7.16	8.00	8.29	5.28	5.04	6.11	5.28	4.88
Total Fixed Costs	26.15	20.79	22.88	25.67	14.96	14.34	17.12	15.00	14.01
Variable Costs									
Power	5.33	15.52	15.52	5.34	4.77	4.77	5.44	5.44	6.30
Absorbent	-	0.14	0.14	0.05	0.35	0.35	0.35	0.35	0.30
KS1-Solvent ($5/kg)	1.02	-	-	-	-	-	-	-	-
Chemicals and Disposal	1.11	-	-	7.09	5.43	4.07	5.43	5.53	0.83
Transportation	3.11	2.99	2.99	3.65	4.36	4.37	3.43	4.63	4.76
Storage	2.53	2.43	2.43	2.97	1.75	1.75	1.75	1.48	1.75
Total Variable Cost	13.09	21.08	21.08	19.10	16.65	15.32	16.39	17.43	13.93
Total O&M	39.24	41.87	43.96	44.78	31.62	29.66	33.52	32.43	27.93

Table 11. Incremental CCS O&M costs for USGC cases.

Cost	Post-C. (USGC)	Pre-C. (USGC)	CLC-4X273 (USGC)	Post-C. - Hi Cap. (USGC)
Fixed Costs				
Direct Labour	1.37	3.82	1.37	1.37
G&A	0.42	1.15	0.42	0.42
Maintenance	6.77	7.21	3.89	3.63
Insurance and P Taxes	4.51	4.80	2.59	2.42
Total Fixed Costs	13.07	16.98	8.27	7.84
Variable Costs				
Power	6.16	6.18	5.52	6.79
Absorbent	-	0.05	0.35	0.30
KS1-Solvent ($5/kg)	1.02	-	-	-
Chemicals and Disposal	1.11	7.09	5.43	0.83
Transportation	3.11	3.65	3.37	3.29
Storage	1.35	1.59	1.47	1.86
Total Variable Cost	12.75	18.56	16.13	13.06
Total O&M	25.82	35.54	24.40	20.91

Table 12 shows the O&M costs for the OTSG cases without CCS. These can be added to the incremental costs of completing CCS to determine the total cost of producing steam from an OTSG with CCS. They were also used to help determine the incremental cost of the CLC cases compared to an OTSG plant without CCS.

Table 13 shows the O&M cost for the chemical looping and high CO_2 capture cases in both Alberta and the USGC.

Table 12. O&M costs for OTSG cases without CCS.

Cost	No CCS	No CCS (USGC)
Fixed Costs		
Direct Labour	0.60	0.60
G&A	0.18	0.18
Maintenance	4.79	2.48
Insurance and P Taxes	3.20	1.66
Total Fixed Costs	8.76	4.92
Variable Costs		
Power	1.91	2.21
Total Variable Cost	1.91	2.21
Total O&M	10.67	7.13

Table 13. Total O&M cost of CLC and high capture OTSG cases in Alberta and USGC.

Cost	CLC-4X273	CLC-3X365	CLC-4X273 Pilot	CLC-4X273 Oxy	Post-C. - Hi Cap.	CLC-4X273 (USGC)	Post-C. - Hi Cap. (USGC)
Fixed Costs							
Direct Labour	1.97	1.97	1.97	1.97	1.97	1.97	1.97
G&A	0.60	0.60	0.60	0.60	0.60	0.60	0.60
Maintenance	12.68	12.30	14.01	12.72	12.12	6.38	6.11
Insurance and P Taxes	8.48	8.24	9.31	8.48	8.08	4.25	4.08
Total Fixed Costs	23.72	23.10	25.89	23.76	22.77	23.72	23.10
Variable Costs							
Power	6.68	6.68	7.35	7.35	8.20	7.73	9.00
Absorbent	0.35	0.35	0.35	0.35	0.30	0.35	0.30
KS1-Solvent ($5/kg)	-	-	-	-	-	-	-
Chemicals and Disposal	5.43	4.07	5.43	5.53	0.83	5.43	.83
Transportation	4.36	4.37	3.43	4.63	4.76	3.37	3.29
Storage	1.75	1.75	1.75	1.48	1.75	1.47	1.86
Total Variable Cost	18.56	17.22	18.30	19.34	15.84	18.33	15.27
Total O&M	42.29	40.33	44.19	43.10	38.61	31.55	28.01

Sensitivity Studies

Figure 6 and Figure 7 show how the cost of capturing CO_2 varies as the price of natural gas changes. All cases except the oxy-fuel cases show an increase in capture cost as the cost of natural gas increases. This is driven by the significant amount of fuel required to produce steam for the capture operation. The relative slopes are related to the fuel usage. The decrease in capture cost for the oxy-fuel case, as natural gas prices rise, is due to the benefit of reduced fuel consumption relative to the reference OTSG. Therefore as the gas price increases, the economic benefit associated with using less fuel increases. If natural gas prices stay low for a significant period of time the benefit is greatly reduced, and this may change investment decisions when life cycle economics are considered.

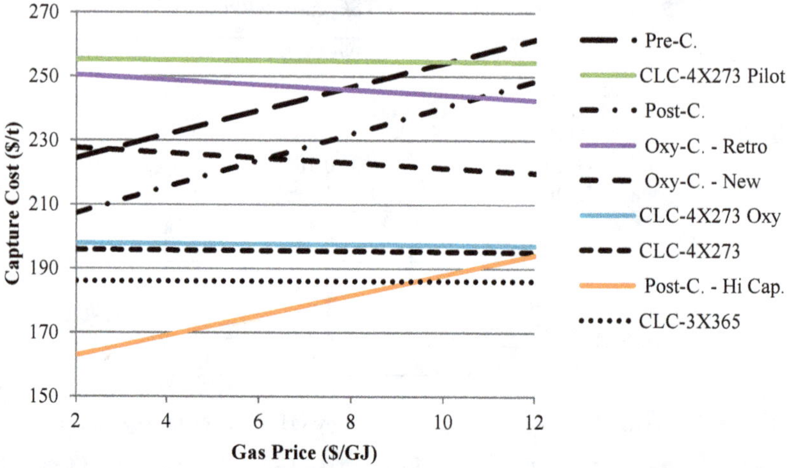

Figure 6. Sensitivity of capture cost vs gas price impacts for Alberta cases

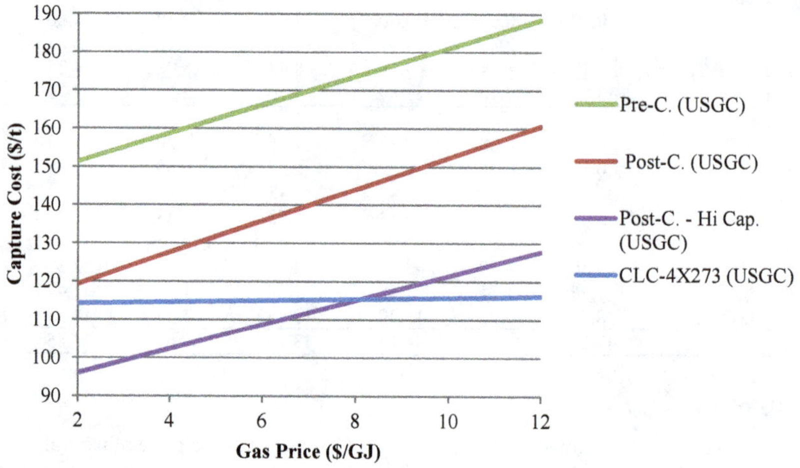

Figure 7. Sensitivity of capture cost vs gas price impacts for USGC cases.

Figure 8 and Figure 9 show how the avoided costs change as gas prices change. As with the graphs above, the oxy-fuel cases show a reduced avoided cost as the price of gas increases.

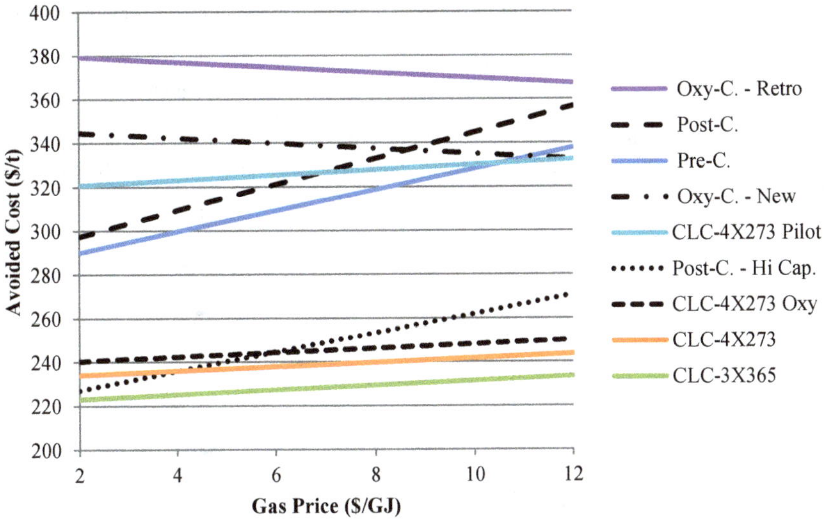

Figure 8. Avoided cost vs gas price for Alberta cases

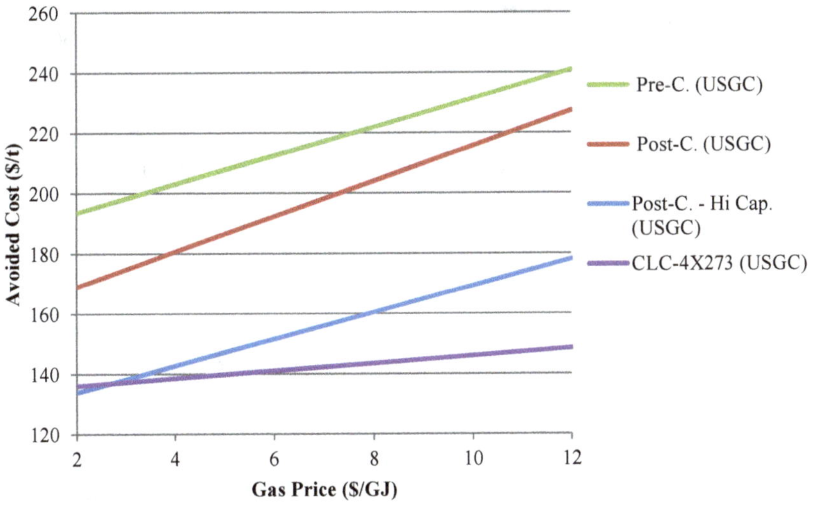

Figure 9. Avoided cost vs gas price for USGC cases.

Figure 10 (Alberta) and Figure 11 (USGC) below show how the avoided cost changes with changes in the GHG emission intensity of the electric power supplied from the grid to the capture operation. As discussed previously, if the GHG emission intensity dips below 0.5 t CO_2/MWh, indicating a gas-fuel-dominated power supply from the grid, the power price becomes a function of the gas price. That accounts for the inflexion in the figures. As the GHG emission intensity increases this indicates a higher reliance on coal. Therefore the price of power in the market may change as a result. Only the avoided cost for the oxy-fuel cases is likely to be significantly impacted by changes in GHG emission intensity of the power employed, since this is the only case for which power is a significant component of the avoided cost.

For the USGC the price of power is 33+5.55 x Gas Price. As the cost of natural gas increases, there is a reasonable expectation that the cost of power will increase and the generation mix in a given area may be influenced more by new coal plants than otherwise would be expected, with all other factors being equal. Figures 10 and 11 indicate which technologies are likely to be sensitive to changes in gas price.

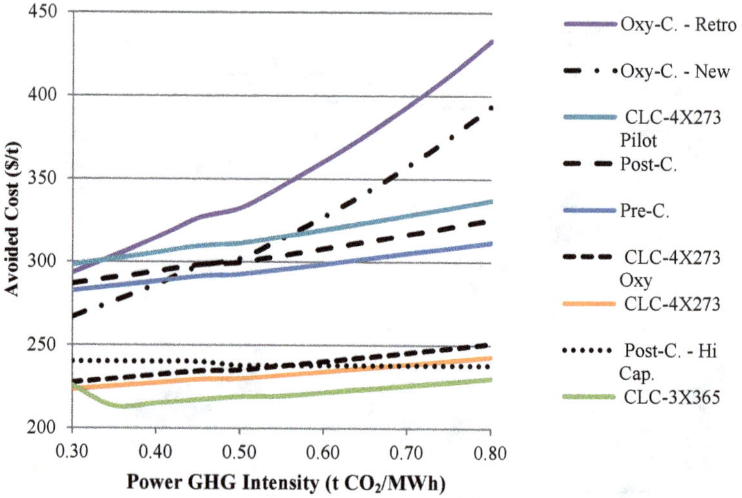

Figure 10. Avoided cost vs power GHG emission intensity for Alberta cases

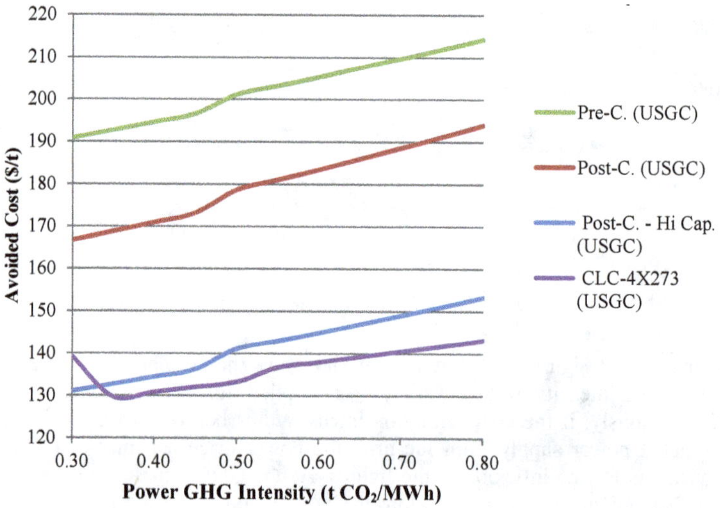

Figure 11. Avoided cost vs power GHG emission intensity for USGC cases.

Figure 12 (Alberta) and Figure 13 (USGC) show how the avoided costs for a project change as the price of power changes. For most cases, power is not a significant cost component of the avoided cost. Therefore, changes in power price have little impact on avoided costs. However, power is a significant cost component for the oxy-fuel cases. Therefore, the avoided cost for these cases is very sensitive to changes in the price of power.

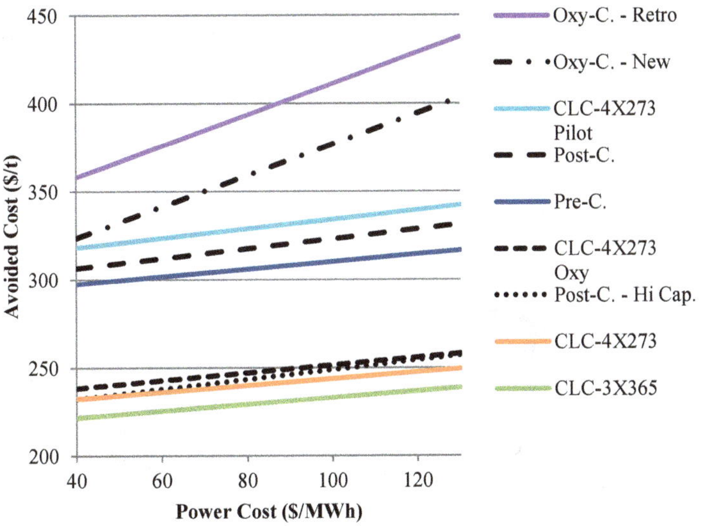

Figure 12. Avoided cost vs power price for Alberta cases.

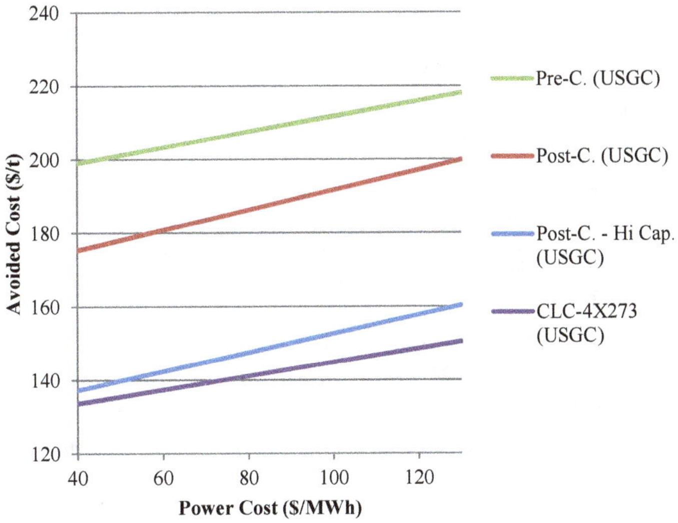

Figure 13. Avoided cost vs power price for USGC cases.

CLC Oxygen Carrier Sesnsitivity

The Amec Foster Wheeler CLC cost estimates were based on using a Ni-based oxygen carrier with a cost of about US \$50/kg. During the INNOCUOUS project a novel carrier was developed with the prospect of a much lower potential cost of US \$6.25/kg. Figure 14 shows base results and a sensitivity of the impact of this lower carrier cost on both the capital cost and fixed operating cost of the CLC-4 X 80 MW project. The base capital cost includes 411 tons of oxygen carrier. The lower cost carrier will reduce this initial capital cost about \$17.9million. The new carrier will however need to be replaced 3 times faster than the Ni-based carrier. Therefore an adjustment was made to the base chemical and disposal cost of \$5.05 million per year: \$5.05 million / (\$US50/t) X (\$US 6.25/t) X 3 = \$1.9 million.

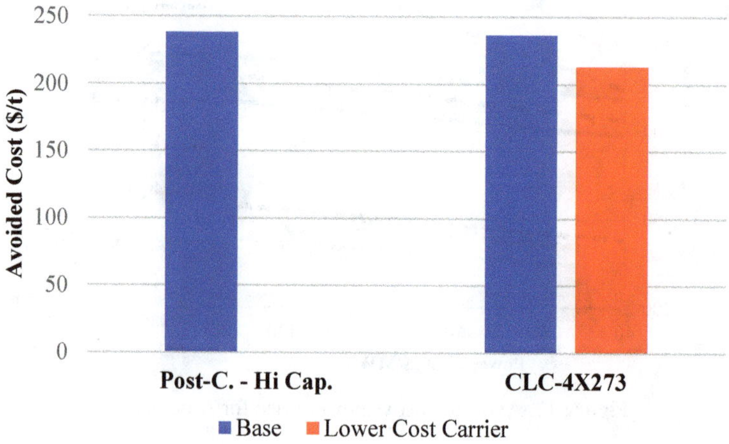

Figure 14. Sensitivity on CLC oxygen carrier.

CONCLUSIONS AND RECOMMENDATIONS

Engineering feasibility and cost analyses have been completed by Praxair and CCP for both a commercial and test scale OTSG operating under air and oxy-fuel (see Chapter 14 of this volume for details). Results of computational fluid dynamic models combined with rigorous heat and mass balance models indicate that oxy-fuel combustion technology with flue gas recirculation can be applied to OTSGs while maintaining key operating parameters (temperature, heat flux) that are similar to operation with air. Therefore, through proper control integration OTSGs can be retrofit for oxy-firing and are capable of either air or oxy-fuel operation.

The comparative cost studies illustrated in Figures 3 and 4 confirmed that CO_2 abatement costs for a typical commercial SAGD OTSG installation located in a low capex environment, such as USGC, could fall in the expected range of \$125 to \$145 \$/tonne on an avoided basis. However, site to site variation and cost reduction opportunities will shift the cost within the range.

The Phase II Oxy-fuel Combustion Test at Cenovus Christina Lake produced data which will be used to fine tune the Phase I results. Phase III, the full-scale CO_2 Capture Demonstration, is well in the future and will require significant capital and resource expenditures. Program requirements and cost estimates for Phase III will be reviewed after the completion of Phase II.

Key results:

- Post-C capture using state-of-the-art solvent is estimated to have lower costs than other capture technologies for this application.
- Pre-C in this application is penalized by relatively low economy of scale, and has a high capture cost. See Chapter 11 for economics of large scale pre-combustion capture.
- Oxy-firing is confirmed as somewhat higher cost option in this application, due to low cost of fuel and high capex (particularly high in Alberta) in this relatively small scale application (see Chapter 7 for larger scale case).
- CLC has promise as a competitive capture technology in the future, especially if lower cost carriers are developed.

Carbon Dioxide Capture for Storage in Deep Geological Formations, Volume 4
Karl F. Gerdes (Editor)

Section 1

CO_2 CAPTURE

Natural Gas Combined Cycle (NGCC)

Carbon Dioxide Capture for Storage in Deep Geological Formations, Volume 4
Karl F. Gerdes (Editor)

Chapter 21

POWER GENERATION FROM NATURAL GAS IN COMBINED CYCLE SYSTEMS WITH INTEGRATED CAPTURE

Mahesh Iyer[1], Richard Beavis[2], Tony Tarrant[3] and Tim Bullen[3]
[1]Shell International Exploration & Production Inc.,
200 N. Dairy Ashford Rd., Houston, TX 77079, USA
[2]BP International Limited, ICBT Chertsey Road, Sunbury-on-Thames, Middlesex, TW16 7LN, United Kingdom
[3]Amec Foster Wheeler, Shinfield Park, Reading, Berkshire, RG2 9FW, United Kingdom

ABSTRACT: The main learning from CCP2 was that deployment of pre-combustion schemes for decarbonising power production from natural gas in the near future was challenging. Hence, the main objective of this phase of study was to focus on post-combustion and oxy-firing technologies. Oxy-fired technology for this scenario (NGCC) continues to remain a challenge, as it requires development of novel gas turbines, so a detailed oxy-firing study was not done for this assessment. CCP3 was able to identify some providers for oxy-firing, so a detailed feasibility analysis is warranted and remains pending.

CCP3 focused on post-combustion technology development for CO_2 capture from a combined cycle power generation facility. For this case study, a 400 MWe NGCC power plant that produces ~1 million tonnes/yr CO_2 was chosen as the basis. The flue gas contains 4 vol% CO_2 and considerable excess oxygen, which presents a challenge. CCP3 selected the MHI-KS1™ solvent technology for the feasibility study after screening several technology providers with capability to provide state-of-the-art deployable technology for this service. MHI offers a single train design for this scale of capture and CO_2 concentration. Integration of the CO_2 capture unit with the power plant includes a flue gas blower and quench (direct contact cooler) before the absorber. In addition, MHI recommends using a rectangular cross-section for the quench and absorber for this scale, since this geometry offers better economies of scale. The solvent regenerator is a conventional cylindrical column. System integration with the NGCC plant can improve overall CO_2 avoidance, as the steam and power utilities for capture are drawn from the power plant with no additional flue gas sources. CO_2 compression and dehydration, for the scale under consideration, is considered a proven technology.

KEYWORDS: CO_2 capture; post-combustion; combined cycle; power generation; NGCC; CCGT

INTRODUCTION

The focus of this study was to understand, from an oil and gas industry perspective, the CO_2 abatement options for power production with natural gas as fuel. In CCP2, the key learning was that pre-combustion schemes for decarbonising power production from natural gas were challenging due to certain technical hurdles for deployment in the near future timeframe of 2020[1]. Hence, the main objective of this phase of study was to focus on post-combustion and oxy-firing technologies for CO_2 capture from natural gas combined cycle (NGCC) power generation. Oxy-fired technology for this scenario continues to remain a challenge as it requires development of novel gas turbines. Several processes based on CO_2 gas turbine cycles and fuel cells are able to operate on natural gas and are examined in detail in the literature (for instance [2]). These approaches are at very early levels of technical development, and none of them have reached the demonstration stage, since most require development of new gas turbines (both combustors and expanders). A feature of CO_2/O_2 gas turbine-based cycles is the theoretical ability to capture nearly 100% of the CO_2 generated. These systems emit very low amounts of other types of pollutants (zero NOx) and do not need external water supply. While CCP3 was able to identify some providers for oxy-firing, a detailed feasibility analysis was felt to be premature for this stage of the project, but is warranted and remains pending. Hence, most of the effort in CCP3 was focused on post-combustion capture from NGCC power generation unit.

TECHNOLOGY SELECTION FOR FEASIBILITY STUDY

The NGCC produces about 4 vol% CO_2 in the flue gas, and capturing 90% CO_2 from this low concentration is a challenge. Hence, the first task was to conduct a feasibility study with the current state-of-the-art post-combustion technology to establish a benchmark for comparison with novel technologies that CCP3 had planned to evaluate in this space. For this purpose, CCP3 in 2010, engaged with Amec Foster Wheeler to conduct a screening of various technology providers available at that time. The screening included technology providers with ammonia, amine and other novel material-based technologies such as:

- Ammonia based solvent technologies
  - Alstom Chilled Ammonia process
  - Powerspan ECO_2 process
- Amine based solvent technologies
  - MHI-KS1™ Hindered Amine process
  - Fluor Econamine FG Plus process
  - BASF, Linde and RWE aMDEA process
  - HTC Pure Energy CCS capture system
  - Aker Clean Carbon "Just catch" process
  - Cansolv Technologies Inc. (acquired by Shell Global Solutions)
- Other novel material based technologies
  - Siemens POSTCAP Amino Acid Salt process
  - Other Solid adsorbent based technologies
  - Other membrane technologies

It is important to note that this technology screening was conducted in early 2010, based on then-current state-of-the-art processes that were sufficiently developed, and for which the technology provider could offer detailed information, for a feasibility assessment for a commercial scale plant that can capture 1 million tonnes/yr of CO_2 from NGCC flue gas. After a preliminary assessment, the MHI-KS1™ technology was selected, as it was considered the state-of-the-art technology with the highest level of development and commercial deployment experience at that time. However, we do recognize that other technologies might have made progress in the recent time period till the publication of this chapter, but are not reflected here. The details of this study including the mass and energy balance and economic evaluations are presented in chapter 22. Based on pilot testing at EERC [3], we have also identified a promising novel solvent from ION Engineering and its economic evaluations, including benchmarking with MHI, are also presented in Chapter 22.

MHI CAPTURE TECHNOLOGY PROCESS SCHEME

MHI and Kansai Electric Power Company developed the proprietary KM-CDR™ (Kansai-Mitsubishi carbon dioxide recovery) process, which has been commercially operating since 1999 since construction of the first plant in Malaysia utilizing the patented KS-1™ solvent. MHI have developed their proprietary KS-1™ solvent based on advanced hindered amine technologies. MHI's carbon capture process consists of three main sections: flue gas cooling, CO_2 absorption and solvent regeneration. Firstly, flue gas is cooled in a direct-contact cooler (water) to an optimum temperature of around 35°C to 45°C before entering the CO_2 absorber. Lowering the flue gas temperature increases the efficiency of the exothermic CO_2 absorption reaction and minimises KS-1^{TM} solvent losses. In addition to cooling the flue gas, the flue gas direct contact cooler is designed to remove other impurities such as NOx, SOx, dust and suspended particulate matter.

The CO_2 absorber has two main sections - the bottom CO_2 absorption section and the top treated flue gas washing section. Upon entering the bottom of the column, the cooled flue gas flows upward through structured metal packing counter current to CO_2-lean KS-1^{TM} solvent, which is distributed evenly at the top of the absorption section onto the packing. CO_2 is absorbed from the flue gas into the KS-1^{TM} solvent upon direct contact. The CO_2 rich KS-1^{TM} solvent is sent to the stripper column for regeneration, and the scrubbed flue gas flows upwards into a treated gas washing section utilizing circulating water to absorb vaporized KS-1^{TM} solvent for recycle back into the CO_2 absorption section. The wash section is provided with cooling to condense water from the exiting flue gas, so a water balance can be maintained within the system.

The rich solvent is pre-heated in a heat exchanger using the hot lean solvent leaving the bottom of the CO_2 stripper (regenerator). The heated rich solvent is then introduced to the top section of the CO_2 stripper where it comes into contact with hot stripping steam at 120°C. The rich solvent is stripped of its CO_2 content, cooled to 40°C and recycled back to the CO_2 absorber as lean solvent, completing the closed loop solvent cycle. The concentrated CO_2 stream (>99.9%) leaves the top of the stripper column and is compressed and dehydrated for transport and sequestration. A schematic of the flow scheme employed by MHI is presented in Figure 1 below.

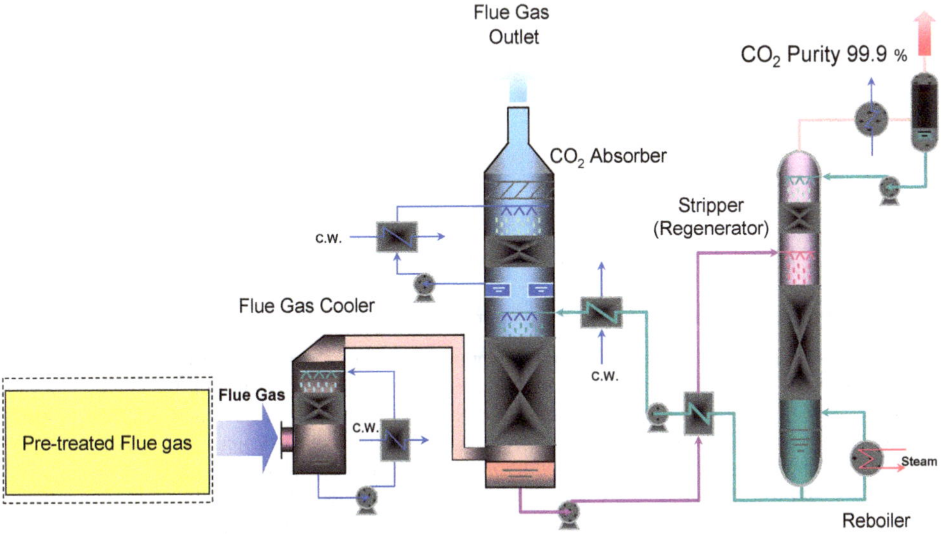

Figure 1. Simplified schematic of the MHI process for CO_2 capture.

Comparing key parameters of the KS-1^{TM} solvent process with those of conventional MEA solvent, the KS-1^{TM} solvent circulation rate is 40% lower, the regeneration energy is 38% lower, and the solvent losses and solvent degradation are both 10% of those experienced with conventional MEA solvent. Reboiler steam usage is 1.1 to 1.2 tonnes for every tonne of CO_2 captured. The regeneration energy for KS-1^{TM} solvent is about 2.6 GJ/ton-CO_2 utilizing 3.0 bar saturated steam in the reboiler.

Based on the numerous commercial plant applications capturing CO_2 from natural gas flue gas streams, MHI now offers large scale commercial single train CO_2 recovery plants. In addition MHI are currently developing even higher efficiency solvents such as KS-2™ and KS-3™ which claim to have lower regeneration energies and lower levels of solvent degradation. MHI are also developing new and improved processes which are to be applied to future plants.

INTEGRATED SCHEME FOR NGCC WITH POST-COMBUSTION CAPTURE

Base case NGCC power plant without capture

For this study, a reference case without any capture is the basis for comparison. The reference NGCC plant is based on a single natural gas fired GE Frame 9FA gas turbine. The gas turbine generates power for export, as well as hot exhaust gas, which is fed to a heat recovery steam generator (HRSG). The steam produced from the HRSG is fed to a single steam turbine which also generates power for export. Following heat recovery in the HRSG, the flue gas is emitted to atmosphere via a downstream stack. The total rated capacity of this power generation system is about 400 MW_e. A simplified schematic of this unit is shown in Figure 2 below. It is important to note that due to excess air firing in the GT, the CO_2 concentration in the flue gas is typically very low, and is about 4 vol% for this specific case.

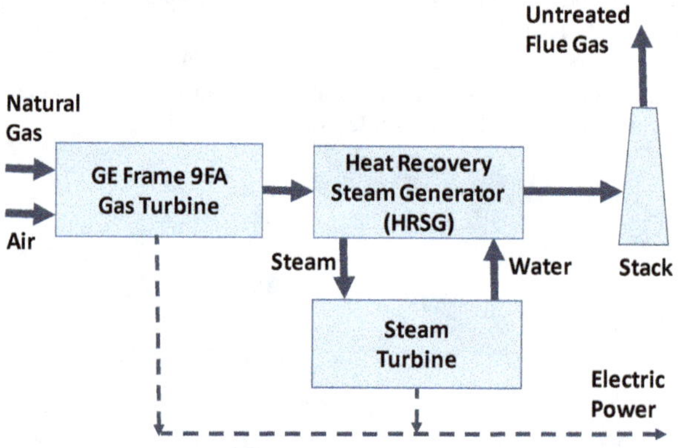

Figure 2. Simplified block diagram of a NGCC plant without capture.

NGCC power plant with post-combustion capture

Flue gas from the HRSG is routed to a CO_2 capture unit in a nearby location of suitable size. It is assumed that 5% of the total flue gas flow bypasses the CO_2 capture plant via the stack. The CO_2 capture unit removes 90% of the CO_2 in the flue gas it receives, with the cleaned flue gas returning to the stack. A single train of CO_2 absorption, stripping, compression and dehydration was considered the most appropriate for the total volume of gas to be treated in the case. Steam from the HRSG or steam turbine supplies the heat required for the stripper reboiler and solvent reclaimer, as well as for regeneration of the dehydration beds. The NGCC also supplies the CO_2 system's electrical power needs. A combination of sea and fresh water cooling is used in the CO_2 capture unit.

Further heat integration between the CO_2 system and the NGCC power island is challenging since the combined cycle power plant is already highly integrated with maximum heat recovery. Efficiency of the overall process may be improved by considering additional low-grade heat use, however, this has not been considered in the scope of this study. Details of the study are presented in Chapter 23 along with the economic evaluations.

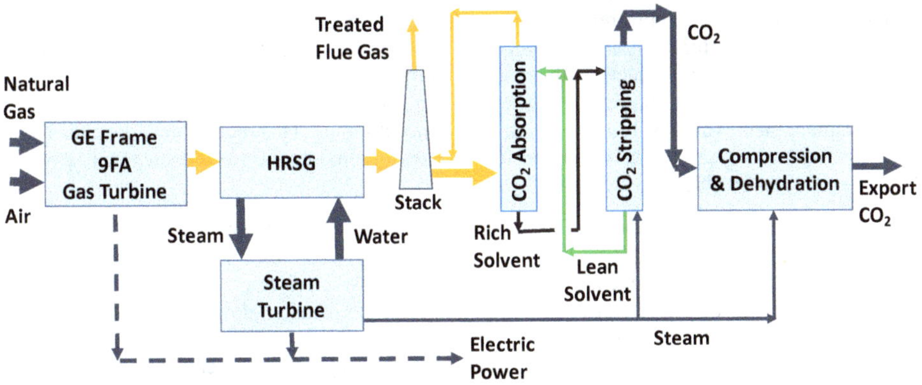

Figure 3. Simplified block diagram of a NGCC plant with capture.

KEY DESIGN CONSIDERATIONS FOR NGCC POST-COMBUSTION SYSTEMS

Integration of the CO_2 capture unit to the power generation unit

The flue gas is extracted upstream from the stack and delivered to the CO_2 capture unit by a flue gas blower. In the event of a failure or trip of the flue gas blower, the flue gas is emitted directly to the atmosphere through the existing stack with no impact on the operation of the power plant. The CO_2 capture unit consists of three main sections; (1) Flue gas pre-treatment section, (2) CO_2 absorption section, and (3) solvent regeneration section. The simplified block flow diagram in Figure 4 demonstrates a typical plant integration configuration.

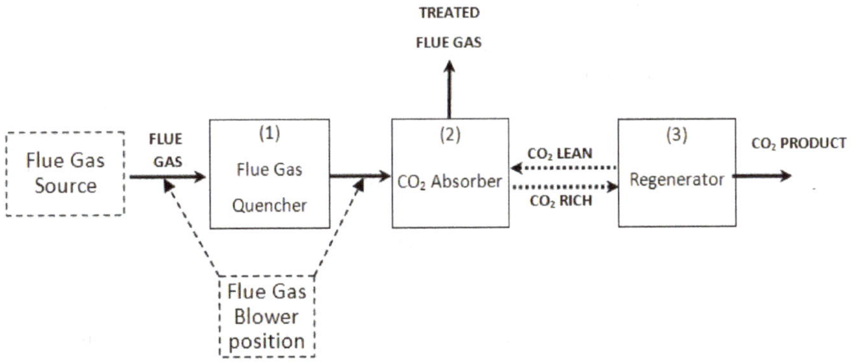

Figure 4. Simplified block diagram of CO_2 capture unit integration with the NGCC plant.

In this case, the flue gas temperature is too high to feed directly into the CO_2 absorber. A low flue gas temperature is preferred for the exothermic reaction between CO_2 gas and KS-1^{TM} solvent. Therefore, the flue gas is cooled in the flue gas quench unit by direct contact with circulating water, upstream of the CO_2 absorber. The circulating quench water is cooled by using cooling water. The flue gas quench also partially removes particulate dust and other impurities.

If the flue gas has a high SOx concentration, the flue gas quencher is also designed to sufficiently minimize SOx in the flue gas entering the CO_2 absorber to ensure efficient, reliable and stable operation of the CO_2 capture unit. In most NGCC cases this would not be an issue as the gas turbine operations would require sulphur removal from the fuel before combustion in the turbine. NO is an inert gas in the CO_2 capture unit and therefore does not affect solvent consumption and requires no pre-treatment. NO_2 partially reacts with amine to form heat stable salts, which are removed by reclaiming. It is preferable to minimise NO_2 by installing selective catalytic reduction (SCR) before the CO_2 capture plant where NO_2 is present in high concentrations. This requirement depends on the gas turbine operations and the solvent used for capture. In some cases, gas turbines that have low NOx burners might be able to avoid this SCR step.

There are two possible positions for flue gas blower (Figure 4) - before or after the quench. A carbon steel flue gas blower, designed to accommodate a high volume of high temperature flue gas can be installed before the quench. Alternatively, a stainless steel flue gas blower, designed to accommodate a lower flow rate and lower temperature can be installed downstream of the quench. The total power requirement of the blower will be increased if it is placed upstream of the quench, due to the higher blower inlet temperature. A site specific study is recommended to determine the optimum blower arrangement. In this study, the blower has been placed before the quench.

Quench and Absorber Tower – design, modularity and construction

Since the NGCC flue gas has very low CO_2 content, the flue gas volume handled per mass of CO_2 captured is very high. Based on extensive experience in the design and construction of large absorption columns, MHI believes that for large scale CO_2 capture plants, such as this example, rectangular quench and absorber towers offer better economics than cylindrical columns. However, in general, there is no specific cut off limit above which a rectangular absorber tower should be selected instead of a cylindrical absorber. This ultimately depends on local site specific factors and construction costs. The Absorber cross sectional area should be increased in accordance with the flue gas flow rate to ensure highly efficient solvent-flue gas contacting inside the CO_2 Absorber. It is also important to design carefully to avoid maldistribution of flue gas and KS-1^{TM} solvent in large diameter gas-liquid contactors.

Cylindrical towers are fabricated in shops and transported to construction area by sea, rail or road. Therefore it is necessary to consider size limitations, due to transport constraints. It is generally cost effective to shop fabricate a complete tower to minimise on-site manpower and time for installation if size permits. Larger towers can be constructed in parts small enough to meet transport limitations, and be assembled at site. For on-site tower construction, it will be necessary to conduct a pressure test according to relevant codes and regulations at site. Rectangular Quench and CO_2 absorber towers are constructed in 2-3 major parts: wall panels, bottom/top cover plates and steel structure items. These are pre-fabricated at assembly yards and transported to the plant site. It is preferable to fabricate panels as large as possible, within the constraints of transportation, to reduce assembly work at site.

Figure 5 shows a typical method for assembling pre-fabricated blocks at site. Assembly and erection of pre-fabricated items, installing internals and connecting associated piping/instrumentation are carried out on the plant site.

For CO_2 capture plants, MHI has developed a standardized design for a single train to capture 3,000 tons per day of CO_2 for natural gas applications (Figure 6). However, MHI is working on even larger single train CO_2 capture plant designs. There is no inherent limitation on train capacity of the Quench or CO_2 Absorber column. A rectangular column can be constructed as a panel plate design, due to operation at atmospheric pressure, with configuration at site. This rectangular design is based on MHI's highly successful flue gas desulphurization (FGD) plant technology which has become the most widely applied FGD technology process in the world. More than 200 commercial units have been deployed worldwide with the largest single train FGD plant treats flue gas with capacity flow rate of 4,384,000 Nm^3/h.

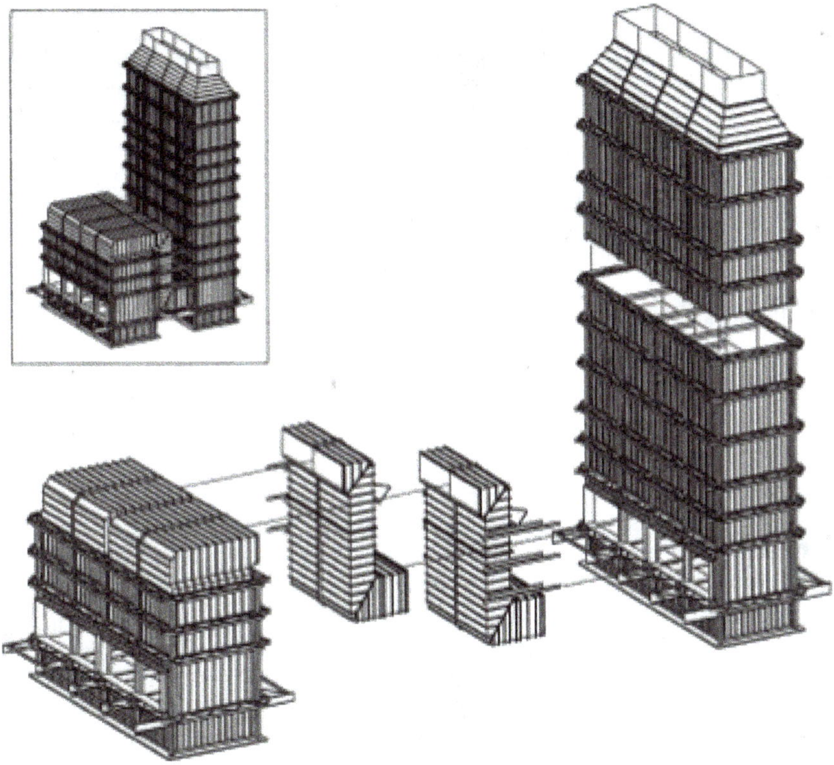

Figure 5. Block construction of CO_2 absorber and quench.

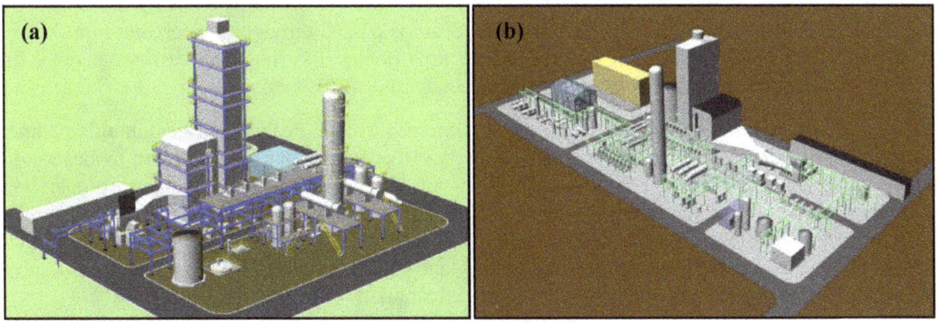

Figure 6. Gas boiler CO_2 capture plant (a) and gas turbine CO_2 capture and compression plant (b).

MHI has gained considerable commercial experience of flue gas CO_2 capture from natural gas-fired boilers and steam reformers. Results from testing at the large scale multi pollutant test plant, which has a flue gas rate equivalent to a 400 MW FGD plant, has been used to tune MHI's computational fluid dynamics (CFD) models used to design larger plants. The major scale up advantage for MHI CO_2 capture plants is based on the flat panel, rectangular design which allows MHI to design large capacity, single-train plants.

The flue gas CO_2 capture process, using amine scrubbing, is widely used throughout the world in various industries. However, a large scale (ie 3,000 tons of CO_2 per day) CO_2 capture plant from NGCC flue gas has not been built as of 2009. This study is based on public information available in 2009 and does not reflect any ongoing commercial planning. Through the successful scale up of several similar environmental plant technologies, such as FGD, MHI is confident in its ability to apply commercially realistic and reliable scale-up methods for large scale commercial CO_2 capture plants. MHI now offers large scale, single train commercial CO_2 capture plants for natural gas fired installations, and has completed the basic engineering and design for a modularized 3,000 tons of CO_2 per day commercial scale plant. These plants are developed for CCS-EOR and CCS projects based on natural gas boilers and gas turbine combined cycle power plants.

Regenerator design and construction

The maximum size of shop fabricated solvent regenerator towers is principally determined by transportation constraints. Since this vessel operates at slightly positive pressure, MHI adopted a cylindrical design, and not a rectangular design used for the absorbers. The maximum diameter of the regenerator tower is limited to around 8-10 m for a single train configuration. A regenerator of this size is capable of producing about 3,000-5,000 tons of CO_2 per day. This single train configuration is adopted for the commercial scale CO_2 capture plant chosen for this study. In addition, train size is limited by reboiler sizing from a mechanical and maintenance point of view. The maximum number of reboilers installed per regenerator tower is 4, as any more than this causes difficulties with respect to piping arrangement and equipment installation.

Regenerators are operated slightly above atmospheric pressure, therefore, once constructed, they must undergo pressure testing. It is possible to fabricate the complete tower at the manufacturer's assembly yard to reduce site work. However there are size limitations for pre-fabricated regenerators due to transportation constraints. Large regenerators may be assembled with pre-fabricated parts at site. Reboilers, auxiliary equipment and piping are fitted to the regenerators at site.

CO_2 compression and dehydration

In this study, the CO_2 compression is carried out in six stages to a final specified pressure of 150 barg. The cooled CO_2 from the regenerator reflux drum is passed through a knock out drum before being fed to the first compression stage. Between each of the first four compression stages is an interstage cooler and a knock out drum to remove condensed water up to a pressure of 40 bara.

The CO_2 is then dried by molecular sieve adsorption to achieve the design specification of moisture. The CO_2 dehydration unit consists of 3 molecular sieve beds, with one in adsorption mode, one in regeneration mode and one on stand-by. The beds are regenerated in a 12 hour cyclic operation using recycled dry CO_2 which is heated against the hot regeneration gas leaving the bed followed by a heater using medium pressure (MP) steam. There are two main types of CO_2 dehydration technology; molecular sieve adsorption and tri-ethylene glycol (TEG). In this study molecular sieve adsorption has been used to achieve the design specification of < 50 ppmv moisture.

The dry CO_2 is then compressed in two final stages, including an intercooler and an after cooler, resulting in a final CO_2 product at specification of 150 barg and 40°C. Oxygen content in the CO_2 is not a major issue, as the post-combustion amine systems are typically very selective for CO_2, resulting in O_2 content in the ppm level.

An electric motor driven, stainless steel CO_2 compressor (and flue gas blower) was assumed in this study. Compression of CO_2 for CCS applications is considered to be achievable with existing, demonstrated, commercial technology (Table 1). Single machines just large enough to compress the total CO_2 flow of the highest flow case considered in this study have been in commercial operation for a number of years. For example, the RG80-8 MAN Turbo machine used at a coal gasification plant in North Dakota has been in operation since 1998 and has characteristics comparable to that needed for this case. A larger machine, the RG100 is now normally recommended by MAN for large-scale CCS applications and is shown in the table below. Also, an MHI CO_2 compressor is in operation at In Salah, Algeria. Similarly, Siemens offers single machines for CCS from its STC-GV series, also shown in Table 1.

Table 1. Large Scale CO_2 compressor comparison.

	RG80-8	RG100	STG-GV	CCP3-NGCC scenario
Inlet Pressure	1.1 bara	Min. 0.4 bara	Not given	1.5 bara
Discharge Pressure	187 bara	Max. 250 bara	Max. 200 bara	151 bara
CO_2 Mass flow Rate	35 kg/s	Not given	Not given	35.7 kg/s
CO_2 Inlet Volume Rate (approximate m^3/h)	100,000	500,000	250,000	51,250

While economies of scale can be realised by using a single machine to compress 100% of the CO_2 from a CO_2 capture unit, in some circumstances opting for 2 x 50% machines arrangement might be preferable. Such circumstances include operations with a high degree of flow variation, such as a load following power station or a process in which high turndown is a frequent occurrence.

System integration for optimal avoidance

Steam consumption is one of the most significant factors influencing the operating cost of the CO_2 capture unit for NGCC systems, as it also affects the net power output of the plant. Therefore, the most effective means for reducing operating costs is to reduce the steam consumption required for solvent regeneration. MHI has developed and commercialised an advanced energy saving system that can be incorporated into the CO_2 capture unit, which includes the provision of additional heat exchangers and pumps (Patent applied by MHI). Through application of the energy saving system, steam requirements can be reduced by approximately 15% compared with the conventional KM CDR Process™. The energy saving system is incorporated in the analysis presented in this study. This system has been developed and proven at the Nanko pilot plant and subsequently applied to a 400 ton per day commercial plant deployed in Abu Dhabi. Due to the additional heat exchanger and pump requirements, there is a slight increase in the initial investment cost but operating cost will be decreased.

MHI typically designs CO_2 recovery plants with an auxiliary boiler for the capture plant routes the boiler flue gas through the CO_2 capture unit in order to improve the overall CO_2 capture efficiency. For this study, the decision was made to have capture plant utilities integral to the NGCC plant, and the design includes integration of capture plant steam and power requirements with the power generation unit. Although no dedicated boiler is needed for the capture plant, the result is power output reduction from the power plant (details in Chapter 23) due to steam extraction from the power cycle. Steam is generally at the crossing point between MP and LP steam turbine sections but there is an optimal position given by an energy analysis.

An emerging opportunity for heat integration to avoid drawing on the power plant it to utilize heat produced at higher temperature compared to conventional equipment by advanced CO_2 compression systems, such as that under development by Ramgen [4,5]. The heat produced by conventional compressors is generally too low in quality for use and is rejected from the process via the cooling system. Further assessment is needed to validate the integration benefits as well as potential capex savings of this concept with the current designs, as the development progresses.

CONCLUSIONS

In this feasibility study CCP3 has focused on post-combustion CO_2 capture from a combined cycle power generation facility with natural gas feed. In a future study, the alternative options, such as CO_2/O_2 gas turbine cycles, should be examined in detail and compared to post-combustion capture, as technology maturity warrants. For this study CCP3 chose a 400 MWe power plant that produces ~1 million tonnes/yr CO_2 in the flue gas containing 4 vol% CO_2. From a screening of several technology providers with capability to provide state-of-the-art deployable technology for this capacity, CCP3 selected the MHI-KS1™ technology for the study. MHI offers a single train design for this scale that can achieve the specified capture rate at the low CO_2 concentration occurring in this application. Integration of the power plant with the CO_2 capture plant includes a flue gas blower and flue gas quench upstream of the absorber. MHI recommends using a rectangular quench and absorber configuration, rather than conventional circular cross section for this size, since it offers better economies of scale. MHI's energy saving systems can reduce steam requirements, and optimized integration with the NGCC power plant can improve overall CO_2 avoidance when steam and power utilities for the capture plant are drawn from the power island. CO_2 compression and dehydration, for the scale under consideration, is considered a proven technology, since there is substantial operating experience, especially in upstream Enhanced Oil Recovery operations.

ACKNOWLEDGEMENTS

CCP3 acknowledges Mitsubishi Heavy Industries (MHI) for providing information for this study.

REFERENCES

1. Miracca I. "CO_2 Capture: Key Findings, Remaining Gaps, Future Prospects", in "Carbon Dioxide Capture for Storage in Deep Geologic Formations – Results from the CO_2 Capture Project" – Volume 3 – pp. 273-275 (Ed. Lars Ingolf Eide), CPL Press (2009).
2. Mathieu P. and Bolland O., "Comparison of costs for natural gas power generation with CO_2 capture" *Energy Procedia* vol. 37 pp. 2406 – 2419 (2013)
3. Brandon Pavlish et al., Partnership for CO_2 Capture – Phase II, Final Report, 2013-EERC-05-03 (May 2013).
4. Kuzdzal, M., and Baldwin, P.; "The Past, Present and Future of CO_2 Compression," *Carbon Capture Journal*, Sept/Oct, 2012
5. Koopman, A.; "Ramgen Supersonic Shock Wave Compression Technology," NETL Capture Technology Meeting, Pittsburgh, PA, July 8-11, 2013.

Carbon Dioxide Capture for Storage in Deep Geological Formations, Volume 4
Karl F. Gerdes (Editor)

Chapter 22

TECHNO-ECONOMIC ASSESSMENT OF POWER GENERATION FROM NATURAL GAS IN COMBINED CYCLE (NGCC) SYSTEMS WITH INTEGRATED CAPTURE

Mahesh Iyer[1], Richard Beavis[2], Tony Tarrant[3] and Tim Bullen[3] and David Butler[4]
[1]Shell International Exploration & Production Inc.,
200 N. Dairy Ashford Rd., Houston, TX 77079, USA
[2]BP International Limited, ICBT Chertsey Road,
Sunbury-on-Thames, Middlesex, TW16 7LN, United Kingdom
[3]Amec Foster Wheeler, Shinfield Park, Reading, Berkshire, RG2 9FW, United Kingdom
[4]Calgary, Alberta, Canada.

ABSTRACT: This study presents techno-economic evaluations of post-combustion solvent CO_2 capture from a natural gas fired combined cycle (NGCC) power plant. A reference plant, based on a GE Frame 9FA gas turbine, producing 400 MW nominal power was integrated with two post-combustion solvent-based capture technologies: MHI-KS1™ a current state-of-art solvent technology and a new solvent technology from ION Engineering. The study was based on a US Gulf Coast location and North West European gas price of $11.5/MMBtu. Integration of post-combustion capture system to the NGCC plant reduced the net power efficiency from 56.4% for the reference plant to 49.8% for MHI and to 49% for ION. The reduction in efficiency is due to power derate caused by integration of the CCS unit with the main power plant, including steam extraction and auxiliary power usage. The net CO_2 avoided in both cases range from 73.7 - 72.6% with the capture percentage very close to 85.5%. The CO_2 avoided cost ranges from $114/tonne for MHI to $125/tonne for the ION Solvent. The addition of CCS increases the reference case cost of power produced by 32% for MHI and 36% for ION. Power derate is the major component of the CO_2 avoided cost, with about 41% of the total cost, while the capex contribution was 22% for ION and 33% for MHI.

The sensitivity of the economic results to changes in gas prices, efficiency and capex are reported. These studies show that efficient integration of the capture process into the power plant is very important. The maturity of the MHI technology enabled significant optimization of system integration, which led to the lower avoided cost for MHI relative to ION, despite the estimated capital costs for ION being lower.

KEYWORDS: combined cycle; power generation; natural gas; NGCC; CO_2 capture; post-combustion; MHI; techno-economics; avoided cost

INTRODUCTION

The primary focus of this work was to understand the CO_2 abatement options and avoided costs for power generation from natural gas - predominantly from post-combustion systems. As mentioned in Chapter 21, both pre-combustion and oxy-firing technologies applied to power production, were not considered for further investigation or detailed economic assessments. Consequently, this chapter presents the results of economic evaluations of post-combustion, solvent-based technologies that can be integrated with a current-state-of-the-art natural gas combined cycle (NGCC) power generation system. As described in Chapter 21, MHI-KS1^{TM} technology was selected, which is considered ready for deployment today, as the base case for benchmarking purposes.

In addition, CCP chose a novel promising solvent from ION Engineering for evaluation, and compared it with the MHI-KS1 technology. The ION solvent performance, upon which the estimates for equipment design, capital cost and operating costs were based, was extrapolated from pilot testing done at the Energy & Environment Research Center – University of North Dakota (EERC) as part of the Partnership for CO_2 Capture Phase II study [1]. A number of solvents were tested for CO_2 capture performance on flue gas from both coal and natural gas combustion. Solvents tested included generic 30 wt % MEA as a reference and several advanced proprietary solvents, including one from ION Engineering, which showed noteworthy improvement on circulation, regeneration energy, and kinetics over reference MEA. The utilization of existing pilot equipment imposed constraints on the testing, so rigorous analysis of the results was not possible. However, comparative testing of the ION solvent relative to MEA at a variety of conditions made possible a semi-quantitative extrapolation of results to commercial conditions. These results, presented here, show that the ION solvent has promise, and testing at more representative conditions is warranted.

BASIS OF DESIGN

For this study the reference plant (without capture) is an NGCC plant based on a single natural gas fired GE Frame 9FA gas turbine. The turbine generates hot exhaust gas which is fed to a heat recovery steam generator (HRSG). The steam produced from the HRSG is fed to a single steam turbine which also generates power for export. The total rated power generation capacity of this reference NGCC plant is 400 MWe. For the abated cases, post-combustion solvent technology is employed to capture CO_2 from exhaust flue gases. The CCP studied two technologies: MHI-KS1 solvent technology and the ION Engineering solvent technology. The preliminary process schemes and block diagrams are explained in Chapter 21.

The post-combustion capture plants are integrated with the NGCC reference plant to provide steam extracted from the low pressure turbine for solvent regeneration and power. The site is assumed to be a green-field location on the US Gulf Coast. The reference conditions for the plant include average ambient temperature of 20°C and pressure of 1.013 bar. A single train of CO_2 absorption, stripping, compression and dehydration was considered the most appropriate for the total volume of gas to be treated in the case. A combination of sea and fresh water cooling is used. Steam and electrical power is supplied over the fence to the CO_2 capture area from the power island.

Streams which cross the CO_2 capture plant battery limits are the following:

- NGCC flue gas from the heat recovery steam generator (HRSG),
- Sea water supply and return,
- Closed loop cooling water supply and return,
- Plant/Raw/Potable/Demineralised/Boiler Feed water,
- Chemicals (including amine),
- Instrument/Plant air, Nitrogen,
- CO_2 product,
- Low and medium pressure steam/condensate.

The composition and conditions of the flue gas have been determined by simulation undertaken by Amec Foster Wheeler, based upon the specified fuel and ambient air and shown in Table 1 below.

Table 1. Simulated exhaust flue gas composition.

Total Flue Gas Flow (kg/hr)	2,287,600
Flue Gas Flow to CO_2 Capture Plant (kg/hr)	2,173,220
Pressure (barg)	0.002
Temper319ature (°C)	88.8
Composition (mol%)	
CO_2	4.17
SO_x	(0.17 ppmv SO_2)*
NO_x	(13.5 ppmv NO, 1.5 ppmv NO_2)*
CO	0.00
N_2	74.40
Ar	0.89
O_2	12.27
H_2O	8.26

*Dry Basis

The fuel specifications are detailed in the table 2 below. The fuel feed conditions are:

- Pressure: 31 bara
- Temperature: 16 °C
- Molecular Weight: 20.74 (kg/kmol)
- LHV: 42.06 MJ/Nm3 (39.88 MJ/Sm3 =11.08 kWh/Sm3, = 45,452 kJ/kg)

Table 2. NGCC fuel properties.

	Composition Vol%		Composition Vol%
CH_4	79.76	n-C_5H_{12}	0.20
C_2H_6	9.68	C_6H_{14}	0.21
C_3H_8	4.45	CO_2	2.92
i-C_4H_{10}	0.73	N_2	0.61
n-C_4H_{10}	1.23	H_2S	5 ppmv
i-C_5H_{12}	0.21		

NGCC CASES: TECHNICAL PERFORMANCE RESULTS

Amec Foster Wheeler, in coordination with CCP, conducted the technical study to develop the process configuration design, simulations to derive mass and heat balances and capital cost estimates. The details of the study methodology are described in Chapter 3. The power island was modelled using GateCycle™ software [2]. For the CO_2 capture technologies, key design and performance information was provided by the two technology providers. The CO_2 capture systems were designed to capture 90% of the CO_2 in the flue gas routed to the CO_2 capture plant. For operational and safety reasons, 5% of the total flue gas bypasses the CO_2 capture plant.

The MHI-KS1 solvent system was selected as the current state-of-the-art post-combustion technology for this assessment, while ION's solvent was selected as a new technology to be evaluated. The MHI technology has undergone a long period of development, and several demonstration plants have been built. In contrast, the ION solvent performance is based on data recently measured in bench and pilot scale testing. Therefore there is significant opportunity to optimize the ION technology and its optimal integration with the NGCC plant. Because of the difference in technical maturity, the ION technology is not directly comparable to MHI technology, and the comparison is only indicative of how the technology may perform if it is commercialized.

Table 3. Overview of the power outputs and efficiency for the NGCC cases.

	Reference plant (unabated)	MHI Post-C. (current)	ION Post-C. (novel)
Gas Turbine (MW)	253	253	253
Steam Turbine (MW)	141	116	112
Gross Power (MW)	394	369	365
Net Power (MW)	390	344	339
Net efficiency LHV	56.4%	49.8%	49.0%
Net efficiency drop	--	6.7%	7.4%

Table 3 above summarizes power output and overall efficiency (gross and net) for the three cases considered in the NGCC scenario study. Note that the gas turbine gross output remains the same in each case, but the steam turbine output changes due to the steam extracted for the capture plant. The overall power derate accounts for auxiliary power consumption in the power plant for all cases, as well as auxiliary power consumption in the capture plant. The net efficiency decrease is relative to the reference (unabated) plant. ION has a slightly larger efficiency loss than the current MHI technology. More detailed assessment of the key metrics is given in Table 4.

The nomenclature used in the tables and figures below are:

- **Ref Case:** Reference unabated case without CCS
- **MHI Post-C.:** NGCC with MHI-KS1 post-combustion capture
- **ION Post-C.:** NGCC with – ION Solvent post-combustion capture

Table 4. Key Metrics of CO_2 Capture Technologies Applied to NGCC system.

Characteristic	Units	Ref. Case	MHI Post-C.	ION Post-C.
Fuel Type		Nat Gas	Nat Gas	Nat Gas
Gross Output	MW	394.0	394.0	394.0
Steam Derate	MW		25.0	29.0
Power Derate	MW		21.0	22.0
Total Derate	MW		46.0	51.0
Auxiliary Power	MW	4.0	4.0	4.0
Net Output	MW	390.0	344.0	339.0
% Derate	%	1.0%	11.7%	12.9%
Net efficiency drop		--	6.7%	7.4%
Heat Rate - LHV	GJ/MWh	6.38	7.23	7.34
Efficiency	%	56.4%	49.8%	49.0%
CO_2 Intensity	t/MWh	0.38	0.06	0.06
CO_2 Generated	Mt/yr	1.16	1.16	1.16
CO_2 Captured	Mt/yr	--	0.99	1.00
CO_2 Emitted	Mt/yr	1.16	0.17	0.17
Reference Emissions - No CCS	Mt/yr	1.16	1.16	1.16
Less: CCS Emissions	Mt/yr		0.17	0.17
Less: Derate Impact	Mt/yr		0.14	0.15
Avoided Emissions	Mt/yr		0.86	0.85

Table 4 summarizes the key performance data developed by Amec Foster Wheeler which was used to derive the economic results. As noted previously, two carbon capture cases were evaluated. The reference NGCC has a net output of 390 MW. This NGCC plant was based on a European design, and a European natural gas price of $11.5/MMBtu (=9.4 €/GJ) was used for the analysis. All results are presented in US dollars. The capital costs were estimated for a US Gulf Coast location, which should also be similar to a North West European location. A reference case without CCS is provided to facilitate the cost comparisons.

Table 4 shows the change in output, the heat rate and efficiency for each capture case, along with the reference case. The steam and power derate, plus the auxiliary power load, are deducted from the gross output to yield the net output. The steam derate is based on the difference between the gross power output for the reference case minus the CCS case and is due to the steam extracted for the capture plant. The total derate is the difference between the net power outputs for the reference and CCS cases, and includes both auxiliary power and steam usage by the capture plant. The CO_2 intensity is based on the net power produced. The net efficiency drop, as mentioned earlier, is the percentage point decrease in efficiency for the entire plant with CCS as compared to the reference unabated plant with a net efficiency of 56.4%.

ECONOMIC ANALYSIS

Table 5 below shows the key financial results for each of the cases.

Table 5. Financial Results.

Characteristic	Units	Ref. Case	MHI Post-C.	ION Post-C.
1st Year Cost of Power	USD/MWh	110.6	146.5	150.0
Increase in Power Cost	USD/MWh		35.9	39.4
% Increase in Power Cost	%		32%	36%
Marginal Cost	USD/MWh	76.7	87.0	88.3
Capital Cost	MUSD	853	1,192	1,098
Incremental Capital	MUSD		339	245
Capex	USD/kW Net	2,188	3,464	3,239
Incremental Capex	USD/kW Net		1,277	1,051
O&M	MUSD/yr	23.8	49.1	61.6
Incremental O&M	MUSD/yr		25.3	35.8
% Capture	%		85.5%	85.7%
% Avoided	%		73.7%	72.6%
Capture Cost	USD/tCO_2		97.9	105.7
Avoided Cost	USD/tCO_2		113.6	124.7
Levelized Capture Cost	USD/tCO_2		114.4	123.5
Levelized Avoided Cost	USD/tCO_2		132.7	145.7
Additionally Yearly Cost	MUSD		97.4	105.4
Levelized Power Cost	USD/MWh	129.3	171.2	175.3
Increase in Levelized Power Cost	USD/MWh		32%	36%

The top part of Table 5 shows the cost of power for each case. The "% avoided" is defined as the mass of CO_2 avoided divided by the mass of CO_2 generated in the base reference case without CCS. The Levelized Capture Cost is based on levelized incremental costs rather than on first year costs (details on calculation methodology are explained in Chapter 4). The Additional Yearly Cost row shows the cost associated with capturing CO_2 each year. It includes an amortized annual cost for capital recovery. It is also important to note that the total Capex of the reference NGCC plant in this study ($2,182/kW), as shown in Table 5 , is higher than typical NGCC plant costs reported in other studies [3-5]. This is primarily due to the use of oil and gas industry design and equipment standards, rather than typical utility standards used elsewhere. Hence, it would not be advisable to compare the percentage increase of the post-combustion capture plants over the reference NGCC plants, but to compare the actual Capex values for a given capture capacity.

Table 5 shows that the CO_2 avoided cost, for the European design, with a natural gas price of $11.5/MMBtu, ranges from $114/tonne for MHI base case to $125/tonne for the ION Solvent case. The levelized avoided cost is higher, ranging from 133-146 $/tonne. The impact of CCS on the power cost (either 1st year or levelized cost), is a 31% increase for MHI and 34% for ION. The % CO_2 avoided in both cases range from 73.7 to 72.6% while the % captured in each case is almost the same at 85.5%.

Figure 1 below shows the breakdown of costs to capture CO_2 for the cases considered; and Figure 2 shows the breakdown for avoided costs. The economic impact of the derate in net power output dominates both the captured and avoided costs. The derate cost is based on the cost to replace generation capacity lost in order to supply steam and power to the carbon capture process, resulting in a lower net export power. Transportation and storage costs for CO_2 are expected to represent a small contribution to the cost.

Both Figures 1 and 2 show that the capex contribution is lower for ION, with 21%, as compared to MHI with 32%. However, for both cases, the contribution of power derate is around 43-44%, which is the largest contributor to the overall capture (or avoided) cost. Hence it is important to focus developments to minimize power derate in the overall scheme. Figure 2 shows the component costs making up the avoided cost of CO_2 for the cases considered.

Table 6 shows the components that contribute to the total capital cost to build the NGCC and to capture CO_2. The top part of the table shows the direct costs for each case. This is followed by indirect OSBL costs. The base capital costs have been adjusted to include escalation from Q1 2009 to the assumed in-service date of Q1 2014. This is followed by estimates for contingencies. To this total, estimates for Owners Costs are added. Finally AFUDC during construction are included. The bottom row shows the total as spent costs expected just before the plant is commissioned.

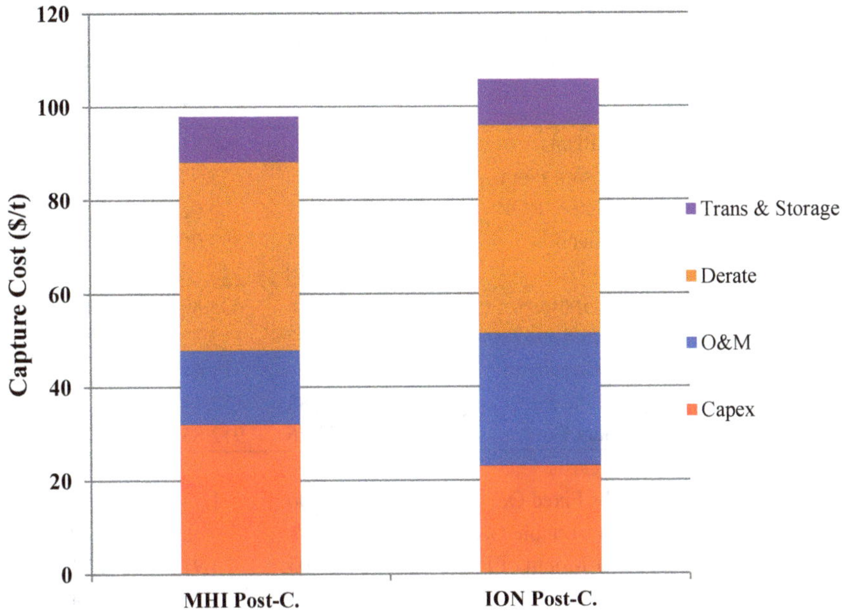

Figure 1. Capture cost breakdown for post-combustion on NGCC system.

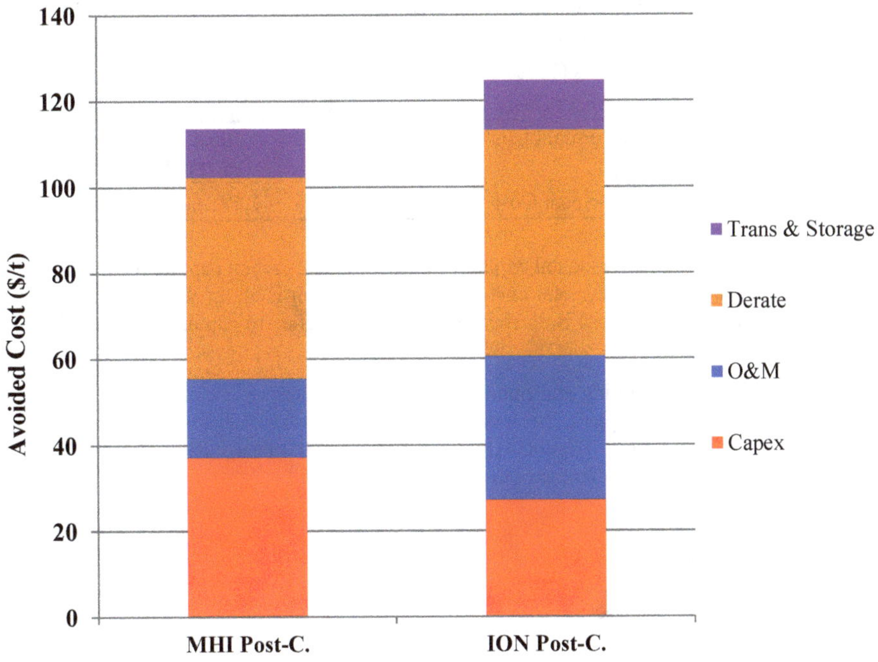

Figure 2. Avoided Cost breakdown for post-combustion on NGCC system.

Table 6. Overall Capital Cost Components for different NGCC cases.

Cost	Ref. Case	MHI Post-C.	ION Post-C.
Capture Plant		96.81	60.43
Compression Plant		35.93	34.44
Steam Gen & BOP		0.91	0.98
Power Island	354.28	353.96	347.71
OSBL (30%)	106.28	146.28	133.07
Direct & Indirect	460.56	633.89	576.63
Escalation from 2009 to 2014	83.82	115.37	104.94
Process Contingency	-	11.44	16.13
Project Contingency (20%)	108.88	152.14	139.54
Total Plant Costs	653.26	912.83	837.23
Owners Costs			
6 Months Fixed O&M	2.00	1.64	2.00
1 Month Variable O&M	-	0.56	1.72
25% of 1 Month of Fuel Costs	2.03	4.91	4.91
2% of TPC	13.07	18.26	16.74
60 Days of Cons.	-	1.12	3.44
Spare Parts (.5% TPC)	3.27	4.56	4.19
Land ($3,000/acre)	0.02	0.02	0.02
Financing Costs (2.7% TPC)	17.64	24.65	22.61
Other Costs (15% of TPC)	97.99	136.92	125.58
Total Owners Costs	136.00	192.65	181.20
Total Overnight Costs	789.27	1,105.47	1,018.43
AFUDC	61.56	86.23	79.44
Total As-Spent Cost	850.83	1,191.69	1,097.86

Table 7 compares the incremental capital costs to complete carbon capture on the NGCC for the two technologies evaluated. The total cost for the MHI system is about $341 million while that of the ION system is $247 million. Both these system are designed to capture about 1 million tonnes per year CO_2.

Table 8 shows the operating and maintenance costs for each case including the cost to operate the NGCC and CO_2 capture plant. The O&M costs have been adjusted for escalation or capacity factor and are reported in $2014. The total O&M cost for the MHI base case is about 2.25 times or 125% higher than the unabated reference case.

Table 7. Incremental Capital Costs for the two post-combustion cases.

Cost	MHI Post-C.	ION Post-C.
Capture Plant	96.81	60.43
Compression Plant	35.93	34.44
Steam Gen & BOP	0.91	0.98
Power Island	-0.32	-6.57
OSBL (30%)	40.00	26.78
Direct & Indirect	173.33	116.06
Escalation from 2009 to 2014	31.54	21.12
Process Contingency	11.44	16.13
Project Contingency (20%)	43.26	30.66
Total Plant Costs	259.57	183.97
Owners Costs		
6 Months Fixed O&M	-0.36	-
1 Month Variable O&M	0.56	1.72
25% of 1 Month of Fuel Costs	2.89	2.89
2% of TPC	5.19	3.68
60 Days of Cons.	1.12	3.44
Spare Parts (.5% TPC)	1.30	0.92
Land ($3,000/acre)	-	-
Financing Costs (2.7% TPC)	7.01	4.97
Other Costs (15% of TPC)	38.93	27.59
Total Owners Costs	56.64	45.20
Total Overnight Costs	316.20	229.16
AFUDC	24.66	17.87
Total As-Spent Cost	340.86	247.03

Table 8. O&M Costs for all the NGCC cases.

Cost	Ref. Case	MHI Post-C.	ION Post-C.
Fixed Costs			
Direct Labour	4.00	4.00	4.00
G&A	0.72	1.21	1.21
Maintenance	12.69	17.47	16.17
Insurance and P Taxes	8.47	11.64	10.78
Total Fixed Costs	25.87	34.32	32.15
Variable Costs			
Capture and compression	-	12.56	25.22
Transportation	-	6.71	6.73
Storage	-	2.92	2.93
Total Variable Cost	-	22.19	34.88
Total O&M	25.87	56.51	67.03

Table 9 shows the incremental capital costs required to capture CO_2. The O&M costs have been adjusted for escalation or capacity factor and are reported in $2014. The incremental O&M cost for the ION case is about 35% higher than the MHI case due to the higher costs of the chemicals assumed for this analysis.

Table 9. Incremental O&M Costs.

Cost	MHI Post-C.	ION Post-C.
Fixed Costs		
Direct Labour	-	-
G&A	0.49	0.49
Maintenance	4.78	3.47
Insurance and P Taxes	3.18	2.32
Total Fixed Costs	8.44	6.28
Variable Costs		
Capture and compression	12.56	25.22
Transportation	6.71	6.73
Storage	2.92	2.93
Total Variable Cost	22.19	34.88
Total O&M	30.63	41.16

Figure 3 shows how the cost of capturing CO_2 varies as the cost of natural gas changes.

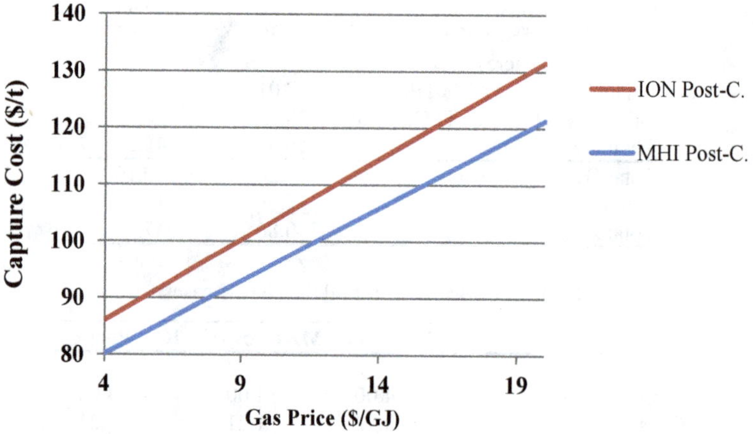

Figure 3. Capture Cost vs Gas Price for the two selected technologies.

Similarly, Figure 4 shows how the avoided costs change as gas prices change. We see from Figure 3 that the base case capture cost for MHI system is about $101/tonne for a gas price of $12/GJ (European price assumption). As the gas price decreases to a $4-5/GJ range, which is currently the case in the United States, the cost of capture then decreases to about $80-85/tonne.

Figure 5 shows the effect of changes in Capex (+25% to -25%) and net power efficiency (+1% to -1%) on the avoided costs for the base case MHI system (keeping all other parameters constant). Increasing the CCS unit Capex by 25% increases the avoided cost from $114/tonne to $122/tonne, which is a 7.9% increase. A capex reduction of 25% reduces the avoided cost to $104/tonne. Reduction in net power efficiency of the system by 1%, from 49.8% to 48.8%, results in an increase in the avoided cost from $114/tonne to $124/tonne, an 8.8% increase. An increase in power efficiency by 1% to 50.8% results in reduction of the avoided cost to $104/tonne. Thus, for this scale (400MW nominal NGCC power), a 1% change in the net power efficiency has almost the same effect as a 25% change in Capex of the CCS system. Hence, efficient system integration for optimal energy efficiency is critical for these applications. The relative sensitivity of avoided cost to capex and efficiency explains why ION Solvent has a slightly higher avoided cost than MHI even though the ION system has about 30% lower capex. ION has about 0.8% lower net power efficiency than MHI and also has 20% higher O&M costs than MHI (Table 8), leading to higher avoided cost. The performance data for the MHI technology has been developed over a long period, and several demonstration plants have been built. The ION performance is based on limited data recently generated in bench and pilot scale testing. Therefore there is significant opportunity to optimize the ION technology and its configuration by proper integration with the NGCC plant.

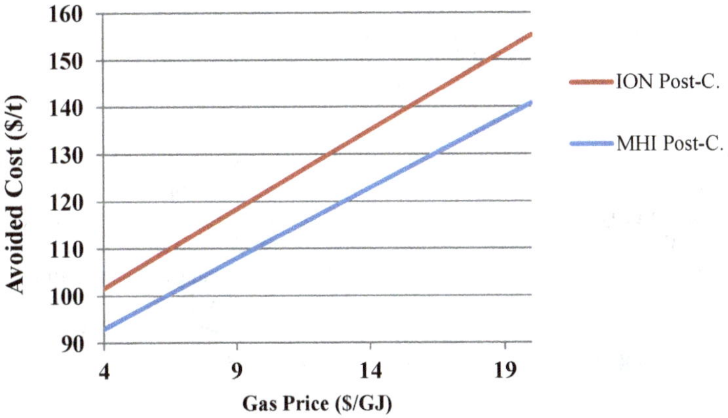

Figure 4. Avoided Cost vs Gas Price for the two selected technologies.

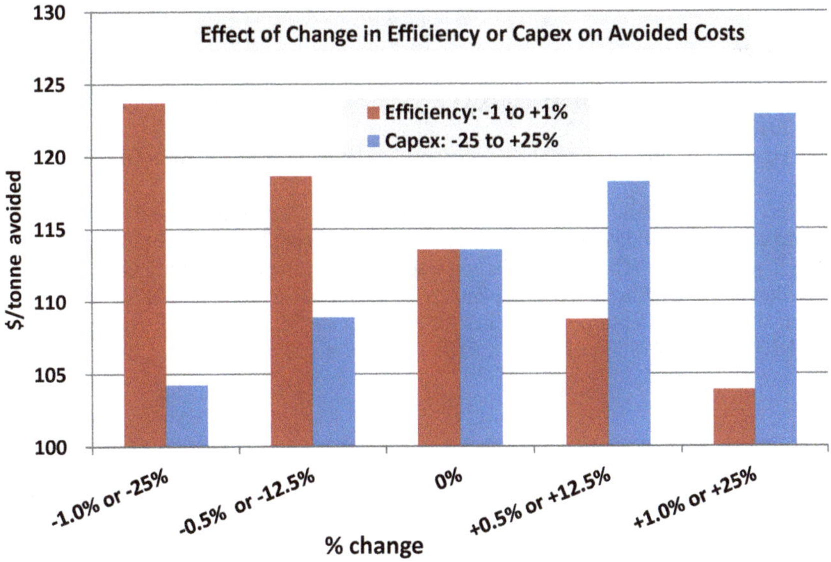

Figure 5. Sensitivity analysis of Efficiency change (± 1%) or Capex change (± 25%) on the CO_2 avoided Costs for the MHI solvent system.

CONCLUSIONS

A techno-economic assessment of solvent-based post-combustion CO_2 capture for a natural gas fired combined cycle power generation system was conducted for two different technologies: MHI-KS1™ , which is a current state-of-art solvent technology, and ION Engineering solvent which is a new technology. Addition of a post-combustion capture system to the NGCC plant reduced the net power efficiency from 56.4% to 49.8% for MHI system and to 49% for ION system. The reduction in efficiency is due to steam extraction and subsequent power derate caused by integration of the CCS unit with the main power plant. The net CO_2 avoided in both cases range from 73.7 - 72.6% with capture very close to 85.5%. The CO_2 avoided cost ranges from $114/tonne for the MH1 base case to $125/tonne for the ION case. The implementation of CCS increases the reference case power cost by 31% for MHI and 34% for ION. Power derate is the major component of the CO_2 avoided cost, accounting for about 44% of the total. Hence, it is important to focus developments on minimizing power derate in the overall scheme.

While the ION system has 30% lower capex than the MHI system, the MHI case has 10% lower CO_2 avoided costs due to better efficiency and O&M costs. Much of this difference can be attributed to the maturity of the MHI system, which has been highly optimized, compared to the ION system. For the scale of 400MW nominal NGCC power facility considered here, a 1% change in the net power efficiency has almost the same effect as a 25% change in Capex of the CCS system. Hence, efficient system integration for optimal energy efficiency is critical for these applications, and expenditure of additional capital to improve efficiency may be warranted. The ION Engineering system looks promising and should be considered in future assessments.

ACKNOWLEDGEMENTS

CCP3 acknowledges Mitsubishi Heavy Industries (MHI) and ION Engineering for working with Amec Foster Wheeler and providing information for this study.

REFERENCES

1. Brandon Pavlish et al., Partnership for CO_2 Capture – Phase II, Final Report, 2013-EERC-05-03 (May 2013).
2. GateCycle™ software for thermal power plant simulation is available from GE Power and Water.
3. Smith N. Miller G., Aandia I., Gadsden R. and Davison J. "Performance and Costs of CO_2 Capture at Gas Fired Power Plants" *Energy Procedia* (37) pp.2443 – 2452 [2013]
4. Rubin E. and Zhai H. "The Cost of Carbon Capture and Storage for Natural Gas Combined Cycle Power Plants" *Environ. Sci. Technol.* Vol. 46, pp. 3076–3084 [2012]
5. Shelton W. "Carbon Capture Approaches for Natural Gas Combined Cycle Systems" DOE/NETL-2011/1470 Final Report [2010].

Carbon Dioxide Capture for Storage in Deep Geological Formations, Volume 4
Karl F. Gerdes (Editor)

Section 1

CO_2 CAPTURE

Capture Conclusions

Carbon Dioxide Capture for Storage in Deep Geological Formations, Volume 4
Karl F. Gerdes (Editor)

Chapter 23

CO_2 CAPTURE – KEY FINDINGS AND FUTURE WORK

Ivano Miracca[1], Raja Jadhav[2] and Jonathan Forsyth[3]
[1]Saipem S.p.A. (Eni) – Via Martiri di Cefalonia, 67 – I-20097 San Donato Milanese, Italy
[2]Chevron Energy Technology Company, 100 Chevron Way, Richmond, CA, USA
[3]BP International Limited, ICBT Chertsey Road, Sunbury-on-Thames, Middlesex, TW16 7LN, United Kingdom

ABSTRACT: The main technical and economic conclusions of CCP3 are summarized. Potential areas for future work are discussed, based on gaps identified.

KEYWORDS: CO_2 Capture; post-combustion; oxy-combustion; oxy-firing; pre-combustion; NGCC; heavy oil; oil sands; OTSG; FCC; refinery

MAIN LESSONS LEARNED

Post-combustion CO_2 capture technology based on aqueous amine solvents has made steady, incremental improvement during the last decade. Commercial technology providers have been able to reduce both energy consumption (mainly the amount of steam used for amine regeneration) and capital cost. These gains have been achieved by optimizing solvent formulations along with the process schemes in which they are used and the design of the equipment and its construction. As a consequence, as the performance offered by the state-of-the-art technology has improved, the reference target for developmental technologies has moved forward as well. As a result new technology performance and cost hurdles are more challenging than they were a decade ago.

Based on the assumptions used for CCP3 techno-economic studies, state-of-the-art, solvent-based post-combustion CO_2 capture technology offers the lowest CO_2 avoided cost for each application scenario, when compared to the oxy-fuel and pre-combustion technology that is available today. However, in many cases there is further opportunity for optimisation and cost reduction of these alternate approaches, since they are less mature than solvent-based post-combution technology. Factors other than new-build cost and performance can change the broad conclusion regarding post-combustion. Site specific factors, such as the number of separate emission sources, scale factors, layout and plot space issues or plant debottlenecking can make the other technologies competitive. The higher additional fuel consumption of post-combustion compared to other techniques is a major factor, although the low cost of natural gas fuel in North America, after the shale gas revolution, serves to minimize this effect at present. Only in a future where there is increased fuel cost, reduced greenhouse gas intensity of imported power, and significant cost improvements in air separation technology, would oxy-firing become competitive with post-combustion. Chemical Looping Combustion (CLC), which is a technology in development that involves separating oxygen from air without the need for a cryogenic air separation unit, showed potential to be competitive with post-combustion and had the lowest avoided cost for the oil sands steam generation application scenario.

It is worth noting that developers of novel capture technology often compare with generic post-combustion capture using monoethanolamine (MEA) as the benchmark in terms of energy consumption. However, established technology providers now licence proven post-combustion technology, already successfully implemented at the scale of more than 100,000 tonnes/year of captured CO_2, with an energy consumption about 30% lower than for generic MEA. These established technology providers also offer commercial performance guarantees for single-train plant capacities of about 1 million tons/year of captured CO_2.

Over the last few years, our industry has seen the emergence of a number of innovative CO_2 capture technologies under development. This presents a rich opportunity for innovation driven by competiton between the technology providers, where care needs to be taken to identify those that have the best prospects for improved performance.

Last, but not least, CCP has found that many new technology developers are targeted at reduction of operating costs (typically energy consumption), often at the expense of higher capital cost. This may be a good strategy for application to power generation, in which a penalty in efficiency has the direct consequence of a lower power output. However, for refinery or oil/gas production applications, since CO_2 capture technology is very capital intensive and, therefore, a large portion of the capture cost is generally attributable to initial investment, research targets could be shifted toward simpler, low capex technology, accepting either a small increase in operating costs or a decrease in the capture rate. This is generally true in current energy price conditions, and particularly in Northern Alberta's oil sands producing region in Canada, where construction costs are particularly high due to the combination of remote location and harsh climate.

The following sections analyse the CCP's key findings in more detail for each application scenario studied. It should be noted that these conclusions are specific to the CCP3 techno-economic study assumptions. Hence, they may not apply to other situations due to the impact of location, utility cost, assumed CO_2 footprint of imported power and other economic assumptions on the CO_2 avoided cost.

REFINERY SCENARIO

Refinery – Fluid Catalytic Cracking Unit (FCC)

Among applications of interest to the CCP, fluid catalytic cracking (FCC) is the most favorable to oxy-firing. The "fuel" being burned in this case is the carbon that is continuously deposited on the FCC catalyst. Oxy-firing is normally a more attractive option when the carbon to hydrogen ratio of the fuel is high, because in this situation, the oxygen that is supplied is used most effectively to create high-purity CO_2, rather than large amounts of water vapour. Another reason for the favourable economics for the oxy-firing process in an FCC is the fact that the FCC is operated slightly above-atmospheric pressure, which avoids any air in-leakage, and so results in a high purity CO_2 undiluted by nitrogen. Even in this favorable case for oxy-fuel, the CO_2 avoided cost of post-combustion was found to be slightly lower than the cost of oxy-firing.

The FCC-oxy-fuel field demonstration run, which has been one of the most important achievements of CCP to date, has confirmed the technical viability of retrofitting an FCC unit to enable CO_2 capture through oxy-firing. The demonstration proved that the FCC unit can work steadily in oxy-firing mode and that oxy-firing can enable a higher throughput (up to 3% without any adverse impact on the process) or allow a switch to processing heavier oil feeds while keeping the same product yield.

Considering the above potential advantage and the typical layout issues of a refinery in which oxy-firing would need much less space close to the FCC unit than post-combustion, oxy-firing may compete with post-combustion for this application.

Research Needs and Future Work

One of the main issues identified in the oxy-FCC demonstration run was the corrosion in the recycle CO_2 compressor due to the formation of sulfuric acid mist. Further work is needed to identify the ways to reduce the presence of oxides of sulfur in the flue gas stream and to avoid the corrosion in the recycle compressor, possibly using corrosion-resistant materials.

The economic study confirmed that using a high purity oxygen stream (99.5 mol%), followed by CO_2 purification using a catalytic de-oxygenation unit has lower costs compared to using a lower purity oxygen steam (97.5 mol%) followed by CO_2 purification using cryogenic processing. Future work should include the demonstration of the catalytic de-oxygenation unit using actual raw CO_2 from an oxy-fired FCC.

Since the cost of cryogenic air separation has a big impact on CO_2 avoided costs for oxy-firing, future work should be directed towards lowering the cost of oxygen production. For example, support might be given to development of improved cryogenic air separation and alternative oxygen production technologies such as oxygen transport membranes or regenerable metal oxides.

Finally, the oxy-firing process needs to have a first commercial demonstration at a larger scale, such as 5,000 BPSD or more, to obtain scale-up confidence for future commercialization of the technology.

Refinery – Heaters & Boilers

Heaters and Boilers (H&Bs) are typically grouped into clusters of a few units and located throughout a refinery, discharging CO_2 to atmosphere via several stacks. A typical refinery might have a total of 20-30 such fired heaters.

The CCP approached this scenario considering the capture of CO_2 from two different sized clusters of H&Bs. The analysis showed that post-combustion is the lowest cost technique, and a strong economy of scale was found, principally because for both cases, a single train of post-combustion capture equipment could be used. This demonstrated an important finding that the size of single train post-combustion plants has a strong effect on specific capture and avoided costs. At the scale of the H&B clusters studied, pre-combustion CO_2 capture and oxy-firing were not economic. However, possible limitations in plot space availability may limit the applicability of post-combustion.

Another approach that CCP studied involved capturing CO_2 from all of the H&Bs in the refinery using a pre-combustion capture approach. The analysis showed that pre-combustion technology might be preferred, since costs were close to those from post-combustion and plot space near the fired equipment is not an issue. The system studied utilized an autothermal reformer (ATR) or steam methane reformer (SMR) located outside the fence, and produced hydrogen to be used as both fuel and feedstock for the whole refinery, while capturing CO_2.

The CCP, with John Zink Company, has demonstrated the technical feasibility of oxy-firing process heaters using existing burners with minimal modifications in a pilot test program. For process heaters, the in-leakage of air caused by sub-atmospheric operating pressure may require further costly purification of CO_2 before transportation and storage.

As a possible improvement for pre-combustion technology, the technical viability of Membrane-enhanced Water Gas Shift (MWGS) technology was evaluated. Palladium alloy-based membrane tubes developed by Pall were tested for more than 1,000 hours in operation with actual syngas, and showed good permeability to hydrogen and no significant performance decay. However, the very high estimated cost of the membrane modules and the number of modules required for a large-capacity plant means that this technology is unlikely to be economically attractive for large scale applications in the near future.

Research Needs and Future Work

The economic analysis confirmed that for the distributed refinery H&Bs, pre-combustion was the most economic CO_2 capture technology for refinery-wide application (though post-combustion remains most attractive for small clusters of H&Bs). For refinery-wide application further work should focus on identifying novel process schemes for lowering the cost of low-carbon/hydrogen-rich fuel production from a large-scale SMR or ATR. Further work should also incorporate study of refinery fuel gas systems to ensure that hydrogen fuel can be used safely without extensive modification. Future pre-combustion technology development work should include conducting pilot and demo runs of hydrogen-firing in a refinery heater.

Refinery Steam Methane Reformer (SMR) Hydrogen Plant

Hydrogen is frequently required in the oil refining process, and it is commonly produced by reforming natural gas in a Steam Methane Reformer (SMR). The CO_2 produced by hydrogen production can account for up to 20% of the total CO_2 emitted by a refinery. CCP evaluated CO_2 capture from a modern SMR, in which the reformer furnace was fired by a mixture of natural gas and tail gas from the hydrogen purification Pressure Swing Adsorption (PSA) unit. Post-combustion CO_2 capture was applied to the reforming furnace. This allowed capture of 90% of the CO_2 generated from a flue gas which had a relatively high CO_2 content of 20%. The high CO_2 content arises because the PSA tail gas fed to the furnace contains all of the CO_2 produced from the reforming and water gas shift reactions used to make hydrogen. This case produced the lowest capture cost of any that was studied in the refinery, due both to the quantity of CO_2 capture and the high CO_2 flue gas.

Research Needs and Future Work

CCP studied only one CO_2 capture configuration for the SMR. Lower capture cost may be achieved from the SMR via pre-combustion capture of only the CO_2 produced from the reforming and water gas shift reactions, because the CO_2-rich process gas stream is at high-pressure and contains no oxygen. This opens up a wider choice of capture technologies (physical solvents, solid adsorbents, membranes and cryo-separation). This approach would lower the overall CO_2 capture rate (ca 50%), and the SMR furnace burners may need adaptation to handle the richer fuel, assuming they were designed for use with recycled PSA tail gas. The assessment of this option is worthy of further study.

The other area for future work will be to integrate the pre-combustion scheme for the distributed refinery H&Bs with the hydrogen plant capture. We have seen in Chapters 8 and 9 that the volume of hydrogen-rich fuel needed for the H&Bs case is easily an order of magnitude higher than the pure hydrogen that is needed for various refinery hydro-processing requirements. Hence, one integrated scheme with pre-combustion capture could be used to produce enough hydrogen for high purity application, replacing the SMR, and also for low carbon/hydrogen-rich fuel production. Details are illustrated in chapter 12.

OIL SANDS/HEAVY OIL EXTRACTION SCENARIO

Oxy-firing

This scenario examines CO_2 capture from the production of steam for injection in oil sands reservoirs in Once Through Steam Generators (OTSGs). OTSGs are typically natural gas fired and deployed in clusters of four, with a firing duty of 250 MMBTU/h each. Considering the relatively small scale of operation and the fuel used, the detailed economic analysis reported in this volume showed that post-combustion was the lowest-cost option. This conclusion is different than previously reported CCP results, which were based on out-dated post-combustion capture technology. Detailed analysis in the current stage of CCP suggested that due to the higher relative construction costs of capital-intensive oxy-firing technology (cryogenic air separation units and their auxiliary systems) in the Northern Alberta oil sands producing region, oxy-firing technology was not economically competitive with the the state-of-the art post-combustion capture technology. Oxy-firing might become of interest in case natural gas is replaced by a heavier fuel, such as a fraction of extracted oil, or alternatively if the cost of natural gas in Canada goes up to the level of 8-10 $/GJ, which is a level previously experienced in the 2004-2008 period.

The CCP co-funded a field demonstration of the technical feasibility of oxy-firing a commercial OTSG unit, operated by Cenovus in their extraction field in Christina Lake (Alberta). The test was a technical success, showing that a commercial OTSG can be retrofitted for oxy-firing and can reliably be operated in either air-firing or oxy-firing mode and safely be transitioned between the two operating modes.

Chemical Looping Combustion

This scenario looks promising for potential future application of Chemical Looping Combustion (CLC), a novel technology which has received development support from CCP beginning in Phase 1. A CLC boiler system, consisting of two reactors, requires more plot space and is taller than a conventional boiler. This is a significant constraint for application to refineries, in which CO_2 capture will mostly be a matter of retrofitting existing equipment. In contrast, the extraction of heavy oil by steam injection (Steam Assisted Gravity Drainage or SAGD) is expanding, and new-build facilities will be located in remote areas where plot space constraints are not likely. CLC could therefore become an attractive solution for next generation capture technology in this application, offering a higher capture rate and better efficiency than solvent-based, post-combustion capture, if challenges of equipment complexity and high capital cost can be overcome.

The recent focus for CCP relating to CLC has been development of novel oxygen carrier materials that may be both cheaper and more environmentally-friendly than currently available Nickel-based carriers. Promising Calcium-Manganese carriers have been developed and evaluated at the pilot plant scale.

Activities in this phase of CCP included developing a detailed design for the next scale unit (10 MW) and for a commercial unit (320 MW) for replacement of OTSGs in conventional SAGD clusters. This information enabled a detailed economic evaluation. Quantified results showed that CLC, compared to state-of-the-art post-combustion, looked slightly more expensive in terms of capex, but had advantages in terms of higher efficiency and CO_2 capture rate, and lower operating costs, resulting in CO_2 avoided costs very close to post-combustion. A valid comparison is difficult, considering that CLC is a technology under development and will not be ready for market before the year 2020. The high capital cost of CLC is also a function of the high location cost factors associated with the Canadian oil sands. CLC would be even more competitive with post-combustion technology in a relatively lower capex/higher fuel cost application.

Research Needs and Future Work

While CLC was identified as the lowest-cost option for this scenario, it is still very expensive (>200 US$/tonne) due to the high cost location factors in Northern Alberta. The capex component of the avoided cost is very high compared to other locations. For example, capex for a US Gulf Coast location is 40% lower than Alberta. Hence, further work is needed to focus on capital cost reduction, even if this results in a compromise on energy efficiency, to arrive at an optimal cost.

Since the CLC technology was favored in the economic assessment, to further reduce its costs, it is recommended to support the development of the technology by demonstrating environmentally-friendly and durable oxygen carriers and the CLC technology at a larger scale (10 MW).

For conventional oxy-firing technology, the current economic gap over the post-combustion and CLC technologies is rather large to overcome by small improvements. To reduce the cost of oxygen production, further effort should go towards improved cryogenic air separation and novel oxygen production technologies with the objective of finding cost reductions.

For this scenario, pre-combustion capture technology was not economically competitive, since the hydrogen was produced from a small scale ATR (80 MMSCFD hydrogen) to supply the fuel to four OTSG units. However, producing low-carbon/hydrogen-rich fuel in a larger SMR or ATR to supply the fuel to a large set of OTSGs may significantly reduce the cost of hydrogen due to the economy of scale and reduce the CO_2 avoided costs. This alternative should be investigated in future work.

NATURAL GAS COMBINED CYCLE (NGCC) POWER STATION

CO_2 capture from a NGCC is a difficult application because of the very low concentration of CO_2 in the flue gas (~4% vol). Post-combustion equipment must be larger and energy consumption per tonne of CO_2 may be larger than for capture from the exhaust of other combustion processes. Previously during CCP2, a thorough investigation of state-of-the-art and future pre-combustion technology was carried out in hopes of identifying lower cost alternatives to post-combustion. Several technology providers collaborated with CCP in the frame of the EU-funded project CACHET. The conclusion was that based on today's technology or on developmental technology that might by deployable in the next decade, pre-combustion is unlikely to challenge post-combustion capture from NGCC, due to lower efficiency and high capital cost. More recent evaluations by CCP have confirmed that commercial state-of-the-art post-combustion technology which has been demonstrated at a relatively large scale (though still one order of magnitude smaller than needed for a 400 MW power station) would cause a drop in efficiency of only ~7 absolute points (from 57% to about 50%). Consequently, in the present phase of work reported in this volume, CCP engaged in a screening process of novel post-combustion technologies through studies, lab and pilot testing. These technologies include novel approaches, like non-aqueous solvents (avoiding water evaporation during regeneration), enzyme-accelerated solvents (enabling the use of low-energy solvents whose kinetics would be very slow in normal conditions) and solid adsorbents. Although some of these approaches are promising and deserve further development, CCP has not identified clear winners at the current stage of development or any that may be considered a breakthrough compared to commercial technology.

Sensitivity studies of the effect on capture cost of efficiency and capex for a 400 MW power station illustrate that a 1% change in the net power efficiency has almost the same effect of a 25% change in capex of the CCS system. As a result, it is clear that the post-combustion technology cost is not only impacted by the solvent performance but also by process integration within a power plant for optimal energy efficiency, an exercise requiring advanced engineering skills often not available in small research organizations. Any future novel solvent should therefore be evaluated on a similar basis after incorporating the possible process integration opportunities in a realistic power plant design.

CCP has also continued to track and evaluate emerging oxy-fuel processes for CO_2 capture, finding that these have potential for improvement but also significant challenges to be overcome which will require large programmes of concerted research and development.

Research Needs and Future Work

In the short term, it is evident that the post-combustion solvent-based technology would be the technology of choice for the NGCC application. Future work, therefore, should focus on identifying and supporting the development of novel solvents, adsorbents, membranes and system integration for energy efficiency that would further reduce costs. Novel process schemes in which the CO_2 concentration in the flue gas is increased, such as membrane sweep processes, should be evaluated further to investigate the technical and economic feasibility.

In the long term, consistent with the earliest published findings from the CCP, breakthrough technologies such as oxy-fuel natural gas fired cycles may have a potential as a low-cost CO_2 capture technology and developments in this area should continue to be monitored.

Carbon Dioxide Capture for Storage in Deep Geological Formations, Volume 4
Karl F. Gerdes (Editor)

Section 2

STORAGE MONITORING & VERIFICATION (SMV)

SMV Overview

Carbon Dioxide Capture for Storage in Deep Geological Formations, Volume 4
Karl F. Gerdes (Editor)

Chapter 24

CCP3-STORAGE MONITORING & VERIFICATION (SMV) OVERVIEW

Scott Imbus[1], Kevin Dodds[2], Andreas Busch[3], Mark Chan[4]
[1]Chevron Energy Technology Co., 1500 Louisiana St., Houston, Texas 77002 USA
[2]Formerly BP Corp. North America Inc. (currently Australian National Low Emissions Coal Research & Development, NFF House, 14-16 Brisbane Avenue, Barton 2600, Australia)
[3]Shell Global Solutions, Kessler Park 1, 2288GS Rijswijk, The Netherlands
[4]Suncor, 150 – 6th Avenue S.W., Calgary, Alberta, Canada T2P 3E3

ABSTRACT: The present chapter outlines the CCP3-SMV program objectives, scope, and key results of the projects conducted by its research partners. It is organized by theme: Subsurface Processes, Monitoring and Verification, Optimization, Field Trialing and Contingencies. Chapter 49 outlines the significance of these modeling / simulation, experimental and field studies in the context of furthering the science and technology of secure and efficient CO_2 storage at commercial scale.

KEYWORDS: CO_2 Storage; subsurface processes; monitoring & verification (M&V); optimization; field trialing (deployment); contingencies (intervention)

INTRODUCTION

At the inception of the CCP3-SMV program (2009/10), the team sought to conduct studies that would provide assurance for CO_2 storage at commercial scale using existing oil and gas industry technology, as well as to develop new technology specific to understanding CO_2 behavior and monitoring its fate in the subsurface. This work aimed to build on technologies developed in the CCP1 (2000-05) and CCP2 (2005-09) SMV programs, as well as addressing emerging issues in CO_2 storage. CCP3-SMV technical work ranged from fundamental R&D (modeling and simulation) to lab experiments to field deployments. Mid-way through the program, the team launched the "Contingencies" initiative, which comprises the first systematic approach to detection, characterization and intervention in CO_2 leakage.

An outline of objectives, findings and significance of the CCP3-SMV project slate is outlined in the subsequent sections of this chapter.

OVERVIEW OF CCP3-SMV PROJECTS BY THEME

Subsurface Processes

For the Subsurface Processes theme, project selection was aimed at challenging current assumptions and providing new insights into physico-chemical processes that impact CO_2 injectivity, migration-trapping and long-term containment of CO_2 in storage systems. Additional, related work was conducted as subprojects as part of the Certification Framework project (Chapter 32) under the Optimization theme.

Challenging assumptions of experimental tests to characterize CO_2 behavior in the subsurface

Two projects sought to qualify the relevance of specific experimental tests conducted to 1) determine the relative permeabilities of CO_2 and brine at different saturations as well as to quantify the proportion of CO_2 that is trapped via capillary forces in the rock pore matrix and 2) assess the reproducibility of testing for capillary entry pressure on sample plugs as a means to assess top seal rock resistance to CO_2 migration.

1. The approach and results of "Relative Permeability" (or "Kr" project) experiments conducted by CoreLabs are not presented as a chapter in this volume, although the following is a brief summary: An ambitious CO_2/brine relative permeability and CO_2 capillary trapping experimental program was developed to assess the sensitivity of inferred flow and capacity to experimental conditions, as well as to determine if commercial laboratories are capable of conducting these experiments on a routine basis. The experimental results were complicated by "anomalous" data, particularly a low apparent capillary trapping number relative to analog fluid runs and literature data. Procedural changes, namely increase in monitoring of CO_2 saturation profiles and adjustment of brine flow rate *to control mass transfer*, eventually yielded residual CO_2 values in the expected (i.e., published) ranges. It seems likely that the initially observed low trapping number was due to an opportunity for CO_2 to dissolve in water under low flow conditions. This begs the question, however, given the range of flow rates expected in the near wellbore and near and distal reservoirs, what is an accurate trapping number for CO_2/brine storage systems?
2. The capillary entry (Pc) project by RWTH Aachen University (Chapter 25) sought to ascertain the effect of small rock inhomogeneities on gas breakthrough for a set of fresh and homogeneous Opalinus (NW Switzerland) top seal facies. Two methods were used – the classical "step by step" gas pressure increase method yielding a value for capillary breakthrough pressure, and the imbibition (or residual pressure method) to record "snap off" pressure on the imbibition pathway. Observed gas breakthrough pressures were highly variable with no clear correlation to permeability. This is attributed to minor inhomogeneities in the rocks, either natural facies variability or artificially induced stress fractures (e.g., de-stressing fractures after core retrieval). Breakthrough (drainage) pressures were, as expected, always higher than snap off pressures on the imbibition path. It is concluded that an improved understanding of how natural and induced variability impacts Pc determinations is needed to rely on this experimental assessment of allowable CO_2 column heights or that statistical distributions of results from multiple samples of a given top seal rocks need to be determined to sufficiently quantify the risk of capillary leakage.

Impact of CO_2 stream impurities on reservoir injectivity, flow, trapping and rock mechanical stability

Over the past several years, it has been speculated that the economics of CCS could be improved if CO_2 could be captured at lower purity and the storage reservoir could accept entrained purities without undue impact on injectivity, flow/trapping and loss of rock mechanical strength (impacts on pipelines, compressors and wells were not considered). The impact of impurities on the subsurface storage systems was addressed (Chapter 26) in terms of impact on 1) migration and trapping behavior and thus, needed changes to pressure management, as well as to the areal extent and timeframe for monitoring, and 2) reactivity, particularly dissolution of rock forming minerals and diagenetic cements and precipitation of minerals near the wellbore, potentially affecting injectivity and fluid flow.

1. Expected ranges of impurities (H_2, Ar, O_2, N_2) from various capture processes were used to conduct original PVT experiments for input into reservoir simulations of CO_2+impurities versus pure CO_2 cases using end-member reservoir type (heterogeneity and depth). In general, these non-compressible gas impurities result in more rapid migration of the gas steam towards the top of the reservoir (via stronger buoyant flow due to lower density) and then more rapid lateral migration. Thus the plume footprint can be markedly increased, although trapping of the plume gases occurs more rapidly (owing to increased exposure to water and rock pores). These effects are expected to be most marked in low heterogeneity, shallower reservoirs compared to the pure CO_2 case. The obvious implication is the need to account for impurities when establishing the area of review for monitoring and pressure management.
2. Autoclave experiments with CO_2+O_2 and pure CO_2 demonstrated that mineral attack (particularly carbonates and FeS_2 [pyrite]) are more intense for the former compared to the latter. In the case of the O_2 / FeS_2 reaction, pH rapidly drops (particularly if carbonates are not present to buffer the system) with subsequent enhanced attack of feldspars. Minor precipitation of minerals as Fe oxides occurs under some conditions but it appears minor volumetrically. The implications are that under some conditions, rock geomechanical strength, and thus, injectivity, may be degraded if pyrite is available and/or rock forming minerals are cemented by carbonates. Subsequent precipitation of Fe oxides may impede fluid flow (e.g., further into the reservoir) if it occurs in pore throats as opposed to pore bodies.

Insights into geomechanics of CO_2 storage using analogs

Modeling of geomechanical concepts was afforded by utilizing data from two well-documented fields: 1) a CO_2 tertiary recovery asset in Saskatchewan (for which a comprehensive data set was gathered for the IEA Weyburn storage research project) (Chapter 27), and 2) a natural gas storage (NGS) facility in Victoria Australia ('look-alike' to the well-documented Otway Basin pilot project) (Chapter 28). The former study examines the concept of "geomechanical hysteresis" through primary depletion, waterflooding and CO_2 recovery. The latter exercise was to test the strength of analogy to more common (than CO_2 storage) NGS sites which have undergone repeated geomechanical stress changes through seasonal gas filling and recovery.

1. Coupled reservoir and geomechanical modeling of thermo- and poro-elastic processes document the relative magnitude of probability of tensile and / or shear stress during subsequent recovery stages for the Weyburn Field. Thermoelasticity dominates during the early waterflood stage, which resulted in minor tensile fracturing of the top seal. That the top seal fracture pressure is reduced prior to CO_2 flood, illustrates the importance of assessing geomechanical hysteresis (as would CO_2 storage "only" post-tertiary recovery).
2. Key findings of the Iona NGS case are that significant horizontal stress changes accompany seasonal fluid pressurization, thereby increasing fault stability (for which the opposite would be shown in classical analyses which do not incorporate the complex poro-elastic phenomena), and that surface deformation would be sufficient during peak charging and discharging to detect using InSAR.

Monitoring & Verification (M&V)

Monitoring and verification is a core process for the long term management and confirmation of the safe storage of CO_2. Since a wide variety of M&V deployments were being undertaken globally in storage pilot projects, for CCP3 it was deemed necessary to identify key areas within this theme that would identify common challenges and demonstrate practical processes and design elements necessary for planning monitoring processes. The M&V theme explored this through establishing design principles for integrating multiple data acquisition types in the space-constrained borehole environment through the Modular Borehole Monitoring (MBM) project (Chapter 29), which was validated within the MBM deployment at Citronelle (Chapter 37). In addition a methodology based on Value of Information (VOI) methodology was illustrated by modelling the decision processes necessary to choose monitoring technology through reference to a the choice of seismic at In Salah (Chapter 30). Finally, processes for defining a site-specific monitoring program are reviewed through reference to design principles for assessing a number of in-reservoir and above-reservoir monitoring methods (Chapter 31). The M&V theme also provided context to the various monitoring activities in the Field Trialing theme.

Modular Borehole Monitoring (MBM)

The MBM project conducted by Lawrence Berkeley National Laboratory (LBNL) (Chapter 29) was conceived to establish design principles and components necessary to build a single completion assembly that can be adapted to site specific monitoring requirements in a single borehole. The chapter describes the integration of diverse specific components such as an inflatable packer, downhole fluid sampler (U-Tube), geophones, multiple single temperature and pressure sensors and fiber optic distributed temperature and acoustic sensing technology. These sensing systems, along with power and data transmission, are conveyed by flat-packed electrical, fiber optic and hydraulic control lines installed using deployment backbones such as sucker rods, tubing and umbilical systems. The CCP3-SMV program funded an assessment of potential improvements to the MBM, including enhanced flexibility, integration of technologies, reliability, pressure control and safety. Detailed ready for deployment engineering designs were developed once these components were defined and their geometric arrangement were defined. A deployment opportunity at Citronelle Dome (Field Trialing, Chapter 37) arose midway through this project, which benefited from the improvements developed up to this time. Development work continued post-deployment with incorporation of lessons learned from the Citronelle Dome deployment experience.

Quantifying the value of monitoring

An approach to valuation of CO_2 storage monitoring based on established oil and gas industry value of information (VOI) processes was developed by the University of Texas (UT) (Chapter 30). This approach was tested against historical technology deployment decision making for the In Salah CO_2 storage project (Algeria). Value of information (VOI) analyses seek to 1) identify activities or technologies that could provide information regarding important uncertainties, 2) quantify the accuracy of this information, 3) understand and quantify how this information would alter our decision making, and 4) assess whether the likely improvement in our decision making is worth the additional cost of the information. In the case of CO_2 sequestration the value against which the monitoring is assessed is the probability of leakage more than a specified design threshold and the impact of the technology for reducing that probability. The study suggested as an objective to be minimized, and demonstrated how to calculate, the 1%-Volume-at-Risk (1%-VaR), which is a volume so high that there is only a 1% chance that more than this amount would be leaked. This type of objective is analogous to that used in monitoring the risk of financial organizations.

Optimizing site-specific monitoring technologies

In this US EPA-funded and CCP3-SMV supported study (Chapter 31) led by the University of Texas – Bureau of Economic Geology (UT-BEG), geological features and CO_2 storage project design were taken into account to assess the relative utility of CO_2 storage monitoring technologies, individually and collectively, based on experience and theory. Technologies selected for this study include time-lapse 3D seismic, above-zone pressure, above-zone temperature, and ground-water field geochemistry. For each of these technologies there is a brief overview of some of the common uses of the technology, and then a small subset of the technology uses was selected to forward model the response in a monitoring deployment, varying the elements of the geologic setting to illustrate the sensitivity of the technology to the environment. The key value of time-lapse seismic is for monitoring CO_2 migration through the reservoir, although it can be used to demonstrate absence (below resolution thresholds) of CO_2 leakage through the top seal. Above zone pressure has the potential advantage over seismic in being able to detect brine migration through the top seal (which may precede CO_2 leakage). Thermal monitoring lacks the resolution of seismic and pressure although it can be used to distinguish between brine and CO_2 leakage, particularly through well casing where wells intersect with faults. Geochemical monitoring is primarily a defensive tool to detect migration of fluids into protected groundwater when other monitoring techniques lack resolution or fluid migration occurs along an unmonitored or unexpected path.

Optimization

In addition to strictly assurance-related project work, the current and prior CCP-SMV programs have supported project work related to efficiency of CO_2 storage and economic offsets such as enhanced oil or gas recovery.

Certification Framework (CF)

Since early in CCP2 (2005), CCP has supported the development of the CF (Chapter 32), the core of which is a leak risk assessment framework for predicting effective CO_2 trapping. The CF has been applied to a number of sites, both to candidate and operating storage sites. A time-lapse CF study of the In Salah storage project (operating 2004-12) used the historical knowledge base at three stages of the project to assess key risks and make comparisons to actual outcomes. Over the course of the CF's development, CCP has conducted modeling and simulation on issues identified through case studies or issues that have been identified by other research groups and the media, including potential for CO_2 leakage along reactivated faults, effect of buoyant fluid in vertical fractures and pressure dissipation in fractures.

CO_2 utilization for shale oil and gas recovery

Stanford University summarized existing knowledge of issues surrounding the use of CO_2 for enhanced unconventional gas and oil recovery and outlined gaps, along with recommended approaches to addressing them (Chapter 33). Topics included: injectivity and storage capacity, multiscale shale characterization, enhanced recovery mechanisms, reservoir/fracture dynamics, fault permeability, fluid dynamics, organic matter desorption/adsorption and interactions with resident water and hydrocarbons.

Field Trialing (Deployment)

The CCP3-SMV program experienced considerable success in accessing, operating and deriving useful (and often remarkable) findings from field trialing of M&V technology. This was accomplished by leveraging third party sites, which have available characterization, reservoir modeling and other M&V data for comparison. Table 1 outlines the field deployment sites along with objectives and key findings.

Table 1. Summary of field trialing (deployment) objectives and outcomes.

Chapter	M&V technology	Site (location)	Type	Objective	Outcome
Well Based					
34	Resistivity (through casing)	Otway (Victoria Australia)	Repeat	Determine if post-CO_2 migration signal can be compared to initial (pre-casing) open hole resistivity logs.	Despite complications with the two contrasting logging approaches and presence of methane (in situ and co-injected), achieved a semi-quantitative measurement of CO_2 saturation which shows that migrating CO_2 is contained above the structural spillpoint.
36	Borehole gravity measurement (BHGM)	Cranfield (Mississippi USA)	Repeat	Test the resolution of low cost BHGM tool to detect pre- versus post-CO_2 migration signal in monitoring wells.	Lithologic boundaries are easily distinguished using the BHGM. Detection of CO_2 passage is an interpretational challenge but there is clear evidence of density partitioning within the reservoir, indicating CO_2 saturation at some level.
37	Modular Borehole Monitoring (MBM)	Citronelle Dome (Alabama USA)	Baseline-(Repeat)	Deployment of improved design (flatpack housing, geophone clamps, DTS) at 3rd party site for baseline and repeat comparison.	Off depth perforations diagnosed using onboard DTS. Follow up deployment of DAS coincidently with 2DVSP indicates comparable resolution and greater flexibility of the latter compared to the former.
Seismic					
35	3D surface vs. 3DVSP	Otway (Victoria Australia)	Repeat	Comparison of the two seismic methods in terms of repeatability, cost and field operations at the Otway pilot site, which has near-surface karst.	3DVSP has superior reservoir imaging capability compared to surface seismic although amplitude migration away from the wellbore must be considered. For this site, permanent buried geophones would combine the repeatability and cost benefits of surface and 3DVSP.

Table 1 continued.

Chapter	M&V technology	Site (location)	Type	Objective	Outcome
Seismic					
NA	Microseismic	Undisclosed	Baseline-(Repeat)	Site identified for deployment, taking advantage of other M&V deployments to analyze results.	The operator would not consent to deployment given potential public sensitivities of recording "earthquakes" associated with the operation.
Remote Sensing					
41	InSAR	Decatur (Illinois USA)	Repeat	Test ability to detect vertical surface uplift with CO_2 injection in complex temperate terrain and mixed use site using stacked radar images tied to existing and installed reflectors.	Essentially no vertical movement was detected which was consistent with BHP data during operations (possibility that permeability of the injection formation was higher than anticipated, thus limiting uplift).
Other					
40	Electomagnetics (EM)	Aquistore (Saskatchewan Canada)	Baseline (-Repeat)	Test of a new BSEM (borehole to surface) configuration entailing introduction of current via reservoir depth electrodes and with transmission through well casing and detection at the surface.	The EM data are consistent with conductivity logs from the newly constructed injection and observation wells. Modeling indicates that the small planned injection of CO_2 (in the 3000m deep reservoir) would be detectable using BSEM method.
38	Soil gas	W. Hastings (Texas USA)	Baseline	Approach to rapid, field-based soil gas analysis through sensing relative humidity, temperature. Pressure, CO_2, O_2 and CH_4 and calculating N_2 as the difference. This would obviate the need for GC instrumentation on site.	Comparison between this method and well-established GC methodology were poor. This may have resulted from less than accurate manufacturer's representation of sensor accuracy and/or complications of field conditions.

Otway CO2CRC: through casing resistivity and RST comparison

Time-lapse wireline logging methods, that are standard practice for petroleum field development in order to locate and manage stranded resources, are routinely applied to monitoring of CO_2 and play a vital role in monitoring injection profiles and detecting gas or water fronts. In this case, fluid saturation logs have been shown as essential in the monitoring portfolio at CO_2 storage sites for regulatory conformance. An opportunity arose at the CO2CRC Otway well to compare the time-lapse response of the traditional open-hole resistivity log with a through-casing resistivity log as a complement to the use of a time-lapse saturation log (Chapter 34). A quantitative result was not achieved due to a variety of factors related to timing of logging runs, low salinity of the formation water, high mud filtrate invasion, the presence of CO_2 inside the borehole and existing residual hydrocarbons in the reservoir. Nevertheless, the more reliable log outputs were evaluated and corrected accordingly in order to produce a semi-quantitative evaluation of saturation post-injection. The results were used to verify that the CO_2 plume is contained above the structural spill point of the storage complex.

Otway CO2CRC: 3DVSP and 3D surface seismic comparison

The objectives of the comprehensive seismic monitoring program at the Otway injection project were expanded by CCP3-SMV to test and benchmark 4D VSP and surface seismic for CO_2 sequestration monitoring according to repeatability, cost and field operation (Chapter 35). Also, the project compared observed differences between the baseline and the later monitoring surveys against the predicted time-lapse signal. Amongst more detailed conclusions on survey geometrical differences and advantages, the project concluded that a 4D VSP is probably a better tool for imaging reservoir changes, while 4D surface seismic is essential for monitoring changes in the overburden.

Cranfield SECARB: Time-lapse borehole gravity evaluation

As part of the Southeast Partnership (SECARB) "early" test of CO_2 injection at Cranfield, Mississippi, time-lapse borehole gravity measurements (BHGM) were collected within two multi-use monitoring wells (Chapter 36). The goals of these BHGM tests were to understand the operational and design aspects of data acquisition, assess the ability of the tool to detect geology and injected CO_2, and to make recommendations for improved future deployment. The time-lapse response from CO_2 injection were less defined than the changes associated with lithological change due to the limited volume of the CO_2. However using this approach, it was observed that the apparent density change from measured gravity data after noise adjustment is largely consistent with the locations predicted from the reservoir simulations. Therefore, while improvements can be identified for future surveys of this type, it is clear from this test project, that the time-lapse borehole gravity data for sequestration monitoring at Cranfield contain meaningful qualitative information about CO_2 movement

Citronelle SECARB: MBM deployment and DAS evaluation

While work was still ongoing within the design phase of the MBM program (Chapter 29), the SECARB Anthropogenic Test in Citronelle, Alabama, USA was identified as a deployment site for the engineered monitoring systems (Chapter 37). The initial step in designing the Citronelle MBM system was to select from the various monitoring tools available to include technologies that were considered essential to the project objectives. Monitoring methods selected included U-tube geochemical sampling, discrete quartz pressure and temperature gauges, an integrated fibre-optic bundle consisting of distributed temperature and heat-pulse sensing, and a sparse string of conventional 3C-geophones. While not originally planned within the initial MBM work scope, the fibre-optic cable was able to also be used for the emergent technology of distributed acoustic sensing (DAS). Results and lessons learned from the Citronelle MBM deployment are addressed along with an example of data collected.

West Hastings BEG: Soil Gas monitoring

A process-based approach to soil gas monitoring at geologic carbon storage sites may provide an accurate, simple, and cost-effective alternative to other soil gas methods that require complex background data collection and analysis (Chapter 38). The technology is lacking for economical, field-deployable, smart data collection of all gas parameters important for a process-based analysis, especially for N_2 concentration. This lack limits the ability to implement this leak monitoring approach on an industrial scale. This project was used to evaluate commercially available, automated sensors that measure CO_2, CH_4, O_2, temperature, relative humidity (RH), and pressure. A method for deriving N_2 by mass balance was developed and field tested at a typical CO_2 enhanced-oil recovery site. The accuracy and precision of the data collected were compared against the current method using gas sampling and chromatography. The results indicate a site-specific bias in the data comprising significant NDIR (non-dispersive infrared) sensor error, and a lesser degree of error from galvanic cell technology. This identifies challenges to improvement of measurement technology performance.

Aquistore PTRC: EM modelling and hardware design downhole EM source

To assess the potential application of electromagnetic monitoring at Aquistore, LBNL and Multi-Phase Technologies collaborated on a two-part study including (1) numerical forward modelling of a time-lapse, controlled-source electromagnetic (CSEM) survey, and (2) an initial engineering study of instrumentation and proposed design for a downhole electric dipole source and electrode sensors (Chapter 39). The study evaluated the optimization of the source location within the borehole. The design of the electromagnetic (EM) transmitter for a borehole dipole source and its ability to provide sufficient current, voltage, and power is also described. There are three major elements to the system design 1) the surface transmitter that provides a carefully controlled current source, 2) the subsurface electrodes and supporting casing, and 3) wires and cables to connect the electrodes to the subsurface.

Aquistore PTRC: Deep EM monitoring using casing connected surface EM source

This project conducted the first ever survey with a new electromagnetic source configuration that takes advantage of the high conductivity path provided by a conventional steel well casing to channel surface generated electric current to reservoir depth (Chapter 40). In contrast to the borehole located source described in Chapter 39, this novel configuration for borehole to surface electromagnetic surveying (BSEM) evaluates deep reservoir electrical properties without needing to put hardware within the borehole, yet still obtains the same results as a borehole located source or receivers. Additionally, a new type of surface electric field sensor (eCube) was deployed to acquire the EM data. The eCube sensor couples capacitively to the electric potential in the ground, removing the need for electrochemical interaction (i.e. ionic exchange) with the ground. This brings considerable operational flexibility and efficiency. The pre-survey modelling, which was in good agreement with conductivity logs for the injection and observation wells, showed that the CO_2 plume would be detectable using this system. The subsurface current flow was detected with very high coherence out to 800 m from the well bore in a reservoir at ~10,000 ft (3030 m). This is considered the deepest confirmed EM imaging of a reservoir. Consequently, a measurement system capable of detecting the projected CO_2 plume signal over a substantial distance from the well, without need to instrument the well has been demonstrated. A follow-up post-injection time-lapse survey is planned within CCP4-SMV.

Decatur MGSC: InSar evaluation

The successful measurement of ground deformation using InSAR at the In Salah CCS project in Algeria spawned a follow-on study at the Illinois Basin Decatur Project to assess the viability of applying advanced InSAR techniques in a temperate climate setting (Chapter 41). While the arid desert environment of the In Salah project was ideal for InSAR, as there were no issues with vegetation or snow cover, the setting is not typical of temperate climates, in which many CCS projects in North America will be sited. The CCP3 consortium funded the TRE Canada-managed InSAR monitoring component of the SMV activities at Decatur to evaluate the density and distribution of measurement points that would be obtained in this setting as well as the impact on the estimation of ground movement. The feasibility analysis led to the design and installation of an artificial reflector network consisting of 21 reflectors placed between the injection well and an observation well. The results of the InSAR monitoring at Decatur indicate that a satisfactory density and distribution of measurement points can be obtained in temperate settings to effectively monitor surface deformation, although actual deformation at the time of this work was insufficient to attribute to CO_2 migration in the reservoir.

Contingencies (Intervention)

Assurance of CO_2 containment is key to gaining stakeholder trust and acceptance, and ultimately, regulatory approval for operating large scale CO_2 storage projects. To the technical experts, comprehensive site characterization, fluid dynamics simulation, and surveillance is sufficient for confidence in such operations. Given unexpected outcomes of some storage pilots and commercial operations, the CCP3-SMV team felt that an executable plan to intervene in such outcome would go a long way in assuring key stakeholders, so the CCP3-SMV team launched the “Contingencies” initiative. This is the first comprehensive effort to develop a systematic approach to addressing detection, characterization and intervention methods for unexpected CO_2 leakage from the storage reservoir. The program includes modeling of storage projects to assess the detectability and characterization of leakage, passive (stop injection) and active (e.g., hydraulic controls, sealant injection) controls and planning to test intervention at the bench-field scale.

Intervention modeling, simulation and experiments

A summary of the Stanford work in leakage detection, characterization and intervention appears in Chapter 42. Figure 1 outlines a recommended workflow and identifies tools associated with each phase.

Individual chapters from the Stanford University studies address:

1. Leakage detection and characterization (Chapter 42) – Detection of leakage using a seismic program designed to monitor CO_2 plume migration through the reservoir system would probably be inadequate unless focused on the area of a suspected leak. Above zone pressure monitoring could be quite sensitive. Leakage characterization (i.e., flow path, rate) can be constrained by fault models (based on dimensions and damage zone models) and speculated effective permeabilities. Parametric studies show that the permeability contrast between the reservoir below and above the top seal is more important than conduit (fault) permeability which in turn is more important than other factors.

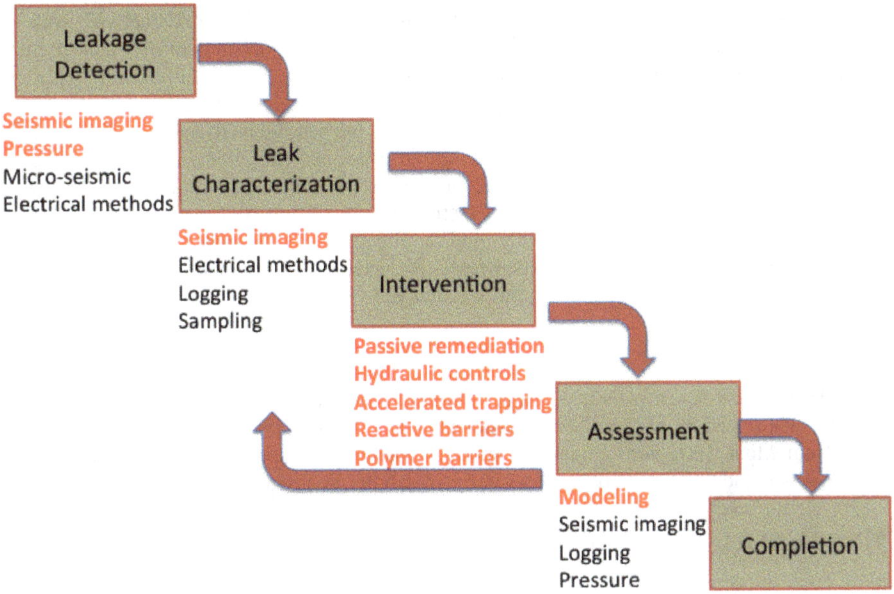

Figure 1. Recommendations for leakage intervention. The items shown in red were conducted as part of the CCP3 Phase I (Agarwal et al., Chapter 42).

2. Remediation (Intervention) (Chapter 43) - Passive control (stopping injection) reduces flow by at least one order of magnitude. Active controls, such as injecting water into the reservoir above the fault and/or extracting water from the injection reservoir near the fault can stop leakage altogether although this operation would need to be continued for an extended period of time.
3. Sealants (Chapter 44) – A commercial organic cross-linked polymer (OCP) used for sealing wells was evaluated for its setting time (as delivered vs. modified) and experimentally tested for its performance for reducing conduit permeability. In addition, a model delivery system for contacting the sealant with a 3 dimensional conduit system was developed. The sealant was found to be very effective at reducing conduit permeability in the laboratory but chemical modification would be needed to achieve useful setting times even with an effective delivery system.
4. Reactive barriers (Chapter 45) – A reactive transport model (RTM) was deployed to assess the leakage sealing capacity of alkaline silica floods. Such floods may be effective if the pH is in a range (i.e., that there is effective mixing between native brine and lower pH brine generated by CO_2 dissolution) that can trigger precipitation of amorphous silica gel. This agent would overcome the rapid setting time problem (i.e., before the agent can be delivered to the affected conduit system) with conventional sealants but is limited by subsurface fluid chemistry and possibly the stability of the precipitated amorphous silica.

Fracture sealing experiment at Mont Terri Underground Laboratory (MT-UGL)

CCP3-SMV contracted GeoScience Ltd. to lead a feasibility study for conducting a "fracture sealing" experiment in the Opalinus Shale at MT-UGL. Chapters 46 and 47 detail the feasibility study and detailed design (respectively) of the experiment at bench-field scale. The experiment entails generating isolated hydraulic fractures from a single packed off "active" well with subsequent drilling of "passive" wells to intersect the fractures, establishment of water circulation followed by injection of up to three different sealants (candidates include Halliburton's H2Zero^{TM}, Montana State University's biofilm (Chapter 48) and LANL's pH triggered smart gel). The experiment would be monitored via pressure of the active and passive wells and near field acoustics. Once the experiment is concluded, the system will be overcored (or cores will be taken in several key locations) to assess the aperture width sealing ability of each sealant. The design document is sufficient to specify fabrication, installation testing and operation of the experiment. However, additional work on the packer assembly and a site test on the near field acoustics were recommended prior to launching the experiment. The experiment could be done by CCP4-SMV starting in 2016 if an appropriate consortium is assembled.

Biomineralization Sealing in Fractured Shale at the Mont Terri Underground Research Facility

CCP3-SMV funded a bench scale test of a biomineralization procedure developed by Montana State University, which had been tested on sandstones. The work described here used Opalinus Shale samples from Mont Terri with extant and induced fractures (Chapter 48). The biomineralization procedure was found to be effective at reducing permeability by up to four orders of magnitude.

SUMMARY AND CONCLUSIONS

The CCP3-SMV program, through the dedicated efforts of team members and reserch partners as well as third party field operators, made important progress in improving confidence in the safety and efficacy of CO_2 storage. This program was built on findings in prior CCP-SMV programs, which in turn will inform the CCP4-SMV program (refer to Chapter 49 for more forward looking information).

Carbon Dioxide Capture for Storage in Deep Geological Formations, Volume 4
Karl F. Gerdes (Editor)

Section 2

STORAGE MONITORING & VERIFICATION (SMV)

Subsurface Processes

Carbon Dioxide Capture for Storage in Deep Geological Formations, Volume 4
Karl F. Gerdes (Editor)

Chapter 25

LABORATORY TESTING PROCEDURE FOR CO_2 CAPILLARY ENTRY PRESSURES ON CAPROCKS

Alexandra Amann-Hildenbrand[1], B.M. Krooss[1], P. Bertier[2], and A. Busch[3]
[1]RWTH Aachen University, Energy & Mineral Resources Group, Institute of Geology and Geochemistry of Petroleum and Coal, Lochnerstr. 4-20, 52056 Aachen, Germany
[2]Clay and Interface Mineralogy, RWTH Aachen University, Bunsenstr. 8, 52072 Aachen, Germany
[3]Shell Global Solutions International, Kessler Park 1, 2288GS Rijswijk, The Netherlands

ABSTRACT: The results presented here are the outcome of a research study comparing two different laboratory testing procedures for the determination of CO_2 capillary pressures of low permeable rock types with permeabilities ranging between $1 \cdot 10^{-21}$ and $6 \cdot 10^{-21} m^2$. A well-characterized core section of the Opalinus Clay (Mont Terri) has been sampled for this study. Laboratory test conditions corresponded to a depth of approximately 1500m, representative of a typical CO_2 storage scenario. The experiments were performed on fully water-saturated samples plugs.

Two gas breakthrough measurement techniques were tested: 1) the classically used step-by-step *gas pressure increase method* yielding a value for the capillary breakthrough pressure, and 2) the *imbibition or residual pressure method* aiming at the detection of the snap-off pressure on the imbibition path. First, imbibition measurements were conducted, followed by the classical step-by-step approach.

Breakthrough pressures of the different sample plugs were strongly variable. The observed gas breakthrough pressures for the core plugs ranged from 3.4 to 12.3MPa for N_2 and from 14 to 17.5MPa for CO_2. This is attributed to the likely occurrence of artificially induced micro fissures or natural inhomogeneities and the poroelastic response to changes in effective stress (development of dilatant pathways). As expected, breakthrough (drainage) pressures were always higher than the snap-off pressures on the imbibition path (ratio 1.6:4). Intrinsic water permeabilities remained unaltered in the course of the gas breakthrough experiments. A simple relationship to more readily obtained parameters, such as permeability, could not be obtained.

The results have implications for the prediction of the seal capacity of CO_2 storage sites. Determination of capillary pressures from a single plug seems insufficient - likely resulting in over/underestimation of the true value. More research is recommended to understand the governing transport and leakage processes and their lateral variability in caprocks.

"Are we using appropriate sample preparation techniques?" "Is it reasonable to use data gained on material after stress-unloading (after coring)?" "Do we use correct wettability data to convert to CO_2/brine conditions?" These are the open questions that need more research focus in order to explain the broad range of critical capillary pressures observed in this study, using drilled twin plugs from a core section only few meters long.

KEYWORDS: gas breakthrough; critical capillary pressure; breakthrough; snap-off; mercury porosimetry (MIP); permeability; porosity; experiments; Opalinus clay

INTRODUCTION

Knowledge of the critical capillary breakthrough pressure is highly important for understanding the performance of geologic systems with regard to fluid confinement. Natural hydrocarbon traps exist due to the abundance of low-permeability capillary sealing layers above reservoirs [1]. In the petroleum industry the amount of hydrocarbons (HC), or specifically the HC-column height, may be estimated based on capillary pressure analyses [2]. The same approach is used when estimating CO_2 storage capacity. Storage sites require caprocks characterized by high capillary breakthrough pressure values to prevent CO_2 from escaping to shallower parts of the earth's crust or the atmosphere [3, 4, 5]. Another risk is associated with mechanical failure of the sealing rock, which would result from high overpressure in the absence of capillary failure.

This study is motivated by the need to ensure that measurement techniques are reproducible not only for non-reactive reference gases, such as nitrogen, but also for reactive gases, such as CO_2. Further, it is industry practice to use mercury porosimetry as an analogue, which requires the extrapolation of the Hg/air petrophysical parameters to the desired reservoir fluids (e.g. CO_2/brine). For comparability of the different measuring techniques, we determined the different critical capillary pressures on a 4m long core section with low spatial variability in mineralogical composition and petrophysical properties. By determining critical capillary pressures for CO_2/brine, N_2/brine and Hg/air, we obtained a comprehensive dataset which was analysed for consistency and to prove (or disprove) if the extrapolation to other systems is feasible.

In the past, different procedures for the determination of the critical capillary pressure were tested and applied:

1. Different studies followed the drainage path by slowly increasing the pressure on one side of the sample until the detection of gas breakthrough (outflow) on the low pressure side [9-12]. This approach can either be achieved by slow continuous gas injection rates controlled by a pump or by a step-wise pressure increase. As gas flow rates right after breakthrough are extremely low the detection of gas bubbles or gas flow rates may become difficult. In general, there exists the risk of "overshooting" the critical pressure when gas pressure is increased too quickly.
2. The imbibition or residual pressure method is conducted by instantaneously creating an initially high gas pressure difference across the sample [13-15]. The subsequent pressure decay is interpreted with respect to flow regime, effective gas permeability and capillary snap-off pressure. In recent years, it has been proven that the snap-off pressure is lower than the breakthrough pressure (approach 1), but this method is somewhat quicker [16-18]. This is due to the creation of an initially high pressure difference which causes more pathways to be drained, and results in higher gas flow rates. The process of gas breakthrough and subsequent imbibition is therefore speeded up, and snap-off pressure is characterized.
3. Egermann *et al.* [19] presented a new procedure, the dynamic approach. Here, the capillary entry pressure is determined by comparing the water outflow rates before and after gas entry. After gas has entered the pore system, the driving pressure gradient for the water phase is reduced by the capillary pressure (i.e. the water flow rate decreases). Assuming that the intrinsic permeability is constant one can solve the equations for the capillary pressure. For this method the pressure difference has to be carefully chosen and sample length must be sufficient in order to measure the water flow rate after gas entry, but before breakthrough.
4. The racking method is similar to the dynamic approach, but instead of a pressure increase on the high pressure side, the pressure is reduced on the low pressure side by extracting water at a constant flow rate [20, 21]. The monitored pressure decay evolution on the low pressure side is a function of water outflow before and after gas entry.

In this study we used the step-by-step pressure increase, method (1) and the imbibition method (2), yielding the capillary breakthrough pressure on the drainage path and the snap-off pressure on the imbibition path. The breakthrough pressure terminates the pressure build-up, defining a maximum gas column height. The snap-off pressure regulates the re-sealing of the system resulting in a residual (minimum) amount of trapped gas. The different capillary pressures (entry, breakthrough, snap-off) as well as their respective relation in a reservoir engineering perspective are further discussed in Busch and Amann-Hildenbrand [17].

The experimental program was subdivided into several work packages. For reference, experiments were first performed with a non-reactive and less soluble gas (N_2). Thereafter, several CO_2 experiments were conducted. We started with the imbibition experiments, which readily provide a value for the snap-off pressure, a "good starting point" for the incremental pressure steps of the drainage experiments. Each gas breakthrough experiment used a new sample plug in order to provide "undisturbed" starting conditions and to test for sample heterogeneity. This way, a pre-test initial gas flow-induced deterioration of the pore system was avoided, which might be caused by either hydrofracturing [12, 22], CO_2/mineral-dissolution effects or simply by the presence of a residual gas conducting backbone within the matrix [23].

The aim of this research study was to obtain/identify:

- The difference between breakthrough and snap-off pressure.
- A correlation between critical pressures and other petrophysical parameters (porosity, permeability).
- Differences in properties before and after gas breakthrough experiments (e.g. water permeability).
- Different flow regimes after gas breakthrough (viscous vs. diffusive flow).

SAMPLE MATERIAL AND SAMPLE CHARACTERISATION

An approximately four meter long core section (BHG-D1_section 4.6-8.6m) from the Mont Terri Test site was provided by NAGRA for this research project.

Table 1. List of sample numbers and their dimensions per experimental method. Material for sample Hg_02* was taken from plug_01 after the flow experiment.

Gas breakthrough and permeability experiments			Hg-porosimetry		Basic Geochemical, mineralogical, petrophysical analysis	
sample no.	length (mm)	diameter (mm)	sample no.	fragments	sample no.	particle size
plug_01	11.55	37.5	Hg_01	7-10 mm	Powder_1	
plug_02	10.55	37.6	Hg_02*	plug_01	Powder_2	
plug_03	14.13	37.65	Hg_03		Powder_3	
plug_04	13.7	37.8	Hg_04		Powder_4	
plug_05	10.25	37.7	Hg_05		Powder_5	~5µm
plug_06	22.4	37.75	Hg_06		Powder_6	
plug_07	17.8	37.7	Hg_07		Powder_7	
plug_08	11.84	37.8	Hg_08	7-10mm	Powder_8	
plug_09	9.4	37.9	Hg_09		Powder_9	
plug_10	6.15	37.9	Hg_10			
plug_11	22.9	37.9	Hg_11			
plug_12	14.0	37.9	Hg_12			
plug_13	10.1	37.85	Hg_13			
plug_14	17.4	37.85				
plug_15	13.85	37.85				

Based on the CT scans we first selected the apparently most homogeneous sections, which were most promising for obtaining good plugs. Plug drilling was performed under axial load and using compressed air for cooling. Plugs and residual material were carefully re-sealed in aluminium foil and PE-foil before further plug preparation and weighing. In order to achieve plane-parallel surfaces plugs were ground carefully in the dry state. Other techniques proved to be less successful. In total, 15 core plugs were drilled. The dimensions of the plugs used for permeability and gas breakthrough tests are listed in Table 1. Also indicated in this Table are the aliquots and sister samples used for mercury injection porosimetry and geochemical, mineralogical and petrophysical analyses.

Geochemical and mineralogical characterisation

The residual material was used for a comprehensive mineralogical and petrophyisical characterisation. Powders were prepared using a McCrone Micronising Mill yielding material with a narrow particle size distribution of about 5μm. In order to minimise strain damage of the samples, particle size was reduced in multiple grinding steps and in ethanol to avoid heat damage. Fragments larger than 7-10mm were kept for mercury porosimetry and He-pycnometry measurements. For all measurements sample material was dried at 105°C in a vacuum oven until constant weight was observed.

For the X-ray diffraction measurements (Table 2) an internal standard (TiO_2, 10 wt.%) was added before milling. All reported mineral compositions relate to the crystalline content of the analysed samples. Mineral quantification was performed on diffraction patterns from random powder specimens, which were prepared by means of a side filling method to minimise preferential orientation. The measurements were done on a Bruker D8 diffractometer using CuK α-radiation produced at 40kV and 40mA. The rotating sample holder was illuminated through a variable divergence slit. The diffracted beam was measured with an energy dispersive detector. Counting time was set to 3 seconds and a step size of 0.02°2θ was used. Diffractograms were recorded from 2° to 92°2θ. Quantitative phase analysis was performed by Rietveld refinement. BGMN software was used, with customised clay mineral structure models [24]. The precision of these measurements, from repetitions, is better than 0.1 m% for phases of which the content is above 2%. The accuracy cannot be determined because of the lack of pure clay mineral standards, but is estimated to be better than 10% (relative).

Table 2. Mineralogical composition determined using X-ray diffraction.

Phase	Quartz	Albite	K-feldspar	Kaolinite	Mus+Illite	Chlorite	Calcite	Dolomite	Siderite	Pyrite	Sphalerite	Anhydrite	Gypsum
Powder_1	19.6	1.9	2.2	21.6	43.8	3.1	3.5	0.8	1.1	0.7	0.0	0.9	0.7
Powder_2	16.2	1.8	2.3	21.2	44.1	8.0	2.9	0.5	0.8	0.8	0.0	0.6	0.7
Powder_3	16.5	1.4	2.0	23.2	46.5	3.7	2.9	0.6	1.2	0.8	0.0	0.6	0.6
Powder_4	15.3	1.4	2.3	20.3	50.4	3.8	2.5	0.5	1.3	0.7	0.0	1.0	0.4
Powder_5	12.5	1.1	1.6	22.2	50.8	4.0	4.6	0.3	0.5	0.7	0.0	1.0	0.8
Powder_6	12.4	1.7	2.1	21.2	47.1	6.6	4.6	0.3	0.9	1.0	0.0	1.1	1.0
Powder_7	12.8	1.6	2.5	22.2	45.1	4.4	7.2	0.3	1.1	1.2	0.0	1.0	0.7
Powder_8	11.6	0.8	1.6	20.5	52.5	3.9	5.3	0.3	0.5	1.5	0.0	0.9	0.6
Powder_9	12.1	0.8	1.6	21.3	52.6	4.4	3.4	0.4	0.3	1.7	0.1	1.0	0.4

Major elemental compositions were determined by energy dispersive X-ray fluorescence spectrometry (XRF, Table 3) with a Spectro XLab2000 spectrometer, equipped with a P_d-tube and Co, Ti and Al as secondary targets. The spectrometer was operated at acceleration voltages between

15 and 53kV and currents between 1.5 and 12.0mA. Major elements were analysed on fused discs (diluted 1:10 with a Li-tetraborate/Li-metaborate mixture, FXX65, Fluxana, Kleve, Germany). Data computation was performed using a fundamental parameter procedure. Loss on ignition was determined by heating the powdered sample to 1000°C for 120min. Samples were dried at 105°C for over 24h prior to determination of loss on ignition. Precisions and accuracy, as determined from repeated measurements on standards, are better than 0.5m%.

Table 3. Major element analysis using X-ray fluorescence spectrometry.

Sample No.	SiO_2 (%)	Fe_2O_3(T) (%)	TiO_2 (%)	Al_2O_3 (%)	MnO (%)	MgO (%)	CaO (%)	Na_2O (%)	K_2O (%)	P_2O_5 (%)	SO_3 (%)	Loi (%)	Total (%)
powder_1	53.38	5.57	0.93	18.50	0.04	1.98	3.96	0.57	2.87	0.21	0.20	10.46	98.66
powder_2	52.14	5.81	1.00	20.09	0.05	2.05	3.43	0.53	3.06	0.19	0.17	10.61	99.13
powder_3	51.17	6.01	0.98	20.02	0.05	2.10	3.43	0.53	2.97	0.18	0.17	10.78	98.40
powder_4	51.55	6.42	1.00	19.85	0.05	2.13	3.50	0.50	2.97	0.17	0.19	11.25	99.59
powder_5	48.03	5.86	0.97	20.61	0.06	2.13	4.81	0.50	3.14	0.15	0.15	12.65	99.05
powder_6	46.79	6.33	0.93	20.23	0.06	2.23	4.87	0.55	3.10	0.18	0.17	12.98	98.41
powder_7	45.96	5.71	0.90	19.44	0.06	2.21	6.66	0.51	2.97	0.38	0.12	13.40	98.32
powder_8	46.45	6.25	0.96	20.33	0.06	2.18	5.59	0.62	3.16	0.24	0.14	12.80	98.77
powder_9	48.23	6.61	0.95	20.66	0.05	2.20	4.03	0.55	3.19	0.17	0.19	11.79	98.62

Cation exchange capacities were analysed by means of the copper(II)-triethylenetetraamine, [Cu(Trien)]2+, method (Table 4). A 0.02M solution of copper(II)-triethylenetetraamine was prepared by mixing 0.1mol triethylenetetraamine (Trien) and 0.1mol copper(II)-sulfate pentahydrate in 100mL deionised water and diluting appropriately. 200mg of sample were added to 20mL of exchange solution. The suspensions were dispersed in an overhead shaker for more than 1h. The supernatant solutions were separated using syringe filters and analysed for their copper(II)-triethylenetetraamine concentration by spectrophotometry (Lambda 11, Perkin Elmer). The adsorption of the supernatant was measured at a wavelength of 576 nm and converted to concentration using a 7 point calibration series (0 – 0.02M). The measured CEC values were corrected for differences between labs and methods by means of a calibration series consisting of 7 standard clays (kindly provided by Dr. S. Kaufhold, BGR, Germany) with CEC values ranging from 30 to 2000meq/kg. After this correction the accuracy of the measurements is better than 10%. Precision was verified by repeated analyses of the same sample and is better than 2%.

Table 4. Cation exchange capacity (CEC).

Sample ID	Corrected CEC (meq/kg)
powder_1	129
powder_2	150
powder_3	147
powder_4	144
powder_5	151
powder_6	152
powder_7	145
powder_8	154
powder_9	157

TOC data were measured with a LECO RC-412 Multiphase Carbon/Hydrogen/Moisture Determinator (Table 5). This instrument operates in a non-isothermal mode with continuous recording of the CO_2 release during oxidation, which permits individual determination of inorganic and organic carbon in a single analytical run and does not require removal of carbonates by acid treatment for TOC measurement. The technique is based on the different decomposition temperatures of organic and inorganic compounds. The total organic carbon content is determined at T< 500 °C and the inorganic carbon content at T> 500°C. Carbon dioxide produced by combustion is quantified by IR absorption. The accuracy and precision of the analyses are better than 4% relative.

Table 5. Total organic and total inorganic carbon contents.

Sample no.	TOC (<500˚C)(%)	TIC (>500˚C)(%)	TC (%)	weight (g)
powder_1	1.2	0.9	2.1	0.1006
powder_2	1.0	0.9	1.9	0.1004
powder_3	1.0	0.9	1.9	0.1001
powder_4	1.3	1.0	2.3	0.1006
powder_5	1.5	1.0	2.5	0.1000
powder_6	1.4	1.1	2.5	0.1008
powder_7	1.2	1.4	2.7	0.1005
powder_8	1.4	1.2	2.6	0.1007
powder_9	1.1	0.9	2.0	0.1001

For the MIP measurements a Micromeritics AutoPore IV 9500 porosimeter was used, yielding information about porosity, density, pore size distribution and critical capillary pressure. Additionally, porosity was determined from the as-received water content, measured on residual material after opening and drilling of the sealed core section.

Basic petrophysical characterisation

In this study, matrix density and porosity was determined by various methods. When referring to one specific measuring technique, we can conclude that the spatial variability of properties in the 4m long core section is small. However, it is evident that values derived from the different measurement methods, like porosity, can be significantly different, because of the nature of the analysis fluid (e.g. Hg versus He) and to the different accessibility of the pore space.

The average matrix density is 2.7±0.03 g/cm³ (He-pycnometry, using a Micromeritics AccuPyc 1330 on powdered material) or 2.6±0.04 g/cm³ (Hg-porosimetry). Densities derived from He-pycnometry are larger likely due to the lower pore accessibility of mercury (r_{pore}>2nm).

Porosities derived from mercury porosimetry (ϕHg) and the as-received water content (ϕWC) range from 8 and 13% (Figure 1). ϕWC was determined on the drilling fragments from the weight loss after vacuum oven drying at 105°C until constant weight is observed, assuming an average water and mineral matrix density of 1 and 2.7g/cm³, respectively. In most cases water content porosities are somewhat higher (average ϕWC = 12.5±1.8%). This is expected, as water is a wetting fluid occupying the entire pore network. We interpret those sections with somewhat lower water content porosities to have partially dried out during sampling, drilling and/or transportation (#7, #11, #12). The average porosity from mercury intrusion experiments is 10.4±0.6% (after surface correction).

The pore size distribution was derived from Hg-porosimetry experiments (Figure 2). Mercury injection, assuming cylindrical pores, a contact angle of 140° and an interfacial tension of 485 mN/m [25], reveals most prominent pores to range between 3 and 5nm.

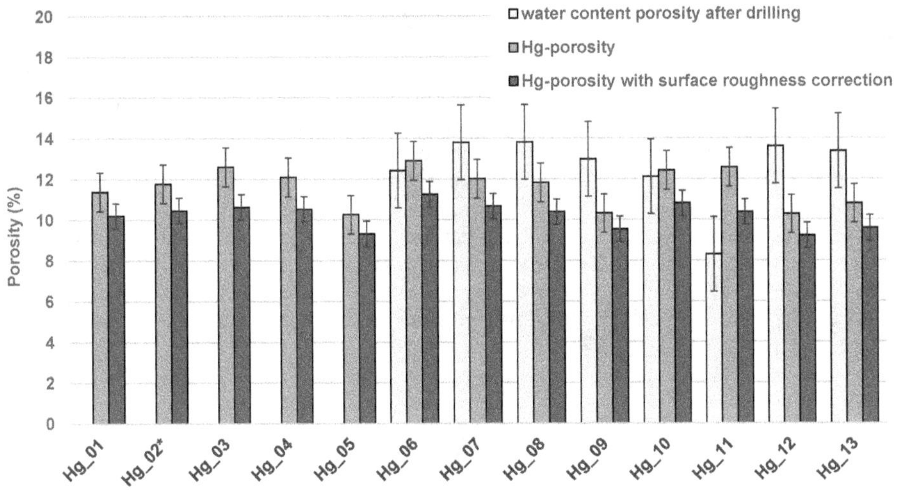

Figure 1. Porosity determined from the prevailing water content (as received) and mercury porosimetry (MIP, with and without surface roughness correction).

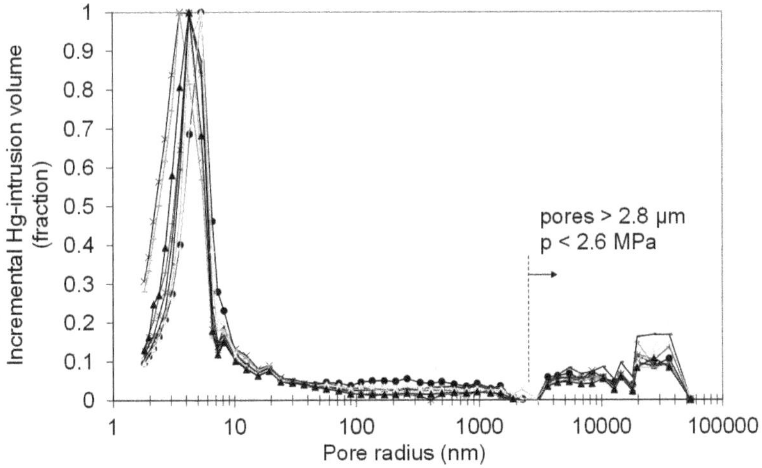

Figure 2. Pore size distribution determined on the sections used for gas breakthrough experiments. The intrusion up to a pressure of 0.2MPa is interpreted as the filling of surface irregularities and small artificial fissures (surface roughness correction).

EXPERIMENTAL

High pressure fluid flow cell

The experimental set-up used in this study has been described in detail in previous studies and the reader is referred to those [e.g. 3, 14, 15]. Here, we will only focus on some of the key aspects of this setup. The triaxial cells for testing flow in low-permeability rocks (e.g. mudrock, shale, coal, tight sandstones) are designed for confining pressures up to 50MPa, axial load of up to 100 kN and maximum temperatures up to 350°C. The sample cells accommodate plug sizes of 30mm length and

28.5 and 38mm in diameter. Basic elements of the set-up are two closed reservoir chambers (V_1, V_2, Figure 3a) on top and bottom of the sample, which are used to determine the mass flux across the sample. Both volumes are connected to pressure transducers (maximum pressure of 25MPa with an error of 0.05% FSO). Prior to the experiments, V_1 and V_2 are calibrated by means of helium expansion tests in the range of 1–6 cm³. The accuracy of the volume calibration is <0.5%. Leak tightness of the system is of uttermost importance, considering the long run time of the experiments. The diffusion tight sleeve system, consisting of a Pb-foil and Al/Cu-tube, is initially squeezed tightly around the sample/piston assembly in order to minimize any bypass along the sample. Ideally, leakage rates are far below the volume transport across the sample, or the leak rate is known with good confidence to account for in the calculations. For this reason leakage rates are determined for each experiment individually prior to the start of the experiments. In both methods, the occurrence of gas breakthrough is marked by a pressure increase on the low pressure side (p_2).

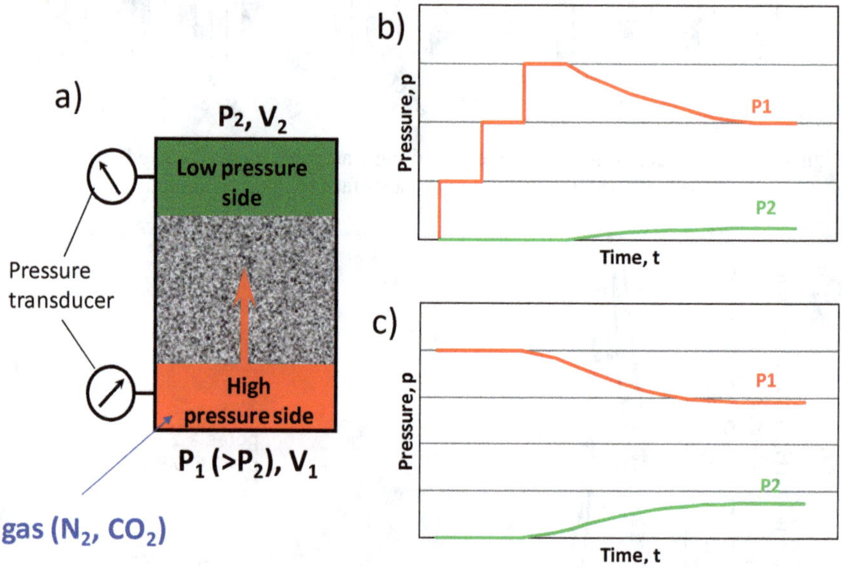

Figure 3. (a) schematic sketch of the experimental set-up for brine permeability and two-phase flow experiments. (b) The classical step-by-step pressure (p_1) increase method and (c) the imbibition or residual pressure method.

Gas/brine flow experiments

In this study a series of non-steady state (nstst) water permeability and gas breakthrough (drainage and imbibition) experiments were performed. Water permeability experiments were always performed prior to gas breakthrough experiments in order to re-establish full water saturation (in case of drying during preparation) and thus to ensure comparable starting conditions. Water permeability experiments after gas breakthrough measurements aimed at the identification of permanent changes of the material caused by the long-term experiments (i.e. creation of micro-fractures or other artefacts due to different loading/unloading cycles).

Intrinsic (water) permeability

The non-steady state (nstst) brine permeability experiments are conducted in a closed system. The high pressure compartment is predominantly filled with brine, which is forced through the sample by an initially applied He-pressure difference, p_1-p_2. The resulting pressure decay within volumes V_1 is converted to the dynamic water volume flux, which is related to the intrinsic (water) permeability. Details are given in [23]. Permeability of the sample, k_{water} (m^2) is derived from Darcy's law for incompressible media:

$$Q = \frac{dV}{dt} = -\frac{k_{water} \cdot A}{\eta}\frac{dp}{dx} \qquad \text{Equation 1}$$

With Q (m^3/s) the Darcy velocity, η (Pa·s) the dynamic viscosity, dp (Pa/m) the pressure gradient, A [m^2] the cross section area, and dx [m] the length of the sample.

Gas breakthrough tests

In contrast to the nstst water permeability experiments, the remaining water is flushed out of the reservoirs in order to ensure rapid contact of the gas phase with the sample for the gas capillary pressure measurements. There are different ways to determine critical capillary pressures in shales, i.e. the pressure where non-wetting (CO_2, N_2) phase is displacing the wetting phase (here brine). Depending on the method used, either the entry, breakthrough or snap-off pressure is measured. There are more ways for determining p_{c_brkth} and an overview is given in [19, 26].

The capillary breakthrough pressure p_{c_brkth} is measured once gas exits the downstream compartment side of the sample and a continuous gas flow is achieved across the sample. For very homogeneous samples and short sample lengths, p_{c_entry} and p_{c_brkth} are very similar. The breakthrough pressure is recorded on the drainage path, defined as the process where a non-wetting phase displaces a wetting phase. Usually, the step-by-step pressure increase method is applied, where the pressure is step-wise increased on one side of the sample (Figure 3b). This is continued until pressure communication between upstream and downstream compartments is observed (Figure 3b). The true breakthrough pressure is somewhere between the last two pressure steps; the accuracy of p_{c_brkth} strictly depends on the pressure step size. Before the occurrence of breakthrough the system is diffusion controlled (dissolved gas in pore water). The step-by-step method has been used earlier in several studies [9-11, 19, 20].

The snap-off pressure $p_{c\_snap\text{-}off}$ is detected on the spontaneous imbibition path (displacement of non-wetting by the wetting phase), and represents the capillary pressure at which the last interconnected gas filled pathway is blocked with the wetting phase. One prerequisite for this type of experiment is that a sufficiently high initial differential pressure is applied, which causes the pore system to drain readily. Due to the pressure communication, Δp decreases leading to the re-imbibition of water, which successively blocks the interconnected gas conducting pore system (Figure 3c). Once the differential pressure Δp has decreased below the snap-off pressure, gas transport is diffusion controlled. A detailed description of the imbibition or residual pressure method with the interpretation of the different flow regimes obtained in this type of measurement is given in [3, 13, 14, 27].

After breakthrough the effective permeability to the gas phase, $k_{eff(gas)}$, can be incrementally calculated from the pressure decay according Darcy's law for compressible media:

$$\frac{dn}{dt} = \frac{V_2 dp_2}{dt} = -\frac{k_{eff(gas)} \cdot A \cdot (p_2^2 - p_1^2)}{\mu \cdot 2 \cdot dx} \qquad \text{Equation 2}$$

Here *dn/dt* (mol/s) is the amount of mass moving through the sample calculated from the pressure increase dp_2/dt (Pa/s) in the calibrated downstream compartment V_2 (m^3). p_1 and p_2 (Pa) are the pressures on the high and low pressure side, $k_{eff(gas)}$ (m^2) is the apparent effective gas permeability at given mean pressure, $(p_1 + p_2)/2$. Data are not Klinkenberg (slip-flow) corrected due to experimental reasons.

Diffusion

From the diffusion controlled flow regimes an apparent effective gas diffusion coefficient can be estimated for each time step according to Fick's first law:

$$J_D = -D_{eff} \cdot \frac{dC_{bulk}}{dx} \qquad \text{Equation 3}$$

J_D (mol/m²/s) is the diffusive flux, D_{eff} (m²/s) is the effective diffusion coefficient, C_{bulk} (mol/m³) is the bulk concentration of the diffusing species and dx (m) the sample length. Here, the bulk concentration is the product of gas concentration within the pore water and sample porosity. Since the time lag, which is the time span required to establish steady state diffusion at constant concentration gradient, is not analysed in this experiment, only an apparent D_{eff}-value is obtained ($D_{eff\_apparent}$).

In addition, the effective diffusion coefficient can be derived from leak tests or the degassing stage after breakthrough experiments. In these experiments, the pressure gradient is zero, so the only transport mechanism is diffusion. D_{eff} is derived from the pressure drop/increase in the closed reservoir volumes according to [28], providing a solution for the diffusion of a dissolved gas species into an infinite plane sheet:

$$\frac{M_t}{M_\infty} = 1 - \sum_{n=1}^{\infty} \frac{2b(1+b)}{1+b+b^2 q_n^2} \exp\left(1 - \frac{D_{eff} q_n^2 t}{dx^2}\right) \qquad \text{Equation 4}$$

M_t is the total amount of solute in the sheet at time t as a fraction of M_∞, the amount of solute after infinite time. Furthermore it is defined that tan(qn)=-b·q_n and b=V_R·V_s-1, which is the ratio of volumes of solution and sheet. V_R corresponds to the reservoir volumes (V_2 + V_2), V_s to the sample volume.

Pore water chemistry

For the water saturation experiments artificial brine was used (Table 6), with its composition taken from [29]. This brine is considered to be close to the composition of the brine in contact with the Opalinus samples used here. It is important to use brine with similar cation compositions so as to not change the swelling states of smectites, which might result in artefacts in the measurements that are difficult to quantify.

Table 6. Pore water composition used during flow tests.

Parameter	**OPA reference [29]**	**APW (this report)**	**Parameter**	**OPA reference [29]**	**APW (this report)**
Temperature	25	25	FeII	0.0524	--
pH	7.203	7.861	FeIII	3.31 10-9	--
Pε	-2.781	--	Si	0.1779	--
Concentration units	mmol/kg	mmol/kg	Cl	160	160
Na	164.4	163.8	SVI	24.72	24
K	2.604	2.551	S-II	12.4 10-9	--
Ca	12.51	11.91	CIV	2.506	0.543
Mg	9.625	9.166	Alkalinity	2.308	0.549
Sr	0.2106	--	P(CO_2) [bar]	10-2.20	10-3.50
			Ionic strength	0.2299	0.2264

MIP capillary pressure prediction – indirect method

Different methodologies to determine the critical capillary pressures from mercury porosimetry data have been reported in the literature [17, 30, 31]. An example of a MIP-pressure curve and the data interpretation applied in this study is given in Figure 4. We define three different values:

The percolation threshold pressure, P_p, is the pressure at which a continuous filament of mercury extends through the sample [32, 33, 34]. This point represents the most prominent pore radius within the sample.

P_d, which is somewhat lower than P_p, is defined as the capillary displacement pressure required to form such a continuous, non-wetting phase-filled pore network. We use the so called tangent-method, for which a tangent is fitted to the inflection point of the cumulative intrusion curve and extrapolated to the logarithmic pressure axis [31].

Often the pressure at 10% Hg-saturation, $P_{(10\%)}$, is taken as valid equivalent of the displacement pressure. This is an average value published by Schowalter [2], who compared the results from several uni-directional nitrogen breakthrough experiments and mercury injection porosimetry measurements.

The evaluation of these critical pressures is performed on the surface roughness corrected data, where Hg-intrusion volumes associated with the filling of fractures and surface irregularities are removed. In this study, Hg-volumes prior to the establishment of the injection plateau in the cumulative injection curve are excluded (Figure 5). This plateau is associated with a final/complete Hg-wetted sample surface but zero injection (i.e. the entry pressure has not been reached at this stage) [30]. In this case, pressures below 0.2MPa were not considered.

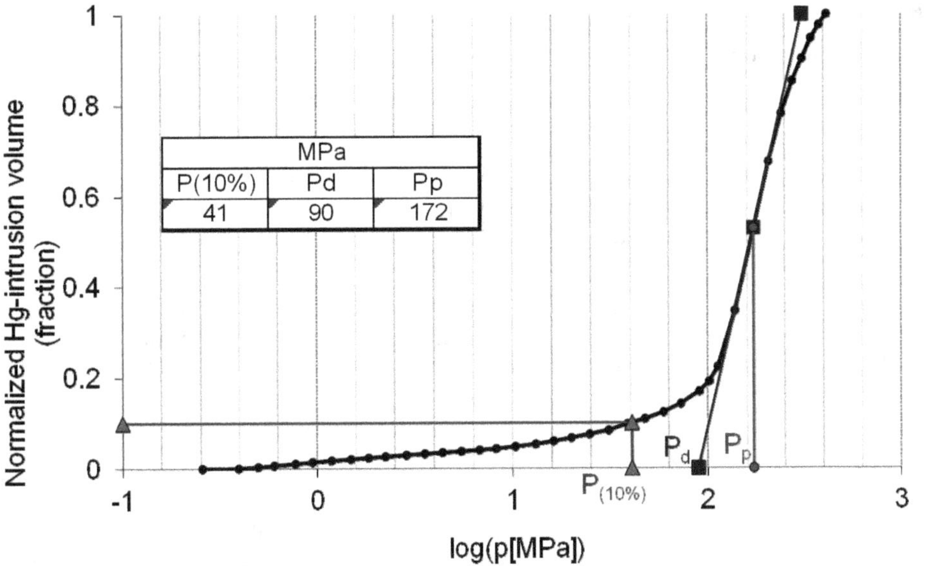

Figure 4. Example of a MIP injection curve and data interpretation.

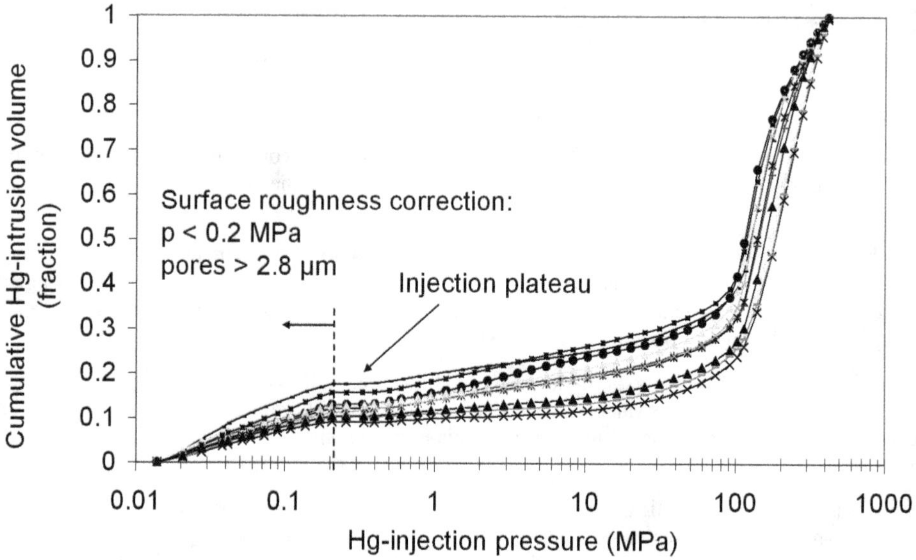

Figure 5. Cumulative MIP intrusion curves determined for the sections from which plugs were prepared for gas breakthrough experiments. The intrusion up to a pressure of 0.2MPa is interpreted as the filling of surface irregularities and small fissures.

RESULTS

Mercury porosimetry

The evaluation of the raw data reveals bulk and matrix densities of approximately 2.3 and 2.6±0.1 g/cm³, respectively (Table 7). The total porosity is 11.6±0.9%. After application of the surface roughness correction by subtracting injection volumes below 0.2MPa, the total porosity reduces to 10.2±0.5%. From the evaluation of the critical Hg-pressures it is obvious that the pressure at 10% saturation ($P_{(10\%)}$ = 28±23MPa) is consistently lower than the displacement pressure. On average, the displacement pressure, P_d, is 85±12MPa. The inflection point is 155±41MPa, yielding values around 5 ±1 nm for the most prominent pore radii.

For further interpretation, Hg-injection data must be converted to the natural fluid system. Based on the Washburn equation and considering the characteristic wetting angles and interfacial tension data of the different systems, the critical capillary pressure of the natural gas/water-system can be calculated:

$$p_{c(gas,brine)} = p_{c(Hg,air)} \frac{\gamma_{gas,brine} \cos(\theta_{gas,brine})}{\gamma_{Hg,air} \cos(\theta_{Hg,air})} \qquad \text{Equation 5}$$

For the system N_2-brine, perfect water wettability ($\theta = 0°$) and an interfacial tension (γ) of 70 mN/m was assumed. The contact angle (θ) in the system CO_2-brine was varied between 0° and 60°, and γ was set to 25 mN/m [35]. The results indicate a N_2 breakthrough pressure of approximately 16MPa, and for CO_2, the capillary pressure was between 2.8 and 5.7MPa (Figure 6).

For the system Hg-air a contact angle of 140° and interfacial tension of 485 mN/m was used.

Table 7. Evaluation of characteristic capillary pressures based on surface roughness corrected Hg-data. $P_{(10\%)}$ = pressure at 10% Hg-saturation, P_d = capillary displacement pressure determined with tangent-method, P_p = inflection point of cumulative intrusion curve, r_{max} = most prominent pore radius at point P_p.

Sample No.	$P_{(10\%)}$	P_d MPa	P_p	r_{max} (P_p) nm
Hg_01	41	90	172	4.3
Hg_02*	31	90	172	4.3
Hg_03	5	72	113	6.6
Hg_04	5	72	113	6.6
Hg_05	59	100	207	3.6
Hg_06	11	76	113	6.6
Hg_07	15	76	138	5.4
Hg_08	20	82	138	5.4
Hg_09	56	94	207	3.6
Hg_10	21	82	138	5.4
Hg_11	14	78	138	5.4
Hg_12	53	102	207	3.6
Hg_13	44	94	207	3.6
Average	28	85	155	5
Standard Deviation	23	12	41	1

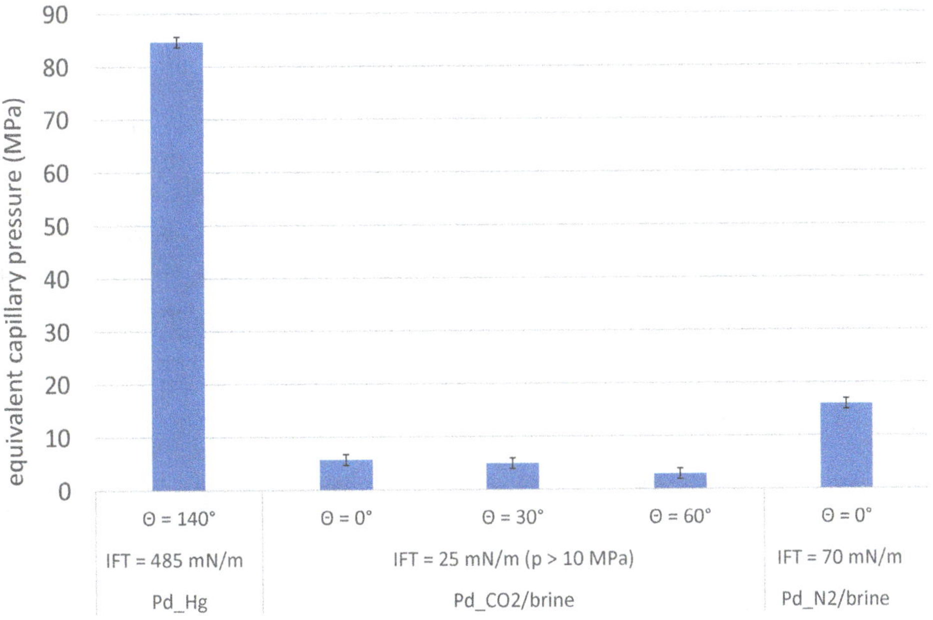

Figure 6. Conversion of Hg-derived displacement pressure to the system CO_2-brine and N_2-brine.

Water permeability

As stated above, single-phase non-steady state water permeability (k_{water}) experiments were conducted prior and after gas flow-through experiments with brine. In these experiments He was used as the driving gas phase. For five samples k_{water} was also derived from the initial phase of the gas (N_2, CO_2) breakthrough experiments. The initial pressure difference was varied from 0.3 to 9.7MPa (average 2.0MPa). For some samples the mean fluid pressure was systematically varied between 0.5 to 12.7MPa in order to test the sensitivity of hydraulic conductivity to effective stress (17-31MPa).

The intrinsic permeability of the entire sample set ranges from 1 to 6 E-21 m². Permeability variations observed over repeated measurements indicate an average standard deviation of 12%. Likely reasons for this are variations in effective stress or flow reversal. However, a clear dependence on mean effective stress is only given for samples plug_02 and plug_12. For these samples a stress release of about 30% led to a permeability increase of approximately 65% (Figure 7).

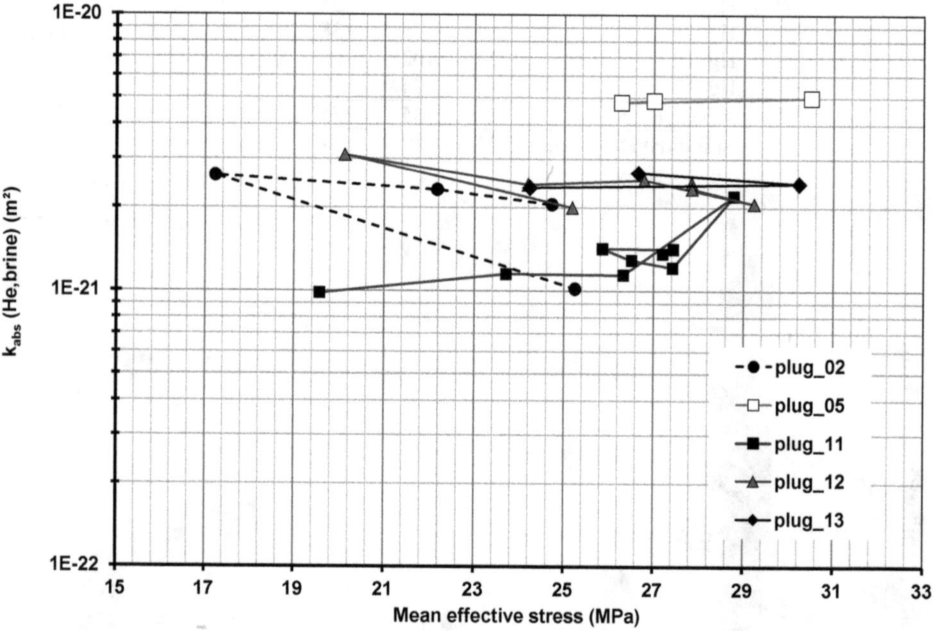

Figure 7. Dependence of intrinsic permeability on changes in effective stress. This was achieved by variations in mean effective fluid pressure (p_{mean} = 0.5 to 12.7MPa).

Two-phase flow experiments

In the following, selected N_2 and CO_2 experiments will be discussed in detail in order to explain the different transport processes involved and observed. We will give explanations with respect to the analysis of the critical capillary pressures, single-phase (water) and two-phase flow regime after breakthrough and transport where diffusion is the dominant transport process.

N_2 *experiments*

plug_01 – N_2 imbibition/residual pressure experiment

Figure 8 shows the pressure decay recorded for the up- and downstream pressure compartment for sample plug_01. The initial pressure drop recorded during the first 10 hours of the experiment corresponds to the pure displacement of residual brine. Fitting of the pressure data to the single-phase flow model yield a water permeability of 7E-21 m². In comparison, the non-steady state experiments before and after the gas breakthrough tests indicated somewhat lower permeability coefficients of 5.9E-21 (stdev 1.6E-21 m², before) and 5.4E-21 m² (afterwards).

At about 10 hours the recorded pressure data deviate from the single-phase flow curves and pressure equilibration is significantly accelerated. This is interpreted as the onset of continuous gas flow across the sample plug. However, this pressure does not strictly correspond to the capillary breakthrough pressure. As this is an imbibition type of experiment, where the initial pressure difference is substantially higher than the capillary breakthrough pressure, the timing of breakthrough strongly depends on the amount of water within the high pressure reservoir and the intrinsic permeability of the sample. Due the non-steady state nature of this experiment, gas permeability is strongly influenced by visco-dynamic time-delay effects. In comparison to steady state flow conditions, gas permeability is somewhat lower during drainage, and somewhat higher during imbibition.

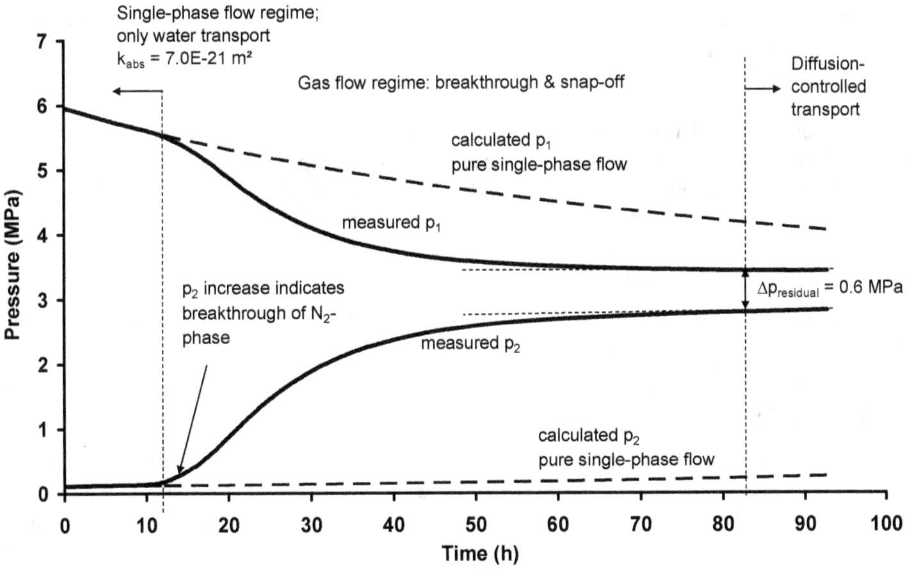

Figure 8. Sample plug_01, results from the N_2-imbibition experiment. The single-phase water flow regime is followed by a clear breakthrough of the gas phase, which is indicated by the pressure increase on low pressure side. The absolute (water) permeability, k_{water}, is 7.0E-21 m². The snap-off pressure after spontaneous imbibition is 0.6 MPa.

Upon gas breakthrough (>10h), the pressure on the low-pressure side increases rapidly. Due to the gas flux, the pressure difference (capillary pressure) decreases and water re-imbibes into the pore system. Permeability decreases until the last interconnected flow-path is shut-off by the water phase, resulting in a constant residual pressure difference (snap-off pressure $p_{snap-off}$ = 0.6 MPa, Figure 9).

At this stage (t>80h) effective gas permeability values derived from the up- and downstream pressure side drop to values in the sub-nDarcy-range (< 1E-21 m²) and transport remains diffusion controlled.

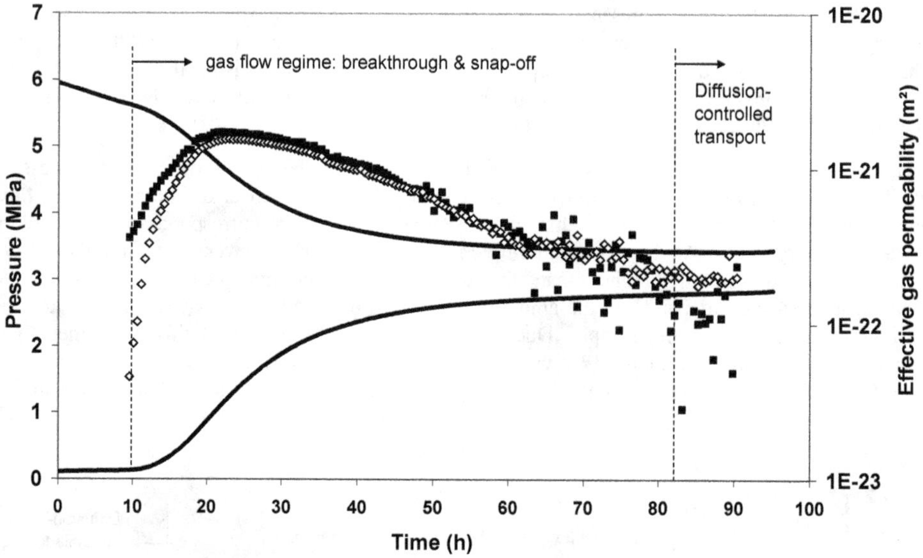

Figure 9. Sample plug_01, results from the N_2-imbibition experiment. The effective permeability to the gas phase, $k_{eff(gas)}$ is incrementally calculated per pressure step, from the pressure drop/increase in the up- and downstream pressure compartment, k_1/k_2, respectively. Gas flow was initiated (after gas came into contact with the sample) at a pressure difference of 5.5 MPa.

plug_02 – N_2 drainage experiment

In contrast to sample plug_01, the experiment on plug_02 can be divided into three different zones, the single-phase water flow regime, capillary sealing and viscous flow regime.

The absolute water permeability determined from the initial phase of the gas breakthrough experiment is 2.5E-21 m² (stdev 3.6E-22 m², Figure 10).

At about 23 days the pressure curves remain almost constant, with small fluctuations due to daily temperature changes only. At this stage, gas is in contact with the sample surface with the pressure difference too low for capillary breakthrough to occur. Flow does not completely stop, but slowly decreases, with a nominal effective gas permeability coefficient of 6E-25 m². It is likely that some residual/remaining water is still displaced from the porous disc until full capillary sealing occurs. We interpret this phase of the experiment to be mainly diffusion controlled.

Further pressure increase (up to approximately 12MPa) ultimately resulted in a gas breakthrough at a pressure difference of about 12.5MPa. Maximum flow rates achieved were still extremely low with maximum effective permeability coefficients of approximately 5E-24 m² (Figure 11). The fact that this "peak permeability" is followed by a permeability drop (>180 hours) is an indication for a viscous controlled imbibition process. The experiment was terminated before final snap-off pressure was achieved.

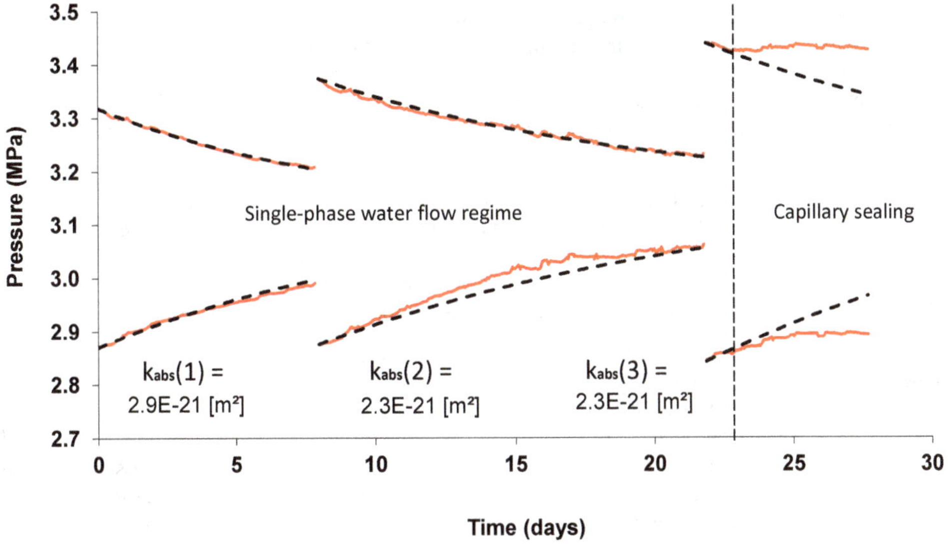

Figure 10. First phase of the gas breakthrough experiment on plug_02. Only water is displaced across the sample. After approximately 23 hours transport is stopped, which is due to capillary sealing (i.e. the gas phase got in contact with the sample surface but pressure difference is still too low to displace water from the pore space).

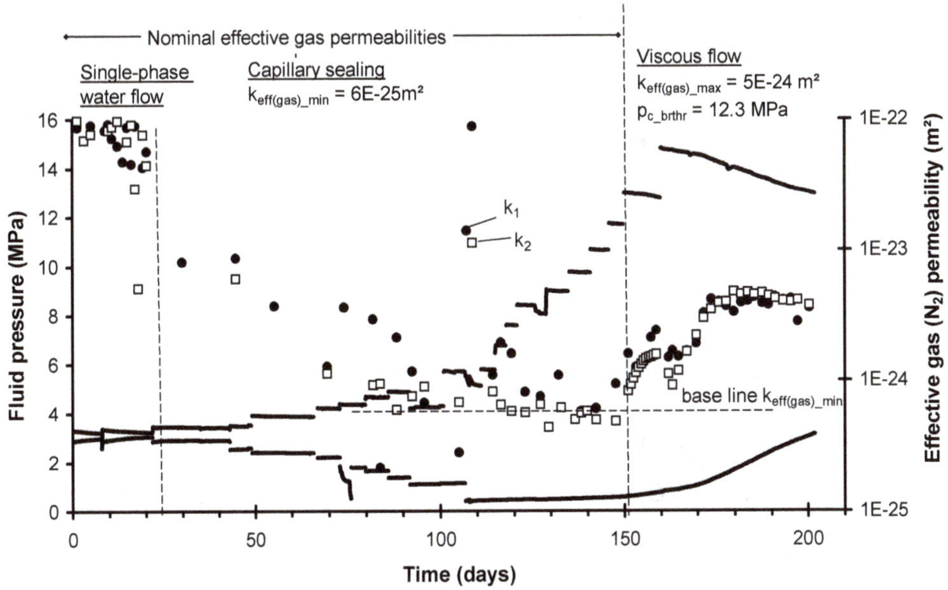

Figure 11. N_2-drainage experiment on sample plug_02. Effective gas permeability coefficients are only nominal (i.e. not necessary related to viscous flow).

Although the single-phase flow properties of sample plug_01 and 02 were relatively similar the resistance to breakthrough was extremely different. To check for any changes of the conducing pore system of sample plug_02 (e.g. compaction/time) we conducted several single-phase flow experiments, before, during and after the gas breakthrough experiment. The interim flow experiments were performed by simply changing the flow direction (~110 days) and recording the brine displacement in the opposite direction.

We observe a slight decrease with time as well as some stress dependence. Data shown in Figure 12 indicate absolute permeability increases with increasing mean fluid pressure. However, the lowest value must be taken with care, as this was the first experiment after gas breakthrough. Low water conductivities could be caused by some remaining N_2 bubbles blocking the pore throats. Without this value the permeability range is small (2 to 3.6 E-21 m²).

The permeability values in the beginning (<10 days) correspond to the initial water displacement phase (k_{water} = 2.5E-21 m²). Thereafter the average nominal effective gas permeability drops to values near 6E-25 m². Due to the absence of a clear pressure increase on p_2 and the large scatter in $k_{eff(gas)}$-values, we interpret this region as being diffusion controlled (i.e. where gas flow is prevented by capillary sealing). At about 150 days (Δp = 11.2-12.5 MPa) the downstream pressure starts to increase, which is interpreted as the onset of viscous gas flow. An even higher pressure difference leads to a further increase in flow rate, resulting in a maximum effective gas permeability of 5E-24m², which is still extremely low.

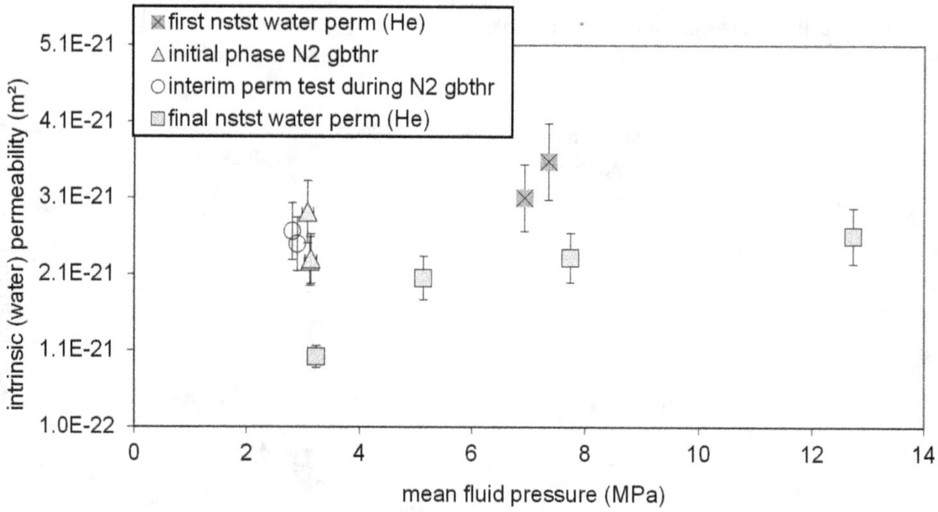

Figure 12. Single-phase brine permeability before, during and after the N_2 gas breakthrough experiment. Permeability data are plotted versus the mean fluid pressure applied during the individual experiments. Error bars represent the average standard deviation of 14%.

CO_2 experiments

In comparison to the N_2 experiments, the execution and evaluation of gas breakthrough tests with CO_2 was more difficult and ambiguous. A clear distinction between the different transport processes (diffusion vs. viscous flow regime) was often not possible, as CO_2 dissolution/degassing in/from the pore water is instantaneous and dominant, which significantly influences the pressure decay.

The evaluation of the pressure decay curves was based on two key-criteria:

- Capillary sealing is clearly established when the initially detected pressure decay (single-phase flow regime) slows greatly. Associated nominal effective gas permeability coefficients drop to values below the sub-nDarcy range ($k_{eff(gas)}$< 1E-23 m²). Here, transport of CO_2 is dominated by diffusion (baseline).
- The onset of gas breakthrough is associated with a significant pressure increase on the low pressure side. Here, viscous effective gas permeability coefficients increase to values above the baseline.

For all incremental pressure steps the solubility of CO_2 in brine was calculated as a function of pressure (p_1, p_2), temperature and salinity [36]. In addition to the calculation of the viscous parameters "absolute water permeability" (k_{water}) and "effective gas permeability" ($k_{eff(gas)}$), an "apparent diffusion coefficient" (D_{eff}) was calculated for all pressure steps. "Apparent" means that for each time step, steady state diffusion was assumed (Fick's first law), even though probably not achieved at that specific time.

Various physical relationships (models) can be applied for the reconstruction of the pressure decline curves or flow rates. However, the model relationship chosen may not always strictly correspond to the actual phenomenon. For example, we use the term "nominal" when assuming another process to be dominant, like "nominal effective gas permeability coefficient" determined in a capillary sealing regime, which is in reality diffusion controlled. These nominal values usually turn out to be unrealistic (too high/low). We consider the maximum effective diffusion coefficient to be 2E-9 m/s (diffusion in pure water).

plug_12 – CO_2 drainage experiment

The single phase permeability based on He-experiments is 2.5E-21m². The experiment ended with capillary sealing, and the flow rate dramatically dropped to values below 1E-22m². The CO_2 gas breakthrough experiment begins under capillary sealing conditions with flow rates below the single-phase flow conditions. However, in terms of permeability, the capillary sealing regime can be divided into two parts: a first one with nominal water permeabilities in the order of 1E-23m² and a second one with even lower values ($k_{eff(gas)}$ = 4E-24m² or D_{eff} = 1E-10m/s). This is due to a successive, step-wise realization of capillary sealing (i.e. that not all pores are capillary blocked at the same time).

With increasing pressure difference, flow rates determined from the pressure drop at the high pressure side (dp_1/dt) increase to 1E-23m². In contrast, mass flow into the downstream compartment remains relatively constant at lower flow rates.

The last two pressure steps are characterised by flow rates above the base line, likely representing the onset of viscous gas flow. This is substantiated by the fact that a maximum permeability is established, followed by a decrease back to the base line values, due to by successive capillary snap-off and closure of pores; both processes are considered to be stress related.

Key-parameters obtained are the capillary breakthrough pressure at a pressure difference of 17.5MPa. The snap-off pressure (taken at the time when flux drops below the base line) is ~10MPa.

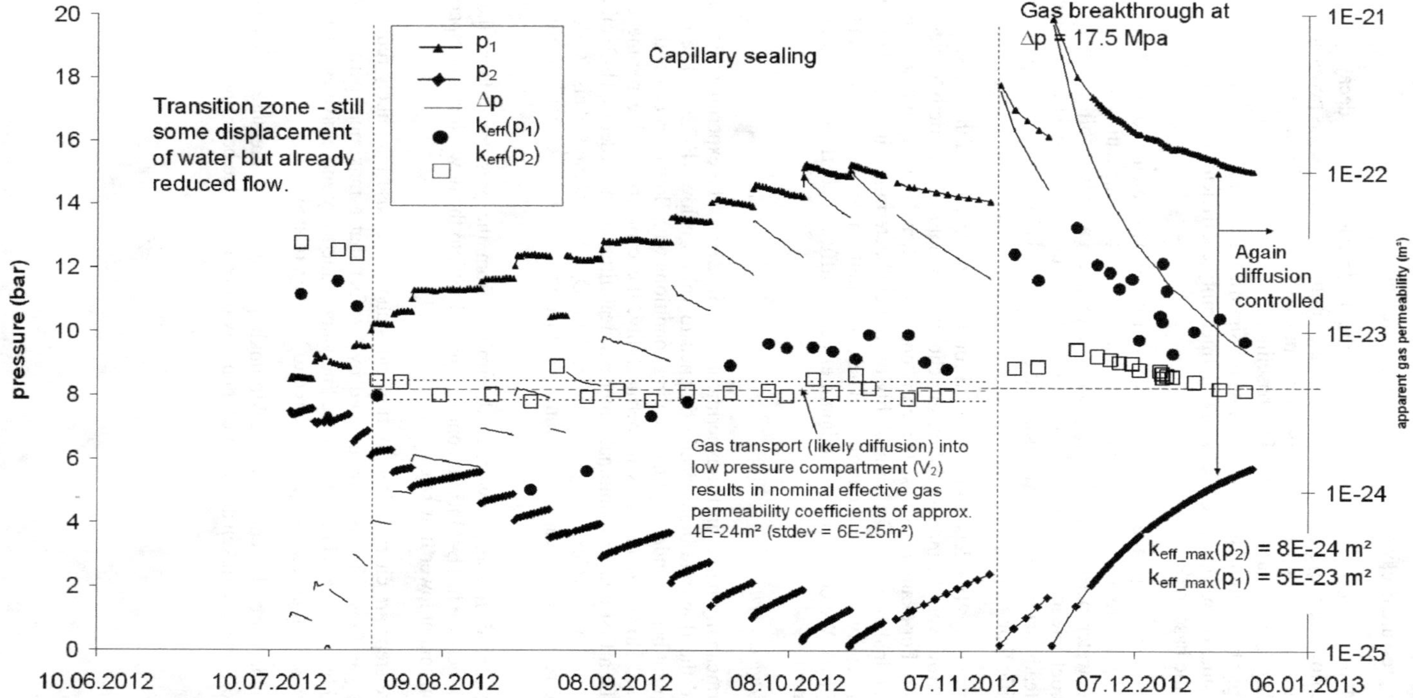

Figure 13. CO_2 gas breakthrough experiment on sample plug_12. Three flow regimes are distinguished, (a) the zone of reduced conductivity, a transition zone, where still some remaining water is displaced through the sample, (b) the capillary sealing regime, with all pores blocked by capillary forces, and (c) the viscous flow regime, which is typically characterized by a peak permeability with successive decrease (imbibition).

plug_13 – CO_2 drainage experiment

Three nstst water permeability tests were initially conducted for sample plug_13. The average water permeability is 2.5E-21 ±1.3E-22 m² for the experiments with He as the drive gas and 1.7E-21 for the CO_2 experiment (Figure 14), which clearly ended with capillary sealing (Figure 15). Accordingly, flow rates at all subsequent CO_2 pressure levels were extremely small, corresponding to apparent water permeabilities of 5E-23 m² (Figure 14) or nominal effective CO_2 permeabilities of 3E-24 m² (Figure 16).

Higher flow rates were detected at differential pressures above 14MPa, with a final/significant increase at a pressure difference of 20MPa (Figure 16). In this last pressure step, the effective gas permeability increased steeply to 1E-21 m². As the pressure difference decreases, permeability drops due to successive imbibition of water. The final pressure difference is 3.5MPa (snap-off pressure).

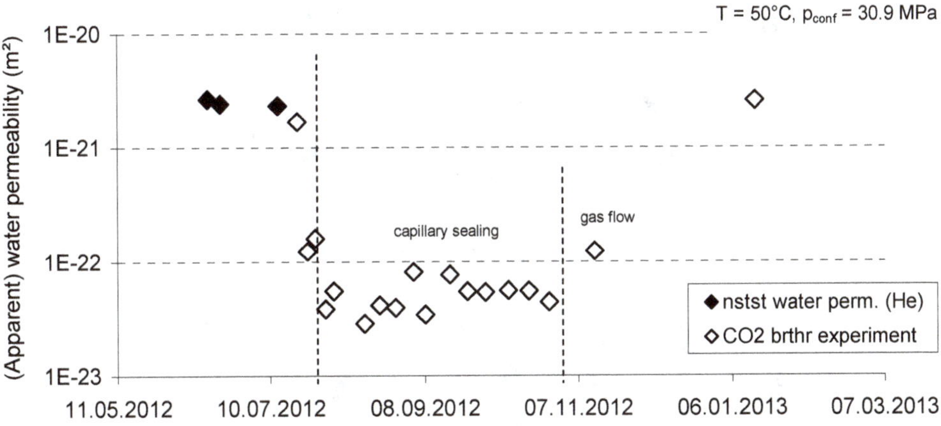

Figure 14. nstst water permeability determined for sample plug_13.

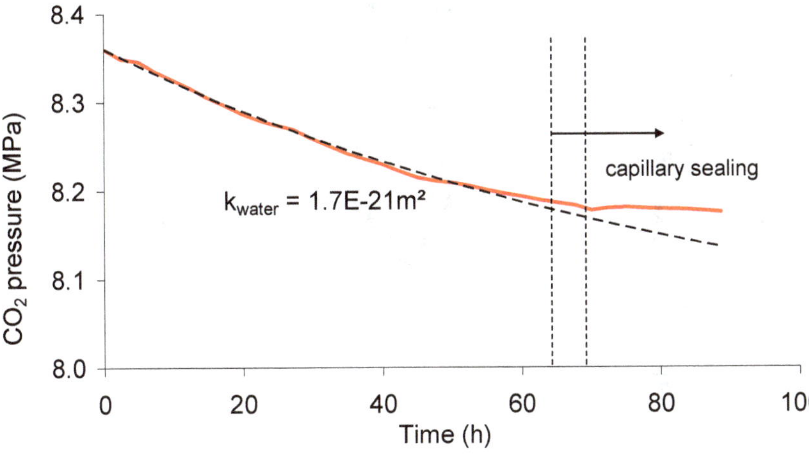

Figure 15. First CO_2 pressure (plug_13).

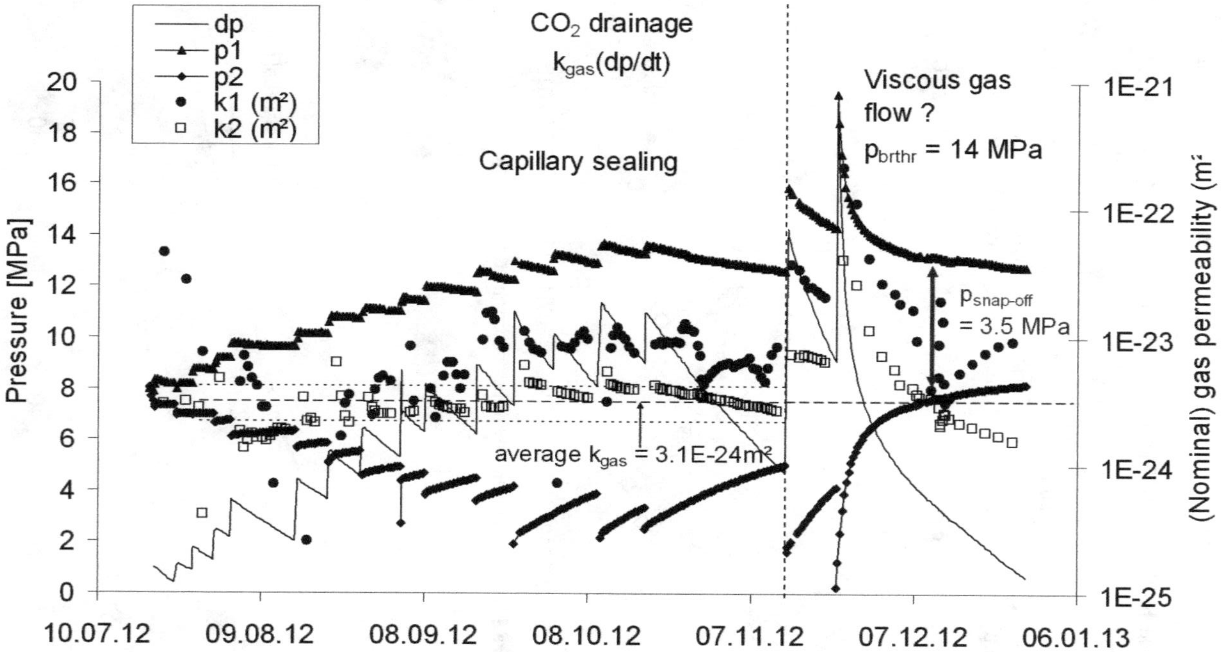

Figure 16. CO_2 gas breakthrough experiment on sample plug_13.

DISCUSSION

Water content (porosity) before and after plug saturation

Porosities determined by mercury porosimetry are much lower than water content porosities determined on the plugs right after flow experiments (ϕ_{WC} = 10-12%, ϕ_{Hg} = 22%, Figure 17). So, which of these values can be considered as the "correct" ones? Experience shows that mercury porosimetry usually yields lower porosity values than water-saturation porosities for very tight rocks. This is because Hg accesses only pores with radii > 2nm, while water as a wetting fluid is able to fill the entire pore space. The much higher water-based porosities in Figure 17 would thus indicate a substantial micropore volume (r<2nm) for most of the samples. The low ϕ(WC)-values in Figure 1 could be due to drying of material during drilling and transportation.

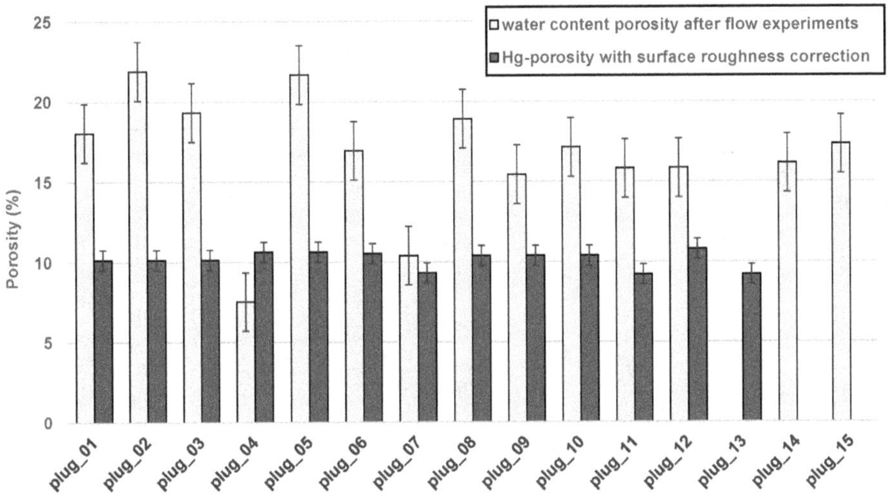

Figure 17. Comparison of Hg-porosimetry data. The water content porosities were determined on sample plugs after fluid flow experiments. It was not possible to determine the weight of sample plug_13 at the end of gas testing.

Critical capillary pressure (breakthrough vs. snap-off)

Substantial differences were observed between gas breakthrough (drainage) and snap-off (imbibition) pressures. Whereas for one sample (plug_01) viscous gas flow occurred immediately at a breakthrough pressure of 6MPa and ended at a snap-off pressure of 0.5MPa, other samples showed a much higher threshold pressure (capillary sealing efficiency). A clear permeability-capillary pressure relationship was not identified (Figure 18). As expected, breakthrough (drainage) pressures were always higher than the snap-off pressures on the imbibition path.

Based on predictions from (omnidirectional) Hg-injection porosimetry (MIP) results, and assuming completely water-wet mineral surfaces, gas breakthrough was expected to occur at approximately 16MPa for N_2 and between 2.8 and 5.7MPa for CO_2. The observed gas breakthrough pressures for the core plugs, however, ranged from 3.4 to 12.3MPa for N_2 and from 14 to 17.5MPa for CO_2.

We conclude that the observed variations in the measured capillary breakthrough and snap-off pressures might be a result of dilatancy (widening of micro fissures or the soft clay matrix, e.g. [8]). In these experiments, a purely capillary-controlled immiscible flow is not considered relevant.

In contrast, mercury injection data are very similar with Hg/air capillary displacement pressures of 85±12 MPa (Table 8). Assuming that the pore network is too tight to allow for unhindered intrusion of mercury, the mineral matrix system would be (partly) compressed. In this case, mercury injection data would only reflect the compressibility of the rock sample.

Table 8. Results from permeability and gas breakthrough experiments. All experiments were conducted as non-steady state experiments. Pressure decay is interpreted with respect to the different fluid flow parameters (single-phase water permeabililty, k_{water}, and critical capillary pressure, breakthrough and snap-off).

	Initial Water Perm.		**Gas Breakthrough Results**							
Sample Plug	**Initial k_{water}**	**Stdev**	**Gas**	**k_{water} (ii) Initial Phase of Gbthr.**	**Stdev**	**k_{water} (iii) After Gbrthr. Exp.**	**Stdev**	**Snap-off (imbibition)**	**Breakthroug h (drainage)**	**Breakthroug h/ snap-off**
	m²				**m²**			**MPa**		
plug_01	5.9E-21	1.3E-21		7.0E-21		5.4E-21		0.6		
plug_02	3.4E-21	1.7E-22	N_2	3.0E-21	2.7E-22	2.0E-21	6.0E-22		12.5	
	3.4E-21	1.7E-22		2. 6E-21	8.3E-22					
plug_0 5	5.7E-21	6.3E-23		5.4E-21	1.2E-22	4.9E-21	8.6E-23	2.1	3.4	1.6
plug_06	1.5E-21	3.8E-22		1.1E-21		1.4E-21		6.4	12.0	1.9
plug_07	1.5E-21	1.2E-22				1.2E-21				
plug_10	3.8E-21	2.2E-22				4.5E-21	1.1E-21			
plug_11	1.4E-21	3.3E-22								
plug_12	2.4E-21	3.3E-22	CO_2			2.6E-21		10.0	17.5	1.8
plug_13	2.5E-21	1.3E-22		1.7E-21		2.6E-21		3.5	14.0	4.0
plug_14	3.3E-21	4.8E-22								
plug_15	2.3E-21	2.7E-22				3.0E-21		3.5		

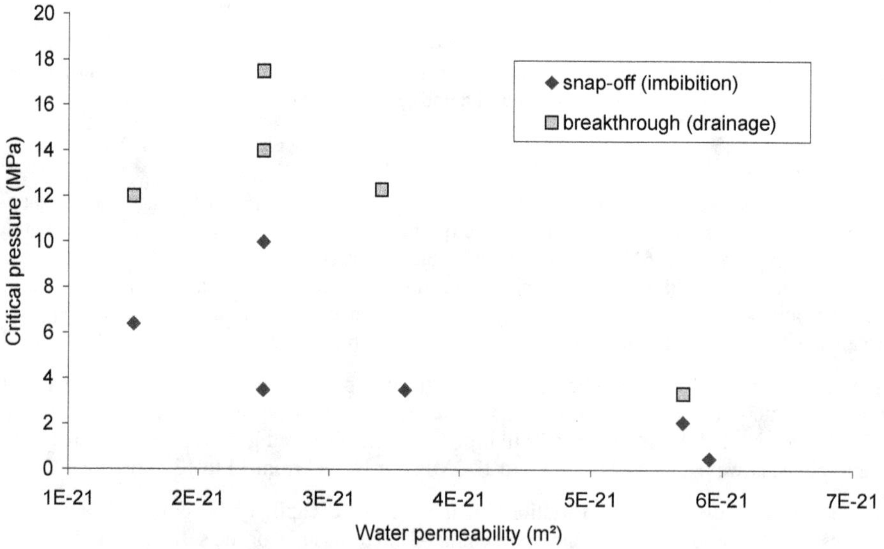

Figure 18. Critical capillary pressures vs. sample permeability.

Viscous flow versus diffusion

Flow rates detected after gas breakthrough are extremely small. Therefore, an unambiguous identification of gas breakthrough is not always possible. Pressure changes upon Δp-increase, uptake and dissolution of CO_2 in/from the pore water (porosity ~10%) result in pressure fluctuations in the reservoirs: When the upstream pressure p_1 is increased, more CO_2 is dissolved in the pore water, resulting in higher apparent flow rates out of the upstream compartment V_1. This results in an overestimation of flow rates. Pressure reduction on the downstream side (p_2) results in degassing of the pore water into the downstream compartment V_2 and, therefore resulting in an overestimation of flow rates. As gas solubility changes significantly when decreasing pressures below 10MPa, degassing effects on the low pressure side are more pronounced. Due to the very low flow rates, a clear differentiation between diffusion, viscous flow and uptake/dissolution processes is difficult or impossible.

Gas fluxes detected in the capillary sealing regimes correspond to nominal effective gas permeabilities below 1E-23 m^2, ranging from 6E-25 m^2 to 7E-24m^2. Corresponding apparent effective diffusion coefficients are 1E-10 m^2/s. From leak and degassing tests, where transport was limited to diffusion (zero pressure gradient), we obtained effective diffusion coefficients ranging from ~5E-11 to ~2E-11 m^2/s.

Conversion MIP to CO_2/brine system

The critical capillary pressure, above which a rock becomes conductive, is often determined from mercury injection experiments (MIP), which provides information on the excess pressure required to displace air at successively decreasing pore throats. At maximum injection pressures of approximately 400MPa, pore radii of roughly 2nm are invaded (wetting angle of 140° measured through denser phase, interfacial tension 485mN/m) [30, 37].

From the cumulative Hg-intrusion curve, different critical capillary pressures may be derived. These are then converted from the Hg-air system to the CO_2-brine-rock system (Equation 5). Literature values for interfacial tension (IFT) and contact angle within the natural system (CO_2-brine) were used for system conversion. While IFT data for CO_2/brine became available recently and seem to be consistent [38, 39], contact angle values seem to vary from study to study and mineral phase to mineral phase. An extensive literature review was recently published by Heath *et al.* [40]. In general, IFT depends on pressure, temperature and brine salinity. At pressures up to the critical CO_2 pressure (~7.4MPa) IFT decreases from approximately 60-70mNm^{-1} to 20-30mNm^{-1}. At pressures above 8-15MPa (at 27-71°C, respectively) a "pseudo-plateau" is reached, where IFT does not further change significantly with pressure. A decrease with increasing temperature is given in this high pressure region. The CO_2-contact angle certainly depends on the mineralogical composition. Contact angles reported for the more wetting minerals (quartz, muscovite, calcite) range between 18° and 60° [35, 41, 42, 43]. Trends indicate that the contact angle increases with increasing salinity and pressure [40].

When using MIP-data, the choice of values for IFT and contact angle has a large influence on the predicted capillary failure prognosis for caprocks. If a perfectly water wet system ($\theta = 0°$, $\cos(\theta) = 1$) were assumed, the capillary sealing efficiency would be strongly overestimated. In contrast, a contact angle of 60° ($\cos(\theta) = 0.5$) would lower the predicted capillary sealing capacity by 50%.

However, as stated above, the "real" breakthrough pressures measured on core plugs are strongly variable and often different from the predicted values. The N_2-drainage breakthrough pressures ranged between 2.8 and 12.3MPa. N_2-snap-off pressures vary between 0.5 and 6.4MPa. For the CO_2 breakthrough experiments, perfect capillary sealing was observed at high differential CO_2 pressure, e.g. 14MPa (plug_13), and 17.5MPa (plug_12).

To compare all the data, we converted all N_2 and MIP data to the system CO_2/brine (using IFT from Li *et al.* [39] and contact angle=0°) and plotted them with the CO_2 data in Figure 19. There is a significant spread in the data for CO_2 breakthrough (1.2 to 17.5MPa) and CO_2 snap-off (0.2 to 10MPa) pressures with a predicted CO_2/brine breakthrough pressure from MIP data of 5.7MPa.

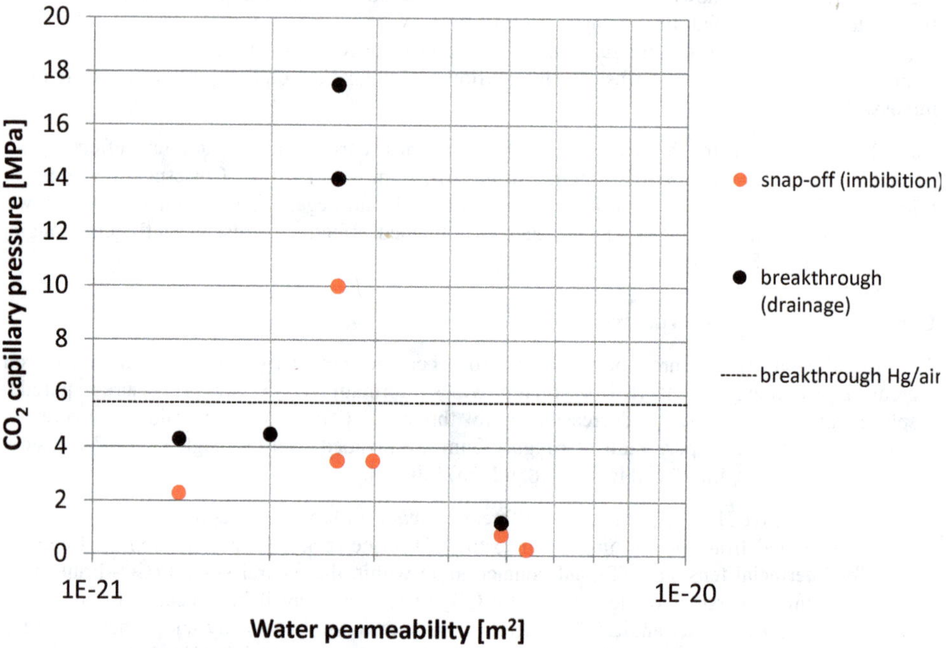

Figure 19. Comparison of critical capillary pressures (breakthrough & snap-off) converted to the CO_2/brine system, using IFT data from literature [39] and assuming contact angle θ = zero.

Permeability before and after gas breakthrough experiments

The intrinsic permeability does not significantly change after the gas breakthrough experiments (Figure 20). Therefore, we conclude that gas breakthrough tests did not alter the pore system on the long term. If pores/fractures were opened because of differential pressure increase, the process seems to be reversible. Fluctuations along the 1:1-line are within the standard deviation of the non-steady state experiments.

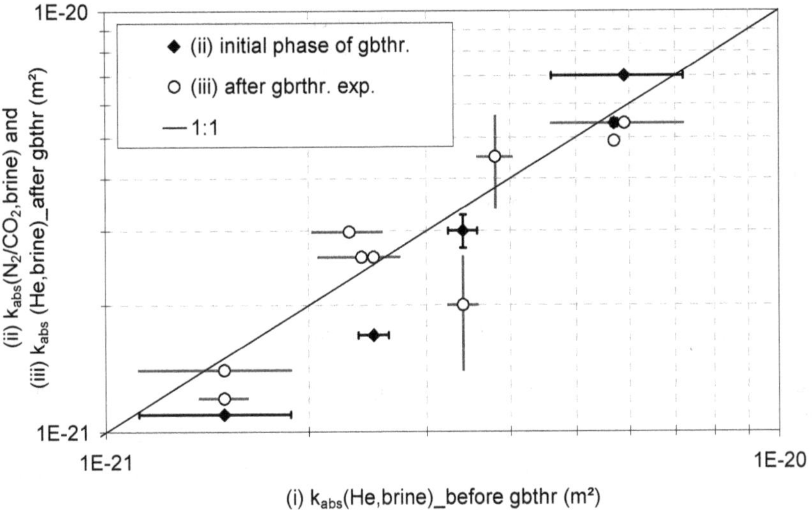

Figure 20. Comparison of permeability values derived from different experimental steps. (i) Before gas breakthrough experiments, thus during the initial re-saturation phase, k_{water} (He,brine). (ii) During the initial phase of gas breakthrough tests where water was still displaced, k_{water} (N_2/ CO_2, brine) before. (iii) And after the gas breakthrough experiments, k_{water} (He,brine)_after.

CONCLUSIONS

We performed a number of drainage and imbibition capillary pressure experiments on Opalinus shale in order to determine capillary breakthrough and snap-off pressures using N_2 and CO_2 on water saturated sample plugs. Measurements were compared to MIP data performed on the same samples and converted from Hg/air to gas/brine data using literature IFT and contact angle data.

Mineralogy and petrophysical data (tested on 9 samples) showed small spatial variations. Porosity values derived from different methods varied, with MIP porosities generally lower (by about 2% absolute) compared to water-content values. This is attributed to differing pore accessibilities of water compared to mercury.

Water permeabilities varied between 1 and 6 E-21 m^2 and no significant differences were observed before and after capillary pressure measurements.

Breakthrough CO_2 capillary pressures varied between 14 and 17.5 MPa for 2 samples, while corresponding snap-off pressures varied between 3.5 and 10.0 MPa for 3 samples. Breakthrough and snap-off pressures for N_2 range between 3.4 to 12.5 MPa (3 samples) and 0.6 to 6.4 MPa (3 samples), respectively. The average displacement pressure from MIP is 85 (±12) MPa, corresponding to N_2 and CO_2 breakthrough pressures of ~16 and 5.7 MPa, respectively (completely water wet system, IFT literature data).

Reasons for the differences are unclear and could be explained by:

1. Micro-fissures from either coring or plug drilling or sample dehydration leading to preferential flow pathways in some of the samples.
2. Heterogeneities in sample pore network (size of largest interconnected capillary) on the plug scale determining capillary pressures
3. Gas transport due to dilatancy
4. CO_2/brine contact angles deviating from perfect water-wet conditions.

In summary it is not straightforward to predict capillary entry pressures of a shale formation from single plug measurements. It is unclear and difficult to resolve if the large variabilities in snap-off and breakthrough pressures observed here are due to formation heterogeneity or sampling.

ACKNOWLEDGEMENTS

We thank P. Marschall for very helpful and constructive comments on the paper and P. Marschall and Chr. Nussbaum for providing Opalinus Shale core material. Thanks also to Gunnar Oeltzschner for his enthusiasm in conducting and discussing the laboratory experiments.

REFERENCES

1. Ibrahim, M.A., Tek, M.R., Katz, D.L., 1970. Threshold pressure in gas storage, Pipeline Research Committee, American Gas Association at the University of Michigan, Michigan.
2. Schowalter, T.T., 1979. Mechanics of secondary hydrocarbon migration and entrapment. AAPG Bulletin 63, 723-760.
3. Amann-Hildenbrand, A., Bertier, P., Busch, A., Krooss, B.M., 2013. Experimental investigation of the sealing capacity of generic clay-rich caprocks. International Journal of Greenhouse Gas Control 123, 20-33.
4. Busch, A., Alles, S., Gensterblum, Y., Prinz, D., Dewhurst, D.N., Raven, M.D., Stanjek, H., Krooss, B.M., 2008. Carbon dioxide storage potential of shales. International Journal of Greenhouse Gas Control 2, 297-308.
5. Li, S., Dong, M., Li, Z., Huang, S., Qing, H., Nickel, E., 2005. Gas breakthrough pressure for hydrocarbon reservoir seal rocks: implications for the security of long-term CO_2 storage in the Weyburn field. Geofluids 5, 326-334.
6. Vivalda, C., Loizzo, M., Lefebvre, Y., 2009. Building CO_2 Storage Risk Profiles With The Help Of Quantitative Simulations. Energy Procedia 1, 2471-2477.
7. Didier, M., 2012. Étude du transfert réactif de l'hydrogène au sein de l'argilite, l'Institut des Sciences de la Terre. Université de Grenoble, Grenoble, p. 275.
8. Marschall, P., Horseman, S., Gimmi, T., 2005. Characterisation of Gas Transport Properties of the Opalinus Clay, a Potential Host Rock Formation for Radioactive Waste Disposal Oil & Gas Science and Technology - Rev. IFP 60, 121-139.
9. Harrington, J.F., Horseman, S.T., 1999. Gas transport properties of clays and mudrocks. Geological Society, London, Special Publications 158, 107-124.
10. Horseman, S.T., Harrington, J., 1994. Migration of repository gases in an overconsolidated clay. British Geological Survey, Keyworth, Nottingham, p. 66.
11. Horseman, S.T., Harrington, J., Sellin, P., 1997. Gas migration in MX80 buffer bentonite. Materials Research Society Symp. Proc. 465, 1003-1010.
12. Horseman, S.T., Harrington, J.F., Sellin, P., 1999. Gas migration in clay barriers. Engeneering Geology 54, 139-149.
13. Hildenbrand, A., Krooss, B.M., 2003. CO_2 migration processes in argillaceous rocks: pressure-driven volume flow and diffusion. Journal of Geochememical Exploration 78–79, 169–172.
14. Hildenbrand, A., Schlömer, S., Krooss, B.M., 2002. Gas breakthrough experiments on fine-grained sedimentary rocks. Geofluids 2, 3-23.
15. Schlömer, 1998. Abdichtungseigenschaften pelitischer Gesteine - Experimentelle Charakterisierung und geologische Relevanz, Institut für Chemie und Dynamik der Geosphäre 4, Jülich, p. 212.
16. Amann-Hildenbrand, A., Ghanizadeh, A., Krooss, B.M., 2012. Transport properties of unconventional gas systems. Marine and Petroleum Geology 31, 90-99.
17. Busch, A., Amann-Hildenbrand, A., 2013. Predicting capillarity of mudrocks. Marine and Petroleum Geology 45, 208-223.

18. Vassenden, F., Sylta, Ø., Zwach, C., 2003. Secondary migration in a 2D visual laboratory model, Proceedings of European Association of Geoscientists & Engineers (EAGE) conference "Fault and Top Seals", Montpellier, France.
19. Egermann, P., Lombard, J.-M., Bretonnier, P., 2006. A fast and accurate method to measure threshold capillary pressure of caprocks under representative conditions, International Symposium of the Society of Core Analysts, Trondheim, Norway, pp. 1-14.
20. Boulin, P.F., Bretonnier, P., Vassil, V., Samouillet, A., Fleury, M., J.M., L., 2011. Entry pressure measurements using three unconventional experimental methods, Society of Core Analysts, Austin, Texas, USA, p. 12.
21. Meyn, V., 1999. Die Bedeutung des Sperrdrucks (threshold pressure) für den Fluidtransport in niedrigstpermeablen Gesteinen: Experimentelle und theoretische Aspekte, DGMK Spring Conference, Celle, Germany, pp. 255-264.
22. Skurtveit, E., Aker, E., Soldal, M., Angeli, M., Wang, Z., 2012. Experimental investigation of CO_2 breakthrough and flow mechanisms in shale. Petroleum Geoscience 18, 3-15.
23. Wollenweber, J., Alles, S., Busch, A., Krooss, B.M., Stanjek, H., Littke, R., 2010. Experimental investigation of the CO_2 sealing efficiency of caprocks. International Journal of Greenhouse Gas Control 4, 231-241.
24. Ufer, K., Stanjek, H., Roth, G., Dohrmann, G., Kleeberg, R., Kaufhold, S., 2008. Quantitative phase analysis of bentonites by the Rietveld method. Clays and Clay Minerals 56, 272-282.
25. Washburn, E.W., 1921. Note on a method of determining the distribution of pore sizes in a porous material, Proceedings of the National Acadamy of Science, pp. 115-116.
26. Busch, A., Muller, N., 2010. Determining CO_2/brine relative permeability and capillary threshold pressures for reservoir rocks and caprocks: Recommendations for development of standard laboratory protocols, Energy Procedia, pp. 6053-6060.
27. Hildenbrand, A., Schlömer, S., Krooss, B.M., Littke, R., 2004. Gas breakthrough experiments on pelitic rocks: comparative study with N_2, CO_2 and CH4. Geofluids 4, 61-80.
28. Crank, J., 1975. The Mathematics of Diffusion. Clarendon Press, Oxford.
29. Mäder, U., 2009. Reference pore water for the Opalinus Clay and "Brown Dogger" for the provisional safety-analysis in the framework of sectoral plan - interim results (SGTZE). Nagra Working Report, Nagra, Wettingen, Switzerland.
30. Dewhurst, D.N., Jones, R.J., Raven, M.D., 2002. Microstructural and petrophysical characterization of Muderong Shale: application to top seal risking. Petroleum Geoscience 8, 371-383.
31. Schlömer, S., Krooss, B.M., 1997. Experimental characterisation of the hydrocarbon sealing efficiency of cap rocks. Marine and Petroleum Geology 14, 565-580.
32. Dullien, F.A.L., 1979. Porous Media: Fluid Transport and Pore Structure, 2nd ed. ed. Academic Press, New York.
33. Katz, A.J., Thompson, A.H., 1986. Quantitative prediction of permeability in porous rock. Physical Review B 34, 8179-8181.
34. Katz, A.J., Thompson, A.H., 1987. Prediction of rock electrical conductivity from mercury injection measurements. Journal of Geophysical Research 92, 599-607.
35. Bikkina, P.K., 2011. Contact angle measurements of CO_2-water-quartz/calcite systems in the perspective of carbon sequestration. International Journal of Greenhouse Gas Control 5, 1259-1271.
36. Duan, Z., Sun, R., 2003. An improved model calculating CO_2 solubility in pure water and aqueous NaCl solutions from 273 to 533 K and from 0 to 2000 bar. Chemical Geology 193, 257-271.
37. Vavra, C.L., Kaldi, J.G., Sneider, R.M., 1992. Geological applications of capillary pressure; a review. AAPG Bulletin 76, 840-850.

38. Bachu, S., Bennion, D.B., 2009. Interfacial tension between CO_2, freshwater, and brine in the range of pressure from (2 to 27)MPa, temperature from (20 to 125) degrees C, and water salinity from (0 to 334 000) mg(.)L(−1). Journal of Chemical and Engineering Data 54, 765-775.
39. Li, X., Boek, E.S., Maitland, G.C., Trusler, J.P.M., 2012. Interfacial Tension of (Brines + CO_2): (0.864 NaCl + 0.136 KCl) at Temperatures between (298 and 448) K, Pressures between (2 and 50)MPa, and Total Molalities of (1 to 5) mol•kg-1. Journal of Chemical & Engineering Data 57, 1078-1088.
40. Heath, J.E., Dewers, T.A., McPherson, B.J.O.L., Nemer, M.B., Kotula, P.G., 2012. Pore-lining phases and capillary breakthrough pressure of mudstone caprocks: Sealing efficiency of geologic CO_2 storage sites. International Journal of Greenhouse Gas Control 11, 204-220.
41. Chiquet, P., Broseta, D., Thibeau, S., 2007. Wettability alteration of caprock minerals by carbon dioxide. Geofluids 7, 1-11.
42. Mills, J., Riazi, M., Sohrabi, M., 2011. Wettability of common rock-forming minerals in a CO_2-brine system at reservoir conditions, International Symposium of the Society of Core Analysts, Texas, USA, 18-21 September 2011, paper number SCA2011-06.
43. Shah, V., Broseta, D., Mouronval, G., 2008. Capillary alteration of caprocks by acid gases, SPE/DOE Improved Oil Recovery Symposium, Tulsa, Oklahoma, USA, 19-23 April 2009, paper number SPE 113353.

SYMBOLS & DEFINITIONS

Abbreviation, symbol	Unit	Definition
p_1 and p_2	Pa	Up- and downstream pressure.
V_1 and V_2	m^3	Calibrated up- and downstream reservoir.
nstst		Non-steady state experiment. Permeability is calculated from the pressure decay in a closed volume.
k_{water}		Absolute permeability determined with brine in a brine saturated sample
D_{eff}		Effective diffusion coefficient derived from the uptake or degassing experiments from the pore space.
$D_{eff\_apparent}$		Estimation of apparent effective diffusion coefficient from pressure increase on low pressure side.
gbthr		Gas breakthrough experiment (nstst).
Drainage experiment		Step-wise pressure increase until breakthrough of gas (pressure increase on low-pressure side).
Imbibition experiment		Starting at a high initial pressure difference and evaluation of the pressure decay until snap-off pressure.
$k_{eff\ max}$		Maximum gas permeability during gas breakthrough experiments.
$k_{eff\_apparent}$		Apparent effective gas permeability. Gas permeability at given mean pressure conditions (not Klinkenberg corrected).
k_1, k_2		Permeability coefficients derived from the pressure changes of p_1 and p_2.
nominal		Process cannot be clearly identified. Detected fluxes can either be related to diffusion or viscous flow.
$P_{10\%}$	Pa	Pressure at 10% Hg-injection volume.
P_d	Pa	Capillary displacement pressure ("percolation threshold"), where an interconnected Hg filament (pathway) across the sample is formed.
P_p	Pa	Capillary threshold pressure at which a continuous Hg-network exists
r_p	μm	Most prominent equivalent pore radius (corresponding to p_p)

Carbon Dioxide Capture for Storage in Deep Geological Formations, Volume 4
Karl F. Gerdes (Editor)

Chapter 26

IMPACT OF CO_2 IMPURITIES ON STORAGE PERFORMANCE AND ASSURANCE

Jean-Philippe Nicot[1], Jiemin Lu[1], Patrick Mickler[1], and Silvia V. Solano[2]
[1]Bureau of Economic Geology, Jackson School of Geosciences,
University of Texas, University Station, Box X, Austin, Texas 78713-8924, USA
[2]previously at the Bureau of Economic Geology

ABSTRACT: The amount of CO_2 in the waste gas stream of various industrial facilities is very variable. Whereas flue gas from conventional power plants involves a capture step before geological storage, some streams, in particular in the chemical industry, consist of >99% CO_2, requiring little or no treatment. However, waste gas streams from power plants that use novel technologies such as oxy-fuel, the focus of this study, can also circumvent the capture step thanks to their CO_2-rich composition (CO_2>90%), but at the expense of stream CO_2 purity. In addition to CO_2, two non-reactive gases make up the bulk of impurities - N_2 and Ar. O_2 is the reactive gas most commonly cited, with molar concentration ranging from <1% to >5%. Subsurface impacts of an impure CO_2 stream could be twofold: (1) complicate flow behavior and reduce static capacity because of density and viscosity differences and (2) undermine reservoir and top seal integrity due to reactions with reactive species (O_2 and minor species such as CO). We approached the study not in absolute terms but by comparing impacts of the neat CO_2 cases to the impure CO_2 cases. We performed (1) numerical flow simulations with varying formation parameters and characteristics to determine impacts on flow behavior and (2) laboratory experiments to determine geochemical impacts by exposing, reservoir rock fragments to CO_2 and to CO_2 and O_2, in an autoclave reactor at reservoir conditions. The analysis of impact of impurities on flow behavior included four gases: CO_2, N_2, Ar, and O_2, whereas the geochemical analysis focused on two reactive gases: CO_2 and O_2.

Overall, an impure stream translates into a larger plume extent - that is, a larger area requiring observation. In addition, there is faster and larger residual trapping, which shortens the period of plume mobility. Both of these processes are amplified at shallower depths. The effect on flow dynamics decreases with depth, as the impure CO_2 stream behaves more and more as neat CO_2. Equally important is that changes in viscosity and density between neat CO_2 and CO_2 mixtures decrease with depth, suggesting that differences in storage capacity are proportionally reduced with depth. Although flow behaviour may impact the larger storage venue reservoir, geochemical impacts are more likely to be restricted to the wellbore environment and adjacent proximal reservoir. Batch experiments conducted in 2 high-pressure, high-temperature autoclave reactors with siliciclastic rocks immersed in synthetic brine and exposed to supercritical CO_2 with and without admixed O_2 show that O_2 is likely to alter the geochemistry of subsurface systems in ways that the pure CO_2 case does not, in particular when ferrous-iron bearing minerals are present. For example, pyrite (FeS_2) is quickly oxidized by O_2 and pH can drop significantly if the system has little or no buffering capacity leading to deeper degradation of feldspars (mostly K and Na silicate). In addition, iron-bearing carbonates (e.g., ferroan calcite, siderite, ankerite) are degraded with CO_2 addition. This mobilizes ferrous Fe, which then precipitates as iron oxides when contacted by O_2. In all 19 of the autoclave experiments, mineral precipitation remained minor because the precursor minerals that supply component ions are not abundant. This suggests that as long as a precursor reactive mineral fraction is a small portion of the rock, O_2 will not have a large geochemical effect on mineral precipitation and therefore on rock stability or fluid flow. The overall results of the study may, therefore, present the CO_2 storage project developer with tradeoffs in capacity, pressure evolution, and monitoring scenarios, with additional costs likely more than offset by reduced capture costs.

KEYWORDS: carbon storage; CO_2 sequestration; impurities; oxygen; oxy-fuel; rock-water interactions; autoclave; reactor; area of review

INTRODUCTION

A key impediment to carbon capture and storage is the cost of CO_2 capture, particularly for conventional power plants whose flue gas is dominated by gases other than CO_2. Opportunities to reduce capture costs provided the incentive to investigate the impact of impurities on reservoir system dynamics and geochemical changes in the injection formation and their consequences on flow and ultimate capacity. Leaving some of the impurities in the injection stream could save both capital and operational costs, possibly without significant economic or HES (health, environment, safety) assurance consequences to the storage part of the project. For example, pipelines and compressors could be re-engineered to handle impurities. Power generation and industrial processes that have CO_2 as a byproduct seldom produce a pure CO_2 stream. When the CO_2 fraction is low (as is the case of flue gas from conventional power plants), an amine-based capture system that will increase the CO_2 fraction (to >99%) and possibly eliminate most of the impurities is needed and highly desirable. However, when the CO_2 fraction in the waste stream is already high (e.g., oxy-firing or gasification), only some level of gas processing might be required, depending on the subsurface characteristics of the strata that are to receive the stream and on other local operational factors (i.e., if compression and transportation are not unduly impacted or can be cost-effectively re-engineered). This chapter explores the impacts of impurities on CO_2 behavior in the subsurface during geological carbon storage and focuses on flue-gas streams from oxy-fired power plants (combustion in ~pure O_2).

Impurities in the flue streams include N_2, Ar, O_2, CO, and, potentially, H_2, SOx, NOx, and other acid species. CO_2 capture experts have not converged yet to a small set of applicable technologies but a range of compositions can be inferred from the literature. Discussions with industry experts, however, suggested that maximum volume fractions are 15% for N_2 (that is, ~10% molar in a binary mixture), 5% for Ar and O_2, 2% for CO, and 0.15% for SOx. Once the impure CO_2 stream is injected into the subsurface, CH_4 and H_2S gases, which are commonly present in brines in a dissolved state in many basins, can form a significant percentage of the mixture as they partition from the brine into the supercritical phase. This phenomenon could occur to some degree even if the injected stream is composed of only pure CO_2.

Impacts can be classified into two types: (1) direct impact on flow dynamic behavior (relevant gases are N_2, Ar, and O_2 to which CH_4 is added) and (2) geochemical impact and indirect impacts on flow, focusing on O_2, the only reactive gas in abundance. Overall, geochemical processes related to reactive species could affect near-field properties such as injectivity and well integrity, whereas larger-scale regional impacts on plume dynamics are likely only an issue with significant mole fractions of non-condensable gases.

FLOW DYNAMIC FINDINGS

Summary of Flow Dynamic Findings

The goal of this part of the study was to understand the impact of major impurities on CO_2 plume dynamics, injectivity, and capacity. As noted above, for CO_2 captured from industrial sources, the main impurities are N_2, O_2, and Ar, to which was added CH_4, ubiquitous in the subsurface of hydrocarbon provinces, which are prime candidates for geological storage. The study considered up to 15% volume for N_2, 5% volume for O_2, and 5% volume for Ar. Other gases such as CO, H_2, and SOx, which could have non-negligible mole fractions, are not considered. Simulations of CO_2 injection floods with selected impurities levels were conducted on generic models broadly representative of major onshore / near-shore clastic systems in North America and Europe (low to moderate permeability). The problem was approached through an extended desktop study using the numerical modeling tool (multiphase flow code CMG-GEM and associated CMG-WINPROP software package). GEM is a compositional multiphase flow code that can accommodate multiple gas components and their interaction with a liquid phase. WINPROP is an allied module useful in determining and tuning equations of state (EOS). In order to work with accurate PVT data (Peng-Robinson EOS), thorough laboratory experiments were performed early in the study to measure viscosity and density of the mixtures. We performed 10 experiments (through an external vendor based in Houston, Texas) to tune EOS parameters for various CO_2 mixtures (incorporating CO_2, N_2, O_2, and Ar) at various temperatures (60, 80, and 100°C) and pressures (13.8, 27.6, and 41.4 MPa). CMG-GEM relies on many empirical mixing rules for density and viscosity calculations that need to be calibrated and tuned. In parallel, a comprehensive literature survey was undertaken to collect information on solubility of those various mixture components into the aqueous phase under various subsurface pressure, temperature, and salinity conditions. We developed binary interaction coefficients between components under a range of pressure (10–50 MPa), temperature (30–120°C), and salinity (0–200,000 mg/L) conditions so as to model dissolution of the mixtures into the brine. The differential partitioning of gas components in the aqueous phase impacts the gas phase composition.

At this point, a comparison between density and viscosity of the main mixture components could be helpful because they vary with depth (Figure 1). In the subsurface, temperature and pressure are positively correlated, and an exploration of the entire P, T space is not needed. Pressure can be assumed to be hydrostatic in most settings, and the temperature gradient varies within a relatively small range (15–35°C/km). Density values of all impurities are significantly less than that of CO_2 by at least a factor of 2, with $CO_2 >> Ar > O_2 > N_2 > CH_4$. Viscosity values follow the same pattern below the depth at which CO_2 mixtures are supercritical. An important observation integral to understanding plume dynamics associated with impurities is that viscosity and density of mixtures are lower than that of neat CO_2 at the same temperature and pressure. Equally important to note, viscosity and density contrasts between mixtures and neat CO_2 decreases with depth.

The numerical geologic models used grow in complexity from simple box-like generic models, to which heterogeneity is added in a second step, to more realistic models constructed from actual U.S. Gulf Coast locations but representative of many sites around the world. The objective was to reproduce end-members of aquifer architecture such as (1) clean homogeneous, medium permeability sand; (2) homogeneous sand/clay, and (3) heterogeneous sand with discontinuous shale partings and continuous baffles. Progressively more complex gas systems, binary, ternary, and beyond, were investigated. The results, normalized by results of the corresponding neat CO_2 case, draw on two metrics, time to hit the top and maximum longitudinal extent (Kumar *et al.*, 2009). These are contrasted for two depths: "shallow" (~1500 m, ~60°C, ~17 MPa, 100,000 mg/L) and "deep" (~3000 m, 125°C, ~30 MPa, 180,000 mg/L). Because O_2, N_2, and Ar have similar properties and behavior, they impact the CO_2 dominated mixtures in a similar way, particularly at the concentration level of a couple percent molar and they can be merged in one unique component with properties of N_2. However, the approximation deviates too much from the "truth" beyond a few percentage points.

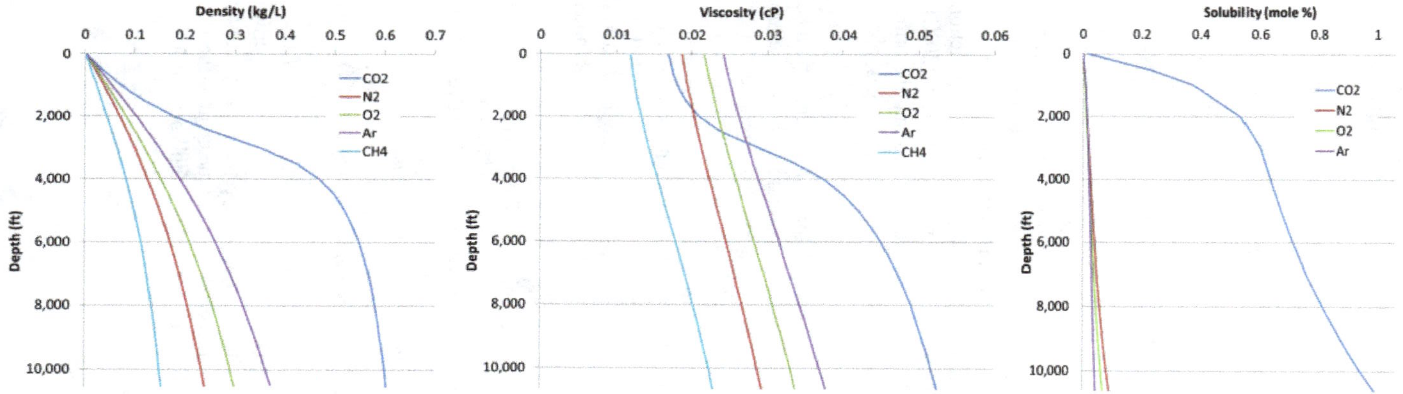

(a) (b) (c)

Note: assumed conditions are hydrostatic pressure and geothermal gradient of 22°C/km. 10,000 ft = 3048 m.

Figure 1. Density (a), viscosity (b), and solubility (at 100,000 mg/L) (c) of a pure component as a function of depth.

Impurities impact density and viscosity of the CO_2-rich mixture. Impurities impact CO_2 capacity not only because of the smaller fraction injected and space needed for storing impurities, but also because of the generally lower density of the mixture with impurities at the same pressure and temperature conditions. An approximate proxy for capacity change owing to impurities is given by the density ratio. The loss in capacity can be as high as >50% at shallow depths (~1000 m, CO_2 and 15% molar N_2) but the difference quickly decreases with depth. Similarly, mass injectivity, which can be represented by the proxy metric of density over viscosity ratio, also shows a decreased value at very shallow depths that quickly recovers with increasing depth.

In terms of plume shape and extent, the impact of impurities is again more marked at shallow depth where the contrast in density and viscosity with neat CO_2 is the largest. The contrast decreases with depth even if the actual extent increases with depth. For example, about 4% mole fraction in a binary system suffices to increase plume length in “shallow” low-dip sloping layers by 25%, whereas 9% to 15%, depending on the component, would cause similar change in a “deep” system. In all cases, plume extent is greater with impurities, although residual trapping occurs faster. This relationship holds for most systems regardless of heterogeneity and complexity. The contrast is most extreme in very simple systems whereas heterogeneity (assuming adequate operational choices) seems to dampen impacts of impurities. This presumably occurs because heterogeneity creates multiple tongues that attenuate the impact of impurities. It also suggests a trade-off between plume extent (area of review with risk of CO_2 leakage) and decreased risk owing to faster trapping. A larger plume translates into a larger area to inspect for leakage pathways such as faults and abandoned wells, but a faster trapping translates into a shorter period of time to monitor the site.

Illustration of Key Flow Dynamic Findings

The general approach consisted of a parametric study and sensitivity analyses of a generic case and of three previously studied sites (two on the U.S. Gulf Coast and one in Alberta, Canada), modified slightly to meet our objectives. Gas composition and range were estimated from various sources (Table 1). We initially used a simple dual generic model: a shallow model reproducing conditions present at the Frio site (southeastern Texas) (Hovorka *et al.*, 2006) and a deep model reproducing conditions prevailing at Cranfield (western Mississippi) (Hovorka *et al.*, 2013). Both sites are in the U.S. Gulf Coast region, but for analysis they were stripped of specific properties, retaining only environmental conditions of pressure, temperature, and salinity (Table 2). In order to focus on the processes of interest and not on specificities associated with an actual site, we developed a generic sloping aquifer and compared results of simulations carried out using various CO_2 mixtures (Nicot *et al.*, 2013).

Table 1. Molar composition of base cases.

Component (mol %)	Neat CO_2	Stream A	Stream C	Single component density at 1 atm and 21°C (kg/m^3)
CO_2	100	96	92	1.834
N_2	—	0.2	1	1.161
O_2	—	2.1	6.5	1.327
Ar	—	1.7	0.5	1.654

Table 2. Characteristics of shallow and deep generic models.

Reservoir Property	Shallow Reservoir Case	Deep Reservoir Case
Model dimensions	11,000 × 4660 × 300 m^3	same
Number of cells x × y × z	120 × 51 × 20	same
Cell dimensions	90 × 90 × 15 m^3	same
Dip in x direction	2°	same
Permeability /kv/kh / porosity	300 md / 0.01 / 0.25	same
Depth at top downdip	1675 m	3040 m
Initial pressure (equilibrium at time 0)	Vert. Equil. ~17.6 MPa at top cell of downdip boundary	Vert. Equil. ~32.4 MPa at top cell of downdip boundary
Temperature	57°C	125°C
Injection rate and period	8.5 m^3/s for 30 years	same
Maximum res. saturation	0.30	same
Boundary	No flow except updip (hydrostatic)	same
Formation water TDS	~100,000 mg/L	~170,000 mg/L
Simulation period	100 yr	same

The only trapping mechanisms simulated in the model were dissolution and residual phase mechanisms. Mineral phase trapping on a meaningful scale is generally understood to require at least hundreds or thousands of years, and was not included. Structural trapping—that is, trapping of CO_2 as would occur in oil and gas accumulations—was not included in the design of the generic model because structural trapping would be of negligible utility in explaining the interplay of all processes and is site specific. The model was large enough (11 km) for the total mass of CO_2 mixtures to be fully trapped as residual saturation before reaching the updip boundary, assuming an injection rate equivalent to 0.5 million tons per year of pure CO_2 for 30 years. Injection occurred at the downdip section of the lower third of the 300-m-thick reservoir (Figure 2). Results are to be understood relative to one another, in particular relative to the base cases, because of numerical and gridding issues. For example, in homogeneous models, plume extent is a function partly of cell size but mostly of cell height (Thibeau and Dutin, 2011; these authors investigated the largest cell thickness that would model CO_2 dissolution correctly in their model and concluded that it is <0.1 m). Scaling the plume extent from various simulation runs to the pure CO_2 base case minimizes this effect.

The metrics used to measure impact on storage consisted of (1) time for the plume to reach the top and, more importantly, (2) extent of the plume at a given time or when all of the injected CO_2 mixture had been immobilized, and (3) time until all CO_2 mixture was immobilized (Kumar *et al.*, 2009). After treating the homogeneous case, we developed reservoir models encompassing a range of heterogeneity:

(1) we handled heterogeneity in a simplistic way by adding four baffles with no porosity, parallel to the formation top and bottom, just upstream of the injection well and short of a few cells, all the way up to the updip boundary and across the whole width of the model (Figure 2a);
(2) we obtained multiple heterogeneous fields through permeability generators (Figure 2b); and
(3) we used models from actual sites.

First, we estimated static capacity, which is especially relevant to the case of structural traps because it relates to the volume occupied by the mixture in the subsurface. Comparison of densities as a function of depth allows for a first-order comparison of capacities. Second, we examined dynamic capacity. Results are consistent with that of a previous study (IEAGHG, 2011; Wang *et al.*, 2011). They addressed topics very similar to those discussed in this document. They focused on the capacity issues and reported that non-condensables such as N_2, O_2, and Ar impact capacity, and that the impact is maximal at a certain pressure under a given temperature. Impurities impact static capacity by causing variations in density and viscosity of the CO_2-rich mixture. A lower density impacts CO_2 capacity not only because of the smaller fraction of CO_2 that can be injected and space taken up by impurities, but also because of the generally lower density of the impurities under the same conditions. An approximate proxy for capacity change owing to impurities is provided by the density ratio. The loss of capacity can be >50% at shallow depths (~1000 m, CO_2 and 15% molar N_2) (Figure 3), but the difference quickly decreases with depth. Similarly, mass injectivity, which measures how much CO_2 can be injected (and which can be represented by the proxy metric of the density:viscosity ratio), also exhibits a value that decreases at shallow depths but recovers with increasing depth.

Dynamic reservoir simulations revealed that, following the pattern of static capacity and for the same reasons, impurities impact CO_2 plume shape (rate of vertical ascent and lateral extent) more markedly at shallow depths where the contrast in density and viscosity with pure CO_2 is at its largest. For example, a 4% mole fraction impurity in a binary system is sufficient to increase plume length in 'shallow' low-dip sloping layers by 25%, whereas a mole fraction of 9 to 15%, depending on the component, is needed to create the same impact in a 'deep' system. Note that pure-CO_2 plume extent is larger at depth than in the shallow case, but that the difference between streams of pure CO_2 and impure CO_2 is smaller in the deep model. In all cases, plume extent is greater when impurities are present, although residual trapping occurs more rapidly. This is generally the case regardless of reservoir heterogeneity and complexity, although heterogeneity tends to moderate the impact of impurities on plume extent because of the multiplicity of smaller plumes (Table 3). Heterogeneity tends to increase plume extent because although CO_2 favors higher-permeability streaks, the contrast between pure CO_2 and CO_2 with impurities is smaller. Overall, a trade-off occurs between larger plume lateral extent owing to the presence of impurities and decreased risk owing to faster trapping (pressure management).

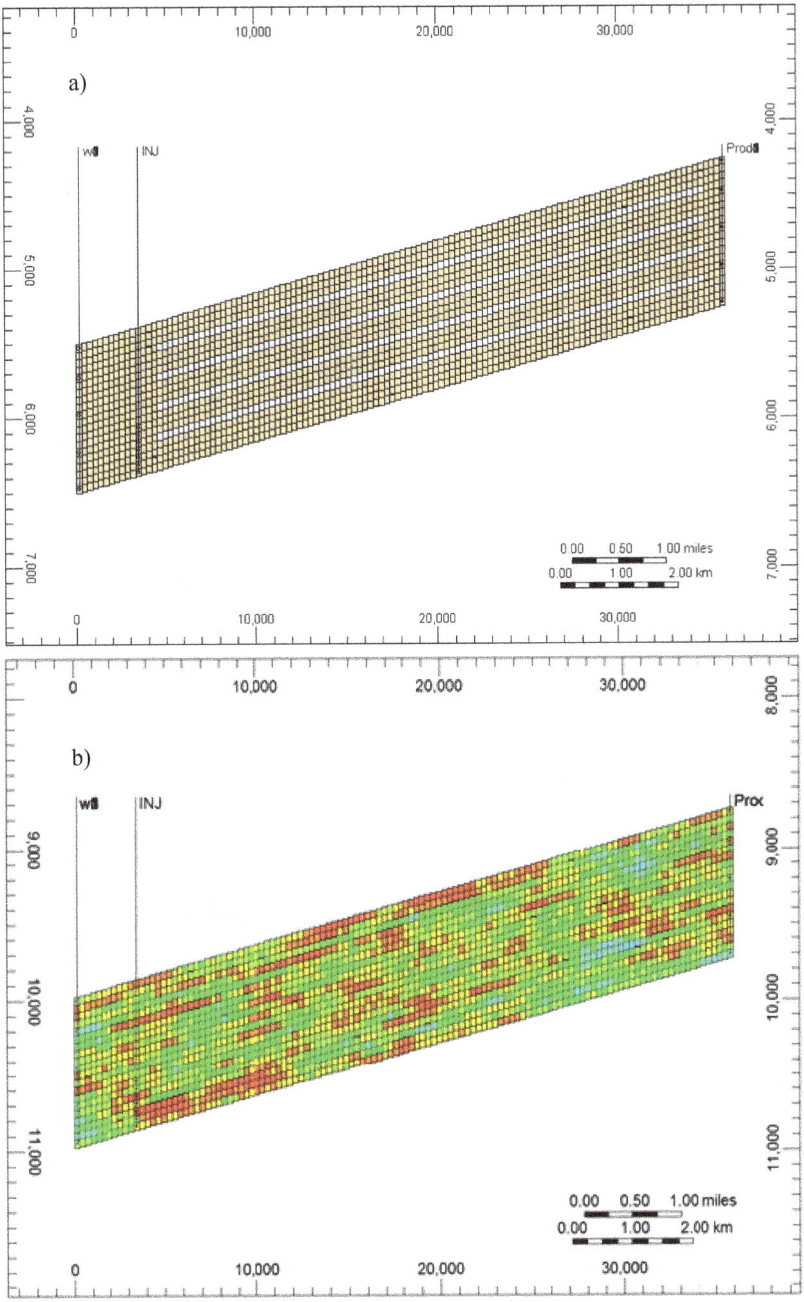

Figure 2. Cross section of generic model displaying homogeneous field with baffles (a) and heterogeneous permeability field (b).

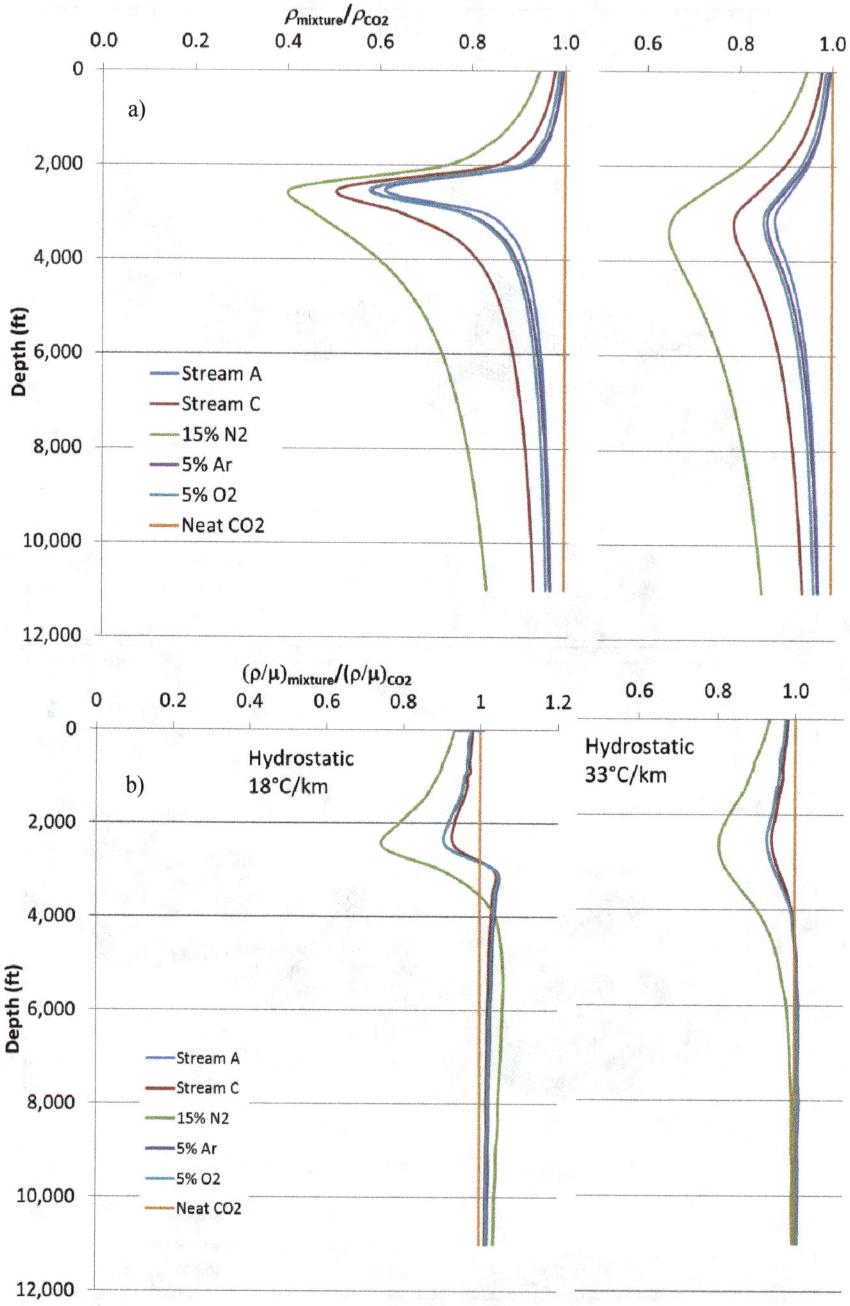

Note: 12,000 ft = 3658 m.

Figure 3. Mixture density (a) and density to viscosity relative to neat CO_2 (b) as a function of depth; hydrostatic conditions and geothermal gradient of 18°C/km and 33°C/km.

Table 3. Comparison of Stream A, stream C, and neat CO_2 (base case) plume extent in various conditions (shallow case): uniform, with baffles, and 5 of several realizations of heterogeneity.

(in m)	Uniform	w/ Baffles	Real.#7	Real.#8	Real.#9	Real.#10	Real.#13
Neat CO_2	2,835	2,560	3,200	2,105	1,555	1,555	2,195
Stream A	3,475	2,835	4,665	2,195	1,645	1,555	2,380
Stream C	4,390	3,290	6,025	2,285	1,645	2,105	2,380
A / neat ratio	1.23	1.11	1.46	1.04	1.06	1.00	1.08
C / neat ratio	1.55	1.29	1.89	1.09	1.06	1.35	1.08

Note: the 5 realizations were computed using a spherical variogram with mean permeability of 255 md and standard deviation of 700 md with a null nugget and a vertical range of 100 ft.

GEOCHEMISTRY FINDINGS

Summary of Geochemistry Findings

The geochemical issue is approached through laboratory autoclave experiments. The autoclave consists of a 250-ml vessel able to sustain temperatures as high as 150°C and pressures as high as 40 MPa, corresponding to reservoir conditions up to a depth of ~3,700 m. A computer automatically regulates pressure and temperature, and the system also allows for water sampling during the experiments. Typically 10 to 15 1-2 ml samples of the solution were taken during the 10-to-16 day course of each of the experiments. Rock samples were exposed to a supercritical mixture of CO_2 and O_2 (in general 3.5% molar) or to pure supercritical CO_2 that filled about half of the reactor cell. The other half consisted in a single core fragment or a few large fragments (~8g total) submerged into ~140 ml of synthetic brine (~1.88 mol NaCl corresponding to a TDS of 110,000 mg/L). The study analyzed three clastic rock samples (Lu *et al.*, 2014a; Lu *et al.*, 2014b): (1) a "dirty sandstone" of Miocene age from a deep well in the shallow offshore off the Texas coast; (2) a relatively clean sandstone from the Cardium Formation of Cretaceous age from Alberta; (3) a chlorite-rich sandstone from the lower Tuscaloosa formation in Mississippi originating from the Cranfield site that BEG has been thoroughly studying for several years. Composition of the samples is presented in Table 4.

Table 4. Summary of sample mineralogical composition.

	Offshore Miocene, TX Well OCS-G-3733 Depth 9205 ft (2805.7 m)	Cardium Sands, AB Well 13-13-048-07W5 Depth 1458 m	Cranfield, MS Tuscaloosa Formation Well CFU31F-3 Depth 10,476.6 ft (3193.3 m)
Quartz	43.5%	75.5%	66.9%
Calcite	11.8%		
Siderite		1%	
Microcline	15.2%	4.2%	
Albite	18.4%	2.5%	1.8%
Chlorite			20.2%
Kaolinite	6.2%	10.4%	7.3%
Illite	5.0%	6.5%	2.0%
Pyrite	Trace++	Trace+++	Trace
Anatase			1.8%
Total	100.1%	100.1%	100%

In addition to quartz, the Miocene sample is dominated by calcite (11.8%) and feldspars (31.6%), the Cardium sample is dominated by clays (16.9%) with some feldspar (6.7%) and siderite (~1%), and the Cranfield sample is dominated by chlorite (20.2%) with some other clays (9.3%). Both the Miocene and Cardium samples show evidence of not uncommon pyrite. The "dirty sandstone" Miocene sample allows for investigating carbonate behavior with and without O_2 whereas the relatively clean and non-reactive Cardium sample is a good candidate to investigate feldspar behavior without the overprint of carbonates. The Cranfield sample with abundant clay minerals dominated by chlorite is even less reactive vis-à-vis CO_2. Minerals potentially sensitive to the presence of O_2 are pyrite (present in the Miocene and Cardium samples), siderite (present in the Cardium sample), and chlorite (abundant in the Cranfield sample). These rocks all contain ferrous iron-bearing minerals.

We performed 19 autoclave experiments. Core segments with pre- and post- reaction rocks were submitted to petrographic analyses (X-ray diffraction –XRD, scanning electron microscope –SEM, and energy dispersive X-ray spectroscopy –EDS) and chemical analyses (TDS, anions, cations, trace elements). The offshore Miocene samples showed dissolution of carbonate (Ca and Mg increase) as well as of feldspars (Ca and K increase, Na concentrations are irrelevant because experiments are done with a NaCl brine). Feldspar dissolution is more intense when O_2 is present. Kaolinite is presumed to form in both cases. The Cardium samples displayed deep attack of carbonates and of some feldspars with kaolinite formation and perhaps very minor authigenic illite. Pyrite and siderite are degraded when O_2 is added and FeOx species precipitate. The Cranfield samples have limited reactivity when exposed to pure CO_2 as they contain little carbonate. When O_2 is added, some chlorite is degraded and FeOx deposits can be observed as well as formation of some authigenic clays.

In terms of release rates, our results confirmed well-known reactions and delivered new observations. In pure-CO_2 cases carbonates were observed to dissolve quickly with a sharp increase in Ca, Mg, and other elements typically present in calcite. Calcite solubility was also observed to decrease with increasing temperature (i.e., reverse solubility). As expected, feldspars showed an increase in solubility with increasing temperature whereas clays, including chlorite, remained unreacted. Adding O_2, however, brought in interesting observations, pyrite framboids were clearly degraded, and thus added H^+ ions to the system with consequent increased carbonate dissolution. The siderite (Cardium sample) is another source of ferrous iron. Both pyrite and siderite attack led to deposition of FeOx on mineral surfaces. Chlorite could also be an important source of ferrous iron but it is mostly stable unless pH drops very low. Feldspar dissolution is enhanced in the presence of O_2 but in an indirect way, though pyrite oxidation and drop in pH. Overall the additional impact of a few percent O_2 when comparing samples reacted with pure CO_2 and with a CO_2+ O_2 mixture is limited. The observation is also true for trace elements released to the brine following mineral dissolution; concentration in some cases increased several-fold, but never at a level high enough to generate a drinking water hazard when diluted with fresh water (relevant if saline waters contacted by CO_2 with impurities were to invade and aquifer). Aqueous concentrations of several elements (V, Mo, As) drop when O_2 is added because they form oxyanions that sorb to precipitating FeOx and clays.

Although in need to be confirmed by flow-through experiments, the preliminary reactive transport modeling done in parallel to the batch experiments suggests that porosity changes due to mineral reactions in siliciclastic material is minor and that mineral precipitation is unlikely to impact fluid flow. Overall, it does not seem that a few percent O_2 in the CO_2 stream has much impact beyond the impact of neat CO_2.

Illustration of Key Geochemistry Findings

The Miocene samples display a more thorough degradation of carbonates and particularly feldspars relative to pure CO_2 case when O_2 is added - the likely impact of minor pyrite oxidation and related pH drop (Figure 4).

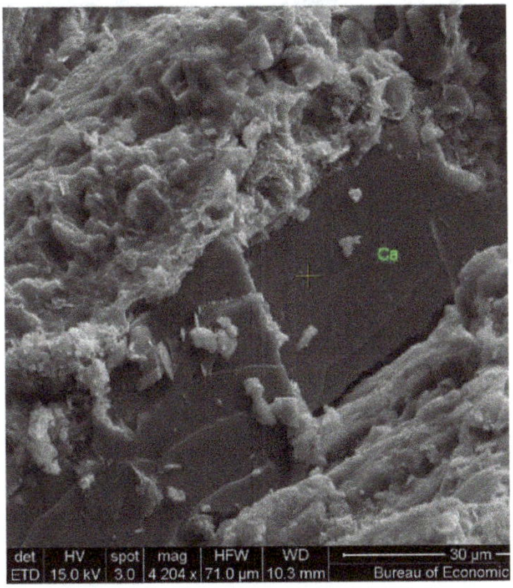

a) Unreacted sample. SEM image showing calcite (Ca) cements with fresh-looking surface.

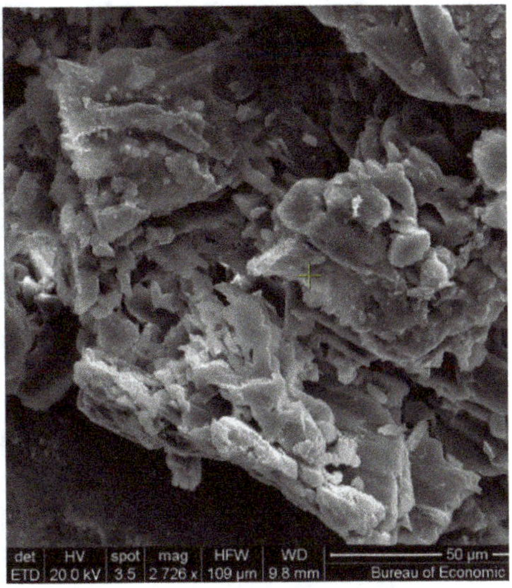

b) Sample surface reacted with brine and CO_2. A strongly corroded plagioclase grain.

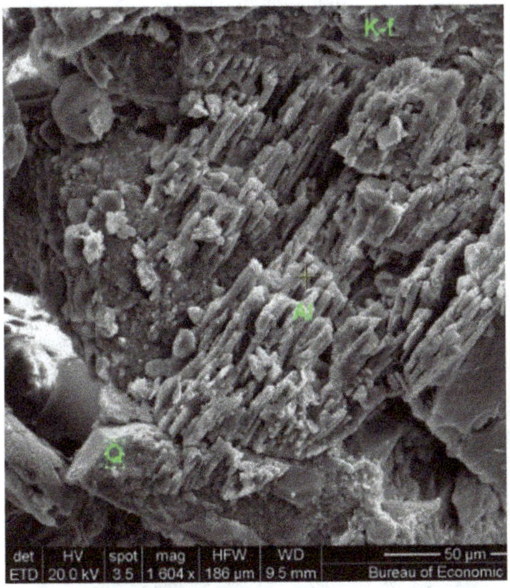

c) Reacted sample with brine and CO_2. A heavily leached albite (Al) grain with only the grain's skeleton left.

d) Reacted sample surface with brine, CO_2 and O_2. Rare remaining of corroded calcite showing severe leaching features.

Figure 4. Unreacted and reacted Offshore Miocene samples.

The Cranfield samples are fairly unreactive under pure CO_2. However, adding O_2 has a clear effect on chlorite crystals (Figure 5) which display some limited degradation dramatically expressed through a dense cover of FeOx precipitates. Very little pyrite, if any, is present in the Cranfield samples (as suggested also by the limited increase in sulfate concentration).

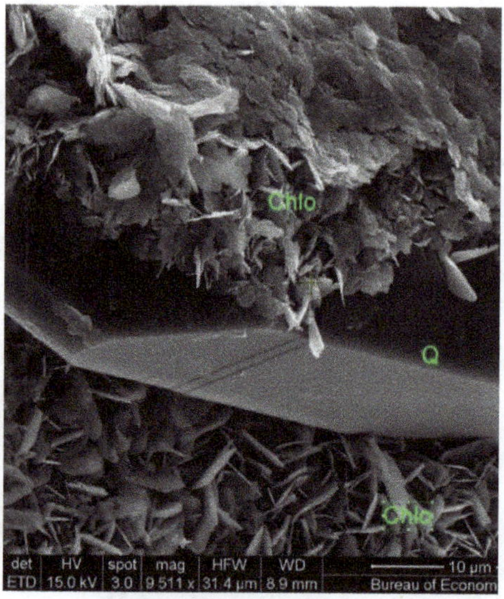

a) Reacted with CO_2. Chlorite flakes surrounding a quartz crystal with no evidence of reaction.

b) Reacted with CO_2 and O_2. Chlorite flakes dotted with iron oxide crystals which may be derived from oxidation of chlorite.

c) Reacted with CO_2 and O_2. Close-up of chlorite covered by iron oxide.

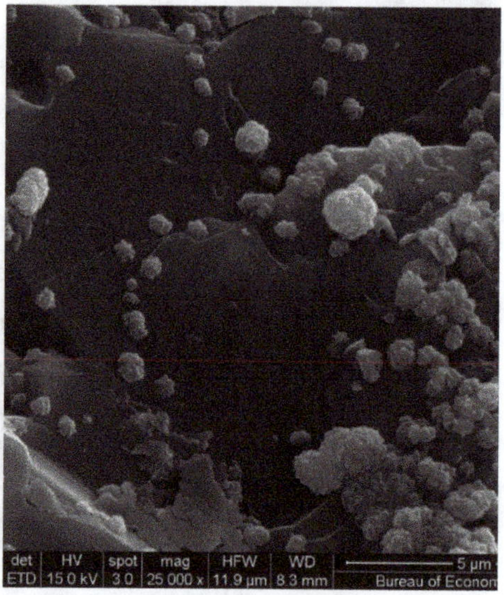

d) Reacted with CO_2 and O_2. Iron oxide buds precipitated on quartz surface.

Figure 5. Unreacted and reacted Cranfield samples.

The Cardium samples (Figure 6) contain two types of ferrous iron-bearing minerals: carbonates (siderite and ankerite) and pyrite. Only the former is corroded with pure CO_2 with ferrous iron released into solution. However, when O_2 is added, pyrite is also oxidized and the soluble ferrous iron is oxidized and precipitated as ferric iron-bearing FeOx minerals.

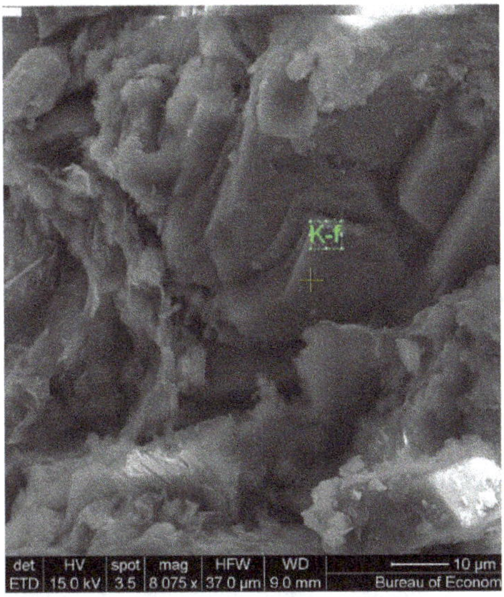

a) Unreacted sample. K-feldspar (K-f) showing clean steps.

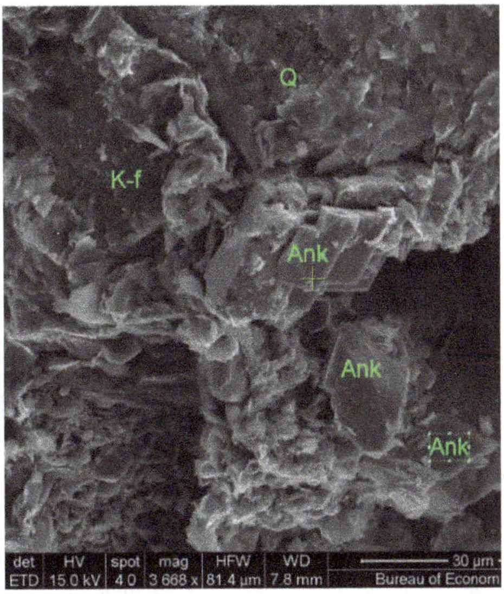

b) Unreacted sample. Euhedral ankerite (Ank) rhombs

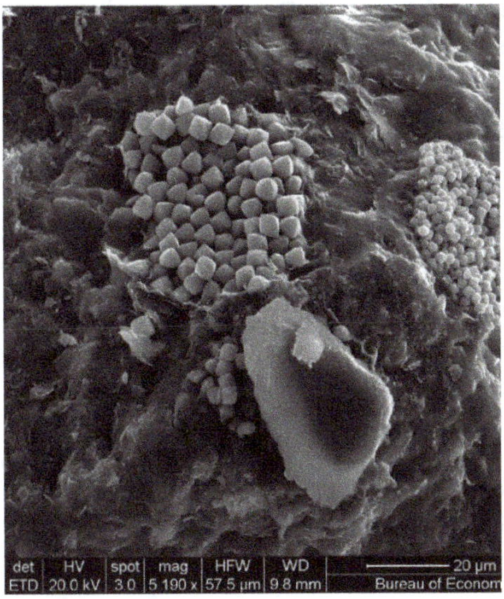

c) Unreacted sample. Small amount of pyrite present as fresh-looking octahedrons.

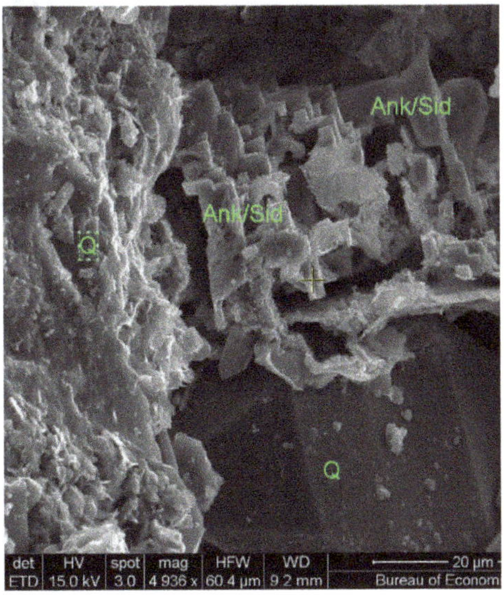

d) Reacted brine and CO_2. Sample showing corroded ankerite (Ank) crystals.

Figure 6. Unreacted and reacted Cardium samples.

We focus on the differences between pure CO_2 and CO_2+O_2 mixtures at different temperatures. An increase in temperature, with no surprise, led to an observed increase in release rates from silicates. O_2 has an impact if there is (1) redox-sensitive mineral species and (2) ferrous iron-bearing minerals (pyrite, chlorite, siderite, ankerite, ferroan dolomite or calcite). The pH drop attending CO_2 injection may be mitigated or reduced by buffering species, such as carbonates or to a lesser extent through dissolution of other species such as feldspars. When reduced species have been mobilized and are in solution, O_2 can oxidize them. A clear example is siderite or pyrite dissolution with iron hydroxides and allied species (FeOx) precipitating. Overall the release rates, although variable, are very fast with asymptotic behavior reached in a few days (Lu *et al.*, 2014a). Clearly this is related to the experimental setup. Columns experiments or field observations will likely reveal a much longer time frame for completion of the reactions. Batch experiments are typically rate-controlled but reactions taking place in column, and especially, field tests are generally diffusion-controlled, a much slower control. These findings are consistent with work by Renard *et al.* (2011) and Shao *et al.* (2013).

To have an impact on permeability by obstructing pores, redox-sensitive minerals must be abundant enough with sufficient oxidizing material (O_2) to impact the system. A reactive transport model using the Cardium sandstones from Alberta as a test case, shows that little material precipitates compared to the pore volume. Permeability change is very sensitive to the exact location in the pore of the authigenic mineral deposits (i.e., pore throats as opposed to pore bodies) but past field experience, such as at the Rousse Field, France (Girard *et al.*, 2013; Monne and Prinet, 2013) or injecting air or flue gas, do not point to this being a major issue.

CONCLUSIONS

The results of this study indicate that the most likely impact of non-compressible CO_2 stream impurities (N_2, Ar, O_2) on clastic reservoirs will result from flow dynamics, particularly at relatively shallow depth (<1200 m). The impact consists in longer plume extents relative to the pure CO_2 case and the additional costs it entails because of the larger Area of Review. Plume shape relative to that of pure CO_2 is more elongated along the bottom of the seal or a barrier. Similarly, because buoyant forces are larger when impurities are present, the plume or plume sections do not widen as much as in the pure CO_2 case in their rise to the top of the formation. However, increasing depth gradually reduces the impact of non-condensable gases because mixtures behave more and more like pure CO_2 due to the decreased density and viscosity contrast with neat CO_2. Reservoir rock heterogeneities tend to dampen the increased plume extent because it is distributed among several tongues (the contrast is most extreme in very simple systems). In addition, trapping occurs faster because the plume is more mobile and thus exposed to greater rock and brine volumes, affording more extensive residual trapping and, possibly, dissolution opportunities. It will also stabilize faster. In contrast, a pure CO_2 plume travels less but stays mobile longer. However, thanks to increased mobility, the impure gas mixture may be more likely to breach a seal via enhanced buoyant pressure and a resultant leak would have to be monitored over a larger area. It also suggests a trade-off between increased plume extent and decreased long-term risk owing to faster trapping.

We addressed capacity of structural traps by comparing density of net CO_2 vs. density of CO_2-rich mixtures as a function of depth. The impact on capacity (with density ratio of mixture to neat CO_2 as a proxy) is a strong function of the geothermal and pressure gradient. For reasonably low and high geothermal gradient, density ratio plots show a large drop in capacity around 1000 m, which is slightly below the accepted depth to keep CO_2 stream in a supercritical state. In other words, the impact of impurities on storage capacity decreases with increasing depth. The effect is the largest for the lower geothermal gradient (only 40% of neat CO_2 capacity at shallow depth) because the density contrast between neat CO_2 and mixture is more pronounced in this case.

Reservoir storage integrity could be impacted if impurities (1) negatively affect the capillary barrier mechanism holding the gas phase in the injection interval or (2) promote permeability of the confining system. An increase in IFT, as it seems would happen with common impurities (N_2, Ar, and O_2), would reinforce the capillary barrier effect making the reservoir more secure if it increases faster than the balancing force due to increased buoyancy. Overall permeability changes resulting from interaction of the seal minerals with reactive gases (mainly O_2) are unclear. On the one hand, precipitation of FeOx may decrease permeability but, on the other hand, oxidation of pyrite, common in shales, a typical seal, may generate a lower pH than with pure CO_2, dissolving carbonates more thoroughly and altering clays in a detrimental way.

Many aspects of impure CO_2 streams remain to be investigated. Because the results depend on viscosity and density models, their sensitivity must be tested with alternate EOS and formulations of the flow parameters. Impact of IFT and wettability changes due to impurities on residual saturation and relative permeability curves also need to be examined. The scope of flow dynamics sensitivity analysis can be further extended to include various formation dip angles, anisotropy ratio, and closed boundary conditions. Other relevant exercises such as monitoring leakage through a (leaking) well located at some distance from the injection well as a function of injection-stream composition could also be performed. Geochemical impacts need to be tested with more reservoir rock types, in particular carbonate rocks. Although batch experiments are very informative, flow-through experiments are needed to better understand potential permeability changes and assess the actual release and reaction rates. Investigative studies of geochemical impact on seals are also needed.

ACKNOWLEDGEMENTS

The authors would like to thank the CO_2 Capture Project Phase 3 for funding this project and, in particular, the project manager, Scott Imbus. We are also grateful to the Computer Modeling Group (CMG), Calgary, Canada, for giving us ~free access to their GEM software and the Alberta Energy Regulator office (AER) for providing the Cardium samples for the autoclave experiments. We are also grateful to the DOE-sponsored SECARB project for providing the Cranfield samples, and to the DOE- and State of Texas-sponsored Mega-Transect Carbon Repository project for providing the Miocene samples.

REFERENCES

1. Kumar, N., Bryant, S. L., and Nicot, J. -P., 2009, Simplified CO_2 plume dynamics for a certification framework for geologic sequestration projects, Energy Procedia, v. 1(1), Proceedings of 9th International Conference on Greenhouse Gas Control Technologies GHGT9, November 16–20, 2008, Washington D.C., p.2549–2556.
2. Hovorka, S. D., Doughty, Christine, Benson, S. M., Freifeld, B. M., Sakurai, Shinichi, Daley, T. M., Kharaka, Y. K., Holtz, M. H., Trautz, R. C., Nance, H. S., Myer, L. R., and Knauss, K. G., 2006, Measuring permanence of CO_2 storage in saline formations: the Frio experiment: Environmental Geosciences, v. 13, no. 2, p.105-121.
3. Hovorka, S. D., Meckel, Timothy, and Treviño, R. H., 2013, Monitoring a large-volume injection at Cranfield, Mississippi--Project design and recommendations: International Journal of Greenhouse Gas Control, v. 18, p.345-360.
4. Nicot, J. -P., Solano, S., Lu, J., Mickler, P., Romanak, K., Yang, C., and Zhang, X., 2013, Potential subsurface impacts of CO_2 stream impurities on geologic carbon storage: Energy Procedia, v. 37, Proceedings of 11th International Conference on Greenhouse Gas Control Technologies GHGT11, November 18-22, 2012, Kyoto, Japan, p.4552–4559.
5. Thibeau, S., and A. Dutin, 2011, Large scale CO_2 storage in unstructured aquifers: Modeling study of the ultimate CO_2 migration distance, Energy Procedia, v. 4, Proceedings of 10th

International Conference on Greenhouse Gas Control Technologies GHGT10, September 19-23, 2010, Amsterdam, The Netherlands, p.4230–4237.

6. IEAGHG, 2011, Effects of Impurities on Geological Storage of CO_2, report 2011/04 prepared by CanmetENERGY, Natural Resources Canada, June, 63 pages + Appendices.
7. Wang, J., 2011, D. Ryan, E. J. Anthony, N. Wildgust, and T. Aiken, 2011, Effects of impurities on CO_2 transport, injection and storage, Energy Procedia, v. 4, Proceedings of 10th International Conference on Greenhouse Gas Control Technologies GHGT10, September 19-23, 2010, Amsterdam, The Netherlands, p.3071–3078.
8. Lu, J., P. J. Mickler, J.-P. Nicot, C. Yang, K.D. Romanak, 2014a, Geochemical impact of oxygen on siliciclastic carbon storage reservoirs: International Journal of Greenhouse Gas Control, v. 21, p. 214-231.
9. Lu, J., P. J. Mickler, and J.-P. Nicot, 2014b, Geochemical impact of oxygen impurity on siliciclastic and carbonate reservoir rocks for carbon storage: Proceedings of 13th International Conference on Greenhouse Gas Control Technologies GHGT12, October 5-9, 2014, Austin, Texas, USA.
10. Renard, S., Sterpenich, J., Pironon, J., Chiquet, P., Lescanne, M., Randi, A., 2011, Geochemical study of the reactivity of a carbonate rock in a geological storage of CO_2: implications of co-injected gases. Energy Procedia, v. 4, Proceedings of 10th International Conference on Greenhouse Gas Control Technologies GHGT10, September 19th 23, Amsterdam, The Netherlands, p. 5364–5369.
11. Shao, H., Kukkadapu, R.K., Krogstad, E.J., Newburn, M.K., Cantrell, K.J., 2013, Leaching of toxic metals from geologic CO_2 sequestration reservoir materials and the impact of oxygen. in: The 12th Annual Conference on Carbon Capture Utilization and Sequestration, May 13–16, Pittsburgh, PA.
12. Girard, J.-P., P. Chiquet, S. Thibeau, M. Lescanne, and C. Prinet, 2013, Geochemical assessment of the injection of CO_2 into Rousse depleted gas reservoir. Part I: Initial mineralogical and geochemical conditions in the Mano reservoir: Energy Procedia, v. 37, Proceedings of 11th International Conference on Greenhouse Gas Control Technologies GHGT11, November 18-22, 2012, Kyoto, Japan, p. 6395–6401.
13. Monne, J. and C. Prinet, 2013, Lacq-Rousse Industrial CCS reference project: Description and operational feedback after two and half years of operation: Energy Procedia, v. 37, Proceedings of 11th International Conference on Greenhouse Gas Control Technologies GHGT11, November 18-22, 2012, Kyoto, Japan, p. 6444–6457.

Carbon Dioxide Capture for Storage in Deep Geological Formations, Volume 4
Karl F. Gerdes (Editor)

Chapter 27

COUPLED RESERVOIR AND GEOMECHANICAL MODELING OF HYSTERESIS EFFECTS ON CAPROCK INTEGRITY FOR CO_2 STORAGE PROJECTS

Somayeh Goodarzi[1] and Dale Walters[2]
[1,2]Taurus Reservoir Solutions, Suite 1060, 1015 – 4th St. S.W., Calgary, Alberta, Canada T2R 1J4

ABSTRACT: A new study on geo-mechanical effects of CO_2 storage was commissioned by the Carbon Capture Project (CCP) through Taurus Reservoir Solutions, Calgary, Canada in 2013, with the objective of better understanding system containment vulnerabilities induced by field development and production operations prior to injection for dedicated storage. The study looked at cumulative reservoir/seal stress changes (via primary and secondary recovery (i.e. waterflood)) that may impact seal resilience during CO_2 injection for storage. A model for an analogue carbonate storage reservoir was developed utilizing publicly available data for the Weyburn Field (Canada). Data included rock and fluid properties, initial stress distribution, production history, initial pressure and temperature, and were used as input to formulate a coupled flow, thermal and geomechanical numerical model to study operational stress path and hysteresis effects in the analogue reservoir. Taurus' propriety GEOSIM simulator was used for this study.

Although the Weyburn field was used as a basis for this case study, there were simplifications which make the observations based on modeling results illustrative in nature, rather than specific to the actual Weyburn field. Thermo- and poro-elastic processes modeled on this field show the relative magnitude of deformation and resultant probability of tensile and/or shear failure of the reservoir and top seal. This study showed that thermoelasticity dominates during the early stage of the waterflood which resulted in tensile fracturing of the reservoir and partial growth of the fracture into the caprock without violating the integrity of the caprock. The maximum injection pressure for avoiding hydraulic fracturing in the reservoir is determined to be around 26 MPa. This is equivalent to the original minimum total stress. Considering the effect of hysteresis on the geomechanical properties and stress changes resulted in lower fracture pressure in the reservoir during CO_2 storage. When hysteresis is included, the maximum injection pressure falls to around 21 MPa which is ~95% of the original fracture pressure. Taurus' results show that the existing caprock in the study area will not be compromised prior and during CO_2 storage. However if the overlying caprock is assumed to have a relatively low rock cohesion (e.g., 3000 kPa), it is expected to undergo shear failure during CO_2 storage. This would increase the chance of CO_2 leakage to upper permeable layers.

KEYWORDS: CO_2 storage; geomechanics; fracturing; thermal

INTRODUCTION

The objective of the proposed scope of work was to evaluate the geomechanical effects of the full operational history of a typical storage reservoir on the seal integrity of that reservoir, specifically during the period of CO_2 storage. Data describing the Weyburn field was used to set up a simplified model system (particularly details of the operational history) for this study. The study focussed on the operational stress path in the caprock over the life of a CO_2 storage reservoir as a measure of the system's containment vulnerabilities. The study included three operational phases: 1) primary production (and associated reservoir compaction and/or shear failures); 2) secondary production by waterflooding (rebound effects due to pressure recovery) and 3) CO_2 storage both below and at fracturing conditions (and associated rebound, potentially even net heave). The stress path was evaluated through this operational history and monitored with respect to the caprock material failure surfaces (shear and tension). Stress levels were used to track the position of the effective stress state to the respective failure surfaces. Hysteresis of the stress-strain behavior of the reservoir material was included based on the material description and used to estimate the potential for permeability enhancement with failure that may impact the seal integrity. Throughout this chapter, results are linked to the Marly and Vuggy formations for convenience. The reader should keep in mind that the modeled results apply to the simplified analgue system and are not meant to reflect expected outcomes in the real Weyburn Field.

BACKGROUND

Taurus' GEOSIM™ software was used to simulate the operational field history to investigate the geomechanical effects imposed on the caprock. GEOSIM™ [1] is Taurus' proprietary coupled reservoir/geomechanics simulator. GEOSIM™ is a modular software system combining a 3D, 3-phase thermal reservoir simulator with a general 3D finite element stress-strain simulator. The modular structure of the GEOSIM™ software allows flexibility in the level of coupling between the geomechanical and reservoir flow/heat systems. The details of the coupling have been described by [2]. In this study two coupling methods of GEOSIM™ were considered:

- Explicitly coupled: Each timestep the reservoir simulator uses the coupling terms from the geomechanical solution of the previous timestep.
- Iteratively coupled: Within each timestep, the reservoir and geomechanical solutions are iterated until convergence with the coupling parameters updated at each iteration.

The explicit coupling mode of GEOSIM™ is typical for most studies and results in reasonable simulation times. In some instances, poroelastic stress changes away from the pressurized zone that result from displacement transfer may require a tighter coupling. By displacement transfer we refer to the deformations induced by the reservoir zone of pressure and temperature change that are transferred to the surrounding non-reservoir material. However, since the stiffness of the reservoir and caprock rock material are large, explicit coupling was assumed to adequately capture these effects and was adopted in this study.

GEOSIM™ simulations were calibrated to historical data for a representative sector model of an inverted nine-spot pattern. The GEOSIM™ model allowed the inclusion of the following coupled processes:

- Reservoir compaction causing changes in porosity and permeability due to a decrease in pressure and increase in effective stress
- Stress transfer to material surrounding the reservoir (including caprock) and associated changes in pressure.
- Increase in porosity and permeability with decreased effective stress, increased shear stress and potential microfracturing around the main tensile fractures
- Creation and propagation of horizontal and vertical fractures.
- Poroelastic pressure changes in adjacent or overlying zones. The displacement field transfer beyond the pressurized storage zone inducing pressure changes in porous, low permeability materials.

This represents a dynamic model of the reservoir level deformations. Once the model is calibrated, it can be used for forward and sensitivity modeling of variations based on the storage operational constraints. This type of model has been used extensively by Taurus for geomechanical simulations of compaction drive reservoirs, waterflood and PWRI fracturing, CO_2 injection, tight gas fracturing and reservoir performance [3-7]. An outline of the study methodology is:

- Build Reservoir and Geomechanical model

  a. Convert case study reservoir model to GEOSIM™ with the purpose of obtaining a representative storage reservoir structure for caprock deformations and stress paths. The GEOSIM™ model captured the reservoir structure and rock property descriptions (porosity, permeability, saturations), but not the full details of well configurations and field development. No history matching of individual well or field performance was completed. Only average pressure levels due to depletion or injection were matched. In order to save computational time and to quickly examine the scale of the pore pressure changes, the reservoir simulations in this section were run in an uncoupled flow only mode. Also since the considered reservoir is not highly compressible, using the uncoupled mode was justified.
  b. Extend reservoir grid to model caprock and surrounding formations.
  c. Build 1D MEM for the intended pilot application based on log data.
  d. Stress Characterization: average gradients by stratigraphic layer of log derived vertical profile.
  e. Geomechanical material characterization: reservoir and caprock.

- Primary Recovery (Primary depletion)

  a. Simulate primary production (generic well locations and controls to represent final average reservoir pressure).
  b. Simulate compaction and calibrate to average field observations.
  c. Evaluate stress path of caprock and monitor for failure using stress ratios.

- Secondary recovery: Waterflooding

  a. Follow primary recovery with the addition/conversion of injectors.
  b. Simulate the waterflood process (below and above fracturing conditions) and pressure recovery of reservoir.
  c. Simulate rebound of reservoir material and associated loading or unloading (thermal effects) to the caprock.
  d. Simulate cooling of the reservoir and caprock with long-term water injection and change in the stress state.

- CO_2 injection and long-term storage

  a. Forecast field abandonment pressure distribution (Secondary depletion).
  b. Forecast CO_2 injection for long-term storage. The GEOSIM™ model will include solubility of CO_2 in the oil, but at this point cannot rigorously model solubility in both the oil and water phases. This effect will be neglected for this study focusing on repressurization levels and rates rather than accuracy of stored volume.
  c. Inject CO_2 below and at fracturing conditions.
  d. Evaluate the rate of pressurization of the field, cooling of the field and the associated stress path in the caprock.
  e. Monitor caprock for failure using stress ratios.
  f. Recommend safe repressurization levels to avoid shear and tensile failure in the caprock.

TECHNICAL APPROACH

The fluid flow model developed in this study is based on the available public data for the Weyburn project. An inverted 9-point element of symmetry grid is constructed with one vertical well at each corner. The size of the model was chosen based on the reported well spacing in Weyburn project (79 acre pattern=682 m well spacing) [8], and later was increased to 782 m to match the original oil in place (from an old Taurus case study) and the average reservoir pressure after primary depletion. The reservoir is composed of two geological layers: dolostone dominated Marly and limestone dominated Vuggy. The input thicknesses of these two layers are in the range reported by Jimenez *et al* [9]. Table 1 shows the model thickness, porosity and permeability for the reservoir and surrounding formations [8-9]. Marly and Vuggy are refined into three and two layers of 1, 2, 4 and 3, 5 m respectively. The caprock is also divided into two layers to predict and capture the potential fracture growth into this layer with better accuracy. The number of grid blocks is 38x38x8 in X, Y and Z directions respectively.

Figure 1 shows the gridding and the location of wells for the inverted 9-point element of symmetry model.

Table 1. Reservoir and Surrounding Formation's Geological Properties

Coarse reservoir simulation layer based on Geology	Thickness (m)	Porosity (fraction)	Initial Horizontal Permeability (md) (Kv/Kh=0.1)
Anhydrite overburden	6	0.04	1E-6
Marly	7	0.26	10
Vuggy (intershoal)	3	0.15	3
Vuggy	5	0.15	50
Frobisher underburden	4	0.04	1E-6

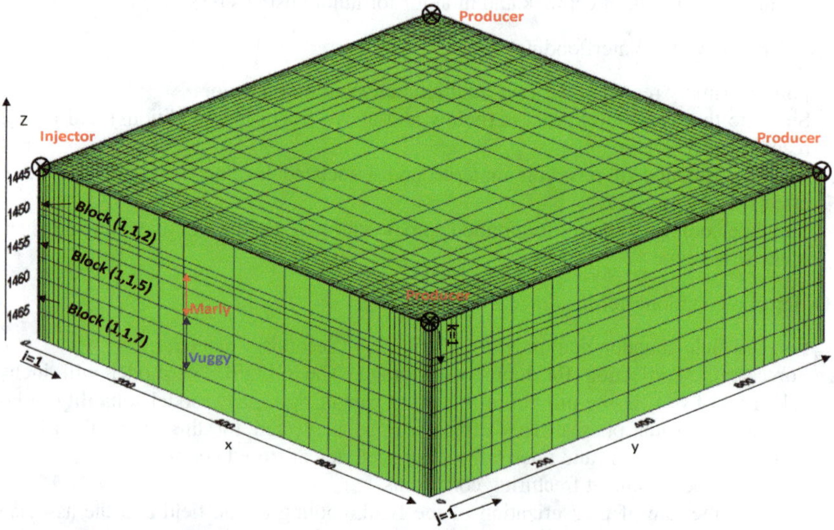

Figure 1. Gridding and the location of the wells for the 9-point element of symmetry model.

The initial pressure, temperature and bubble point pressure are 14,500 kPa, 60 °C and 6,900 kPa respectively [9]. Figure 2 shows the input relative permeability curves derived from an old Taurus case study in the Weyburn area. This function could be re-generated based on a Corey function.

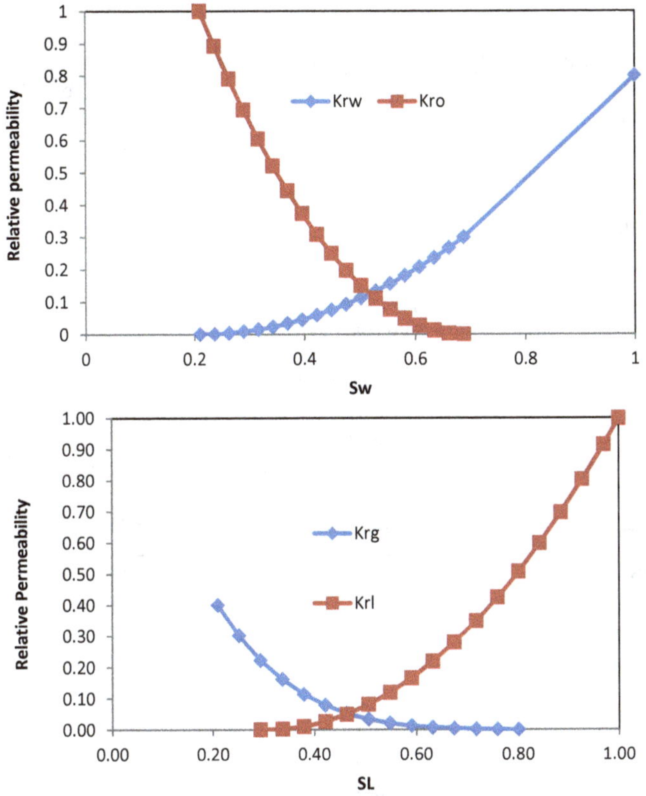

Figure 2. Input Relative Permeability. Krw, Kro, Krg and Krl are water, oil, gas and liquid (water+oil) permeability.

The PVT data generated for the reservoir fluid below bubble point was originally based on the reported fluid properties and later was extended to include CO_2 swelling for gas injection scenarios. An oil API of 32 which is in the range of the measurements reported was assumed[10}. Vasquez and Beggs correlation [11] was used for creating oil formation volume factor and solution gas oil ratio input. Gas formation volume factor was estimated from Z-factor which was derived from Hall and Yarborough correlation [12] with a gas specific gravity of 0.6. Carr *et al* correlation [13] was used for generating viscosity.

Undersaturated oil compressibility was assumed as 7.5E-7 1/kPa based on the reported average oil bulk modulus [14]. The saturated oil viscosity was generated from the Beggs and Robinson correlation and a slope of 4E-6 cP/psi was assumed for the undersaturated oil viscosity.

Results-Uncoupled flow only model

As the first step, the fluid flow model was run in an uncoupled fashion and was compared against the production data from an old study in the area. The uncoupled model was chosen in order to save computational time and to quickly examine the scale of the pore pressure changes. The other reason for using this approach was to validate the model based on the results of an old Weyburn study using a similar fluid-flow-only model. The cumulative production by the end of primary depletion was about 50,000 m^3.

COUPLED GEOMECHANICAL MODEL

The geomechanical grid is usually extended beyond the reservoir in order to eliminate the effects of the boundary conditions. In this case, since the modeling was completed for a 9-point element of symmetry grid, the lateral extent of the geomodel remained the same as the flow model in order to account for the potential full field pressure effect from surrounding wells. The model was however extended vertically to the ground surface above the reservoir and 800 m below the reservoir grid. The overburden and underburden zones were refined into layers with smaller thicknesses to eliminate/lower the effect of boundary conditions on the solution and correctly capture the arching of those zones.

Table 2 lists the input rock mechanical properties used for the target zone and layers surrounding it. Young's Modulus and Poisson's ratio were derived from the reported measured data in [9]. Grain Modulus for the Marly and Vuggy reservoir was taken from [8]. Typical values were assumed for grain Modulus for other layers based on published data [15]. Thermal expansion coefficient was assumed as 1E-5 $^{o}C^{-1}$ for all geomechanical layers [9]. It should be noted that this is a reasonable assumed value, and not based on physical measurements on rocks from the Vuggy or Marly reservoirs. Linear elasticity was chosen for the constitutive models of the rock in this study.

Table 2. Input Geomechanical properties for the reservoir and surrounding layers

Bounding Layers	Thickness (m)	Young's Modulus (GPa)	Poisson's Ratio	Grain Modulus (GPa)
Surface-Ratcliffe beds	1444	11.52	0.33	26
Midale, Anhydrite	6	22.7	0.26	125
Marly, Dolomite	7	10.3	0.29	83
Vuggy,Limestone	8	16.85	0.3	72
Frobisher, Evaporite	4	15.8	0.313	125
Frobisher beds, added underburden	800	15.8	0.313	125

Figure 3 shows the profile of initial pressure, horizontal and vertical stresses with depth in the study area [9]. Based on Jimenez *et al* [9], the azimuth of the maximum horizontal stress is around 40-50°, whereas the minimum horizontal stress has an azimuth of 130-140°. The geo grid in this study was aligned with these directions to avoid having initial non-zero shear stresses.

The boundary constraints for the geomechanical model are as follows. The right and left sides of the model are fixed in the x-direction and the front and back sides of the model are fixed in the y-direction. The bottom side of the model is fixed in the vertical direction and the top of the model is free to move in all directions.

In order to account for the post-production compaction effect, a stress dependant porosity multiplier function was incorporated in this model based on equations (1-4).

$$\Phi^{mult} = \frac{\Phi^{*n+1}}{\Phi^{o}} = \frac{\frac{V_p^{n+1}}{V_b^{o}}}{\Phi^{o}} = \frac{\frac{V_b^{n+1}-V_s^{n+1}}{V_b^{o}}}{\Phi^{o}} \qquad \text{Equation 1}$$

$$V_b^{n+1} = V_b^{n} - V_b^{o}(\Delta\varepsilon_v) \qquad \text{Equation 2}$$

$$\varepsilon_v^{n+1} = \varepsilon_v^{n} + C_b\Delta\sigma'_{avg} \qquad \text{Equation 3}$$

$$V_s^{n+1} = V_s^{n} - V_s^{n}C_s(\Delta P) \qquad \text{Equation 4}$$

where, Φ is the porosity, V_b is the bulk volume, V_s is the grain volume, V_b is the pore volume, ε_v is the volumetric strain, σ'_{avg} is the average effective stress, C_b is the bulk compressibility, C_s is the grain compressibility and P is pressure. The superscripts are: *mult* is for multiplier, * is for apparent porosity, o is for original, *n* and *n*+1 is for time levels of *n* and *n*+1.

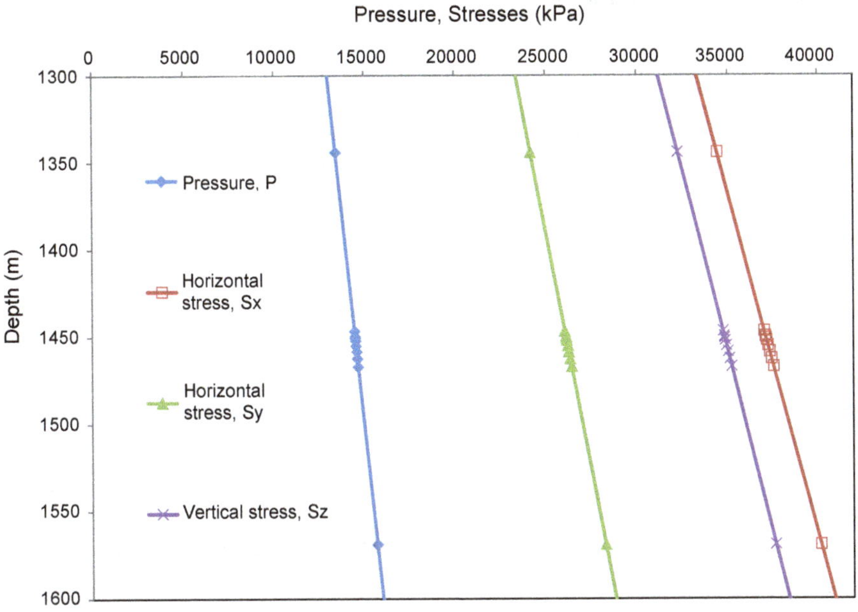

Figure 3. Initial horizontal and vertical stresses (S) by depth.
Horizontal stresses are in x and y direction and vertical stress is in z direction.

With inclusion of this function, the compaction effect is considered by dynamic incorporation of porosity as pressure is reduced due to reservoir depletion.

MODELING NATURAL FRACTURES

The first set of coupled geomechanical models were run with input mechanical and stress properties provided in Table 2 and Figure 3. The results showed that the Vuggy layer would fracture during the waterflood under an injection pressure which was much lower than expected based on the observed fracture pressure in the field and Taurus's old studies in the area. This has been linked to the fact that input moduli might have been overestimated and resulted in much larger thermal induced stresses which causes a lower fracture pressure. This speculation was supported by a University of Alberta geomechanical lab study which reported a Young's modulus of 2E6 and 4E6 GPa for the Marly and Vuggy cores [16]. Later it was reported that the core was damaged and the measured moduli did not represent the intact rock (matrix) moduli. However, these softer moduli values were assumed to be representative of the reservoir materials (Marly and Vuggy) with microfractures present – as is expected in the field.

STRESS DEPENDANT MODULI FUNCTION

The reported existence of natural fractures in the reservoir indicates there may be softening of moduli occurring with opening of natural fractures during waterflood [17]. The idea of incorporating natural fractures in the model was also supported by the observed anisotropy in the horizontal permeabilities from past history match experiences [18-19]. Mclellan *et al* [17] has reported the mean fracture azimuth as 51° which is very close to the maximum horizontal stress direction (59°). This suggests that the natural fractures are perpendicular to the minimum horizontal stress direction. Figure 4 shows the assumed natural fracture direction with respect to direction of stresses in the field.

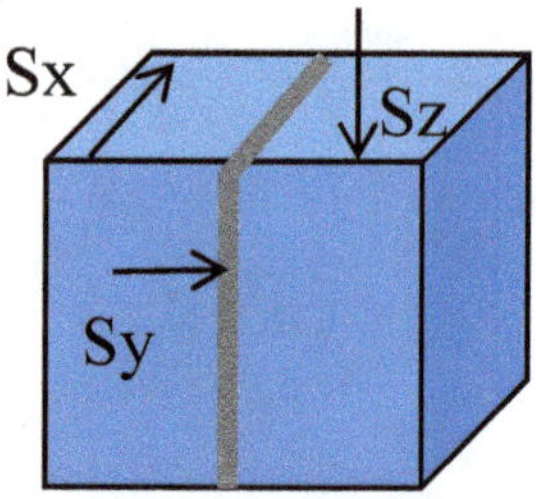

Figure 4. Schematic of the simulation grid block containing natural fracture.

Churcher and Edmunds carried out a fracture study for PanCanadian in 1994, where core from over 150 vertical wells was described, along with five FMS logs from five horizontal wells and 51 Repeat Formation Tests (RFT). Their results showed a vertical to sub-vertical set of fractures oriented NE-SW, which controls the directional permeability of the field. A fracture spacing of about 3 m for the Marly, 0.3 m for the intershoal Vuggy, and 2.5 m for the shoal Vuggy was identified in their results. Fracture heights of 28 cm and 47 cm with average apertures of 0.06 to 0.4 mm were measured for the Marly and Vuggy in the Midale field, respectively. Fracture aperture is reported to vary between 0.004 mm and 0.2 mm for the Marly and Vuggy in Weyburn field [9].

In this study we have assumed an average fracture spacing of 3 and 1 m for the Marly and Vuggy reservoir layers respectively. The maximum fracture aperture for the Marly and Vuggy is taken from the reported range as 0.09 and 0.3 mm respectively.

A Barton-Bandis constitutive model [20] for joints was used for modeling the mechanical behavior of the fracture elements (Equations 5-9).

$$\sigma_n = \frac{K_{ni} V_j}{1 - \frac{V_j}{V_{jmax}}}$$ Equation 5

$$v_j = \frac{\sigma_n}{k_{ni} - \frac{\sigma_n}{v_{jmax}}}$$ Equation 6

$$\varepsilon_j = \frac{v_j}{s}$$ Equation 7

$$K_n = \frac{\partial \sigma_n}{\partial V_j} = K_{ni}\left[1 - \frac{\sigma_n}{V_j K_n + \sigma_n}\right]$$ Equation 8

$$\frac{1}{E_t} = \frac{1}{E_m} + \frac{1}{S * K_n}$$ Equation 9

The K_{ni} was back calculated based on the fracture spacing and the measured moduli for the damaged core for the Marly and Vuggy.

Figure 5 shows the input stress dependent moduli function for the Marly and Vuggy layers.

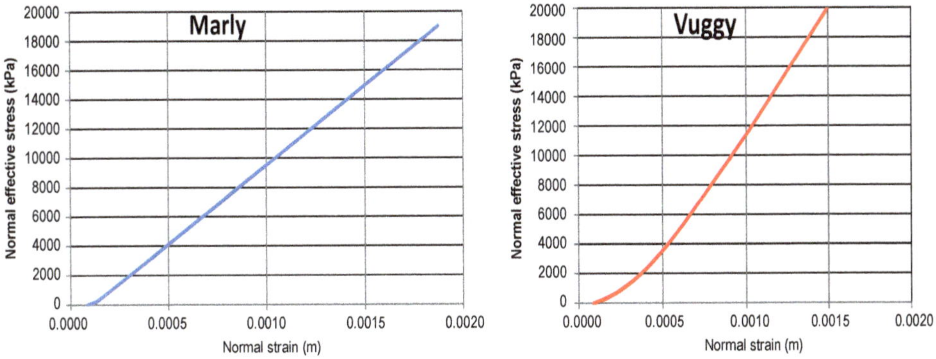

Figure 5. Input stress dependant moduli function for the Marly and Vuggy.

STRESS DEPENDENT PERMEABILITY FUNCTION

During changes in effective stress, natural fractures will open or close potentially changing the permeability of the reservoir dramatically. Therefore a dynamic permeability function based on the displacement of fracture walls derived from Equation (6) as a function of minimum effective stress was used in the model. The algorithm for deriving the two stress dependent functions is described in Figure 6, and Figure 7 shows the resulting function for the Marly and Vuggy layers.

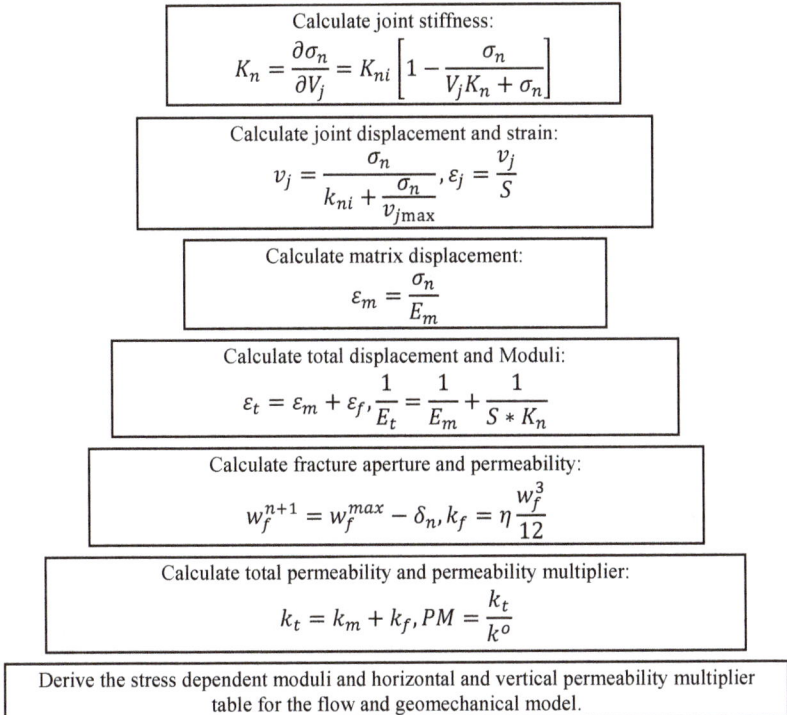

Calculate joint stiffness:

$$K_n = \frac{\partial \sigma_n}{\partial V_j} = K_{ni}\left[1 - \frac{\sigma_n}{V_j K_n + \sigma_n}\right]$$

Calculate joint displacement and strain:

$$v_j = \frac{\sigma_n}{k_{ni} + \frac{\sigma_n}{v_{j\max}}}, \varepsilon_j = \frac{v_j}{S}$$

Calculate matrix displacement:

$$\varepsilon_m = \frac{\sigma_n}{E_m}$$

Calculate total displacement and Moduli:

$$\varepsilon_t = \varepsilon_m + \varepsilon_f, \frac{1}{E_t} = \frac{1}{E_m} + \frac{1}{S * K_n}$$

Calculate fracture aperture and permeability:

$$w_f^{n+1} = w_f^{max} - \delta_n, k_f = \eta \frac{w_f^3}{12}$$

Calculate total permeability and permeability multiplier:

$$k_t = k_m + k_f, PM = \frac{k_t}{k^o}$$

Derive the stress dependent moduli and horizontal and vertical permeability multiplier table for the flow and geomechanical model.

Figure 6. Deriving the stress dependent moduli and permeability multiplier function based on minimum effective stress.

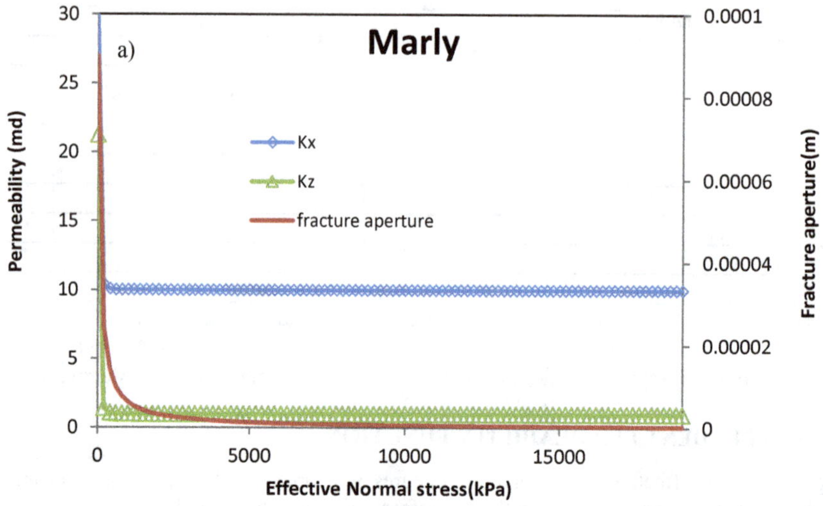

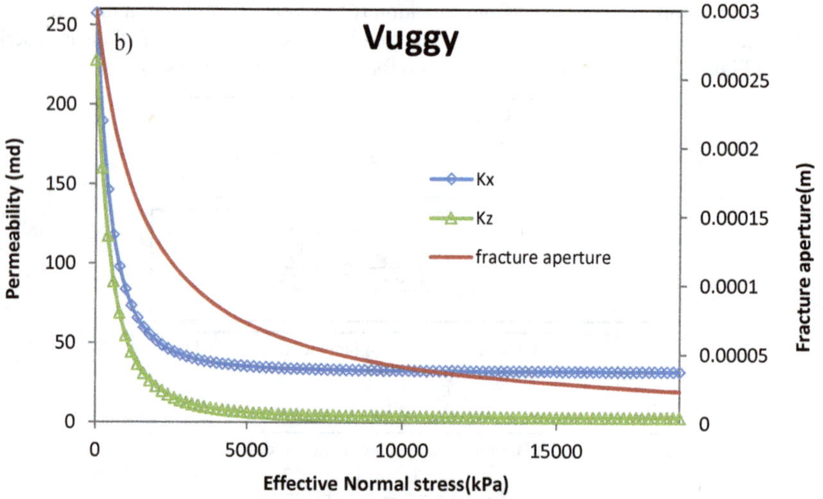

Figure 7. Permeability function for the a) Marly and b) Vuggy layers as a function of minimum effective stress.

DYNAMIC FRACTURE PROPAGATION MODELING

If the minimum effective stress falls below the tensile strength of the rock, tensile fracturing is expected in the rock. This study includes a strong thermoelastic effect (due to cooling) and, therefore, hydraulic fracturing is likely to occur in the reservoir. This was considered in the Geosim model by defining a separate effective stress dependent multiplier to account for the dynamic hydraulic fracture propagation. Figure 8 shows the induced fracture plane with respect to stress directions. This figure shows a section which is cut from the full geomechanical model with the injection well located at the front corner of this grid. Blue and yellow layers in this picture show the reservoir and surrounding layers respectively.

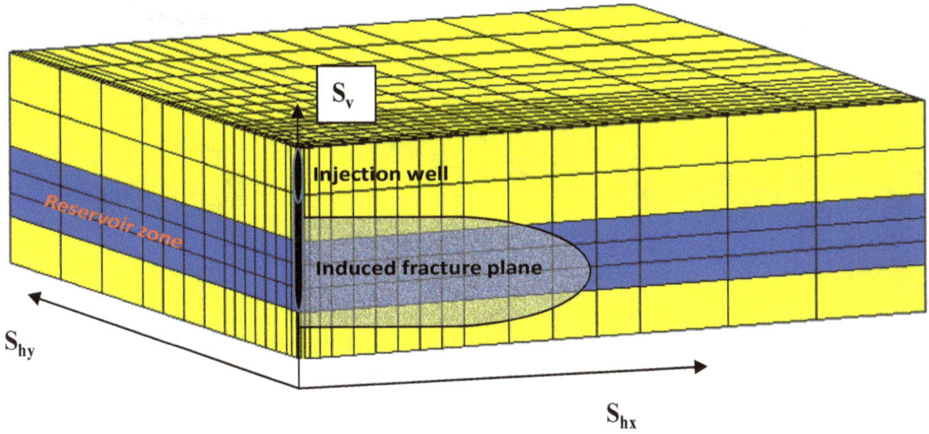

Figure 8. Schematic showing an arbitrary induced fracture plane direction with respect to stress directions: S_{hx}: horizontal stress in x-direction, S_{hy}: horizontal stress in y-direction, S_v: vertical stress in z-direction.

In order to model the dynamic fracture propagation, a permeability multiplier table was incorporated in the reservoir module. It is important to note that an initial value of the fracture half height is used to derive the transmissibility function but the actual fracture height is dynamically changing as a function of minimum effective stress. The entries can be calculated based on the estimation of the width of a 2D crack in a cross-section [21] as a function of the pressure in the fracture as described in equations (10-12). H_f in Equation (11) is the estimated fracture half-height based on the 2-D Perkins-Kern geometry assumption of vertical fracture with smooth closure at the top and bottom [22].

$$K_f = R_{fa}\frac{w_f^2}{12} \qquad \text{Equation 10}$$

$$w_f = \frac{\Delta P H_f}{\bar{E}} = \frac{\Delta P H_f * 4(1-v^2)}{E} = \frac{(P-P_{foc})\mathrm{H}_f * 4(1-v^2)}{E} \qquad \text{Equation 11}$$

$$T_r = \frac{K_m A_m + K_f A_f}{K_m A_m} = 1 + R_{fa}\frac{{w_f}^3}{12 K_m w} = f(P - P_{foc}) \qquad \text{Equation 12}$$

Since P_{foc} in Equation 12 is equal to the stress perpendicular to the fracture plane, ΔP can be also replaced by negative value of effective stress with $\alpha = 1$. The Transmissibility Multiplier table can be incorporated in the model either as a function of net pressure or effective minimum stress. In order to calculate the multiplier function (12), a fracture half height of 7.5 m (equal to half-height of the reservoir) has been considered and the rest of the data are taken from the mechanical properties of the injection zone. Figure 9 shows the incorporated permeability multiplier for the grid blocks along the designated fracture plane in the x and z direction and the coupling algorithm. If the fracture height exceeds the assumed fracture height (7.5 m), the actual multipliers would be higher, but this representation is still valid. This assumed the effect of increased permeability due to fracturing on fluid flow is negligible after the permeability has reached a significantly high magnitude that corresponds to an initial assumed fracture height. When fracture height is extended, the permeability multiplier is also increased further from an already very high multiplier. However the consequent effect on fluid flow is negligible and the assumption remains valid.

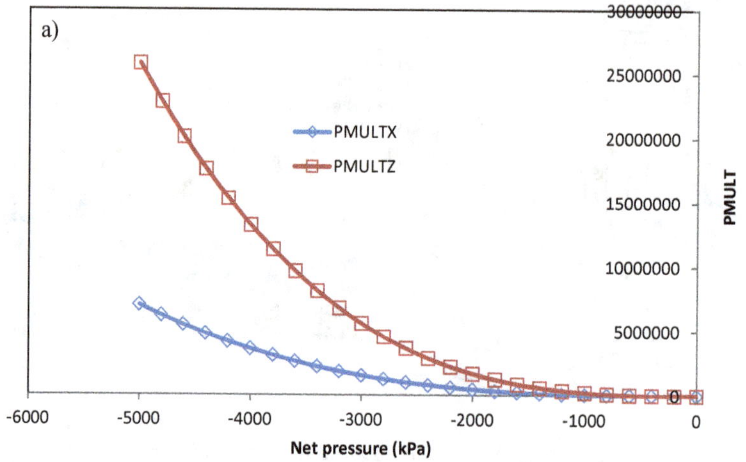

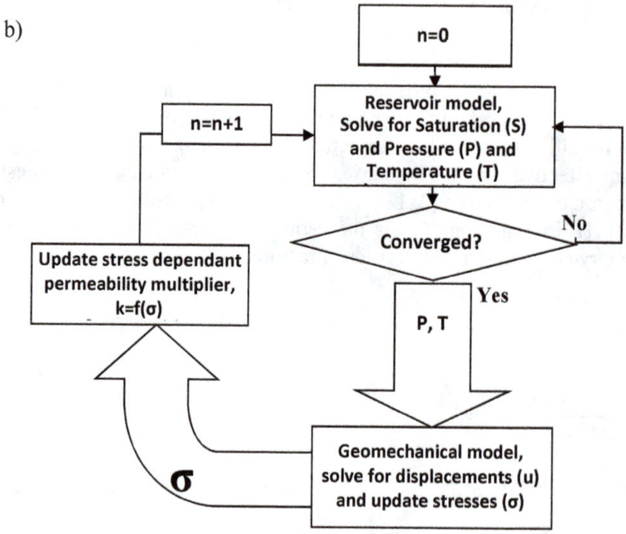

Figure 9. a) Permeability multiplier functions for the grid blocks along the along the designated fracture plane in the X and Z direction in the X direction and b) the coupling algorithm.

RESULTS-ISOTHERMAL WATERFLOOD

The oil production at the end of the primary depletion reached a cumulative amount of 49.2 Mm^3 (3.7% recovery). The primary production was then followed by waterflooding through injector well (Figure 1) with an average injection rate of 40.6 m^3/day for ~36 years. Figure 10 and Figure 11 show the water saturation for the bottom Vuggy layer and the XZ section of the reservoir along the wellbore by the end of waterflooding. As expected, since the bottom Vuggy layer has the highest permeability, most of the injected water is flowing through this layer and by the end of the operation; the injected water has swept a large portion of the Vuggy zone within the reservoir. The cumulative oil production by the end of the water injection is ~298 Mm^3 (22.5 % recovery). The vertical displacement at surface at the end of waterflooding is shown in Figure 12.

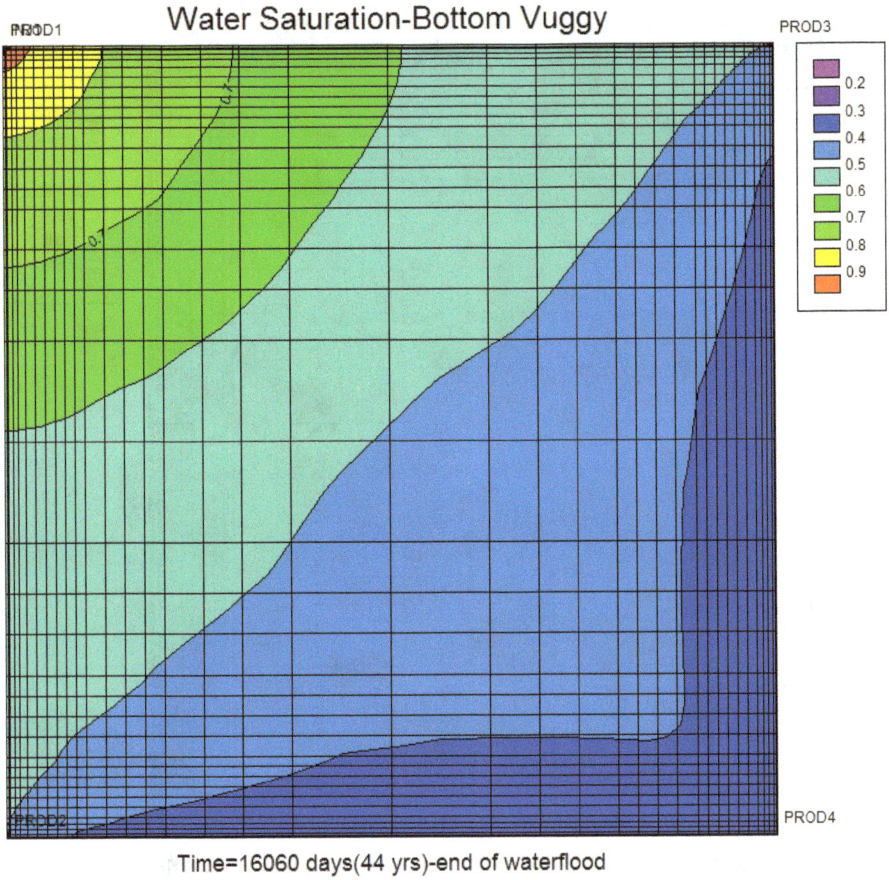

Figure 10. Water saturation for the bottom Vuggy layer by the end of the waterflood- map view.

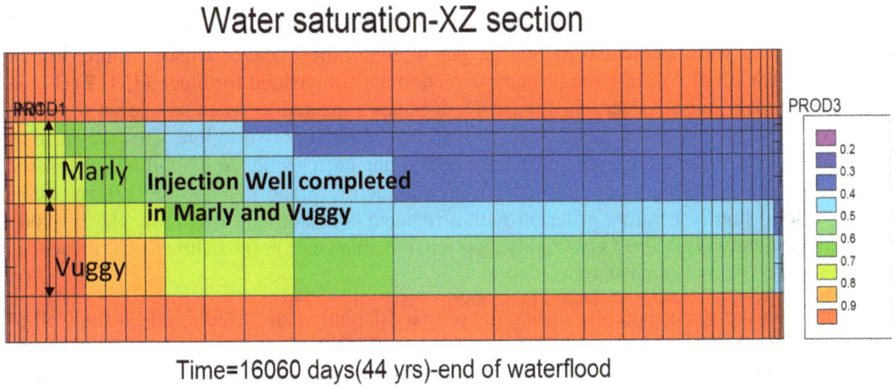

Figure 11. Water saturation for the bottom Vuggy layer by the end of the waterflood-XZ section of the reservoir along the wellbore in z-direction.

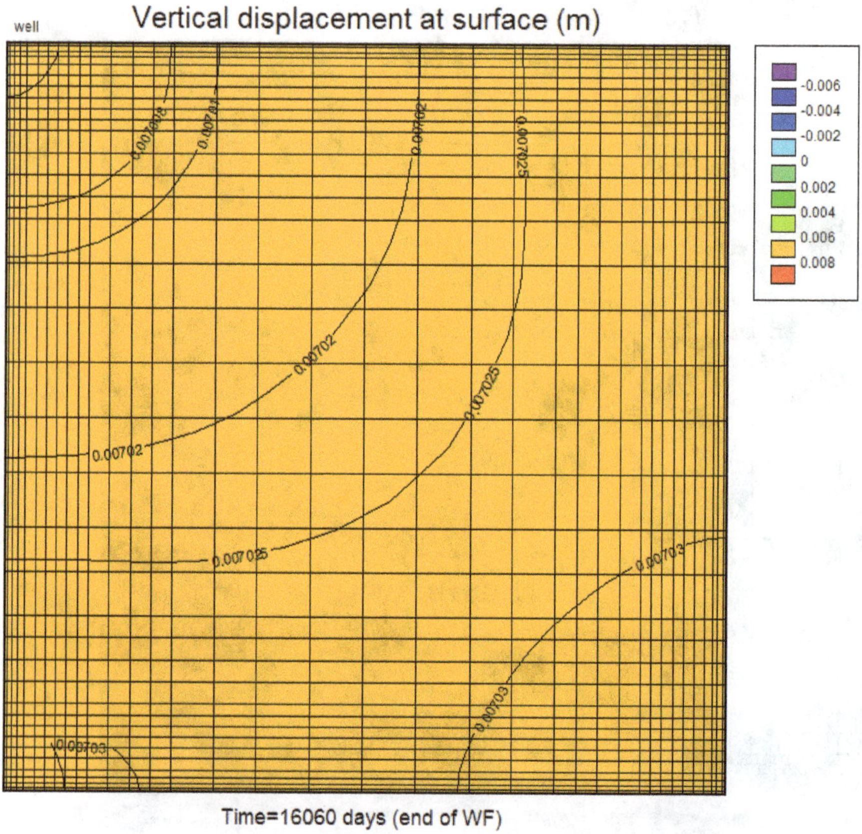

Figure 12. The vertical displacement at surface at the end of waterflooding.

SHEAR OR TENSILE FAILURE DURING PRIMARY DEPLETION AND WATERFLOOD

The stress path followed within several reservoir and caprock gridblocks during primary depletion and waterflooding were studied to assess the potential for shear and/or tensile failure. Figure 13 shows the stress path plot during the primary depletion and waterflood for block (1, 1, 2) (Figure 1) in the first caprock layer. As seen in Figure 13, it is not expected to reach tensile or shear failure during primary depletion and waterflood in the caprock. There are two failure lines drawn in stress path plots for the caprock in this study. The red dotted line corresponds to the failure criteria for the Midale Anhydrite assuming a cohesion and friction angle of 18,150 kPa and 45° [9], respectively. The purple solid line is a failure criterion with a reduced cohesion and friction angle of 3000 kPa and 30°, respectively. This line is plotted to examine the chance of shear failure in the presence of a less stiff, weaker caprock such as weak shale.

Figure 14 shows the stress path plot during the primary depletion and waterflood for the well block in the lowest layers in the Marly and Vuggy reservoirs (blocks (1, 1, 5) and (1, 1, 7) in Figure 1). Two failure criteria lines plotted have the same friction angle of 30°. The red line assumes a cohesion of 3500 kPa which corresponds to the Weyburn reservoir [9] and given the presence of natural fractures within the reservoir [17], a different failure criteria was plotted with an arbitrary lower cohesion of 3000 kPa to account for their effect on average rock cohesion.

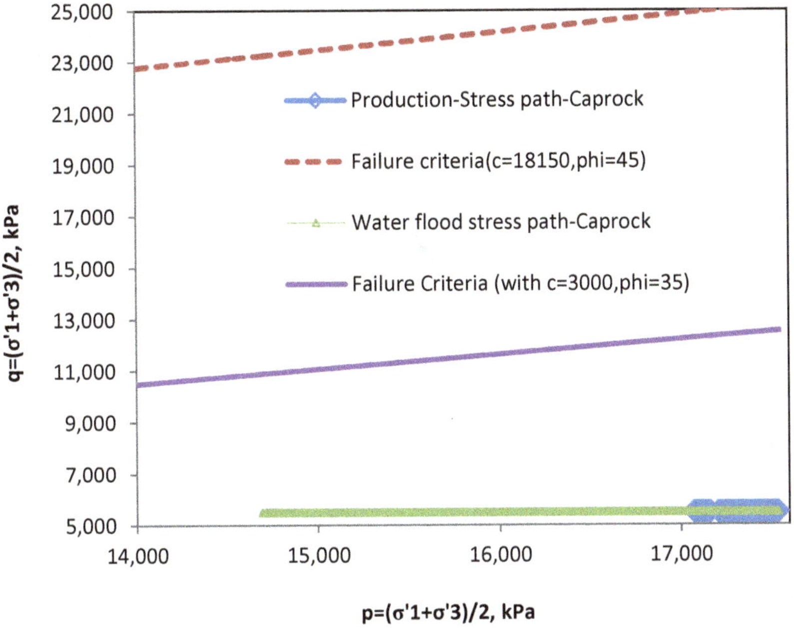

Figure 13. Stress path for block (1, 1, 2) (Figure 1) in the first caprock layer during the primary depletion and waterflood for the caprock for the isothermal model.

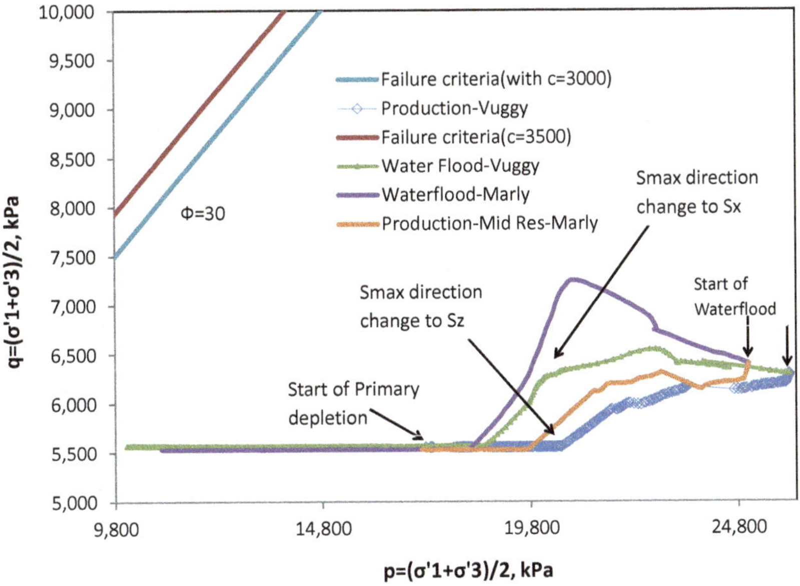

Figure 14. Stress path for the well block in the lowest layers in the Marly and Vuggy reservoirs (blocks (1, 1, 5) and (1, 1, 7) in Figure 1) during the primary depletion and waterflood for the isothermal model.

When primary depletion starts, with reduction of reservoir pressure, average effective stress increases. As seen in Figure 14, at some point during the primary depletion, the direction of the maximum stress changes from the original x-direction to vertical. Since the boundary condition at surface is free displacement, the vertical stress does not change significantly. Therefore, when vertical stress becomes the maximum stress, the difference between the maximum and minimum stress (q) increases. This is illustrated in Figure 15.

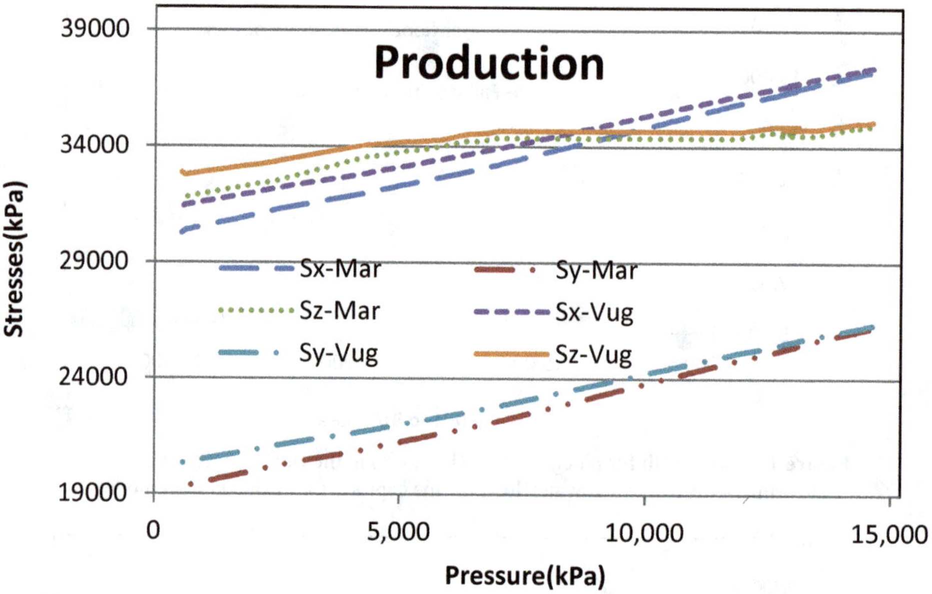

Figure 15. Variation of stresses (S) with pressure (P) for the well block in the lowest layers in the Marly and Vuggy reservoirs (blocks (1, 1, 5) and (1, 1, 7) in Figure 1) during primary depletion for the isothermal model. Where S_x=horizontal stress in x-direction, S_y=horizontal stress in y-direction, S_z=vertical stress in Z-direction, Mar=Marly and Vug=Vuggy.

When waterflooding starts, effective stresses will decrease due to the increase in reservoir pressure. The rate of pressure change and pressure gradient in the reservoir during the primary depletion and waterflood are different, therefore the stress path for the two operations does not fall on the same curve.

Since horizontal stresses increase during the waterflood, at some point, the maximum stress direction changes back to the x-direction and therefore the deviatoric stress (q) decreases. This is well illustrated in Figure 16.

For the wide range of assumptions adopted in this model, the simulation results infer that, under isothermal production/injection conditions associated with primary production and waterflood secondary recovery operations, shear and/or tensile failure are unlikely to occur within the reservoir or caprock.

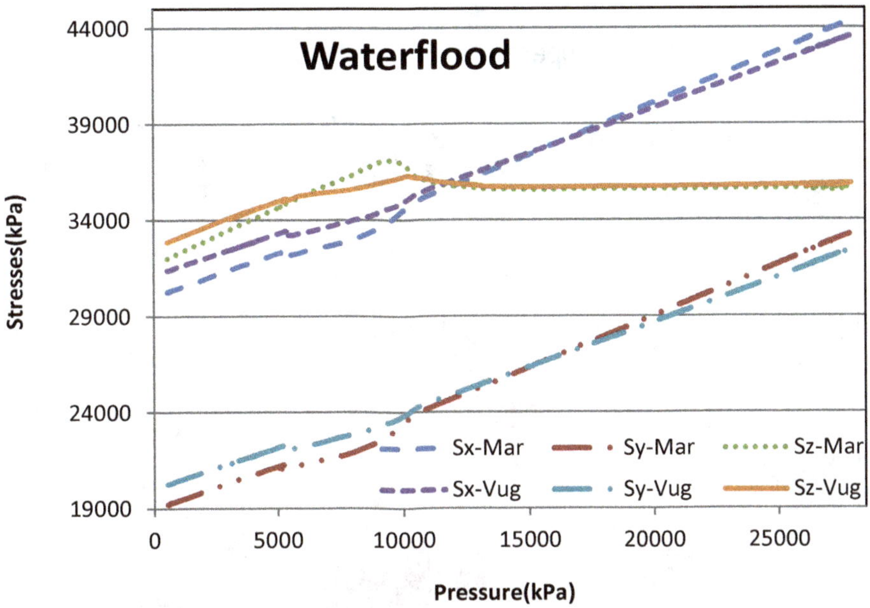

Figure 16. Variation of stresses (S) with pressure (P) for the well block in the lowest layers in the Marly and Vuggy reservoirs (blocks (1, 1, 5) and (1, 1, 7) in Figure 1) during waterflooding for the isothermal model. where, S_x=horizontal stress in x-direction, S_y=horizontal stress in y-direction, S_z=vertical stress in Z-direction, Mar=Marly and Vug=Vuggy

THERMAL EFFECTS OF INJECTION

In the next set of models, thermal effects of injection are incorporated by coupling the energy conservation equation with the mass balance and geomechanical system of equations. The water injection temperature is assumed as 35°C and the initial reservoir temperature is 60 °C. The linear thermal expansion coefficient for all geomechanical layers was assumed as 1E-5 $1°C^{-1}$. This is equivalent to a bulk thermal expansion coefficient of 3E-5 $1°C^{-1}$. The selected thermal expansion coefficient is within the range published for different rocks [23]. This value is also close to the measured bulk thermal expansion coefficient measured for a Devonian limestone sample (3.9E-5 $1°C^{-1}$) [24]. Figure 17 shows the temperature distribution in the bottom Vuggy reservoir by the end of the waterflood. Lower temperature results in lower minimum effective stress that, in turn, results in opening of natural fractures and higher conductivity for the cooled part of the reservoir. As seen in Figure 7, the fracture aperture and the permeability increase as the effective stresses decrease for the reservoir. This results in softening of the moduli and increasing of the grid block permeability.

Figure 18 shows the stress path for the block (1, 1, 2) (Figure 1) in the first caprock layer during the primary depletion and waterflood with cooling effects (thermal model). Due to the thermal conduction between the reservoir and caprock, horizontal stresses decrease. This occurs due to thermoelastic effects of reduction of temperature in the caprock. As a result vertical stress becomes the maximum of the three stresses. Since the vertical stress does not change very much (due to the free displacement boundary condition at surface), deviatoric stress (q) increases. However after the temperature of the grid block in the caprock drops to the minimum (35 °C), the effect of pore pressure changes (poroelasticity) dominate the stress solution and result in increasing horizontal stresses and decreasing deviatoric stress.

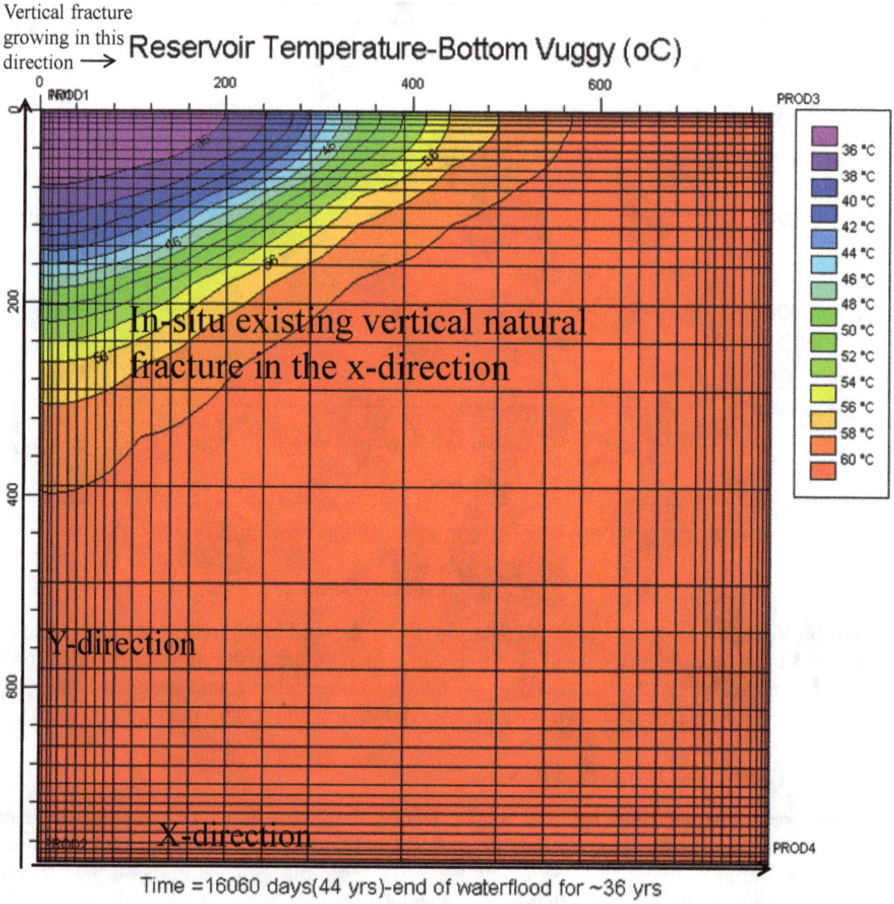

Figure 17. Temperature distribution (map view) in the bottom Vuggy reservoir by the end of the waterflood.

Stress and pressure history for the block (1, 1, 2) (Figure 1) in the first caprock layer from the beginning of the field operation is shown in Figure 19. As seen, the early thermoelastic effects reduce the horizontal stresses significantly whereas vertical stress drops slightly due to the local arching effects and then retreats back due to the free displacement boundary at surface and spreading the cooled zone reducing the localization effect. When the thermal effects have lowered the maximum horizontal stress to the point that it becomes smaller than the vertical stress, vertical stress becomes the maximum stress. Since maximum stress does not largely change (free displacement boundary at surface), deviatoric stress rises. Thermoelastic effects are mostly important during the early injection times. As illustrated in Figure 20, the block temperature drops very quickly to the minimum right after injection due to thermal conduction between reservoir and caprock. However the increase in pressure and consequent poroelastic effects occur later due to the slow pressure transfer between reservoir and caprock (low permeability). The induced reduction in horizontal stresses in the caprock has also been observed by Preisig and Prevost [25] and Gor *et al* [26]. Contrary to these studies, Vilarrasa *et al* [27] concluded that thermal effects of cold CO_2 injection increases the total horizontal stresses. Since the induced temperature change profile was not provided by Vilarrasa *et al* [27], it is hard to find a reason for reaching an opposite conclusion. However one speculation is that the stress transfer effect of the induced volumetric strain caused by thermal contraction in the reservoir dominated the effect of thermal contraction in the caprock.

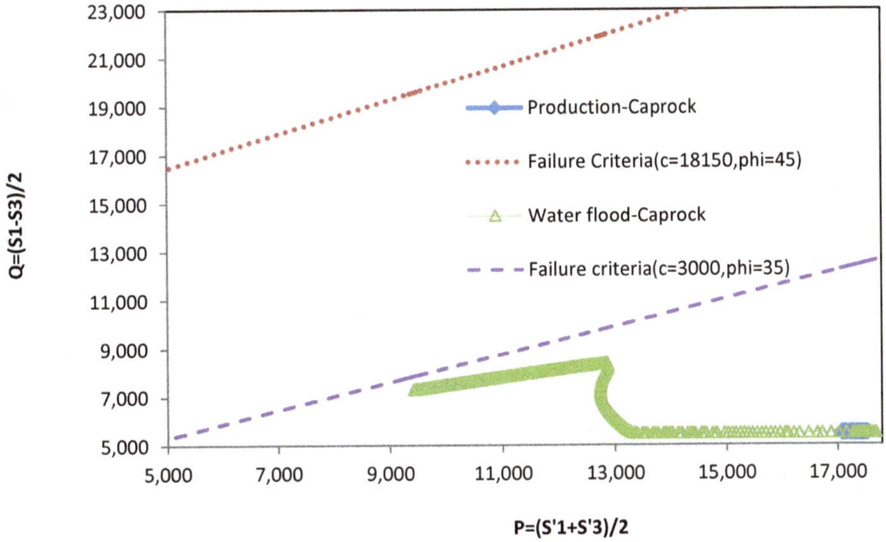

Figure 18. Stress path for block (1, 1, 2) (Figure 1) in the first caprock layer during the primary depletion and waterflood with cooling effect (thermal model).

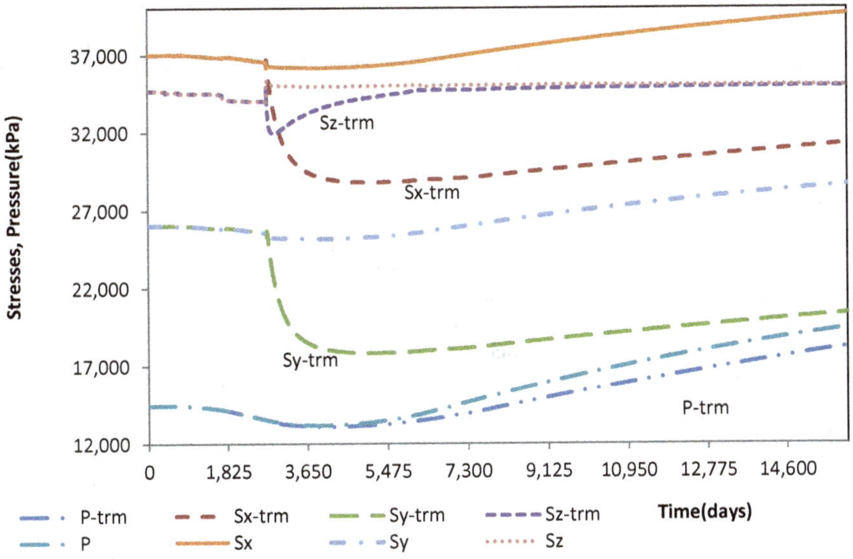

Figure 19. Stress(S) and pressure (P) history since the start of the field operation for the block (1, 1, 2) (Figure 1) in the first caprock layer in the caprock. Trm=thermal.

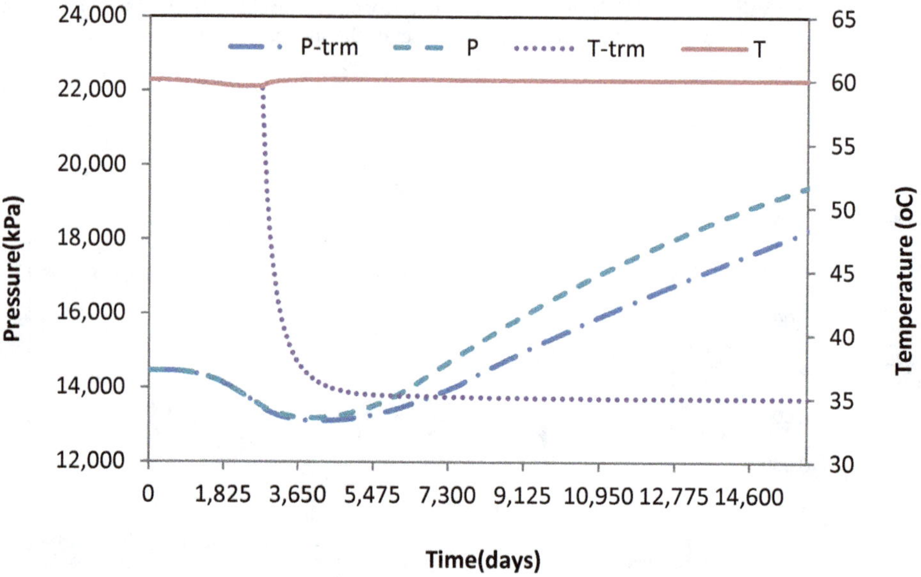

Figure 20. Pressure (P) and temperature (T) history since the start of the field operation for the block (1, 1, 2) (Figure 1) in the first caprock layer, trm=thermal.

By the end of the waterflood, the failure criteria line for the Midale Anhydrite and the assumed shale caprock is not crossed. Based on this, shear failure in the caprock is not expected to occur by the end of the waterflood.

Figure 21 shows the stress path for the well block in the lowest layers in the Marly and Vuggy reservoirs (blocks (1, 1, 5) and (1, 1, 7) in Figure 1) during the primary depletion and waterflood with cooling effects. By the end of primary depletion, vertical stress is the maximum stress. When waterflood is started, the pressure increases and temperature decreases in the reservoir. By the time temperature has dropped to the point that thermoelastic effects (early time effect) have reduced the horizontal stresses in the reservoir, the deviatoric stress increases. Once the grid block is cooled to the minimum temperature, poroelastic effects dominate and the pressure increase results in higher horizontal stresses. This reduces the deviatoric stress.

Figure 22 shows the stress and pressure history for well block in the lowest layer in the Vuggy reservoir (block (1, 1, 7) in Figure 1) for the primary depletion and waterflooding. Figures 23 shows the pressure and temperature history for the well block in the lowest layer in the Marly reservoir (block (1, 1, 5) in Figure 1) for the primary depletion and waterflooding As seen in Figure 23, the block temperature very quickly drops to injection temperature. This is followed by an increase in the pore pressure (local pressure and average pressure of reservoir zones) and the domination of poroelasticity over thermoelasticity and the consequent increase of the horizontal stresses.

The stress path for the Vuggy reservoir crosses the lower failure criteria line (shale) (Figure 21). This means that the Vuggy reservoir is expected to fail in shear during the waterflood, if the effect of the natural pre-existing fractures is accounted for by lowering the cohesion of the reservoir rock. As seen in the figure, when the stress path crosses the failure line, the minimum effective stress is still positive. This means that the reservoir rock has failed in shear before any tensile fracturing. Tensile fracturing is expected to occur later after about 6200 days since the start of field operation.

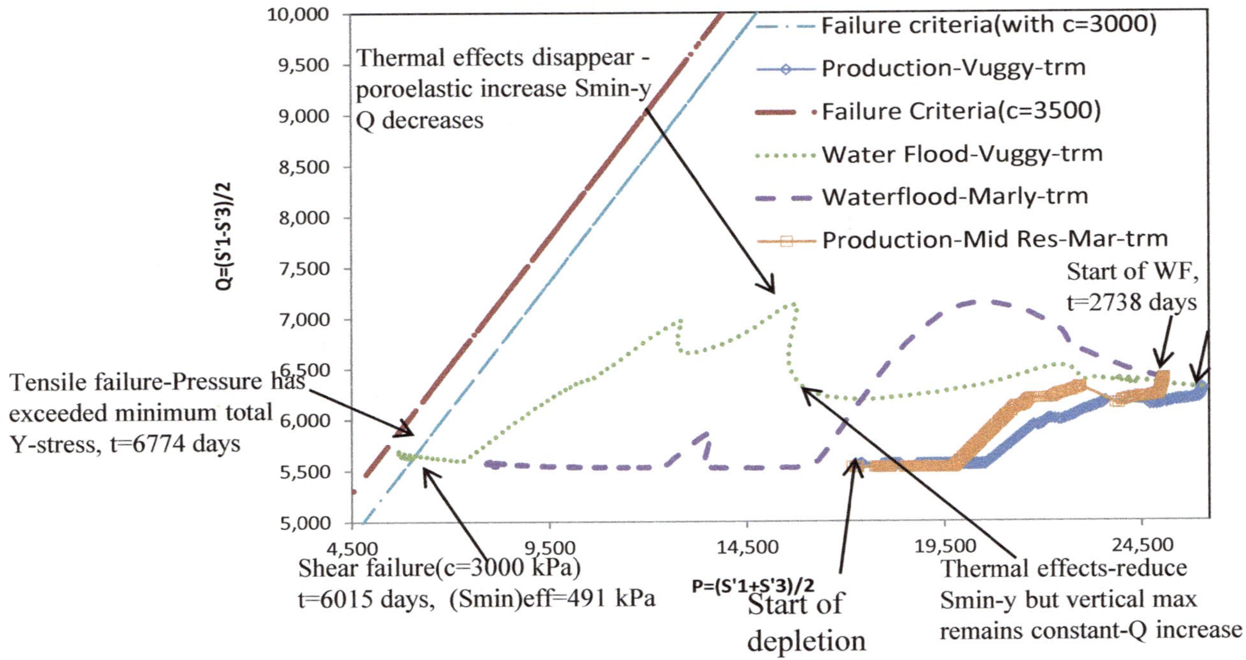

Figure 21. Stress path for the well block in the lowest layers in the Marly and Vuggy reservoirs (blocks (1, 1, 5) and (1, 1, 7) in Figure 1) during the primary depletion and waterflood with cooling effects

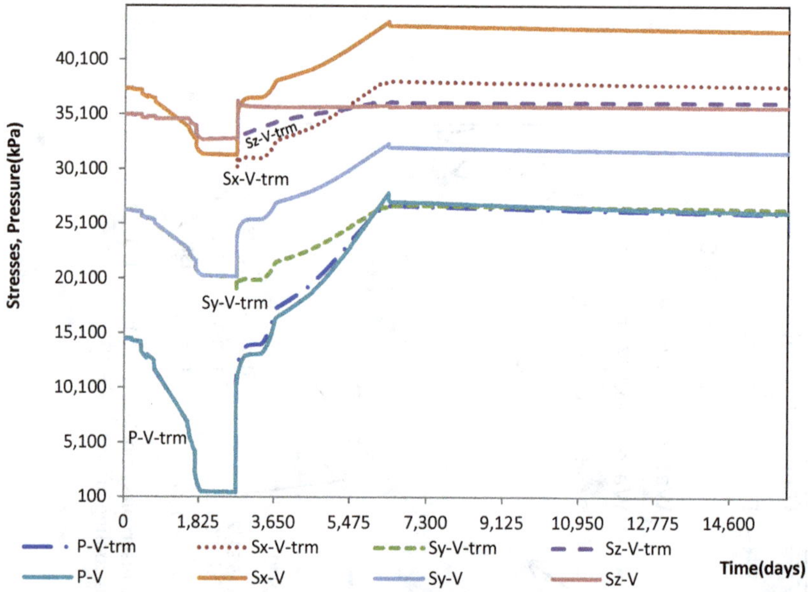

Figure 22. Stress (S) and pressure (P) history for the well block in the lowest layer in the Vuggy reservoirs (block (1, 1, 7) in Figure 1) for the primary depletion and waterflooding, V=Vuggy, trm=thermal.

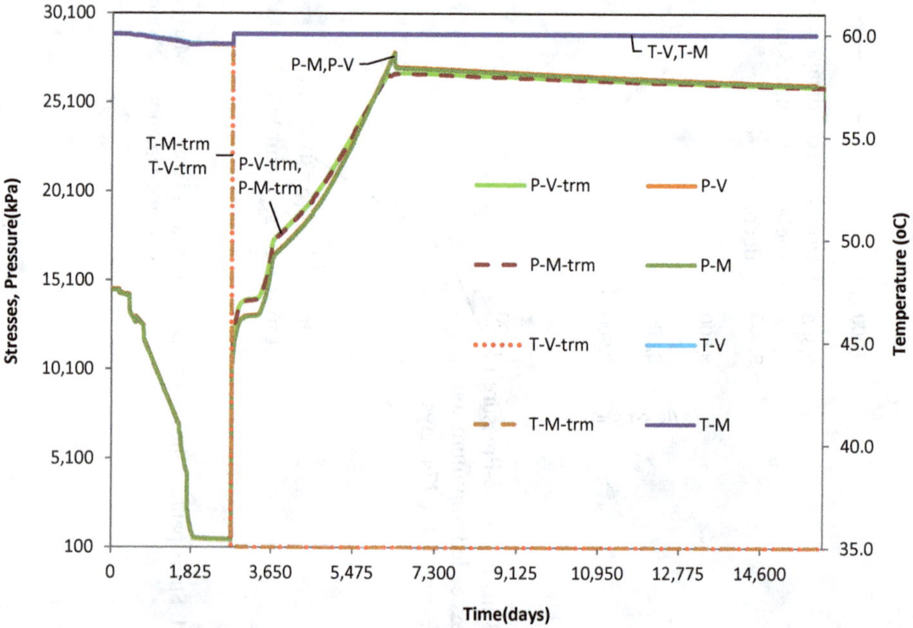

Figure 23. Pressure (P) and temperature (T) history for the well block in the lowest layers in the Marly and Vuggy reservoirs (blocks (1, 1, 5) and (1, 1, 7) in Figure 1) for the primary depletion and waterflooding, M=Marly, V=Vuggy, trm=thermal.

As seen in Figure 22, the block pressure in the reservoir is slightly higher for the thermal model than the isothermal model before tensile fracturing while Figure 20 shows a lower block pressure in the caprock for the thermal model. This is speculated to be the effect of reduced water mobility due to reduced water viscosity at lower temperature. Therefore for injecting the same volume of water, the pressure rise would be larger for the model with less phase mobility. Also since the phase mobility is reduced, a smaller quantity of water has diffused from reservoir to caprock and therefore, the pressure in the caprock is lower for the thermal model.

Figure 24 shows the maximum extent of the fracture during waterflooding. While there is no fracturing predicted with the isothermal model, the thermally reduced minimum horizontal stress results in a lower (relative to isothermal model) fracture pressure of about 26 MPa for the reservoir. This is about 20% reduction in the fracture pressure compared to the isothermal model with a fracture pressure of 32 MPa. This is in contrast with isothermal model that does not fail in tension at such injection pressure.

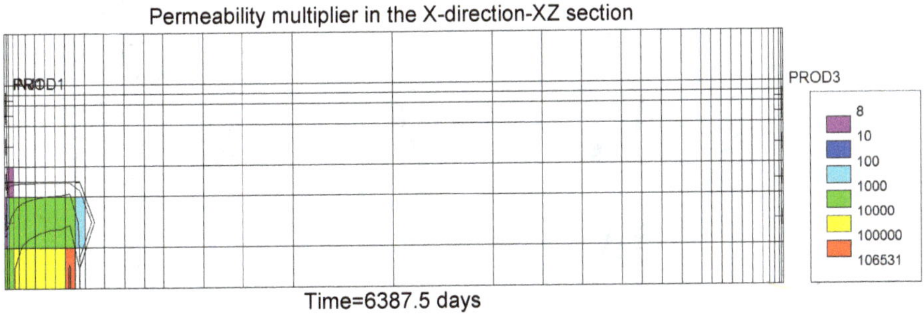

Figure 24. Permeability multiplier in XZ section of reservoir and the immediate surrounding layers - shows the maximum fracture extent during waterflooding. The starting multiplier in the legend is about the maximum multiplier for existing natural fractures at zero stress.

SECONDARY DEPLETION AND CO_2 FLOOD

The oil production after about 36.5 years with an average rate of 40.6 m^3/day of waterflooding reaches a cumulative amount of ~298 Mm^3 (Figure 26). In order to lower the reservoir pressure in preparation for the CO_2 flood/disposal, a secondary depletion phase was started by stopping the injection well (upper left corner, Figure 1) and continuing the production for about 2.5 years from the three corner wells which reached a cumulative production of 309.2 Mm^3. Following the secondary depletion, CO_2 was injected through the upper left corner well (Figure 1) with a rate of 12000 m^3/day (8.7 Mt/yr) for about 18 years which resulted in a stored CO_2 volume of ~58800 Mm^3 (116.4 Mt) (before shear or tensile fracturing of the caprock, see Figures 27-29). During injection the other three corner wells were under production with a limiting water cut of 0.9 as the Shut-In criteria. During CO_2 injection, two of the producers met this critetia and were forced to shut in. This is the reason for the increased slope of the bottomhole pressure history during CO_2 storage (Figure 25). Although the process does not follow a conventional CO_2 storage scenario as oil and gas production is still occurring during CO_2 injection, the intention is to create voidage for storing larger volume of CO_2 by continually removing oil. Also this model scenario is similar to current operations at Weyburn, which is currently undergoing WAG or straight CO_2 flooding. The injection pressure is increased significantly during injection to match Weyburn operations that attempt to ensure flood pressures are above the miscibility pressure of about 17 – 18 MPa to optimize the flood performance. In this simulation the injection pressure was allowed to continue to increase to investigate the maximum safe pressure in the case of a true storage scenario.

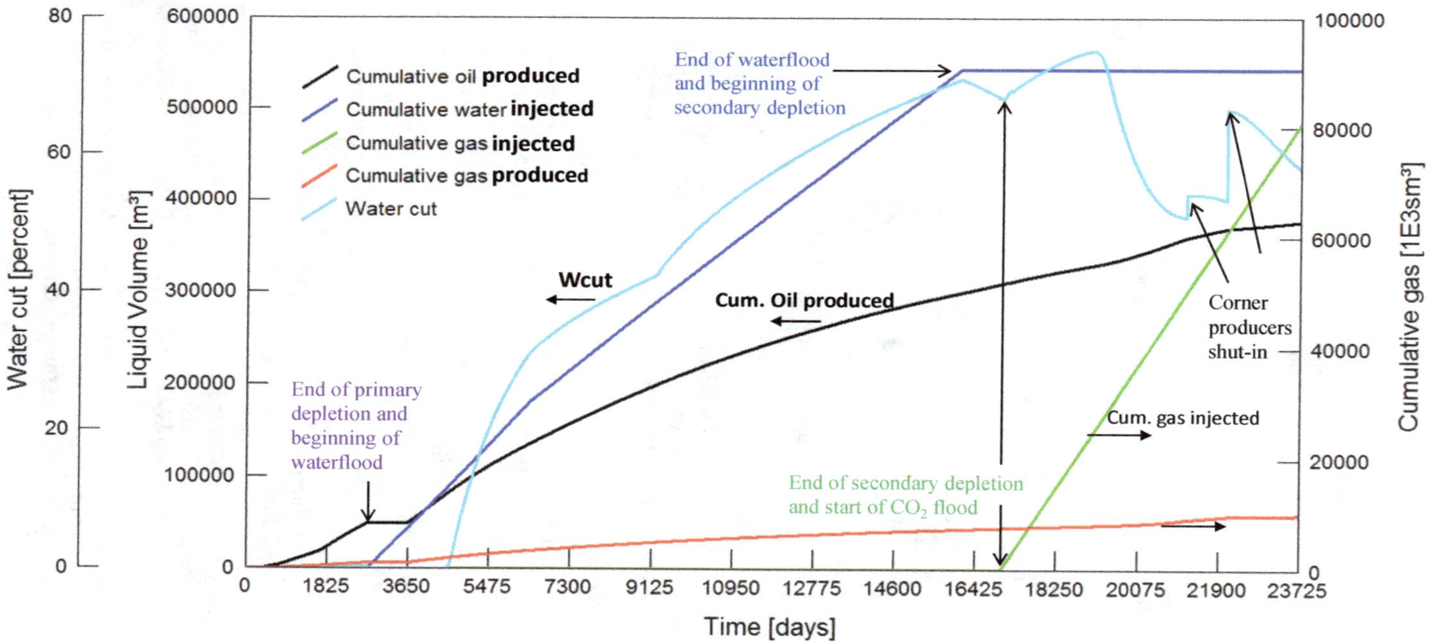

Figure 25. Cumulative oil and gas production, cumulative water and gas injection and watercut history during the primary depletion, waterflooding, and secondary depletion and CO_2 storage.

One of the most important concerns with CO_2 storage is the integrity of the caprock, which could be endangered by potential shear or tensile failure in the reservoir or caprock during CO_2 injection. In order to examine the possibility of reaching this state in the reservoir or caprock, the stress path of the reservoir and caprock for the secondary depletion and CO_2 flood was studied. Figure 27 shows the stress path for the block (1, 1, 2) (Figure 1) in the first caprock layer for all field operations in this study. The stress path of the caprock crosses the failure line for the arbitrary shale caprock (Figure 27). This occurs after about 16.9 years of CO_2 injection and is followed shortly after (17.6 years of injection) by tensile fracturing of the caprock. If CO_2 injection is continued beyond this point, the injected CO_2 can diffuse and flow through the created conductive channels in the caprock and reach upper permeable layers.

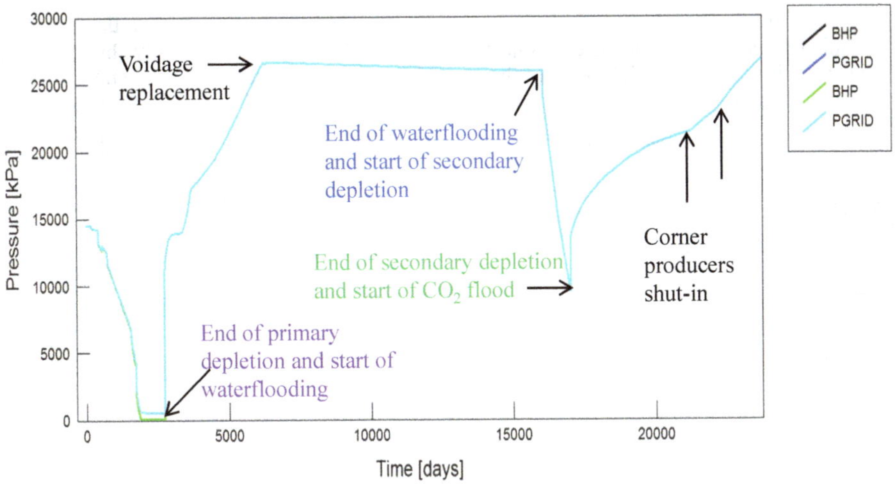

Figure 26. Injection well pressure during the primary depletion, waterflooding, secondary depletion and CO_2 storage.

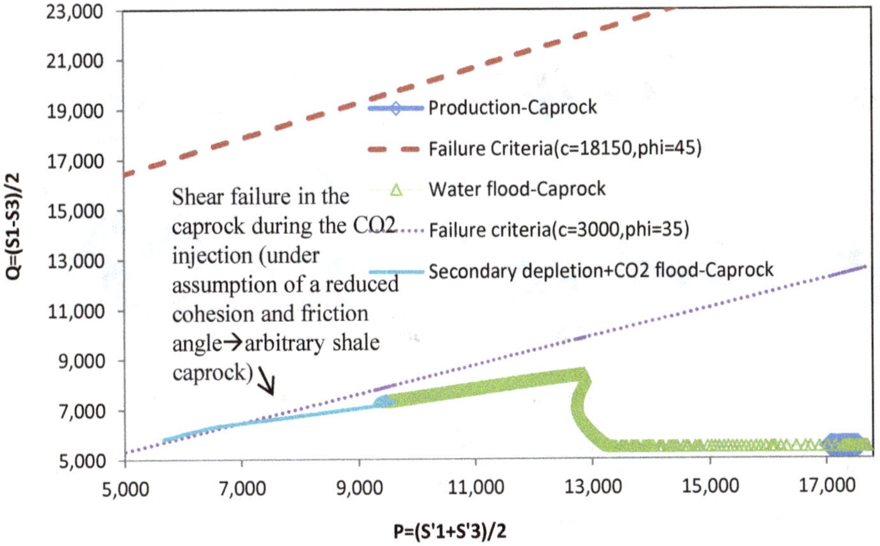

Figure 27. Stress path for the block (1, 1, 2) (Figure 1) in the first caprock layer for all field operations in this study.

Figure 28 shows the maximum extent of the fracture by the end of CO_2 injection. As seen, the caprock is fractured even though the Marly is still not showing any tensile failure. This is due to the large thermoelastic effects resulting from the thermal conduction between the reservoir and caprock and large stiffness of the caprock. The reason for the lack of any tensile failure in the Marly reservoir (while caprock is getting fractured) is the softening effect of the Marly rock with opening of existing natural fractures after injection. This softening reduces the thermal stress changes during cooling.

Figure 29 shows the gas saturation for the XZ-section of the reservoir and the immediate surrounding layers by the end of CO_2 injection. As seen, CO_2 is flowing mostly through Vuggy reservoir but it does not breakthrough to the producer well by the end of CO_2 injection. Also it is important to note that with fracturing of the caprock, CO_2 has slightly diffused into the caprock.

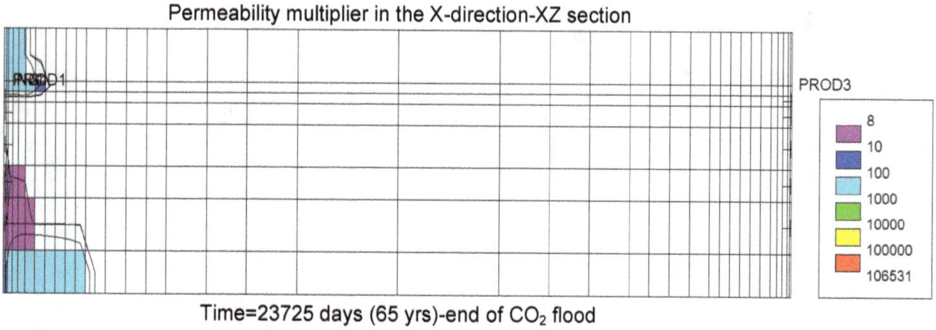

Figure 28. Permeability multiplier in XZ section of reservoir and the immediate surrounding layers - shows the maximum fracture extent during CO_2 storage. The starting multiplier in the legend is about the maximum multiplier for existing natural fractures at zero stress.

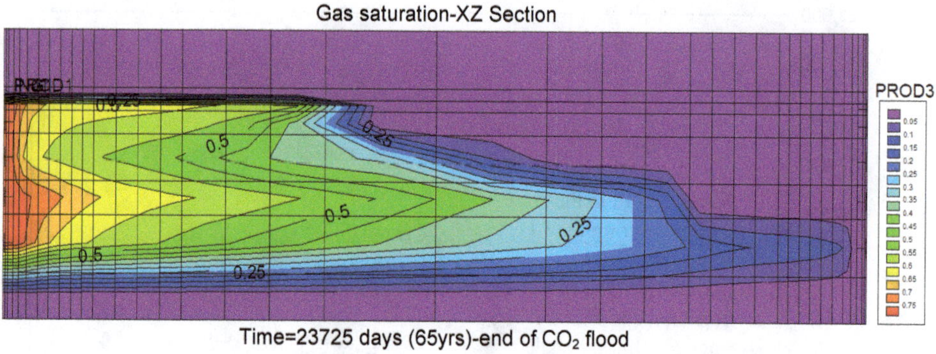

Figure 29. Gas saturation for the XZ-section of the reservoir and the immediate surrounding layers by the end of CO_2 injection.

The stress paths for the Marly and Vuggy reservoir have shown the occurrence of shear failure during the waterflood. During the secondary depletion, the maximum stress direction change to vertical and therefore with continuation of depletion, the deviatoric stress increases (Figure 30). When CO_2 injection is started and minimum horizontal stress is increased due to poroelastic effects, the difference between the maximum (vertical stress) and minimum stress (q) decreases. This would continue until one of the horizontal stresses has increased beyond the vertical stress so it becomes the maximum stress. Following that, the deviatoric stress remains almost constant but the average effective stress decreases due to injection. With continuation of CO_2 injection and reduction of the average effective stress, shear failure line for an arbitrary shale caprock is reached.

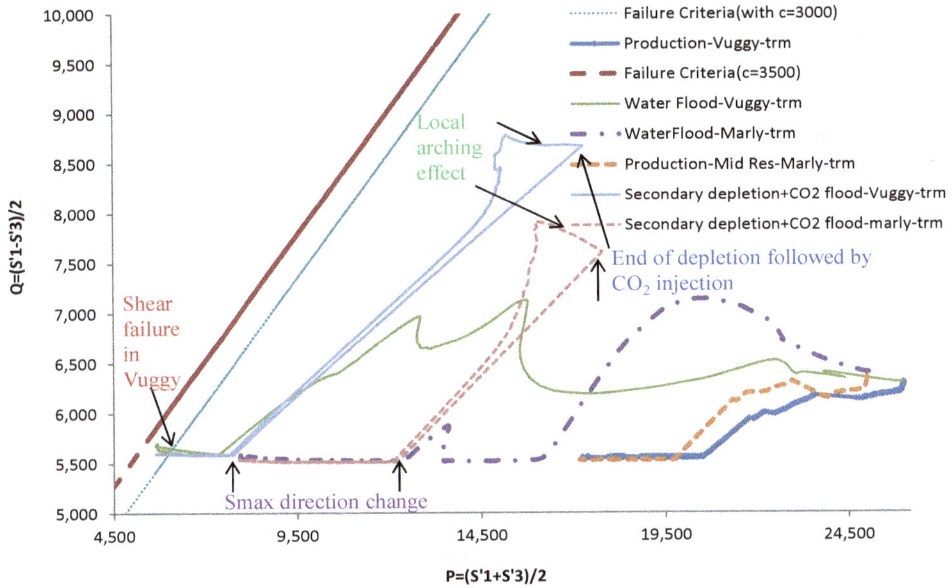

Figure 30. Stress path for the well block in the lowest layers in the Marly and Vuggy reservoirs (blocks (1, 1, 5) and (1, 1, 7) in Figure 1) for all field operations in this study.

EFFECT OF HYSTERESIS

Rock mechanical behavior is stress path dependent and, so, as rock undergoes pressure and temperature induced stress changes, the mechanical properties may vary so that under an equal induced stress change, the magnitude of deformation will be different depending on the past geomechanical history of the formation.

In this section the effect of hysteresis is studied by including the effect for both the fracture and matrix elements. Figures 31-34 show the input hysteric stress-strain curves for Marly and Vuggy reservoir for different operations in this study. The hysteresis for the fracture elements arises from crushing of the joints contacts/asperities from past operations. This will result in reduction of maximum closure of the fractures.

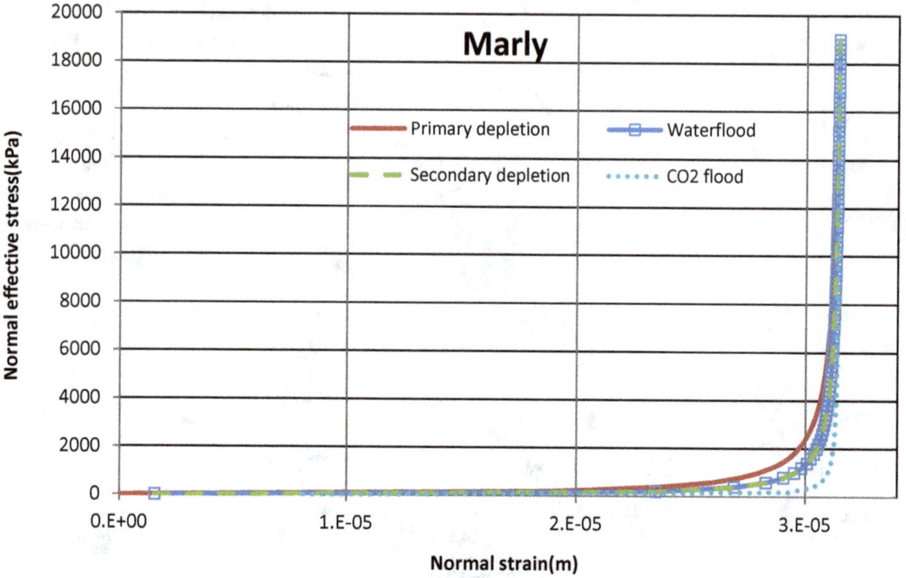

Figure 31. Input hysteric stress-strain curves for fractures in the Marly reservoir for different operations in this study.

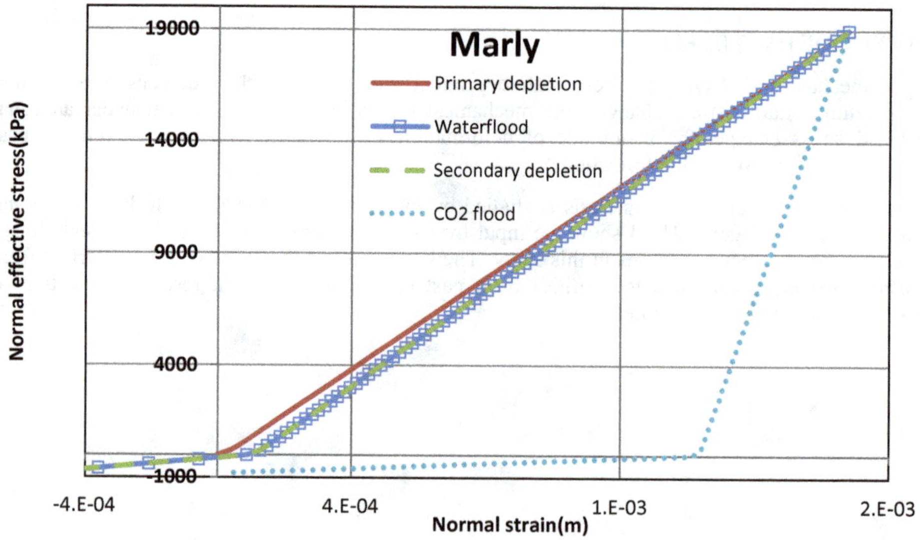

Figure 32. Input hysteric stress-strain curves for composite matrix grid block that contains fracture elements in the Marly reservoir for different operations in this study.

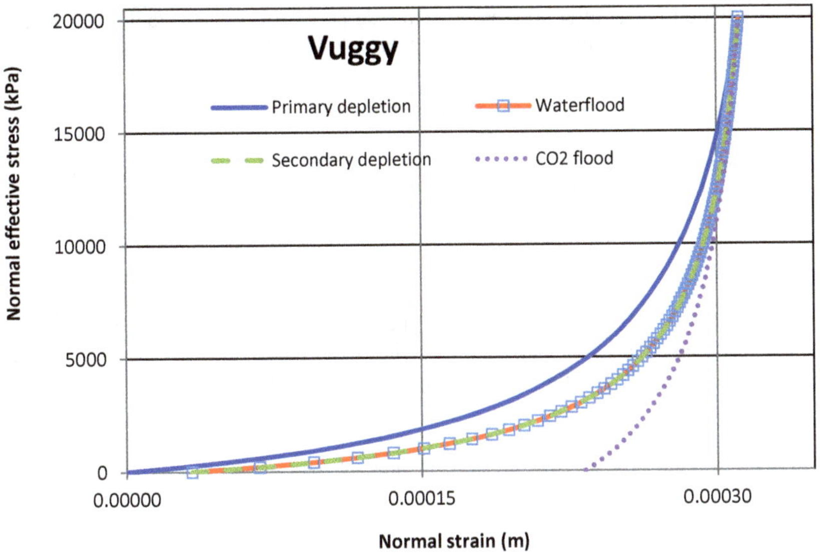

Figure 33. Input hysteric stress-strain curves for fractures in the Vuggy reservoir for different operations in this study.

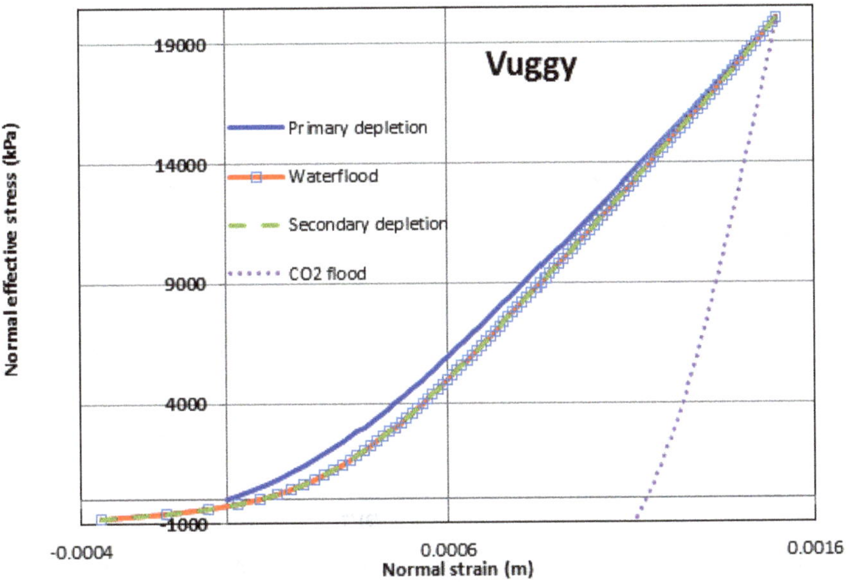

Figure 34. Input hysteric stress-strain curves for composite matrix grid block that contains fracture elements in the Vuggy reservoir for different operations in this study.

The input hysteric behavior for joint is based on Bandis *et al* [28]. The Young's modulus of the matrix elements is assumed to remain unchanged until the secondary depletion is completed. After that, the Young's modulus of the matrix elements is assumed to increase by a factor of 3. The idea behind this assumption is that after all pressurization, de-pressurization and cooling of the field; the rock material will be harder compared to the original state of the formation. An equivalent moduli was then calculated based on Equation 9.

Figure 35 shows the stress path for the Marly and Vuggy reservoir, when hysteresis effects are included in the model. As seen in the graph, the reservoir rock takes a different stress path during CO_2 injection compared to the stress path shown in Figure 30. This resulted in an early crossing of shear failure lines for the rock with both the original and lowered cohesion (to account for the existence of natural fractures). The reason for this behavior is that since with the hysteresis the moduli of the rock is larger, the poroelasticity constant (equation 13) becomes smaller. Hence the poroelastic effect on minimum stress is reduced (equation 14) and therefore difference between the maximum (vertical) and minimum stress drops faster. With the same cause, the average effective stress also decreases faster compared to the original model. It should be noted that since past water injection has lowered the temperature around the well block to the injection temperature, poroelasticity is the only active dominating geomechanical mechanism during the CO_2 storage.

$$\eta = \alpha \frac{1-2\upsilon}{1-\upsilon} = \left(1 - \frac{K_b}{K_s}\right)\frac{1-2\upsilon}{1-\upsilon} = \left(1 - \frac{\frac{E}{3(1-2\upsilon)}}{K_s}\right)\frac{1-2\upsilon}{1-\upsilon} \qquad \text{Equation 13}$$

$$\Delta S_h = \eta \Delta P \qquad \text{Equation 14}$$

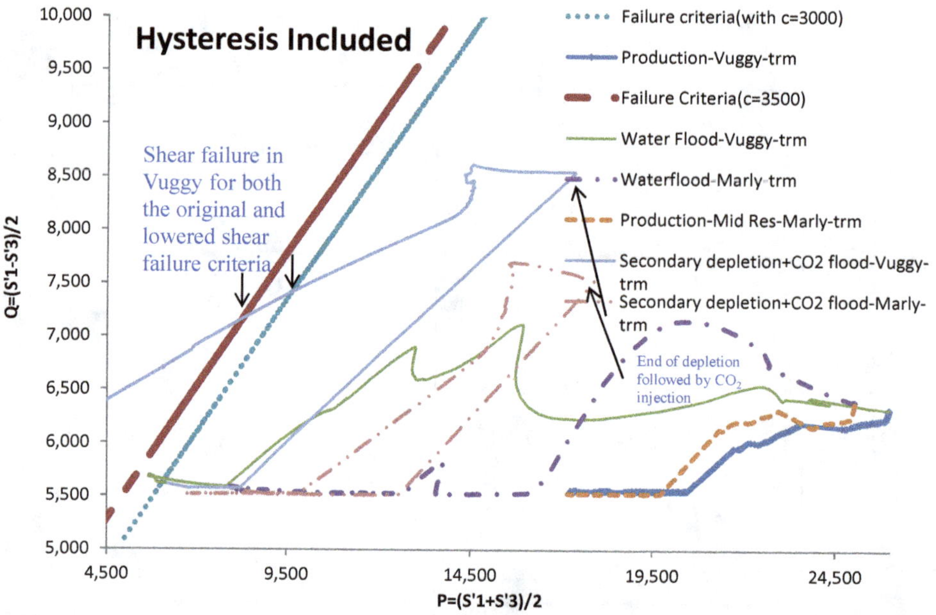

Figure 35. Stress path for the Marly and Vuggy reservoir when hysteresis effects are included in the model.

Hysteresis also affects the fracture pressure. When hysteresis effects are included, the minimum stress change with pressure becomes smaller (equation 14). Therefore with increasing pressure during CO_2 injection, the minimum stress will be met at lower fracture pressure. This effect is illustrated in Figure 36. This is a very important effect of hysteresis as it lowers the maximum operating pressure if the injection well is designed to operate below fracture pressure.

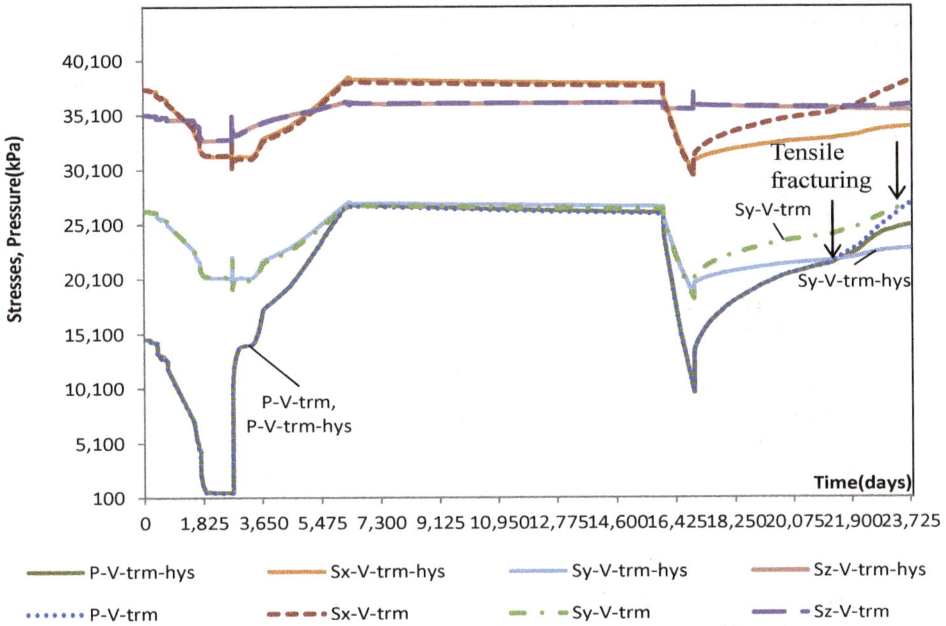

Figure 36. Stress (S) and pressure (P) history for the well block in the lowest layers in the Vuggy reservoirs (block (1, 1, 7) in Figure 1) for the original and hysteric model, V=Vuggy, trm=thermal.

MAIN CONCLUSIONS

In this study a coupled fluid flow, thermal and geomechanical model for a typical oil reservoir was formulated to evaluate the geomechanical effects of the full operational history on the seal integrity of that reservoir, specifically during the period of CO_2 storage. Field operations were started with primary depletion and then followed by waterflooding, secondary depletion and CO_2 injection for storage and tertiary recovery purposes. The cumulative oil production and stored CO_2 volume by the end of safe CO_2 injection (no leakage) period reach a maximum of 368.2 Mm^3 and the stored CO_2 volume of 58800 Mm^3 (116.4 Mt).

The model has incorporated the effect of existing natural fractures in the reservoir (in the reservoir and geomechanics module) and the potential for tensile fracturing of the reservoir and caprock. It also has examined the effect of induced stress-strain hysteresis from past operation such as depletion and waterflood on the fracture pressure during CO_2 storage. The following summarizes the main conclusions drawn from this study:

- The occurrence of shear or tensile failure was not predicted during primary depletion.
- Vertical displacement at surface is quite small due to the depth and stiffness of the reservoir formation. Surface deformations reach a maximum of ~ -7, +6 and +8 mm by the end of depletion, waterflood and CO_2 injection respectively.
- Thermoelasticity dominates during the early stage of the waterflood, which resulted in tensile fracturing of the reservoir and partial growth of the fracture into the caprock.
- Studying the stress paths has shown that depending on the scenarios considered, poroelastic and thermoelastic effects could lead to maximum stress direction changes, tensile and/or shear failure in the reservoir and/or caprock.

- If the effect of existing natural fractures in the reservoir is accounted for by lowering the rock cohesion, the Vuggy reservoir is expected to fail in shear around the well block.
- If the overlying caprock is assumed to consist of a material with a relatively low rock cohesion (such as 3000 kPa), the caprock is expected to undergo shear failure during CO_2 storage. This would increase the chance of CO_2 leakage to upper permeable layers.
- The hydraulic fracture (created during waterflooding) closes after the start of secondary depletion but then re-opens during the CO_2 injection in the Vuggy at a similar fracturing pressure (when there's no hysteresis effect). Therefore when there's no hysteresis effect expected for the reservoir and caprocks, fracturing pressures observed in waterflooding may be useful for designing Maximum operating pressure (MOP) during CO_2 injection.
- Safe injection pressure for CO_2 storage can be estimated based on avoiding the initiation of tensile and shear failure in the caprock. If the operation is designed for avoiding hydraulic fracturing in the reservoir, the maximum injection pressure should be around 26 MPa for the model storage unit described in this study. This is determined considering the cooling effect of CO_2 injection and is equivalent to the initial horizontal minimum stress in the reservoir. Considering the effect of hysteresis on the geomechanical properties and stress changes has resulted in lower fracture pressure in the reservoir during CO_2 storage. When hysteresis is considered, this pressure falls to around 21 MPa. This is 95% of the original fracture pressure.

RECOMMENDATIONS FOR FUTURE WORK

- As geomechanical hysteresis effect (from past operations before CO_2 storage) has proven to be significant (in terms of the estimated fracture pressure and the potential for shear fracturing), we suggest incorporating the observed hysteresis properties in the lab in the coupled geomechanical model.
- Due to the presence of existing natural fractures in the reservoir, it is recommended to build a dual porosity- dual permeability model for the fluid flow model and couple to the equivalent continuum geomechanical model, which tracks the matrix and joint strains separately.
- Also, including joint elements in the geomechanical model explicitly could be useful in analyzing the shear and potential shearing of major joints or faults to better evaluate the risk of permeability enhancement and containment loss.
- In the geomechanical model used for this study, the effect of opening of existing natural fractures is accounted for by softening the moduli in an isotropic fashion. This results in larger than expected deformations in the two directions aligned with the plane of fracture. We suggest improving this model by either:
  - incorporating Barton-Bandis joint model (to be able to define separate constitutive model for the joint and matrix elements);
  - using the equivalent continuum constitutive model that accounts for joint set orientation resulting in a dynamic, potentially nonsymmetric, modification to the rock stiffness matrix.
- It is recommended to model shear failure of the matrix and/or fractures by including an elasto-plastic constitutive model to account for the stress changes when shear failure occurs in the reservoir and caprock (with inclusion of associated shear induced permeability enhancement). That is very important especially when the objective is to determine the fracture pressure/max operating pressure. Regardless of the extent of the failed region, the correct modeling of stresses and displacements will affect the estimated fracture pressure since failure tended to occur near the well and propagate outward.
- The dissolution of CO_2 in water is ignored in the black oil model used in this study. This is important especially due to waterflooding before CO_2 injection. Therefore the results shown are conservative in terms of the amount of CO_2 that can be injected before fracturing. It is recommended to modify the source code of the flow model to allow gas solubility in the water

phase. In this way the model can account for the dissolution of CO_2 in water without the large computational requirement of a compositional model.

- Alternate reservoir descriptions could be investigated in a similar manner.
  - Sandstone reservoirs tend to have much more dominant geomechanical effects in depletion (such as compaction and associated permeability and porosity changes) as a result of compressive failure. Hysteresis is guaranteed for such failure. Also, the much larger compactive deformations during depletion can cause much larger arching stresses to develop in the caprock, which may impact the containment capability of the caprock during high pressure CO_2 storage.
  - Faulted reservoirs also pose a problem somewhat similar to the fractures modeled in the Weyburn field. Fault failure and permeability enhancement could significantly impact CO_2 storage characteristics.

REFERENCES

1. Taurus. 2013. www.TaurusRS.com
2. Settari, A. and Walters, D.A. Advances in Coupled Geomechanical and Reservoir Modeling With Applications to Reservoir Compaction. SPE Journal, Vol. 6, No. 3, Sept. 2001, pp. 334-342.
3. Settari, A. Al-Ruwaili and V. Sen, Upscaling of Geomechanics in Heterogeneous Compacting Reservoirs, presented at the 2013 SPE Reservoir Simulation Symposium, Feb 18 - 20, 2013 2013, The Woodlands, TX, USA, SPE 163641.
4. Goodarzi, S., Settari, A. Zoback and D. Keith, Thermal Aspects of Geomechanics and Induced Fracturing in CO_2 Injection With Application to CO_2 Sequestration in Ohio River Valley, Paper SPE 139706, presented at the SPE International Conference on CO_2 Capture, Storage, and Utilization held in New Orleans, Louisiana, USA, 10–12 November 2010.
5. Bachman, R., Harding, T., Settari A. and D. Walters .Coupled Simulation of Reservoir Flow, Geomechanics, and Formation Plugging With Application to High-Rate Produced Water Reinjection. Paper SPE 79695, SPE Reservoir Simulation Symposium, Houston, TX. Feb. 3-5, 2003
6. Settari, A. R.B. Sullivan, D.A. Walters and P.A. Wawrzynek. 3-D Analysis and Prediction of Microseismicity in Fracturing by Coupled Geomechanical Modeling. Paper SPE 75714, presented at the 2002 SPE Gas Technology Symposium held in Calgary, Alberta, Canada, 30 April–2 May 2002.
7. Settari, A., R.B. Sullivan, and R.C. Bachman. The Modeling of the Effect of Water Blockage and Geomechanics in Waterfracs. Paper SPE 77600, presented at the Annual Techn. Conf. of SPE, San Antonio, TX. Sept. 29 – Oct. 2, 2002.
8. Yamamoto, H. 2004. Using Time-Lapse Seismic Measurements to Improve Flow Modeling of CO_2 Injection In the Weyburn Field: A Naturally Fractured, Layered Reservoir, A thesis submitted to the Faculty and the Board of Trustees of the Colorado School of Mines in partial fulfillment of the requirements for the degree of Doctor of Philosophy (Petroleum Engineering).
9. Jimenez Gomez, J., A. 2006. Geomechanical Performance Assessment of CO_2 – EOR Geological Storage Projects. A thesis submitted to the Faculty of Graduate Studies and Research in Partial Fulfillment of the Requirements for the Degree of Doctor of Philosophy in Geotechnical Engineering Department of Civil and Environmental Engineering Edmonton, Alberta.
10. Srivastava, R.K., Huang, S.S., Laboratory Investigation of Weyburn CO_2 Miscible Flooding, Saskatchewan Research Council, Was Presented at the Seventh Petroleum Conference of the South Saskatchewan Section, the Petroleum Society of CIM, Held in Regina October 19-22, 1997.

11. Vazquez, M.E., and Beggs, H.D., Correlations for Fluid Physical Property Prediction; Journal of Petroleum Technology pp. 968-970, June 1980.
12. Hall, K.R. and Yarborough, L. 1973. A New Equation of State for Z-Factor Calculations. Oil & Gas J. 71 (25): 82.
13. Carr, N.L., Kobayashi, R., and Burrows, D.B. 1954. Viscosity of Hydrocarbon Gases under Pressure. J Pet Technol 6 (10): 47-55. SPE-297-G. http://dx.doi.org/10.2118/297-G
14. Brown, L.T. 2002. Integration of rock physics and reservoir simulation for the interpretation of time-lapse seismic data at Weyburn Field, Saskatchewan, M.Sc. Thesis: Golden, CO, Colorado School of Mines.
15. Havens, J. 2011. Mechanical Properties of the BAKKEN Formation, A thesis submitted to the Faculty and the Board of Trustees of the Colorado School of Mines in partial fulfilment of the requirements for the degree of Master of Science (Geophysics)
16. Kashib, T. 2013. Personal communication
17. McLellan, P.J., Lawrence, K., and Cormier K. 1992. A multiple-zone acid stimulation treatment of a horizontal well, Midale, Saskatchewen. Journal of Canadian Petroleum Technology, v. 31, no. 4, p. 71-82.
18. Galas, C.M.F., Churcher, P.L., Tottrup, P. Predictions of Horizontal Well Performance in a Mature Waterflood, Weyburn Unit, Southeastern Saskatchewan, JCPT, November 1994. Volume 33, No. 9
19. Elsayed, S.A., Baker, R., Churcher, P.L. Edmunds, A.C. Multidisciplinary Reservoir Characterization and Simulation Study of the Weyburn Unit. JPT, Volume 45, Number 10, October 1993
20. Settari, A., 2007. Reservoir Geomechanics, course note, Department of Chemical and Petroleum Engineering, University of Calgary, Chapter 6, p. 9-12
21. Sneddon, I.N., Lowengrub, M., 1969. Crack Problems in the Classical Theory of Elasticity. John Wiley & Sons Inc, New York, p. 19.
22. Perkins, T.K. and L.R. Kern. 1961. Widths of hydraulic fractures, Journal of Petroleum Technology. Vol, 13, No.9, 937-949.
23. Ramey, H.J., Kruger, P., Miller, F.G., Horne, R.N., Brigham, E.E., and Gudmundsson, J.S. Proceedings ninth workshop geothermal reservoir engineering. December 13-15, 1983
24. Kosar, K., "Geotechnical Properties of Oil Sands and Related Strata", A thesis submitted to the faculty of graduate studies and research in partial fulfilment of the requirement for the degree of doctor of philosophy in geotechnique", 1989, Department of Civil Engineering, University of Alberta, p. 167
25. Preisig, M., Prévost, J.H., 2011. Coupled multi-phase thermo-poromechanical effects. Case study: CO_2 injection at In Salah, Algeria. International Journal of Green-house Gas Control 5 (4), 1055–1064.
26. Gor, Y.G., Elliot, T.R., Prévost, J.H., 2013. Effects of thermal stresses on caprockintegrity during CO_2 storage. International Journal of Greenhouse Gas Control 12, 300–309.
27. Vilarrasa, V., Olivella, S., Carrera, J., Rutqvist, J., Long term impacts of cold CO_2 injection on the caprock integrity, International Journal of Greenhouse Gas Control, Volume 24, May 2014, Pages 1-13, ISSN 1750-5836, http://dx.doi.org/10.1016/j.ijggc.2014.02.016.
28. Bandis, S.C., Lumsden, A.C., Barton, N.R. 1983. Fundamentals of rock join deformation, Int. J. Rock Mech.

Carbon Dioxide Capture for Storage in Deep Geological Formations, Volume 4
Karl F. Gerdes (Ed.)

Chapter 28

GEOMECHANICAL MODELLING STUDY OF THE IONA GAS STORAGE FIELD

E. Tenthorey, G. Backe, R. Puspitasari, Z.J. Pallikathekathil,
S. Vidal-Gilbert, B. Maney and D. Dewhurst
GPO Box 46314 – 16 Brisbane Avenue, Barton, Canberra, ACT

ABSTRACT: We present a 3D geomechanical modelling study of the Iona gas storage facility in the state of Victoria, Australia. The results provide important information pertaining to gas storage, which can then be used to understand certain geomechanical aspects of CO_2 storage. A key finding in this paper is that significant changes to the horizontal stress magnitudes are imparted by changes to the fluid pressure due to gas injection or withdrawal. This effect, known as the reservoir stress path, significantly influences fault stability by counteracting the changes to effective stress. In the case of Iona, pressurisation of the field results in a stress path which is parallel to the failure criterion rather than towards it, as would be expected in a classical treatment which does not incorporate complex poro-elastic effects. Another output of interest relates to reservoir deformation, which would be manifested at the ground surface as heave or subsidence. During periods of peak gas withdrawal and injection, surface ground movement is predicted to be on the order of −9 mm and +2.5 mm, respectively. These numbers are similar to the surface deformation observed at the In Salah CO_2 injection project, but much smaller than the subsidence observed in some producing hydrocarbon fields.

KEYWORDS: Iona gas field; CO_2 storage; geomechanics; reservoir stress path; fluid pressure; ground deformation; fault stability

INTRODUCTION

Injection or withdrawal of gas into the subsurface will invariably lead to displacement of *in situ* fluids and result in a concomitant change in the pore fluid pressure. Fluctuations to fluid pressure are important as they change the effective stress state of the reservoir and in many cases may alter the local *in situ* stress field via complex poro-elastic phenomena. These changes are of interest to geomechanicists as they have important implications for issues such as well integrity, fault stability, cap rock integrity and surface deformation; all of which are prime considerations during storage of CO_2 in the subsurface. Unfortunately, there is little publicly available research on the complex 3-dimensional effects of gas injection or withdrawal. Most of the work that has been done using modern finite element techniques is proprietary and has remained in the possession of industry.

Fortunately, EnergyAustralia (formerly TRUenergy), a major gas retailer in Victoria, has agreed to share data from the Iona field so that a detailed study could be performed characterising the geomechanical effects of methane gas injection and withdrawal into the subsurface. The Iona field is currently a methane gas storage facility located in southern Victoria, approximately 20 km east of the CO2CRC Otway Project. The formation being used for gas storage is the Cretaceous Waarre C Formation, which was the same unit that was used for CO_2 storage during Stage 1 of the Otway Project. Therefore, in addition to providing important geomechanical constraints which can be applied broadly, Iona is also useful in that it may provide information that can be applied directly to the Otway Project operations. The Iona facility is unique in that there is a long injection/withdrawal history together with accompanying pressure data, from which history matched static and dynamic flow models have been generated. In this study, we use these history matched models as the foundation for generating a 3-D geomechanical model describing the physical evolution of the field since its conversion to a gas storage facility in 2000. Construction and execution of a 3D geomechanical model involves a number of steps and quality control measures involving data from different sources. The workflow implemented in this modelling study is shown in Figure 1.

There are several main outputs of interest from this modelling study. First, all the geomechanical data from Iona and regionally were assembled in order to determine the initial stress condition at Iona prior to initial production in 1992. This critical information was input into the dynamic geomechanical model, which was run from 1992 to 2007 using the actual Iona injection and withdrawal history. The geomechanical model provides information on the reservoir stress changes as a result of production and injection at selected time steps. Time steps generally correlated with periods of peak overpressure and under-pressure, as these are the periods of greatest interest geomechanically. The information gained regarding stress changes directly allows the magnitude of the reservoir stress path to be determined, which is an important consideration for any gas storage field. Stress and strain outputs from the model also provide information regarding cap rock integrity and fault stability. Finally, outputs from the model also allow some statement to be made regarding the magnitude of possible ground movement during peak periods of injection and withdrawal.

STUDY AREA

The Iona gas storage facility is located in the Port Campbell region of the state of Victoria, Australia and is operated by EnergyAustralia. Figure 2 shows the Iona gas storage facility in relation to the CO2CRC Otway Project. Following the Longford gas incident in 1998, in which gas supply was cut to Victoria for several days, it was decided that an additional gas storage facility was needed to ensure adequate supply. Conversion from a producing gas field to a storage facility was completed in 2000. Iona is located approximately 20 km ESE of the CO2CRC's Otway Project in the Naylor Field. At Iona, methane gas is piped from offshore gas fields and stored in the Upper Cretaceous Waarre C Formation, approximately 1300 m below ground level. Gas injection and withdrawal are carried out via 5 wells, with two observation wells on the edge of the field to ensure that the reservoir is not filled to spill (Figure 3).

Structurally, the Iona field is a tilted anticline, which is bound to the north and south by major faults. The surface shown in Figure 3 is the top of the Waarre C Formation. The south fault behaves as a juxtaposition seal to the gas column, while the seal capacity of the north fault is uncertain, as it has barely been exposed to gas. The Waarre C sandstone at Iona is a high permeability sandstone and is divided into three units which are intercalated with thin shale units. The Waarre sandstone is of late Cretaceous age (Turonian) with a significant rift tectonic influence associated with Australia-Antarctica breakup. Although a number of different depositional models have been proposed, current models now suggest a stacked series of fluvial dominated, low sinuosity channels frequently influenced by tidal processes and marine storm surges. Some previous work suggests that the intraformational shales at Iona are thicker and more laterally continuous than at the Otway Project (Spencer and La Pedalina, 2006). This difference may be a result of various tectonic influences or proximity differences to the basin depocentre. Continuing work involving data from Iona will hopefully provide greater constraints for depositional models of the Waarre formation. The Waarre reservoir rocks are capped by the Belfast Mudstone, a thick (150 m) regional seal to most of the gas-bearing reservoirs in the Otway Basin.

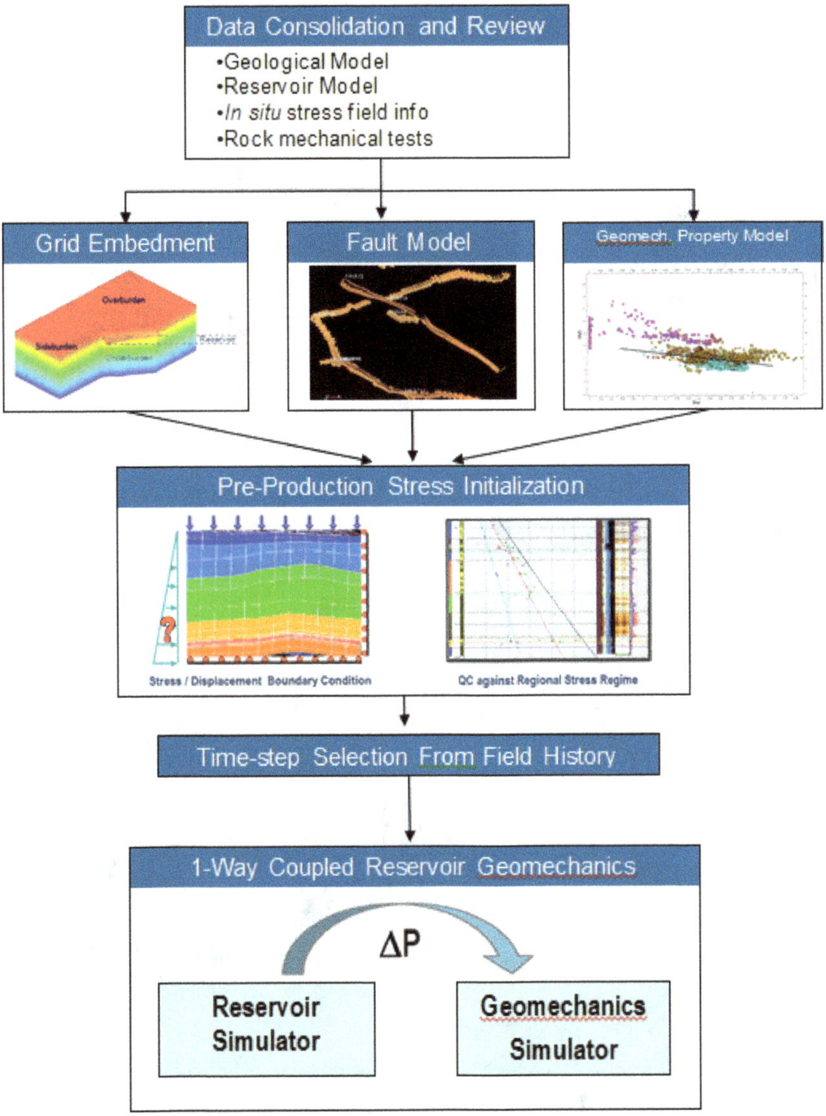

Figure 1. General workflow used in this study.

In this work, the dynamic simulation is 1-way coupled meaning that the changes to permeability and porosity caused by pressure changes are not fed back in to the static model as in a fully coupled model. Fully coupled models are computationally more difficult and in the simple structural case of Iona would likely be similar to the 1-way coupled scenario.

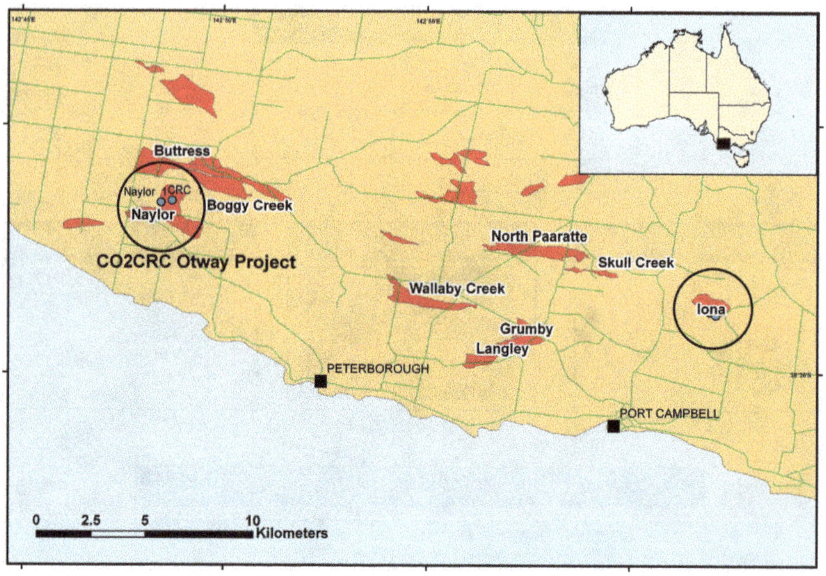

Figure 2. Study area.

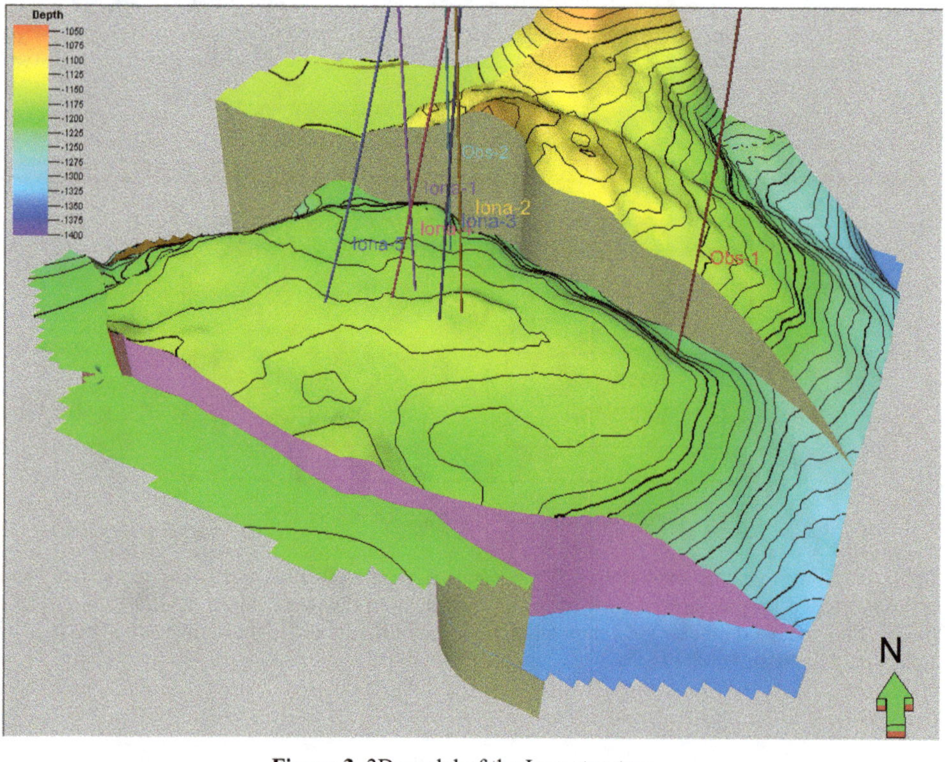

Figure 3. 3D model of the Iona structure.

DATA CONSOLIDATION AND REVIEW

Previous Modelling

Development of a 3D geomechanical is contingent upon the availability of a static geological model, a dynamic flow model and sufficient information on the local *in situ* stress tensor. All of these have either been provided by EnergyAustralia or are available from previous work conducted at Iona or at the Otway Project site. Additionally, a number of rock mechanical tests were performed on core from the Otway Project which has led to the development of a transform which allows the sonic logs to be used as a measure of elastic rock properties.

The static geological model used for the dynamic was constructed in Petrel previously by RISC consultants (IonaWellLogs.pet). It was based on 3D seismic data and information from 7 wells, which included significant core. The static model was also upscaled by RISC and exported to Eclipse for dynamic flow simulation. The Eclipse model size is 160,000 cells (79x78x26), which equates to a horizontal and vertical resolution of 50m and 2m, respectively. There are 5 injector/producer wells, 2 observation wells and 10 faults, all of which are considered to be non-sealing. The history match analysis was performed from Dec 1, 1992 to Jan 1, 2007 (~14 years). Figure 4 shows the overall injection and withdrawal history at Iona, together with the resulting changes to reservoir pressure. The field was variably produced until 2000 at which point it was converted to a gas storage facility. The largest oscillations in pressure are observed at around this point in time, thereby representing a window which is of great interest geomechanically.

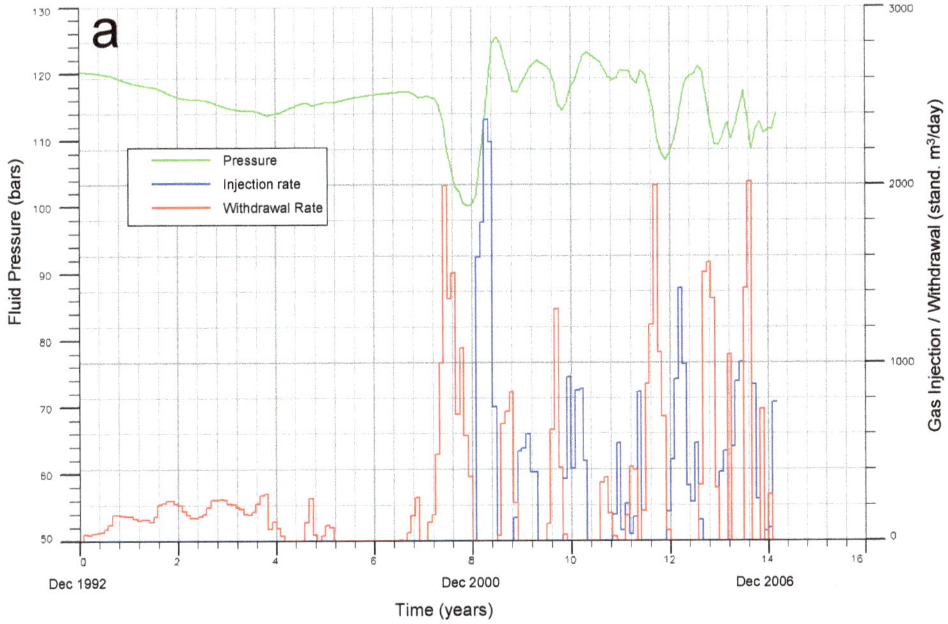

Figure 4. Fluid pressure evolution of the Iona field together with injection and withdrawal history.

In situ Stress Field

Information for the *in situ* stress tensor at Iona was acquired using data presented in van Ruth *et al.* (2007), Berard *et al.* (2008), Vidal-Gilbert *et al.* (2009a), Tenthorey *et al.* (2010) and Lawrence

(2011). These reports document *in situ* stress data from wells throughout the Otway Basin, as well as some specific information on the Otway Project and Iona. The key stress results from these reports are summarised in Table 1. Perhaps the most valuable information pertaining to the horizontal stress magnitudes are contained within Tenthorey *et al.* (2010). This report outlines the results of a mini-frac test programme following the drilling of the CRC-2 well at the Otway project. Contained within this report is a refined *in situ* stress estimation for the Otway Project site. Due to the lack of significant stress information at Iona, we have used the refined stress model for the Otway Project from Tenthorey *et al.* (2010) for use in the Iona modelling presented here. We therefore assume that the Iona Field is in a normal faulting regime with a vertical stress gradient (Sv) of 21.4 kPa/m, maximum horizontal stress (Shmax) of 17 kPa/m and minimum horizontal stress (Shmin) of 14 kPa/m. The maximum horizontal stress direction is taken to be 142°.

Table 1. Summary of *in situ* stress data available.

Document	Sv (kPa/m)	Shmax (kPa/m)	Shmin (kPa/m)	Shmax Azimuth	Notes
Berard *et al.* 2008	22.01	18.13	15.98	141° ±9°	Normal stress regime, horizontal stress magnitude calculated by SonicScanner
Vidal-Gilbert *et al.* 2009a	21.45	37.0	18.5	142° ±5°	Strike-slip stress regime based on various Otway Basin well data
Tenthorey *et al.* 2010	21.40	16.16-18.43	13.86-14.77	142°	Normal stress regime, estimated from ELOT and MDT mini-frac data
Lawrence, 2010	-	-	-	NW-SE	FMI resistivity image study

Rock Mechanical Testing

To populate the static geomechanical model with rock mechanical properties, it is critical to have some mechanical tests available which can be used to calibrate the sonic logs for properties such as Young's modulus, Poisson's ratio and unconfined compressive strength (UCS). These 1D, log-based mechanical and elastic properties can then be used in conjunction with other associated properties such as porosity or V-shale to populate the 3D model with rock mechanical properties. Unfortunately, rock mechanical testing of Iona core proved to be impossible, mainly due to the poor quality and friable nature of the core plugs. This made jacketing of the specimens prior to triaxial testing very difficult. As a result, rock mechanical tests conducted on core from the CO2CRC Otway Project's Naylor field were used, as the mechanical properties of the core are assumed to be similar to Iona which is located only 20 km away. Following the drilling of the CRC-1 well, two specimens were tested mechanically, one from the Waarre C Formation and one from the overlying Belfast Mudstone which is the primary sealing lithology to most of the gas fields in the onshore Otway Basin. During the drilling of the CRC-2 well in early 2010, significant core was extracted from which 5 specimens were selected for rock mechanical testing. Four of the samples were from the Paaratte Formation, while the other was from the Pember Mudstone. These tests were used successfully to develop transforms which could be applied to the sonic logs from the Iona wells. The transforms used and results of the sonic log calibrations are presented below.

Rock mechanics testing was performed at CSIRO Earth Science and Resource Engineering laboratories in Perth. The tests were performed on an autonomous triaxial cell capable of imposing 70 MPa confining and pore pressures and up to 400 MPa axial load on a 38 mm diameter sample.

The inner diameter of the cell is considerably larger than the sample, allowing the installation of internal instruments such as a load cell, axial displacement transducers, acoustic transducers and temperature sensors.

Samples from CRC-2, nominally of 76 mm length and 38 mm diameter, were deformed in a standard triaxial cell with full independent control of cell pressure, pore pressure and axial load. Where sample length/diameter ratios were <2, a correction was applied to the differential stress as per standard ASTM testing procedures. Due to the small amounts of samples available, multi-stage triaxial tests were run (Fjær *et al.*, 2008). Samples were deformed to within a few MPa of peak strength (5-10%) at a set confining pressure then unloaded. Confining pressure was then increased followed by further application of axial load until again close to failure. This cycle was repeated until the desired number of steps (usually four or five, ranging between 8 and 21 MPa for the CRC-2 samples) had been reached. The final cycle was taken through to failure and residual strength. A pore pressure of 3 MPa was applied to both sandstone and shale samples.

STATIC 3D GEOMECHANICAL MODEL

Development of a robust static geomechanical model requires not only a detailed geomechanical characterisation of the reservoir interval, but also of the rock volume enclosing it (Vidal-Gilbert *et al.*, 2009a, 2009b; Rutqvist *et al.*, 2010). A number of different methods must be used to assign poromechanical properties to the reservoir and surrounding formations. The best and most direct manner of assigning rock mechanical properties is through rock mechanical testing of the different rock types. In such a case, the static moduli of interest (Poisson's ratio and Young's modulus), can be used directly in the model. However, in many cases rock mechanical measurements will not be available and sonic logs must be used to assign poro-mechanical properties to the reservoir and cap rock. These sonic logs can be used to calculate the undrained dynamic moduli, which can then be converted to drained dynamic moduli using the Biot-Gassmann (Gassmann, 1951) equation or similar. Finally, drained dynamic moduli are converted to static moduli either by calibration to rock mechanical testing or using empirical relationships (Wang, 2000). It is these transforms that are used in conjunction with other properties in the static geological model, to generate a 3D model of poro-elastic properties. Also critical is a quantitative understanding of the brittle failure parameters such as unconfined compressive strength and friction angle, as these properties will control whether the reservoir and cap rock will be able to support the changes in pore fluid pressure, or whether new fault and fracture networks form.

As discussed above, rock mechanical tests from the Otway Project's CRC-2 well were used to develop the transforms for Young's modulus, Poisson's ratio and unconfined compressive strength (UCS). This process is shown graphically for Young's modulus in Figure 5. The first step in this process is to convert the compressional (DTCO) and shear slowness (DTSM) logs to compressional and shear velocities, Vp and Vs, respectively, which is straight forward. The dynamic Young's modulus, E_{dyn}, a measure of rock stiffness, was then calculated using:

$$E_{dyn} = \rho V_s \left(\frac{3V_P{}^2 - 4V_S{}^2}{V_P{}^2 - V_S{}^2} \right) / 10$$

where E_{dyn} is in units of GPa and ρ is density in g/cc. Results for Young's modulus acquired from rock mechanical testing were then plotted up against the corresponding dynamic value calculated above to yield an empirical relationship that could be used to calibrate the entire dynamic Young's log. The resulting log for static Young's modulus is shown on Figure 5 together with the results from rock mechanical testing. One can see that the empirical relationship used is able to closely match the log to the mechanical test data.

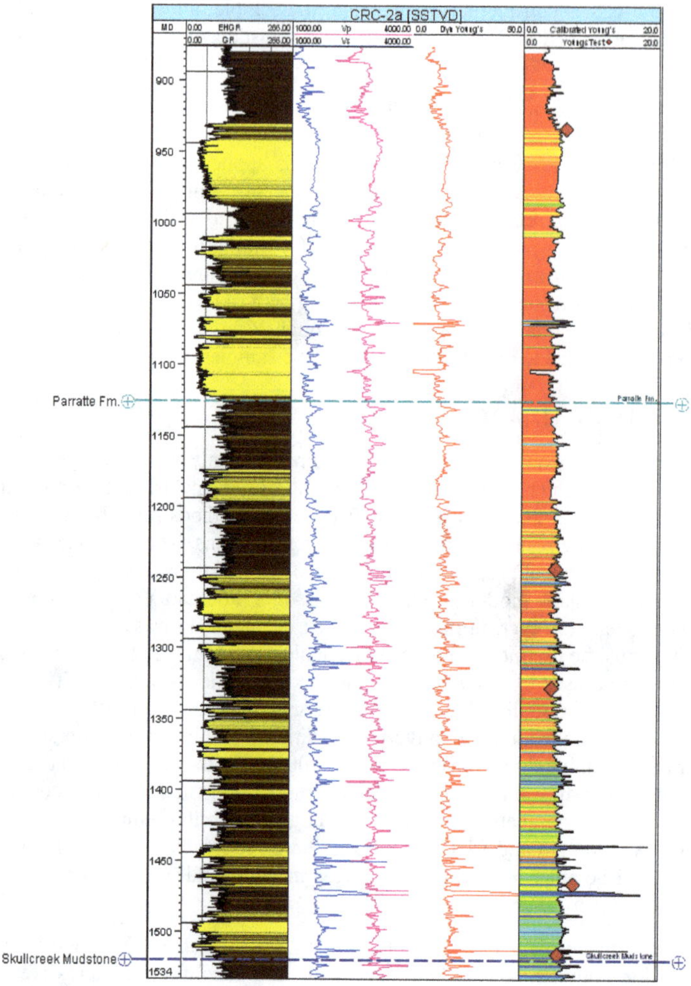

Figure 5. Well section for the CRC-2 well.

Similarly, the dynamic Poisson's ratio, ν_{dyn}, the ratio of transverse strain to axial strain, is calculated by the following:

$$\nu_{dyn} = \frac{V_P{}^2 - 2V_S{}^2}{2(V_P{}^2 - V_S{}^2)}$$

For the case of Poisson's ratio, the calculated values using the above relationship and the results from mechanical testing matched very closely, and we therefore assume that $\nu_{dyn} = \nu_{stat}$.

The unconfined compressive strength (UCS) of the specimens also correlated well with the dynamic Young's modulus allowing the following relationship to be used as a continuous measure of UCS:

$$UCS = 7.76e^{0.0388E_{dyn}}$$

The transforms developed above, using rock mechanical tests from CRC-2 core, were then extended deeper down to the level of the Belfast Mudstone and Waarre C Formation to ensure that they were consistent with the mechanical tests conducted during Stage 1 of the Otway Project. This extrapolation resulted in a very good match and therefore validates use of the transform for the geomechanical modelling. The vertical stress gradient was calculated by integrating the density logs over the length of the section. The minimum and maximum horizontal stress magnitudes were calculated using the static Young's modulus and Poisson's ratio in the relationships presented in Thiercelin and Plumb (1994).

Following the successful calibration of the sonic logs from top to bottom of the Iona section, it was necessary to grid the reservoir together with the over-, under-, and side-burdens. For the reservoir, gridding from the ECLIPSE model was used (79 x 78 x 26; 160k cells). However the complete embedded model was significantly larger, measuring 747k cells (101 x 100 x 74), and is shown in Figure 6. The over-burden in the model was extended to the surface, while the under-burden was extended to a depth of 16000 m. The side-burdens were extended to 3 times the reservoir length on each side. It is assumed that the faults present in the geological model do not extend into the over-burden or under-burden. While this assumption does not characterize the exact structure of the field, it is necessary for practical modelling purposes. Furthermore, as no fault properties are available from the dataset, values for the various key properties were estimated based on average reservoir stiffness and average fault spacing. Normal and shear stiffness were assumed to be 300,000 kPa/m and 150,000 kPa/m, respectively, with cohesion of 1 kPa, friction angle of 20° and angle of dilation of 10°.

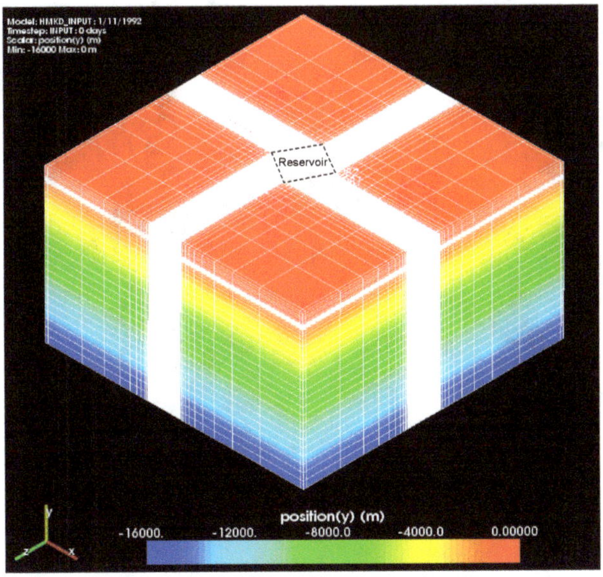

Figure 6. 3D geomechanical model grid.

The final step in the development of the static geomechanical model is population of the 3D grid with the required mechanical properties such as Young's modulus, Poisson's ratio, UCS and density. Originally, it was planned that the grid would be populated in a layer-cake manner based on the calibrated well data. However, it was later decided that correlations would be sought between the mechanical properties and various lithological properties in the different wells, so that the grid could be populated more accurately. The only property which shows some degree of correlation is porosity, which correlates to some degree with Young's modulus and density (Figure 7).

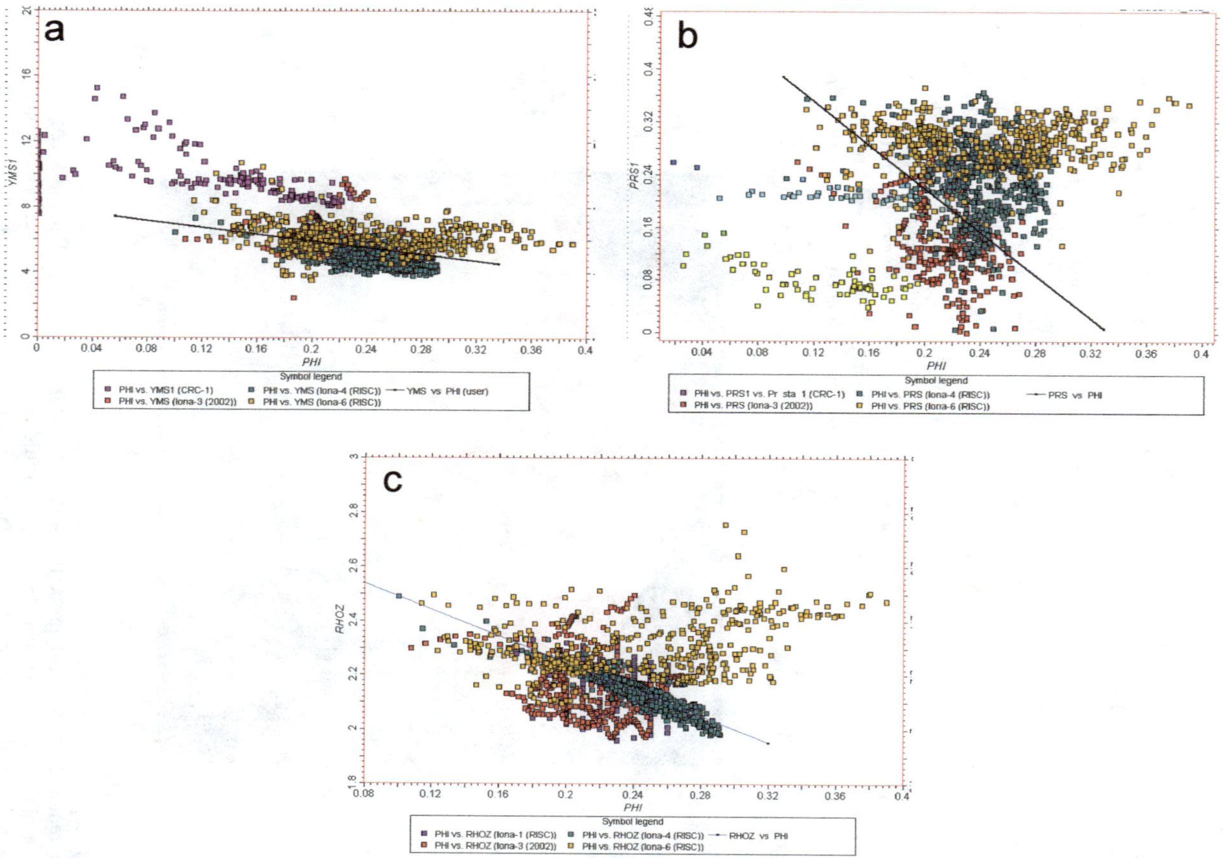

Figure 7. Co-variograms for Young's modulus (a), Poisson's ratio (b) and density (c). The different colours on each figure represent data from different wells in the Iona field.

The variograms of Figure 7 are graphed against porosity for various well data. Young's modulus and density exhibit moderate correlation with porosity; correlations which were used to populate the 3D mechanical grid.

In contrast, Poisson's ratio does not show any correlation with porosity and was therefore estimated based on a correlation with Young's modulus, as was UCS. The over-burden and under-burden were populated as layer cake based on the upscaled value from the calculated rock property log. Meanwhile, the side-burden mechanical properties were populated based on the mean value of the reservoir mechanical property. A 3D visualization of the key results from the property modelling is shown in Figure 8.

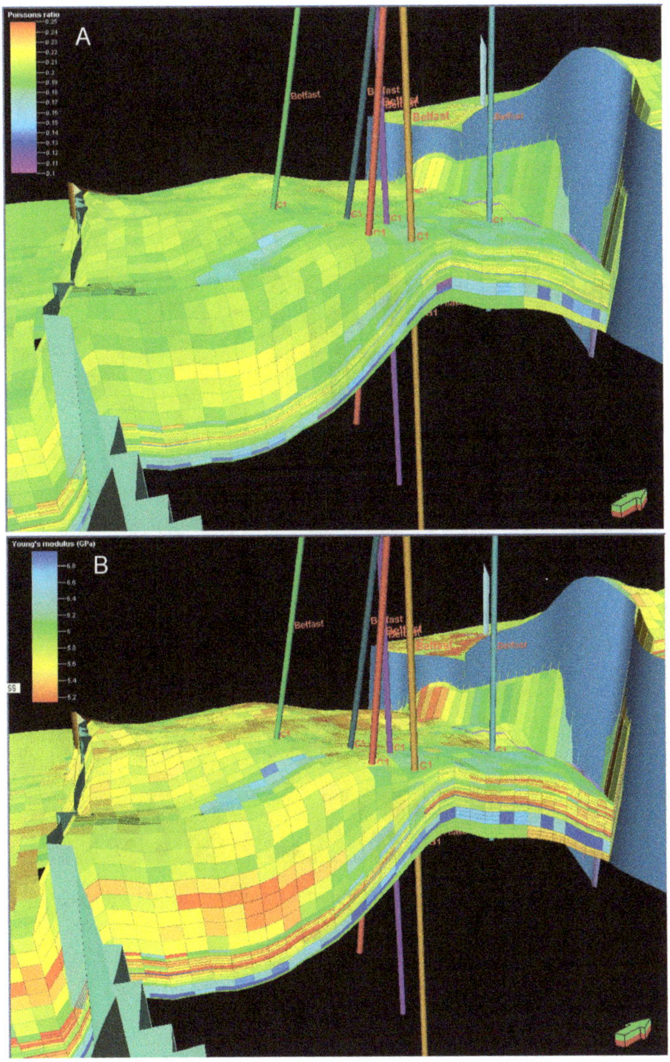

Figure 8. 3D illustration of the Iona field showing the model input variations for Poisson's ratio (A) and Young's modulus (B).

DYNAMIC GEOMECHANICAL SIMULATION

Pre-production Stress Initialization

Prior to running the dynamic geomechanical simulation, it is necessary to determine the boundary stress conditions at time zero. For the horizontal stresses, the initial stress condition is applied only at the edge of the model, with the stresses inside the model being controlled by the elastic properties of the rock. The magnitudes of the horizontal boundary stresses were determined so as to satisfy the stress gradients presented in Tenthorey *et al.* (2010). The vertical stress within each model element is calculated using the density data in the 3D grid cells, as presented in the previous section. It is interesting to note that at shallow depths (<500m), the stress regime is strike-slip and transitions to normal faulting at greater depths. This is consistent with the 1D mechanical earth model for the CRC-1 well, and is likely a result of the rocks having greatly reduced stiffness and/or density at shallow depths.

The assumed pre-production pore pressure is based on the ECLIPSE results for fluid-active cells. Where the cells are inactive and for the embedding cells, the pore pressure is assumed to follow a normal hydrostatic gradient. A 3D visualization of the three principal stress magnitudes, pre-production is shown in Figure 9 through Figure 11.

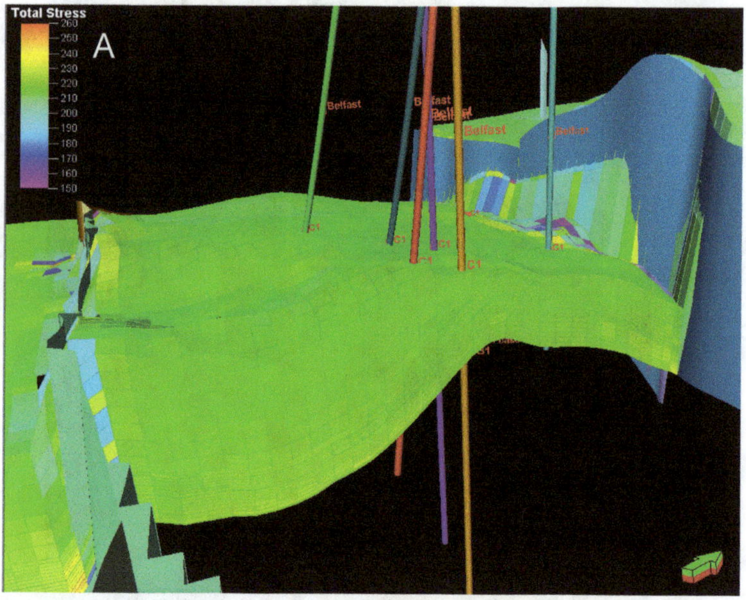

Figure 9. 3D illustration of the geomechanical model (minimum horizontal stress).

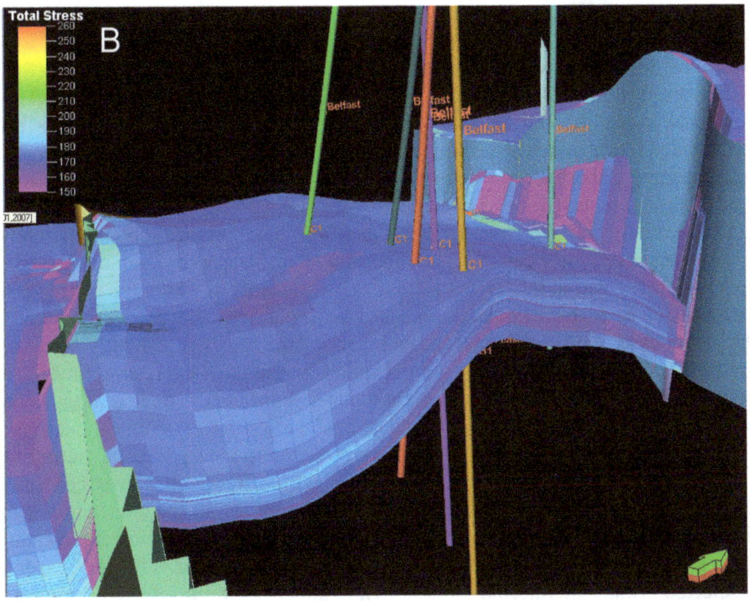

Figure 10. 3D illustration of the geomechanical model (maximum horizontal stress).

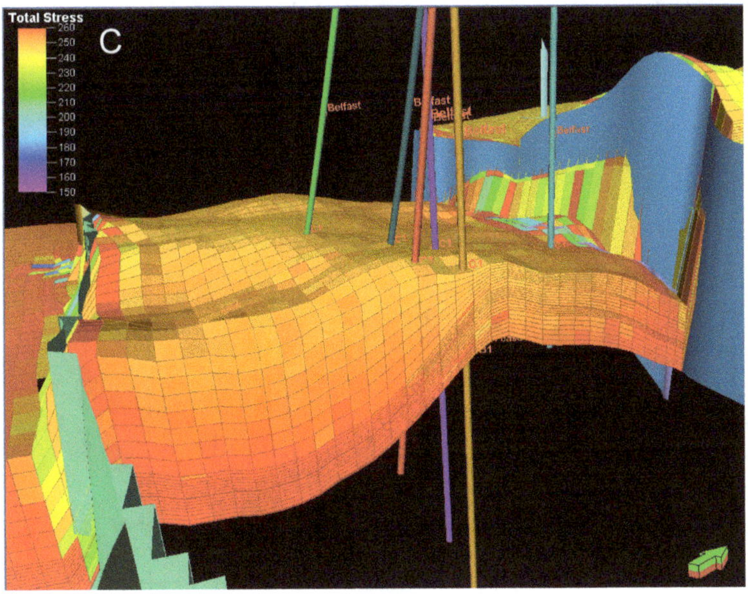

Figure 11. 3D illustration of the geomechanical model (vertical stress).

Time-Step Selection and Model Runs

The geomechanical modelling results presented in this report were conducted by the Schlumberger geomechanics team in Perth, Australia. The software used was Schlumberger's VISAGE package. 3D geomechanical modelling can be fully coupled or partially coupled (see Vidal-Gilbert *et al.*, 2009b, for a detailed description). The fully coupled approach simultaneously solves the whole set of equations that govern the hydromechanical problem, meaning that changes to permeability and porosity caused by geomechanical phenomena, are continuously fed back into the reservoir simulator. In the partially coupled case, the stress and flow equations are solved separately, although information is transferred between the geomechanical and reservoir simulators. The partially coupled modelling presented in this report is "one way coupled", in which the pore pressure history from a conventional reservoir simulation is used as input into the geomechanical equilibrium equation. In such a case, the geomechanical behaviour of the reservoir can only be assessed at key times during the history of the reservoir. Geomechanically, the most interesting time steps are those in which pore fluid pressure increases or decreases most significantly, as it is at these times that *in situ* stresses, deformation and ground movements will vary most. For this modelling study, six different time steps were chosen. The time steps were chosen to incorporate the period in which the Iona field was a producing field, as well as the later period in which it was converted to a gas storage facility. The exact timeline for injection and withdrawal and the dates chosen for the time steps are shown in Figure 12. As is clear from the figure, the greatest and most rapid changes to pore pressure occur around the time of conversion to a gas storage facility. It is for this reason that 3 time steps were selected around this period. Another two time steps were selected during a significant injection/withdrawal phase at around year 12. The final time step was chosen at the end of the model. Although this final time step did not occur during a significant pressurisation phase, it was selected due to the fact that some of the geomechanical phenomena (such as fault reactivation) are dependent on cumulative changes in a parameter.

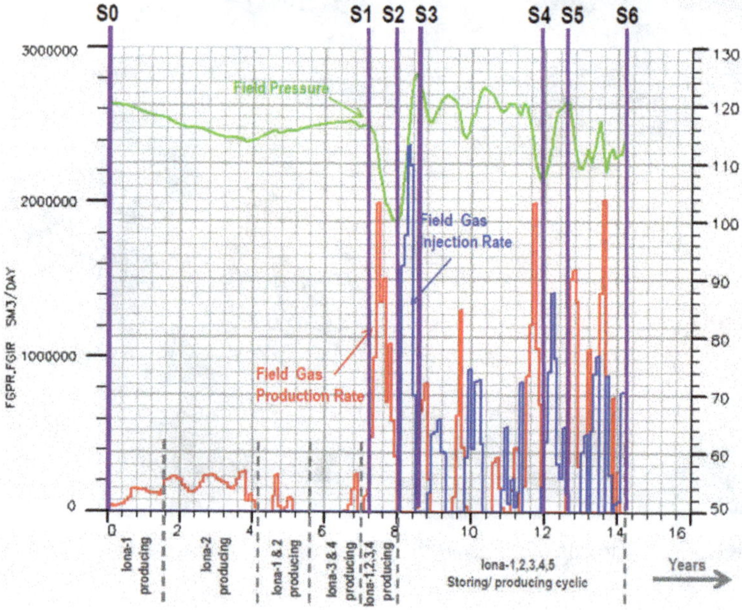

Step	Date
S0	Nov 01, 1992
S1	Dec 01, 1999
S2	Dec 01, 2000
S3	May 01, 2001
S4	Oct 01, 2001
S5	Jun 01, 2005
S6	Jan 01, 2007

Figure 12. Time-step selections for the dynamic geomechanical modelling.

Reservoir Stress Path

In general, simplified geomechanical models usually rely solely on the effective stress concept to characterise the impact of pore fluid changes on reservoir integrity and fault stability. The effective stress concept was first developed by Terzaghi (1943) for soil systems, where an increase in pore fluid pressure results in an equally reduced effective stress on the rock mass, and vice versa. The effective stress concept implies that fluid pressure changes can alter the effective stresses felt by a body of rock, but cannot alter the absolute magnitudes of the principal *in situ* stresses themselves.

By contrast, the reservoir stress path (sometimes referred to as pore pressure-stress coupling) is another important geomechanical phenomenon that can greatly impact on a geomechanical response by imparting changes to the *in situ* stress field. The reservoir stress path generally describes the degree to which the minimum horizontal stress changes in response to perturbations of pore fluid pressure, either in an injection or withdrawal scenario (eg. Hillis 2000). The reservoir stress path and its effect on the reservoir may be understood conceptually as follows. As a liquid or gas is injected into a reservoir, pore fluid pressure builds up and the reservoir tries to expand in all directions. As the vertical direction is bound by a free surface (the Earth's surface), there is no change to the vertical stress. However, as the reservoir tries to expand laterally, there is a counteracting force that is imparted into the reservoir which causes the minimum horizontal stress to increase. In a normal faulting or strike slip faulting environment, an increase in the minimum horizontal stress does not favour reactivation of faults, but rather "stabilises" existing faults by reducing the shear stress/normal stress ratio on the fault surface.

The magnitude of pore pressure stress coupling can be expressed as follows (Engelder & Fischer, 1994):

$$\frac{\Delta\sigma_h}{\Delta P} = \alpha\left(\frac{1-2\nu}{1-\nu}\right)$$

where $\Delta\sigma_h$ is minimum horizontal stress, ΔP is pore pressure, ν is Poisson's ratio and α is the Biot coefficient (which in turn is a function of the rock and mineral compressibility). The magnitudes observed for pore pressure coupling generally range from about 0.5 to 0.8.

Figures 13-16 show the change in pore pressure and the three principal stresses at time step 2. S2 is the time of maximum depletion and minimum fluid pressure.

During S2, pore fluid pressure decreases in the permeable part of the reservoir by approximately 18 bars, on average. As discussed above, if the reservoir stress path effect was non-existent, then there would be no change in the magnitudes of the horizontal stresses. However, in Figure 14 and Figure 15, the minimum and maximum horizontal stresses in the reservoir section decrease by about 13 bars on average. This yields a value for the reservoir stress path of 0.72. Above the reservoir section, in the Belfast Formation immediately above, horizontal stresses actually increase slightly, sometimes by as much as 2 bars. This is an example of stress partitioning which leads to a non-linear, and somewhat complex fracture gradient. In Figure 16 the stress change for the vertical stress is also shown. As expected, most of the reservoir section does not exhibit any change in the vertical stress. However, intra-reservoir, low permeability baffles do exhibit some significant increases in the vertical stress, as high as 6 bars. This change is not a result of the reservoir stress path, but rather a result of stress arching effects.

During S3 (Figures 17-20), pore fluid pressure increases by about 5.8 bars in the main reservoir section, while the concomitant horizontal stresses increase by about 4.5 bars. This yields a reservoir stress path effect upon re-pressurisation of 0.77, which is a rough, integrated value for the whole reservoir section. As was the case for depletion during S2, vertical stress does not change significantly throughout the reservoir section. However, during pressurisation, low permeability baffles do undergo a reduction in the vertical stress. This is caused by elastic expansion in the surrounding rocks which is a pressure relief mechanism for such static cells during pressurisation of the reservoir, and is again reflective of stress arching processes.

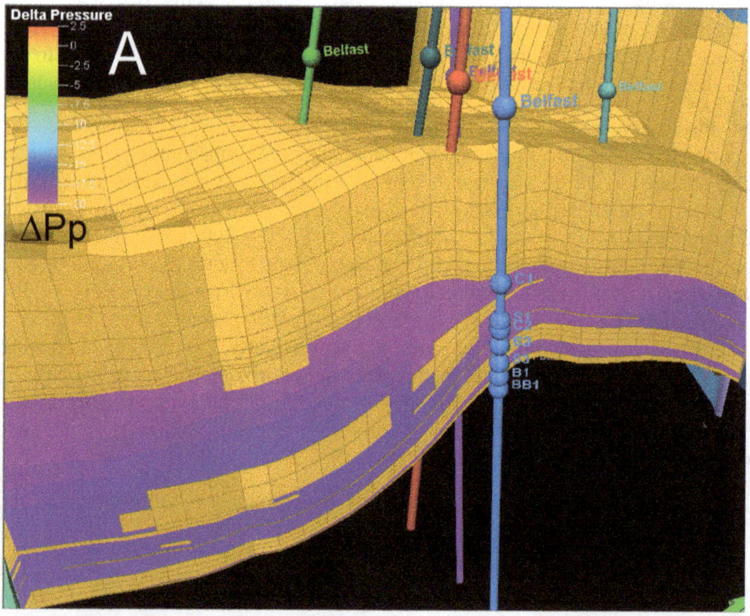

Figure 13. Change in the pore pressure during S2 (relative to S0).

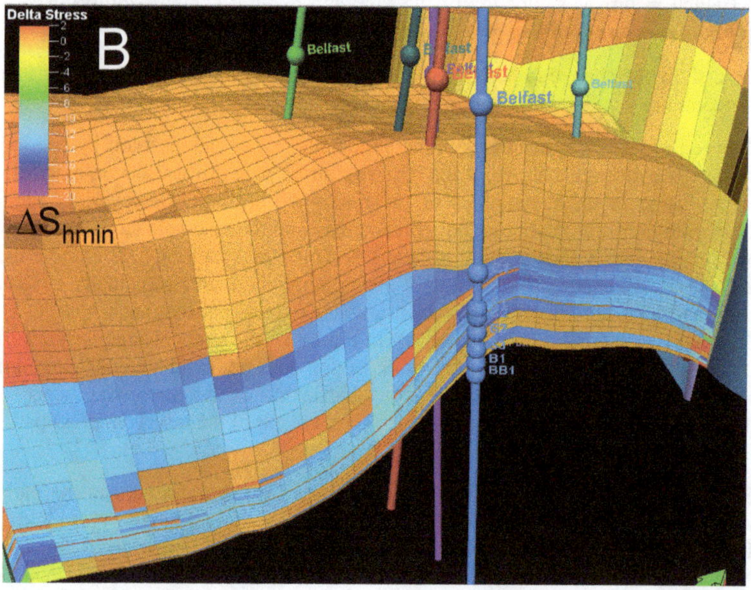

Figure 14. Change in the minimum horizontal stress during S2 (relative to S0).

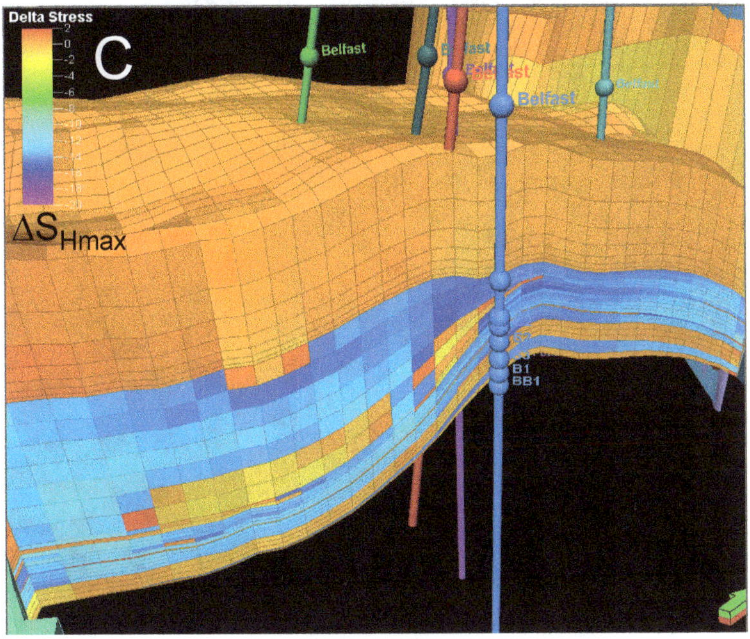

Figure 15. Change in the maximum horizontal stress during S2 (relative to S0).

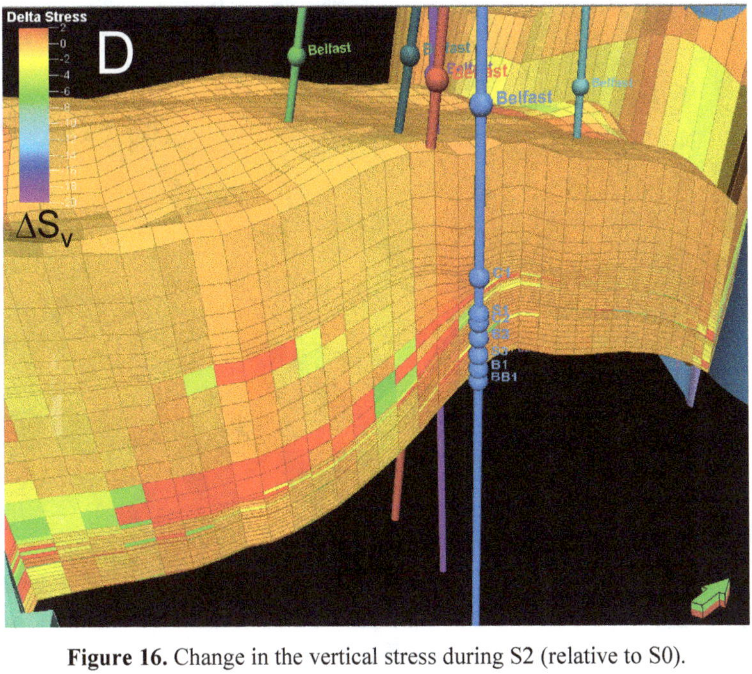

Figure 16. Change in the vertical stress during S2 (relative to S0).

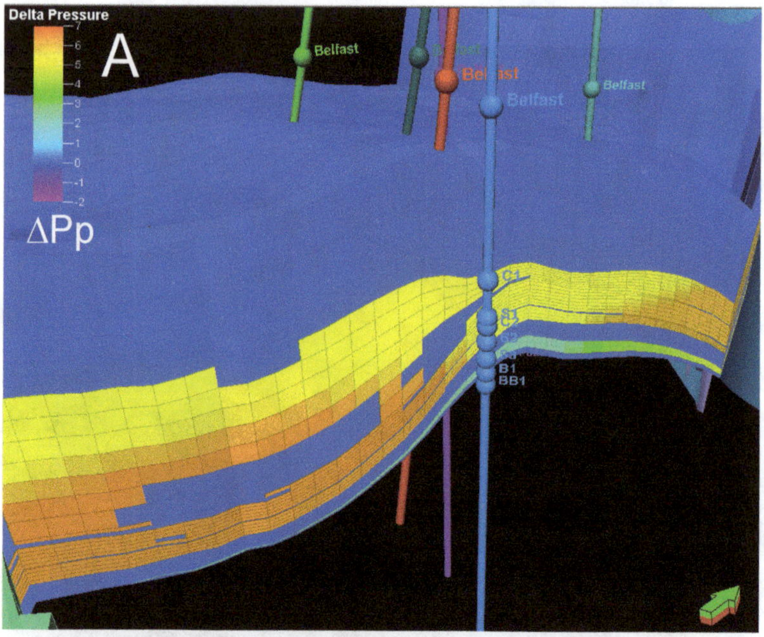

Figure 17. Change in the pore pressure during S3 (relative to S0).

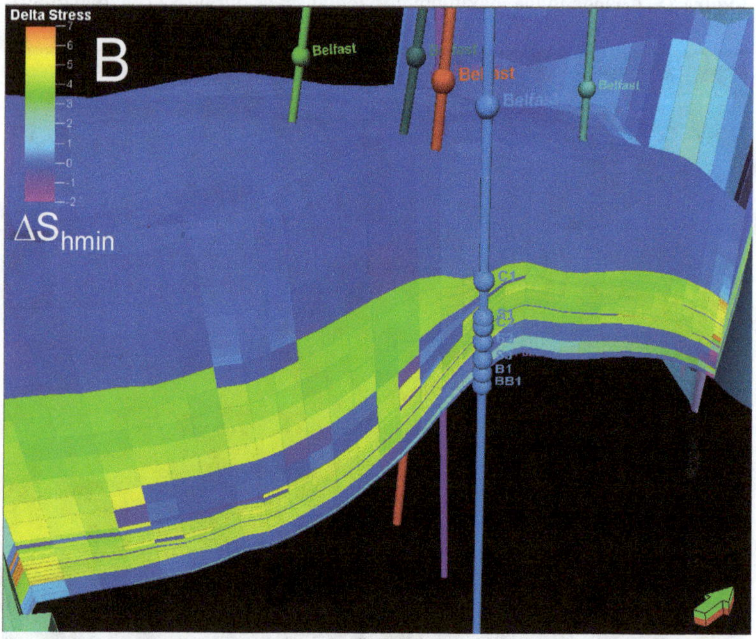

Figure 18. Change in the minimum horizontal stress pore during S3 (relative to S0).

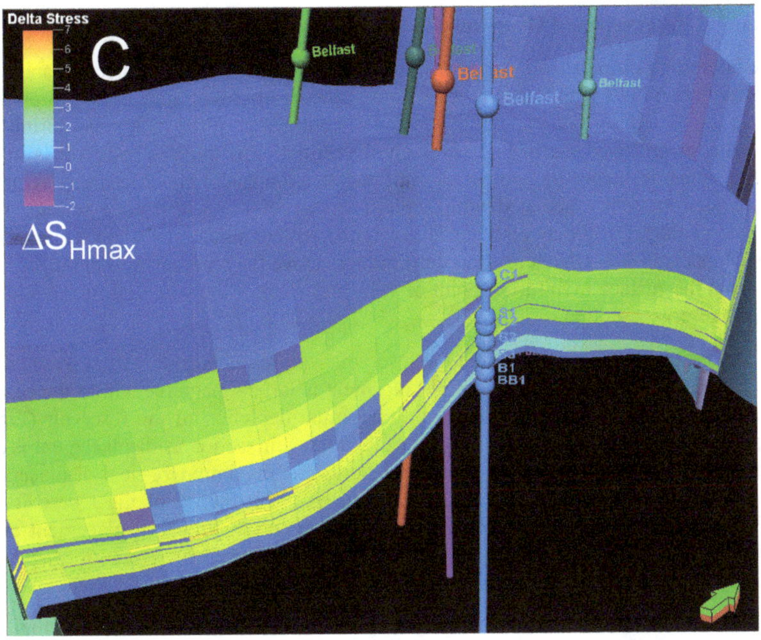

Figure 19. Change in the maximum horizontal stress during S3 (relative to S0).

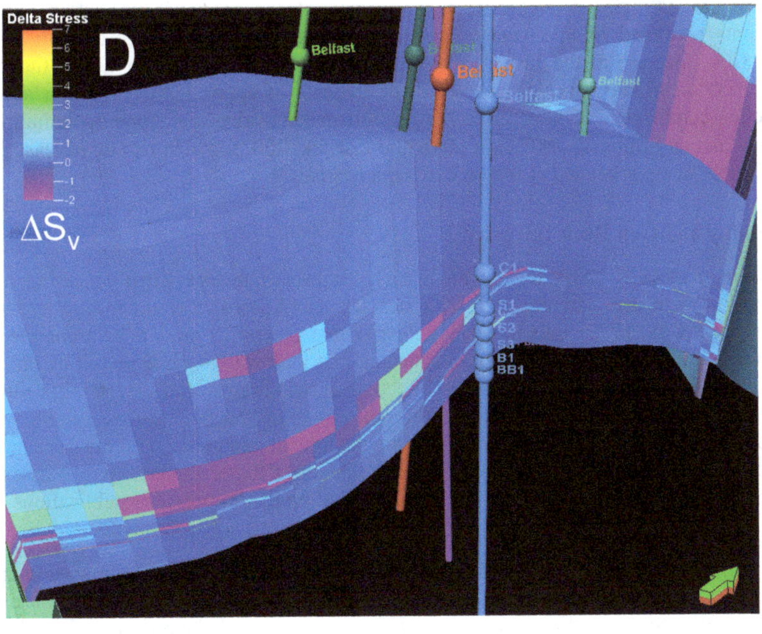

Figure 20. Change in the vertical stress during S3 (relative to S0).

Deformation of Reservoir and Over-burden

One of the interesting outputs from the geomechanical modelling relates to injection and withdrawal induced deformation both within the reservoir and in the overburden. Such deformation occurs due to the elastic (or sometimes non-elastic) expansion or contraction of the rock mass during periods of over-pressuring or under-pressuring, respectively. Deformation within the reservoir is important as it may have some influence on potential fault reactivation and may affect cap rock integrity. Furthermore, deformation associated with depleting reservoirs can also be a serious problem in terms of casing collapse and may also introduce significant complications when it comes to drilling later wells (Zoback, 2007). Characterising deformation above the reservoir and at the surface is also important from the surface monitoring point of view and also for the potential reactivation of faults outside of the reservoir section.

As expected, vertical displacement within the reservoir with respect to S0, was at its greatest during time steps S2 and S3. During S2, vertical displacement is largely negative within the reservoir due to compactional processes being active during maximum depletion of the reservoir (Figure 21a). The vertical displacement is greatest at the top of the reservoir section as displacement is a function of the cumulative displacements of all the cells below. The maximum vertical displacement at the top of the reservoir is approximately 12 mm, which is about 0.05% of the reservoir thickness. During S3, the period of maximum pore fluid pressure increase, vertical displacement exhibits the opposite behaviour, as expected (Figure 21b). Maximum predicted vertical displacement is approximately 4 mm in an upward direction.

The translation of this reservoir deformation to the surface is shown in Figure 22 and Figure 23. The dashed lines represent the projection of the two bounding faults to the surface. The maximum subsidence during S2 relative to S0 is 9mm and the maximum heave at S3 is 2.5 mm. While the greatest surface displacements are expected directly above the reservoir, some displacement is expected as far as 2.5 km away from the centre of the reservoir projection. The surface displacements predicted by the geomechanical model are slightly smaller than those observed following CO_2 injection at the In Salah project (Rutqvist *et al.*, 2010), in which Interferometric Synthetic Aperture Radar revealed surface displacements of approximately 15 mm several years after injection commenced. Compared to the documented subsidence at some depleting oil and gas fields (eg. Chan and Zoback, 2007; Mallman and Zoback, 2007), which can be at least 25 cm, the surface displacements predicted at Iona are extremely small.

Fault Stability and Cap Rock Integrity

Fault stability and cap rock integrity are clearly important factors when assessing a field for potential gas storage use. Rupture of fault zones are well known to result in an increase in fault permeability (eg. Cox, 1995), thus increasing the risk that gases may migrate up along the fault zone to shallower levels during induced seismic events. Often fault stability is assessed by modelling the expected changes in fluid pressure during injection or withdrawal and determining the resulting changes to the shear/normal stress ratio at different points on the fault. Using assumed values for fault cohesion and friction, one can then gauge the theoretical maximum fluid pressure increase that the fault zone can sustain before undergoing slip. This was the technique used to assess fracture stability of the three main faults at the CO2CRC Otway project in previous publications (van Ruth *et al.*, 2007; Vidal-Gilbert *et al.*, 2010).

The approach taken in this study is different in that it looks at the plastic strain accumulated in the different cells as a measure of how close the system is to fault rupture or cap rock failure. Based on years of experience modelling and observing geomechanical behaviour of various fields, Schlumberger typically uses a value of 1% strain to delineate the point at which the risk of rock failure becomes elevated. Unlike elastic strain, plastic strain is irreversible and only increases during the life of a field. Figure 24 shows the predicted accumulated plastic strain near the main faults (within the Waarre C Formation) at Iona. Plastic strain is predicted to be greatest in the northern bounding fault, with values up to 0.09% in some cells, at the end of the model run. Even the cells that exhibit these maxima in plastic strain are well below the critical value at which fault reactivation or rupture becomes a significant risk.

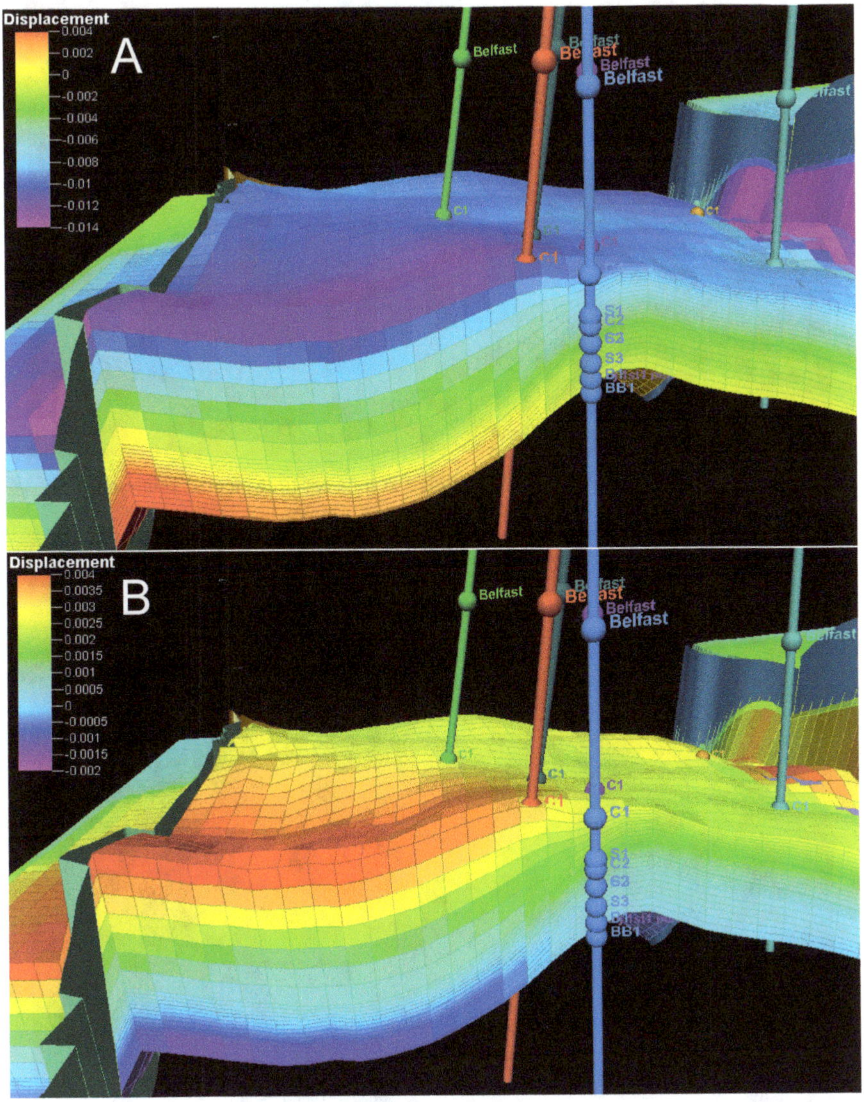

Figure 21. Vertical displacement at A) minimum fluid pressure, S2 and B) maximum fluid pressure condition, S3.

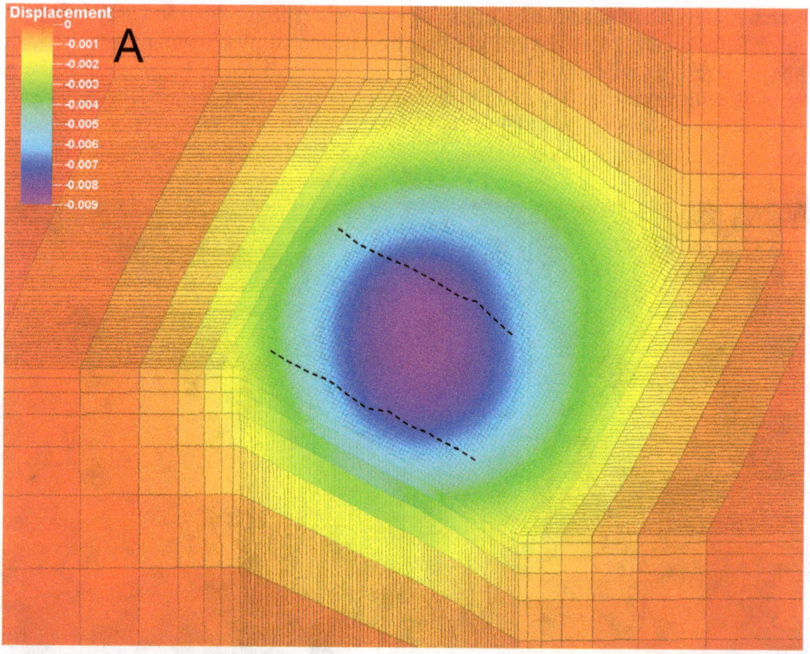

Figure 22. Surface displacements predicted at Iona for time step S2 relative to S0.

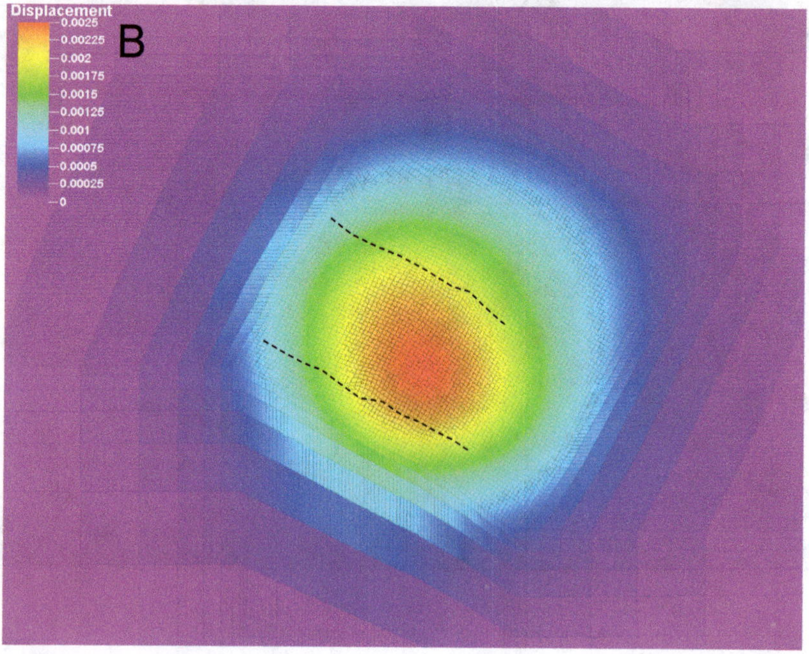

Figure 23. Surface displacements predicted at Iona for time step S3 relative to S0.

The same plastic strain metric was used to evaluate cap rock integrity. In the model, the Belfast Mudstone , which measures approximately 150 m in thickness, was modelled as overburden layers 10 to 17. Layer 17 is deepest and is immediately overlying the Waarre C Fm. The maximum change in plastic strain within the Belfast Mudstone at the end of the history match is 0.135%, and occurs in the vicinity of the northern bounding fault. This value is well below the 1% benchmark and therefore suggests that there is minimum risk of cap rock integrity failure. However, it should be noted that the fault model data is only available in the reservoir section. Should the faults extend into the overburden, then a higher potential for cap rock integrity failure is possible.

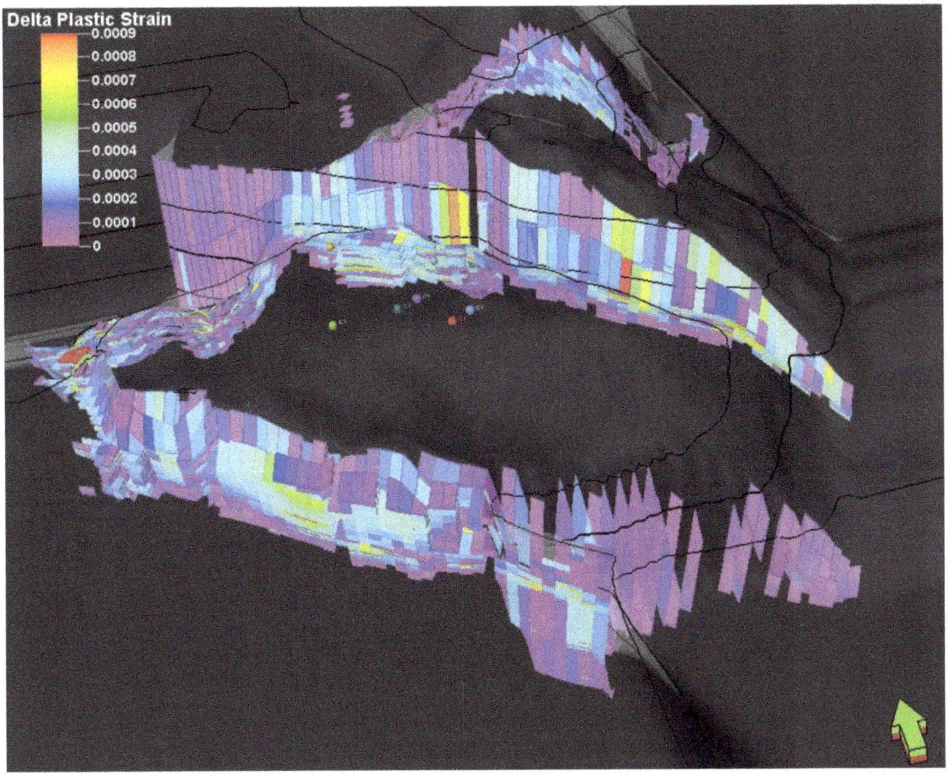

Figure 24. Overview of the predicted cumulative plastic strain near the main faults, at the end of the model run (S6).

SUMMARY AND CONCLUSIONS

The contents of this report provide valuable information, on two fronts, which can be applied to our understanding of gas storage in general. Firstly, the report itself describes the step by step workflow for building a detailed 3 dimensional dynamic geomechanical model for a potential gas storage facility. This includes calibration of sonic well data for rock mechanical properties using mechanical test data, population of the static model with mechanical properties, pre-production stress initialisation and execution of the dynamic geomechanical model using history matched flow models. Although many geomechanical models such as these have likely been constructed, there is a surprising paucity of information in the public domain regarding the construction and execution of such models.

Secondly, this model study is not only relevant to the Iona gas storage facility. The information gained from the modelling can be applied to other potential gas storage facilities around the world, be they methane storage facilities (such as Iona) or CO_2 storage sites. The results help us to understand various stress related phenomena, such as the reservoir stress path, surface ground movement and fault and cap rock integrity. A clear understanding of these issues together with some predictive capability will facilitate development of future gas storage facilities and also help minimize geomechanical risk.

The key outcomes of this study are as follows:

- Geomechanical properties have been populated in the 3D static model based on CRC-2 rock property correlations and correlations with reservoir porosity. This approach captured the reservoir heterogeneity more closely than a layer cake approach.
- CRC-2 MDT mini-frac tests and extended leak-off tests were used as a guide to determine the initial stress conditions in Iona field. VISAGE pre-production stress simulation shows a consistent stress regime in the Iona field and CRC-1 well. The stress regime observed is normal regime which changes to strike slip regime towards the surface.
- One-way reservoir geomechanical coupling was conducted in 6 selected time-steps, representing the evolution of field pressure changes due to production and injection. The changes in field stresses were calculated based on the changes in reservoir pressure.
- Horizontal stress changes within the reservoir during gas injection/withdrawal suggest a relatively strong reservoir stress path. In a normal faulting regime, such as at Iona, this large pore pressure stress coupling would serve to stabilise faults by increasing the normal stress on the fault plane.
- There is minimal vertical displacement observed, translating to minimal reservoir compaction/ expansion and surface subsidence/ heave. The maximum predicted subsidence and heave are 9 mm and 2.5 mm, respectively.
- There is minimum plastic strain change observed, translating to minimum risk of fault instability and cap rock integrity failure. The maximum plastic strain change is predicted in the vicinity of the northern bounding fault and does not exceed 0.14% which is well below the critical 1% threshold.

REFERENCES

1. Berard, T., Sinha, B.K., van Ruth, P., Dance, T., John, Z., Tan, C., 2008. Stress estimation at the Otway CO_2 storage site, Australia. In: SPE Asia Pacific Oil and Gas Conference and Exhibition, SPE 116422, Perth, Australia, October 20–22.
2. Chan, AW, and Zoback, MD, 2007. The role of hydrocarbon production on land subsidence and fault reactivation in the Louisiana coastal zone. Journal of Coastal Research, 23(3), 771-786.
3. Cox, SF, 1995. Faulting processes at high fluid pressures: An example of fault valve behavior from the Wattle Gully Fault, Victoria, Australia, J. Geophys. Res. 100, 12841-12859.
4. Engelder, T. and Fischer, M.P., 1994, Influence of poroelastic behavior on the magnitude of minimum horizontal stress, Sh, in overpressured parts of sedimentary basins. Geology v22. 949-952.
5. Fjær E., Holt, R.M., Horsrud, P., Raaen, A.M. and Risnes, R., 2008, Petroleum Related Rock Mechanics, 2nd Edition. Developments in Petroleum Science, 53, Elsevier, Amsterdam, 514 pp.

6. Gassmann, F, 1951. Ueber die Elastizitat poroser Medien. Vierteljahrschrift der Naturforschenden Gesellschaft in Zurich, 96: 1-23.
7. Hillis, R, 2000. Pore pressure/stress coupling and its implications for seismicity. Exploration Geophysics v31, 448-454.
8. Lawrence, M 2011. Structural Analysis of Borehole Images from Otway Wells CRC-1 and CRC-2. Cooperative Research Centre for Greenhouse Gas Technologies, Canberra, Australia, CO2CRC Publication Number RPT11-2833.
9. Mallman, EP and Zoback, MD, 2007. Subsidence in the Louisiana coastal zone due to hydrocarbon production. Journal of Coastal Research, Special Issue No. 50, pp. 443–449.
10. Rutqvist, J, Vasco, DW, and Myer, L, 2010. Coupled reservoir-geomechanical analysis of CO_2 injection and ground deformations at In Salah, Algeria. International Journal of Greenhouse Gas Control v4, 225–230.
11. Spencer, L and La Pedalina, F, 2006. Otway Basin Pilot Project Naylor Field, Waarre Formation Unit C; Reservoir Static Models. Cooperative Research Center for Greenhouse Gas Technologies, Canberra, Australia, CO2CRC Publication Number RPT05-0123. 250pp. DOI:10.5341/RPT05-0123
12. Tenthorey, E, John, Z and Nguyen, D, 2010. CRC-2 Extended Leak-Off and Mini-Frac Tests: Results and Implications. Cooperative Research Centre for Greenhouse Gas Technologies, Canberra, Australia, CO2CRC Publication Number RPT10-2228. 23pp.
13. Terzaghi, K, 1943. Theoretical soil mechanics. John Wiley and Sons, New York, 528 pp.
14. Thiercelin, MJ and Plumb RA, 1994. A core based prediction of lithologic stress contrasts in East Texas Formations. SPE 21847.
15. van Ruth, P, Tenthorey, E and Vidal-Gilbert, S, 2007. Geomechanical Analysis of the Naylor Structure, Otway Basin, Australia. Pre-Injection. Cooperative Research Centre for Greenhouse Gas Technologies, Canberra, Australia, CO2CRC Publication Number RPT07-0966. 27pp.
16. Vidal-Gilbert, S, Tenthorey, E and Dewhurst, D, 2009a. Description of the geomechanical model for the Iona field, Otway Basin, Australia. Cooperative Research Centre for Greenhouse Gas Technologies, Canberra, Australia, CO2CRC Publication Number RPT09-1730.
17. Vidal-Gilbert, S, Nauroy, J-F and Brosse, E, 2009b. 3D geomechanical modelling for CO_2 geologic storage in the Dogger carbonates of the Paris Basin. International Journal of Greenhouse Gas Control v3, 288-299.
18. Vidal-Gilbert, S, Tenthorey, E, Dewhurst, D, Ennis-King, J, van Ruth, P and Hillis R, 2010. Geomechanical analysis of the Naylor Field, Otway Basin, Australia: Implications for CO_2 injection and storage. International Journal of Greenhouse Gas Control v4, 827–839.
19. Wang, Z., 2000. Dynamic versus static elastic properties. In: Seismic and Acoustic Velocities in Reservoir Rocks. SEG Geophysics Reprint Series, No. 19.
20. Zoback, M, 2007. Reservoir Geomechanics. Cambridge University Press, Cambridge, 449 pp.

Carbon Dioxide Capture for Storage in Deep Geological Formations, Volume 4
Karl F. Gerdes (Editor)

Section 2

STORAGE MONITORING & VERIFICATION (SMV)

Monitoring & Verification

Carbon Dioxide Capture for Storage in Deep Geological Formations, Volume 4
Karl F. Gerdes (Editor)

Chapter 29

MODULAR BOREHOLE MONITORING AN INTEGRATED DEPLOYMENT PACKAGE DEVELOPMENT

Barry M. Freifeld, Thomas M. Daley and Paul Cook
Lawrence Berkeley National Laboratory, One Cyclotron Rd, Berkeley, Ca, USA

ABSTRACT: The Modular Borehole Monitoring (MBM) Program, funded by the CO_2 Capture Project, was a three year research and development effort by Lawrence Berkeley National Laboratory to develop a next generation CO_2 sequestration integrated well-based monitoring system. As a key participant in numerous field demonstration projects around the world, Berkeley Lab's team of scientists and engineers have experience in the design and operation of a wide variety of monitoring systems to understand the behavior of geologically sequestered carbon dioxide (CO_2). The MBM program initially consisted of four tasks: (1) perform a review of technologies and tools applicable to well-based monitoring of CO_2 sequestration, (2) identify a subset of critical technologies and perform a conceptual engineering design of an integrated monitoring platform, (3) move the conceptual engineering design into detailed engineering and identify suitable vendors, and (4) support the development of a prototype package. When an opportunity presented itself to rapidly shift to a field demonstration of the MBM system, two additional tasks were added to the initial program: (5) design, fabricate and install an MBM System as part of SECARB's Citronelle Dome Storage Program, and (6) operate the MBM system for diagnostic purposes. Following an initial test of fiber optic seismic acquisition with the Citronelle MBM system, a specific test incorporating comparison of fiber-optic and geophone seismic data from vertical seismic profile acquisition was added (Task 7). We here summarize this work and present deployment, operation and DAS testing in Chapter 36 of this volume.

The review of existing technologies (Task 1) identified several monitoring technologies that were deemed of critical importance to incorporate in all well-based storage monitoring programs (Task 2), including our MBM deployment at Citronelle Dome (Tasks 3-5). These technologies include reservoir pressure and temperature, fluid sampling, integrated fiber-optic bundles, and the ability to facilitate wireline logging. Task 6 consisted of the field deployment at Citronelle, and the engineering details and baseline data collected is presented in Chapter 36. Seismic monitoring is also considered to be of high-importance and two different technologies were incorporated in the Citronelle MBM: (1) a semi-permanent, tubing-deployed, 3C-geophone array with a unique cable and clamping design and (2) a recent technology development, fiber-optic distributed acoustic sensing (DAS). The incorporation of DAS in the Citronelle MBM package provided an ideal opportunity to compare conventional geophones to DAS (Task 7).

KEYWORDS: monitoring and verification; borehole monitoring; downhole instrumentation

OBJECTIVE OF A BOREHOLE MONITORING PROGRAM

The overarching objective of monitoring geologically sequestered CO_2 is to demonstrate the safe and effective long-term storage and integrity in the target reservoir. This is accomplished through a multi-faceted monitoring program by which data is acquired that: (1) assures the public and regulators that the reservoir is behaving as intended, (2) validates conceptual models developed for reservoir engineering and storage management, and (3) demonstrates protection of drinking water and the greater environment. The modular borehole monitoring (MBM) concept aims to provide a suite of monitoring tools that can be deployed cost effectively in a flexible, robust package at geosequestration sites (or other sites requiring dedicated monitoring wells). It incorporates many of the technologies considered most desirable for CO_2 plume characterization, such as pressure/temperature sensing, fluid sampling and seismic monitoring, in a way that maximizes the data collected from a single wellbore. Furthermore, by performing the engineering needed for successful integration of these different functions, it lowers the costs and risks that accompany the installation of advanced monitoring tools.

This report summarizes the results of a project to bring the MBM concept to reality, including the following topics: a summary of techniques and tools available to include in a MBM system; a review of their application at CO_2 storage field sites (summarizing an earlier report); a review of the Citronelle MBM design and deployment; a description of the Citronelle system commissioning and baseline data collection, including well completion diagnostics; a report on dedicated test comparison of MBM geophone data and MBM DAS data; a summary of lessons learned; a discussion of conclusions including a high-level discussion of cost-benefit analysis of the MBM approach. Attached in an appendix section is a detailed description of U-tube operation, and also the specifications for a conventional geophone string.

TASK 1: IDENTIFICATION OF MONITORING TOOLS AND REVIEW OF USAGE

ENVIRONMENTAL CONDITIONS AND CHALLENGES

The geologic reservoirs targeted for geosequestration are normally at pressures and temperatures above the critical point of CO_2: 31.1 °C and 73.8 bar. These conditions are typically found at depths greater than ~750 m, and sequestration pilots have typically been carried out at 2-3 km depth. There are many engineering challenges associated with the environmental conditions encountered, which include elevated pressures, temperatures, and aggressive groundwater chemistries, to name a few.

The depth of the target reservoir and the corresponding hydrostatic pressure provides a significant challenge to the design and survivability of complex monitoring instruments. A 3000 meter deep well will have a static pressure at bottom of around 300 bar, or even greater, depending on the salinity of the fluid.

In addition to pressure, elevated temperatures present additional engineering challenges for monitoring, verification and accounting (MVA) tool design. For example, temperature impacts MVA because downhole electronics experience increasing rates of failure at temperatures above 100°C, with lifetimes of downhole electronic circuitry decreasing exponentially with increased temperatures.

A study conducted by Quartzdyne, Inc. a major manufacturer and OEM supplier of quartz crystal and electronic circuit boards for permanent pressure/temperature gauges found that surface mounted electronics could be used reliably at up to 150°C, with lifetimes of 5 years at 125°C. Hybrid electric circuitry assemblies can last up to two years at 200°C or five years at 180°C [1].

Similarly, fiber-optics also suffer from degradation at elevated temperatures. Standard acrylate coated optical fibers are rated for use up to 85°C, with high temperature acrylate fibers acceptable for extended usage at 150°C. Polyimide coatings are used at temperatures up to 300°C, while difficult-to-manufacture fibers with metallic coatings are available beyond this temperature. Two of the challenges with metallic-coated fibers face are the reliable fabrication of long lengths and the difficulty in recoating after splicing. Optical fibers in general suffer a condition known as hydrogen darkening at elevated temperatures, where hydrogen diffuses into the fiber and degrades the optical characteristics. In high temperature boreholes (> 200°C) with hydrocarbons present, the diffusion of hydrogen into fibers can be severe and seriously degrade the life of a fiber-optic cable in the timespan of several months.

Corrosion and chemical resistance of the materials selected for downhole use in the MBM system is an important consideration and is more difficult to achieve at higher temperature because of the exponential dependence of reaction rates on temperature. Deep sedimentary aquifers, often rich in dissolved salts, are considered the largest potential targets of CO_2 sequestration. Monitoring in wells used for fluid sampling means equipment exposure to CO_2-rich fluids. CO_2 dissolved in formation waters will form carbonic acid, with the resulting acidity determined by the buffering capability of the host formation. Acidic waters form a hostile environment to most ferritic materials commonly used in well completion [2].

Processes and properties to monitor

Deciding what to monitor for future geologic sequestration will undoubtedly be driven by regulatory oversight requirements. U.S. EPA guidelines under the Safe Drinking Water Act for geosequestration are aimed at the protection of USDW. The EPA final rule for Federal requirements under the underground injection control (UIC) program for carbon dioxide (CO_2) geologic sequestration (GS) wells prescribes that site characterization and monitoring has to address the following issues:

- Geologic site characterization to ensure that sequestration wells are appropriately sited;
- Requirements to construct wells with injectate-compatible materials and in a manner that prevents fluid movement into unintended zones;
- Periodic re-evaluation of the area of review around the injection well to incorporate monitoring and operational data and verify that the CO_2 is moving as predicted within the subsurface;
- Testing of the mechanical integrity of the injection well, ground water monitoring, and tracking of the location of the injected CO_2 to ensure protection of underground sources of drinking water;
- Extended post-injection monitoring and site care to track the location of the injected CO_2 and monitor subsurface pressures.

Many of the MVA tools can be adapted from current environmental and oil and gas practices to meet the above objectives. These tools fall broadly into the categories of geophysical, hydrological, geochemical, and geomechanical monitoring methods, discussed here in general terms.

The types of geophysical processes and properties that can be monitored using well-based methods are wide ranging. The term geophysical typically implies 'remote sensing' type measurements as compared to wireline geophysical methods with tools that monitor only the near-borehole environment. In the context of this project, we are focusing on seismic methods as the most likely to be deployed for monitoring CO_2 sequestration. More broadly, geophysical includes seismic, electric and gravity techniques.

Hydrologic monitoring is accomplished through a program of continuous measurements of flowrates, pressures, and temperatures within the zones of interest. A comprehensive monitoring program may also monitor parameters in formations above the primary storage reservoir for indication of leakage through the cap rock.

Geochemical monitoring of sequestration sites can help demonstrate the performance and long-term integrity of the geologic containment system. It can also identify potential problems and threats to USDW. A combination of field-acquired fluid samples, and laboratory analysis of geologic cores can provide input parameters for geochemical models that can estimate rates of mineral trapping and identify threats to cap rock stability when exposed to CO_2-rich fluids.

Geomechanical monitoring is focused on the long period (weeks to years) variation in ground deformation and associated changes in stress. This is apart from the short period (milliseconds) deformation caused by seismic waves, and relates to structural changes such as layer thickness or fracture opening induced by subsurface pressure changes or tectonic forces.

CURRENT STATE OF THE ART TECHNOLOGIES

Well completion technology

Permanent deployment: cemented outside casing

A standard technique within the oil and gas industry is to instrument the outside of a well casing with control lines that are cemented in place. The installation of distributed temperature sensing (DTS) cables outside of the casing provides a real-time and continuous evaluation of cement operations, allowing the concentration of cement to be assessed by its exothermic curing process. Other instrumentation can be deployed on casing as part of an MVA effort. Many MVA tools, including electrical resistivity tomography (ERT), seismic sensors, and samplers have been installed using casing deployment in demonstration programs such as the CO2SINK Project in Ketzin Germany, and SECARB's Cranfield distributed acoustic sensing (DAS) test in Cranfield, Mississippi. There are several significant benefits to deployments of instruments behind casing, which includes leaving the wellbore available for wireline logging and other temporary tool deployments and better coupling to the formation for seismic or electrical sensors. Deployments behind casing are usually assumed to increase the risk of leakage pathways to the surface that can be difficult and expensive to remediate, but this is more based on anecdotal information rather than a comprehensive review of installations. The entire deployment of instrumentation on casing requires the use of specialized subcontractors that have experience in completion operations that are modified to accommodate the physical presence of the instrumentation.

While casing deployment is similar in many ways to tubing deployment, as spooling units and control line protectors are also used, there are numerous complexities that arise with casing deployment that are not encountered with tubing deployment. The cementing operation of the casing has to take into consideration the damage that could occur during casing movement which is used to improve the cement job. Rotation of the casing is not permitted, however reciprocation can usually still be performed. Perforation needs to be performed in such a way as to mitigate the risk of the perforation charges damaging the instruments. One way to do this is to install behind casing charges which are aimed away from the instruments. This method has most frequently been used for the installation of behind casing pressure/temperature sensors. If the perforation will be performed after cementing then some method for oriented perforating, as well as "blast shields" or other protective housings placed over critical instruments, are usually employed.

Semi-permanent deployment

Tubing

In many ways tubing instrumentation deployments are operationally similar to electric submersible pump (ESP) deployments, as the specialized equipment to protect and run-in-hole with instrumentation control lines are identical. Specialized vendors are required to oversee the installation and operation of their particular instruments and a spooling operator coordinates with the rig floor workers for the installation of mandrels, clamps, and bands during the installation. The wellhead will need to accommodate control lines feeding through the tubing hanger and out through the tubing head adapter flange. Tubing deployment of instruments is more common than installation outside of casing, and the variety of vendors and service organizations with familiarity with the process is greater. However tubing deployment lacks the benefit of behind casing sampling for sensors that require close contact to the formation, particularly seismic and electrical sensors.

Coiled tube

A coiled tubing (CT) rig is potentially more economical than a standard workover rig used for conventional tubing deployment. Deployment is more rapid because joints don't have to be made up and there are no control line protectors to be positioned on each joint. However the engineering for instrumented deployments using coiled tubing is far less mature than for convention tubing deployment, and the availability of CT rigs and specialized personal are considerably lower, leading to large variability in the ability to perform instrumented CT deployments. An example of a service provider offering instrumented CT is Precise Downhole Services Ltd., located in Nisku, Alberta, Canada. To date there has not been a CO_2 monitoring well completed with instrumented coil tubing, although a temporary seismic hydrophone cable was deployed at Weyburn with CT.

Wireline/Umbilical

An umbilical system, as used in subsea applications that runs from platform to wellhead, could bridge the gap between flatpack coiled tubing and standard wireline deployment. CJS Production Technologies, Calgary Alberta, Canada, have been commercializing an umbilical style flat-pack. They have modified a conventional CT rig to use rectangular shaped push blocks that can grip and deploy a rectangular umbilical. More significantly, they have worked on methodologies for performing pressure control, which is one of the significant engineering challenges in an umbilical style deployment. The flat-pack at Citronelle dome is really a hybridization of a conventional tubing deployment with a flat-pack encapsulated instrumentation bundle. Challenges that CJS Production Technologies have encountered include leakage between the encapsulant material and the instrumentation lines, as well as the need to engineer highly customized wellhead components.

Deployment Pressure control issues

For both casing and tubing deployment, pressure control is critical. Pressure control must be maintained at all times in open hole casing deployment and for tubing deployment in a perforated well. For completed wells this means having the previously mentioned zonal isolation at some depth above the perforations (such as a packer or seal bore) or a well head with a gate valve. All such zonal isolation requires more engineering when monitoring control lines need to be passed through seals. While running-in-hole, 'kill-fluid' (high density drilling mud) serves as primary well control, with secondary control devices such as a hydril, blind ram or shear ram as part of a BOP stack.

Measurement Technology

Seismic

Measuring the ground motion (displacement, velocity or acceleration) requires mechanical coupling of a sensor to the ground. For fluid or air-filled wells, the ground motion is acquired with mechanical clamping of velocity sensors (geophones) or accelerometers. For fluid filled wells, this can be accomplished by fluid-coupled hydrophones, which measure pressure changes induced in the borehole fluid by the ground motion of a seismic wave.

Electrical / Electromagnetic (EM)

Measuring the electrical conductivity of formations is recognized as a fundamental measurement in well-logging. However, achieving the tomographic imaging of conductivity between wells is a much more difficult problem. Electrical borehole measurements have the fundamental problem of highly conductive steel casing limiting the measurement. Recent advances in crosswell electromagnetic (EM) acquisition and electrical resistivity tomography (ERT) are significant, but still optimal only with non-conductive casing. Because the electrical wire 'backbone' of a modular borehole monitoring package could be easily adapted to electrodes or EM sensors, we include consideration of this technology.

Pressure/Temperature

Permanently deployed discrete pressure/temperature gauges are commercially mature products. They span a range of measurement methods, with the deep well environment dominated by piezoresistive and quartz gauge technology. Resonating quartz cells are considered the most stable and accurate, with resolution as high as 1×10^{-8} and accuracy of 0.01% of full scale (FS).

Data from permanent gauges are typically read out at the surface through single-conductor, tubing encased conductors (TEC). Alternatively, memory gauges can be installed in side pocket mandrels and retrieved periodically to download data and replace batteries.

Fluid Sampling

There are numerous methods for obtaining subsurface fluid samples, including wireline samplers, formation testers, gas lift systems, and U-tube samplers [3]. For fluid samples from two-phase reservoirs, such as exist in mixed brine CO_2 systems, methods that preserve the relative ratio of the separate phases are preferred as they provide information deemed important to understanding the state of the reservoir. Electrical pumps and gas lift significantly distort the composition of the fluid, and hence, downhole wireline and U-tube samplers are the preferred techniques for monitoring CO_2 sequestration reservoirs. A comparison of all of these sampling methods was conducted at the Citronelle field site by a team led by Yousif Kharaka, USGS Menlo Park. Unpublished results showed that the wireline and U-tube samples provided the least disturbed dissolved gas chemistry, resulting in more representative samples than from submersible pumps and gas lifting fluids.

Additional tools have been developed by major oilfield service provides for sampling fluids through casing. This involves creating a hole, extracting fluid, and repairing the hole. As expected, these tools are highly specialized and carry significant costs to mobilize and use. They however can provide one of the few methods by which suspected leakage above zone can be investigated.

If it is known in advance that fluid samples are required to be collected above the reservoir, there are a couple of different experimental methods by which a permanent sampling system can be installed outside of the casing. As part of the PTRC Aquistore Project, a cement diverter has been installed with a U-tube sampling port and fluid sampling lines cemented outside of casing. To date, the performance of the system is unknown, as it has not been function tested since installation, which occurred shortly before the writing of this report.

An alternative method is to deploy a U-tube as part of a behind casing perforation system. Behind casing perforation systems have been used to couple discrete pressure/temperature gauges to the formation. This works by installing a hollow perforation charge carrier connected through capillary tube to the pressure sensor. The perforations create a fluid pathway between the formation and the pressure gauge. This type of device has been marketed by several companies including Promore, Houston, TX and Sage Rider, Rosharon TX. Alternatively this same deployment method can be used to couple the formation to a U-tube fluid sampler.

Fiber Optic Technologies: State-of Sensor Technology

Fiber optic based sensor systems are either distributed, based upon Raman or Brillouin scatter, or discrete or multi-point, based upon Fabry-Perot cavities or Fiber Bragg Gratings (FBGs). Distributed temperature sensing is by far the most widely adopted well monitoring technique, having been first developed in the early 1980s at the Southampton University in England. The technique was commercialized initially by York Sensors Ltd. Several other companies, including Sensortran, Sensornet, LIOS Technology and APSensing (a spin-off from Agilent Systems), have since developed commercial products. Performance specifications for RAMAN based DTS systems are usually a function of the overall cable length and the integration period for each measurement cycle, with spatial resolutions typically 15 cm to 1 m and temperature resolution as high as 0.01 °C.

Brillouin optical scattering based temperature monitoring systems typically have lower measurement resolution and accuracy than Raman optical scattering Systems. However, since strain induced variations in optical properties can be decoupled from the temperature measurements, the Brillouin technique is less susceptible to noise induced by strain on the cables.

Because the Brillouin technique uses low loss single-mode fiber it can be operated at ranges as long as 100 km. Brillouin measurements use single mode fiber in comparison to the multimode fiber employed for Raman based temperature measurement. Brillouin sensing is also used for monitoring fiber-strain. Typical sensitivity limits for stain are from 2 $\mu\varepsilon$ to 10 $\mu\varepsilon$ up to as high as 4% strain depending on the cable material. One difficulty in monitoring strain is the challenge of transferring environmental strain onto the cable in a way that accurately transfers the strain but does not degrade the environmental integrity of the fiber-optic cable encapsulation, which needs to still resist the elevated pressures of the deep subsurface environment. This is still an area of active research. FBG strain sensors are more commonly deployed to monitor strain at discrete locations because of the difficulty of imparting strain onto a continuous fiber. Baker Hughes and Shell jointly developed an FBG based real-time compaction imaging system to monitor sand screen deformation and casing shape which used FBG strain sensors.

A technology that is more recent than DTS, but has rapidly evolved in only a few years, is distributed acoustic sensing (DAS). Discrete fiber-optic based geophone sensors have been marketed for many years based on FBG technology. However, there was little commercial uptake of the technology, as the advantage over conventional copper wire based geophone sensors was not significant enough to overcome the price for utilizing the fiber-optic technology. DAS uses commercial grade single-mode telecom fibers to monitor with high spatial resolution (up to 1 m) to provide truly distributed sensing over kilometers of cable.

Fiber-optic DTS monitoring specifically for CO_2 sequestration has been deployed at the CO2SINK site at Ketzin, Germany [4], the CO2CRC Otway Project and the SECARB Cranfield Site, in Mississippi [5] and at the Quest project in Alberta, Canada [6]. Both the CO2SINK and Otway Project sites deployed a variant of passive DTS monitoring, referred to as heat-pulse monitoring which provides for the creation of a thermal pulse to investigate the thermophysical setting of the near wellbore environment [7].

Other: gravity, tilt, strain

Many technologies have been developed for borehole deployment as stand-alone measurements. We considered these to the extent they could possibly be integrated into a modular deployment. A good example is strain measurement. Current fiber optic technology, typically used for distributed temperature sensing, is being applied to strain measurements. Current measurement sensitivity is sufficient for sensing casing damage.

MONITORING TOOLS AND CONVEYANCE EXAMPLES

Numerous sequestration pilots have deployed some type of semi-permanent borehole monitoring instrumentation, from relatively straight forward completions, like pressure/temperature gauges at the Nagoaka Project, to complex multifunction deployments such as the SECARB Cranfield, which incorporated seismic electrical and fiber-optic monitoring capabilities as well as behind casing deployments of pressure/temperature gauges [8].

Table 1 shows a summary of some CO_2 sequestration sites, their properties, the types of wells, measurements and sensors used and a comparison of conveyance systems.

Table 1. Storage pilot properties and borehole monitoring deployments.

	Frio II	**Otway Stage 1**	**Cranfield**	**Ketzin**	**Pennwest**	**Weyburn**	**Nagaoka**
Configuration	Injector/ Observation	Observation	Two observation	Injector/ 2 Observation	Multiple well (EOR) single observation	Multiple well (EOR) horizontal injector/producers Vertical water injectors Single vertical observation well	1 Injector/ 3 Observation
Formation	sand/brine	methane gas/brine	sand/brine	sand/brine	sand/hydro-carbon	carbonate/ hydrocarbon	sand/brine
Depth (ft)	5,000	5,100	10,500	2,300	5,250	4,600	3,300
CO_2 Quantity	250 tons	50,000 tons	7,500 tons (locally)	30,000 tons	>250,000 tons	18 million tons	10,400 tons
Conveyance	tubing	sucker rods	tubing, casing cemented	casing cemented	cemented inside casing	coiled tubing	wireline
Separation (ft) (from injector)	100	984	227, 367	164, 328	~1600 from nearest injector	164 from nearest horizontal injector	~130, 200, 400 (nonlinear)
Sensors	Injector: dual piezo sources Observation: hydrophones P/T gauges, U-tube samplers both wells	Walkaway one component, Microseismic triaxial, Geophones – High resolution TT, Hydrophones - High resolution TT, P/T, 3 U-Tubes	Casing conveyed: ERT, DTS and heater Tubing conveyed: single U-tube ea. P/T digital ea. dual piezo sources obs 1 Hydrophones obs 2	Gas membrane sensor (CO_2 gas dispersed in Argon), ERT, DTPS room for wireline VSP	Geophone array outside casing 3 pair P/T sensors 8 3c geophones 2 downhole fluid samplers Casing deployed U-Tube	8 level 3C geophone	Resistivity, induction RST Nuclear (neutron. Gamma) X-well seismic bottom hole PT sonic

TASKS 2, 3 AND 4

After the initial review for CCP, the MBM project tasks were as follows:

- Task 2: Design a modular deployment package
- Task 3: Detailed Engineering and procurement support
- Task 4: Field site identification and testing

These three tasks were addressed as part of an extended work scope which included focused development at a field site of opportunity and which became Task 5: Design, fabricate, install and operate a multifunction borehole monitoring (MBM) system at the SECARB Citronelle Dome CO_2 storage site. This work included integration of the following monitoring technologies: geophone array, U-tube fluid sampling, distributed heat pulse and in-zone pressure/temperature monitoring. The results of original tasks 2, 3 and 4 are incorporated into the following section.

TASK 5: CITRONELLE DOME PROJECT

PROJECT BACKGROUND

The Citronelle project is formally known as the Phase III Anthropogenic CO_2 Injection Field Test, operated by the Southeast Regional Carbon Sequestration Partnership (SECARB). The following summary text is taken from the project 'fact sheet' (http://www.secarbon.org/files/anthropogenic-test.pdf accessed Feb 13, 2013).

Past work by Southeast Regional Carbon Sequestration Partnership's (SECARB) has identified that a series of thick, regionally extensive saline formations with high-quality seals exist within the Gulf Coastal Region. These saline formations have the potential to hold large volumes of carbon dioxide. One such formation, the Cretaceous age Paluxy Formation sandstone, is the target for the SECARB Anthropogenic CO_2 storage test.

The Anthropogenic CO_2 storage field test is being performed in southwest Alabama near the town of Citronelle in northern Mobile County (Figure 1).

The CO_2 source for the test is a newly constructed post-combustion CO_2 capture facility at Alabama Power's existing 2,657 MW Barry Electric Generating Plant (Plant Barry). A small amount of flue gas from Plant Barry (equivalent to the amount produced when generating 25 MW of electricity) will be diverted from the plant and captured using a process developed by Mitsubishi Heavy Industries to produce highly pure CO_2.

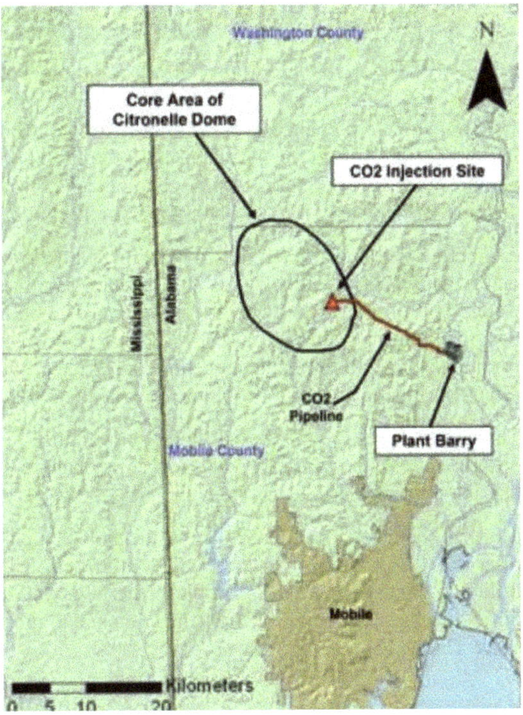

Figure 1. Location map showing Citronelle Dome, the injection site and the Plant Barry (CO_2 source) site.

Plant Barry is a coal- and natural gas-fired electrical generation facility located in Bucks, Mobile County, Alabama, (Figure 2) (Alabama Power is a subsidiary of Southern Company).

Figure 2. CO_2 capture facility at Plant Barry.

The CO_2 storage site is located within the Citronelle Dome geologic structure. The Citronelle Dome, which provides secure four-way closure free of faults or fracture zones, is located approximately 15 kilometers west of Plant Barry. A geological cross-section is shown in Figure 3 showing the location of the Paluxy Formation targeted for storing of captured CO_2.

A pipeline was constructed in 2011 to link the CO_2 capture system with the Paluxy Formation, a major reservoir containing saline water (i.e. water that is too deep and salty to serve as a drinking water supply). The Paluxy occurs at a depth of 3,000 to 3,400 meters and is overlain by multiple geologic confining units that serve as vertical flow barriers and will prevent CO_2 from escaping from the storage reservoir. Lateral flow is expected to be confined within the Paluxy due to residual trapping of the relatively low volume injection (small footprint).

Three new wells will be drilled during the test; a reservoir characterization well, an observation/backup injection well, and a dedicated CO_2 injection well. The characterization well, the first deep well drilled at Citronelle since the 1980s, was completed in January 2011.

Modern characterization data were collected on the injection zone, confining zones and the oil reservoir. The primary injection well was drilled in December 2011 and the backup injector was drilled in January 2012. In addition to the new wells, the project will utilize several existing idle oilfield wells surrounding the CO_2 injection site to monitor injection operations and to ensure public safety (Figure 4).

System	Series	Stratigraphic Unit	Major Sub Units	Potential Reservoirs and Confining Zones
Tertiary	Plio-Pliocene		Citronelle Formation	Freshwater Aquifer
Tertiary	Miocene	Undifferentiated		Freshwater Aquifer
Tertiary	Oligocene		Chickasawhay Fm.	Base of USDW
Tertiary	Oligocene	Vicksburg Group	Bucatunna Clay	Local Confining Unit
Tertiary	Eocene	Jackson Group		Minor Saline Reservoir
Tertiary	Eocene	Claiborne Group	Talahatta Fm.	Saline Reservoir
Tertiary	Eocene	Wilcox Group	Hatchetigbee Sand	Saline Reservoir
Tertiary	Paleocene		Bashi Marl	Saline Reservoir
Tertiary	Paleocene		Salt Mountain LS	Saline Reservoir
Tertiary	Paleocene	Midway Group	Porters Creek Clay	Confining Unit
Cretaceous	Upper	Selma Group		Confining Unit
Cretaceous	Upper	Eutaw Formation		Minor Saline Reservoir
Cretaceous	Upper	Tuscaloosa Group	Upper Tusc.	Minor Saline Reservoir
Cretaceous	Upper	Tuscaloosa Group	Mid. Tusc.: Marine Shale	Confining Unit
Cretaceous	Upper	Tuscaloosa Group	Lower Tusc.: Pilot Sand, Massive sand	Saline Reservoir
Cretaceous	Lower	Washita-Fredericksburg	Dantzler sand	Saline Reservoir
Cretaceous	Lower	Washita-Fredericksburg	Basal Shale	Primary Confining Unit
Cretaceous	Lower	Paluxy Formation	'Upper' 'Middle' 'Lower'	Injection Zone
Cretaceous	Lower	Mooringsport Formation		Confining Unit
Cretaceous	Lower	Ferry Lake Anhydrite		Confining Unit
Cretaceous	Lower	Donovan Sand	Rodessa Fm. Upper'	Oil Reservoir
Cretaceous	Lower	Donovan Sand	'Middle'	Minor Saline Reservoir
Cretaceous	Lower	Donovan Sand	'Lower'	Oil Reservoir

Figure 3. Stratigraphic column of the Citronelle Dome Site, modified from Pashin *et al.* [9], showing the location of the Paluxy CO_2 storage formation above the oil bearing Donovan Sand.

Beginning in 2012, between 100,000 and 150,000 metric tons of CO_2 captured from the pilot facility at Plant Barry will be transported to the storage site over a period of two to three years. During the injection period, multiple CO_2 monitoring technologies will be deployed to track the CO_2 plume, to measure the pressure front, to understand CO_2 trapping mechanisms of the Paluxy saline formation, and to monitor for potential leakage. Three years of post-injection monitoring are planned, with site closure expected to occur in 2017. The wells will either be plugged and abandoned per state regulations or repermitted for CO_2-enhanced oil recovery operations in a deeper mineral formation.

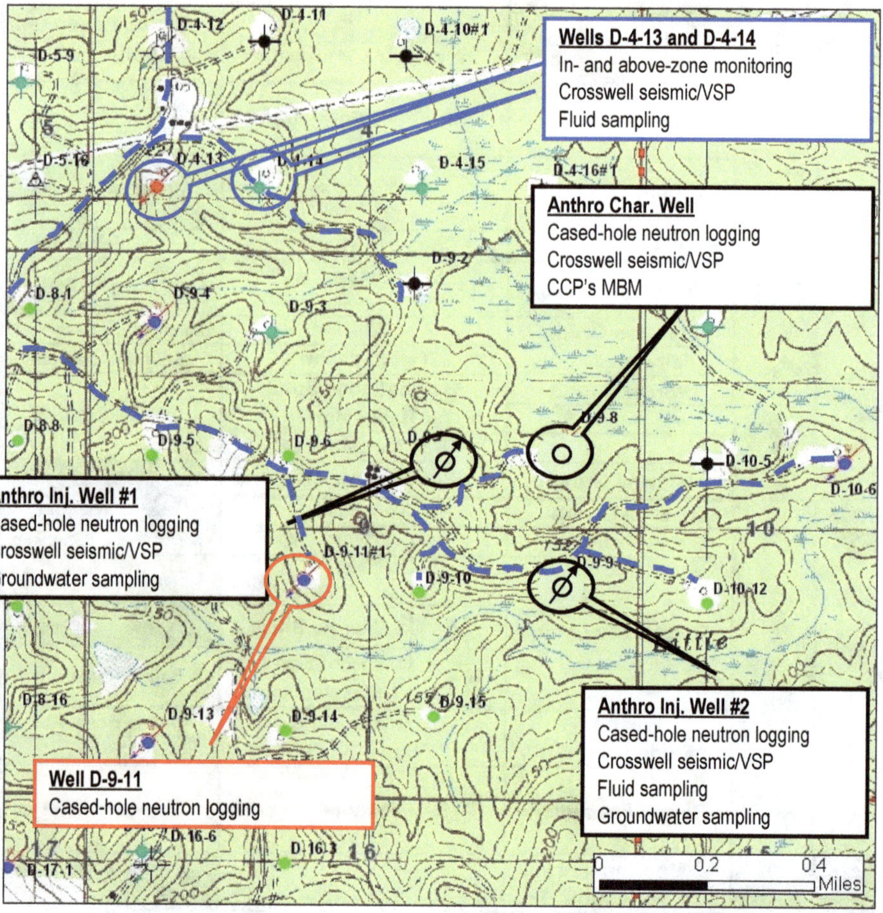

Figure 4. Location Map of wells used for the Anthropogenic test, with proposed monitoring use, along with map of other existing wells. Figure courtesy of ARI.

Project Timeline

Initial discussions for LBNL/CCP involvement in the Citronelle project began in late 2010, with scoping studies completed in early 2011. In parallel to the execution of the new Task V, deployment of an MBM at the Citronelle Dome site, detailed engineering and design work had begun, with an initial go/no-go decision scheduled for mid-June, for a potential deployment in late 2011. Delays in drilling the D9-8 observation well and preparing the well site allowed for further design and fabrication time. Deployment of the MBM package in the Citronelle monitoring well began on March 19, 2012 and downhole operations concluded on March 24, 2012, with surface connection and initial testing work continuing through March 26, 2012. In all, less than two years separate conception and deployment, including engineering design. Given what the project team learned, a similar MBM package could be designed and deployed in less time for future applications.

SELECTION OF MONITORING TECHNOLOGIES

The technologies selected for incorporation in the MBM were chosen to support data gathering to meet the Federal Requirements under the UIC Program for CO_2 Geologic Sequestration (GS) Wells, 40 CFR §124, 144, 145, 146, and 147. The technologies were down selected from a longer list of instruments and methods that have been used at other CO_2 sequestration sites, based on a careful review of objectives, costs and performance. The criteria considered for selecting a particular monitoring technology included:

- The EPA requirements of Class VI Well Monitoring
- Feasibility from an engineering and cost perspective to be integrated in a single well-based installation
- Complimentary with additional monitoring tasks performed outside of the CCP MBM scope
- Long-term potential for advancing the state-of-the-art in subsurface monitoring

The technologies chosen for the MBM include (1) discrete digital quartz pressure/temperature gauges, (2) integrated hybrid copper/fiber-optic cable incorporating distributed acoustic, temperature and heat-pulse features, (3) short-string geophone array, and (4) U-tube geochemical sampling system. Tubing deployment was chosen to provide access to the reservoir and facilitate periodic wireline logging.

Some monitoring technologies that are in use, but were not chosen, include in-well chemistry and/or pH monitoring, electrodes for single-well EM, or ERT, and borehole gravimetry. U-tube fluid sampling was chosen over in-well monitoring since U-tube sampling has been proven to provide representative fluid samples of high quality at numerous other CO_2 sequestration sites. In-well sensors tend to have finite lifetimes within which the data is calibrated, and the replacement of the sensors would be at a significant cost. The electrical methods weren't chosen, as the D9-8 well was cased through the reservoir using steel casing, rendering electrical methods impractical. Fiberglass reinforced composite casing would be needed to effectively use electrical methods. While the short MBM geophone string could be used for microseismic monitoring, it was beyond the scope of the CCP MBM program to install and operate a recording system, and the site operator made the decision to only periodically operate the geophone short string for VSP monitoring, and not continuously operate it for microseismic recording.

CONCEPTUAL DESIGN

Operational design parameters (based on Citronelle Site)

The Paluxy Formation at the Citronelle Site is a fluvially deposited coarsening upward sequence of interbedded sands and shales that spans 9,400 ft to 10,500 ft deep [10]. Because of the coarsening sequence the primary target sand is also the shallowest, referred to as "9460" with a porosity of 21.5% and a permeability estimated from a porosity-permeability cross plot at 450 mD [11]. Temperatures at this depth are 108 °C at a pressure of 298 bar.

String design with packer

Figure 5 shows the D9-8 completion schematic tubing and primary component layout as-built. The primary tubing is 2 7/8" 6.5 ppf L-80 RTS-8 with an internal coating of Tuboscope TK-805. The instrumentation layout is shown in Figure 6. To provide pressure control a dual-mandrel hydraulic set packer is used. The dual-mandrel is asymmetric with the long-string connection used to attach to the 2 7/8" tubing, and the short string connection used to feedthrough the instrumentation lines in metal compression fittings. Above the packer a flat-pack cable is used to convey all of the instrumentation lines up to the wellhead. The instrumentation lines are broken out from the flatpack just above the packer, so that they can run through the short-string mandrel.

COMPANY	FIELD	WELL NAME			
DENBURY	CITRONELLE	D-9-8 # 2			

COMPANY REP		CASING	SIZE	WEIGHT	GRADE	THREAD	TYPE COMPLETION FLUID IN CASING		
STEPHEN BUMGARDNER			7	26	L-80		4% KCl w/ Packer Fluid		
COUNTY	STATE	LINER	SIZE	WEIGHT	GRADE	THREAD	TUBING WT.ON	LOWER	MIDDLE	UPPER
MOBILE	AL.							Neutral		

TUBING	SIZE	WEIGHT	GRADE	THREAD			
UPPER	2 7/8	6.5	L-80	NTS-8	TYPE LATCH	LOWER	MIDDLE	UPPER
MIDDLE				TK805	OPERATOR'S NAME: Kelvin O. Smith	MSO: 14434	
LOWER					OFFICE: LAUREL	DATE: 3/24/12	

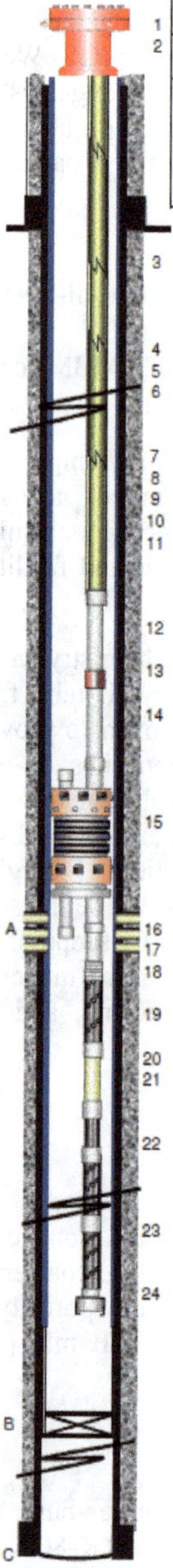

ITEM	DEPTH	LENGTH	JTS	DESCRIPTION	OD	ID
KB	0.00	18.00		ELEVATION		
1	18.00	0.75		TUBING HANGER		
2	18.75	0.68		2 7/8" DOUBLE PIN SUB, 2 7/8" EUE X NTS-8, TK805 IPC	3.462	2.347
3	19.43	6092.52	200	2 7/8" TUBING 6.5 ppf, L-80, NTS-8, TK-805 IPC	2.875	2.347
4	6111.95	3.80		2 7/8" X 4' PUP JT 6.5 ppf, L-80, NTS-8, TK-805 IPC	2.875	2.347
5	6115.75	61.55	2	2 7/8" TUBING 6.5 ppf, L-80, NTS-8, TK-805 IPC	2.875	2.347
6	6177.30	7.81		2 7/8" X 8' PUP JT 6.5 ppf, L-80, NTS-8, TK-805 IPC	2.875	2.347
7	6185.11	215.73	7	2 7/8" TUBING 6.5 ppf, L-80, NTS-8, TK-805 IPC	2.875	2.347
8	6400.84	3.81		2 7/8" X 4' PUP JT 6.5 ppf, L-80, NTS-8, TK-805 IPC	2.875	2.347
9	6404.65	277.28	9	2 7/8" TUBING 6.5 ppf, L-80, NTS-8, TK-805 IPC	2.875	2.347
10	6681.93	3.81		2 7/8" X 4' PUP JT 6.5 ppf, L-80, NTS-8, TK-805 IPC	2.875	2.347
11	6685.74	2686.27	88	2 7/8" TUBING 6.5 ppf, L-80, NTS-8, TK-805 IPC	2.875	2.347
12	9372.01	5.80		2 7/8" X 6' PUP JT 6.5 ppf, L-80, NTS-8, TK-805 IPC	2.875	2.347
13	9377.81	0.48		COMBO COLLAR- 2 7/8" NTS-8 BOX x 2 7/8" EUE BOX, L-80 NPC	3.250	2.371
14	9378.29	5.67		2 7/8" X 6' PUP JT 6.5 #, L-80, EUE, NPC	2.875	2.441
15	9383.96	5.32		HYDRO II DUAL HYDRAULIC SET PKR, 7", 26#	5.938	P - 2.441
				2 7/8" EUE BOX x PIN PRIMARY, 1.9" EUE BOX x PIN		S - 1.610
				SECONDARY, LONG STRING SET, HNBR ELASTOMER, NPC		
16	9389.28	0.33		2 7/8" TUBING ADAPTER EUE NPC	3.250	
17	9389.61	10.10		2 7/8" X 10' PUP JT 6.5 ppf, 13Cr, EUE BOX x PIN	3.462	2.441
18	9399.71	1.00		2 7/8" LANDING NIPPLE, RN - OTIS TYPE PROFILE	3.470	2.188
				2 7/8" EUE BOX x JFE BEAR PIN, 17-4 ss	NO-GO	2.010
19	9400.71	31.62	1	2 7/8" JFE BEAR 13Cr, 6.4 ppf, PERFORATED TUBING	2.875	2.441
20	9432.33	0.50		2 7/8" 13 Cr X-over, JFE BEAR BOX x 2 7/8" EUE PIN	3.310	2.441
21	9432.83	8.61		EASTERN SERIES 150 OVERSHOT NPC, 2 7/8" EUE BOX UP	4.700	2.500
	9441.44	9.00	6.65	OVERSHOT SWALLOWS 6.65' OF CUTOFF JOINT		
22		15.65		15' CUT-OFF JT, 2 7/8", 6.4 ppf, JFE BEAR 13 Cr		
23	9450.44	346.55	11	2 7/8" JFE BEAR, 13 Cr PERFORATED TBG, 6.4 ppf	2.875	2.441
24	9796.99	0.35		2 7/8" WIRELINE RE-ENTRY GUIDE, 13 Cr, JFE BEAR BOX	3.310	2.441
	9797.34			####END OF TUBING####		
A	EST. 9394' - 9424'			PERFORATIONS 2 JSPF, JUMBO CHARGES		
B	10777.00			CIBP w/ CMT ON TOP		
C	11817.00			TD OF WELL		
				NOTES		
				1.) TUBING WAS TORQUE TURNED IN THE HOLE (PWR)		
				2.) GEOPHONE STRING FROM 5952.83' - 6852.83'		

Taken from Pryor Packers
Completion sheet
S. Bumgardner
Advanced Resources Int'l.

PAGE 1 OF 1

Figure 5. As-built completion design detailing casing, tubing, and specialized completion materials.

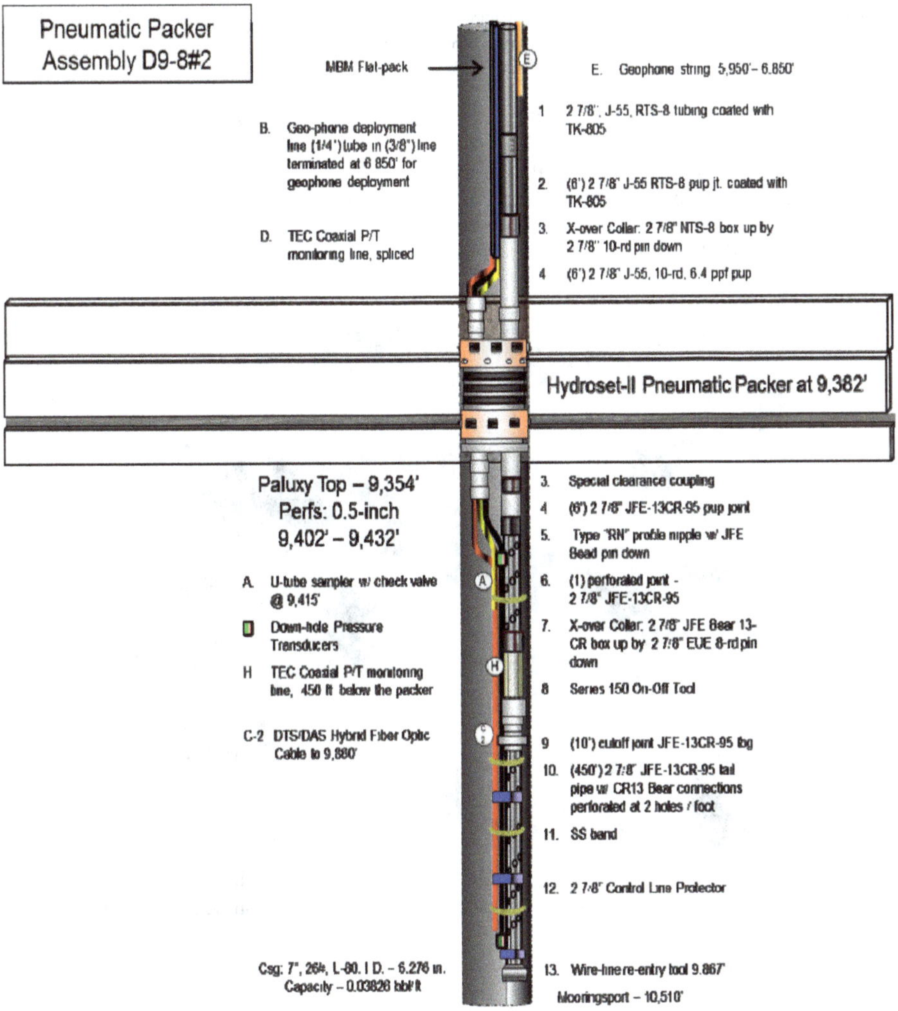

Figure 6. Instrumentation and bottom-hole assembly for the D9-8 completion.

DETAILED MBM ENGINEERING

Packer and overshot design

In considering zonal isolation for the bottom hole assembly (BHA), the project team considered inflatable and hydraulic set packer options. Inflatable packers are generally considered not as reliable since any slight leak that develops in the gland or seals can lead to deflation, and the multi-year life required of the completion string required the highest dependable installation possible. Mechanical set packers require twisting of the string which is not permitted at the packer because of the three control lines that pass through the seal location. For Citronelle, a hydraulic set packer coupled with an overshot to connect the tailpiece to the packer was selected for coupling the BHA to the support string based upon recommendations by Denbury Resources and experience they have in long-life installations.

The packer selected was a D&L Hydroset II Packer, which is a hydraulic set, mechanically held dual string packer with asymmetric short and long string connections. The 2 7/8" long string connection was used for the production tubing while the smaller 1.900 EUE facilitates pass-throughs for the fiber-optic, pressure/temperature gauge, and U-tube sampling lines. Figure 7 shows the dual-mandrel packer with an inset picture highlighting the pass-throughs that penetrate the short string coupling. The overshot selected to couple the tailpipe to the packer is a Logan Series 150 basket grapple overshot with a high-pressure packer-off assembly.

Figure 7. The D&L Hydroset II hydraulic set packer. The inset photograph shows the pass-throughs for the fiber-optic, pressure/temperature, and U-tube control lines.

Fiber-Optic Instrumentation

The fiber-optic monitoring cable is a custom-fabricated, hybrid copper/fiber-optic design that incorporates six copper conductors surrounding a fiber-in-metal-tube (FIMT) containing four high-temperature acrylate fibers – two single mode fibers (SMF) and two multimode fibers (MMF). The identical design was previously used at the CO2CRC Otway Project Stage IIb Residual Gas Test. The cable was fabricated by Draka Cableteq as part of the multi-function flat-pack. Figure 8 shows the cable as constructed. The entire assembly is rated for 150°C with an external collapse pressure of 20,000 PSI. The copper conductors are 20 AWG FEP insulated, tin-coated copper. The FIMT is a gel-filled 304 SS tube with an OD of 1.8 mm. The assembly of six electrical conductors surrounding the FIMT is wrapped with PTFE tape encapsulated in a polypropylene jacket and then further encapsulated in a 3/8" OD x 0.035" wall 316L SS tube. The hybrid fiber-optic cable was fabricated by Draka Cableteq of North Dighton, MA.

SIX 20 AWG CONDUCTORS & FOUR FIBER FIMT STAINLESS STEEL TUBE

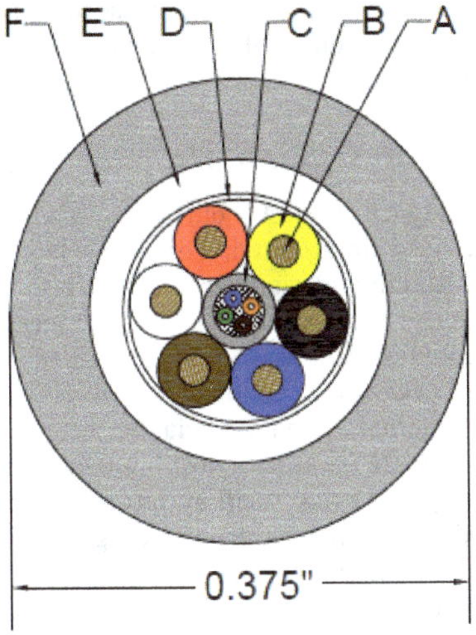

Components

A:	6 x 20 AWG 7/28 Tin Coated Copper; O.D.: 0.96 mm (0.037") Nominal
B:	Colored T-01 (FEP); O.D.: 1.73 mm (0.068") Nominal;
C:	316L FIMT containing gel and 2 x 50/125 & 2 x SM HT Acrylate Coated Fibers; O.D.; 1.8 mm
D:	PTFE Tape (0.003" Thickness) Wrap over Cabled Core
E:	White P-06; O.D.: 7.75 mm (0.305") Nominal
F:	316L Stainless Steel Tube; Wall Thickness: 0.89 mm (0.035"); O.D.: 9.53 mm (0.375") Nominal

Physical Characteristics

System Pressure	: 10,000 psi
External Collapse Pressure	: 20,000 psi
Cable Weight, kg/km (lbs/1000 ft)	: 286 (192)
Temperature Rating	: 150°C

Electrical / Optical Characteristics

DC Resistance at 20°C Center Cond.	34.8 ohms/km (10.6 ohms/1000')
Voltage Rating	1000 volts d.c.
Insulation Resistance at 20°C	1524 megohm-km; (5,000 megohm-kft)

Figure 8. Hybrid copper/fiber-optic cable with multifunction fiber-optic bundle containing two SMF and two MMF. Six 20 AWG conductors allow for heat-pulse monitoring.

U-tube Operation

The U-tube sampling system used in the Citronelle Dome monitoring system is similar to those used previously at other CO_2 injection monitoring sites [3] [12], but is described in detail in Appendix 1 because of the importance to the MBM operation. U-tube sampling relies upon high pressure N_2 gas to force fluid up to the surface from a loop of tubing. The fluid enters the tubing through a downhole check valve that closes when the N_2 drive pressure is applied. Because the U-tube is part of the permanent installation, it provides for cost-effective repeat sampling events that does not require the mobilization of additional equipment out to the field. The only consumable is the compressed N_2 used to drive the samples to the surface.

SEISMIC SYSTEM DESIGN AND SPECIFICATIONS

The seismic system was designed to test both active and passive monitoring. For passive monitoring, 3-component sensors are required. For active monitoring both 3-component and 1-component sensors are used. An initial design decision was use of geophone sensors (other options considered were hydrophones and accelerometers). Hydrophones were considered because they are fluid coupled and therefore could be deployed in the fluid filled annulus between tubing and casing without special clamping. The selection of geophones required that a clamping mechanism be used to provide coupling to the external formation, via the cemented casing. The total number of geophone channels was limited to 24 because of cost limitations and to allow inexpensive recording with a 24-channel acquisition system (very small by seismic standards). The 24 channels were used for 15 vertical geophones and 3 3-component geophones, giving a total of 18 geophone 'pods'. The 3-component sensors were placed at the top, bottom and middle of the array. A spacing of 50 feet (15 m) was chosen between pods. A key design decision was to use a geophone cable that was fully steel encased with no seals. Stainless steel tubing-encased conductor (TEC) cable was specified with 48 wires (24 geophones x 2 wire each), and the TEC was welded to each geophone pod. Previous deployments had failed at the connections between cable and pod or at downhole connections made to allow packer pass through. The use of a non-rotating packer attachment in the MBM, combined with welded connections between geophone cable and pods, allowed removal of all seals and connectors from the seismic system. Specifications used to purchase the MBM Geophones are given in Appendix 2.

Figure 9 shows a single geophone pod attached to the TEC cable. Identical pods were used for both the vertical and 3-componenet geophones. Figure 10 shows the entire reel of geophones and TEC as delivered to the Citronelle well site and set up for installation.

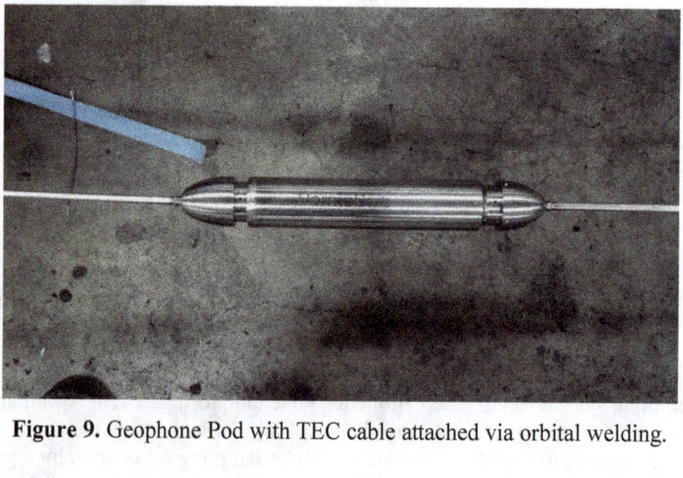

Figure 9. Geophone Pod with TEC cable attached via orbital welding.

Figure 10. Geophone TEC cable on spool at Citronelle well site.

Clamping mechanisms: review and MBM design

Review of borehole geophone clamping

The MBM system is designed to carry geophone pods that need to be seismically coupled to the borehole casing. An ideal geophone couple is one that has no free (not coupled to ground) mass of its own and provides a seismically contiguous connection to the surrounding ground. The first barrier to an ideal system is the casing itself that has its own considerable mass that, on its own; is not representative of natural ground. This can be mitigated by having a good cement bond, making the casing and ground seismically contiguous. For cemented in place geophones (inside or outside of casing) this is where the clamping problem ends. For non-cemented geophones deployed inside of casing there is the inherent problem of allowing the conveyance to proceed upon deployment but then making sure the geophones are set and coupled properly when deployed. Usually this requires some kind of mechanism or clamp to carry the geophone. The clamping mechanism will strive to have minimal free mass of its own and have a solid, stable contact with the inside of the casing. Also for practical reasons an effort is usually made to design a clamp that maximizes the clamping force when finally deployed but that allows for easy deployment by minimizing contact while running in hole.

A common type of mechanism is the bowspring. When mounted on a bowspring, a geophone can be deployed on tubing wireline sucker rod, for example. It will then contact the inside of the casing immediately with full clamping force as soon as the deployment begins thereby forcing full contact all the way down the casing. A balance therefore has to be obtained between the amount of clamping force and the nuisance and wear caused by having full contact while running in hole. For shallow deployments this can be tolerated to some degree. For deep deployments of more than a few thousand feet there is concern that excessive wear on the contact points will diminish the efficiency of the eventual seismic coupling. There may also be interference issues if sufficient buildup of casing material scrapes off and gathers in front of the contact points causing debris and even entanglement. Bowspring designs can utilize a sprung or unsprung method. In the sprung design, the spring resides between the geophone package and the deployment string. In the unsprung design the geophone is mounted solidly to the deployment string and the spring provides the seismic contact to the borehole. One advantage the first bowspring design has even over some active coupling methods is the partial decoupling of the geophone from the deployment structure (such as tubing) mass due to having an elastic element situated between the geophone and the mass. However this is the same elasticity that has to provide the coupling force for the geophones to the ground so the free mass of the tubing can never be separated entirely from the geophone. This second design is even less desirable because the geophone will be better coupled to the mass of the string than to the borehole.

Active methods of clamping whereby a mechanism is activated upon reaching depth have been utilized instead of, or in addition to springs. By keeping the geophones close to the tubing during run-in, the problem with full contact is solved. However, if the active mechanisms move the geophones from the tubing to contact the casing when setting, then the problem of engaging the free mass of the tubing is actually worsened since there is now no isolation at all between the tubing and the geophone. This problem can be somewhat mitigated by putting a spring between the move out mechanism and the geophone, thus bringing it back to the balance between contact force and tubing coupling.

The two common means of actively setting or actuating geophone clamps are electrical and hydraulic (Figure 11).

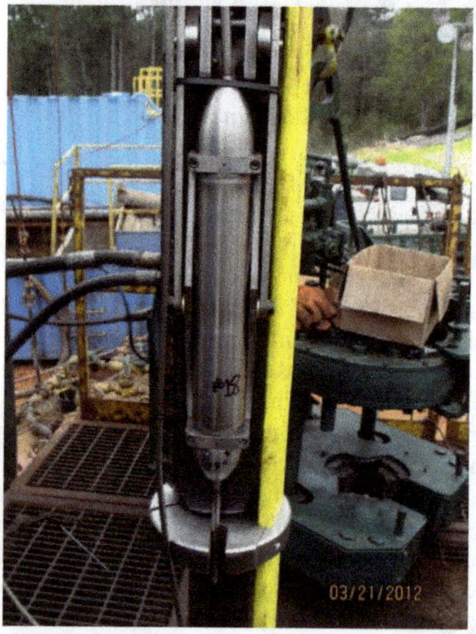

Figure 11. Geophone pod with hydraulic actuated clamp, including flatpack clamped to production tubing.

In wireline applications there are electrical conductors available in the deployment systems that can be utilized, for example, to power electric motors to drive worm screws to provide sufficient mechanical advantage to a locking arm mechanism. In tubing deployment there is no a priori need for electrical service so the power is often supplied hydraulically. The tubing itself can be the hydraulic conduit and can be pressured up to set clamps. The tubing in this case has to be dedicated to just this (and perhaps other simultaneous hydraulic tasks such as packer setting) or has to utilize a removable set plug below the actuation levels to allow for other uses of the tubing later. In other cases there may be available small diameter control lines external to the tubing that can supply hydraulic pressure.

MBM Geophone Clamp Design

For the Citronelle MBM system, we decided to design a specialized clamp for tubing-deployment. The MBM clamp design attempts to completely decouple the conveyance and the geophones mechanically so that no springs are needed and the tubing mass in not in contact with the geophones once they are actively clamped. This is in addition to resolving the problem of casing contact during run-in. An active hydraulic setting mechanism is used, taking advantage of MBM control lines for hydraulic supply, thereby leaving the tubing free for other types of wireline logging tools to be used without retrieving a temporary set plug.

The MBM clamping system takes advantage of the high pressures obtainable in typical hydraulic systems and multiplies this mechanically to obtain very high clamping forces. Combined with a fully floating (from tubing) design, this clamping force approaches the ideal of having a solid couple with casing and no contact elsewhere. There is of course a slight coupling due to the geophone cabling itself but this is very flexible compared to the locking effect of the clamps.

To accomplish their task, the MBM clamps incorporate a unique combination of features. The foremost of these is the tubular design of the support frame that completely surrounds the deployment tubing with an annular and a top and bottom gap that allow the frame to float with a full six degrees of freedom around the tubing. The floating frame is mounted as two circular halfshells and loosely trapped top and bottom between two protective collars. The collars limit the tubing's sideways excursions in the casing so that it cannot trap the clamp frame between it and the casing. Figure 12 and Figure 13 show schematic representation of the two halfshells with geophone pod and clamp components.

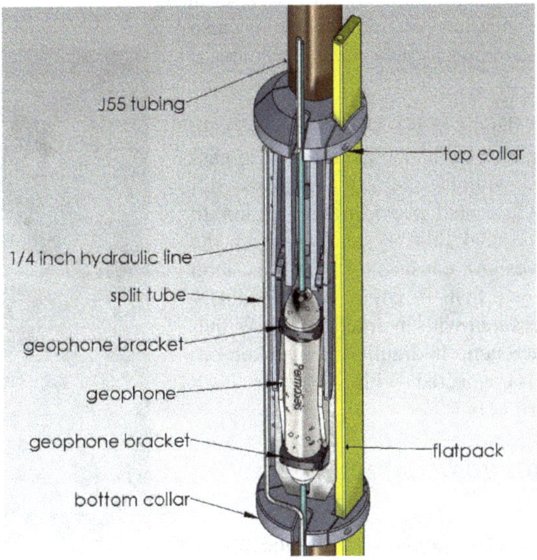

Figure 12. Schematic drawing of geophone clamp assembly with geopohone pod (front).

To gain mechanical advantage, the hydraulic actuator uses compound lever arms to multiply the force. Towards the end of arm travel the force multiplication goes up many times. This multiplication can actually reach a tipping point or over-center geometry wherein the multiplication coefficient becomes theoretically infinite, limited only by the stiffness of the system. Driven past the tipping point, the arm will actually not come back on its own even if the hydraulic pressure is released. The arm is then at a local optimum, slightly on the other side of the maximum excursion. The MBM system is designed to take advantage of this phenomenon to lock itself mechanically. After locking in this manner the force is no longer dependent on the applied hydraulic pressure. The clamps therefore do not require sustained function of the hydraulic system, which is beneficial because with many hydraulic seals the system can lose sealing if any corrosion or seal degradation (always a problem with polymers) is present on the sealing surfaces.

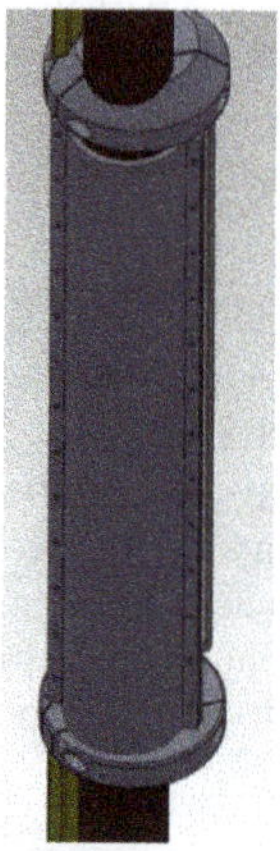

Figure 13. Schematic drawing of geophone clamp assembly (rear).

After locking, the cylinders are designed to drop out of the mechanism so as to allow release when the MBM system is pulled from the hole. To accomplish this task a dedicated mechanism is added to the clamps to kick out the lock. Special kick-out yolks utilize an upwards jar from the tubing string to bump the locks free and let the geophones fall back (horizontally) to the tubing when the system is ready to be retrieved. To optimize this clamping method the excursion of the clamps is adjustable via eccentrics (micro-adjustments) and pivot pins (large adjustments) to make sure the over-center locking action coincides precisely with the contact point for any particular casing, i.e. the clamps are adjustable for planned tubing and casing diameters. In reality this setpoint is set well outside the casing thereby taking advantage of the slight elasticity in the overall mechanism to provide some strain in the lock, ensuring they stay set under various conditions of movement and thermal cycling. The amount of over-center is also adjustable to allow fine tuning for any given deployment situation.

Other important design details of the MBM clamp system allow for both the geophone cable and the flatpack to pass by each geophone pod without interfering with the full floating design and while also providing protection to those components. The hydraulic system (part of the integrated flat pack design) consists of a closed loop system (tube-in-tube) so that it could be completely purged of air when pressured with hydraulic fluid. Within the geophone array, the clamp hydraulic line is taken out of the flat pack and connected with a 'T' to each geophone clamp (Figure 14).

Figure 14. Geophone clamp hydraulic line 'T', along with flatpack attached to production tubing with steel banding.

FLATPACK AND GEOPHONE CABLE PROTECTORS

To protect the flatpack and geophone TEC and attach them to the deployment tubing, cast steel control line protectors from Gulf Coast Downhole Technologies were utilized. One protector was installed at each tubing joint.

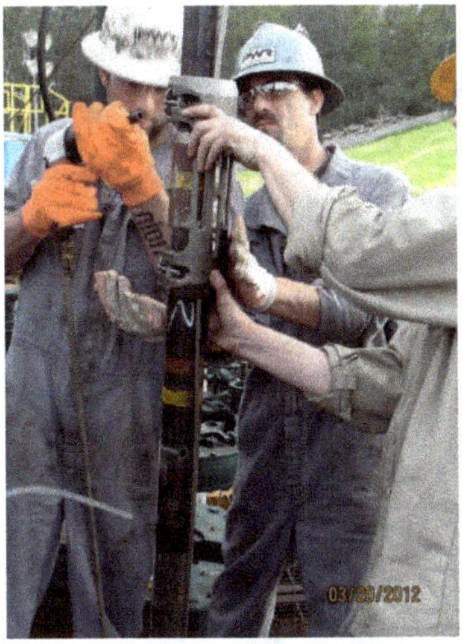

Figure 15. Control line clamp and protector being attached to perforated tubing below packer.

For the CR-13-95 tail pipe (450 feet) below the packer, the control lines were not in a flatpack. For this tail pipe section (expected to be in a more corrosive environment), 15 protectors fabricated from Super Duplex corrosion resistant alloy clamped the separate TEC, P/T gauge, and tubing encased fiber/heater to the tail pipe (Figure 15).

Flatpack Design

The flatpack utilized for the Citronelle MBM serves as the MBM monitoring backbone, transmitting electrical power and signals, fiber optic light pulses and fluids from surface to reservoir. Figure 16 shows a picture of the four-element flatpack as well as a schematic of the MBM control lines. Ideally the geophone cable would have been integrated into the flatpack, but delivery time constraints did not allow for the additional engineering. The key difficulty is separation of geophone pod sections from the flatpack. The design did include incorporation of the geophone clamp hydraulic line in the flatpack. This line was 'broken out' of the flat pack above the top geophone and below the bottom geophone, with short segments connecting the geophone pods (50 ft spacing).

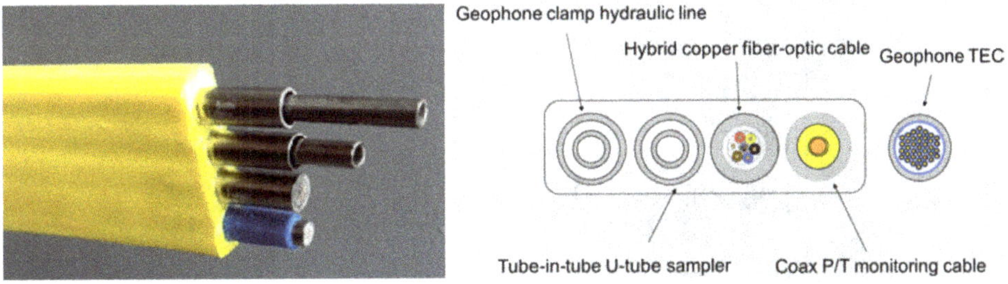

Figure 16. Flatpack with DAS cable photo (left) and schematic (right). Note that the geophone tubing encapsulated conductor (TEC) cable was deployed separate from the flatback.

MBM MONITORING WELLHEAD ENGINEERING

Use of an MBM system requires control and instrument lines to be routed through the well head with full sealing. Therefore the design and deployment of an MBM system requires close cooperation with wellhead design. It is notable that the wellhead design could be reused if the MBM system is consistent. The wellhead for D9-8 was supplied by GE Oil & Gas, (formerly Wood Group).

Figure 17 shows the ported tubing hanger and tubing head adapter, designed to accommodate the MBM control lines which would penetrate the wellhead. The tubing hanger incorporates test ports, so that the pressure integrity of each passthrough could be verified.

Figure 18 shows further details on the layout of the tubing head adapter and tubing head.

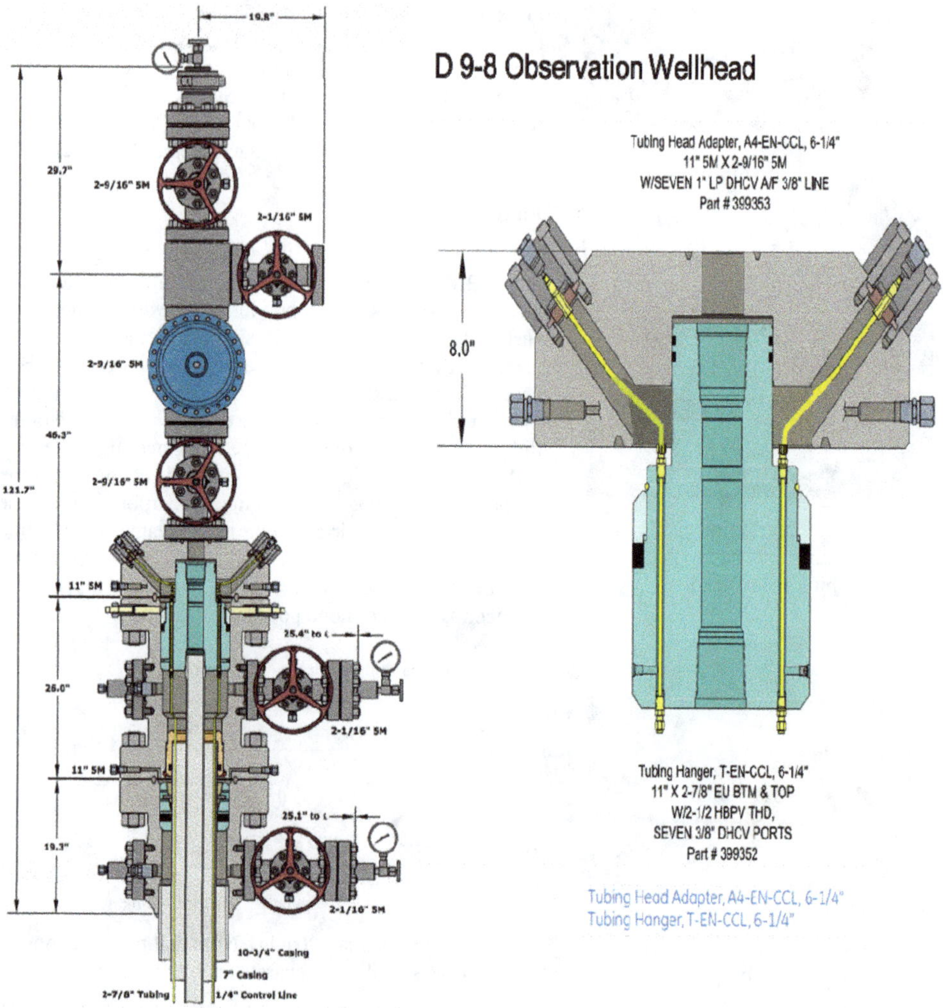

Figure 17. Wellhead tree/ported tubing adapter used for the D9-8 MBM deployment.

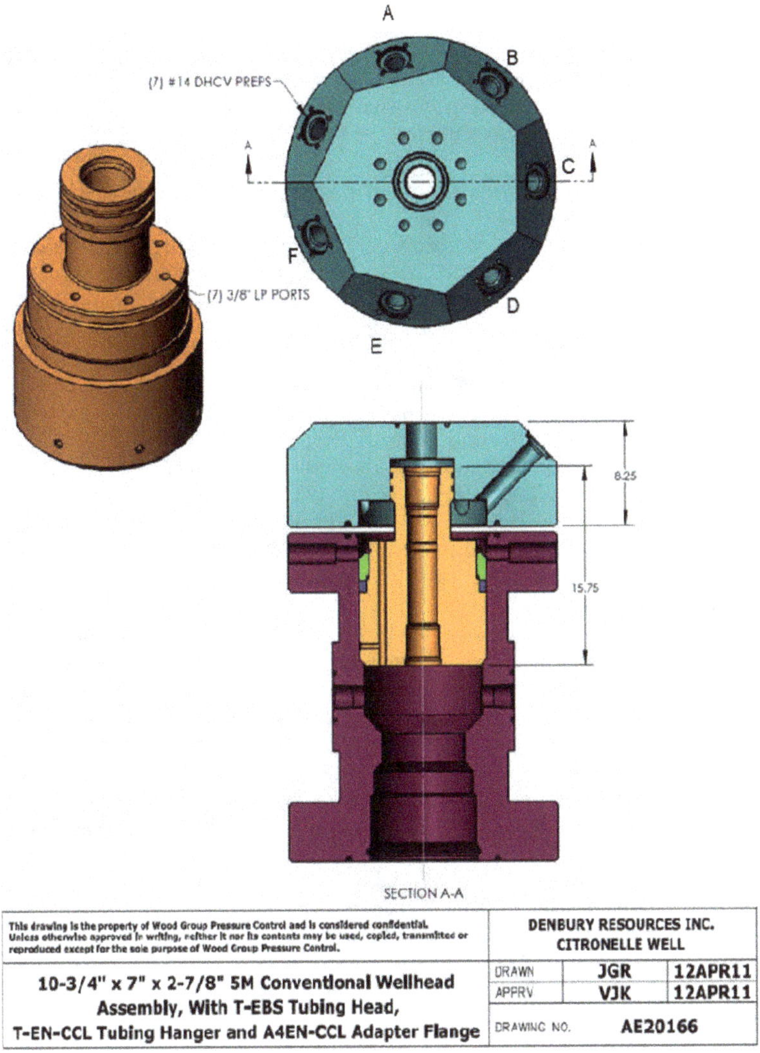

Figure 18. Tubing adapter flange details and tubing head for the D9-8 well. From Wood Group.

ACKNOWLEDGEMENTS

We would like to specifically thank Kevin Dodds of the CO_2 Capture Project for his unwavering support of the MBM concept, development and deployment. We thank the SECARB team, including Jerry Hill of SSEB, Rob Trautz of EPRI, George Koperna and Dave Reasonberg of ARI and Gary Dittmar of Denbury. Acquisition of seismic data (geophone and DAS) was assisted by Dale Adessi of SR2020 and Michelle Roertson of LBNL. This work was supported by the CO_2 Capture Project, and performed at Lawrence Berkeley National Laboratory under U.S. DOE Contract No. DE-AC02-05CH11231.

REFERENCES

1. Watts, M. 2003, High temperature circuit reliability testing, Quartzdyne, Inc. accessed Sept 21, 2013 http://www.quartzdyne.com/pdfs/CktRelPaper.pdf.
2. Ueda, M. and Ikeda, A., Effect of Microstructure and Cr Content in Steel on CO_2 Corrosion, Sumitomo Metal Ind. Ltd. Source CORROSION 96, March 24 - 29, 1996, Denver, Co NACE International.
3. Freifeld, B. M., 2009, The U-tube: a new paradigm in borehole fluid sampling, *Scientific Drilling*, 8, doi:10.2204/iodp.sd.8.07.2009.
4. Giese, R., Henninges, J., Lüth, S., Morozowa, D., Schmidt-Hattenberger, C., Würdemann, H., Zimmer, M., Cosma, C., Juhlin, C., CO2SINK Group (2009): Monitoring at the CO2SINK site: A concept integrating geophysics, geochemistry, and microbiology, 9th International Conference on Greenhouse Gas Control Techniques (GHGT-9) (Washington DC, USA 2008), 2251-2259.
5. Núñez-López, Vanessa, 2011, Temperature monitoring at SECARB Cranfield Phase 3 site using distributed temperature sensing (DTS) technology: presented at the 10th Annual NETL Carbon Capture & Sequestration Conference, Pittsburgh, Pennsylvania, May 2-5, 2011.
6. SECARB Phase III Geophysical Monitoring: Quarterly Reports July-December 2009, Lawrence Berkeley National Laboratory, Berkeley, CA.
7. Freifeld, B. M., Finsterle, S., Onstott, T. C., Toole P., and Pratt, L. M., 2008, Ground surface temperature reconstructions: using in situ estimates for thermal conductivity acquired with a fiber-optic distributed thermal perturbation sensor, *Geophys. Res. Lett.* 35, L14309, doi:10.1029/2008GL034762, LBNL-1755E.
8. Hovorka, S.D., Timothy A. Meckel, Ramon H. Trevino, Jiemin Lu, Jean-Philippe Nicot, Jong-Won Choi, David Freeman, Paul Cook, Thomas M. Daley, Jonathan B. Ajo-Franklin, Barry M. Freifeild, Christine Doughty, Charles R. Carrigan, Doug La Brecque, Yousif K. Kharaka, James J. Thordsen, Tommy J. Phelps, Changbing Yang, Katherine D. Romanak, Tongwei Zhang, Robert M. Holt, Jeffery S. Lindler, Robert J. Butsch, Monitoring a large volume CO_2 injection: Year two results from SECARB project at Denbury's Cranfield, Mississippi, USA, *Energy Procedia, Volume 4, 2011, Pages 3478-3485.*
9. Pashin, J. C., McIntyre, M. R., Grace, R. L. B., and Hills, D. J, 2008, Southeastern Regional Carbon Sequestration Partnership (SECARB) Phase III: Final Report prepared for Advanced Resources International, 57 p.
10. Esposito, R., Rudy, R., Trautz, R., Koprena, G., and Hill, G., Integrating carbon capture with transportation and storage, *Energy Procedia*,4 (2011) 5512 – 5519, doi:10.1016/j.egypro.2011.02.537.
11. Riestenberg, D., 2012, SECARB Citronelle Geologic Characterization, SECARB Joint Project Review Meeting, June 14, 2012.
12. Freifeld, B.M., Daley, T.M., Hovorka, S.D., Henninges, J., Underschultz, J. & Sharm, S. Recent advances in well-based monitoring of CO_2 sequestration. *Energy Procedia*, 2009, Elsevier.
13. Freifeld, B.M., Trautz, R.C., Yousif K.K., Phelps, T.J., Myer, L.R., Hovorka, S.D., and Collins, D., 2005, The U-Tube: A novel system for acquiring borehole fluid samples from a deep geologic CO_2 sequestration experiment, *J. Geophys. Res.*, 110, B10203, doi:10.1029/2005JB003735.

APPENDIX 1: U-TUBE OPERATION

The operation of the U-tube sampling system requires high pressure nitrogen (N_2) gas. N_2 from standard industrial K-size cylinders is supplied at atmospheric pressure to the inlet of a gas compressor (Bauer Compressors, Norfolk, VA) which can boost the pressure up to 413 bar. The compressor supplies a manifold in which reside four, 413 bar rated 43.3 liter capacity N_2 gas cylinders used in parallel to provide reserve capacity. The U-tube consists of the patent pending tube-in-tube design (provisional US patent 20130220594 A1), where a ¼" x 0.035" 316L SS tube is encapsulated within a 3/8" x 0.03" 316L SS tube. The distal end of the U-tube terminates above the packer, and connects using a swaglok union to a check valve. Using perforations in the ¼" tube, the tube-in-tube forms a "tee" within the connector so as to function identically to the traditional U-tube sampler that is composed of two separate lines [13]. The formation fluid enters through a 40 μ SS sintered metallic inlet filter and then is directed through a Hastelloy check valve to the tube-in-tube tee. Hastelloy is used for greater corrosion resistance.

The check valve opens in the direction that allows well fluid to enter through the filter and into the U tubing when the valves at the surface of the tubing are open to atmosphere. To operate the U-tube sampler N_2 is allowed to enter the outer conduit of the tube-in-tube assembly, the upstream (drive) leg of the tubing, where it will push any fluid residing in the tubing up the inner conduit and up to the surface. The check valve in the well prevents any fluid from re-entering the well while under N_2 pressure.

Once at surface the fluid and chasing N_2 flows through a 10 micron stainless steel filter and then a valve to an atmospheric pressure waste accumulation tank. This fluid can alternately be redirected right after the filter to an analysis manifold (containing temperature, electrical conductivity and pH probes) in which pressure is maintained using a back-pressure regulator (BPR) in line with the exit of the fluid to the aforementioned waste tank. Small portable fluid samples at BPR pressure and also at regulated reduced pressure can be obtained from this manifold for shipment for further analysis. While fluid is flowing through the BPR it can be branched off before the regulator by a valve into the bottom of a 3 liter vertically oriented sample vessel where it will fill this volume with fluid at the same pressure as is set by the BPR. Once the vessel is full, fluid can be bled off the top of the vessel through a control valve, then a bleed valve and a set of regulators to an atmospheric pressure phase separator. Liquid phase is allowed to drop out the bottom of the separator and gasses can emanate from the top. The gasses from the separator are then sealed and stored as samples at atmospheric conditions and/or analyzed for constituents.

After the sampling/analysis is obtained, the sample vessel can purged of fluid using the N_2 supply as fed through a pressure reducing regulator set at 55.1 bar or below. This N_2 can blow down the vessel through another valve on the top of the vessel and exit through a drain valve on the bottom of the vessel to the waste tank. The fluid in the waste tank, if gaseous, will dissipate through a flue into the air outside the working space while any liquid will, at a preset level, be pumped into a storage tank onsite. Any remaining fluid in the U-tube will be driven out through the non-regulated exit valve by continued N_2 pressure until the U is clear of well fluid. At this point the exit valve can be closed thereby maintaining N_2 pressure in the U-tube. When the next well sample is desired, the atmospheric (non-regulated) exit valves at both ends of the U are opened to atmosphere to vent the N_2 out and to let the well fluid flood the U tubing thereby obtaining another fluid sample ready for retrieval. The time at which this last activity occurs is the time stamp of the next sample.

For the operation at the Citronelle Dome site there is an extra stage to the operation of the U-tube controller. The depth of the monitoring well (3000 m) and the salinity of the formation fluid (200g/L TDS Na-Ca-Cl brine) giving a fluid hydrostatic pressure gradient of 0.117 bar per meter combine to make the needed pressure at the bottom of the U-tube about 351 bar to close the check valve and deliver the sample to the surface. This pressure is roughly the same as the practical maximum delivery pressure of the N_2 from the compressor, and therefor above the average delivery pressure for a cycle performed at a reasonable rate. The extra operational stage consists of a secondary fluid drive to retrieve the sample using freshwater. Using a high pressure air driven fluid pump, freshwater is pumped into the drive leg of the U-tube to push out the well fluid sample. The combined head of the freshwater with the extra pressure delivered by the fresh water pump can easily provide the needed pressure to close the check valve and drive out the sample. In addition the sample can be collected at any desired pressure on surface up to the formation pressure thus preserving the pressure on the sample as it is drawn and keeping the chemistry of the sample intact. There is very little mixing of the freshwater and the sample because the area of contact between the two in the tube is miniscule compared to the volume and length of the samples. This keeps any diffusive mixing to a trivial amount. In addition, the formation sample is taken from the volume corresponding to the bottom of the U-tube, which is very remote from the drive water interface in the drive leg of the U-tube and any trickle back of freshwater in the sample leg from a previous sampling cycle. The lack of mixing was confirmed by salinity testing on delivered samples which gave results corresponding exactly to the delivery quantities expected of the formation water and the freshwater drive. After the sample has been collected, nitrogen N_2 is used to blow out the freshwater, drive fluid from the U-tube and purge the liquid from the U-tube system. The freshwater is significantly lighter than the formation water requiring an easily sustainable 293 bar from the nitrogen N_2 supply and therefore much easier to flush from the tubing.

As described above, the U-tube at the Citronelle site utilized a tube-in-tube design that was integrated into the flatpack for deployment. The coaxial design of the tube-in-tube allows the system to use just one control line, making a significant contribution to the ease of installation and the minimization of space needed in the flatpack and for feed-throughs at the packer and wellhead. In addition, by using the tube-in-tube configuration, the actual U of the U-tube is simplified so that only one regular pass through fitting (which takes less space and is potentially more robust that the conventional T fitting) is needed to provide the loopback for the U.

APPENDIX 2: MBM GEOPHONE SPECIFICATIONS

The specifications given to the vendors in a request for quotations were as follows:

1. Cable must be stainless steel tubing enclosing at least 24 pair of 28 AWG wire, i.e. tubing encased conductors (TEC). The TEC should be no larger than 3/8" diameter. Cable total length is 10,000 feet. Cable should be rated for at least 5,000 psi and 250° F.
2. Cable will have 18 borehole geophones welded in-line at 50 foot spacing. The exact geophone locations are to be determined, but will be approximately 7000-8000 feet from surface end of cable
3. The 18 geophone sections are 15 vertical component dual geophone sections and 3 3-component dual geophone sections. The dual geophones will be wired to give double output voltage. The geophone section should be no more than 2.0" outer diameter and rated for at least 5,000 psi and 250° F.
4. The geophone section should allow for mounting a clamping mechanism, e.g. bolt holes, specification to be determined. The vendor will be asked to interact with LBNL on the design of mounting specifications.
5. LBNL will conduct an acceptance test of the final product before shipment to the field site where it will be deployed.
6. Delivery time is a factor, ideally 12 weeks or less.
7. Vendor should demonstrate experience with fabrication of borehole geophone systems.

Two companies responded to a request for quotation and the selection was made based on price and experience. The winning bid was from Paulsson Inc, with the following specifications:

1. Borehole Geophone Array Design: Fully welded, hermetically sealed and pressure tested design consisting of the following main components:
   a. Stainless steel geophone pods with an outside diameter of 1.90" using a fully welded design with a pressure rating exceeding 10,000 psi. This pressure rating will be verified with finite element analysis of the final design as well as physical tests of prototypes in a pressure and temperature vessel.
   b. Cable consisting of 30 twisted pairs using #28 copper solid strand magnet wires rated to 400F/200C. The cable will be jacketed with a 300F/150C material. The cable will be supplied by a leading US cable manufacturer with a long history of successful manufacturing of specialized cables to the geophysical industry.
   c. 0.312" (5/16") stainless steel tubing with 0.049" wall encasing the geophysical cable. The pressure rating for this tube exceeds 15,000 psi. A leading US stainless steel tubing manufacturer will encase the Paulsson supplied cable.
2. The seismic array is comprised of a total of 18 levels in the following configuration:
   a. Three x 3C geophone pods using dual geophones for all components. The geophones are mounted inside heavy duty stainless steel geophone pods. The 3C geophone pods will be placed in position 1, 9 and 18 in the 18 level array
   b. 15 x 1C geophone pods using dual geophones for the axial (vertical) component mounted inside heavy duty stainless steel geophone pods. The 1C geophone pods will be placed in positions 2 – 8 and 10 – 17.
   c. Dual Geophones will be used for all pods and all components. The dual geophones will be wired to give double output voltage.
3. Geophone pod spacing center to center: 50 ft +/- ¼"
4. Geophones: Oyo Geospace Omni 2400, 15 Hz omni-directional geophones. Temperature rating; 400F/200C
5. The Paulsson geophone pod design includes "hardpoint" mounting holes for the LBL clamping mechanism. The mounting design will be finalized in collaboration with LBL to fit LBL's clamping design.
6. Total length of borehole geophone array including lead in section: up to 10,000 ft.
7. Delivery time for the system: 12 weeks.

Carbon Dioxide Capture for Storage in Deep Geological Formations, Volume 4
Karl F. Gerdes (Editor)

Chapter 30

QUANTIFYING THE BENEFIT OF CCS MONITORING AND VERIFICATION TECHNOLOGIES

J. Eric Bickel
Graduate Program Operations Research & Industrial Engineering
The University of Texas at Austin

ABSTRACT: In this chapter, we develop a procedure to quantify the benefit of monitoring technologies in carbon capture and storage projects, which we demonstrate within the context of the In Salah Gas CO_2 Project operated by the In Salah Gas JV, a joint venture between Sonatrach, BP and Statoil. The procedure we develop is based on the decision-analytic concept of Value of Information (VOI). Using previously conducted probabilistic risk analyses and a new analytic model we quantify the value of the 2010 seismic survey conducted at In Salah. As part of this effort, we illustrate the importance of choosing the correct objective. Previously, the In Salah Joint Industry Project (JIP) chose to minimize the risked CO_2 leakage volume. As we demonstrate here, this objective understates the risk of significant leakage volumes and will tend to minimize the value of monitoring. In its place, we recommend minimizing the Volume-at-Risk, which we define as the volume such that there is a 1% chance of leaking more than this amount. Under this objective, we demonstrate that the 2010 seismic survey significantly reduced the risk at In Salah. We also argue that VOI methods should be used to value future information-gathering activities.

KEYWORDS: Value of Information; value of monitoring; decision analysis; Carbon Capture and Storage

INTRODUCTION

All decisions, including decisions regarding the injection of carbon dioxide (CO_2) into a saline aquifer, are composed of three elements:

- Alternatives (what we can do).
- Information (what we know or, more importantly, what we are uncertain about).
- Preferences (what we want or are trying to maximize).

Value of information (VOI) analyses seek to perform the following [1,2]:

1. Identify activities or technologies that could provide information regarding important uncertainties.
2. Quantify the accuracy of this information.
3. Understand and quantify how this information would alter decision-making.
4. Assess whether the likely improvement in our decision making is worth the additional cost of the information.

For example, within the context of CO_2 injection monitoring, a VOI analysis would ask questions such as the following:

1. What risks most affect our objective (i.e. the long-term storage of CO_2 away from the atmosphere)?
2. What technologies could we deploy to reduce our uncertainty regarding these risks, and how accurate are they?
3. What actions could we take if we learned that our objective was at risk?
4. Given the answers to 1 through 3, what is the benefit of each technology?

This last question deserves particular emphasis: VOI analyses seek to quantify the value of information, rather than qualify it using verbal labels such as 'high benefit with medium cost'. If possible, value is quantified in monetary terms. This is helpful if monetary resources are being expended to gather the information (e.g. the information is worth $100, but costs $125). However, 'value' may include benefits other than monetary ones. For example, monitoring technologies can be ranked based on how much they increase the probability the project will meet its objectives (e.g. the probability that CO_2 is sequestered for a given length of time). The decision would then be, 'what set of technologies maximize our objective given a particular monitoring budget?'

In order to apply this VOI methodology at In Salah, we re-examined a decision to acquire seismic information. This decision was taken to better manage In Salah and reduce the probability of CO_2 migrating outside the licensed containment area. In this chapter, we refer to this scenario as loss of containment or leakage. It is important to stress that leakage does not refer to the loss of CO_2 to the atmosphere. We constructed a VOI model based on the 2008 and 2010 quantitative risk assessments developed by URS Corporation. Based on this VOI analysis, as described in this chapter, we recommend that the JIP make a substantial change in objective, from minimizing risked leakage and the Total Containment Risk, to one that minimizes the probability of leaking more than a given amount, which we call 'Volume at Risk (VaR).' This is similar to objectives used in the financial industry to measure exposure to low-probability, highly-consequential events.

In this chapter, we detail our application of VOI methods to the management of In Salah Gas JV's In Salah CO_2 injection project. We begin in the next section with a summary of VOI concepts and methods. In following section, we explain the VOI model that we constructed. This model is based on the Quantitative Risk Analysis (QRA) performed by URS in 2008 and 2010. In the penultimate section, we apply our model and the VOI methodology to quantify the benefit of acquiring seismic information in 2008. The final section demonstrates how to couple VOI techniques with an analytic CO_2 migration model developed by the BP Institute. We hope that this work will serve as a foundation and roadmap for how monitoring programs can be valued.

VALUE OF INFORMATION BACKGROUND

When considering the acquisition of information, we must consider two uncertainties [3]:

1. The uncertainty that is important to us, but that we cannot directly observe. For example, we want to know the location of any faults that may be present, but we cannot directly observe this subsurface feature.
2. The result of the test we may conduct in order to provide information regarding the uncertainty we cannot observe. Continuing the example above, we may consider conducting a seismic survey to image the subsurface. This image (test result) provides imperfect information regarding the presence of faults.

We represent and communicate the relationship between the uncertainty that is important to us (e.g. the presence of faults) and what we can observe (e.g. seismic results) through the use of an ‘influence’ diagram [3]. Figure 1 displays an influence diagram and an example probability tree for a simple information-gathering situation related to the presence of faults. We begin with a prior probability distribution that quantifies our uncertainty regarding the presence of faults before we gather any further information. In this example, we believe there is a 25% chance that faults are present based on what we currently know. The arrow between the two nodes in Figure 1 represents the relevance of (or the dependence between) the true state of faults in the subsurface and what the seismic test will report. The strength or accuracy of the relationship is codified by the likelihood function. If faults are truly present (Fault Present = Yes) then we believe there is a 90% chance that seismic will report this (Seismic Report = “Fault”) and a 10% chance the seismic will miss the faults (Seismic Report = “No Fault”). Conversely, if faults are not present (Fault Present = No), we believe there is a 90% that the seismic will report that they are absent and a 10% chance seismic will incorrectly report that faults are present.

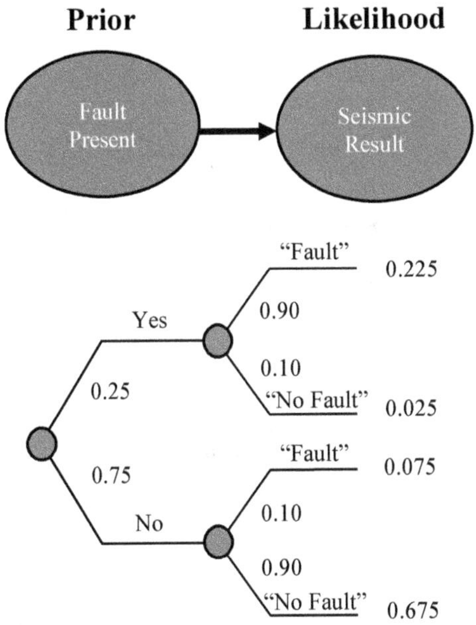

Figure 1. Information diagram in assessed form.

These assessments imply that the probability a fault is present and the seismic reports “Fault” is 0.225 (0.25x0.90). This probability is the product of two events and is referred to as a joint probability. The other joint probabilities are 0.025, 0.075, and 0.675. Notice that the joint probabilities sum to 1.

The influence diagram in Figure 1 is in *assessed form*, which is generally how the reliability or accuracy of the testing procedure is assessed (i.e. we assess how likely the test is to report different outcomes for a known subsurface configuration). While the reliability of the test is assessed in this order, this is not the order in which the information is revealed in the real application. In the actual testing situation, we will first observe the test and then update our beliefs regarding the presence of faults. For example, we would like to know the probability of a fault being present given that the seismic reports that a fault is present. This is depicted in the influence diagram shown in Figure 2, which we refer to as being in *inferential form*, as an inference is being drawn from an observation.

The probabilities depicted in Figure 2 are simply calculated from those in Figure 1. For example, from Figure 1 we know that the probability of a "Fault" test result and a fault actually being present is still 0.225. The other joint probabilities are obtained similarly. Once we know the joint probabilities, the other probabilities follow directly. For example, the probability of a "Fault" test result is 0.30 (0.225 + 0.075). The probability of a fault being present *given* a "Fault" test result is 0.75 (0.225/0.30). It is important to note that once the influence diagram in Figure 1 has been fully specified and the probabilities assessed, the probabilities in Figure 2 are a logical consequence, not an opinion.

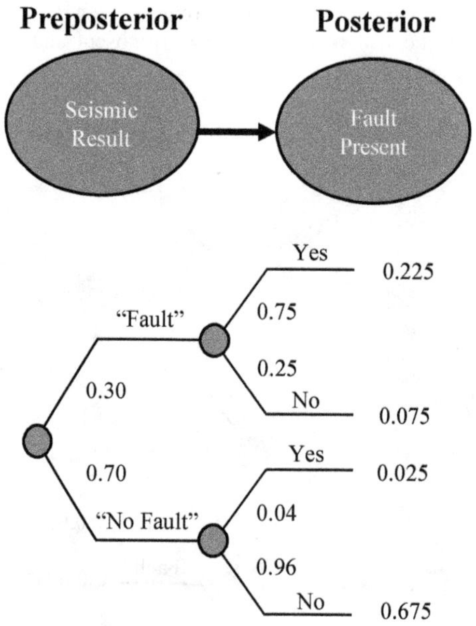

Figure 2. Information diagram in inferential form.

Once the prior distribution has been updated based on the test result, it is referred to as the 'posterior.' The 'pre-posterior' encodes the probability of observing different test results before the test is actually completed. The reversal of the arrow between the assessed and inferential forms (i.e. flipping the order of the two uncertainties) represents the application of Bayes' theorem. For an introduction to Bayesian calculations, refer to Clemen and Reilly [4].

Based on the inferential form of the tree, if the test result comes back "Fault," then the probability of a fault is actually present is revised from 0.25 to 0.75. On the other hand, if the test result comes back "No Fault," the probability of a fault is present is revised from 0.25 to 0.04. It is necessary that all testing situations include cases where the posterior probability is greater than or less than the prior. It is not possible for the posterior probability to be greater than the prior in all cases, or else the prior would have to be larger before the test.

Calculating the Value of Information

The VOI is defined as the most the decision maker should pay for additional information on the uncertainty of interest. If the decision maker is risk neutral, as assumed in this chapter, then:

$$VOI = \begin{bmatrix} \text{Expected value } \textit{with} \\ \text{additional information} \end{bmatrix} - \begin{bmatrix} \text{Expected value } \textit{without} \\ \text{additional information} \end{bmatrix} \qquad \text{Equation 1}$$

If the additional information is perfect (i.e. the information provides perfect knowledge of the state of the world), then the VOI is the value of perfect information (VOPI), which places an upper bound on any information-gathering activity. No test, no matter how sophisticated, is worth more than the VOPI. The VOI can only be greater than the value without information if some decision may be affected by the revelation of the information. If the decision maker would make the same decision no matter how the test comes back, then VOI = 0, and the test is worthless.

IN SALAH QUANTIFIED RISK ASSESSMENTS

In order to apply the above VOI methods to the In Salah project, we analysed a previous decision to acquire seismic information. We based this analysis upon the 2008 and 2010 QRAs conducted by URS [5,6]. We were provided with URS reports, which included their assessments and analysis. The underlying model used by URS was not available. Using the URS report and inputs, we constructed a new QRA model for use in this project. As discussed below, our model results are nearly identical to those of URS. Where our results differ, the discrepancies are minor.

In Salah Background

The In Salah Gas Krechba field development, which has been in operation since 2004, included an industrial-scale CCS project. As of 2010, this project had captured, transported, and injected over 3 MMt of CO_2 in a deep saline formation down-dip of the Krechba Carboniferous gas producing horizon [7]. The planned storage capacity of this facility is 17 MMt of CO_2.

1. A Joint Industry Project (JIP) was set up by the In Salah Gas Joint Venture as research platform to test and demonstrate CCS monitoring technology. The U.S. Department of Energy, and EU DG Research also contributed to JIP funding. The JIP had the following high-level objectives:Provide assurance that secure geological storage of CO_2 can be cost-effectively verified and that long-term assurance can be provided by short-term monitoring.
2. Demonstrate that CCS is a viable strategy to mitigate greenhouse gas (GHG) emissions.
3. Set precedents for the regulation and verification of CCS projects, including eligibility for GHG credits.

Injection of CO_2 was accomplished via three horizontal (1,500 m – 1,800 m) injection wells, KB-501, KB-502, and KB-503, shown in Figure 3, which injected a combined total of typically 50 MM standard cubic feet per day. As of 2010, approximately 75% of the 3 MMt of injected CO_2 had been injected through the two northern injection wells: KB-502 and KB-503. The injection wells have performed as expected, but data suggests that they have been subjected to injection pressures above the reservoir fracture pressure [8]. CO_2 breakthrough was detected at KB-5 in 2007. Tracer analysis confirmed that this 10CO_2 was from KB-502 [8].

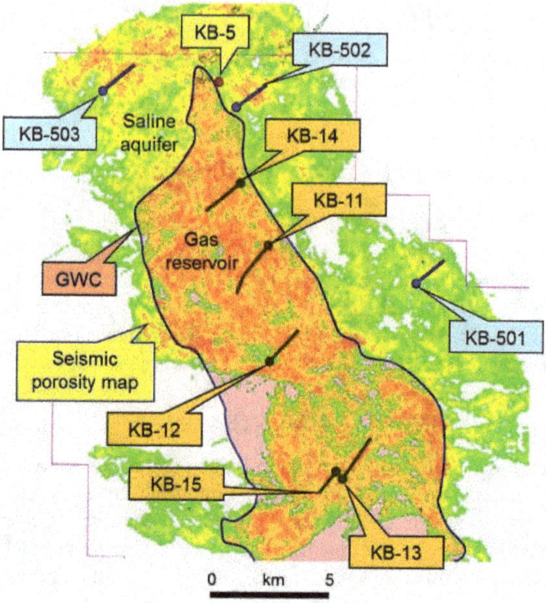

Figure 3. Krechba field layout.

JIP Objective

The value of obtaining further information depends critically upon the objective we seek to optimize. The In Salah JIP set a target that there be an "80% confidence [probability] that 99% of the CO_2 would be retained for 1,000 years" [8]. This was interpreted to mean that the risked (or mean) leakage should not exceed 0.2% ([1-80%] x [1-99%]) of the total injected volume. Since the intended storage volume was 17 MMt, or 17,000 kilotonnes (17,000 kt), the JIP set a leakage threshold of 34 kt (0.2% · 17,000). Again, as stated in the Introduction, within the context of In Salah, "leakage" may mean vertical movement of CO_2 out of the reservoir, or lateral migration beyond the production licence boundary.

As discussed more fully below, setting a target based on risked leakage, rather on low-probability high-consequence events can understate the risk. This could bias monitoring towards technologies that lessen the chance of more common events, rather than on those that are less likely, but more consequential. Another, and we believe better, interpretation would be that "the probability that more than 170 kt (1% of planned storage) of CO_2 escapes the storage facility over the life of the project must be less than 20%."

These two objectives are not the same. The risked leakage can be calculated using the tree in Figure 4. The risked leakage is equal to the probability of a leak, p, multiplied by the risked volume lost if there is a leak, x. Based on the JIP's objective, they set p equal to 0.2 and x equal to 170 kt, yielding a risked leakage of $p{\cdot}x$ = 34 kt. However, there are an infinite number of combinations of p and x that satisfy the 34 kt target. Some of these might be quite concerning. For example, a 1% chance of leaking 3,400 kt (20% of the planned storage volume) would also satisfy the JIP's stated objective. Or, a 1-in-500 chance of leaking 17,000 kt, the entire planned storage volume, would also result in a risked leakage of only 34 kt. The problem is that the mean of a distribution is only a measure of central tendency. It does not provide much information regarding tail risks (low-probability high-consequence events).

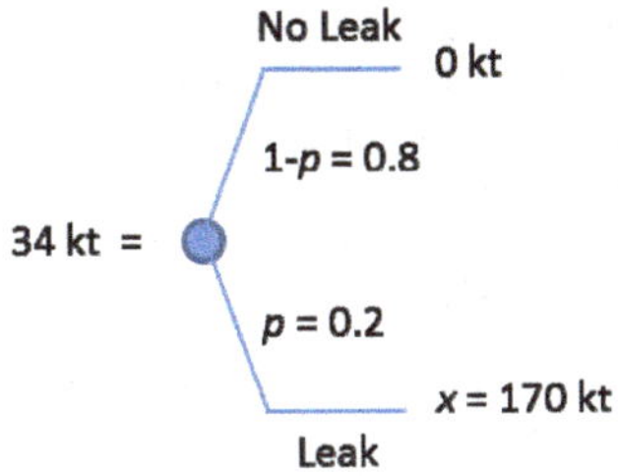

Figure 4. Probability tree illustrating JIP's objective.

Instead, we would recommend a probability threshold similar to that previously described. Probability-threshold goals are quite common. For example, in the financial industry, regulators require companies to hold capital such that they could cover a loss so large that there is only a 5% chance they could experience a loss larger than this. This target is referred to as the 5% Value at Risk. A 1% Value at Risk is also common. Later, we will use the financial analogy and discuss the 5% and 1% 'Volume at Risk' (5%-VaR and 1%-VaR), which we think are a better measure of leakage risk.

Modifications to URS Inputs

For both the 2008 and 2010 QRAs, URS modelled the potential leakage for specific events, as shown in Figure 5. Specifically, URS determined the total mass lost from the storage complex by taking the product of the number of possible items that might be impacted (e.g. four injection wells), the potential flow rate per item (kt/yr), and the duration of the event. These three items yield the mass lost if there was an occurrence. URS multiplied this number by the probability the event will occur, which yields the mean or risked leakage (RL).

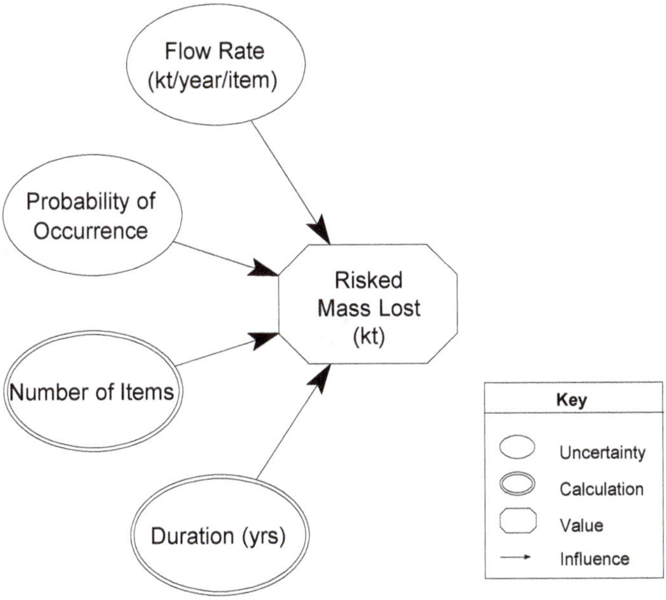

Figure 5. URS QRA model structure.

As shown in Figure 5, URS assumed that the flow rate and the probability of occurrence were uncertain, while the number of items affected and the duration of the event were deterministic. We adopt these assumptions, given our goal of demonstrating the use of VOI methods. We do note, however, that assuming one is uncertain about probability poses some philosophical challenges [9,10]. Rather than thinking that we are uncertain about the chance of occurrence, we view the range considered by URS as a sensitivity analysis.

URS considered eight different possible events: permeable zones in seal; leakage via an undetected fault; leakage via wells; regional-scale over pressurization; exceeding spill point; earthquake induced fractures; and migration direction. URS assumed that both the probability of occurrence and the flow rate for each event were distributed lognormally. Since the lognormal distribution is fully described by two parameters, URS needed only to assess two points to specify each distribution. In the case of probability of occurrence, they assessed the 1st (P1) and 99th (P99) percentiles of the lognormal distribution. The rate uncertainty was encoded by assessing the P50 and the P95 percentiles.

For the VOI analysis, we prefer to work with the P10, P50, and the P90 points, as is common practice in the oil and gas industry. We then calculated these values based on the inputs provided by URS. In the case of the flow rate, where the P50 and P95 were assessed, we determined the mean μ and standard deviation σ of the lognormal distribution as:

$$\mu = \ln P50 \qquad \text{Equation 2}$$

$$\sigma = \frac{\ln P95 - \mu}{\Phi^{-1}(0.95)} \qquad \text{Equation 3}$$

where Φ^{-1} is the inverse of the standard normal cumulative distribution function. For the probability of occurrence, we determined the mean and standard deviation of the lognormal distribution, respectively, as:

$$\mu = \frac{\dfrac{\ln \text{P99}}{\Phi^{-1}(.99)} - \dfrac{\ln \text{P1}}{\Phi^{-1}(.01)}}{\dfrac{1}{\Phi^{-1}(.99)} - \dfrac{1}{\Phi^{-1}(.01)}} \qquad \text{Equation 4}$$

$$\sigma = \frac{\ln P99 - \mu}{\Phi^{-1}(0.99)} \qquad \text{Equation 5}$$

Given the mean and standard deviation, the P10, P50, and P90 may be obtained as follows:

$$P10 = Exp\left[\mu + \Phi^{-1}(0.10)\sigma\right] \approx Exp\left[\mu - 1.28\sigma\right] \qquad \text{Equation 6}$$

$$\text{P50} = Exp\left[\mu + \Phi^{-1}(0.50)\sigma\right] = Exp\left[\mu\right] \qquad \text{Equation 7}$$

$$\text{P90} = Exp\left[\mu + \Phi^{-1}(0.90)\sigma\right] \approx Exp\left[\mu + 1.28\sigma\right] \qquad \text{Equation 8}$$

2008 QRA Results

The 2008 QRA inputs are given in Figure 6. These assessments cover the probability of occurrence, the number of events, the flow rate, and the duration. As mentioned previously, URS assumed each uncertain input was distributed lognormally and specified this distribution using two percentiles. For example, for the uncertainty "Permeable zones in seal," URS assumed that the flow rate should this event occur is lognormally distributed (see the "LN Dist" in Figure 6 for this uncertainty) with a P50 of 15 tons per year (see the CL50% in Figure 6 for this uncertainty in the Rate column) and a P95 of 60 tons per year (see the CL95% in Figure 6 for this uncertainty in the cell to the right of the P50). The implied P10, Mean, and P90 for the probability of occurrence and the short-/long-term flow rates are given in Table 1.

Table 1. Modified URS 2008 QRA data.

Probability of Leakage	**P10**	**Mean**	**P90**
Permeable Zones in Seal	2.81E-06	1.63E-05	3.56E-05
Undetected Faults	2.81E-06	1.63E-05	3.56E-05
Leakage Near Wells	1.68E-04	3.57E-04	5.96E-04
Leakage Devonian Wells	1.68E-02	3.57E-02	5.96E-02
Carboniferous Wells	1.68E-04	3.57E-04	5.96E-04
Chemical Alteration	2.81E-06	1.63E-05	3.56E-05
Regional-Scale Over Pressurization	2.81E-06	1.63E-05	3.56E-05
Leak Near Well Pressurization	2.81E-04	1.63E-03	3.56E-03
Exceeding Spillpoint	2.81E-06	1.63E-05	3.56E-05
Earthquake Induced Fractures	2.81E-07	1.63E-06	3.56E-06
Migration Detection	1.68E-02	3.57E-02	5.96E-02
Short-term Leakage Rate (ktpa)	**P10**	**Mean**	**P90**
Leakage Near Wells	0.06	0.11	0.17
Leakage Devonian Wells	0.06	0.11	0.17
Carboniferous Wells	0.06	0.11	0.17
Regional-Scale Over Pressurization	0.05	0.21	0.44
Leak Near Well Pressurization	23.98	30.46	37.54
Exceeding Spill Point	384.70	510.57	649.86
Earthquake Induced Fractures	0.17	2.66	6.01
Migration Direction	168.08	201.85	237.98
Long-term Leakage Rate (ktpa)	**P10**	**Mean**	**P90**
Permeable Zones in Seal	0.01	0.02	0.04
Undetected Faults	0.06	0.11	0.17
Chemical Alteration	0.05	0.21	0.44
Regional-Scale Over Pressurization	0.01	0.02	0.04
Leak Near Well Pressurization - Surface	0.01	0.02	0.04
Earthquake Induced Fractures	0.02	0.03	0.05

Combining the assessments in Figure 6 and Table 1, we can calculate the risked leakage (RL) for each event, which is shown in Figure 7. The total containment risk (TCR) is 36.2 kt, which exceeds the JIP target of 34 kt. Nearly 100% of the TCR is comprised of migration direction, in which case CO_2 migrates out of the Krechba lease. The other events, such as leakage through an undetected fault, are much smaller. This is somewhat difficult to see because of the log scale used in Figure 7. For example, the RL through an undetected fault was estimated to be only 0.18 kt. This value stemmed from an assumption that the chance of such an event was 1.6E-5 (0.0016%). If it did occur, the total mass lost would be approximately 11 kt.

INITIATING EVENTS	Annual prob occurring	Prob of presence	Number of items		Rate (t/year/item)	Duration (yrs)	Mass lost (t)	Risk Quotient
Permeable zones in seal		1.E-05	1		15.00	1,000	15,000	0.1500
		1.0E-05			15			
	LN dist	1.E-06	1.E-04	LN dist	15	60		
		CL 01%	CL 99%		CL50%	CL95%		
Leakage - Undetected fault		1.E-05	1		100	100	10,000	0.10
		1.0E-05			100			
	LN dist	1.E-06	1.E-04	LN dist	100	200		
		CL 01%	CL 99%		CL50%	CL95%		
Leakage - Wells								
							4.E-02	
Leakage - Injection and near wells		3.E-04	4		100	0.1	33	731
		3.0E-04			100			
	LN dist	1.E-04	1.E-03	LN dist	100	200		
		CL 01%	CL 99%		CL50%	CL95%		
							36.75	
Leakage - Devonian		3.E-02	7		100	0.3	175	
		3.0E-02			100			
	LN dist	1.E-02	1.E-01	LN dist	100	200		
		CL 01%	CL 99%		CL50%	CL95%		
							0.27	
Leakage - Carboniferous		3.E-04	6		100	0.3	150	
		3.0E-04			100			
	LN dist	1.E-04	1.E-03	LN dist	100	200		
		CL 01%	CL 99%		CL50%	CL95%		
							693.6	
Leakage - chemical alteration		1.E-05	10		150	1	1,500	0.015
		1.0E-05	1.E-05	Guide midpoi	150			
	LN dist	1.E-06	1.E-04	LN dist	150	600		
		CL 01%	CL 99%		CL50%	CL95%		

Figure 6. URS 2008 QRA data.

Regional-scale overpressurization								
Short term leakage - during injection		1.E-05	10		150	1.00	1,500	0.015
		1.0E-05			150			
	LN dist	1.E-06	1.E-04	LN dist	150	600		
		CL 01%	CL 99%		CL50%	CL95%		
Long term leakage - post injection		1.E-05	3		15	1,000	22,500	
					15			
				LN dist	15	60		
					CL50%	CL95%		
Local-scale overpressurization								
Short term leakage		1.E-03	3		30,000	0.08	7,500	30.000
		1.0E-03			30000			
	LN dist	1.E-04	1.E-02	LN dist	30000	40000		
		CL 01%	CL 99%		CL50%	CL95%		
Long term leakage		1.E-03	3		15	1,000	22,500	
				LN dist	15			
					15	60		
					CL50%	CL95%		
Exceeding spillpoint		1.E-05	1		500,000	1.00	500,000	5
		1.0E-05			500000			
	LN dist	1.E-06	1.E-04	LN dist	500000	700000		
		CL 01%	CL 99%		CL50%	CL95%		
Earthquake induced fractures								
Short term leakage	1.0.E-06	1.E-03	1		1,000	0.01	10	0.010
	1.0E-06				1000			
LN dist	1.E-07	1.E-05	LN dist	LN dist	1000	10000		
	CL 01%	CL 99%			CL50%	CL95%		
Long term leakage	1.E-06	1.E-03	1		30	0	0	
					30			
				LN dist	30	60		
					CL50%	CL95%		
Migration direction		3.E-02	1		200,000	5	1,000,000	30,000
		3.0E-02	3.E-02 Guide midpoi		200000			
	LN dist	1.E-02	1.E-01	LN dist	200000	250000		
		CL 01%	CL 99%		CL50%	CL95%		

Figure 6 continued.

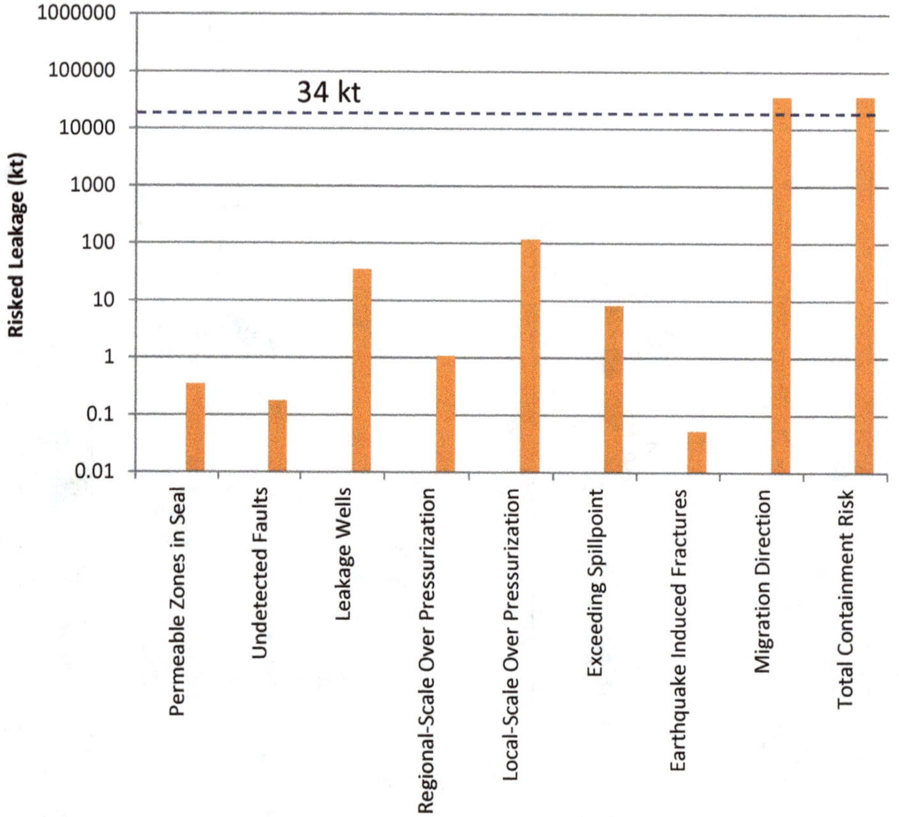

Figure 7. 2008 Risked leakage for each event.

Figure 8 presents a 'tornado diagram,' which analyses the individual drivers of the containment risk. This diagram is centred on the base-case estimate of 36.2 kt, which is the value one obtains if all inputs are set to their mean values (shown in Table 1). Each bar in the tornado diagram represents an 80% probability interval. For example, the low value, which is the P10, for the probability of leakage due to migration was 1.7E-02 (1.7%). If this was the case then the RL would be 17.1 kt. If, instead, the probability of migration leakage was the high value, or the P90, of 6.0E-02 (6.0%) then the RL would be 60.3 kt. The second most important risk is the migration leakage rate. The other uncertainties contribute relatively little to the total containment risk.

While 36.2 kt exceeds the JIP's target, one may conclude that risk is still rather small, since the TCR is ~ 0.2% of the planned storage volume. However, focusing on the risked losses is misleading in this case because it is dominated by the large probability of leaking nothing. Figure 9 presents an excess probability distribution for the 2008 QRA. The horizontal scale is the volume leaked and the vertical scale is the probability of leaking that volume or more. In this case, there is about a 92.7% chance of leaking nothing and a 7.3% chance of leaking something. If a leakage does occur, the conditional risked leakage (i.e., the risked leakage given a leak has occurred) is around 489 kt. Thus, the true risk at In Salah, as it stood in 2008, may be better depicted as shown in Figure 10, where there is a 92.7% chance of no leakage and a 7.3% chance of leaking 501 kt. The probability of leaking more than the design criteria of 170 kt is 3.1%.

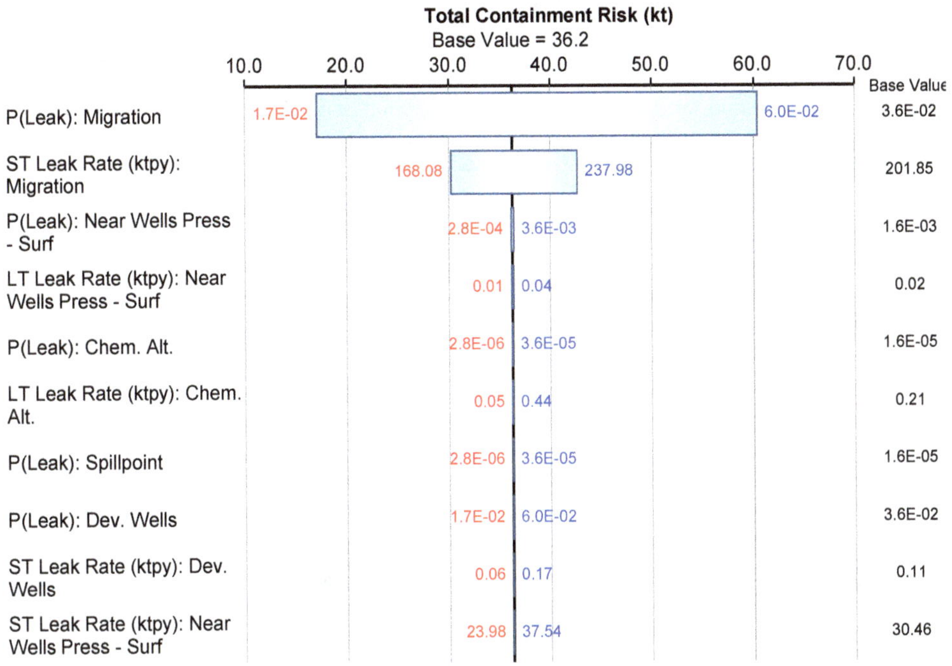

Figure 8. Tornado diagram for 2008 total containment risk.

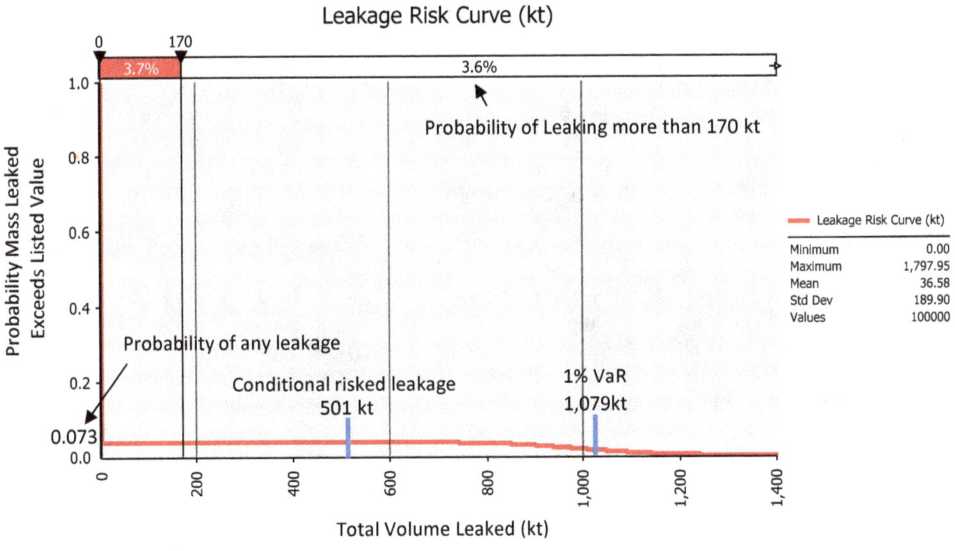

Figure 9. 2008 leakage risk curve.

To provide a more accurate depiction of these risks at In Salah, we now introduce the 1% Volume at Risk (1%-VaR). This is a leakage volume so large that it is estimated there is only a 1% of leaking more than this amount. For the 2008 QRA the 1%-VAR is 1,079 kt, which is about 6.4% of the planned storage volume.[1] Thus, given the 2008 assessments, it is believed there is a 1% chance that more than 6.4% of the planned storage volume will eventually be leaked. Whether or not this risk is acceptable, is a question for the JIP management. The point, however, is that focusing on the RL fails to properly consider risks that may be the most concerning: those that are unlikely, but extremely consequential. This will play an important role in our VOI analysis.

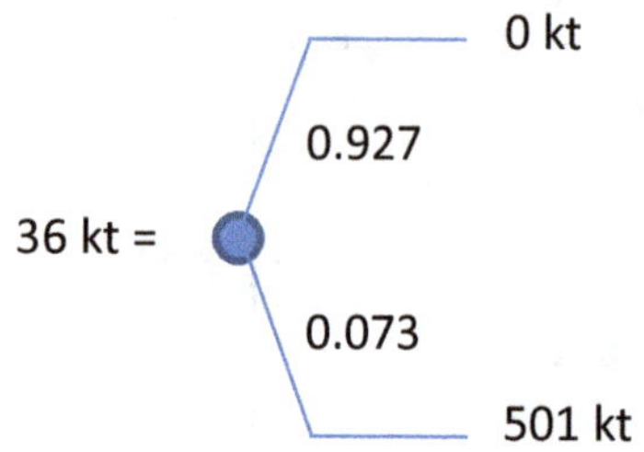

Figure 10. Probability tree for In Salah 2008 QRA.

2010 QRA Results

In 2010, the In Salah Gas JIP commissioned an update of the 2008 QRA for the JV. This update was based on the acquisition and interpretation of a new 3D seismic survey and new geo-mechanical modelling. This new information and interpretation indicated the existence of faults through the overburden near KB-5 to KB-502 and KB-503. Further, substantial uncertainty regarding the structural spill point remained.

Based on this data URS updated its 2008 QRA. Figure 11 presents the 2010 URS assessments. Our modification of these assessments, to yield the P10-P90 range, is given in Table 2.

The most significant change relative to the 2008 QRA is the inclusion of fractures for KB-501, KB-502, and KB-503. The base case estimate is now that there is a 10.7% chance of a fracture at KB-501, a 10.7% chance of a fracture at KB-502, and a 3.6% chance of a fracture at KB-503. This is a significant increase relative to the 2008 QRA, which assessed the chance of an undetected fault at 0.0016%. If these fractures do exist, the mean short-term flow rates for KB-501, KB-502, and KB-503 are 175 ktpa, 320 ktpa, and 38 ktpa, respectively.

The RLs for each of the events identified in the 2010 QRA are given in Figure 12. In this case three events, NW Fracture KB-502, NW Fracture KB-501, and Migration Direction, each exceed the 34 kt threshold, and NW Facture KB-503 is not far behind. The TCR is now about 691 kt. The sensitivity of this amount to uncertainty in our inputs is shown in the tornado diagram presented in Figure 13. The largest risk driver is now the probability of leakage at KB-502. For example, if the probability of leakage was 18%, instead of 11%, the RL would be over 950 kt. The uncertainty regarding the presence of fractures at KB-501 and KB-502 is now a significant uncertainty.

---

[1] Based on 100,000 Monte Carlo trials.

Table 2. Modified URS 2010 QRA data.

Probability of Leakage	P10	Mean	P90
Permeable Zones in Seal	2.81E-06	1.63E-05	3.56E-05
Undetected Faults	2.81E-06	1.63E-05	3.56E-05
NW Fracture KB502	5.03E-02	1.07E-01	1.79E-01
NW Fracture KB501	5.03E-02	1.07E-01	1.79E-01
NW Fracture KB503	1.68E-02	3.57E-02	5.96E-02
Leakage Old Wells - Surface	1.68E-02	3.57E-02	5.96E-02
Leakage Old Wells - Aquifer	1.68E-04	3.57E-04	5.96E-04
Leakage New Wells - Surface	2.81E-04	1.63E-03	3.56E-03
Leakage New Wells - Aquifer	2.81E-06	1.63E-05	3.56E-05
Leak Near Well Press. - Surface	2.81E-07	1.63E-06	3.56E-06
Leak Near Well Press. - Aquifer	2.81E-07	1.63E-06	3.56E-06
Exceeding Spill Point	2.81E-06	1.63E-05	3.56E-05
Earthquake Induced Fractures	2.81E-07	1.63E-06	3.56E-06
Migration Detection	1.68E-02	3.57E-02	5.96E-02
Short-term Leakage Rate (ktpa)	**P10**	**Mean**	**P90**
NW Fracture KB-502	305.24	320.22	335.48
NW Fracture KB-501	160.85	175.38	190.39
NW Fracture KB-503	11.05	38.49	76.08
Leakage Old Wells - Surface	0.02	0.27	0.60
Leakage Old Wells - Aquifer	0.02	0.27	0.60
Leakage New Wells - Surface	0.02	0.27	0.60
Leakage New Wells - Aquifer	0.02	0.27	0.60
Leak Near Well Pressurization - Surface	0.02	0.27	0.60
Leak Near Well Pressurization - Aquifer	0.02	0.27	0.60
Exceeding Spill Point	384.70	510.57	649.86
Earthquake Induced Fractures	0.17	2.66	6.01
Migration Direction	168.08	201.85	237.98
Long-term Leakage Rate (ktpy)	**P10**	**Mean**	**P90**
Permeable Zones in Seal	0.01	0.02	0.04
Undetected Faults	0.06	0.32	0.70
NW Fracture KB-502	20.15	31.48	44.67
NW Fracture KB-501	11.50	18.47	26.63
NW Fracture KB-503	2.01	3.15	4.47
Leakage Old Wells - Surface	0.02	0.27	0.60
Leakage Old Wells - Aquifer	0.02	0.27	0.60
Leakage New Wells - Surface	0.02	0.27	0.60
Leakage New Wells - Aquifer	0.02	0.27	0.60
Leak Near Well Pressurization - Aquifer	0.02	0.27	0.60
Earthquake Induced Fractures	0.02	0.03	0.05

RISKS	INITIATING EVENTS	Annual prob occurring	Prob of presence	Number of items		Rate (t/year/item)	Duration (yrs)	Mass lost (t)	Risk Quotient
Subsurface CO2 escape	Permeable zones in seal		1.E-05	1		15.00	1,000	15,000	0.1500
			1.0E-05	1.E-05	Guide midpoint	15			
		LN dist	1.E-06	1.E-04	LN dist	15	60		
			CL 01%	CL 99%		CL50%	CL95%		
	NW Fracture corridors KB-502		3.E-03	1		200	1,000.00	200,000	6.E+02
			3.0E-03	3.E-03	Guide midpoint	200	1000	Duration	
		LN dist	1.E-03	1.E-02	LN dist	200	1000	Short Term	
			CL 01%	CL 99%		CL50%	CL95%		
						0.01	10	Duration	
						0.01	0.011	Long Term	
						CL50%	CL95%		
	NW Fracture corridors KB-501		3.E-02	1		100,000	1.00	1,275,000	4.E+04
			3.0E-02	3.E-02	Guide midpoint	100000	11	Duration	
		LN dist	1.E-02	1.E-01	LN dist	100000	120000	Short Term	
			CL 01%	CL 99%		CL50%	CL95%		
						17500	10	Duration	
						17500	30000	Long Term	
						CL50%	CL95%		
	NW Fracture corridors KB-503		3.E-03	1		29,000	1.00	349,000	1.E+03
			3.0E-03	3.E-03	Guide midpoint	29000	11	Duration	
		LN dist	1.E-03	1.E-02	LN dist	29000	100000	Short Term	
			CL 01%	CL 99%		CL50%	CL95%		
						3000	10	Duration	
						3000	5000	Long Term	
						CL50%	CL95%		
	Leakage - Undetected fault		1.E-05	1		200	100	20,000	0.20
			1.0E-05	1.E-05	Guide midpoint	200			
		LN dist	1.E-06	1.E-04	LN dist	200	1000		
			CL 01%	CL 99%		CL50%	CL95%		

Figure 11. URS 2010 QRA data.

	Leakage - Wells								
Well survey, remediation	Leakage - Old wells								
Leakage - Devonian	Leakage to surface		3.E-02	4		100	1.0	400	151
			3.0E-02	3.E-02 Guide midpoint		100			
		LN dist	1.E-02	1.E-01	LN dist	100	1000		
			CL 01%	CL 99%		CL50%	CL95%		
	Leakage to potable aquifer		3.E-04	4		100	1,000.0	400,000	
			3.0E-04	3.E-04 Guide midpoint		100			
		LN dist	1.E-04	1.E-03	LN dist	100	1000		
			CL 01%	CL 99%		CL50%	CL95%		
	Leakage - New wells								
	Leakage to surface		1.E-03	17		100	1.0	1,700	
			1.0E-03	1.E-03 Guide midpoint		100			
		LN dist	1.E-04	1.E-02	LN dist	100	1000		
			CL 01%	CL 99%		CL50%	CL95%		
	Leakage to potable aquifer		1.E-05	17		100	1,000.0	1,700,000	
			1.0E-05	1.E-05 Guide midpoint		100			
		LN dist	1.E-06	1.E-04	LN dist	100	1000		
			CL 01%	CL 99%		CL50%	CL95%		
	Near well pressurization								
	Leakage to surface		1.E-06	3		10	1.00	30	0.015
			1.0E-06	1.E-06 Guide midpoint		10			
		LN dist	1.E-07	1.E-05	LN dist	10	100		
			CL 01%	CL 99%		CL50%	CL95%		
	Leakage to potable aquifer		1.E-06	3		10	1,000	15,000	
					LN dist	10			
						10	100		
						CL50%	CL95%		
Drill 2x more wells	Drill 2x more wells		1.0E-04	1.E-04 Guide midpoint		30000			
		LN dist	1.E-05	1.E-03	LN dist	30000	40000		
			CL 01%	CL 99%		CL50%	CL95%		
	Long term leakage		0.E+00	0		15	1,000	0	
					LN dist	15			
						15	60		
						CL50%	CL95%		

Figure 11 continued.

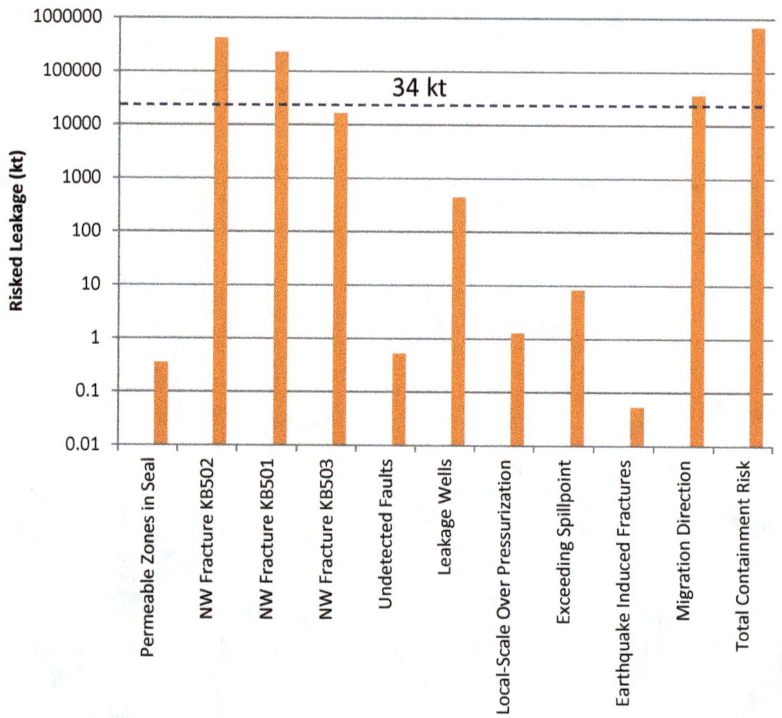

Figure 12. 2010 risked leakage for each event.

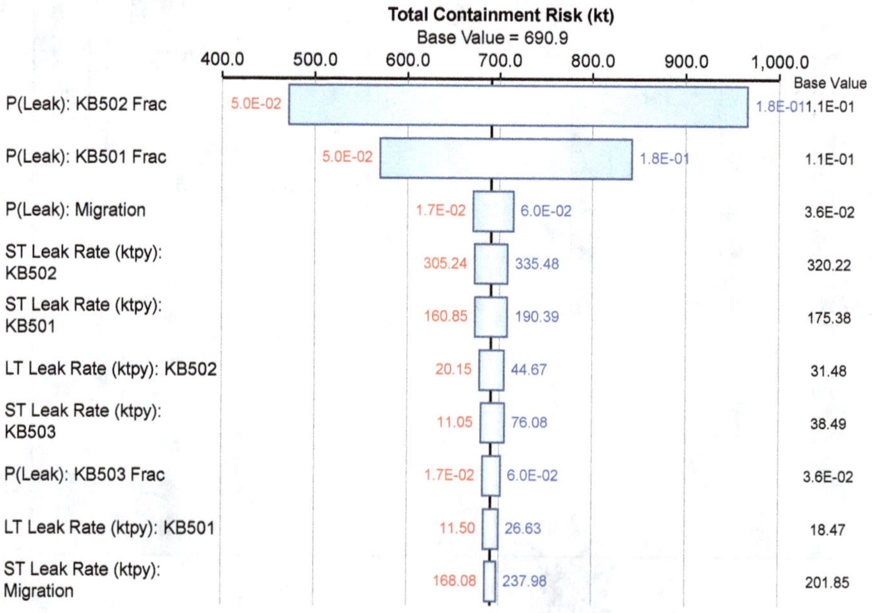

Figure 13. Tornado diagram for 2010 total containment risk.

Figure 14 presents the updated leakage risk curve. The probability of any leakage has increased from 7.3% in 2008 to 27.8%. If there is a leak, the RL is 2,462 kt, or almost 15% of the planned storage volume. This alternate way of looking at the RL is given in Figure 15. The probability of leaking more than 170 kt is 24.9%. The 1%-VaR is 5,750 kt (34% of planned storage volume). Thus, we see the importance of looking at the extremes of the distribution.

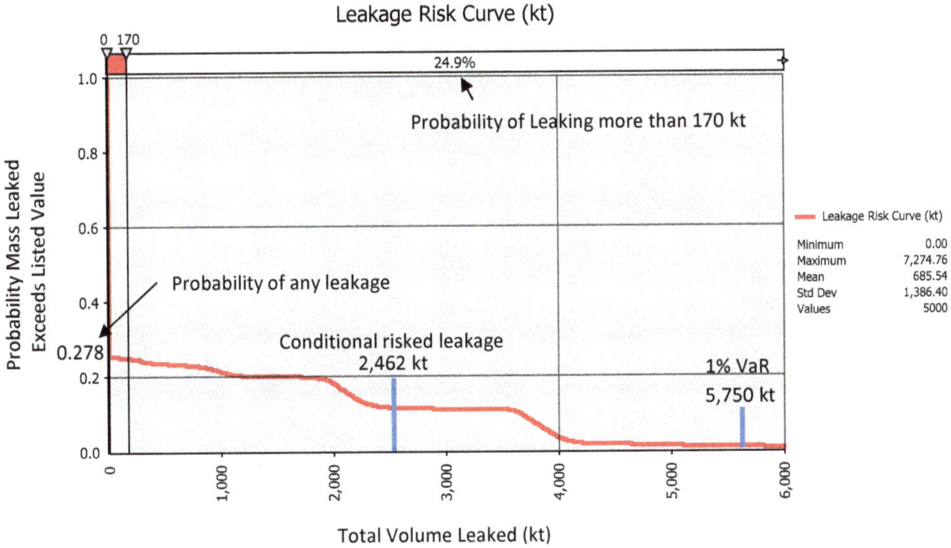

Figure 14. 2010 leakage risk curve.

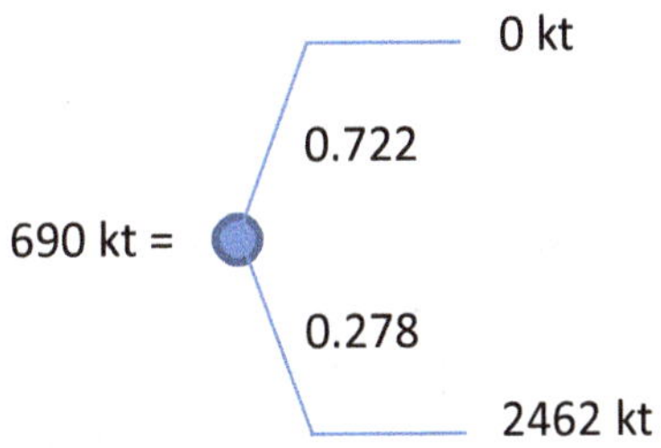

Figure 15. 2010 probability tree for In Salah.

2010 QRA Results - Post-Mitigation Strategy

Based in part on the previous analysis, the JV decided to suspend injection at KB-502, reduce injection at KB-501, and reduce injection at KB-503. The predicted impact of each of these actions is as follows:

1. Suspending the injection at KB-502 reduces the probability of leakage through that corridor to between 1.7E-3 (P10) and 6E-3 (P90). The flow rate then lowers to between 0.6 ktpa (P10) and 0.70 ktpa (P90). However, this flow rate would persist for 1,000 years. Suspending KB-502 also reduces the flow rate to between 2.9 ktpa (P10) and 35 ktpa (P90) in the event of a flow-over from the spill point.

2. Reducing the injection rate at KB-501 reduces the likelihood of leakage to between 1.7E-2 (P10) and 6E-2 (P90). The flow rate then lowers to between 87 ktpa (P10) and 115 ktpa (P90). The duration would not be affected.
3. Reducing injection at KB-503 reduces the likelihood of leakage to between 1.7E-3 (P10) and 6E-3 (P90). The flow rate and duration would not be affected.

The risked leakage for both the pre- and post-mitigation strategy is shown in Figure 16. The TCR still exceeds the target of 34 kt. However, the risk of exceeding 34 kt at KB-502, KB-501, and KB-503 is reduced significantly. The post-mitigation strategy tornado diagram is shown in Figure 17. The most significant risk is the leakage risk at KB-501.

Figure 18 presents the post-mitigation strategy leakage risk curve. Note: the vertical scale ranges from 0 to 0.1, instead of 0 to 1.

There is now about an 11% chance of experiencing some leakage. If a leakage does occur, the average amount leaked is ~478 kt. The probability of leaking more than 170 kt is 4.7%. These performance parameters are on par with that of the 2008 QRA and are a substantial reduction from the pre-mitigation strategy performance. The 1%-VaR is 1,375 kt (or ~8% of the planned storage volume). This is lower than the pre-mitigation strategy 1%-VaR of 5,750 kt and the 2008 estimate of 1,699 kt. As we will discuss in the next section, this demonstrates the ex post value of the seismic survey. The survey results have enabled the JV to gain a better understanding of the project and adjust the field management strategy, which is believed to reduce the risk.

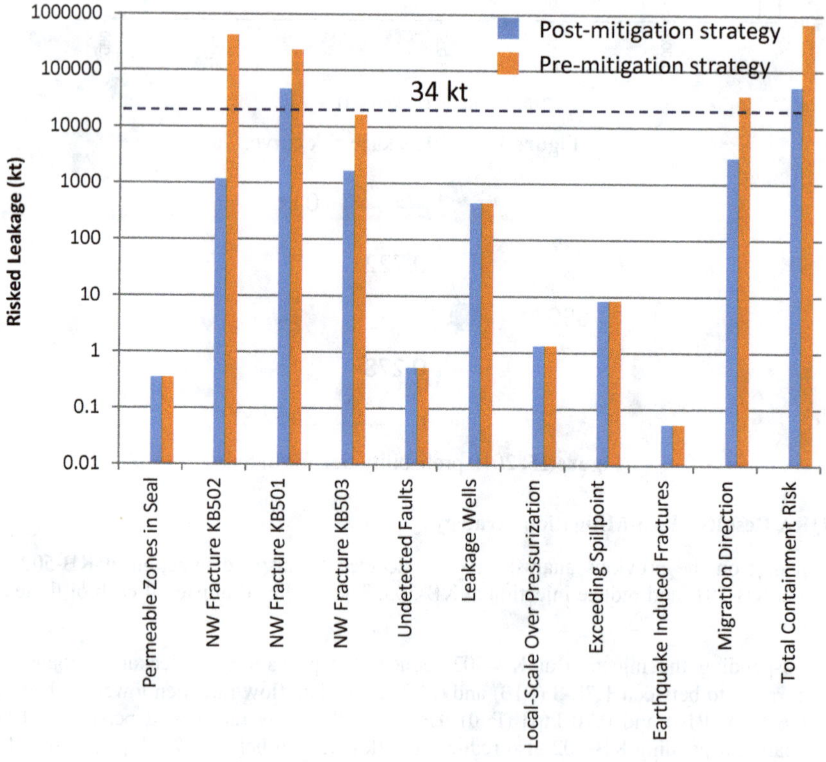

Figure 16. 2010 risked leakage - post-strategy.

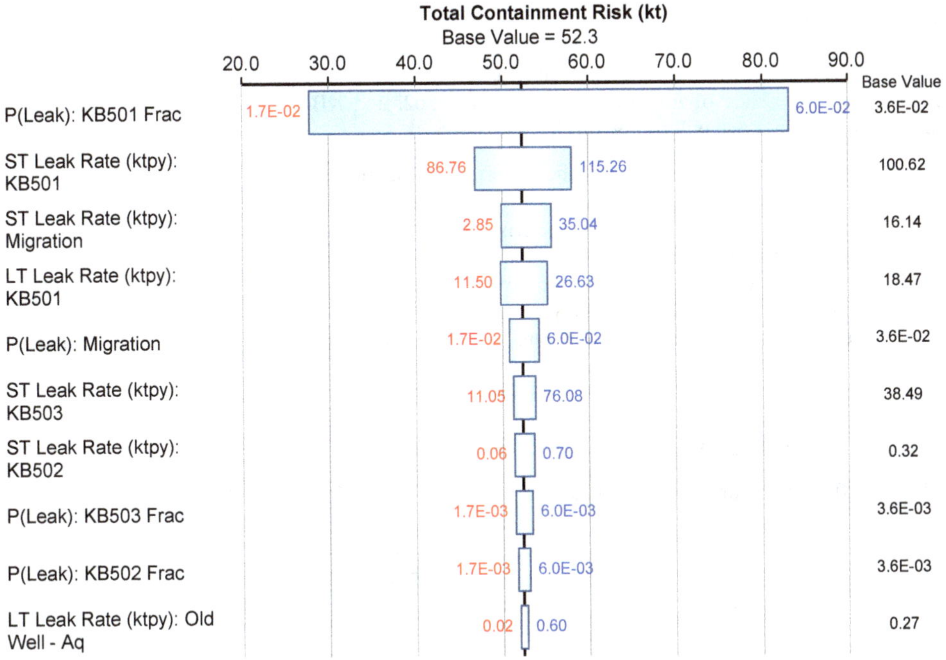

Figure 17. Tornado diagram for 2010 total containment risk - post-mitigation strategy.

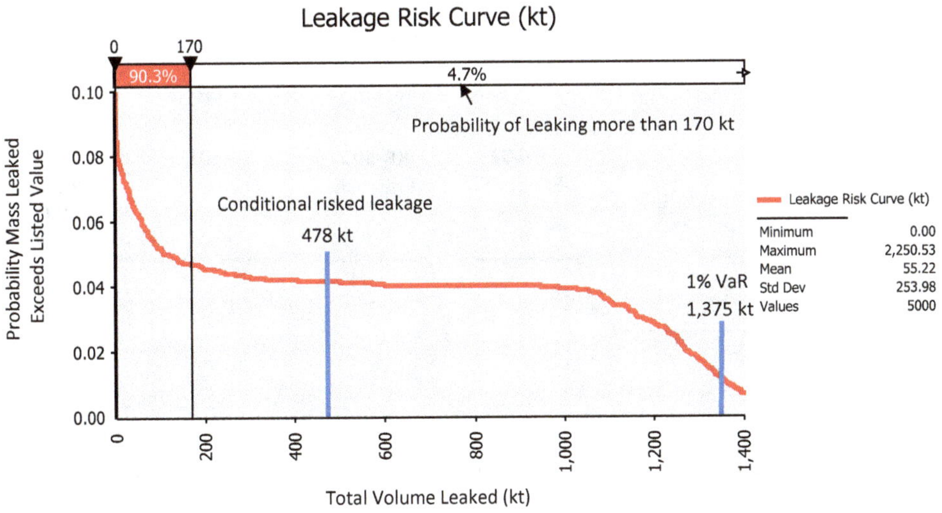

Figure 18. 2010 leakage risk curve – post-mitigation strategy.

VALUE OF INFORMATION

In this section, we demonstrate how to use the QRA model developed in the previous section to understand the value of conducting a new seismic survey at In Salah. We imagine we are in 2008 and are trying to decide if we should acquire a new seismic survey that would provide information regarding the presence of fractures at KB-501, KB-502, and KB-503. To determine the value of seismic information, we will follow this VOI Roadmap:

1. Specify prior distributions on the uncertainties of interest about which seismic would provide information.
2. Determine the performance characteristics of In Salah assuming that we do not collect any further information.
3. Specify the accuracy of the seismic survey.
4. Calculate the posterior probabilities on the uncertainties of interest.
5. Determine how we would operate In Salah given each of the possible seismic interpretations.
6. Determine the performance characteristics of In Salah conditional on the seismic information.
7. Compare (2) and (6) to determine the benefit of gathering the seismic information.

This workflow is captured in the schematic decision tree shown in Figure 19.

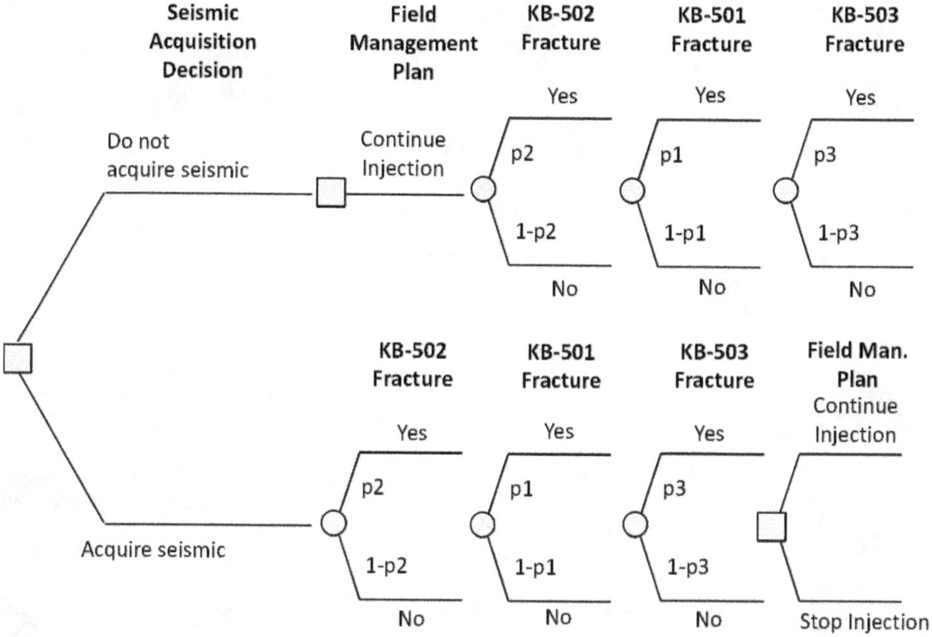

Figure 19. Schematic decision tree for 2008 seismic acquisition decision.

We first face a decision about whether or not to acquire seismic. If we choose not to acquire seismic, we then need to decide upon the field management plan. We assume that we will continue injection as there will be no reason to do otherwise. At this point, after we have made our choices, we learn whether or not fracture corridors exist at the injection wells. The probability of these events is given by the probabilities p1, p2, and p3. If fractures do exist, then we assume that CO_2 leakage will occur. In actuality, fractures may be present but may still not leak. This additional uncertainty could be included, but we do not do so here in the interest of easing our explanation and demonstrating the VOI workflow. If fractures do not exist, then leakage will not occur via these pathways. However, leakage could still occur via the other mechanisms given in Table 1.

If we choose to acquire seismic, then we assume that the existence of fractures will be revealed and then we make our field management decision, which is to either continue or stop injection. The tree in Figure 19 assumes that the information is perfect in that we gain clairvoyance on the presence of fractures before we make our field management decision. We will handle the case of imperfect information later.

There is likely to be strong dependence in the existence of fractures because they would be the result of a common operating procedure (e.g. operating the injectors above the fracture pressure of the reservoir). Due to this dependence and the need to reduce the complexity, we will assume there is a single uncertainty called "fracture corridors." If fracture corridors exist then they exist at all injectors. This simplified tree is shown Figure 20.

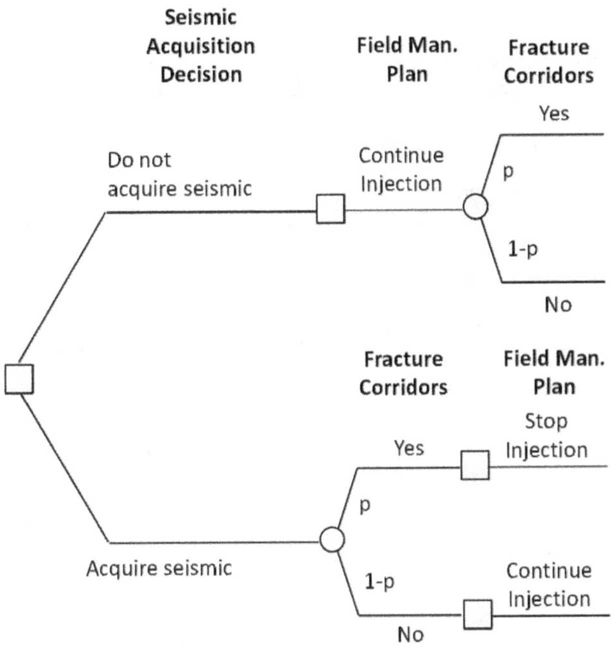

Figure 20. Simplified decision tree for 2008 seismic acquisition decision.

We now illustrate the VOI Roadmap by analysing the decision in 2008 to acquire seismic information. It is a challenge to assume only the 2008 data after knowing 2010 survey results. For example, the 2010 QRA explicitly considered the possibility of fractures KB-501, KB-502, and KB-503, whereas the 2008 QRA considered a single uncertainty for 'undetected faults.' This single uncertainty was meant to capture the possibility of faults at KB-501, KB-502, and KB-503. In order to make use of both the 2008 and 2010 QRAs, we assume that in 2008, the JIP explicitly considered fracture corridors at KB-501, KB-502, and KB-503.

Prior Distributions

The uncertainty of interest is whether or not fracture corridors exist for KB-501, KB-502, and KB-503. As discussed above, the JIP did not assess the probability of these specific events in 2008. Rather, they assumed that the probability of undetected faults was 1E-5 (1 in 100,000). We assume this is the prior probability p assigned to fracture corridors as shown in Figure 20.

We must also assess probability distributions on the volume that would be leaked if these fracture corridors existed, allowed leakage and the In Salah Gas JV did not change its operation of the field. For this, we adopt the 2010 QRA assessments given in Table 2. Specifically, the P10, Mean, and P90 flow-rate (ktpa) assessments (lognormal distribution) for KB-501, KB-502, and KB-503 are as follows:

- KB-501: 160.85, 175.38, 190.39
- KB-502: 305.24, 320.22, 335.22
- KB-503: 11.05, 38.49, 76.08

While these assessments were provided in 2010, we are assuming they are what would have been assessed had the 2008 QRA explicitly considered these risks.

Performance Characteristics without Further Information

We now determine the performance characteristics of In Salah without acquiring seismic information. This is illustrated in the upper branch of Figure 20, where we average over the uncertainty regarding the presence of fracture corridors. The chance of fracture corridors and the flow rate uncertainties were defined previously. All other uncertainties are defined according to the 2008 QRA given in Table 2. Given these assessments, the TCR is 30.7 kt, which is slightly below the target of 34 kt (Table 3). The probability of leaking more than 170 kt (1% of the planned storage) is 3%. More concerning is the fact that there is a 1% chance of leaking more than 1,061 kt (6% of the planned storage volume).

Table 3. In Salah performance characteristics given seismic is not acquired.

TCR (kt)	P(Leak > 170kt)	1% VaR (kt)
30.7	3.0%	1,060.7

Seismic Accuracy

If we decide to acquire seismic, we need to specify its accuracy. By accuracy, we mean how likely seismic is to report a fracture corridor if one exists and not report one if one does not exist. We show this graphically in the probability trees given in Figure 21 and Figure 22. Figure 21, which is in assessed form, is a case of perfect information. If a fracture does exist, then seismic will report "Fracture" with probability 1.0; if a fracture does not exist, then seismic will report "No Fracture" with probability 1.0. Figure 22 shows a case in which we would say seismic is 80% accurate. As discussed earlier, these trees do not have to be symmetric as assumed in this report.

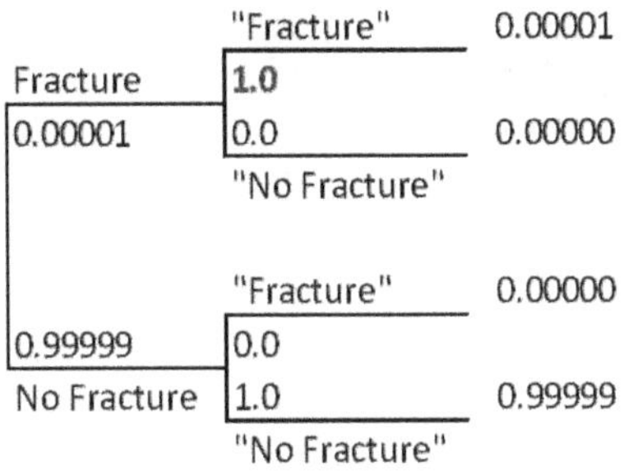

Figure 21. Seismic accuracy in assessed form: perfect information.

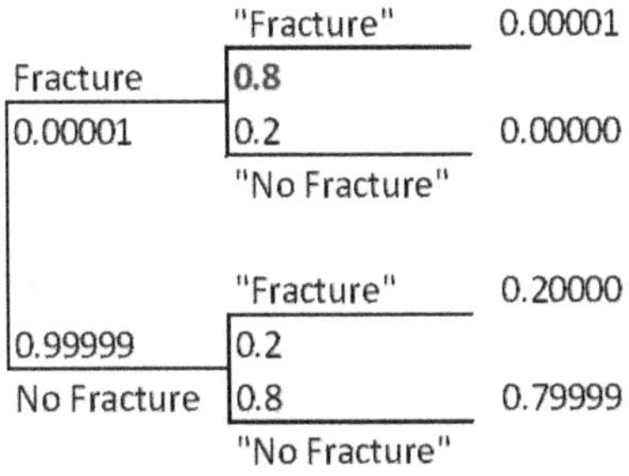

Figure 22. Seismic accuracy in assessed form: 80% accuracy.

Posterior Distributions

Using the trees in Figure 21 and Figure 22, we determine the posterior distributions by placing the trees in inferential order. That is, we 'flip' the tree or perform Bayes Rule. This is executed in Figure 24.

As before, Figure 23 is a case of perfect information. If seismic is perfect then the probability of Fracture or No Fracture is certain if the test reports "Fracture" or "No Fracture," respectively. In addition, the probability that we must assign to the survey reporting "Fracture" is 1E-5, which the prior probability assigned to "Fracture."

Figure 24 corresponds to the case of an 80% accurate test. In this case, there is only a 4E-5 (.00004) chance that a fracture will be present even if the seismic survey reports "Fracture." The posterior probability is four times larger than the prior, but is still low in an absolute sense. This occurs because the prior on a fracture being present was so low to begin with. Essentially, the prior belief is that it is nearly impossible that a fracture corridor could exit. The prior assessment implies that, out of 10,000 In Salah-like situations, only one would have a fracture corridor. Later, we will test sensitivity to this assessment. This is an important example, however, of just how important test accuracy can be. If the test is perfect, we are certain a fracture exists if the test reports fracture. However, if the test is 80% accurate we are almost certain a fracture is not present given the same report - we are not 80% certain. If the 80% accurate seismic reports "No Fracture," then the report confirms our initial beliefs.

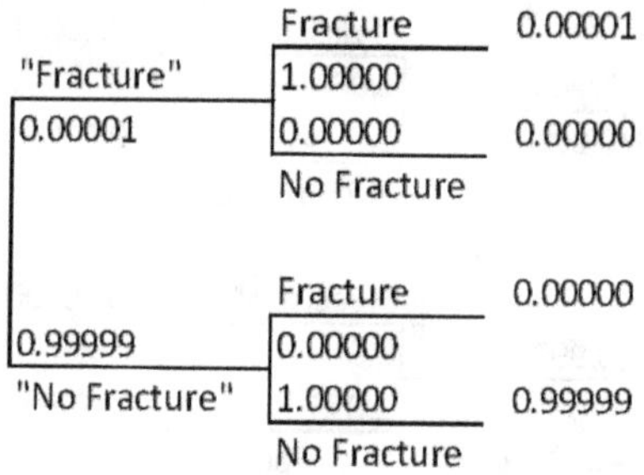

Figure 23. Seismic accuracy in inferential form: perfect information.

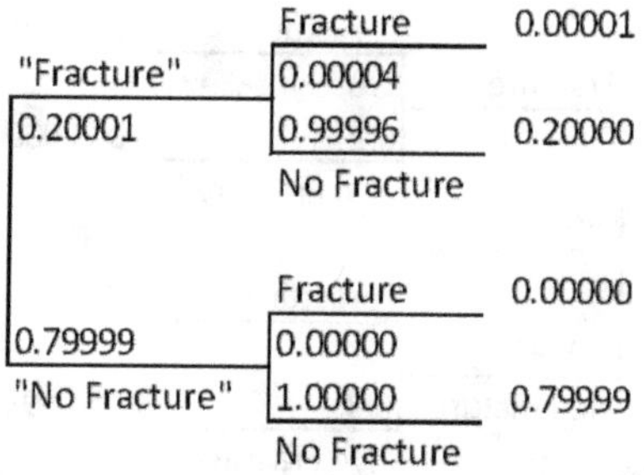

Figure 24. Seismic accuracy in inferential form: 80% accuracy.

Performance Characteristics Given Seismic Information

We now determine the performance characteristics of In Salah under two conditions: the seismic reports "Fracture" and the seismic reports "No Fracture." If the seismic reports that a fracture corridor is present, we assume that we stop all injection. The flow rates for this situation were provided in the 2010 QRA (Figure 11). If the seismic determines that a fracture is not present then we assume that we continue injection. Table 4 presents the performance characteristics for each seismic result. If the test reports that a fracture is present then the TCR increases from 30.7 kt to 972 kt; the probability of leaking more than 170 kt would increase to almost 100%; and, the 1% VaR would be 3,612.1 kt. If, instead, the test reports "No Fracture" then the performance characteristics would be very similar to those with the case of no seismic information (Table 3) because that case only assumes a 1-in-10,000 chance of a fracture.

Table 4. In Salah performance characteristics given seismic result (perfect seismic).

Seismic Result	TCR (kt)	P (Leak > 170kt)	1% VaR (kt)
"Fracture"	972	99.5%	3,612.1
"No Fracture"	30.6	3.0%	1060.7

Comparing the Cases without Further Information and with Information

Since we are uncertain whether or not seismic will report "Fracture" or "No Fracture," we need to probability-weight these two scenarios, or calculate their expectation. In the case of perfect information, the probability of "Fracture" is 0.00001 and "No Fracture" is 0.99999. The risked loss given we acquire seismic is then 0.00001x972 + 0.99999x30.6 ≈ 30.6 kt. Likewise, the probability of leaking more than 170 kt is 3.0%. In the case of the 1% VaR, taking the expectation of the VaRs in the case of a fracture and with no fracture is not theoretically correct.[2] However, it is very close in this case and is easier to explain. Doing so yields a 1%-VaR if we acquire seismic of 1,060.6 kt.

We now summarize the cases we have examined above by displaying them in Figure 25. We see that, because the performance metrics between the two alternatives are almost identical, it is believed that acquiring seismic will not significantly improve the performance characteristics of the project. This is true even though the seismic was assumed to be perfect. This conclusion is reached due to the prior assumption that the probability of a fracture corridor was 1E-5. We have assumed that we already know that no fractures are present, therefore there is no need to obtain seismic.

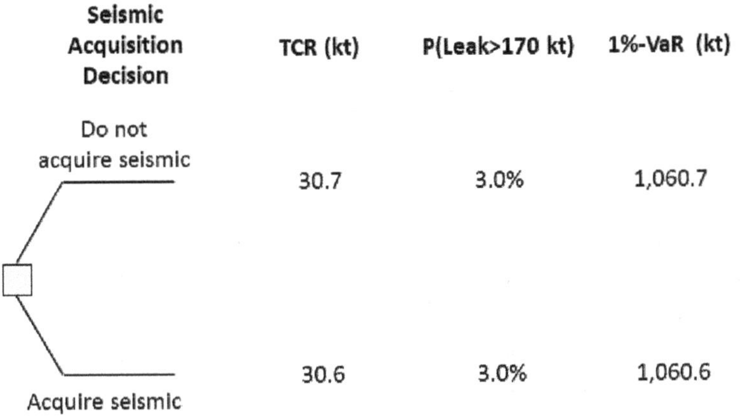

Figure 25. Seismic acquisition decision tree assuming perfect information (prior = 1e-5).

The result that seismic will not add value will not hold up under less certain prior assumptions. For example, assume that we believe is the probability of fracture corridors is 0.01. In this case, the TCR increases to 94.5 kt if we do not acquire seismic (refer to Figure 26), the probability of leaking more than 170 kt is 5.8% and the 1%-VaR is almost 3,510 kt, or 21% of the planned storage volume, and the performance statistics are improved significantly by acquiring seismic. For example, seismic lowers the TCR by almost 55 kt and the 1%-VaR by over 2,400 kt. This occurs because we expect there is a 1% chance the seismic will report a fracture is present and in that case we stop injection. Even this low probability of taking a remedial action results in large risk reductions because continuing to inject in the presence of a fracture corridor produces large losses.

We believe that Figure 26 presents the 'bottom-line' value of seismic in a clear and compelling manner. Without seismic, there a 1% chance more than 21% of the planned storage volume (3,510 kt) will be leaked. Acquiring seismic and stopping injecting, should seismic identify a fracture, reduces this loss to 6% of the planned storage volume (1,086 kt). Whether or not this risk reduction is worth the cost of the seismic survey is a judgment the JV management must make, but this information provides a consistent and convenient framework within which to have that discussion. If a dollar-for-dollar comparison was desired, a monetary damage amount would need to be applied to the leakage volumes.

[2] For example, the expectation of 1st percentiles is not the 1st percentile of the mixture of distributions.

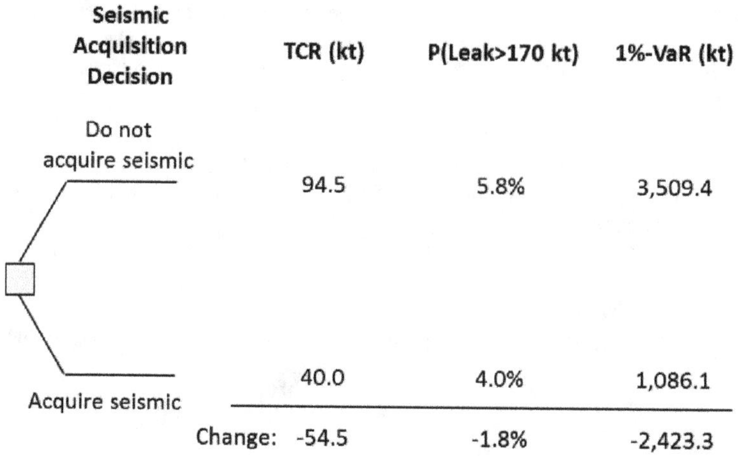

Figure 26. Seismic acquisition decision tree assuming perfect information (prior = 0.01).

Figure 27, Figure 28, and Figure 29 display the TCR, probability of leaking more than 170 kt, and the 1%-VaR, respectively, as a function of the prior probability of a fracture corridor. Focusing on Figure 27, if the prior probability is very low, on the order of 1E-5, then seismic is not expected to produce much benefit. However, as the prior probability increases, the TCR, given that seismic is not obtained, increases significantly. Conversely, if seismic is obtained, the TCR increases slowly, because we can decide to stop injection if the seismic survey identifies a fracture (i.e. the decision to purchase seismic provides the option of stopping injection once one has learned about the structure of the reservoir).

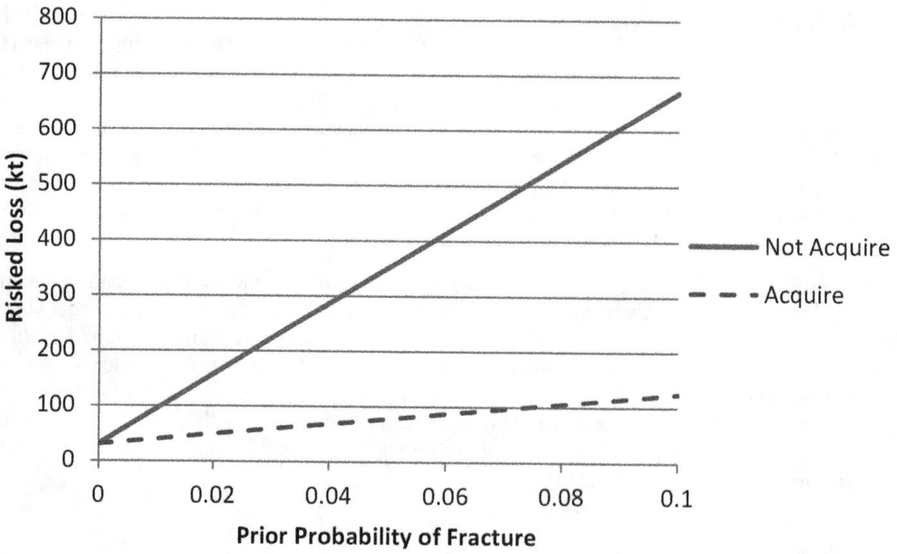

Figure 27. Total containment risk as a function of prior probability of fracture corridors.

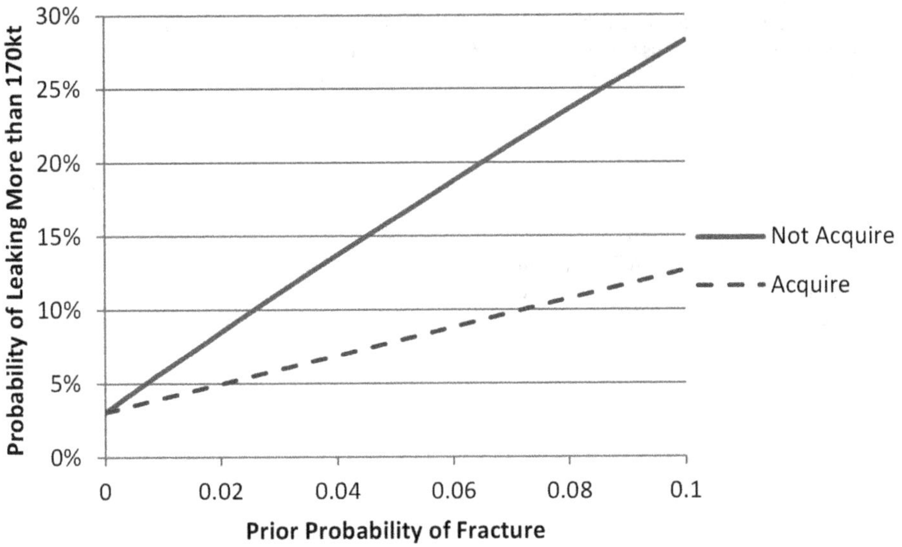

Figure 28. Probability of leaking more than 170 kt as a function of prior probability of fracture corridors.

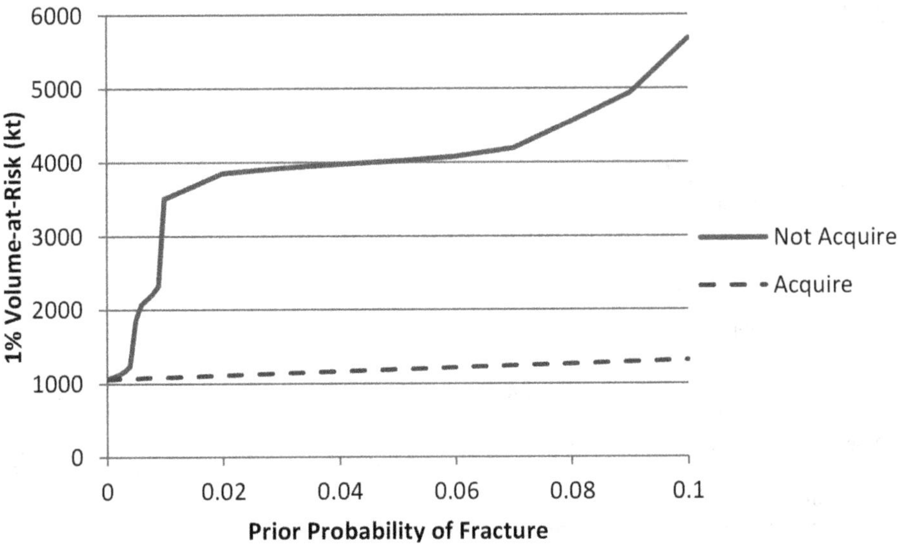

Figure 29. 1%-VaR as a function of prior probability of fracture corridors.

The other performance metrics display similar trends. The 1%-VaR is particularly striking. This metric increases non-linearly and rapidly with the prior probability of a fracture. However, if seismic is acquired, this tail event is reduced dramatically because we can stop injection if a fracture is detected.

The preceding analysis assumed that the seismic information was perfect. While this is rarely the case, it does serve as a useful upper bound (i.e. no technology can offer more than perfect information). In the next section, we consider the case of imperfect information.

Imperfect Information

As illustrated in Figure 22 and Figure 24, if seismic is not perfect, we will be uncertain about the presence of a fracture corridor, even if the survey indicates that a fracture is present. Figure 30 shows the posterior probability of a fracture corridor being present. This assumes that the seismic survey has reported "Fracture" as a function of the prior probability of such a fracture, with accuracies of 80%, 90% and 100%. The case of 100% accuracy is the perfect information case we analysed earlier. As long as the prior probability is not 0, the posterior will always be 100% if the test is perfect. Alternatively, if the test is 90% accurate and the prior is 0.1, the posterior would be 50%. If the accuracy was 80% the posterior would be about 31% in this case.

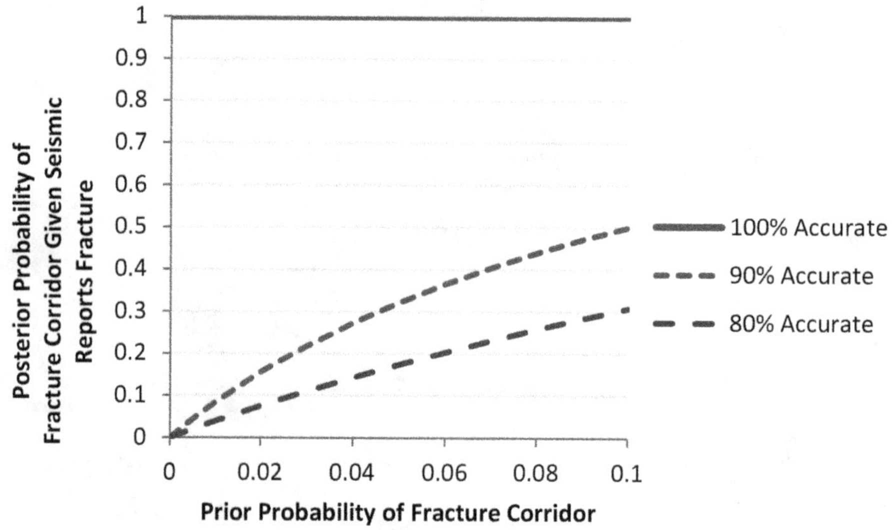

Figure 30. Posterior probability of a fracture corridor being present given seismic reports "fracture."

The procedure to determine the benefit of acquiring seismic is the same as outlined in the previous section. The only difference is the calculation of posterior probabilities is a bit more complex than the case of perfect information. Figure 31 displays the TCR as a function of the prior probability of a fracture for differing seismic accuracies. The lines labelled "Not Acquire" and "Acquire (100%)" are identical to the two cases considered in Figure 27. We see the benefit of the seismic decreases along with its accuracy. However, in this case, absolute benefit of seismic is still quite robust.

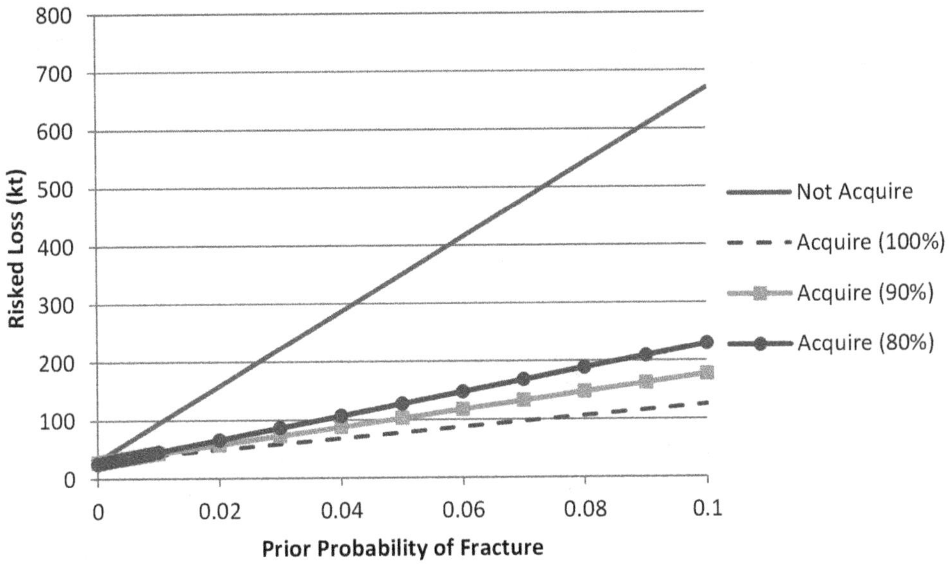

Figure 31. Total containment risk for differing levels of seismic accuracy.

ANALYTIC VALUE OF INFORMATION MODEL

In the previous section, we demonstrated how to estimate the benefit of obtaining more information on fracture corridors at In Salah using the URS assessments. In this section, we demonstrate how to use an analytic model of leakage rates to estimate the value of information over differing time windows.

The analytic model used in this report was developed by Prof. Andy Woods (BP Institute, Cambridge) and is loosely based on In Salah. Given estimates of reservoir parameters and fault properties, the model determines the cumulative volume leaked over any time frame. Inputs to the model are listed in Table 5. The model is deterministic in that it produces a single estimate of the total volume leaked for a given set of inputs. However, by linking this model with a Monte Carlo software package (e.g. @Risk) we are able to produce a probability distribution regarding the volume leaked over differing time frames.

Table 5 also displays the uncertainty surrounding each input, as provided by Prof. Woods. The P10 is a value for the input that is so low that there is only a 10% chance the true value is less than this. The P90 is a value so high such that there is 90% chance that the true value will be less than this. These two assessments provide two points on the underlying probability distribution. Prof. Woods also specified the probability distribution family for each input, as shown in Table 5. Each input was assumed to be either normally or lognormally distributed. Since these distributions are completely specified by two parameters (e.g. their mean and standard deviation), the P10 and P90 assessments are sufficient to define the underlying distributions. Table 5 also includes the mean and standard deviation that are implied by the P10-P90 assessments.

Table 5. Inputs to analytic model.

Variable	Units	P10	P90	dist type	mean	Std dev
Reservoir						
Permeability of Reservoir	m^2	1.00E-15	1.00E-14	lognormal	4.73E-15	5.27E-15
Porosity	fraction	0.10	0.15	lognormal	0.12	0.02
Thickness Aquifer	m	15	25	normal	20	3.9
Permeability	m^2	4.00E-13	1.00E-13	lognormal	2.32E-13	1.35E-13
Residual Saturation	fraction	0.05	0.25	normal	0.15	0.08
Fracture						
Distance to Fracture	m	1000	3000	normal	2000	780
Width of Fracture	m	1.00	0.01	lognormal	0.5	2.5
Permeability of Fracture	m^2	1.00E-12	1.00E-10	lognormal	5.02E-11	2.47E-10
Angle of Slope	degrees	1.0	3.0	normal	2.0	0.8

The *subsurface* single-well injection rate was assumed to be 86.40 m^3/day or 0.247 MMt of CO_2 per year.[3] This rate is nominally based on the single-well subsurface injection rate at In Salah. The planned injection time was assumed to be 40 years, yielding a modelled single-well injection volume of about 9.89 MMt of CO_2. Using the JIP target of not leaking more than 1% of the injected volume, the leakage target is 0.0989 MMt.

Figure 32 presents the 1,000-year leakage risk curves for three different assumptions regarding the distance to the fault (or fracture): 1,000 m, 2,000 m, and 3,000 m. The closer the fault is to the injection location, the more likely it is that we will exceed a particular level of leakage. For example, if the fault is at 1,000 m, there is a 90% chance we will leak at least 0.57 million m^3 within 1,000 years. If the fault is at 2,000 m or 3,000 m, there is a 90% chance of leaking at least 6.95 MMt and 5.91 MMt, respectively.

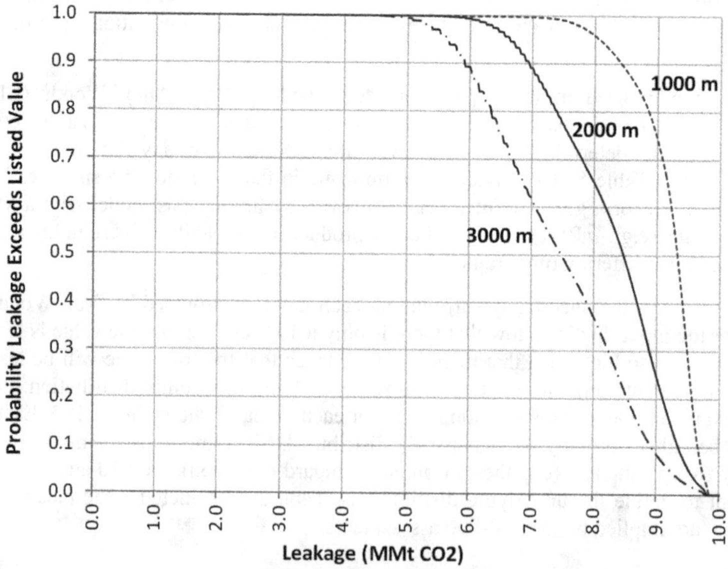

Figure 32. 1,000-year leakage risk curve.

[3] Based on conversion rate of 1.2755 m^3 per ton of CO_2 at a depth of 2 km under hydrostatic pressure.

Table 6 summarizes the performance characteristics for the three fault locations. The TCR (the risked leakage volume) decreases from 9.17 MMt to 7.45 MMt as the fault is moved from 1,000 m to 3,000 m. These numbers are large, being between 75% and 92% of the total modelled injected volume. The probability of exceeding the designed leakage threshold of 0.0989 MMt is 100% for all three fault locations. Given the assumptions above, it is not possible to prevent less than 1% of the modelled storage volume from leaking, assuming we keep injecting for 40 years. The 1%-VaRs are also quite large, being nearly 100% of the modelled injected volume.

Table 6. 1,000-year performance characteristics.

Distance to Fault	TCR (MMt)	P(Leak > Target)	1% VaR (MMt)
1,000 m	9.17	100%	9.83
2,000 m	8.28	100%	9.78
3,000 m	7.45	100%	9.69

Figure 33 presents the 50-year leakage risk curves. While the risk of leakage is reduced over the next 50 years, relative to the 1000-year risk, it is still substantial. Table 7 summarizes the performance characteristics. The TCRs exceed the target by a large margin. The 1%-VaRs are also quite large, being at least 73% of the modelled injected volume.

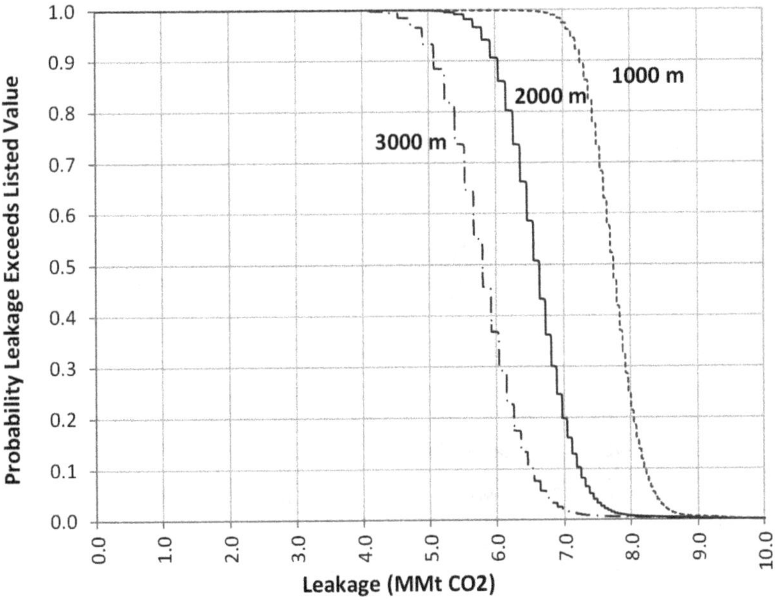

Figure 33. 50-year leakage risk curve.

Table 7. 50-year performance characteristics.

Distance to Fault	TCR (MMt)	P(Leak > Target)	1% VaR (MMt)
1,000 m	7.74	100%	8.69
2,000 m	6.61	100%	7.84
3,000 m	5.81	100%	7.25

We now analyse a case where we assume that we have been injecting for 3 years, but then stop because we believe a fault is present. To simplify the analysis and our presentation of the results, we will assume that the fault is located at 2,000 m.

Figure 34 shows the 50-year and 1,000-year leakage risk curves if we stop injecting after 3 years. In this case, the risk is almost eliminated within the next 50 years. However, over the next 1,000 years, we would still expect a substantial amount of CO_2 leakage. The performance characteristics for these two cases are given in Table 8. If we only consider the next 50 years, the project would be just within our performance limits: the probability of exceeding the targeted leakage is 13%. However, if taking the next 1,000 years into account, even limiting the injection period to only 3 years incurs a probability of 85% that we will exceed the target.

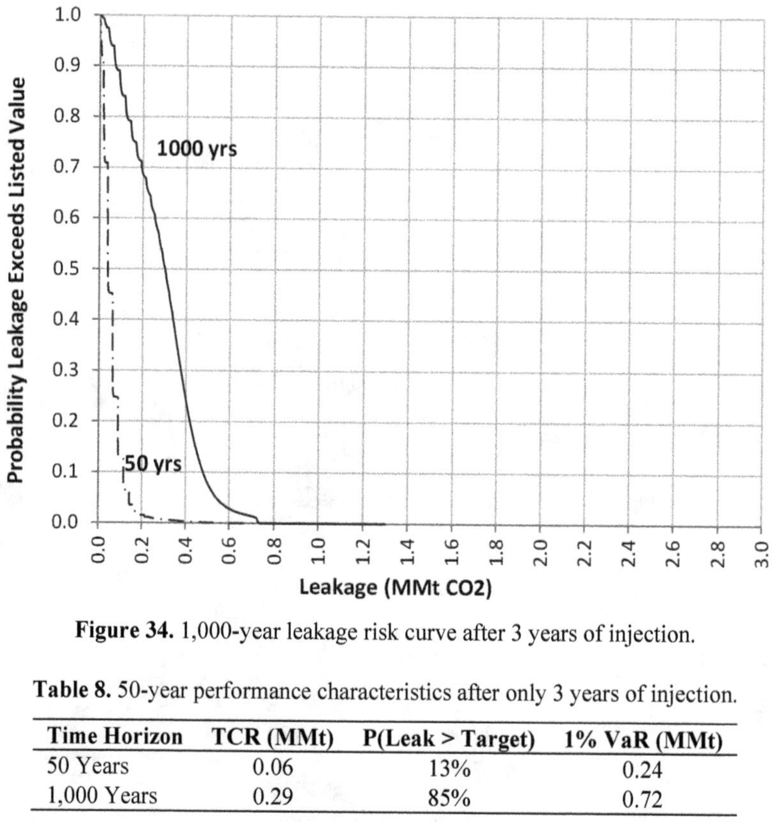

Figure 34. 1,000-year leakage risk curve after 3 years of injection.

Table 8. 50-year performance characteristics after only 3 years of injection.

Time Horizon	TCR (MMt)	P(Leak > Target)	1% VaR (MMt)
50 Years	0.06	13%	0.24
1,000 Years	0.29	85%	0.72

To quantify the benefit of gathering seismic data, we developed decision trees (Figure 35). We have included a tree based on a 50-year time horizon and another based on a 1,000-year time horizon. We assume the probability of a fault (or fracture) is p.

If we do not acquire seismic, then we will continue to inject for 40 years. If a fracture is present, we are certain to leak more than our target and fail to achieve our objectives for the project. If a fracture is not present then no leakage will occur.

If we do acquire seismic, then we assume that seismic will perfectly reveal whether or not a fracture exists. Since the test is perfect, the probability the seismic will report "Fracture" is *p*. We assume that if we learn that a fracture is present, then we will stop injection and only the 3-years of injected volume is subject to leakage. By stopping injection, we reduce the probability of not meeting our objective to 13% over 50 year and 85% over 1,000 years.

Figure 36 plots the probability of meeting our target objective over 50 and 1,000 years if we either acquire or do not acquire seismic. Over 50 years, acquiring seismic significantly reduces the risk of not meeting our objectives. Over 1,000 years, acquiring seismic reduces, but cannot eliminate risk and still leaves an unacceptably large risk of leaking more than the target. This occurs because we inject 7.5% of the total injected volume – almost 0.75 MMt – over 3 years. This volume exceeds the target leakage volume of 0.0989 MMt. If a fault is present, the analytic model forecasts that almost the entire injected volume will be lost within 1,000 years. Thus, while seismic is still helpful here, we are already in a situation where the project objectives cannot be achieved.

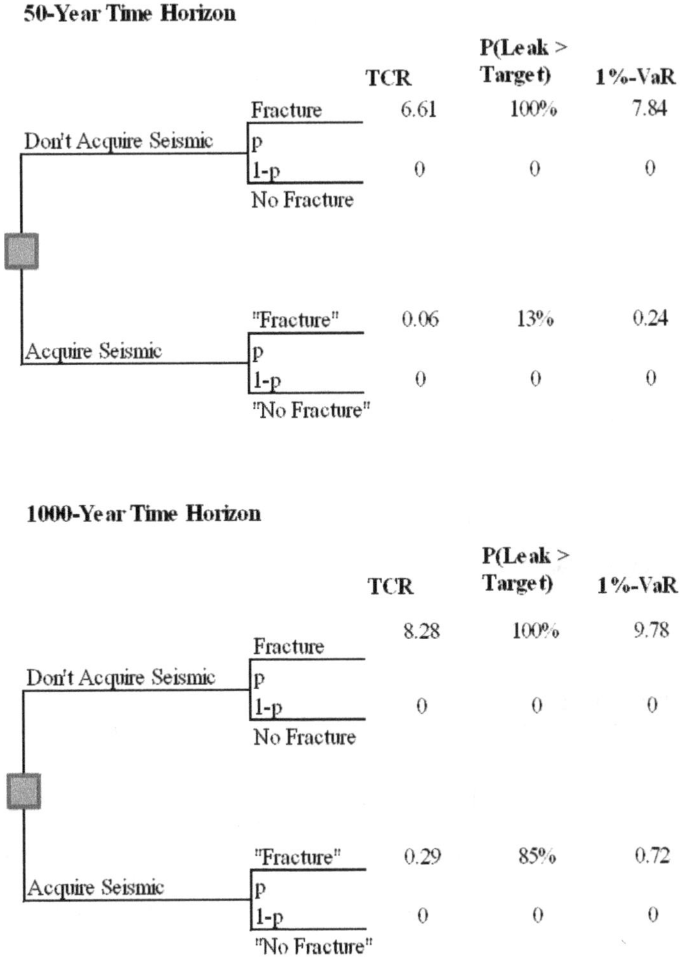

Figure 35. Decision tree to obtain seismic for 50-year and 1,000-year time horizons.

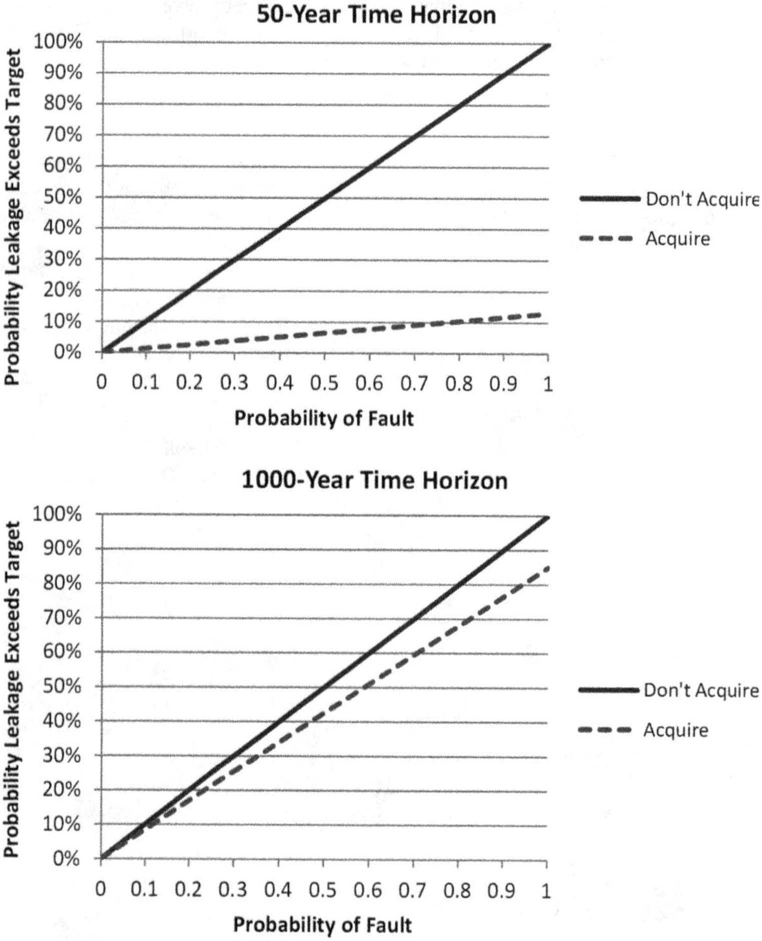

Figure 36. Ability of seismic to help us achieve project objectives.

CONCLUSION

We have demonstrated how the concept of value of information (VOI) can be applied to assess the benefit of monitoring programs in carbon capture and storage (CCS) projects. As shown, the value of these programs depends critically upon the optimization goal and prior beliefs regarding the reservoir.

As suggested in this report, the objective of minimizing the risked leakage is unlikely to quantify properly the risk in CCS projects, where the operator and the public may be concerned with low-probability but high-consequence events. We recommended computing the probability that the project will leak more than its design threshold or defining a probabilistic limit, such as the 1%-Volume-at-Risk (1%-VaR). This is a volume for which there is only a 1% chance that more than this amount would be leaked.

We also stressed the importance of developing quality preliminary assessments and models. The value of any information-gathering program will depend critically upon our beliefs regarding the reservoir before the information gathering program is launched. For example, if we assume that the probability of fractures is very low or zero, then we will not attribute much or any value to learning more about fractures. Learning about the presence of fractures does not add any value when we have already assumed we are sure they do not exist.

Finally, we demonstrated how to use either probabilistic risk assessment or analytic-based modelling as part of our VOI roadmap. Expert elicitation is useful when analytic models are infeasible or costly to build. On the other hand, the analytic model we used in this research was able to generate many interesting insights and allows analysts to perform sensitivity analysis and answer many questions that are likely to be posed by decision makers.

The general structure of any VOI analysis is the same. First, one must specify the activities or technologies that could provide information regarding important uncertainties. Second, the accuracy of this information must be quantified. Third, the decisions that would be altered by the information must be modelled. Therefore, the JV should be able to use the framework detailed here to value new monitoring activities. This could include detection of faults, the monitoring of the CO_2 plume, or any other information that would be helpful in managing the effectiveness and mitigating the risk at In Salah.

LIST OF ACRONYMS

CCS	Carbon Capture and Storage
CO_2	Carbon Dioxide
GHG	Greenhouse Gas Emissions
JIP	Joint Industry Project
MMt	Million tonnes
QRA	Quantitative Risk Analysis
RL	Risked Leakage
TCR	Total Containment Risk
VaR	Volume at Risk
VOI	Value of Information
VOII	Value of Imperfect Information
VOPI	Value of Perfect Information

REFERENCES

1. Bickel, J. Eric, Richard L. Gibson, Duane A. McVay et al. 2008. Quantifying the reliability and value 3d land seismic. Reservoir Evaluation and Engineering 11(5) 832-841.
2. Bratvold, Reidar B., J. Eric Bickel and Hans Petter Lohne. 2009. Value of information in the oil and gas industry: Past, present, and future. SPE Reservoir Evaluation & Engineering 12(4) 630-638.
3. Howard, Ronald A. 1988. Uncertainty about probability: A decision analysis perspective. Risk Analysis 8(1) 91-98.
4. Clemen, Robert T. and Terence Reilly. 2001. Making hard decisions. Duxbury, Pacific Grove, CA.

5. URS. 2008. CO_2 containment and effectiveness risk assessment: unpublished In Salah JIP internal report.
6. URS. 2010. Salah Kretchba project: Geologic storage risk review 2010.
7. Mathieson, Allan, John Midgely, Kevin Dodds et al. 2010. In Salah CO_2 storage JIP: CO_2 sequestration monitoring and verification technologies applied at Krechba, Algeria. The Leading Edge 29(2) 216–222.
8. In Salah CO_2 Storage: Joint Industry Project phase 1 (2006-2010); unpublished In Salah JIP internal report
9. Howard, Ronald H. 2005. Decision analysis manuscript. unpublished.
10. Cox, Louis A., Gerald G. Brown and Stephen M. Pollock. 2008. When is uncertainty about uncertainty worth characterizing. Interfaces 38(6) 465-468.

Carbon Dioxide Capture for Storage in Deep Geological Formations, Volume 4
Karl F. Gerdes (Editor)

Chapter 31

SITE-SPECIFIC OPTIMIZATION OF SELECTION OF MONITORING TECHNOLOGIES

Susan D. Hovorka, Mehdi Zeidouni, Diana Sava, Randy L. Remington, and Changbing Yang
Bureau of Economic Geology, Jackson School of Geosciences, The University of Texas at Austin

ABSTRACT: A diverse suite of monitoring techniques is available for application to CO_2 storage. However, the processes for selecting the suite of techniques and tools for commercial sites are immature. This study explores parts of the technical basis for decision-making processes by which some techniques are selected and others are rejected or placed in reserve through two types of decision-making processes and an initial goal-setting activity. A technology selection activity follows. The goal-setting activity is critical because it determines the expectations to be met by the technology. It is important to recognize that monitoring goals may differ between projects, and that such differences have a major effect on the technologies selected.

The technology selection activities are the focus of this study. The geological conditions at a site have a profound effect on the feasibility and value of each monitoring technology. While the site-specific effect is widely recognized by geologic storage researchers, proponents, and regulators, no systematic overview of how site-specific conditions affect different types of monitoring is available. This study begins to fill the gap.

This study evaluates a small subset of the possible site-specific variability on a small number of possible technologies to illustrate the method of intersecting the geologic variability with monitoring technologies. Technologies evaluated include time lapse 3D seismic detection of CO_2 substitution for brine in sandstone, both pressure and thermal detection of migration out of the injection zone into an above-zone monitoring interval, and geochemical changes for detection of leakage out of the injection zone into fresh groundwater.

KEYWORDS: CO_2 storage monitoring; site-specific design

INTRODUCTION

Geologic storage of captured CO_2 as part of a carbon capture and storage (CCS) project lessens the rate of increase of CO_2 in the atmosphere [1]. Long retention times for CO_2 in the ocean-atmosphere system result in the need for a high standard of retention in storage [2], commonly described as 'assurance of storage permanence.' Storage permanence is created by selecting a natural geologic site that limits migration, and by injecting the CO_2 in such a way that the natural limits to migration are retained or augmented [3,4,5]. Design of injection facilities and an operational plan that is suitable for the site uses one of a number of numerical modelling approaches [6]. Together, the site characterization, facilities, and operational design form the key part of a permit to inject.

In addition to characterization and operational data, authorities may require activities that confirm that injection is operating as planned [7]. These activities are generally referred to as 'monitoring,' and include a number of acronyms with subtle differences in meaning (e.g. MMV;MRV;MVA. Examples of monitoring expectations are provided in [8,9,10,11,12,13].

Most geologic storage regulations and protocols recognize and require that characterization and monitoring plans identify and adapt to geologic variability. One technology selection tool specifies different technology choices for onshore and offshore [14]. However, no examples recommend how to adapt to other variability in injection zone, overburden, and the near surface settings.

SETTING GOALS AND MONITORING

One major variable that impacts the selection of monitoring technologies are the goals that are to be achieved. Regulations may prescribe the technologies and standards under which monitoring is conducted, in which case the topics explored in this study have minimal relevance. However, many evolving geologic storage regulations avoid prescriptive approaches for data collection [e.g.10 and 11]. Accounting frameworks in development also avoid strongly prescriptive protocols for monitoring, leaving open the question of selection of techniques appropriate to the site [9,12,15,16,17].

Risk assessment has been used as a method for optimization with respect to site-specific characteristics. Risk assessment considers the high-level project goals, and through a formal process explores features, events, and processes that might cause the goals not to be attained. For examples, refer to references [18], [19], and [20]. Site-specific characteristics have a strong impact on risk assessment. For example, the possibility that CO_2 could escape from storage might be linked to lateral migration at one site, but to the possibility that a fault is transmissive at another site.

Most guidance is under-constrained in terms of specifying exactly how the risk assessment links to monitoring. Some regulations and protocols discuss matching the observed response of the plume to the modelled response of the plume, and if miss-matched, correcting the model. However, monitoring leads to management and mitigation only if models have been created that delimit the acceptable variability of the system, and define clearly how monitoring observations show unacceptable response is occurring (Table 1). For example, the average saturated thickness of the plume in the injection zone can be observed to be thinner or less continuous at early times than predicted, so that the plume area will be larger than expected. This occurrence can then be mitigated by preparing for a large plume or by changing injection strategy to force the plume to access other parts of the injection formation and become thicker.

This approach only works if the monitoring program is designed to detect unacceptable outcomes, or preferably trends toward unacceptable outcomes before any damage occurs. Modelling that compares acceptable with unacceptable outcomes allows identification and quantification of the thresholds that lead to the unacceptable outcomes. The monitoring plan is then designed to be sensitive to detection of these thresholds. Design of monitoring, using quantitative modelled linkage to risk, has the important advantage of systematically eliminating risk factors, by formally documenting that the design has been effective in eliminating the issue of concern. Reduction of risk and uncertainty then builds a case toward closure of the site.

Examples of risk assessments linked to monitoring are listed in Table 1.

To connect monitoring and risk assessment, a second and less widely discussed assessment must be made. Would the technology shown to be effective in reducing risk at geologic setting C be equally effective in reducing the same risk at geologic setting D? Project organizers and regulators have a tendency to repeat a previous success. However, if the technology is not properly sensitive to the signal, costs of deploying the technology will be lost, the planned assurance will not be obtained, and unacceptable outcomes may not be mitigated in a timely manner. Because geologic variability among geologic settings is high, assessment of the range of conditions over which monitoring is effective must be made.

Table 1. Example cases of risk process linked to monitoring through modelling.

Process	Example Case
Identification of risk.	Flaws in confining system such that fluid will leak upward at unacceptable rate.
Modelling range of acceptable outcomes. Set thresholds that indicate unacceptable performance.	More than a prescribed amount of fluid leaks into an overlying aquifer per year.
Design monitoring to detect key indicators of acceptable and unacceptable response.	Select above-zone monitoring interval and set well spacing and pressure gage sensitivity such that this leakage rate could be detected.
Observe trend toward unacceptable response – mitigate.	Increase of pressure greater than threshold observed. Locate and mitigates flaws in confinement, or set surveillance higher to test attenuation.
Observe no unacceptable trend – risk concern eliminated.	Once confining system integrity is shown, monitoring can stop and site be closed.

PURPOSE OF THIS STUDY

This study considers a few common monitoring technologies in order to evaluate, as quantitatively as possible, how site-specific variability impacts the ability of the monitoring tools to detect a signal. This study cannot fully test either the range of variability among sites or the range of possible monitoring technologies. Rather, this study provides example cases of how the sensitivity of monitoring technologies can be tested against a range of geologic settings. Monitoring plan designers, regulators, and other stakeholders can use the example cases to engage technology specialists to forward-model the sensitivity of the technology array to the signal to be detected.

In order to communicate clearly with the intended audience and bring to the foreground the need to conduct a forward-modelling assessment, the example cases are basic but important. This study advocates for (but is not to be used as a substitute for) detailed site-specific characterization followed by detailed forward-analysis and testing of sensitivity by technology specialists in each candidate monitoring technology prior to selection as part of the monitoring program.

METHODS

Technologies selected for this study include: 1) time-lapse 3D seismic; 2) above zone pressure; 3) above zone-temperature; and 4) ground-water field geochemistry. A small subset of each technology was selected to forward-model the response to illustrate the sensitivity of the technology to the environment. The non-geologic parameters that effect sensitivity are not reviewed.

This study is unique in that it focuses on illustrating the limitations of technologies in order to help guide selections among competing technologies. Optimization approaches (e.g. to combine and jointly invert two technologies) can be used to overcome, to some extent, the limitations shown. All of these technologies have many additional uses as well as potential for advancement through research and development beyond those considered. The case studies can also serve as a model for evaluation of such advanced or optimized technology.

RESULTS

Time-lapse 3D Seismic

Overview of Technology

The major purpose of time-lapse 3D seismic in a geologic storage monitoring program is to detect the change resulting from the replacement of brine in pore spaces by CO_2 and use this signal to interpret the distribution of the injected CO_2 in a rock volume. A baseline survey is conducted prior to emplacement of CO_2. After a period of injection, the survey is repeated as exactly as possible. Subtracting the difference between the surveys then shows changes, which, in the absence of other significant changes, can be attributed to injection. Substitution of CO_2 for brine is one major change caused by injection; change in seismic properties resulting from pressure change may be co-mingled. Further repetitions of the same survey add value in statistical confidence in seismic response and show evolution of the plume. Sonic well logging and vertical seismic profiling are used in concert with a 3D survey to improve time-depth interpretations and rock physics models. Testing and forward-modelling are used commercially to optimize source and receiver selection and placement and minimize noise and other interferences.

The principle values of time-lapse 3D seismic in terms of meeting monitoring goals include 1) mapping the area of the main mass of CO_2, to show that it is behaving as predicted by models; and 2) identifying areas where no change above background shows that measureable amounts of CO_2 have not migrated into these areas. This second use is sometimes thought of as mapping the edges of the detectable plume, as well as showing that no detectable anomalies that are potentially indicative of leakage are developing in zones above the intended injection zone.

Seismic data collection, processing, and interpretation have had heavy investment over many decades, so that the technology tool kit is well developed and highly flexible [21]. Many types of optimization are conducted both in commercial best practices and active research areas. The flexibility and potential for optimization makes assessment of the site-specific sensitivity of time-lapse 3D seismic difficult. However, in the context of this study we illustrate the impact of a few site-specific parameters on seismic detection of changes resulting from injection in a simple case. Similar forward modelling can be used to assess improvements by advanced techniques.

Time-lapse 3D seismic has been highly successful in imaging the CO_2 plume in the Sleipner project beneath the Norwegian North Sea [21]. However, this success brings a risk that stakeholders will expect similar successes to be obtained at other geologic storage sites. The Sleipner project is a near-ideal site for imaging fluid substitution in that:

1. The setting is sub-sea, so that error and noise resulting from on-land collection can be minimized.
2. The plume is shallow, thick, and in a poorly consolidated, high porosity sandstone, so that the velocity contrast where CO_2 has migrated is large.

Another well-documented time-lapse 3D seismic success story is reported from Weyburn field, Saskatchewan, Canada where injected CO_2 was imaged [23,24,25]. Predicted areas of fluid and pressure change along horizontal wells are imaged, as well as some preferential flow paths between the horizontal wells.

A number of other geologic storage projects have attempted to collect time-lapse 3D seismic, but met with less success. Projects in the public domain falling in this category include West Pearl Queen and Nagaoka, Ketzin, In Salah and Cranfield [26,27,28]. In these surveys, a change in time-lapse signals is found where change is expected as result of injection. However, the interpreted signal is not simple to statistically separate from noise. Some projects have created a map of the

CO_2 plume through making a series of assumptions. However, these assumptions typically mean that an unexpected, small, isolated CO_2 accumulations would be discounted as noise, which fails to achieve the second goal of the monitoring.

It is difficult to determine the reason for less successful mapping of the plume, in part because the less successful surveys are not widely or comprehensively reported. End member possibilities for reduced success in clearly separating plume from non-plume include:

- Less-than optimal data collection (human limit to technology).
- Less-than optimal data processing (human limit to technology).
- Change signal too weak to be successful detected above noise (site-specific geological limit).

In the first two cases, it is reasonable for a regulator or significant stakeholder to require that the project optimize techniques to meet the project goals. However, in the third case, such a demand is futile, and the project developer would need to seek alternative technologies (possibly including improved seismic methods) for accomplishing the project goals. The rapid advancement of seismic technologies has combined with the research nature of most existing geologic storage projects to confuse the issues limiting seismic detection. It is always possible to collect additional data or conduct additional processing to improve a time lapse survey [29]. However, it may not be possible with existing technologies to improve sufficiently to meet the monitoring goals.

One project that formally explained why the change in seismic velocity resulting from CO_2 injection was too weak to be detected is the Otway project's first injection [30,31]. In this project, gas saturation existing in the reservoir before placement of CO_2 was identified as a major factor in weakening the time-lapse signal. Non-repeatability and noise were also formally assessed [32,33].

In the site-specific rock physics assessment of seismic attribute sensitivity to CO_2 saturation conducted for this study, a small set of idealized sites is modelled to determine the factors leading to diminished signal from placement of CO_2 in a brine-filed reservoir considering both change in P-wave amplitude and P-P reflectivity contrast between the reservoir and the overlying lowest part of the confining system (referred to as 'caprock' or 'seal'). In design of a seismic survey, a much more comprehensive study should be conducted to determine the change in signal that could be extracted from the change of fluids in the reservoir. This would include optimization of the sources and receiver placement and characteristics and consideration of a wide variety of advanced processing options. An assessment of noise and sources of error can be used to assure that the predicted signal can be detected above noise on the site and repeatability is suitable for time lapse processing. If noise is too high or repeatability too low, a number of approaches can be considered to improve them.

Site-specific Assessment

The objective is to evaluate the seismic attribute sensitivity to CO_2 saturation in geologically realistic scenarios. Quantitative links are built from rock physics theories for elastic properties of porous media partially saturated with CO_2.

Sonic logs of injection zone-caprock pairs from real sites which have been injected with CO_2 were modelled to determine sensitivity to seismic P-wave velocity change and PP reflectivity change from 0 to 20% CO_2 in the pores. This provided confidence that the modelling approaches are reasonable. Then, three artificial cases were developed with different compressibility to assess attenuation of seismic P-wave velocity change and PP reflectivity change with compressibility as it decreases with depth. The third set of experiments varied the thickness of the CO_2 plume in the injection zone.

Well-log Derived Models

The sensitivity of seismic P-wave velocities and PP reflectivity to CO_2 saturation was assessed using well-log data from a shallow, poorly consolidated sandstone (Frio sandstone of South Liberty field, Texas) and a deep, well-consolidated sandstone (Cranfield Field, Mississippi).

A simplified model with average elastic properties derived from the available well-log data was created for each example. Because the initial elastic properties of the sequestering formation (P and S-wave velocities, bulk density, porosity, bulk modulus of the mineral) and the pore fluid properties (bulk modulus and density) at reservoir conditions are known, the Gassmann theory can calculate the elastic properties of the rock containing various ratios of fluids whose elastic properties are also known [34}. This theory assumes that the shear modulus of the rock does not change with fluid property change, only the bulk modulus - that is, fluids do not support shear strength. In this study we concentrate on the P-wave velocity examples; other types of analysis considering the effect of fluid substitution on S-wave velocity are possible. The fine scale mixing assumption between brine and CO_2 and Reuss (1929) average to determine the bulk modulus of the two-phase fluid mixture are used in order to predict the P and S-wave velocities for variable CO_2 saturation [35].

Case 1: Representative Real Sandstone at shallow depth (Frio Formation)

Figure 1 shows well log data for the Frio Sandstone acquired before the CO_2 injection, so the sandstone is fully brine saturated. The injection zone between 5,045 ft and 5,115 ft. The relatively low P and S-wave velocities and large V_P/V_S ratios indicate a poorly consolidated sandstone. The caprocks are shales; sandstone and shale caprocks have similar elastic properties. Well-log data was used in creation of a simplified two-layer model of the caprock and the Frio sandstone. Mineralogy, porosity and reservoir conditions for case 1 is from previous work by Sakurai *et al*., Hovorka *et al*., and Daley *et al*. [36,37,38].

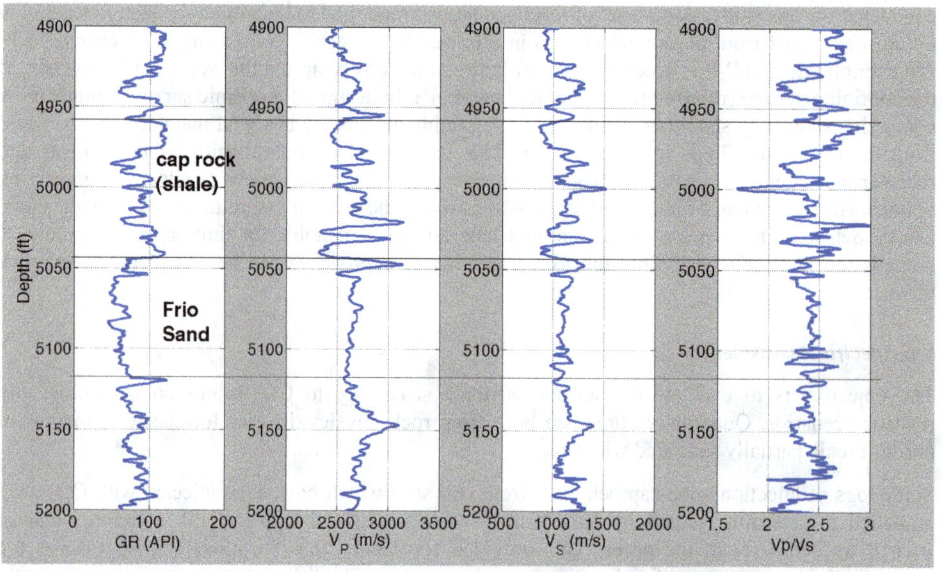

Figure 1. Well-log data (brine –saturated conditions) from Frio Sand: Gamma Ray (API), P-wave velocity (m/s), S-wave velocity (m/s), VP/VS ratio.

The elastic properties needed for fluid sensitivity analysis for the Frio sandstone are presented in Table 2, Case 1. Frio sandstone has orthoclase minerals in its composition, which make Kmin larger than the one corresponding to pure quartz. Gassmann theory is used to predict seismic velocities for different CO_2 saturation conditions [34]. The in-situ elastic properties of the two fluid phases, brine and CO_2, are presented in Table 3, Case 1.

Table 2. Average sandstone properties used in model.

Case	Model from	V_P (m/s)	V_S (m/s)	Bulk density (kg/m^3)	Porosity (%)	Kmin (GPa)
1	Frio-log based	2,663	1,114	2,143	29	44
2	Cranfield	3,486	1,991	2,185	23	23
3	Compressible sandstone (theoretical)	2,280	901	2,165	30	37
4	Medium compressible sandstone (theoretical)	2,776	1,443	2,164	30	37
5	Less compressible sandstone (theoretical)	3,226	1,850	2,164	30	37

Table 3. Sandstone bulk modulus (GPa) and density (kg/m^3) for brine and CO_2 at *in situ* 55°C and 15 MPa used in models

Case	Model from	Brine Bulk modulus (GPa)	Brine Density (kg/m^3)	CO_2 Bulk modulus (GPa)	CO_2 Density (kg/m^3)
1	Frio log based	2.73	1,033	0.143	651.85
2	Cranfield	2.86	1,035	0.383	826
3	Compressible sandstone (theoretical)	2.67	1,028	0.1776	699.85
4	Medium compressible sandstone (theoretical)	2.76	1,023	0.244	736.7
5	Less compressible sandstone (theoretical)	2.81	1,020	0.33	789.7

Results for the fluid substitution calculations for P-wave velocity are presented in Figure 2. The fine-scale mixing assumption (Reuss average) implies that just a few percent CO_2 in the pores can significantly lower the P-wave velocity of the sequestering formation if the sandstone is poorly consolidated and has a large fluid sensitivity. For 20% CO_2 saturation in pores, the P-wave velocity dropped from 2,663 m/s to 2,196 m/s, equivalent to 467 m/s change. This is a large decrease in P-wave velocity that can be monitored seismically, and fits the observed response in the Frio test injection.

The seismic PP reflectivity from the top of Frio sandstone can be calculated using Zoeppritz (1919) equations, assuming a two-layer model [39]. The PP reflectivity represents an interface attribute and depends on both the elastic properties of the sequestering formation (Frio sandstone) and on the elastic properties of the caprock. The caprock properties required for Zoeppritz calculations are extracted as average values from the well-log data (Figure 1) and are presented in Table 4, Case 1.

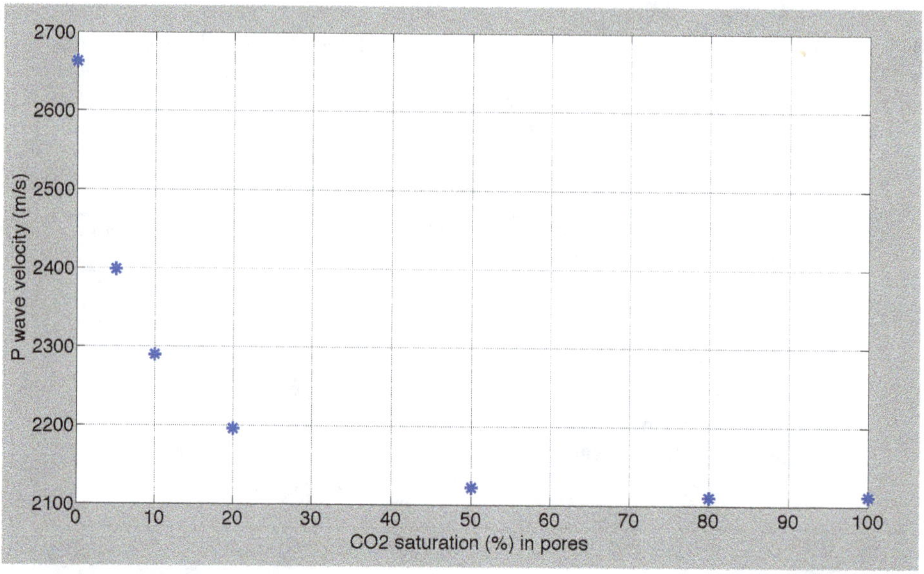

Figure 2. P-wave velocity calculations as a function of CO_2 saturation in pores using Gassmann Equation for the Frio Sandstone model (Table 2, case 1, Table 3, case 1).

Table 4. Caprock properties for models

Case	Model from	V_P (m/s)	V_S (m/s)	Bulk density (kg/m^3)
1	Frio Log based	2,611	1,077	2,224
2	Cranfield deep sandstone	4,299	2,455	2,334
3	Compressible sandstone (theoretical)	2,349	979	2,187
4	Medium compressible sandstone(theoretical)	2,859	1,429	2,186
5	Incompressible sandstone(theoretical)	3,322	1,846	2,185

Figure 3 shows the PP reflectivity at top Frio sandstone as a function of incidence angle for different CO_2 saturations in pores.

The elastic properties of the Frio sandstone and those of the caprock are very similar (compare Table 2 with Table 4). Therefore, the PP reflectivity for the 100% brine saturated Frio sandstone model is very small (0.0087). However, due to the large fluid sensitivity of the Frio sandstone, just a few percentages of CO_2 in pores decrease the reflectivity at the top Frio sandstone significantly. For 20% CO_2 saturation in pores, the PP reflectivity at normal incidence (0 degrees angle of incidence) drops to -0.1098, which is more than one order of magnitude smaller than the PP reflectivity for 100% brine saturated Frio sandstone. This decrease can be easily detectable with seismic technology, as shown in Figure 4.

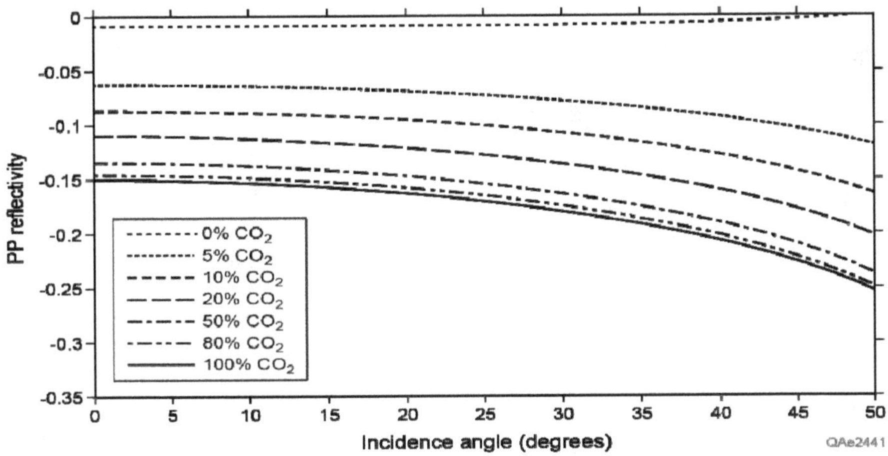

Figure 3. PP reflectivity as a function of incidence angle for different CO_2 saturations (Frio sandstone model). Calculations are performed using Gassmann Equation for fluid substitution and Zoeppritz Equations for PP reflectivity. The caprock properties are kept fixed (Table 4, Case 1).

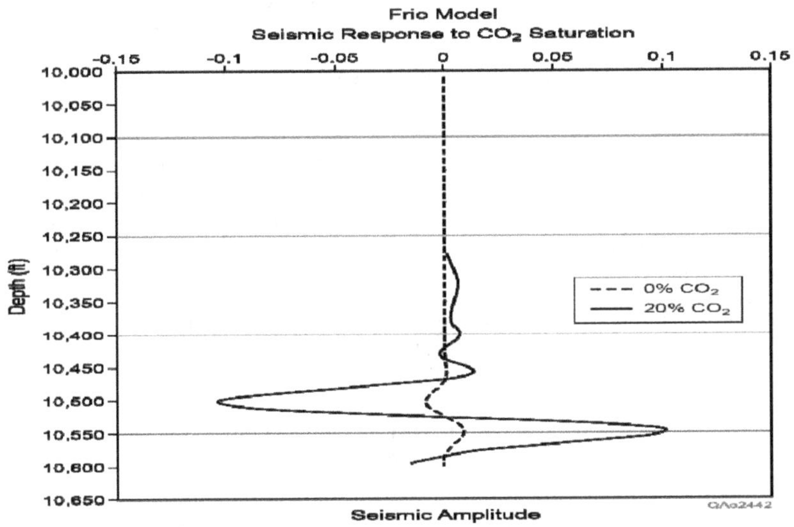

Figure 4. 1D Synthetic model showing seismic amplitude for different CO_2 saturations (Frio sandstone model).

Case 2: Representative Deep Sandstone (Tuscaloosa Formation)

The second example is a 10,500 ft deep sandstone from the Tuscaloosa Formation at Cranfield Field, Mississippi. This sandstone is particularly interesting because it has a large average porosity (23%) for its deep burial depth. Figure 5 presents the well-log data from this site for 100% brine saturated rocks. P and S-wave velocities are larger than those of the Frio sandstone, while the V_P/V_S ratio is smaller, indicating a consolidated sandstone. In this case, the caprocks have larger P and S-wave velocities than the sequestering formation.

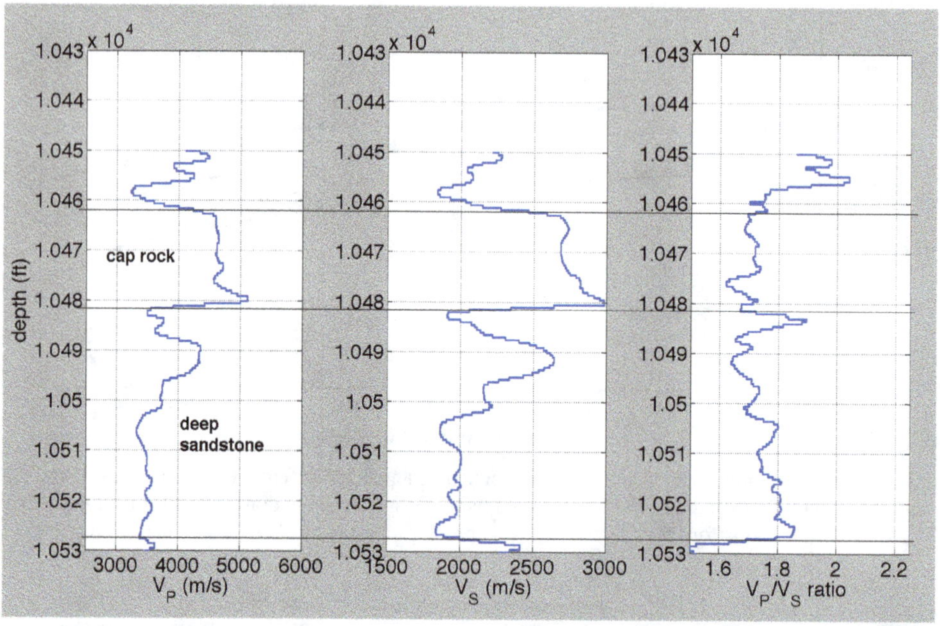

Figure 5. Well-log data from the deep well-consolidated sandstone: P-wave velocity (m/s), S-wave velocity (m/s) and VP/VS ratio.

The elastic properties needed for fluid sensitivity analysis of the deep sandstone are presented in Table 2 Case 2. This sandstone has clay minerals in its composition, and the bulk modulus is smaller than that of pure quartz. Gassmann theory is again used to predict seismic velocities for different CO_2 saturation conditions [34]. The in-situ elastic properties of the two fluid phases, brine and CO_2, are presented in Table 3 Case 2.

Results for the fluid substitution calculations for P-wave velocity for the deep sandstone are presented in Figure 6. For 20% CO_2 saturation in pores, the P-wave velocity dropped from 3,486 m/s to 3,313 m/s, a 173 m/s change. This decrease is significantly smaller than the equivalent P-wave velocity decrease of 467m/s for Frio sandstone. The deeper sandstone has smaller porosity, is better consolidated, and less compressible than the Frio sandstone. Therefore, seismic methods are less successful in monitoring CO_2 migration for the deep sandstone than for the Frio sandstone. However, 173 m/change can still be detectable with good quality seismic data.

We can also compute the seismic PP reflectivity from the top of the deep sandstone using Zoeppritz equations, assuming a two-layer model. However, Zoeppritz results are representative for thick intervals, and they do not take into account the thin-layers effects relevant to this site. Nevertheless, we use Zoeppritz equations to understand what the PP reflectivity would look like if the deep sandstone was thicker. The caprock properties required for Zoeppritz calculations are extracted as average values from the well-log data (Figure 5) and are presented in Table 4 Case 2.

Figure 7 shows the PP reflectivity at top of the deep sandstone as a function of incidence angle for different CO_2 saturations in pores.

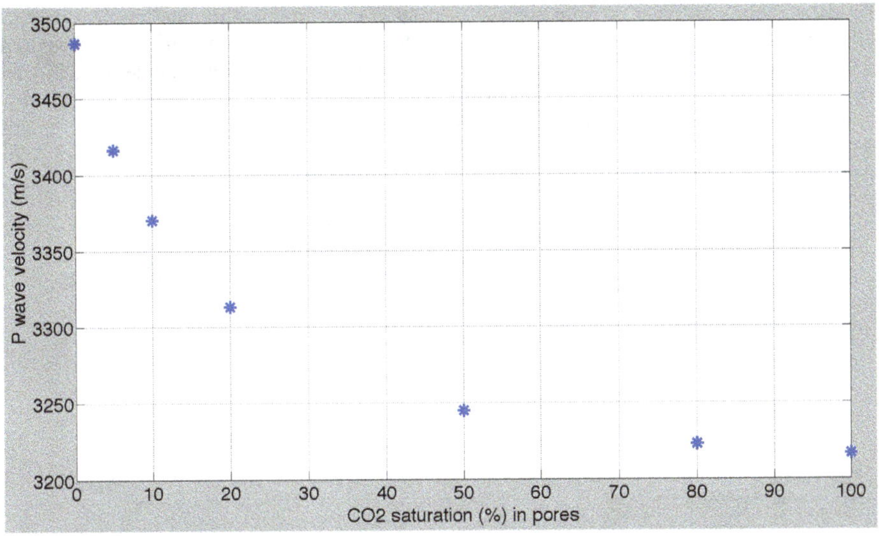

Figure 6. P-wave velocity calculations as a function of CO_2 saturation in pores using Gassmann Equation (deep sandstone model).

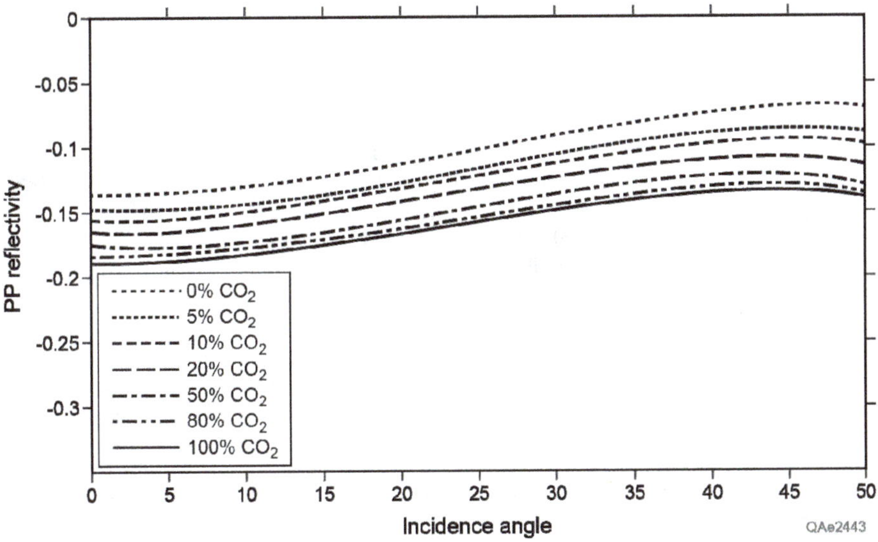

Figure 7. PP reflectivity as a function of incidence angle for different CO_2 saturations (deep sandstone model). Calculations are performed using Gassmann Equation for fluid substitution and Zoeppritz Equations for PP reflectivity. The caprock properties are kept fixed (Table 4, Case 2).

In this example, the caprocks are significantly stiffer than the sandstone, and the PP reflectivity for the 100% brine-saturated sandstone is much lower than 0. At normal incidence (0 degrees angle of incidence), this PP reflectivity is - 0.137. Once the CO_2 is injected in the pores, the PP reflectivity decreases. However, since this sandstone has lower porosity than the Frio sandstone, is well consolidated, and is not very compressible, the decrease in PP reflectivity due to CO_2 is not large. For example, for 20% CO_2 saturation in pores, the PP reflectivity at normal incidence has dropped from -0.137 to only -0.1635. This change in reflectivity is relatively subtle to be detected seismically. In addition, there is also the challenge related to the thickness of the sandstone, which is smaller than the typical seismic wavelengths. The 1D synthetic model showing seismic amplitude shown in Figure 8 illustrates the smaller amplitude to change as a result of replacement of brine by 20% CO_2.

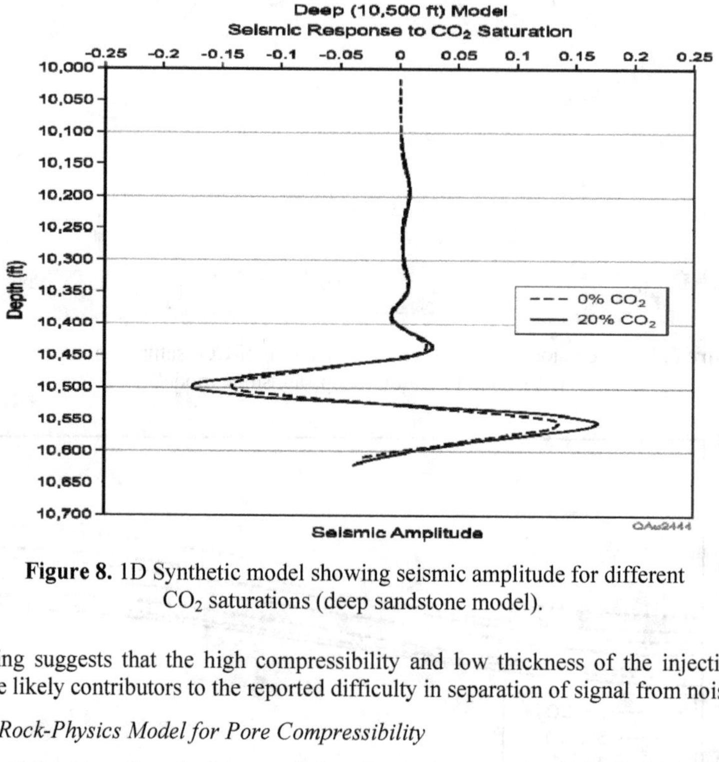

Figure 8. 1D Synthetic model showing seismic amplitude for different CO_2 saturations (deep sandstone model).

The modelling suggests that the high compressibility and low thickness of the injection zone at Cranfield are likely contributors to the reported difficulty in separation of signal from noise.

Theoretical Rock-Physics Model for Pore Compressibility

To further widely extrapolate the impact of the effective pressure (and depth), as well as of pore compressibility on seismic fluid sensitivity, a theoretical rock-physics model was developed using the self-consistent approximation [40]. This model allows us to account for the changes that occur within the dry mineral frame due to variation in effective pressure. The model assumes idealistic ellipsoidal pore shapes with different aspect ratios. Pores with small aspect ratios are considered soft or compressible and are easily closed when applying pressure. Those with large aspect ratios are considered incompressible. At lower effective pressure, the compressible, crack-like pores are open, making the rocks softer and more fluid-sensitive to seismic waves.

To model a rock, a volumetric distribution of ellipsoidal pores with various aspect ratios is assumed. The distribution of the aspect ratios can be determined by calibrating the velocity-pressure dependence with laboratory measurements on rock samples from the sequestering formation. When such measurements are not available, analogues can be used from laboratory data on velocity versus effective pressure. Figure 9 presents P-wave laboratory measurements performed on a dry sample as a function of effective pressure.

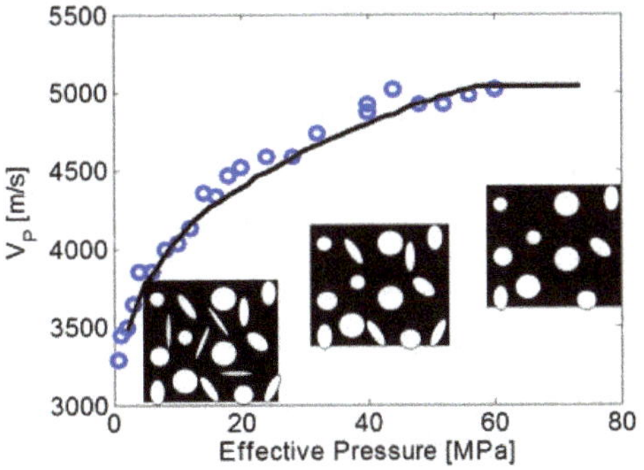

Figure 9. Laboratory measurements performed on dry core sample (circles): P-wave velocity as a function of effective pressure.

Data courtesy of Chris Purcell and William Harbert of University of Pittsburgh. Superimposed as a curve are the results for P-wave velocity of the dry mineral frame as a function of effective pressure calculated using Berryman (1995) self-consistent approach. The boxes represent schematically the dry mineral frame: the black regions represent the mineral and the white ellipsoids represent the pores with various aspect ratios.

The illustrations in Figure 9 represent schematically the dry mineral frame: the black regions correspond to the mineral and the white ellipsoids correspond to the pores with various aspect ratios. The P-wave velocity increases with effective pressure due to the stiffening of the mineral frame, caused by closing of the crack-like pores (pores with low aspect ratios). The shape of the velocity pressure dependence is related to the shape (or aspect ratio) of the cracks that close at a given effective pressure. The fraction of porosity that is pressure dependent is very small, only a few percentages of the total porosity. However, this fraction has a strong impact on velocity-pressure dependence and on the seismic fluid sensitivity.

Superimposed on the data (circles) in Figure 9 are the modelling results (curve) for P-wave velocity as a function of effective pressure, using Berryman self-consistent approach. A good agreement can be observed between the modelling results and the laboratory measurements. This example suggests that, even though the model assumes idealized pore shapes, it can capture the physical behaviour of the P-wave velocity as a function of effective pressure.

Theoretical Examples: 30% Porosity Sandstone

The theoretical model used in this study represents a quartz sandstone with 30% porosity. The model started with a uniform distribution of aspect ratio of the pores that would close gradually with the increasing effective pressure from around 18 MPa to 36 MPa. These effective pressure values may correspond to different depths ranges, depending on the specific environments. However, a rough estimate may be from around 5,000 ft for the low effective pressure value, up to about 10,000 ft for the high effective pressure value. This depth range is used based on the well-log data examples presented in the previous section.

Using Berryman self-consistent approximation, the P-wave velocity pressure dependence for the theoretical 30% porosity sandstone is calculated. The modelled distribution of compressible pores is displayed in Figure 10. As the crack-like pores are closing with the increasing effective pressure, the P-wave velocity increases as well. This increase is more than 1,000 m/s over the whole effective pressure range. Based on this example, three different cases were chosen and are graphically represented with symbols on the P-wave pressure dependence curve (Figure 11).

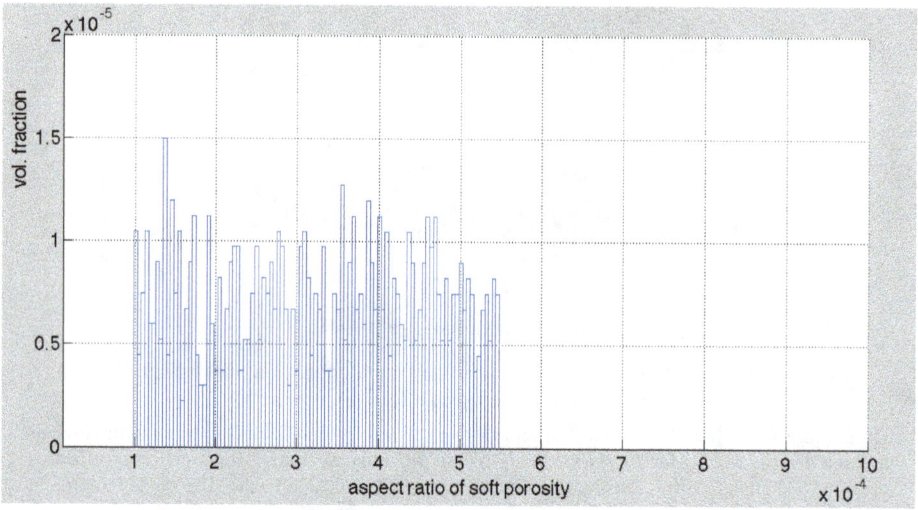

Figure 10. Initial distribution for the volumetric fractions of aspect ratios for the compressible, pressure-dependent pore space of the theoretical sandstone model.

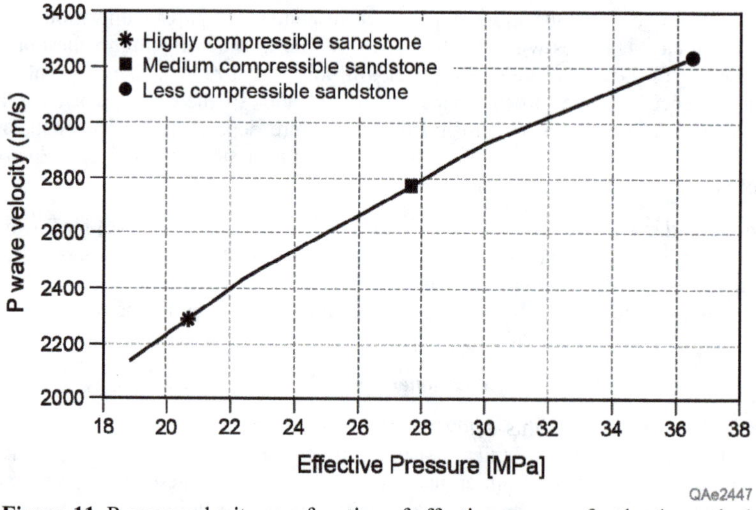

Figure 11. P-wave velocity as a function of effective pressure for the theoretical 30% porosity sandstone for three cases using self-consistent approximation. The theoretical sample is 100% brine saturated.

Case 3: Theoretical Highly Compressible Sandstone

At low effective pressure (star symbol in Figure 11) the 30% porosity sandstone has low velocity because most of the compressible pore space is open. This sample is poorly consolidated, and like the Frio sandstone presented in the previous section, has high seismic sensitivity to fluids. The Gassmann theory using the inputs on Table 2, Case 3 and Table 3, Case 3 quantifies this fluid sensitivity.

Results for the fluid substitution calculations for P-wave velocity from the highly compressible sand are presented in Figure 12. For 20% CO_2 saturation in pores, the P-wave velocity dropped from 2,280 m/s to 1,725 m/s, equivalent to 555 m/s change. This is a large decrease in P-wave velocity that can be successfully monitored seismically.

We also analyse the PP reflectivity response at top of the highly compressible sandstone, assuming a caprock with the elastic properties presented in Table 3, Case 3. Zoeppritz Equations are used for modelling the PP reflectivity response with the assumption of half-space modelling (both caprock and sequestering formation are thick in comparison with the seismic wavelength). We consider two cases. Figure 13 shows that the PP reflectivity for the 100% brine saturated sandstone is small, equal to 0.0198 at normal incidence (0 degrees angle of incidence).

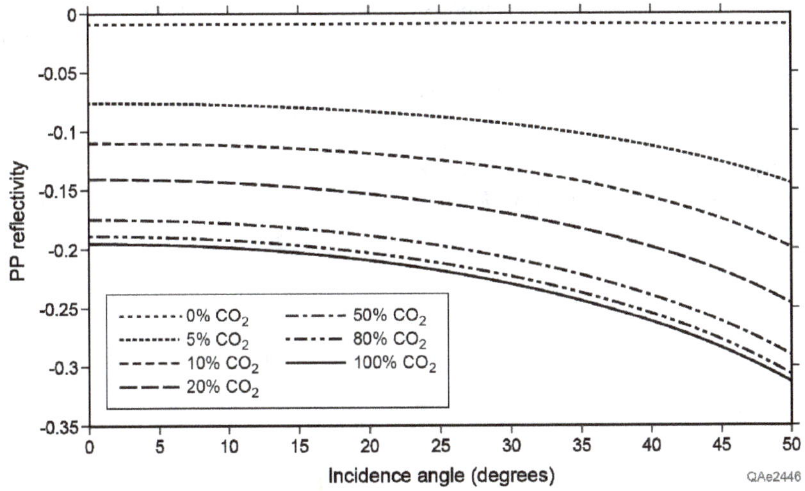

Figure 12. P-wave velocity calculations as a function of CO_2 saturation in pores using Gassmann Equation for the theoretical model of highly compressible sand (Table 2, Case 3).

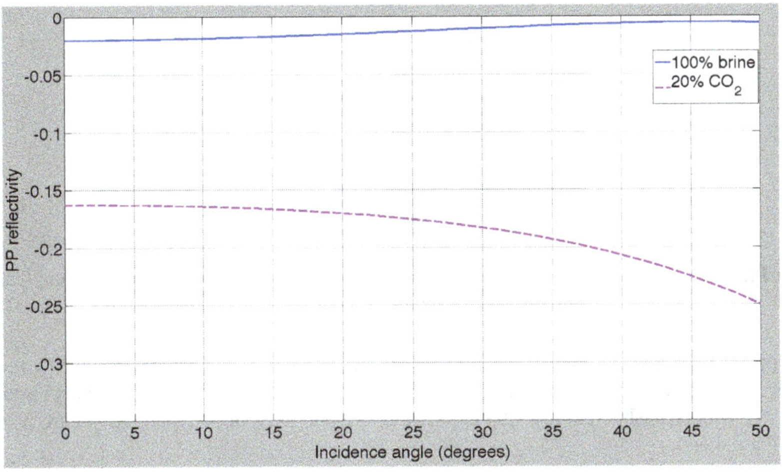

Figure 13. PP reflectivity results at top of the highly compressible sandstone.

For 20% CO_2 saturation in pores, the PP reflectivity at normal incidence drops to -0.163, which an order of magnitude smaller than the PP reflectivity for 100% brine saturated sandstone. This decrease can be easily detected seismically.

Case 4: Theoretical Medium Compressible Sandstone

The medium compressible sandstone has a smaller fraction of compressible pores, and the aspect ratios of the ones left are larger, since the smallest aspect ratio pores have been closed due to increasing effective pressure. Therefore, the P-wave velocity of the sandstone is larger. We again use Gassmann fluid substitution to derive the P-wave velocity of the sandstone partially saturated with CO_2.

The elastic properties of the medium compressible sandstone are presented in Table 2, Case 4 and bulk modulus and density in Table 3, Case 4. As expected, the results of Gassmann fluid substitution show the P-wave velocity decrease with CO_2 saturation, as displayed in Figure 14. For the medium compressible sandstone, this decrease is smaller than for the highly compressible sandstone, as expected. P-wave velocity dropped from 2,776 m/s for 100% brine saturated sandstone to 2,429 m/s for 20% CO_2 saturation in pores. This is equivalent to 346 m/s P-wave velocity change, which is seismically detectable. Therefore, CO_2 migration can be successfully imaged using seismic methods in the 30% porosity sandstone with medium compressibility.

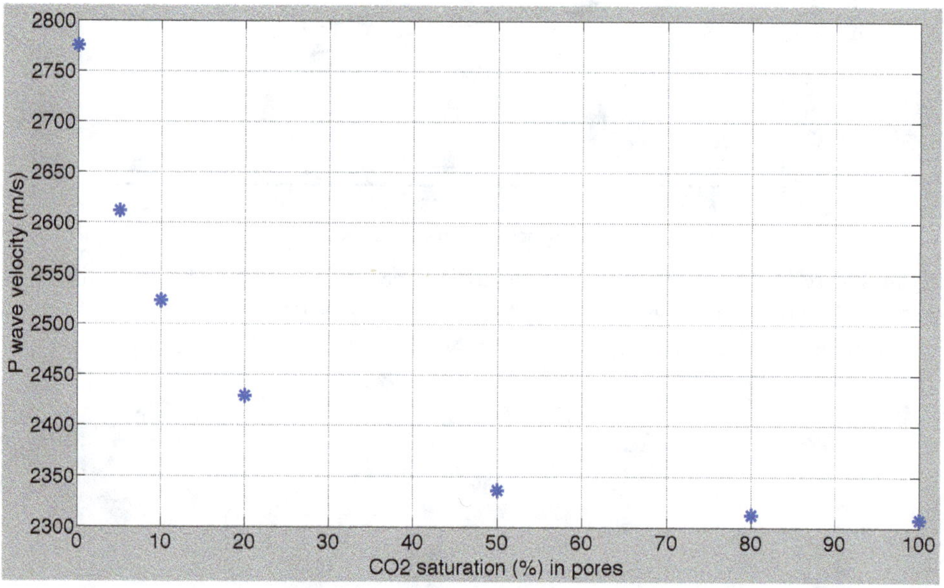

Figure 14. P-wave velocity calculations as a function of CO_2 saturation in pores using Gassmann Equation for the theoretical model of medium compressible sand (Table 4, Case 4).

The Zoeppritz Equation results for seismic PP reflectivity at the top of the medium compressible sandstone are presented in Figure 15, both for 100% brine saturated rock (solid curve) and for 20% CO_2 saturated rock (dashed curve). The caprock properties used in the calculations are presented in Table 5, Case 4.

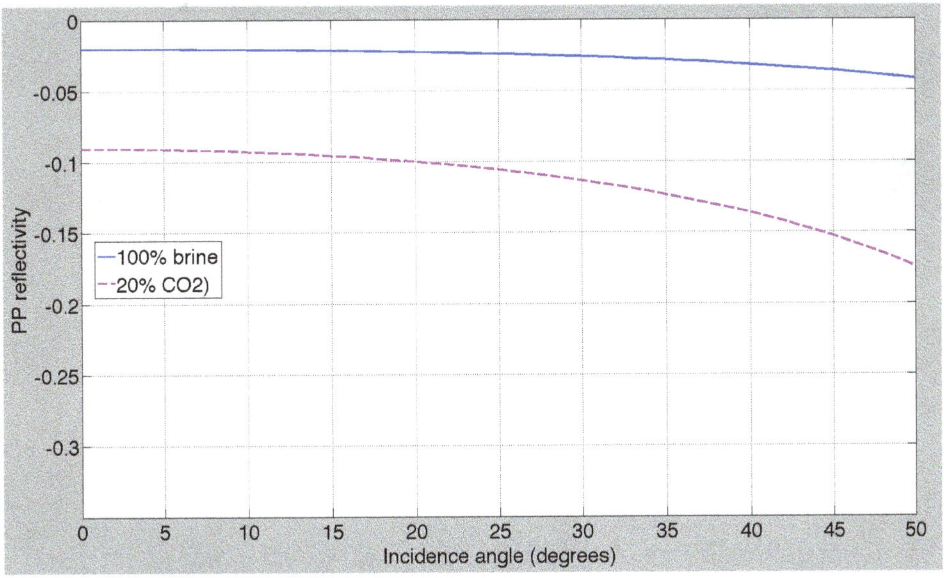

Figure 15. PP reflectivity results at top of the medium compressible sandstone. Caprock properties are kept fixed (Table 5, Case 4).

The PP reflectivity at normal incidence dropped from -0.0198 for 100% brine saturated case to -0.09 for 20% CO_2 saturated sandstone. Even though the decrease in PP reflectivity due to the presence of CO_2 is not as large as for the highly compressible sandstone, it is still seismically detectable.

Case 5: Theoretical Less Compressible Sandstone

The less compressible sandstone has most of the highly compressible pores closed due to the increasing effective pressure (greatest depth). Therefore, the P-wave velocity of the sandstone is the largest of the three cases considered in this analysis. Similar to cases 3 and 4, calculated elastic properties of the less compressible sandstone are and presented in Table 2, Case 5 and the elastic properties of the fluids are presented in Table 3, Case 5.

Figure 16 shows that as expected, for the less compressible sandstone the decrease in P-wave velocity due to the presence of CO_2 is not large. With 20% CO_2 saturation in pores, the P-wave velocity dropped only 218 m/s, from 3,226 m/s to 3,008 m/s. Nevertheless, this decrease may still be seismically detectable with excellent quality seismic data.

The results from the Zoeppritz Equation for seismic PP reflectivity at the top of the less compressible sandstone at 0 and 20% CO_2 saturation are presented in Figure 17. The caprock properties used in the calculations are presented in Table 5, Case 5. The PP reflectivity at normal incidence dropped from -0.0198 for the 100% brine saturated case to -0.0577 for the 20% CO_2 saturated sandstone. This decrease in PP reflectivity due to a relatively small presence of CO_2 is relatively small and more challenging to detect. However, excellent quality seismic data may still be able to map fluid changes using PP reflectivity for this high porosity, but less compressible sandstone.

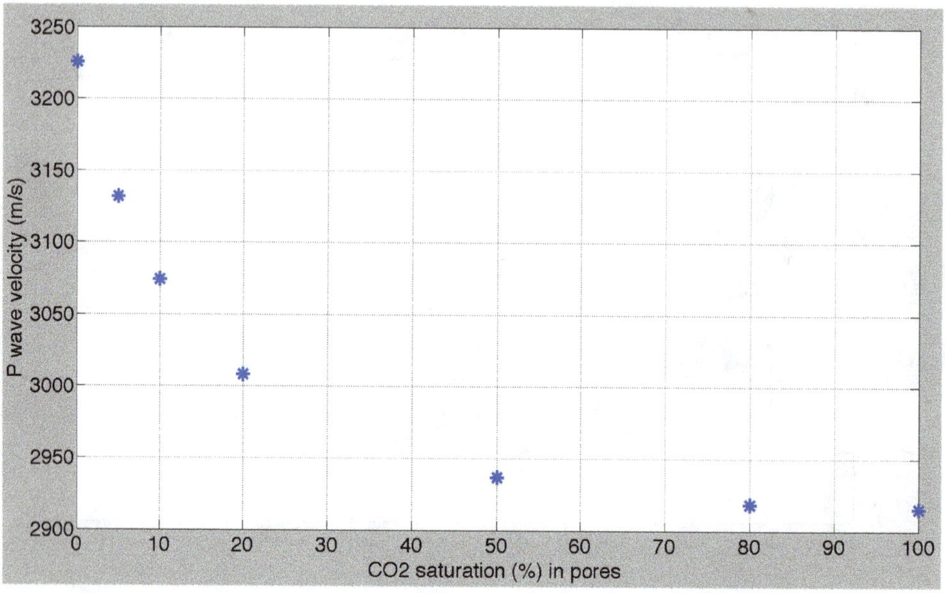

Figure 16. P-wave velocity calculations as a function of CO_2 saturation in pores using Gassmann Equation for the theoretical model of less compressible sand (Table 4, Case 5).

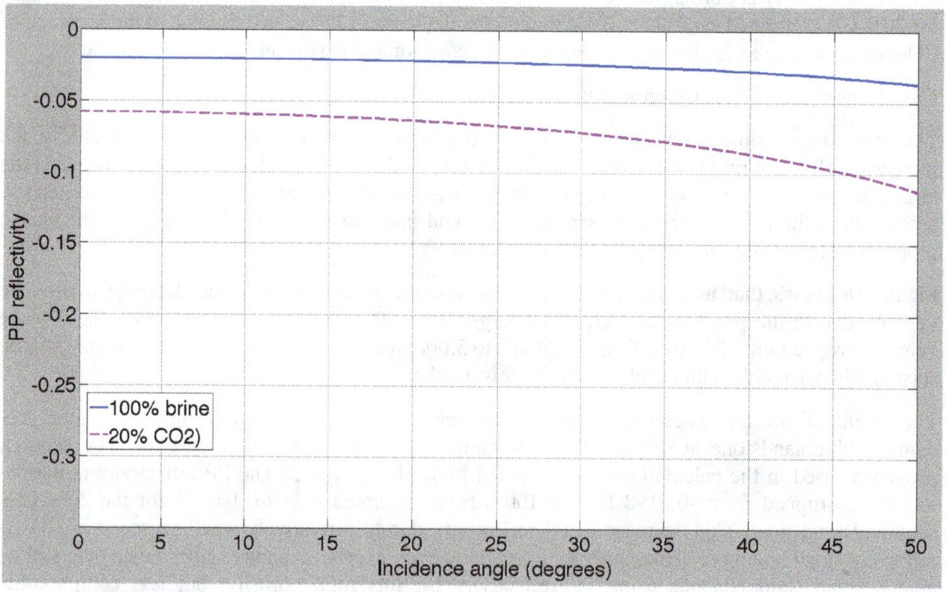

Figure 17. PP reflectivity results at top of the less compressible sandstone. Caprock properties are kept fixed (Table 5, Case 5).

Comparison of the sandstone models with different degrees of pore compressibility

The theoretical modelling based on the self-consistent approximation shows that the compressibility of the pore space plays a very important role on seismic fluid sensitivity. The model considered in this analysis has a large porosity equal to 30%. However, the ability of seismic methods to monitor CO_2 migration in the sequestering formation varies with the degree of pore compressibility. At low effective pressures (and shallower depths) the seismic fluid sensitivity is very large. However, as the effective pressure increases, the crack-like pores with low aspect ratio close and the rock frame stiffens, such that the fluid sensitivity decreases significantly. Since this example uses a model with large porosity, the absolute change in P-wave velocity even for the less compressible sandstone is likely still detectable with excellent quality seismic data, however the risk of signal being obscured by noise increases.

Seismic sensitivity to time-lapse changes in CO_2 saturation at various reservoir thicknesses

The next element of modelling using the rock physics model to predict the limits of detection focused on thickness of the injection zone. The series of models with decreasing bed thickness were conducted using data from the petrophysical properties from Case 1 Frio model shown in in Table 6.

Table 6. Properties of the Frio sandstone model.

100% Brine sandstone (0% CO_2)	**20% CO_2**	**Caprock**
V_P_0%CO_2=8,737.41 ft/s	V_P_20%CO_2=7,206.13 ft/s	V_P_cap=8,567 ft/s
V_s_0%CO_2=3,654.06 ft/s	V_s_20%CO_2=3,673.05 ft/s	V_s_cap=3,533.88 ft/s
Density0%CO_2=2.14 g/cc	Density20%CO_2=2.12 g/cc	Density_cap=2.22 g/cc

The original model was run using a reservoir thickness of 50 ft. Logs were edited to test seismic response to doubled thickness (100 ft) as well as decreased thickness to 3 ft and a 1D forward model run for each log. Note the decrease in amplitude scale with decreasing thickness on Figure 18.

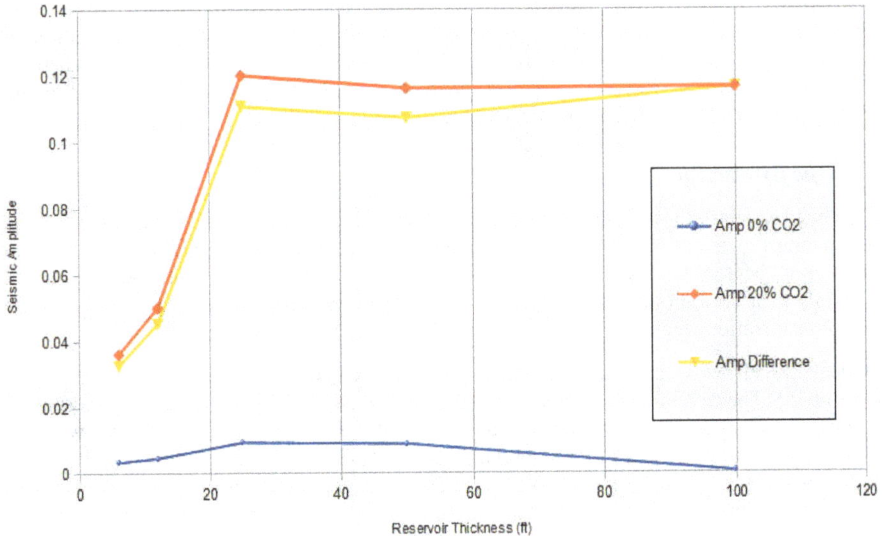

Figure 18. Maximum amplitude response for varying reservoir thickness.
For moderate and thick reservoirs, the difference between saturation conditions shows that the detectability increases with thickness.

In cases of reservoir thickness greater than 20 ft, using the Frio sandstone properties, increase in CO_2 saturation is easily detected by seismic. Below 20 ft, seismic amplitude response drops dramatically, decreasing the difference in brine filled and CO_2 saturated reservoir seismic response.

Above-Zone Pressure Monitoring (AZMI)

Overview of Technology

The purpose of collecting pressure data above the intended injection zone is to test the function of the confining system in limiting vertical fluid migration. The confining system is a sequence of layered strata that overlay the injection zone. Low permeability layers within the confining system limit vertical flow. Transmissive strata within the confining system correct for any minor flaws, such as microannuli through well cement sheaths that penetrate the lower part of the confining system, because upward fluid migration of higher pressure brine or buoyant, low viscosity CO_2 that contacts transmissive strata will bleed off into that layer [41]. In this model, vertical flow will only be able to continue upward by building pressure and accumulating CO_2 saturation in each zone within the confining system. By making pressure measurements at one or more of these transmissive strata designated as above-zone monitoring intervals (AZMI), the attenuation function of the confining system can be assed to determine if it is adequate. AZMI monitoring has been previously deployed at gas storage sites; however, it has only sparsely been tested for geologic storage.

Characterization of the prospective AZMI is important to establish that the pressure signal obtained will meet the fluid migration detection goal. The analysis conducted considers the impact of thickness, permeability, and distance above the injection zone, focusing on leakage occurring as flow focused along a flow path such as a flawed well completion or a transmissive fault.

The hydrologic connectivity between the leakage path and the AZMI, and between the AZMI and the pressure gage are other factors that need to be considered. If the leakage path is hydrologically isolated (e.g. leakage is occurring inside a well casing that is intact across the AZMI) the signal will be isolated and the monitoring ineffective. Additional study is needed to define other types of pressure signals, such as diffuse flow across low permeability units, which over a large area could be significant [42], and the hydrologic response of the AZMI to a geomechanical signal from the injection zone [43]. The extent to which the boundary conditions of an AZMI are hydrologically open or closed has a strong impact on its hydrologic function.

Site-specific Assessment

Above-zone pressure monitoring will be successful in detecting leakage only if the pressure change corresponding to leakage exceeds a measurement threshold. The leakage rate may be too small or the leak may be too far away causing pressure change at the observation location to not be successfully measured. For this assessment, we define a detectable pressure change as 0.1 psi. Site-specific considerations including noise and trend can change this threshold.

Analytical models were used to evaluate the pressure change in response to leakage into the AZMI [44,45]. The model is simplified to single-phase flow, as it is intended to probe sensitivity. The situation assessed is where the leakage into AZMI is outside the two-phase region, which is the case most useful for early detection. Should a pressure change be detected, additional assessment is needed which can explore cases of two-phase flow. The model assumes the most general case of infinite-acting boundary conditions to the aquifer. Based on this model, leakage through a leak located at distance R from the injector and distance ρ from the observation well causes leakage rate and pressure change given by:

$$P_D = \frac{1}{(1+T_D)}\left(-\frac{\gamma}{2}-\frac{1}{2}\ln\left(\frac{R_D^2}{4t_D}\right)\right)-\frac{q_{lD}}{2T_D}\left(k+\ln\left(\frac{\rho_D^2}{\eta_D}\right)\right)$$

$$q_{lD} = \frac{1}{1+1/T_D}\left(1-C(t_D)\left(k+\ln(R_D^2)\right)\right)$$

where:

$$C(t_D) = \frac{1}{(k-2\gamma+\ln(4t_D))}-\frac{\gamma}{(k-2\gamma+\ln(4t_D))^2}+\frac{\gamma^2-\frac{\pi^2}{6}}{(k-2\gamma+\ln(4t_D))^3}$$

$$-\frac{\gamma^3-\frac{\pi^2\gamma}{2}+2\xi(3)}{(k-2\gamma+\ln(4t_D))^4}+\frac{\gamma^4-\pi^2\gamma^2+\frac{\pi^4}{60}+8\gamma\xi(3)}{(k-2\gamma+\ln(4t_D))^5}$$

$$k = \frac{2T_D}{\alpha(T_D+1)}+\ln\left(\frac{\eta_D^{\frac{1}{(T_D+1)}}}{r_{lD}^2}\right)$$

The subscript D stands for dimensionless parameters which are related to the dimensional terms based on the equations below:

$$R_D = \frac{R}{r_w}, \rho_D = \frac{\rho}{r_w}, L_D = \frac{L}{r_w}, \alpha = \frac{r_l^2 k_l}{2khh_l}, T_D = \frac{k_a h_a}{kh},$$

$$\eta_D = \frac{\eta_a}{\eta}, P_D = \frac{2\pi kh}{q\mu B}(P_i - P), t_D = \frac{\eta t}{r_w^2}, q_{lD} = \frac{q_l}{q}$$

The subscript α stands for the properties of "above-zone." The parameter (referred to as leakage coefficient) is a measure of the leak transmissibility and is related to r_l, k_l, h_l, which are the leak radius, leak permeability, and leakage interval respectively.

For the sake of generalizing the results of this work to various fields, it would be useful to present the detectability in terms of dimensionless terms. However, the pressure change threshold is dimensionally defined. Constant pressure threshold (P_i-P=0.1 psi) converts to different dimensionless pressure threshold (P_D) values for different formations. Therefore, the results should be given for a range of dimensionless pressure thresholds (P_D) considering a range of operational and field parameters. The range of the parameters is given in 7. Based on this table, P_D values corresponding to 0.1 psi pressure threshold may vary from 0.00005 to 0.05. Therefore, the results are plotted for this range of P_D values.

Table 7. Parameter definition for dimensionless curves for AZMI leakage detection.

Parameter	Lower Limit	Upper Limit
k, injection formation permeability (Darcy)	0.1	3
h, injection formation thickness (m)	10	50
$\dot{m}$, mass injection rate (Mt/year)	0.5	1.5
m, initial reservoir fluid viscosity (cp)	0.5	1
B, CO_2 formation volume factor (rm^3/sm^3)	0.0025	0.004

The iso-P_D (dimensionless pressure) curves are presented for leak dimensionless parameters versus dimensionless transmissivity and dimensionless diffusivity. When varying each of the leak parameters, the rest of the leak parameters are fixed to reference values. For simplicity, the leak of interest is considered to be halfway between the injector and observation well (i.e. R_D=ρ_D). Therefore the only leak parameters to vary are α and R_D (or ρ_D) the reference values for which are 0.00001 and 1,000 respectively.

The distances are given constant values: R_D=ρ_D=1,000. From these figures, it is observable that the detectable region spans a larger area as time increases and therefore the leak detectability increases with time. This is because larger volume of fluids will leak and larger pressure response will be induced as time advances. For instance, for T_D= 0.001, and η_D= 0.5, and considering the pressure threshold P_D=0.05, the smallest detectable leakage coefficient is 10^{-5} at dimensionless time of 10^8. This value decreases by one order of magnitude to 10^{-6} when dimensionless time is increased to 10^{13}.

Increasing T_D deteriorates the detectability significantly. As T_D increases (with fixed η_D), AZMI becomes more receptive to leakage fluids and less leakage-induced pressure change will be observed in AZMI. For instance, for t_D= 10^{11},η_D=1, P_D= 5×10^{-5} the minimum detectable leakage coefficient increase by four orders of magnitude (from 10^{-9} to 10^{-5}) as a result of increasing T_D by four orders of magnitude (from 10^{-3} to 10). Increasing η_D implies faster dissipation of the leakage fluids in AZMI and reduced resistance to leakage. Therefore, increasing η_D leads to increased detectability (i.e. smaller leaks will be made detectable as η_D increases). However, η_D has much less effect on leak detectability compared to T_D. The minimum detectable leakage coefficient reduces from 10^{-5} to 7×10^{-6} as a result of increasing η_D by two orders of magnitude from 0.01 to 1 (for t_D= 10^{13},T_D=10, P_D= 5×10^{-5}).

By increasing the pressure monitoring period from dimensionless time 10^8 to 10^{13}, the maximum dimensionless leak distance increases from 10^3 to 10^5 (considering T_D= 0.001, η_D= 0.5, P_D=0.05). Increasing T_D from 10^{-3} to 10, (for t_D = 10^{11}, η_D =1, P_D = 5×10^{-5}), the maximum dimensionless distance decreases by three orders of magnitude (from 2×10^5 to 400). And when increasing η_D from 0.01 to 1 the maximum distance increases from 1,000 to 3,000 (for t_D= 10^{13},T_D=10, P_D= 5×10^{-5}).

The method for using Figure 19 through Figure 20 is as follows:

1. Obtain the dimensionless properties h_D and T_D based on field specifications.
2. Pick a pressure change threshold value and translate to dimensionless pressure (P_D).
3. Pick a time after which the leak is to be detectable and translate to dimensionless time (t_D).
4. Refer to the graphs to find the detectable leak properties.
5. Translate the dimensionless leak properties to dimensional values based on field specifications.

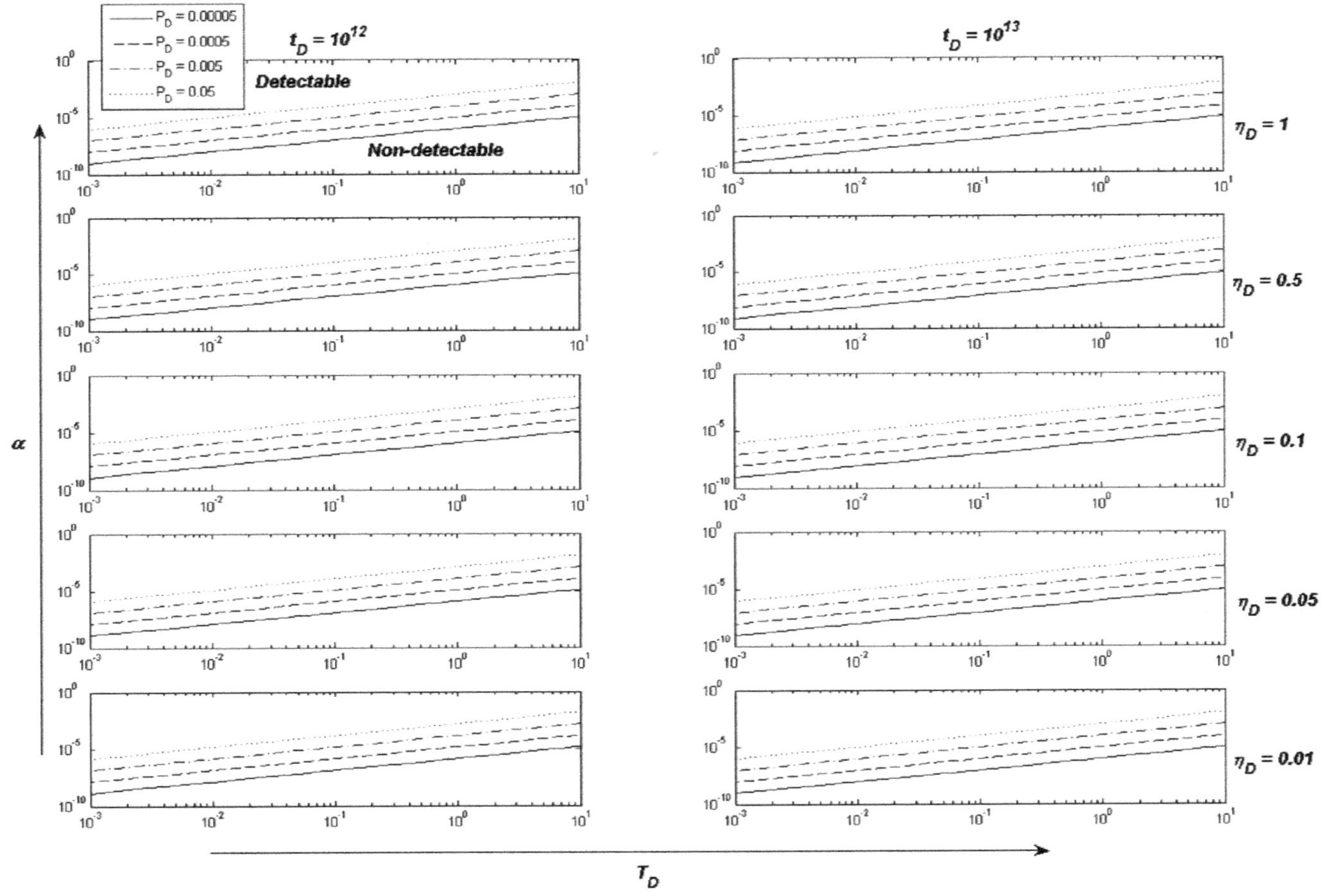

Figure 19. Dimensionless curves for calculating the leak transmissibility considering $t_D = 10^{12}$ and 10^{13}.

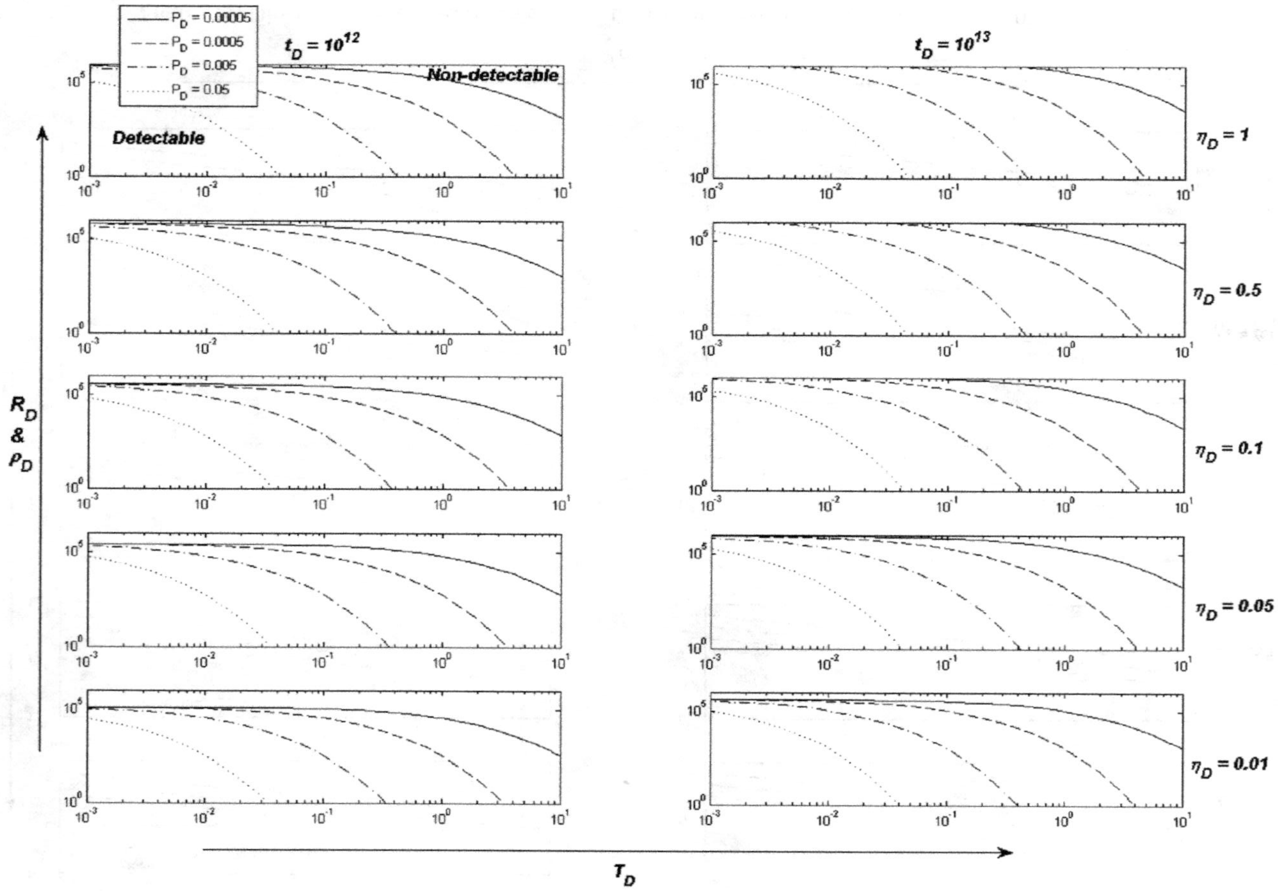

Figure 20. Dimensionless curves for calculating the leak distance considering tD =10^{12} and 10^{13}.

An Example: Cranfield Injection Zone and AZMI

In this section the above developed graphs are used to find the detectability limits of the Cranfield project. The average properties of AZMI and Tuscaloosa formations are obtained based on the well testing program conducted at Cranfield (Table 8). Based on these properties the dimensionless diffusivity and transmissivity are: η_D=0.5, T_D=0.28.

Table 8. Average properties of AZMI and Tuscaloosa formations are obtained based on well tests at of Cranfield.

Parameter	**Value**
m, brine viscosity (Pa/s)	1×10^{-3}
Tuscaloosa formation compressibility (1/Pa)	3×10^{-10}
Tuscaloosa formation porosity (fraction)	0.22
k, Tuscaloosa formation permeability (m^2)	2×10^{-13}
h, Tuscaloosa formation thickness (m)	18
h, diffusivity (m^2/s)	3.03
AZMI compressibility (1/Pa)	3×10^{-10}
AZMI porosity (fraction)	0.22
k_a, AZMI permeability (m^2)	1×10^{-13}
h_a, AZMI thickness (m)	10
h_a, AZMI diffusivity (m^2/s)	1.51
q, injection rate (Mt/year)	1
$r_{w,}$ injection well radius (m)	0.1
$_{rl}$, Leak radius (m)	0.1
h_l, leakage interval (m)	90
B, CO_2 formation volume factor (Rm^3/stm^3)	0.00259

Also, the following relationships are obtained to relate the time and pressure change to the dimensionless terms: $t_D \approx 1{,}010$ t and $P_D \approx 0.0033\ \Delta P$ where t is time in years and ΔP is the AZMI pressure change in psi. Taking the threshold pressure change of $\Delta P = 0.15$ psi results in $P_D = 0.0005$. Based on the above graphs, after t = 1 year (t_D = 1,010) of injection, a leak with parameters $\alpha = 6.5 \times 10^{-6}$ and $R_D = \rho_D = 2{,}474$ is detectable. The relationships calculated for leak properties are:

$$\alpha = 1.54 \times 10^7 k_l,\ R_D = 10R, \rho_D = 10\rho,$$

Therefore, the detectable leak permeability and distance are $k_l = 4.2 \times 10^{-13} m^2$ and R=ρ=250 m. The reference values of leak parameters α and R_D (or ρD) translate to leak permeability of 650 m_D and distances of 100 m. In summary, for Cranfield project, the AZMI pressure monitoring can detect leaky wells with permeability of 420 m_D or larger if they are located within 100 m distance from the injector/monitor. Also, if the monitoring well is within 250 m from the leak, any leaky well with permeability of 650 m_D or larger can be detected.

Above-Zone Temperature

Overview of Technology

Temperature monitoring is a mature technique used for detection of upward fluid leakage, typically along the rock-casing annulus of wells. A temperature survey run inside a well casing can be used to identify fluids migrating from deeper horizons to shallower zones, because the deeper fluids are hotter than the ambient fluids and cause a deviation in temperature gradient. A thermal analysis method was used to eliminate the possibility that brine flow along the well as a result of pressure increase in the injection was the cause of a pressure increase at the SECARB Cranfield phase II observation well [47]. This study adds to the assessment by considering the interactions between supercritical CO_2 and brine in leakage scenarios, where heating because of upward transport and cooling because of expansion of CO_2 are both occurring. In addition, this study considers feasibility of a novel application, using thermal monitoring to determine if a fault is vertically transmissive or sealing.

Site-specific Assessment

CO_2 injection induces temperature changes as a result of processes such as Joule-Thompson (JT) cooling, endothermic water vaporization, and exothermic CO_2 dissolution. CO_2 leaking from the injection zone, in addition to initial temperature contrast due to the geothermal gradient, undergoes similar processes, causing temperature changes when it migrates into a shallower zone. Numerical simulation tools were used to evaluate temperature changes associated with CO_2 leakage from the storage aquifer to an above-zone monitoring interval and to assess the monitorability of CO_2 leakage on the basis of temperature data. Specific details of the numerical model as well as the outcomes of many tests are provided by Zeidouni *et al.* [47].

In analysing temperature response in the injection zone, Han *et al.* (2012) identified the following as key parameters affecting temperature change [48]: porosity, vertical and horizontal permeability, injection rate, and injected CO_2 temperature. These parameters were varied for the AZMI and investigate their effect on temperature pulse. Other important parameters are leak permeability and rock-heat capacity.

One important finding is that under all conditions, the area of thermal change (radius of the area covered) is small, so that site-specific variations in tool effectiveness are modest. For example, Figure 21 shows the effect of increasing the permeability of the leakage pathway by an order of magnitude only increases the area over which it can be detected two times, no matter how long the leakage continues. Other site–specific parameters tested such as AZMI porosity and horizontal and vertical permeablities showed small sensitivity [47].

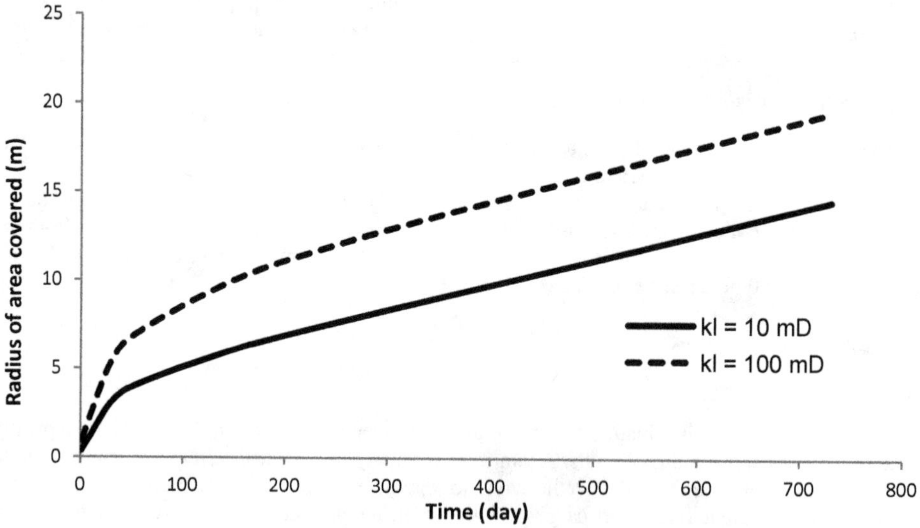

Figure 21. Extent of temperature change in AZMI versus the permeabilities of the leakage pathway.

Injected CO_2 may be either cooler (e.g. at Cranfield) or warmer (e.g. CO_2SINK) than the brine-saturated injection zone. Non-isothermal modelling of the wellbore is required to determine the temperature of CO_2 at the sandface, where the CO_2 leaves the well and enters the geologic formation. Lu and Connell (2008) and Han *et al.* (2010) studied the non-isothermal behaviour of CO_2 injection wells by solving coupled heat [50,51], mass, and momentum equations using various fluid and thermodynamic properties. They have shown that CO_2 generally undergoes cooling, followed by heating as the depth increases. As a result, the sandface temperature at the injection well may be warmer or cooler than that of the injection zone.

Running a temperature log or DTS system along suspected wells is the conventional use of thermal monitoring, and it has been shown to be effective in leak detection under certain conditions. However, in spite of possibly high temperature perturbation at the location of the leaking well completion, the thermal perturbation is localized, and it will be difficult to detect well leakage based on temperature data at monitoring wells distant from the leaking well. Likewise, monitoring a fault for leakage requires placement of monitoring wells where they intersect the fault to be assessed. If fault leakage is distributed along much of the fault surface, wells that intersect the fault plane (dipping fault and vertical wells or slanted wells and vertical faults) may be worth testing. If flow on faults is focused in small areas, the geometry becomes similar to wells and the value of monitoring is decreased. Additional evaluation is needed.

Geochemical Leakage Detection

Overview of Technology

The purpose most commonly given for monitoring groundwater to detect leakage of CO_2 is to assure that groundwater is protected from damage for use as drinking water. This perspective comes from the Safe Drinking Water Act Underground Injection Control Program [7], where underground injection of hazardous, as well as non-hazardous, fluids are regulated. An additional element considered in some monitoring plans is the extent to which monitoring groundwater can be used to document CO_2 retention in the deep subsurface and effective isolation from the atmosphere. In some areas, groundwater flow across the site provides a potentially useful wide-area sampling system that could be sensitive to any leakage from depth.

Data collection optimization is needed to assure that the monitoring meets the goals for which it is intended. Optimization techniques are mature and widely used for contaminated site studies (for example Sharma and Reddy, 2004 [52]). Characterization is needed prior to well placement to determine sources of ambient variably such as hydrologic gradient, recharge and discharge, aquifer heterogeneity, aquifer isolation, and cross-aquifer flow.

The impact of past usage is also needed. Placing wells with respect to these features are as applicable to geological storage monitoring as they are to contaminated sites (e.g. sampling upgradient of contamination to measure changes in regional properties and down gradient to assess the impact of contamination on groundwater geochemistry).

Vertical placement of screened intervals of wells is as important in geologic storage as it is in contaminated sites where contaminants are of different density than groundwater. It is important to place screens strategically to sample sufficient parts of the aquifer to provide the designed monitoring. However, leaving too much of a well screened will allow preferential flow from high permeability zones to dominate, leaving sampling of other zones diluted or unrepresented.

The selection of constituents to be analysed is a critical step that is often neglected or insufficiently considered in geologic storage protocols. Reactive transport modelling validated with laboratory and field testing is a recommended approach to selection of constituents to be analysed. This process will determine which geochemical parameters are most indicative of CO_2 interaction with the rock-fluid system. Our study of site specific sensitivity presented in the following section serves as an example of how this can be done.

Once the geochemical parameters to be assessed have been selected, appropriate sampling methods, analytical protocols, and a match between sensitivity of the methods and the leakage signal goal to detect can be designed. Dissolved CO_2 in groundwater can be affected by sampling methods, including turbulence, outgassing, and temperature change. Other dissolved constituents such as minerals near saturation or that sorb also can be perturbed during sampling. It is therefore important to provide documentation that the sampling and analytical methods designed are appropriate. A post-sampling analysis is likely to be needed. One set of analyses will be a statistical assessment of trend, used to determine if change greater then noise has been detected. If a change is detected, a second set of analyses, including fluid flow and reactive transport modelling, may be needed to determine if the change can be attributed to CO_2 or brine leakage, and to help constrain any follow-up testing needed for leakage accounting or mitigation.

Questions of risk to groundwater in terms of liberation of cations have been the subject of a large amount of current research (e.g. Birkholzer *et al.*, 2008 [53]). The studies completed so far have found changes in cations to be small. Stability of the cations changes as indicators of CO_2 rock water interaction need to be further assessed to determine if they survive transport.

Site-specific Assessment

The environmental factors that may affect sensitivity of detection have been classified into chemical factors and physical factors. The chemical factors are related to geochemical processes after CO_2 is leaked into the aquifer, such as mineralogy in aquifer sediments, and initial groundwater chemistry, which are the focus of this analysis. The physical factors are related to CO_2 migration or transport processes, and variations include confined or unconfined aquifers, variable groundwater velocity, groundwater recharge, extraction, aquifer heterogeneity, and monitoring location and depth. The following groundwater parameters have been selected as primary indicators of leakage of CO_2 into groundwater to be studied: pH, DIC, alkalinity, and HCO_3-. The site-specific sensitivity of the response to leakage was tested against reactive minerals in the aquifer sediments and initial aquifer chemistry. The detailed methodology and data are included in Yang *et al.* [54].

Three of the most common minerals (quartz, albite, and calcite) were selected and simulated in the generic model. For the first set of models, aquifer chemistry is held constant, with a composition of the fresh water aquifer (Dockum and Ogallala formations) above the SACROC aquifer in Scurry County, Texas [55]. As a follow on, the influence of aquifer water composition was compared using the Cranfield shallow aquifer in Natchez, Mississippi [56], and a shallow aquifer in Montana [57].

The results of this study show that the presence of carbonate in the monitored aquifer has an important impact on groundwater monitoring for leakage. Models of aquifers with nonreactive mineralogy such as quartz exhibit a leakage response to CO_2 as negative shifts in pH, positive shifts in total inorganic carbon, and negligible changes in alkalinity [56], which is replicated in this study (Figure 22).

Responses of geochemical parameters to CO_2 leakage simulated in the carbonate-poor aquifer with some silicate (quartz + 5% albite) show slightly different responses of geochemical parameters to CO_2 leakage than in the carbonate-poor aquifer (quartz only) because albite dissolution can slightly buffer pH. This is due to the dissolution of CO_2 in groundwater. However, in a short time after CO_2 is leaked into the aquifer, responses of geochemical parameters to CO_2 leakage in the aquifer containing only quartz are very similar to the responses of geochemical parameters to CO_2 leakage in the aquifer in the aquifer containing quartz and albite because albite dissolution is relatively slow.

Groundwater pH calculated in the carbonate-rich aquifer is buffered compared to groundwater pH in the carbonate-poor aquifer (Figure 22a). Alkalinity show very different responses to CO_2 leakage in the carbonate-poor aquifer compared to that in the carbonate-rich aquifer (Figure 22b). Alkalinity is almost unchanging in response to leakage into the carbonate-poor aquifer while alkalinity increases in the carbonate-rich aquifer as increase in CO_2 leakage rate (Figure 22b). As expected, HCO_3- shows very similar behaviour as alkalinity when CO_2 leakage is simulated (Figure 22d).

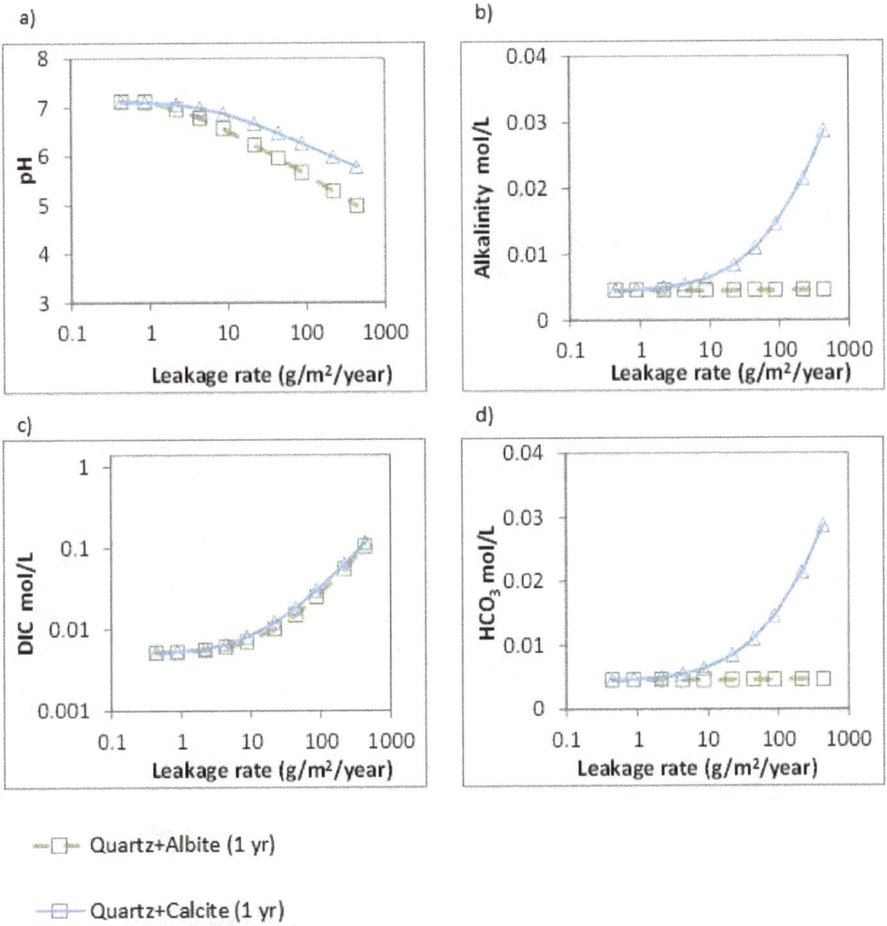

Figure 22. Modelled responses of groundwater pH, Alkalinity, DIC, and HCO^{3-} to CO_2 leakage rate comparing calcite –bearing and non-calcite bearing aquifer rocks.

It is interesting to note that responses of DIC and dissolved CO_2 (not shown) in groundwater to CO_2 leakage rate appear to be independent of aquifer mineralogy, although DIC could be slightly higher in the carbonate-rich aquifer than in the carbonate-poor aquifer (Figure 22c). It also can be seen that responses of geochemical parameters shown in all parameters are similar in the carbonate-rich and carbonate-poor aquifers if CO_2 leakage rate is very small, suggesting that small CO_2 leakage rate may not be detectable by monitoring geochemical parameters.

A series of assessments of the impacts of different freshwater aquifer chemical compositions on detection of leakage were conducted. We conclude that that impacts of initial groundwater chemistry in the aquifer has minor impacts on sensitivity of geochemical parameters to CO_2 leakage. Details are presented by Yang *et al.* [54].

DISCUSSION AND CONCLUSIONS

Each of the four technologies assessed in this study has a different role to play in monitoring. Time-lapse 3D seismic has a role in validation of fluid flow models if there is a reasonable match between the area where CO_2 migration was predicted and area where it was observed as a change in seismic response. In addition, where the change signal is sufficiently above noise, the time-lapse 3D seismic survey can be interpreted to show no CO_2 migration out of the intended area or into overlying strata.

Pressure is a classic reservoir monitoring tool. In the injection zone, pressure change in response to injection is a key element in demonstrating that the fluid flow model properly represents the aquifer properties. Use of pressure monitoring in AZMI is a relatively new approach, and should be a powerful tool in determining the extent to which the confining system is functioning as predicted. Unlike seismic, AZMI pressure signal is not specific to which fluid has migrated, brine or CO_2. This can be an advantage because advance warning of flaws in confinement can be identified prior to any CO_2 leakage.

Thermal monitoring lacks the value of seismic and pressure, which can show a signal far from the leakage path. The role for thermal monitoring falls more in the category of diagnostic data, to separate brine leakage from CO_2 leakage, for example. It is also being tested as a through-casing tool for assessing vertical leakage along faults intersected by many wells.

Geochemical monitoring is undertaken primarily to document no damage to usable quality fresh water, and may mostly be needed if other tools show concern about fluid migration above the confining system. Monitoring in freshwater also may be needed if leakage paths bypass deeper zones of the confining system (e.g. leakage occurs up the centre of improperly plugged wells and exits the well in the near surface).

Due to these different roles, the technologies are difficult to compare directly. For example, there are different outcomes to utilizing pressure monitoring instead of time-lapse 3D surveys and further difference in combining the two. Similarly, it is difficult to assess the value or lack of value to a monitoring program of collecting these different types of data. This study therefore does not accomplish a full value of information assessment on which to base decisions. However, the analysis readily shows that where site-specific variability in key properties is present, the value of each technology is quite different in different projects. This should be of substantive use to project developers when comparing different projects, to determine if the monitoring selections made at one site should be duplicated or substantive modified for application at another site.

In all four technologies assessed, noise and ambient long term trends are significant limits on the sensitivity of the method. Many protocols recommend collection of a pre-injection "baseline" with the implication that the monitoring trend can be subtracted from the baseline to show if a response to injection is observed. Consideration of the site-specific characteristics creates a different perspective on this need. Characterization and perhaps test data inputs are needed to forward model the magnitude of the signal that would be required to meet the monitoring goal. Further, data on the ambient variability of the signal is needed to determine if the modelled injection response signal is above the noise and ambient trends. Without such investment, monitoring results that find no leakage signal are indeterminate. This is especially an issue where absence of monitoring response is the signal expected for a properly performing site with good permanence. That is, no change in seismic response above the injection zone, no change in AZMI pressure and temperature, and no change in groundwater chemistry is the expected indicator. The lack of signal could be a result of no leakage. However, in a monitoring plan without proper design, the lack of signal could just as convincingly be a result of insensitivity of the monitoring array to leakage.

Rock physics models show why time-lapse 3D seismic response is such a valued monitoring tool. Modelled substitution of CO_2 for brine leads to a strong change in P-wave velocity, which can be measured over a wide range of rock types both by differencing velocity and by measuring PP reflectivity change between a pre-injection survey and following during- or post-injection survey(s). Models show the change in velocity with saturation produces a steep curve at low saturation. Although some very low saturations could be non-detectable, even a few percent change in CO_2 saturation produces a strong effect. The well-known levelling off of the rate of change at higher saturations limits the feasibility of using seismic velocity response alone to quantify the mass of stored CO_2, as it is difficult to quantify the difference between moderate saturations of 20% and high saturations of 50%. In addition, this levelling off with increasing saturation limits the use of time-lapse 3D seismic in reservoirs where gasses (e.g. methane) are trapped or present as a residual phase. For the increase in saturation when CO_2 is added to existent gas, the change in seismic response may occur in the flat part of the curve and not be detectable above noise. The shape of the curve of velocity change vs. saturation is also sensitive to the model used, therefore the conclusion presented here are subject to revision in cases where a different velocity model is applicable.

Models show a moderate decrease in velocity change with decreasing compressibility, and a strong decrease in velocity change as the injection zone becomes thin. This means that deeply buried and stiffer rocks, even if they have high porosity, lead to a weaker change in velocity. Therefore, for the projects in which detection of the CO_2 plume was near the limits of resolution, a thin plume is a likely cause of weak change in velocity and poor detectability above noise.

For the pilot-scale projects for which seismic response was not far above noise, available data suggests that parts of the plumes away from the injection point may be thin. Thinning is driven by the buoyancy of CO_2 relative to formation brine, also by lower viscosity of CO_2 relative to formation brine and by heterogeneity of reservoirs. Plume thinning is therefore to be expected away from the injection point and as time passes. A corollary of this finding is that it is unlikely that ordinary time-lapse 3D seismic can be used to detect and map thin edges of plumes. If such features are identified in the risk assessment as significant, the monitoring approach should be adapted. Research in thin-bed and fracture-zone detection using seismic is active and suitable technology improvements may be available.

A set of issues not yet assessed in this study to evaluate the suitability of time-lapse 3D seismic for monitoring include noise, factors that attenuate signal, and repeatability of signal between surveys. It is expected that these issues will show strong site-specific variability.

Planners and reviewers of monitoring should balance the strengths of time-lapse 3D seismic with its limitations. Investment in well-constrained rock-physics models to guide the expectations of change that can be detected are needed to increase the probability that the monitoring goals will be met. A wide spectrum of optimization technologies can be considered.

Above-zone pressure monitoring is shown to be a simple and relatively direct measure of out-of zone leakage. The hydrologic properties of the rocks in terms of thickness and permeability have a strong effect on detectable signal. This analysis of the leakage pathway considered the injection zone to be linked to only one overlying formation. Nordbotten *et al.* show that in presence of multiple overlying zones (assuming similar petrophysical properties) [41], most of the leakage occurs at the zone immediately overlying the injection zone. Therefore, even in presence of multiple overlying permeable zones, it is likely that the results are applicable if the lowest zone is developed as AZMI.

Pressure monitoring sensitivity can be simply estimated using a set of dimensionless type-curves. Leak detectability increases with time as the induced pressure increases; however a long duration of leakage may be unacceptable in terms of the risk reduction goals. Placing AZMI wells closer together is the most effective way to increase sensitivity [59, 60]. When comparing well spacing from one site to another, the dimensionless type-curves show that a high transmissivity ratio of AZMI to injection zone significantly reduces the detectability; that is, a thicker AZMI is less sensitive. The diffusivity ratio of AZMI to injection zone also changes detectability.

One pressure effect that requires further assessment is pressure 'noise.' A variety of pressure effects are found in subsurface environments. Periodic signal such as tidal effects can likely be processed out. However, transients introduced by geomechanical effects, distant pumping or injection, or equilibration after past perturbation may mask leakage signal and require assessment prior to deployment of an AZMI pressure based monitoring system.

Thermal techniques for leakage detection are known and applicable in locations very near the leakage path. Thermal techniques are also shown to be useful in the negative (i.e. for eliminating a suspect location as a significant leakage pathway because the thermal signal is lacking). However, slow leakage has low sensitivity, because the fluid is thermally equilibrated with the flow path during slow migration. Analytical methods are useful in making the determination that slow or no leakage is occurring. Modelling of a mixed brine and CO_2 system shows an interesting and potentially diagnostic signal. Thermal monitoring along a leakage pathway initially shows warming, as deeper brine, warmer because of the geothermal gradient, migrates during initial plume development. As CO_2 leakage begins, the JT cooling in response to pressure drop create the reverse trend, cooling. So a warming then cooling trend may show response to change in the type of fluid leaking. Thermal perturbations are localized, and therefore may be less subject to artefacts than pressure signal. However, collection of thermal signal is perturbed by well drilling and workover, and the presence of the steel and fluid-filled well bore may provide artefacts or noise near the temperature measurement point.

Chemical techniques for monitoring leakage of CO_2 into brine requires assessment of the nature of rock-water interaction with introduced CO_2. Aquifer geochemistry has a minor effect on leakage detectability, however the presence of even minor amounts of carbonate in the aquifer rock matrix have a major effect. When this difference is put into a reactive transport model, the impact of minor carbonate will become even more profound, as it will quickly buffer and consume the leaked CO_2. Forward modelling the impact of CO_2 on aquifer chemistry as conducted for this study can be validated by laboratory and field testing [54]. In addition, it is important to assess the ambient variability and any trend in the aquifer system, to determine if the predicted signal is above noise. Factors such as land use or climate change should be considered, including positive trends such as aquifer recovery from overuse or past contamination.

REFERENCES

1. IPCC (2005) IPCC Special Report on Carbon Dioxide Capture and Storage, prepared by Working Group III of the Intergovernmental Panel on Climate Change. Metz, B., O. Davidson, H. C. de Coninck, M. Loos, and L. A. Meyer (eds.), New York: Cambridge University Press.
2. Lindeberg, Eric (2002) The Quality of a CO_2 Repository: What is the Sufficient Retention Time of CO_2 Stored Underground, Proceedings of 6th International Conference on Greenhouse Gas Control Technologies, 255-260.
3. World Resources Institute (WRI) (2008) Guidelines for Carbon Dioxide Capture, Transport, and Storage; http://www.wri.org/publication/ccs-guidelines

4. U.S. Department of Energy, National Energy Technology Laboratory (NETL) (2010) Best practices for: Site screening, selection, and initial characterization for storage of CO_2 in deep geologic formations. p. 110.
5. Carpenter, M., Kvien, K., Aarnes, J. (2011) The CO2QUALSTORE guideline for selection, characterisation and qualification of sites and projects for geological storage of CO_2. *International Journal of Greenhouse Gas Control*, v. 5, p. 942-951.
6. IEA Greenhouse Gas R&D Programme (IEA GHG) (2009) 1st Geological Storage modelling network meeting, Report no. 2009/05 http://cdn.globalccsinstitute.com/sites/default/files/publications/106391/1st-CO2-geological-storage-modelling-network-meeting.pdf
7. US Environmental Protection Agency (EPA) (2012) Protecting drinking water though underground injection control, Office of Water (4606M), EPA 816-K-10-004, http://water.epa.gov/type/groundwater/uic/upload/pocketguide_uic_protecting_dw_thru_uic.pdf
8. Interstate Oil and Gas Compact Commission (IOGCC) (2007) Storage of Carbon Dioxide in Geologic Structures: A Legal and Regulatory Guide for States and Provinces, Oklahoma City, OK: The Interstate Oil and Gas Compact Commission, http://iogcc.publishpath.com/Websites/iogcc/pdfs/Road-to-a-Greener-Energy-Future.pdf
9. Det Norske Veritas (2009) CO_2QAULSTOR guidelines and qualification of sites and projects for geological storage of CO_2: *DNV Report* No. 2009-1425.
10. Official Journal of the European Union (2009) EU Directive on the Geological Storage of Carbon Dioxide; http://eur-lex.europa.eu/LexUriServ/LexUriServ.do?uri=OJ:L:2009:140:0114:0135:EN:PDF
11. US Environmental Protection Agency (EPA) (2010) Federal Requirements under the Underground Injection Control (UIC) Program for Carbon Dioxide (CO_2) Geologic Sequestration (GS*) Wells 40 CFR* Parts 124, 144, 145, 146, and 147 77230 *Federal Register*; v. 75, no. 237; accessed Friday, December 10, 2010; http://www.gpo.gov/fdsys/pkg/FR-2010-12-10/pdf/2010-29954.pdf
12. United Nations Framework Convention on Climate Change (2011) Draft decision-/CMP.7: Modalities and procedures for carbon dioxide capture and storage in geological formations as clean development mechanism project activities; http://unfccc.int/files/meetings/durban_nov_2011/decisions/application/pdf/cmp7_carbon_storage_.pdf
13. U.S. Department of Energy, National Energy Technology Laboratory (NETL) (2009) Monitoring verification, and accounting of CO_2 stored in deep geologic formations: DOE/NETL-311/081508.
14. British Geological Survey (BGS) (2006) *Interactive Design of Monitoring Programmes for the Geological Storage of* CO_2, released October 18, 2006, http://www.CO2captureandstorage.info/CO2tool_v2.1beta/index.php
15. U.S. Environmental Protection Agency (EPA) (2009) Clean Air Act Part 98—Mandatory Greenhouse Gas Reporting, *Federal Register*, Washington, DC.
16. International Energy Agency (2010) Model regulatory framework information paper, http://www.iea.org/publications/freepublications/publication/model_framework.pdf
17. McCormick, Mike (2012) A greenhouse gas accounting framework for carbon capture and storage, Center for Climate and Energy Solution http://www.c2es.org/docUploads/CCS-framework.pdf.
18. U.S. Department of Energy, National Energy Technology Laboratory (NETL) (2011) Risk analysis and simulation for geologic storage of CO_2, DOE/NETL-2011/1459.

19. Det Norske Veritas (2013) CO_2RISKMAN Guidance on effective risk management of safety and environmental major accident hazards from CCS CO_2 handling systems. *DNV Report* No.: I3IJLJW-2 http://www.dnv.com/binaries/CO2riskman%20guidance%20-%20level%201%20rev%201b_tcm4-536360.pdf
20. Quintessa (2013) CO_2 FEP database, http://www.quintessa.org/CO2fepdb/.
21. Graebner, W.A., Hardage, B .A, and Schneider, W. A. (2001) 3D Seismic Exploration, Geophysics Reprint Series, Society of Exploration Geophysisicists, 857 p.
22. Chadwick, Andy, Williams, Gareth, Delepine, Nicolas, Clochard, Vincent, Labat, Karine Sturton, Susan, Buddensiek, Maike-L, Dillen, Menno, Nickel, Michael, Lima, Anne Louise, Arts, Rob, Neele, Filip, Rossi, Giuliana (2010) Quantitative analysis of time-lapse seismic monitoring data at the Sleipner CO_2 storage operation *The Leading Edge*, February 2010, v. 29, p. 170-177.
23. Wilson, M., Monea, M. (2004) IEA GHG Weyburn CO_2 monitoring & storage project. Summary report 2000-2004, 7th International Conference on Greenhouse Gas Control Technologies, Petroleum Technology Research Centre, Vancouver, BC, p. 273.
24. Petroleum Technologies Research Center (2012) Best Practices for Validating CO_2 Geological Storage: Observations and Guidance from the IEAGHG Weyburn-Midale CO_2 Monitoring and Storage Project, Geoscience Publishing, http://www.geosciencepublishing.ca/.
25. White, Don (2013), Seismic characterization and time-lapse imaging during seven years of CO_2 flood in the Weyburn field, Saskatchewan, Canada, *International Journal of Greenhouse Gas Control*, v. 16S, S95-S102.
26. Pawar, Rajesh J., Warpinski, Norm R. Lorenz, John C., Benson, Robert D. , Grigg, Reid B., Stubbs, Bruce A., Stauffer, Philip H, . Krumhansl, James L. Cooper, Scott P., Svec, Robert K, 2006, Overview of a sequestration field test in the West Pearl Queen reservoir, New Mexico, Environmental Geoscience , v. 13, pp. 163 – 180 http://www.es.ucsc.edu/~phlip/LDRD_ER/Pawar_EnvironmentalGeoscience.pdf
27. Ditkof, J., Zeng, H., Meckel, T.A., Hovorka, S.D. (2011) Time lapse seismic response (4D) related to industrial-scale CO_2 injection at an EOR and CCS site, Cranfield, MS. GCCC Digital Publication Series #11-19.
28. Zhang, Rui, Ghosh, Ranjana, Sen, Mrnal, and Srinvatsan, Sanjeay (2012) Time-lapse surface seismic inversion with thin bed resolution for monitoring CO_2 sequestration: A case study from Cranfield, Mississippi. Int. J. Greenhouse Gas Control, http://dx.doi.org/10.1016/j.ijggc.2012.08.015
29. Lumley, D. (2010) 4D seismic monitoring of CO_2 sequestration, The Leading Edge, v. 29, no. 2, p. 150–155 . doi: 10.1190/1.3304817.
30. Urosevic, M., Pevzner, R., Shulakova, V. ,Kepic, A. Caspari, E., and Sharma, S. (2011) Seismic monitoring of CO_2 injection into a depleted gas reservoir – Otway Basin Pilot Project, Australia, Energy Procedia, v.4, p. 3550-3557.
31. Jenkins, C., P. J. Cook, J. Ennis-King, J. Underschultz, C. Boreham, T. Dance, P. de Caritat, D. Etheridge, D. Freifeld, B., Hortle, A., Kiresti, Dirk, Patterson, L., Pevzner, R. Schacht, U, Sharma, S., Stalker, L. Urosevic, M. (2011) Safe storage and effective monitoring of CO_2 in depleted gas fields; Proceedings of the National Academy of Sciences of the United States of America, v.109 , no. 2: p. E36-E41. Doi: 10.1073/pnas.1107255108.
32. Al-Jabri, Y., and Urosevic, M. (2010) Assessing the repeatability of reflection seismic data in the presence of complex near-surface conditions, CO_2CRC Otway Project, Victoria, Australia: *Exploration Geophysics* v. 41, no. 1, p. 24-30.
33. Pevzner, R., Shulakova, V. Kepic, A. and Urosevic, M. (2011) Repeatability analysis of land time-lapse seismic data: CO_2CRC Otway pilot project case study: *Geophysical Prospecting*, v. 59, P. 66-77.

34. Gassmann, F. (1951) On the elasticity of porous media, Veirteljahrsschrift der Naturforschenden Gesellschaft in Zurich, v. 96, p. 1-23.
35. Reuss A. (1929) Berechnung der Fliessgrenzen von Mischkristallen auf Grund der Plastizitatsbedingung fur Einkristalle, *Zeitschrift fur Angewandte Mathematik und Mechanick*, v. 9, p. 49-58.
36. Sakurai, Shinichi, Hovorka, S. D., Ramakrishnan, T. S., Boyd, A., and Mueller, N. (2005) Monitoring saturation changes for CO_2 sequestration: petrophysical support of the Frio Brine Pilot Experiment, in SPWLA 46th Annual Logging Symposium: Society of Petrophysicists and Well Log Analysts, Paper No. 2005-YY.
37. Hovorka, S. D., Doughty, Christine, Benson, S. M., Freifeld, B. M., Sakurai, Shinichi, Daley, T. M., Kharaka, Y. K., Holtz, M. H., Trautz, R. C., Nance, H. S., Myer, L. R., and Knauss, K. G. (2006) Measuring permanence of CO_2 storage in saline formations: the Frio experiment: *Environmental Geosciences*, v. 13, no. 2, p. 105–121.
38. Daley, T.M., R.D. Solbau, J.B. Ajo-Franklin, and S.M. Benson (2007) Continuous active-source monitoring of CO_2 injection in a brine aquifer, *Geophys.*, 72(5): A57–A61, doi:10.1190/1.2754716.
39. Zoeppritz, K. (1919) Erdbebenwellen VIIIB, On the reflection and propagation of seismic waves. Gottinger Nachrichten, v. I, p. 66-84.
40. Berryman, J. G. (1995) Mixture theories for rocks. *In Rock Physics and Phase Relations: a Handbook of Physical Constants*, ed. T. J. Ahrens. Washington, DC: American Geophysical Union.
41. Nordbotten, J., Celia, M., Bachu, S. (2004) Analytical Solutions for Leakage Rates through Abandoned Wells: Water Resources Research, v. 40, no.4, W04204.
42. Manzocchi, T. and Childs, C. (2013) Quantification of hydrodynamic effects on capillary seal capacity, *Petroleum Geoscience*, v. 19, p. 105–121, doi: 10.1144/petgeo2012-005.
43. Kim, S. H. and Hosseini, S.A. (2013). Above-zone pressure monitoring and geomechanical analyses for a field-scale CO_2 injection project, Cranfield, MS. Greenhouse Gases: Science and Technology. DOI: 10.1002/ghg.1388.
44. Zeidouni, M., Pooladi-Darvish, M., and Keith, D. (2010) Leakage detection and characterization through pressure monitoring, in Proceedings of 10th International Conference on Greenhouse Gas Control Technologies, September 19□23, Amsterdam.
45. Zeidouni, M. (2012) Analytical model of leakage through fault to overlying formations, Water Resources Research, 48, W00N02, doi:10.1029/2012WR012582.
46. Tao, Q., Bryant, S.L., Meckel, T.A. (2013) Above-zone measurements of pressure and temperature for monitoring CCS Sites, *International Journal of Greenhouse Gas Control*, http://dx.doi.org/10.1016/j.ijggc.2012.08.011
47. Zeidouni, Mehdi, Nicot, J-P. Hovorka, S. D. (in review 2014) Monitoring above-zone temperature variations associated with CO_2 and brine leakage from the storage aquifer.
48. Han, W. S, Kim, K.-Y, Park, E., McPherson, B., Lee, S.-Y Park, M.-H. (2012) Modelling of spatiotemporal thermal response to CO_2 injection in saline formations: Interpretation for monitoring: *Transport in Porous Media* v.93, no.3p. 381-399.
49. Oldenburg, C.M and Pruess, Karsten (1998) Layered Thermohaline Convection in Hypersaline Geothermal Systems, *Transport in Porous Media*, ISSN 0169-3913, v. 33, no. 1, p. 29 – 63.
50. Lu, M. and Connell, L.D. (2008) Non-isothermal flow of carbon dioxide in injection wells during geological storage: *International Journal of Greenhouse Gas Control* v. 2 , no. 2 p.248-258. doi: 10.1016/S1750-5836(07)00114-4.
51. Han, W. S., G. A. Stillman, M. Lu, C. Lu, B. J. McPherson, E. Park (2010) Evaluation of potential non-isothermal processes and heat transport during CO_2 sequestration, *J. Geophys. Res.*, 115, B07209, doi:10.1029/2009JB006745.

52. Sharma, H. D. and Reddy, K. R (2004) Geoenvironmental engineering: site remediation, waste containment, and emerging waste management technologies, John Wiley and Sons, Hoboken NJ, 968 p.
53. Birkholzer, J.T., Apps, J.A., Zheng, L., Zhang, Y., Xu, T., Tsang, C.-F. (2008) Research Project on CO_2 Geological Storage and Groundwater Resources: Water Quality Effects Caused by CO_2 Intrusion into Shallow Groundwater, Berkeley, CA.
54. Yang, Changbing (in preparation) Sensitivity of Groundwater Geochemical Parameters to Potential CO_2 Leakage in *Shallow Aquifers: A Generic Modeling Approach.*
55. Romanak, K. D., Smyth, R. C., Yang, C., Hovorka, S. D., Rearick, M., and Lu, J. (2012) Sensitivity of groundwater systems to CO_2: application of a site-specific analysis of carbonate monitoring parameters at the SACROC CO_2-enhanced oil field: *International Journal of Greenhouse Gas Control*, v. 5, no. 1, p. 142-M.
56. Yang, Changbing, Mickler, P. J., Reedy, Robert, Scanlon, B. R., Romanak, K. D., Nicot, J.-P., Hovorka, S. d., Trevino, R. R., and Larson, Toti (2013) Single-well push–pull test for assessing potential impacts of CO_2 leak-age on groundwater quality in a shallow Gulf Coast aquifer in Cranfield, Mississippi. *International Journal of Greenhouse Gas Control.* http://dx.doi.org/10.1016/j.ijggc.2012.12.030
57. Wilkin, R.T., DiGiulio, D.C. (2010) Geochemical Impacts to Groundwater from Geologic Carbon Sequestration: Controls on pH and Inorganic Carbon Concentrations from Reaction Path and Kinetic Modeling. *Environmental Science & Technology*, v.44 no.12, p. 4821-4827.
58. Yang, Changbing, Hovorka, S., Young, M, Trevino, R., 2013, Geochemical sensitivity of aquifers to CO_2 leakage: detection in potable aquifers at CO_2 sequestration sites: Greenhouse Gas Science and Technology, Wiley Online Library (wileyonlinelibrary.com). DOI: 10.1002/ghg.1406
59. Sun, A. Y., and Nicot, J. -P. (2012) Inversion of pressure anomaly data for detecting leakage at geologic carbon sequestration sites: Advances in Water Resources, v. 44, p. 20-29.
60. Sun, A. Y., Zeidouni, M., Nicot, J. -P., Lu, Zhiming, and Zhang, D. (2013) Assessing leakage detectability at geologic CO_2 sequestration sites using the probabilistic collocation method: Advances in Water Resources, v. 56, p. 49–60.

Carbon Dioxide Capture for Storage in Deep Geological Formations, Volume 4
Karl F. Gerdes (Editor)

Section 2

STORAGE MONITORING & VERIFICATION (SMV)

Optimization

Carbon Dioxide Capture for Storage in Deep Geological Formations, Volume 4
Karl F Gerdes (Editor)

Chapter 32

HEALTH, SAFETY, AND ENVIRONMENTAL RISK ASSESSMENT OF GEOLOGIC CARBON SEQUESTRATION: OVERVIEW OF THE CERTIFICATION FRAMEWORK, EXAMPLE APPLICATION AND SELECTED SPECIAL STUDIES 2010-2014

C. M. Oldenburg[1], J.-P. Nicot[2], P. D. Jordan[1], Y. Zhang[1], L. Pan[1], J. E. Houseworth[1], T. A. Meckel[2], D. L. Carr[2], and S. L. Bryant[3]
[1]Earth Sciences Division, Lawrence Berkeley National Laboratory, Berkeley
[2]Bureau of Economic Geology, Jackson School of Geosciences, University of Texas, Austin, Texas
[3]Department of Chemical and Petroleum Engineering, Schulich School of Engineering, University of Calgary, Calgary, AB Canada

ABSTRACT: The Certification Framework (CF) is a Geologic Carbon Sequestration (GCS) leakage risk assessment framework developed around the concept of effective trapping. Since the time of its development in 2009, we have carried out CF applications and related special studies. Here we summarize a selected few of these related studies. For example, we applied the CF to the In Salah CO_2 storage project at three different stages in the state of knowledge of the project: (1) at the pre-injection stage, using data available just prior to injection around mid-2004; (2) after four years of injection (September 2008), and (3) between 2008 and 2010. We found that increasing knowledge about the condition of wells and injection pressures led to an increase in the assessment of likelihood of well and fault leakage, while the assessment of consequences remained the same. With the benefit of hindsight now in 2014, we believe the CF analysis of 2010 has been validated because our analysis correctly pointed out the hazard of higher-than-planned injection pressures in the reservoir, and the increased likelihood for fracturing to extend upward partly into the cap rock immediately overlying the reservoir.

In a special study, the comparison of fault displacement and seal thickness distribution in a thick section of the Texas Gulf Coast led to the conclusion that fault offset following a single event should not be of concern for cap-rock leakage. The applicability of the results elsewhere remains an open question, but it should be noted that the Gulf Coast Basin contains a significant fraction of the USA's CO_2 storage capacity. In another special study, we have carried out numerical experiments in simple model systems to calculate static pressure and pressure evolution in large vertical fractures filled with buoyant (less dense) CO_2. We found that buoyant fluid can be expected to generate high fracture-tip pressure differences, and these differences may grow with fracture height, provided there is a continuous supply of fluid injected at the fracture base and a continuous connected fluid column. As for pressure dissipation in such fractures, we observed that pressure rises rapidly with time in the fracture for both water-filled and CO_2-filled fractures, that dissipation of pressure by flow into adjacent cap-rock is negligible, and that pressure dissipation by flow into the matrix along the fracture planes is not an effective process for mitigating vertical fracture propagation over time scales of hydraulic fracturing.

KEYWORDS: geologic carbon sequestration; risk assessment; surface leakage; In Salah; induced seismicity

INTRODUCTION

The Certification Framework (CF) is a leakage risk assessment framework developed around the concept of effective trapping, where effective trapping implies that the product of likelihood and consequences of leakage are low (i.e., leakage risk is low). The CF has been fully described in a previous publication [1]. Since the time of development of the CF, we have applied it to several sites and carried out numerous special studies related to refining risk assessment through better understanding of the behaviour of deep subsurface systems in response to CO_2 injection. The purpose of the special studies is to investigate specific leakage-risk-related issues to inform estimates of likelihood and impact severity that can be integrated into future development and applications of the CF. In this chapter, we present a brief overview of the CF, and then we summarize a case study application and three selected special studies carried out since 2010.

OVERVIEW OF THE CF

The purpose of the CF is to provide a framework for project proponents, regulators, and the public to analyze the risks of geologic CO_2 storage in a simple and transparent way to certify startup and decommissioning of geologic CO_2 storage sites. The CF currently emphasizes leakage risk associated with subsurface processes and excludes compression, transportation, and injection-well leakage risks. The CF is designed to be simple by (1) using proxy concentrations or fluxes for quantifying impact rather than complicated exposure functions, (2) using a catalog of pre-computed CO_2 injection results, and (3) using a simple framework for calculating leakage risk. For transparency, the CF endeavors to be clear and precise in terminology in order to communicate to the full spectrum of stakeholders. Definitions used in the CF are as follows:

- ***Effective Trapping*** is the proposed overarching requirement for safety and effectiveness.
- ***Storage Region*** is the 3D volume of the subsurface intended to contain injected CO_2.
- ***Leakage*** is migration across the boundary of the Storage Region.
- ***Compartment*** is a region containing vulnerable entities (e.g., environment and resources).
- ***Impact*** is a consequence to a compartment, evaluated by proxy concentrations or fluxes.
- ***Risk*** is the product of probability and consequence (impact).
- ***CO_2 Leakage Risk*** is the probability that negative impacts will occur to compartments due to CO_2 migration.
- ***Effective Trapping*** implies that CO_2 Leakage Risk is below agreed-upon thresholds.

In the CF, impacts occur to compartments, while wells and faults are the potential leakage pathways. Figure 1a shows how the CF conceptualizes the system into source, conduits (wells and faults), and compartments HMR, HS, USDW, NSE, and ECA, defined as:

- ***ECA*** = Emission Credits and Atmosphere
- ***HS*** = Health and Safety
- ***NSE*** = Near-Surface Environment
- ***USDW*** = Underground Source of Drinking Water (= potable water)
- ***HMR*** = Hydrocarbon and Mineral Resource

Figure 1b shows a flow chart of the general CF logic and inputs and outputs. Further details can be found in [1]. A cumulative list of CF-related publications is presented in Table 1.

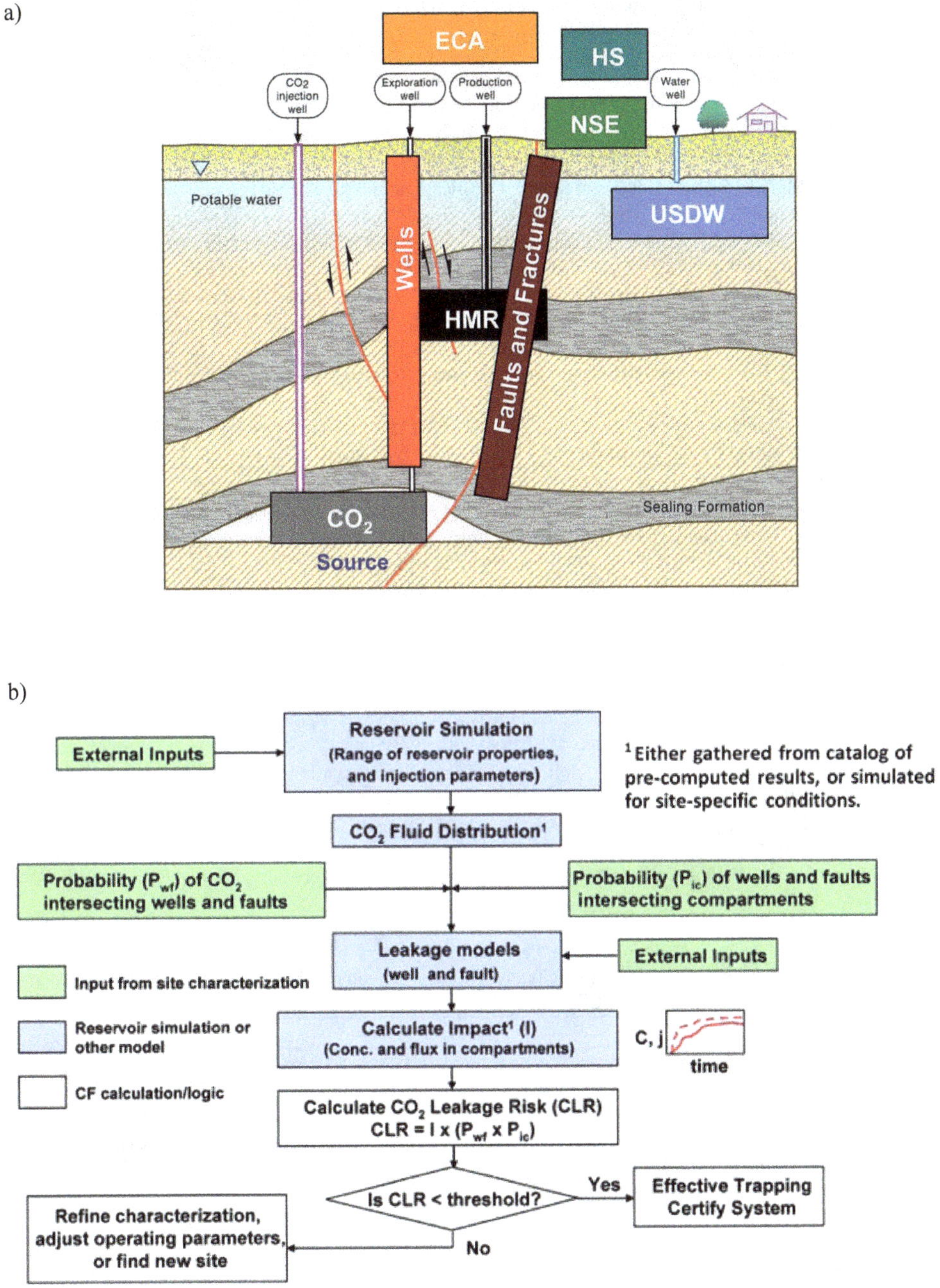

Figure 1. (a) Generic schematic of compartments and conduits in the CF, and (b) flow chart of the CF approach.

Table 1. Cumulative list of publications on CF-related research grouped by publication type.

Journal articles
Jordan, P.D., C.M. Oldenburg, J.-P. Nicot (2013). Measuring and modeling fault density for CO_2 storage plume-fault encounter probability estimation, *AAPG Bulletin*, 97(4), 597-618.
Jordan, P.D., C.M. Oldenburg, J.-P. Nicot (2011). Estimating the probability of carbon dioxide plumes encountering faults. *Greenhouse Gases: Sci. and Tech.*, 1(2), 160–174. *LBNL-5284E.*
Oldenburg, C.M., S.L. Bryant, J.-P. Nicot (2009). Certification framework based on effective trapping for geologic carbon sequestration, *Int. J. Greenhouse Gas Control, 3*, 444–457. *LBNL-1549E.*
Pan, L., C.M. Oldenburg, Y.-S. Wu, K. Pruess (2011). Transient CO_2 leakage and injection in wellbore-reservoir systems for geologic carbon sequestration, *Greenhouse Gases: Sci. and Tech.*, 1(4), 335-350.
Zhang, Y., C.M. Oldenburg, S. Finsterle (2009). Percolation-theory and fuzzy rule-based probability estimation of fault leakage at geologic carbon sequestration sites, *Env. Earth Sci.*, *59*(7), 1447-1459. *LBNL-2172E.*
Zhang, Y. (2011). Using the Choquet integral for screening geological CO_2 storage sites, *Greenhouse Gases: Sci. and Tech.*, 1(2): 175-179.
Nicot, J.-P., C.M. Oldenburg, J. Houseworth, J.-W. Choi (2013). Analysis of potential leakage pathways at the Cranfield, MS, U.S.A., CO_2 sequestration site: *Int. J. of Greenhouse Gas Control*, 18, 388-400.
Book Chapters and Reports
Oldenburg, C.M., S.L. Bryant, J.-P. Nicot, N. Kumar, Y. Zhang, P. Jordan, L. Pan, P. Granvold, F.K. Chow (2009). Model Components of the Certification Framework for Geologic Carbon Sequestration Risk Assessment, in *Carbon Dioxide Capture for Storage in Deep Geological Formations, Volume 3*, L.I. Eide (Ed.), CPL Press and BP. *LBNL-2038E.*
Houseworth, J.E., C.M. Oldenburg, A. Mazzoldi, A.K. Gupta, J.-P. Nicot, S.L. Bryant (2011). Leakage Risk Assessment for a Potential CO_2 Storage Project in Saskatchewan, Canada. Lawrence Berkeley Laboratory Report *LBNL-4915E.*
Nicot, J.-P., J.E. Houseworth, C.M. Oldenburg, J.-W. Choi, H.R. Lashgari, S. Coleman, T.A. Meckel, P.Jordan, A. Mazzoldi (2011). Certification Framework: Case Study V, Leakage Risk Assessment for the SECARB Phase III CO_2 Storage Project at a Mississippi EOR Site.
Zhang, Y., C.M. Oldenburg (2011). A Simplified 1-D Model for Calculating CO_2 Leakage through Conduits, Lawrence Berkeley National Laboratory Report *LBNL-3266E.*
Oldenburg, C.M., S.L. Bryant, J.-P. Nicot, M.J. Coombs, C. Doughty, P.D. Jordan, N. Kumar, J. Wagoner (2008). Preliminary risk assessment for WESTCARB's Phase 3 CO_2 injection at the Kimberlina Power Plant, San Joaquin Valley, California, Lawrence Berkeley National Laboratory, Internal Report, April, 2008.
Nicot, J.-P., Meckel, T.A., Carr, D.L., Costley, R., Zeidouni, M., Oldenburg, C.M., Fifariz, R., Osmond, J. (2013). Critical Topics in Geologic Carbon Sequestration. Topic 1.1.1: Induced Seismicity and Topic 2.1.1: Storage Capacity: The University of Texas at Austin, Bureau of Economic Geology, contract report prepared for CO_2 Capture Project (CCP) Phase III, 35 p.
Proceedings
Oldenburg, C.M., P.D. Jordan, J.-P. Nicot, A. Mazzoldi, A.K. Gupta, S.L. Bryant, Leakage risk assessment of the In Salah CO_2 storage project: Applying the Certification Framework in a dynamic context. *LBNL-4278E. Energy Procedia, 2*, Elsevier, GHGT-10, Amsterdam, The Netherlands, Sept. 19–23, 2010.
Nicot, J.-P., C.M. Oldenburg, S.L. Bryant, S.D. Hovorka, Pressure perturbations from geologic carbon sequestration: Area-of-review boundaries and borehole leakage driving forces, *Energy Procedia, 1*(1), 47-54, February 2009, *LBNL-3064E.*
Oldenburg, C.M., J.-P. Nicot, S.L. Bryant, Case studies of the application of the Certification Framework to two geologic carbon sequestration sites, *Energy Procedia*, GHGT9 conference, Nov. 16-20, 2008, Washington DC. *LBNL-1421E.*
Zhang, Y., C.M. Oldenburg, P.D. Jordan, S. Finsterle, K. Zhang, Fuzzy Rule-Based Probability Estimation of Fault Leakage at Geologic Carbon Sequestration Sites, *Energy Procedia*, GHGT9 conference, Nov. 16-20, 2008, Washington DC. *LBNL-1415E.*

EXAMPLE CF APPLICATION: DYNAMIC LEAKAGE RISK AT IN SALAH

Introduction

From 2004-2011, the In Salah CO_2 storage project in Algeria injected CO_2 stripped from produced natural gas into the water leg of the Carboniferous C10.2 gas reservoir at Krechba for long-term storage (e.g., [2,3]). Over the years prior to and during injection, the joint industry operators have carried out extensive analyses of the Krechba system including three risk assessment efforts - one before injection started, one carried out in September 2008 and one undertaken in July 2010. The long history of injection at Krechba, and the accompanying characterization, modeling, and performance data provided a unique opportunity to test and evaluate risk assessment approaches. In 2009-2010, we applied the CF to the In Salah CO_2 storage project at three different stages in the state of knowledge of the project: (1) at the pre-injection stage, using data available just prior to injection around mid-2004; (2) after four years of injection (September 2008) to be comparable to the other risk assessments, and (3) from 2008-2010. A complete description of the study summarized here is presented in [4].

Overview of the In Salah CO_2 storage project

The Krechba gas reservoir structure is a 20 m-thick sandstone anticline at a depth of approximately 1,800 m with gas-cap footprint of dimension 20 km (NS) by 8 km (EW), located in the Tademait Plateau of Algeria. Laterally extensive and thick (total thickness of 950 m) sealing mudstones provide effective seals that have trapped natural gas in a closed anticline. The unconformably overlying Cretaceous Continental Intercalaire (CI) aquifer is confined at Krechba and resides in multiple layers of mudstone, sands, and gravels creating a complex aquifer system ([5,6,7]).

The net thickness of water-producing zones (upper and lower units) is ~400 m and the total or gross thickness of the aquifer complex is ~900 m. The local water-table depth is approximately 150 m bgs (below ground surface) within the Upper Cretaceous. Water quality is good, with less than 1,000 mg/L TDS to a depth of more than 500 m.

Seventeen wells comprising appraisal (oldest), gas production, and CO_2 injection wells (newest) penetrate the Carboniferous age target CO_2 storage and gas reservoir as shown in Figure 2. The five Carboniferous gas production wells and three CO_2 injection wells have long horizontal sections within the reservoir for enhanced production and injection, respectively.

The conceptual geologic model of Krechba is that of a broad two-way plunging anticline influenced by underlying strike-slip faulting with a component of reverse displacement [8]. These faults are reactivated and were formerly extensional faults. Predominantly east-west faults on the west limb and some subtler faults on the east limb are associated with these deeper faults. Drilling mud losses have occurred at various depths in the Viséan suggesting fractures are present.

For the purposes of the CF definition of leakage, the lateral boundaries of the storage region were defined in two different ways to honor the temporal aspects of the system relative to production of the natural gas resource. Specifically, the storage region for the next 20 years was defined to include (1) a boundary defined by the edge of the gas cap, and (2) the lease boundary. After 20 years, the lateral boundary of the storage region will be the lease boundary all around the reservoir because the natural gas resource will be considered depleted. The bottom boundary is a deep Devonian age unit. The upper boundary to the system is defined by a seismically mappable "hot shale" unit with a fairly distinct gamma ray response in well logs within the 950m sealing Carboniferous mudstones. This is a reasonable choice for the effective cap rock because it is predominantly mudstone within which no drilling mud loss events occurred. With these choices of boundaries for the storage region, the only CO_2 leakage fluxes that need to be considered are (1) upward to USDW or higher compartments through wells and faults, and (2) laterally for the next 20 years into the Krechba gas resource (potential impact to HMR).

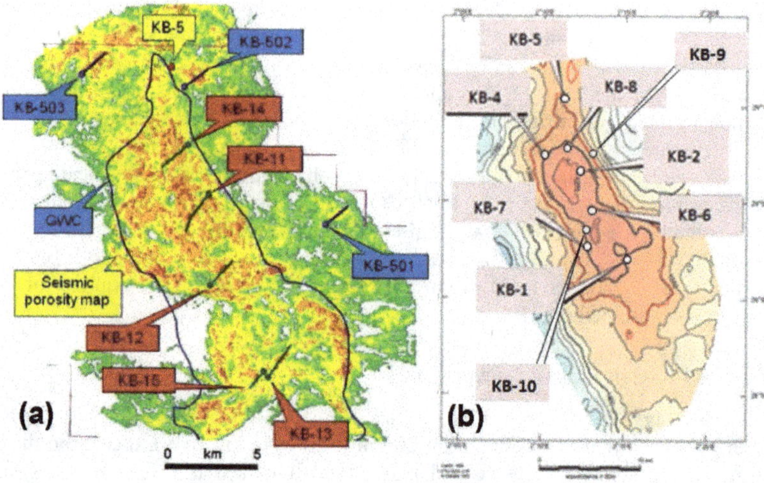

Figure 2. (a) Production wells (KB-11 – KB-15) and injection wells (KB-501 – KB-503), and (b) appraisal wells.

Pre-injection (Pre-2004) leakage risk

Wells

Statistical data on uncontrolled flow from wells from oil and gas operations in general can be used to estimate a likelihood of approximately 1% that a CO_2 injector well will fail in the project lifetime [9]. The non-operational wells at In Salah were not expected to encounter CO_2 within many decades with the exception of KB-5 (which later had a very small leakage amounting to less than one tonne of CO_2) and possibly KB-4 which are relatively close to injectors KB-502 and KB-503. Statistics for non-operational wells in general suggested that an uncontrolled flow from such wells was improbable, however, impact severity is potentially high, so we recommended that the integrity of these wells should be evaluated and checked at appropriate intervals, and if necessary, full decommissioning should be carried out at an appropriate time. Depending on the abandonment procedure, well leakage of CO_2 would then become unlikely to improbable, rendering the well leakage risk low. The corresponding brine leakage risk was considered very low because of the small driving force for brine flow. The CO_2 leakage risk to USDW assessment for this period is shown in Figure 3. Definitions and more discussion of likelihood and severity for this case study are given in [4].

Faults and Fractures

With respect to leakage through faults and fractures, the static and post-closure periods were not considered a concern due to the record of gas accumulation and the long-term pressure decline expected in the system due to gas production. However, the injection period may produce overpressures in the injection reservoir that could be of concern. In particular, the injection pressure range proposed was such that it could re-activate fractures in the storage reservoir, and possibly in thin mudstones above to enhance injectivity, but that would not open fractures in the overlying cap rock. The upper limit of the bottom hole injection pressure was variously set at 24 MPa or 29 MPa (3,500 psi or 4,200 psi). The former is equivalent to a gradient just above 1.3 SG (SG = specific gravity of water) and the latter to 1.45 SG given the reservoir depth. Given that injection may re-activate fractures in the mudstone above the reservoir but below the thick cap rock, these fractures could become filled with CO_2, and extend relatively unattenuated to the base of the cap rock. By defining the upper storage region boundary at a unit within the cap rock, such upward migration is not considered leakage. However, further fracture propagation upward through the cap rock could not be ruled out. See [4] for further discussion.

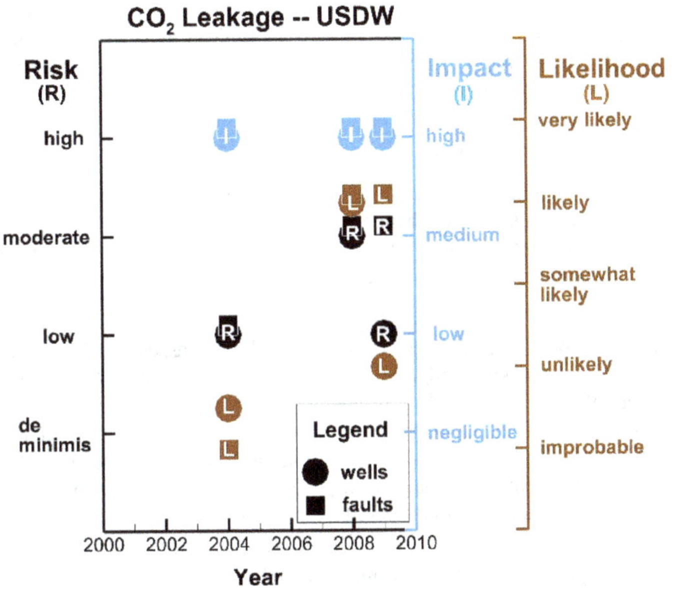

Figure 3. Assessment of CO_2 leakage to USDW likelihood (L), impact (I) and risk (R), the qualitative combination of likelihood and impact) for existing wells and hypothetical faults and fractures shown as they were assessed at different times using data available up to that time. Note that the scales are arbitrary and qualitative, and L and I do not necessarily combine to produce an R between L and I. As discussed in the text, the assessment of impact of leakage of CO_2 to USDW does not change with additional data, while the assessment of likelihood does change.

Therefore we assessed in 2010 based on the pre-injection state of knowledge (pre-2004) that fault/fracture CO_2 leakage was improbable, but when combined with high potential impacts of such leakage on USDW, the results of fault/fracture leakage risk to USDW had to be considered low rather than de minimis (see Figure 3). The corresponding brine leakage risk was considered de minimis through faults and fractures. Note that we considered hypothetical rather than known faults and fractures in this analysis because there are no mapped faults which penetrate the Carboniferous caprock through to the Continental Intercalaire.

Prior to September 2008 leakage risk

Cement bond log (CBL) surveys of wells at Krechba were carried out by various operators during the exploration and appraisal phase, results from which were made known in 2006 (after start-up of injection). CBL results revealed that several intervals in several appraisal wells had poor cement bonds. Modeling of leakage fluxes up injection wells with good cement bonds produced fluxes of CO_2 too small to have measurable impact. Other injection wells with poor cement bonds provide a higher probability of subsurface blowout.

KB-5 experienced a minor wellhead leak of CO_2 in 2007, following the arrival of CO_2 from KB-502 much faster than expected [2]. The flow rate of gas through this opening was small and the well is located in an unpopulated area, so the consequence of this wellhead leak was negligible. By 2010, KB-5 was fully decommissioned, but the occurrence of KB-5 well leakage raised the expected probability of well leakage to likely from improbable-unlikely as observations revealed faster CO_2 transport than was modelled, cement bonds may be poor, and a flange was found to have been removed by persons unknown from the wellhead [2]. While ECA impacts were still estimated to be low from such events, USDW impacts were assessed to be high resulting in an assessment that such leakage risk was moderate (see Figure 3).

October 2008-2010 leakage risk update.

Additional information became available on the (CI), and new water wells were constructed, but neither of these developments led to changes in our risk assessment. Analysis of InSAR data between 2008 and 2010 suggested some pressurization may have occurred at the base of the Viséan but still within the storage region, a topic of research at that time using approaches such as those of [10] and [11].

A summary of the likelihood, consequences, and combined (risk) assessed at each of the three times based on the existing state of knowledge at those times is presented in Figure 3. As shown, the assessment of potential consequences did not change with greater knowledge in this case, but the assessment of likelihood of the leakage scenario changed depending on available data and mitigation activities. As for the latter, the plugging and abandonment of KB-5 lowered our assessment of likelihood of leakage through wells in 2009, resulting in the lowering of risk of CO_2 leakage through wells from moderate in 2008 to low in 2009.

Overall Recommendations from our CF analysis in 2010

Given the occurrence of faster-than-expected flow of CO_2 from KB-502 to KB-5 along the preferred orientation of faults [2], we noted in 2010 the potential for accelerated transport of CO_2 from KB-503 to KB-4, and recommend that the integrity of legacy wells KB-2, 4 and 8 should be evaluated and checked at appropriate intervals, and decommissioned if necessary [4]. We also recommended in 2010 [4] that work should be undertaken to determine if InSAR data inversion can discriminate brine pressurization in the base of the Viséan alone from upward movement of CO_2 via numerous fractures in addition to such pressurization. In addition, we stated in [4] that continued injection at the pressures maintained up to that date created a significant opportunity for leakage out of the storage region that would propagate upward toward USDW. It was recommended that modeling of such an occurrence should be undertaken to determine the flow rates that might occur if such an event occurred. We also stated in [4] that a more prudent course of action would be to limit the pressures in all three injection wells to the lower end of the fracture gradient range for the base of the Viséan (a course of action that was in fact followed prior to our findings unbeknownst to us at that time).

Post-2010 conclusions on the dynamic CF analysis of In Salah

A broad program of monitoring and modeling focused on understanding various surface deformation (e.g., [12]), CO_2 migration [2], and seismic survey [13] observations at In Salah was carried out, the findings of which have been coming out in the literature over the last five years including some very recent publications (e.g., [10,11,14,15]). The general conclusions from this body of work are that (1) a fracture or fracture zone formed in the shale immediately above the reservoir, (2) this fracture or fracture zone may have allowed relatively fast migration of CO_2 from KB-502 to KB-5 within the injection horizon, and (3) this fracture or fracture zone does not span vertically through the entire 950m-thick cap rock, but rather is limited to the deepest layers. As preliminary results from these studies were being discussed, the joint industry operators in 2011 made the decision to stop injecting CO_2.

Looking back on our work reported in 2010 and summarized above, we believe the CF analysis approach was validated. In particular, without the benefit of 3D seismic data or advanced interpretations of the double-lobe uplift observed in InSAR data that came out later, our analysis correctly pointed to the hazard of higher-than-planned injection pressures in the storage reservoir, and the likelihood for fracturing to extend upward into the cap rock from these high injection pressures (see section *Prior to 2008 leakage risk* (above)).

SELECTED STUDY (1): SIZE OF FAULTS ACTIVATED BY INDUCED SEISMICITY

Introduction

Some recent publications have suggested that large-scale CO_2 injection will trigger earthquakes and that even small- to moderate-sized earthquakes may threaten the seal integrity of the injection zone [16]. For example, it was stated in [17] that "*Because of the critically stressed nature of the crust, fluid injection in deep wells can trigger earthquakes when the injection increases pore pressure in the vicinity of pre-existing potentially active faults.*" In the following, we assess such impact on near-offshore sediments along the Gulf of Mexico (see Figure 4). We do not discuss the validity of this claim for Gulf Coast sediments, which are somewhat unconsolidated and where regional tectonic stresses are currently low and extensional. Rather, we simply assume a scenario in which earthquakes do occur, and we assess their impact on seals. Induced seismic events of various origins in Texas have been described in several recent publications [18,19,20,21]. The slip generated by a single seismic event may have two major impacts on a CO_2 repository: (1) depending on the seal thickness and the fault size, it may allow CO_2 leakage through a newly created sand-on-sand pathway across the fault plane (assuming a sufficient fault transverse permeability); and (2) depending on the relative thickness of "sand" and "shale" intervals, it may create a vertical buoyant pathway along the fault plane. This section focuses on the first impact.

Three main types of faults are represented in the Gulf Coast area as shown in Figure 4: syndepositional growth faults, radial faults associated with shale or salt piercement structures, and, less commonly, regional post-depositional faults. Faults are generally envisioned mostly as features that act as seals but that would let fluids through in pulses [22]. Growth faults are generally thought to originate from sediment loading due to extensive sedimentary input. Growth faults are organized in systems with major faults that extend along strike 20 to 50 km [23]. Growth faults are the focus of this section. Throw of radial faults associated with shale or salt piercement structures can be important but is quickly attenuated away from the dome. It is currently thought that there is little salt diapiric activity in the Gulf Coast.

The objective of this special study was to determine fault-size distribution and inferred throw and compare it to the distribution of seal/cap-rock thickness, assuming that fault displacement is due to a single induced seismic event and is related to current fault size. Many peer-reviewed journal articles and reports have documented that fault instantaneous and cumulative displacement is related to fault size. For example, Mazzoldi et al. [24, p. 438], citing [25], suggested an average displacement over trace length ratio of 10^{-4} for single events. Another example [25, p. 118], citing [26], stated that a single-event ratio would range from 10^{-3} to 10^{-5}.

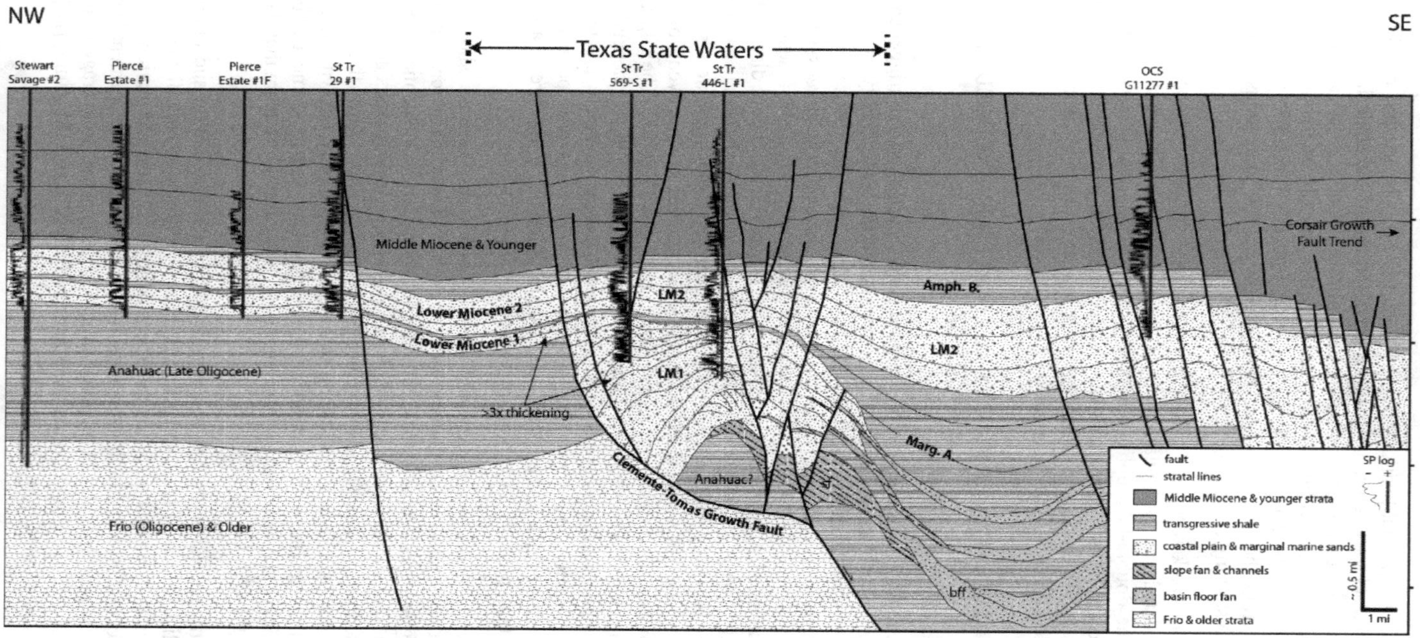

Figure 4. NW-SE cross section in the vicinity of the study area showing relationship between potential injection intervals, seals, and growth faults. Source: [28, Figure 2].

This study focuses on a 10×50 km^2offshore area of the Texas Gulf Coast. The Gulf Coast has been described as having a large potential capacity because of the thick sedimentary package [27]. We used a data set recently collected in the Texas offshore state waters, a 10-mile-wide track of submerged lands paralleling the Texas coastline between Houston and Victoria/Port Lavaca east of Matagorda and Brazoria counties where information on both seal thickness and fault size distributions is available. Seal-thickness distribution was obtained from compilation of individual well logs, whereas fault-size distribution was acquired through analysis of seismic information.

The vertical section analyzed is of Miocene age above the regional Anahuac seal at the bottom and mostly below the *Amphistegina* B (*Amph.* B) seal at the top. The rock volume contains two major intervals delimited by two regressive horizons, LM1 and LM2 (Lower Miocene 1 and 2) [28, p. 4]. LM2 is marked by the top of *Amph.* B, whereas the top of *Marginulina* A (*Marg.* A) denotes the top of LM1. The two thick regional shale layers (*Amph.* B and Anahuac) are relatively easy to correlate between wells. However, many smaller seals that may provide primary seals to CO_2 storage sites are not easily correlated across wells because of (1) the commonly large change in thickness of the growth fault across the fault line and (2) the deltaic depositional system favoring shaly intervals limited in horizontal extent and possibly of dimensions smaller than the interwell distance. The Anahuac Formation is the secondary and ultimate seal for injection in the very thick Frio Formation (not studied here). Similarly, the *Amph.* B. shale is the secondary and ultimate seal for all injection intervals in the Lower Miocene. Given the thickness of both the Anahuac and the *Amph.* B shale, no realistic seismic event or small series of seismic events will offset them significantly. However, this may not be true of smaller local primary seals, as we analyze next.

Data

We relied on two types of data regarding the same rock volume: well logs and seismic. The well logs are from historical wells drilled in the area with resistivity and spontaneous potential (SP) tracks from which we extracted shale thickness information. The 3D seismic package covers the same area, but its resolution is generally too low to allow for shale layer correlation, except for the largest such as Anahuac and *Amph.* B.

Methodology

Overall, the methodology to assess the induced seismicity risk is (1) to extract information about seal thickness and compile statistics; (2) to extract information about fault length and compile statistics; (3) to derive single-event displacement from fault length; and (4) to compare seal thickness and single-event displacement. We assume that all displacements are vertical.

Results

We analyzed geophysical logs from a total of 71 wells. The net shale values of the studied interval increase from the shoreline seaward. Overall, from well logs, shale abundance is ~68%, in agreement with similar observations elsewhere in the Gulf Coast area. Only shale intervals having a thickness >10 ft were retained as viable seals. A detailed analysis of shale and sand thickness shows that many sand (not much capacity) or shale (not much of a seal) intervals are thin and that their distribution can be fit to a lognormal distribution (2,871 of each sand and shale intervals). Suitable injection intervals would consist of thick sands (e.g., >50 ft, about 6% of all sand intervals) overlain by a good seal (e.g., >20 ft, about 40% of all shale intervals). No attempt was made to count these suitable combinations.

We collected data from the seismic survey at five horizons. We represented the 2D fault traces by their midpoint. Each fault segment was then assigned five data pairs: midpoint location and length for each of the five horizons. The analysis was done in a time domain not converted to depth, but this small imprecision should not have a large impact on approximately horizontal trace measurements. Length statistics are given in Table 2. Length distribution of fault segments (723 segments measured) is lognormal with a long tail with many segments measured at <2,000 ft, some representing major growth faults longer than 50,000 ft.

Table 2. Fault length count.

Fault length (km)	Top Miocene	Intra-Miocene	*Amph.* B	Lower Miocene	Base Lower Miocene
0–1.5	8	20	45	121	74
1.5–4	4	16	41	79	42
4.0–7.0	8	24	32	52	19
7.0–13.0	9	15	24	20	19
13.0–31.4	5	11	13	11	4
Total	**34**	**86**	**155**	**283**	**158**

We then performed a simple Monte-Carlo analysis comparing shale thickness to fault displacement. Fault displacement was estimated as the product of the fault length times an uncertain multiplier coefficient. The multiplier coefficient is assumed to follow a triangular distribution with a 10^{-3} to 10^{-5} range and a mode of 8×10^{-5} [25]. Fault length and seal thickness were not fit to a specific distribution; rather, the data from the study area were used. We achieved stable results after ~200,000 trials. In the vast majority of cases, displacement is much smaller than seal/shale thickness (Figure 5). Only 1.8% of thickness/displacement pairs display a displacement greater than 20% of the seal thickness. Only 0.05% of thickness/displacement pairs result in a clear seal rupture, and only 0.26% of the pairs result in a displacement of half the seal thickness.

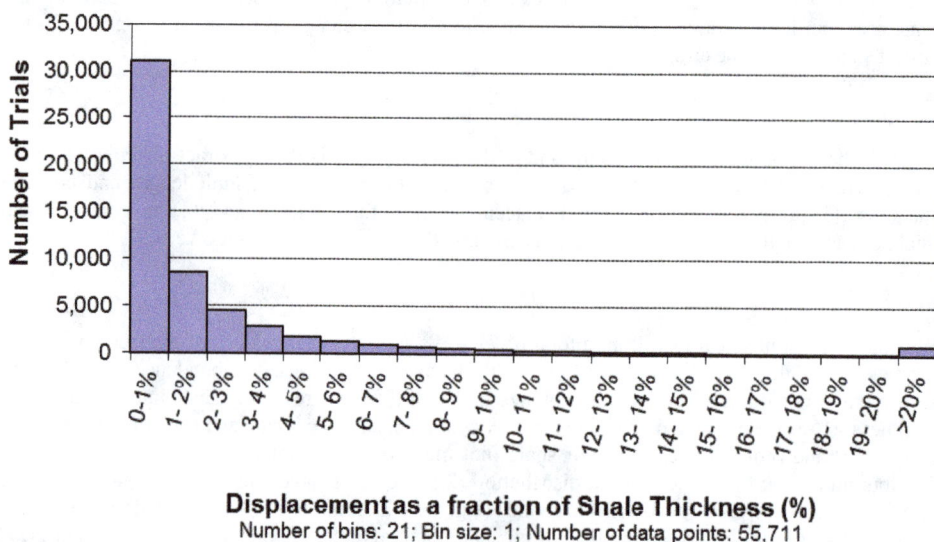

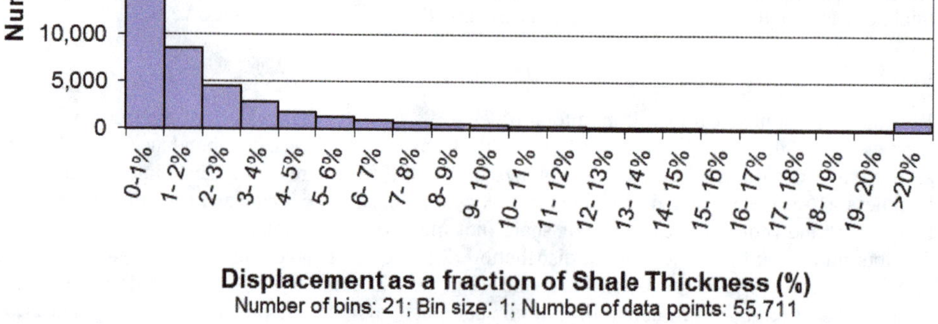

Figure 5. Histogram of single-event maximum displacement as a function of shale thickness (55,711 trials).

Discussion

The results suggest that induced seismicity is not an issue in terms of storage integrity in the Gulf Coast area. A more accurate calculation would include the fact that seals may have already been offset by several previous events. Maximum fault trace versus maximum fault displacement data for 297 faults from the LM2 horizon shows good agreement with the fault-growth models (*F'* = 3 GPa shear modulus) of Walsh and Watterson [29]. Clearly if the shale interval has already been fully offset, it may not act as a seal any longer. A quick analysis reveals that in ~75% of cases the fault throw (historical cumulative impact of individual single seismic events) is larger than the shale interval thickness. This, however, does not change the statistics previously presented. Rather, it could limit the capacity of the entire Lower Miocene section, that is, keeping the statistics valid but for a smaller number of sites.

We now compare the seismic event magnitude M that would be generated by the displacements calculated above. The magnitude is given by:

$$M = \frac{2}{3}[\log(M_0) - 9.1] \qquad \text{Equation 1}$$

with $M_0 = F'D_SA$, where F' is the shear modulus (F' = 3 GPa), D_s is the single-event displacement, and A is the area of rupture (as quoted in [24]). The only unknown and uncharacterized parameter in this equation is the area A. For smaller faults, A is generally understood as a disk with the trace length defining the diameter. However, such an approach is unrealistic for faults having larger trace lengths. In this case, we defined the height of area A, assumed to be 1,000 ft, as a conservative estimate of the vertical section of the fault exposed to some degree of overpressure following CO_2 injection. The resulting magnitude for the 476 cases with full seal offset ranges from 4.7 to 5.8 with an average of 5.2. Reasoning *a contrario*, such high magnitudes are not typical of injection-related seismic events, suggesting that the area A is probably overestimated. Decreasing vertical length to 500 ft yields a range of 4.5 to 5.6 with an average of 5.1. The National Research Council [30, Table S.1] reported that the maximum magnitude of felt events was 4.9 for waterfloods and 4.8 for disposal wells, in approximate agreement with the results of this CF special study.

SELECTED STUDY (2): VERTICAL GROWTH OF CAP-ROCK FRACTURE FILLED WITH BUOYANT FLUID

Overview

Large vertical fractures through cap rock are potential pathways for fluid leakage into valuable groundwater resources in geologic CO_2 sequestration and shale hydrocarbon production systems. We have carried out numerical experiments in simple model systems to calculate static pressure and pressure evolution in vertical fractures. In the first experiment, we examined static pressures in 1 km-high vertical fractures filled with a buoyant (less dense) fluid such as CO_2. We found that buoyant fluid can be expected to generate high fracture-tip pressures. This result suggests that fractures that are initiated at depth due to overpressure created during injection of CO_2 and which grow vertically will have a growing pressure-driving force with fracture height, provided there is a continuous supply of fluid injected at the fracture base.

Introduction

GCS sites will typically be favored in subsurface environments with thick cap-rock seals capable of trapping buoyant CO_2 for millennia or longer. During the injection process, pressures in the storage zone may locally exceed the fracturing pressure of the cap rock if safe operating procedures are not followed, or injected CO_2 may fill existing open vertical fractures or faults. In either case, buoyant CO_2 could fill the fracture leading to high fracture pressures because of (1) direct connection to the injection zone, (2) high permeability of the fracture, and (3) buoyancy of the CO_2. These high fracture pressures could promote upward fracture growth and potentially cause the fracture to propagate farther into or across the thickness of the cap rock. Such fracturing is a concern for CO_2 leakage from the storage region into USDW (e.g., [1]). Similar concerns for groundwater contamination by means of fluid flow upward through vertical fractures exist for shale-gas and shale-oil production enhanced by hydraulic fracturing.

Here we present results of numerical experiments to calculate pressure in fractures containing buoyant CO_2, and dissipation of pressure due to flow of CO_2 or water into the fracture wall rock. These results serve to enhance understanding of potential fracture propagation and growth that can be incorporated into future application and development of the CF. The simple numerical experiments do not include geomechanical coupling that would allow simulation of the propagation of the fracture. Instead we present results in terms of the pressure-driven tendency for fracture growth. These analyses provide readers with the basis for intuition about pressure in gas-filled fractures and related pressure dissipation. Our simple non-geomechanical analyses are relevant to early stages of risk-assessment during which hazards are being analyzed at a high level by simple conceptual analyses and thinking. Fully coupled hydro-geomechanical modeling would be the next step if actual impact assessment due to fracture propagation and fluid leakage fluxes were warranted at the site. While we are motivated primarily by interest in GCS, the results are relevant also to hydrocarbon production from shale where hydraulic fracturing is carried out intentionally, and vertical fractures may become filled with buoyant natural gas or oil.

Prior Work

Fluid injection-induced fracture propagation and growth is normally studied with consideration of coupled hydro-geomechanical processes and models (e.g., [31,32,14,10]). These modeling studies and analyses have addressed key issues related to propagation of fractures through cap rock, to fluid leakage through faults and fractures, and to fault re-activation for studies of induced seismicity. These sophisticated modeling studies are capable of providing a detailed understanding of coupled hydro-geomechanical processes on the one hand, and the most defensible estimates of likelihood of induced seismicity and upward fluid leakage fluxes (needed for quantitative risk assessment studies at sites where considerable investments are being made) on the other hand. Stepping back a bit from the complexity of coupled hydro-geomechanics and the potential dependence of related processes on potentially poorly known properties of the system, e.g., stress state and fracture stiffness, we have taken the view in this special study that there is a need for more basic intuition around the effects and processes related to hydraulic fracturing. The simple modeling reported here can serve as the basis for a better understanding of the first-order behavior of buoyant-fluid-filled fractures in cap rock, and serve the needs of much simpler analyses and thinking that need to be done quickly and economically at multiple sites in the early stages of risk assessment.

Methods

We carried out static and dynamic analyses of buoyant pressure rise and pressure dissipation in fractures using TOUGH2/ECO2H [33], a numerical simulator of multiphase and multiple component fluid and heat flow in porous or fractured media. TOUGH2/ECO2H calculates thermophysical properties of water, salt, and CO_2 under varied pressure and temperature conditions typical of deep subsurface environments. In the static analysis, we modeled an open vertical fracture through cap rock as a one-dimensional pressure distribution problem. The fracture is filled with the injection fluid and it is closed everywhere except at its base where it is exposed to the injection-induced reservoir pressure. In the dynamic analysis, we used a two-dimensional domain to describe the open fracture and flow into surrounding permeable wall rock. The transient fluid flow along the fracture was simulated numerically along with the flow through the permeable wall rock in response to buoyancy and injection-based pressure driving forces.

Static analysis of pressure due to injection overpressure and buoyant column height

In this first numerical experiment, we investigated the potential for vertical growth of a cap rock fracture filled with buoyant fluid by calculating the pressure in the fracture relative to hydrostatic pressure in the cap rock. The pressure at the tip of a vertical fracture is a function of the injection overpressure at the base of the fracture and the fluid column height above the base of the fracture. Different fluids within the fracture relative to formation fluids may have different effects on fracture tip pressure because of different fluid density. Our static analysis arbitrarily assumed a 1,000 m-high fracture had already formed at a given depth because pressure exceeded the local fracturing pressure, and that the pressure within the vertical fracture has reached an equilibrium state (i.e., no flow). The three unconnected model systems are shown in Figure 6. The fracture was discretized into 20 m equispaced grid blocks in the vertical direction. The bottom boundary is held at constant pressure (e.g., the fracturing pressure of the given depth). The top boundary and the side boundaries are closed. We present results for two sets of cases, namely all-water-filled and all-CO_2-filled fracture cases. Each case is investigated for the fractures starting at three different depths (2,000, 3,000, and 4,000 m) to calculate the pressure distributions as a function of depth. The bottom boundary pressure is 31.7, 47.5, and 63.3 MPa for depths of 2,000, 3,000, and 4,000 m, respectively. Although isothermal simulations are used (temperature does not change with time), the temperature varies from 15 °C at the ground surface (depth = 0 m) to 115 °C at 4 km depth assuming a geothermal gradient of 25 °C/km.

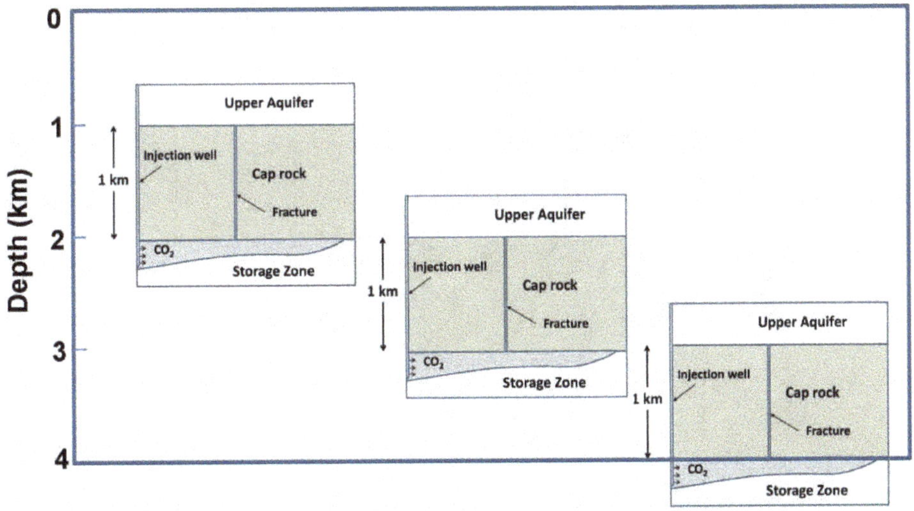

Figure 6. Three depths of the idealized model system for analysis of pressure in a large vertical fracture filled with water and with CO_2 at three different depths.

As shown by the results in Figure 7a, the all-water cases (red lines) have slopes that are the same as the hydrostatic pressure gradient because water density is not very sensitive to the pressure. For the all-CO_2 cases (blue dashed lines), the pressure increases as a function of depth in the fractures less than that in the all-water case, because the density of CO_2 is smaller than water at the same pressure and temperature. Since the entire pressure range of the cases examined is above the critical pressure of CO_2, the effects due to the sharp change in CO_2 density, as it transitions from super critical conditions to gaseous conditions on the pressure distribution, does not occur in these cases.

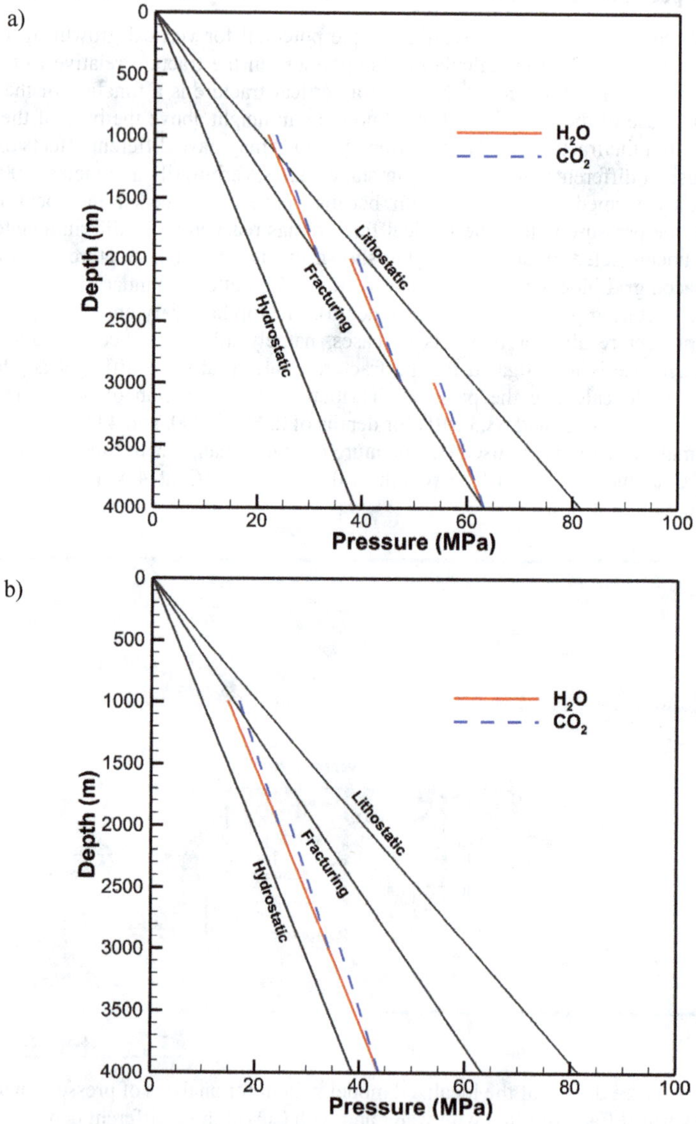

Figure 7. Pressure distributions in vertical fractures if the injection-induced pressure reaches (a) the fracturing pressure), and (b) 5 MPa above the hydrostatic pressure at the given injection depths (2,000, 3,000, and 4,000 m). The red lines are the all-water cases whereas the blue lines are the all-CO_2 cases. The gray lines (as labeled) are hydrostatic, fracturing, and lithostatic pressure distributions, respectively.

As a result of pressurized fluid filling the open fracture, fracture-tip pressure could exceed lithostatic pressure as the fracture extends upward from the shallow injection point (2,000 m) because the lithostatic pressure decreases as the fracture becomes shallower much more quickly than the hydrostatic pressure does. As our 1 km-high fracture is much larger than we would expect at actual field sites, and GCS is usually targeted at deep reservoirs, exceeding lithostatic pressure is not considered likely.

Figure 7b shows the results of calculations similar to those in Figure 7a but with injection-induced pressure of 5 MPa above the hydrostatic pressure at the given depths (2,000, 3,000, and 4,000 m, respectively). In this scenario, the injection does not fracture the formation at the injection point, thus we are examining the pressure profile in the case that injected water or CO_2 fills a pre-existing fracture. As expected, the all-water cases of pressure in 1,000 m-high (pre-existing) hypothetical fractures form a hydrostatic pressure profile shifted by 5 MPa relative to the non-injection gradient. In contrast, for the all-CO_2 cases, the lower density of CO_2 causes a slower decrease of pressure in the fracture as the fracture becomes shallower (Figure 7b). For the same 1,000 m length of the fractures, the pressures at a hypothetical fracture tip could exceed the fracturing pressure for the all-CO_2 case, but not for all-water case. But neither case exceeds the fracturing pressure if the injection depth is below 3,000 m.

Figure 8a shows the fracturing driving force (the fluid pressure in vertical fractures minus the fracturing pressure at the given depth (Pl - Pf) and the dissipation driving force (the fluid pressure in vertical fractures minus the hydrostatic pressure at the given depth (Pl - Ph) for the same cases as those shown in Figure 7a. The fracturing driving forces are positive at all depths and increase from the base to the tip of the fractures. Such increases are more profound in the all-CO_2 cases than in the all-water cases because of the buoyant effects of CO_2. For the same fracture length of 1,000 m, the all-CO_2 cases have about 1.51, 1.38, and 1.27 (MPa) more fracturing driving pressure than the all-water case if the depths of the fracture base are at 2,000, 3,000, and 4,000 m, respectively. Similarly, the dissipation driving force is almost constant in each vertical fracture in the all-water cases (< 0.5%) but it increases significantly from the base to the tip of each fracture in the all-CO_2 cases. The magnitude of such increase can reach 1.46, 1.30, and 1.15 (MPa) for the depth of the fracture base of 2,000, 3,000, and 4,000 m, respectively.

For the same injection-induced pressure perturbation of 5 MPa shown in Figure 7b, we present in Figure 8b the fracturing and dissipation driving forces. Note in Figure 8b that the fracturing driving force is negative (i.e., no fracture propagation will occur) in the deep (depth > 2,000 m) cases for both the all-water and all-CO_2 cases. At shallower depths (e.g., 1,000 m), the fracturing driving force becomes positive for the all-CO_2 case. For the same fracture height of 1,000 m, the all-CO_2 cases get about 2.11, 2.21, and 2.22 MPa more fracture driving pressure than the all-water cases if the depths of the fracture base are at 2,000, 3,000, and 4,000 m, respectively. The corresponding magnitudes of increase in dissipation driving force due to buoyant effects of CO_2 are 2.09, 2.19, and 2.19 MPa, respectively.

Simple hydrostatic analyses suggest that fractures initiated at depth by injection of a buoyant (less dense) fluid and filled with that buoyant fluid such as CO_2, can be expected to generate anomalously high fracture-tip pressures. We also found through these simple analyses that if the pressure is elevated by injection, e.g., by 5 MPa, but not above the fracture pressure, the fluid pressure at the fracture tip of a high vertical fracture at shallow depths can exceed the fracture pressure if the fracture is filled with CO_2. This result illustrates the containment problems that can arise when long connected fluid column heights exist in high-permeability fractures or fault zones. When the injection-induced pressure exceeds the fracturing pressure at the base, the pressure dissipation driving forces are larger for the deeper fracture cases, even though the fracturing driving forces are about the same at each depth. When the injection-induced pressure exceeds hydrostatic pressure but not fracturing pressure at each depth, the pressure dissipation driving force increases as the fracture becomes shallower for CO_2-filled fractures. In this case, fracturing pressure is only exceeded for shallow fractures. This study did not consider the effects of pressure dissipation, considered in the next study.

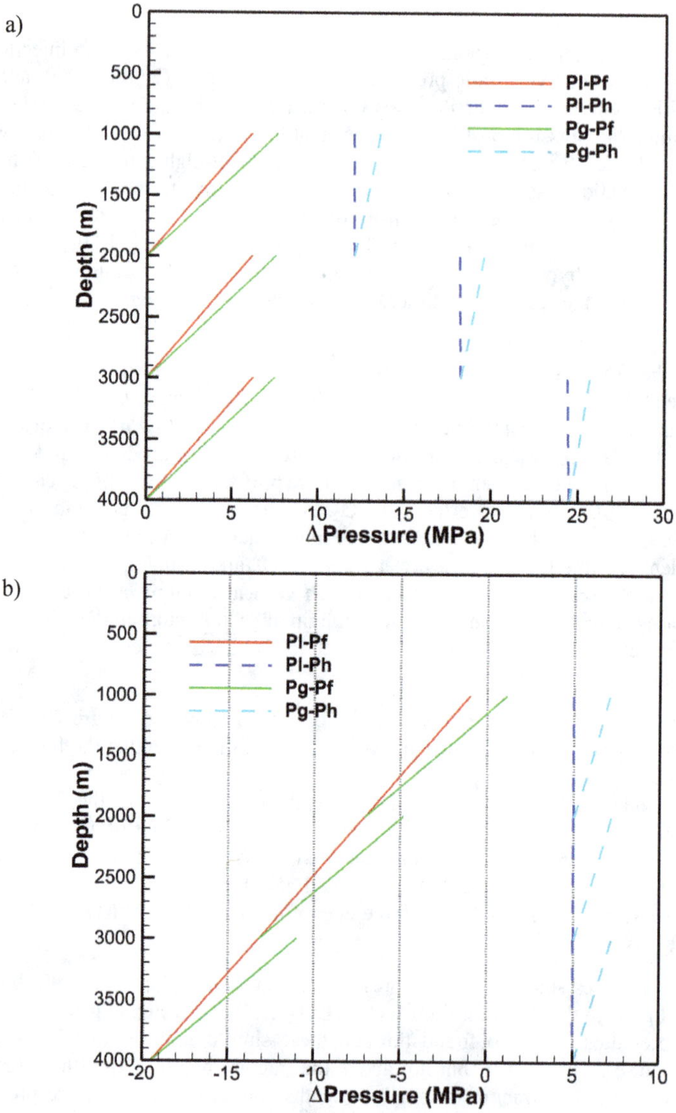

Figure 8. Pressure difference in vertical fractures plotted for the fracturing driving force ($Pl - Pf$, or $Pg - Pf$, solid lines) and the dissipation driving force ($Pl - Ph$, or $Pg - Ph$, dashed lines) for the cases that the injection-induced pressure reaches (a) the fracturing pressure, and (b) 5 MPa above the hydrostatic pressure at the given depth. Pl = pressure of water; Pf = fracturing pressure; Ph = hydrostatic pressure; Pg = pressure of CO_2

SELECTED STUDY (3): PRESSURE DISSIPATION IN FRACTURES

Effects of permeable matrix wall rock on the dissipation of pressure in fracture

In this set of numerical experiments, we developed a simple test problem in a 2D Cartesian geometry and examined the dissipation of pressure by explicitly modeling fluid flow into the fracture wall rock. This pressure dissipation plays off against buoyancy in the fracture and over pressure from the reservoir to control fracture tip pressure. We observed that pressure rises rapidly with time in the fracture for both water-filled and CO_2-filled fractures even with permeable wall rock. The dissipation of pressure by flow into the wall rock is negligible unless the matrix permeability is of the order 10^{-2} Darcy, a high value for cap rock. We conclude from this experiment that pressure dissipation by flow into wall rock is not an effective process for mitigating vertical fracture propagation in cap rock over time scales of hydraulic fracturing.

The simplified model of a single fracture (between the depths of -2,000 m and -3,000 m) with surrounding matrix (wall rock) is shown in Figure 9. The simulations are assumed to be isothermal with a constant geothermal gradient (25 °C/km and 15 °C at surface). The system is initially under hydrostatic pressure and filled with pure water. At time zero, an instantaneous pressure perturbation of 100% water or CO_2 saturation at fixed pressure was imposed at the bottom of the fracture and maintained thereafter. Each case is simulated for 4,000 time steps (~26 hours). Properties of the system are given in Table 3. We used T2Well/ECO2H [34] with the wellbore capability unengaged to simulate the flow processes. We note that geomechanical coupling is not considered here as we focus on pressure evolution in an existing fracture.

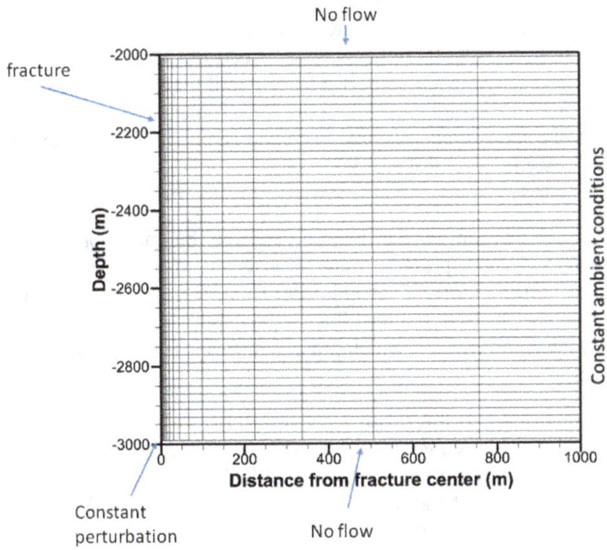

Figure 9. Numerical mesh (nodes and connections) and boundary conditions used in the analysis. The simulated domain is filled by matrix except for the left-most side, which is the fracture which has an aperture of 2 mm (one-half of this is represented in the mirror-plane symmetric grid. The top and bottom boundaries of the domain are no-flow except for the bottom of the fracture where a constant fluid perturbation condition (e.g., pressure and saturation) is maintained. On the far right-hand side (1,000 m away), constant ambient conditions (i.e., hydrostatic, all water, and ambient temperature) are maintained. The horizontal nodal distance varies from 1 mm near the fracture to about 200 m in far field whereas the vertical nodal distance is 20 m uniformly.

Table 3. Properties of fracture and matrix (cap rock)

Parameter	Fracture	Matrix	Notes
Permeability	10^{-6} m^2	10^{-16} m^2, ± 2 orders	
Porosity	0.50	0.50	
Parameters for relative permeability:			Liquid relative permeability using van Genuchten-Mualem model [35] and (for matrix) gas relative permeability using Corey model [36]
Residual gas saturation	0.00	0.04 or 0.14	
m_{VG}	0.20	0.20	
Residual liquid saturation	0.027	0.27	
Saturated liquid saturation	1.0	1.0	
Parameters for capillary pressure:			Capillary pressure using van Genuchten-Mualem model [35]
Residual liquid saturation	0.025	0.25	
m_{VG}	0.20	0.20	
Characteristic capillary pressure	11.90 Pa	1190.48 or 4961.91 Pa	
Maximum capillary pressure	1×10^2 Pa	1×10^5 Pa or 1×10^7 Pa	
Saturated liquid saturation	1.0	1.0	
Pore compressibility	10^{-10} Pa^{-1}	10^{-10} Pa^{-1}	

We present in Figure 10a the profile with depth of the pressure-difference (P - $P_{fracturing}$) in the fracture for various combinations of fluid and matrix properties at the end of the simulation. The base case is one with only the fracture; no matrix is present. In general, as the permeability of the matrix increases, the pressure in the fracture should decrease. However, we observed that such pressure dissipation effects are smaller than the buoyant effects of CO_2 if the fracture is long and the cap rock permeability is low, as we would expect it to be. Until the matrix permeability becomes as large as 10^{-14} m^2, the change in fracture pressure in the test problem was not significant in both the water-filled fracture and the CO_2-filled fracture. With the same small permeability of the matrix (10^{-16} m^2), the fracture pressure was also not very sensitive to the residual gas saturation or the capillary pressure of the matrix.

The effects of pressure dissipation to the matrix seem to be more profound at early time (Figure 10b), which implies that the pressure dissipation could play a more important role in the case of a growing fracture. Furthermore, if second-order small fractures develop in the neighboring matrix along the main fracture, the bulk permeability in the neighborhood of the main fracture could change tremendously which could result in much more significant effects of the pressure dissipation than those simulated here.

Note that the system has not reached the steady state for all of the cases shown in Figure 10b. For example, for the case "P14_CO2" (i.e., the most pressure-dissipated case), the ending time of 4,000 time steps is 25.841 hours and the pressure profile is still developing. The pressure gradient has been established near the fracture (at the left-hand side), and CO_2 has leaked into the matrix but remains within a limited distance of the fracture.

Dynamic simulations of fluid filling a fracture showed that pressure rises rapidly for both water- and CO_2-filled fractures. The dissipation of pressure was negligible over time scales on the order of 10 days unless the matrix permeability was on the order of 10^{-2} Darcy or higher, not typical of cap rock. The implication of this result is that pressure dissipation due to fluid leak-off into wall rock matrix of cap-rock fractures in water-saturated systems cannot be counted on as a means of reducing fracture-driving forces. In short, once a fracture is initiated and filled with buoyant fluid, high fracture tip pressures will persist if there is a continuous supply (pressure maintenance) of fluid from below.

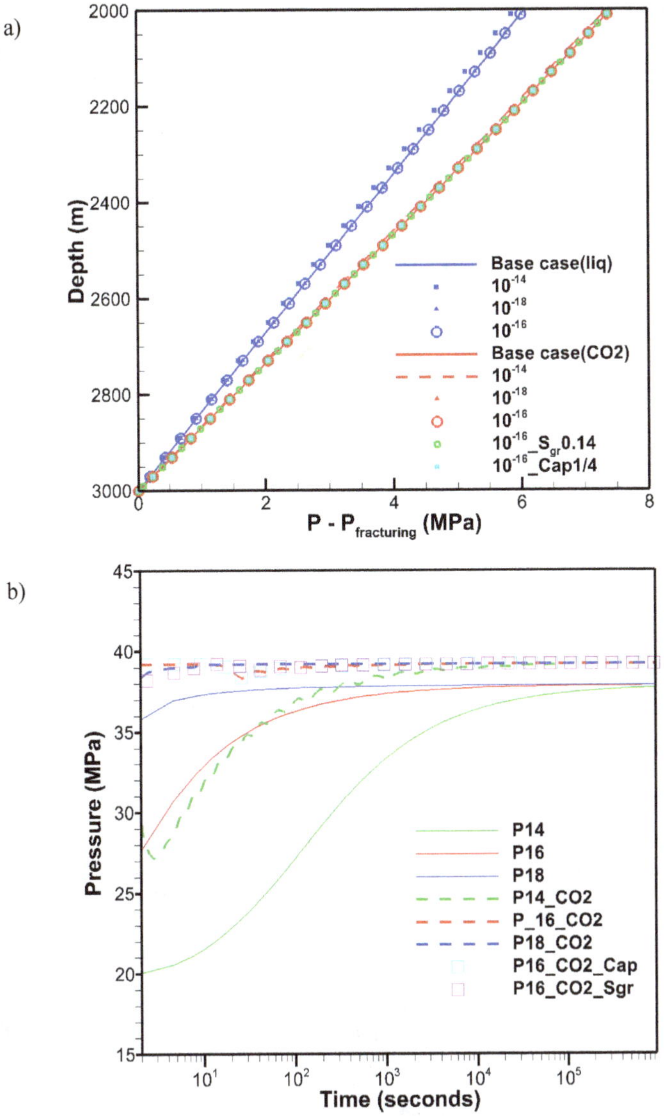

Figure 10. (a) Simulated profile of pressure with depth in the fracture for different fluids and matrix permeabilities. "Base case" is the fracture-only case (static pressure without permeable matrix). The number labels indicate the corresponding permeability of the matrix. "Sgr 0.14"indicates that the residual gas saturation is 0.14 (the other cases are 0.04) and "Cap1/4" indicates that the characteristic capillary pressure is 4961.91 Pa (the other cases are 1190.48 Pa) and the maximum capillary pressure is 10^7 Pa (the other cases are 10^5 Pa) for the matrix. The properties of the fracture stay the same in all cases. "(liq)" indicates water only whereas "(CO_2)" indicates CO_2 only cases. (b) Simulated pressure at the fracture tip as a function of time for various cases. "P14" indicates permeability of 10^{-14} m^2 and similarly for P16 and P18. "CO_2" indicates the CO_2-filled fracture case (default is water-filled fracture).

CONCLUSIONS

During the period 2010-2014, we built upon the CF through applications to sites and through special studies of various topical GCS risks. In this chapter, we summarized a selected short-list of four studies: (1) the application of the CF to the In Salah dynamic risk assessment; (2) potential impacts on leakage of induced seismicity; (3) effect of buoyant fluid in vertical fractures; and (4) pressure dissipation in fractures.

ACKNOWLEDGEMENTS

We thank Quanlin Zhou (LBNL) and an external reviewer for constructive comments and suggestions. This work was supported by the CO_2 Capture Project (LBNL Award No. WF009343, Sponsor Award No. SMV-031, and UT award No. 61798, Sponsor Award No. SMV-032BP). Additional support came from Lawrence Berkeley National Laboratory through the U.S. Department of Energy under Contract No. DE-AC02-05CH11231.

REFERENCES

1. Oldenburg C.M., Bryant S.L., Nicot J.-P. (2009) Certification framework based on effective trapping for geologic carbon sequestration. *International Journal of Greenhouse Gas Control* 3(4): 444-457.
2. Ringrose P., Atbi M., Mason D., Espinassous M., Myhrer Ø., Iding M., Mathieson A., Wright I. (2009) Plume development around well KB-502 at the In Salah CO_2 storage site. *First Break* 27, 49-53.
3. Dodds K. (2009) In Salah CO_2 JIP: status and overview. Presentation to the *5th IEA Monitoring Meeting*, Tokyo, Japan, June 2-3, 2009.
4. Oldenburg C.M., Jordan P.D., Nicot J.-P., Mazzoldi A., Gupta A.K., Bryant S.L. (2010) Leakage risk assessment of the In Salah CO_2 storage project: Applying the Certification Framework in a dynamic context. LBNL-4278E. *Energy Procedia*, 2, Elsevier, GHGT-10, Amsterdam, The Netherlands, Sept. 19-23, 2010.
5. Edmunds W.M., Guendouzb A.H., Mamouc A., Moullab A., Shanda P., Zouarid K. (2003) Groundwater evolution in the Continental Intercalaire aquifer of southern Algeria and Tunisia: trace element and isotopic indicators, Applied Geochemistry, 18, p.805-822.
6. Guendouz A., Michelot J.-L. (2006) Chlorine-36 dating of deep groundwater from northern Sahara, Journal of Hydrology, 328, 572– 580.
7. Ould Baba Sy M. (2005) Recharge et paleorecharge du systeme aquifère du sahara septentrional, PhD Dissertation, University of Tunis, Tunisia, 261p. [in French].
8. Ringrose P.S., Roberts D.M., Gibson-Poole C.M., Bond C., Wightman R., Taylor M., Raikes S., Iding M., Østmo, S. (2011) Characterisation of the Krechba CO_2 storage site: Critical elements controlling injection performance." *Energy Procedia* 4, 4672-4679.
9. Jordan P.D., Benson S.M. (2008) Well blowout rates and consequences in California oil and gas district 4 from 1991 to 2005: implications for geological storage of carbon dioxide, *Environmental Geology*.
10. Rutqvist J., Vasco D.W., Myer L. (2010) Coupled reservoir-geomechanical analysis of CO_2 injection and ground deformations at In Salah, Algeria. *International Journal of Greenhouse Gas Control* 4(2): 225-230.
11. Iding M., Ringrose P. (2010) Evaluating the impact of fractures on the performance of the In Salah CO_2 storage site. *International Journal of Greenhouse Gas Control* 4.2, 242-248.

12. Vasco D.W., Rucci A., Ferretti A., Novali F., Bissell R.C., Ringrose P.S., Mathieson A.S., Wright I.W. (2010) Satellite-based measurements of surface deformation reveal fluid flow associated with the geological storage of carbon dioxide. *Geophysical Research Letters* 37(3), L03303.
13. Gibson-Poole C.M., Raikes S. (2010) Enhanced understanding of CO_2 storage at Krechba from 3D seismic. In: *9th Annual Conference on Carbon Capture and Sequestration*, Pittsburgh, PA, May 10-13, 2010.
14. Rinaldi A.P., Rutqvist J. (2013) Modeling of deep fracture zone opening and transient ground surface uplift at KB-502 CO_2 injection well, In Salah, Algeria. *International Journal of Greenhouse Gas Control* 12, 155-167.
15. White J.A., Chiaramonte L., Ezzedine S., Foxall W., Hao Y., Ramirez A., McNab W. (2014) Geomechanical behavior of the reservoir and caprock system at the In Salah CO_2 storage project. *Proceedings of the National Academy of Sciences*, 111(24): 8747–8752.
16. Verdon, J.P., Kendall J.-M., Stork A.L., Chadwick R.A., White D.J., Bissell R.C. (2013) Comparison of geomechanical deformation induced by megatonne-scale CO_2 storage at Sleipner, Weyburn, and In Salah. *Proceedings of the National Academy of Sciences* 110.30: E2762-E2771.
17. Zoback M.D., Gorelick S.M. (2012) Earthquake triggering and large-scale geologic storage of carbon dioxide, *Proceedings of the National Academy of Sciences of the United States of America*, 109(26): 10164–10168.
18. Frohlich C. (2012) Two-year survey comparing earthquake activity and injection-well locations in the Barnett Shale, Texas, *Proceedings of the National Academy of Sciences of the United States of America*, 109(35): 13934–13938.
19. Frohlich C., Glidewell J., Brunt M. (2012) Location and felt reports for the 25 April 2010 m(bLg) 3.9 earthquake near Alice, Texas: was it induced by petroleum production? *Bulletin of the Seismological Society of America*, 102(2): 457–466.
20. Gan W., Frohlich C. (2013) Gas injection may have triggered earthquakes in the Cogdell oil field, Texas, *Proceedings of the National Academy of Sciences*, 110(47): 18786–18791.
21. Justinic A.H., Stump B., Hayward C., Frohlich C. (2013) Analysis of the Cleburne, Texas, earthquake sequence from June 2009 to June 2010, *Bulletin of the Seismological Society of America*, 103(6): 3083-3093.
22. Sibson R.H. (1992) Implications of fault-valve behavior for rupture nucleation and recurrence, *Tectonophysics*, 211, 283-293.
23. Nicot J.-P., Hovorka S.D. (2009) Chapter 17. Leakage pathways from potential CO_2 storage sites and importance of open traps: case of the Texas Gulf Coast, *in* Grobe, M., J.C. Pashin, and R.L. Dodge, eds., Carbon dioxide sequestration in geological media—state of the science, *American Association of Petroleum Geologists Studies in Geology*, 59, 321–334.
24. Mazzoldi A., Rinaldi A.P., Borgia A., Rutqvist J. (2012) Induced seismicity within geological carbon sequestration projects: Maximum earthquake magnitude and leakage potential from undetected faults, *International Journal of Greenhouse Gas Control*, 10, 434–442.
25. Ferrill D.A., Smart K.J., Necsoiu M. (2008) Displacement–length scaling for single-event fault ruptures: insights from Newberry Springs Fault Zone and implications for fault zone structure, *in* Wibberley C.A.J., W. Kurz, J. Imber, R.E. Holdsworth, and C. Collettini eds., *The Internal Structure of Fault Zones: Implications for Mechanical and Fluid-Flow Properties*: The Geological Society of London, London, http://dx.doi.org/10.1144/SP299.7
26. Wells D.L., Coppersmith K.J. (1994) New empirical relationships among magnitude, rupture length, rupture width, rupture area, and surface displacement, *Bulletin of the Seismological Society of America*, 84(4), 974-1002,A1001-A1004,B1001-B1011,C1001-C1049.

27. NETL (2012) The 2012 United States carbon utilization and storage atlas—fourth edition (Atlas IV): 129 p., http://www.netl.doe.gov/technologies/carbon_seq/refshelf/atlas/, accessed November 2013.
28. Nicholson A.J. (2012) Empirical analysis of fault seal capacity for CO_2 sequestration, Lower Miocene, Texas Gulf Coast: The University of Texas at Austin, Department of Geological Sciences, M.S. thesis, 78 p.
29. Walsh J.J., Watterson J. (1988) Analysis of the relationship between displacements and dimensions of faults, *Journal of Structural Geology*, 10(3): 239–247.
30. NAS (National Research Council) (2012) Induced seismicity potential in energy technologies, *The National Academies Press*, Washington, D.C., 225 p.
31. Rutqvist J., Birkholzer J.T., Tsang C.-F. (2008) Coupled reservoir-geomechanical analysis of the potential for tensile and shear failure associated with CO_2 injection in multilayered reservoir-caprock systems. *International Journal of Rock Mechanics and Mining Sciences* 45(2): 132-143.
32. Cappa F., Rutqvist R. (2011) Modeling of coupled deformation and permeability evolution during fault reactivation induced by deep underground injection of CO_2. *International journal of Greenhouse Gas Control* 5(2), 336-346.
33. Pruess K., Oldenburg C.M., Moridis G.J. (2012) TOUGH2 User's Guide Version 2. Lawrence Berkeley National Laboratory Report LBNL-43134 (revised).
34. Pan L., Oldenburg C.M. (2013) T2Well-An integrated wellbore-reservoir simulator. *Computers & Geosciences* 65, 46-55.
35. Van Genuchten M. Th. (1980) A closed-form equation for predicting the hydraulic conductivity of unsaturated soils. *Soil Science Society of America Journal* 44(5): 892-898.
36. Corey A.T. (1954) The interrelation between gas and oil relative permeabilities. *Producers monthly* 19(1): 38-41.

Carbon Dioxide Capture for Storage in Deep Geological Formations, Volume 4
Karl F. Gerdes (Editor)

Chapter 33

ASSESSMENT OF CO_2 UTILIZATION FOR SHALE GAS AND OIL

Anshul Agarwal and Dawn Geatches
Department of Energy Resources Engineering, Stanford University

ABSTRACT: It is anticipated that over the next two decades, tens of thousands of wells will be drilled in the 23 states in which organic-rich shale gas deposits are found. This review investigates the feasibility of using CO_2 enhanced recovery techniques in these formations. If feasible, the number of sites where CO_2 can be utilized increases dramatically. The feasibility of utilizing CO_2 in shale beds depends on many factors at multiple length-scales. For example, there are potential interactions of CO_2 with shale components at the nano- and micro- scales. Such questions can be addressed by a combination of experiments and simulations, where characterization experiments identify the components and composition of shale, providing structural input necessary to build robust models for simulations. In this way, modeling and simulations can be used to investigate both chemical and non-chemical interactions, e.g., the conversion of CO_2 to carbonates and CO_2 adsorption and transport within the shale matrix. Modeling such interactions is a powerful means of accessing information that can be difficult, if not impossible, to obtain from experiments alone. To capture this full range of anticipated interactions requires a variety of modeling techniques, with each having its own strengths and limitations. For example, modeling the possible chemical conversion of CO_2 to carbonates requires quantum mechanical modeling at the sub-nanometer scale. This is computationally expensive, which limits the size of the models to a few hundred atoms. Modeling the physical aspects of CO_2 interacting with a shale environment, such as transport and adsorption of CO_2 through, and within the shale matrix, requires both molecular (i.e., nano- to micro-) scale modeling, as well experiments. Although not limited so much by the size of the model, there are limitations on the time period that can feasibly be simulated, as well as the complexity of the model and its data availability.

KEYWORDS: CO_2 adsorption; enhanced oil recovery (EOR); enhanced gas recovery (EGR); clay swelling; shale gas; CO_2 storage; storage potential; recovery factor; fracking

INTRODUCTION

A literature review was conducted on the reported uses of CO_2 in shale gas and shale oil reservoirs, from both theoretical and field demonstration points of view, in order to evaluate various aspects such as (i) primary and secondary enhanced gas & oil recovery (EGR, EOR) with CO_2; (ii) CO_2 sequestration; and (iii) use of CO_2 as a fracking fluid. There are four main focus areas: I. Physical and chemical aspects of CO_2/shale interactions at the pore scale, II. Transport processes of critical state CO_2 in hydrofracs, natural fractures and pores, III. Chemical interactions with resident water, and IV. Trap and seal mechanisms of CO_2 in shale gas and oil reservoirs. These four areas address the principal scientific questions that arise with CO_2 utilization in shale, which are to determine how the physical and chemical processes associated with CO_2 interaction with organic-rich shales affects CO_2 injectivity over long periods of time, and the ability of the shale to store CO_2 as both a free phase and an adsorbed phase for thousands of years. The primary focus of the analysis in this chapter is the pore-scale interactions. While hydraulic fracturing would be beneficial and is required for the overall improved recovery, it is important to understand what goes on at the smallest scale. We develop a roadmap to address knowledge and technology gaps, as well as serve a preliminary guide to address complications such as organic matter swelling, clay stability, flow back and production.

CO_2 ENHANCED OIL AND GAS RECOVERY

Enhanced oil and gas recovery from fractured low permeability reservoir rock is very challenging. In many cases, primary production has already taken place, creating continuous gas channels for delivering a gas for secondary recovery. The effectiveness of the secondary recovery by gas injection depends upon various factors, important ones being resident gas saturation, injection pressure, and flow conditions. An example of a low permeability hydrocarbon-containing formations is the Monterey Shale formation in California, USA. At larger depths, due to diagenetic processes, the rock loses permeability, and various siliceous shales such as the Brown and Antelope exist [1]. These deeper shales require gas injection rather than hot water or steam injection in order to improve their recovery factors. Enhanced recovery mechanisms when CO_2 is below minimum miscibility pressure (MMP) with the oil include oil-phase swelling, viscosity reduction, and gas-oil displacement. Above the MMP, oil and CO_2 miscibility leads to a zero capillary entry pressure barrier for CO_2, and the displacement efficiency is higher.

The CO_2 MMP is significantly lower than other gases such as nitrogen, flue gas or natural gas, making it advantageous for enhanced recovery. However, due to the higher mobility ratio, the displacement is unstable with lower volumetric sweep efficiency. Kovscek *et al* [1] quantified the recovery potential of fractured low permeability reservoir rock by experiments using both miscible and immiscible CO_2 gas injection, and X-ray CT imaging. The core samples used were 1.5 inches in diameter and roughly 3 inches long with 1.3 mD permeability and 35% porosity. They discovered that for the immiscible tests, the incremental oil recovery ranged from 0-10% for countercurrent injection flow mode and between 18-25% for the cocurrent mode. The oil recovery potential is lower because of low permeability, rock heterogeneity, and existing oil and gas distribution in the core. At miscible conditions, they discovered that countercurrent injection recovery was up to 25% of the oil in place and only 10% for concurrent case. Vega *et al* [2] followed up the experiments above MMP, and also simulated the compositional model with diffusion & convection controlled flow that supported the recovery efficiency percentages. Combining both countercurrent and then concurrent modes led up to 93% oil recovery due to multi-contact miscibility regime.

The use of CO_2 for enhanced oil and gas recovery (EOR & EGR) in tight oil reservoirs such as shale plays, is a relatively new concept. Mohanty *et al* [3] studied miscible CO_2 huff ‘n’ puff in a shale matrix typical of the Bakken formation. A compositional simulator was used to represent multiple vertical fractures along a horizontal well in a formation with matrix permeability and porosity of 0.01 mD and 8% respectively. CO_2 injection was found to outperform primary production for the case of a heterogeneous reservoir. Tovar *et al* (2014) evaluated CO_2 EOR in unconventional liquid reservoirs experimentally, using a layer of glass beads along with the core to simulate the presence of hydraulic fractures. The core soaked with CO_2 was tracked for changes in saturation using x-ray computed tomography and revealed an oil recovery in the range of 18 to 55% of OOIP. There is potential to not only store large volumes of CO_2, but also be able to utilize it for producing more incremental oil and gas.

The Energy & Environmental Research Center (EERC) has conducted laboratory and modeling activities to examine the potential for CO_2 storage and EOR in the Bakken (2012 – May, 2014) [4][5][6]. They suggest that CO_2 may be effective in enhancing the productivity of oil from the Bakken by as much as 50%, as well as possibly store between 120Mt and 3.2 Gt of CO_2. The effectiveness of EOR has been obtained using single porosity-single permeability and dual permeability simulation models, with a double injection well and single producer well scenario. Capillary pressure data has been neglected. Diffusion plays an important role in the movement of CO_2, as indicated in the laboratory and modeling results and the delayed improvement in oil production from the Burning Tree injection test in the Bakken shale. However, there is no clear answer as to what would be the most effective approach for using CO_2 to improve the oil productivity of storage capacity of the Bakken. It depends on the fracture networks and interaction of CO_2 with the resident rock matrix and fluids.

Challenges for Recovery Factors

Recent total oil in place estimates for the Bakken Petroleum System range from 100 Bbbl to over 900 Bbbl. Most estimates for primary recovery range from 3% to 6%, depending on reservoir characteristics. When considering these low primary recovery factors in the context of such a large resource, it is clear that just small improvements in productivity could increase technically recoverable oil in the Bakken by billions of barrels. The challenges of EOR within the Bakken have to do with the mobility of traditional fluids (i.e., reservoir fluids and injected water vs. CO_2, polymers, or surfactants) through natural or induced fractures relative to very low matrix permeability and the aversion of exposing swelling clays to water, which can reduce permeability and damage the formation. Further, the oil-wet nature of much of the Bakken system will dramatically minimize the effectiveness and utility of water flooding. With these issues in mind, the use of CO_2 as a fluid for EOR in the Bakken may be effective.

The use of CO_2 for EOR in conventional reservoirs began in West Texas in the 1970s and has since been applied at locations around the world [7]. However, its use for EOR in tight oil reservoirs is a relatively new concept. In conventional reservoirs, vertical heterogeneity, wettability, gravity, and relative permeability characteristics can have a significant effect on the effectiveness of an EOR scheme, and fracture networks could be detrimental to EOR operations [7]. However, tight oil reservoirs, such as the Bakken, rely on natural and hydraulically induced fracture networks for their productivity. Because of the tight matrix, dominance of fractures, and oil-wet nature of the Bakken, the conventional notion of positive and negative attributes of a candidate injection reservoir may or may not apply. With respect to CO_2, fracture networks are the primary means of movement throughout the reservoir, and their characteristics control the contact time that CO_2 has with the oil in the reservoir.

Evaluation of Pilot-Scale CO_2 Injection in Bakken

The Elm Coulee area in Richland County, Montana, is one of the first areas to see prolific oil production from the middle member of the Bakken Formation. With the use of horizontal drilling combined with hydraulic fracturing technologies, oil production was economically viable. In 2009, three companies (Continental Resources, Enerplus, and XTO) jointly conducted a pilot-scale CO_2 injection test in the Burning Tree-State 36-2H well (or the Burning Tree well), primarily to evaluate EOR. The Burning Tree well was drilled and completed in March and April 2000. The top of the Bakken in the Elm Coulee area at the Burning Tree well location is approximately 9740 ft. The horizontal leg of the well was drilled into the middle member of the Bakken, which at that location is approximately 40 ft thick and dominated by sandy to silty dolostones. The well was drilled with a horizontal lateral leg 1592 ft in length, completed with 5.5 inches of cemented production casing, perforating approximately 813 ft of the wellbore, and stimulated using a single-stage, hydraulic fracturing operation - with 20-40 mesh sand as proppant. The initial oil production of the well was 196 bbl/day of 40.1 API gravity crude, but after producing for about 8 years it dropped to between 30 and 40 bbl/day.

In a huff 'n' puff scheme, CO_2 is injected in a single well (huff), after which it is allowed to soak while the well is closed, and then it is produced again (puff). These tests can be an effective means of evaluating the response of a reservoir to CO_2, both with respect to EOR and CO_2 storage. Previously, huff 'n' puff experiments with CO_2 have been done on conventional reservoirs [8], rather than tight oil formations such as the Bakken shale. In this regard, the Burning Tree CO_2 huff 'n' puff test was a pioneering effort.

Over the course of a 45-day period in early 2009, approximately 45,000 mcf (2570 tons) of CO_2 were injected into the Burning Tree well. The maximum injection pressure was 1848 psi BHP. While the average daily injection rate was approximately 1000 mcf/day, the actual injection operation was intermittent, with injection rates ranging from 0 to 3000 mcf/day. After injection, the well was capped and the CO_2 was allowed to soak for 64 days, after which the well was opened for production. Daily oil, water, and gas production data for the Burning Tree well for a period of about six months were examined by EERC. After rapidly climbing to a peak oil production of over 160 bbl/day 8 days after the well was brought back into production, the oil production settled into an average of about 20 bbl/day during the first 30 days after the end of the soak period.

By the end of two months, the well was no longer flowing and was put on pump. The average oil production over the following 3 months rose slightly to about 22 bbl/day, with a range of about 15 to 25 bbl/day. By the end of five months, the range of daily oil production was from 20 to 30 bbl/day, which was still below the pre-injection range of 30 to 40 bbl/day. By the end of 2009, average daily oil production had risen to nearly 28 bbl/day. Oil production continued to slowly rise in early 2010, reaching a peak post-injection high approaching 44 bbl/day in March 2010, which is a higher rate of production than was achieved during any of the 14 months immediately prior to the injection test. While oil production from the Burning Tree well remained above 35 bbl/day throughout the summer of 2010, by November of that year, it was back down to less than 30 bbl/day and has continued to decline. By the end of November 2013 (the latest month for which data were available) production had declined to slightly less than 15 bbl/day. Figure 1 shows the monthly oil production history of the Burning Tree well from June 2000 to November 2013. Figure 2 shows the monthly oil production from February 2008 to November 2013.

Typical "successful" CO_2 huff 'n' puff operations in conventional wells see a dramatic improvement in oil production immediately following the soak that often takes several weeks, or even months, to return to pre-injection rates [8]. When compared to conventional huff 'n' puff tests, particularly when looking at the first 6 months of data after the soak, the Burning Tree huff 'n' puff test might not be considered successful. However, the Bakken is an unconventional play. In the Elm Coulee area, the porosity of the middle member of the Bakken commonly ranges from 4% to 6% and the permeability typically ranges from 0.06 to 0.12 mD [9], and the play, therefore, requires the artificial generation of fracture networks to enable hydrocarbons to flow to wells. In a conventional reservoir, a highly fractured rock with a tight matrix is not an ideal candidate for any type of CO_2-based EOR due to CO_2 escaping. The first few days saw an initial spike in fluid production that was likely the result of pressure build-up as opposed to any miscibility-related effects of CO_2, the Burning Tree well did not see a dramatic increase in oil production. But CO_2 was successfully injected and reservoir fluids produced from this tight, unconventional formation. When a longer view is taken, we see a gradual increase in oil productivity, and though it was delayed and certainly not dramatic, this improved productivity, which lasted for another several months through the summer of 2010, might be attributable to the injection of CO_2 or some other operational factors not related to CO_2. An analysis of the CO_2 monitoring data would have been useful, along with a mass balance on the produced CO_2.

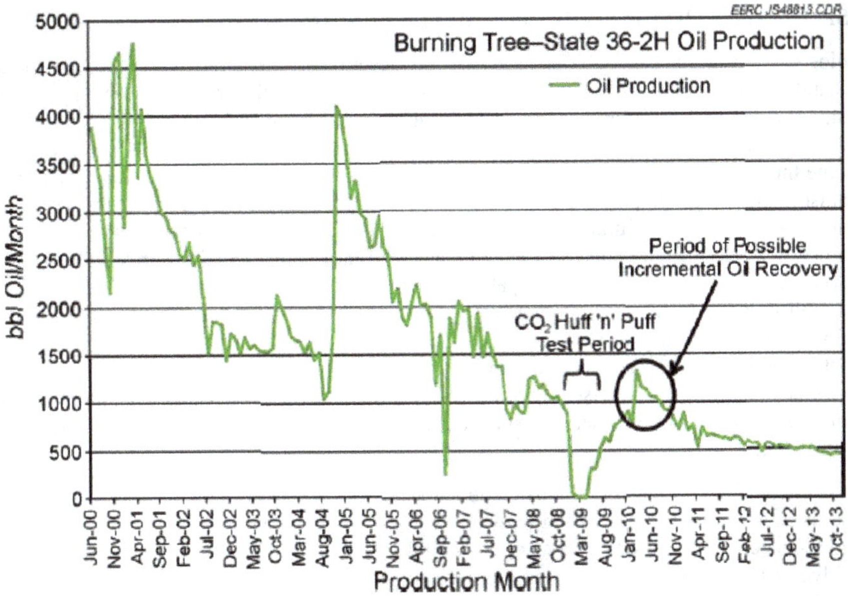

Figure 1. Monthly oil production history of the Burning Tree well from June 2000 to November 2013. Courtesy of: Final Report on CO_2 Storage and Enhanced Bakken Recovery Research Program, by Sorensen *et al* [9].

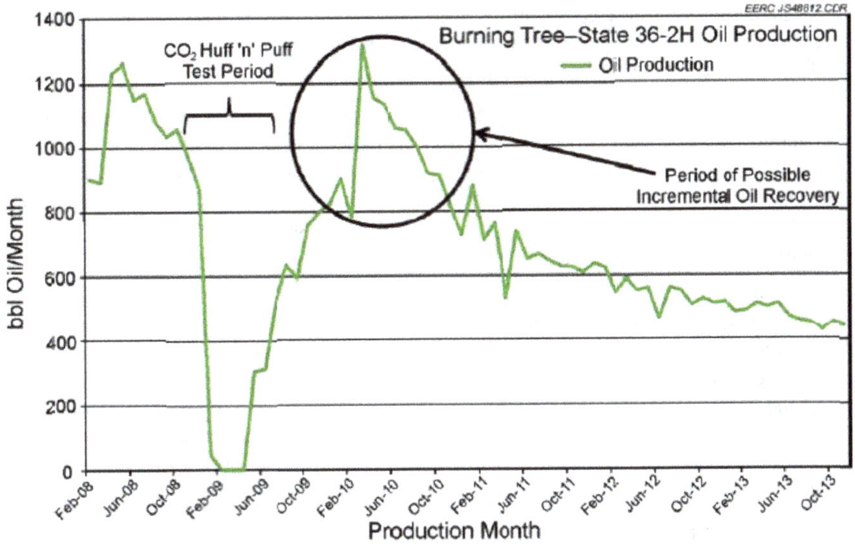

Figure 2. Monthly oil production from February 2008 to November 2013. Courtesy of: Final Report on CO_2 Storage and Enhanced Bakken Recovery Research Program, by Sorensen *et al* [9].

CO_2 SEQUESTRATION IN DEPLETED SHALE FORMATIONS

A geological formation can be considered for sequestering Carbon dioxide (CO_2) provided we understand its storage attributes. Godec (2013) studied the potential implications on gas production from shales and coal for geologic storage of CO_2 in a report to the IEAGHG[1]. A high pore volume formation like a sedimentary rock is a great choice due to good injectivity of CO_2, provided we overcome the leakage risks. Shale beds offer a suitable storage environment enhancing the physical mechanisms such as gas trapping, adsorption and potentially dissolution, if there is free phase water present. Coalbeds and naturally occurring gas hydrate reservoirs have been proposed as economically feasible choices due to the twin reasons of the possibility of preferential adsorption of CO_2 over methane as well as storing CO_2 at the same time. On the other hand, this might be difficult to accomplish due to injectivity problems associated with CO_2 and the low permeability formations typically prevalent in shale. CO_2 storage in saline reservoirs is only considered at depths exceeding 800 meters in order to ensure that the hydrostatic pressure is sufficient to keep CO_2 in a dense phase (supercritical phase). However shale formations are typically much deeper and there is carbon containing material; so even when pressure depletes, it is expected that the CO_2 will be adsorbed onto this material.

In the context of non-shale formations, sedimentary rocks are rarely homogenous and thus there may be risks of heterogeneities within seals providing leakage pathways from storage reservoirs. Of particular concern is the risk of exceeding the capillary breakthrough pressure, potentially allowing undissolved CO_2 to migrate through the caprock and possibly, of diffusive loss of dissolved CO_2 through the water saturated caprock [10]. Capillary breakthrough pressure, also known as threshold pressure or entry pressure, is the capillary pressure at which the non-wetting CO_2 first begins to flow through a porous medium (i.e. caprock) saturated with a wetting fluid. Thus capillary sealing is effective as long as the capillary breakthrough pressure exceeds the buoyancy force created by the density contrast between the CO_2 and the displaced reservoir fluid. Capillary sealing is related to fluid rock interactions, caprock permeability, wettability and pore size distribution. Both numerical and experimental results indicate that most caprocks have the potential to accommodate column heights of CO_2 on the tens to hundreds of meters.

Rates of diffusion can vary significantly over geologic time and are strongly dependent on the pore structure, and to the extent that effective stress may change the porosity of the caprock. A number of studies have examined rates of diffusive loss [11]. These studies agree that diffusion takes place at geologic time scales and that even thin caprock seals (~10m) seem to be secure for thousands of years. When chemical reactions and potential CO_2 sorption by clay minerals are considered, risk of diffusive leakage through caprocks remain low [11].

We will now discuss the potential of storing CO_2 in shale formations. A significant amount of literature that exists is based on laboratory studies of storage capacity of shale formations. Shale sediments with the potential for natural gas production are generally rich in organic matter, also known as kerogen. Table 1 shows the total organic content (TOC) in shale as a weight % of the rock in some of the North American plays. Even though these numbers are small in terms of mass, they are greater by volume, with this organic material providing sites where most of the natural gas is adsorbed. These TOC values form up to 70% of the pore volume for Barnett samples, and therefore are a promising avenue for sequestering CO_2.

Ambrose *et al* [12] reported the results of an analysis involving hundreds of 2-D images and their recombined 3-D shale segmentations using Barnett shale samples from different locations and depths. The digital segment analysis revealed kerogen- and pore-networks. They concluded that:

[1] http://www.ieaghg.org/docs/General_Docs/Reports/2013-10.pdf

(i) organic-rich shale matrices consist of both organic and inorganic materials that could be dispersed in one another, bicontinuous or intertwined; (ii) a major part of the porosity is contained in the organic matter; and therefore (iii) much of the gas storage capacity is associated with the organic fraction of shale. Gas sorption experiments have shown that pore volume compressibility has a very minor effect, and that the primary mode of storage in the organic matter is that due to adsorption. The density for the adsorbed phase is close to the liquid density of the chemical species used during the experiment.

Table 1. Typical TOC of North American shale gas plays (from [12]).

Shale Type	Average TOC (weight %)
Barnett	4
Marcellus	1-10
Haynesville	0-8
Horn River	3
Woodford	5

EOR Mechanisms in Shale

In shale oil and gas reservoirs, most of the resource is contained in the low-permeability matrix, which needs to be produced through the high permeability fractures. Moreover, each type of shale has a different combination of permeability and pore size distribution along with varying organic vs clay content. It can be assumed that the injection of CO_2 will recover oil from shale, however, the physical mechanisms associated with production are not evident. The injected CO_2 would provide energy owing to its compressibility, as well as dissolve in the oil and decrease its viscosity. Additionally, CO_2 miscible flooding reduces the oil and gas capillary pressure leading to better recovery [1][2].

It is unknown whether a cyclic injection - a huff 'n' puff scheme, where CO_2 is allowed to soak through and pressurize the reservoir above the minimum miscibility pressure - would be a better strategy, as compared to a continuous injection of CO_2. The best choice would depend upon the role of diffusion and slip flow in nano- and micro-scale pores in the matrix, and the time required to notice the effects. In order to understand the relative importance and sequence of these mechanisms, we need more experiments with in-situ visualization at various scales to learn how diffusion and slip flow translate to Darcy flow. Additionally, once the primary flow regimes are established in various parts of the shale matrix, it is important to devise methods to combine these scales and develop upscaling techniques for reservoir simulation. This is not a straightforward procedure due to the different time scales involved with the slower diffusion and slip flow as opposed to the faster Darcy flow.

An added complexity in modeling CO_2-based EOR/EGR in shale systems is the CO_2 phase behavior through the variable pore network. Since a significant number of the pores have a size very close to that of a CO_2 molecule, it is essential to understand if there would be any interfacial tension affects that would cause changes or reduction in the bubble point pressure, gas bubble nucleation and coalescence, as well as subsequent mobilization. Honarpour *et al.* [13] showed that the bubble point pressure is suppressed significantly by high capillary pressure in nano pore rock and that high critical gas saturation results in delayed gas mobilization and reduction in producing gas oil ratio (GOR). Alharthy *et al.* [14] concluded that hydrocarbon fluids can move through low permeability shale reservoirs due to favorable phase envelope shift of hydrocarbon mixtures in the nano- and meso-scale pores in gas-condensate and bubble-point systems. They also noted that when the phase envelope is crossed in gas condensate systems, a large gas-to-oil volume split in the nano, meso, and macro-pores plays a crucial role in hydrocarbon recovery during depletion. For the bubble-point oil region, the low viscosity of the liquid phase and the delay in gas bubble evolution appears as the main reason for favorable oil production.

Challenges for CO_2 Storage

The obvious primary challenge of using any tight oil formation as a target for large-scale storage of CO_2 is the characteristic low porosity and low permeability of the formation. The tight nature of the shale formation presents challenges to both CO_2 injectivity and storage capacity. Furthermore, the presence of complex, heterogeneous lithologies (including organic-rich, oil-saturated shales) complicates the ability to understand and predict the effectiveness of various mechanisms (e.g., diffusion, sorption, dissolution, etc.) that act on CO_2 mobility and storage.

Previously, some work has been published on the potential storage capacity of tight, natural gas-rich shale formations, including studies on gas shales in Kentucky [15], Texas [16], and the Appalachian region [17]. The authors of those studies assumed that the CO_2 storage, and subsequent methane recovery, in organic-rich gas shales is controlled by similar adsorption and desorption mechanisms as CO_2 storage and methane recovery in coal seams. In those cases, the sorptive capacity of the organic content in the shales plays a prominent role in estimating their potential CO_2 storage capacity. Unfortunately, those approaches may have limited applicability to other shale plays due to the composition that consists of a combination of organic-rich shales, tight carbonates, and clastics; and, whether the formation is saturated with oil and brine as opposed to gas in coal. The diversity of lithology and presence of oil and brine may substantially limit the effects of sorptive mechanisms on CO_2 storage as compared to the gas shale. To assess accurately the potential for tight oil formations to store CO_2, it is necessary to develop a better understanding of the fundamental mechanisms and unique formation properties (e.g., tight matrix, microfractures, high organic carbon content, etc.) controlling the interactions between CO_2 and the rocks and fluids of those tight oil formations. EERC developed a first-order, reconnaissance-level estimate of the potential CO_2 storage capacity of the Bakken Formation in North Dakota.

Estimation of Potential CO_2 Storage Capacity in Bakken

Most of the Bakken Formation is not organic-rich shale but, rather, oil- and brine-saturated tight carbonates and clastics, as discussed in the section above. Thus, the typical basis for storage capacity estimation in natural-gas-rich shales noted in the previous section, does not apply. With these characteristics in mind, published methods to estimate the storage capacity of oil reservoirs may be more applicable to estimating the potential storage capacity of the Bakken.

To develop first-order CO_2 storage capacity estimates for the Bakken in North Dakota, an approach was used that estimates the amount of CO_2 needed for EOR in the Bakken. Specifically, the methodologies for estimating CO_2 storage capacity in oil formations based on production and volumetrics as presented in the *Carbon Sequestration Atlas of the United States and Canada* [18] were applied to the Bakken Formation in North Dakota.

The first method, referred to as the volumetrics method, is largely based on estimating the OOIP of the Bakken according to known reservoir properties. Specifically, the product of the area, net thickness, average effective porosity, original hydrocarbon saturation (1-initial water saturation, expressed as a fraction), and the initial oil (or gas) formation volume factor yield the OOIP. The storage efficiency factor is derived from local CO_2 EOR experience or reservoir simulation as standard volume of CO_2 per volume of OOIP. Using OOIP data from Nordeng *et al.* [19] for North Dakota, an estimate of a 4% increase in oil recovery (4% of OOIP) and two utilization factors, the mass of CO_2 needed for a Bakken EOR effort (i.e., the potential CO_2 storage capacity of the Bakken in North Dakota) ranges from 1.9 to 3.2 billion tons.

A second approach, generally applied to mature oil fields or those for which key reservoir property data are unavailable, to determining OOIP is to use cumulative production divided by a recovery factor (e.g., 36%). In the case of the Bakken in North Dakota, a recovery factor of 7% was used along with a cumulative production of 732,000,000 bbl. This approach results in a predicted OOIP of 10.5 billion bbl and a corresponding CO_2 storage capacity for the Bakken ranging from 121 to 194 Mt. The estimates using the reservoir property-based OOIP approach are likely too high because the U.S. Department of Energy (DOE) method was developed based on knowledge derived from decades of studies and experience related to CO_2 injection, utilization, and storage in conventional oil reservoirs. While the OOIP of the Bakken is known to be high [20][21][22], the extremely tight nature of the formation may adversely affect injectivity and storage efficiency and thus reduce the storage capacity estimates.

It is evident that more data from laboratory- and field-based research efforts are required to develop improved CO_2 storage capacity estimates for tight oil formations in order to better understand the pore-scale transport processes. Future evaluations of CO_2 storage potential in tight oil formations like the Bakken may consider using a hybrid method that combines some elements of the shale gas capacity methods with elements of the oilfield methods.

CO_2 UTILIZATION AS A FRACKING FLUID

Oil and gas bearing reservoirs benefit from hydraulic stimulation through increased production rates. Hydraulic fracturing is a routine process in many reservoirs which is achieved by pumping liquids at high pressure, and rates into the reservoir rock with sufficient energy to break or "hydraulically fracture" or "hydraulically stimulate" the formation to generate cracks or "fractures" through which gases and liquids can flow more rapidly to the well. This increase in production rate can significantly enhance the economics and make production in many wells commercial. In order to break the rock, the stimulation treatments often require energy both to overcome its tensile strength and also to overcome any additional tectonic forces which may be present. Both of these forces must be exceeded to generate a hydraulic fracture.

The tectonic forces present can close the fracture when the hydraulic pressure is removed at the end of the pumping operation. Therefore, proppants are frequently used to keep these fractures from closing completely. There are several considerations regarding proppant selection relative to its strength and size. Sand, which has been carefully sized and cleaned, is the least expensive proppant and is generally used. The proppant is transported into the fractures by the pumped liquid, generally with enhanced properties to improve its fracturing and transport characteristics.

Water-based fracturing liquids are the most commonly used. Generally, chemical additives are mixed with the water to improve its ability to transport the proppant. This is achieved through the addition of gels to increase the viscosity and also to reduce fluid loss from the fracture by temporarily plugging or bridging the natural permeability of the reservoir rock. Once the pumping is completed, fractures are created when the chemical additives (gel) transform, and these fractures serve as channels for oil and gas flow to the well. The use of hydraulic stimulations and chemicals to enhance oil and gas production has been very successful, since first used in the 1950's [23]. The gels sometimes do not completely break and there is always a residue following its decomposition. These materials, along with liquid blockage, could damage the reservoir.

An advantage of using CO_2 is to avoid formation damage. CO_2 can be pumped as a liquid and then it vaporizes and flows from the reservoir, leaving no liquid or chemical damage. This process can also transport proppant in limited volumes but requires a proper mixing of the liquid CO_2 with proppant. With the CO_2/sand dry-frac stimulation process, formation damage is reduced, well clean-up is fast, and the economics significantly improved. . However, if formation water is present, scaling can be a long-term problem when the CO_2-formation water chemistry is unfavorable. The process is best applied in tighter (less permeable), lower pressure, dry gas reservoirs where stimulation liquids are foreign to the formation and reduce its permeability to gas, such as shale beds. The initial cost is generally more expensive than conventional because CO_2 is more expensive than the water and chemicals. Moreover, supply of CO_2 to the injection site might not be easy and offer logistic issues such as risk of transporting the compressible gas. Secondly, after injection when it converts back to gaseous form, it might escape from wells and would need to be captured for reinjection or flaring. The concept of fracturing with 100% CO_2 as the fracturing fluid and proppant carrying fluid was first introduced in the early 80's [24][25]. The method was pioneered by a Canadian service company called Fracmaster who performed fracture stimulations on thousands of wells in Canada via CO_2 sand fracturing (100% CO_2 and proppant) with great success. At the time, the company proved that it could produce more oil and natural gas than using water because of higher fracking pressures achieved with CO_2 [26][27].

The viscosity of CO_2 is low, about one-tenth that of water, and much less than water with gels and other chemicals as typically used in conventional treatments. Proppant transport relationships with liquid CO_2 have not been found in the literature. Additionally, because of the low viscosity of liquid CO_2 and the absence of gels, it can readily escape from the fracture into the surrounding rock, thereby reducing its ability to transport proppant. We need to establish an understanding of this behavior in order to develop stimulation models. To understand and reduce the likelihood of damage and leakage is an incentive for researchers. However, we still do not know the degree to which slip along faults in the caprock creates potentially permeable pathways along which fluids could leak. Further, we want to learn whether certain caprock compositions are more vulnerable to leakage than others, probably such as clay-lean, if triggered fault slip occurred. Additionally, we do not know, how and to what extent pre-adsorbed water and CO_2 affect the reactivation and flow through fractures.

The fundamental understanding of fracture flow of sorbing and non-sorbing gases and the response of transport properties to changing stress conditions can aid in assessing issues such as caprock integrity for CO_2 storage applications or use of CO_2 for enhanced gas recovery from unconventional gas or liquid-rich reservoirs. Therefore, it is important to understand the evolution of fault permeability and the development of a fault damage zone in shaly rocks during shear deformation and under various effective normal stress conditions.

SHALE AND THE ROLE OF CLAY MINERALS

Shale is a mix of mudstone and organic matter, and the mudstone is a mix of silica, carbonates and clay minerals with the latter comprising up to 50% of the overall shale contents, [28]. The most common clay minerals within gas shale are kaolinite, smectites (i.e., swelling clays such as montmorillonite), chlorite and illite. Their contributions vary depending on the shale play. For example, the Fayetteville shale contains 25 to 30% clay, which is primarily illite with 20 to 30% chlorite [29]; the Barnett shale contains 5 to 40% clay, primarily illite with some smectite [30]; and the Marcellus shale play contains 10 to 35% illite and a variety of mixed-layer clays plus 1 to 10% chlorite [31]. The total surface areas of these clays varies from approximately $85m^2/g$, [32] for illite, $350m^2/g$ for montmorillonite, and between 10 and $20m^2/g$ for kaolinite. Given the volume contribution of clay minerals to shale's composition, together with their large surface areas, it is reasonable to conclude that clay minerals comprise the majority of the surface area within shale. Therefore, it is highly probable that CO_2 pumped into a depleted shale bed will come into contact with clay mineral surfaces, and these clay minerals interface with organic matter, and all components of mudstone including other clay minerals.

Organic matter within shale is holds methane (natural gas), the extraction of which, following hydraulic fracturing, will create empty volume space within the organic matter. On depletion of the natural gas within a shale play, (if not during EOR) the injected CO_2 could enter this empty space as well as readily migrate through the many micro-, (<2nm) meso-, (<50nm) and macro- (> 50nm) pores and induced fractures within the shale matrix. Given the large contribution of clay minerals to the total surface area, and also that they interface with all components of shale, then some of the pore spaces and fractures could comprise a mixture of clay minerals and/or organic matter, silicates and carbonates. This suggests that CO_2 could experience a variety of different surfaces, dominated by the presence of clay minerals, and undergo a range of interactions from the chemical conversion of CO_2 to carbonates, to the transport and adsorption of CO_2 to the surfaces of, and through, (predominantly) clay-lined pores. The role of clay minerals within shale is important. There has been work demonstrating the sorptive capacity of clay minerals, which is discussed later.

Chemical interactions with resident water

At the *in situ* temperatures and pressures in shale beds, the clay mineral surfaces are hydrated, which means that a robust model representing clay minerals within shale must include water. The exploration of chemical interactions requires quantum mechanical modeling methods such as density functional theory (DFT). There are a few DFT studies involving hydrated kaolinite [33][34], more investigating hydrated smectites [35][36], and only one focusing on the hydration of illite surfaces [37].

DFT investigations may be needed to determine whether carbonate formation occurs with the addition of CO_2 on hydrated clay mineral surfaces. Currently there appears to be no work published in this field, although some experimental studies recently presented at the 50th Anniversary of the Clay Minerals Society Annual Meeting (October 2013), found that carbonate formation depended on the particular clay. For example, experimental studies of CO_2 on mica and mica/quartz substrates found both bicarbonates and carbonates are formed [38] (illite is a mica), whereas another experimental study found no carbonate species intercalated into montmorillonite [39] on exposure to CO_2.

Physical aspects of CO_2/shale interactions

Physical interactions between CO_2 and shale at the micro- and meso-scale include transport through the shale matrix, and adsorption within its various components. From an atomistic modeling and simulation perspective these processes are addressed using classical modeling techniques such as molecular dynamics and Monte Carlo as these methods are able to access larger scale phenomena than those occurring during chemical interactions. For example, molecular dynamics uses classical force fields to simulate CO_2 transport through micro- to meso-scale pores, and Monte Carlo methods simulate the adsorption of CO_2 molecules within the (clay) pores. In both methods the reservoir temperature and pressure are included, with the latter encapsulating the Monte Carlo chemical potential, thus making a direct connection between the macroscopic reservoir conditions and the environment at the molecular level in micro- to meso-scale pores.

As previously mentioned, at the *in situ* temperatures and pressure experienced by shale beds, the clay mineral surfaces will be somewhat hydrated. As the fracking fluid is injected and fissures are created throughout the shale, the clay minerals will be prone to further hydration, and the smectites could swell. The swelling nature of smectites lends itself to investigation using Monte Carlo and molecular dynamics methods, and there are many studies in this field although they are not limited to clay minerals within shale, and they include swelling inhibition. The development of the molecular dynamics force field – CLAYFF - specifically designed for clay minerals, [40] facilitated many of these studies.

Clay Swelling

At the 50th Anniversary of the Clay Minerals Society Annual Meeting (October 2013) recent experimental and theoretical studies were presented concerning clay minerals found in shale and caprock[2] and their interactions with CO_2. Topics ranged from the experimental characterization of smectite-water-natural organic matter interfaces to the impact of clays and their properties on oil recovery, (reports online)[41] to a summary of the role of clay in shale formations [42]. Molecular dynamics modeling studies, include creating edge sites and cation adsorption sites, investigations into hydrated illite surfaces, as well as *ab initio* molecular dynamics[3] modeling of the intercalation of CO_2 in montmorillonite [43] found that the degree of swelling caused by CO_2 depends on the initial water content of the interlayer space. The experimental studies presented corroborated these simulation results.

These results on different source materials such as shale, clay minerals and caprock, were in agreement that a certain degree of hydration of the smectite interlayer cations (sodium, potassium, calcium) is required for maximal expansion of smectites during CO_2 uptake [39]. Furthermore, illites adsorb the least amount of CO_2, followed by kaolinite and the most, smectites, and finally, that there is a cooperative effect of CO_2 adsorption - it is easier to adsorb more CO_2 into the clay interlayers when some is already adsorbed.

The effect on the surrounding environment of swelling smectites was also addressed at the conference. One study showed that the increase in interlayer space due to CO_2 uptake leads to immobilization and decreasing reservoir pressure, and that CO_2 uptake can lead to both fracture closure and fracture activation due to the effect of the swelling on the *in situ* stress [44][45]. There is a wealth of relevant information available from other research areas in addition to the aforementioned interactions of clay minerals with CO_2. These include molecular dynamics studies of humic acid and clay minerals in the context of CO_2 sequestration, DFT investigations of CO_2 adsorption on carbon materials, and research on carbon-bearing fluids at interfaces of nanoscale materials including clays. Considering the specific shale environment, comprising multiple-sized pores and heterogeneous surfaces, the next stage of modeling is to focus on adsorption and transport through clay, clay/organic matter, clay/quartz and clay/carbonate pores. Combining these results with those obtained from the cited studies focusing on clay interlayer swelling, would begin to bridge the length-scale gap between experimental and simulated data. Some work on clay pores and transport has already been done, such as experiments and molecular dynamics investigation of illite/mica slit pores,[4] the adsorption of both CH_4 and CO_2 in clay-like slit pores, and transport through swelling clays.

The presence of natural gas (mostly CH_4) within shale and how it interacts with clay minerals, CO_2, water and how hydraulic fracturing fluids affect shale's components need to be considered to gain real understanding of the long-term effects of injecting CO_2 into depleted shale beds. This is undoubtedly a highly topical issue involving past, current and future research on diverse topics, such as the structure of shale components, how they respond during the hydraulic fracturing process, and how they interact with CO_2, CH_4, H_2O etc. By combining the results of these studies it should be possible to create a realistic model of shale and how CO_2 interacts with all of its components.

---

[2] Caprock in this document refers to the clay-containing rock overlying and plugging geological, CO_2 storage features such as aquifers.

[3] *Ab initio* molecular dynamics involves both DFT and molecular dynamics in one code, with the advantage of describing both chemical and dynamic interactions.

[4] A slit pore is modeled as a rectangular cavity.

CHALLENGES, NEEDS AND RECOMMENDATIONS

The benefits from injection of CO_2 into shale formations are potentially large and diverse in nature. The current low recovery rates of oil (~5%) and gas (~25%) from shale formations might benefit greatly from CO_2-enhanced production. The enormous volume of shale formations tapped by tens of thousands of wells could be used for CO_2 storage, and CO_2 could be used as an alternative to water as a fracking fluid. However, while the opportunity is large, so are the challenges! Shale science and engineering is still in its infancy. Shales compared to conventional reservoir rocks are characterized by pore spaces measured in nanometers instead of micrometers, and by permeabilities ranging from nanodarcies to microdarcies as compared to millidaricies to darcies. Additionally, geochemical reactivity and geomechanical properties are potentially much more variable. All of these factors challenge conventional approaches for site characterization, hydraulic fracturing, reservoir engineering, and production engineering. Assessing the potential and realizing the beneficial use of CO_2 in shale formations requires significant progress across a number of fronts. This report summarizes the current activities focused on understanding beneficial uses of CO_2 in shale formations based on a review of the existing literature. From this survey of the current state-of-the-science, we identify key challenges and opportunities for research activities to address them.

- ***Injectivity and storage capacity of the low permeability/low porosity shale rocks:*** The low permeability and potential swelling behavior due to CO_2/rock/organic interactions is the single largest challenge for CO_2-enhanced recovery and CO_2 storage in shale. A forward simulation study of CO_2 injection into a shale reservoir with a different resident fluid and sensitivity on permeability/porosity combination is required to understand the EOR/EGR and storage ability of shale.
- ***EOR mechanisms in Shale:*** Since the transport models (convection/sorption/diffusion) are not completely understood for such a varied system of scales encountered in shale rocks, experimental investigation using CT imaging in the lab are required for understanding CO_2 pore volume occupancy and CO_2/shale interactions.
- ***Enhanced shale characterization for CO_2 injection:*** Develop methods for characterizing the natural microfractures and organic material distributions that provide transport pathways for CO_2. Develop mathematical models that provide reliable predictions of CO_2 and multiphase transport in these rocks.
- ***Reservoir/fracture dynamics during CO_2 injection:*** How do the fracture networks behave as reservoir pressures are depleted during primary production and then repressurized due to CO_2 injection, and how does that behavior impact reservoir permeability? How would CO_2 phase behavior at nano scale affect recovery? Would this affect production or storage? How would cold CO_2 injection affect the thermal stresses and favor fracturing? What is the time frame for these dynamic effects?
- ***Scale Characterization & Upscaling:*** It is evident that we encounter many differing scales in the shale rock, from nano meter sized pores to a few meter sized fractures to fracture zones extending hundreds of meters. We need upscaling techniques such that they could be applied to the whole spectrum of pore sizes and flow regimes.
- ***Fault permeability:*** Does a pre-existing fault in shale seal itself during shear slip or does it open up a pathway for potential CO_2 leakage? How does fault permeability in shale evolve during shearing? How does rock clay content affect the fault permeability?
- ***Fluid dynamics:*** How does the mineral composition of shales affect the poro-elastic and fluid dynamic properties? Does CO_2 and CO_2+water change the friction properties and the transport properties of shale/caprocks?
- ***Clay smear:*** What kind of influence is contributed by the clay smear itself and how does it differ between the different types of clay? How does the permeability change during sliding and does it matter which gas is permeating the fracture?

- ***Clay pore models:*** Build a series of clay pore models using DFT at a range of length scales representative of micro to meso-pore with opposing clay walls comprising kaolinite/kaolinite, kaolinite/smectite, kaolinite/illite, smectite/smectite, smectite/illite and illite/illite. Use advanced synchrotron techniques such as small angle x-ray scattering to identify the interfaces between the various components of shale and what lies between them.
- ***Simulate organic matter for CO_2 adsorption:*** Based on clay pore models, build mixed clay/organic matter pores using DFT where, in the first instance the organic matter is simulated by sheets of graphene. Use these models in grand canonical Monte Carlo simulations to determine CO_2 adsorption isotherms at a range of temperatures and pressures.
- ***Interaction with resident water/methane:*** Add complexity to the above models by adding water molecules and methane based on expected water saturation of CO_2, and experimentally predicted quantity of methane in the same sized pores.

ACKNOWLEDGEMENTS

This project was supported by CCP3; a joint industry project sponsored by BP, Chevron, Eni, Petrobras, Shell, and Suncor.

ACRONYMS

Bbbl	Billions of barrels
GOR	Gas oil ratio
Gt	Gigatons
Mcf	Million cubic feet
OOIP	Original oil in place
TOC	Total organic content

REFERENCES

1. Kovscek, A.R., Tang, G.Q., and Vega, B (2008). Experimental Investigation of Oil Recovery from Siliceous Shale by CO_2 Injection, SPE 115679 presented at the SPE Annual Technical Conference and Exhibition in Denver, Colorado, USA, 21-24 September.
2. Vega, B., O'Brien, W.J., and Kovscek, A.R. (2010). Experimental Investigation of Oil Recovery from Siliceous Shale by Miscible CO_2 Injection, SPE 135627 presented at the SPE Annual Technical Conference and Exhibition in Florence, Italy, 19-22 September.
3. Mohanty, K., Chen, C., and Balhoff, M. (2013). Effect of Reservoir Heterogeneity on Improved Shale Oil Recovery by CO_2 Huff-n-Puff. SPE 164553 presented at the SPE Unconventional Resources Conference, The Woodlands, Texas, USA.
4. Hawthorne, Steven B., Charles D. Gorecki, SPE, James A. Sorensen, SPE, Edward N. Steadman, John A. Harju, Energy & Environmental Research Center; Steven Melzer, Melzer Consulting (2013). Hydrocarbon Mobilization Mechanisms from Upper, Middle, and Lower Bakken Reservoir Rocks Exposed to CO_2. SPE-167200-MS, presented at the SPE Unconventional Resources Conference-Canada held in Calgary, Alberta, Canada, 5–7 November.
5. Kurtoglu, Basak, Marathon Oil Company; James A. Sorensen, Jason Braunberger, Steven Smith, Energy & Environmental Research Center; Hossein Kazemi, Colorado School of Mines (2013). Geologic Characterization of a Bakken Reservoir for Potential CO_2 EOR, URTeC 1619698, presented at the Unconventional Resources Technology Conference held in Denver, Colorado, USA, 12-14 August.

6. Tovar, Francisco D., Oyvind Eide, Arne Graue and David S. Schechter (2014). Experimental Investigation of Enhanced Recovery in Unconventional Liquid Reservoirs using CO_2: A Look Ahead to the Future of Unconventional EOR, SPE 169022, presented at the SPE Unconventional Resources Conference, The Woodlands, Texas, USA 1-3 April.
7. Jarrell, P.M., Fox, C.E., Stein, M.H., and Webb, S.L., (2002), Practical aspects of CO_2 flooding: SPE Monograph v. 22, Henry L. Doherty Series, Richardson, Texas, 220 p.
8. Mohammed-Singh, L., Singhai, A.K., and Sim, S., (2006). Screening criteria for carbon dioxide huff 'n' puff operations: SPE 100044.
9. Sorensen *et al*, (2014), Final Report prepared for NDIC, titled "CO_2 Storage and Enhanced Bakken Recovery Research Program".
10. Busch, A., S. Alles, B. M. Krooss, H. Stanjek, and D. Dewhurst (2009), Effects of physical sorption and chemical reactions of CO_2 in shaly caprocks, *Energy Procedia, 1*(1), 3229-3235.
11. Busch, A., S. Alles, Y. Gensterblum, D. Prinz, D. N. Dewhurst, M. D. Raven, H. Stanjek, and B. M. Krooss (2008), Carbon dioxide storage potential of shales, *International Journal of Greenhouse Gas Control, 2*(3), 297-308.
12. Ambrose, R.J., Hartman, R.C., Diaz-Campos, M., Akkutlu, I.Y., and Sondergeld, C.H. (2010). New Pore-scale Considerations in Shale Gas in-place Calculations. SPE 131772, presented at the SPE Unconventional Gas Conference held in Pittsburgh, Pennsylvania, February 23-25.
13. Honarpour, M.M., Nagarajan, N., Orangi, A., Arasteh, F and Yao, Z. (2012). Characterization of Critical Fluid, Rock & Rock-Fluid Properties Impact on Reservoir Performance of Liquid-Rich Shales. SPE 158042, presented at the SPE Annual Technical Conference & Exhibition, San Antonio, October 8-10.
14. Alharthy, N.S., Thanh N. Nguyen, Tadesse W. Teklu, Hossein Kazemi, and Ramona M. Graves. (2013). Multiphase Compositional Modeling in Small-Scale Pores of Unconventional Shale Reservoirs. SPE 166306 presented at the SPE Annual Technical Conference and Exhibition held in New Orleans, Louisiana, USA, 30 September–2 October.
15. Nuttall, B.C., Eble, C.F., Drahovzal, J.A., and Bustin, M.R., (2005). Analysis of Devonian black shales in Kentucky for potential carbon dioxide sequestration and enhanced natural gas production: Kentucky Geological Survey Final Report to U.S. Department of Energy, 120 p.
16. Uzoh, C., Han, J., Hu, L.W., Siripatrachai, N., Osholake, T., and Chen, X., 2010, Economic optimization analysis of the development process on a field in the Barnett Shale Formation: EME 580 Final Report.
17. Godec, M., Kuuskraa, V., Van Leeuwen, T., Melzer, L.S., and Wildgust, N., (2011). CO_2 storage in depleted oil fields—the worldwide potential for carbon dioxide enhanced oil recovery:Energy Procedia, v. 4, p. 2162–69.
18. U.S. Department of Energy, 2007, Carbon sequestration atlas of the United States and Canada: U.S. Department of Energy Office of Fossil Energy, March, www.precaution.org/lib/carbon_sequestration_atlas.070601.pdf.
19. Nordeng, S.H., LeFever, J.A., Anderson, F.J., Bingle-Davis, M., and Johnson, E.H., (2010). An examination of the factors that impact oil production from the middle member of the Bakken Formation in Mountrail County, North Dakota: North Dakota Geological Survey, RI-109.
20. LeFever, J., and Helms, L., (2008). Bakken Formation reserve estimates: North Dakota Geological Survey white paper, Bismarck, North Dakota, North Dakota Geological Survey, March, p. 6.
21. Nordeng, S.H., and Helms, L.D., (2010). Bakken source system – Three Forks Formation assessment: North Dakota Department of Mineral Resources, April.
22. Continental Resources, Inc., (2012). Bakken and Three Forks: www.contres.com/operations/bakken-and-three-forks (accessed May 30, 2013).

23. Mazza, R.L. Petroleum Consulting Services, Liquid-Free Stimulations - CO_2\Sand Dry-Frac: http://www.mde.state.md.us/programs/Land/mining/marcellus/Documents/Liquid_free_stimilations.pdf
24. Greenhorn, R., and Li, E., (1985). Investigation of high-phase volume liquid-CO_2 fracturing fluids, Paper no. 85-36-34, presented at the 36th Annual Technical meeting of the Petroleum society of CIM, June 2.
25. Lancaster, G., Barrientos, C., Li, E., and Greenhorn, R., (1987). High-phase- volume liquid-CO_2 fracturing fluids, Paper no. 87-38-71, presented at the 38th Annual Technical meeting of the Petroleum society of CIM, Calgary, June-7-10.
26. Yost, A.B., Mazza, R.L., Gehr, J.B., (1993). CO_2/sand fracturing in Devonian shales. SPE-26925, 1993 Eastern Regional Conference& Exhibition, Pittsburgh, PA, U.S.A., 2-4 November.
27. Ribeiro L. H., Sharma, M.M., (2013). Fluid selection for energized fracture treatments, SPE 163867, Hydraulic Fracturing Technology Conference, The Woodlands, Texas, USA, 4-6 February.
28. Środoń, J. (2009). "Quantification of illite and smectite and their layer charges in sandstone and shales from shallow burial depth." Clay Minerals **44**: 421-434.
29. Bai, B., M. Elgmati, H. Zhang and M. Wei (2013). "Rock characterization of Fayetteville shale gas plays." Fuel 105: 645-652.
30. Jarvie, D. M., R. J. Hill, R. T. E. and P. R. M. (2007). "Unconventional shale-gas systems: The Mississippian Barnett Shale of north-central Texas as one model for thermogenic shale-gas assessment." AAPG Bulletin **91**: 475-499.
31. Bruner, K. R. and R. Smosna (2011). A Comparative Study of the Mississippian Barnett Shale, Fort Worth Basin, and Devonian Marcellus Shale, Appalachian Basin. U. S. D. o. Energy. **DOE/NETL-2011/1478**.
32. Macht, F., K. Eusterhues, G. J. Pronk and K. U. Totsche (2011). "Specific surface area of clay minerals: Comparison between atomic force microscopy measurements and bulk-gas (N-2) and -liquid (EGME) adsorption methods." Applied Clay Science **53**(1): 20-26.
33. Geatches, D. L., A. Jacquet, S. J. Clark and H. C. Greenwell (2012). Monomer Adsorption on Kaolinite: Modelling the Essential Ingredients. J. Phys. Chem.
34. Wang, J., S.-W. Xia and L.-M. Yu (2014). "Adsorption mechanism of hydrated Pb(OH)(+) on the kaolinite (001) surface." Acta Physico-Chimica Sinica **30**: 829.
35. Chatterjee, A. (2005). "Application of localized reactivity index in combination with periodic DFT calculation to rationalize the swelling mechansim of clay type inorganic material." J. Chemical Sciences **117**: 533-539.
36. Mignon, P., P. Ugliengo, M. Sodupe and E. R. Hernandez (2010). "*Ab initio* molecular dynamics study of the hydration of Li+, Na+ and K+ in a montmorillonite model. Influence of isomorphic substitution." Phys. Chem. Chem. Phys. **12**: 688-697.
37. Suehara, S. and H. Yamada (2013). "Cesium stability in a typical mica structure in dry and wet environments from first-principles." Geochimica Et Cosmochimica Acta **109**: 62-73.
38. Li, Q., B. Lee, A. Fernandez-Martinez, G. A. Waychunas and Y.-S. Jun (2013). Calcium carbonate polymorphs on mica/quartz substrates and their interfacial free energies: Implications for mineral trapping mechanisms in geologic CO_2 sequestration. Dept. of Energy, Environmental and Chemical Engineering, Washington University, St. Louis, MO, qingyun.li@wustl.edu; ysjun@seas.wustl.edu.
39. Bertier, P., A. Busch, Y. Gensterblum, G. Rother, P. Weniger and H. Stanjek (2013). CO_2-clay interaction: Results from in-situ neutron diffraction and supercritical CO_2 sorption measurements. Clay and Interface Mineralogy, Energy and Mineral Resources Group, RWTH-Aachen University, D-52072 Aachen, Germany, Pieter.Bertier@emr.rwth-aachen.de.

40. Cygan, R. T., J. J. Liang and A. G. Kalinichev (2004). "Molecular models of hydroxide, oxyhydroxide, and clay phases and the development of a general force field." J. Phys. Chem. B **108**: 1255-1266.
41. Hughes, R. and B. Seyler (2013). Differences in pore-filling clay mineral compositions in Illinois reservoirs and their impact on the recovery of oil. Illinois State Geological Survey, Prairie Research Institute, University of Illinois, 615 E. Peabody Dr., Champaign, IL, r-hughes@illinois.edu.
42. Daniels, E. (2013). Importance of clays and shale properties for shale reservoir hydraulic stimulation and groundwater protection. Chevron Energy Technology Company, San Ramon, ericdaniels@chevron.com.
43. Cygan, R. T., E. M. Myshakin, V. N. Romanov, W. A. Saidi and K. D. Jordan (2013). Molecular models of CO_2 intercalation in montmorllonite. Geoscience Research and Applications Group, Sandia National Laboratories, Albuquerque, NM, rtcygan@sandia.gov.
44. Busch, A. (2013). CO_2/clay (physical) interactions: Why do they matter? Shell Global Solutions International, 2288GS Rijswijk, Netherlands.
45. Zhang, M., C. J. Spiers, S. M. de Jong, A. Busch and R. Wentinck (2013). Swelling stress development Faculty of Geosciences, Utrecht University, 3584 CD Utrecht, Netherlands, m.zhang@uu.nl.

Carbon Dioxide Capture for Storage in Deep Geological Formations, Volume 4
Karl F. Gerdes (Editor)

Section 2

STORAGE MONITORING & VERIFICATION (SMV)

Field Trialing

Carbon Dioxide Capture for Storage in Deep Geological Formations, Volume 4
Karl F. Gerdes (Editor)

Chapter 34

MONITORING CO_2 SATURATION FROM TIME-LAPSE PULSED NEUTRON AND CASED-HOLE RESISTIVITY LOGS

Tess Dance[1] and A. Datey[2]
[1]CO2CRC/CSIRO, Earth Science and Resource Engineering,
PO Box 1130, Bentley, W.A., 6102, Australia, Email: tess.dance@csiro.au
[2]Schlumberger Data Services, Melbourne, Victoria, Australia

ABSTRACT: Time-lapse well logging has long been a valuable petroleum reservoir management technique for monitoring relative changes in near-well bore hydrocarbons and formation fluid. As interest grows in the monitoring and accounting of Carbon Dioxide (CO_2) for enhanced recovery and sequestration, techniques such as pulsed neutron and cased-hole resistivity logging have been put to the test in the quantitative evaluation of CO_2. Despite this being beyond the original design purpose of the tools, and a lack of calibration specific to CO_2 injection conditions, results from demonstration projects and storage sites around the world have shown promise. In this chapter a case study is presented from the CO2CRC Otway project, Australia, where time-lapse well logging was applied to monitoring of CO_2 storage in a depleted gas field. Not all of the interpreted products for the quantitative characterisation of CO_2 saturation were as definitive as hoped. This was due to a variety of factors related to timing of logging runs, low salinity of the formation water, high mud filtrate invasion, the presence of CO_2 inside the borehole, and existing residual hydrocarbons in the reservoir. Nevertheless, the more reliable log outputs were evaluated and corrected accordingly in order to produce a semi-quantitative evaluation of saturation post-injection. The results were used to verify that the CO_2 plume is contained above the structural spill point of the storage complex. The lessons learned from Otway show that these logging techniques can be used effectively as part of a monitoring portfolio at CO_2 storage sites provided the execution is carefully controlled and variables are well understood.

KEYWORDS: time-lapse monitoring; carbon storage; pulsed neutron logging; cased-hole resistivity; residual CO_2 saturation; CO2CRC Otway project

INTRODUCTION

Time-lapse wireline logging methods, that are standard practice for petroleum field development in order to locate and manage stranded resources, are routinely applied to monitoring of CO_2 and play a vital role in monitoring injection profiles, detecting gas or water fronts, and for fine-tuning formation evaluations. Despite having a limited depth of penetration, well logs provide a high vertical resolution profile (<0.2 m) of the formation. This data can then be used to calibrate between the finer scale direct measurements from cores and fluid samples, and the coarser scale geophysical and pressure data from the reservoir. Thus fluid saturation logs are seen as essential in the monitoring portfolio at CO_2 storage sites for regulatory conformance, CO_2 accounting, model validation, and storage integrity assurance [1].

Two examples of logging methods that are commonly used for this purpose are: i) cased-hole resistivity which differentiates hydrocarbons from formation water on the basis of changes in resistivity [2]; and ii) pulsed neutron logging which is used to record the carbon-oxygen ratio and Sigma log (denoted by Σ), from which formation fluids may be distinguished on the basis of how thermal neutrons interact with various atoms in the formation [3, 4].

These logging methods have proven successful in well based monitoring of CO_2 saturation at a number of storage sites around the world [1, 5, 6]. For example, at the Frio Brine Project in Texas, it was demonstrated that time-lapse pulsed neutron logging is an appropriate method to monitor near-well CO_2 saturation changes in highly saline reservoirs due to a large sigma (Σ) contrast between the saline formation waters (high Σ) and the injected CO_2 (low Σ) [7, 8]. At the Nagaoka site in Japan, it was shown that by combining time-lapse induction and pulsed neutron logging techniques [9, 10], with time-lapse cross-well seismic tomography [11], it is possible to observe timing of breakthrough and migration direction of a plume over time [12], which in turn provides indication of the dissolution of the injected plume, displacement of water, and residual trapping potential within the reservoir [13]. At the Cranfield site in Mississippi, both techniques were used in an integrated monitoring study which revealed complex fluid flow in the sub-surface [14]; and at Svelvik Ridge Norway, down-hole electrical resistivity methods were used to monitor a controlled leak experiment [15]. Pulsed neutron logging has also recently been applied to monitor CO_2 flooding during an enhanced oil recovery project in the Middle East [5], and assessed in the context of enhanced gas recovery at the Altmark site in Germany [16].

The examples listed above uphold these methods as promising technologies for CO_2 monitoring at field scale. However, a gap in the literature exists surrounding the pitfalls and uncertainty that may be encountered during execution and interpretation of the data products when acquired under less than ideal conditions. This chapter presents a case study from the CO2CRC Otway site, Australia, where time-lapse logging was conducted during stage 1 of the project. Specifically this stage was aimed at demonstrating geological storage of over 65,000 tonne of CO_2 rich gas in an onshore depleted natural gas field and testing an array of conventional and novel monitoring techniques [17-19]. The post-injection conditions were logged at the injection well with a Slim Cased-hole Formation Resistivity Tool (SFRT) 14 months after injection ceased. The results were compared to the open-hole laterolog data to assess any changes in resistivity that may be attributed to CO_2. Pulsed neutron logging with a Reservior Saturation Tool (RST) was also run (using Inelastic Capture and Sigma mode) 23 months after injection ceased and compared to a base line RST acquired shortly after the well was cased. The aim of this paper is to document the complications that arose during the quantitative interpretation of CO_2 saturation from these methods. The lessons learned can be applied to improve the design of similar logging programs and interpretation workflows for CO_2 storage monitoring in time-lapse mode.

OVERVIEW OF LOGGING TOOLS USED AT OTWAY

The following provides details of the tools specific to the Otway Project monitoring. To this end the authors feel it is necessary to use vendor specific language. However, it must be acknowledged that there are equivalent tools available through other service providers that work on similar principles.

Cased-hole resistivity with the SFRT

The Cased-hole Formation Resistivity tool (CHFR) provides deep-reading measurements for an estimate of the formation resistivity behind steel casing. The slim-hole version of the formation resistivity tool (SFRT) fits in a casing as small in diameter as 2 7/8 inches (~73 mm). An overview of the development of the cased-hole logging tools can be found in Ferraris, [20]. In essence the tools work by inducing a current into the casing, which acts as a focusing electrode to force the current deep into the formation, past the zone of invasion. This current returns to a surface electrode. Most of the injected current will flow back to the surface within the casing thickness but a small fraction of it will leak into the formation. At a given depth the amount of leaking current is proportional to the formation/fluid conductivity. Some of the limitations of the tool as outlined in Aulia *et al.* [21], include loss of data in areas of the casing collars, or due to scale build-up which prevents good electrical contact between electrodes and the casing. Similarly, heavy casing or tubing can limit its use. However, a significant advantage of the SFRT is the depth of investigation which is between 2 and 10 m. This is more than an order of magnitude deeper than nuclear measurements and provides reliable results from beyond the invaded zone. Vertical resolution is a function of voltage spacing and station reading and is in the order of 1.2 m for bedding features, and fluid contacts can be identified +/- ~30 cm. In principal the SFRT measurement is comparable to that recorded by a laterolog, for example the HALS (or High resolution Azimuthal Laterolog Sonde) which passes current into the formation through electrodes that are in contact with the open bore-hole. Thus interpretations of formation saturation based on resistivity changes (e.g. high resistivity gas replacing low resistivity formation brine) often use the open-hole logs as a baseline.

Pulsed neutron logging with the RST

Pulsed neutron logging tools work by emitting bursts, or pulses, of neutrons into the formation. As the neutrons interact with various elements in the borehole, rock matrix and formation fluids, gamma rays are generated that are measured by the tool. These gamma rays are recorded and analysed to interpret fluid saturation. The RST (reservoir saturation tool) can operate in both Sigma and inelastic capture mode. Sigma (Σ) is the neutron capture cross section recorded with a "dual burst" pattern: short burst of neutrons followed by a long burst. The short burst is influenced by properties of the near well environment, and the long burst provides information from the formation. The depth of investigation is approximately 0.25 m. The neutron capture cross section is heavily influenced by chlorine and hydrogen; hence the response is largely determined by salinity and molecules like methane and water that contain hydrogen. Examples of the application of pulsed neutron capture logging for reservoir monitoring are provided by Morris *et al.* [22]. Inelastic capture measurements are made using a single long burst, and period where the generator is turned off. The gamma-ray energy resulting from inelastic scattering and capture is used to produce a spectrum from which the carbon-oxygen ratio (C:O) can be derived. Although it is less accurate and depth of penetration is limited to ~ 0.15 m, this method has the advantage of being independent of formation salinity.

As part of the CO_2 Capture Project Phase 2, a series of experiments were conducted at Schlumberger's Environmental Effects Calibration Facility to evaluate the RST's ability to detect CO_2 in a sandstone formation under controlled conditions [23]. It was concluded that sigma mode measurements show the best promise when the formation is saturated with highly saline brine prior to injection. This is because of the large capture cross-section difference between CO_2 and brine. However, when the formation water is relatively fresh (<20-50 ppk) the difference between the sigma of CO_2 and formation water is not adequate. To compute saturation from the C:O ratio method requires a calibration database in the same completion, formation and borehole fluids as the acquired log. Since such a database does not exist for CO_2 in formation and borehole means that the inelastic-capture measurement should only be used in qualitative interpretations in order to provide a method for detecting CO_2 rather than estimate saturation or quantify changes.

An ancillary measurement of the RST is thermal decay porosity (TPHI). TPHI is a measure of the hydrogen index (HI) obtained from the ratio of the "near to far" detector capture count rates. Changes in the HI can be used to interpret changes in saturation, for example CO_2 with HI ≈ 0 replacing water with HI ≈ 1. The large difference between the HI of formation water and CO_2 makes TPHI an attractive measurement to consider especially in low salinity formations as the difference between Sigma of fresh formation water and CO_2 is not significant. Complications can arise, however, in the case where the tool is surrounded by CO_2 in the borehole, logging injection wells for example. Furthermore, TPHI is not a calibrated porosity and the output from the various passes needs to be matched over intervals where no change of formation fluids is expected. For this purpose shales or highly cemented sandstones above and below the reservoir, can serve as useful lithological calibration markers.

When compared to the SFRT outlined above, the RST has significant limitation in the depth of investigation. This means that measurements may be completely invalid in formations with high mud filtrate invasion. However, the tool can provide a detailed fluid saturation profile along the borehole, with approximately 0.15 m vertical resolution– i.e. measurements are taken at 15 cm depth intervals (Adolph *et al.,* 1994). Also RST measurements can be recorded through dual casing and tubing strings while the SRFT can only be recorded through a single casing string.

MONITORING AT OTWAY

Project background and aims

The CO2CRC Otway project, stage 1, was conducted in the depleted Naylor gas field, located in the onshore Otway Basin, south western Victoria, Australia. Over the course of 18 months, between March 2008 and August 2009, 65,445 tonnes of CO_2 rich gas with a composition of 80% CO_2 / 20% CH_4 (molar) was injected into the field. The existing production well, Naylor-1, was recompleted for the purpose of monitoring and a new well, CRC-1, was drilled down-dip for use as an injector targeting the 25 m thick Waarre C Formation reservoir (Figure 1). The reservoir comprises heterolithic sandstone and mudstone. The porosity of the reservoir sandstones is between 18 % and 29 %, and permeability is in the order of 1.5 Darcy. Salinity of the formation water is approximately 20 ppk TDS (total dissolved solids), and pressure and temperature at the beginning of injection was 85 °C and 17.4 MPa respectively. Containment at the field is via mudstone seal juxtaposed to reservoir in a three way structural dip closure providing a spill point at a depth of around 2015 m below mean sea level. For containment assurance it is essential the injected CO_2 does not exceed the spill point which corresponds to the pre-production gas water contact. At the time of injection residual methane (average 19 % S_{gr}) was present throughout the storage reservoir as well as small gas cap at the top of the structure (Figure 1).

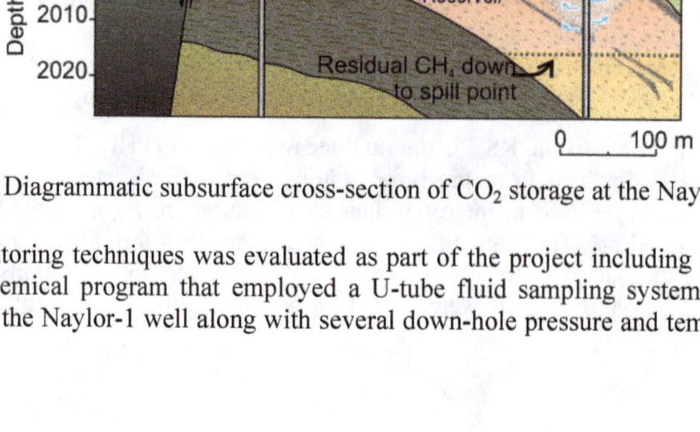

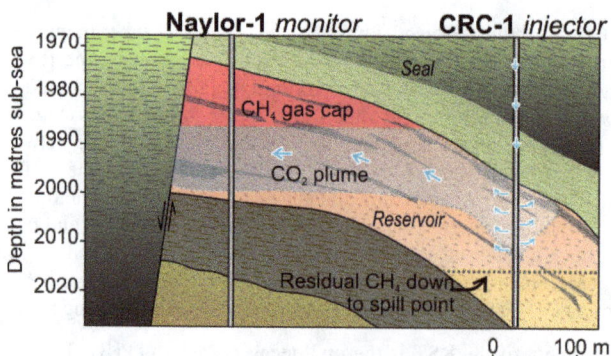

Figure 1. Diagrammatic subsurface cross-section of CO_2 storage at the Naylor Field.

An array of monitoring techniques was evaluated as part of the project including a comprehensive down-hole geochemical program that employed a U-tube fluid sampling system [24, 25]. These were deployed at the Naylor-1 well along with several down-hole pressure and temperature gauges, geophones and sensors in order to observe the rate of CO_2 migration from the injector up-dip to the monitoring well and the dynamic chemical changes that occur as the plume filled the structure. For a detailed background and overview of the project see Cook [26]. Further details of the project's operation and planning can be found in Sharma *et al.* [27], site characterisation is in Dance [28], a technical overview and initial results of the monitoring in Underschultz *et al.* [18], and overall research implications and impacts in Jenkins *et al.* [19].

Time-Lapse well logging

The permanent installations in the Naylor-1 monitoring well meant it was prohibitive to access the borehole for wire line logging at any stage of the experiment. Instead the CRC-1 injector was used for this purpose. CRC-1 is a steel cased, 4½ inch (11.43 cm), vertical monobore, perforated over 11 meters in the top half of the target reservoir. At four stages of the project wireline logs were acquired at CRC-1. A summary of the timeline of logging and other key events at the well is given in table 1. Baseline for the resistivity monitoring was provided by the HALS laterologs acquired in the open hole soon after drilling. Six months later baseline RST in sigma and inelastic capture modes were acquired in the cased-hole. The well was perforated and a well injection test was attempted using water with Potassium Chloride (KCl) added to prevent the swelling of clays in the reservoir. The test revealed blocked perforations so the perforations were extended a few meters down and the formation water allowed to back flow into the well. The well test was not repeated after this, nevertheless it is acceptable to assume during this stage fluid salinity changes occurred in the near-well bore region due to the saline injection fluid and then from the "fresher" formation water flushing the mud invasion zone. Unfortunately RST or SFRT logging was not run after this critical step to observe the effects of the exchange of fluids on formation salinity, radioactivity response and porosity in the invaded zone. Furthermore, this would have provided a more appropriate baseline of logs in the cased-hole prior to injection.

Post-injection logging was performed first with the SFRT 14 months after injection ceased. Then the RST monitoring logs were run 9 months after that (i.e. nearly two years after the end of injection) (Figures 2 and 3).

Table 1. CRC-1 logging timeline.

Logging stage	Date	Event
	Feb 2007	Well spudded
Baseline	March 2007	Open hole well logs acquired
Baseline	Sept 2007	RST: sigma & inelastic capture logs run in cased-hole
	Dec 2007	Well perforated & well test attempted with KCl brine
	Feb 2008	Well perforations extended & well test not repeated
	March 2008 to Aug 2009	**CO_2 rich gas injected**
Monitoring	October 2010	Slim cased-hole formation log (SFRT) acquired
Monitoring	July 2011	RST: Sigma & inelastic capture logs acquired

Predicted response

The two measurements that were used to invert for saturation were Sigma (Σ) and TPHI. In order to estimate the sensitivity margin of each, the predicted change in the specific reservoir conditions was computed using a commercial nuclear parameter code (SNUPAR Schlumberger's Nuclear Parameter Code [29]).

Figure 2. Logging of the Otway Project CRC-1 injection well with RST in July 2011.

Figure 3. Close up view of the RST logging at the well head.

Figure 4 shows the estimated change in Σ and Figure 5 is the change in THPI, as the mixture fluid (80% CO_2 and 20% CH_4 fluid at formation temperature and pressure) replaces water for a clastic sandstone with 18 % porosity and formation water salinity around 20000 mg/L TDS (20 ppk). The quoted accuracy for Σ logs from the RST is 1 capture units (1 cu). Although salinity in the formation is low, the change in Σ when CO_2 (with Σ of ~0.03 cu) replaces formation water was expected to be visible on the logs compared to the precision of the measurement. The change in TPHI is less affected by salinity and therefore this output from the tool is considered more reliable for CO_2 saturation interpretation.

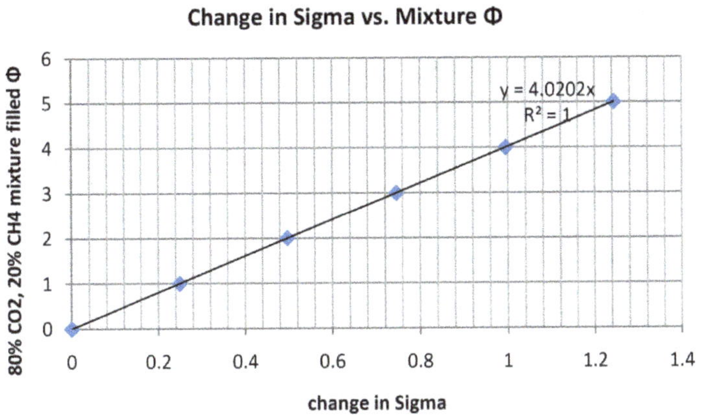

Figure 4. The estimated change in Sigma with CO_2 saturation.

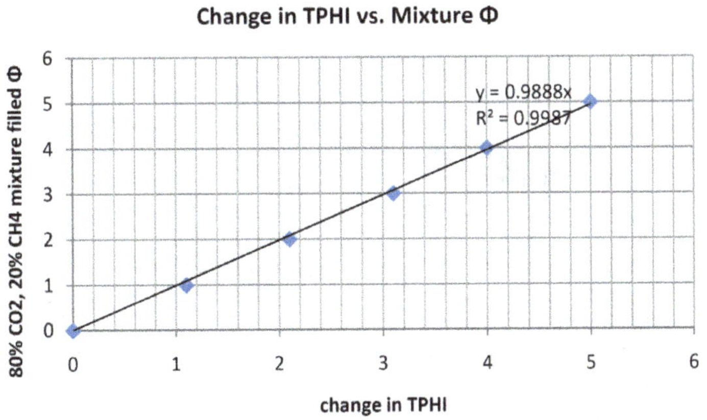

Figure 5. The estimated change in TPHI when the CO_2 rich mixture replaces formation water.

RESULTS AND INTERPRETATION

A composite display of all logging outputs is shown in Figure 6 (depth is in meters measured from rotary table). The Waarre C Formation is the interval from 2052 m to 2082 m. The interval 2052 m to 2062 m represents the perforated sand across which the CO_2 rich gas was injected. CO_2 saturation is quantitatively interpreted from the sigma logs and the thermal decay porosity logs (TPHI). Due to low salinity of the formation brine (~20 ppk), the change in Σ is not significant and can be prone to statistical errors. Thus TPHI is the preferred inversion method. These interpretations, as well as the results from each of the other logging methods, are discussed in more detail below.

Gamma ray

Four sets of Gamma Ray (GR) logs were recorded:

1. GR open hole: recorded with open hole Platform Express logs.
2. GR preInj RST: recorded in cased-hole with RST before the injection.
3. GR postInj SFRT: recorded in cased-hole with SFRT after the injection.
4. GR postInj RST: recorded in cased-hole with RST after the injection.

In the first track in Figure 6 the open hole GR is presented from scale 0-250 gAPI and cased-hole GRs are presented from scale 0-150 gAPI, so that the open hole and cased-hole GR log can be compared when presented on the same track. The log indicates that GR open hole, pre-injection and post-injection match across all intervals except over 2052 m to 2064 m and 2071 m to 2078 m, which corresponds to the permeable sandstones. The GR from post-injection acquisition for both these intervals are higher than the GR from pre-injection acquisition. A possible explanation is that the potassium in the KCl mud cake has precipitated out and increased radioactivity over this interval [30]. Another source of increased gamma can be from radioactive scale precipitation on the well casing [31]. However, without independent verification, no reasonable interpretation is offered for the source of increased GR at CRC-1.

Resistivity

Results of the resistivity measurements are shown in the second track in Figure 6. The pre-injection, open hole logs comprise laterolog (HART), deep resistivity (HLLD) and shallow (HLLS). The post-injection SFRT logs (RTCH_K) were calibrated to the open hole logs over shale markers. The zones where cased-hole resistivity is greater than the open hole measurements are shaded yellow and are apparent throughout the formation (from 2055 m to 2083 m). In principal this difference may be used to compute the change in water saturation (S_w) between the timing of the logs. The assumption being the increased resistivity is attributed to the resistive CO_2 rich gas displacing the conductive formation water. However, the observations did not match the expected response over some intervals. For example, the greatest change in resistivity was in the sands below the perforations. It appeared from the interpretation that S_w was reduced from nearly 100% to 60-70% and the implication was that CO_2 had displaced the water. Furthermore, increases in resistivity were also observed in sandstone beds several hundreds of meters above the injection zone. From the simulation studies, it was highly unlikely CO_2 would migrate to sandstones beneath the perforations and even less likely to be above the seal. Before jumping to the conclusion that CO_2 has displaced water in the overlying aquifers, thus implying there is a breach of containment, we must consider the multiple phenomenon that may be attributed to the source of increased resistivity in the time between the open hole logs and the post-injection logs.

Firstly, during drilling, the low salinity *in situ* formation brine was highly invaded and mixed with saline fluid from the KCl drill mud. This was intentional as the well was drilled slightly overbalanced to improve well bore stability. The contrast is even more pronounced in the relatively fresh Paaratte Formation above that has salinity of 2-8 ppk and permeability in the order of 2 Darcy. Evidence from core plug analysis shows there was substantial invasion, as well as the failed well injectivity test in an otherwise highly permeable sandstone also supports the interpretation that mud invasion was high. Furthermore, the test itself was performed with KCl brine after the open hole logging and may or may not have contributed to lowering the resistivity. Unfortunately, due to poor injectivity, it is unknown to what degree the well test fluids entered the formation. However, it is certain that the process of back flowing water from the formation into the well at this time would result in additional changes to the resistivity response. The implication is that the low conductivity "fresh" formation fluid flushed the high conductivity KCl solution from the near well region. Therefore using the open-hole laterologs as a baseline would artificially set the starting point for resistivity too low in the region of the perforations resulting in an overestimation of gas saturation in this zone. What is puzzling is that the sands below the perforated interval show a greater difference in resistivity than the sandstones that actually received CO_2 during injection. As a result the resistivity logs are inconclusive in determining reservoir saturation changes.

The lesson here is to log a pre-injection cased-hole SFRT measurement and compare it to the post-injection cased-hole SFRT measurement. This will eliminate any interpretation anomalies and add assurance that any resistivity changes would be a result of injection. When acquiring SFRT, some bad measurement points will be unavoidable due to poor electrode contact, casing joints, mud-cake in the annulus, and non-conductive cement. This can also result in a mismatch between the open-hole data and cased-hole data. The addition of a pre-injection cased-hole baseline SFRT measurement would reduce these uncertainties in the subsequent interpretation.

Capture cross-section (Σ)

The logs from the RST run in Σ mode are used to compute saturation (track 4 in Figure 6). The SIGM postInj is less than SIGM preInj across the interval from 2052 m to 2064 m adjacent to the perforations. This decrease is attributed to the injected CO_2, which has low capture cross section, replacing water with higher capture cross section across this zone. The interpreted CO_2 saturation is between 10-30 pu. Σ is unchanged over the rest of the interval. Due to the low salinity of the formation brine, the Σ response is not very sensitive to the change in water / CO_2 saturation

Thermal Decay Porosity (TPHI)

Thermal Decay Porosity is one of products of RST Sigma log acquisition. Two sets of RST TPHI were recorded (Figure 6, tracks 6 & 7).

1. TPHI preInj: recorded in cased-hole with RST before the Injection
2. TPHI postInj: recorded in cased-hole with RST after the Injection

TPHI is not a calibrated porosity and the TPHI from the two passes need to be matched over the intervals where no change of formation fluids is expected. Thus a bulk shift of -0.01pu is applied to TPHI postInj before comparing it to TPHI preInj. The TPHI postInj is less than TPHI preInj across the perforated interval 2052 m to 2064 m and is largely unchanged over rest of the interval. This indicates that CO_2, which has Hydrogen Index ≈ 0, has replaced water with Hydrogen Index ≈ 1, across the interval 2052 m to 2064 m. The TPHI logs were used to compute CO_2 saturation and at the time of logging CO_2 saturation was in the order of 15-20 pu. The estimated structural spill point for the Naylor field at a minimum depth of ~2015 m TDV SS, translates to ~2066 m MD in CRC-1. An important observation is that no changes can be observed in the TPHI logs below the lowest perforation at 2064 m, adding assurance the injected gas has not filled downwards to the estimated structural spill point of the reservoir.

At the time of this study, the RST was not calibrated for operating with CO_2 in well bores. At CRC-1 the post-injection logs were acquired nearly two years after the end of injection. However, it is likely that CO_2 was still present in the borehole and annulus. There are available methods that can account for this effect during the analysis. Pressure information is first used to identify the various well bore fluid interfaces. Then a shift is applied to the TPHI log over the intervals where CO_2 was in the borehole to account for the changed response in the near-tool region. Similarly, for the Σ output processing, the Thermal Decay Time-Like processing (SIGM TDTL) can be used for evaluation as its computation is less affected by CO_2 in the wellbore than the standard processing for Σ. However, these adjustments bring with them an additional layer of uncertainty.

The Carbon:Oxygen logs

Two sets of RST Inelastic Capture (IC) log were recorded.

1. IC preInj: recorded in cased-hole with RST before the Injection
2. IC postInj: recorded in cased-hole with RST after the Injection

The Far Carbon to Oxygen (C:O) ratio was computed from the log. FCOR:CO ratio is computed using spectral data from far detector. FCOR has good accuracy but has poor statistics. Acquiring reasonable statistics for FCOR would have required equivalent logging speed of ~ 6 ft/hr which is currently not possible. FWCO:CO ratio is computed using windows method from the far detector. This method obtains the ratio of C:O by placing broad windows or "bins" over the carbon and oxygen spectral peaks [3]. The windows method has good statistics and therefore is more precise but is often less accurate. Hence the FWCO from the two passes need to be matched over the intervals where no change of formation fluids is expected. A shift of +0.05 units is applied to FWCO postInj before comparing it to FWCO preInj. Comparing the preInj and postInj FWCO clearly indicates CO_2 replacing formation water across the perforated interval. The FWCO logs are picking up the increased levels of CO_2 at the injection interval (and above) 2052 m to 2063 m, and this is interpreted to be a result of the injected gas. When the RST Inelastic Capture logs were evaluated in Schlumberger's Environmental Effects Calibration Facility for CO_2 in formation, the response was difficult to model [28]. Thus FCOR or FWCO cannot yet be converted to CO_2 saturation but can be used qualitatively. A specific method of interpreting the carbon isotope data at Otway with CO_2 in the borehole is outlined in Quinlan *et al.* [6]. The inelastic capture ratio count rates (IRAT), which characterise near well and bore hole fluids, is used to correct for late capture ratios (TRAT), which see deeper in the formation.

Discussion

Current petrophysical logging techniques do not provide a direct measurement of the property we ultimately want to solve for (i.e. saturation). Tools measure the physical response from various elements and one must make assumptions to derive the end output. One complication for quantitative interpretation of the logs from the Otway site is the presence of residual methane in the baseline data. Prior to injection, the formation fluid is a mixture of water and gas. In all likelihood the injection fluid could be displacing existing water and gas from the formation. The assumption for the resistivity, sigma and TPHI interpretation is that the injection fluid is only displacing water. One measurement is being made with the RST tool (Sigma and TPHI are non-unique responses). One measurement can solve of only one variable. Thus change in volume of only one fluid can be computed when comparing the baseline measurement and post-injection measurement. If a change in Sigma or TPHI response is due to a change in fluid salinity, CH_4 volume and CO_2 volume, a unique solution cannot be determined. The assumption in the workflow is that only CO_2 volume has changed. TPHI logs combined with core corrected/gas corrected NMR porosity logs, coupled with the density logs may be used as a combination to identify the methane in the reservoir.

In time-lapse mode, the interpretations are further complicated by physical or geochemical changes that may have occurred during well operations. These include changes as a result of casing, patches, collars, the drill mud itself, well test fluids, and well bore scale. Using oil based muds will minimise the issues associated with induced apparent salinity changes. But in order to record the most accurate baseline characterisation, logs would need to be acquired at each stage of the well work over up to the point of injection.

Similarly, post-injection logging should be properly timed to optimise results. At CRC-1 the post-injection monitoring logs were acquired long after injection stopped (i.e. >12 months for resistivity, and nearly 2 years for the RST). Simulation results predicted that the free gas would migrate up-dip from the injection well, and a strong aquifer drive would enhance the imbibition process leaving only residually trapped CO_2 behind. Dissolution would further reduce the percentage of CO_2 remaining. This coupled with the difficulty that there was already 10-20% residual methane in the reservoir at the time of the baseline logs, complicates the log interpretation further. With this in mind, timing of the post-injection monitoring as soon as feasible after injection may improve the likelihood of detecting higher saturation and as a consequence enhance the contrast between logs.

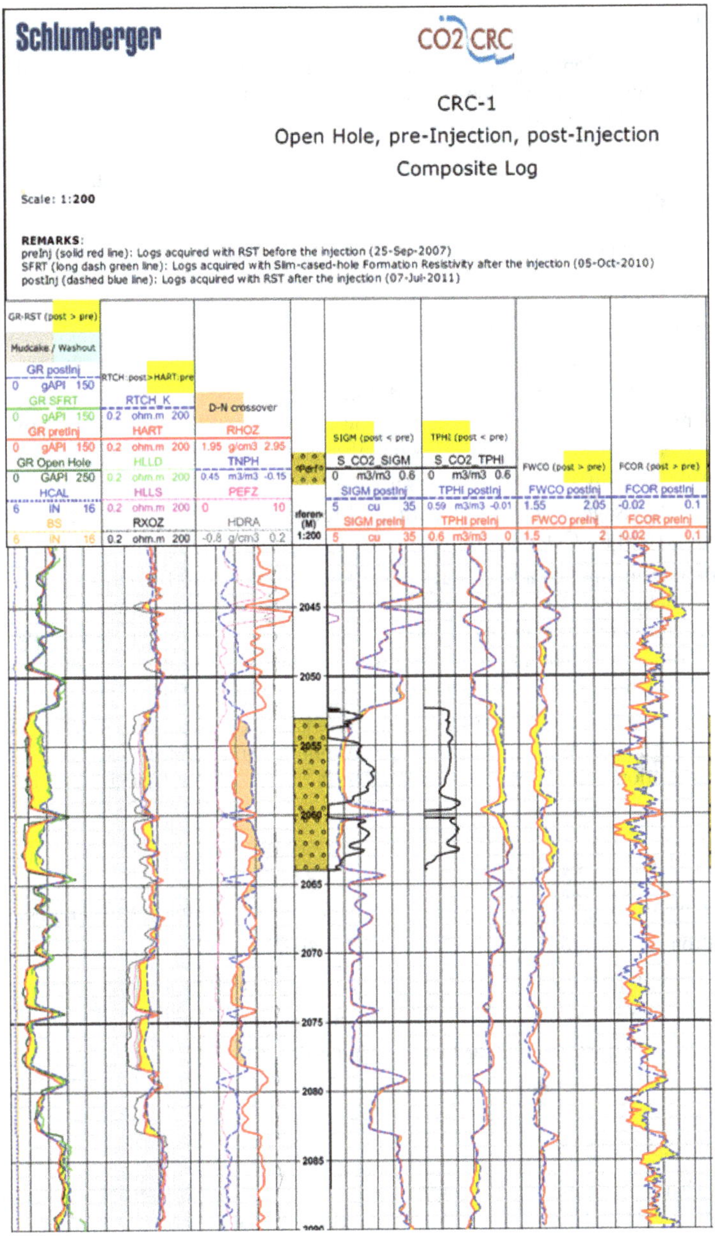

Figure 6. The integrated display of all logging outputs acquired in the Waarre C Formation (2052 m to 2082 m). From left to right: gamma ray; resistivity from the open hole (HART, HLLD - deep, and HLLS - shallow) and the cased-hole resistivity (RTCH_K); the density-neutron cross over used as a proxy for lithology; the perforated section; the derivation of CO_2 saturation from Sigma logs; and from the Thermal Decay Porosity (TPHI) logs; the carbon-oxygen ratio computed using the windows method (FWCO); and from the spectral data from the far detector (FCOR).

CONCLUSIONS

Monitoring CO_2 saturation from time-lapse pulsed neutron and cased-hole resistivity logs can deliver useful qualitative information that can be helpful in managing the CO_2 reservoir. At the CO2CRC Otway Project all logging products were cross-evaluated to detect petrophysical changes that were attributed to the injected gas. However, the timing of logs at Otway and conditions in which they were acquired were not ideal for reliable quantitative interpretation of CO_2 saturation. The summary of findings includes:

1. Although the reduced conductivity in the zone below the perforations is unexplained, both the Sigma and TPHI logs, as well as the C:O logs independently confirm that injected gas did not encroach the sands below the perforations, which was useful for containment assurance and may encourage further use of this combination for cross validation purposes.
2. The gamma ray and the resistivity from the SFRT are prone to unexplainable uncertainties which in hindsight may have been avoided if logs were run after every event. A baseline log run is highly recommended. **The baseline log must be as representative as possible of pre-injection conditions with no other variables changing. Open hole logs are not a good quantitative baseline for cased hole logs.**
3. In order for the usage of RST to be optimised for CO_2 related operations, the tool needs to be fully characterised for CO_2 conditions.
4. This study has highlighted a few issues in the interpretation workflow that need to be addressed if RST is to be used for CO_2 estimation in depleted fields with mixed fluids. TPHI logs combined with core corrected/gas corrected porosity curve with the density logs can help to account for the methane in the reservoir. This not so critical for saline aquifer storage monitoring which is expected to be more commonly used for CO_2 injection.
5. The timing of logs with respect to well work overs is an important consideration that needs to be made at storage sites using this technology to monitor saturation in CO_2 injection wells. The same complication will not apply when logging in dedicated monitoring bores that are not used for injection.

NOMENCLATURE

PNC = Pulsed neutron capture

SIGM = Neutron capture cross section, cu

TPHI = PNC neutron porosity, pu

TPHI = Thermal neutron porosity

TRAT = Thermal count rate ratio (near/far)

TDT = Thermal Decay Time tool

SIGM TDL = Sigma with TDT-Like processing

Σ = Sigma

cu = Capture cross section units, 10^{-3} cm^{-1}

s.u. = Saturation units

pu = Porosity units

ACKNOWLEDGEMENTS

The authors wish to acknowledge the support given by the CO_2 Capture Project for acquiring the valuable logging data at CRC-1 which made this study possible. The CO2CRC acknowledge the funding provided by the Australian government through its CRC program to support the Otway research project. Financial assistance is provided through Australian National Low Emissions Coal Research and Development (ANLEC R&D). ANLEC R&D is supported by Australian Coal Association Low Emissions Technology Limited. Special thanks to Tom McDonald for analysis of the cased-hole resistivity logs. Thanks also to Rajindar Singh, Lincoln Paterson, Tom Daley, and Kevin Dodds of the CO_2 Capture Project for their expert technical input and advice.

REFERENCES

1. Freifeld, B.M., Daley, T.M., Hovorka, S.D., Henninges, J., Underschultz, J., and Sharma, S. (2009). *Recent advances in well-based monitoring of CO_2 sequestration.* Energy Procedia **1**: pp. 2277-2284.
2. Zhou, Q., Julander, D., and Penley, L. (2002). *Experiences with cased-hole resistivity logging for reservoir monitoring*, Society of Petrophysicists and Well-Log Analysts, 8 p.
3. Albertin, I., Darling, H., Mahdavi, M., Plasek, R., Cedeno, I., Hemingway, J., Richter, P., Markley, M., Olesen, J.-R., Roscoe, B., and Zeng, W. (1996). *Many facets of pulsed neutron cased-hole logging.* Oilfield Review. **8**(2): p. 28-41.
4. Adolph, B., Stoller, C., Brady, J., Flaum, C., Melcher, C., Roscoe, B., Vittachi, A., and Schnorr, D. (1994). *Saturation monitoring with the RST reservoir saturation tool.* Oilfield Review. **6**(1): p. 29-39.
5. Al Arayni, F., Obeidi, A., Brahmakulam, J., and Ramamoorthy, R. (2013). *Pulsed-neutron monitoring of the first CO_2 enhanced-oil-recovery pilot in the middle east.* SPE 141490-PA, SPE Reservoir Evaluation & Engineering, **16**(1): p. 72-84.
6. Quinlan, T.M., Sibbit, A.M., Rose, D.A., Brahmakulam, J.V., Zhou, T., Fitzgerald, J.B., and Kimminau, S.J. (2012). *Evaluation of the carbon dioxide response on pulsed neutron logs*, SPE 159448, presented at the SPE Annual Technical Conference and Exhibition held in San Antonio, Texas, USA, 8-10 October.
7. Sakurai, S., Ramakrishnan, T.S., Boyd, A., Mueller, N., and Hovorka, S. (2006). *Monitoring saturation changes for CO_2 sequestration: Petrophysical support of the Frio Brine pilot experiment.* Petrophysics. **47**(6): p. 483-496.
8. Müller, N., Ramakrishnan, T.S., Boyd, A., and Sakruai, S. (2007). *Time-lapse carbon dioxide monitoring with pulsed neutron logging.* Int. J. Greenhouse Gas Control. **1**(4): p. 456-472.
9. Xue, Z., Tanase, D., and Watanabe, J. (2006). *Estimation of CO_2 saturation from time-lapse CO_2 well logging in an onshore aquifer, Nagaoka, Japan.* Exploration Geophysics. **37**(1): p. 19-29.
10. Nakatsuka, Y., Xue, Z., Garcia, H., and Matsuoka, T. (2010). *Experimental study on CO_2 monitoring and quantification of stored CO_2 in saline formations using resistivity measurements.* International Journal of Greenhouse Gas Control. **4**(2): p. 209-216.
11. Saito, H., Nobuoka, d., Azuma, H., Xue, Z., and Tanase, D. (2006). *Time-lapse crosswell seismic tomography for monitoring injected CO_2 in an onshore aquifer, Nagaoka, Japan.* Exploration Geophysics. **37**: p. 30-36.
12. Mito, S. and Xue, Z. (2011). *Post-injection monitoring of stored CO_2 at the Nagaoka pilot site: 5 years time-lapse well logging results.* Energy Procedia. **4**: p. 3284-3289.
13. Nakajima, T. and Xue, Z. (2013). *Evaluation of CO_2 saturation at Nagaoka pilot-scale injection site derived from the time-lapse well logging data.* Energy Procedia. **37**: p. 4166-4173.
14. Butsch, R., Brown, A.L., Bryans, B., Kolb, C., and Hovorka, S. (2013). *Integration of well-based subsurface monitoring technologies: Lessons learned at SECARB study, Cranfield, MS.* International Journal of Greenhouse Gas Control. **18**: p. 409-420.
15. Denchik, N., Pezard, P.A., Neyens, D., Lofi, J., Gal, F., Girard, J.-F., and Levannier, A. (2014). *Near-surface CO_2 leak detection monitoring from downhole electrical resistivity at the CO_2 field laboratory, Svelvik Ridge (Norway).* International Journal of Greenhouse Gas Control. **28**: p. 275-282.
16. Baumann, G. and Henninges, J. (2012). *Sensitivity study of pulsed neutron-gamma saturation monitoring at the Altmark site in the context of CO_2 storage.* Environmental Earth Sciences. **67**(2): p. 463-471.

17. Dodds, K., Daley, T., Freifeld, B., Urosevic, M., Kepic, A., and Sharma, S. (2009). *Developing a monitoring and verification plan with reference to the Australian Otway CO_2 pilot project.* The Leading Edge. **28**(7): p. 812-818.
18. Underschultz, J., Boreham, C., Dance, T., Stalker, L., Freifeld, B., Kirste, D., and Ennis-King, J. (2011). *CO_2 storage in a depleted gas field: An overview of the CO2CRC Otway Project and initial results.* International Journal of Greenhouse Gas Control. **5**(4): p. 922-932.
19. Jenkins, C.R., Cook, P.J., Ennis-King, J., Undershultz, J., Boreham, C., Dance, T., de Caritat, P., Etheridge, D.M., Freifeld, B.M., Hortle, A., Kirste, D., Paterson, L., Pevzner, R., Schacht, U., Sharma, S., Stalker, L., and Urosevic, M., (2012). *Safe storage and effective monitoring of CO_2 in depleted gas fields.* Proceedings of the National Academy of Sciences. **109**(2): p. E35-E41.
20. Ferraris, P. (2002). *Current affairs.* Middle East & Asia Reservoir Review. **3**: p. 33-45.
21. Aulia, K., Poernomo, B., Richmond, W.C., Wicaksono, A.H., Béguin, P., Benimeli, D., Dubourg, I., Rouault, G., VanderWal, P., and Boyd, A. (2001). *Resistivity behind casing.* Oilfield Review. **13**(1): p. 2-25.
22. Morris, C.W., Morris, F., Quinlan, T.M., and Aswad, T.A. (2005). *Reservoir monitoring with pulsed neutron capture logs*, SPE 93889, presented at the SPE Western Regional Meeting held in Irvine, CA, USA, 30 March - 1 April.
23. Climent, H. (2009). *CO_2 detection - response testing of rst in sandstone formation tank containing CO_2 and water-based fluid.*, in *Carbon dioxide capture for storage in deep geologic formations - results from the CO_2 capture project. Volume 3: Advances in CO_2 capture and storage technology results (2004-2009)*, Editor: L.I. Eide, CPL Press.
24. Boreham, C., Underschultz, J., Stalker, L., Kirste, D., Freifeld, B., Jenkins, C., and Ennis-King, J. (2011). *Monitoring of CO_2 storage in a depleted natural gas reservoir: Gas geochemistry from the CO2CRC Otway Project, Australia.* International Journal of Greenhouse Gas Control. **5**(4): p. 139-154.
25. Freifeld, B.M., Trautz, R.C., Kharaka, Y.K., Phelps, T.J., Myer, L.R., Hovorka, S.D., and Collins, D.J. (2005). *The U-Tube; a novel system for acquiring borehole fluid samples from a deep geologic CO_2 sequestration experiment.* J. Geophys. Res. **110**(B10): p. B10203.
26. Cook, P.J.E., (2014). *Geologically storing carbon: Learning from the Otway Project experience.*, P.J.E. Cook, Editor, Wiley CSIRO Publishing: Melbourne. p. 400.
27. Sharma, S., Cook, P., Robinson, S., and Anderson, C. (2007). *Regulatory challenges and managing public perception in planning a geological storage pilot project in Australia.* Int. J. Greenhouse Gas Control. **1**(2): p. 247-252.
28. Dance, T. (2013). *Assessment and geological characterisation of the CO2CRC Otway Project CO_2 storage demonstration site: From prefeasibility to injection.* Marine and Petroleum Geology. **46**: p. 251-269.
29. McKeon, D.C. and Scott, H.D. (1989). *SNUPAR-a nuclear parameter code for nuclear geophysics applications.* Nuclear Science, IEEE Transactions on. **36**(1): p. 1215-1219.
30. Fertl, W.H. and Frost Jr., E. (1982). *Experiences with Natural Gamma Ray Spectral Logging in North America*, SPE 11145, presented at the Society of Petroleum Engineers Annual Technical Conference and Exhibition, 26-29 September, New Orleans, Louisiana.
31. Crabtree, M., Eslinger, D., Fletcher, P., Miller, M., Johnson, A., and King, G. (1999) *Fighting scale—removal and prevention.* Oilfield Review. **11**(3): p. 30-45.

Carbon Dioxide Capture for Storage in Deep Geological Formations, Volume 4
Karl F. Gerdes (Editor)

Chapter 35

4D VSP MONITORING OF CO_2 SEQUESTRATION INTO A DEPLETED GAS RESERVOIR, CO2CRC OTWAY PROJECT

Roman Pevzner[1], E. Caspari[1], M. Urosevic[1], V. Shulakova[2] and B. Gurevich[3]
[1]Curtin University and CO2CRC
[2]CSIRO and CO2CRC
[3]Curtin University, CSIRO and CO2CRC

ABSTRACT: The Otway Project is Australia's first demonstration of the deep geological storage of carbon dioxide. The first stage of the project included injection of a CO_2-rich gas mixture into the depleted gas reservoir of the Naylor gas field. This field is located onshore in Victoria, about 300 km west of Melbourne. It was discovered in 2000 after acquisition of the large 3D seismic survey (Curdie Vale 3D) and subsequently produced in 2002-2004 using the single Naylor-1 well. During the first stage of the injection project, 66,000 tonnes of 80/20 % CO_2/CH_4 gas mixture were injected through the dedicated, newly-drilled injector well (CRC-1) located about 300 m away from Naylor-1 and drilled in the down dip direction of the gas reservoir. A comprehensive seismic monitoring program is part of the project. The Stage I monitoring program objectives were: (1) to ensure detection of possible gas leakages out of the reservoir into other formations, (2) to attempt detection of changes in seismic response due to CO_2 injection into the reservoir and (3) to test and benchmark various seismic monitoring techniques. In this chapter we compare applicability of 4D VSP and surface seismic for CO_2 sequestration monitoring based on repeatability, cost and field operation. Also, we compare observed differences between the baseline and the monitor surveys against the predicted time-lapse signal.

KEYWORDS: seismic monitoring; VSP; Otway project; depleted gas reservoir

INTRODUCTION

Stage I of the CO2CRC Otway Project consisted of the injection of 66,000 tonnes of a 80/20 % CO_2/CH_4 mixture into the depleted Naylor gas reservoir, located at a depth of about 2 km. A dedicated injection well (CRC-1) was drilled for the project. The seismic monitoring program included two simultaneously acquired pairs of 3D surface seismic and 3D VSP surveys before and after the injection.

The first pair of seismic surveys was acquired in December 2007-January 2008 (baseline survey) and the second in January, 2010. An intermediate 3D surface seismic survey was acquired in January, 2009; however, it was not combined with 3D VSP acquisition.

Due to limited availability of the Schlumberger VSP crew and equipment during the baseline survey acquisition, only half of the shot points used for surface seismic were also used for the 3D VSP. A full set of source points was used for both the surface seismic and the 3D VSP survey in 2010.

Both the baseline and monitoring 3D VSP surveys were processed by Schlumberger [1, 2]. Surface seismic data was processed by Curtin University and CSIRO. This chapter describes the cross equalisation of 4D VSP data and compares the repeatability of borehole and surface seismic surveys. A seismic modelling study was performed in order to predict the time-lapse signal and compare it to the achieved level of repeatability. This report also compares technical and operational advantages and disadvantages of both methods.

OTWAY SURFACE AND BOREHOLE TIME-LAPSE SEISMIC: ACQUISITION

Acquisition parameters of VSP and surface seismic surveys are presented in Table 1.

Table 1. Acquisition parameters.

	Survey I	Survey II	Survey III
Date	VSP: December, 2007 Surface seismic: December 2007 – January 2010	January 2009	January, 2010
Source	Weight drop, Hurricane 10, 750 kg, free fall from 1.2 m 4 stacks	IVI minibuggy, Linear sweep 12 s, 10-150 Hz, 4 sweeps per source point	
Number of source lines	VSP: 15 Surface seismic: 29	29	
Source line source spacing	200 m	100 m	
Source spacing	20 m	20 m	
Number of source points	VSP: 1139 Surface seismic: 2181	VSP: 0 Surface seismic: 2223	2223
Receiver Parameters: VSP			
VSP tool	3C Schlumberger VSI, 10 level	N/A	3C Schlumberger VSI, 8 level
Acquisition interval	1485-1620 m		1500-1605 m
Receiver spacing along borehole	15 m		15 m
Receiver Parameters: Surface Seismic			
Recorder	Seistronix, EX-6	Seistronix, EX-6	
Receiver type	10 Hz, single geophones	10 Hz, single geophones	
Number of active channels	440	873	
Number of receiver lines	10	10	
Receiver line spacing	100 m	100 m	
Receiver point spacing	10 m	10 m	
Recording pattern	Orthogonal cross-spread pattern, odd source lines were recorded by the first 5 receiver lines, even source lines by 6-10 receiver lines	Orthogonal cross-spread pattern, all receiver lines were active for all shots	

The baseline (2007-2008) survey was acquired in two stages. The 3D VSP and the first part of the surface 3D seismic data were shot in December 2007. Only odd-numbered source lines and the first five surface receiver lines were used (Figure 1 and Figure 2). The second stage (January 2008) consisted of only surface seismic data being acquired using receiver lines 6-10.

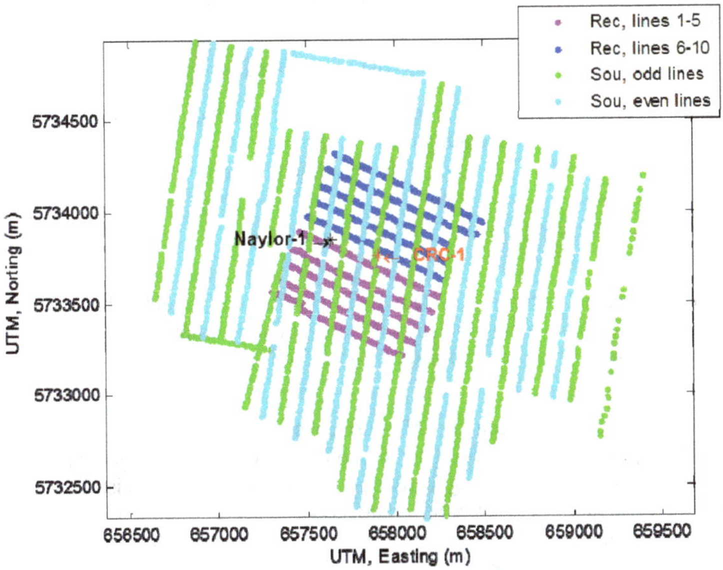

Figure 1. Surface seismic acquisition scheme.

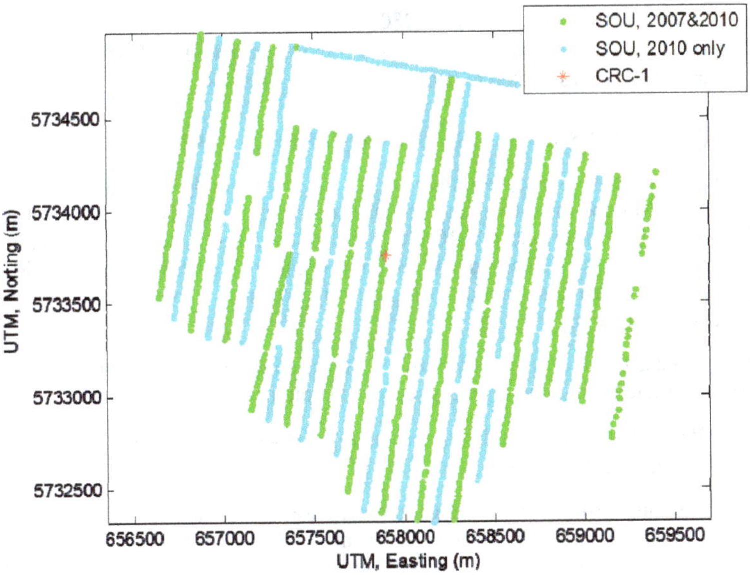

Figure 2. 4D VSP Source point locations.

During the repeat surveys (2009 and 2010), all surface seismic receiver lines were active for all shot points. The repeat 3D VSP survey acquired in 2010 was also shot using all 29 source lines.

Each of these (baseline and monitor) surveys took approximately 14 working days to acquire. 3D surface seismic surveys took an additional 4-5 days to deploy and remove the recording spread (deploy geophones and cables and retrieve them after the survey).

The seismic crew which participated full time in field operations consists of the following members:

- Crew that operated the seismic source (~2 persons in the baseline survey operating the weight drop vs. 3-4 persons operating the IVI minibuggy in the monitoring surveys)
- A person that operated the GPS to provide real-time positioning of the seismic source
- VSP crew: 1 person
- Surface seismic crew: 3-4 persons
- Support personnel: 1 person

To operate the EX-6 seismic recording system each working day, about 1 hour was required to distribute batteries before commencing the shooting and about 30 minutes to retrieve and recharge them. This means that the surface seismic acquisition effectively took 10-15% more time than the 3D VSP.

Another important difference between the 3D surface seismic and 3D VSP acquisition is the influence of various types of surface noise. The borehole seismic is not too sensitive to surface noise sources (e.g. wind, rain, trucks, farming activities), provided that the seismic source is more powerful than any individual noise source. Several delays in surface seismic operations were caused by strong wind/rain and local traffic. Even when weather conditions were less severe and allowed the survey to proceed, they still affected the data quality.

Estimates of the acquisition costs are:

- 3D VSP acquisition: Schlumberger, A$280K
- 3D Surface Seismic : Curtin University (Non-commercial rates) – A$200K
- Miscellaneous expenses related to landowner access. ~A$10K

4D SEISMIC DATA CROSS-EQUALISATION AND REPEATABILITY ANALYSIS

The surface seismic baseline survey acquired in 2007/8 was initially processed by a contractor, DECO Geophysical (Moscow, Russia). Later the datasets acquired in 2007/8, 2009 and 2010 were reprocessed in-house by Curtin and CSIRO/CO2CRC researchers. After the acquisition of each repeat survey, all the datasets available to date were reprocessed together in order to achieve a better level of repeatability. The average time of each processing round is about 5 months using 1.5 full-time equivalent staff members.

Due to changes in the acquisition geometry between the baseline and both monitor surveys, the final processing was completed using two different sets of 3D volumes:

- 2007/8+2009+2010, using acquisition geometry matched to the baseline survey
- 2009+2010, using full datasets

4D VSP data processing (using only the Z component) was performed by Schlumberger [2]. The same processing methodology as for 3D surface seismic was employed: the baseline and monitor surveys were processed together using only those source points that were occupied in both surveys and also, the monitoring surveys were processed using all source points.

The cross-equalisation of seismic and VSP volumes was performed independently between the two methods. The main reasons for that are summarised as follows:

1. Different illumination of the Waarre C horizons using down-hole and surface receivers and differences in the velocity model used for imaging resulted in noticeable differences in the topography of the reflectors.
2. Incident angles of the reflected waves that contributed to the VSP images are different from those which contributed to the surface seismic image. This means that we could expect to have differences between the two images due to dependence of the reflection coefficient on incidence angle.
3. Amplitude correction approaches utilised in VSP and surface seismic processing were also different.

Cross-equalisation and repeatability analysis of the Otway surface time-lapse seismic data were discussed in "Repeatability analysis of land time-lapse seismic data: CO2CRC Otway pilot project case study" [5]. Here we summarise cross-equalisation of the 4D VSP dataset.

Cross-equalisation Flow Overview

Migrated 3D VSP volumes created by Schlumberger are listed in Table 2. Only datasets from the first two groups share the same acquisition geometry. One of the usual approaches to control the VSP migration aperture is to define the anticipated dip of the reflectors and allowed deviation from this dip (see, for instance, Dillon [3]). We found that volumes migrated with 1 degree aperture contain significant migration artefacts, thus they were not used for analysis.

Table 2. List of data sets.

Baseline, 7° aperture, 0° central dip Baseline, 7° aperture, 5° central dip Baseline, 1° aperture, 5° central dip	Baseline survey migration from coincident sources – group 1
Repeat, 7° aperture, 0° central dip, limited bandwidth Repeat, 7° aperture, 5° central dip, limited bandwidth Repeat, 1° aperture, 5° central dip, limited bandwidth	Repeat survey migration from coincident sources, 8-90Hz – group 2
Repeat, 7° aperture, 0° central dip, full bandwidth Repeat, 7° aperture, 5° central dip, full bandwidth Repeat, 1° aperture, 5° central dip, full bandwidth	Repeat survey migration from all sources, 10-150 Hz – group 3

These considerations reduced the selection of available VSP volumes to following two valid pairs that could be cross equalised:

- Pair 1: Baseline and repeat volumes, 7° aperture, 0° central dip, limited bandwidth
- Pair 2: Baseline and repeat volumes, 7° aperture, 5° central dip, limited bandwidth

The cross-equalisation workflow consisted of the following steps:

1. **Post-stack static shifts analysis between 2007 and 2010 data**
   Due to the limited vertical extent of the image (especially above the reservoir), the entire trace length was selected to compute cross-correlations between corresponding pairs of traces. In principle, we expected to see virtually no shifts between the surveys after the model-based static correction was applied during the processing. Figure 3 shows the distribution of time shift magnitudes between baseline and repeat 3D VSP volumes (7° aperture, 0° central dip); the standard deviation is about 0.55 ms. Spatial distribution of

these shifts is presented in Figure 4. Large values are only found at the edges of the image area. The inner part of the area has shift values of less than 0.5 ms. The other pair of migrated volumes (with a central dip equal to 5°) demonstrates similar behaviour (Figure 5 and Figure 6). A slightly larger migration grid was used for these volumes, which explains the visual difference between Figure 4 and Figure 6.

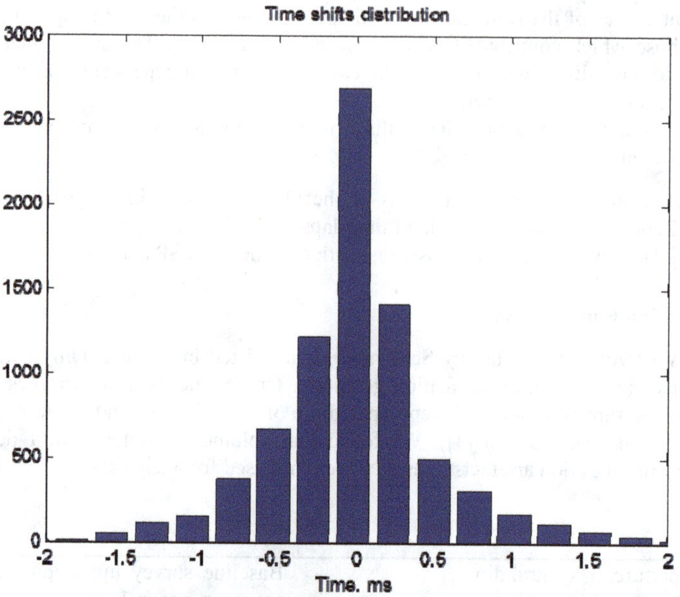

Figure 3. Histogram of static shifts between baseline and repeat 3D VSP volumes (7° aperture, 0° central dip).

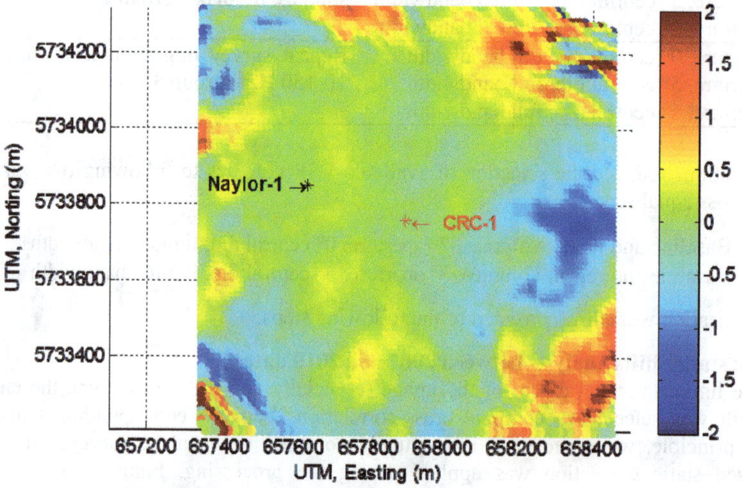

Figure 4. Static shifts between baseline and repeat 3D VSP volumes (7° aperture, 0° central dip, colour bar units are milliseconds).

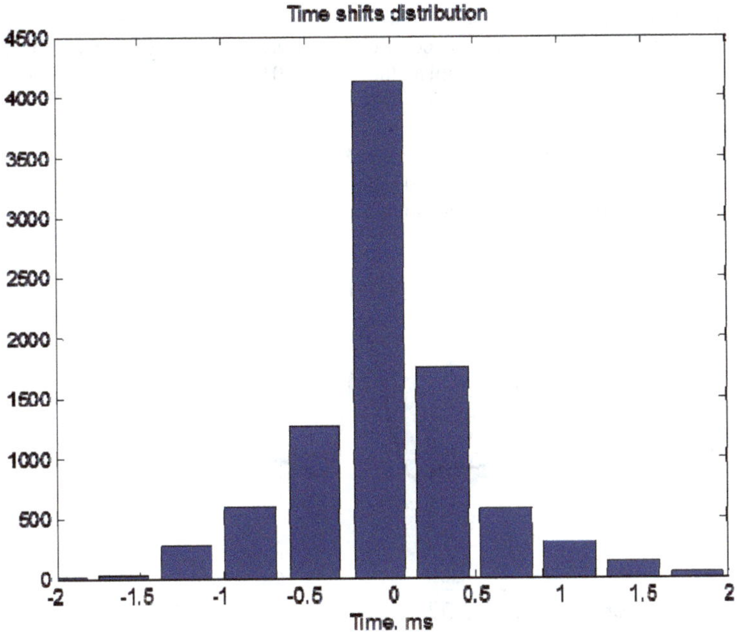

Figure 5. Histogram of static shifts between baseline and repeat 3D VSP volumes (7° aperture, 5° central dip).

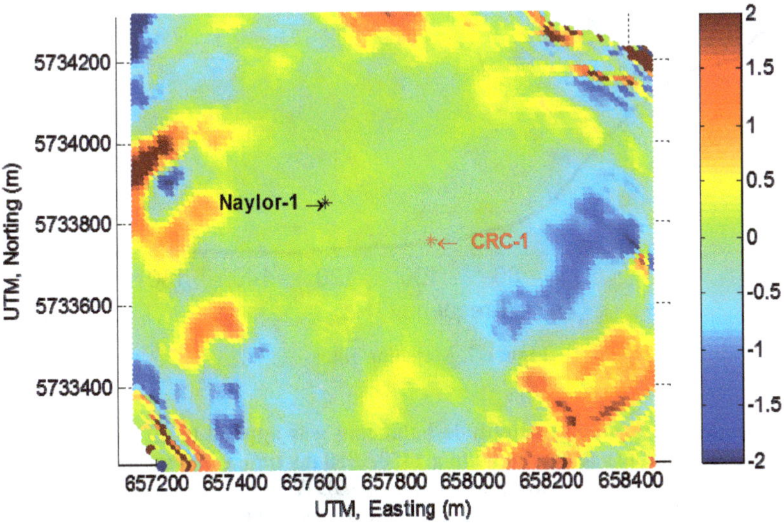

Figure 6. Static shifts between baseline and repeat 3D VSP volumes (7° aperture, 5° central dip, colour bar units are milliseconds).

2. **Single matching filters were derived and applied to match 2010 to 2007**
For all pairs of traces with a correlation coefficient of 0.8 and greater we computed a 200 ms matching Wiener filter with white noise factor of 0.01. These filters were averaged to achieve a single filter, which was applied to the 2010 datasets to match the amplitude and phase spectra of the wavelet of the 2007 data. The resulting filter is shown in Figure 7. One can see that the filter is almost perfectly zero-phase, which means that the de-convolution applied during the processing to both 2007 and 2010 data produced wavelets with a very similar phase despite the use of different source types.

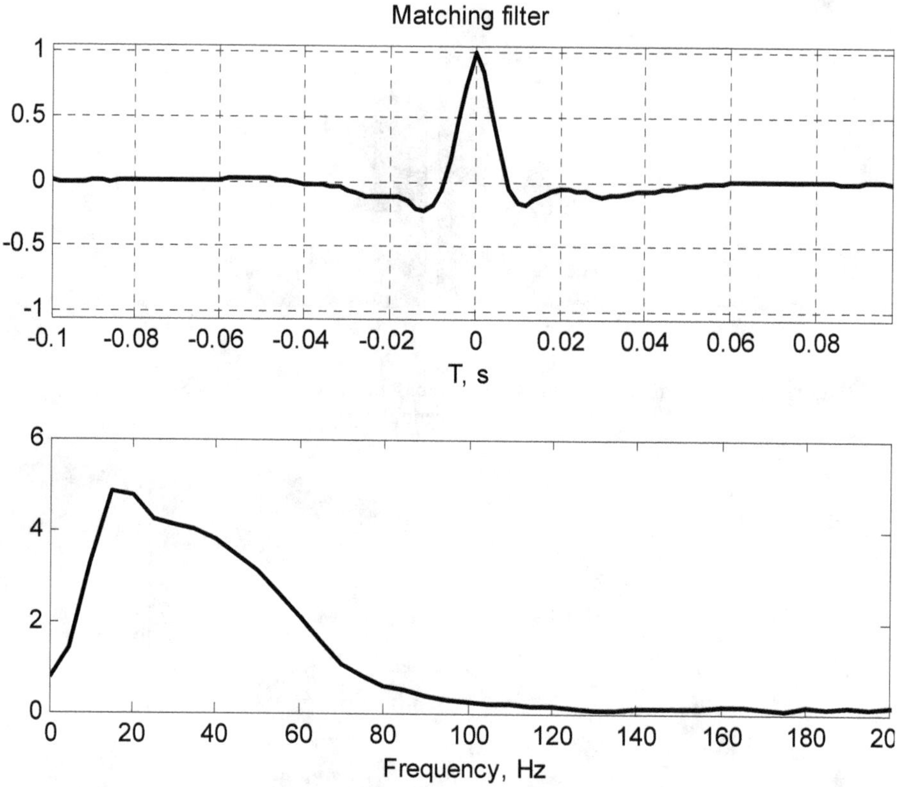

Figure 7. Matching filter.

3. **Single scalar (one per volume) amplitude balancing was applied to all volumes using the RMS absolute amplitude computed in a 1450-1850 ms time window.**
The result of the Schlumberger processed time-lapse 3D VSP data and their difference is shown in Figure 8, top (central dip 0°). The corresponding results after the cross-equalisation are presented in Figure 8, bottom. It is clear that cross-equalisation has significantly improved the repeatability of time-lapse VSP data.

4. **3D VSP data sets were interpolated to match a 3D surface seismic binning grid**
Unfortunately the grid used for the migration of 3D VSP data was chosen without considering orientation of the surface seismic binning grid or orientation of the source lines (Figure 9). In order to share the picked horizons or do trace-by-trace comparison we interpolated 3D VSP data onto a surface 3D seismic binning grid. Different velocities used for the migration of surface seismic and VSP data together with different migration aperture resulted in a difference of the observed morphology and the dip of the target horizon (Waarre C., Figure 10). This made further cross-equalisation between the surface seismic and VSP data impossible to carry out in practice. However, we can conclude that nearly all events identified in the 3D VSP data can be traced to 3D surface seismic and vice versa (Figure 11).

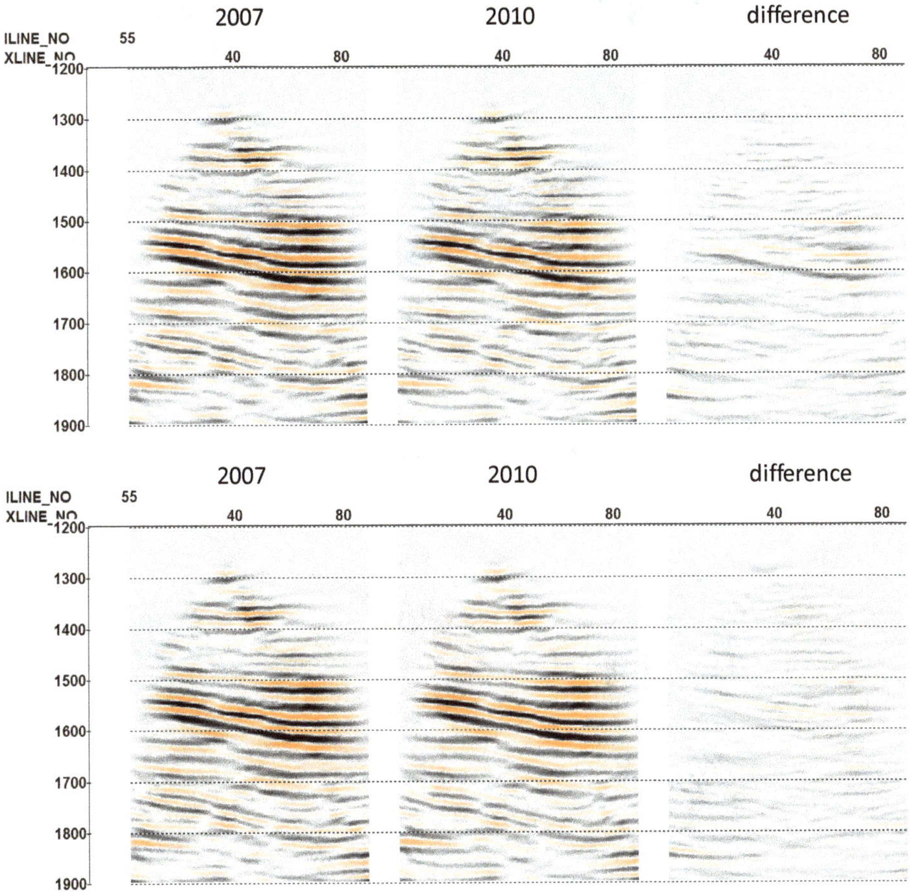

Figure 8. 4D VSP processing result before (top) and after (bottom) cross-equalisation.

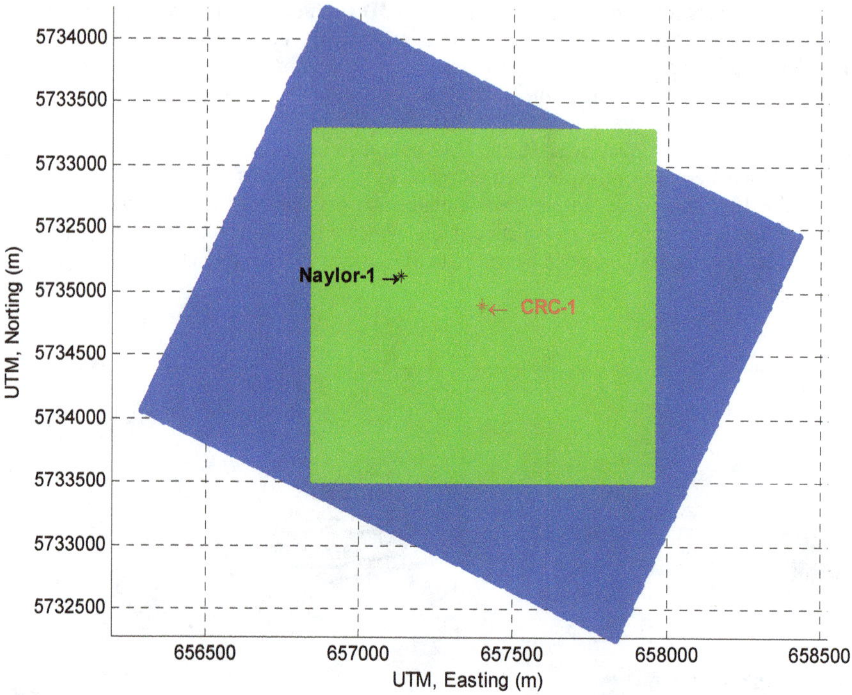

Figure 9. Comparison of the surface seismic binning grid (blue) and 3D VSP migration grid (green).

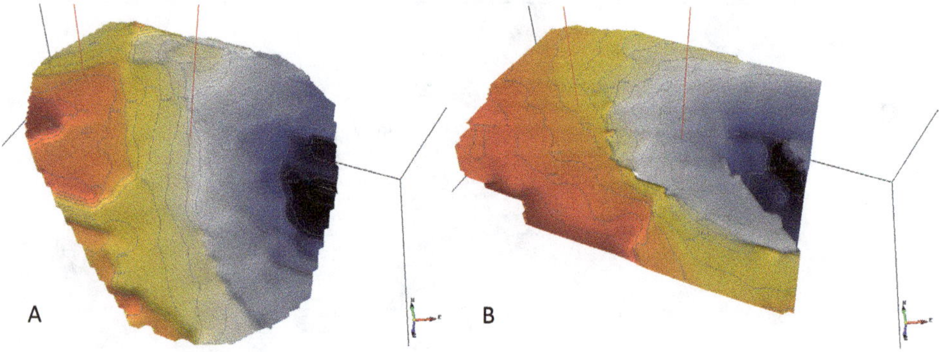

Figure 10. Waarre C horizon picked on 3D surface seismic (A) and 3D (0° central dip) data (B).

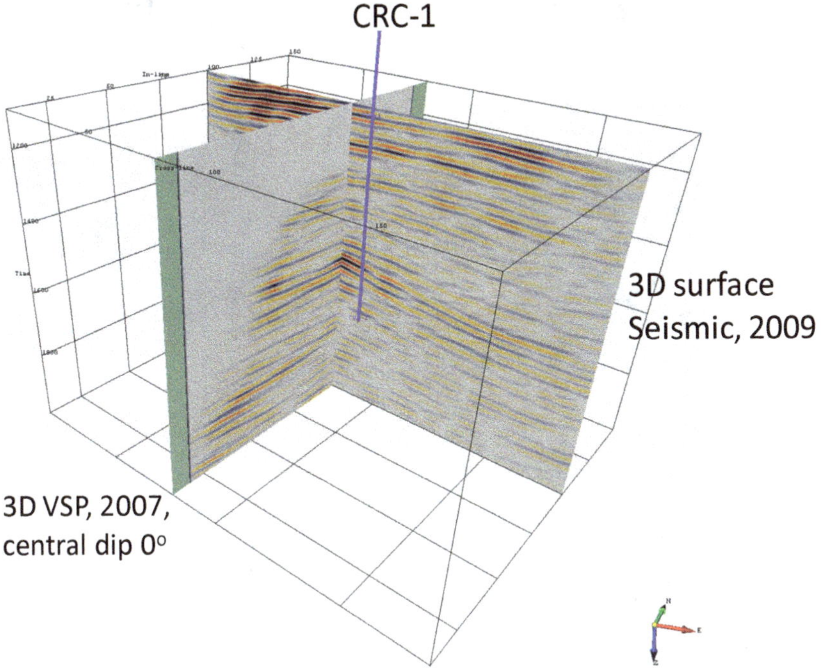

Figure 11. Comparison of 3D surface and 3D VSP data.

Repeatability Analysis

To analyse repeatability of 3D VSP data, the NRMS volume is computed using cross-equalised pairs of surveys. The same parameters of computations (60 ms sliding window) are used in Pevzner *et al.*, [6], so the results can be compared across different types of surveys.

Figure 12 compares the achieved repeatability of both 3D surface seismic and 3D VSP surveys, computed for the correspondent pairs of surveys. Time lapse VSP surveys demonstrate the outstanding level of repeatability with a median NRMS value equal to 16 percent.

The seismic amplitudes at the Waarre C horizon measured in 2007 and 2010 and the different 3D VSP volumes are shown in Figure 13. One can see that the raw amplitude has an obvious dependence on distance from the CRC-1 borehole and is not sensitive to other factors. This can be explained by a range of factors including the influence of the migration aperture (varying with source offset) and features of a particular VSP migration software, such as amplitude and phase corrections applied to the signal during summation, directivity of the receiver (these volumes were produced from the Z-component only), improper amplitude decay compensation and probably AVO effects (image points located further away from the borehole will have greater influence from bigger incidence angles). None of these factors were compensated for during the processing. However, the amplitude of the difference volume does not have such a radial pattern and all amplitude variations follow geological features of the reservoir (with absolute changes of no more than 10%). It is important to note that these amplitude variations do not exceed the background repeatability level (i.e. the amplitude differences observed elsewhere in the difference volume). To determine the ability to interpret these amplitude variations as a time-lapse signal related to CO_2 injection, a synthetic modelling study is performed in the next section.

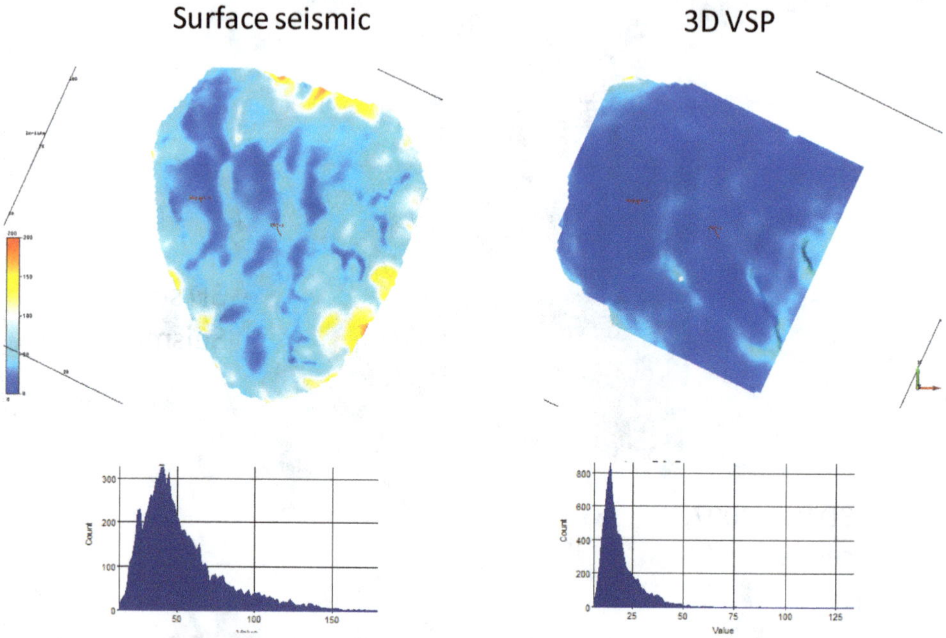

Figure 12. 2007 and 2010, NRMS value computed over the Waarre C horizon in 60ms window for surface seismic (left) and 3D VSP (right) surveys. Histograms of NRMS values are shown below the maps.

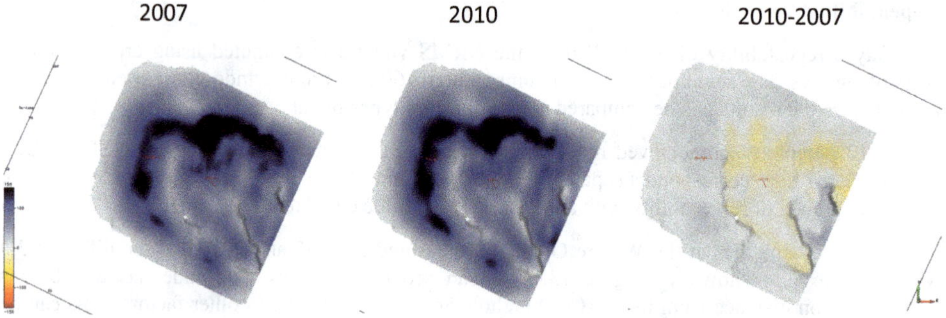

Figure 13. 2007 and 2010 seismic amplitude on Waarre C horizon, central dip 0°.

TIME-LAPSE SIGNAL FROM CO_2 INJECTION TO WAARRE C RESERVOIR

The aim of this modelling is to predict the time lapse seismic response in 2009 and 2010 from flow simulation [4] and acoustic impedance (AI) inversion results of the seismic baseline data in 2008. Therefore, a small reservoir model (Figure 14) is built based on the seismic time horizons, log data and the acoustic impedance volume. This model was then altered using the flow simulation results.

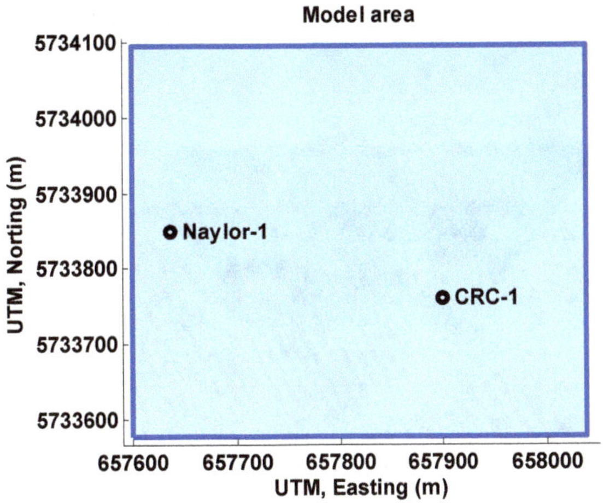

Figure 14. Model area.

Reservoir Model in ChronoSeis

RokDoc ChronoSeis (IconScience) built models from seismic time horizons (Top Clifton, Top Dillwyn, Top Sherbrook, Top Paaratte, Top Skull Creek, Top Belfast Formation, Top Waarre C, Top Waarre B, Top Waarre A and Top Eumeralla) and depth of the horizons. The horizon depths are calculated by a layer cake method and fixed to the well tops in CRC-1 and Naylor-1. To match the flow simulation grid with the depth model, the depth horizons of the top Waarre C and top Waarre B are replaced by depth surfaces (Top and Bottom Waarre C) of the flow simulation grid. These depth surfaces are limited to the area in which the AI inversion and the flow simulation results are available (Figure 15, Figure 16).

Two models with different reservoir properties were developed. For the first model (Model-1) the porosity model is taken from the flow simulation results. The acoustic impedance in the reservoir is created by interpolating AI logs (for 20% gas saturation) between CRC-1 and Naylor-1 using the Collocated Kriging method with the porosity model as a soft property. The same is done for the grain P-wave modulus, which is calculated in Naylor-1 and CRC-1 from the clay and quartz volumes using the average of the lower and upper Hashin-Shtrikman bounds. Afterwards, fluid substitution is performed to resemble the gas saturation in 2008 using the flow simulation results. Outside the reservoir, the AI model is built from the AI inversion volume of the seismic baseline data.

For the second model (Model-2), the AI is taken from the AI inversion results in 2008. Since the porosity model (flow simulation) is much more detailed (heterogeneous) than the AI inversion result, the porosity model is smoothed using a mean filter (Figure 17). For the grain properties, the properties of quartz are used.

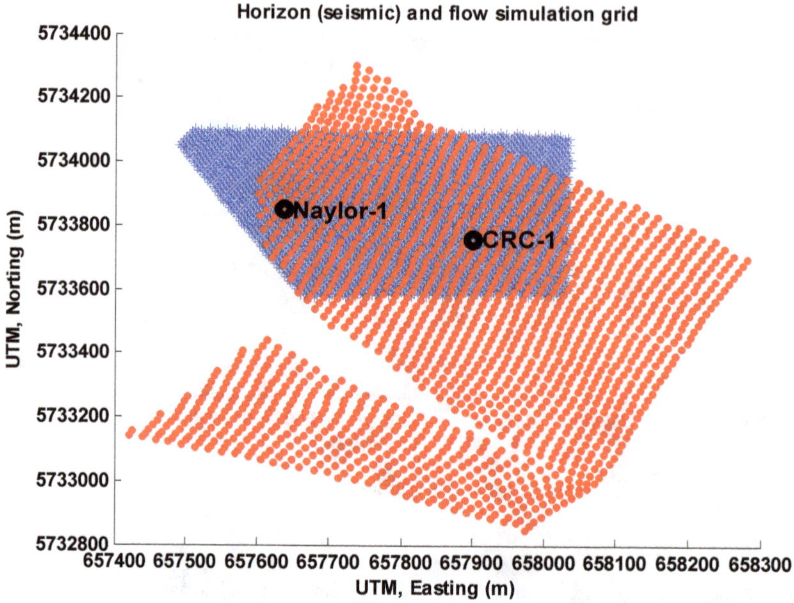

Figure 15. Seismic time horizon (top Waarre C).

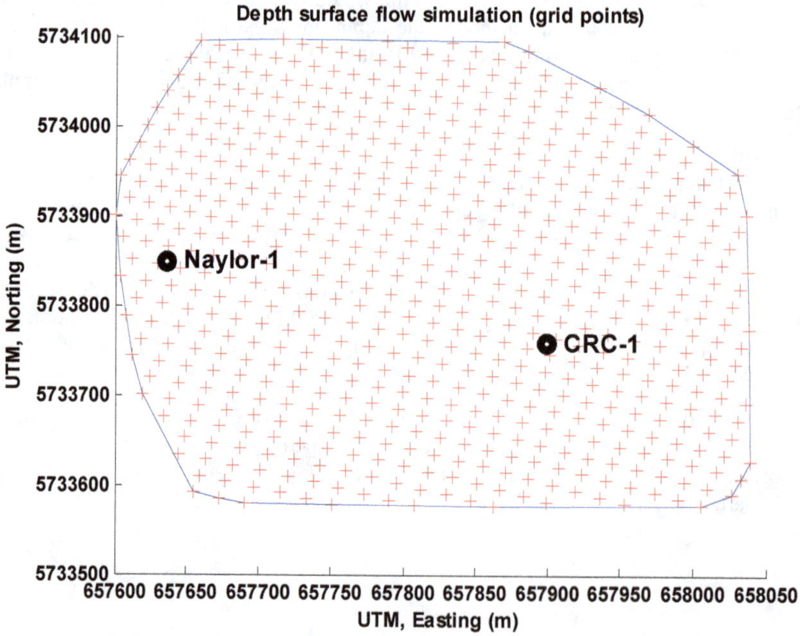

Figure 16. Area between time horizon and flow simulation grid.

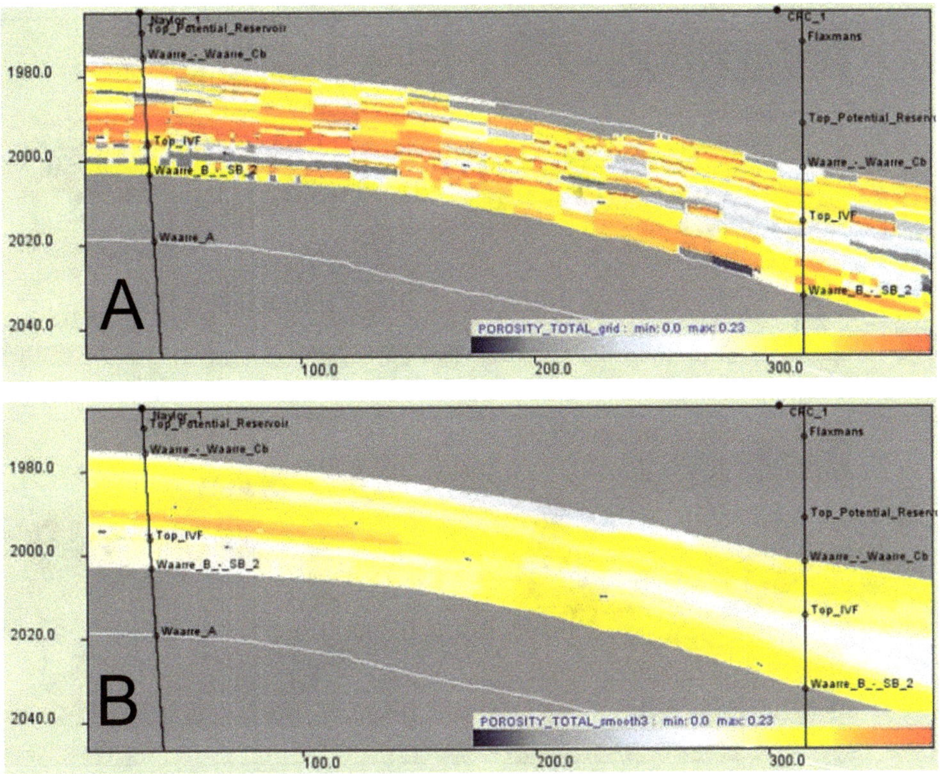

Figure 17. Slice through the unsmoothed (A) and smoothed (B) porosity model on a line between Naylor-1 and CRC-1.

Figure 18 shows a comparison between the AI of Model-1 and Model-2. We see that Model 1 is much more detailed than Model 2, but average AI values of Model-2 and the AI values of Model-1 have the same order of magnitude. In summary, Model-1 honours the static model of the flow simulation and Model-2 the inversion result of the seismic data.

Fluid Substitution Modelling

Fluid substitution modelling was performed for the conditions of pre-injection (in 2008) and an injection of a CO_2/CH_4 mixture of 35 kt (in 2009) and 65 kt (in 2010). The gas saturations and gas properties are taken from the flow simulation results (Figure 19). The in-situ brine properties are computed by the empirical formula of Batzle and Wang,[7]. Since the S-wave velocity is unknown, an approximate method for solving Gassmann's equation with Wood's formula based on the P-wave modulus is applied [8]. For Model 1 no fluid substitution is performed in areas where the porosity is below six percent. For Model 2 (smoothed porosity), the porosity is above six percent everywhere.

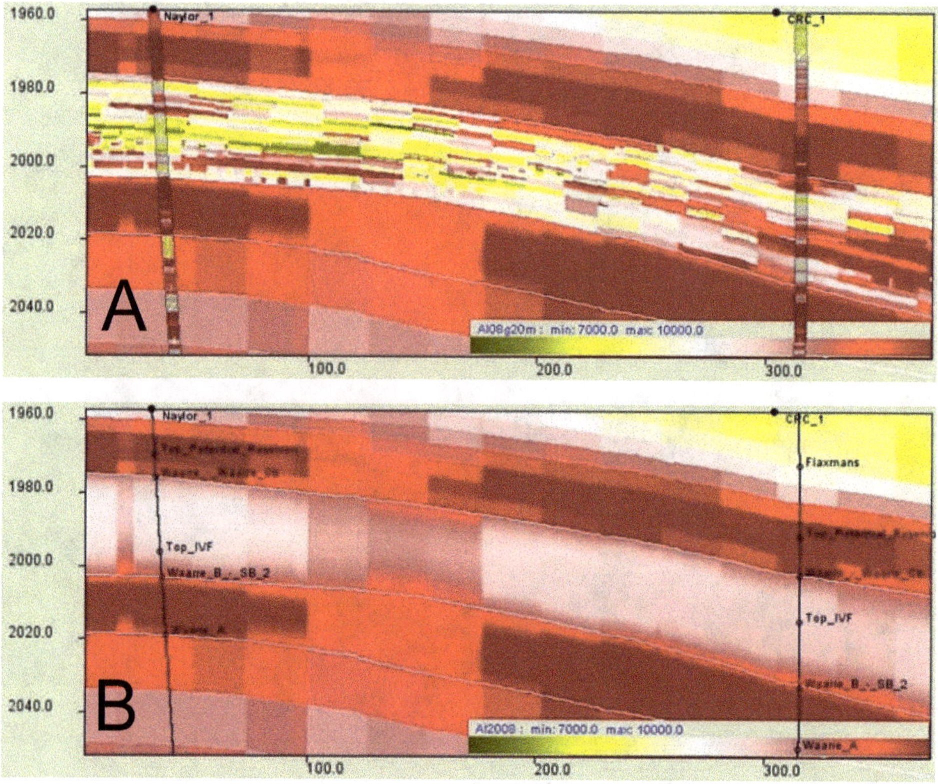

Figure 18. Slice through AI for Model-1 (A) and Model-2 (B) on a line between Naylor-1 and CRC-1.

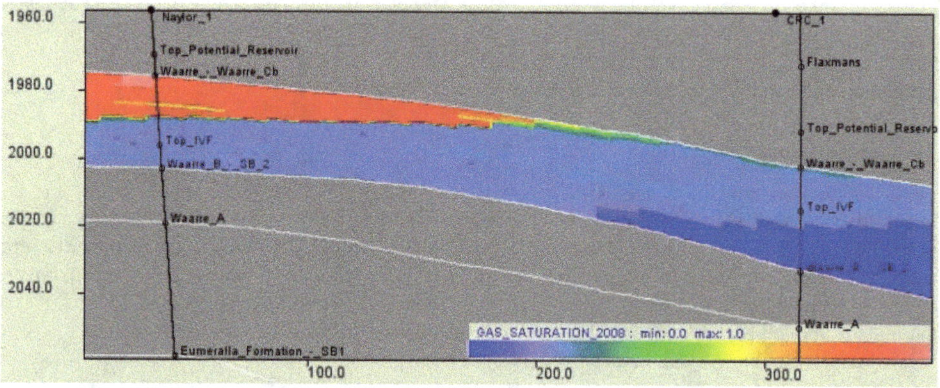

Figure 19. Gas saturation in 2008.

The results show that the acoustic impedance in the reservoir decreases when brine is replaced by the CO_2/CH_4 mixture. The absolute differences in AI between 2008 and 2010 (including 2009), are shown in Figure 20 for 2009-2008 and in Figure 21 for 2010-2008. These differences are of the same order of magnitude for both models. As expected, the differences reflect the CO_2 mass fractions in 2009 and 2010, respectively. Furthermore, it shows that the largest change in AI occurs halfway between Naylor-1 and CRC-1.

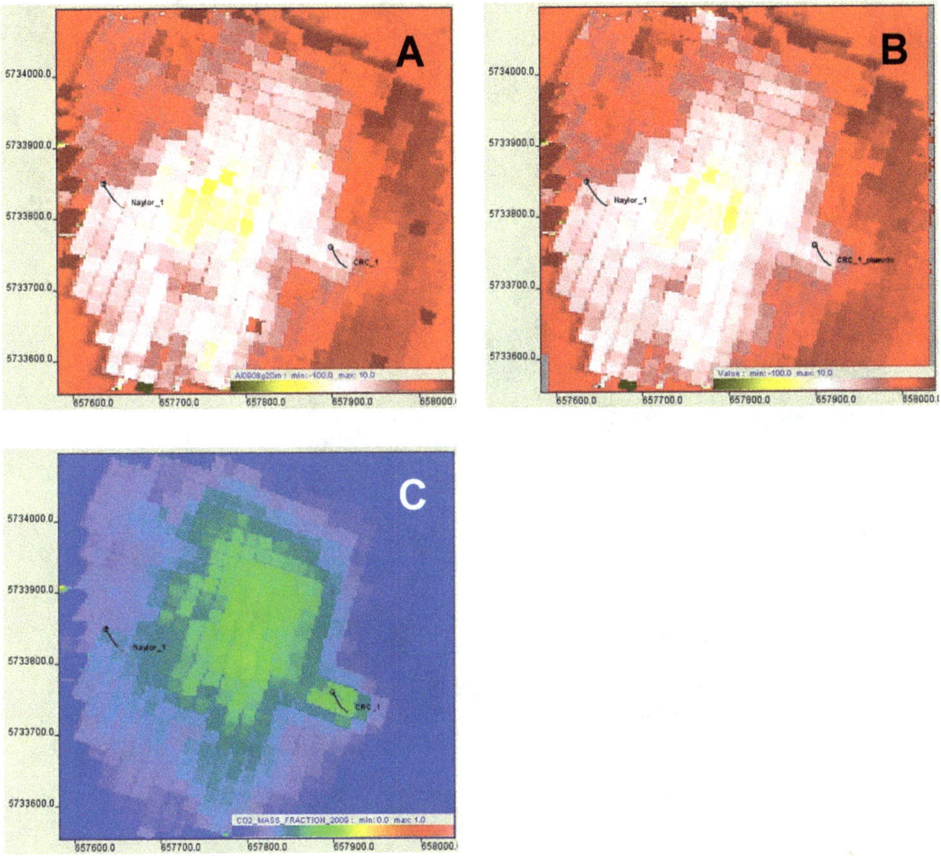

Figure 20. Absolute difference of AI (m/s*g/cm^3) between 2009 and 2008 for Model-1 (A) and Model-2 (B) - averaged over the depth interval of the reservoir zone; C:CO_2 mass fractions for 2009 - averaged over the depth interval of the reservoir zone.

Synthetic Seismograms

Zero-offset synthetics are computed from the fluid substitution results using a statistical wavelet from the 2008 seismic data, as shown in Figure 22.

One of the inlines, extracted from the baseline seismic volume is presented in Figure 23.

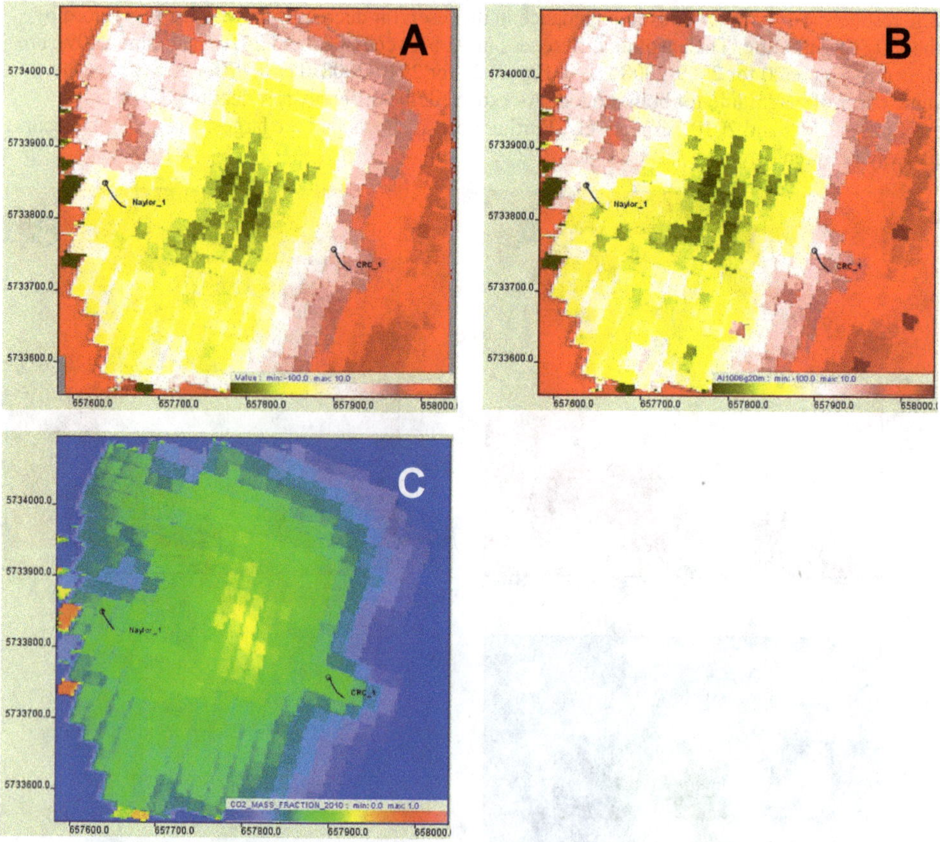

Figure 21. Absolute difference of AI (m/s g/cm3) between 2010 and 2008 for Model-1 (A) and Model-2 (B) - averaged over the depth interval of the reservoir zone; C:CO_2 mass fractions for 2010 - averaged over the depth interval of the reservoir zone.

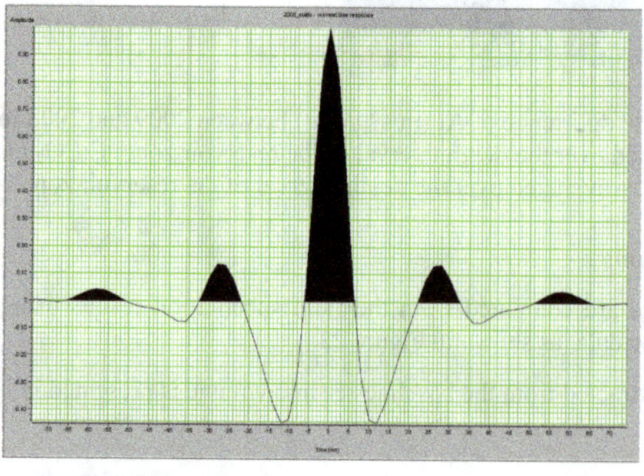

Figure 22. Statistical wave from the seismic data, 2008.

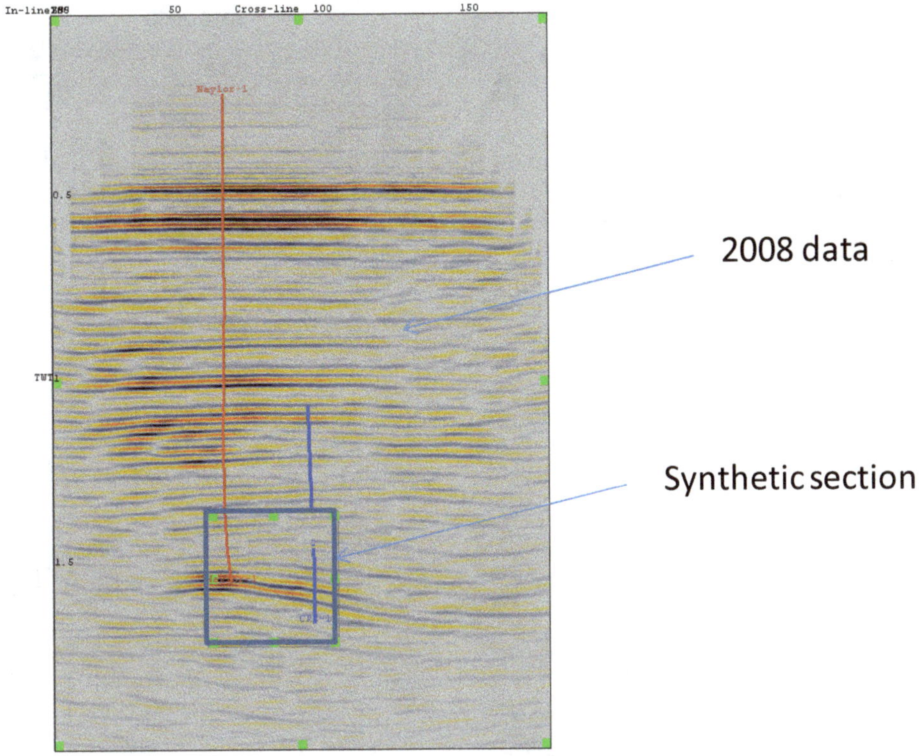

Figure 23. Inline 87, baseline volume, portion of reconstructed synthetic data inserted around the reservoir.

DISCUSSION

The comparison of the predicted difference between the baseline and monitor surveys is presented in Figure 24. The left column contains synthetic section (A) compared to the real data (C) (both of them represent baseline survey, 2008). Minor differences between amplitude distributions along the Waarre C horizon are related to the averaging of the properties required for fluid substitution (in the current workflow). The right column shows predicted time-lapse response (B) and the real difference obtained from the data (D). Unfortunately, the signal level is significantly smaller compared to time-lapse noise.

The map of predicted the NRMS difference for 2010-2008 (i.e. computed from the synthetic volumes with all of the differences related to changes in the reservoir level) is shown in Figure 25. When compared to Figure 13, this shows that the repeatability of the surface seismic data (even if it should be considered as relatively high, especially for the land time-lapse data) is too low to detect the signal. For the time-lapse 3D VSP data, the achieved NRMS difference is also below the level required to detect the signal robustly. However, in this case the predicted signal is at least of similar order to the magnitude as time-lapse noise.

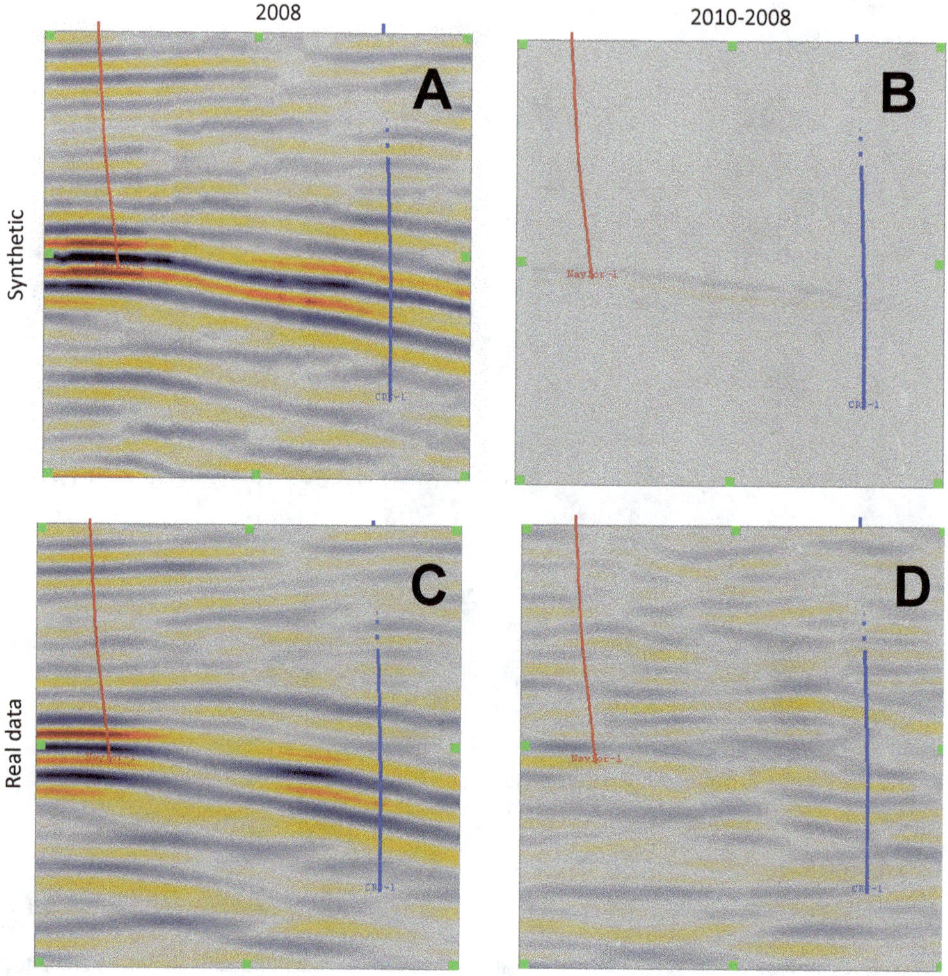

Figure 24. Inline 87, field vs synthetic data.

Our previous repeatability studies performed using surface seismic data show significant increase in repeatability between the first (2008-2009) and the last (2009-2010) pair of surveys due to the increase in the CMP fold and power of the seismic source (Figure 26). We could expect to have similar improvements in repeatability of the time-lapse 3D VSP data had both monitor and baseline surveys been acquired with the acquisition parameters of the monitoring (2010) surveys. This allows us to speculate that in such a case 4D VSP could probably be applicable to monitoring of the injection of CO_2 sequestration, even in depleted gas reservoirs.

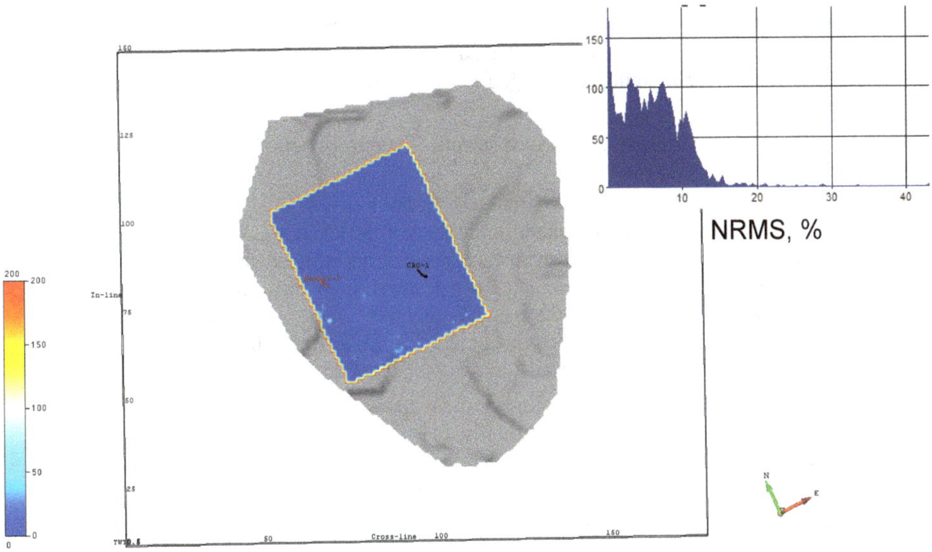

Figure 25. Predicted NRMS difference for 2010-2008 (computed from synthetic data).

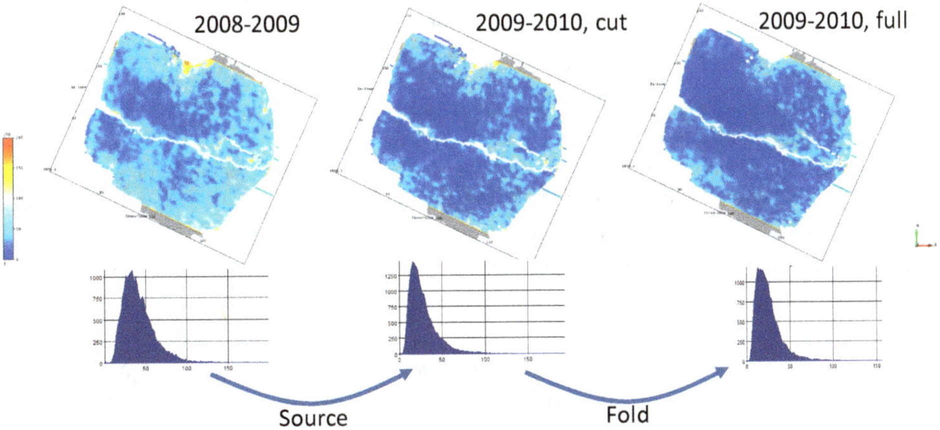

Figure 26. NRMS maps and histograms computed for one of the horizons in the Paaratte formation (~1 s TWTT) showing improvement in repeatability due to use of more powerful source and increasing the fold. From left to right: 2008-2009, reduced acquisition geometry (440 active channels, different source types); 2009-2010, reduced acquisition geometry (440 active channels, same source); 2009-2010, full acquisition geometry (873 active channels, same source).

To compare borehole (4D VSP) and surface 4D seismic methods we use several criteria.

1. **Coverage.** Processed surface seismic data provides an image from the top part of the section to the target horizon, while the 3D VSP can be used to generate an image only below the top receiver in the downhole array. In principle, it is possible to use down-going PS waves [9], but these techniques are not well developed.

2. **Repeatability.** Borehole seismic methods demonstrate superior repeatability in comparison to the surface seismic data. In fact, all time-lapse seismic methodologies used in Otway basin could be arranged in the following order according to the level of repeatability [2, 10]:

   2D surface seismic << 3D surface seismic ~= zero-offset/offset VSP < 3D VSP

3. **Field operations.**

   a. **Acquisition time/performance.** In the case of the CO2CRC Otway Project, 3D VSP has been more rapid than surface seismic, has a much shorter mobilisation time and better performance due to a less required daily maintenance (e.g., distribution of charged batteries to receiver lines, replacement of dead geophones). The borehole seismic is much less sensitive to the surface noise and weather conditions, which results in a higher acquisition speed.

   b. **Cost.** In our case, the acquisition cost for the 4D seismic data is less than the cost of the 4D VSP. However, this is due to the fact that surface seismic was acquired by the Curtin University crew and not a contractor. Most likely 4D VSP surveys will be the cheaper option if commercial rates are used.

   c. **Disruptions to farming activity.** Borehole seismic surveys do not require deploying receiver cables and geophones. This means that interaction between the seismic crew and land users in this case will be minimal.

4. **Processing.** The processing of 3D VSP data is significantly less time consuming due to the smaller amount of data and generally higher signal/noise ratio.

Other benefits of 3D/3C borehole seismic methods for monitoring purposes are:

- Higher temporal and (most likely) spatial resolution
- Possibility to use shear (mode-converted) waves for imaging
- Principal possibility to evaluate changes in local anisotropy parameters and their changes due to potential changes in stress fields [2, 11, 12].

The principal limitations of the 4D VSP are limited coverage (mentioned above) and strong influence of the AVO effect on the obtained image.

CONCLUSIONS

4D VSP data acquired within the Otway Basin Pilot Project in 2007 (baseline) and 2010 (monitor) were processed to migrated volumes and subsequently cross-equalised using exactly the same technique as developed for surface Otway 3D time-lapse seismic data. The main conclusions can be summarised as follows:

- The 4D VSP data demonstrates a superb level of repeatability in comparison to all other time-lapse datasets acquired in the survey area (with median NRMS value on the target horizon of ~16%).
- The amplitudes of the seismic events in the migrated volume show a strong variation with distance from the borehole, which can be explained by the influence of such factors as directivity of receivers and/or migration aperture. During quantitative interpretation of the 4D VSP survey these factors must be taken into account and/or the processing flow should be improved.
- In order to compare the migrated volumes obtained from 3D surface seismic and 3D VSP surveys, the velocity models used for migration should be matched.

- 4D VSP is probably a better tool for imaging reservoir changes while 4D surface seismic is essential for monitoring changes in the overburden (e.g., possible leakages).
- Permanent near-surface installations should be considered as a part of the Otway CO_2 sequestration monitoring program because they could combine many advantages of the borehole seismic with those of surface seismic.
- The cost of acquisition of 4D VSP data can be reduced if cheaper hardware, (e.g., hydrophone string) is used. Additional field tests are required.
- More effort should be spent on Otway 4D/3C VSP data analysis as not all potentially useful information has been extracted yet.

ACKNOWLEDGEMENTS

The Australian Commonwealth Government through the CO2CRC and the CO_2 Capture Project (CCP) sponsored this body of work. The authors thank CO2CRC colleagues, and in particular, Tess Dance, Jonathan Ennis-King, Martin Leahy, Putri Wisman, Anton Kepic, Christian Dupuis and Andrej Bona for their contribution to this study. The support of Peter Cook, Sandeep Sharma, Kevin Dodds (of the CO_2 Capture Project) and Charles Jenkins is gratefully acknowledged. Thanks also to Schlumberger for their extensive cooperation in the acquisition and processing of the borehole seismic data.

REFERENCES

1. Campbell, A. and Shujaat, A., 2010, Borehole Seismic Processing Report, 3D VSP Survey, CO2CRC, CRC-1 3D VSP, Baseline and First Repeat Survey, Otway Basin, Australia. pp. 69. CO2CRC, Schlumberger.
2. Campbell, A., Nutt, L., Ali, S., Dodds, K., Urosevic, M., Pevzner, R. & Sharma, S. 2011. An Early Look at a Time-Lapse 3D VSP. In: EAGE Borehole Geophysics Workshop - Emphasis on 3D VSP, pp. BG16. EAGE.
3. Dillon, P.B. 1990. A comparison bewteen Kirchhoff and GRT migration on VSP data. *Geophysical Prospecting* **38**, 757-777.
4. Ennis-King, J., Dance, T., Xu, J., Boreham, C., Freifeld, B., Jenkins, C., Paterson, L., Sharma, S., Stalker, L., and Underschultz, J., 2011, The role of heterogeneity in CO_2 storage in a depleted gas field: History matching of simulation models to field data for the CO2CRC Otway Project, Australia: Energy Procedia, v. 4, p. 3494-3501.
5. Pevzner, R., Shulakova, V., Kepic, A. and Urosevic, M., 2011, Repeatability analysis of land time-lapse seismic data: CO2CRC Otway pilot project case study. Geophysical Prospecting, doi: 10.1111/j.1365-2478.2010.00907.x.
6. Pevzner, R., Bona, A., Gurevich, B., Yavuz, I., Shaiban, A. and Urosevic, M., 2010a, Seismic anisotropy estimation from VSP data: CO2CRC Otway project case study. SEG Technical Program Expanded Abstracts 29, 353-357.
7. Batzle, M. and Wang, Z., 1992, Seismic properties of pore fluids, Geophysics, 57, 1396-1408.
8. Mavko, G., Chan, C., and Mukerji, T., 1995, Fluid substitution: estimating changes in Vp without knowing Vs. Geophysics, 60, 1750-1755.
9. Luo, Y., Liu, Q., Wang, Y.E. and M.N. AlFaraj, M.N., 2006, Imaging reflection-blind areas using transmitted PS-waves. Geophysics 71, S241-S250.
10. Pevzner, R., Urosevic, M. and Nakanishi, S., 2010b, Applicability of Zero-offset and Offset VSP for Time-lapse monitoring – CO2CRC Otway Project Case Study, 72nd EAGE Conference & Exhibition incorporating SPE EUROPEC 2010 Barcelona, Spain, P284.
11. Urosevic, M., Pevzner, R. and Vidal-Gilbert, S. 2009, CO_2 monitoring with time-lapse seismic anisotropy, SEG 2009 Summer Research Workshop CO_2 Sequestration Geophysics, pp. Abstracts.

12. Vidal, S., Longuemare, P., Huguet, F. and Mechler, P., 2002, Reservoir parameters quantification from seismic monitoring integrating geomechanics. Oil & Gas Science and Technology-Revue De L Institut Francais Du Petrole 57, 555-568.

Carbon Dioxide Capture for Storage in Deep Geological Formations, Volume 4
Karl F. Gerdes (Editor)

Chapter 36

EVALUATING TIME-LAPSE BOREHOLE GRAVITY FOR CO_2 PLUME DETECTION AT SECARB CRANFIELD

Kevin Dodds[1], Richard Krahenbuhl[2], Anya Reitz[2], Yaoguo Li[2] and Susan Hovorka[3]
[1]Formerly BP Corp North America Inc (Australian National Low Emissions Coal Research & Development, NFF House, 14-16 Brisbane Avenue, Barton 2600, Australia)
[2]Center for Gravity, Electrical & Magnetic Studies, Colorado School of Mines, 1500 Illinois St., Golden, CO 80401
[3]Bureau of Economic Geology, Jackson School of Geosciences, University of Texas, University Station, Austin, Texas 78713-8924

ABSTRACT: Monitoring of CO_2 storage processes may be enabled by adapting Oil and Gas monitoring technology to provide measurements required of operators by regulatory authorities and the public. These monitoring requirements include providing cost-effective measurements that will support an understanding of safe containment of the CO_2 plume, and provide confirmation that injected CO_2 remains in the storage complex for the purposes of greenhouse gas accounting. Such monitoring demands prompt an assessment of deep reading measurements besides seismic, such as gravitational and electromagnetic technologies, which are sensitive to density and electrical properties of the fluids, and are less costly than seismic. In particular this paper explores the operational and interpretational challenges of borehole microgravity measurements in a CO_2 time lapse survey in two wells at the SECARB Cranfield injection site.

The Southeast Partnership (SECARB) test at Cranfield, Mississippi, was the first of the commercial scale projects comprising a staged array of field deployments testing key issues of capacity and best methods for assuring storage permanence. More than 3 million metric tons of injected CO_2 will have been injected since the start of injection in July of 2008. The site is a historic oilfield at a depth of 2500 m in the Cretaceous fluvial Tuscaloosa Formation, developed by Denbury Onshore LLC. The project was unique in deploying a range of geophysical measurements, including cross-well seismic and electrical resistivity tomography, from two monitoring wells positioned in line from the injector and several hundred feet down-dip in the Tuscaloosa. The CO_2 Capture Project (CCP) took advantage of the multiple field acquisitions of borehole data during the injection phase of this project to acquire time-lapse borehole gravity in the two monitoring wells. The objective was twofold; the first to understand the operational aspects and design of the acquisition, while the second was to assess the ability of the surveys to detect the injected CO_2. The baseline acquisition occurred in October 2009 and the repeat survey in September 2010.

KEYWORDS: CO_2 storage; borehole gravity; time-lapse; operations; SECARB; Cranfield

INTRODUCTION

Gravity measurements provide one of three fundamental deep reading geophysical techniques: seismic, electromagnetic sensing and gravity sensing. Precise measurements of the acceleration due to gravity have formed one of the earliest exploration tools available to geophysicists [1]. The variations in subsurface density can be measured on the surface using highly sensitive gravimeters. These instruments are sensitive to variations in acceleration of the order of 10s of microgals to milligals (1 mgal = 10^{-3} cm/sec^2). The measurements are subject to environmental noise, often exceeding the desired signal in size, but which can be removed by reference to their properties, such as cyclical earth-tides, influence of nearby masses, precision location measurements, air pressure effects, vibration and electronic noise. Some instruments are based on acceleration of a weight on a spring, and produce relative measurements, as the mechanical spring constants can drift, requiring re-calibration. Other instruments are based on the acceleration of a falling mass, and produce absolute measurements since they are independent of mechanical properties and don't drift with time [1]. The surface measurements have been able to provide information on the distribution of density associated with ore bodies, depth of basins, subsurface salt structures [2]. Surface gravimeters have also been used to measure the more subtle response of water floods in a gas reservoir [3], where they monitor the density difference between the gas replaced by water. Also the time-lapse effects of CO_2 and water have been measured using sub-marine gravity [4].

The potential to measure density changes associated with an injected CO_2 migration has been recognised as a means to evaluate the total amount of injected CO_2. This would supplement the areal definition provided by seismic, where the seismic is unable to determine the volume or saturation of the plume [5]. The demands of sensing the density differences between oil and water and CO_2 and water, which are of the order of 0.02 - 0.03 gm/cc, place demands on the sensitivity of the instruments, and the depth to which the difference can be observed. In most instances, the variations in density due to water floods or CO_2 plumes are not detectable from the surface. Consequently borehole deployed measurements have been proposed and modelled to assess their viability [6,7]. There are relatively few borehole deployable gravimeters in the world and hence, operating experience is limited, although the capability has existed since the late 1950s. In the late 60s Lacoste and Romberg adapted their surface gravimeter to the borehole. This tool was able to read from 5-10 μgals. This tool is limited to casings greater than 5 inches, vertical deviations less than 14 degrees, and since it is thermally sensitive, it requires a Dewer flask to make measurements above 125 °C. This is due to the complexity of the mechanical systems that have been reduced in size to borehole dimensions and to their relative fragility.

We used one of these borehole tools for our acquisition of borehole gravity data. More recently there has been interest in building borehole gravimeters with wider operating parameters, such as a slim-hole tool by Scintrex [8] which allows higher vertical deviations but has limited temperature range. In light of potential breakthroughs in borehole technology [9], we took the opportunity to test borehole gravity performance within the Cranfield project and compare its response to the range of other measurements that were acquired. This was intended to evaluate their operating requirements, to test their sensitivity to the injected plume and to develop operational procedures which would be useful for future surveys. We would refer the reader to Alixant [10] for an excellent discussion on survey design and operational constraints for mapping residual oil saturation. Due to time and budget constraints we were limited in scope for this survey. On the other hand, our survey represented an example of what could be done with reasonable care and time.

Operational Planning

It is important in all geophysical surveys and especially for borehole gravity, that the expected response is modelled. The simulated amplitude would provide confidence of detectability while the time-lapse difference response with depth would help in selection of the depth and spacing of the stationary measurement levels. Ultimately modelling provides not only improved survey design, but a means to manage the length of the operation and hence, its costs. Similar response modelling had been previously undertaken [6,7], although that exercise had no corresponding field data with which to compare.

Since we simulated the effect of the CO_2 plume, we only needed to model the effect of the difference in fluid density post-injection in the reservoir interval. All the other lithological densities are unchanged and hence, would be subtracted out. The model consisted of two slabs, one 15 feet thick and a second 7 feet thick separated by 23 feet, corresponding to two simplified plumes in an upper and lower zone in the Tuscaloosa formation. Each slab was assigned a post-injection density difference of -0.027 g/cc, equivalent to 30% saturated CO_2 in a formation of 0.2 porosity units (p.u.). The slabs were 400 feet on a side. A modified Talwani 2.5D gravity modelling program was used to simulate three scenarios; one where the slab was equi-centred about the well, one where the slab's edge was touching the well and one where the slab was offset from the well at 200 feet (Figure 1).

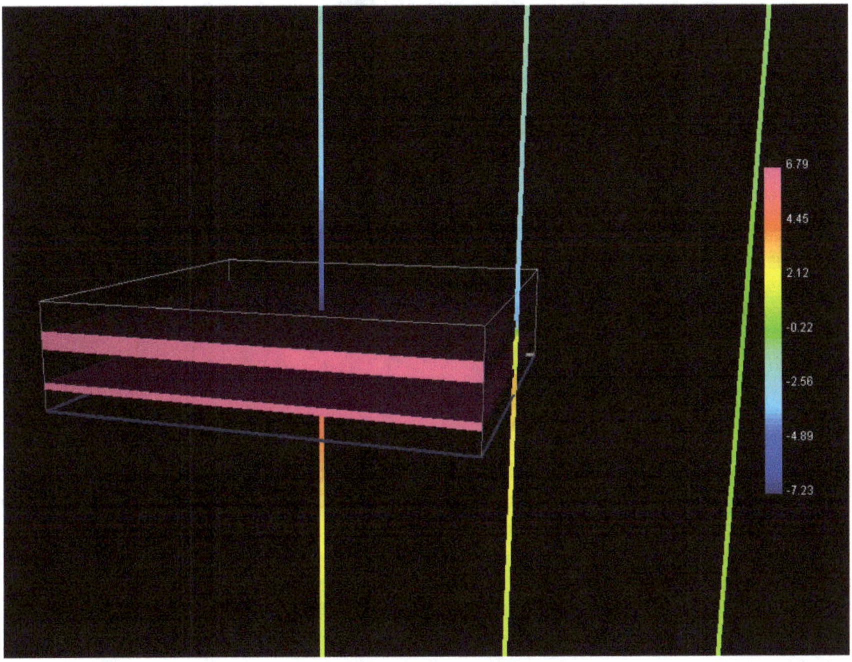

Figure 1. 3D slab model built by CSM GMRC showing the monitoring well symmetrically located in the middle, asymmetrically on it edge and offset from the slab by half its length.

The modelled time-lapse response exhibited a characteristic increasing negative response as the measurement point approached the slab from above (Figure 2). This is because the density difference between CO_2 and water is negative. The response shows an inflection point at the top of the slab and begins moving towards positive due to the influence of CO_2 above and below the measurement point as it moves down the interval. A further extreme positive inflection point occurs at the base of the lower slab, with this response decreasing asymptotically to zero as the measurement point moved away from the slab. The gap between the slabs gave rise to two smaller inflection points in the switch over region, which may or may not be observable in the data. It is important to note that, the modelled gravity response is a continuous curve. However since the measurement is made at discrete depths, sample points need to be chosen to sufficiently identify these changes, especially over the inflection points. This means that the spacing would be closer over the interval of interest and wider above and below to define the trend. There is an increasing error in determining the apparent density with decreased distance between the two stations due to the error in determining the depth of each station. With standard care of depth measurement and correlation one can expect more than a 0.02 g/cm^3 density error and increasing for station spacing less than 10 feet for a 3 μgal sensitivity gravimeter. Improved instrument sensitivity (1 μgal) and precise depth positioning can improve that by about 0.01 gm/cc [10]. A method for calculating densities using an inversion technique has been described [11]. Using this method, useful densities can be recovered at station spacings as small as three feet. The inversion results have excellent correlation with gamma-gamma density logs, and an uncertainty of ± 0.02 g/cm^3 or better.

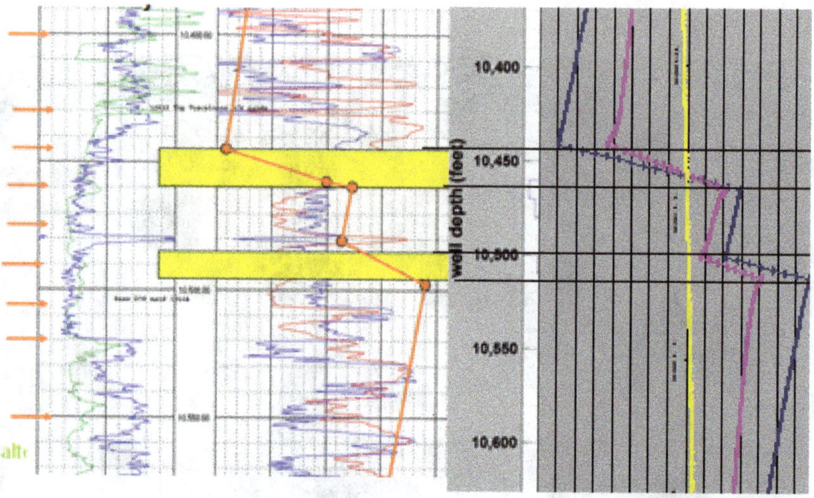

Figure 2. Simulated response on the right track where dark blue is the symmetrical case, light blue is the non-symmetrical response, and the yellow is the response of the offset slab. The extreme offsets are ±7 μgals either side of the zero line (yellow curve). The symmetrical response (red) has been plotted over the expected CO_2 plumes in upper and lower interval (yellow boxes) adjacent to the target interval SP log response. The possible station levels have been plotted on the left of the GR and SP logs on the left to show possible sampling interval choice.

A borehole survey consists of a series of stationary measurements. Time must be taken to allow the instrument to stabilize at each measurement point. Each of the levels has also to be re-occupied up to four times to reduce errors from measurement noise statistics and average depth errors between stations. Given that the errors in apparent density increase for decreasing station spacing (positioning errors), care had to be exercised to select measurement points of sufficient resolution to properly describe the shape and yet be optimal in terms of survey expense and timing. The inflection points were also critical to define, as they indicated the top and bottom of the slabs. The maximum excursion at the inflection points was predicted to be of the order of 7 microgals. The final choice of measurement levels was 14 stations with 8 central levels spaced at 10 feet and the remaining upper and lower levels spaced between 30 and 60 feet. This ensured we could capture the long tails, and still have resolution over the interval of interest for both wells.

Operational Considerations

With prior selection of target depth levels, we started the operation with a safety briefing for all involved to discuss anticipated actions and possible HSSE considerations. The operation was initiated by running-in the tool to a position below the target formation. Several methods, including stretch calculations, were carried out to ensure proper depth control. Once below the expected interval, a pass is made recording with high definition casing collar locator and formation gamma ray, keeping close attention to the depth systems, and then compared to previous runs of the corresponding logs. This was carried out while logging up to set the correct stretch to the cable, to maintain repeatability.

The casing collar locator (CCL) is a high sample rate device, consequently repeat passes can be quite precise. The original CCL, usually acquired with a cement bond log, uses a much lower precision 6 inch sample rate. For an accurate gravity log, depth errors have to be less than half an inch over 10 feet, as relocation error of 1.5 inches can give rise to a gravity error of 6 microgals [12]. In our case there was a large non-magnetic fiberglass section of casing, which could have been a problem, (absence of collars to serve as markers). However, there were pin-point radioactive markers in this section (so that the perforations could be oriented away from the resistivity cabling). These were ideal for confident precision of depth positioning. Using all these techniques, we believe we were able to establish the depth repeatability for positioning the tool to well less than an inch by correlating against a radioactive pill in the completion. Calibration of the sensor was accomplished by sampling the response over several 10s of minutes for a range of gimbal orientations, which when curve fitted, established the true vertical position for the sensor. Each of the middle stations were occupied 4 times, while the outer levels were sampled three times, which meant a repetition of correlations and depth positioning and then stabilization time. The average measurement rate was about 5 levels per hour. All of these activities added up to 24 hours acquisition for each well - a considerable investment in rig time. We were surprised during the calibration survey to see large excursions of the signal of greater than 100 μgals, which when we checked on the internet NEIC site we found was due to a level 6 earthquake in Venezuela. We had to wait for over an hour for the reverberations to die down before resuming calibrations. The baseline survey was acquired in September 2009, and the subsequent follow-up survey was acquired in October 2010. The repeat survey was marred by high instrument drift compared to the baseline survey. This was depth related and mechanical in nature related to temperature, since it was repeatable rather than intermittent. We decided to complete the survey assuming that, although high, the drift rate was linear with time and consequently could be compensated in the subsequent evaluation of the data. However it considerably complicated the analysis of the data.

The data shows for the baseline a decrease of 600 microgals from top to bottom of the survey due to the change in overburden density and depth. There was a positive change of slope over the reservoir of 200 microgals (blue box in Figure 3). The upper and lower inflection points correspond precisely with the upper and lower bounds of the reservoir formations as shown by the correlative SP and neutron density porosities. The shape, although similar to the time-lapse difference simulated in Figure 2, is in fact related to the gross lithological apparent density changes and not the time-lapse difference. The statistical variation of the measurements at a particular depth ranged from 20 to 50 μgals, with the largest variations associated with the 2010 survey. The response should be exactly the same above and below the reservoir, with any differences due to the CO_2 in the repeat pass being identified as departures from the baseline. As shown in Figure 2, the expected magnitude of the variation from the base line is on the order of ±7 μgals. Given the drift bias in the repeat survey and the very small expected differences, it is difficult to see the difference directly. There is a significant shift however in the second pass for F02 associated with the lower interval although not observable in the furthest well from injection F03. This shift does not appear to be an artefact of the measurement uncertainty or the drift bias, since the shift seems consistent over several measurement levels, even with the uncertainty. There are similar variations, although small, in the F03 well in the reservoir interval, while the trend is distinctly parallel and straight above and below this interval, further hinting at response. This change appears to be responding to the CO_2. The signal to noise presented a large challenge in assessing significant quantifiable detection.

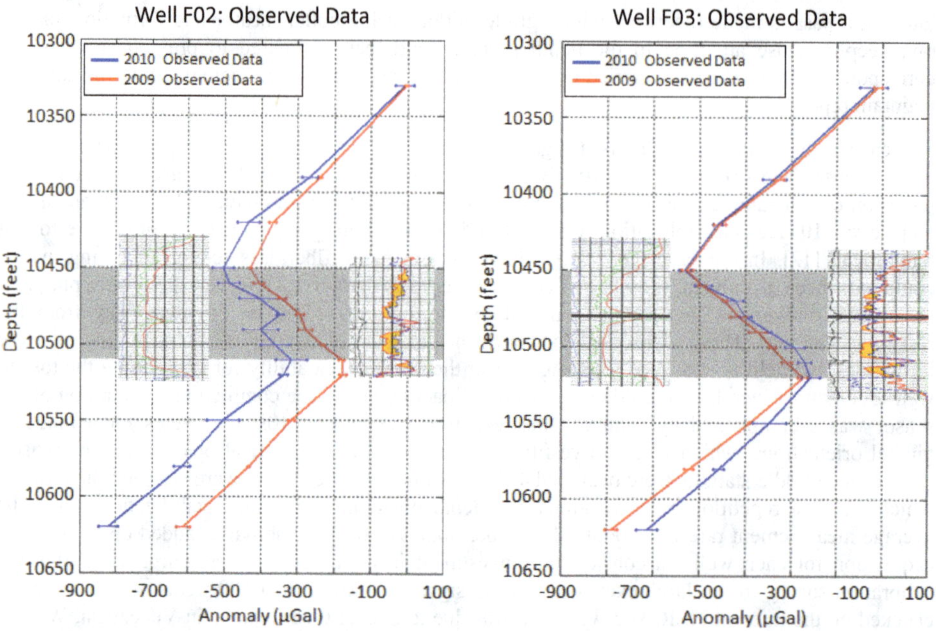

Figure 3. Observed data for both wells with red denoting the 2009 baseline passes and blue denoting 2010 repeat passes. Horizontal lines denote the statistical variation of repeated measurements at each level. The increasing separation from top to bottom is due to the anomalous instrumental drift observed on the repeat passes.

POST-ACQUISITION ANALYSIS

The time-lapse borehole gravity data collected at Cranfield were evaluated in two ways. The first is a comparison of the field data with simulated reference data calculated from detailed reservoir simulation models. The second is the analyses of apparent density information from the borehole data directly to identify the best approach to interpreting borehole gravity data as currently available, and to make recommendations for future time-lapse borehole surveys.

Simulation Model

We first examine a set of simulated reference data calculated at the borehole measurement locations using detailed time-lapse density models from simulations. The primary model, Figure 4 is a time-lapse density contrast model of the field built from the local geology, porosity, and fluid density weighted by predicted CO_2 and brine saturations. The majority of the field is predicted to have an average density change of approximately -0.02 g/cm^3. Given the reservoir depth near 3 km, a surface gravity survey will not detect the gravity response due to such a small density change, and therefore, is not a viable tool for monitoring. Thus, an alternative detection method, such as borehole gravity, is needed.

The time-lapse density model (Figure 4) is the foundation for creating the reference data to simulate the predicted time-lapse responses in the boreholes, the regions of influence around each borehole contributing to these data, and the differences predicted in the gravity response at the two borehole locations. In Figure 4, the black well is the CO_2 injector, and the observation wells are the blue (F02) and red (F03). The measurement locations within each well are identified with black points.

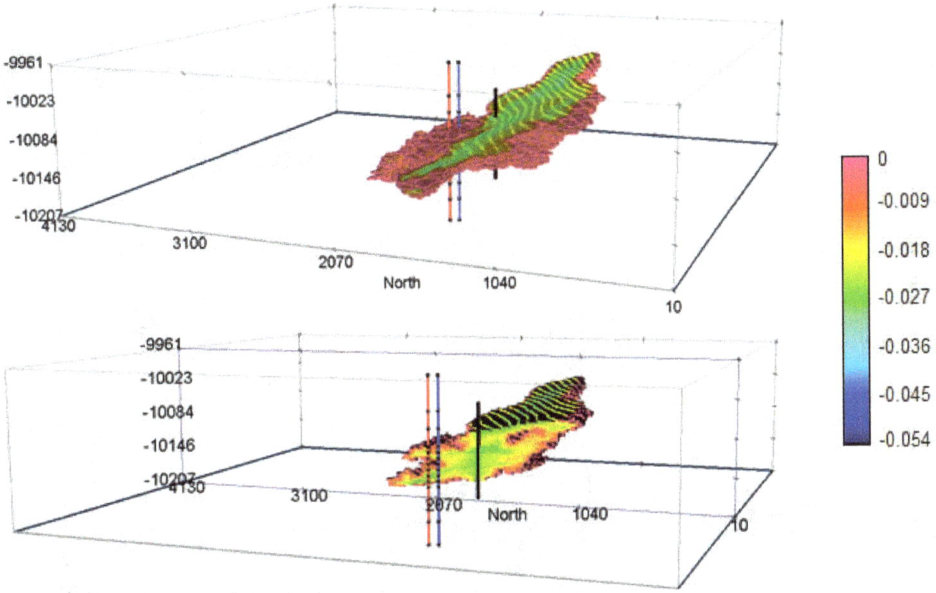

Figure 4. 3D representation of the simulation model showing the full plume (upper) and the plume sliced along the plane of the injection and monitoring wells (lower). The right hand well is the injection well with the two monitoring wells F02 and F03 to the left. Colours define saturation.

Simulation Data

The simulated time-lapse gravity responses predicted in the two boreholes are shown in Figures 5 and 6. The data are calculated for a CO_2 plume expanding from the injector into the porous formations where although the monitoring wells were downdip, the simulation shows that the plume still expands past these wells. The data in Figure 5 identify the depths of the predicted peak responses within wells F02 and F03 around 10450-ft (peak negative) and 10520-ft (peak positive). The peak anomalies are predicted to have a magnitude of ± 10 to 11 μGal. A negative response above the field and positive response beneath is consistent with a negative apparent density change from CO_2 injection. There is also no apparent distinct changes associated with two distinct formations as expected from the pre-survey modelling in Figure 2.

We next evaluate the peak responses (both positive and negative) of the reference data (Figure 5) within each borehole from CO_2 plume expansion. The simulations predict a time-lapse gravity response in both monitoring wells above the detection limit of the borehole gravimeter deployed in this survey. Additionally, the simulations predict a stronger time-lapse response in the lower zones of the reservoir versus the upper zones for both borehole datasets. The south-western well (F02) should have an earlier and larger final time-lapse gravity response than well F03, since the former is closer to the injector well and located in the thicker portion of the reservoir. A CO_2 plume expansion beyond approximately 600-ft from the injector is not predicted to affect the borehole gravity response in either monitoring well. Beyond that radius, the time-lapse responses in both monitoring wells reach their respective asymptotic values governed primarily by the density changes in the immediate vicinities of the well.

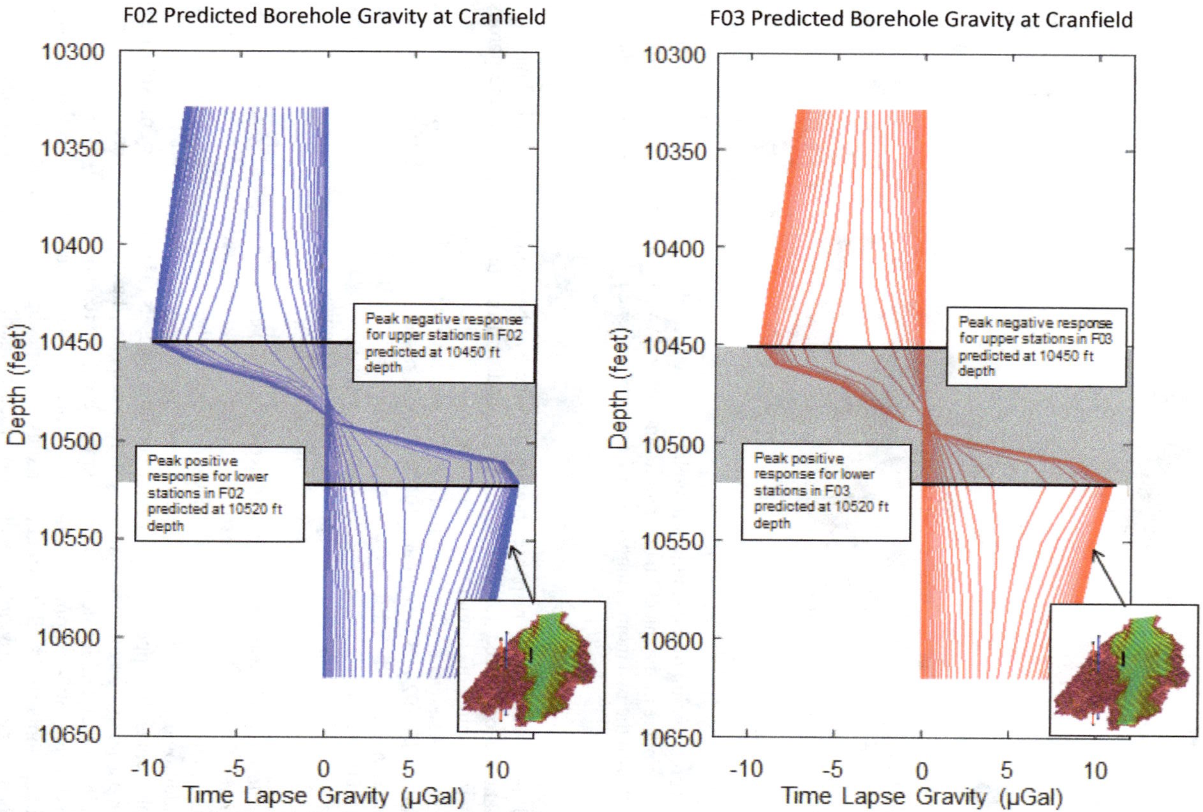

Figure 5. A more detailed time-lapse response using the simulation model showing an increasing upper and lower lobes as the plume builds out from the injection well.

In Figure 6, the stages of the plume growth are indicated from right to left at the bottom. The upper and lower curves plot the extremes of the upper and lower lobes in Figure 5 blue F02 and red F03. The vertical axis is gravity response and the horizontal curve is growth of plume in terms of radius of expansion. This plot shows the buildup in response if we were able to have permanently installed sensors with continuous read out available. In our case we are restricted to one position in time, and hence at one point along the x axis where the curves had reached maximum departure.

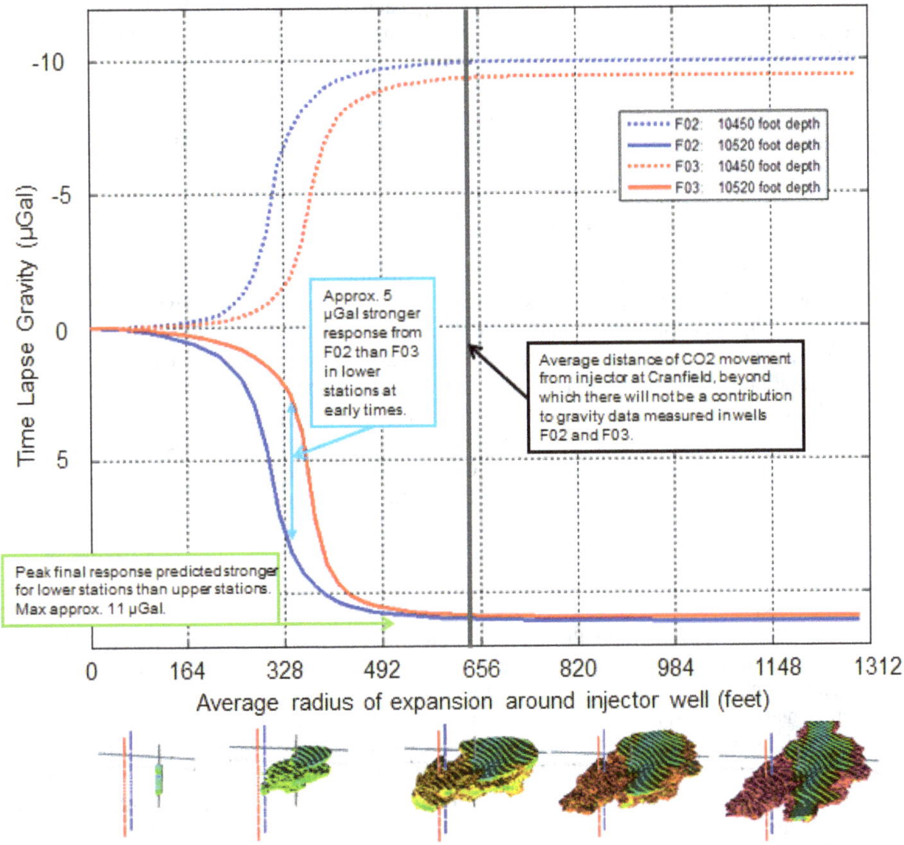

Figure 6. Predicted Peak Gravity Locations Based on Simulated Cranfield Parameters and Well/Station Locations.

CHOICE OF INTERPRETATION METHODS

Interpretation of the borehole gravity data can be performed in one of two ways. The first is inversion of the time-lapse anomaly data calculated from the measured fields before CO_2 injection and at a later time. The advantages of the time-lapse inversion are that it provides direct information on locations of density change occurring within a reservoir from CO_2 movement, and the results can be compared to reservoir simulations and additional monitoring data such as seismic and well-logs.

As borehole gravity technology continues to advance and is further tested for reservoir monitoring applications, we are likewise seeing advances in inversion technologies formulated for these time-lapse problems [13]. We note that for this test project, the data are too sparse for practical inversion, with coarse sampling in two boreholes, and we require an alternative approach to evaluate the instrument's sensitivity to CO_2 movement here.

The second approach is curve matching of the measured data with reservoir simulations or additional monitoring techniques. While this approach does not provide the same location information as the inversion, it does offer an objective means to address two of primary goals of the SECARB Cranfield injection study with borehole gravity: understanding the interpretational challenges of borehole microgravity for time-lapse sequestration monitoring, and assessing the ability of the instrument to detect geology and the injected CO_2. Given the sparse data sampling here, we opt to follow this approach over direct inversion of time-lapse data to address these underlying project goals.

APPARENT DENSITY CALCULATION

The instrumental drift observed for the bore hole gravity tool injects additional uncertainties in the direct subtraction of the two passes. As a solution, we evaluate the sensitivity of the borehole gravimeter to CO_2 injection by calculating an alternative data form, apparent densities [14], such that the time lapse difference should not change significantly above or below the reservoir. Apparent density calculation from borehole gravity data is a well-established interpretation tool that has been successfully developed in parallel to field instrumentation [14-17], and implemented successfully for hydrocarbon discoveries [18]. It is calculated from the difference between two gravity observations (vertical gradient) separated vertically within a borehole. The most common form of this formula to determine rock density between two stations is given by LaFehr [14] as:

$$\rho_a = \frac{F}{4\pi\gamma} - \frac{\Delta g/\Delta z}{4\pi\gamma} \qquad \text{Equation 1}$$

Where ρ_a is the apparent density, F is the free-air gradient between the stations, γ is the gravitational constant, Δg is the difference in gravity readings for the two stations, and Δz is the vertical separation between the two measurements. For the case of horizontal bedding extending infinitely with constant thickness and density, the apparent density derived from Equation 1 is equal to the formation density. If the layered earth assumptions are not met, as is the case in this study, the apparent densities from Equation 1 serve as a valuable form to represent the borehole gravity data, which can be directly compared with independent density information such as density logs. The importance of apparent density lies in its ability to express the gravity measurement, which senses deep into the formation from the borehole and can detect 3D density variations [14].

The advantages of using apparent densities for the current datasets are twofold. First, the unknown drift levels in the data do not have as significant an impact on calculated apparent densities as it does on gravity data themselves. Second, we can objectively correct, to an extent, for the influence of the unknown drift of gravimeter data by requiring that the apparent density changes far above and below the reservoir be zero, since the formation densities in those regions do not change significantly over time. Only apparent density changes within or near reservoir boundaries should exist for time-lapse problems.

Figure 7 shows that there is a distinct change between the top and bottom of the reservoir (shaded boxes). The variations between the 2009 and 2010 data are more pronounced and consistent in the reservoir interval than above and below. The shaded boxes correspond to two porous intervals within the reservoir, with an impermeable interval in the middle.

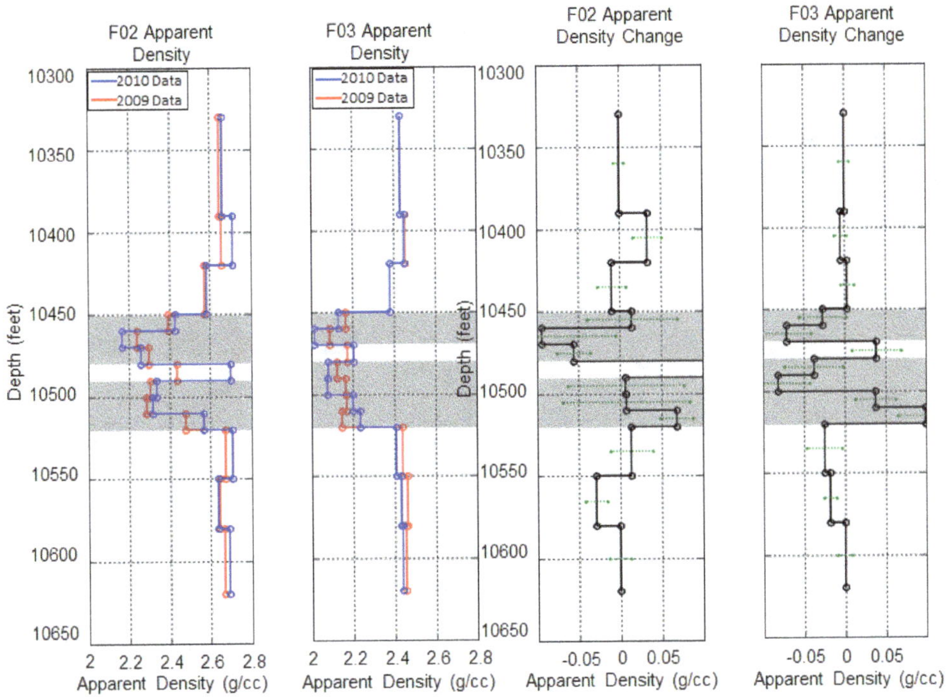

Figure 7. Apparent density data for both wells and corresponding time lapse differences on the right.

We note that Equation 1 involves the differentiation of an already noisy data set. This problem is compounded by the time-lapse calculation of apparent density change from multiple data sets. However, given the large station spacing, the resultant apparent density values appear reasonable and capable of facilitating the interpretation here. The change in apparent density provides the most objective means to evaluate the time-lapse borehole gravity data for the SECARB Cranfield injection study given the limited data sampling and lack of quiet reference points.

The simulated apparent densities are highlighted by the red box in Figure 8. The Blue shade in the observed apparent densities show a negative apparent density within the interval. There are correlatable logs on each side showing the porous interval and the impermeable break. The simulation plume is shown at lower left for reference to the zones.

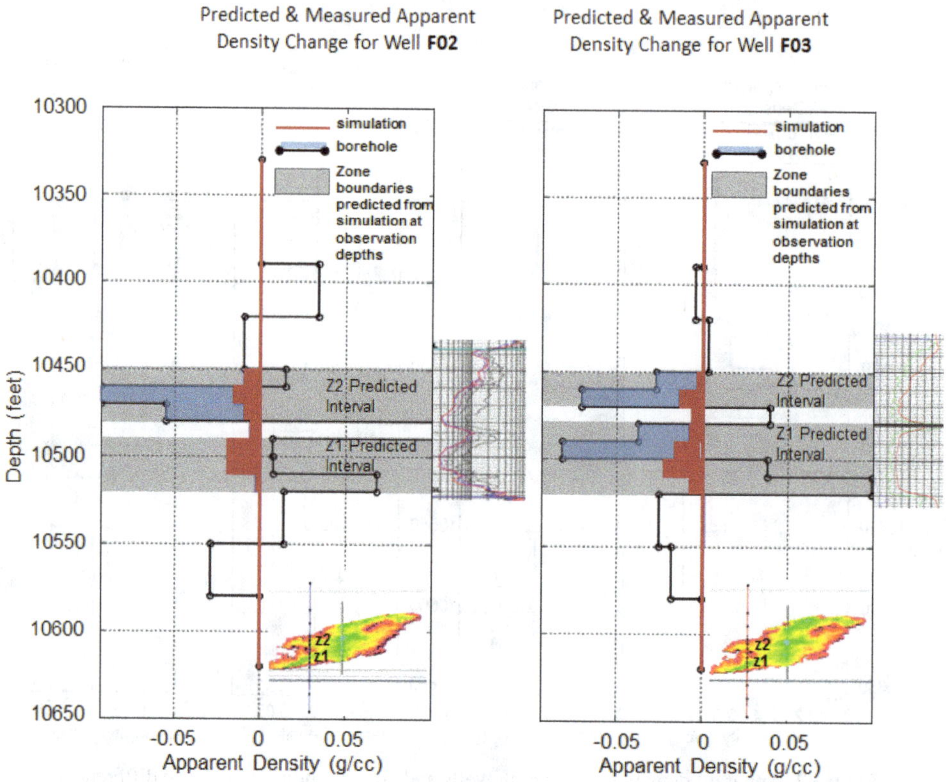

Figure 8. Differenced apparent densities.

DISCUSSION

As part of the Southeast Partnership (SECARB) "early" test of CO_2 injection at Cranfield, Mississippi, time-lapse borehole gravity measurements (BHGM) were collected within two multi-use monitoring wells. The project is the first commercial scale study designed to evaluate operational and interpretational challenges of an array of geophysical tools with the goals of understanding their ability to monitor storage performance. The goals of these BHGM tests were to understand the operational and design aspects of data acquisition, assess the ability of the tool to detect geology and injected CO_2, and to make recommendations for improved future deployment.

The borehole gravimeter data were evaluated on two levels: sensitivity to the larger geologic response and the smaller time-lapse signal from injected CO_2. All four data sets, two for well F02 and two for F03, demonstrated distinct Poisson jumps at the boundaries of the Cranfield reservoir and are clearly seeing geology of the site. The positions of these boundaries are consistent with known geology, represented by the 3D reservoir model and simulated reference data for comparison. Time-lapse changes within the reservoir zones are likewise consistent with the 3D problem, where data are seeing deep within a reservoir that is neither constant density or represented by horizontally continuous layers.

The time-lapse response from CO_2 injection is less defined than the changes associated with lithological change. This is understandable since there are several influences on the data acquisition and analysis that are comparable in size and obscure the density differences we would like to observe. These influences include the mechanical (temperature) related drift that although of low frequency and assumed to be linear, we were not able to identify the actual origin and hence assess its full consequence. Additionally the very small volume of the plume and displacement of the water means that the observable signal is comparable to the noise generated by differencing the measurements between levels. This noise is twofold. The first is derived from any small errors in repeating the depths within the survey and between the two time-lapse surveys. The second is derived from differences in the tool response at each level between surveys and being very close to the limit of resolution of the tool. Our numerical experiments indicate that working with apparent density enabled us to deal with some of the influence of noise in the Cranfield borehole gravity data, including a large drift in each data set. Using the approach, we observed that the apparent density change from measured gravity data after adjustment is largely consistent with the locations predicted from the reservoir simulations. Therefore, while improvements can be identified for future surveys of this type, it is clear from this test project, even with unknown instrument drift, the time-lapse borehole gravity data for sequestration monitoring at Cranfield contain meaningful but qualitative information about CO_2 movement.

Despite these challenges in quantitatively assessing the saturation, it was deemed important to report the experience, acquisition and analysis methods. Borehole gravity has a long history [1] in providing unique formation density for depth of investigations well beyond other geophysical logs. It is complementary to technologies including VSP and deep reading resistivity and in the case of Cranfield injection, provides an independent comparison to the crosswell seismic and resistivity tomography acquisitions. The future of borehole gravity technology is with development and production surveillance where despite the challenges of time-lapse illustrated within this paper, deep reading sensing of saturations, flood fronts (where the volumes provide a significant density difference) provide the promise of intrawell sensing with possibility of optimising production and injection processes [12].

The anomalous drift issues observed between the two sets of surveys arose from the age of the technology (unchanged since the 1960s) and its highly precise mechanical system to achieve the required resolution (~ 10^{-9} of the acceleration of gravity). New technology is in development [9] that will reduce the diameter of the tool and increase its well deviation range.

Motivation for the principle author to pursue this survey was not only to evaluate its potential for CO_2 monitoring but also because of a BP sponsored development of a break-through robust MEMS triaxial gravity sensor that is now in the process of commercialisation. When this technology is available, it will easily address all deviations and casing sizes and, moreover, is adaptable to multiple levels - essentially eliminating the need for depth precision between levels - and deployable both on wireline and permanently.

ACKNOWLEDGEMENTS

This work was supported by CO_2 Capture Project (CCP) and its participating companies, BP Chevron, ENI, Petrobras and Shell We would also acknowledge the generosity of Denbury Resources, access to their field and wells, as well as the South East Regional Carbon Consortium (SECARB) and the researchers in the Gulf Coast Carbon Center, in particular Susan Hovorka, Seyyed Hosseini and JP Nicot.

REFERENCES

1. Nabighian M., Ander M., Grauch V., Hanson O., LaFehr T., Li Y., Pearson W., Phillips J., Ruder M. Historical development of the gravity method in exploration. *Geophysics* Vol 70, No. 6, 2005
2. Huston, D. C., H. H. Huston, and E. Johnson, 2004, Geostatistical integration of velocity cube and log data to constrain 3D gravity modeling, deepwater Gulf of Mexico; *The Leading Edge*, 23, 842–846
3. Hare J.L., Ferguson J.F., Aiken C.L.V., and Brady J.L., The 4-D microgravity method for waterflood surveillance: A model study for the Prudhoe Bay reservoir, Alaska; *Geophysics*, Vol. 64, No.1 (1999).
4. Nooner S., Eiken O., Hermanrud C., Sasagawa G., Stenvold T., Zumberge M., Constraints on the in situ density of CO_2 within the Utsira formation from time-lapse seafloor gravity measurements; *International Journal of Greenhouse Gas Control* (2007)
5. Arts R., R. Elsayed, L. van der Meer, O. Eiken, S. Ostmo, A. Chadwick, Estimation of the mass of injected CO_2 at Sleipner using time-lapse seismic data , 64th European Association of Geoscientists and Engineers (EAGE) Meeting, Florence, Paper H016 (2002).
6. Gasperikova E. and G. M. Hoversten, G.M. Gravity monitoring of CO_2 movement during sequestration: Model studies High-resolution density from borehole gravity data; *Geophysics* Vol. 73, No. 6, 2008.
7. Sherlock, D., Toomey A., Hoversten M., Gasperikova E., Dodds K., Gravity monitoring of CO_2 storage in a depleted gas field: A sensitivity study; *Exploration Geophysics* (2006) 37, 37-43.
8. Nind C., Seigel H., Chouteau M., Giroux B., Development of a borehole gravimeter for mining applications. *First Break*, 25:71–77, 2007
9. Loermans, T., Kelder, O., Intelligent Monitoring? Add Borehole Gravity Measurements!, SPE 99554, 2006
10. Alixant J-L, Mann E., In-Situ Residual Oil Saturation to Gas from Time-Lapse Borehole Gravity SPE 30609 Annual Technical Conference and Exhibition held, Dallas, 22-25 October 1995.
11. MacQueen J., Micro-g LaCoste, High-resolution density from borehole gravity data, SEG Annual Meeting, San Antonio, 2007
12. Beyer L.A., Borehole gravity surveys: Theory, Mechanics and Nature of Measurements, ISGS 1983.
13. Krahenbuhl R.A., and Yaoguo Li, Time-lapse gravity: A numerical demonstration using robust inversion and joint interpretation of 4D surface and borehole data; *Geophysics*, Vol. 77, No. 2 2012
14. LaFehr, T. R., 1983, Rock density from borehole gravity surveys; *Geophysics*, Vol 48, No. 3, 341-356.
15. Hammer, S., 1950, Density determination by underground gravity measurements; *Geophysics*, 15, 637 – 652.
16. Smith, N.J., 1950, The case for gravity data from boreholes: Geophysics, 15, 605 – 636.
17. McCulloh, T.H., 1965, A confirmation by gravity measurements of an underground density profile based on core densities. *Geophysics*, Vol 30, 1108-1132.
18. Bradley, J. W., 1974, The commercial application and interpretation of the borehole gravimeter: Contemporary Geophysical Interpretation Symposium, Geophysical Society of Houston, *Proceedings*.

Carbon Dioxide Capture for Storage in Deep Geological Formations, Volume 4
Karl F. Gerdes (Editor)

Chapter 37

MODULAR BOREHOLE MONITORING: DEPLOYMENT AND TESTING

Barry M. Freifeld[1], Thomas M. Daley[1], Paul Cook[1] and Douglas E. Miller[2]
[1]Lawrence Berkeley Laboratory, 1 Cyclotron Rd, Berkeley, Ca, USA
[2]Silixa Ltd., Elstree, UK

ABSTRACT: The Modular Borehole Monitoring (MBM) Program, funded by the CO_2 Capture Project, was a three year research and development effort by Lawrence Berkeley National Laboratory to develop a next generation, integrated, well-based monitoring system for CO_2 sequestration.

The previous MBM chapter (Chapter 29 in this volume) covered our review of existing technologies, which identified several monitoring technologies that were deemed of critical importance to incorporate in all well-based storage monitoring programs, including our MBM design for Citronelle Dome. The critical monitoring technologies include reservoir pressure and temperature, fluid sampling, integrated fiber-optic bundles, and the ability to facilitate wireline logging.

This chapter describes the field deployment at Citronelle Field, Citronelle, Alabama. Seismic monitoring is also considered to be of high-importance, and two different technologies were incorporated in the Citronelle MBM: (1) a semi-permanent, tubing-deployed, 3C-geophone array with a unique cable and clamping design and (2) a recent technology development, fiber-optic distributed acoustic sensing (DAS). The incorporation of DAS in the Citronelle MBM package provided an ideal opportunity to compare conventional geophones to DAS (Task 7).

The Citronelle Dome MBM system was successfully deployed in March 2012 in the 9400 ft deep D9-8 monitoring well. All monitoring systems passed functionality testing upon installation, and have continued to work for the first 18 months of operation (as of this writing). The integrated fiber-optic bundle incorporated a heat-pulse distributed temperature sensing (DTS) system, which proved invaluable in diagnosing off-depth perforations. The surface read-out pressure/temperature data continues to show that CO_2 has yet to arrive at the D9-8 well. An initial MBM-DAS VSP recorded in May 2012 provided proof-of-concept data showing MBM-DAS acquisition potential. This result lead to a designed DAS VSP test, conducted in May 2013, which provided quality data, quantified the relative sensitivities of MBM tubing-deployed geophones and DAS, and demonstrated the potential of DAS for future applications.

The MBM deployment benefited greatly from lessons learned from prior deployments. In particular, material selection was guided by failures of polymers and rubber seals when exposed to supercritical CO_2. Several new features incorporated into Citronelle Dome's MBM are considered noteworthy including: (1) a hydraulic clamping system for coupling a tubing-deployed, welded TEC geophone string against the well casing, (2) an integrated fiber-optic bundle incorporating DTS, DAS and heat-pulse sensors, and (3) an integrated flatpack deployment system that simplifies the run-in-hole by combining numerous control lines into one bundle. The MBM deployment achieved its objectives of incorporating a suite of high priority monitoring methods in a package that limited the overall risk profile of a complex well completion. Future deployments should benefit from our project teams extensive documentation and dissemination of the engineering of the Citronelle MBM package.

KEYWORDS: monitoring; boreholes; instrumentation; deployment; field testing

DEPLOYMENT

The deployment of the MBM in the D9-8 well was preceded by numerous planning teleconferences and two pre-job meetings, which resulted in the development of a detailed Well Completion Procedure to guide the installation. The effort was led by the Citronelle Project manager from Advanced Resources International, Inc. (ARI), with participation by program managers and completion service and equipment providers including EPRI, LBNL, Denbury Onshore Resources LLC., Eastern Tools, Paulsson Inc. and GE Oil & Gas (formerly Wood Group).

The installation procedure included the following high-level steps: clean-up the well, assemble and lower the BHA components with control lines terminated above the packer, connect to main MBM assembly (flatpack), run-in-hole, set the well-head, and commission the instrumentation. In this section we give an overview of the installation. The detailed well completion schematic can be found in Chapter 29: "Modular borehole monitoring – an integrated deployment package development." Further details on the integrated fiber-optic cable and flatpack cable can also be found in Chapter 29. In particular, Figure 16 in Chapter 29 is a picture and schematic of the flatpack configuration.

Figure 1. An overview of the worksite showing the placement of the spooling units, pipe stands and mud tanks. Light poles facilitated 24-hour operations.

The D9-8 was already perforated prior to commencement of the workover operation. The clean-up followed standard procedures with a scraper run and circulation of fluid until the well produced clean fluid. Next the control lines were prepared. Figure 1 shows an overview of the worksite showing the location of the spooling units, pipe racks, and mud tanks. Light stands permitted 24-hour operations.

The fiber-optic cable was the first control line installed, starting at the bottom of the bottomhole assembly (BHA). The 3/8" stainless steel tube was run through the packer, which was pre-assembled with pup-joints and placed in the rig pipe stand prior to the run-in-hole (Figure 2). Prior to shipping out to the field, the fiber-optic cable was terminated with a welded end to ensure a hermetic seal.

Figure 2. Packer (wrapped in protective cover) and placed vertical in pipe stand prior to starting the run-in-hole of the BHA. The worker has just started feeding the fiber-optic cable through a packer feedthrough and the end can be seen dangling in the air.

Figure 3 shows a GCDT Inc. control line protector being installed to cover the termination of the fiber-optic line. Note that one of the 3/8" perforations in the 450' 2 7/8" tubing, which allows fluid to equilibrate between the inside and outside of the tubing, is visible in this image,. The tailpiece serves to keep any logging instruments free from the fiber-optic cable and pressure/temperature sensors and organizes the sensors and protectors during installation. Figure 4 shows how the flatpack cable was delivered without encapsulation covering 450 ft of control lines, which would be passed through the feedthroughs in the packer and attached to the tailpipe.

Figure 5 shows the locations of the sheaves in the derrick to run the flatpack and seismic cable. Because of the stiffness of the flatpack, setting the sheave as high up in the derrick as practical is important in order to use the weight of the flatpack to straighten itself out from spooling curvature. The seismic sheave needed to be lowered while the seismic pods were installed, allowing manual control so that the 2" OD seismic pod would not hang up going through the sheave, damaging the cable. The drawback to the lower sheave is the hazard of the cable running through the middle of the rig platform, limiting the rig crew's mobility and requiring more careful operations. Also, pup joints were needed to keep any geophone pod from landing too near a joint (due to the mismatch of 50 ft pod spacing and 30 ft joints). As soon as the last of the 18 pods were installed, the sheave was raised up again for the remainder of the run-in-hole operation. Stainless steel bands were used to organize the cables (Figure 6).

Figure 3. Installing a GCDT Inc. control line protector over the fiber-optic cable termination to protect it while running-in-hole.

Figure 4. Location where the polypropylene flatpack encapsulation ended, allowing 450' of exposed control lines to be available for passing through the packer.

Figure 5. Location of the sheave position high up on the rig derrick for running the flat-pack. The seismic cable sheave was set below the monkey board, but lowered to work-platform height when running geophone pods over the rollers.

Once the MBM was landed at the proper depth (using pup joints for adjusting the final spacing), the control lines were broken outside of the flatpack. This proved to be a very time consuming operation, as it was originally envisioned that a 200 ft length of the fiber-optic cable would be broken out of the flatpack (removing the polypropelene encapsulation) for running into the data acquisition office container. This was designed to avoid a fiber splice. A good quality splice in the SMF should have minimal impact on DAS data, given the large optical power budget and relatively short length of SMF, however the quality of the splices for the MMF are critical so as to not degrade the performance of the DTS system. Given the depth of the well at over 9000 ft., we wanted to minimize fiber signal loss due to splices, as reduced losses along the fiber means a lower the noise level in the data collected within the reservoir.

Figure 6. Installing stainless steel bands to organize the cables. Small bands were used at mid-joint locations and above and beneath each geophone pod.

However, after an 8 hour delay for mechanically cutting off the polypropylene housing and/or using a belt sander, it was decided to save rig time and cut the fiber cable and splice to surface running lines. The impact of the thickness of the polypropylene and its toughness in cold weather was not appreciated during planning. Whereas polypropylene around a single control line can often be scored along the tubing with a pull knife or box cutter and removed as one piece, the massiveness of the plastic encapsulation on the flatpack meant that virtually all of the material removed would need to be cut or sanded away. Since the cable has a special termination, there was no length with which to practice removal of the encapsulation prior to mobilization to the field.

After deciding to splice to surface lines, the cut lines were passed through the tubing hangar and wellhead, and the wellhead was completed. Figure 7 shows the control lines being fed through the valve spool adapter after landing the tubing hanger. Figure 8 shows the completed wellhead with the control line protectors installed over the port collars.

Figure 7. Feeding the control lines through the valve spool adapter after landing the tubing hanger.

Figure 8. D9-8 wellhead as completed with control lines penetrating through port collars and collar sleeves.

Figure 9. Setting of the hydraulic system using a small air driven piston pump. Hydraulic oil is injected down the ¼" center tube in the tube-in-tube and returns up the outer 3/8" tube.

Baseline Data Collection and Commissioning of System

Following deployment, the MBM 'system' components were individually tested. This included testing of the U-tube sampling by using compressed N_2 to verify functionality of the downhole check valve. The fiber-optic heat-pulse cable was tested after splicing pigtails onto the ends of the fibers. The electrical conductors used for heat-pulse measurements were Hi-Pot tested to ensure that there was no loss of integrity to the wire insulation. In addition the resistance of each of the three loops of conductors was measured. After initial 'state of health' testing, the system components were used in baseline data acquisition as part of a system commissioning.

Seismic Acquisition: Geophone and Fiber Optic VSP

As part of initial testing of the MBM components, the geophones needed to be clamped in place. To lock the geophones against the casing, the hydraulic cylinders in the geophone clamps were actuated using hydraulic oil circulated in a tube-in-tube control line. Hydraulic oil was injected using an air driven piston pump down the center ¼" control line, and then ported through the downhole tee's to each hydraulic clamp. The 3/8" outer control line in the tube-in-tube was used to allow hydraulic fluid to circulate back out to the surface. This arrangement meant that practically all of the air could be circulated out from the hydraulic system. Figure 9 shows the setting of the geophone clamps using the hydraulic system.

The initial test recordings were simply background noise recording during well testing and 'tap tests' on the wellhead, which indicated that most of the geophones were operational. The horizontal components of the 3C geophones apparently failed during deployment, which was later attributed to a correctable design problem, while the vertical channels all tested okay. The initial seismic source testing was conducted by SR2020, who are contracted for the vertical seismic profile (VSP) monitoring program at Citronelle. The test was in fact the initial walkaway VSP data recording. As part of the testing, we also utilized the single mode fiber cable deployed as part of the MBM package for novel distributed acoustic sensing (DAS) using a surface recorder provided by Silixa, LLC. This innovative test is described in the following text taken from a paper published in the *The Leading Edge* (Society of Exploration Geophysicists) [1].

DAS Background

Distributed acoustic sensing (DAS) is a relatively recent development in the use of fiber optic cable for measurement of ground motion. Discrete fiber optic sensors, typically using a Bragg diffraction grating, have been in R&D and field testing for over 15 years with geophysical applications at least 12 years old [2][3]. However developments in recent years have sought to remove the need for point sensors by using the fiber cable itself as a sensor [4][5][6][7].

Through Rayleigh scattering, light transmitted down the cable will continuously backscatter or 'echo' light so that it can be sensed. Since light in an optical fiber travels at about 0.2 m/ns, a 10-nanosecond pulse of light occupies about 2 meters in the fiber as it propagates. The potential of DAS is that each 10 nanoseconds of time in the optical echo-response can be associated with reflections coming from a 1-meter portion of the fiber (two-way time of 10 ns). By generating a repeated pulse every 100 μsec and continuously processing the returned optical signal, one can, in principle, interrogate each meter of up to 10 km fiber at a 10 kHz sample rate. Local changes in the optical backscatter due to changes in the environment of the fiber can thus become the basis for using the fiber as a continuous array of sensors with nearly continuous sampling in both space and time.

Since the technology for deploying fiber optic cable in boreholes is well developed for thermal sensing (distributed temperature sensing, or DTS), a DAS system has the potential of having thousands of sensors permanently deployed in the subsurface, at relatively low cost. Currently DAS systems use single-mode fiber, as opposed to the multi-mode fiber typically used for DTS, but the type of fiber does not affect deployment, and multiple fibers are easily deployed in a single capillary tube.

Recent advances in opto-electronics and associated signal processing have enabled the development of a commercial distributed acoustic sensor (DAS) that actualizes much of this potential [6]. Unlike disturbance sensors, [8], the DAS measures the strain on the fiber to characterize the full acoustic signal. Unlike systems relying on discrete optical sensors [2][3][9]) the distributed system does not rely upon manufactured sensors and is not limited by a need for multiple fibers or optical multiplexing to avoid optical crosstalk between interferometers.

The responsiveness of the optical fiber to seismic energy within the flatpack warrants discussion and further consideration. While we went through considerable effort to ensure strong physical coupling between our conventional geophones and the borehole walls, the DAS measurements were not planned until the MBM was already installed. Transmission of seismic energy is through fluid coupling and as noted in Daley *et al* [1], a cable cemented external to casing provides better signal-to-noise than a cable floating freely in fluid. The fact that a P-wave arrival can be detected in the fluid coupled fiber is significant and is the subject of continued research into the exact physical mechanism by which dynamic strain is measured through changes in Rayleigh scattering. Future projects need to consider if the added benefits in data quality attributable to installing a fiber-optic cable external to the casing warrants the potential increase in drill bit size and protective equipment required for protecting and centralizing the casing string. The benefit is a clear improvement in DAS signal-to-noise, resulting in less seismic source effort needed for performing surveys. Drawbacks include the impacts on the casing completion program and consideration of future fluid migration scenarios along the cable pathway.

Initial Field Test

The DAS fiber was a 'fiber in metal tube' (FIMT) which was itself part of a multi-conductor cable inside a molded 'flatpack' (Figure 16, Chapter 29) which was clamped to the production tubing, in the fluid –filled annulus between tubing and casing. The MBM flatpack was deployed to almost 3 km depth.

The DAS seismic data acquisition at Citronelle was a walk-away vertical seismic profile (VSP) recorded with an early version of the Silixa iDAS system. Data from initial DAS test, using a ~35,000 lb force vibroseis truck, were processed with a synthetic linear 16 s sweep from 10 to 160 Hz. From 4 to 6 sweeps were recorded at each source point (Figure 10). A strong tube-wave is observed along the entire length of the cable.

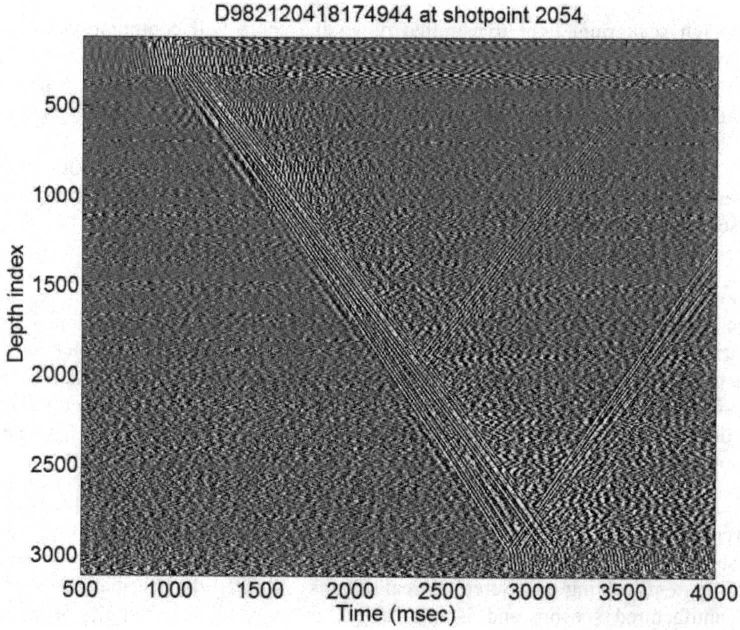

Figure 10. DAS data from tubing-deployed MBM flatpack for a shot point about 100 ft from the well. There are two observed wave speeds, 1.4 km/s and 1.3 km/s, this is likely from two modes of tube waves related to the presence of a fluid filled annulus [10]. Depth index is in meters. Analysis courtesy of D. Miller (Silixa), data collection courtesy of SECRAB, EPRI and ARI.

We were encouraged by the observation that DAS data does record seismic energy, however the DAS recordings do not have sufficient S/N for observing P-waves below about 1600 m (the 2.7 km/s event in Figure 11), while P-wave energy is easily seen on the clamped geophones at 6000 ft (1.8 km) to 7000 ft (2.1 km) (Figure 12). We felt this result, while not useful for seismic monitoring of the ~2.9 km deep reservoir at Citronelle, was sufficiently successful to move forward with work on improving acquisition and planning for a future repeat field test. We plan to return to the Citronelle site for further testing, where the MBM package remains installed and serves as an example of multiple instrument deployment and a test site with geophones and DAS co-deployed.

The MBM geophones were used for a full walkaway VSP (Figure 12 is data from one shot point). The full data set was processed into migrated reflection images by SR2020.

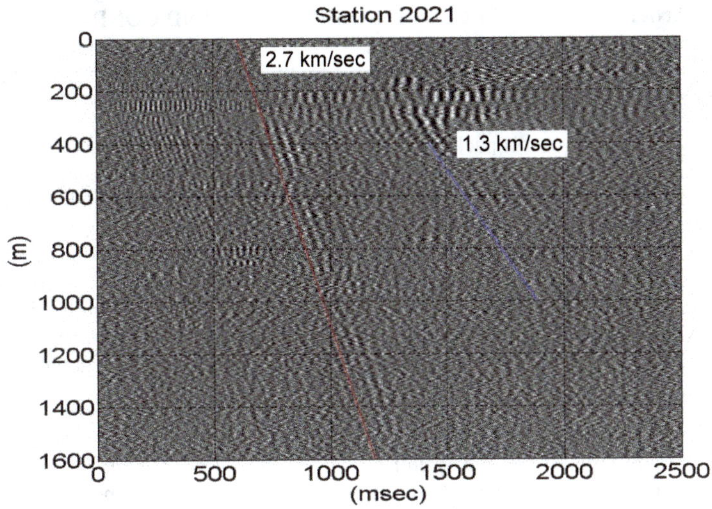

Figure 11. DAS data from source station 2021 at Citronelle, approximately 700 feet offset from the D-9-8 sensor borehole. Estimated wave speeds for two events (red and blue lines) are labeled in km/s. Analysis courtesy of D. Miller (Silixa). Data collection courtesy of SECRAB, EPRI and ARI.

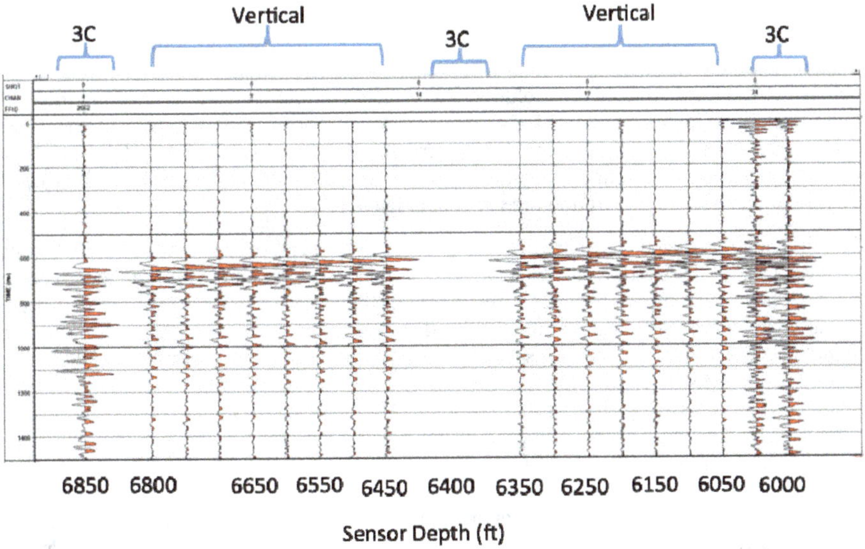

Figure 12. MBM tubing-deployed, clamped geophone data (50-foot interval between geophones) from Source Station 2021 (approximately 700 foot offset) with 60 Hz Notch Filter and removal of bad traces. Vertical and 3-component geophones are labeled with most of the 3C channels removed. A clear P-wave arrival is seen between 500 and 600 ms.

DIAGNOSTIC ANALYSIS OF WELLBORE COMPLETION AND CONDITIONS

Heat-Pulse Test

Objective of D9-8#2 Heat-Pulse Testing

During the setting of the Hydroset II packer in the D9-8#2 well it was observed that the annulus was not isolated since when pressured up, the annulus pressure bled down. To investigate the cause of fluid moving past the packer we utilized the tools that were installed for monitoring CO_2 in the reservoir – namely the two PANEX pressure/temperature sensors, and the hybrid fiber-optic heating cable. Two goals were achieved related to understanding the well completion. First, the response of the fiber-optic distributed temperature sensor (DTS) to a distributed heat-pulse was used to identify the location of the packer. Subsequently, flowing D9-8#2 by opening the annulus provided thermal signatures that would indicate the depth of the perforations and any casing leakage points, if they exist, as cooler formation fluid would move past the DTS cable, removing the heat pulse.

Testing Methodology

The heat-pulse thermal monitoring data is collected using the DRAKA hybrid copper-fiber-optic control line located in the flatpack. The DRAKA cable end is terminated 5 ft above the reentry guide located at 9797 ft KB (per initial tubing deployment depth calculations). The DRAKA hybrid cable consists of six 20 AWG electrical conductors which are shorted together approximately 1 ft above the physical end of the cable. They form three looped conductors with approximately 220 Ohm resistance per leg. When 1000 VAC 3-phase power is applied to the conductors they heat the cable at 6.06 W/m.

The fiber optic DTS cable, which is centered within the six helical wound conductors, registers the temperature along the cable using a Silixa Ultima DTS unit. The spatial resolution of the Silixa unit is 13 cm (0.43 ft). The temperature resolution is dependent on the integration time and the distance along the fiber (with increasing noise for greater distances). For the 5 minute measurements shown here the noise level at the level of the reservoir is approximately ±0.2 °C. The noise is predominantly Gaussian, so longer integration periods or spatial averaging results in a reduction in noise by the square root of integration time or spatial length.

The length along the fiber is measured by using the known speed of the light to time when pulses are reflected back to the instruments detector. The distance from the Silixa unit to the point of entry of the fiber-optic cable into the wellhead (the length of the surface cable run) was determined to be 160' by applying a heat pulse right at the wellhead and noting its location in the temperature traces. Depth in the well is referenced from this point. There are two sources of uncertainty in measurement of fiber depth (the length along the fiber): (1) excess fiber length (EFL) and (2) deviation from a perfectly straight line as the DRAKA flatpack is strapped onto the tubing. The EFL is caused by the manufacturer purposely installing extra fiber in the small metallic housing that encapsulates the fibers and gel to limit fiber strain. EFL is typically around 0.15% extra – or about 15 ft over the length of fiber in Well D9-8. The deviation from straight can be attributed to the cable both spiraling around the tubing and also taking bends around the joints and geophone clamps. This error is unknown, but thought to be around the same magnitude as the EFL. Together, the uncertainty in the position along the fiber is around 0.3% or around 30 ft in 10,000 ft, and both errors make the fiber depth greater than true distance along the borehole. These uncertainties will cause the tubing tally lengths to differ from the measurement length along the fiber. However, since the main points of interest for our testing occur at the bottom of the well, we can use the known termination point of the fiber near the bottom of the tubing to calculate the short distance back up the well. Since the distance from Reentry guide to the top of the packer, according to the as-built drawing, is 413' in length we can then measure up from the fiber to estimate packer location. This has less uncertainty than measuring from the top of the well down (about 1 ft vs 30 ft).

Testing Sequence

Two series of heating/flow tests were conducted in Well D9-8#2. Initial testing was performed on April 10, 2012. After acquiring baseline temperature and pressure data (no heating or flowing the well), the heater was turned on at 5:32 PM. At 8:32 PM a valve on the annulus was opened and fluid was produced at 10 GPM. The flow was shut in at 9:32 PM and the recovery was logged until 10:24 PM. Figure 13 shows the pressure and temperature data as recorded on the PANEX pressure gauges. Gauge RGA5109 and RGA5108 are located 21.01 ft and 45.93 ft below the top of the packer respectively. During the subsequent vertical seismic profile (VSP) acquisition, additional data was collected which allowed for additional temperature trace averaging and obtained a lower noise floor in the data which clarified the observations obtained earlier but did not change the initial interpretations of well conditions. The PANEX gauge data collected during this second set of heating and flow tests are shown in Figure 14. Heating commenced on April 17 at 2:20 PM. The heater was turned off on April 18 at 8:00 AM to allow for the VSP acquisition without electrical interference. Following seismic acquisition, heating restarted on April 18 at 7:30 PM followed by flowing the well at 8.7 GPM at 7:45 PM. The pressure response indicated that overnight, the well stopped flowing on its own, as the more dense formation fluid filled in the annulus and the less dense workover fluid was produced from the annulus.

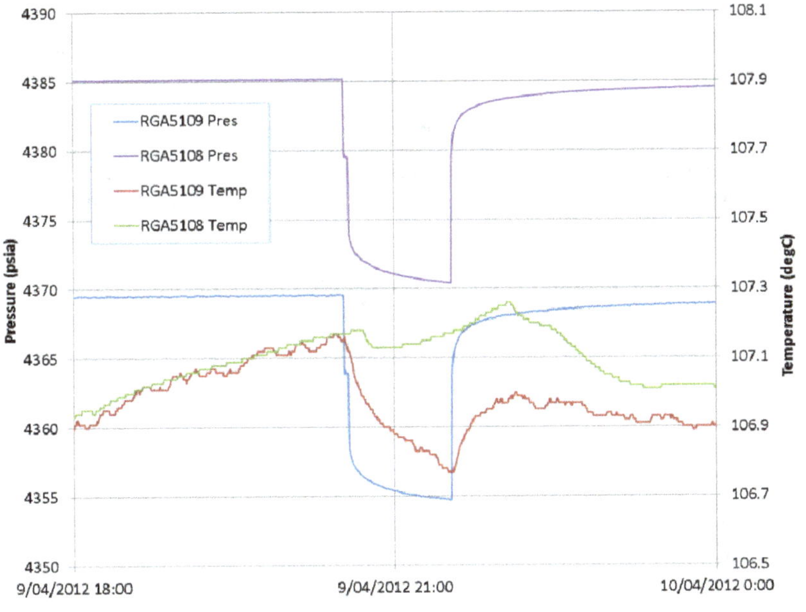

Figure 13. Pressure and temperature as recorded on the PANEX gauges in Well D9-8 during the heating and subsequent pumping test, April 9, 2012.

Figure 15 shows the baseline thermal profile (purple) as well as the thermal profile with the heater turned on (red). The step in the blue baseline curve from 9425 ft to 9450 ft is a remnant of the cold fluid circulated before packer setting. Another step is seen where the polypropylene encapsulation of the flatpack acts as a thermal insulator, retaining more of the added heat, with the end of the flatpack identified by the drop in temperature (red curve) at a depth of about 9370 ft.

Figure 16 shows the same temperature profiles shown in Figure 15, but highlights the region from 9300 ft to 9850 ft. The fiber depth has also been modified for comparison to the Kelly bushing (KB) depth used in the D9-8#2 as-built drawing by adding 18 ft to the overall fiber length that was previously adjusted so that the 0 length was at the top of tubing hanger.

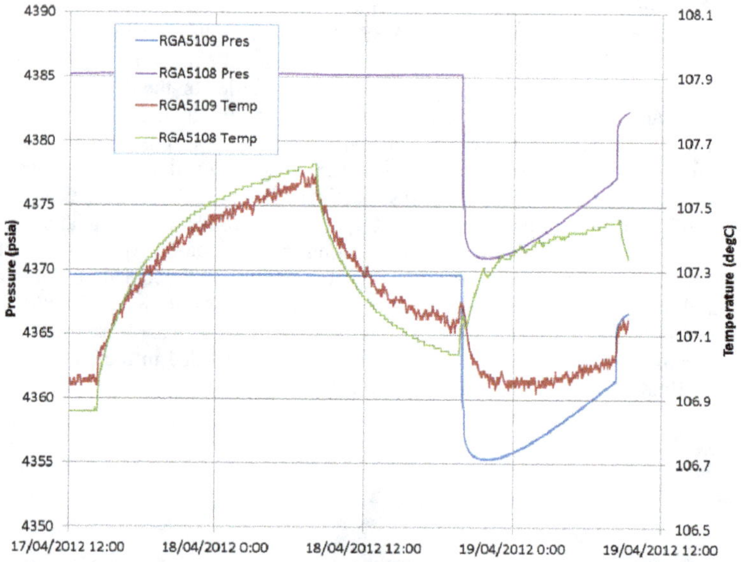

Figure 14. Pressure and temperature as recorded on the PANEX gauges in Well D9-8 for the second series of heating and flow tests conducted during the VSP acquisition campaign.

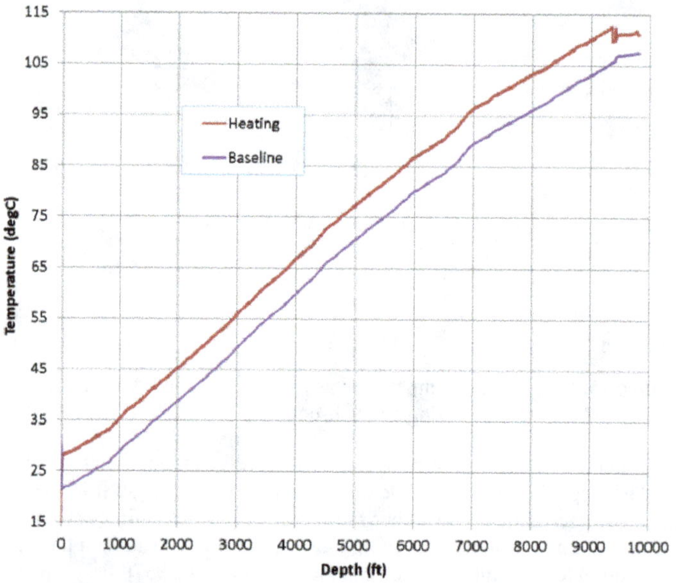

Figure 15. The purple temperature profile is the baseline conditions from the wellhead down to the end of the fiber. The depth is the length along the fiber. The red line is the temperature profile with heating at 6.06 W/m.

To identify the location of the packer, we used the anomalous thermal profile associated with the fiber-optic line running within the packer. While the packer body is steel, the fiber enters a Swagelok compression fitting at the top of the packer and is guided through a 1.25" diameter water-filled tube to where it exits the bottom of the packer with no seal. This water filled tube does not permit advection of fluid since it is sealed only at the top – hence Figure 16 identifies the packer as a short section at a depth of ~9430 ft that is warmer than the surrounding fiber with the stagnant water in the packer tube acting as an insulator. The measured fiber distance from the bottom of the well to the base of the packer is 402 ft, which agrees with the as-built expected distance (400 ft). The very small discrepancy in distance can be attributed to the fiber wrapping around the tubing and taking a slightly longer path because it is not stretched taut.

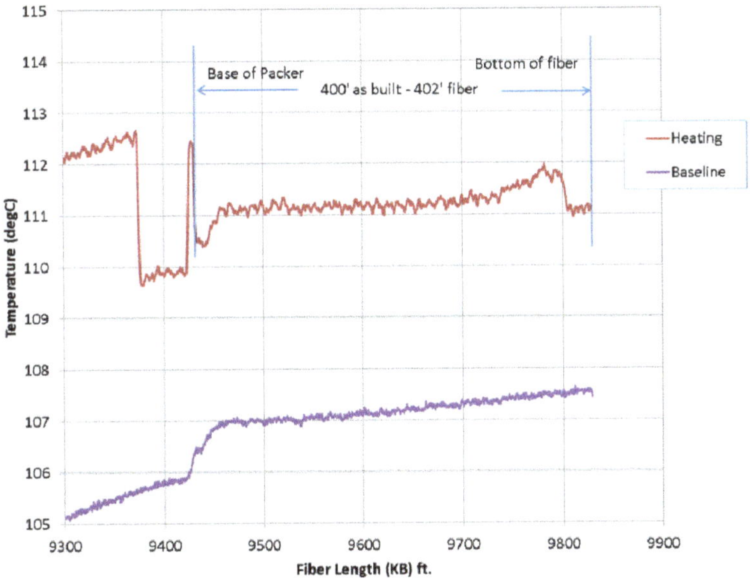

Figure 16. The purple line is the initial baseline profile from the depth of 9300 ft to 9900 ft. The red curve is the thermal profile with heating. Depth is referenced to KB.

What appears as a rather regular noise pattern in the heating temperature profile (see red thermal profile, Figure 16), with cyclical temperatures with an approximate period of 10 ft, is not noise but results from the regular banding of the fiber-optic cable to the production tubing. Where stainless steel bands are located, the fiber-optic cable or flat-pack is in good thermal contact, and hence each band appears as a broad cool spot (with the steel conducting heat away). This also occurs at each joint protector. Away from the stainless steel bands the cable sits a few millimetres away from the tubing, and hence is warmer. This cyclical pattern is evident whenever the fiber-optic cable is heated, but disappears when collecting DTS data without heating.

Figure 17 shows the difference (green) between the heating temperature profile (red) and the profile when both heating and producing the well at 8.7 GPM (blue). The shift in the difference curve (green) below 0 at the bottom of the well is attributed to drift in the fiber-optic temperature measurement unit, and not a real reduction in temperature. The cooler temperatures observed both above and below the packer are attributed to fluid entering the wellbore, presumably through the perforations. The fact that the temperatures above and below the packer are significantly different is an indication that the packer is properly set. While this result does not preclude some behind-casing leakage past the packer through cement, it is more likely that the fluid entering below the packer is flowing up the borehole, then out into the formation around the packer, and back into the wellbore above the packer. The much cooler temperatures above the packer than below indicate the in-flow is predominantly above the packer.

The large decrease in temperature change (green curve) observed at the top of the packer is not confidently explained in our initial interpretation, but may indicate that the heated fiber above the packer was not initially subject to advective cooling (because the packer body blocks the formation of convection cells), until the well was produced and water from the formation was forced across the heated-fiber, cooling it approximately two degrees.

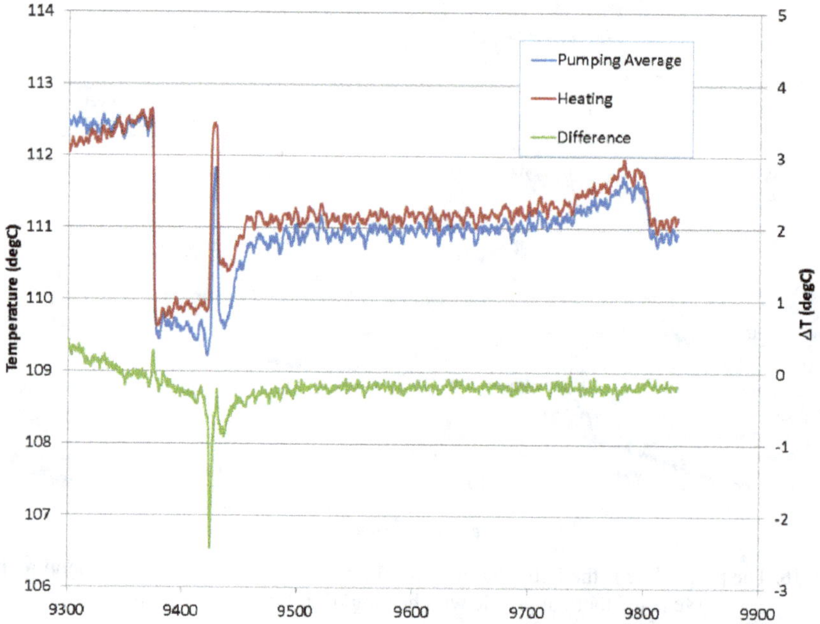

Figure 17. The difference between the heating temperature profile and the profile while heating and producing the D9-8#2 well at 8.7 GPM are highlighted with the difference profile (green). Depths are reference to KB.

Summary of Heat Pulse Testing

Figure 18 highlights the region around the packer, and shows our best estimates of the perforation and packer locations. The perforations are identified by distinct cooling noted from a depth of about 9421±5 ft. to 9451±5 ft. KB (measured via length along the fiber from bottom of fiber). The most likely location of the packer is given as 9426±1 ft to 9432±1 ft.

Given that the rubber seals on the packer are located approximately 2 ft. below the top of the packer and that the thermal profiles indicate flow both above and below the packer, there is a strong likelihood that the packer was set within the perforated interval.

These results demonstrate that operation of the MBM system for heat-pulse testing was successful in two separate tests and has proved very useful in diagnosing well flow zones and packer location, in both a relative and absolute sense.

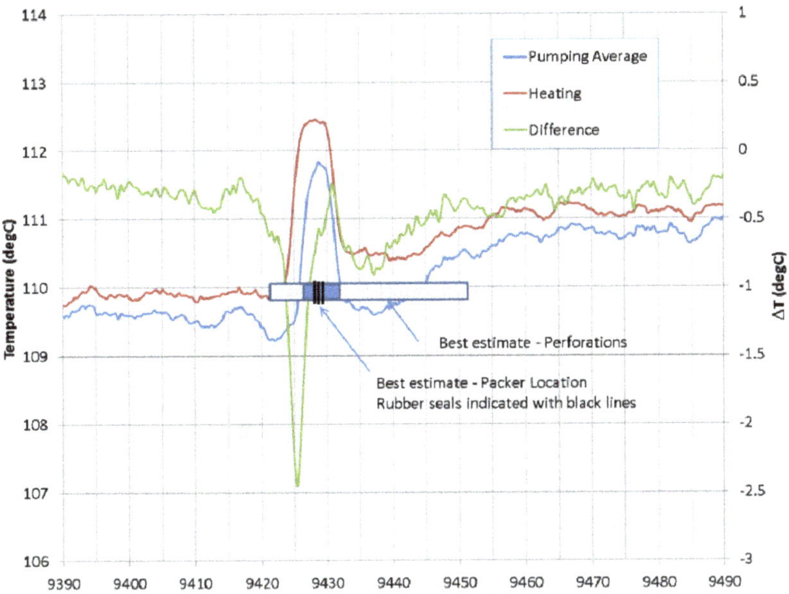

Figure 18. Thermal data from D9-8#2 showing the best estimate for the packer location and the most likely location for the 30 ft. length of perforations. Depths are referenced to KB.

Pressure and temperature monitoring data

The Ranger Gauge Systems digital quartz pressure/temperature gauges were installed at the top and middle of the first sand interval. Figure 19 shows approximately one year of monitoring the injection of CO_2 in the D9-7. The injection was started and halted numerous times, leading to the repeated build-up and fall-off in the pressure data. The large temperature spikes correspond to when the heater was turned on for conducting fiber-optic heat-pulse studies. To date no changes have been seen in the thermal response to indicate the arrival of CO_2 in the D9-8 well.

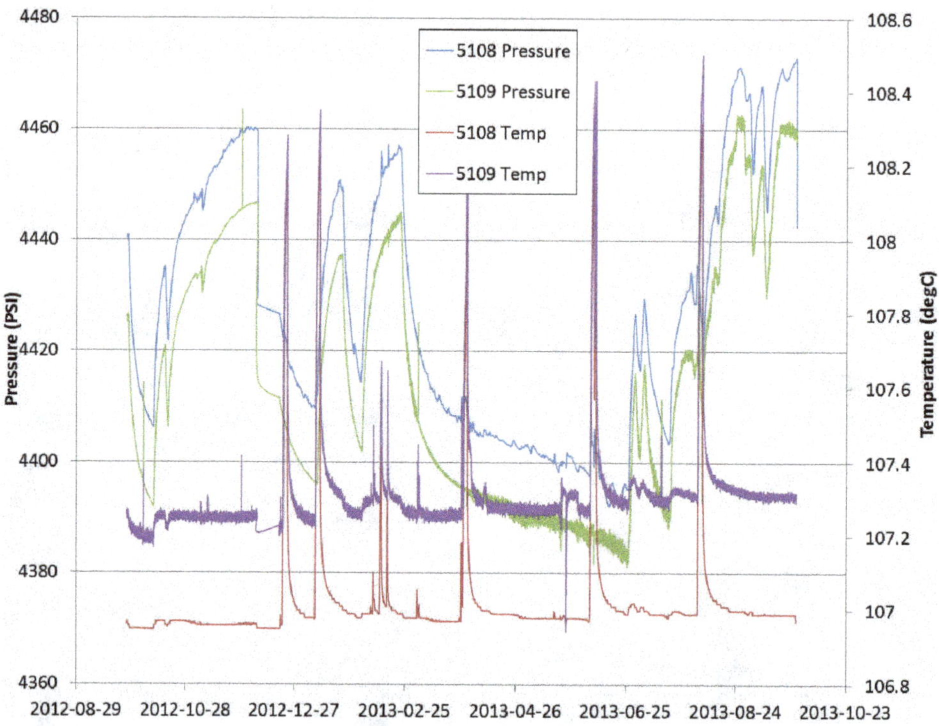

Figure 19. Pressure and temperature at the D9-8 reservoir level as recorded using two digital quartz gauges.

DEDICATED ACQUISITION OF DAS AND COMPARISON WITH GEOPHONES

Background

A novel opportunity to acquire distributed acoustic seismic (DAS) data with the Citronelle MBM system became available, utilizing recent advances in DAS technology. Using fibers deployed in the MBM, we tested the distributed acoustic sensing (DAS) system currently under development by Silixa, Ltd (Silixa's iDAS) and compared iDAS to the clamped geophones deployed as part of the MBM system. DAS is a new technology with limited testing, but it holds good promise to simplify borehole seismic monitoring. Silixa, as a collaborator, processed the DAS data.

Initial DAS testing was completed with processing and analysis as reported in baseline data, above, however the initial results indicated that problems with the data acquisition compromised the data quality and the ability to process the field data with available resources. This initial DAS data was acquired as a "piggyback" to a walkaway VSP survey run with parameters set for efficient geophone recording. Issues related to the vibroseis source truck led to uncertainty in the sweep parameters used. Because of these issues and the limited number of sweeps used, we felt confident that significantly better iDAS results could be obtained. Despite the limitations of the initial test, conducted in March 2012, the results were encouraging and demonstrated that DAS data could be acquired with the MBM package fibers at Citronelle. The results from the 2012 test were reported in Daley *et al* [1]. Following discussions with project team members, we proposed to work with the Citronelle sequestration project team and Silixa to acquire new DAS data with the MBM system.

The MBM project was extended with a task of acquiring new iDAS data and determining the source effort (number of sweeps) needed to obtain signal-to-noise levels using Silixa's iDAS unit which would be comparable to the MBM geophone data.

Based on the results from the initial DAS survey, we designed the work to focus on a large source effort (64 sweeps per source point rather than the standard 4) at a limited number of locations (maximum of 4). The acquisition was designed to leverage a planned repeat of vertical seismic profile (VSP) acquisition using the MBM geophones. Following planning coordination between LBNL, Silixa, SR2020 (the VSP acquisition contractor), and the SECARB-Citronelle team, the data acquisition was conducted in August 2013.

Data Acquisition

The primary focus for testing on SP 2021 (Figure 20), which showed better data quality in the 2012 test (Figure 21).

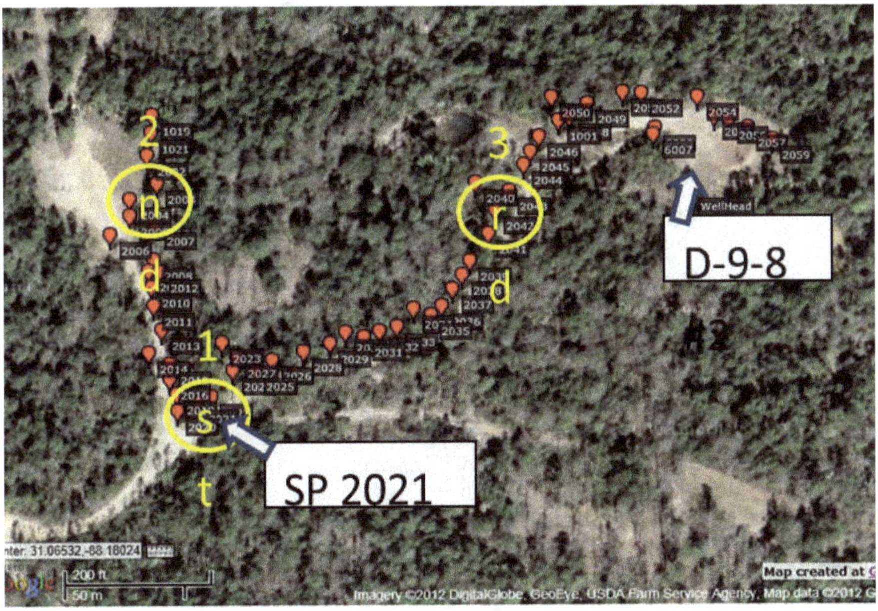

Figure 20. Location map of Citronelle MBM DAS testing. Well with MBM is D-9-8 #2. Shot points for VSP are shown in red, with three DAS test focus points shown in yellow labels 1, 2 and 3. SP 2021 was the initial primary test location (yellow circled 1).

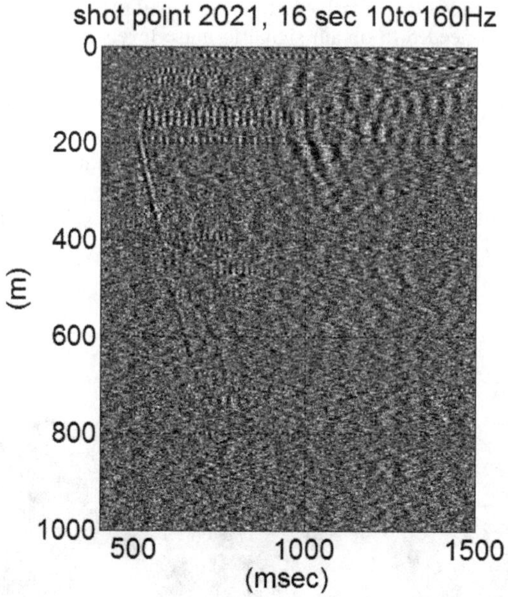

Figure 21. DAS data from SP 2021 in March 2012 survey. Note that only the top 1 km of data is shown since the deeper data had no observable body waves.

We planned that the maximum number of sweeps at each SP would be determined in the field, based on near real-time analysis of stacked data by Silixa. In the 2012 test, real time analysis was not possible because the iDAS was operated in continuous recording mode, with GPS timing used to 'cut' the vibroseis data out of the continuous records in post-processing. For the 2013 survey the iDAS system had a triggering box, and SR2020 provided a +5V, 20ms width TTL pulse trigger to the Silixa iDAS system. Silixa also provided a GPS antenna and timing module for the iDAS, while SR2020 has GPS timing hardwired in the doghouse, and on the vibroseis truck (a Mertz model 35 with about 35,000 lbs force).

In triggering mode, the iDAS required a 5-second 'wait time' between records in order to be ready for the next incoming trigger. The remote triggering of the iDAS system appeared to work well and in-field analysis of stacked data by Silixa was achieved. Note that data files were not supplied to LBNL in the field, and were not in a standard format such as SEG-Y.

For best performance, the iDAS system records at a sample-rate higher than the output geophysical records (typically x10, i.e. 10 kHz sampling for 1 kHz output data). While newer iDAS systems can perform the down sample step in real time at acquisition, the one that was available for this survey, did not. This meant that only the first 20 or so records were available for on-site analysis.

Silixa chose to test four internal iDAS settings ("D50", "D30", A50", "A30") with a single Vibroseis sweep (12-110 Hz, 16 sec linear) used throughout. These sweep parameters are somewhat different than those used for the storage monitoring geophone VSPs, however, all sweeps used for iDAS testing were recorded on the MBM geophone system. The MBM geophone data was recorded on a DAQlink III recording system, made by Seismic Source. The iDAS data acquired is summarized in Table 1.

Because of the number of sweeps recorded at each SP, the vibroseis truck had to move slightly to prevent road damage or coupling issues from too deep of a 'footprint' from the vibrator baseplate. The vibe moved up one pad width for each set of 20 or 24 sweeps, making a 3x1 grid at each SP. For a second set of 64 sweeps at the same SP, the vibe moved and lined up another 3x1 grid parallel to the first. For SP2021, the vibe made a 3x4 grid of padmarks (256 sweeps) from one side of the road to the other. For SP2003, the vibe made a 3x2 grid (128 sweeps). For the third shotpoint the vibe made a 3x1 grid (64 sweeps). Figure 22 shows the relative location of baseplate footprints from SP 2021.

Table 1. Number of sweeps at each shot point.

Shot Point	D30	D50	A30	A50	Total Sweeps
2021	64	88	64	64	280
2003	129				129
2040	20				20
2041	44				44

Figure 22. Effects of vibroseis sweeps at SP 2021. Multiple sets of sweeps were recorded together with the first and last set 'pad marks' labeled with yellow arrows. Each pad mark had 20 or 24 sweeps.

Initial Data Analysis (2013)

Figure 23 shows DAS data from ~2740 sensor segments, each ~1 m (3.3 ft) long, from well head to reservoir. We see that the data is much improved from the 2012 acquisition (Figure 21). The known differences are certainty of triggering, verified sweep parameters with a narrower frequency range, increased number of stacks (20 vs 4) and a newer iDAS recording system. Figure 23 demonstrates that the MBM deployed fiber can obtain useful VSP data.

A primary attribute of DAS data acquisition, as compared to traditional geophones, is the large spatial sampling at small intervals. For comparison, the MBM geophones (18 sensors spaced 15 m (50 ft) between 6000 and 7000 ft depth) are shown in Figure 24 at the same scale as the DAS data in Figure 23.

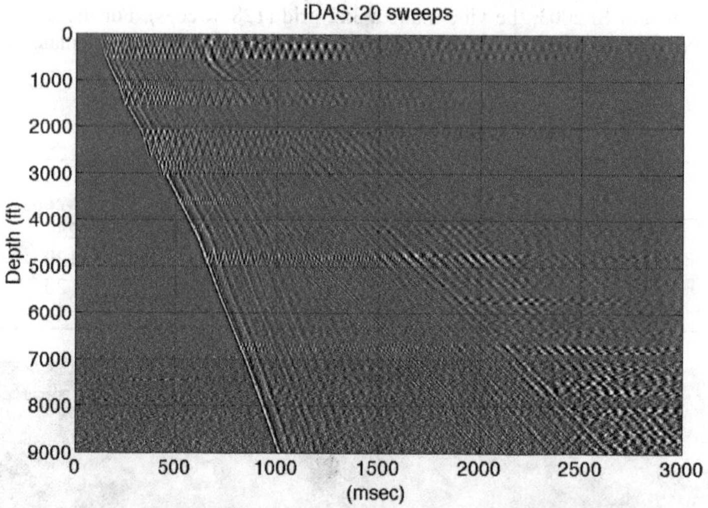

Figure 23. A stack of 20 sweeps of iDAS data from SP 2021.

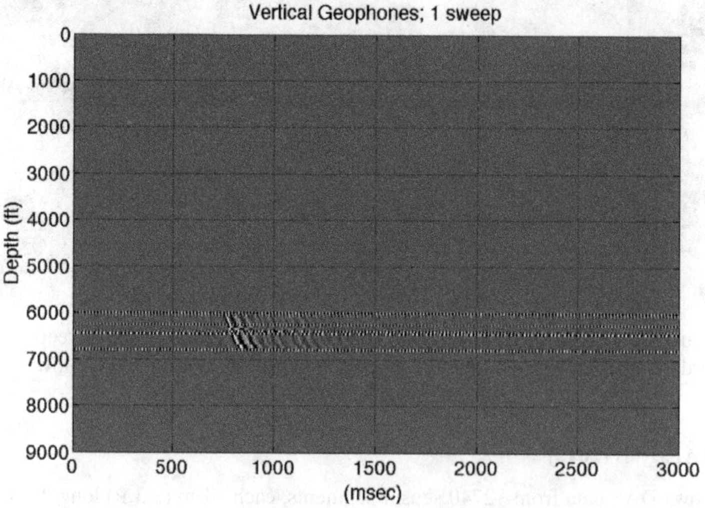

Figure 24. A single sweep of geophone data for SP 2021.
Geophone data displayed at same scale as DAS data in Figure 23, for comparison.

One advantage of the small spatial sampling is the ability to average over larger intervals to improve signal-to-noise ratios (SNR). Figure 25 shows that while a single DAS channel has poor SNR compared to a geophone, averaging DAS data over the +/- 50 ft spacing between geophones gives comparable SNR. Similarly, SNR for individual DAS channels is improved by source stacking, as shown in Figure 26 for all channels and Figure 27 comparing a single channel to a geophone.

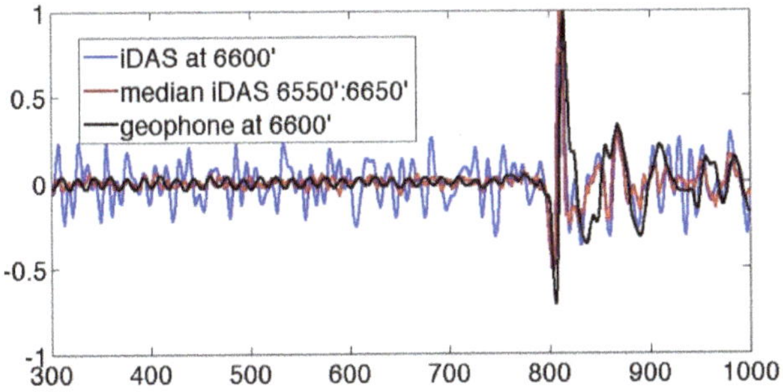

Figure 25. Comparison of a single iDAS channel (blue) with a single geophone (black) and the median average of the iDAS channels spanning the 50 ft interval between geophones (red).

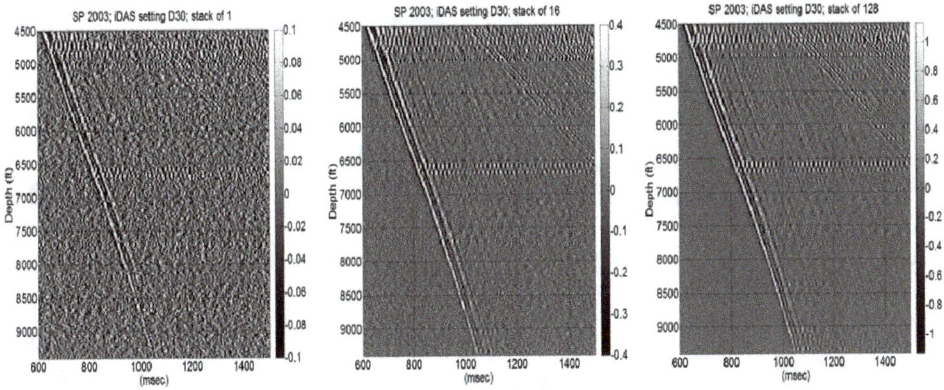

Figure 26. Comparison DAS stacking for SP 2003 VSP.
Shown are 1, 16 and 128 stacks for SP 2003.

Figure 27 indicates that a stack of 129 sweeps on the DAS system is close to the SNR of 24 sweeps into the MBM geophone. This gives an important conclusion that 5-6 times the source effort is needed for DAS VSP to equal geophone VSP data, within the MBM system at Citronelle.

A key issue is phase and frequency response of DAS compared to conventional geophones. Figure 28 shows the frequency response of a 20 sweep stack of a single geophone and the scaled filtered DAS recording after application of a bandpass filter. A single DAS channel matches well the spectral response of a geophone following a bandpass filter to match the approximate sweep parameters.

The similarity of phase response can be seen in Figure 29 which shows the time series for single channels of DAS and a geophone. This similarity is indicating that the DAS system is measuring velocity (or, more specifically, strain rate along its axis).

The overall quality of iDAS recording can be seen in Figure 30 which shows the entire ~9700 ft data set for the 128 stacked sweeps at SP 2003. Many interesting features can be observed in Figure 30. Notable are multiple zones of 'ringing' (reverberant events trapped between two depths, like a wave guide). We hypothesize that these events are related to waves propagating in the steel casing and may indicate zones of reduced cement bond. For example, between 2145 ft and 3140 ft, in Figure 30, the waves can be seen to have initial downgoing segment with faster apparent velocity than the P-wave.

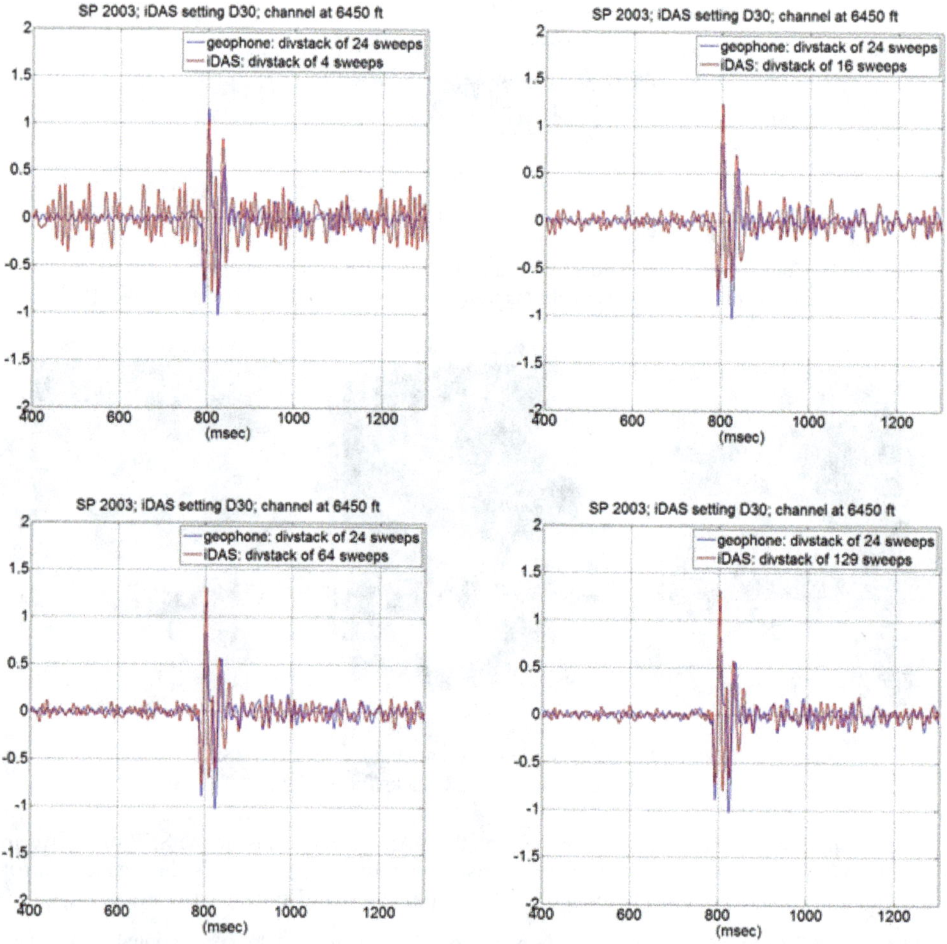

Figure 27. Comparison of a single vertical geophone stack of 24 sweeps (blue) with a single DAS channel at the same depth (6450 ft). The four panels show DAS stacks of 4, 16, 64 and 129 sweeps, as labeled. All data collected with the source at SP 2003 (SP number 2 on Figure 20).

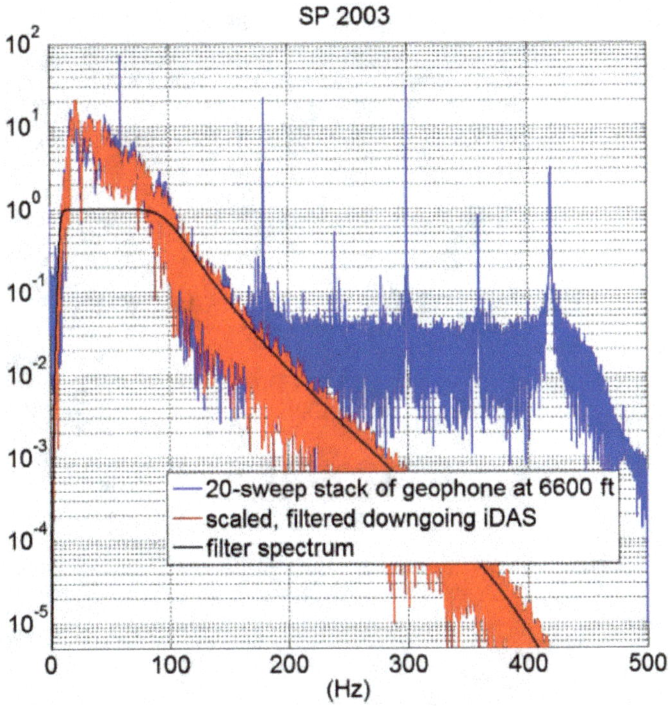

Figure 28. Spectral response of a 20 sweep stack of a single geophone (blue), iDAS channel with bandpass filter (red) and the filter (black).

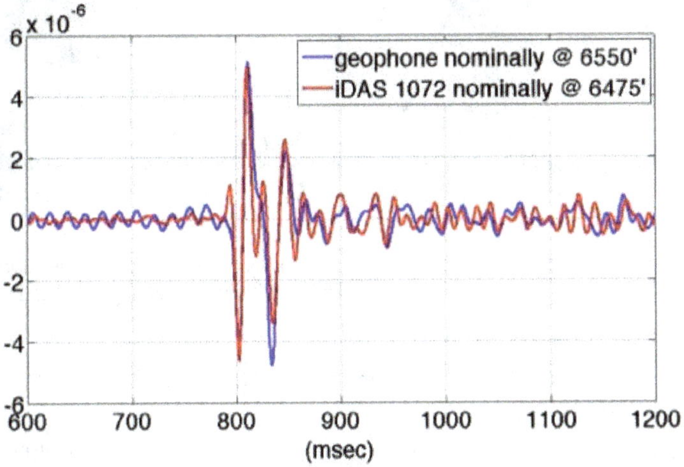

Figure 29. Comparison of a single geophone at 6550 ft depth and the iDAS channel most closely matching (nominally 6475 ft depth).

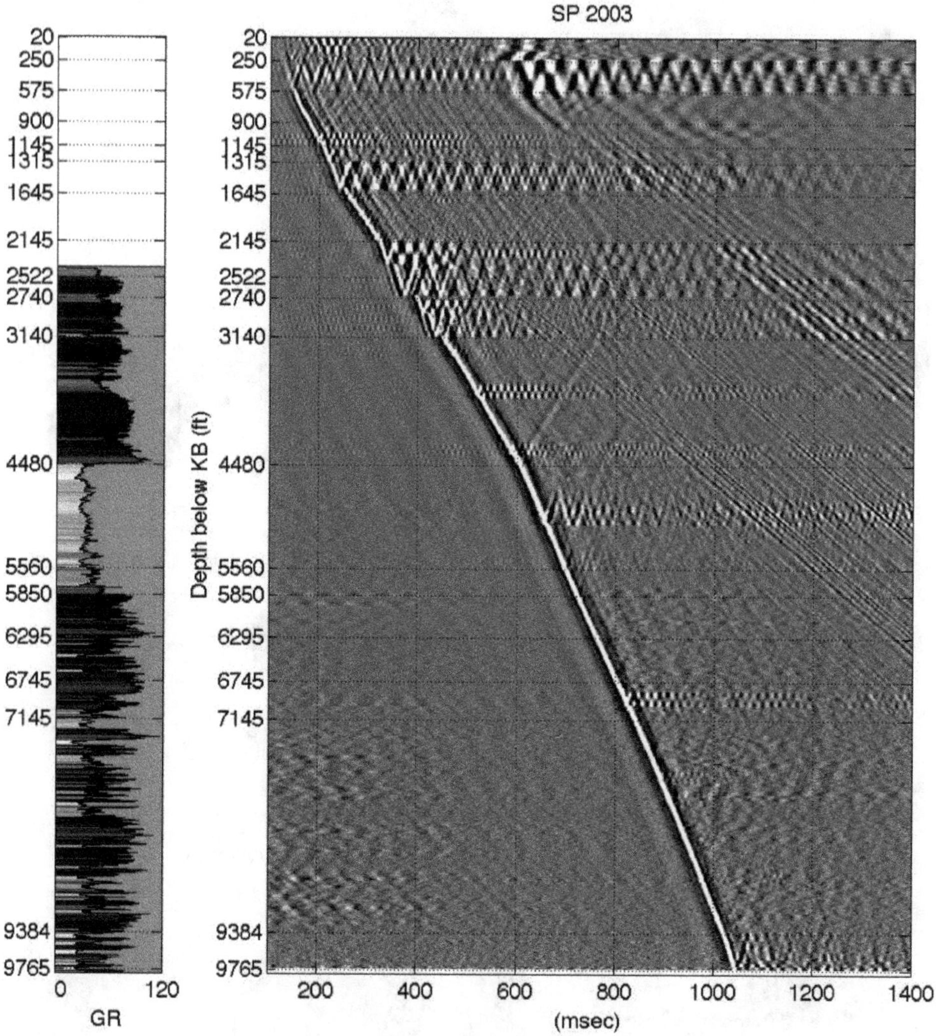

Figure 30. DAS data from SP 2003 (right) for stack of 128 source sweeps, along with gamma ray log data (left), and depths associated with velocity changes in the DAS data (lines and labeled depths on Gamma Log).

At the end of the initial DAS data processing and analysis extension of MBM work, several conclusions were reached:

- Excellent VSP results were obtained from three test shotpoints and different iDAS internal settings. Improvement from the 2012 test was clear and unambiguous.
- Standard improvement from repeated stacking of the source was observed (i.e. DAS noise is largely random and the signal is repeatable). SNR comparable to the MBM geophones was obtained with at least 6 times the source effort (24 to 64 sweeps for Citronelle). This is a key conclusion of our test.

- Source downgoing signature varies with source location (as typically observed). The spatial sampling of iDAS is ideal for deconvolution design using the downgoing wavefield. This is an important advantage of DAS over geophone arrays with limited length (such as the MBM array).
- With our acquisition parameters, 64 sweeps can be done in 30 min (or about 2 SP per hour).

Updated DAS Analysis (2014)

Following completion of the MBM DAS project with the initial data processing described above, work on the Citronelle DAS data continued by Silixa Ltd and LBNL. For completeness, the following updated analysis reports on the work performed outside of and beyond the scope of the original MBM project.

Signal and Noise

The processing and analysis of distributed sensor data is fundamentally different than point sensors such as geophones. Upon detailed study of the Citronelle data set, we realized that the experimental goal stated above, to compare source effort needed for comparable SNR for DAS and Geophones, is not the same as comparing two types of point sensors. Furthermore, since the physical property measured by a DAS system is different from a geophone, the properties of the 'noise' in the SNR are also different for DAS and geophones.

Most borehole seismic tools currently are constructed using geophones (sensors of current generated by the motion of a coil in a magnetic field) that are idealized as sensitive to components of the local particle velocity of the medium at the point where the tool is clamped. In contrast, the iDAS is effectively continuous spatially as well as temporally. In its native format (as recorded at Citronelle) the iDAS is designed to produce an output such that each sample, indexed by axial location along a cable's fiber core (the sampled 'channel') and recording time interval (the sampled 'time'), represents the average change in fiber elongation during the corresponding time interval between points that are a reference distance apart and centered at that channel location. Thus if $u(z,t)$ represents the dynamic displacement from rest position of the fiber at axial location z and time t, the iDAS output is an estimate of:

$$\left[u\left(z+\frac{dz}{2},t+dt\right)-u\left(z-\frac{dz}{2},t+dt\right)\right]-\left[u\left(z+\frac{dz}{2},t\right)-u\left(z-\frac{dz}{2},t\right)\right]$$

where dz and dt are the reference distance and sample rate respectively.

As such, the iDAS output represents an estimate of the fiber strain-rate $\frac{\partial\partial u}{\partial z\partial t}$ as calculated by difference operators applied in time and axial dimensions.

For comparison with geophone data, this may also be viewed as the running average (in both space and time) difference between point velocity measurements at positions separated by the reference distance. This appears to be a common design feature of DAS systems [7][11].

When comparing signal and noise for data recorded by the iDAS unit, the useable signal captured from the native output is typically limited by broadband noise that is inherent in the optical scattering process upon which the system depends. Because the system response is linear and coherent in the dynamic local strain, repeated stacking of iDAS traces over repeated shots is expected to result in a SNR improvement following the inverse square root relationship between SNR and number of stacks. However, different from geophone sensors, detailed analysis of the DAS optical scattering statistics has shown that simple stacking is far from optimal in recovering weak signal in the presence of this noise.

In its native format, the iDAS strain-rate measurement is limited by broadband speckle noise that is neither spectrally white nor uniform in its distribution with respect to channel index or recording time. Using conventional descriptions of optical scattering and proprietary knowledge of the iDAS opto-acoustic demodulation carried out within the iDAS, Silixa scientists have recently developed a noise-adaptive rebalancing method that combines an optimally weighted averaging over repeat measurements with a rebalancing of the temporal spectrum to create an output signal that, to good approximation, represents dimensionless axial strain (not the native strain-rate) of the fiber core plus broadband noise that has a flat temporal spectrum and a uniform power distribution with respect to channel index and recording time.

Figure 31 shows the result of applying this rebalanced stacking (followed by correlation with the sweep) for a representative subset of the Citronelle data and compares it with the result of a simple stack and correlate processing.

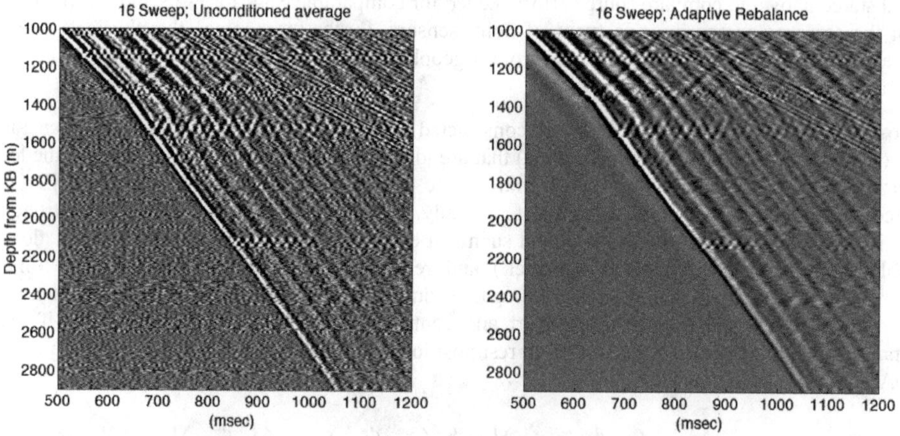

Figure 31. A 16 sweep stack of DAS for SP 2021, comparing (left) stacking of native DAS data with (right) DAS data with adaptive noise reduction, rebalancing and stack.

Figure 32 compares this rebalanced result with data from the 2012 survey (and processing) at the same location. The data quality in Figure 32 is greatly improved from the 2012 DAS acquisition shown in the inset (and in Daley *et al* [1]). In addition to the rebalancing operator, the new data benefitted from certainty of triggering, verified sweep parameters with a narrower frequency range, increased number of stacks (16 vs 4) and a newer, improved iDAS recording system.

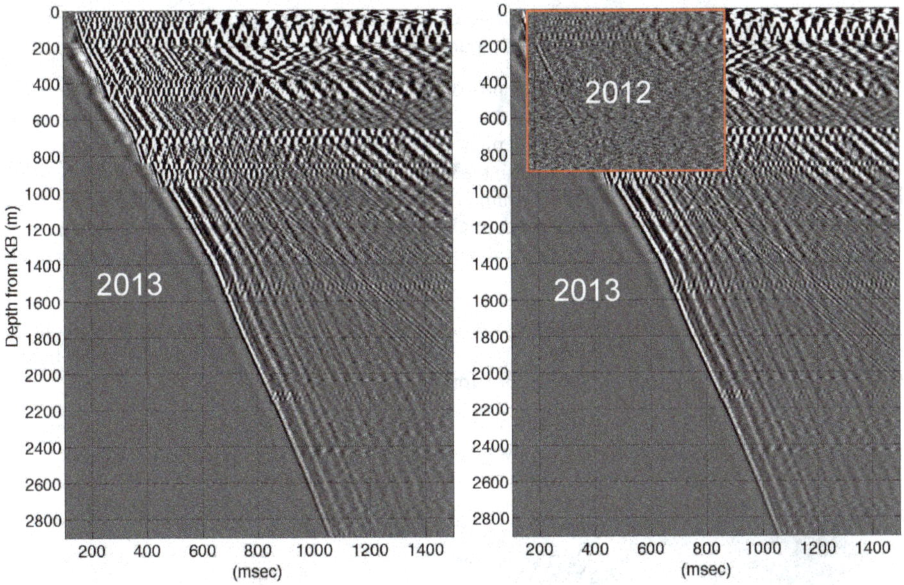

Figure 32. Correlated DAS data for SP 2021, comparing 4-sweep stack from the April 2012 survey (right inset) with a 16-sweep optimized stack from the August 2013 survey (left).

Conversion to Geophone Equivalent Signal

An important issue for use of DAS is relative response to industry 'standard' geophones. The MBM deployment provides a platform for direct comparison, with caveats for the clamping difference described above and the fundamental difference of distributed and point sensors. Since the rebalanced DAS recording is a local strain, comparison to a geophone (which measures local particle velocity) requires conversion.

First, consider propagating seismic signals, such as a harmonic plane wave. Strain, displacement and particle velocity are related as follows (e.g. [12]).

For ε_{zz} = extensional strain in the z direction, and u_z = displacement in the z direction with velocity c, where $u_z = U\, e^{-i\omega\,(t-z/c)}$ and $v_z = du_z/dt = U\,(-i\omega)\, e^{-i\omega\,(t-z/c)}$ is the axial particle velocity, it follows that $\varepsilon_{zz} = du_z/dz = v_z/c$.

However the relationship is more general and applies to any propagating disturbance with a stable phase function. Again writing $u(z,t)$ for the dynamic fiber displacement, a stably coupled propagating disturbance will be self-similar under suitable translation in space and time. That is, it will take the form $u(z,t) = u(\varphi)$ where $\varphi = (t_0 + t \pm z/c)$ is a characteristic phase function with propagation speed c.

Differentiating with respect to time and distance respectively, we obtain the fiber particle velocity

$$v = \frac{\partial u}{\partial t} = \frac{\partial u}{\partial \varphi} \text{ and the fiber strain } \varepsilon = \frac{\partial u}{\partial z} = \pm 1/c\, \frac{\partial u}{\partial \varphi},$$

Comparing these equations, it is evident that $c\,\varepsilon = \pm v$. That is, the ratio between fiber particle velocity and fiber strain is given by the propagation speed along the fiber cable (apparent velocity) with a sign determined by direction of propagation. In general, the total fiber displacement, velocity and strain may be the superposition of multiple events and the propagation may be dispersive (i.e. propagation speed may depend on temporal frequency).

In-situ coupling of the fiber cable to waves propagating in the earth is also an important factor that can affect scaling DAS data to earth movement. It is beyond the scope of the present paper to investigate the details of how to combine plane wave decomposition and models of mechanical coupling to rescale data from complex fiber installations. In our case the data appear to be consistent with the simple assumption that the fiber strain and the geophone coil-to-casing velocity are faithful transducers of the corresponding environmental strain and velocity. With this assumption, the rebalanced DAS signal is converted to equivalent geophone signal by multiplying the dimensionless strain by the local propagation speed.

In our VSP data the vertical propagation speed across the zone covered by the geophones (1829 m to 2088 m) is approximately 3500 m/s. We have used that value to rescale our noise-reduced, rebalanced iDAS strain values to velocity units for the uncorrelated data.

Following this velocity conversion the Citronelle DAS data and SNR can be directly compared to the Citronelle geophone data. Figure 33 shows the uncorrelated DAS-geophone comparison.

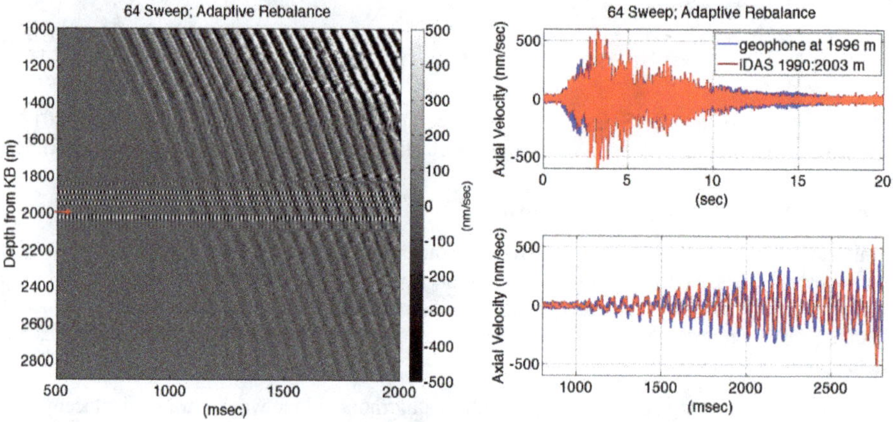

Figure 33. (A) (left) SP 2021 DAS data uncorrelated, noise-reduced, rebalanced and velocity-converted for a stack of 64 sweeps, shown with ~2 m channel spacing. In the zone covered by the geophone array (1829m to 2088m), the DAS data is overlain by geophone records. The geophone data have a stack of 4 sweeps, and the 60 Hz electrical noise from some geophones is easily seen. (B) (top right) shows a full 20 s of data for a single geophone at 1996 m (blue) and the DAS data summed over a 13 m interval centered at the geophone. (C) (lower right) Same as (B) but with the data having a zoom view of 500-3000 ms. Note that the DAS and geophone data have been independently converted to true velocity units (nm/s) and only normalized by the number of sweeps.

Correlated data is shown in Figure 34, along with spectral analysis of signal and noise for the noise-reduced, rebalanced, velocity-converted iDAS data. We display uncorrelated data in true velocity units (nm/s); while for correlated data, following industry convention, we have normalized the correlated data. (Note: for a sweep of amplitude A and length T the correlated amplitude is $A^2T/2$, but this is typically not removed as many sweeps are correlated with a synthetic pilot signal.) For our data, we have set the sweep autocorrelation equal to 1 and then divided by 3500 m/s, yielding units for correlated data that are dimensionless nanostrain.

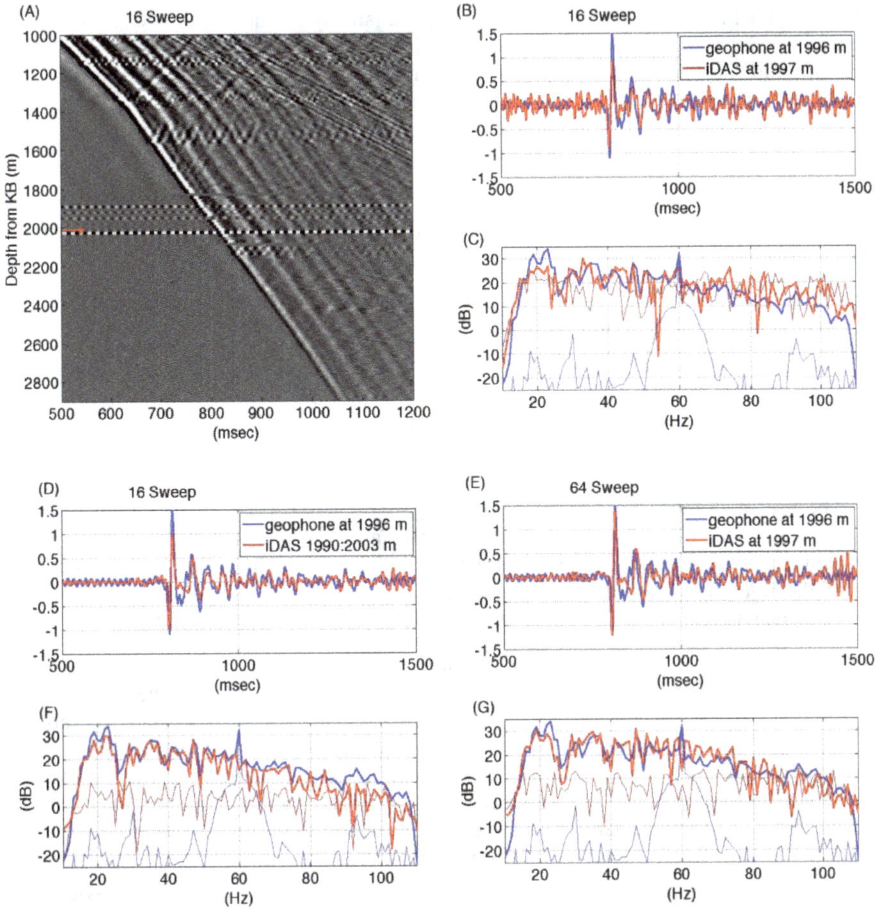

Figure 34. All panels show DAS data noise-reduced, rebalanced and correlated with a normalized sweep. The geophone data is a stack of 4 sweeps and has been correlated with the same normalized sweep and then divided by 3500 m/sec to give comparable dimensionless nanostrain. Panels show DAS data with a stack of either 16 sweeps (A:D, F) or 64 sweeps (E, G), and correlated geophone data with a stack of 4 sweeps. (A) Comparison of about 1900 m (950 channels) of DAS data with about 260 m (18 channels) of geophone data inserted at their depths. (B) Data for a single geophone at 1996 m depth (indicated with red arrow in (A)) and the DAS data from the nearest 2 m long channel (center at 1997 m). (C) Spectra for DAS and geophone data of (B) for signal (dark lines) and noise (light lines). (D) and (E) show a comparison of single geophone to (D) DAS summed over a 14 m interval centered at the geophone and (E) data from the single nearest DAS channel. (F) and (G) show spectra for the data in (D) and (E) respectively. For the spectral plots, 'signal' in dark curves is calculated from data in a 500-1500 ms time window (peak seismic wave amplitudes), while 'noise' (light red and light blue curves) is from data in a time window of 8000-9000 ms (after seismic waves have decayed to minimal amplitudes).

By calculating the mean RMS noise levels we can quantitatively evaluate DAS SNR as a function of stacking fold. Using uncorrelated data, and a time window of 200-7000 ms for depths 2000-2800 m we find the following: 4 sweep have RMS noise of 186 nm/sec, 16 sweep have rms noise of 89 nm/sec and 64 sweep have RMS noise of 43 nm/sec. This is about 6.4 dB for each factor of 4 in sweeps (with a 'theoretical' expectation of 6 dB). We believe the additional 0.4 dB comes from the adaptive rebalancing using larger number of data sets (sweeps).

Similarly, DAS noise is reduced by sampling larger spatial intervals. We find the decrease in mean rms noise in the 16 sweep data upon resampling from 2 to 8 m output is 5.5 dB. (from 89 to 47 nm/sec nrms).

The data of Figure 34, which compares a vertical geophone to DAS data converted to axial velocity, demonstrates a number of comparison observations.

1. DAS data can match the spectral response of a geophone, within the source bandwidth used at Citronelle (see Figure 34C).
2. With independent processing and nominal gains, there is clear similarity of amplitude and phase response (i.e. the time series) between geophone and DAS (see Figure 34B, D and E).
3. The use of DAS spatial sampling to average over larger intervals to improve SNR. Improvements in the signal processing discussed above have enabled us to achieve SNR results that are effectively equivalent to the geophone data using a small portion (15 m) of the DAS data. Alternatively, DAS data with fine spatial sampling (~2 m) can be equivalent to geophone data with extra source effort (see Figure 34D and E).

An attribute of DAS seen in Figure 34, as compared to geophones, is the lack of sensitivity to electrical noise. The noise in the geophone data is dominated by 60 Hz (power line) electrical noise, most visible at depths where geophone wiring has electrical leakage to ground. We also have observed a difference in sensitivity to tube-waves (borehole interface waves) such that the geophones which are actively clamped to casing and decoupled from the tubing string show reduced tube-wave sensitivity compared to the fiber cable in the flatpack which is strapped continuously to the tubing. While geophone clamping is known to reduce tube-wave sensitivity, the uncertain coupling of the fiber cable (described above) makes quantitative comparative analysis difficult. Clearly, coupling of the fiber cable should be actively considered in DAS survey design, just as geophone coupling is considered. As pointed out in Daley *et al* [1] the observation of two distinct tubewave speeds indicates a tubewave coupling between the fluid-filled annulus and the central portion of the tubing.

VSP Analysis

We can now compare data in the context of the goals of a typical VSP survey – imaging the subsurface. Imaging is improved with increasing spatial coverage and sampling. A primary attribute of DAS VSP data, as compared to traditional geophones, is the large spatial coverage at small sampling intervals. Figure 32 demonstrated that the MBM tubing-deployed fiber can obtain useful VSP data over nearly 3 km of borehole from a reasonable source effort (16 vibroseis sweeps). For comparison of DAS to the MBM geophones (18 sensors spaced ~15 m (50 ft) between 6000 and 7000 ft depth), the vertical geophone data is inserted in a DAS data plot (Figure 34). Figure 34 shows both the match of data phase and SNR described above and the much greater spatial sampling achieved by DAS from a single source stack. While the MBM tubing deployment has many fewer sensors than a typical temporary wireline-deployment geophone VSP, semi-permanent geophone arrays are often limited to 10-20 levels as in the MBM.

Depth Estimation

Both precision and accuracy in sensor depth are important for VSP analysis. DAS depths are measured via the speed of light in the fiber (e.g. [1]). However, DAS data has inherent depth uncertainty due to extra fiber length (EFL) installed in the fiber encapsulation to prevent fiber breaking due to differential stretch of glass fiber and steel encapsulation. Additionally, the fiber cable inevitably has some spiraling on the tubing, while the tubing itself has some spiraling during deployment, both of which add length and reduce certainty with regard to the physical depth increment per channel. Moreover, the accuracy of the assigned depths depends upon knowing one or more physical locations where the corresponding DAS channel is confidently determined.

For Citronelle we used two reference depths for calibration – the surface well head and the packer. The packer depth below surface was determined via measurement of tubing joint lengths and its location relative to the fiber was confirmed with DTS and heat-pulse analysis at 2873 m. Previous use of the MBM for distributed heat-pulse studies [13] had located well bore completion components (such as the top of packer and the end of the flatpack) with a precision of about 0.25 m. The distributed temperature sensing (DTS) system used in this study has higher spatial resolution than DAS (about 0.15 m) while still measuring the fiber length. By assigning the packer depth to the DAS channel observed to have tube-wave reflections coming from the packer, we fixed a deep DAS channel at a known depth. A shallow DAS channel was fixed by a 'tap test' on the wellhead (just above ground level). Dividing the depth difference with the number of channels, we were able to estimate the distance per channel and compare that result to the true fiber length from optical measurement. The result was an EFL of 0.74%, giving a DAS channel spacing of 2.033 m rather than the nominal (straight) length of 2.048 m. Similarly, the DAS data depth can be calibrated by matching observed DAS reflection depths with well log measured property change, such as a reflection from a sonic velocity change at ~1360 m. Thus, the complete sampling of the well with fine spacing allows the DAS data set to fully address the problem of depth matching for DAS channels.

Deconvolution

A standard component of VSP processing is designing a deconvolution operator based on the downgoing wavefield [14]. The fact that DAS VSP data will typically cover the entire well increases the precision of the downgoing decon operator design, leading to improved quality of deconvolved data. We have applied a downgoing deconvolution based on [15]. The overall quality of deconvolved DAS VSP data is shown in Figure 35 which has nearly the entire ~2900 m data set. Comparison of Figure 32 (or Figure 34A) with Figure 35 shows the effect of the downgoing deconvolution in removing multiples. Note that the deconvolved wavelet is zero-phase and, with the reduced DAS noise level, has visible side lobes before the first arrival.

Interesting features can be observed in Figure 35. Notable are zones of 'ringing' (reverberant events trapped between two depths, like a wave guide). We hypothesize that these events are related to waves propagating in the steel casing and may be related to a lack of cement bond at these locations. For example, between 653 and 836 m, in Figure 35, the waves can be seen to have initial downgoing segment with faster apparent velocity than the P-wave, indicating propagation at least partly in non-formation material (likely steel or cement). Another feature demonstrated in Figure 35 is the depth match between well-log measured interfaces and the reflected events observed in the VSP, e.g. the reflection event at 1360 m.

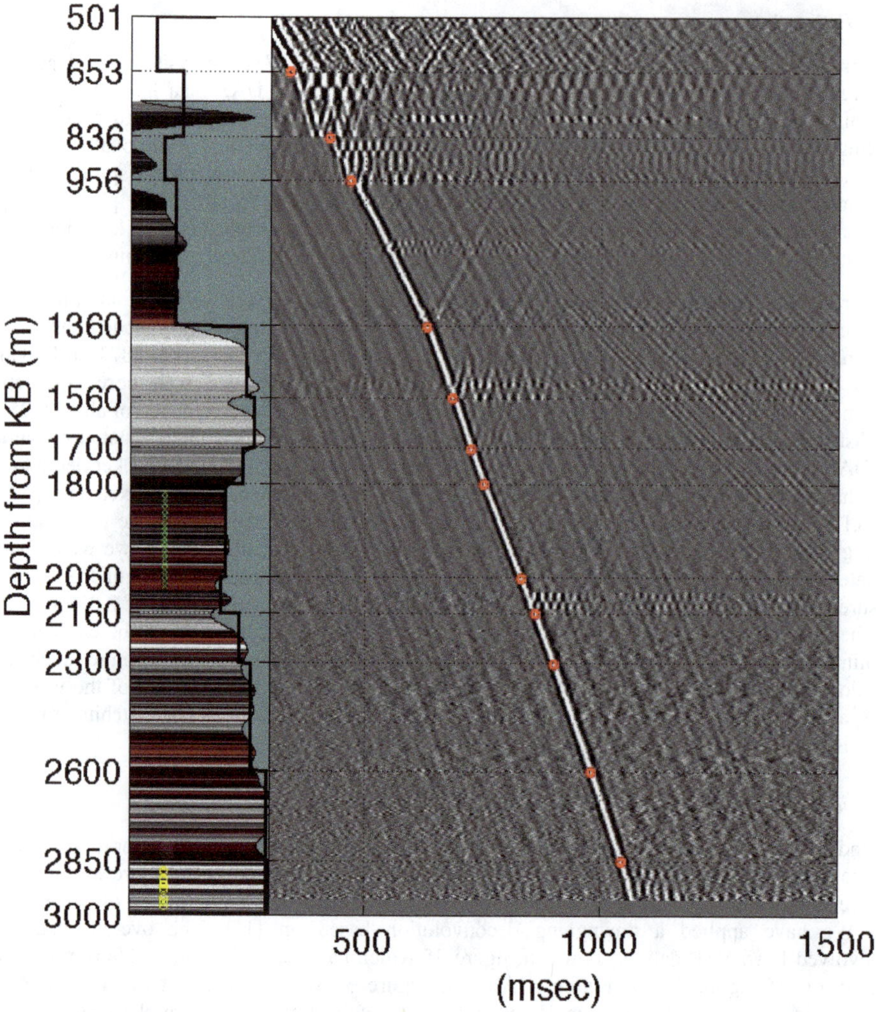

Figure 35. Deconvolved DAS data for a stack of 16 source sweeps, along with sonic velocity log (left side) colored by gamma-ray, and a blocked velocity model from the DAS first arrival times (solid black line), and depths associated with major velocity changes in the DAS data (red dots).

VSP Reflections

Following conventional VSP processing, including deconvolution, the DAS data can provide upgoing reflectivity, which is typically one primary goal of a VSP survey. The processing for reflections included the following operations: deconvolution, time shifting to two-way travel time, smoothing with a median filter and bandpass filter. These operations were applied to both DAS and MBM geophone data, and the results are compared in Figure 36. Increased coherency of reflections are observed in the DAS data. Also shown in Figure 36 is a corridor stack [14] and well log data. Increased vertical extent of DAS reflection data, above the geophones, is seen by comparing the two corridor stacks in Figure 36.

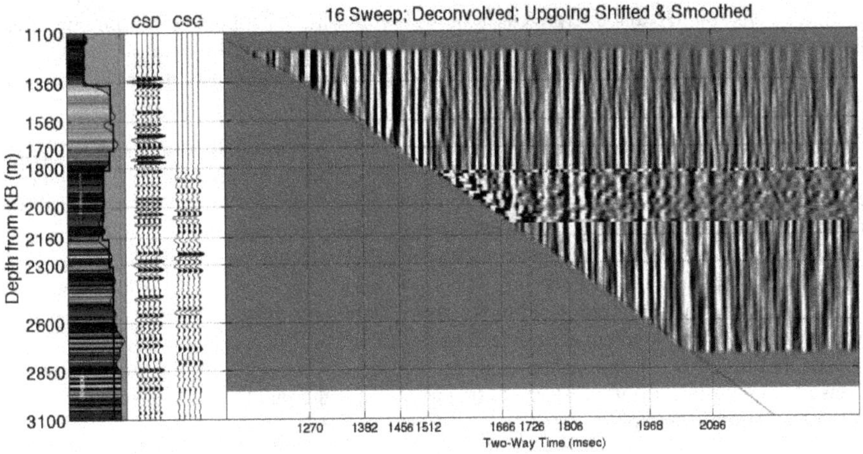

Figure 36. Upgoing reflections from DAS VSP (right) with MBM geophone data inserted, along with corridor stacks for both DAS (labeled CSD) and geophones (labeled CSG). The far left panel shows sonic log data colored by gamma-ray, and a blocked velocity model from the DAS first arrival times. The reflection panel shows depths associated with major velocity changes in DAS data indicated by red dots and DAS picked travel times (red line).

Summary

A modular borehole monitoring system was designed and successfully deployed in a 2900 m well for CO_2 monitoring. The tubing-deployed MBM system provided a platform for simultaneous acquisition of clamped geophone and DAS VSP data, allowing direct comparison of two sensor types. Excellent VSP results were obtained from three test source points. Improvement in DAS data quality from the initial 2012 test shown in Daley *et al* [1] was clear and unambiguous. Improvement was seen initially due to improved recording procedures and DAS acquisition hardware. Further improvement was gained from DAS processing.

We have described the native measurement in the iDAS unit as localized strain rate. Following a processing flow of adaptive noise-reduction and rebalancing the signal to dimensionless strain, standard improvement from repeated stacking of the source was observed (i.e. the remaining DAS noise is temporally flat and uniform between channels while the signal is repeatable). Conversion of the rebalanced signal to equivalent velocity units allows direct comparison of DAS and geophone data. We see a very good match of uncorrelated time series in both amplitude and phase, demonstrating that velocity-converted DAS data can be analyzed equivalent to vertical geophones. Comparison of SNR between distributed sensors and point sensors (i.e. DAS and geophones) can be done in various ways. For a single ~2 m DAS channel, we find SNR comparable to the MBM geophones is obtained with about 16 times the source effort (4 vs 64 sweeps), implying about 24 dB greater sensitivity for the casing-clamped geophone than the tubing-clamped fiber in flatpack. However, using a 15 m section of DAS cable (the distance between geophones) centered at a geophone, comparable SNR is obtained with about 4 times the source effort (4 vs 16 sweeps), or about 12 dB greater sensitivity for the clamped geophone. These are key conclusions of our test.

The DAS recordings were processed for VSP reflectivity, including downgoing wavefield deconvolution and generation of corridor stacks, both of which are improved by the fiber cable's large spatial coverage. This is an advantage of DAS over geophone arrays with limited length (such as the Citronelle MBM array). Following depth corrections, the reflectivity was shown to have very good correlation to intervals interpreted from well log data. The DAS recordings also appear sensitive to well completion, with zones of trapped energy interpreted as related to casing bond. This observation requires further dedicated study and has potential use for wellbore integrity studies.

We have described many fundamental attributes of DAS VSP data and compared data quality of tubing-coupled fiber cable to casing-clamped geophones. During testing at the Citronelle site, we have seen improvement in the specific DAS recording system used (the iDAS) and expect further improvements to DAS technology. An important observation is that the cost and effort of the clamped geophone deployment was far more than the fiber cable deployment. Therefore, the extra seismic source effort currently needed for tubing-deployed DAS seems to be well-compensated by the benefit of extra spatial sampling and lower deployment cost. Further study and improvement in DAS technology and deployment should lead to further gains in a cost-to-benefit ratio. Given that the proven lifetime of fiber-optic installations can be 50 years or more, for the application of long-term seismic monitoring of carbon sequestration as well as other applications, DAS VSP appears to be a useful and promising tool.

One caveat to the expected long-life of optical fibers is that hydrogen darkening, which occurs at elevated temperatures when hydrogen can rapidly permeate the protective stainless steel tubing encapsulating the optical fibers, can substantially diminish their useful lifetime. The amount of hydrogen expected in the downhole environment needs to be taken into consideration for deep completions in hydrogen-rich hydrocarbon bearing units. Darkening of the fibers will over time lead to a decrease in the signal-to-noise ratio for all optical measurements. The impact of optical fiber degradation is more severe in DTS measurements, where the optical budget is less than for DAS, and darkening leads to not only an increase in noise but increased inaccuracy in the measurement estimates.

LESSONS LEARNED AND CONCLUSIONS

An overarching goal of this MBM project was development of a robust monitoring package. The fact that all of the instruments installed are still functioning as intended is a testimony to the robustness of the Citronelle deployment package. We consider the Citronelle MBM system to be a successful prototype and 'blueprint' for CO_2 storage monitoring. Still with each installation there are improvements to be made and lessons learned.

While the overall installation of the MBM proceeded smoothly without incident, there were some aspects of the installation that would benefit by modification and reengineering. Because most completion materials are "off-the-shelf," duplicating this installation would not require a repeat of the design engineering, which was an objective of the MBM concept. However, being off-the-shelf does not mean that all of the completion materials are stock items with minimal lead times for ordering. For example the control line joint protectors are manufactured specifically for the tubing used and the control lines being run. They can be considered semi-custom, because the actual overall design is standard, but some of the minor details are modified for each job. Since there are numerous vendors that can provide quotations and fabricate the equipment as part of their business operations, we consider these items to be off-the-shelf within the oil and gas industry. The tubing-deployed geophone clamp is one item that is not off-the-shelf, as it was designed at LBNL. Given our success in the operation of them, we are hoping that through technology transfer, the basic concept and design can be implemented commercially so that improvements to our initial design can be realized. Similarly, the geophone string was specified by LBNL and special ordered, but with the information given here multiple vendors could be approached to provide the product.

Some refinements which would improve the installation are quite minor, and resulted from doing things for the first time. The flatpack packaging of the multiple control lines was a success that greatly sped up the handling of the control lines and reduced the chances of a stray line being caught in the slips. However, it was part of the original work plan to remove approximately 150 ft of the polypropylene encapsulant so as to avoid having to perform fiber-optic splices before the recording equipment. This proved a task too difficult, as the flat-pack jacket resisted easy removal by cutting, grinding or any other method we tried. In retrospect we should have planned to only remove enough to safely go through the wellhead, with enough extra fiber-optic cable to permit multiple attempts to splice onto a surface conductor cable. Once the wellhead is in place and secure, more time is available for work on cables.

Another lesson is preparation for post-deployment wellhead cable slack. While we did not initially have a robust splice enclosure box to leave at the wellhead, a trip to the hardware store allowed us to procure an outdoor rated electrical enclosure that served that purpose. With planning, the outdoor enclosure would contain both the necessary electrical junctions as well as a standard fiber-optic splice box. Leaving an extra 2 m of coiled optical fibers would ensure that any future use would allow multiple re-terminations.

A more significant learning relates to deploying the geophone pods and clamps. The hydraulic clamped geophone mounts required a significant amount of time (rig time) to install due to numerous screws for the two half shells. We knew ahead of the field operation that it would be slow going with the clamps, but the timeline of the MBM program limited the amount of design engineering and testing before proceeding to final fabrication. Many ways to couple the half clamps together exist and should be considered before the next deployment. Also the stainless steel control lines (hydraulic and geophone wiring) need to be better managed at the clamp as they tended to extend out far enough so that they could contact the casing during the installation.

A further lesson learned relates to potential early use of the MBM system. At Citronelle, the final depth of landing the MBM and the location of the perforations did not match the original plan by the site operator. While it was determined that the perforations were off depth, it would have been possible to identify using diagnostic techniques, similar to those performed after installing the wellhead, to learn about the off-depth issue. Given that this style packer can only be set once, the idea of adding 12 hours of diagnostics time before setting the packer could have been considered. During all of our discussion on risk planning, off depth installation was not a risk considered. Even had the risk been identified, it would still have been difficult to plan ahead to perform the necessary diagnostics. There would have been a large crew mobilized, and the cost and difficultly of the rapid interpretation would have to be weighed against the potential benefit. For the MBM installation the necessary monitoring instruments were not in place yet, so the testing could not have been performed. However one could imagine having a temporary set-up containing a DTS unit, and power-supply for heat-pulse operations ready to go. All that would be needed would be to temporarily terminate the fiber-optic heat pulse cable, and then a short injection test would provide the relative separation between the completion string and the perforations. While it is easy to consider such an operation, the reality is more challenging, as the weather is variable, fiber-optic splices are best done when things are not rushed, and the well-pad is a difficult environment in which to operate because of all of the other equipment.

Finally, the successful testing of DAS VSP provided a useful lesson in flexibility of design. The use of fiber-optic cables for DAS testing was not in the early plans. Having multipurpose fiber packing (both multi-mode and single-mode) allowed us to take advantage of a DAS testing opportunity. Similarly, the tube-in-tube design of U-tube sampling was put to use for hydraulic activation of the geophone clamps. This type of monitoring flexibility was a goal of the MBM project.

In conclusion, the MBM deployment is considered a success. The pressure-temperature gauges are providing high quality data. The fiber-optic cable has been used for passive DTS, active DAS and active heat-pulse monitoring. The short geophone string is also working, although some of the 3C pod channels are not providing good data. Three WVSP surveys have been acquired for monitoring with the MBM geophones. The geophone string itself has since been redesigned by the manufacturer to eliminate the problem that led to the loss of channels. Finally, the U-tube is providing samples as intended from the reservoir that can be used to positively confirm the arrival of CO_2 and tracers. We expect that future deployments will certainly benefit from our efforts, particularly through the use of a highly engineered instrumentation flat-pack that can replace a large number of independent control lines.

ACKNOWLEDGEMENTS

We would like to specifically thank Kevin Dodds of the CO_2 Capture Project for his unwavering support of the MBM concept, development and deployment. We thank the SECARB team, including Jerry Hill of SSEB, Rob Trautz of EPRI, George Koperna and Dave Riestenberg of ARI and Gary Dittmar of Denbury. Acquisition of seismic data (geophone and DAS) was assisted by Dale Adessi of SR2020 and Michelle Robertson of LBNL. Thanks to Bjorn Paulsson and John Thornburg of Paulsson, Inc for fabrication and deployment support of MBM geophones. This work was supported by the CO_2 Capture Project, and performed by Lawrence Berkeley National Laboratory under Contract No. DE-AC02-05CH11231.

REFERENCES

1. Daley, T. M., Barry M. Freifeld, Jonathan Ajo-Franklin, Shan Dou, Roman Pevzner, Valeriya Shulakova, Sudhendu Kashikar, Douglas E. Miller, Julia Goetz, Jan Henninges, Stefan Lueth, 2013, Field testing of fiber-optic distributed acoustic sensing (DAS) for subsurface seismic monitoring, The Leading Edge 32, 6(2013); pp. 699-706, http://dx.doi.org/10.1190/tle32060699.1
2. Bostick, F., 2000, Field experimental results of three-component fiberoptic seismic sensors: 65th Annual International Meeting, SEG, Expanded Abstracts, http://dx.doi.org/10.1190/1.1815889
3. Keul, P. R., E. Mastin, J. Blanco, M. Maguérez, T. Bostick, and S. Knudsen, 2005, Using a fiber-optic seismic array for well monitoring: The Leading Edge, 24, no. 1, 68–70, http://dx.doi.org/10.1190/1.1859704
4. Mestayer, J., S. Grandi Karam B. Cox P. Wills A. Mateeva J. Lopez, D. Hill & A. Lewis, 2012, Distributed Acoustic Sensing for Geophysical Monitoring, 74th EAGE Conference & Exhibition incorporating SPE EUROPEC 2012 Copenhagen, Denmark, 4-7 June 2012
5. Miller, D., T. Parker, S. Kashikar, M. Todorov, and T. Bostick, 2012, Vertical Seismic Profiling Using a Fiber-optic Cable as a Distributed Acoustic Sensor, 74th EAGE Conference & Exhibition incorporating SPE EUROPEC 2012 Copenhagen, Denmark, 4-7 June 2012
6. Farhadiroushan, M., Parker, T.R., and Shatalin, S. [2009] *Method And Apparatus For Optical Sensing*. WO2010136810A2
7. Mateeva, A., J. Mestayer, B. Cox, D. Kiyashchenko, P. Wills, J. Lopez, S. Grandi, K. Hornman, P. Lumens, A. Franzen, D. Hill, J. Roy, 2012, Advances in Distributed Acoustic Sensing (DAS) for VSP, Society of Exploration Geophysicists Annual Meeting, Las Vegas, DOI http://dx.doi.org/10.1190/segam2012-0739.1
8. Shatalin, S. V., V. N. Treschikov, and A. J. Rogers, 1998, Interferometric optical time-domain reflectometry for distributed optical fiber sensing: Applied Optics, 37, 5600-5604
9. Hornby, B., F. Bostick III, B. Williams, K. Lewis, and P. Garossino, 2005, Field test of a permanent in-well fiber-optic seismic system: Geophysics, 70, no. 4, E11–E19

10. Marzetta, T. and M. Schoenberg, 1985, Tube waves in cased boreholes: 55th Annual International Meeting, SEG, Expanded Abstracts, 34–36, http://dx.doi.org/10.1190/1.1892647
11. Hartog, A. H.; Kotov, O. & Liokumovich, L.: The Optics of Distributed Vibration Sensing Second EAGE Workshop on Permanent Reservoir Monitoring 2013 – Current and Future Trends, 2013
12. Aki, K., and P. G. Richards, 2002, Quantitative Seismology, University Science Books, Sausalito, CA, cited page 635
13. Daley, T.M., Freifeld, B.M., Cook, P., Trautz, R., Dodds, K., 2013, Design and Deployment of a Modular Borehole Monitoring System at SECARB's Citronelle Sequestration Site, Twelfth Annual Conference on Carbon Capture, Utilization & Sequestration May 13 – 16, Pittsburgh, Pennsylvania
14. Hardage, B.A., 1985, *Vertical Seismic Profiling*, 2nd ed. Elsevier
15. Haldorsen , J. B. U., D. E. Miller, J. J. Walsh, 1994, Multichannel Wiener deconvolution of vertical seismic profiles, Geophysics, v 59, n10, p1500-1511, DOI:10.1190/1.1443540

Carbon Dioxide Capture for Storage in Deep Geological Formations, Volume 4
Karl F. Gerdes (Editor)

Chapter 38

FIELD PRACTICAL GUIDE TO ENVIRONMENTAL AND LEAK CHARACTERIZATION USING A PROCESS-BASED SOIL GAS MONITORING METHOD

K.D. Romanak[1], G.L. Womack, D.S. Bomse, and K. Dodds
[1]Corresponding author: Bureau of Economic Geology, The University of Texas at Austin University Station, Box X Austin, TX 78713-8924
[2]Formerly BP Corp. North America Inc. (currently Australian National Low Emissions Coal Research & Development, NFF House, 14-16 Brisbane Avenue, Barton 2600, Australia)

ABSTRACT: A process-based approach to soil gas monitoring at geologic carbon storage sites may provide an accurate, simple, and cost-effective alternative to other soil gas methods that require complex background data collection and analysis. This approach uses ratios of coexisting gases (CO_2, O_2, N_2, and CH_4) to promptly and simply identify a leakage signal in the near-surface rather than using complex comparison of pre- and post-injection CO_2 concentrations. However, the technology for economical field-deployable smart data collection of all gas concentrations important for a process-based analysis, especially N_2, is lacking, limiting the ability to implement this leak monitoring approach on an industrial scale. A new method using commercially-available, automated sensors that measure CO_2, CH_4, O_2, temperature, relative humidity (RH), and pressure, followed by deriving N_2 as the remaining gas balance was developed and field tested at a typical CO_2 enhanced-oil recovery site. The accuracy and precision of the data collected were compared against the current method using gas chromatography. The results indicate a site-specific bias in the data comprising significant non-dispersive infrared (NDIR) sensor error, and a lesser degree of error from galvanic cell technology. A method for processing sensor data to acquire gas concentrations necessary for a process-based analysis is possible. However, neither NDIR nor galvanic cell technologies presently available consistently produced data in the field with the necessary quality to perform process-based monitoring at geologic carbon storage sites. No obvious reason for the bias was identified, although similar results have been reported in other research.

KEYWORDS: process-based; CCS; monitoring; soil-gas; carbon storage; sensor; NDIR; CO_2

INTRODUCTION

Soil-gas Monitoring

Soil gas geochemistry is an accepted technique for monitoring terrestrial geologic carbon storage sites for near-surface CO_2 that may have leaked from a deep storage formation [1,2,3,4]. As soil gas measurements are point measurements, risk assessment commonly informs and directs soil gas monitoring to the most likely areas of CO_2 migration (e.g. near faults, fractures, or wellbores) [5,6,7]. Such a targeted monitoring approach focuses on obtaining measurements within several compact (e.g. 100 m x 100 m) areas of interest within the larger area of review. If anomalous CO_2 gas concentrations (i.e. concentrations outside of the defined natural variation) are detected, they are assessed as either resulting from natural variation within the environment or the arrival of a leakage signal into the near-surface. If a leakage signal is identified, reliable and accurate means of quantifying the leakage would be needed to supply an optimized method for accounting of credits, assessing environmental impacts, and predicting cost of remediation if needed.

A 'concentration-based' approach is a popular method for attributing anomalous soil gas CO_2 concentrations to either leakage or natural variation. This approach relies on measuring soil gas CO_2 before injection begins to establish a baseline range. Any statistically significant increase in CO_2 concentration above baseline detected during the lifetime of a project could signal a storage formation leak. One challenge with this soil gas monitoring protocol is that it relies on lengthy collection (one to three years) of baseline soil gas CO_2 and supporting weather data (e.g. rainfall, temperature, barometric pressure) and requires complex statistical comparisons of these pre- and post-injection parameters to identify a potential leakage signal (e.g. Schloemer *et al*, 2013 [8]). The method also relies on the assumption that the baseline condition defined at the beginning of a project is static and applicable as baseline over the life of a project (including the post-injection monitoring phase). This assumption may not be valid given the influence of seasonal and climatic variations on soil gas concentrations, the dynamic nature of these influences over space and time, and the unpredictability of weather and climate conditions in the face of global climate change.

To address the complexities and uncertainties of current concentration-based methods, a process-based approach is being applied to near-surface soil gas monitoring, verification, and accounting (MVA). This approach was developed at a West Texas, US, playa lake, a natural CO_2-rich analogue site [9], and applied at a CO_2-enhanced oil recovery site where exogenous CH_4 in the near-surface was oxidized to CO_2 [10]. The method was also demonstrated at the ZERT controlled release site in Bozeman, Montana [11], and applied at the Kerr farm in Saskatchewan, Canada, where landowners claimed CO_2 leakage affected their property [12,13].

The process-based approach is powerful, yet simple, compared to the more complex existing protocols. It can promptly identify a leakage signal using three simple relationships among coexisting gases (CO_2, N_2, O_2, CH_4) to distinguish processes acting in the near-surface (Figure 1). Such geochemical signatures can indicate natural processes such as the following:

- Biologic respiration
- CO_2 dissolution and reaction with soil carbonate
- CH_4 oxidation
- Dilution of soil gas through atmospheric mixing

It may also indicate a potential leakage signal as an influx of exogenous gas into the near-surface. If the presence of an exogenous gas is identified, further assessment is necessary but CO_2 leakage from the reservoir may be indicated.

Implementing a process-based monitoring approach requires periodically sampling discrete locations for analysis with a gas chromatograph (GC). For MVA, GC analysis is relatively inefficient, requires use of a consumable carrier gas, and does not directly measure Argon (Ar), which is estimated based on N_2 concentrations, or water vapour, which is assumed under all conditions to be 2.3% (the vapour-saturated condition at 20°C and 100 kPa). Developing the capability for automated data collection and continuous measurement of all gases necessary for a process-based analysis will be a significant improvement over the current application. Assuming that areas of highest leakage potential into the near-surface are correctly identified, continuous monitoring within targeted sites will provide the temporal data density needed for assurance that no leakage occurs. The complete data set afforded by continuous monitoring may be important to alleviate public concerns about the near-surface environment and will increase defensibility in case of litigation. In the event that a CO_2 leak is detected, continuous monitoring of any on-going CO_2 flux will support updates to CO_2 inventory reporting and will also help evaluate the effectiveness of any mitigation measures designed to control leakage.

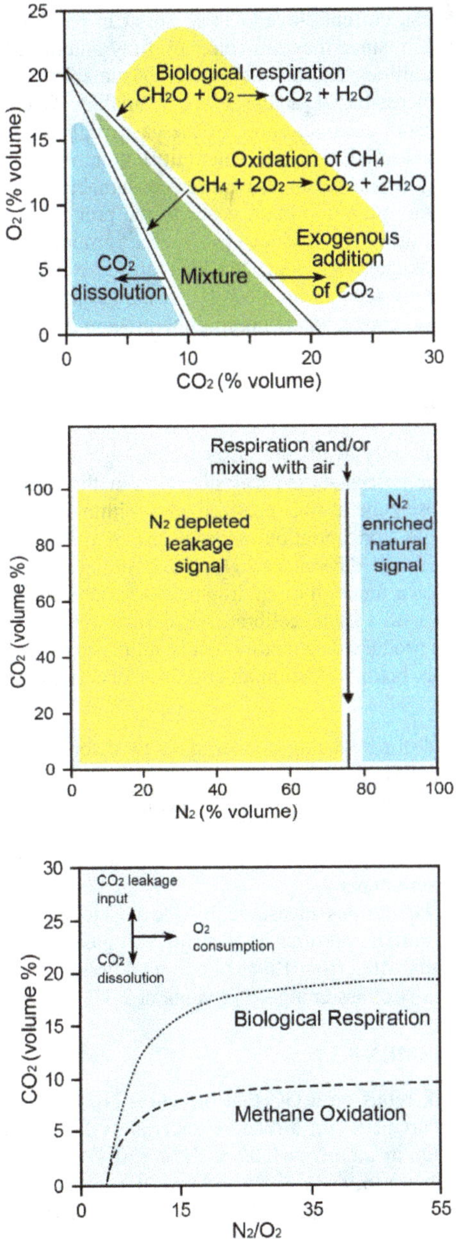

Figure 1. Three simple ratios used in a process-based analysis to distinguish leakage signal from natural variation in the near-surface above geologic carbon storage sites. For more information on the various processes represented by the graphs see [10,12].

For selected gases of interest, currently-available sensors can provide continuous monitoring for a concentration-based approach, since they are compact, fully automated, can be dedicated to a site, have data transmission capabilities, and do not require consumable supplies. However, not all of the sensors needed for a process-based analysis are available. Whereas compact cost-effective commercially-available sensors exist for some of the gases important to a process-based analysis (O_2, CO_2, CH_4, and H_2O measured as relative humidity), none exist for N_2 measurement. Existing N_2 measurement technology is cumbersome, requires consumable supplies, or is incompatible with O_2 analysis. N_2 is one of the most important gases to the process-based method because it gives strong evidence for whether gases are produced in-situ (background) or whether they migrate from a remote area (potential leakage). N_2 concentrations are also used to accurately estimate Ar concentrations which generally account for close 1% of the gas mixture. Without the technology for reliable continuous N_2 measurement, implementing the process-based method at an industrial scale project may be hindered.

OBJECTIVES AND APPROACH

In order to understand if currently-available sensor technology could ultimately provide continuous smart data collection for the process-based leak detection method, small commercial sensors were field-tested within the vadose zone at two localized sites within a typical Gulf Coast, USA oilfield. Continuous simultaneous gas concentration were measured using both the GC and commercial sensors over a period of several days at each localized site. The response of each measurement device was evaluated both as a function of environmental variability during baseline conditions and while 'spiking' the system with various calibration gas mixtures. The aim was to investigate if an approach using sensors can produce data quality that is similar to or better than the current approach to process-based monitoring. Field validation and performance assessment of the two methods were accomplished by:

1. Devising a method for reducing sensor data to determine parameters not directly measured by sensors (Ar, H_2O, N_2)
2. Comparing the precision of the two measurement types as a function of environmental variability
3. Assessing the accuracy and error of sensor measurements relative to "true values" defined as GC measurements
4. Comparing N_2 concentrations measured by the GC to N_2 concentrations derived by difference after the partial pressures of all major soil gases (CO_2, O_2, and CH_4 and H_2O) are measured and subtracted from 100% (1 atm total pressure)
5. Addressing the validity of assuming water vapor at 2.3%

DATA QUALITY ASSESSMENT

Current process-based work relies on a GC that measures N_2, O_2, CH_4, and CO_2 with accuracies equal to ±2% of each measurement (i.e. a relative uncertainty). A 2% relative uncertainty of a GC measurement of ambient O_2 in air gives 0.02 x 21% = 0.4% error in the actual (absolute) O_2 concentration, which rounds up to 0.5%. A 2% relative uncertainty of a GC measurement of CO_2 measured at the high end of the sensor operating range (0 to 20% which also is comparable to concentrations of 18% seen at the West Texas playa lake natural analog site) yields an error in the actual (absolute) CO_2 concentration that is also 0.4% (which rounds up to 0.5%). With H_2O concentrations of about 2.3% and Ar of about 1%, absolute uncertainties of these gas components are negligible. Assuming CH_4 concentrations far less than 20% at CCS sites (except in rare circumstances), these errors are also relatively small and can be assumed negligible. Manufacturers' specifications for several NDIR and galvanic cell sensors meet these requirements; therefore, these sensors should be capable of providing data quality that can match that of the GC. Thus, 0.5% absolute is the target for assessing if data quality of each individual component measured by sensors is comparable to the GC.

With individual gas component accuracies defined, the statistical relationship between multiple variables and their standard deviations can be used to identify the acceptable error for an N_2 concentration that is calculated as the remaining balance of partial pressures in the gas mixture. In this case, measurement uncertainties are defined as uncorrelated; thus the combined uncertainty is calculated as the square root of the sum of the squares as follows:

$$\sigma_{N_2} = \sqrt{\Delta_{CO_2}{}^2 + \Delta_{O_2}{}^2 + \Delta_{CH_4}{}^2 + \Delta_{Ar}{}^2\, \Delta_{H_2O}{}^2} \qquad \text{Equation 1}$$

Where σ_{N_2} *is* the total error propagated by the addition of the uncorrelated measured variables for each gas component and $\Delta_{CO_2}, \Delta_{O_2}, \Delta_{CH_4}, \Delta_{Ar}, \Delta_{H_2O}$ are the absolute standard deviations of the respective gas concentration measurements [14]. The simplified assessment assuming negligible error in H_2O, Ar and CH_4 is as follows:

$$\sigma_{N_2} = \sqrt{\Delta_{CO_2}{}^2 + \Delta_{O_2}{}^2} \qquad \text{Equation 2}$$

Using 0.5% as an acceptable level for absolute uncertainty for concentrations of O_2 and CO_2 leads to an uncertainty sum of 0.7%. This assessment shows that the uncertainty in N_2 concentrations derived by difference should be significantly below the 2% relative uncertainty of a GC measurement of N_2. If sensors function as well in the field as reported by sensor manufacturers, the data quality of N_2 determined using sensor data should be an improvement on current methods.

SENSOR FUNCTIONALITY

Screening a wide range of existing sensor capabilities (e.g. NDIR sensors, differential optical absorbance sensors, galvanic) against the functional requirements (e.g. range, accuracy, sensitivity, environmental tolerance, sampling interval) for a continuous process-based system indicates NDIR and galvanic cell technologies as the most suitable, cost-effective, low-maintenance sensor technologies for our purposes (Table 1). These sensors have continuous monitoring and data-logging capabilities with programmable data collection frequencies as fast as every 10 seconds. Sensor lifetimes exceed one year and most manufacturers claim greater than five years. Components are compact and can fit into a 13 cm borehole. When sensors function as a coordinated array, the pressure and temperature ranges common to all sensors will define the relevant operating environmental conditions: 0°C-45°C and 0% to 85% relative humidity (RH). These conditions are theoretically suitable for borehole monitoring.

Because there are typically two functional ranges for NDIR CO_2 sensors (0-2000 ppm and 0-20%) two CO_2 sensors are needed to ensure adequate accuracy over the full range of concentrations observed in natural settings. For example, at a natural West Texas playa where the process-based method was developed, CO_2 produced by biologic respiration was found to range from atmospheric concentrations to 18% [10] and at the Kerr Farm in Saskatchewan, Canada, biologic CO_2 as high as 11% was measured. At the IEAGHG Weyburn-Midale CO_2 Storage and Monitoring site, soil CO_2 reached just above 12% [5]. Some geologic CO_2 storage sites such as In Salah, Algeria, however, show CO_2 concentrations only as high as 0.27% due to a relative lack of biologic respiration [15].

When biologic respiration in vadose-zone sediments outpaces oxygen influx into the near-surface, O_2 concentrations can fluctuate between 21% (atmospheric concentration) and 0%. For O_2 measurement, galvanic cell sensors typically have auto-ranging capability among three range settings (0-5% O_2, 10% O_2, or 25% O_2) which is optimal for measuring O_2 concentrations in the vadose zone. Auto-ranging ensures 1% accuracy over the entire range of concentrations (0-21%) expected at geologic carbon storage sites.

Table 1. Factory-derived sensor specifications.

Sensor Type	Range	Accuracy	Response Time	Environmental Limits	Technology	Manufacturer
CO_2						
GMT 222	0 to 2000 ppm	± 1.5% of range + 2% of reading	30 secs	-5°C to +45°C 0 to 85% RH	NDIR	Vaisala
GMT 221	0% to 20%	± 1.5% of range+2% of reading	20 secs	0°C to 50°C 90% RH	NDIR	Vaisala
O_2						
Series 2000	0% to 25%	±1%	<30 secs	0°C to 50°C no RH spec	Electro-chemical	Alpha Omega
CH_4						
MSH-P-HR	to 10% 10 to 100%	100 ppm 1000 ppm	30 secs	-25 to 50°C 0 to 95% RH	NDIR	Dynament
Humidity and Temperature						
HUMICAP180R Humidity	0 to 100% RH	1.5% RH	8 secs	-40 to +80°C	Capacitive thin-film polymer	Vaisala
Pt1000RTD Temperature	-40 to +80°C	0.4°C	8 secs	-40 to +80°C		Vaisala
Pressure						
PTB110	500 to 1100 hPa	0.3 hPa	500 ms	-40 to 60°C	Micromechanical	Vaisala

NDIR sensor technology is the most suitable for CH_4 detection; however, sensitivity for the CH_4 NDIR sensors is 100 parts per million (ppm) or 0.01 volume %. This level of sensitivity means these sensors do not respond to ambient atmospheric CH_4 background concentration of 1.8 ppm nor can they detect typical soil gas concentrations which were, for example, measured at 1 ppm to 12 ppm at the Kerr site [10,16]. Instead, NDIR devices are effective for detecting CH_4 concentrations that are large enough to have an impact on the total volume of the gas mixture. This dilution effect should be characterized accurately when N_2 concentrations are inferred by difference.

Measurement Technology

Gas Chromatography

For GC analysis, a known volume of a gas mixture is introduced into a controlled-temperature column using carrier gas. Individual gas components are separated in the column and successively measured by detectors as they elute, one by one, from the column. Avoiding cross interference among components is accomplished by choosing the correct methodology for analysis with respect to the column material, operating temperature, carrier gas flow rate and detector type. Detector response is represented by GC software as a peak where the area beneath the peak is proportional to the concentration of the gas component as defined by calibration standards. A flame ionization detector (FID) provides sensitive measurement of CH_4, with a lower detection limit of 1 ppm. Using a methanizer to convert CO_2 to CH_4 provides CO_2 detection at near-atmospheric concentrations (0.04%). A thermal conductivity detector (TCD) provides detection of per cent concentrations of CH_4 and CO_2 in addition to O_2 and N_2. Temperature and volume effects are avoided using a fixed-volume sample loop that injects gas into the temperature-controlled column; however, slight differences in pressure can occur at the inlet valve of the sample loop. These effects are small and easily corrected by normalizing gas concentrations to 1 atm pressure or 100 volume% as described in [10, 12].

Galvanic Cell Technology

Electrochemical (galvanic) sensors that measure O_2 operate using a sensing electrode (cathode) and a counter electrode (anode) separated by electrolyte. A hydrophobic membrane prevents electrolyte leakage but enables gas diffusion and subsequent reaction with the sensor electrode, producing an electrical signal that is proportional to the concentration of the gas. Electrochemical sensors generally have linear output with good selectivity, repeatability, and accuracy. Cross sensitivity does not affect O_2 measurement using galvanic cell technology. Instead, the greatest potential challenge to data quality is from the presence of toxic gases which may degrade sensor components. Water vapour (or humidity) does not directly interfere with galvanic sensor measurements; however, extremely dry or humid environments can affect the water content of the electrolyte solution and destabilize the sensor. Also, because galvanic technology relies on diffusion of gas across a hydrophobic membrane, condensation on the membrane will inhibit gas diffusion into the cell, degrading response time and accuracy. These sensors are not directly affected by pressures within normal operating ranges; however, pressure fluctuations can result in disturbance of the diffusion gradient across the hydrophobic membrane leading to false increases in concentration. Since these sensors are based on fuel cell technology, their output is dependent on temperature as defined by the Nernst equation describing electrochemical output. However, temperature compensation is generally a part of the sensor component (e.g. Rammoorthy *et al*, 2003 [17]).

Humidity Sensors

Humidity sensors use a capacitive thin-film polymer operating principle for relative humidity (RH) measurements. Thin-film polymer is placed between two electrodes and absorbs or releases water vapour as the RH fluctuates. Absorbing and releasing water causes the dielectric properties of the polymer to change and thus the capacitance of the sensor. This is converted into a humidity reading. The micromechanical sensor for pressure measurement has a silicon membrane which bends with pressure fluctuations. These fluctuations increase or decrease the height of the vacuum in the sensor. Because the two sides of the vacuum act as electrodes, the changing height produces a change in the capacitance of the sensor. This is converted into a pressure reading. The response of a RH sensor can be non-linear, complex, and sensitive to many parameters including sensor temperature, hysteresis, contamination, and condensation [18,19,20].

NDIR Technology

NDIR technology is commonly used for accurate CO_2 and CH_4 measurement and is available as small components suitable for borehole use. These sensors use NDIR technology with a single-beam infrared (IR) source and a dual wavelength detector. IR light is absorbed by the sample gas at its specific wavelength and optical filters isolate additional portions of the IR wavelength region. The sensor signal is determined by recording changes in the amount of light that is ultimately transmitted; the strength of the resulting light signal indicates gas concentration. The use of a second "reference" channel, operating at a different wavelength compensates for any interfering environmental conditions such as sensor aging or contamination, ensuring sensor stability over time.

Interferences from co-existing gases with overlapping absorption spectra are a potential hindrance to NDIR technology. Manufacturers of NDIR CO_2 sensors generally report that there are no interferences from other gases, including water vapour, in the CO_2 absorption spectra (wavelength of 4.2 to 4.35 μm). This lack of interference from water vapour is also stated by Gibson and MacGregor [21] and inspection of the HITRAN database [22] indicates only extreme concentrations (5000 ppm) of phosphine (PH_3) would result in false positive reading during CO_2 measurement. However, conflicting information is reported by Herber *et al* [23] and Shrestha and Maxwell [24,25,26,27]. It is also implied by the US Environmental Protection Agency which provides guidance on assessing the interference from humidity on NDIR CO_2 vehicle emissions testing under rule 40 CFR parts 1065.350 and 1065.355 [28]. These studies report an "effect" from humidity but do not specify that the cause is from interference in absorption spectra.

In contrast, CH_4 NDIR sensor manufacturers report strong cross sensitivity to other gases including most hydrocarbons and ethanol (Figure 2).

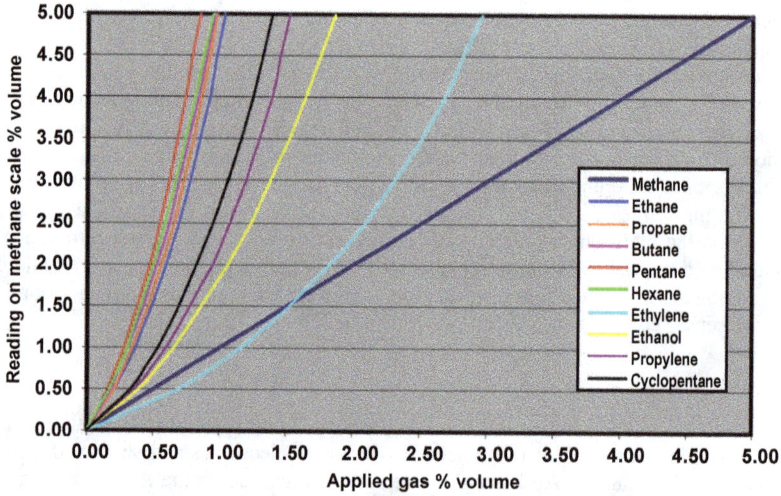

Figure 2. Cross-sensitivity of hydrocarbons and ethanol with methane from a typical commercial NDIR sensor. (Dynament, Ltd., Technical Data Sheet TDS0068; http://www.dynament.com/infrared-sensor-data/tds0068.pdf).

Sensor response to propane (a possible component at a CO_2-enhanced oil recovery geologic CO_2 storage site) is five-fold larger than to CH_4. In other words, a concentration of 1000 ppm of propane (0.1%) would produce a false positive 5,000 ppm (0.5%) concentration reading for methane. Avoiding cross-sensitivity problems and/or obtaining accurate measurement of CH_4 concentrations below 100 ppm requires a laser-based analyser or flame ionization detection. Flame ionization detectors are not a practical alternative. They require consumable gases and in fact are the same sensors used for hydrocarbon detection on the GC, therefore, these have no advantage over automated gas chromatographs.

The response of NDIR sensors is a function of the compressibility of gases and therefore NDIR output can vary with pressure [29,30,31,32,33]. Sensors generally have automatic and/or manual pressure and temperature compensation; sensor manufacturers may provide correction factors for small amounts of bias caused by pressure and temperature fluctuations within the sensor's operating range. These correction factors can be applied when temperature and pressure data are available.

Fortunately, precise and accurate pressure and temperature sensors are available with miniature components suitable for borehole measurement.

Despite manufacturer instructions for pressure and temperature compensation, NDIR sensors are commonly reported in the literature to require additional application- and/or site-specific correction for bias beyond what is reported by the manufacturer [24,33,34,35,36,37,38,39]. Correction of bias is complicated by the fact that compact NDIR analysers have short and constant path lengths and therefore exhibit non-linear absorption especially at high gas concentrations [34,40]. Although using two or more filter channels to analyse for one gas decreases this effect, the conflict between dynamic range and linearity cannot be completely avoided [33]. Bias is therefore commonly propagated at higher concentrations [41,42]. Hysteresis (the degree to which a sensor produces a different output when concentrations are increasing versus decreasing) is also documented in the literature but rarely indicated by the sensor manufacturer [24, 26, 27] and may add to bias. Sun *et al.* [34] suggest that the pressure of water vapour, rather than spectral interference, may be the origin of bias above and beyond what manufacturers report for CO_2 measurement.

Determination of H_2O Vapour Content

Commercially-available sensors appropriate for borehole monitoring cannot measure H_2O vapour in volume % directly. However, this parameter can be indirectly calculated from RH by first assessing the relationship between RH and water vapour pressure as follows:

$$RH = \frac{Pw}{Pws} \times 100 \qquad \text{Equation 3}$$

Where RH is the % relative humidity, Pw is the partial pressure of water vapor in the gas mixture (in kPa), and Pws is the saturated water vapour pressure (in kPa) at 100% RH at a given temperature and pressure. Pws is estimated using an experimentally-defined polynomial equation, the Goff–Gratch equation [43,44], which has been updated and refined throughout the years by various researchers. We use the polynomial fitting reported by Schlatter and Baker (1981) (Annex I) to calculate Pws at the various temperatures and pressures measured during the study [45].

Given RH from sensor measurements and Pws calculated using the Goff-Gratch equation with the temperatures and pressures measured during the study, the partial pressure of water can be calculated.

MATERIALS AND METHODS

Field Tests

A total of nine continuous monitoring tests were conducted within two 1.3 m deep hand-augured boreholes within a typical Gulf Coast oil field. The two borehole sites are located near oilfield infrastructure in brown/black clay sediments that grade from clay to silty clay and finally to sandy clay at 1 m depth. Localized red oxidized zones (2 cm to 5 cm in diameter) and powdery calcareous zones were found in both boreholes' sediments. Soil gas in the vicinity of each test site was sampled in March 2013, about four months before sensor testing began. At that time, areas near each test site (referred to as Site 1 and Site 2) exhibited soil gas concentrations that differed from each other. In March 2013 near Site 1, CO_2 in the upper 2 m of sediments averaged 2% with trace amounts of CH_4. In March 2013 near Site 2, CO_2 in the upper 2 m of sediments averaged 13% with CH_4 as high as 3.0%. No additional hydrocarbons were detected at either site.

Seven commercially-available sensors comprised the monitoring array for the field tests: Vaisala Carbocap Carbon Dioxide GMT 220 series (0 to 2000 ppm and 0 to 20%), the Dynament MSH-P-HR High Resolution Methane Sensor, Alpha Omega Series 2000 Percent

Oxygen Analyzer, the Vaisala Humicap Humidity and Temperature HMT130 sensors, and the Vaisala Barocap Barometer PTB110. At each location, sensor arrays were installed in 13 cm diameter boreholes cased with PVC pipe screened throughout the bottom 76 cm interval to allow diffusion of gas from sediment to sensors (Figure 3).

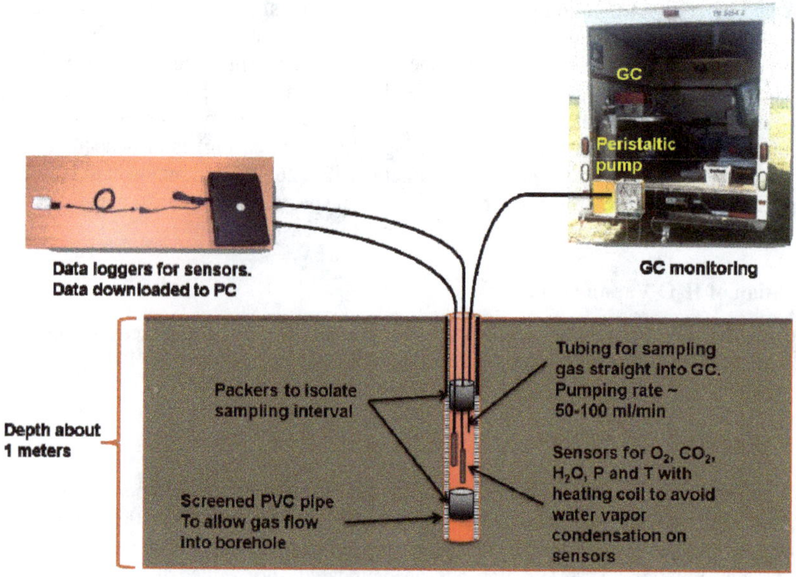

Figure 3. Schematic representation of the field deployment of sensors and the gas chromatograph for continuous monitoring. Not to scale.

The annulus of the screened interval was backfilled with sand to enhance permeability of gas from sediments into the monitoring interval while the remainder of the annulus was backfilled with wetted bentonite in order to inhibit communication with atmosphere through the borehole annulus. The sensor array was deployed using a custom-made packer system designed to isolate a 20 cm interval within the near-surface for soil gas monitoring (Figure 4). The system consisted of an aluminium rod with two sets of paired circular plates enclosing foam pads as packers. A spacer rod with a circular mounting plate separated the upper and lower packers. The circular mounting plate had customized slots to accommodate each sensor and two additional screw rods for attaching and stabilizing the sensors. Tubing and wires connecting sensors to data loggers at ground surface fit through notches in the upper packer system. A clamp mechanism at the surface was used to compress the foam around the wires and expand it into the borehole, isolating the monitoring interval from communication with atmosphere.

The GC column temperature was maintained isothermal at 50˚C. For real-time analysis, gas was pumped through a 3 mm diameter stainless steel tube into the GC at a flow rate of 50-100 mls/min during monitoring, and the GC was automated to sample every four minutes. Because sensors used in this test are designed for installation in heating and air conditioning ducts, sensors can tolerate flow rates of many liters per minute; however, the small flow rate used in the study was the best way of matching the sampling and response times of the various measurement methods. Although each measurement device sampled at different time resolutions, all sensor data used in our analysis corresponded to the sample resolution of the GC (every four minutes) for comparison. Corrections were made for travel time to the GC to match GC measurements with concurrent downhole in-situ sensor measurements.

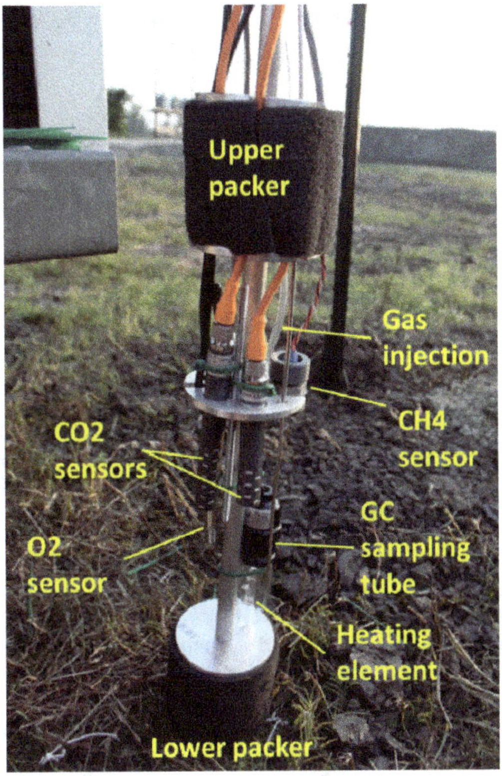

Figure 4. Sensor array and packer system before deployment into borehole.

To prevent condensation, which could be detrimental to sensor functioning, the sensor platform was warmed with a programmable heating element. Adding a heating element downhole guaranteed relative humidity below 85% which is the upper range for environmental functionality of all sensors. Warming the samples does not change the concentration relationships among the gases. The necessary use of the heating element hindered investigation of one of the goals of the study which was to evaluate the natural water vapor content of the sediments; however, for a short time period, the output of the heating coil was adjusted to observe the effect of temperature on RH and total pressure. This was done during a 'temperature test' that ran for nearly 24 hours at Site 1 beginning at 8.30 on 3 July, 2013. After heating the monitoring interval to a temperature of 49.9°C for approximately one hour, the heating element intensity was decreased until 19.40 on 3 July when it was again increased. The effects on total pressure and RH were subsequently observed for the duration of the test.

Gas concentration tests consisted of two periods of initial background monitoring and seven 'injection' or 'spike' tests. Injection tests involved introducing different gas mixtures into a 20 cm monitoring interval at 1.3 m depth to determine the response of sensors and the GC to changing gas concentrations. Each injection test consisted of a period of background monitoring followed by introduction of air and/or gas mixtures into the borehole. The tests performed and the gas mixes involved are shown in Table 2 and Table 3.

Table 2. Field tests performed during the study

Site 1	Temperature test
	Background monitoring
	Air injection
	Gas mix 4 injection
	Gas mix 6 injection
	Gas mix 5 injection
Site 2	Background monitoring
	Air injection
	Gas mix 4 injection
	Gas mix 6 injection
	Gas mix 6 injection-static
	Atmospheric monitoring

Table 3. Gas concentrations of gas mixes used for injection tests.

Gas Mix	**CO_2%**	**O_2%**	**CH_4%**	**N_2%**
Air	0	21	0	78
Mix 4	5	1	0	91
Mix 5	0	0	4	96
Mix 6	15	5	1	79

Tests began at Site 1 at 16.14 on 9 July 2013 and ran until 15.54 on 10 July 2013. A short period of background monitoring (14 minutes) was followed by four injection tests using ambient atmosphere, Mix 4, Mix 6, and Mix 5. Tests began at Site 2 at 21.30 on 10 July 2013 and ran until 11 July 2013 at 20.30. At Site 2, a relatively longer period of background monitoring (10 hours and 36 minutes) was followed by three injection tests using ambient air, Mix 4, and Mix 6. A Mix 5 test could not be performed because of a malfunction of the pump on 11 July 2013 at 20.30, which permanently disabled gas input into the GC.

To perform air injection tests, the peristaltic pump was disconnected from the GC and flow was reversed in order to inject air into the monitoring interval. During air injection, sensors continued to collect data but no samples reached the GC. The GC was then used to monitor the recovery of the system after air injection. Introduction of gas mixtures was accomplished via the pressure sensor tubing which was disconnected briefly during injection. This procedure was justified by a pressure test that confirmed that total pressures in the borehole were not sensitive to gas injection. Subsequent injection tests did not begin until system recovery and stabilization of gas concentrations in the borehole had occurred.

Sensors were factory calibrated and their calibrations checked before and after field testing. The GC was calibrated regularly throughout testing using ambient air and certified low and high standard gas mixtures spanning expected nominal concentration ranges before, during, and after each day of sampling. All calibration curves were forced through the origin. The GC retained stable calibrations over the duration of the study. Four-point CO_2 calibration curves had 98.9% confidence levels, three-point O_2 calibrations showed 98.6% confidence, two-point N_2 calibration curves showed 95.8% confidence and four-point methane calibrations showed 99.2% confidence levels (Figure 5).

After field testing, the O_2 sensor calibration was checked under ambient conditions and the CH_4, and CO_2 sensor calibrations were checked in the laboratory using a calibration mix having 2.5% CO_2 and CH_4 each. The results showed sensors did not significantly lose their calibrations during field testing.

The readings were as follows:

- The O_2 registered a constant reading of 20.9% in laboratory air.
- The CO_2 sensor measured an average concentration of 2.5% ± 0.03% for 60 readings of 2.5% CO_2.
- The CH_4 sensor measured an average concentration of 2.8% ± 0.07% for 55 readings at 2.5% CH_4.

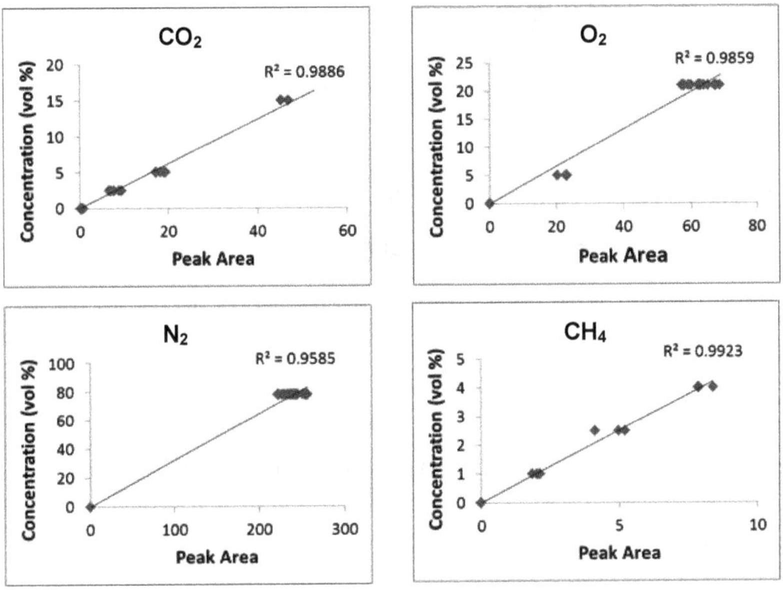

Figure 5. Gas chromatography calibrations over the duration of the study show high confidence levels.

RESULTS AND DISCUSSION

Temperature, Pressure, and Humidity

Temperature remained stable during the study due to the heating element within the boreholes. Even though the heating element was held at the same intensity setting during tests at both sites, temperature averaged about 10 degrees higher at Site 1 (mean = 39.1°C ± 0.3°C) compared to Site 2 (mean = 29.3°C ± 0.0°C). Average pressures were similar between the two sites (101.7 kPa at Site 1 and 101.9 kPa at Site 2). However, pressures were more stable at Site 1 (standard deviation of 0.2 kPa) compared to Site 2 (standard deviation of 1.0 kPa). There is no apparent correlation between weather systems and the difference in pressure between the sites. The cause of the difference in pressures is unknown but may be related to subtle differences in sediments or borehole construction between the two sites. RH was significantly lower at Site 1 compared to Site 2 with Site 1 RH averaging 58.5% ± 4.5% and the latter averaging 90.2% ± 1.6% RH.

Temperature Test

The presence of the down-hole heating element to prevent condensation on the sensors limited efforts towards one of the study goals, which was to test the validity of assuming 2.3% H_2O which represents water vapour saturated conditions in near-surface sediments. Controlling the temperature obscured the natural variation in water vapour content; however, by varying the temperature in a controlled test, the co-dependence of temperature, pressure and humidity could be observed.

During the temperature test, temperature varied from 30.1°C to 44.9°C and RH varied from 45.5% to 89.3%, translating to a range in H_2O vapour from 3.7% to 4.4% (Figure 6, Figure 7).

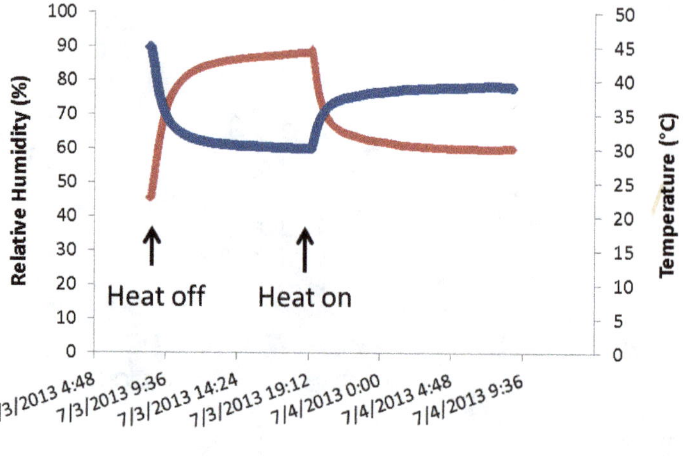

Figure 6. Variation in relative humidity (red) and temperature (blue) during the temperature test.

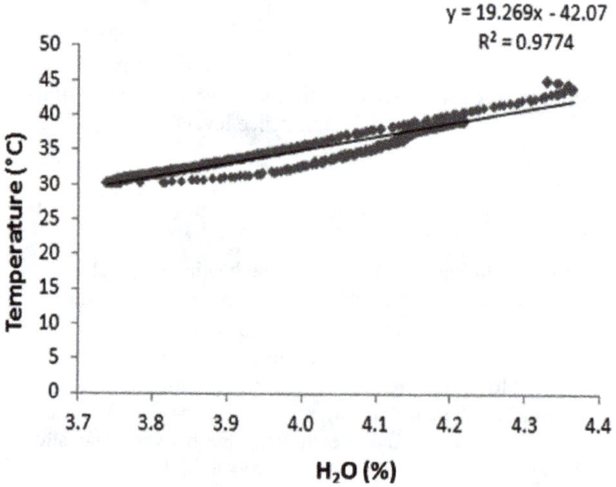

Figure 7. Correlation between the temperature in the borehole and the % H_2O calculated from relative humidity measurements.

The results of the test indicate a direct correlation between H_2O vapour concentration and temperature with a R^2 of 0.98. Also evident is a hysteresis effect on the RH sensor that is no greater than 0.1% (absolute) of the total % H_2O over the temperature range of the test. The results suggest that even under highly fluctuating moisture conditions the bias from hysteresis of RH sensors would not significantly contribute to error in measuring H_2O concentrations or estimating N_2 values by difference. Note also that an increase in H_2O vapour was not correlated with changes in pressure and that pressure remained relatively stable at 101.5 kPa during the test. This result suggests that changes in water vapour pressure do not perceptively contribute to changes in total pressure, an outcome that could have implications for assessing whether pressures asserted from water vapour might interfere with sensor functioning as suggested by Sun *et al* [34].

After shut-down of the heating element, temperature in the borehole equilibrated to 30.1°C. This temperature is 10 degrees higher than the temperature at which H_2O vapour can be assumed to be 2.3% (at 20°C). The higher temperature in the borehole caused H_2O vapour concentrations as high as 3.7%. This value is higher than the historically-assumed value of 2.3% but slightly lower than the standard saturated water vapour pressure at 30°C, which is listed on published tables as being 4.2 kPa or 4.2%. H_2O vapour concentrations observed at the highest temperature measured during the test (44.9°C) was 4.4%, again lower than the corresponding saturated water vapour pressure at 45°C, which is listed in published tables as 9.5 kPa or 9.5%. The effect of prolonged heating of the borehole sediments before the temperature test is unknown; however, the results suggest that H_2O vapour may, in some cases, be undersaturated and that temperatures in soils may vary enough to make the 2.3% H_2O vapour assumption invalid. Due to the unknown potential effect of borehole heating on results, more testing is recommended to confirm these preliminary conclusions.

Gas Concentration Data Reduction

The analytical capabilities and thus the data collected by sensors are different than those of the GC. Sensor data therefore requires an amended approach for its reduction and analysis from the standard GC approach (Table 4). An approach for sensor data reduction was devised during this study and data collected during the injection tests were reduced so as to represent each gas concentration's composition for a process-based analysis depending on the analytical method employed (either sensors or GC).

Table 4. Comparison of traditional GC and proposed sensor methods for process-based approach.

	Current Approach	Proposed Approach
Analysed	CO_2, O_2, N_2, CH_4	CO_2, O_2, CH_4 P, T, RH
Not Analysed	H_2O (assumed at 2.3%), P, T	N_2 calculated by difference
Analytical Tool	Gas chromatography	NDIR, Electrochemical sensors

GC Data Reduction Method

The standard GC approach to data reduction and analysis for process-based interpretation uses a GC to analyse O_2, CO_2, CH_4, and N_2. As discussed previously, H_2O vapour is assumed to be 2.3 kPa or 2.3% by volume. Another estimated value is Ar concentration. During GC analysis, O_2 and Ar are not separated but are detected and measured simultaneously; therefore, Ar must be estimated and its value subtracted from the measured 'O_2' concentration. Due to the overall non-reactivity of both Ar and N_2 in most sediments, the Ar/N_2 ratio can be assumed the same as that found in the atmosphere or 1/83. This assumption holds true except in rare cases of extreme denitrification in sediments [46,47]. Using the GC data reduction method, Ar is therefore estimated as 1/83 the N_2 concentration and subtracted from the O_2 measurement. Finally, when all components have been either measured or estimated, gas concentrations are then normalized to 100 kPa total pressures or 100% [10].

Sensor Data Reduction Method for Ar and N_2

One of the main objectives of the study is to assess the validity of estimating N_2 concentrations by difference given measurement of all other major parameters. Current economical field-deployable sensor technology does not exist for measuring Ar or N_2. Since estimating one of these parameters depends on knowing the other, the process of estimating N_2 and Ar concentrations is slightly iterative. Dalton's law of partial pressures states that the total pressure of a gas mixture is the sum of its partial pressures. Since H_2O, CO_2, N_2, CH_4, O_2, and Ar account for all the major soil gas components, the partial pressures of N_2 and Ar can be assigned the remaining gas balance.

Ar is then calculated as 1/83 of N_2 + Ar. This procedure slightly overestimates Ar. Calculated Ar is then subtracted from Ar + N_2 to yield the final N_2 value. It has been calculated that given a range of N_2 values from 50 to 100%, the maximum possible error using this method of estimation is negligible, on the order of 0.01% absolute (Table 5). Such an insignificant error validates this method as accurate for determining N_2 and Ar for a process-based approach using sensor data.

Table 5. Calculation of the error involved in estimating Ar from N_2 concentrations derived by difference.

GC Concentration		Concentration derived by difference	Estimated Concentration		Error of Estimation	
N_2 (vol%)	Ar (vol%)	Ar + N_2	N_2 (vol%)	Ar (vol%)	Δ (Absolute) Ar	Δ (Absolute) N_2
50.00	0.60	50.60	49.99	0.61	0.01	-0.01
100.00	1.20	101.20	99.99	1.22	0.01	-0.01

Data Quality Assessment

In order to best compare the data quality of a method that utilizes sensors to a method that utilizes a GC, it is necessary to evaluate the uncertainty inherent in each measurement type. A formal assessment of uncertainty (ISO Guide to the Expression of Uncertainty in Measurement and the corresponding American National Standard ANSI/NCSL Z540-2) is a complex process beyond the scope of this study; however a simple sensitivity analysis that quantifies and assesses the error in sensor measurement relative to GC measurement is necessary for comparing the two methods tested in the study.

The main components of our analysis are accuracy (defined as the agreement of a measured value with its true value), and precision (defined as the degree of closeness in independent measurements made under the same conditions). For the purposes of this study, we define the "measured value" as the sensor data and the "true value" as the GC data. We also consider the two data collection methods to be independent measurements made under the same conditions; thus both accuracy and precision are represented as absolute error (also termed absolute uncertainty) and reported as the difference (denoted as Δ) in gas concentrations measured by each of the methods.

Environmental Variability

During continuous testing, all sensors detected concentrations within their ranges with the exception of the 0-2000 ppm CO_2 sensor. Even with the injection of air, CO_2 concentrations in the boreholes were well above this sensor's operating range at both sites (>1%). Future investigations will be necessary to test the response of this sensor which may have limited use at geologic carbon storage sites which commonly exhibit natural soil CO_2 concentrations above 2000 ppm.

Maximum, minimum and average functional concentrations for all measurements after data reduction are summarized in Table 6 and Table 7. Overall gas concentration ranges measured during the study by the GC and sensors are dependent on baseline concentrations but also heavily dependent on the gas concentrations of the Mix being injected. Baseline sensor gas concentrations at Site 1 averaged 2.6% for CO_2 and 19.2% for O_2 with no detectable CH_4, whereas baseline sensor gas concentration at Site 2 were somewhat different, averaging 3.1% CO_2 and 17.6% O_2 with CH_4 also at non-detectable levels.

Table 6. Summary of the maximum, minimum and average functional measurements for a process-based analysis after data reduction (Site 1).

Parameter	Sensor Max	Sensor Min	Sensor Avg	GC Max	GC Min	GC Avg	Δ Max	Δ Min	Δ Avg
Background Site 1									
Pt (kPa)	101.8	101.8	101.8						
RH (%)	60.5	59.9	60.2						
Temp (°C)	39.3	39.1	39.2						
H_2O (%)	4.2	4.2	4.2	4.2	4.1	4.1			
CH_4 (%)									
CO_2 (%)	3.2	3.1	3.1	1.3	1.3	1.3	1.9	1.8	1.9
O_2 (%)	19.4	19	19.1	18.9	18.9	18.9	0.5	0.1	0.2
N_2 (%)	72.5	72.1	72.4	74.6	74.5	74.5	2.5	2.0	2.1
Air Test Site 1									
Pt (kPa)	102.7	101.7	102.1						
RH (%)	61.7	58.7	60.9						
Temp (°C)	39.3	38.5	38.7						
H_2O (%)	4.2	4.0	4.1	4.2	3.9	4.0			
CH_4 (%)									
CO_2 (%)	3.3	0.6	2.2	1.3	0.4	1.0	2.0	0.4	1.3
O_2 (%)	21.0	19.3	20.1	20.0	18.0	19.0	2.1	0.4	1.1
N_2 (%)	73.2	72.0	72.4	75.6	74.5	74.8	3.4	1.5	2.4
Mix 4 Site 1									
Pt (kPa)	101.7	101	101.6						
RH (%)	60.9	44.4	59.5						
Temp (°C)	39.4	38.8	39.2						
H_2O (%)	4.2	3.1	4.1	4.3	3.0	4.1			
CH_4 (%)									
CO_2 (%)	3.9	2.6	2.9	1.7	1.2	1.3	2.3	1.4	1.6
O_2 (%)	19.3	9.9	18.4	19.0	8.8	18.1	2.4	0.1	0.2
N_2 (%)	81.8	72.3	73.4	85.2	74.4	75.3	4.2	1.5	1.9
Mix 6 Site 1									
Pt (kPa)	101.8	101.5	101.7						
RH (%)	61.0	38.4	57.1						
Temp (°C)	40.4	38.8	39.2						
H_2O (%)	4.2	2.8	4.0	4.8	3.2	4.4			
CH_4 (%)	0.8	0.0	0.2	0.5	0.0	0.2	0.5	0.0	0.2
CO_2 (%)	12.4	2.8	5.1	5.9	1.0	2.7	6.9	0.0	2.4
O_2 (%)	19.5	6.4	16.0	18.9	4.7	13.9	5.4	0.5	2.1
N_2 (%)	76.4	72.3	73.5	84.3	74.2	77.6	8.0	1.8	4.1
Mix 5 Site 1									
Pt (kPa)	102.7	101.1	101.6						
RH (%)	60.3	39.2	55.2						
Temp (°C)	39.7	38.6	39.0						
H_2O (%)	4.1	2.8	3.8	4.8	3.0	4.0			
CH_4 (%)	3.0	1.1	2.0	1.6	0.0	0.6	2.1	0.0	0.8
CO_2 (%)	3.2	0.6	2.1	2.3	0.0	0.9	3.2	0.1	1.1
O_2 (%)	19.5	2.9	12.0	20.4	0.0	10.7	13.4	0.0	3.2
N_2 (%)	89.6	72.3	79.7	92.1	74.3	81.8	92.1	74.2	81.8

Table 7. Summary of the maximum, minimum and average functional measurements for a process based analysis after data reduction (Site 2).

Parameter	Sensor Max	Sensor Min	Sensor Avg	GC Max	GC Min	GC Avg	Δ Max	Δ Min	Δ Avg
Background Site 2									
Pt (kPa)	104.3	102.0	102.3						
RH (%)	90.9	89.9	90.6						
Temp (°C)	29.3	29.3	29.3						
H_2O (%)	3.6	3.6	3.6	3.9	3.3	3.6			
CH_4 (%)									
CO_2 (%)	5.8	1.8	3.6	3.1	1.0	2.0	2.8	0.7	1.6
O_2 (%)	18.6	15.3	17.6	19.5	15.5	18.2	1.2	0.2	0.6
N_2 (%)	75.0	73.4	74.1	76.9	74.6	75.1	2.7	0.0	1.0
Air Test Site 2									
Parameter	Sensor Max	Sensor Min	Sensor Avg	GC Max	GC Min	GC Avg	Δ Max	Δ Min	Δ Avg
Pt (kPa)	105.8	101.4	102.9						
RH (%)	92.0	90.8	91.3						
Temp (°C)	29.4	29.3	29.3						
H_2O (%)	3.7	3.5	3.6	4.8	3.5	3.9			
CH_4 (%)									
CO_2 (%)	2.9	0.2	1.4	1.4	0.2	0.8	1.5	0.1	0.8
O_2 (%)	20.2	17.8	19.0	20.8	18.4	19.3	1.5	0.0	0.6
N_2 (%)	75.1	74.1	74.8	75.8	73.9	74.9	1.2	0.0	0.5
Mix 4 Site 2									
Parameter	Sensor Max	Sensor Min	Sensor Avg	GC Max	GC Min	GC Avg	Δ Max	Δ Min	Δ Avg
Pt (kPa)	104.8	100	101.4						
RH (%)	90.8	85.0	90.1						
Temp (°C)	29.3	29.2	29.3						
H_2O (%)	3.7	3.4	3.6	4.2	3.6	3.9			
CH_4 (%)									
CO_2 (%)	4.7	3.2	3.8	2.5	1.8	2.1	2.2	1.2	1.6
O_2 (%)	18.1	2.5	12.5	17.5	0.1	11.0	6.3	0.2	1.9
N_2 (%)	87.8	73.8	79.0	92.0	75.5	81.7	5.2	0.2	3.0
Mix 6 Site 2									
Parameter	Sensor Max	Sensor Min	Sensor Avg	GC Max	GC Min	GC Avg	Δ Max	Δ Min	Δ Avg
Pt (kPa)	101.3	99.3	100.7						
RH (%)	90.6	81.7	88.6						
Temp (°C)	29.4	29.2	29.3						
H_2O (%)	3.7	3.3	3.6	4.1	3.5	3.9			
CH_4 (%)	0.9	0.0	0.4	0.6	0.0	0.2	0.4	0.0	0.2
CO_2 (%)	13.2	3.3	7.7	7.5	1.9	4.5	5.7	1.4	3.2
O_2 (%)	17.4	5.4	12.7	16.9	3.7	11.3	3.1	0.1	1.6
N_2 (%)	76.2	72.3	74.4	82.9	75.3	78.8	7.3	1.4	4.4

Environmental variability is represented by the test in two ways:

1. Spatially, by the different conditions between each test site (i.e. baseline gas concentrations, RH and temperature)
2. Temporally as a function of the injection tests

Precision of the GC and sensor measurements as a function of this environmental variability is also assessed in two ways: visually through time series analysis (Figure 8 through Figure 12), and statistically by error and regression analyses (Figure 13).

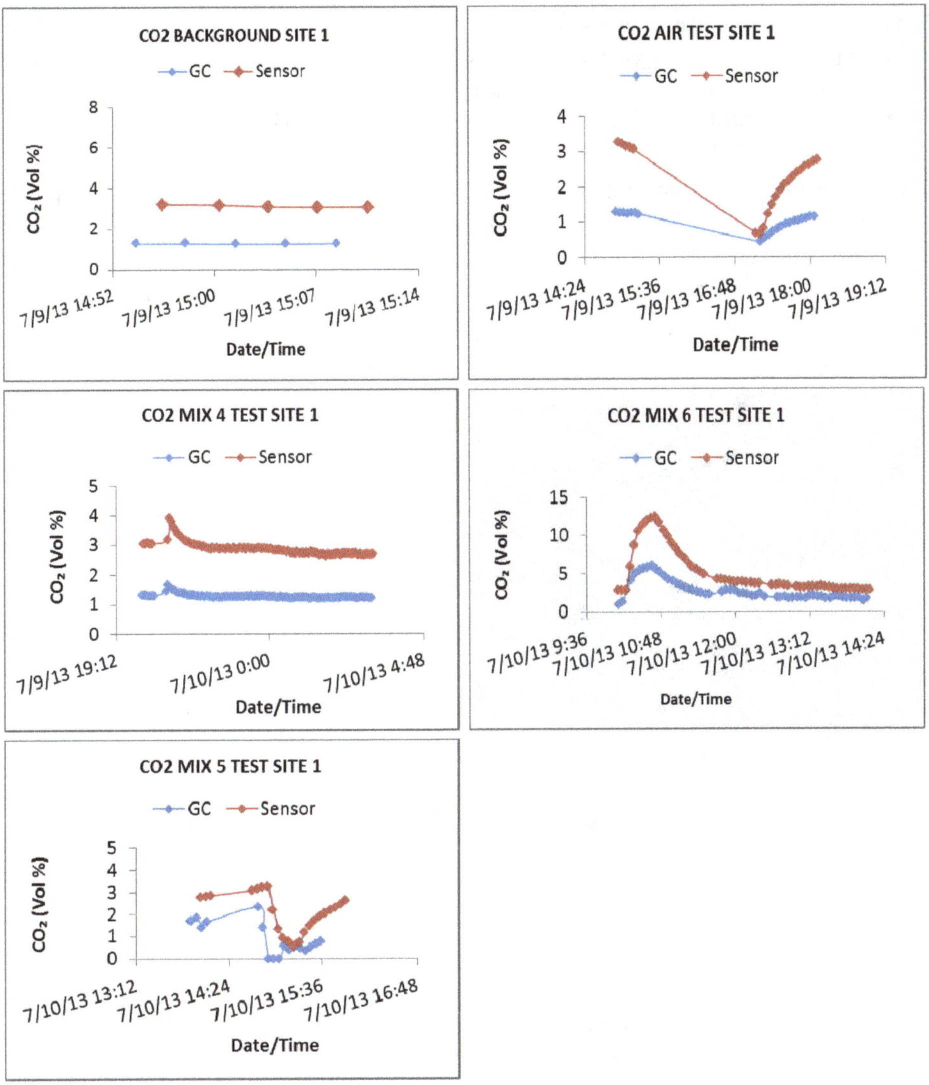

Figure 8. CO_2 data measured by the Vaisala NDIR sensor (red) and the GC (blue) during the various injection tests at Site 1.

Based on the figures, overall response times of the sensors and GC are comparable. However, sensor measurements conducted with NDIR technology consistently measured significantly higher concentrations of CO_2 and CH_4 compared to the GC. This effect is less pronounced with O_2, which was measured using galvanic cell technology. Subsequent lab testing of the sensors showed factory calibrations remained valid after field tests ruling out sensor malfunction as the source of the discrepancy.

Lab testing consisted of the following:

- 54 measurements of 2.5% CH_4 which yielded an average value of 2.8% ± 0.07%, giving an absolute error of 0.3% for the CH_4 sensor
- 59 measurements of 2.5% CO_2 which yielded an average value of 2.5 ± 0.03% for an absolute error of 0.0%.

The post-field laboratory testing illustrates the sensors retained their calibrations and discrepancies between sensor and GC measurements are not due to measurement drift during field tests.

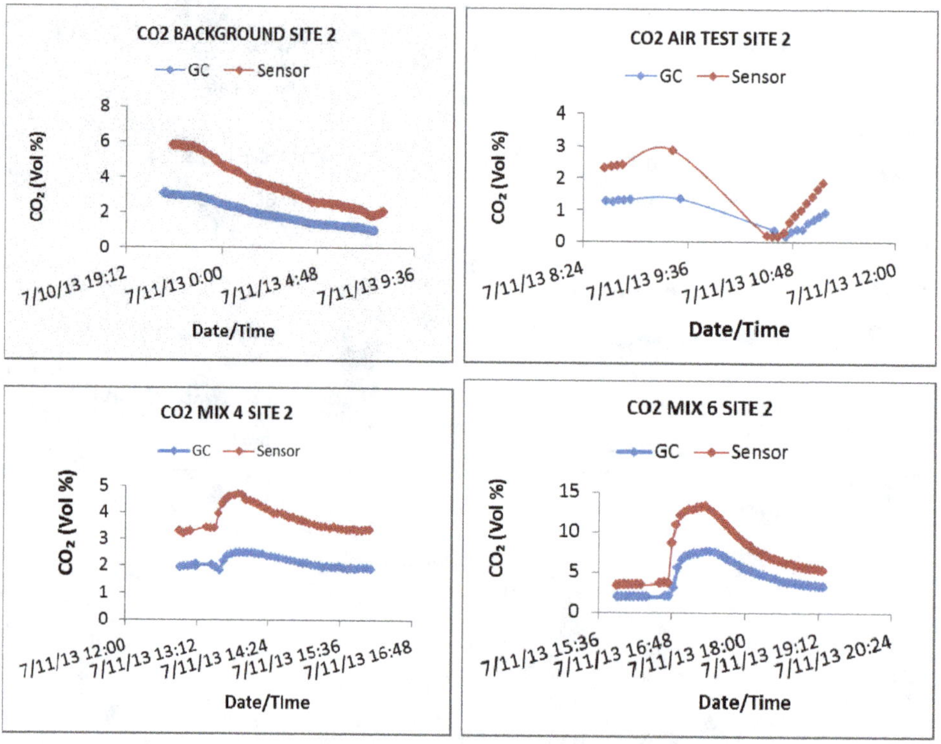

Figure 9. CO_2 data measured by the Vaisala NDIR sensor (red) and the GC (blue) during the various injection tests at Site 2.

The discrepancy between GC and sensor data during injection tests is represented by the absolute standard deviations about the mean for each gas concentration (Table 8). Although baseline conditions and thus starting gas concentrations differ between sites, the injection of similar amounts of gas during injection tests will theoretically create the same magnitude of sensor response at each site. A different sensor response is expected only if environmental conditions cause variability in sensor functionality. For the air injection test, CO_2 at Site 1 averaged 2.2% compared to 1.4% at Site 2. However, their absolute responses differed only by 0.1%. Sensor response to O_2 during the air test differed between sites only by 0.4%. Responses of CO_2, O_2 and CH_4 sensors during the Mix 6 injection tests were virtually identical between sites only differing by as much as 0.3%. The only discrepancy of note is the behaviour of the O_2 sensor during the Mix 4 tests which was significantly more sensitive to gas injection at Site 2 (4.9% absolute standard deviation) than at

Site 1 (1.5% absolute standard deviation). The reason for the outlier response from O_2 during the Mix 4 test is unknown. Taken together, however, both time-series and absolute response data show an overall similarity in both sensor and GC response to injection tests at Sites 1 and 2 suggesting that variation in site conditions will not add complexity or barriers to implementing sensors for process-based monitoring at a number of localized sites within an area of review.

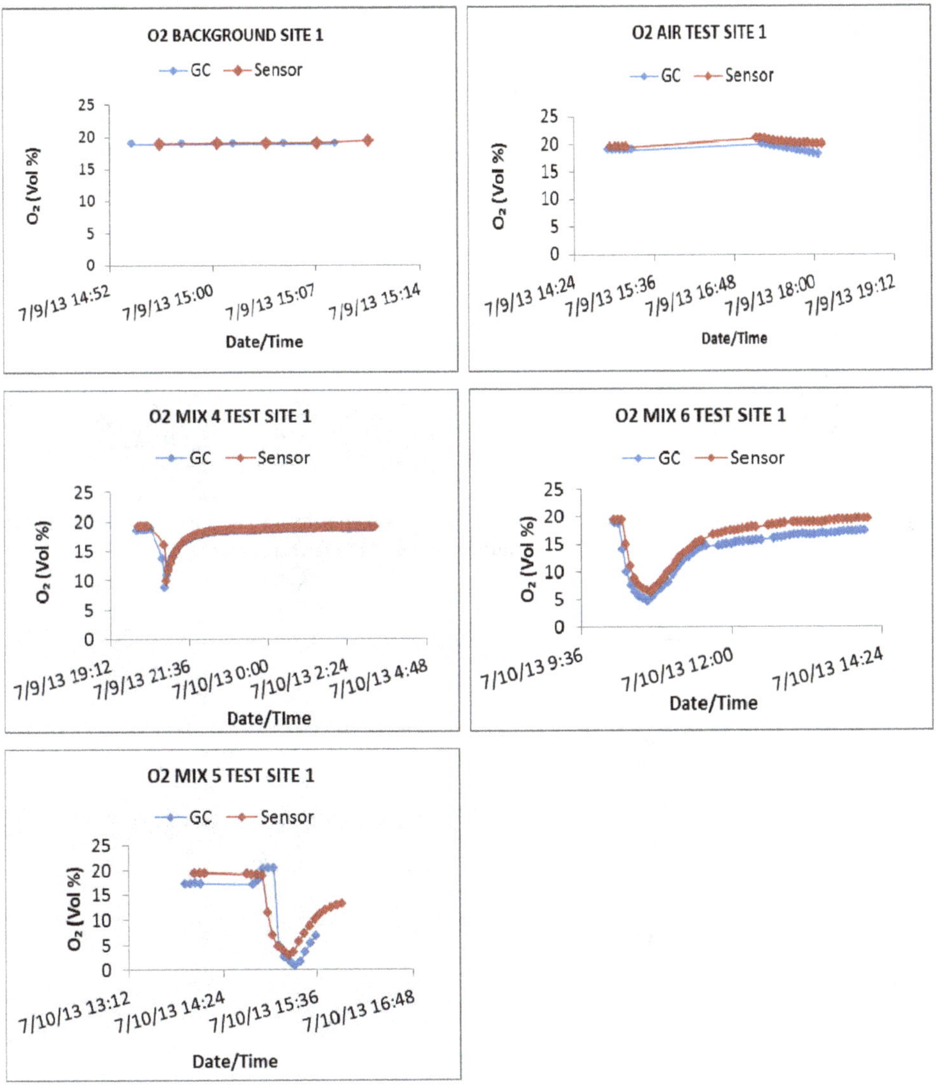

Figure 10. O_2 data measured by the Alpha Omega galvanic cell sensor (red) and the GC (blue) during the various injection tests at Site 1.

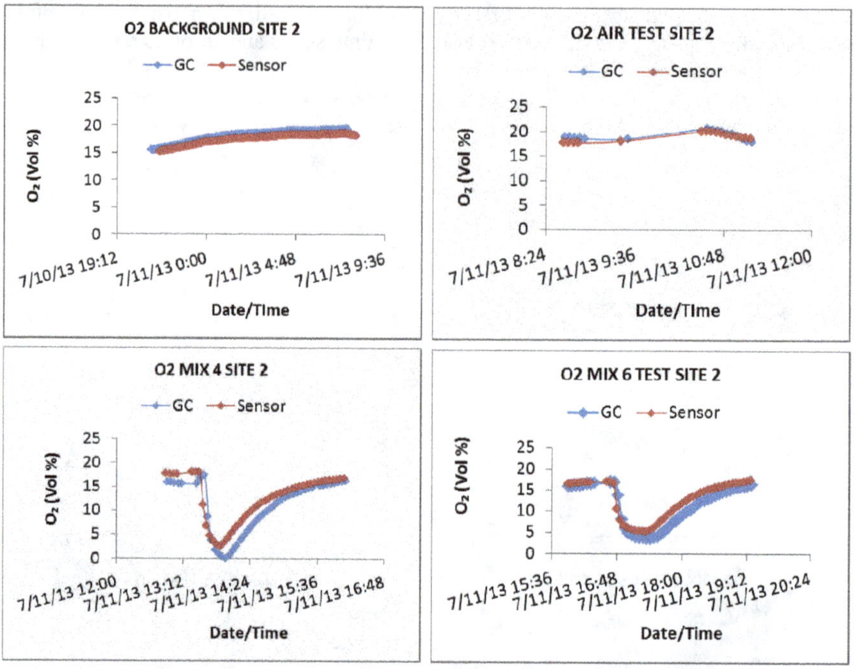

Figure 11. O_2 data measured by the Alpha Omega galvanic cell sensor (gray) and the GC (black) at Sites 1 (left) and Site 2 (right) during the various injection tests (Site 2).

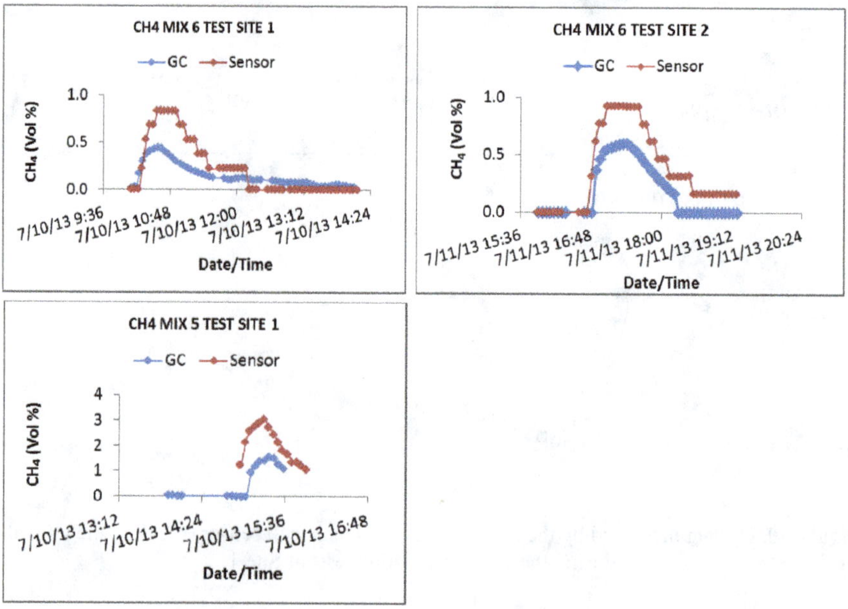

Figure 12. CH_4 data measured by the Dynament NDIR sensor (red) and the GC (blue) during the various injection tests at Site 1 and Site 2.

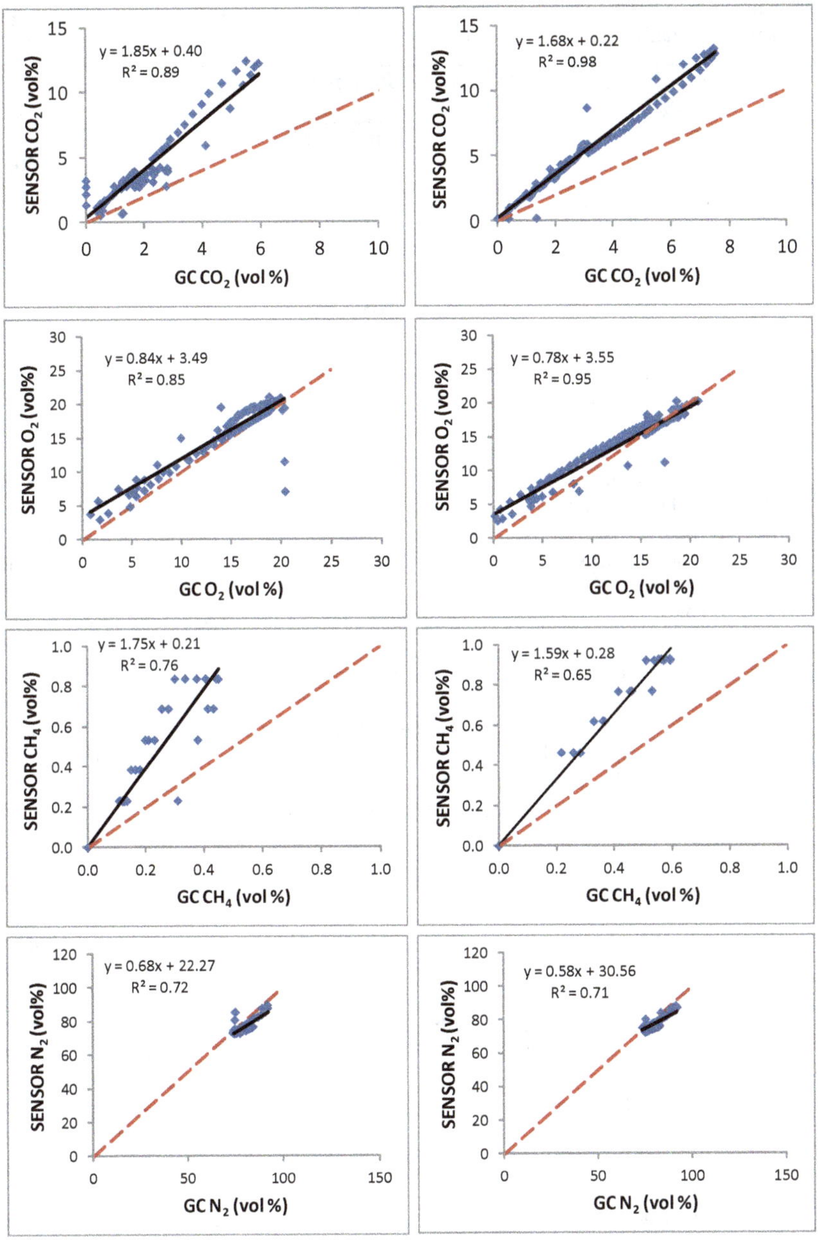

Figure 13. Regression analysis comparing sensor data and GC data for Site 1 (left set) and Site 2 (right set). Perfect match between the two data sets would yield a regression with a slope of 1. The data show that galvanic cell measurements are more accurate than NDIR. Discrepancies between NDIR and GC data increase at higher concentrations. Discrepancies between galvanic cell and GC data increase at lower concentrations.

Table 8. The discrepancy between GC and sensor data during injection tests as represented by the absolute standard deviations about the mean for each gas parameter at each test site.

Parameter	Sensor Avg (%)	Absolute Response (%)	Parameter	Sensor Avg (%)	Absolute Response (%)
Air Test Site 1			**Air Test Site 2**		
CO_2	2.2	0.8	CO_2	1.4	0.9
O_2	20.1	0.5	O_2	19	0.9
Mix 4 Site 1			**Mix 4 Site 2**		
CO_2	2.9	0.2	CO_2	3.8	0.5
O_2	18.4	1.5	O_2	12.5	4.9
Mix 6 Site 1			**Mix 6 Site 2**		
CH_4	0.2	0.3	CH_4	0.4	0.3
CO_2	5.1	3.0	CO_2	7.7	3.3
O_2	16	4.2	O_2	12.7	4.5

To further investigate the discrepancy between sensor and GC measurements, data from each site were also compared and statistically evaluated according to the method employed by by Von Bobrutzki et al [48]. This method compares the accuracy of data measured by the sensors to that of the GC measurements. Perfect agreement between GC and sensor measurements would yield a regression through the origin with a slope of 1. Regressions for each site (Figure 13) indicate no major differences in trends between the two sites; however regressions of NDIR measurements (both CO_2 and CH_4) are statistically different from galvanic measurements. Regressions of NDIR data show slopes between 1.37 and 1.85 with y intercepts between 0.13 and 0.4 and R2 between 0.98 and 0.76. These statistics support evidence from the time series data; namely, that the error of NDIR sensor measurements when defining the GC measurement as the 'true' value increases with increasing concentration. Galvanic cell measurements show the opposite trend, with regression slopes which are closer to 1.0 (specifically 0.78 and 0.84) and therefore can be determined overall to be more accurate than NDIR measurements; however, with y intercepts between 3.49 and 3.55, O_2 measurements are more precise at higher concentrations within the concentration range expected for O_2 (0 - 25%) at geologic CO_2 storage sites.

Accuracy and Precision Assessment

Because a process-based analysis relies on ratios of several gas compositions, it is important to assess the accuracy and precision of each individual gas component measurement. Equally important is to assess the accuracy and precision of N_2 values derived by difference. A comparison of the error in each sensor measurement to the 2% relative uncertainty of the corresponding GC measurement shows that virtually none of the CO_2 measurements and only 12% of the CH_4 measurements are within the acceptable accuracy range of a GC. The performance of the galvanic O_2 sensor is more accurate and precise than the NDIR sensors, however, only 31% of O_2 measurements fall within the accuracy range of a GC. Application of manufacturers correction factors to the data result in small and insignificant changes that do not sufficiently correct data quality. Only 33% of the calculated N_2 values fit the requirement for 2% relative uncertainty of the corresponding GC measurement. Similar to other reports in the literature, sensor data quality during field deployment does not match the quality expected from sensor accuracies determined in the laboratory and reported by manufacturers. Most of the error for NDIR sensors occurs in the higher concentration ranges whereas for galvanic measurement error is concentrated in the low concentration ranges.

CONCLUSIONS AND RECOMMENDATIONS

Commercially-available sensors for CH_4, O_2, RH, and CO_2 measurements in soil gas were screened and found to possess factory-derived accuracies that meet the desired specifications for implementing the process-based method in the field. However, field testing of sensors yielded data with insufficient accuracy and precision to match the current GC method. Overall, sensors responded quickly to changes in concentration. However, significant sensor error was identified. NDIR sensor error propagated at high CO_2 and CH_4 concentrations, and a lesser degree of galvanic cell sensor error propagated at low O_2 concentrations. Discrepancies between sensor and GC measurements are likely the result of field conditions during deployment. However, there was no apparent effect of total pressure, RH, or temperature on sensor functioning. None of the CO_2 measurements, 12% of the CH_4, and 31% of O_2 measurements fell within the acceptable accuracy range of a GC. Application of manufacturers' correction factors to the data result in small changes that do not significantly correct data quality. This phenomenon of site-specific bias has been reported in the literature by other researchers, but the source of the bias remains unknown. A method for reduction of sensor data was devised and assessed and seems feasible; however, because sensors failed to collect sufficient quality data, full assessment of estimating N_2 and Ar by subtraction of all major gas components from 100% was not possible. Assessment of water vapour contents in soils was hindered by the need to warm the borehole to prevent condensation on sensors but preliminary results indicate the following:

1. Soils may often be undersaturated with respect to water vapour.
2. Soil temperature may be an important parameter to measure for assessing water vapour content.

The authors recommend further testing of commercial sensors in order to understand the discrepancy of manufacturers reported accuracies and those observed in the field. At the same time, continuation of development of field-deployable sensor technology should continue.

ACKNOWLEDGEMENTS

This work was funded by the CO_2 Capture Project under contract SMV-037 BP Agreement Reference: 61798. Many thanks go to Mark Crombie, Stephen Bourne, and Michael Young for helpful editorial review. Appreciation is also extended to Michael Young, Associate Director and Scott Tinker, Director of the Bureau of Economic Geology and to Susan Hovorka, Principal Investigator the Gulf Coast Carbon Center for their unending support.

REFERENCES

1. European Commission (2009), Directive 2009/31/EC of the European Parliament and of the Council of 23 April 2009 on the geological storage of carbon dioxide, Off. J. Eur. Union, 140, 114–135
2. U.S. Environmental Protection Agency (2010a), General technical support document for injection and geologic sequestration of carbon dioxide: Subparts RR and UU, report, Washington, D. C.
3. U.S. Environmental Protection Agency (2010b), Mandatory reporting of greenhouse gases: Injection and geologic sequestration of carbon dioxide, Fed. Regist., 75, 75,060–75,089.
4. U.S. Department of Energy, National Energy Technology Laboratory, 2012, Best practices for monitoring, verification, and accounting of CO_2 stored in deep geologic formations- 2012 update.

5. Riding, J. B., and C. A. Rochelle (2009), Subsurface characterization and geological monitoring of the CO2 injection operation at Weyburn, Saskatchewan, Canada, in Underground Gas Storage: Worldwide Experiences.
6. Strazisar, N. R., A. W. Wells, J. R. Diehl, R. W. Hammack, and G. A. Veloski (2009), Near-surface monitoring for the ZERT shallow CO2 injection project, Int. J. Greenh. Gas Control, 3(6), 736–744, doi:10.1016/j.ijggc.2009.07.005.
7. Furche, M., S. Schlömer, E. Faber, and I. Dumke (2010), One year continuous vadose zone gas monitoring above an EGR test site, Geophys. Res. Abstr., 12, Abstract EGU2010-3095-1.
8. Schloemer S, Furche M, Dumke I, Poggenburg J, Bahr A, Seeger C, Vidal A, Faber E. A, 2013, review of continuous soil gas monitoring related to CCS – Technical advances and lessons learned. Applied Geochemistry Volume 30, Pages 148–160.
9. Romanak, K.D, 1997, Vadose-Zone Geochemistry of Playa Wetlands, High Plains, Texas, PhD dissertation, 273 pp., The University of Texas at Austin.
10. Romanak, K.D., Bennett, P.C., Yang, C, Hovorka, S.D., 2012, Process-based Approach to CO2 Leakage Detection by Vadose Zone Gas Monitoring at Geologic CO2 Storage Sites: Geophysical Research Letters, v. 39, L15405, doi:10.1029/2012GL052426.
11. Romanak, K.D., Dobeck, L., Dixon, T.E., Spangler, L., 2013a, Potential for a process-based monitoring method above geologic carbon storage sites using dissolved gases in freshwater aquifers, Procedia Earth and Planetary Science vol. 7, pp. 746-749, doi: 10.1016/j.proeps.2013.03.122
12. Romanak, K. D., Wolaver, B., Yang, C., Sherk, G. W., Dale, J., Dobeck, L. M., & Spangler, L. H. (2014). Process-based soil gas leakage assessment at the Kerr Farm: Comparison of results to leakage proxies at ZERT and Mt. Etna. International Journal of Greenhouse Gas Control, 30, 42-57.
13. Romanak, K.D., Sherk, G. W., Hovorka, S.D., and Yang, C., 2013b, Assessment of alleged CO2 leakage at the Kerr farm using a simple process-based soil gas technique: Implications for carbon capture, utilization, and storage (CCUS) monitoring, Energy Procedia Volume 37, 2013, Pages 4242–4248.
14. Ku, H., 1966, Notes on the Use of Propagation of Error Formulas, J Research of National Bureau of Standards-C. Engineering and Instrumentation, Vol. 70C, No.4, pp. 263-273.
15. Jones, D. G., Lister, T. R., Smith, D. J., West, J. M., Coombs, P., Gadalia, A., Brach, M., Annunziatellis, A., and Lombardi, S. (2011). In Salah gas CO_2 storage JIP: Surface gas and biological monitoring. *Energy Procedia, 4,* 3566-3573.
16. Beaubien, S.E., Jones, D.G., Gal, F, Barkwith, A.K.A.P. , Braibant, G. , Baubron, J.-C. , Ciotoli, G, Graziani, S, Lister, T.R., Lombardi, S., Michel, K., Quattrocchi, F., and M.H. Strutt, 2013, Monitoring of near-surface gas geochemistry at the Weyburn, Canada, CO2-EOR site, 2001–2011, International Journal of Greenhouse Gas Control, Volume 16, Supplement 1, June 2013, Pages S236-S262, ISSN 1750-5836, http://dx.doi.org/10.1016/j.ijggc.2013.01.013
17. Rammoorthy, K., Dutta, P.K., Akabar, S.A., 2003, Oxygen sensors: Materials, methods, designs and applications ,Journal of Materials Science vol. 38, 4271 – 4282.
18. Benyon, R.; Lovell-Smith, J.W.; Mason, R.S.; Vicente, T., 2001, State of the art calibration of relative humidity sensors in Proceedings of TEMPMEKO 2001, 8th International Symposium on Temperature and Thermal Measurements in Industry and Science, Berlin, pp. 1003-1008.
19. Lovell-Smith, J.W.; Benyon, R,. 2002, Immersion error in relative humidity probes, Papers from the 4th International Symposium on Humidity and Moisture ISHM2002, Taipei, pp. 389-396.
20. Lovell-Smith, J., and Neilsen, J., 2004, Calibration equations for humidity applications, Proceedings of TEMPMEKO 2004, 9th International Symposium on Temperature and Thermal Measurements in Industry and Science, Cavtat, M., pp. 645-950.

21. Gibson, D, and MacGregor, C, 2013, A Novel Solid State Non-Dispersive Infrared CO2 Gas Sensor Compatible with Wireless and Portable Deployment, Sensors volume 13, no 6, 7079-7103; doi:10.3390/s130607079
22. Rothman, L.S., et al., 2009, The HITRAN 2008 molecular spectroscopic database, Journal of Quantitative Spectroscopy & Radiative Transfer 110: p. 533-572.
23. Herber, S., Bomer, J., Olthuis, W., Bergveld, P., & Berg, A. V., 2006, A miniaturized carbon dioxide gas sensor based on sensing of pH-sensitive hydrogel swelling with a pressure sensor. Biomed Microdevices, 7(3), 197–204. doi:10.1007/s10544-005-3026-5.
24. Shrestha, S. S., & Maxwell, G. M., 2009, LO-09-44 An Experimental evaluation of HVAC-grade carbon dioxide sensors—Part I: Test and Evaluation Procedure. ASHRAE Transactions, 115(2).
25. Shrestha, S. S., 2009, Performance evaluation of carbon-dioxide sensors used in building HVAC applications, Graduate Thesis, Iowa State University, Paper 10507, 94 pp.
26. Shrestha, S. S., & Maxwell, 2010, Experimental evaluation of HVAC-grade carbon-dioxide Sensors: Part 2, Performance test results. ASHRAE Transactions, 116(1).
27. Shrestha, S. S., & Maxwell, 2010, Experimental evaluation of HVAC-grade carbon-dioxide Sensors: Part 3, humidity, temperature and pressure sensitivity test results, ASHRAE Transactions, 116(1).
28. U.S. Environmental Protection Agency, 2012, 40 CFR 89, Subpart D - Emission Test Equipment Provisions CFR 1065.350 - H2O interference verification for CO2 NDIR analyzers.
29. Park, J.S., Cho, H.C., Yi, S.H., 2010, NDIR CO2 gas sensor with improved temperature compensation, Proc. Eurosens., vol 5, 303–306.
30. Mukhopadhyay, Subhas Chandra, 2013, Intelligent Sensing, Instrumentation and Measurements. Dordrecht: Springer, 175 pp.
31. Chen, S.; Yamaguchi, T.; Watanabe, K., 2002, A Simple, Low-Cost Non-Dispersive Infrared CO2 Monitor. In Proceedings of the ISA/IEEE Sensor for Industry Conference, Houston, TX, USA, 19–21 November 2002; pp. 107–110.
32. Mizoguchi, Y. and Ohtani, Y, 2005, Comparison of response characteristics of small CO2 sensors and an improved method based on the sensor response. J. Agric. Meteorol. Vol. 61, 217–228.
33. Yasuda, T., Yonemura, S, and Tani. A, 2012, Comparison of the Characteristics of Small Commercial NDIR CO2 Sensor Models and Development of a Portable CO2 Measurement Device, Sensors, 12(3), pp. 3641-3655; doi:10.3390/s120303641
34. Sun, Y. W., Liu, C, Chan, K. L., Xie, P. H., Liu, W. Q., Zeng, Q, Wang, S. M. Huang, S.H., Chen, J.,Wang, Y.P., and F. Q. Si, 2013, Stack emission monitoring using non-dispersive infrared spectroscopy with an optimized nonlinear absorption cross interference correction algorithm, Atmos. Meas. Tech. Discuss., 6, 2009–2053, doi:10.5194/amtd-6-2009-2013
35. Sayed, A. M. M. and Mohamed, H. A., 2010, Gas analyzer for continuous monitoring of carbon dioxide in gas streams, Sensors and Actuators B: Chemical, 145, pp. 398–404
36. Bingham, D. and Burton, C. H., 1984, Analysis of multi-component gas mixtures by correlation of infrared spectra, Appl. Spectrosc., 5, pp. 705–709.
37. Tyson, L., Ling, Y. C., and Charles K. M., 1984, Simultaneous Multi-component Quantitative Analysis by Infrared Absorption Spectroscopy, Appl. Spectrosc., 5, pp. 38–56.
38. Lopez F, and de Frutos J. 1993, Multispectral interference filters and their application to the design of compact non-dispersive infrared gas analyzers for pollution control, Sensors Actuat A 1993;37:pp. 502– 506.
39. Kuske T, Jenkins C, Zegelin S, Mahabubur M, and Feitz A, 2013, Atmospheric tomography as a tool for quantification of CO_2 emissions from potential surface leaks: Signal processing workflow for a low accuracy sensor array, Energy Procedia 37, pp. 4065-4076.

40. Andre, G., Gerard, F., and Pierre, C., 1985 Gas concentration measurement by spectral correlation: rejection of interferent species, Appl. Opt., 14, pp. 2127–2132.
41. Dinsmore, K. J., Billett, M. F., and Moore, T. R, 2009, Transfer of carbon dioxide and methane through the soil-water-atmosphere system at Mer Bleue peatland, Canada, Hydrol. Process. 23, 330–341. DOI: 10.1002/hyp.7158
42. Pandey, S.K., and Kim, K.H., 2007, The relative performance of NDIR-based sensors in the near real-time analysis of CO2 in air. Sensors, 7, 1683–1696.
43. Goff, J. A., and S. Gratch, 1946, Low-pressure properties of water from −160 to 212 °F, in Transactions of the American Society of Heating and Ventilating Engineers, pp 95–122, presented at the 52nd annual meeting of the American Society of Heating and Ventilating Engineers, New York, 1946.
44. Goff, J. A., 1957, Saturation pressure of water on the new Kelvin temperature scale, Transactions of the American Society of Heating and Ventilating Engineers, pp 347–354, presented at the semi-annual meeting of the American Society of Heating and Ventilating Engineers, Murray Bay, Que. Canada.
45. Schlatter T.W., and Baker D.V, 1981, Algorithms for thermodynamic calculations, NOAA/ERL PROFS Program Office, Boulder, CO, 34 pp.
46. Martin, G. E., Snow, D.D., Kim, E., and R. F. Spalding, 1995, Simultaneous determination of Argon and Nitrogen, Ground Water, 33, 781-785.
47. Smith, K. A., and J. R. M. Arah, 1991, Gas chromatographic analysis of the soil atmosphere, in Soil Analysis, edited by K. Smith, pp. 505–546, Marcel Dekker, New York
48. Von Bobrutzki, K., Braban, C.F., Famulari, D., Jones, S.K., Blackall, T., Smith, T.E.L., Blom, M., Coe, H., Gallagher, M., Ghalaieny, M., McGillen, M.R., Percival, C.J., Whitehead, J.D., Ellis, R., Murphy, J., Mohacsi, A., Pogany, A., Junninen, H., Rantanen, S., Sutton, M.A. and Nemitz, E., 2010, Field Inter-comparison of Eleven Atmospheric ammonia measurement techniques, Atmos Measurement Techniques, vol. 3, pp. 91-112.

ANNEXES

Annex I

Parameters for calculating saturation pressure of water vapor from the Smithsonian Tables, 1984, after Goff and Gratch [43].

Es = saturation pressure of water vapour = Eso/p^8
Eso = saturation vapour pressure over liquid water at at 0˚C

Eso = 6.1078
p = (c0+T*(c1+T*(c2+T*(c3+T*(c4+T*(c5+T*(c6+T*(c7+T*(c8+T*(c9))))))))))
T = temperature, deg C
c0 = 0.99999683
c1 = -0.90826951*10-2
c2 = 0.78736169*10-4
c3 = -0.61117958*10-6
c4 = 0.43884187*10-8
c5 = -0.29883885*10-10
c6 = 0.21874425*10-12
c7 = -0.17892321*10-14
c8 = 0.11112018*10-16
c9 = -0.30994571*10-19

+T*(c_1+T*(c_2+T*(c_3+T*(c_4+T*(c_5+T*(c_6+T*(c_7+T*(c_8+T*(c_9))))))))))

Carbon Dioxide Capture for Storage in Deep Geological Formations, Volume 4
Karl F. Gerdes (Editor)

Chapter 39

BOREHOLE EM MONITORING AT AQUISTORE WITH A DOWNHOLE SOURCE

Thomas M. Daley[1], J. Torquil Smith[1], John Henry Beyer[1] and Douglas LaBrecque[2]
Lawrence Berkeley National Laboratory [1]
Multi-Phase Technologies, LLC [2]

ABSTRACT: To assess the potential application of electromagnetic monitoring at the Aquistore CO_2 storage project, Lawrence Berkeley National Laboratory and Multi-Phase Technologies collaborated on a two-part study including (1) numerical forward modelling of a time-lapse, controlled-source electromagnetic (CSEM) survey, and (2) an initial engineering study of instrumentation and proposed design for a borehole electric dipole source and electrode sensors. In section I, we present the background model development, including CO_2 and CSEM-source properties, the model results, including response to varying injection volumes. We also present a study of optimizing the source location within the borehole and a consideration of using borehole sensor electrodes (single-well measurement). The overall conclusion of this study is that the CSEM method with borehole source dipoles could be a useful monitoring tool for the planned Aquistore sequestration pilot.

In section II, we study the design of the electromagnetic (EM) transmitter for a borehole dipole source and its ability to provide sufficient current, voltage, and power as described in Section I. There are three major elements to the system design:

1. The surface transmitter that provides a carefully controlled current source,
2. The subsurface electrodes and supporting casing, and
3. Wires and cables to connect the electrodes to the subsurface.

Our suggestion is to make use of an existing, off-the-shelf, transmitter. We also consider the costs of two separate electrode and cable designs. The overall, installed cost of both designs will be nearly equivalent. Design 2 is less dependent on rigorous clamping of the cable, which will be an extra expense. In addition, Design 2 is superior both mechanically and electrically.

KEYWORDS: geologic CO_2 storage; electromagnetic monitoring; Aquistore; controlled source electromagnetics (CSEM); borehole-to-surface electromagnetics (BSEM); borehole electrode design

INTRODUCTION

Geologic carbon sequestration (GCS) is a technology whose goal is to prevent atmospheric release of greenhouse gases via injection of carbon dioxide (CO_2) into an underground reservoir for long term storage. GCS is typically part of a program of carbon capture and storage (CCS) that captures CO_2 from point sources such as power plants, transports the CO_2 to a storage site, and operates an injection facility. One recent CCS pilot project is the Aquistore CO_2 sequestration project, near Estevan, Saskatchewan, Canada. The Aquistore project is managed by the Petroleum Technology Research Centre (PTRC) and will be one of the first integrated CCS projects storing CO_2 in a deep saline aquifer from a coal fired power plant (PTRC, 2011). Aquistore is expected to store 500,000 tons of CO_2 during its lifetime (Ministry of Environment, 2012).

Assuring the long-term, safe storage of CO_2 requires the development of effective monitoring strategies. As part of the geophysical monitoring effort at Aquistore, there were initial plans for deployment of borehole electrodes for electrical or electromagnetic measurements to monitor CO_2 within the reservoir. The injected CO_2 displaces saline brine in the reservoir, and because CO_2 has a high resistivity compared to brine, the overall resistivity of the formation increases and can be monitored by measuring electric or magnetic fields. Previous unpublished work by Lawrence Berkeley National Laboratory (LBNL) had indicated that borehole-to-surface electromagnetic monitoring, using an electric dipole source near the bottom of a well penetrating the reservoir, could detect the resistivity change induced by GCS.

To assess the potential application of electromagnetic monitoring at Aquistore, Lawrence Berkeley National Laboratory and Multi-Phase Technologies collaborated on a two-part study including (1) numerical forward modelling of a time-lapse, controlled-source electromagnetic (CSEM) monitoring survey, and (2) an initial engineering study of instrumentation and proposed design for a borehole electric dipole source and electrode sensors.

This study addresses a specific geologic/geophysical model for the Aquistore site and the engineering design limitations imposed by the specific Aquistore monitoring well deployment. As such, this is a site-specific study and not a generic analysis. The modelling work was focused on determining the potential detection of injected CO_2 with a borehole electric dipole source and, as such, it was a forward-modelling detection study designed to guide the acquisition geometry and give a go/no-go decision point. We did not undertake a plume-mapping study, which would need to include an inversion with sensitivity analysis to understand resolution. However, we did look at estimated noise levels to assess likely detectability of the plume in field data for various plume sizes.

This report summarizes the results of the study in two sections. In Section I, we will present the background model development, including CO_2 and CSEM source properties. The model results, including response to varying injection volumes, are included in Section 1, along with a study of optimizing the source location within the borehole. Also included is a consideration of using borehole sensor electrodes (single-well measurement) with analysis of the effect of cable crosstalk for the proposed Aquistore deployment cables, as considered by the engineering study.

In Section II, we present an engineering design study specifically focused on the planned monitoring well for Aquistore. The engineering design section builds on the numerical results of the first section to consider the design of the electromagnetic (EM) transmitter and its ability to provide sufficient current, voltage, and power. The specific hardware considered is that which is best suited for the specific project constraints known at the time of the work, including factors such as borehole diameter, depth, time to completion, and delivery time of fabricated components. The hardware in the engineering design includes borehole electrodes for sources and sensors and the wire/cables which would run to surface.

Work on the two sections was carried out largely in parallel due to the short time available for possible deployment at Aquistore. Therefore, there are some differences, notably in the placement of the source electrodes (at the top of reservoir for modelling, below and above the reservoir for engineering).

SECTION I: NUMERICAL MODELING OF BOREHOLE SOURCE CSEM AT AQUISTORE

EM response to partial brine displacement by a resistive fluid has been modelled before. Hoversten and Gasperikova (2006) report changes in modelled in-line surface electric fields due to the presence or absence of CO_2 on the order of 1.5% for injection of ~4Mt (megatonnes) (~7×10^6 m^3 *in situ*) of CO_2 in sands at 1,100m to 1,400m depth, for a roving surface horizontal electric dipole source (2006). Andréis and MacGregor (2010) found measurable seafloor electric field sensitivity to the withdrawal of $4\times10^5 m^3$ ($\sim3\times10^5 m^3$ *in situ*) of natural gas from 1,100m to 1,400m below seafloor instruments (2010). However, source fields typically fall off as $1/r^3$ away from a (low frequency) dipole source, and anomalous fields due to localized targets fall off as $1/r^3$ (e.g. dipole fields), or as $1/r^2$ (e.g. magnetic fields due to secondary currents of electric dipole moment of target), making imaging of localized zones at depths much over 2km below sources and receivers fairly difficult. Thus, single-well, cross-well, or wellbore-to-surface measurements can be more sensitive to the injection or withdrawal of gas at large depths. For example, Swanepoel *et al* (2012) show 2% to 100% anomalies in synthetic single-well electric field data 300m from 0.1kt to 10kt (kilotonnes) CO_2 injected in a resistive layer in a homogeneous background.

Background Resistivity Model Design

Our simulated injection of CO_2 is into the lower Deadwood formation from 3,370m to 3,400m, as is anticipated for the Aquistore Esteban CO_2 injection well. We assume a temperature of 107°C, consistent with well bottom temperatures in the area and a vertical geothermal gradient of 0.04°C/m. We assume 350grams/litre (g/L) (350,000 ppm) salinity, consistent with values for the Deadwood formation at a similar depth in a plotted cross section of the Weyburn field slightly to the north of the Esteban drill site (from Whittacker *et al*, 2009). This gives an estimated brine conductivity of 70 siemens/m (resistivity of 0.014 ohm-m) (at 100°C). To estimate *in situ* resistivity, we assume hydrostatic pressure and a porosity of 5%, and we use Archie's law with parameters a = 0.62 and m = 1.95, typical for well-cemented Palaeozoic sediments, and well log resistivities from the nearby Imperial Halkett Well 15 7 3 8 (unique well number 101150700308W200_1524_MD_L1_MD) adjusted for differences between micro laterolog resistivities and long and short-normal resistivities. For the depths injected with CO_2, we assume a gaseous (supercritical) saturation of 0.3, and a slight amount in aqueous solution (extrapolated from solubility values at 90°C, 300 bars and 0, 20, 50, 80, and 100 g/l salinity given in Doughty, 2009).

In the area around the Aquistore site, deep horizons dip less than 1% (see cross section in Whittacker *et al*, 2009), so we use a one-dimensional (1D) resistivity model. We average well log resistivities from the Imperial Halkett well into a 1D anisotropic model with vertical and horizontal resistivities given by series and parallel averages of the well log resistivities. Boundaries between layers were selected sequentially at the maxima of a Komolgorov-Smirnov statistic (for example, Press *et al*, 1986) on the distribution of well log resistivities normalized by layer mean resistivities. The number of layers were kept to 15 (plus a basement half-space), which models the gross variation in well log resistivities with depth while retaining an overall simplicity. Where available, 1.5 times the long-normal resistivity less 0.5 times the short-normal resistivity was used to represent the formation resistivity. In the lower parts of the well, only micro laterolog values were available, which, in the upper part of the well, were consistently higher than short and long-normal resistivities, so they were scaled using scale factors estimated from higher in the well. Resistivity values for the Prairie evaporate formation (salt) were replaced with a 100 ohm-m value (depth 2700m?), because while salt-saline surfaces at a well are conductive, salt in bulk is resistive. The depths of the model resistivity layers were increased slightly to put the Precambrian basement at 3,400m, anticipated at the Esteban drill site. Based on a comparison of basement resistivities from other wells in the area, the measured 90 ohm-m basement resistivity was replaced with a 400 ohm-m value. The resistivity model resulting from the above analysis is shown in Figure 1.

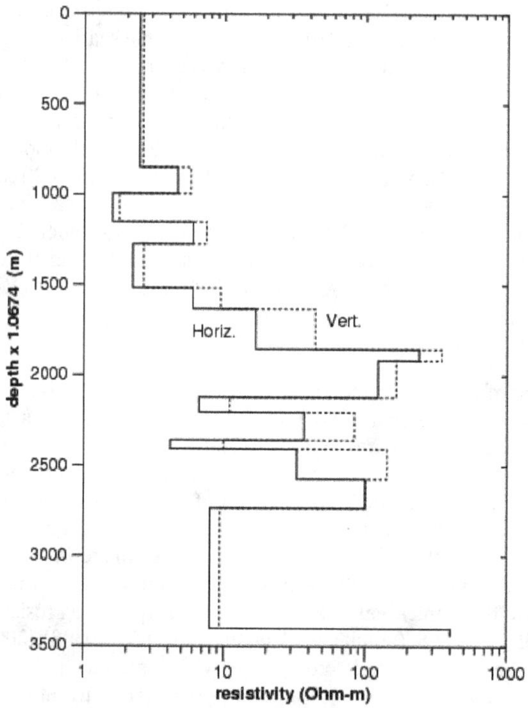

Figure 1. 1D resistivity model used in this study, taken from averaged resistivity analysis from Imperial Halkett Well 15 7 3 8.

CO_2 Plume and EM Source Model

For the study, CO_2 was modelled as injected into a 30m thick reservoir layer from 3,370m to 3,400m depth. Within the background resistivity model, the reservoir layer is not distinguished from layers above it, as the Komolgorov-Smirnov statistic at its top is smaller than at the boundaries of the 15 layer model. The injected CO_2 is modelled as a 30m high inverted (upside down) spherical cap of the requisite volume, centred laterally at the injection well. The injection well is 100m from the observation well in the x direction. A down-hole vertical electric dipole source was modelled at the top of the reservoir layer in the observation well, as shown in Figure 2. Based on the well logs, an 8.6 ohm-m average resistivity is anticipated in the reservoir before CO_2 injection, and this value was used for calculating the anticipated dipole current available. In this dipole, two electrodes 2m long x 0.2m diameter are separated 30m and have 4.1 ohms contact resistance. Connecting each to four 3,400m long #20 (American) gauge wires in parallel adds 65.2 ohms. Against 69.3 ohms, a 10 kW source produces 12.0 Amps at 832 V, yielding a source dipole moment of 360 Amp-m. A configuration with the source dipole at the top of the injection horizon enables inducing some horizontal electric fields in the CO_2, and assures good electric contact with the injection horizon. A detail of the source dipole and CO_2 injection configuration is shown in Figure 2, together with model vertical conductivity values after discretization on the finite difference grid used in modelling, for the case of 10kt injected CO_2.

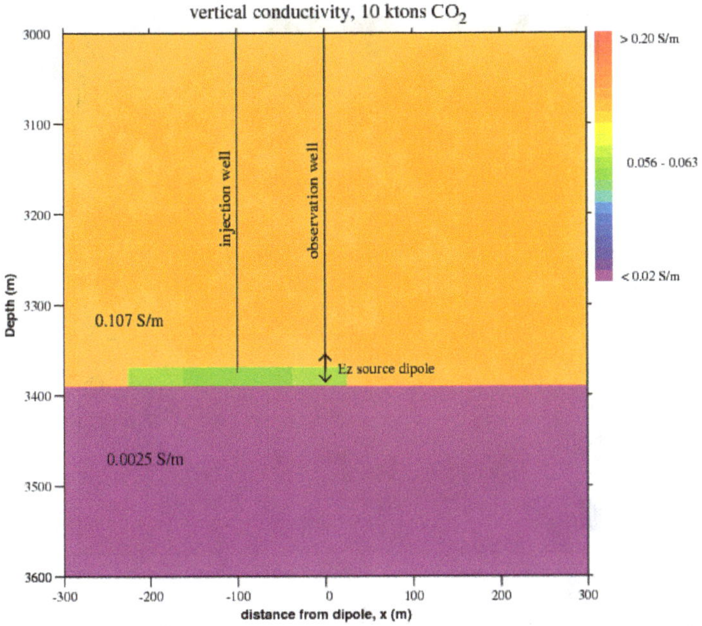

Figure 2. Detail of vertical conductivity model on finite-difference gridding, with 10kt CO_2 injected, showing wells and source dipole location.

Modelling Results

We model fields between ±15,840 m in x and y, and from -14,321 to 14,066m in z (down) using the code of Commer and Newman (2008) in a forward modelling mode using a secondary field formulation with primary fields determined by the 1D model section at the injection well prior to injection (2008). The fields were modelled on a 105 by 105 by 73 grid, with 250m grid spacing in x and y over most of the central ±6,115m, decreasing to 62.5m near the well, with 20m vertical spacing in the 260m surrounding the injection zone, increasing to 300m spacing by 1,265m to 1,865m depth, and decreasing to 193m vertical spacing near the Earth's surface.

Anomalous Surface Fields

To view the effect of CO_2 injection, we calculate the anomalous fields (i.e. the difference in a given EM field component between the model with CO_2 injected and the baseline model without CO_2). Anomalous horizontal electric, horizontal magnetic, and vertical magnetic field magnitudes (at 1 Hz) at the Earth's surface are shown in Figures 3, Figure 4 and Figure 5, respectively, for the case of 10kt injected CO_2, scaled for a 360 Amp-m source. These fields resemble those of a horizontal electric dipole at depth aligned in the direction of offset (between wells) from injection well to transmitter dipole (x direction in Figure 3, Figure 4, Figure 5 and Figure 6). Such a dipole has magnetic field lines in rings about the dipole axis, and electric field lines emanating from one pole and returning to the other. This resemblance is consistent with the fields from a horizontal electric dipole moment induced in the CO_2 being reinforced by an image horizontal electric dipole in the resistive basement, because a no-current flow boundary condition holds approximately there, with fields from any similarly induced vertical electric dipole moment being effectively cancelled by its image in the basement which opposes it.

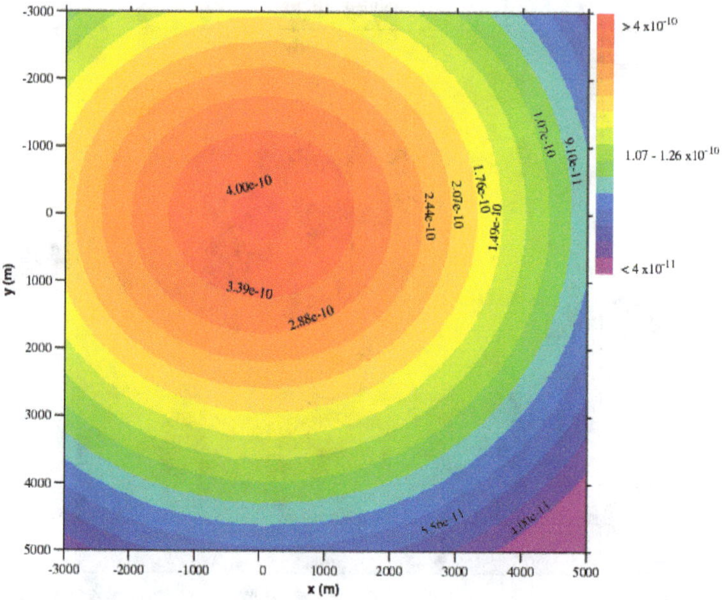

Figure 3. Surface anomalous horizontal electric field (Eh) for 10kt CO_2 with 360 Amp-m, 1 Hz downhole source. Source well at 0,0.

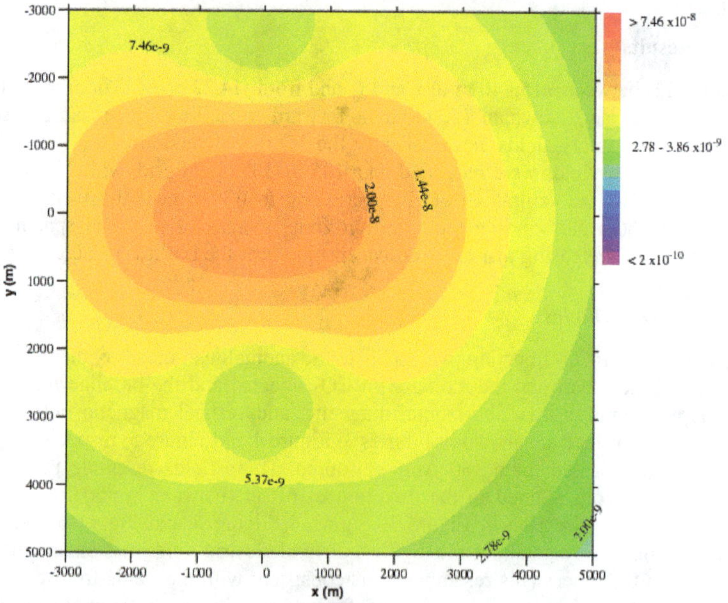

Figure 4. Surface anomalous horizontal magnetic field (Bh) for 10kt CO_2 with 360 Amp-m, 1 Hz downhole source.

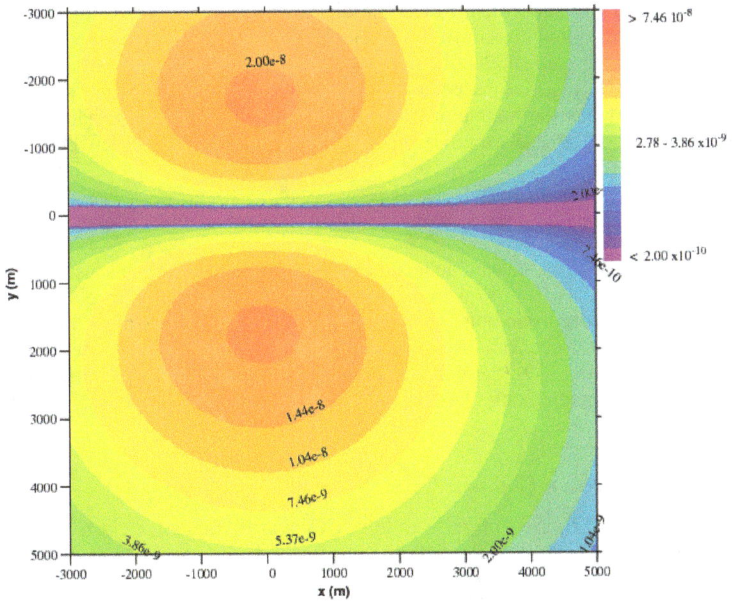

Figure 5. Surface anomalous vertical field, 360 Amp-m, 1 Hz downhole source.

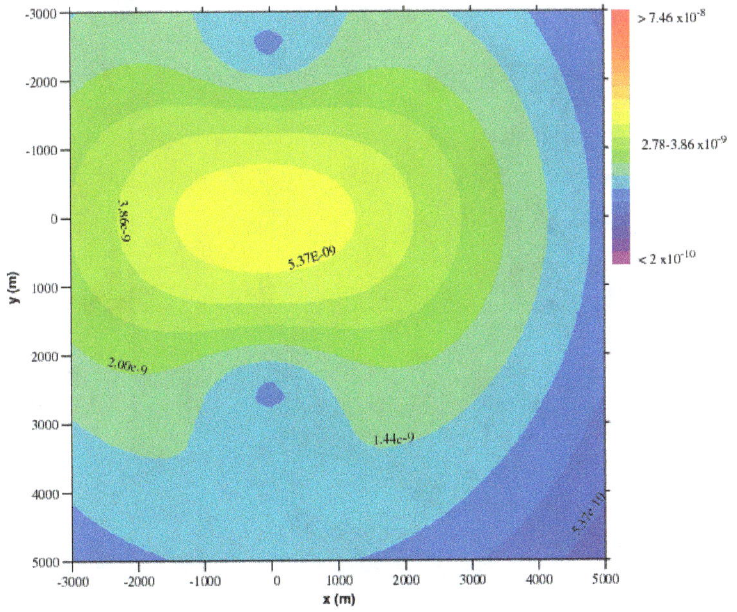

Figure 6. Surface anomalous horizontal magnetic field (Bh) for 2.5kt CO_2 with 360 Amp-m, 1 Hz downhole source.

In general, the use of a downhole vertical electric dipole source is important for producing vertical electric fields (the current flow direction) in the vicinity of a flat-lying resistive CO_2 plume, which results in significant anomalous electric and magnetic fields. However, the existence at the Aquistore site of a highly resistive basement immediately below the CO_2 reservoir and the source dipole significantly reduces the anomalous fields. This highlights the importance of site-specific geologic conditions and the use of modelling to assess appropriate survey designs for CSEM monitoring.

Surface Measurement S/N

With 30 minutes of stacking (a reasonable acquisition time for each surface sensor site), anticipated noise levels at 1Hz are 2.8×10^{-10} V/m in E_h, with copper-copper sulphate electrodes and 100m electrode lines (based on electrode noise measurements by Petiau and Dupis, 1980), and 2×10^{-10} T (Tesla) in B_h (assuming 95% cancellation of natural signals using a remote reference). Thus, for a 360 Amp-m source and 10kt CO_2, the maximum electric field anomaly raises a factor of 1.4 above electrode noise, whereas magnetic field anomalies attain signal-to-noise levels of 135 and 105 in B_h and B_z, respectively.

Response to Varying Injection Volumes

Surface anomalous E_h and B_h fields for 1kt to 20kt CO_2 and B_z fields for 1kt to 1Mt CO_2 are very similar in form to the 10kt fields, while differing in amplitude. For example, surface anomalous B_h fields for 2.5kt of CO_2 are shown in Figure 6. These are above the anticipated noise level throughout the plotted region. From 50kt to 250kt, surface anomalous B_h field shape changes continuously towards the fields at 250kt shown in Figure 7 (with a different colour scale). These are similar in form to the sum of fields from opposing horizontal electric dipoles induced on the CO_2 plume on either side of the source, with greater moment in the +x direction (away from the injection well).

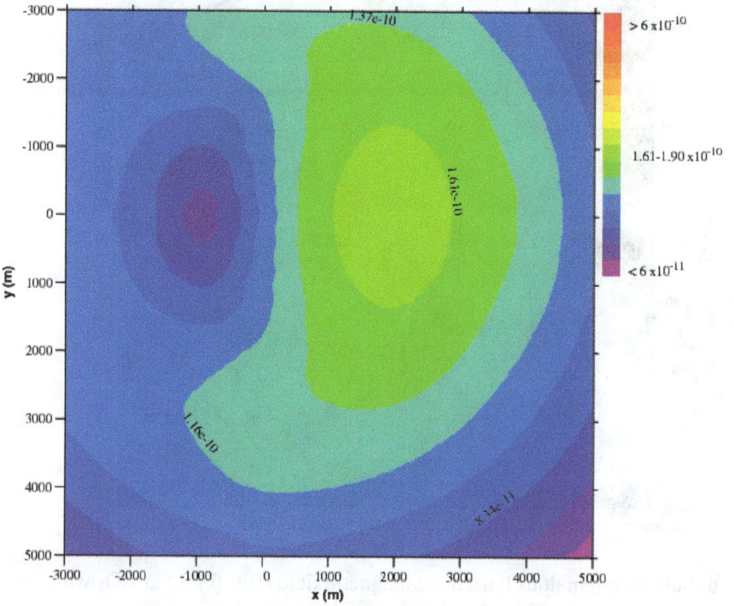

Figure 7. Surface anomalous horizontal magnetic (Bh) field for 250kt CO_2 with 360 Amp-m, 1 Hz downhole source.

While difficult to measure, for more than 250kt the surface anomalous B_h field shape continues to change, towards fields with magnitudes which are symmetric on reflection about the observation well in both x and y directions (not shown).

For a 360 Amp-m source dipole at 1Hz, Figure 8 shows the maximum changes in horizontal and vertical magnetic fields and in horizontal electric fields at the Earth's surface as a function of the mass of CO_2 injected. The maximum anomalies are observed for 10kt injected CO_2 (272m horizontal span at top), with magnitudes 4×10-10 V/m (E_h), 2.7×10-8 T (B_h), and 2.1×10-8 T (B_z). Thus, maximum magnetic field anomalies are expected to rise above the noise level with injection of about 0.8kt of CO_2, increase to a maximum at about 10kt, and decrease to noise levels by 170kt (for B_z) and by 270kt (for B_h) corresponding to spherical caps with 710m and 770m horizontal spans. Thus, the proposed experiment design can detect and monitor initial CO_2 injection with magnetic field anomalies, but may lose sensitivity for the planned 500,000-tonne injection. For the given geometry, the detectability appears to decrease as the plume grows and the edges of the plume move to far field.

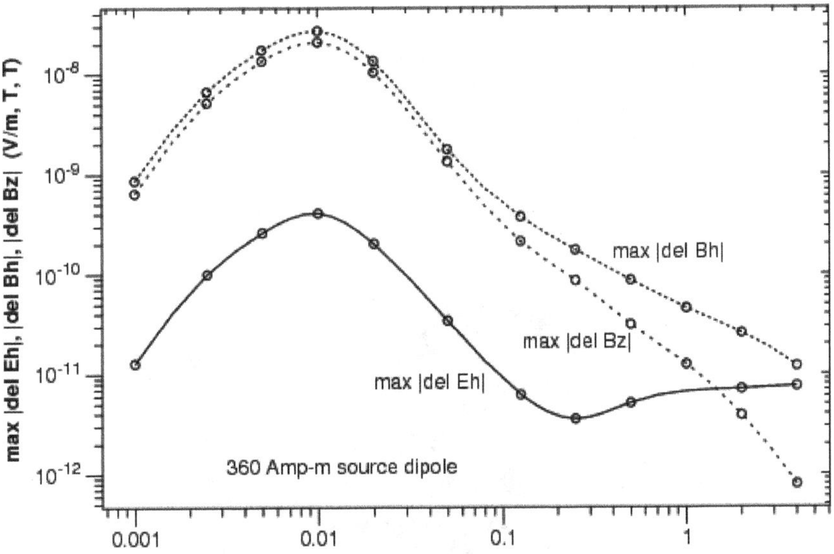

Figure 8. Maximum changes in surface horizontal and vertical magnetic fields and in horizontal electric fields at Earth's surface, for 360 Amp-m dipole at 1 Hz, as a function of CO_2 mass injected in million tonnes.

Placement of Surface Sensor Stations

Regarding instrument location for sensitivity to various masses of injected CO_2 (i.e. survey area for monitoring CO_2 injection), we consider at what injection mass the anomalous field magnitudes are some multiple of the anticipated noise level. Figure 9 shows the mass of CO_2 needed for a 4:1 signal to anticipated noise level in a surface anomalous horizontal magnetic field. This shows B_h sensitivity to 1kt of CO_2 only above the source transmitter location, with sensitivity to larger masses of CO_2 at greater distance from the source well location. Figure 10 is a similar plot for B_z sensitivity to CO_2 mass. Beyond about 7km radius from the source well, the anomalous B_z magnitude drops below a 4:1 signal to anticipated noise ratio for all injected CO_2 masses modelled, whereas anomalous B_h does not fall below this ratio until a distance of 8.5km from the source well.

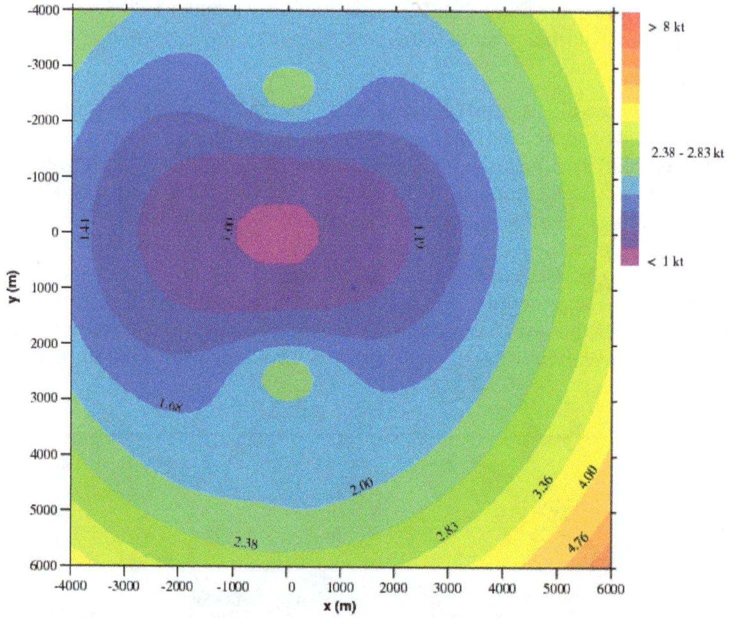

Figure 9. Mass (kt) of CO_2 needed for 4:1 signal to anticipated noise level in surface anomalous Bh, 360 Amp-m, 1Hz downhole source.

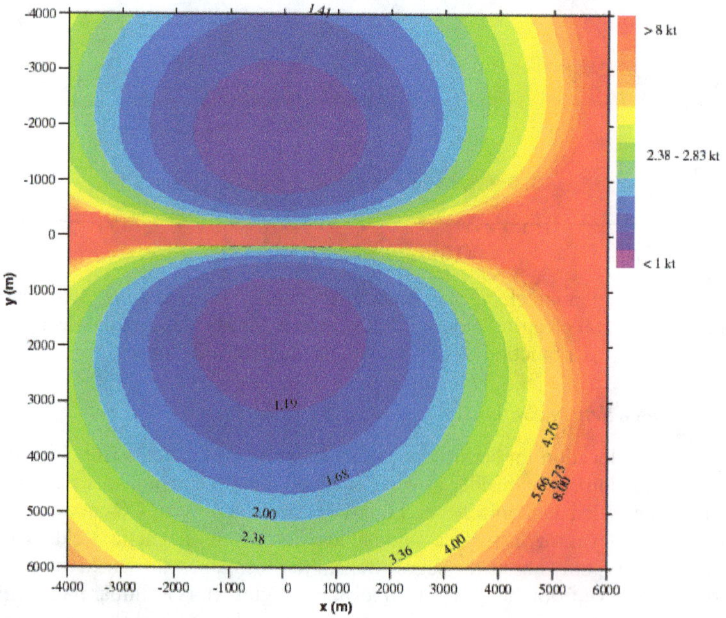

Figure 10. Mass (kt) of CO_2 needed for 4:1 signal to anticipated noise level in surface anomalous Bz 360 Amp-m, 1Hz downhole source.

Optimizing Source Location

The above results have been computed for a point dipole source at the top of the reservoir in which the CO_2 is injected (i.e. at top of inverted spherical cap of CO_2). Because the CO_2 plume is just above a resistive basement, surface measurements are mostly sensitive to the horizontal electric-dipole moments induced in the CO_2. This suggests using a vertical electric-dipole source centred a height of half the lateral offset to the CO_2 centre, above the CO_2 centre (e.g. 50m above the CO_2 centre) to maximize the horizontal electric field at the plume centre. For the smallest CO_2 masses, that would double the signal levels above those reported here. Or, taking into account that the actual source is a finite length dipole rather than a point dipole, the electrode placement can be optimized to maximize the horizontal electric field at the plume centre. In the zero frequency limit, for an electrode at distance x' laterally and height z_0' above the plume centre, which itself is a height b above the resistive basement, and approximating the structure above basement as a uniform half-space, the horizontal electric field at the plume centre is proportional to:

$$E_h \propto \frac{x'}{(x'^2 + z_0'^2)^{3/2}} + \frac{x'}{(x'^2 + (z_0' + 2b)^2)^{3/2}}$$

including effects of an image in the resistive basement. Assuming $|z_0'| \ll |x'|$, the horizontal electric field at the plume centre is maximized for the electrode at:

$$z_0' \approx \frac{-2b}{1 + \left(\frac{1 + 4b^2}{x'^2}\right)^{5/2}}$$

for b = 15 m and x' = 100m, z_0' = -13.4 m, suggesting a lower electrode 1.6m above basement. Adding a second electrode at z_1' the horizontal electric field at the plume centre is proportional to:

$$E_h^{(t)} \propto \frac{x'}{(x'^2 + z_0'^2)^{3/2}} - \frac{x'}{(x'^2 + z_1'^2)^{3/2}} + \frac{x'}{(x'^2 + (z_0' + 2b)^2)^{3/2}} - \frac{x'}{(x'^2 + (z_1' + 2b)^2)^{3/2}}$$

Directly above the electrodes at the height of the Earth's surface the (zero frequency) horizontal magnetic field is proportional to:

$$B_h^{(surf)} \propto \frac{1}{d - z_1'} - \frac{1}{d - z_0'}$$

where d is the distance of the plume centre below the Earth's surface. Using z_0 from Eq. (2) and using Equation 3 and Equation 4, we maximize the ratio of $|E_h^{(t)}/B_h^{(surf)}|$ by searching the interval $[z_0', d]$ in small steps for the value z_1 giving its largest value. For the same values of b, x' and z_0', we have z_1'=72.4m, suggesting placing the upper source electrode 87.4m above basement. Such an electrode configuration (1.6 m and 87.4 m above basement) would increase the horizontal electric fields at the plume centre by a factor of 3.1 over those due to the point dipole considered above, so for the smallest CO_2 masses considered, the measured anomalous fields at the surface should increase by this factor.

Use of Borehole Electrode Sensors in Single Well Monitoring

Since Aquistore was considering borehole electrodes in the same well as a possible borehole dipole source, it is interesting to think of the potential single-well measurements for monitoring. For the single well geometry, we have considered the sensitivity of downhole measurements of vertical electric field strengths as a function of depth and mass of injected CO_2. Figure 11 shows anomalous vertical electric fields in the observation well plotted as a function of depth below surface (at 25m intervals) at 1 Hz, for the 360 Amp-m vertical electric dipole source at 3,370m depth (as before), for 1, 2.5, 5, and 10kt of injected CO_2. Amplitude increases and spikes at the source dipole depth as CO_2 mass approach 10kt. (For actual finite length receiver electrode arrays, the measured signals are the receiver electrode separation times the field strengths plotted in Figure 11.) For the smallest CO_2 mass considered (1kt, 79 m span), anomalous E_z evidently changes sign within the plotted interval, but not for the greater CO_2 masses considered.

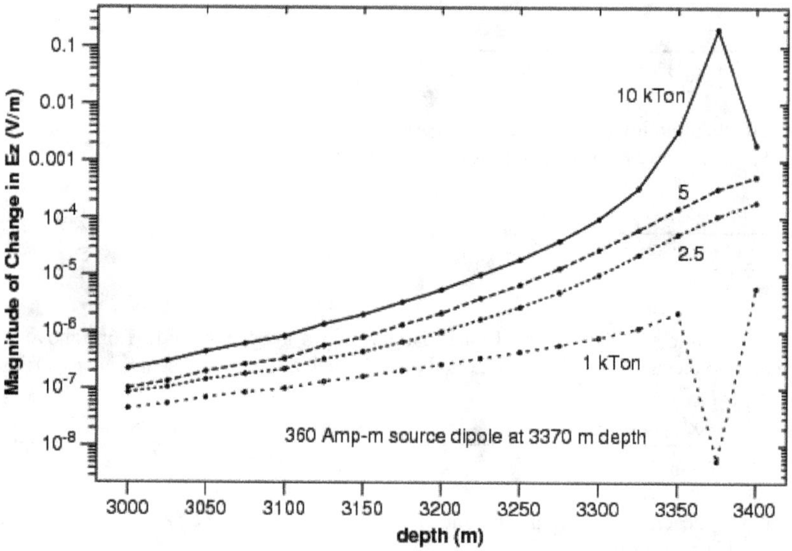

Figure 11. Anomalous vertical electric fields (Ez) measured at the observation well, as a function of depth for various amounts of CO_2.

Anomalous vertical electric field magnitudes, relative to vertical electric fields without the injected CO_2, are plotted in Figure 12. We see increased relative strength of the anomalous vertical electric field away from the source dipole where source fields are weaker, with maximum relative anomalies of 0.014, 0.027, 0.04 and 0.12 for 1, 2.5, 5, and 10kt CO_2, respectively. These anomalous fields are much smaller than the previously mentioned results of Swanepoel *et al* (2012), presumably due to the destructive interference of image dipole fields in the resistive basement immediately below the CO_2 mass (as discussed above).

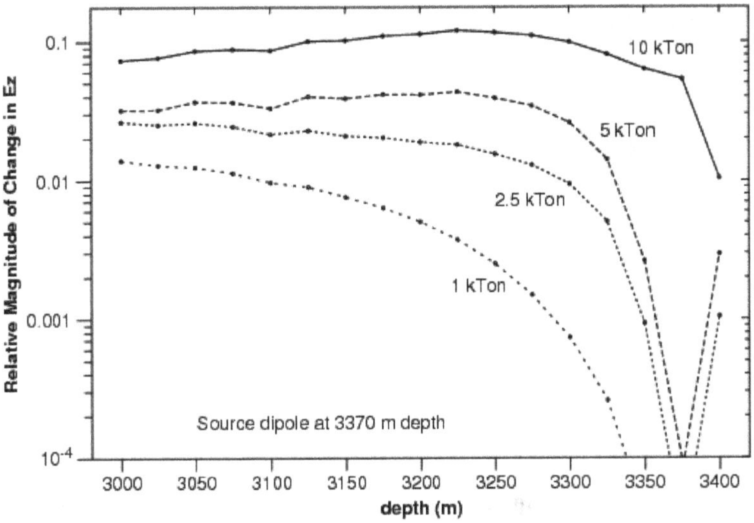

Figure 12. Relative anomalous vertical electric field (Ez) in observation well, as a function of depth for various amounts of CO_2.

Cable Crosstalk in Single Well Monitoring

When using a monitoring well for a downhole dipole source and downhole magnetic and/or electric field sensors, an important issue is potential crosstalk between the cables going to the source and receivers. For the Aquistore monitoring well, there were constraints on the cables that could be deployed to carry current to the source electrodes. We therefore considered the effects of non-optimal cable design on the electrical crosstalk between source and receiver wires. A 12-conductor cable was considered for powering the source electrodes (see Cable Design 1 in Section II below). The source cable design contained a set of three conductors 0.0029m from the cable centre and 120° from each other, connected to one electrode; and a similar set of three conductors rotated 40° from the first set, connected to the other electrode. A second 12-conductor cable with conductors separated by 0.00185m would be used for leads to the receiver electrodes. The two cables were proposed to be deployed outside the well casing on opposite sides.

For comparison with the above single-well anomalous fields, Figure 13 plots the expected level of inductive crosstalk at 1Hz between source dipole cables and downhole receiver dipole leads in a separate cable as a function of receiver lead/transmitter cable separation, for parallel transmitter and receiver cables and a 12-amp source dipole current. Since each of the cables consist of several wires, the relative positions of wires in one cable vary with respect to those in the other with rotation of either cable about its entire length. Therefore, inductive crosstalk depends on the rotation of each of the cables with respect to the other. The two inductive crosstalk values plotted are for the pair of cable rotations giving the largest magnitude crosstalk, and for the crosstalk averaged over all possible cable rotations. This assumes no twisting of cables, or identical twisting of source and receiver cables. Differing twists in source and receiver cables would give less inductive crosstalk. With some differences in twist randomizing orientations along the cables, the actual crosstalk is expected to be closer to the averaged crosstalk than to the worst case. For these cable configurations, crosstalk varies approximately inversely with the cube of the cable separation (and varies proportionally with frequency). A typical casing size for monitoring wells is 5.5" or 7" (14cm or 18cm) outer diameter, giving the maximum cable separation.

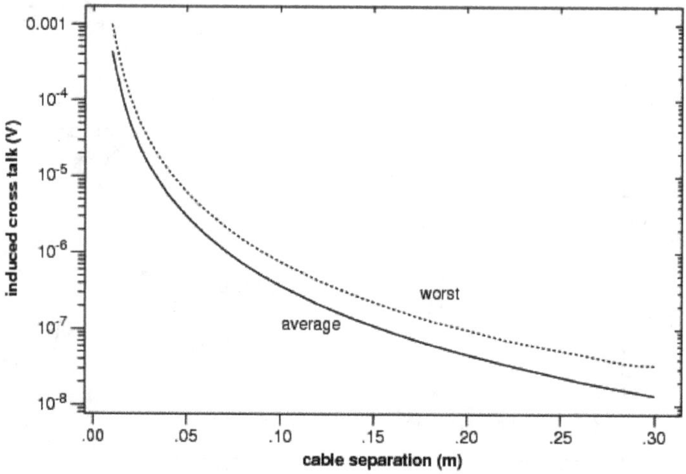

Figure 13. Expected inductive crosstalk between the source electrode cable and downhole receiver electrode leads as a function of the separation between the source cable and receiver leads.

For a 1cm separation between source cables and receiver leads, the expected inductive crosstalk at 1Hz is 4×10^{-5} V; for 15cm separation it is 9×10^{-8} V. The electrode noise levels expected at 1Hz with half an hour of stacking is 4×10^{-8}V (with steel or brass electrodes). The magnitudes of average crosstalk noise and electrode noise are comparable at 21cm cable separation, which is greater than the likely casing diameter. For 1cm cable separation, crosstalk is much larger than electrode noise. So, for 1cm receiver/transmitter cable separation (i.e. cables next to each other), with 10m receiver electrode pair separation, for 1kt injected CO_2, the magnitudes of all but one of the sample points of change in E_z plotted in Figure 11 are below that of the expected inductive crosstalk. For 2.5kt injected CO_2, downhole E_z anomalies are above crosstalk in the bottom 125m plotted. For 15cm receiver/transmitter cable separation (e.g., for casing deployed cables), expected inductive crosstalk is below the E_z anomaly expected in the observation well due to CO_2 injections from 1kt to 10kt, for receiver electrodes spaced 3 m or more, for all but one of the points plotted in Figure 11. Thus, we see that crosstalk and cable separation are important variables to consider over the possible range of cable separations.

Use of Borehole Magnetometers

We have also computed downhole magnetic field responses. For a 1D background conductivity, with the injection/observation well separation in the x axis direction and a vertical electric dipole source in the observation well (as we have been considering), the only magnetic fields at the observation well are due to the inhomogeneity (the CO_2) and are in the y direction. Anomalous horizontal (By) responses at 1 Hz, for the same 360 Amp-m vertical electric dipole source, for 1, 2.5, 5, and 10kt of injected CO_2 are shown in Figure 14. Using simple 1,000 turn 0.075×0.75m (air core) windings as inductive sensors gives 56.25m^2 effective coil area, and coil outputs of 56.25 V/T, so the lowest signal levels of Figure 14, would correspond to $2.3\times10\text{-}5$, $2.8\times10\text{-}4$, $8\times10\text{-}4$ and $1.3\times10\text{-}3$ V, respectively, for the four curves. For coils with no downhole amplification, crosstalk between transmitter cable and receiver cable is expected to be by far the largest source of noise. Assuming 10-15cm transmitter/receiver cable separation, these signal levels are well above the expected crosstalk. Therefore, a borehole magnetic field sensor, possibly wireline deployed, is a feasible monitoring tool.

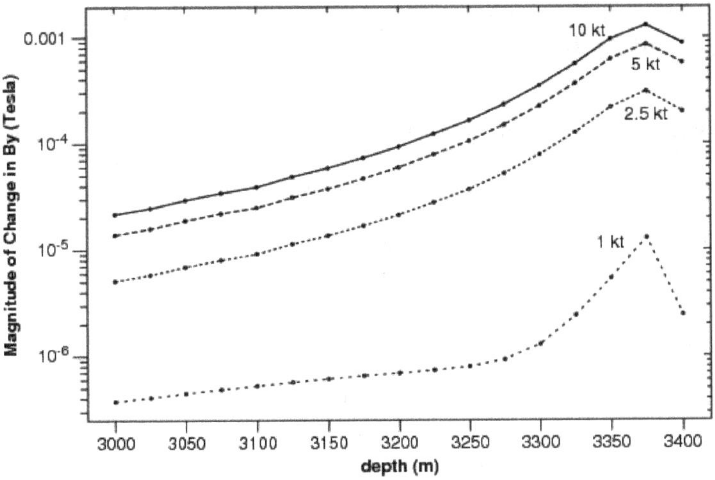

Figure 14. Anomalous horizontal magnetic field (By), as a function of depth, for various amounts of CO_2.

MODELING SUMMARY AND CONCLUSIONS

We have built a 1D resistivity model of the Aquistore sequestration site. For this model, we have considered borehole-to-surface and single-well (i.e., borehole to shallower E-field measurements in the same well) EM monitoring using a downhole vertical electric dipole source. We have modelled a source with realizable moments, located in a monitoring well 100m from the injection well, similar to the Aquistore site. The borehole-to-surface measurements are sensitive to injected CO_2 masses between 0.8kt and 270kt, with peak sensitivity at 10kt. Downhole vertical electric field measurements (the single-well configuration) are similarly sensitive to injected CO_2, but require separation of source and receiver cables for signals to rise above the expected crosstalk level induced in the sensor cable. Crosstalk is a function of cable design and location. We considered non-optimal cables due to specific project limitations and found the single-well CSEM method could still provide CO_2 detectability with adequate cable separation (e.g. 7cm for 1kt CO_2, 100m well separation, and untwisted cables at 1Hz).

An important conclusion is that the sensitivity of CSEM measurement is site specific. For the Aquistore geologic model of CO_2 injection just above resistive basement, the basement has a strong effect on the modelled EM fields. Also, there is a clear peak in sensitivity versus volume of injected CO_2, a result which again is specific to the site and transmitter/receiver geometry. We have shown that for a given model and well spacing, we can find an optimal source electrode location within the proposed EM source well.

The overall conclusion of this study is that the CSEM method with borehole source dipoles could be a useful monitoring tool for the planned Aquistore sequestration pilot.

SECTION II: DESIGN ENGINEERING FOR AQUISTORE BOREHOLE EM MONITORING

Design Overview

This section discusses the design of the electromagnetic (EM) transmitter for a borehole dipole source and its ability to provide sufficient current, voltage, and power as described in Section I. There are several constraints to be considered when designing the EM transmitter. The constraints and design options to overcome the constraints are also discussed. Major design constraints include: (1) the total cable length of 3,400 m (including drilling 50m into the basement) (2) instrumentation of the bottom 200 m of the borehole, and (3) the maximum transmitted (Tx) voltage of 1,000 volts. There are three major elements to the system design:

1. The surface transmitter that provides a carefully controlled current source.
2. The subsurface electrodes and supporting casing.
3. Wires and cables to connect the electrodes to the subsurface.

Our suggestion is to make use of an existing, off-the-shelf, transmitter. The transmitter needs to provide high power, high voltage, and a current-controlled waveform with an accurate phase reference that can be used with one or more phase-locked receivers on the surface. Table 1 lists three possible transmitters. All three companies have significant track records in supplying this instrumentation. Two of the transmitters, the TXU-30 and the GGT-30, have almost identical specifications, including a maximum voltage of 1,000V and a maximum transmitted power of 30kW. The third system, the IRIS VIP10000 has a much higher maximum voltage, 3,000V, but much lower maximum power, 10kW.

Table 1. Companies that can provide high power, high voltage and current controlled waveform transmitters.

Transmitter	Maximum Power	Maximum Voltage	Maximum Current
Iris Instruments VIP 10000	10 Kw	3,000 volts	20 Amps
Phoenix Geophysics	30 Kw	1,000 volts	45 Amps
Zonge Engineering	30 Kw	1,000 volts	45 Amps

The design for the Aquistore monitoring well, at the time of our planning, called for the deployment of transmitting electrodes on the outside of the casing. A critical part of the design is to make certain that there are no conductive electrical pathways near the electrodes. Possible electrical pathways include:

- The casing itself.
- Tubing inside the casing.
- Armoured electrical cable.
- Metal tubing along the outside of the casing for fibre-optic lines, electrical cables, fluid sampling, or other purposes.
- Electrical instrument housings with a common ground.

If any of these pathways come in direct, electrical contact with an electrode or multiple electrodes, they can completely short circuit the source dipole or provide a current path from one electrode to shallower depths. Either of these cases would decrease the effectiveness of the source dipole and/or create interference effects.

Even if the pathways discussed above do not directly contact the electrodes, they can still create significant effects if they are in electrical contact with the formation near the electrodes. To minimize these effects, it is important whenever possible to:

- Use electrically nonconductive materials such a fiberglass for casings.
- Use polymer encapsulated tubing for instrumentation.
- Insulate housings of instruments with polymer resins.
- Isolate the grounds of down-hole instruments.

When it is not possible to completely insulate all of the metal surfaces from the formation, it is important to:

- Keep the exposed metal surface as far from the electrode as possible; the impact of such a leakage path generally will decrease dramatically with increased distance.
- Keep the exposed metal surface as small as possible, thus increasing its effective contact resistance.
- Avoid having dissimilar metals in contact with each other and the formation.

It is important to note that leakage effects are difficult to quantify and model. The surface impedance of the metal depends on the geometry of the surfaces, the types of metals, and the composition of the pore waters or cement in contact with the metal. It is strongly recommended that the electrodes are placed on fiberglass casing. Electrodes have been successfully deployed on fiberglass wrapped casing of well sites in the 800m to 1,000m range (Tøndel *et al*, 2011, Schmidt-Hattenburber *et al*, 2011). The site at Ketzin Germany used Ryt-Wrap™ (Schmidt-Hattenburger *et al*, 2011), and the site described by Tøndel *et al* (2011) used a proprietary method. We are not aware of any deployments to depths of 3,000m using fiberglass-wrapped casing. The primary issue with fiberglass-wrapped steel casing is the damage that can occur at the collars when making up the casing string and presence of centralizers and cable and instrument protectors.

Standard oil-field tooling invariably damages the fiberglass coating near the collars. The methods of dealing with this problem, ranked from the most to least practical, are:

- Keep the electrodes as far from the damaged areas as possible by placing them near the mid joints.
- Patch the damaged areas by painting them with quick-set epoxy or similar polymer. This is not difficult to do, but the value of the coating is questionable, particularly if it is applied in cold or wet conditions.
- Place rubber sheeting or prefabricated covers over the joints. Often this is not compatible with other design aspects, such as centralizers and cable protectors.
- Develop tooling that will not damage the fiberglass coatings. This is likely beyond the scope of this project.

Although there are a number of designs for casing centralizers and cable protectors used to secure the centralizers to the casing, the design invariably involves either set screws or metal teeth designed to bite into the metal casing. This pierces the electrically insulated fiberglass layer, and the entire surface of the centralizer/protector becomes a leakage path. There are a couple of work-arounds for this problem. The first work-around is to keep the electrodes as far from the centralizer/protector as possible. If the centralizers/protectors can be kept near the collars, then this work-around becomes feasible. The second work-around would be to use centralizers that do not pierce the fiberglass insulation. These types of centralizers were successfully implemented in the Ketzin study. Their centralizers used large, flat clamping surfaces, so the total static frictional force was sufficient to counteract the expected shear forces on the centralizers.

Electrode Design

Figure 15 shows the conceptual design for a 40" (roughly 1m) long electrode used for monitoring of electric fields. The electrode was designed assuming a fiberglass-wrapped steel casing, but could easily be adapted to a fiberglass casing. Primary design considerations for the electrodes are to provide effective electrical insulation between the steel casing and the electrode, and make certain the electrode does not move on the casing during well completion. The proposed design uses a pair of half-cylinder shells of stainless steel (type 316) that are bolted together. The inside of the electrode is covered with a layer of fiberglass to provide an additional layer of insulation and is attached from the coated metal casing using a circular metal block that is welded to the casing. The metal block is covered with a layer of fiberglass, and then covered by a cup-shaped cap of PEEK or similar plastic. The electrode is designed to fit over the metal block locking it into place on the casing.

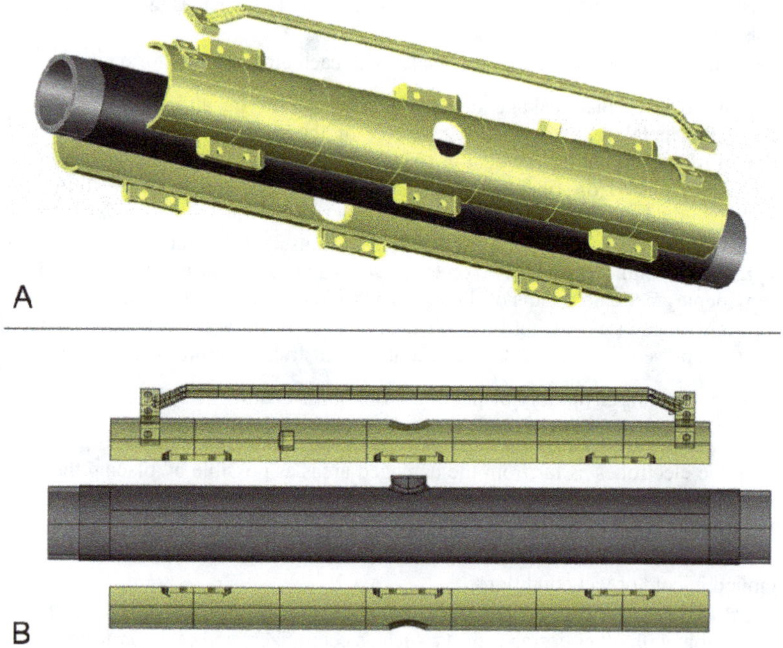

Figure 15. Proposed electrode design to prevent the electrode from moving upon well completion and keeping the electrode completely isolated from the steel well casing: (A) isometric view and (B) side view of assembly.

The attachment point for the electrode is critical to protect the relatively delicate copper conductors from both physical damage and corrosion. It is important that all of the exposed metal surfaces are created from the same type of metal (type 316 stainless steel).

Figure 16 shows a possible takeout. The connection to the copper cable is encapsulated in polymer resin and brazed into the end of a piece of encapsulated stainless steel tube. The tubing acts as the final conductor and is attached to the electrode via standard, high pressure threaded couplings.

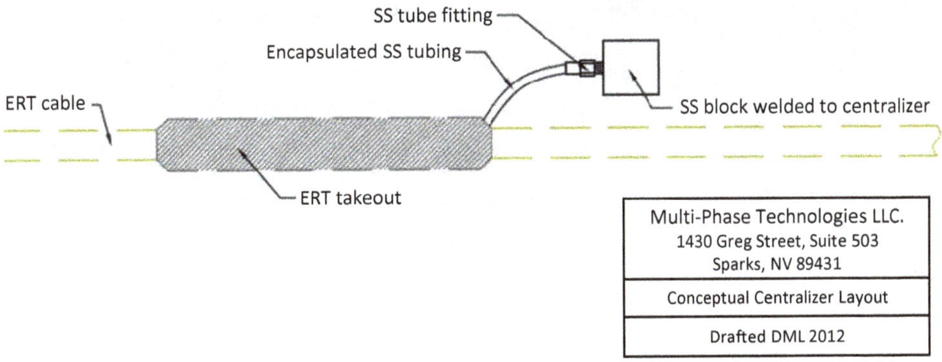

Figure 16. Drawing of a possible take out to prevent leakage along the casing.

Two additional factors to be considered in the higher current EM source electrodes are the overall contact resistance of the electrodes and the ohmic heating of the electrodes. Virtually all of the electrical energy transmitted into the ground will be converted into heat. The fraction of the total energy dissipated near the electrodes is of particular concern. In Appendix A, the approximate heat (temperature) generated by the electric current flow near the electrodes is derived. The increase in heat for a continuous current flow can be summarized as:

The constant A is dependent only on the geometry of the electrodes, and k is the thermal conductivity of the formation. The thermal conductivity, k, depends on rock type and formation age, and is less variable than electrical resistivity. For a nominal thermal conductivity value of 3 (W/m/C), a casing diameter of 4.5 inches and lengths of 1m, 2m and 3m, the value of A/k is approximately equal to 0.020, 0.086 and 0.0050 (C/VA/m), respectively. Table 2 shows the median values of resistivity near the Precambrian base rock of four well logs near the Aquistore site bedrock (White, 2012). For a resistivity value of 3 ohm-m, which corresponds to the median value above the Precambrian in Well *12106200613W200*, a current flow of 20 Amps would generate temperature changes of 24°C for the shortest (1m) electrode. The increase in temperature is likely to damage the insulation on the cables. Increasing the length of the electrode to 3m would drop this value to about 6°C, which would be considered acceptable. For a background value of 21 ohm-m, the same 3m electrode with a current of 20 Amps would produce a temperature increase of 42°C, which could damage the insulation on the cables. In this case, the current flow would need to be reduced to below 10 Amps to keep the temperature increase below 10°C.

Table 2. The average resistivity from well log data in four wells near the Aquistore site. The resistivity values are given near the Precambrian bedrock.

Well UWI	Top Precambrian (MD)	Median Resistivity100 m Above Precambrian (Ohm-m)	Median Resistivity Below Precambrian (Ohm-m)	Median Resistivity Bottom 10 m
101150700308W200	3169.08	21.0	26.5	568
12106200613W200	2905.52	3.0	23.4	149
101132400203W200	2961.08	4.2	1,630	1,940
33023001710000W	3589.01	14.5	4,510	21,300

Due to the high resistivity values, there is a high likelihood that the current flow within the Precambrian will be less than 10 Amps and possibly small fractions of an ampere. Due to the uncertainty in the resistivity near the Precambrian, it is important to place at least 3 source electrodes with two of the electrodes above the Precambrian interface. Although the configuration is less sensitive to the CO_2 plume in the reservoir, it is far more likely to produce sufficient current flow. It is important that the source electrodes be as long as possible. Unfortunately, the need to keep the electrodes away from the casing joints limits the length of the electrodes to something substantially less than 13m, the nominal length of the joints. This, and the difficulties in handling extremely long electrodes, limits the length to 2m or 3m.

The second factor closely related to heating is the contact resistance of the electrodes. Both the power and the voltage of the sources are limited; the resistance of the cables and electrodes may also limit the maximum current flow and need to be considered in the design. For a cylindrical electrode the approximate contact resistance can be found from the single driven rod formula of Tagg (1964). Tagg's approximation for a rod at the surface is based on image theory, but one can back out the whole-space formula as:

$$R \cong \frac{\rho}{2\pi l} ln \left(\frac{2L}{d} \right)$$

where R is the approximate resistance of a grounded cylindrical rod, ρ is the background resistivity, L is the length of the rod, and d is the diameter of the rod. For the 3m long electrode and a background resistivity of 21 ohm-m the electrode contact resistance is still only 5.18 ohm. At these resistances, the voltage drop at 20 Amps would be 103.6 volts and the power loss about 2,072 watts. Both the voltage drop and power lows are substantially lower than the capabilities of the transmitters and are not limiting factors in the current flow, even when the effects of two electrodes are added together.

Cable Design

Since the depth of the site is over 3,300m, designing a durable cable is a key obstacle. The present design is to have separate cables for the EM source cable and ERT receiver cable(s). As discussed in the modelling section, separate transmitting and receiving cables are required if measurements are made with the receiver and transmitter in the same hole. An early borehole-to-surface EM study called for a single pair of source electrodes placed below the reservoir (Smith and Beyer, 2011); however, for Aquistore, we feel that three electrodes are needed for the source: two below and one above the reservoir. This gives flexibility for increasing the signal-to-noise ratio and still leaves a source in the event that an electrode is damaged. The cable can be designed for two likely possible construction scenarios: (1) a combination of standard wire line cable and custom cable or (2) a continuous custom cable.

Design 1

Design 1 would use 3,100m of standard armoured wire-line cable. The suggested cable is the Rochester Cable 12-H-464. This cable has 12 conductors of 20 gauge wire. This cable consists of 12 wires plus filler material wrapped with tape and covered with two layers of steel armour. This design would use identical 12 conductor cables for both EM and ERT. For ERT, each individual conductor would attach to a single electrode. Each wire will have about 120 ohms resistance.

The EM cable will use three conductors together to reduce the resistance to about 40 ohms. The remaining wires in the EM bundle will connect to three additional ERT/style electrodes to give a total of 15 ERT electrodes, plus three EM source electrodes. The transition at the bottom of the borehole will need to have a very high-strength clamp design to pick up the force on the cable. The cable will need to be clamped periodically along the length of the casing.

The advantages of the armoured wire-line cable are:

- Durability: the steel armour is designed for deep borehole use and has a high probability of surviving to the required depth.
- Shielding: for the EM technique the armour would act as a cable shield, giving some electrical separation between the transmitter and receiver bundles. Without the shield, it is not possible to conduct single-well EM.
- Price (refer to Table 3): If the lowest temperature range (135°C continuous) is used, the cable is about $2.25 per foot (0.3048 m). This puts the cost of the cable at about $25,000 per cable, (which is lower than initially estimated as part of the Aquistore planning). The next highest temperature range cable increases the price to $3.25 per foot (0.3048 m) or about $35,000. (Note that this does not include the bottom 200 m of cable, electrodes, fabrication, and clamps, which could be substantial costs). Table 3 provides the estimated costs for the armoured wire line cable.

Table 3. Estimated Costs for the armoured wire line cable (Design 1).

Item	Est. Cost
Armoured ERT cables	$71,500
Armoured Source Cables	$34,375
Polymer ERT Cables	$8,550
Polymer Source Cable	$4,275
Transition	$15,000
Fabricating Cable Takeouts	$9,000
Electrodes	$18,000
Total	$160,700

- Testing: Since this is a stock cable, samples could be received quickly for clamp designs, connection tests, etc.
- Scheduling: The ERT arrays can be built in parallel with the cable manufacture and spliced on at the end. Although a special cable run is needed, it would be a short run. Most of the companies that build out small-run custom cables are not equipped to handle 3,400 m runs of cable.
- Redundancy: We would have two cables with somewhat overlapping capabilities; as with any deep installation, there is a chance of failure, and having two cables makes it likely that at least some functionality would be present.

The disadvantages are:

- Transition: We will need a transition/splice at ~3,100m to a custom cable type, because the armoured cable will short circuit the ERT electrodes. Stripping the armour away for the bottom 200m is not an option either, because of the way the cable is designed. If the armour were stripped, then the remaining loose wire would never survive. The transition is a major reliability issue and was likely the source of problems at the Cranfield site (La Brecque, 2012). Since there will be a transition from high tensile strength cable (18,000 lbf) to low tensile strength (probably 1,000 to 2,000 lbf), there will need to be a clamp system that can pick up the difference, so that any tension on the upper 10,000ft (3,048 m) of the cable doesn't tear the ERT section apart. At Cranfield, the clamps were pre-welded to the casing. The pre-welded clamps created several issues. First, they had to do the splice on the rig floor. Since the cables had to be trimmed to the correct length, this meant they could not "pot up" the splice, and it apparently leaked water over time. Also, one of the transitions became damaged in handling of the casing; the order of the wires was mixed up during the splice process.

- Bottom Cable: We need a run of custom cable to build out the electrode arrays. The manufacture time may be as long as fabricating the main cables.
- Electrical Leakage: The armour is a potential electrical leakage path. If there is a pinhole anywhere in the 11,000ft (3,353m) of cable, water will go into the cable (water in the cable is inevitable over time) and will create a leakage path from the wire to the armour. We will have the company add a silicon water block to the cable to help reduce this risk; however there is a fairly good chance of one or more conductors having leaks over long time periods.

Design 2

Design 2 would use custom cable for both the ERT sensor electrodes and EM source electrodes. Design 2 would use custom-built metal-tubing encapsulated wires. A total of three encapsulated tubes would be included: two ¼-inch tubes each containing seven 20 gauge conductors, and one 5/16-inch tube containing three 14 gauge conductors. To allow connection to the electrodes, the "cables" would be custom-manufactured without the metal tubing, but with polymer encapsulation on the bottom 200m of the cables.

The advantages of this design are:

- Robustness: Using tubing-encapsulated cable is the most robust method of building the cable and has a high likelihood of surviving to depth.
- Electrical Isolation: This would likely be a better cable with a greatly reduced chance of leakage over Design 1. This is likely the most important consideration, since the underlying reason for the cable is to collect good electrical data.
- Source wire resistance: With the custom cable, we can get the single source wire resistance lower than the armoured cable, down to about 30 ohms. Unfortunately, this wire resistance is too high to achieve a current flow of 20 Amps with any of the transmitters listed in Table 1. The total wire resistance for both electrodes in a transmitting dipole would be 60 ohm. At 20 Amps, this would result in a voltage drop of 1,200 volts, too large for the Phoenix or Zonge transmitters, and a power loss of 24,000 watts, too large for the Iris transmitter. However, if the formation resistance is low enough, it should be possible to achieve current flows of 15 Amps.
- Transition: The design of the transition would be essentially eliminated. Although the cable costs are substantially higher than the standard cables, much of this cost would be offset by reducing the need for highly specialized clamps along the casing. Table 4 shows the estimated costs for the custom-built cable.

The disadvantages of this cable design are:

- Cost: The initial cost of this cable is significantly higher than the standard logging cable. Refer to Table 4.
- Testing: No samples are likely available; some of design may have to wait for the cable to be complete.
- Time: Delivery times are likely to be longer for the custom cable.

Table 4. Estimated Costs for the custom built cable (Design 2).

Item	Est. Cost
ERT cables	$134,420
Source Cables	$66,660
Misc.	$3,444
Fabricating Cable Takeouts	$9,000
Electrodes	$18,000
Total	$231,524

The bulk of the costs listed in Tables 3 and 4 for both designs are the cost of the cables themselves. Not included in the costs are the insulating coating, modification of the cables, cables protectors, clamps, and clamp blocks. Cable costs for Design 1 are cheaper than those for Design 2. However, Design 2 offers a more robust design and will not need as robust of a clamping system as Design 1. Design 1 requires robust custom-designed clamps to maintain tension on the cables throughout the installation, whereas Design 2 could simply use industry-standard Cannon type clamps along the upper portions of the casing.

ENGINEERING DESIGN CONCLUSIONS/ RECOMMENDATIONS

Although the primary costs for the cables and electrodes are higher for Design 2, the overall, installed cost of both designs are going to be nearly equivalent. Design 2 is less dependent on rigorous clamping of the cable, which will be an extra expense. In addition, Design 2 is superior both mechanically and electrically. Unfortunately, a number of costs could not be fully evaluated at the time of design, since they require a more complete design of the well, more communication with groups involved in designing and completing casing, and those deploying other instrumentation on the casing. Both of the proposed designs would require insulating not only the casing, but the tubing or cables running to other instruments in the deployment zone. Also, both designs will require modification to the casing to allow robust clamping of the electrodes and cables to the casing. The proposed clamping solutions incorporate welding blocks to the casing, then covering them with fiberglass insulation. This will require either a long lead time or performing fabrication on the well site. One additional advantage to Design 2 is that it is possible, with careful design, to construct a clamping system for the electrodes that would not require the welding of blocks to the casing. A similar system was deployed in the injection well at Ketzin (C. Schmidt-Hattenberger, 2012). Although that deployment was a shallow site, the proposed electrodes are much longer, more robust, and could be designed with much higher clamping force achieving static friction forces approaching the tensile failure forces of the casing.

ANNEX A CABLE HEATING

Nearly all of the electrical energy transmitted into the borehole will be converted to heat in the cables, electrodes, and the formation. Although the overall energy loss in the cable can be quite large, the heat will dissipate over a very large volume and will not cause a significant temperature rise in the cable itself. However, on and near the electrodes, energy is dissipated over a fairly small area, possibly increasing the temperature to the point of damaging the cable or the electrical coating on the cables. This increase in temperature is discussed below.

For a given current flow, formula for electrical contact resistance can be used to estimate the total steady-state heat dissipation within the formation as:

$$WT = I^2 \cdot R \qquad \text{Equation A1}$$

where WT is the total dissipated power, I is the current flow, and R is the electrode contact resistance.

The temperature increase near an electrode

To calculate the temperature increase near the electrode, we assume both the current flow and heat flow to be radially outward. This should be a good approximation near the centre of the electrode for radial distance much less than the length of the electrode.

The volumetric Joule heat flow is given by:

$$q = e \cdot j = |j|^2 \rho \qquad \text{Equation A2}$$

where q is heat flow, ρ is the background resistivity, e is the electric field and j is current density. For the case where the current flow is radially outward from a circular rod, the current density, j, at radius r, is given by:

$$j = \frac{I}{L}\frac{1}{2\pi r} \qquad \text{Equation A3}$$

where L is the length of the rod. So total dissipated heat between the surface of the rod, r_1, and some radius r_2 would be:

$$q = \int_0^L \int_0^{2\pi} \int_{r_1}^{r_2} \rho \left(\frac{I}{L}\frac{1}{2\pi r}\right)^2 r\partial r\partial\theta\partial z = \frac{\rho I^2}{2\pi L} \ln\left(\frac{r_2}{r_1}\right) \qquad \text{Equation A4}$$

If the heat flow, q, were entirely dissipated on the casing surface, radius r_1, then the temperature increase, ΔT, for the same volume would be given by:

$$\Delta T = \frac{q_1}{2\pi Lk} \ln\left(\frac{r_2}{r_1}\right) \qquad \text{Equation A5}$$

where k is the thermal conductivity and q_1 is the heat dissipated at radius r_1.

Simply combining equations A4 and A5 overestimates the ΔT, since the heat is dissipated throughout the volume. The incremental heat flow in the region r_1 to r_2 near the electrode can be found by substituting r for r_1 and taking the derivative of the equation (A4) with respect to r:

$$q'(r) = \frac{\rho I^2}{2\pi L}\left(\frac{1}{r}\right) \qquad \text{Equation A6}$$

Combining equations A5, A6, and integrating gives:

$$\Delta T = \int_{r_1}^{r_2} \frac{\rho I^2}{2\pi L}\left(\frac{1}{r}\right) \frac{1}{2\pi Lk} \ln\left(\frac{r_2}{r}\right) \partial r = \frac{\rho I^2}{8\pi^2 L^2 k}\left(\ln\left(\frac{r_2}{r_1}\right)\right)^2 \qquad \text{Equation A7}$$

This formula is only valid near the electrode, where the electric current flow pattern and density is similar to that of a two-dimensional rod. The approximation breaks down for a radius greater than about ½ L. Substituting $r_2 = ½\, L$ gives the final equation:

$$\Delta T \approx \frac{\rho I^2}{8\pi^2 L^2 k}\left(\ln\left(\frac{0.5L}{r_1}\right)\right)^2 \qquad \text{Equation A8}$$

ACKNOWLEDGEMENTS

This work was supported by the CO_2 Capture Project (CCP). We acknowledge the support of the Aquistore Project, operated by the Petroleum Technology Research Center. We thank K. Dodds of the CCP for support and technical discussion of this work. G.M. Hoversten provided logs for Imperial Halkett Well 15 7 3 8. W. Zaluski provided additional neighbouring logs. We thank D. White for project technical support.

Lawrence Berkeley National Laboratory is supported by the U.S. Department of Energy under Contract No. DE-AC02-05CH11231.

REFERENCES

1. Andreis, D. L. and L.M. MacGregor (2010) Using CSEM to Monitor the Production of a Complex 3D Gas Reservoir - A Synthetic Case Study. *EAGE*, June 2010, 14-17.
2. Commer, M. and G.A. Newman (2008) New advances in controlled source electromagnetic inversion. *Geophysical Journal International*, vol. 172, 513-535.
3. Doughty, C. (2009) Investigation of CO_2 plume behavior for a large-scale pilot test of geologic carbon storage in a saline formation, *Transport in Porous Media*, doi: 10.1007/S112423-009-9396-z.
4. Hoversten, G.M., and E. Gasperikova (2006) Investigation of novel geophysical techniques for monitoring CO_2 movement during sequestration. *Report to Carbon Capture Project, Lawrence Berkeley National Laboratory*, publication no. 56822.
5. LaBrecque, D., Pers comm.,2012.
6. Ministry of Environment (2012) *Go green Saskatchewan*: Aquistore, =English.
7. Press, W.H., B.P. Flannery, S. A. Teukolosky, and W.T. Vetterling (1986) *Numerical Recipes, The Art of Scientific Computing*, Cambridge Univ. Press, Cambridge, Great Britain, 1986, pp 472-473.
8. Petiau, G. and Dupis (1980) A Noise, temperature coefficient, and long time stability of electrodes for telluric observations, *Geophysical Prosp*ecting 28(5), 792-804.
9. Petroleum Technology Research Center (PTRC) (2011) *Aquistore: Saskatchewan's deep saline CO_2 storage research project*, http://www.environment.gov.sk.ca/adx/aspx/adxgetmedia.aspx?MediaID=001689ad-a72e-4a96-b99b-cf45cf01b7db&Filename=Aquistore%20Fact%20Sheet%202012.pdf.
10. Schmidt-Hattenberger, C., personal conversation, 2012.
11. Schmidt-Hattenberger, C., P. Bergmann, D. Kielbling, K. Kruger, C. Rucker, H. Schutt, Ketzin Group (2011) Application of a vertical electrical resistivity array (VERA) for monitoring CO_2 migration at the Ketzin site: First performance evaluation, *Energy Procedia*, 4, 3363-3370.
12. Smith, J. T.,and J.H. Beyer (2011) Monitoring CO_2 Sequestration using a down-well electric dipole and surface receivers, *LBNL*, Unpublished manuscript.
13. Swanepoel, R., B Harris, and A. Pethick (2012) Three dimensional modeling for time-lapse cross-well CSEM monitoring of CO_2 injection into brine filled reservoirs, *Eleventh meeting of the Australian Society for Exploration Geophysics* (Abstract).
14. Tagg, G. F. (1964) Earth Resistances, George Newness Ltd, London.
15. Tøndel, Richard, J. Ingham, D. LaBrecque, H. Schütt, D. McCormick, R. Godfrey, J. Rivero, S. Dingwall, and A. Williams (2011) Reservoir monitoring in oil sands: Developing a permanent cross-well system: *Society of Exploration Geophysicists, Expanded Abstracts*, 30(1), 4077-4081.
16. White, D., personal conversation, Jun 14, 2012.

17. Whittacker, S., K. Worth, and C. Preston (2009) Aquistore: CO_2 storage in deep brines in Saskatchewan Canada. *Presentation at Institut Francaise Petrolier, Rueil-Malmaison, France, May 2009* http://www.ifpenergiesnouvelles.com/content/download/67983/1473837/file/26_Whittaker.pdf, accessed March 2012.

Carbon Dioxide Capture for Storage in Deep Geological Formations, Volume 4
Karl F. Gerdes (Editor)

Chapter 40

TEST OF A NEW BSEM CONFIGURATION AT AQUISTORE, AND ITS APPLICATION TO MAPPING INJECTED CO_2

Andrew D. Hibbs (deceased)[1]
GroundMetrics, San Diego, California

ABSTRACT: The goal of this project was to apply new electromagnetic sensor and source technology to image the carbon dioxide (CO_2) injected into the ground by the Aquistore R&D program. This report describes use of a survey of the site prior to the first injection of CO_2 and use of the acquired data and sophisticated modeling to assess whether the system would be able to detect a signal from the injected CO_2. This project conducted the first ever survey with a new electromagnetic source configuration that takes advantage of the high conductivity path provided by a conventional steel well casing to channel electric current to reservoir depth. The new data were in good agreement with conductivity logs for the injection and observation wells, and the model showed that the CO_2 plume would be detectable using the researchers' system. The next step in this research will be to conduct a survey after injection and establish the system's capability to image the plume.

KEYWORDS: CO_2 sequestration electromagnetic survey; resistivity; electromagnetic sensor; electromagnetic source

INTRODUCTION

A series of surveys in China over approximately the last eight years [1], and a 2011 pilot study in Saudi Arabia [2] have reported that the borehole to surface electromagnetic (BSEM) survey configuration has the capability to map hydrocarbon reservoirs at 2 km depth. Very recently TechnoImaging, LLC, working under a GroundMetrics, Inc. (GMI) led DOE project [3], has calculated via 3D modeling and inversion that BSEM can image an injected CO_2 reservoir at the 3 km Kevin Dome formation in Montana [4].

The BSEM method introduces, via reservoir depth electrodes, an electric current that flows to a counter electrode at the surface, adjacent to the well. The distribution of electric current in the ground is measured by a surface array of electric field sensors oriented radial to the well. One of the primary benefits of BSEM is that, in addition to probing to reservoir depth, signals can be measured out to a lateral distance of 2-4 km, providing a mapped area of approximately 25 km^2. In contrast, crosswell EM, the only other available EM technique at reservoir depth, has a range of no more than 1 km and can only map the region between two boreholes, corresponding to a mapped surface area of no more than ~0.5 km^2.

For this test program to further demonstrate use of BSEM for CO_2 monitoring, GMI made the following technology improvements:

1. Using a new type of electric field sensor to acquire the surface EM data. The sensor couples capacitively to the electric potential in the ground, removing the need for electrochemical interaction (i.e. ionic exchange) with the ground. The resulting eQube™

[1] Contact Paul Esposito - Business Development Manager - GroundMetrics Inc.

sensors have significantly higher measurement precision than conventional technology, and the inherent capability for permanent deployment for long-term monitoring.

2. Applying a new configuration for the source that takes advantage of the high conductivity path provided by a conventional steel well casing to channel electric current to reservoir depth.
3. Applying a new methodology to model the well casing. Present finite element, finite difference and integral equation methods are not well suited to a source that directly includes the well casing as part of the current path, owing to the very large conductivity contrast (>10^6:1) and scale length ratio (1 cm vs. 3 km). In collaboration with Berkeley Geophysics Associates (BGA), we have developed and tested a code that explicitly solves the DC field of a casing in a layered Earth model.

The goal of this program is to apply the new sensor and source technology to image the carbon dioxide (CO_2) injected into the ground by the Aquistore R&D program. This chapter describes a survey of the site done before the first injection of CO_2, with the intent that, if models project a detectable signal from the injected CO_2, then a second survey will be conducted after injection (planned for 2014).

DESCRIPTION OF THE WORK COMPLETED

This project was commissioned by the Carbon Capture Project (CCP) to evaluate the performance of borehole to surface electromagnetic (BSEM) survey technology for imaging injected CO_2 plumes at the Aquistore site. Aquistore is an independent research and monitoring project that intends to demonstrate that storing injected CO_2 deep underground (in a brine and sandstone water formation) is a safe, workable solution to reduce greenhouse gases. Aquistore is located in south-eastern Saskatchewan, Canada near the community of Estevan.

A significant question for an EM survey near a developed site is the level of cultural (manmade) EM interference. As a first check GMI used a small, man-portable B-field sensor to gauge the environmental magnetic field at Aquistore. The results of this were favourable, and so two parallel 34 m electric field dipoles were set up in the eastern portion of the site at the nearest location to the Boundary Dam Power Station. The spectra of the recorded E-field are shown in Figure 1. This recording was 15 minutes in duration and so only limited signal averaging was possible. The expected Schumann resonances are just visible at 8 Hz, 14 Hz, and 20 Hz and above. At 1 Hz the total environmental noise level is about 10 nV/m√Hz. At this noise level there is no visible cultural interference. The maximum time domain output voltage was 70 mV. When scaled to the 80 m dipole length used for the actual survey, this peak voltage scales to about 150 mV, which is easily within the dynamic range of the system.

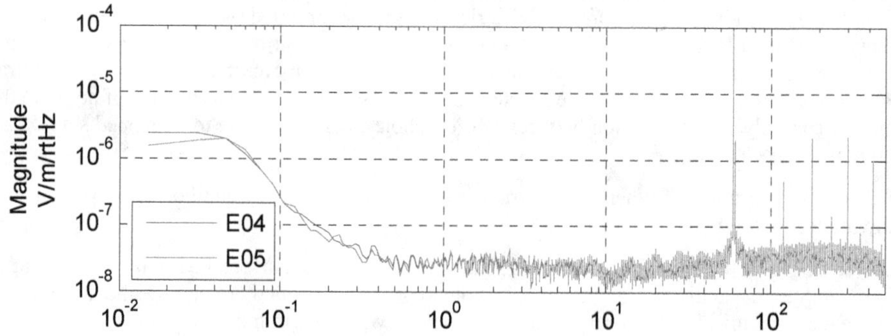

Figure 1. Spectra of the voltage closest to the power station.

The voltage is measured by two E-field sensors deployed in the eastern portion of the site. The total environmental noise at 1 Hz and above is approximately 10 nV/m√Hz.

Survey Execution

Finalizing the Survey Design

GMI had Dr. Clifford Schenkel of Berkeley Geophysics Associates (BGA) calculate the predicted CO_2 signal using a mathematical approach first published in the early 1990s via a code he developed for GMI [5,6]. This new code is a simplified model for DC only, and so for an electrical source it cannot include the inductive response of the target (i.e. the magnetic field caused by the target's redirection of the subsurface current). However, for electric current sources, the low-frequency asymptote of the full EM solution produces the maximum change in amplitude and phase to subsurface changes. For electric sources and electric dipole receivers this essentially means that DC modeling of the response at the surface for resistivity inhomogeneities at depth yields a good first order solution to the surface and subsurface fields. In May 2013, GMI conducted a preliminary test of the casing source at an oilfield in Montana. An example of the agreement between the measured surface fields and those calculated by BGA is given in Figure 2.

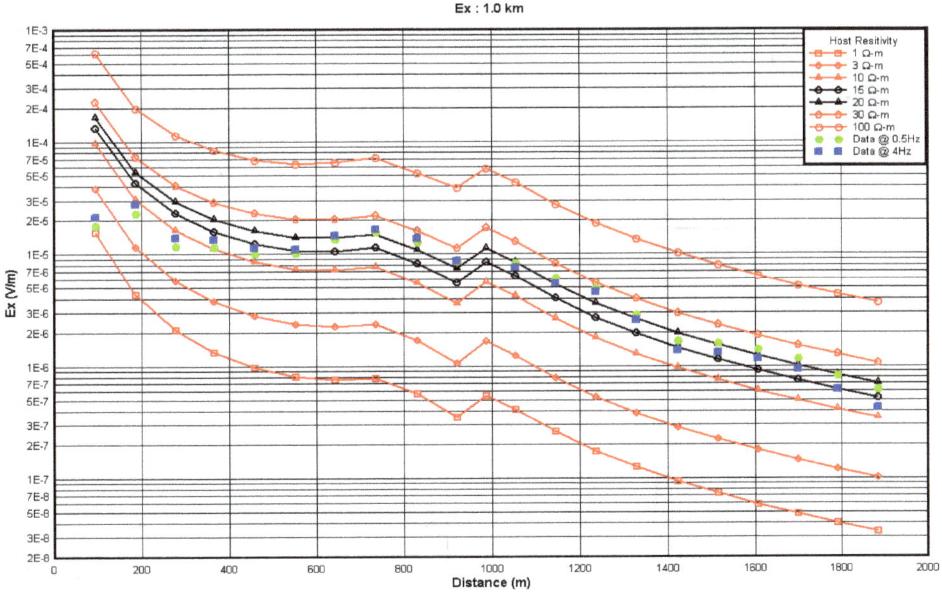

Figure 2. Comparison of the BGA model results (red lines) with measured data (green circles, blue squares) from a preliminary source test.

The average subsurface resistivity down to the depth of the well was 15 Ω-m - 20 Ω-m.

The projected surface field and change in that field due to CO_2 injection at Aquistore are shown in Figure 3. The top graph shows surface field vs. distance for eight discrete surface electrodes with 20 A total current. The bottom graph shows the projected change in the surface field due to injection of CO_2. The plume is modelled as a thickness of 30 m and radius of 150 m. At 800 m the projected change in surface field is 5 nV/m.

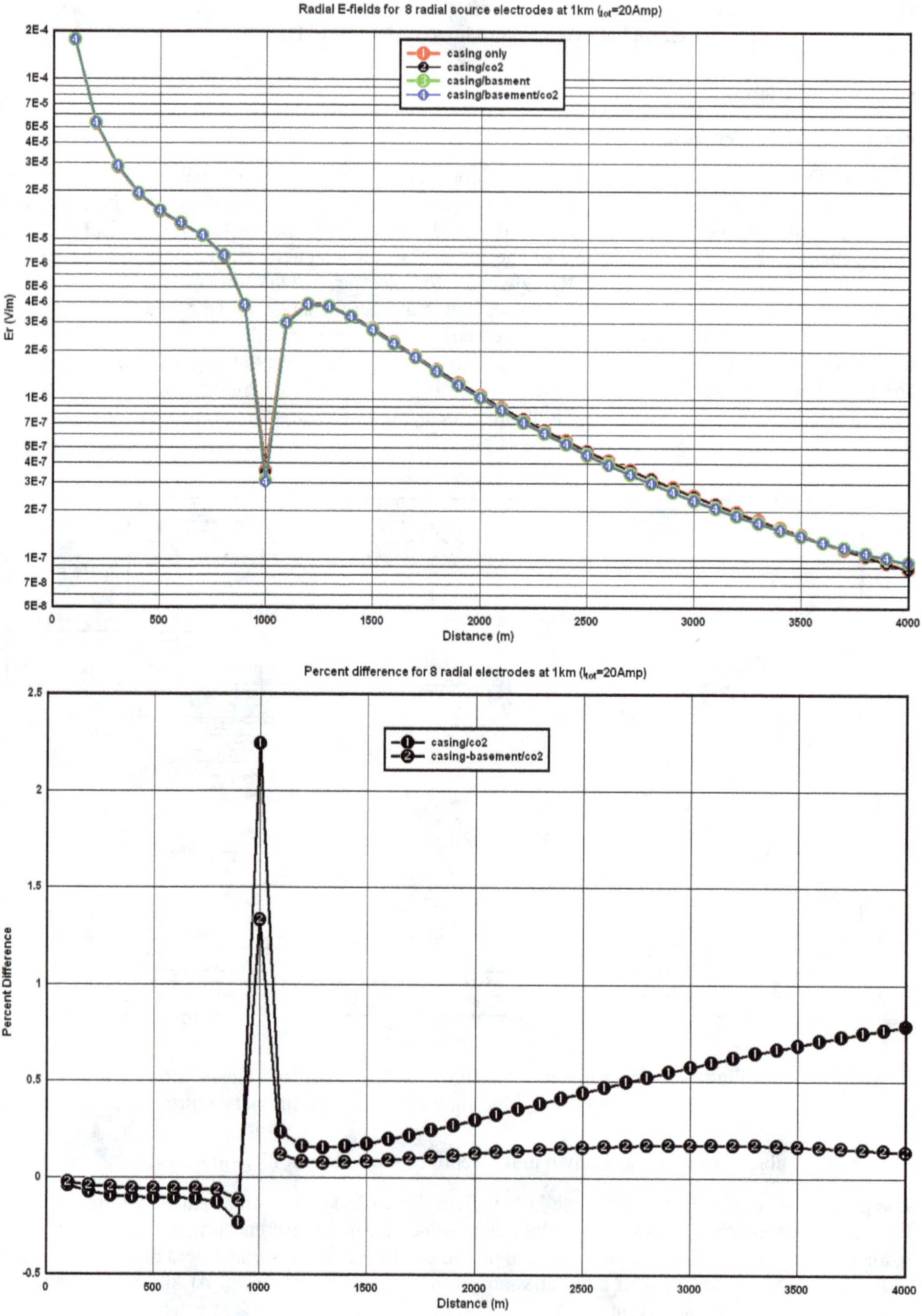

Figure 3. BGA's model results for the CO_2 signal at Aquistore.

The model calculation in Figure 3 is based on DC. Solutions for a casing in a conducting medium show that to first order the effect of attenuation at non zero frequency can be approximated by scaling by the skin depth of the host medium [7]. In a uniform medium of resistivity ρ, the skin depth δ at frequency f is approximately given by $\delta = 500\ (\rho/f)^{1/2}$. For a layered medium, ρ can be estimated by using a weighted average resistivity over the depth of the well. For Aquistore the weighted average resistivity is 2.6 Ω-m, which is very low, and results in a skin depth at 1 Hz of only 800 m. The formation depth is 3.4 km (~4 skin depths), which produces a signal attenuation at 1 Hz of 50 x, i.e. only 2 % of the signal remains.

The results of the modeling had the following implications for the survey:

1. There is little projected signal beyond 800 m offset.
2. The transmitted current should be increased and/or the measurement time should be extended.
3. The fundamental frequency should be as low as possible.
4. At least some data should be taken for a single electrode carrying all the current (24 A).
5. Telluric cancellation should be used to cancel atmospheric noise.

Conducting the DSEM Survey

The source current used at Aquistore was provided by a Zonge ZTG30 transmitter powered from a trailer mounted generator (Figure 4, left). GroundMetrics eQube™ electric-field sensors were deployed along two lines 1040 m and 1120 m long as shown in Figure 5. Individual sensor locations were determined using a Real Time Kinematic Global Positioning System (RTK-GPS). The quoted accuracy is ±1 cm and measurements made on emplaced sensors indicated the sensors were emplaced to the RTK-GPS markers within a total error of ±1.5 cm. A custom fixture that was used to align each eQube™ electric field sensor to the survey marker is shown in Figure 4.

Figure 4. Left: Zonge Generator. Right: Emplacement of two sensors relative to a survey marker.

For each receiver position (see Figure 5), current was transmitted for durations of at least one hour. However, sometimes a transmission was interrupted to allow the team to access part of the test site, and in some cases transmission comprised two or more time intervals that added up to one hour.

Spectra for two typical receiver data and the applied current are shown in Figure 6. The excitation waveform is a simple square wave of fundamental frequency 0.5 Hz. The harmonics (e.g. 1.5 Hz, 2.5 Hz, and 3.5 Hz) are clearly visible in Figure 6 well above 100 Hz. Even though these sensors were close to the end of Line 1 (800 m from the well), the spectra are so similar that the only region where one can be differentiated from the other is near the minimum at 0.24 Hz.

Figure 5. Aquistore receiver (blue) locations.

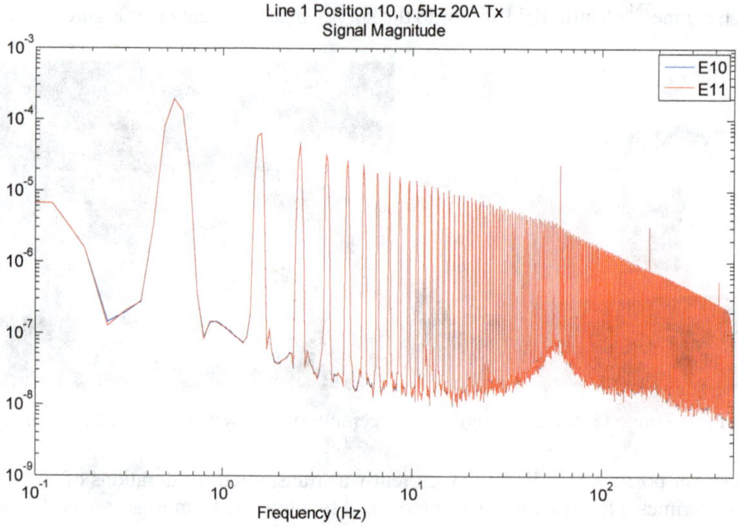

Figure 6. Spectra of two side-by-side parallel E-field sensors at 800 m distance.

Figure 7 shows the spectrum for one of the E-field sensors in Figure 6 and the spectrum of the current waveform recorded at the same time. The small details in the current waveform spectrum (e.g. the small subpeak just to the right of the base of the 1.5 Hz peak) are clearly recorded in the surface E-field spectrum. The vertical axis unit is V/m√Hz.

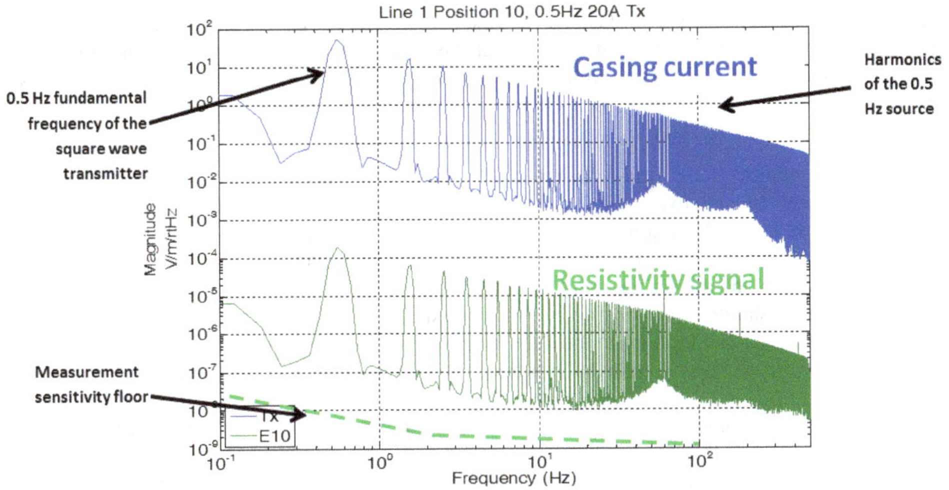

Figure 7. Spectra of the current in the casing and the E-field at 800 m distance recorded at the same time.

The coherence between the current waveform and the surface E-field waveform is given in Figure 8. The coherence is so high that it is visually indistinguishable from 1 at all harmonics of the transmitter.

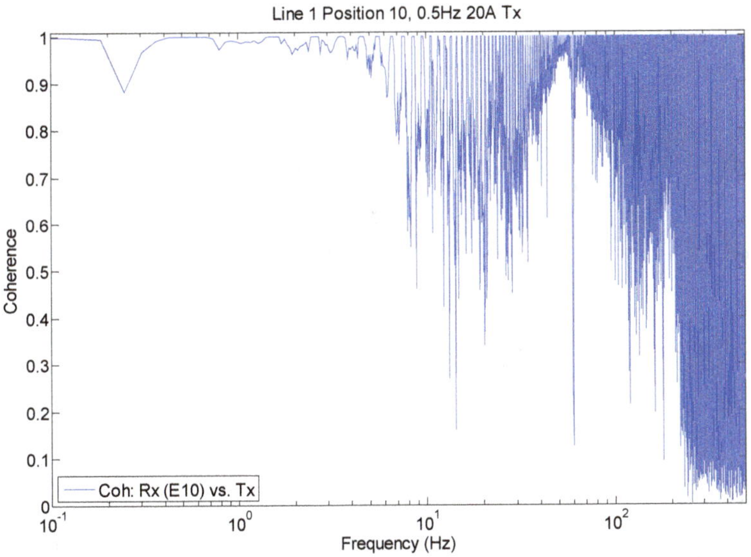

Figure 8. Coherence between the transmitted current and measured E-field at 800 m.

To make the coherence value easier to view we have plotted the logarithm of 1-coherence in Figure 9. A value of 10^{-6} in Figure 9 is equal to 99.9999 %. The coherence between the current and E-field is thus above 99.9999% at the spectral peaks (i.e. minima in Figure 7) of the transmitted current. This residual incoherent energy is a combination of local environmental noise and E-field sensor internal noise. It is important to note that the noise in the output current of the transmitter travels through the Earth as signal to the E-field sensor. Thus although the transmitter produces a level of broadband energy at frequencies between its harmonics, this energy has no negative impact on the EM measurement. There are two points to note:

1. At harmonics of the transmitted waveform, the coherence is very high (1-coherence ~10^{-8}) and equal between the two E-field recordings (E - E) and between one E-field and the current (E - I) recording at low frequency. At higher frequency (>3 Hz) the E - I coherence (red) is less than the E - E coherence (blue). Given that the spectral peaks are produced by the transmitter and measured at the transmitter with no path attenuation, E-I coherence should always be greater than or equal to E - E coherence. This unexpected result may be due to noise in the data recorder used to acquire the current waveform, or may be due to noise conducted into the current monitoring unit by the source electrode wires.
2. Above 3 Hz the coherence at frequencies at which the source is not emitting power is higher for E- E than for E -I. This is expected because parallel E-field sensors measure the same environmental noise and this is coherent above 3 Hz.

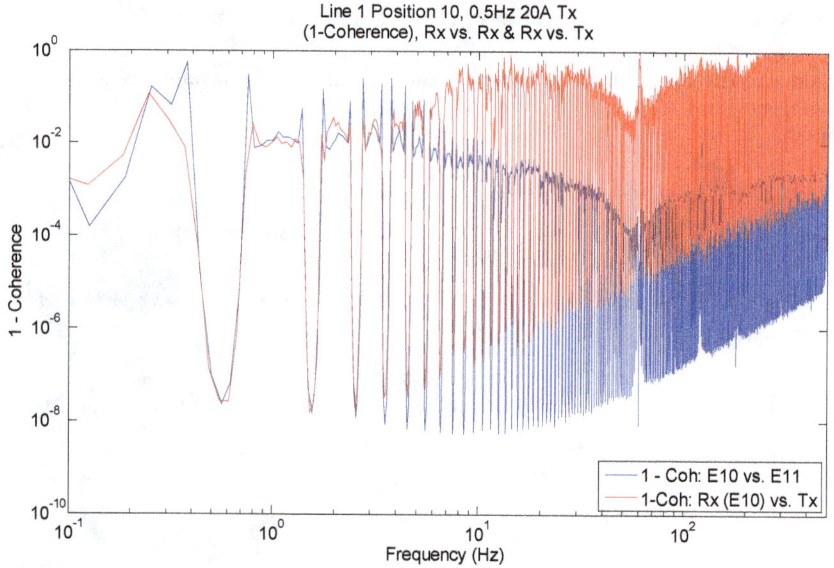

Figure 9. Data of Figures 6 and 7 plotted as 1-coherence to illustrate the very high coherence.

Data Processing

A typical recording of the surface E-field is shown in Figure 10, showing the transient airwave signal and the beginning and end of the square wave. The large overshoot present in the waveform is known as the airwave. It is caused by a magnetic field produced by the current flow in the wires between the transmitter and source electrodes. This magnetic field induces currents in the ground that are detected by the E-field sensors. The amplitude of the induced current is proportional to the rate of change of the current in the wire, and so is at its maximum at the beginning and end of each square wave.

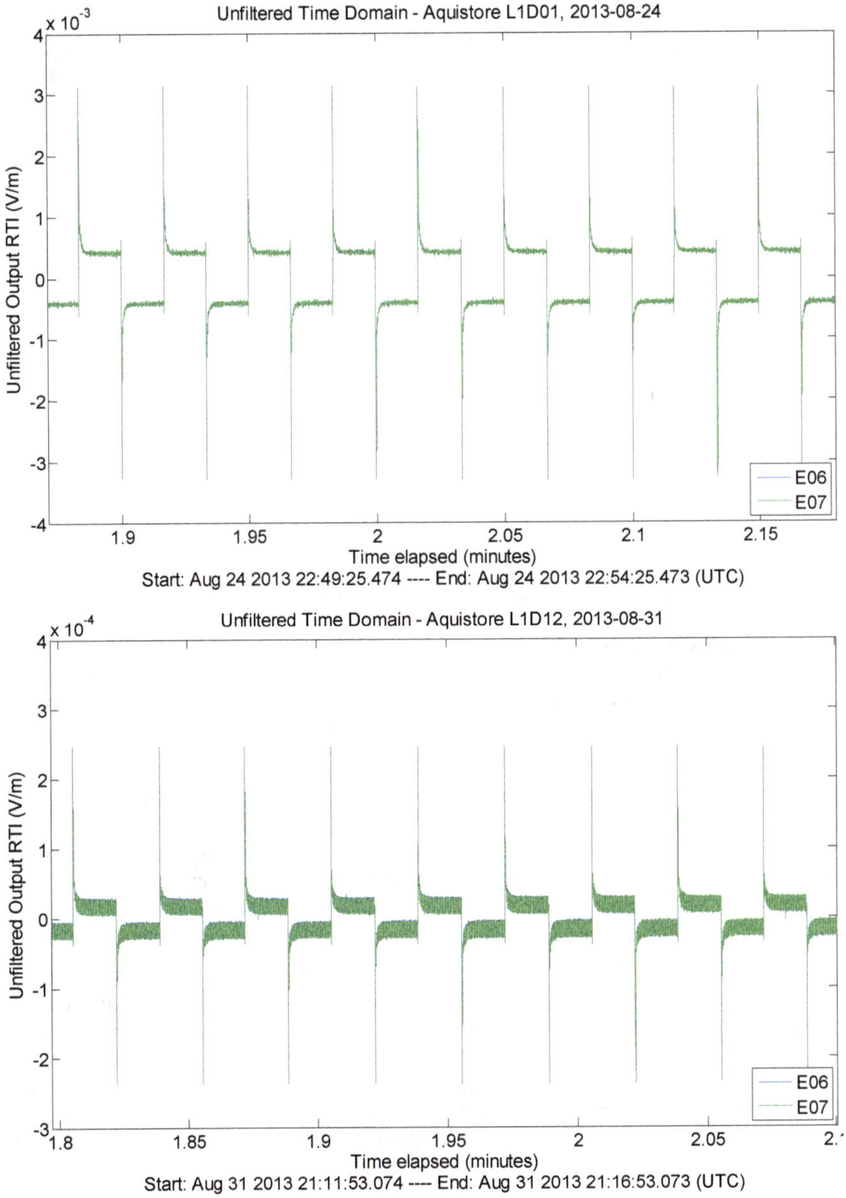

Figure 10. Typical E-field waveforms.

Figure 11 shows a typical recorded E-field waveform on an expanded time axis to highlight the part of the wave that is deleted. The first 12 data points (up to the blue dashed line) are set to zero in data processing.

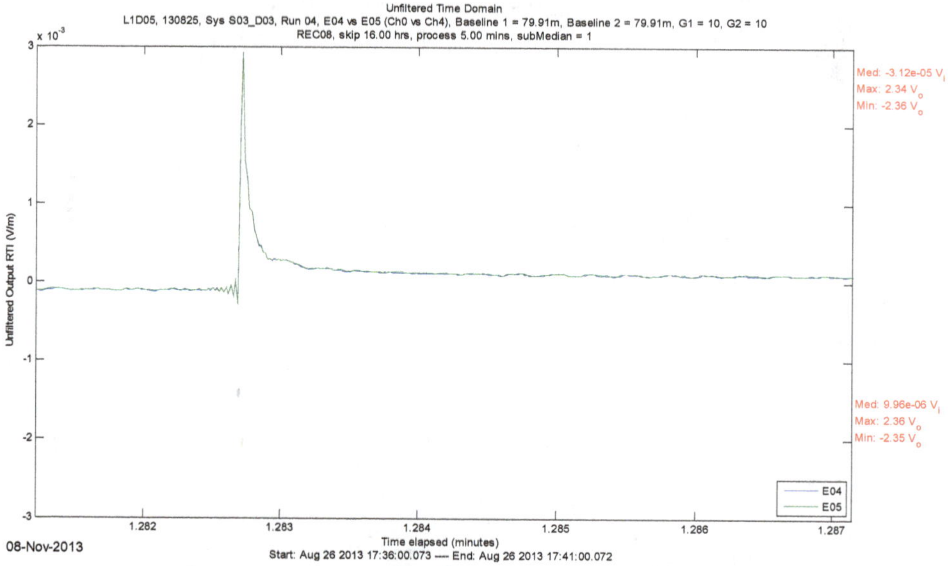

Figure 11. Transient airwave signal on an expanded time access.

The airwave component must travel from the wires to the sensors and so varies with distance across the survey area. The airwave occurs at the same frequency as the fundamental of the square wave and so its presence will affect the amplitude of all spectral components of the field. To remove the airwave signal from the data acquired at Aquistore we set the first 12 data points of each square wave to the average value of the E-field data (i.e. zero).

The standard way to process the EM data is to plot the transfer function between the recorded E-field and transmitter current at each survey location. This transfer function (TF) is calculated using the part of the E-field signal that is correlated with the transmit-current waveform. Thus, the TF measure excludes E-field noise that is present in the environment or produced in the sensor. However, as noted above, noise in the output of the transmitter that reaches the E-field sensor is included when determining the transfer function (i.e. such transmitter noise is counted as signal).

Figure 12 through Figure 15 show the DSEM signal. Our original intention for quantifying the E-field measurement noise floor was to inspect frequencies just below and just above the source signal frequency. However, as noted above there is broadband energy produced by the transmitter (Figure 10 and Figure 11) that is recorded by the E-field sensors at a higher level than the sum of their internal noise and the local environmental noise. Thus, to quantify the SNR of the survey we must look at intervals when the transmitter is off. Figure 12 and Figure 13 show E-field spectra acquired at the end of the day and in the morning prior to running the transmitter, respectively. The noise floor at 0.5 Hz is 10 - 15 nV/m√Hz in each spectrum.

Schumann resonances due to atmospheric noise are clearly evident in both spectra in Figure 12 and Figure 13. Parallel E-field data were acquired, and resulting spectra for coherent cancellation of the environmental noise in one sensor using a measurement of that noise by the second sensor are shown in Figure 14 and Figure 15. The cancelled noise in Figure 14 corresponds to an E-field noise of 5 nV/m√Hz, while that of Figure 15 is 12 nV/m√Hz. In comparison, the average noise level at 0.5 Hz over a 10-hr run was 636 nV/√Hz, corresponding to 7.95 nV/m/√Hz for eQube™ sensors separated by 80 m. This is substantially higher than the internal noise level of the eQube™ sensors of 100 nV/√Hz.

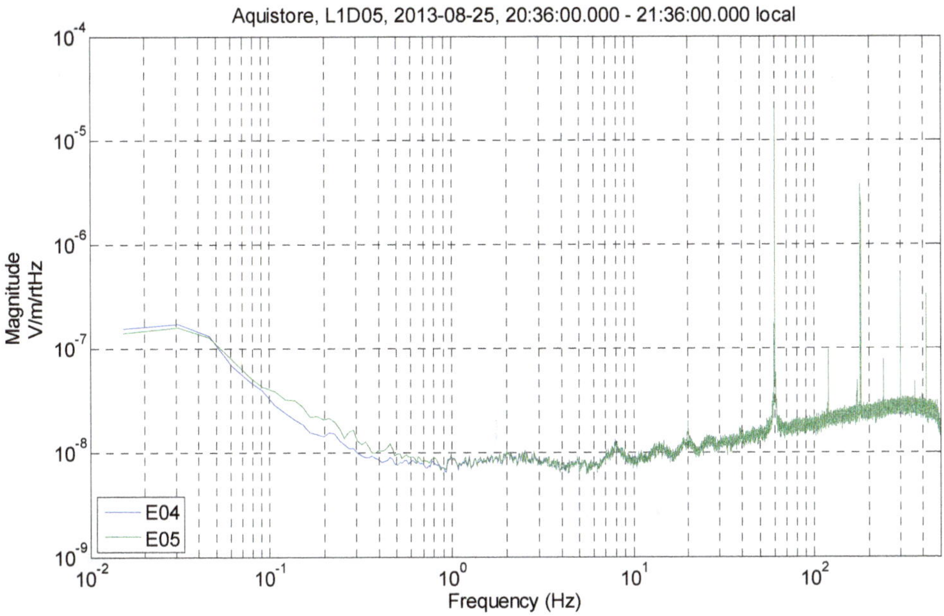

Figure 12. Spectra of parallel E-field sensors at the end of a day. The transmitter is not operating.

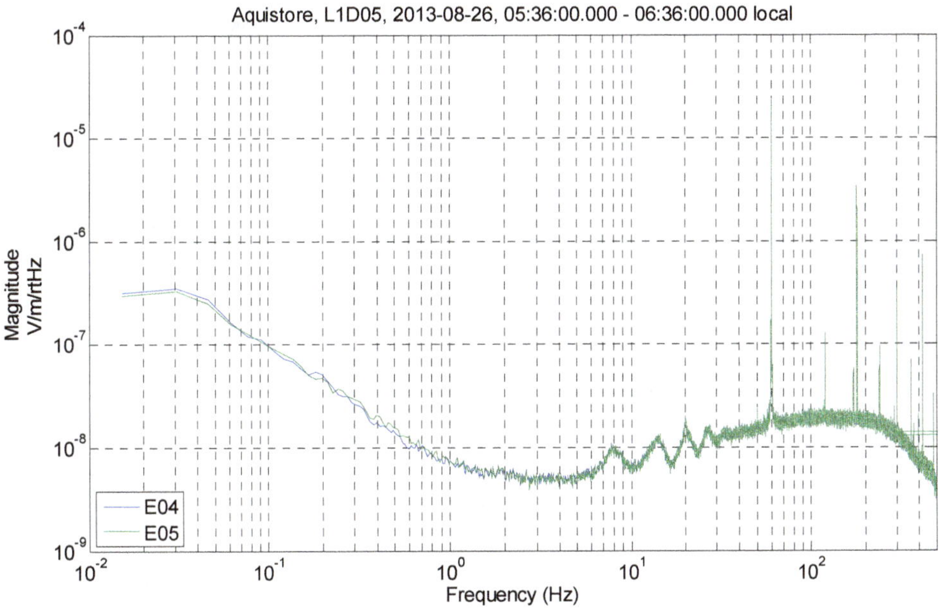

Figure 13. Spectra of parallel E-field sensors in the early morning prior to operating the transmitter.

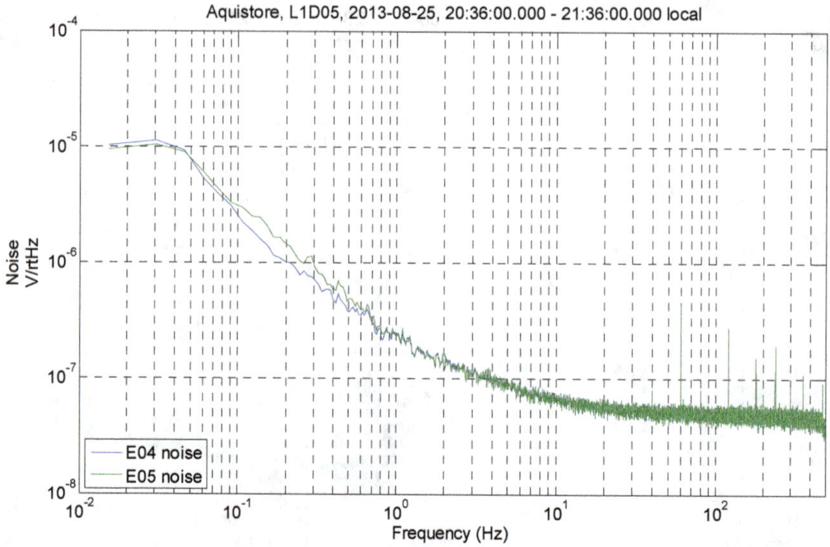

Figure 14. Voltage noise spectra of parallel E-field sensors after cancellation of coherent environmental noise.

The original spectra are in Figure 12. The transmitter is not operating. The voltage noise level at 0.5 Hz is 400 nV/√Hz, corresponding to an E-field noise of 5 nV/m√Hz.

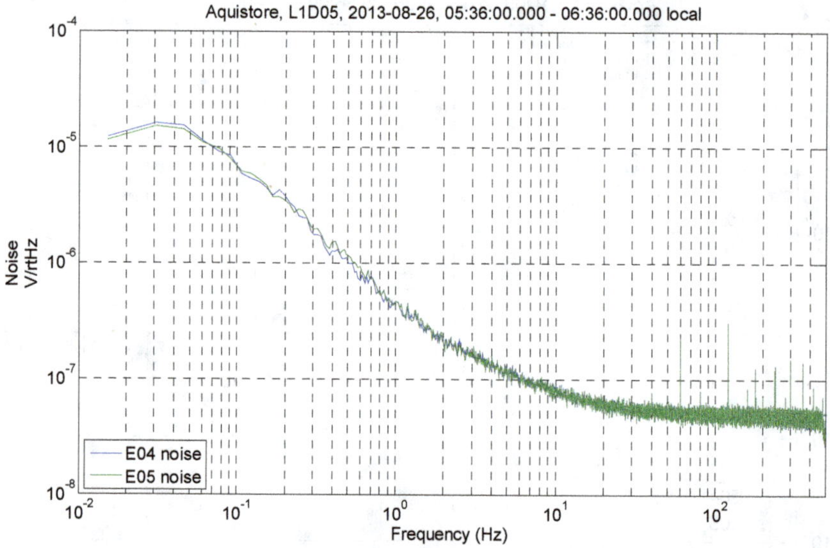

Figure 15. Voltage noise spectra of parallel E-field sensors after cancellation of coherent environmental noise.

The original spectra are in Figure 13. The transmitter is not operating. The voltage noise level at 0.5 Hz is 1 μV/√Hz, corresponding to an E-field noise of 12 nV/m√Hz.

We can project the SNR along a survey line as a function of distance in the following way. First we calculate the surface signal along a survey line using BGA's modeling code for a cased well. Once the surface E-field is projected it can then be divided by the average measurement noise to give an estimate of the SNR at each location. Figure 16 gives the result of this calculation along Line 1 extended out to 4 km with a 24 A amplitude square wave. We have taken the noise at 0.5 Hz to be 8 nV/m√Hz. We see that for this configuration an SNR of >10,000 is achievable out to 1 km, and >1,000 out to 2.5 km.

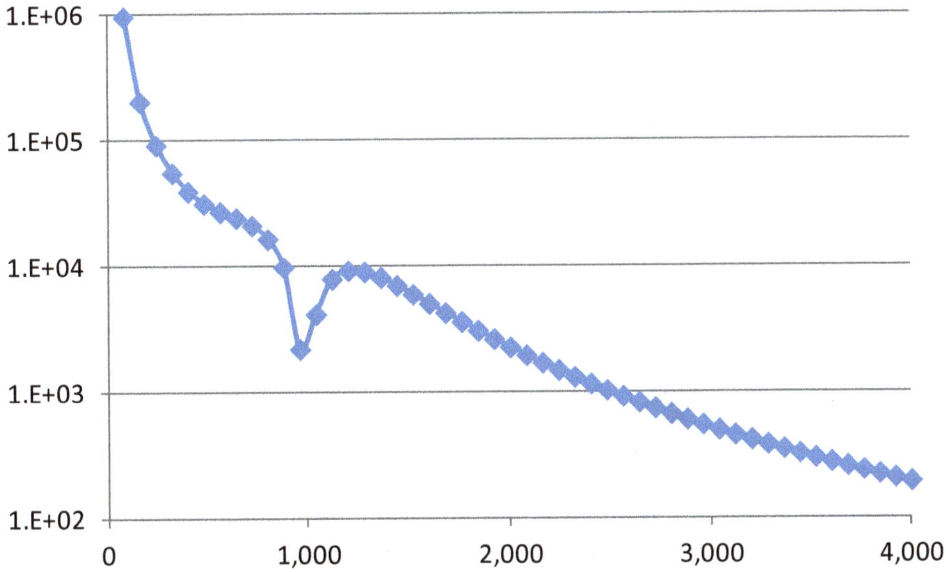

Figure 16. Projection of the achievable signal-to-noise ratio for transmission of 24 A current.

We can assess performance of the new GMI source by comparing the measured fields with those projected by BGA's model. Such a comparison provides a high level way to confirm that the source is producing subsurface currents as expected. Figure 17 and Figure 18 show a comparison for Lines 1 and 2. In both figures, distance is measured in meters (m) and:

- Purple: 0.1 S/m.
- Blue: 0.2 S/m.
- Brown: 0.5 S/m.
- Green: Measured surface E-field.

We see in Figure 17 and Figure 18 that the measured data generally agree best with a uniform surface conductivity of around 0.5 S/m to 0.2 S/m (equal to a resistivity of 2 Ω-m - 5 Ω-m). Figure 19 shows that this range is in good agreement with the conductivity log for the Aquistore wells down to a depth of approximately 1800 m. A value of 0.2 S/m was used in the SNR projection shown in Figure 20.

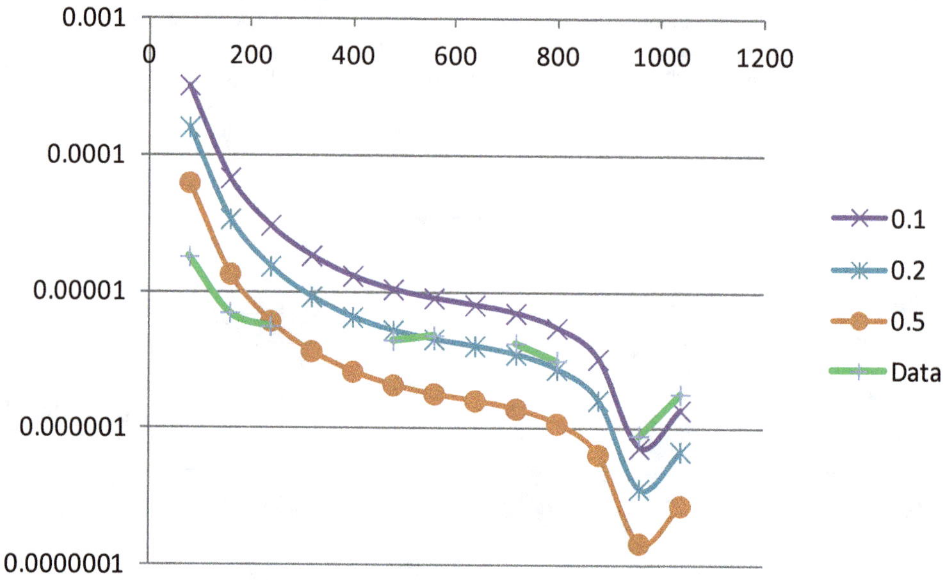

Figure 17. Comparison of the BGA DC model for uniform Earth conductivity with the measured surface E-field data (V/mA) along Line 1.

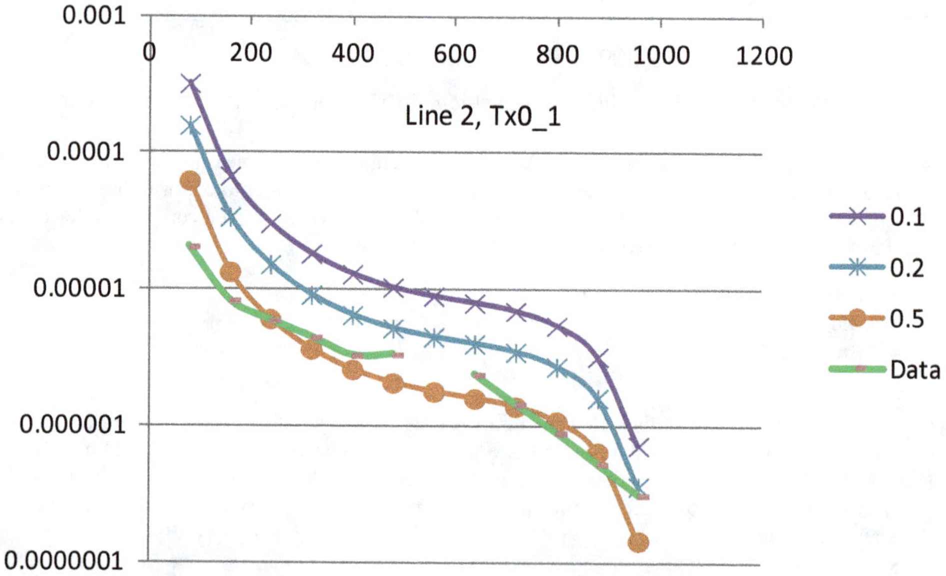

Figure 18. Comparison of the BGA DC model for uniform Earth conductivity with the measured surface E-field data (V/mA) along Line 2.

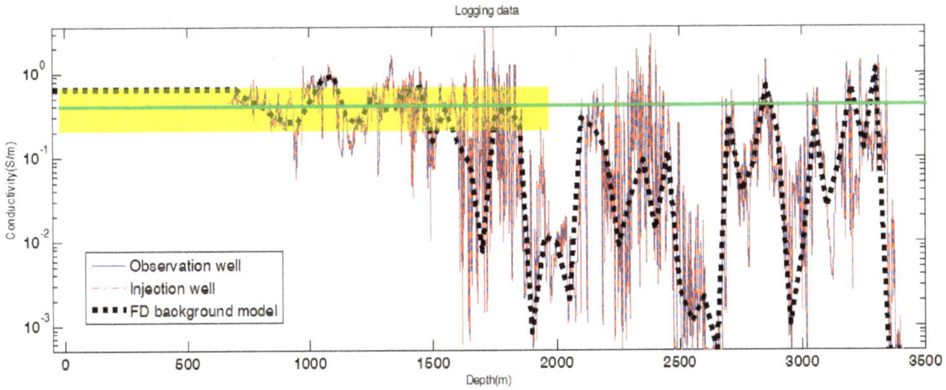

Figure 19. Conductivity logging data for the injection well and observation well at Aquistore.[2]

The dashed line is the conductivity profile used by LBNL in their modeling of the Aquistore site, described elsewhere in this volume. The yellow region is the range of values estimated from comparing the BGA solutions for a uniform Earth to measured data. The green line corresponds to a conductivity of 0.4 S/m (2.6 Ω-m).

For all three resistivity values in Figures 17 and 18, the agreement of the projected and measured data near the well is poor. However, we note that down to a few hundred meters the well casing will be surrounded by a layer of high quality concrete. The resistivity of this concrete will be 200 Ω-m or higher. Also the casing will in general be of larger diameter at the top [8]. Both of these effects will increase the current flowing down the casing and decrease the current flowing laterally in the Earth near to the casing, and therefore make the projected fields smaller near the well.

In Figure 18, the agreement for Line 2 is poor at 600 m compared to the Line 1 data in Figure 17, but we note that Line 2 crosses a waterway around this distance from the well and considerable distortion in the surface field is likely. Line 9 approached the waterway at around 900 m and data were not acquired at these receiver locations.

To address the question of detecting CO_2 we must rely on Figure 3. Fortunately the uniform background conductivity of ~0.4 S/m (2.6 Ω-m) used in our model is very close to the conductivity at reservoir depth (3200 -3400 m) indicated by the well logs (Figure 19). In Figure 3 we see the projected change in surface field is 0.1 % to 0.2 %. However, the BGA model is based on DC, and the first approximation to scale the model to 0.5 Hz excitation is to calculate attenuation based on a skin depth approach. At 0.5 Hz the skin depth in a uniform medium of resistivity 2.6 Ω-m is 1140 m. The casing is thus almost exactly 3 skin depths deep resulting in only 5% of the DC current reaching its end. This means the CO_2 contrast is decreased to approximately 0.01%.

A surface field change of 0.01% is detectable in time-lapse monitoring using the survey technology described in this report. Multiplying the SNR plot of Figure 16 by 10^{-4} gives the result in Figure 20. We see that the 150 m radius plume produces a signal with an SNR ≥ 10 out to 240 m lateral offset from the well and a detectable signal (SNR ≥ 2) out to about 600 m. This is very promising because we could increase the transmit current and/or extend the signal averaging time to further increase the SNR. We could also place additional receivers close to the well at intermediate offsets (40 m, 120 m, 200 m) to acquire additional data points. Thus, very close to the well it is possible we could

[2] Courtesy of Lawrence Berkeley National Laboratory

see a CO_2 plume one tenth of its maximum projected size. However, considering the approximations we have made to apply the DC model to our data, the best assessment we can make is to say the full size CO_2 plume is detectable via the new source and receiver technology.

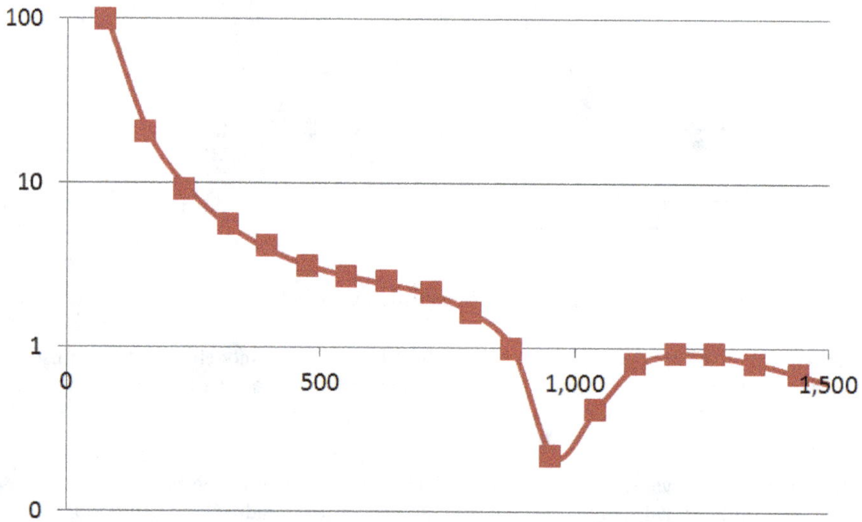

Figure 20. Projected SNR vs. radial distance (m) from the well for measurement of a 150 m diameter x 30 m thick CO_2 plume.

CONCLUSIONS

Under this project GMI has conducted the first ever survey with a new EM source configuration that takes advantage of the high conductivity path provided by a conventional steel well casing to channel electric current to reservoir depth. By providing a current path directly to formation depth, this new source has the potential to extend EM methods to reservoir monitoring even in very deep formations such as at Aquistore. Furthermore, the eQube^{TM}, a new type of electric field sensor that does not require galvanic contact to the ground, was used to acquire all data. The eQube^{TM} opens the door to precision long-term monitoring, and if desired, permanent installation of E-field sensors.

This project has produced the following important results:

1. The subsurface current flow was detected with very high coherence, >99.99999% at 800 m (Figure 9). The measurement SNR corresponding to the coherence in Figure 9 is 10,000, which is compatible with detecting a 0.01% change in signal due to CO_2 injection. Closer to the well the SNR was >5 x higher. Thus we have demonstrated a measurement system capable of detecting the projected CO_2 plume signal over a substantial distance from the well.
2. The data were compatible with a simple DC model of the casing in a uniform Earth background. The optimum fit was for an Earth conductivity of 0.2 S/m to 0.5 S/m. This range was in good agreement with conductivity logs for the injection and observation wells.
3. The CO_2 plume is detectable under the conditions of this first pilot study. Possibly the nearest four sensors will have an SNR of greater than 5, with the nearest three greater

than 10. A longer survey time and more powerful source could be used in a practical monitoring situation.
4. We have demonstrated the new survey technology at a geographically challenging full scale test site. The well depth and low Earth resistivity make Aquistore an extreme case and most oil reservoirs, for example, are less demanding. Further, the surface field recorded at 1140 m from the well was only seven (7) times smaller than the field at 80 m, opening the door to true wide area reservoir monitoring.

EM methods offer new ways to image fluid displacements in reservoirs and fluid injection into formations. The primary application for Aquistore, and in many hydrocarbon reservoirs, is time-lapse monitoring. To that end, all sensors were deployed at Aquistore via RTK-GPS using markers flush to the ground to allow data to be acquired in exactly the same location in a subsequent survey. Once CO_2 injection is complete in March 2014, Aquistore will offer a challenging but technically relevant test site for further exploration of EM imaging.

ACKNOWLEDGEMENTS

This work was supported by the CO_2 Capture Project (CCP). We acknowledge the support of the Aquistore Project, operated by the Petroleum Technology Research Center. We thank K. Dodds of the CCP for support and technical discussion of this work. In addition, we thank Dr. H. Frank Morrison of Berkeley Geophysics Associates for his advice and expertise.

REFERENCES

1. He, Z., Zhao, Z., Liu, H., and Qin, J. TFEM for oil detection: Case studies. 2012. The Leading Edge. v. 31, no. 5, p. 518-521. May.
2. Marsala A.F., Al-Buali M., Ali Z., Ma S. M., He Z., Biyan T., He T. and Zhao G. First Pilot of Borehole to Surface Electromagnetic in Saudi Arabia: a New Technology to Enhance Reservoir Mapping & Monitoring. 2011. EAGE I005 extended abstract presented at the 73rd EAGE Conference & Exhibition. Vienna, Austria, 23-26.
3. Department of Energy grant DE-SC0008264, Permanent Electromagnetic Monitoring of CO_2 Sequestration in Deep Reservoirs.
4. Zhdanov, M.S., Endo, M., Black N., Spangler, L, Fairweather, S. Hibbs, A., Eiskamp, G.A., and Will, R. 2013. Electromagnetic monitoring of CO_2 sequestration in deep reservoirs: First Break, Vol 31, February, p. 85-92.
5. Schenkel, C., Morrison, H.F. Numerical Study on Measuring Electrical Resistivity Through Casing in a Layered Medium. 60th Annual International Meeting Expanded Abstracts, Society of Exploration Geophysicists, 1990.
6. Schenkel, C. J. and Morrison, H. F. 1994. Electrical resistivity measurement through metal casing: Geophysics, 59, no. 7, 1072-1082.
7. Wait, J.R., and Hill, D.A. Theory of transmission of electromagnetic waves along a drill rod in conducting rock. 1979. IEEE Transactions on Geoscience Electronics, GE-17(2):21–24, Apr.
8. Spragg R, Castro J, Nantung TE, Parades M, Weiss J. "Variability Analysis of the Bulk Resistivity Measured Using Concrete Cylinders." Advances in Civil Engineering Materials. Publication FHWA/IN/JTRP-2011/21. Joint Transportation Research Program, Indiana Department of Transportation and Purdue University, West Lafayette, Indiana, 2011. doi: 10.5703/1288284314646.

Carbon Dioxide Capture for Storage in Deep Geological Formations, Volume 4
Karl F. Gerdes (Editor)

Chapter 41

INSAR MONITORING OF GROUND DEFORMATION AT THE ILLINOIS BASIN DECATUR PROJECT

Giacomo Falorni, Vicky Hsiao, JeanPascal Iannacone,
Jessica Morgan and Jean-Simon Michaud
TRE Canada Inc. #410 – 475 West Georgia St., Vancouver, BC, V6B 4M9, Canada

ABSTRACT: Advanced InSAR techniques were used at the Illinois Basin Decatur Project to monitor and measure ground deformation. The motivation was to assess the viability of applying InSAR techniques in a temperate climate setting following the successful deployment of the technique at the In Salah CCS project in Algeria. While the arid desert environment of the In Salah project was ideal for InSAR, as there were no issues with vegetation or snow cover, the setting is not typical of temperate climates, in which many CCS projects in North America will be sited. The CCP3 consortium funded the InSAR monitoring component of the MMV activities at Decatur to evaluate the density and distribution of measurement points that would be obtained in this setting as well as the impact on the estimation of ground movement. The InSAR investigations, performed using an advanced data processing algorithm, SqueeSAR, to process a stack of radar images acquired over the site comprised a feasibility study to provide an initial assessment of point distribution, a baseline analysis and ongoing monitoring during CO_2 injection. The feasibility analysis led to the design and installation of an artificial reflector network consisting of 21 reflectors placed between the injection well and an observation well. The surface deformation monitoring carried out during CO_2 injection highlighted an almost complete lack of ground movement response to injection activities, which is in good agreement with other monitoring data, such as bottomhole pressure and the geology of the site. The results of the InSAR monitoring at Decatur indicate that a satisfactory density and distribution of measurement points can be obtained in temperate settings to effectively monitor surface deformation.

KEYWORDS: remote sensing; InSAR; SqueeSAR; monitoring; ground deformation; Decatur; CCS; carbon capture and sequestration

INTRODUCTION

Carbon capture and storage (CCS) has drawn the attention of governments, scientists and industries worldwide for its potential to mitigate climate change by preventing CO_2 from being released into the atmosphere. The basic principle is that the CO_2 captured from the flue gas of large point sources is then sequestered underground, where it will remain indefinitely. A key aspect of the storage of CO_2 is the necessity to ensure that it is secured permanently and safely, with no leaks into overlying aquifers or the atmosphere [1]. Large research projects are being carried out to investigate many different aspects associated with the storage of CO_2, including the use of numerical modelling to predict the effects of injected gas on the environment, plume migration with time, surface deformation linked to the injection of high pressure fluids into a reservoir, etc. [2, 3, 4, 5, and 6]. In all projects monitoring activities play a key role as they need to ensure the fluid behaves as predicted once it has been injected into the geological reservoir.

There are many different monitoring techniques in use today at CCS projects around the world. The pioneer In Salah CCS project had as one of its main objectives the assessment of the effectiveness of the different monitoring technologies available, including geochemical, geophysical and remote sensing approaches [7]. One of the relevant outcomes of the evaluation was the usefulness of surface deformation monitoring by means of satellite remote sensing and, in particular, interferometric synthetic aperture radar (InSAR). The precise measurement of the ground deformation that occurred following the start of CO_2 injection was instrumental following an observed leak, as it contributed to the reconstruction of a preferential pathway for CO_2 fluid migration [8]. It also allowed the identification of a fault opening at depth around one of the injection wells that eventually led to the suspension of fluid injection [9]. The surface deformation data has since been used extensively to characterize plume development around the three injectors at In Salah and to constrain geomechanical models [10, 11, and 12].

In light of the results obtained at In Salah, the CCP3 consortium funded an InSAR monitoring program at the Illinois Basin Decatur Project (IBDP) with the objective of evaluating the application of the technology in a temperate climate setting. In Salah was located in an arid desert environment that is particularly suited for InSAR as there is no vegetation or snow cover to attenuate the satellite signal. The IBDP project is set on the outskirts of the city of Decatur, Illinois where land cover consists of a mix of urban and industrial infrastructure, agricultural fields and forested areas. An initial pre-injection feasibility study was performed to assess the distribution of measurement points and, based on the findings, 21 artificial reflectors (ARs) were installed between the injection well and an observation well.

An advanced form of InSAR processing, known as SqueeSAR [13], was applied to assess surface deformation at the IBDP. This approach processes a stack of radar images to identify radar targets and assess their displacement over time. The technique uses advanced algorithms to reduce the contribution of atmospheric and orbital disturbances, which often leads to a millimetre-scale precision of the measurements [14]. Following the pre-injection InSAR feasibility study, satellite image acquisitions started in July 2011 to establish the surface deformation baseline.

Injection operations at the IBDP started in November 2011 and CO_2 was injected into the Mount Simon Sandstone, which extends in depth from the top of the pre-Cambrian zone (~2.140 m – 7020 ft) to the base of the Eau Claire Shale (1689 m – 5541 ft) [15]. The Eau Claire Shale forms a 96 m (315 ft) thick confining top unit of the reservoir. Compared to In Salah the target reservoir is much thicker and has a higher permeability [12, 16].

STUDY SITE

The study site is located northeast of the city of Decatur, Illinois and covers an area of 25.25 square kilometres (9.75 square miles; Figure 1). The topography is mainly flat or gently sloping towards Lake Decatur and land cover consists of industrial sites, farmland, forest, and residential areas. Designing a monitoring program for this type of mixed environment is not trivial [17] and resulted in an irregular distribution of measurement points with significantly higher densities over urban areas and reduced densities over agricultural fields and forests (Figure 1). The absence of points in the field northwest of the injection well, where deformation measurements were required, led to the installation of a grid of artificial reflectors (Figure 1).

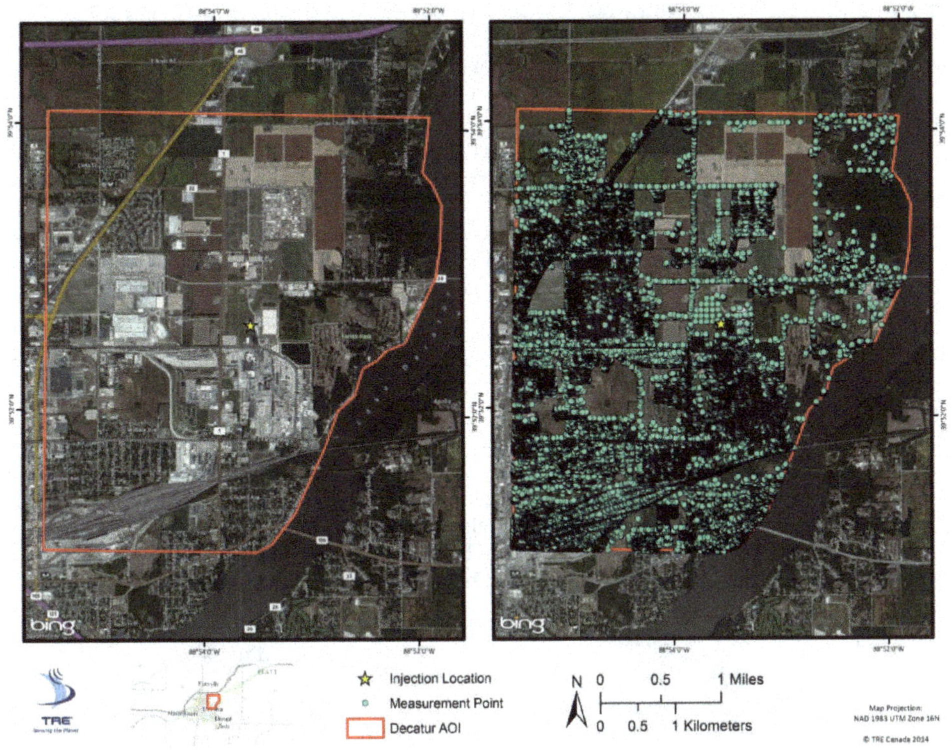

Figure 1. The Decatur study site.

INSAR

Ground displacement information from radar imagery is obtained by a process known as interferometric synthetic aperture radar (InSAR), which compares the transmitted and received signals from radar sensors onboard a satellite. When objects on the ground surface move, the distance between the sensor and the object changes, thereby producing a corresponding shift in the signal phase between successive images. Radar satellites view the ground at an oblique angle that can vary between 20 and 55 degrees off-nadir, depending on the satellite and the beam mode adopted. Movement is measured as motion towards or away from the satellite along the viewing angle. This is known as a line-of-sight (LOS) measurement.

The most basic form of InSAR, known as Differential InSAR or DInSAR, produces an interferogram, which is the representation of shifts in the phase of the received signal between two radar images. The qualitative information in interferograms can usually be converted to quantitative information by means of deformation maps. These provide a spatial representation of ground movement between two images. As only two images are utilized, movement can only be represented as linear in time and any accelerations or cyclical fluctuations cannot be captured. A limitation of interferograms is a lack of coherence over areas with vegetation. Coherence is a measure of the quality of an interferogram and it is typically low in vegetated areas as the reflected radar signal changes too much over time to produce displacement measurements. Furthermore, DInSAR measurements have an associated precision in the centimeter range, which is considerably less than the millimeter precision of more advanced approaches.

Advanced InSAR

Advanced forms of InSAR use the information within multiple interferograms derived from stacks of radar imagery to determine displacement over time. Measurements are based on stable targets identified within every radar image within a stack, and since a series of radar images is processed, it is possible to estimate atmospheric and orbital errors in the imagery and remove them. This increases measurement precision to millimeter scale [14, 18, 19], which can be important for the measurement of slow deformation such as could be expected from the injection and storage of CO_2. The PSInSAR technique developed by Ferretti et al. [19] focuses on dominant reflectors of the radar signal, referred to as Permanent or Persistent Scatterers (PS), to measure surface motion from targets on the ground. PS correspond to objects located within individual pixels such as buildings, transmission towers, pipelines, or rock outcrops.

The SqueeSAR algorithm, an improvement of PSInSAR, extends the detection of targets to non-urban contexts by addressing the lower density of PS in rural settings. Through the detection and analysis of radar signals reflected from large homogeneous areas of ground surface called Distributed Scatterers (DS - Figure 2). This technique produces a significant increase in the density and spatial coverage of InSAR-based measurements [13]. Distributed Scatterers are often hundreds to thousands of square meters in size and are often identified from fallow fields, rangeland, sparsely vegetated areas, scree or bare earth. SqueeSAR also incorporates the PSInSAR algorithm, meaning both PS- and DS-type targets are identified. The algorithm is computationally intensive as it involves several advanced statistical analyses to be performed over areas that can be thousands of square kilometers.

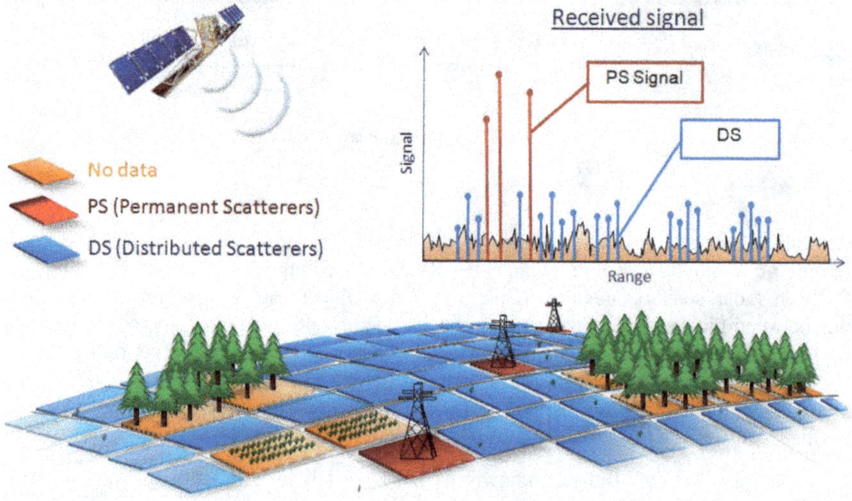

Figure 2. Identification of Permanent Scatterers (PS) and Distributed Scatterers (DS) with the SqueeSAR algorithm.

Artificial Reflectors

At sites where measurement points are required at certain locations, or where vegetation or snow cover reduces the number of natural reflectors, artificial reflectors (AR) can be installed to provide additional measurement points. Artificial reflectors act as strong, stable targets that produce reliable measurements over time without being influenced by weather events, including significant snowfall.

To complement the natural targets identified in the initial feasibility study at Decatur, the Illinois State Geological Survey installed 21 ARs in an area between the injection well and a nearby observation well. The ARs, based on a design provided by TRE, are double geometry structures, meaning they are visible from both ascending and descending satellite orbits. The reflectors were installed on posts drilled into the ground.

The 21 ARs were installed with a spacing of 75 meters (246 ft) by the Illinois State Geological Survey (ISGS) in March and April, 2011. After installation AR visibility to the satellite was assessed in June 2011 by analyzing the amplitude of the reflected signal in four satellite images: all twenty-one ARs were shown to have strong, stable reflectivity.

Figure 3. Artificial reflectors installed at the Decatur site.

DATA

A stack comprising 72 radar images acquired by the Cosmo-SkyMed (CSK) constellation was used in this analysis. The imagery spans the period from 8 July 2011 to 19 June 2013. Figure 4 shows the temporal distribution of the imagery. All images were acquired from an ascending orbit (satellite travelling from south to north and imaging to the east) with a viewing angle of 33.5° off the vertical.

CO_2 injection rate (tonnes/hour) and cumulative volume corresponding to the date and time of each CSK image acquisition were provided by the IBDP. This information, coupled with the average daily downhole pressure, also provided by the IBDP, was compared with the measured ground deformation from SqueeSAR.

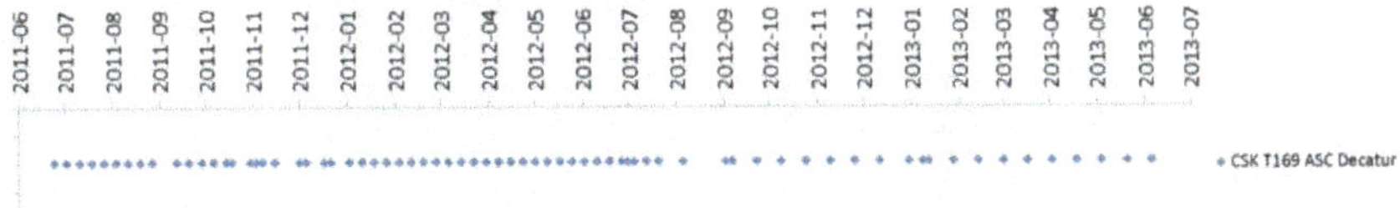

Figure 4. Temporal distribution of the radar imagery acquired over the Decatur site.

RESULTS

Natural Radar Targets

A total of 108,895 measurement points were obtained over the site, leading to a density of 4,313 points per square kilometre (11167 pts/sq.mi., Figure 5). Dense clusters of points are located on buildings, infrastructure and along roads, particularly in the southwestern portion of the site. A close-up of the results in the injection well area is displayed in Figure 6, which shows a cluster of measurement points on the injection well pad and the surrounding area, including the artificial reflectors. Each measurement contains a time series of deformation spanning July 2011 to June 2013, consisting of 14 measurements prior to the start of injection and 58 after the start of injection.

The average displacement of all measurement points located on the well pad indicates an uplift rate of approximately 2 mm/yr (0.08 in./yr) along the LOS over the two-year period, as compared to an average rate of 0.3 mm/yr (0.01 in./yr) for the entire study site. The time series of two measurement points located on or near the injection wellhead are shown in Figure 7, where the x axis indicates the image acquisition date, the y axis denotes ground deformation in millimetres, and the vertical blue line specifies the date of the injection. Both points indicate LOS uplift of 3.5 and 3.7 mm/yr (0.14 in./yr), respectively. This uplift is confined to these two points as other points farther away do not have comparable behaviour. It is interesting to observe that the time series become significantly noisier after the start of injection, which is likely caused by vibrations induced by the high fluid injection pressures. There is little or no ground movement that appears to be correlated with injection activities beyond the two points mentioned previously. No concentric patterns of uplift focused around the injection well are visible in the results.

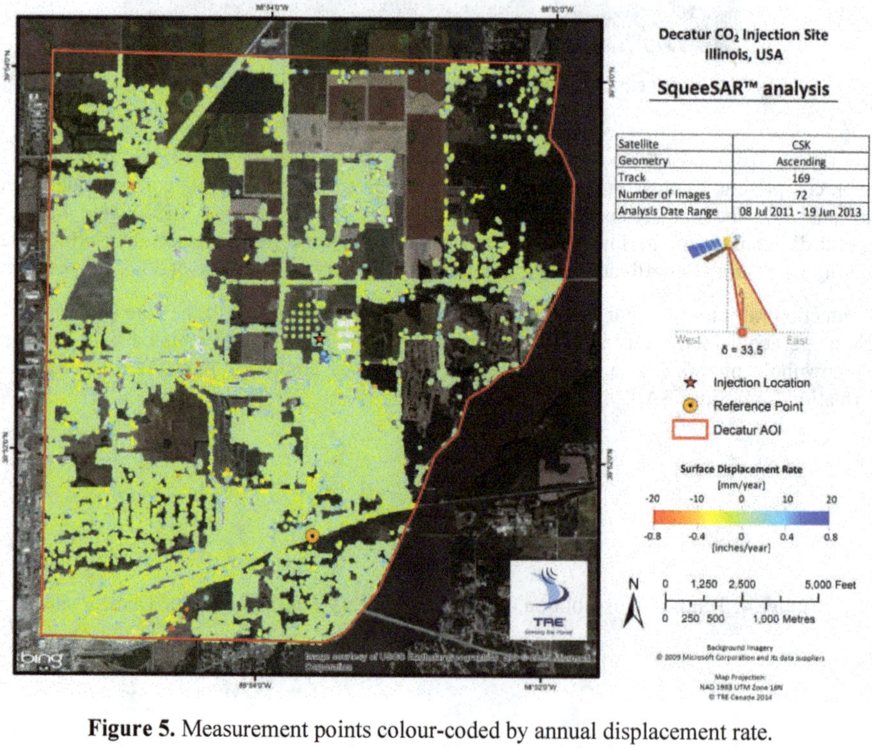

Figure 5. Measurement points colour-coded by annual displacement rate.

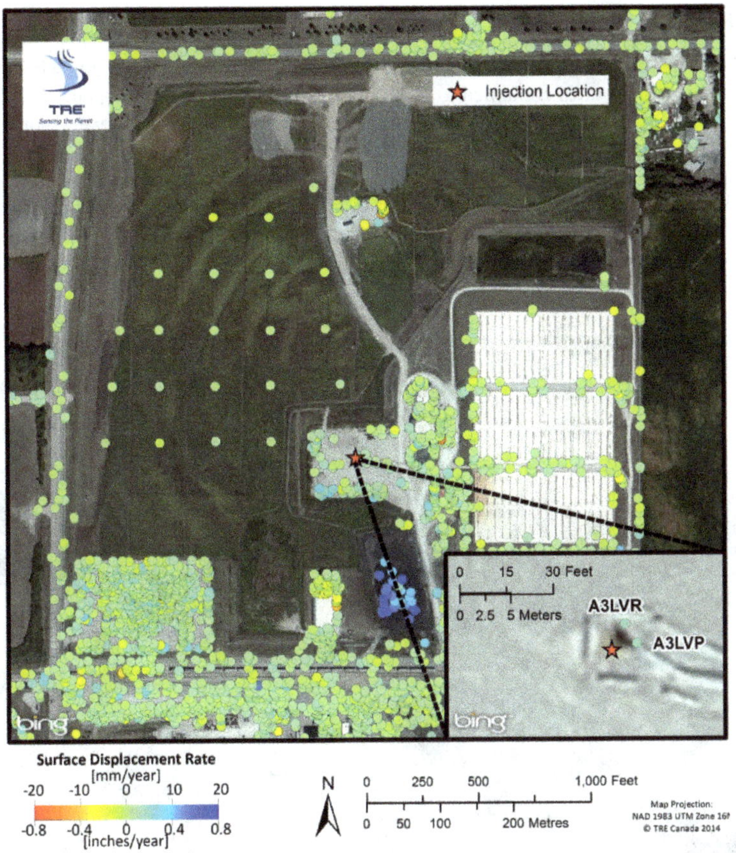

Figure 6. Deformation results in the vicinity of the injection pad. Inset shows the location of the two points for which the time series are shown in Figure 7.

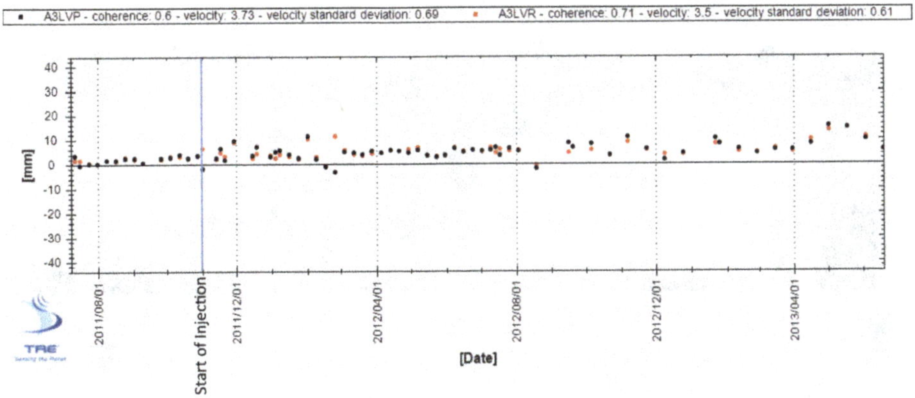

Figure 7. Time series of two measurement points in the vicinity of the injection well. The vertical blue line indicates the start of injection.

Artificial Reflector Analysis

The movement recorded by the ARs over the two-year period was marginal, with all reflectors exhibiting displacement rates contained to -4 ± 0.5 mm/year (-0.16 ± .02 in./yr) to +2 ± 0.5 mm/year (+0.08 ± .02 in./yr) (Figure 8). An average time series of deformation for all twenty-one reflectors is shown in Figure 9. No deformation can be observed and the start of injection does not appear to have an impact. It is interesting to note that ARs 15, 20, and 21, located in the north-eastern corner of the field subside slightly. However, the subsidence appears to occur over a brief period in the summer of 2012 (Figure 10). AR 5, the closest reflector to the injection well, after an initial period of subsidence prior to injection measures approximately 8 mm (0.31 in.) of uplift after the start of injection operations (Figure 11).

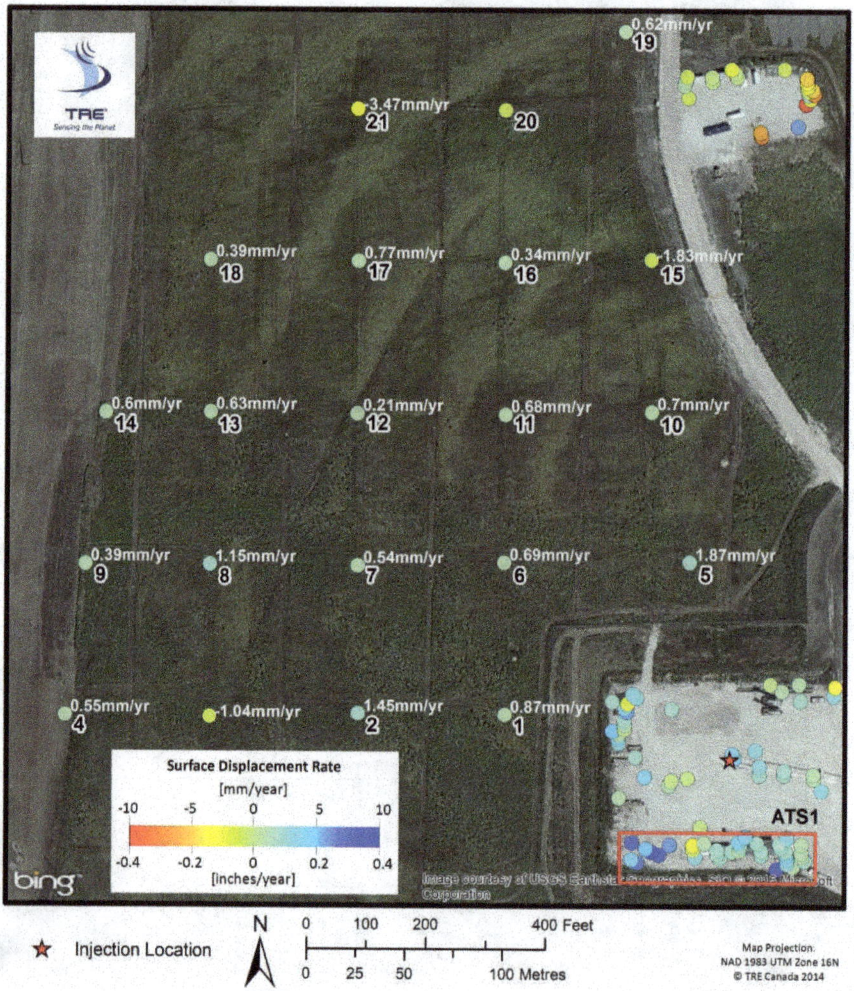

Figure 8. Annual displacement rate of the 21 artificial reflectors and of other points in the area of the wellpad. The polygon labelled as ATS1 indicates the points used to calculate the average time series shown in Figures 12 and 13.

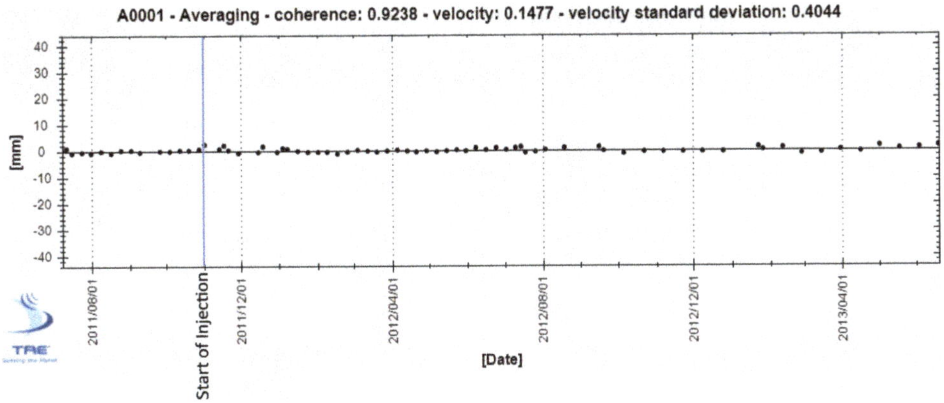

Figure 9. Average time series of all 21 reflectors.

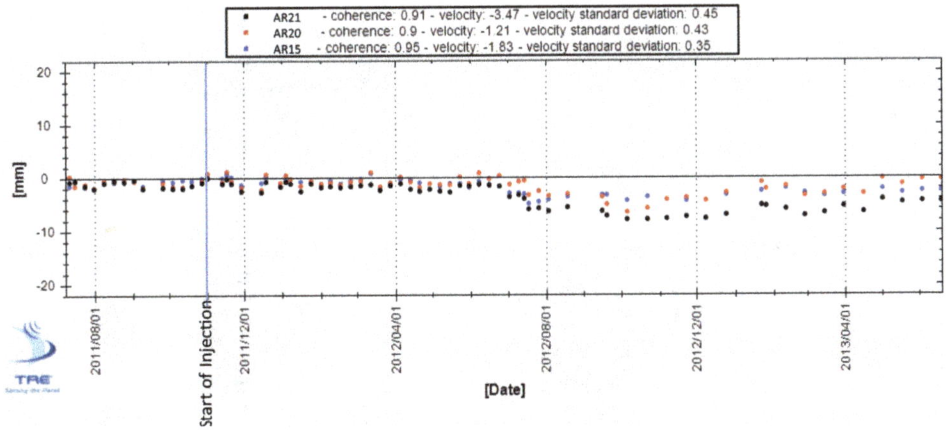

Figure 10. Time series of AR 15, AR 20 and AR 21.

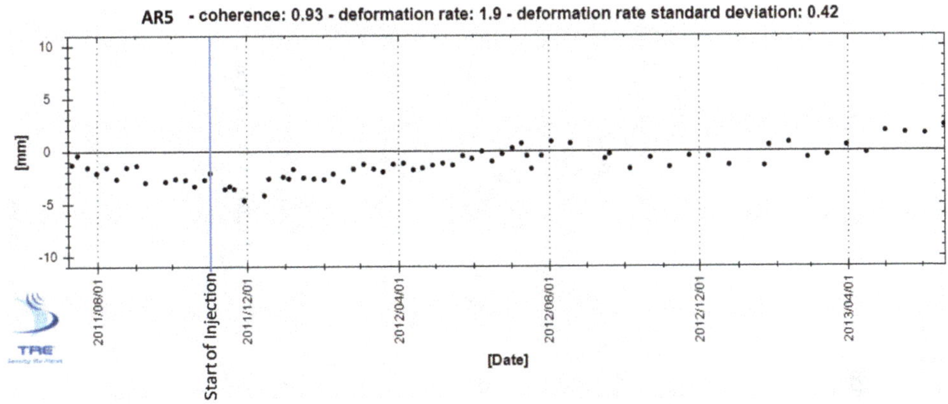

Figure 11. Time series of AR 5.

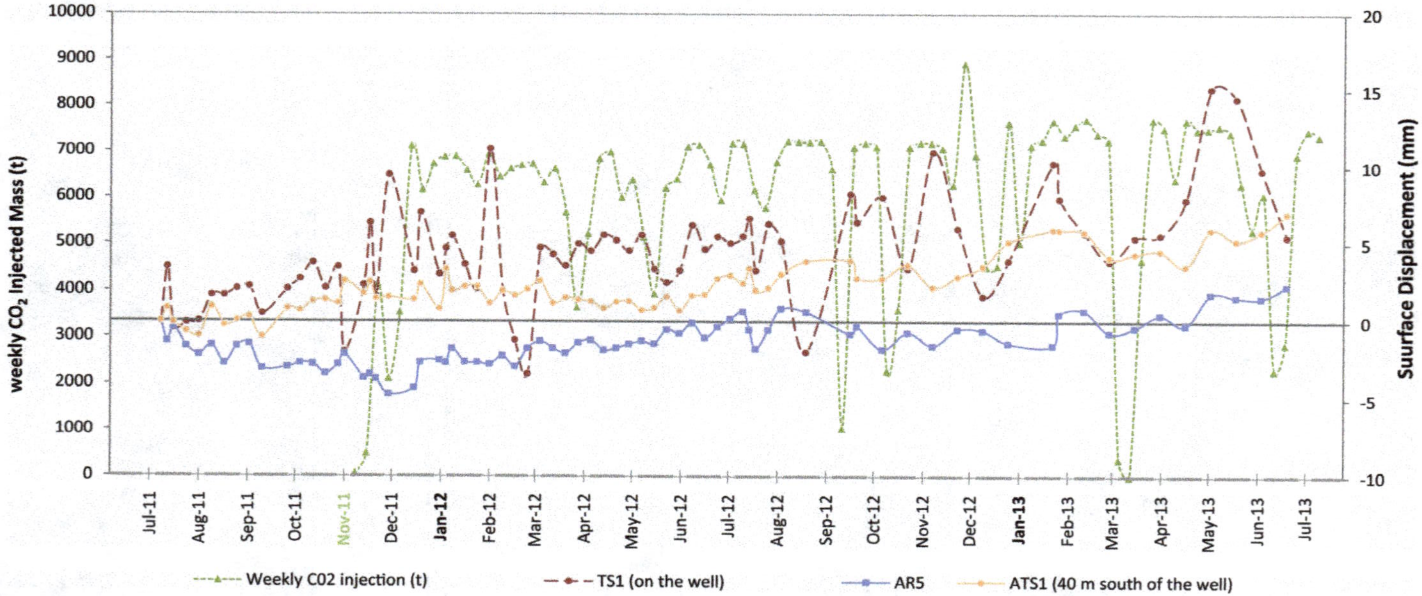

Figure 12. Comparison of time series of A3LVP (on the injection well), average time series of ATS1, and the time series of AR 5, with weekly injected CO_2.

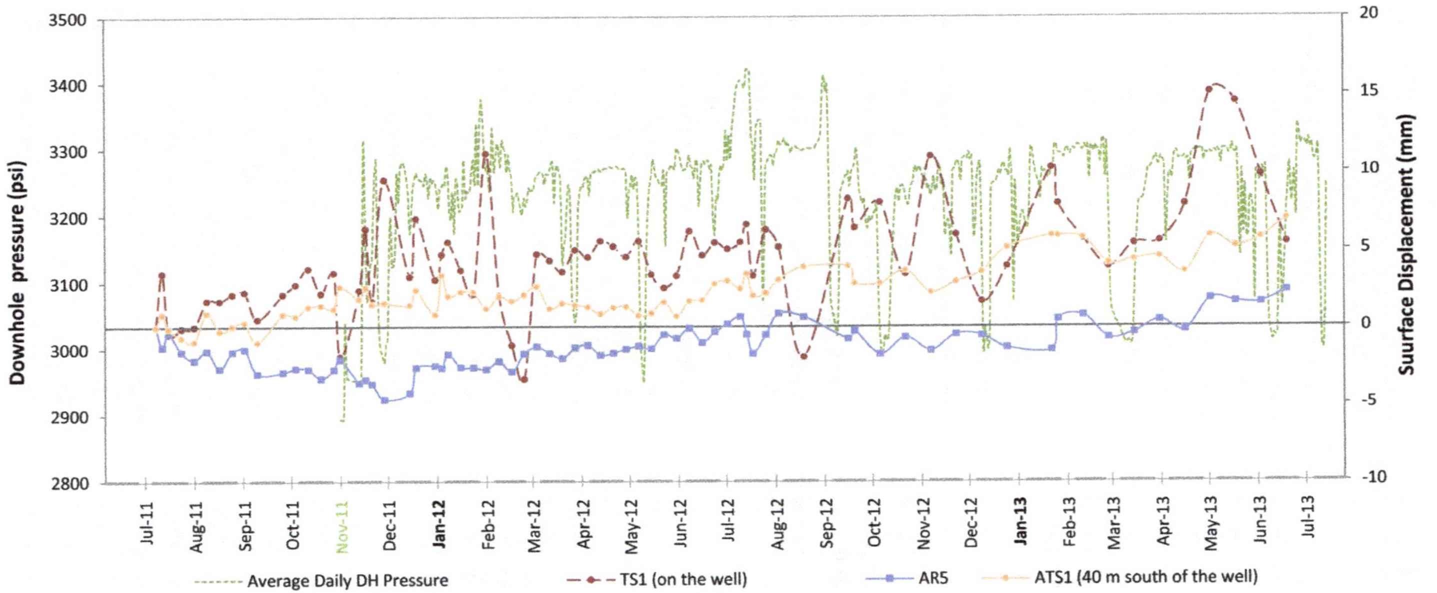

Figure 13. Time series of A3LVP (on the injection well), average time series of measurements 40 metres south of the injection well, and average time series of 21 ARs, along with average daily downhole pressure.

Ground Deformation Correlation with Injected CO_2 and Bottomhole Pressure

The ground deformation detected near the injection well was compared to both the weekly injected CO_2 volume and the average daily downhole pressure to identify possible correlations. The comparison involved one of the measurement points on the injection well (A3LVP in Figure 6), the average time series of the cluster of measurement points south of the injection well (ATS1 in Figure 8), and the time series of AR 5 against weekly injected CO_2 volumes (Figure 12), and the average daily downhole pressure (Figure 13). The results indicate a possible weak correlation between ground deformation and the injection of CO_2, as point A3LVP on the wellhead, AR5 and ATS1 all gradually uplift over time from the start of injection. It also worth noting that, following the start of injection, point A3LVP has much stronger oscillations compared to the pre-injection period.

DISCUSSION AND CONLUSIONS

The results of the advanced InSAR analysis carried out over the Decatur area indicate that it is possible to obtain a large number of measurement points in a temperate climate setting in which vegetation and snow cover can introduce adverse conditions for the identification of natural radar targets. The distribution of the points is not uniform, as significant clustering occurs near manmade structures such as buildings and roads. In areas where it is necessary to have the certainty of obtaining measurement points it is possible to install artificial reflectors. Twenty one were installed between the injection well and an observation well to ensure measurements in this area.

The results of the surface deformation measurements indicate that little ground movement has occurred in response to injection and that this is largely limited to the area of the well pad and the nearest artificial reflector. Any points that are located farther than 100 m (328 ft) from the well did not show any response to the injection of CO_2. These results are in line with expectations as the formation into which the injection is occurring has a high permeability, is very thick, and has shown little change in bottomhole pressure since the start of injection. This lack of response in the fluid pressure field is in marked contrast with the In Salah experience, in which the pressure field responded rapidly to CO_2 injection [10] and where surface deformation approaching 5 mm (0.2 in.) per year was measured. The absence of a surface expression at the IBDP is a positive result, albeit an unexciting one, as it indicates that over the two years of observations there was no unexpected behavior associated with CO_2 injection.

The monitoring of surface deformation by means of InSAR is one of the very few methods available to provide early warning of potential CO_2 leaks. Other approaches such as geochemical analyses aimed at detecting CO_2 in the soil or in aquifers necessarily identify the leak after it has already occurred, perhaps weeks, months or even years after the fluid actually escaped from the reservoir. Ground deformation responds to changes in the pressure field at reservoir level, which can occur over a very short time frame. This was the case, for example at In Salah, where the characteristics of the reservoir (i.e. low permeability, limited thickness) led to rapid increases in subsurface pressure and to a rapid response in terms of surface expression. InSAR detected deformation within a month of the start of injection [8, 12]. Furthermore in cases where the subsurface pressure increases noticeably, surface deformation measurements provide a means to indirectly image the subsurface plume migration [12], including the elongation in agreement with regional stress patterns or, even more importantly, propagation along vertical fractures within the overburden [9]. When leaks do occur InSAR can be used in conjunction with other methods as a forensic tool to reproduce the development of the CO_2 plume leading to the breakthrough.

ACKNOWLEDGEMENTS

The authors would like to thank the CCP3 consortium for providing funding for the present research and the IBDP for supporting the research and providing relevant data. In particular, the authors acknowledge Craig Hartline, Mark Crombie, Rob Finley, Kevin Dodds and Randy Locke for making this work possible and their contributions.

REFERENCES

1. Verdon J.P., Kendall J.-M., Stork A.L., Chadwick R.A., White D.J., and Bissell R.C., *Comparison of geomechanical deformation induced by megatonne-scale CO_2 storage at Sleipner, Weyburn, and In Sala*h (2013) Proceedings of the National Academy of Sciences, 110 (30), pp E2762-E2771NAS 2013.
2. Said A., and Hagrey, A., (2011) *2D Model Study of CO_2 Plumes in Saline Reservoirs by Borehole Resistivity Tomography*. International Journal of Geophysics, 2011, Article ID 805059.
3. Mazzoldi A., Hill T., and Colls J.J., (2011) *Assessing the risk for CO_2 transportation within CCS projects, CFD modelling*. International Journal of Greenhouse Gas Control, 5(4), pp 816-825.
4. Peixue J., Xiaolu L., Ruina X., Yongsheng W., Maoshan C., Heming W., and Binglu R., (2014) *Thermal modeling of CO_2 in the injection well and reservoir at the Ordos CCS demonstration project, China*. International Journal of Greenhouse Gas Control. 23, pp 135–146.
5. Mackay E.J., (2013) *Modelling the injectivity, migration and trapping of CO_2 in carbon capture and storage (CCS),* In: Geological Storage of Carbon Dioxide (CO2). (Ed. by Gluyas J., and Mathias S., pp 45-67,68e-70e, Woodhead Publishing.
6. Martens S., Liebscher A., Möller F., Henninges J., Kempka T., Lüth S., Norden B., Prevedel B., Szizybalski A., Zimmer M., Kühn M., (2013) *Ketzin Group, CO_2 Storage at the Ketzin Pilot Site, Germany: Fourth Year of Injection, Monitoring, Modelling and Verification.* Energy Procedia, 37, pp 6434-6443.
7. Vasco, D.W., Rucci A., Ferretti A., Novali F., Bissell R., Ringrose P., Mathieson A., Wright I. (2010) *Satellite-based measurements of surface deformation reveal fluid flow associated with the geological storage*. Geophysical Research Letters. 37, L03303.
8. Mathieson, A., Wright, I., Roberts, D., Ringrose, P., (2009) *Satellite imaging to monitor CO_2 movement at Krechba, Algeria*. Energy Procedia, 1, pp 2201-2209
9. Rucci, A, Vasco, DW, Novali, F. (2010) *Monitoring the geologic storage of carbon dioxide using multicomponent SAR interferometry*. Geophys J Int. 2013;193(1):197–209.
10. Rutqvist, J., Vasco, D.W., Myer, L. (2009) *Coupled reservoir-geomechanical analysis of CO_2 injection and ground deformations at In Salah Algeria*. Energy Procedia, 1(1), pp 225–230
11. Vasco, D.W., Ferretti, A., Novali, F., (2008) *Estimating permeability from quasi-static deformation: temporal variations and arrival time inversion*. Geophysics 73(6), pp. O37–O52.
12. Ringrose, P., Atbi, M., Mason, D., Espinassous, M., Myhrer, Ø., Iding, M., Mathieson, A.,Wright, I., (2009) *Plume development around well KB-502 at the In Salah CO_2 storage site*. First Break, 27, pp. 85–89.
13. Ferretti, A., Fumagalli, A., Novali, F., Prati, C., Rocca F., and Rucci, A., (2011) *A new algorithm for processing interferometric data-stacks: SqueeSAR™*. IEEE Transactions on Geoscience and Remote Sensing, 99 pp. 1-11.
14. Ferretti, A., Savio, G., Barzaghi, R., Borghi, A., Musazzi, S., Novali, F., Prati, C., and Rocca F., (2007) *Submillimeter accuracy of InSAR time series: Experimental validation*. IEEE Transactions on Geoscience and Remote Sensing, 45, pp. 1142-10153.

15. Senel O., Chugunov N., (2013) *CO_2 Injection in a Saline Formation: Pre-Injection Reservoir Modeling and Uncertainty Analysis for Illinois Basin – Decatur Project*, Energy Procedia, 37, pp. 4598-4611.
16. Locke, R.I., Larssen, D., Salden, Walter, Patterson, C, Kirksey, J., Iranmanesh, A., Wimmer B., Krapac, I. (2013) *Preinjection Reservoir Fluid Characterization at a CCS Demonstration Site: Illinois Basin – Decatur Project, USA*. Energy Procedia, 37, 6424-6433.
17. Couëslan, M.L., Leetaru, H.E., Brice, T., Scott Leaney, W., McBride, J.H. (2009) *Designing a seismic program for an industrial CCS site*. Energy Procedia, 1, 2193-2200
18. Bürgmann, Roland, George Hilley, Alessandro Ferretti, and Fabrizio Novali. (2006) *Resolving Vertical Tectonics in the San Francisco Bay Area from Permanent Scatterer InSAR and GPS Analysis.* Geology 34 (3): 221–24. doi:10.1130/G22064.1.
19. Ferretti, A., Prati, C., and Rocca, F., 2000. *Non-linear Subsidence Rate Estimation Using Permanent Scatterers in Differential SAR Interferometry*. IEEE Transactions on Geoscience and Remote Sensing, 38 p. 2202-2212.

Carbon Dioxide Capture for Storage in Deep Geological Formations, Volume 4
Karl F. Gerdes (Editor)

Section 2

STORAGE MONITORING & VERIFICATION (SMV)

Contingencies

Carbon Dioxide Capture for Storage in Deep Geological Formations, Volume 4
Karl F. Gerdes (Editor)

Chapter 42

OVERVIEW OF ASSESSMENT OF LEAKAGE DETECTION AND INTERVENTION SCENARIOS FOR CO_2 SEQUESTRATION

Anshul Agarwal[1], Tom Aird[1], Sally Benson[1], David Cameron[1], Jenny Druhan[2], Jerry Harris[3], Kate Maher[2], Julia Reece[3], Stéphanie Vialle[2], Christopher Zahasky[1], Sergio Zarantonello[4] and Mark Zoback[3]
[1]Department of Energy Resources Engineering, Stanford University
[2]Department of Geological and Environmental Sciences, Stanford University
[3]Department of Geophysics, Stanford University
[4]Visiting Researcher, Department of Geophysics, Stanford University

ABSTRACT: CO_2 storage security, monitoring and verification are of utmost importance for the long term implementation of CCS for the purpose of reducing global emissions of carbon dioxide. There are several mechanisms which have the possibility of compromising the security of super critical CO_2 stored in deep saline aquifers or depleted oil and gas reservoirs. Of particular interest is the risk of leakage via faults and fractures which could provide fluid migration pathways from the storage reservoir to overlying aquifers or even to the earth's surface. A focus on the investigation and development of methods for dealing with CO_2 leakage is urgently needed for industry and regulators alike to be prepared for managing such events should they arise. A quick response to unforeseen events will reduce the risks and costs associated with them, and increase public confidence in CCS as a climate mitigation option. This chapter introduces and summarizes conclusions from the three chapters that follow dealing with detection and remediation.

KEYWORDS: carbon storage; leakage intervention; sealant; pressure monitoring; seismic monitoring; hydraulic barriers; fault characterization

PROJECT INTRODUCTION

The objective of this work was to develop and quantitatively evaluate intervention options for responding to unforeseen fluid leakage or lack of project conformance in CO_2 storage projects. Although CO_2 storage has to date been shown to be safe and effective, the possibility exists for unforeseen CO_2 leakage via faults and fractures, active or abandoned wells, or unexpected plume migration, or for brine displacement into drinking water aquifers. Being prepared to address these residual risks with a set of options for stopping leakage and controlling plume migration within the storage reservoir, will eliminate or minimize environmental damage to drinking water or to the near surface environment. This will provide an additional measure of confidence for regulators and the public alike, thus facilitating deployment of CCS. Additionally, both the European CCS Directive and the US EPA CCS regulations require the development of contingency plans for each stage of the CCS project.

A focus on the investigation and development of methods for dealing with CO_2 leakage is urgently needed for industry and regulators alike to be prepared for managing such events should they arise. A quick response to unforeseen events will reduce the risks and costs associated with them, and increase public confidence in CCS as a climate mitigation option. Specifically, options for managing each step in the chain of events illustrated in Figure 1, from leak detection to completion of the intervention, are needed.

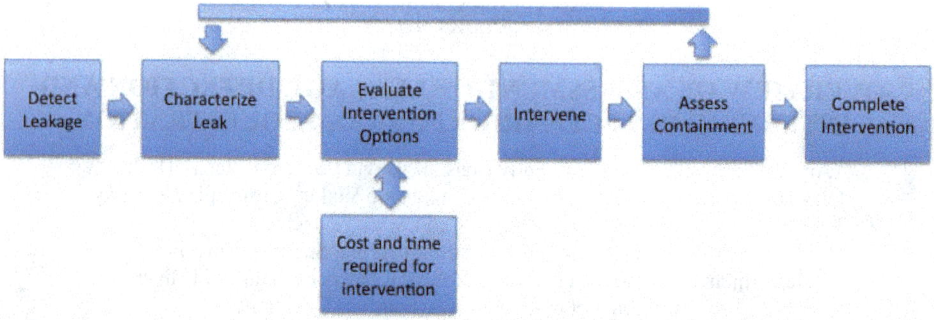

Figure 1. Work-flow for leakage intervention showing the chain of events expected over the intervention activity.

We reviewed the existing literature and summarized the state-of-the-science for leakage intervention activities in 2013. In some cases a wealth of information is known. For example, there are a number of monitoring approaches available for leak detection, with field demonstrations of a variety of techniques confirming detection of small quantities of CO_2 in the subsurface. In others areas, such as intervention options, only a handful of papers exist. Moreover, all of the existing studies on intervention are based solely on numerical simulation, lacking any sort of field experimentation or validation. Based on that review, we made the following recommendations:

1. **Real-Time Leak Detection:** Develop and test real-time methods for leak detection and conformance monitoring to provide early warning that the project is not performing as expected.
2. **Leak Characterization:** Develop and test methods for providing detailed leak characterization once it has been detected. Important information includes the cause for the leak, precise location, the CO_2 saturation distribution, and the quantity leaked. Additionally, characterization of the hydraulic properties of the fracture is also needed. These methods will also be useful for tracking progress toward leak containment/mitigation and eventually be needed to cease the intervention activities.
3. **Intervention:** Perform numerical simulations to assess the potential of passive remediation, and hydraulic controls (both injection and extraction, or a combination of both) for a wide variety of scenarios, using realistic fault architectures and permeabilities and heterogeneous rocks. These studies should include geomechanical deformation of the fractures and stress dependent permeability. Approaches for optimizing well locations and flow rates are also needed.
4. **Seal Placement Location:** Perform numerical simulations to evaluate seal placement location for realistic fault geometries. These simulations can be done, even without simulation of the emplacement process itself. These will be useful for determining the spatial extent, geometry and transport properties of effective seals.
5. **Sealant Delivery:** Perform numerical simulations to assess the potential for delivering sealants to faults – and for identifying the optimal properties of seal materials (e.g. initial viscosity, set times, etc.). Model enhancements to accomplish this are needed to realistically model seal emplacement. Laboratory tests of seal performance will be needed to provide model inputs.

6. **Assessment and Completion:** Develop a test bed where a variety of techniques for leakage management, from leak detection to completion of intervention, can be tested and evaluated. Having prior experience with leakage intervention techniques will be of great value to the industry when the need for deployment arises.

Based on this assessment and a workshop held by CCP3, we identified a work program that would address many of these issues. A schematic showing the options for addressing the issues is shown in Figure 2. The items indicated in red were explicitly addressed by this project. This chapter provides an overview of the major aspects of the project.

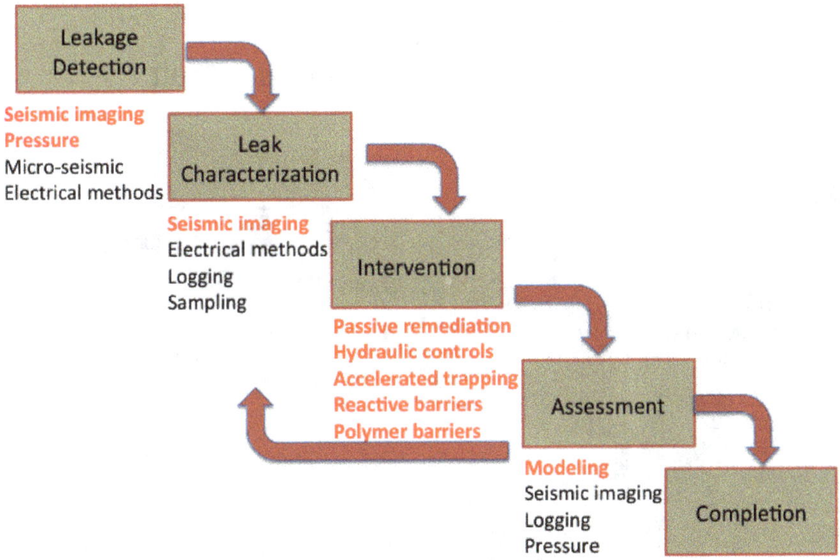

Figure 2. Recommendations for leakage intervention.
The items shown in red were conducted as part of the CCP3 project.

PROJECT PLAN AND STRATEGY

A scenario-based approach was used to carry out and integrate the results of this project. We began by simulating leakage in a typical storage project in a saline aquifer. The model was based on a deep section of the Powder River Basin in Wyoming, consisting of a thick sequence of sandstone, shale, and anhydrite. Leakage rates, pressure build-ups, and saturations from this model were used for calculating seismic responses and assessing how soon after injection began would it be possible to detect leakage. Based on these simulations we concluded that the leak could be detected with a few years after the project started, when a total of about 5000 tonnes of CO_2 had leaked out of the storage reservoir and accumulated in an overlying aquifer above the seal. These simulations set the stage for higher resolution simulations, which addressed the following issues:

- Influence of fault and reservoir properties on leakage;
- Effectiveness of stopping injection on leak mitigation;
- Effectiveness and permanence of hydraulic controls such as fluid extraction and brine injection on leak mitigation;
- Potential for accelerated trapping of CO_2 in the overlying aquifer;

- Effectiveness of polymer and reactive barriers (e.g. amorphous silica) on stopping leakage;
- Barrier emplacement strategies; and
- Evolution of fault permeability due to plugging with reactive materials.

In addition to these simulation studies, we also performed laboratory experiments to test the most promising commercially available sealant, H_2Zero^{TM} by Halliburton. A barrier emplacement strategy, based on measured gel times was also developed. Figure 3 illustrates the components of this program and how they fit together.

A summary of the elements of the project are described briefly the following sections.

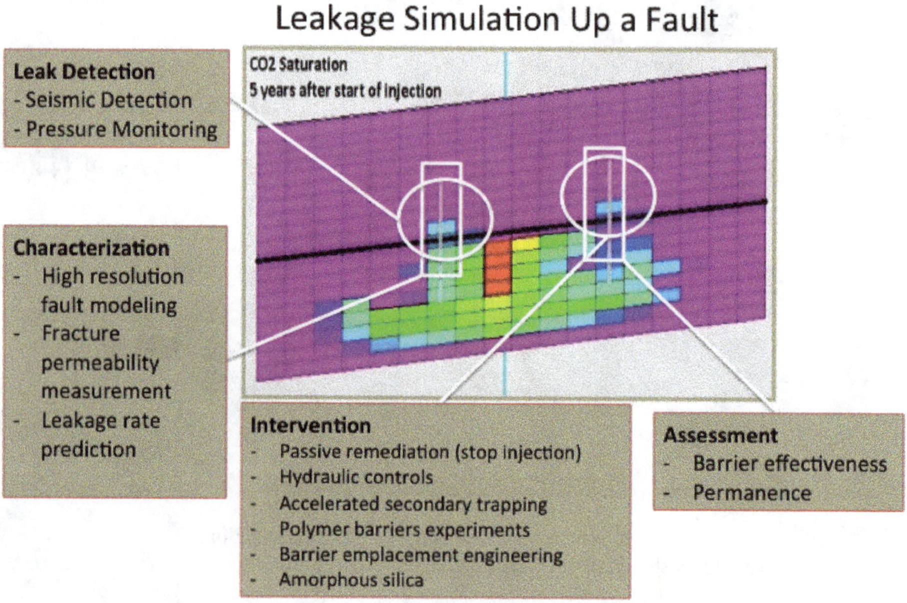

Figure 3. Schematic showing the major elements of this project and how they fit together.

LEAKAGE DETECTION

Pressure Monitoring

Because pressure responses in the subsurface propagate quickly, pressure data from monitoring and injection wells in the storage formation and/or overlying aquifer may be useful in the early detection of leaks. Previously, the location and transmissibility of an abandoned well have been determined using pressure data collected over 50 days from nearby injection and monitoring wells. These analyses ignore geologic uncertainty and use an oversimplified aquifer model, but it shows that leakage detection from pressure data may be possible. The detectability of leaks through abandoned wells and fractures using pressure-transient data from wells in the overlying aquifer was analyzed by Chabora & Benson [1]. They found that this data can indicate CO_2 breakthrough in some cases, though their tests were performed on simplified aquifer models with known geology. Sun & Nicot [2] confirmed that pressure anomalies indicating leakage are detectable in the presence of measurement error and spatial heterogeneity. More recently, Sun *et al.* [3] used a probabilistic collocation method to determine when pressure data might be used to detect leakage. The method considers the signal-to-noise ratio of pressure anomaly data compared to background noise. This approach provides an effective means for detecting when a leak exists. The inverse problem of determining the size and position of leaks via data assimilation was not addressed. Furthermore, the detectability of multiple leaks using pressure data remains to be investigated.

In this work we describe a 3-D, heterogeneous, synthetic aquifer model, where CO_2 may leak through the cap rock into an overlying aquifer from any number of potential leakage locations. We analyze the degree of CO_2 leakage in the model with respect to leak location, leak permeability and other geological features. We present a history matching algorithm for detecting leaks in the cap rock, and provide results for different leakage scenarios using various amounts of observation data. Our overall analysis allows us to estimate the detectability of leakage conduits using pressure data from injectors and monitoring wells. A data assimilation (history matching) procedure, which uses monitoring well pressure data and bottomhole pressure data, was developed with the goal of characterizing leaks. Aquifer geologies for the storage formation and overlying aquifer were represented using the Karhunen-Loève (K-L) expansion, which significantly reduces the number of geologic variables to be determined in the history match. Particle Swarm Optimization (PSO) was applied for the minimizations required for the data assimilation procedure. A measure for determining the efficacy of history matching for a set of models was also developed.

Data assimilation results for a number of single-leak and multi-leak cases using both a data-rich scenario and data-scarce scenario provided insight into the detectability of leaks from pressure data. Both the data-rich and data-scarce scenarios resulted in reasonable matches of leak location and CO_2 leakage fraction for the single-leak cases that were tested. For multi-leak cases, the CO_2 leakage fraction and the location of the largest leaks were matched with some degree of accuracy in the data-rich scenario, whereas the data-scarce scenario provided less accurate predictions. Overall, the most useful data appears to be pressure monitoring data from the overlying aquifer, followed by pressure monitoring data from the storage formation. Bottomhole pressure data from the injection wells does not appear to be useful in locating leaks.

Seismic Monitoring

The goal of seismic monitoring is to assess safe containment of CO_2 and detect leaks from the storage reservoir. Even though there is no established geophysical method proven or prescribed for "early" detection of CO_2 leaks, the basis for early detection by a seismic method is the sensitivity to changes in saturation and pressure in areas above a container seal or in and around leak paths. Successful leak detection will depend on the repeatable use of seismic data acquisition methodologies, data processing, and interpretation. Additionally, early leak detection will depend on the frequency of the geophysical measurements. The frequency of surveys and sensitivity are related because our ability to distinguish small signal changes from noise depends on survey frequency as well as the degree to which measurements can be repeated. Conceptually speaking, true 4-D seismic imaging should replace 3-D much like the way 3-D imaging has replaced 2-D. True 4-D seismic methods can be very powerful for leak detection when implemented continuously or frequently. Moreover, successful early detection of leaks will no doubt depend on the quantity of CO_2 leaked, size and spatial distribution of the leak, and the proximity of the leak to data gathering operations.

The most successful monitoring approaches will likely include (1) reconnaissance scale monitoring to identify a potential leak, in combination with (2) high-resolution targeted monitoring for delineation of a leak. Reconnaissance monitoring can be surface-based (e.g., low-resolution) in order to cover a large and expanding containment volume. Delineation methods may include borehole-based as well as high-resolution surface-based methods. Such complementary methods and techniques might be implemented sequentially in time or simultaneously depending on reservoir conditions, environmental considerations, and risk assessment.

We developed a complete time-lapse workflow, including flow simulation, seismic rock properties, and seismic image simulation. We conclude from the simulations that surface seismic monitoring can be effective in detecting saturation changes caused by a leak in the primary storage reservoir. This result is no surprise. The change in saturation associated with the leak must be of significant dimensions both vertically and horizontally to present a gradient contrast large enough to generate a reflection response. Detectability is entirely site dependent and depends, of course, on survey geometry and signal frequency. For the case study investigated here, leaks of 5000 tonnes could be detected a few years after injection started.

Conventional surface seismic is not likely to be effective at early detection of pressure changes above the seal. This negative conclusion comes because the changes in seismic properties due to pressure alone are small. Moreover, the diffuse nature of pressure change above the seal is not conducive to reflectivity imaging from the surface seismic geometry. Such diffuse velocity changes may be detectable with borehole seismic methods.

LEAK RATE ASSESSMENT

The first step in understanding potential fluid migration through faults or fractures is to first characterize the fault zone size, geometry and architecture. Sub-seismic faults, which could easily go undetected during site characterization, are the focus of this study. While the sub-seismic threshold can vary based on fault properties and observation techniques, it is generally considered to include faults with displacements or offsets of less than 10 meters. Based on this initial constraint of fault displacement, published correlations between fault displace and length, and fault displacement and fault width, can be used to establish an appropriate fault geometry. Additionally, the permeability structure of sub-seismic fault zones in sedimentary rocks is explored in order to estimate a range of potential permeability approximations for simulation studies.

A systematic set of sensitivity studies were carried out using TOUGH2 (available from Lawrence Berkeley National Laboratory) to identify the properties of the reservoir/fault/aquifer system that control leakage rates. Surprisingly, we found that the leakage rates are as strongly affected by the permeability of the storage reservoir and overlying aquifer as they were to the fault itself. Based on this observation we developed a simple analytical model that can be used to bound the rate of leakage. This will be extremely useful for assessing risks and designing remediation plans. Details of this work are provided in Chapter 42 of this volume on Remediation [Zahasky & Benson].

FAULT CHARACTERIZATION

Understanding the permeability distribution in a fault zone is key to designing an effective leakage intervention plan. Often, the complexity of actual faults is neglected during simulation of leakage and intervention. Here we review the dominant features of fault zones, with an emphasis on the relation between initial mineralogy, rock properties (porosity, elastic and viscous properties), chemical and mechanical evolution and structure of the resulting damage zone system. In the modeling work, we provide a physical representation of a fractured system that may be encountered in a seal layer of a CO_2 storage site. This physical representation is less simplified than a single high permeability vertical fracture, and guided by both field observations and indirect geophysical interpretations. An example of a fault-damage zone system is a fault core of low permeability and a damage zone of variable permeability field with preferential flow along the fault plane. For a typical representation of such a system, we developed an upscaling relationship for the permeability field in the damage zone that is an analytical expression of permeability as a function of fracture aperture and fracture density. By applying the expression for permeability in each grid block of the domain, we transform the simplified discrete fracture network into a grid-based continuum model. Leakage scenarios of CO_2 through this fault damage zone were modeled with the simulator TOUGH2.

We demonstrated the importance of the physical heterogeneities for the geochemical and hydrologic evolution of a fault damage zone that is interacting with a CO_2 injection. Our results highlight the necessity to improve the characterization of fractured zones and their hydraulic behavior. Some of the fractures surrounding a fault can be below the resolution of the current geophysical methods and are thus not detected during the site characterization. And even if they are detected, some may be hydraulically active and some not. There are very few studies that have obtained the hydraulic properties of a damage zone from back analyses of hydraulic tests in boreholes set across faults [4,5], mainly because of the heavy equipment that must be installed *in situ*.

We also investigated different scenarios of CO_2 leakage through a fractured damage zone having a heterogeneous permeability field, details of which are discussed in Chapter 44 on Simulation of Reactive Barrier Emplacement [Druhan & Maher]. One scenario considered the migration of dry CO_2 in the caprock. Though this is likely to happen only close to the injection well, the results obtained in these simulations are important to help design mitigation strategies. Here some precipitation reactions are induced in the caprock, leading to a lateral migration and decrease of the leak over time. This can be seen as an analog of the placement of a Silica Polymer Initiator (SPI-CO_2) gel, a polymer that remains a low viscosity fluid until triggered by contact with CO_2, unlike current technology where gel setting is a function of time, regardless of location. If the polymer is strategically placed in the main CO_2 flow path(s), the leak will migrate to smaller fractures as well as simultaneously decrease, since the permeability is reduced in smaller fractures.

INTERVENTION AND REMEDIATION

In this study, we evaluated four different approaches for looking or stopping leaks: stopping injection; hydraulic controls through extraction and injection of brine; gelling polymers; and a reactive barrier (amorphous silica). The major results are described briefly below.

Passive Remediation

The fastest and lowest cost option available is passive remediation - that is, simply stopping injection. The effectiveness of this method is highly dependent on two factors: how soon the leak is detected, and the residual CO_2 saturation in the storage reservoir. For leaks that are detected quickly and injection stopped immediately thereafter, and for cases where the residual CO_2 saturation is 20% or greater, leakage will stop shortly after injection stops. However, when the residual CO_2 saturation is lower or for larger leaks (where the CO_2 plume has migrated far beyond the location of the fracture), stopping injection with slow the rate of leakage significantly, but not stop it completely.

Hydraulic Barriers

Usually injection shutoff alone is not able to completely stop leakage. However, in many cases the quantities of CO_2 leaked may be far below levels that would be of concern, typically less than 5% of the initial leakage. If passive remediation is not sufficient, various hydraulic controls such as water injection [6] or reservoir fluid production can be employed to further reduce leakage rates and trap leaked CO_2. Water injection is a very effective remediation technique because it: (1) increases the pressure in the overlying aquifer relative to the base of the fault which can quickly stop leakage, (2) is able to push some amount of CO_2 back down the fault, and (3) is able to dissolve large quantities of CO_2 in the vicinity of the water injection. In some cases, water injection alone may be enough to stop current and future CO_2 leakage, especially when the leak is detected early, before a large amount of CO_2 has accumulated below and above the fault. Simulations were run using a range of injection rates and well configurations (horizontal and vertical wells). To reduce the risk of reoccurrence of leakage in the future, brine can also be injected into the reservoir below the fault. This reduces the mobility of CO_2, accelerating residual gas trapping in the portion of the storage reservoir underlying the fault. Specific scenarios are demonstrated for accomplishing this involving sequential and simultaneous brine injection above and below the fault. Finally, scenarios are evaluated where brine is also extracted from the reservoir, providing even greater assurance of permanent leak containment, with the additional benefit of providing a source of brine. Overall these studies demonstrate that hydraulic controls can quickly stop leakage, and correctly implemented, provide a permanent remedy for stopping the leak.

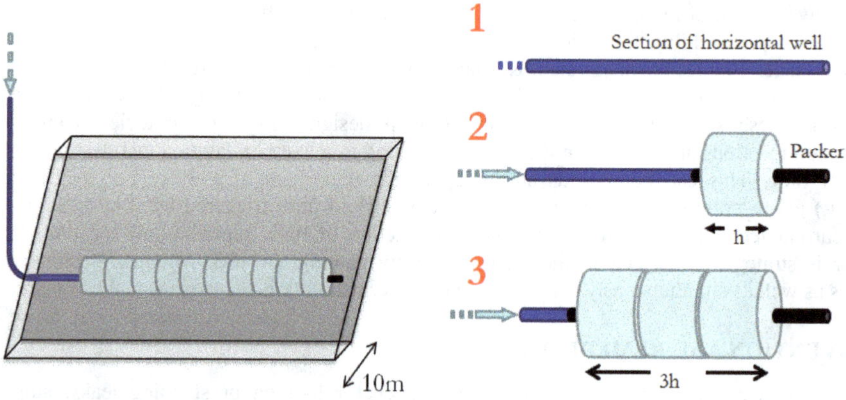

Figure 4. Diagrams of horizontal injection scheme into a 10m vertical fracture plane (left) and a three step process diagram of a horizontal well with sealant injection intervals (right).

Sealants

An alternative mitigation strategy is the introduction of a physical barrier in the form of injected sealant material. An organically crosslinked polymer (OCP) system was investigated as a potential sealant material for leaks from CO_2 reservoirs. This OCP system was developed by Halliburton Energy Services and is currently marketed as H_2Zero^{TM} Service. The OCP system possesses many advantages over other sealant methods. The relatively low initial viscosity (30 cP) enables greater penetration depth into matrix formation than that of traditional squeeze cements. It possesses long-term thermal stability that enables it to keep its structure even at elevated temperatures. This system has also been proven to be unaffected by formation fluids, lithology, or heavy metals [10]. The formulation of the original OCP system is designed to work in a temperature range from 49 to 127°C. Two other formulations were developed from this original system: one for lower temperatures (16-60°C) and one for higher temperatures (127-177°C). The lower temperature OCP system uses a polyacrylamide base polymer, while the high temperature OCP system adds a water soluble carbonate retarder to the original formulation.

A key to successful implementation of gelling polymers it to delay the gel time for as long as possible, so as to be able to emplace the polymer in a large fracture. To evaluate gel times, we tested a variety of formulations by making viscosity measurements in the temperature range from 40-60°C. We found satisfactory formulations that would increase gel times up to 8 hours.

Given the relatively short gel time, we developed an emplacement scheme using a horizontal well drilled through the middle of the fracture plan. Through the use of packers to isolate individual zones, it would be possible to seal a long fault by a series of gel emplacements (see Figure 4). More details are discussed in Chapter 43 on Evaluation of organically-crosslinked polymer system for leakage remediation [Aird & Benson].

REACTIVE SEALANT DELIVERY AND ITS EFFECTIVENESS

Physical and chemical trapping mechanisms are both necessary for long-term storage of CO_2. Initially, the most important of these are structural and hydrodynamic entrapments, which inherently require the presence of low permeability beds to act as seals above the injection formation. This physical trapping can subsequently lead to chemical trapping processes over longer periods of time, including dissolution, sorption and mineralization [11]. The time period over which chemical mechanisms become effective is frequently long enough that physical trapping offers the first defence against CO_2 loss due to plume migration and buoyancy, particularly while primary injection of the supercritical phase is still active.

While the chemical mechanisms for CO_2 stabilization may require on the order of tens to thousands of years to become a substantial component for secure storage [12], perturbations to the geochemical composition of formation water in response to the introduction of concentrated CO_2 can be quite rapid. As a result, should CO_2 intrusion into an overlying reservoir occur due to a defect in the caprock structure, this migration could be associated with abrupt chemical changes as a function of the initial composition of the aquifer [13]. These changes, including increased dissolved inorganic carbon (DIC), lowered pH, trace metal dissolution and shifts in the stable isotope ratios of carbon, oxygen and hydrogen, present a potential means of chemically identifying and characterizing a newly formed leak. Furthermore, the distinct chemical composition of the CO_2 intrusion relative to the surrounding formation water may be leveraged for engineered intervention strategies. Here we introduce the expected geochemical changes associated with a newly formed CO_2 leak, provide examples from both engineered and natural sites, and review the properties of compounds available as potential chemical sealants for a CO_2 leak.

Effectively emplacing a barrier in a leaking fracture is challenging due to the complex geological, geometrical configuration, and fault heterogeneity. In light of this, it is also attractive to identify sealants that would react with the leaking CO_2 itself to form a precipitate that would plug the pore spaces of the rock. However, effectively designing emplacement schemes for reactive barriers is also challenging due to multi-phase flow effects, which make it difficult to deliver a sufficiently large mass of the reactants to fully plug the pore space of the rock. Numerical simulations of reactive barrier emplacement (amorphous silica as a model system) were carried out to investigate the feasibility of reactive barrier emplacement, including location and rates of reactant injection wells. These studies, as discussed in Chapter 44, highlight the challenges of reactive barrier emplacement and provide a good starting point for additional investigations.

CONCLUSION

This multi-faceted study forms the basis for improved understanding of the options for intervention measures to stop or eliminate CO_2 leaks. Overall these results are encouraging, showing that it is possible to detect CO_2 leakage relatively early, when only a very small fraction of CO_2 has leaked from the reservoir. Once detected, leakage can be quickly slowed by a number of measures investigated here. The work provides the foundation for follow-on studies – in particular, field tests to evaluate and refine the effectiveness of the approaches presented here.

ACKNOWLEDGEMENTS

This project was supported by CCP3; a joint industry project sponsored by BP, Chevron, Eni, Petrobras, Shell, and Suncor.

REFERENCES

1. Chabora, E. and Benson, S. (2009) Brine displacement and leakage detection using pressure measurements in aquifers overlying CO_2 storage reservoirs. Energy Procedia, 1(1), pp 2405-2412.
2. Sun, A. and Nicot, J. (2012) Inversion of pressure anomaly data for detecting leakage at geologic carbon sequestration sites. Advances in Water Resources, 44, pp 20-29.
3. Sun, A. and Zeidouni, M. and Nicot, J. and Lu, Z. and Zhang, D. (2013) Assessing leakage detectability at geologic CO_2 sequestration sites using the probabilistic collocation method. Advances in Water Resources, 56(1), pp 49-60.
4. Medeiros, W.E., do Nascimento A.F., Alves da Silva F.C., Destro N., and Deme´trio J.G.A., (2010) Evidence of hydraulic connectivity across deformation bands from field pumping tests: Two examples from Tucano Basin, NE Brazil. Journal of Structural Geology 32, 1553–1864.
5. Jeanne, P., Guglielmi Y., and Cappa F., (2013) Dissimilar properties within a carbonate reservoir's small fault zone, and their impact on the pressurization and leakage associated with CO_2 injection. Journal of Structural Geology 47, 25-35, doi:10.1016/j.jsg.2012.10.010.
6. Reveillere A, Rohmer J, Manceau J-C. (2012) Hydraulic barrier design and applicability for managing the risk of CO_2 leakage from deep saline aquifers. Int J Greenhouse Gas Control 2012; 9:62-71.
7. Benson SM, Hepple RP. (2005) Detection and Options for Remediation of Leakage from Underground CO_2 Storage Projects. In: Wilson M, Morris T, Gale J, Thambimuthu K, editors. Proceedings of the 7th International Conference on Greenhouse Gas Control Technologies, Oxford: Elsevier; 2005, Vol. II, Part 1, p. 1329-1338.
8. Esposito A, Benson SM. (2012) Evaluation and development of options for remediation of leakage into groundwater aquifers from CO_2 storage reservoirs. Int J Greenhouse Gas Control 2012; 7:62-72.
9. Imbus, S.W., Dodds K., Otto C.J., Trautz R.C., Christopher C. A., Agarwal A., and Benson S.M., (2012) CO_2 Storage Contingencies Initiative: Detection, Intervention and Remediation of Unexpected CO_2 Migration, presented at the GHGT-11 conference, 18- 22 November 2012, Kyoto, Japan.
10. Vasquez, Julio, and Larry Eoff (2010) "Laboratory Development and Successful Field Application of a Conformance Polymer System for Low-, Medium-, and High-Temperature Applications." SPE Latin American and Caribbean Petroleum Engineering Conference. 2010.
11. IPCC, (2005) - Bert Metz, Ogunlade Davidson, Heleen de Coninck, Manuela Loos and Leo Meyer (Eds.) *Special Report on Carbon Dioxide Capture and Storage*, Cambridge University Press, UK.
12. Kharaka, Y.K. and Cole, D.R. (2011) Geochemistry of geologic sequestration of carbon dioxide. In Harmon, R. (ed.), Frontiers in Geochemistry: Contribution of Geochemistry to the Study of the Earth, chapter 8, Wiley-Blackwell, pp. 135 – 169.
13. Little, M.G. and Jackson, R.B. (2010) Potential impacts of leakage from deep CO_2 geosequestration on overlying freshwater aquifers. Environmental Science & Technology, 44, 9225 – 9232.

Carbon Dioxide Capture for Storage in Deep Geological Formations, Volume 4
Karl F. Gerdes (Editor)

Chapter 43

QUANTIFICATION AND REMEDIATION OF POTENTIAL CO_2 LEAKAGE FROM CARBON STORAGE RESERVOIRS

Christopher Zahasky and Sally Benson
Stanford University Department of Energy Resources Engineering, Stanford, USA

ABSTRACT: One of the greatest concerns associated with large-scale adoption of CCS technology is the risk of carbon dioxide leakage from sequestration reservoirs, which creates the need to develop intervention and remediation strategies. Here we first investigate the effectiveness of passive remediation (stopping CO_2 injection), and then determine its effectiveness in combination with a variety of different hydraulic controls such as injection of water in the overlying aquifer, injection of water in the injection reservoir near the fault, and production of brine in the lower reservoir. Regardless of when the leak is detected, simulation results show that passive remediation almost immediately reduces the CO_2 leakage rate by an order of magnitude. In order to completely stop leakage in a homogenous reservoir, the implementation of hydraulic controls is necessary. Water injection into the overlying aquifer directly above the fault is able to completely terminate leakage for as long as injection continues. The most effective hydraulic controls for stopping CO_2 leakage combine brine extraction from the storage reservoir with water injection into the overlying aquifer near the fault. This study demonstrates that temporally limited, multi-stage intervention strategies, such as hydraulic barriers, can permanently terminate CO_2 leakage while having the additional benefit of dissolving most of the CO_2 in the overlying aquifer into the resident brine.

KEYWORDS: CO_2 sequestration; leakage; remediation; hydraulic barriers; intervention; simulation; numerical model

INTRODUCTION

There are two main forces that could drive CO_2 and brine leakage from a storage reservoir. The first is the pressure build up in the storage reservoir due to CO_2 injection. Carbon dioxide injection will often increase the fluid potential in the reservoir to levels higher than that of the overlying aquifer(s). If a permeable flow path exists in the seal between the injection reservoir and overlying aquifer (e.g. abandoned well or fault zone), fluid will migrate from the areas of high fluid potential in the injection reservoir to areas of lower fluid potential in overlying aquifer(s). The second driver of CO_2 leakage is the buoyancy of CO_2 relative to the resident reservoir fluid. Typically, supercritical CO_2 is less dense than the surrounding reservoir fluid; this gravity driven buoyancy force is capable of driving CO_2 from the storage reservoir.

If a leak in a sequestration reservoir is identified, it is important to evaluate the risks, operational and economic costs, and environmental impacts of the leakage in order to determine if remediation is necessary and/or feasible. Several remediation strategies have been proposed, which generally fall into one of four categories:

1. Hydraulic barriers and pressure management [1-3],
2. Production and removal of CO_2 from the storage reservoir [4],
3. Biological barriers [5], and
4. Sealants and other physical barriers (e.g. [6]).

Hydraulic barriers rely on the manipulation of pressure fields in the overlying aquifer and storage reservoir by either injecting water (sourced from the injection reservoir) or producing reservoir fluids in order to stop and possibly reverse leakage. In this report, water injection is differentiated from brine injection, in that water injected into the subsurface is required to be treated in many places around the world. Production of CO_2, likely from the original injection well, removes much of the mobile CO_2 in the storage reservoir. This reduces long term leakage significantly but can have huge financial consequences if the CO_2 must be re-injected in another storage reservoir or if penalties are incurred for venting the CO_2 into the atmosphere. Biological barriers, sealants and other physical barriers attempt to alter the permeability of the fault zone by introducing gels, grouts or other materials that have the ability to fill the pore spaces through which CO_2 is escaping from the storage reservoir.

In this study, simulation models are developed in order to analyze and characterize the behavior of CO_2 leaking from a storage reservoir through a subseismic fault. In this scenario, CO_2 is injected into a lower reservoir capped by an impermeable seal. Above the seal is another smaller saline aquifer. In this chapter, 'overlying aquifer' will refer to this saline aquifer located directly above the seal of the injection reservoir. At some injection sites this overlying aquifer may be considered an upper portion of the injection reservoir that is sealed from the injection interval by some impermeable barrier (i.e. caprock). As CO_2 is injected, the CO_2 plume migrates through the injection reservoir. The plume reaches a fault zone which provides a pathway for fluid connectivity between the reservoir and the overlying aquifer.

After developing a detailed understanding of the leakage process, timing, pressure and saturation conditions, different intervention options are evaluated. These options include CO_2 injection termination and hydraulic controls such as water injection in the overlying aquifer above the fault, injection of water in the lower reservoir below the fault, and reservoir fluid production away from the CO_2 plume. While these remediation techniques were developed and tested for a limited range of conceptual models, additional work looking at the effect of additional system complexities such as heterogeneity is ongoing.

METHODS AND MODEL DEVELOPMENT

Simulation model development

All simulations are performed with TOUGH2, a fully implicit numerical simulator designed to model non-isothermal, multiphase, multicomponent flow in porous and fractured media [7]. TOUGH2 was run with the fluid property module ECO2N [8]. In order to properly model the simultaneous drainage and imbibition processes that take place during remediation activities, a specialized TOUGH2 source code was used in order to properly model hysteresis. Details of the code can be found in [9].

The simulation model used for this analysis, is 2.5 km by 2.5 km by 100 m in extent, with structured grid geometry (Figure 1). The grid is locally refined around the injection well and fault zone. The smallest grid cells are one meter wide in the fault and three meters wide at the injection well. The largest grid cells are 500 m wide near the model boundaries. In order to model a basin scale reservoir, the model boundary cells have volume factors of 10^{50} such that the model is effectively of infinite areal extent. The lower reservoir has a thickness of 68 m, the caprock is 12 m thick, and the overlying aquifer is 20 m thick. The storage reservoir and overlying aquifer both have permeabilities of 28 mD and porosities of 10%. The caprock has a permeability of 0.2 nanodarcy and a porosity of 5%. The fault zone geometry was constrained using published scaling relationships starting with the initial assumption that the fault zone is subseismic and therefore has a displacement less than 10m. The fault is three meters wide and 500 m long. For simplicity, the permeability structure that typically exists in fault zones in sedimentary rocks (e.g. [10]) is ignored and a single permeability value is assigned to the fault zone.

The aquifer and reservoir porosity, permeability and characteristic capillary pressure curves are modeled after the Arqov sandstone as described by Pini *et al* [11]. The characteristic capillary pressure curves for the fault and caprock are determined by scaling the Arqov capillary behavior using the Leverett function. The characteristic relative permeability curves are Corey's Curves with S_{lr}=0.2 and S_{gr}=0. The reservoir temperature and pressure conditions are established based on typical values for a reservoir located at a depth of roughly 1600 m below the water table. Prior to remediation initiation, CO_2 is injected at a constant rate of 7.9 kg/s (0.25Mt/yr) into the lower reservoir over a completion interval of 0 m to 55 m from the bottom of the reservoir.

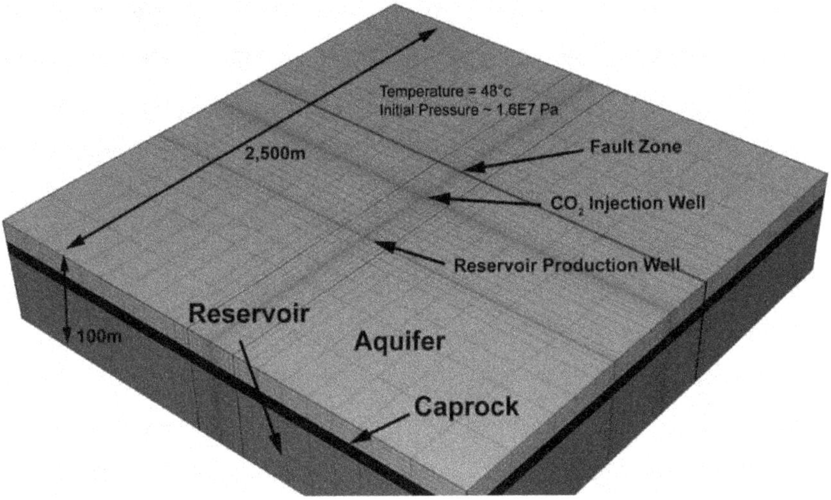

Figure 1. Grid geometry model used for TOUGH2 simulations.
Note the height of the system is scaled 10x greater than the x and y directions.

Residual trapping

An important factor in plume stabilization and thus, remediation effectiveness, is the extent of CO_2 residual trapping. To evaluate the influence of residual trapping, the commonly used Land trapping method [12] is implemented in the ECO2 simulation module.

Trapping coefficient and residual trapping measurements have been measured in lab and field settings [13]. Krevor *et al.* [14] measured Land trapping coefficients in four cores from different sandstone formations and found that core-averaged CO_{2rmax} (here referred to as S_{nmax}) values ranged from 0.21 to 0.33, with typical values around 0.3. Other laboratory experimental studies (e.g. [15,16]) used a variety of methods for relating initial and residual saturation values; these studies yielded similar results. Residual trapping values were also measured in a carbon storage reservoir as a part of the CO2CRC Otway demonstration project [17]. Results from this project estimated that the residual CO_2 saturation in the reservoir was between 11-20% [18]. Typical gas saturation values of 20-50% in the reservoir, and a S_{nmax} trapping coefficient of 0.3, leads to residual trapping values between 13-25%. Based on these studies, a S_{nmax} value equal to 0.3 was assigned for all of the remediation model scenarios in the following sections. Details of the Land trapping implementation are described in [9].

BASE LEAKAGE SCENARIO

Prior to the development of intervention methods, it is useful to examine the character of leakage through a fault zone as a function of time using the model described above. Carbon dioxide is injected at a constant rate of 7.9 kg/s and initially there is no leakage of CO_2 through the fault. However as CO_2 is injected into the reservoir, reservoir fluid/brine is displaced, driving brine up the fault. This flux of brine into the overlying aquifer will result in a pressure increase in the overlying aquifer which has important implications for leakage detectability. The aquifer pore volume, permeability, and distance of the leak from the monitoring well will determine if the pressure increase from the brine flowing into the aquifer is large enough to be detectable [19]. While it is unlikely that a monitoring well would be located in such a convenient location, Figure 2 illustrates the pressure response at the top of the fault zone and CO_2 leakage rate through the fault zone. In this model scenario, the pressure increase is on the order of hundreds of kilopascals up to several hundred meters from the point of leakage, which is likely to be detectable with current measurement tools [19,20]. The pressure increase would be lower if the overlying aquifer had a much large pore volume (i.e. higher porosity or greater thickness).

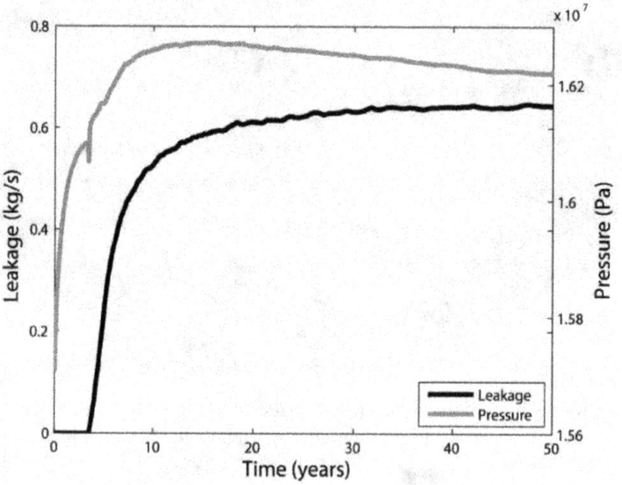

Figure 2. CO_2 leakage rate (black line) through the fault zone and pressure at the top of the fault (gray line) resulting from both brine and CO_2.

Following this initial period of brine leakage via the fault, the CO_2 injection plume reaches the fault after 3.4 years in this model scenario. The pressure drops briefly when CO_2 begins to migrate up the fault. This occurs because the relative permeability of water/brine drops precipitously during drainage, causing the cumulative fluid flow (i.e. brine and CO_2) to drop. As the CO_2 plume continues to advance in the lower reservoir, the rate of leakage continues to increase as does the pressure at the top of the fault. Leakage increases because the length of fault exposed to the plume increases until the CO_2 plume covers the entire base of the fault zone and because the thickness of the plume underneath the fault continues to increase for many years, increasing the buoyancy force driving CO_2 up the fault zone. During this time, the pressure at the top of the fault decreases slightly as the gas saturation in the overlying aquifer continues to increase, increasing the mobility of the CO_2 into the overlying aquifer.

The exact timing and rates of potential CO_2 leakage could vary tremendously based on the geology of the injection reservoir, caprock, fault, and overlying aquifer properties. Previous work has shown that the reservoir and aquifer permeability are as important as the fault permeability in controlling leakage into the overlying aquifer [21]. For example, when the overlying aquifer has a very low permeability relative to the injection reservoir, the reservoir injection pressure may be too low to overcome the capillary entry pressure of a low permeability overlying aquifer. Alternatively, if the overlying aquifer permeability is high relative to the reservoir permeability then the leakage rate is only limited by the fault permeability. If the fault permeability is also high relative to the injection reservoir permeability, the leakage rates are likely to be larger than those shown in Figure 2. These factors will also influence the efficacy of the different remediation options.

The leakage rate reaches a steady state flow rate roughly 30 years after injection begins in this model system. Timing and rates change dramatically depending on reservoir, fault and aquifer conditions but the key stages of leakage are: (1) brine leakage, (2) CO_2 leakage increasing at a decreasing rate and (3) steady-state leakage. These stages are observed in all cases when the fault is not in the immediate vicinity of the injection well, when the permeability of the fault zone is not influenced by pressure-driven geomechanical or poroelastic effects caused by injection activities, and when boundary conditions are nearly infinite.

LEAKAGE INTERVENTION

The most immediate action that can be taken after a leak is identified is to stop CO_2 injection in the well(s) nearest the fault. If this is not sufficient to stop leakage or reduce leakage rates below acceptable levels, additional measures such as using hydraulic barriers or producing CO_2 from the injection well can be implemented. Through injection and production activities, it is possible to increase dissolution of CO_2, change the direction of fluid flow in the fault zone, push CO_2 away from the fault zone at the base of the caprock, and manipulate the CO_2 plume geometry throughout the injection reservoir. The extent and duration of leakage intervention will be dependent on the fault, reservoir and overlying aquifer properties. Many leakage scenarios may only require minimal remediation, such as injection shutoff to terminate leakage, while some leakage scenarios may require extended remediation and monitoring for many years. The extent of remediation and tolerable long-term leakage rates will depend on the location of the storage site (e.g. offshore or onshore), the vicinity of the storage reservoir to sources of underground drinking water, and the economic and regulatory framework that is incentivizing the CCS project.

All of the intervention options presented in this study have been examined through a lens of practicality and feasibility of implementation. As a result, active remediation strategies (i.e. injection and production) are limited to a decade. Pressure build up from fluid injection was evaluated, and basic cost analysis ruled out the use of elaborate well injection and production patterns that have been suggested by other authors (e.g. [3,4]). However, detailed geomechanical and financial analysis is beyond the scope of this study.

Since the effectiveness of remediation is likely to be highly dependent on when the leak is detected, two initial leakage scenarios were examined. The first scenario is that detection occurs after five years of CO_2 injection, almost immediately after leakage begins. In the homogenous base model described earlier, just over 5,000 tons of CO_2 leaked into the upper aquifer, or roughly 0.5% of the total CO_2 injected into the system. This includes CO_2 in both the supercritical and aqueous phase. The extent of the leakage plume in the overlying aquifer is a 100 m wide and 300 m long plume of supercritical CO_2. In the 10 year intervention scenario, intervention begins after 10 years of CO_2 injection. The total amount of CO_2 leaked in this scenario is 76,000 tons, or roughly 3% of the total CO_2 injected. The leakage plume in the aquifer above the storage reservoir is 500 m wide and 800 m long.

Injection shutoff

Shutting off CO_2 injection removes the source of the pressure build up in the storage reservoir, allowing the increased pressure in the reservoir to dissipate over time, which leads to decreasing leakage rates. Figure 3 shows the dramatic decrease in the amount leakage after injection shutoff for the two different detection scenarios compared with no injection shutoff. In the 5 year detection scenario, injection shutoff reduces the mass of CO_2 leaked after 50 years by 99.8% and in the 10 year detection scenario the reduction in total CO_2 leaked is 98.7%.

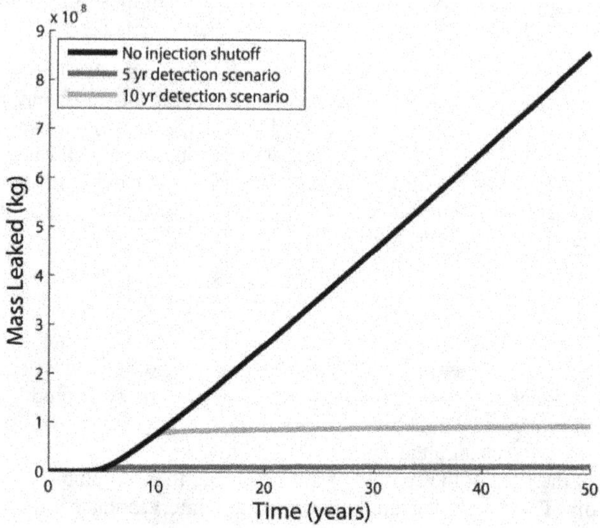

Figure 3. Comparison of injection shutoff at 5 years (gray line), shutoff at 10 years (light gray line), and continued injection on the cumulative mass of CO_2 leaked into the upper aquifer (black line).

Hydraulic Controls

If stopping injection fails to reduce CO_2 leakage below the required threshold then additional intervention methods can be implemented. This section will provide detailed procedures and results for implementing hydraulic controls. All intervention methods are assumed to begin after 5 years of CO_2 injection, and CO_2 injection is terminated prior to beginning any further remediation.

Water Injection into an Overlying Aquifer

Water injection into the overlying aquifer creates a region of higher pressure at the top of the fault. This zone of higher pressure can reduce or reverse the hydraulic gradient causing leakage. As a result, this method is able to completely stop and, in some cases, reverse the CO_2 leakage in the fault. The other benefit of injecting water into the overlying aquifer is that it may be possible to dissolve much of the CO_2 that has leaked into the aquifer, thereby reducing the mobility and concentration of supercritical CO_2.

The length of the fault will determine the well geometry (vertical or horizontal) necessary to create a region of high pressure above the fault. A vertical well will concentrate the pressure build up to a single point. Horizontal wells are able to extend the pressure build up over a larger area, reducing the pressure increase at any one point and thus reducing the risk of hydraulic fracturing or induced seismicity. All of the models examined in this study use vertical injection wells completed only in the top 5 meters of the overlying aquifer. The hydraulic and geometric properties of the aquifer will also influence the well orientation and injection rates. Larger aquifer pore volumes, aquifer permeability, and aquifer height may require horizontal wells and would likely require larger injection rates to produce a more uniform region of elevated pressure in the region around the top of the fault.

In order to show the influence of water injection rate on leakage through the fault zone, a number of simulations were run with injection rates varying from 1 kg/s to 5 kg/s at the top of the aquifer, directly above the fault in the homogeneous base model (Figure 1). In all cases water injection is terminated after 10 years from the initiation of hydraulic controls. Figure 4 compares the impact of four different injection rates on the total leakage percent of CO_2 in aquifer and the percent of CO_2 in the aquifer that is dissolved in water/brine. This figure shows that water injection not only has the potential to decrease the cumulative amount of CO_2 in the overlying aquifer, but even at the lowest injection rates over 90% of the CO_2 in the overlying aquifer becomes dissolved in water. It is important to note that for simplicity, the salinity of the storage reservoir and overlying aquifer was set to zero in these simulations. Increasing salinity decreases the capacity for CO_2 dissolution in the system.

Remediation is more effective at higher injection rates and thus injection rates should often be set as high as possible in order to maximize the efficacy and reduce the duration of remediation. However, caution is required when choosing an injection rate because the geomechanical effects must also be taken into account. The rate of pressure build up in the vicinity of the well will be highly dependent on aquifer properties such as permeability, porosity, and aquifer height. In this model setup, with a vertical injection well, an aquifer permeability of 28 mD and an aquifer height of only 20 m, injection rates higher than 2 kg/s cause aquifer pressure to exceed 150% of initial pore pressure. Prior to water injection, it will be necessary to characterize the fault zone and the regional stress regime in order to estimate the injection threshold above which induced seismicity or slip along the fault would become a significant risk.

Leakage in the homogeneous system can be quickly stopped by injecting water into the overlying aquifer as shown by the examples in Figure 4.

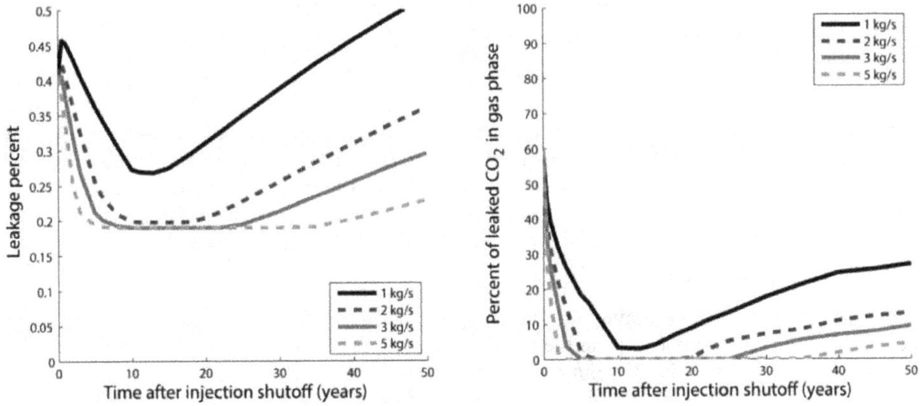

Figure 4. a) Comparison of the influence of different water injection rates on percent of CO_2 in the upper aquifer. The leakage percent is calculated by dividing the total mass of CO_2 in the aquifer by the total mass of CO_2 injected into the system. In all cases, the water injection rate is constant for the first 10 years, after which injection stops. b) Percent of gas phase CO_2 in the overlying aquifer after 5000 tons of CO_2 have leaked into the overlying aquifer.

Leakage eventually resumes after water injection ends regardless of the injection rate. If the size of the overlying aquifer is much larger, the efficacy of this method of remediation could be diminished. The timing and rates of water injection will have to be optimized based on the system properties and characterization. Overall, injection of water in the aquifer will usually act as a temporary containment method to stop leakage while the risks and benefits of other options are being evaluated.

Reservoir Brine Production

A major limitation of water injection in the overlying aquifer is that it has minimal influence on the CO_2 plume in the injection reservoir. As a result, gravity driven equilibration of the injection plume continues to drive CO_2 up the fault zone after water injection is terminated. One option to reduce buoyancy driven leakage is to produce reservoir brine in order to pull the CO_2 plume away from the fault zone.

The first step in fluid production remediation is to determine the ideal placement of a production well in the injection reservoir. Early scoping studies, using a number of different gridding techniques to test different production locations, found that producing brine anywhere on the fault-side of the CO_2 injection plume increased the long term CO_2 leakage. Production in this region pulled more CO_2 under the fault while decreasing the pressure below the fault. This was good in the short term for preventing leakage up the fault but after the cessation of remediation there remained a large column of CO_2 below the fault that subsequently leaked up the fault in the following decades.

The results from this scoping work showed that the most effective production well location was on the opposite side of the plume from the fault. This production well could either be a horizontal or vertical well located at the bottom of the injection reservoir. Figure 5 illustrates the reduction in leakage of CO_2 using water injection activities with (dotted lines) and without reservoir fluid production (solid lines). In these simulations, brine production is balanced with water injection (above the fault in the overlying aquifer) which was set to 2 kg/s for 10 years in the cases shown in Figure 5. The production well was placed 750 m away from the original CO_2 injection well, far enough from the plume such that CO_2 was never extracted from the storage reservoir during brine production. While production alone is insufficient to stop leakage into the overlying aquifer, it contributes to the long term CO_2 leakage reduction.

Simultaneous injection and production addresses some of the financial and logistical concerns associated with hydraulic remediation strategies. In order to limit the costs of water storage and multiple water sources, all of the cases examined in this study have total production rates equal to total injection rates. This assumption relies on the ability to inject reservoir brine into the overlying aquifer, which may not always be possible—or may require filtration or treatment—depending on the salinity, chemistry, and current and future utility of the reservoir and aquifer fluids.

Water injection below the caprock

The second option to prevent gravity driven leakage following remediation termination is to inject water below the fault in order to push the CO_2 plume away from the vicinity of the fault zone and reduce the mobility of CO_2 by dissolution into the injected water. The initiation of water injection below the fault must take place sometime after the injection above the fault has stopped the leakage and re-saturated the fault with brine. This prevents water injection in the reservoir from displacing more CO_2 up the fault zone. In these simulations, water is first injected above the fault with brine produced at the same rate as injection (2 kg/s). After two years of injection above the fault, injection below the fault begins. The water injection well in the storage reservoir is located directly below the

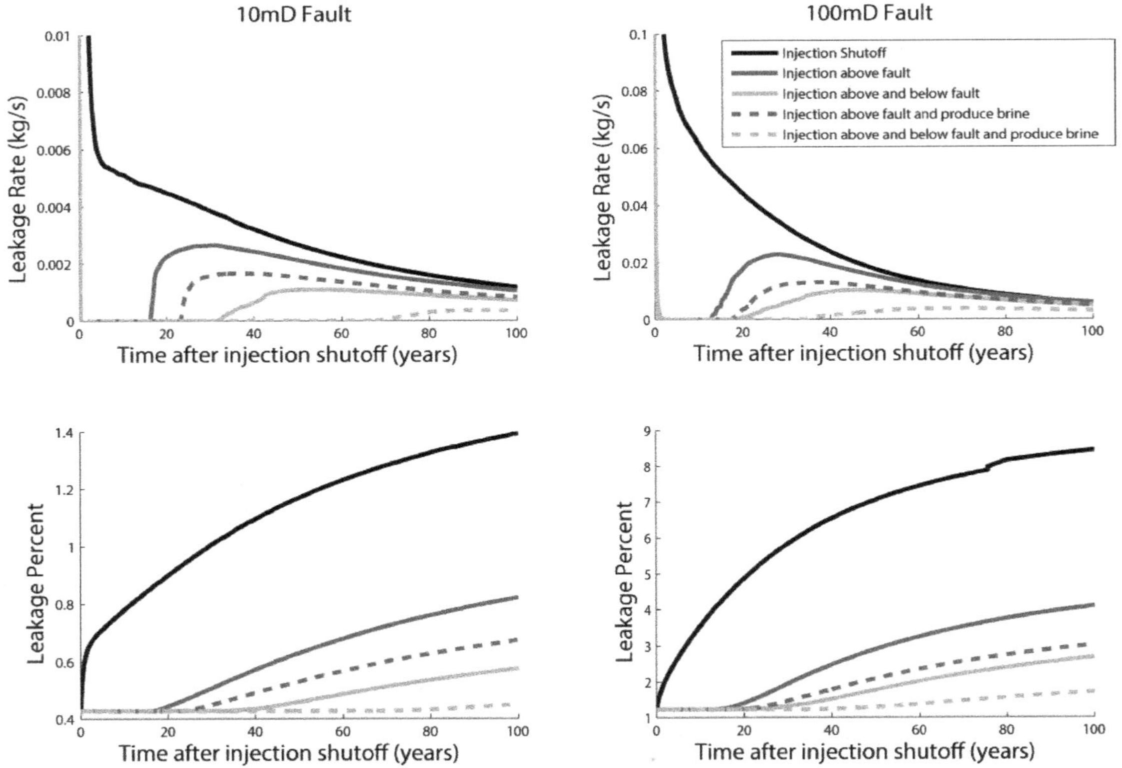

Figure 5. Comparison of different stages of remediation for a 10 mD fault zone (left figures) and a 100 mD fault zone (right figures), showing both leakage rates (top figures) and leakage percent values as a function of time (bottom figures). Note the y-axis limits on the leakage rate figures are an order of magnitude larger in the 100 mD fault scenario. The medium grey solid line corresponds to water injection of 2 kg/s above the fault only, the dashed medium grey line corresponds to water injection and reservoir fluid production of 2 kg/s. The light grey line corresponds to water injection above the fault and below the fault at a rate of 2 kg/s in each well. The dashed light grey line shows the leakage of CO_2 in the case of injection of water below the fault, above the fault, and fluid production from the reservoir.

water injection well in the overlying aquifer. This placement allows both injection methods to be performed using only one well perforated at two different intervals, one below the caprock and one above the caprock. Water injection below the fault is set to the same rate as water injection above the fault (in this case 2 kg/s). When water injection below the caprock begins, the rate of fluid production increases to 4 kg/s to match the combined water injection rates below and above the caprock. After ten years, all injection and production is terminated.

The combination of water injection above and below the fault with concurrent reservoir fluid production, makes it possible to terminate leakage in the homogeneous model for decades. Figure 5 compares the efficacy of different stages of remediation for different fault permeability scenarios. The character of the CO_2 leakage is very similar for the different fault permeabilities. However, the leakage rates are higher when the permeability of the fault is larger, as would be expected. It is important to point out that for the 10 mD fault, when only injection shutoff was implemented, the leakage rates after 100 years are roughly 30 tons per year relative to the 19,000 tons per year leaking prior to the initiation of remediation. For the 100 mD fault, the highest leakage rates are roughly 300 tons per year after 100 years with a cumulative leakage percent of roughly 8% of the total CO_2 injected into the system. Thus, while leakage is still occurring following remediation termination, it is very small and decreasing in an exponential fashion. The implementation of hydraulic barriers manage to drop the percent of total CO_2 leaked from the storage reservoir to less than 2% with the 100 mD fault and less than 0.5% in the 10 mD fault case after 100 years.

CONCLUSION

Leakage intervention and remediation requires a balance of engineering efficacy, geomechanical assessment and cost-benefit analysis. Injection shutoff or passive remediation is the fastest, easiest, and may often be the cheapest method to quickly reduce leakage from the storage reservoir, depending on the economic conditions which may incentivize CCS in the future. Usually injection shutoff alone is not able to completely stop leakage. However, in many cases the quantities of CO_2 leaked may be far below levels that would be of concern, typically less than 5% of the initial leakage. If passive remediation is not sufficient, various hydraulic controls such as water injection or reservoir fluid production can be employed to further reduce leakage rates and trap leaked CO_2. Water injection is a very effective remediation technique because it: (1) increases the pressure in the overlying aquifer relative to the base of the fault which can quickly stop leakage, (2) is able to push some amount of CO_2 back down the fault, and (3) is able to dissolve large quantities of CO_2 in the vicinity of the water injection. In some cases, water injection alone may be enough to stop current and future CO_2 leakage, especially when the leak is detected early, before a large amount of CO_2 has accumulated below and above the fault. Additional hydraulic controls such as reservoir fluid production and water injection below the caprock increase rates of dissolution, and thus, trapping of CO_2 in the storage reservoir. These findings should provide assurances to industry, policy makers, and the public that intervention measures can quickly and effectively mitigate potential leakage from carbon sequestration reservoirs.

ACKNOWLEDGEMENTS

This project was supported by the Stanford University Department of Energy Resources Engineering and the Stanford Center for Carbon Storage. This work is part of a larger project, "Assessment of Leakage Detection and Intervention Scenarios for CO_2 Sequestration," supported by the CCP3; a joint industry project sponsored by BP, Chevron, Eni, Petrobras, Shell, and Suncor.

REFERENCES

1. Réveillère, A., Rohmer, J., Manceau, J.C., 2012. Hydraulic barrier design and applicability for managing the risk of CO_2 leakage from deep saline aquifers. *International Journal of Greenhouse Gas Control*, 9, 62–71.
2. Le Guénan, T., Rohmer, J., 2011. Corrective measures based on pressure control strategies for CO_2 geological storage in deep aquifers. *International Journal of Greenhouse Gas Control*, 5, 571–578.
3. Buscheck, T.A., Sun, Y., Chen, M., *et al.*, 2012. Active CO_2 reservoir management for carbon storage: Analysis of operational strategies to relieve pressure buildup and improve injectivity. *International Journal of Greenhouse Gas Control*, 6, 230–245.
4. Esposito A., Benson S.M., 2012. Evaluation and development of options for remediation of CO_2 leakage into groundwater aquifers from geologic carbon storage. *International Journal of Greenhouse Gas Control*, (7) 62-73.
5. Cunningham, A.B., Gerlach, R., Spangler, L., Mitchell, A.C., 2009. Microbially enhanced geologic containment of sequestered supercritical CO_2. *Energy Procedia*, 1, 3245–3252.
6. Ito, T., Xu, T., Tanaka, H., Taniuchi, Y., Okamoto, A., 2014. Possibility to remedy CO_2 leakage from geological reservoir using CO2 reactive grout. *International Journal of Greenhouse Gas Control*, 20, 310–323.
7. Pruess, K., Oldenburg, C., Moridis, G. 1999. TOUGH2 user's guide, Version 2.0. Tech. Rep., Lawrence Berkeley National Laboratory, Earth Sciences Division, LBNL-43134
8. Pruess, K., 2005. ECO2M: A TOUGH2 Fluid Property Module for Mixtures of Water, NaCl, and CO_2, Including Super- and Sub-Critical Conditions, and Phase Change between Liquid and Gaseous CO_2. User Manual.
9. Patterson, C.G., Falta, R.W., 2012. PROCEEDINGS, TOUGH Symposium 2012 Lawrence Berkeley National Laboratory, Berkeley, California, September 17-19, 2012.
10. Caine, J.S., Evans, J.P., Forster, C.B., 1996. Fault zone architecture and permeability structure. *Geology*, 24, 1025–1028.
11. Pini, R., Krevor, S.C.M., Benson, S.M., 2012. Capillary pressure and heterogeneity for the CO_2/water system in sandstone rocks at reservoir conditions. *Advances in Water Resources*, 38, 48–59.
12. Land, C., 1968. Calculation of Imbibition Relative Permeability for Two- and Three-Phase Flow from Rock Properties. *SPE Journal*, 8(2).
13. Burnside, N.M., Naylor, M., 2014. Review and implications of relative permeability of CO_2/brine systems and residual trapping of CO_2. *International Journal of Greenhouse Gas Control*, 23, 1–11.
14. Krevor, S.C.M., Pini, R., Zuo, L., Benson, S.M., 2012. Relative permeability and trapping of CO_2 and water in sandstone rocks at reservoir conditions. *Water Resources Research*, 48, 1-16.
15. Holtz, M.H., 2002. Residual Gas Saturation to Aquifer Influx: A Calculation Method for 3D Computer Reservoir Model Construction. *Proceedings* of SPE Gas Technology Symposium.
16. Bennion, B., Bachu, S., 2005. Relative Permeability Characteristics for Supercritical CO_2 Displacing Water in a Variety of Potential Sequestration Zones in the Western Canada Sedimentary Basin. SPE 95547
17. Paterson, L., Boreham, C., Bunch, M. *et al.*, 2013. Overview of the CO2CRC Otway residual saturation and dissolution test. *Energy Procedia*, 37, 6140-6148
18. Laforce, T., Ennis-King, J., Boreham, C., Paterson, L., 2014. Residual CO_2 saturation estimate using noble gas tracers in a single-well field test: The CO2CRC Otway project. *International Journal of Greenhouse Gas Control*, 26, 9–21.
19. Chabora, E.R., 2009. The utility of above-zone pressure measurements in monitoring geologically stored carbon dioxide. Master's Thesis

20. Meckel, T.A., Zeidouni, M., Hovorka, S.D., Hosseini, S.A., 2013. Assessing sensitivity to well leakage from three years of continuous reservoir pressure monitoring during CO_2 injection at Cranfield, MS, US. *International Journal of Greenhouse Gas Control.* 18, 439-448.
21. Zahasky, C., Benson, S.M., 2014. A simple approximate semi-analytical solution for estimating leakage of carbon dioxide through faults. International Conference for Greenhouse Gas Technology-12, Austin, Texas, October 6-9, 2014.

Carbon Dioxide Capture for Storage in Deep Geological Formations, Volume 4
Karl F. Gerdes (Editor)

Chapter 44

A POTENTIAL SEALANT FOR CONTROLLING LEAKS FROM CARBON STORAGE RESERVOIRS

Thomas Aird and Sally Benson
Stanford University Department of Energy Resources Engineering, Stanford, USA

ABSTRACT: A possible mitigation strategy for leaks from CO_2 reservoirs is the introduction of a physical barrier in the form of injected sealant material. An organically crosslinked polymer (OCP) system was investigated as a potential sealant material for leaks from CO_2 reservoirs. The OCP system has been extensively used in the oil industry to address conformance issues related to fracture shutoff, gravel pack isolation, waterconing, casing leak repair, and high permeability streaks. The initial viscosity properties of the sealant were tested first for later use in modeling and core flood experiments. Viscosity measurements demonstrate that gel times for this polymer could be altered by changing the concentration of the crosslinker or by changing the gel setting temperature. Sandstone core flood experiments were performed to determine changes in permeability when the polymer gel was placed within artificial fractures. Core flood tests were carried out at reservoir conditions of 50° C and 7 MPa. Permeability tests revealed a reduction of 100% in fracture permeability after sealant emplacement for two rock core samples (Core 1 and 2). Further reductions in overall core permeability were measured after CO_2 exposure. Finally, CT scans of these sealed rock cores were carried out in order to visually assess the location of the sealant within the fracture. These scans revealed that the aperture and distribution of the sealant in the fracture varied along the length of the cores. The results from the experimental tests were used to create a basic injection scheme that modelled sealant injection via a horizontal well to a target fracture plane. The high initial viscosity was shown to create large differential pressures during injection. Formation penetration was limited by this sealant property and the short gel time. Despite these limitations, under ideal conditions, a physical barrier with an approximate radius of 5 m could be created using this polymer sealant system. These tests performed on the OCP system affirmed its potential as a candidate for controlling leaks from CO_2 reservoirs. Field scale trials of this sealant should be performed to gain insight beyond that of these modelling and laboratory tests.

KEYWORDS: CO_2 sequestration; leakage; remediation; sealants; barrier; polymer; gel

INTRODUCTION

Carbon sequestration is a promising method to combat increasing levels of atmospheric carbon dioxide. However, the storage of large volumes of CO_2 in the subsurface creates a range of potential risks. One of the primary risks is the leakage and subsequent vertical migration of CO_2 from these storage reservoirs. A possible mitigation strategy for controlling leaks is the introduction of a physical barrier in the form of injected sealant material. An organically crosslinked polymer (OCP) system (H_2Zero^{TM} by Halliburton) was investigated as a potential sealant material for leaks from CO_2 reservoirs. The OCP system possesses many advantages over other sealant methods. The relatively low initial viscosity (30 cP) enables greater penetration depth into the matrix formation than that of traditional squeeze cements. It possesses long-term thermal stability that enables it to keep its structure even at elevated temperatures. This system has also been shown to be unaffected by formation fluids, lithology, or heavy metals [1].

Sealant description

The main components of the OCP system are a base polymer (copolymer of acrylamide and t-butyl acrylate, PAtBA), a crosslinker (polyethyleneimine, PEI), and a mixing brine (2% KCl) [1]. The crosslinker and the base polymer are easily diluted in the mixing brine. The polymer system is initiated when the amine groups on the PEI react with ester/amide groups on the PAtBA. Multiple groups of these moities on the crosslinker and the base polymer molecules allow for crosslinked three-dimensional structures to form. This polymerization process is thermally activated, and the OCP blend is adjusted to be initiated by the formation temperature, thus allowing for in-situ gelation. The formulation of the original OCP system is designed to work in a temperature range from 49 to 127°C. Two other formulations were developed from this original system: one for lower temperatures (16-60°C) and one for higher temperatures (127-177°C). The lower temperature OCP system uses a polyacrylamide base polymer, while the high temperature OCP system adds a water soluble carbonate retarder to the original formulation. The low temperature OCP system was selected for testing in this study due to its favourable temperature range for application to containment of leakages from geological storage reservoirs. Temperatures used in viscosity measurements varied from 40-60°C. This temperature range reflects possible formation temperatures found above carbon storage reservoirs.

EXPERIMENTAL RESULTS

Sealant Viscosity Measurements

The objective of the viscosity experiments was to measure the initial properties and gel time dependence of the sealant for a range of compositions and temperatures. The initial viscosity properties need to be fully understood in order to develop an injection model of the sealant. Likewise, gel time behaviour in regards to composition and temperature must be known to accurately model sealant injection.

The effects of crosslinker concentration on gel time were investigated by preparing samples with 1.0%, 3.0%, and 6.0% PEI concentrations. The concentrations of base polymer and KCL were kept constant at 25.0% and 2.0% respectively. Each sample was prepared and immediately introduced to the viscometer with the attached water bath. A temperature of 50° C was held constant for each concentration.

From Figure 1, the last measurable viscosity (28,000 cP) for the 6% PEI concentration occurred at 160 minutes. Similarly, the last measurable viscosity occurred at 332 minutes for the 3.0% PEI concentration (with μ=29,745 cP), and 478 minutes for the 1.0% PEI concentration (with μ=29,000 cP). These results indicate that crosslinker concentration is inversely correlated with gel time; or longer gel times require smaller PEI crosslinker concentrations. Ultimate viscosities exceeding 2M cP were measured after the samples had been held at 50° C for approximately 24 hours. The effect of temperature on gel time was also investigated by testing samples with the same composition at 40, 50, and 60° C. The concentration of PEI crosslinker was set at 6.0% to minimize overall gel time. The samples exposed to 60° C were quickest to gel with a final viscosity of 29,761 cP measured at 132 minutes. This compares to 160 and 194 minutes for the samples at 50 and 40° C respectively.

Several important properties of the sealant were determined from these viscosity experiments. First, the initial viscosity of the sealant is consistently around 30 cP, regardless of temperature or composition. This is a relatively high viscosity that will limit any injection scheme. The final viscosities of all sealant samples were sufficiently high to consider them immobile if allowed to set in pore spaces. Another important parameter is the maximum gel time of approximately 8 hours.

This will be the maximum time window for sealant injection before gelation occurs. Low crosslinker concentrations must be used to achieve this maximum gel time. Temperature is less impactful on gel time than crosslinker concentrations. Higher formation temperatures will cause gelation to occur at a faster rate.

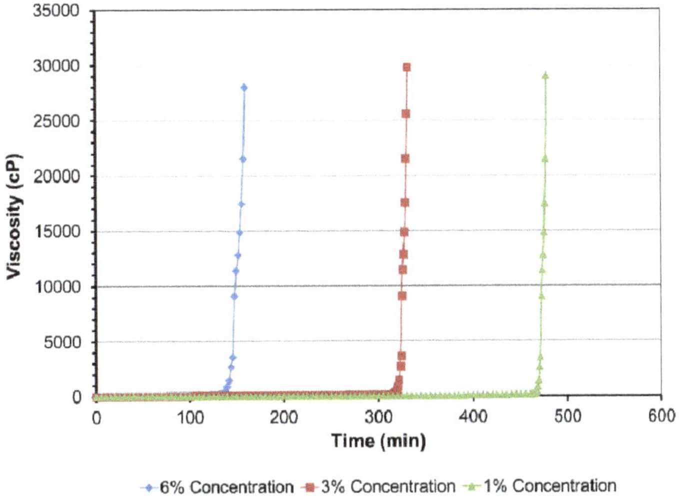

Figure 1. Viscosity measurements over time for gel formulations with different PEI crosslinker concentrations at 50°C.

Core Flood Experiments

Two Berea sandstone samples were chosen for the core flood experiments. The approximate permeability of each sample was known based on previous core flood experiments of cores from the same shipments. In order to achieve a high level of contrast, one core was selected to have a low permeability (~20 mD) while the other a higher permeability (~200 mD). For convenience, the low permeability core will subsequently be referred to as Core 1, and the high permeability core will be referred to as Core 2.

A sequence of core flood experiments was performed to characterize the effect of the sealant on permeability. The first core floods established a base-line permeability of each core. A second set of core floods measured the permeability after each core was artificially fractured by sawing the rock into two parts. The third set of tests re-measured the cores after the fractures were sealed with the polymer system. The final set of core floods measured permeability after the cores were left in a supercritical CO_2 environment for 48 hours. The results of these tests provide an initial assessment of how the polymer system performs under possible injection conditions. Additional testing would be needed to evaluate the performance of the seal material over the longer time periods.

Figure 2 depicts the core flood experiment data for Core 1. The diamond accented line represents the unfractured core before any alterations were done. The square accented line represents the core after it had been artificially fractured in half. The triangle accented line is the core once the sealant had been introduced. Not only does the permeability decrease, it is reduced below that of the original core. The cross accented line is the final core flood experiment after the core had been exposed to supercritical CO_2. This final line represents a further decrease in permeability with Core 1.

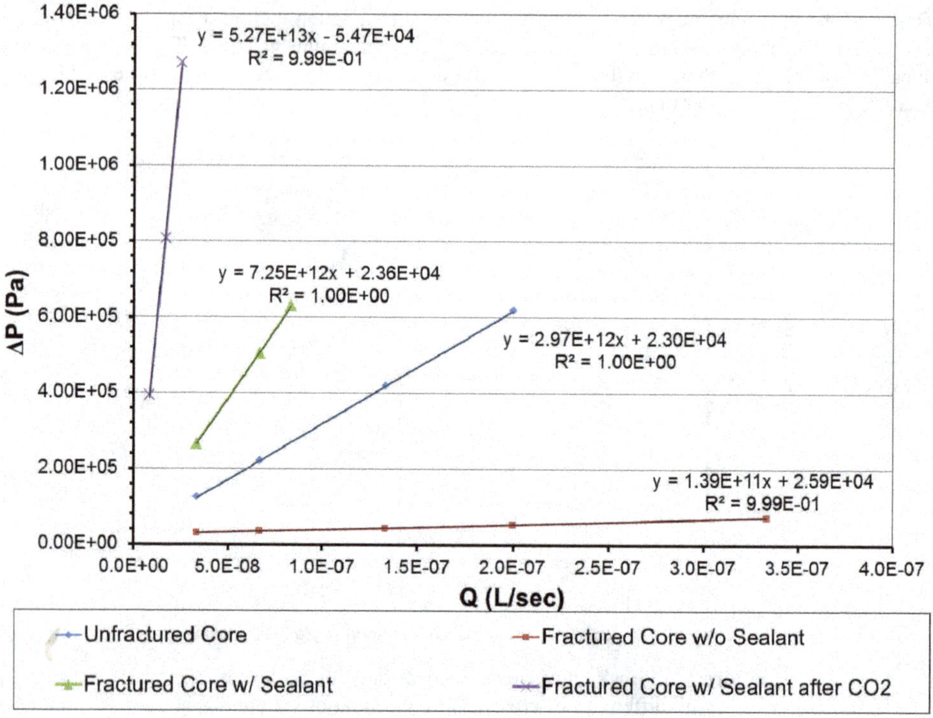

Figure 2. Pressure change vs. flow rate for Core 1.

Figure 3 compares the permeability values for Core 1 and Core 2. The values are plotted on a logarithmic scale. Both Cores 1 and 2 exhibit the same trend in permeability values. Both cores experience a one order of magnitude reduction in permeability when the sealant is introduced. After the sealant is emplaced into the fracture the permeability decreases below the original permeability of the intact core. Another order of magnitude reduction is seen after prolonged exposure to supercritical CO_2. This final measured permeability value is the smallest recorded for both cores.

The data can also be used to calculate the influence of sealant introduction on the permeability of the fracture itself. The initial permeabilities of the fractures were estimated to be 11.7 and 214.5 Darcy for Core 1 and 2 respectively, using a fracture width of 1 mm for each core. The fracture permeabilities were reduced by at least 3 orders of magnitude, and below the level of detection for this experimental setup. These core flood experiments demonstrate the effectiveness of the polymer system as a sealing agent. Upon introduction to the fracture, the OCP system decreased the overall permeability of each core by two orders of magnitude. Prolonged exposure to supercritical CO_2 caused the permeabilities to decrease even further. The drastic reductions in permeability suggest extensive pore space penetration by the sealant or even sealant swelling *in situ*. X-ray computed tomography (CT) scans were then carried out in order to probe these permeability reductions.

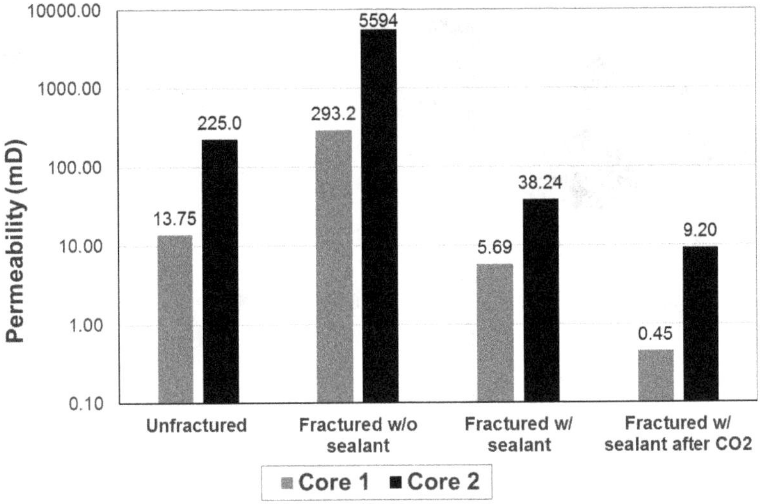

Figure 3. Calculated permeability values for Core 1 and Core 2.

X-Ray CT Imaging

The primary objective of the CT measurements was to measure density changes along the length of the fracture for each core. Any deviations in width and composition might be evidence of the sealant's influence on cores. Conclusions derived from these measurements are limited by the lack of any scans prior to the emplacement of the polymer sealant. Furthermore, each sealed core had been the subject of several core flood experiments prior to these CT measurements. Therefore, the data are used only to qualitatively evaluate the location of the sealant within the fractures and rock matrix.

Figure 4 is comprised of CT images of Core 1 cut along its length with the fracture running perpendicular to the horizontal plane. In all of these core images, the outlet is the outward facing end. These images show the density differences between the inlet, center, and outlet of the core. Both the inlet and outlet fracture zones have small reductions in density compared to that of the core away from the fracture. In the middle of the core, the fracture opens up to a greater extent. The low CT values in the center of the core suggest a decrease in sealant material within this region. These scans reveal a large variation in density along the fracture, thus proving an uneven distribution of sealant material within this region. There is no obvious penetration of the sealant into the rock matrix (which would be evident from higher rock matrix densities adjacent to the fracture).

These CT measurements allowed for a basic understanding of sealant distribution within the fracture of each core. For Core 1, large differences between the wet and dry scans suggest the absence of polymer within the center of the core. The narrowing of the fracture width near the outlet may also attest to sealing properties of the crosslinking polymer. The scans of Core 2 revealed the presence of sealant material within the fracture. Some density differences between the wet and dry scans are identifiable, but no clear patterns existed. Although the sealant material was not uniformly distributed within the fracture, the overall fracture permeability was drastically reduced upon sealant emplacement. These scans further highlight the effective sealing properties of the OCP system despite incomplete sealant distribution within the fracture zone.

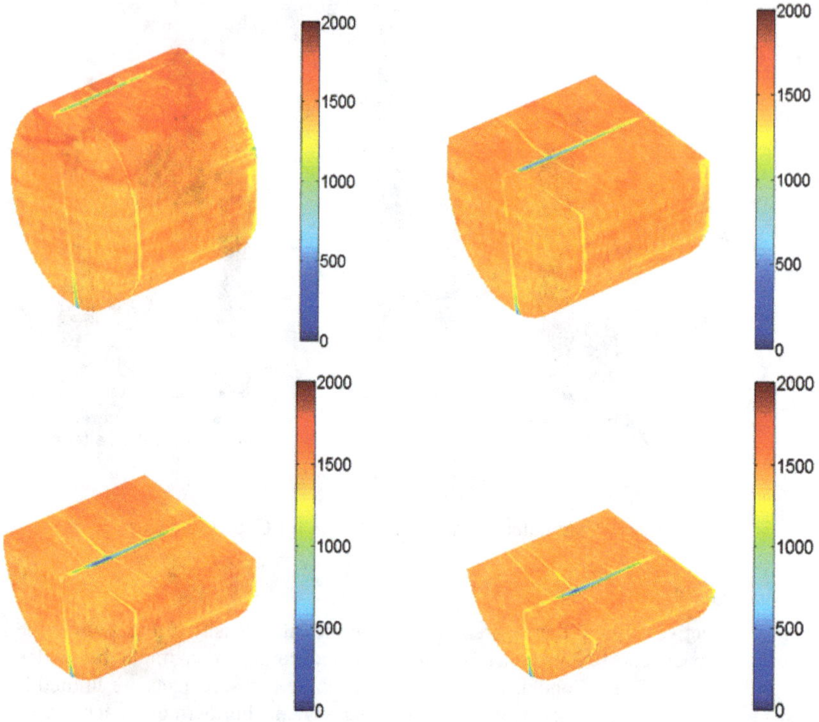

Figure 4. Dry CT image cross sections of Core 1 with CT numbers expressed in Hounsfield Units.

SEALANT INJECTION

A plan for injecting the OCP system was developed in order to assess the sealant's ability to create a physical barrier for migrating CO_2 leaks. Any injection scheme would be limited by the initial viscosity and gel time of the sealant. The high initial viscosity (~30 cP) causes large pressure changes during injection while the gel time allows for a limited injection time window. These challenges require special consideration when designing an effective barrier emplacement scheme. For example, a series of these injections would be required in order to create a barrier for a fault zone with a length of 100s of meters or more. This could be accomplished with multiple vertical wells or a single horizontal well drilled to the desired injection site. The seal could be emplaced either in the fault zone itself, directly below the fault zone in the storage reservoir, or even directly above the fault in the formation into which the CO_2 is leaking.

Here we provide an example of injecting the sealant directly into the fault zone. We assume that the width of the fault zone is 10 m. Figure 5 provides a simple schematic of injection using a horizontal well drilled into the fault plane. (Note that drilling a horizontal well in the middle of a fault zone presents significant technical challenges, calling into question the feasibility of creating a barrier within the fracture). After the well is completed, the OCP system could then be injected over a series of intervals using packers. This series of injections would create a long horizontal barrier. This figure illustrates the process of injecting over a defined interval 'h', and then through the use of packers, injecting multiple times in sequence along the horizontal well. Once the sealant is allowed to gel within the rock pores, the fault plane would become impervious to leakage up the fault.

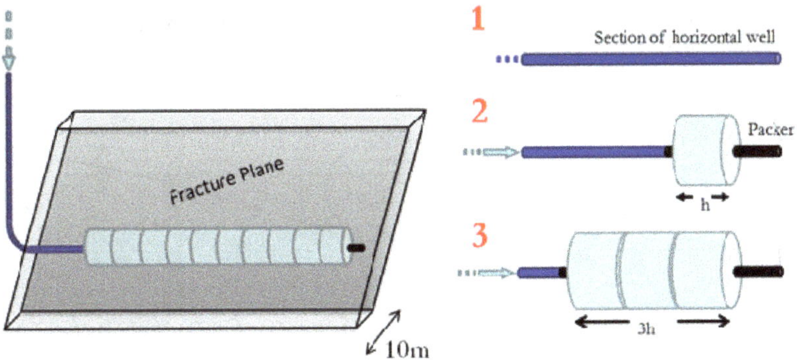

Figure 5. Diagrams of horizontal injection scheme into a 10m vertical fracture plane (left) and a three step process diagram of a horizontal well with sealant injection intervals (right).

The pressure build-up during injection of the sealant is an important parameter to consider when designing injection schemes because the formation rock could fracture if the injection pressure approaches the least principle stress in the formation. We assume this can be avoided if the injection pressure does not exceed 50% of the initial hydrostatic pressure. (Note that the actual fracture pressure will be site specific and could deviate significantly from the value selected here for example only). Assuming radial flow with complete displacement, the following equation can be applied:

$$\Delta P(t) = \frac{Q_{inj}\mu_{ocp}}{2\pi kh} ln \frac{r_f}{r_w} \qquad \text{Equation 1}$$

This equation calculates the pressure increase at the injection well assuming that the pressure build-up at the "front" of the OCP solution is negligible compared to the pressure build-up within the flooded region. In this equation, μ_{ocp} is the viscosity of the sealant, r_f is the radius of the front, k is permeability, and r_w is the well radius. Importantly, we assume that the entire sealant volume is emplaced before the polymer begins to gel. Therefore, the viscosity is assumed to be constant and equal to the viscosity of the pre-gelled mixture. Using this equation, a set of three injection profiles (injection rate per unit injection length) was used in order to model injection behaviour: 1.77, 2.65, and 5.3 $cm^3/cm/s$. The highest of these rates corresponds to injecting 0.0053 m^3/s into a 10 m long interval. In the following calculation, the fracture permeability was set at 250 and 1000 mD. An average initial viscosity of 30 cP was chosen based upon prior measurements.

Figure 6 depicts the expected change in pressure at the injection well during injection of the sealant in order to achieve a 5 m radius using the three injection profiles under two permeability conditions. For example, with the 250 mD fracture zone, the injection profile with the highest rate per length interval produces the largest change in pressure (37 MPa). If the injection depth is about 1 km, this high pressure would exceed the lithostatic pressure (~23 MPa). In order to keep the pressure below 50% of the original *in situ* fluid pressure (6 MPa limit assuming approximately 12 MPa original *in situ* pressure), the injection rate per length interval would have to be less than 2.65 $cm^3/cm/s$. However, it cannot be too low. For example, if it is as low as 1.77 $cm^3/cm/s$, the sealant would only reach a radius of about 4 m after 8 hours of injection, short of the fracture ½ width of 5 m. If the permeability of the fault zone is as high as 1000 mD, all injection profiles at approximately 2.65 $cm^3/cm/s$, or lower, can be injected without exceeding 50% overpressure. Under these conditions, the injection rate of 2.65 $cm^3/cm/s$ should be chosen since it can reach a radius of 5 m in approximately 7 hours. This would allow for the greatest formation penetration since the OCP would not gel for another 1-2 hours.

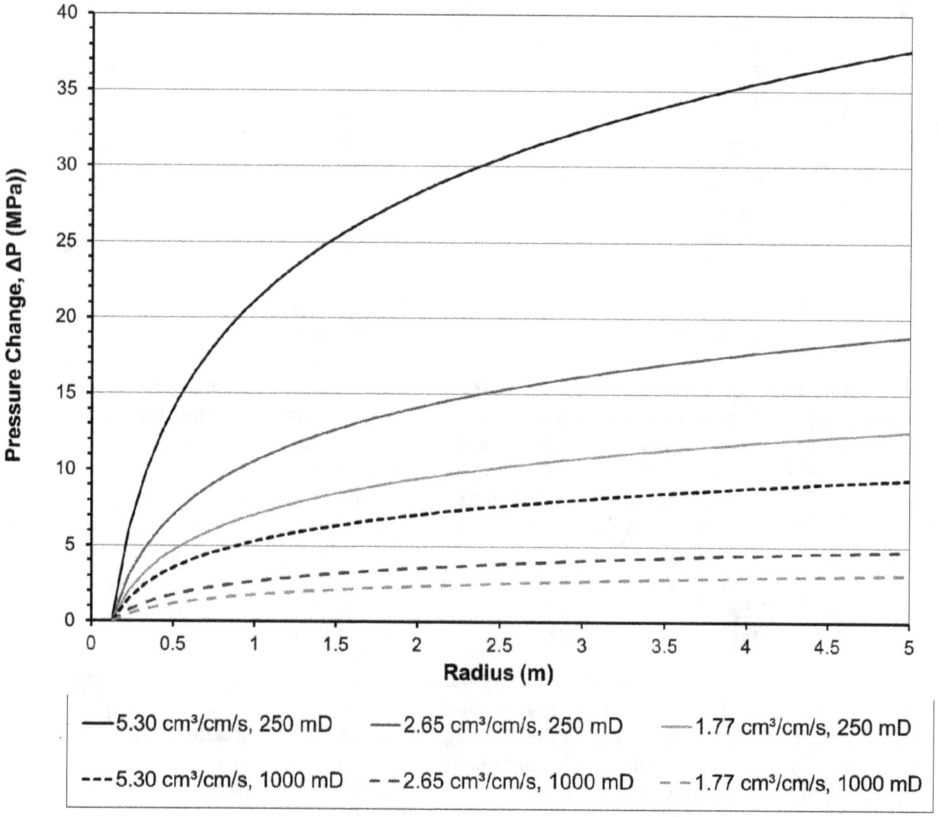

Figure 6. Graph of pressure change vs. barrier radius for three injection profiles.

The results of these calculations reveal one of the main limitations of the sealant as a barrier to CO_2 leaks. For low permeability conditions, the sealant must be injected over a large interval and at a corresponding small rate to limit the bottom hole pressure change. For high permeability conditions, the sealant can be injected at a greater rate per length interval since the change in pressure is smaller than that of the low permeability conditions. This injection rate could create a barrier with a 5 m radius in less time than the sealant requires for gelation. Additional injection time would further increase the size of barrier thus sealing more pore volumes within the fracture zone.

One of the main limitation of this sealant system as an injected physical barrier is the high initial viscosity of 30 cP. This property of the sealant is a main factor behind the high projected pressure build-up upon sealant injection. Despite this limitation, these modelling calculations indicate that a physical barrier with a diameter of approximately 10 m could be created with this sealant system under some circumstances. The barrier would have to be placed with a high degree of precision in order to plug pore spaces found in a targeted fault plane. Upon gelation in these pore spaces, this sealant system would be impervious to CO_2 as suggested by the core flood experiments. If the initial viscosity of the sealant could be reduced by a factor of three, then barrier emplacement in a 250 mD zone would also be achievable. This suggests that reducing the initial viscosity should be an important design goal for seals used in this application.

Other limitations may also exist which were not addressed here. For example, the fault zone is likely to be highly heterogeneous, resulting in uneven emplacement of the sealant. Tongwa and others [3] show that sealing fractures with large diameters (>0.25 mm) is challenging with polymers. Also, if CO_2 injection continues and further movement along the fault occurs, fracture movement could compromise the integrity of the seal. Finally, we have not addressed the issue of the mechanical strength of the sealed fault zone and how this would influence the efficacy of the seal.

CONCLUSION

A general assessment of a crosslinking polymer system has been presented here. The initial experimental tests provided an understanding of the sealant's viscosity properties and behaviour. Core flood experiments were then performed to assess the sealant's effects on fracture permeability under reservoir conditions. Distribution of the sealant within the rock fractures was confirmed with CT imaging. A horizontal injection well scheme was modelled to assess sealant delivery to a target fracture plane.

Viscosity measurements proved that gel times for this polymer system could be altered by changing the concentration of the PEI crosslinker or through temperature changes. Maximum gel times of 8.5 hours could be achieved by lowering the crosslinker concentration to 1.0%. The ultimate viscosity of the sealant remains unaffected by these changes to crosslinker concentration.

The effectiveness of the sealant for stopping leakage of CO_2 through fractures was tested by two core flood experiments. In both high and low permeability rock samples, the sealant reduced the overall permeability upon introduction to an induced fracture. After reaching its final state, the sealant fully plugged the fracture and reduced the permeability of the core below its initial (unfractured) value. The overall permeability decreased by 59% and 83% from the original permeabilities for Cores 1 and 2 respectively. The fracture permeability was decreased by 100% upon sealant emplacement for both Cores 1 and 2. The integrity of the sealed fracture was not compromised after exposure to supercritical CO_2 at reservoir conditions for 48 hours. The absolute permeabilities of the cores were further reduced below their initial values after exposure to supercritical CO_2. This reduction seems to suggest that the CO_2 caused the sealants to swell and penetrate some distance into the rock matrix. These initial tests confirm the sealant's tolerance to a CO_2 environment at least in the short term. Further tests should be undertaken to confirm the sealant's long-term resistance to such environments.

X-ray CT scans of the two sealed rock cores showed that the effectiveness of sealant placement varied within the fracture. Large changes in density occur randomly along the length of the fracture suggesting an incompletely sealed fracture. Despite both fractures appearing partially sealed, prior core flood experiments showed reduction of fracture permeability to less than the detection limit upon sealant emplacement.

Based on experimental demonstration of the effectiveness of the sealant at reducing the permeability, and using data derived from the viscosity measurements, a horizontal injection scheme was developed for emplacing the sealant in a fault plane at a depth of 1 km. A simplified model was developed to calculate pressure build-up at the injection well and to provide guidelines for injection rates and injection intervals. As a conservative assumption, the injection pressure must remain below 50% of the hydrostatic pressure in order to prevent fracturing the cap rock. For a vertical fracture plane with a 10 m width, a horizontal well could be drilled through or near this potential CO_2 leak zone. A fracture filling horizontal barrier could be created by injecting over a set of defined intervals along the length of the well. Assuming complete displacement, three injection profiles were developed to model sealant injection rate over a given injection interval. Injecting at a high rate per length interval results in excessive pressure changes under low permeability conditions (~250 mD). These pressure changes exceed the lithostatic pressure of 23 MPa at the given depth (1 km). Under high permeability conditions (~1000 mD), pressure changes of 4-9 MPa could be expected in order to create a 10 m diameter barrier. These pressures are near or below the 50% hydrostatic pressure threshold for most depths below 1 km. The high initial sealant viscosity (30 cP) was a limiting factor in all these pressure calculations.

Pilot scale field trials are the next step in the assessment of this polymer sealant system. The purpose of these tests would be to assess the behaviour of sealant's injection and propagation under real world conditions. Results from this experimental work have already been requested in the initial design of these projects.

The OCP sealant system potentially allows for the creation a rapid physical barrier to mitigate leaks from CO_2 sequestration reservoirs. Core flood experiments affirmed the ability of the sealant to reduce overall permeability. The sealant's high initial viscosity and gel times limit its use for specific applications. Leaks through abandoned wells or tight fault planes may warrant its application. Further reductions in the initial viscosity of the sealant system would alleviate the excessive injection pressures presently estimated using the current sealant formulation. These pressures are a limiting factor in the application of this sealant system. The investigation of this polymer sealant system reflects why it is important to consider all options when planning for possible leakage scenarios for CO_2 sequestration projects.

ACKNOWLEDGEMENTS

We could like to thank Halliburton for providing samples of the sealant for testing. This project was supported by the CCP3.

REFERENCES

1. Vasquez, Julio, and Larry Eoff. "Laboratory Development and Successful Field Application of a Conformance Polymer System for Low-, Medium-, and High-Temperature Applications." SPE Latin American and Caribbean Petroleum Engineering Conference. (2010).
2. Chiaramonte, Laura, Mark D. Zoback, Julio Friedmann, and Vicki Stamp. "Seal integrity and feasibility of CO_2 sequestration in the Teapot Dome EOR pilot: geomechanical site characterization." *Environmental Geology* 54, no. 8 (2008): 1667-1675.
3. Tongwa, P., Nygaard, R., Blue, A., & Bai, B. (2013). Evaluation of potential fracture-sealing materials for remediating CO_2 leakage pathways during CO_2 sequestration. *International Journal of Greenhouse Gas Control, 18*, 128-138.

Carbon Dioxide Capture for Storage in Deep Geological Formations, Volume 4
Karl F. Gerdes (Editor)

Chapter 45

NUMERICAL SIMULATION OF REACTIVE BARRIER EMPLACEMENT TO CONTROL CO_2 MIGRATION

Jennifer L. Druhan[1], Stéphanie Vialle[1,2], Kate Maher[1], Sally Benson[3]
[1]Department of Geological and Environmental Sciences, Stanford University, Stanford, CA, USA
[2]Department of Exploration Geophysics, Curtin University, Australia
[3]Department of Energy Resource Engineering, Stanford University, Stanford, CA, USA

ABSTRACT: Long-term storage of anthropogenic CO_2 in the subsurface generally requires that caprock formations will serve as physical barriers to the upward migration of CO_2. As a result, geological carbon storage (GCS) projects require reliable techniques to monitor for newly formed leaks, and the ability to rapidly deploy mitigation measures should leaks occur. Here, we develop a two-dimensional reactive transport simulation to analyze the hydrogeochemical characteristics of a newly formed CO_2 leak entering an overlying reservoir and emplacement of a hypothetical pH-dependent sealant in the vicinity of the leak. Simulations are conducted using the TOUGHREACT multi-component reactive transport code, focusing on the comparatively short time period of days to months following formation of the leak. The simulations are used to evaluate (1) geochemical shifts in formation water indicative of the leak, (2) hydrodynamics of pumping wells in the vicinity of the leak, and (3) delivery of a sealant to the leak through an adjacent well bore.

KEYWORDS: caprock defect; CO_2 leak; sealant; reactive transport modeling

INTRODUCTION

Physical and chemical trapping mechanisms are both associated with long-term storage of CO_2. Initially, structural and hydrodynamic barriers are the primary forms of containment, and require low permeability units, or caprocks, to act as seals above the injection reservoir. Efficient physical trapping will lead to chemical trapping processes over longer periods of time, including dissolution, sorption and mineralization. The time period over which chemical processes become dominant is long enough that physical trapping offers the first defence against CO_2 migration, particularly while primary injection of CO_2 is still active [1].

Solubility trapping requires dissolution of CO_2 into the aqueous phase and is thus a function of pressure, temperature and ionic strength of the fluid [2]. While the majority of total DIC in solution exists as $CO_{2(aq)}$, the dissociation of carbonic acid governs the shift in pH and the stability of solid phases. The combined effect of these factors leads to a general expectation that the pore water immediately adjacent to a newly formed CO_2 leak will exhibit an initial drop in pH. This chemical shift is significant because it may precede the breakthrough of a supercritical CO_2 plume in a down-gradient monitoring well. Experimental evidence of pH shifts prior to the arrival of CO_2 were demonstrated within the primary injection reservoir during the Frio carbon capture and storage (CCS) project [3]. Detecting such a shift requires that the chemical and hydrodynamic conditions allow for pore water in contact with the leak to subsequently mix into the surrounding fluid and advect down-gradient in advance of the CO_2 plume.

Such observations from both natural and anthropogenic systems highlight two important points. First, characteristic responses in pore water chemistry, namely decreasing pH and elevated alkalinity, may be detected before the breakthrough of CO_2. These indicators may offer a means of advanced warning where measurements are feasible. Second, these shifts distinguish the geochemical signature of the CO_2 leak from the surrounding reservoir, and thus offer a potential avenue for engineered interventions such as the introduction of a sealant with pH- or CO_2-dependent solubility.

One of the key issues in mitigating extended leakage zones (e.g., fractured systems in connected wells or in fault/damage zone systems) with current sealant technology is that the initial viscosity and setting time of most sealants will not allow sufficient lateral penetration across a fracture system. For example, hydraulic cements have high viscosity (up to 1000's of cP) and a short setting time [4]. Currently only a few sealants reported in the literature have an initial viscosity low enough to support delivery to large damage zones. These include: a water-like viscosity *in situ* generated polymer (IGP) [5], and a CO_2-activated silicate polymer initiator (SPI-CO_2) [DOE project DE-FE0005958]. The SPI-CO_2 gelation time is highly dependent upon the pH of the solvent, but can remain in solution for up to several days [6]. The high pH required in order to maintain an SPI in solution reflects the pH dependence of silica saturation. $SiO_{2(aq)}$ will readily precipitate as amorphous $SiO_{2(am)}$ at low pH, and similarly, silicic acid will form a hydrated gel. This pH dependence suggests the potential to develop of a class of sealants that are not dictated by a setting time, but rather remain in solution until an acidic CO_2 plume is contacted.

In general, effective sealant delivery to a CO_2 leakage zone requires extensive study. Conformance issues present a significant challenge, mainly because a comprehensive rock/fluid property database and operational experience are still lacking [7], and because rock formations are heterogeneous at all scales and cannot be fully characterized by geophysical techniques. An additional limitation is that current reactive transport simulators, which should be used to guide intervention strategies, do not yet have capabilities for accurately modeling the chemical and transport properties of many sealant classes, and for many of these sealants the properties required to develop even basic simulations are currently unavailable. The purpose of the present chapter is to initiate this effort by demonstrating the current capabilities and limitations of a reactive transport modeling approach to describe emplacement of a reactive barrier to mitigate a CO_2 leak. Here we present a conceptual model of CO_2 infiltrating into a confined aquifer above the primary storage reservoir and attempt to deliver a hypothetical pH-dependent sealant to the damage zone through a nearby injection well. All parameters are referenced from CCS literature and provide a generalized abstraction for both application to specific site conditions and introduction of alternate sealant classes.

MODEL DEVELOPMENT

Simulations are conducted using the TOUGHREACT non-isothermal multicomponent reactive transport code [8,9] with the ECO2N thermophysical property module for H_2O-NaCl-CO_2 mixtures in the range of temperatures and pressures appropriate for CO_2 sequestration [10]. Mathematical formulations for the sequential iteration approach [11] utilized in TOUGHREACT to solve the basic mass and energy conservation equations are described in detail elsewhere [10,12]. The TOUGHREACT code has been used previously to simulate the hydrogeochemistry of CO_2 storage in saline aquifers [13-16], and more recently to consider both the geochemical behaviour of a CO_2 leak into an overlying aquifer [17-19] and the associated consequences for drinking water quality [20-22]. To the authors' knowledge, the current study is the first to report the application of a reactive transport code to the simulation of sealant delivery for remediation of a CO_2 leak into an overlying aquifer.

Model domain and thermophysical conditions

The current study focuses on the short term geochemical response to a newly formed CO_2 leak into an overlying aquifer. As a result, the model domain is restricted to the area immediately adjacent to the leak, comprising a 50 m vertical and 2 km lateral extent and simplified to two dimensions (2-D) (Figure 1). The upper boundary of the domain is held to a no-flow condition, representing an upper confining unit above the aquifer. A highly refined grid of $1m^2$ blocks is used to discretize the domain from the left boundary to a distance of 450 m laterally, at which point the grid blocks increase exponentially in area with further distance to yield a quasi-infinite boundary condition on the right side of the domain. The left boundary of the domain is specified as no-flow and may be conceptualized as a fault or sedimentary basin margin. Grid cells in this left boundary are specified with a low injection rate, resulting in a net flux of 1 cm/day from left to right in the aquifer to represent regional groundwater flow. The bottom boundary of the domain constitutes the upper section of the primary CO_2 injection reservoir, and is separated from the overlying aquifer by a thin, horizontal caprock. The initial thermophysical conditions of the system are shown in Table 1.

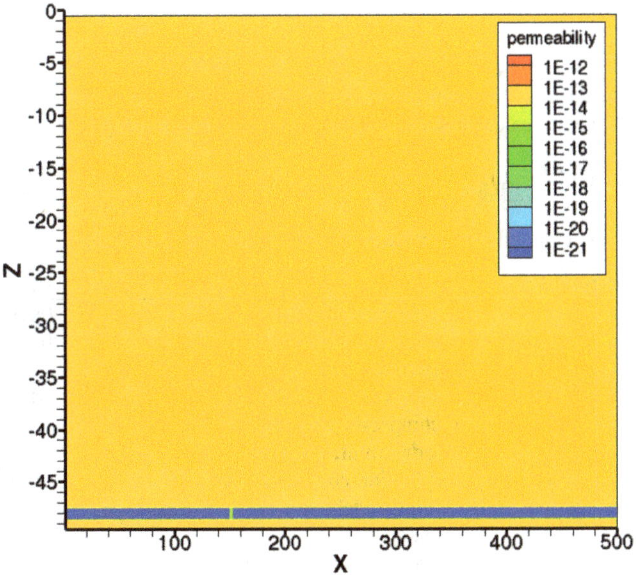

Figure 1. z-permeability of 2D domain with caprock separating upper portion of primary CO_2 injection reservoir from overlying aquifer. A 2 m wide defect in the caprock centered at x=150 m allows flow between the CO_2 reservoir and the upper aquifer. The starting z-permeability of this defect is varied in model simulations. Grid cells comprising the domain are 1 m x 1 m, such that the figure x:z ratio is scaled 10:1. The model domain continues in the x-direction to a final length of 2 km with increasing grid cell size to generate a quasi-infinite boundary condition.

Prior to initialization, a pressure of 17 MPa and a temperature of 55 °C are specified across the domain. These values fall within the range bounded by the hydrostatic and lithostatic gradients and correspond to an approximate depth of 1 – 1.5 km below land surface. To separate the injection reservoir from the overlying aquifer, permeability in the caprock is set to 1×10^{-20} m^2, eliminating any flow through this portion of the domain. Two adjacent grid cells within the caprock unit centered 150 m from the left boundary of the domain are assigned an initial z-permeability less than that of the surrounding caprock to create a 2 m wide 'defect' in the otherwise impermeable unit.

Two starting values of z-permeability for this defect of 1.0 x 10^{-15} and 1.0 x 10^{-17} m^2 were tested in model simulations.

At the specified temperature and pressure range of these simulations, two liquid phases are present as saline H_2O and supercritical CO_2. Relative permeability of the H_2O liquid phase (k_{rl}) is calculated based on H_2O saturation (S_l) using a van Genuchten relation [23] for a specified irreducible water saturation as (all parameters defined in Table 1):

$$k_{rl} = \sqrt{S^*}\left\{1 - \left(1 - [S^*]^{1/m}\right)^m\right\}^2 \qquad \text{equation 1}$$

$$where \quad S^* = \frac{S_l - S_{lr}}{1 - S_{lr}}$$

while relative permeability of the CO_2 phase (k_{rc}) is calculated using a Corey relation [24] based on H_2O saturation, irreducible water and CO_2 saturation (Table 1) as:

$$k_{rc} = (1 - S')^2(1 - (S')^2) \qquad \text{equation 2}$$

$$where \quad S' = \frac{S_l - S_{lr}}{S_l - S_{lr} - S_{gr}}$$

The capillary pressure (p_{cap}) necessary to overcome interfacial tension between H_2O and CO_2 phases in the porous media is also calculated using a van Genuchten relationship [23] as:

$$p_{cap} = -P_0\left([S^*]^{-1/m} - 1\right)^{1-m} \qquad \text{equation 3}$$

Table 1. Hydrogeologic parameters for the confined aquifer, based on values used in Xu *et al.* [8,9].

model parameter		value
k_x	x permeability (m^2)	1.0x10^{-12}
k_z	z permeability (m^2)	1.0x10^{-13}
ϕ	porosity	0.15
τ	tortuosity	0.5
c	compressibility (Pa^{-1})	1x10^{-9}
T	temperature (°C)	55
D	diffusion coefficient (m^2s^{-1})	1x10^{-9}
relative permeability – van Genuchten		
m	exponent	0.457
S_{lr}	irreducible water saturation	0.3
S_{gr}	irreducible gas saturation	0.05
capillary pressure – van Genuchten		
m	exponent	0.457
S_{lr}	irreducible water saturation	0.0
P_0	strength coefficient (kPa)	19.61

Geochemical conditions

The short time span of the current simulation relative to the chemical trapping mechanisms described in section 1 supports the use of simplified geochemical conditions, as the extent of metal-bearing primary silicate dissolution and secondary clay accumulation expected to occur in this interim is extremely limited. The initial reservoir composition was thus simplified to 80% quartz, 20% feldspar. The feldspar composition is a 20% anorthite, 80% albite solid solution ($Ca_{0.2}Na_{0.8}Al_{1.1}Si_{2.8}O_{7.8}$) with temperature-dependent equilibrium constants calculated from

Arnórsson and Stefásson [25] to account for non-ideal mixing. The regression coefficients necessary to obtain this temperature dependence were refit from the Arnórsson and Stefásson [25] values to match the equation used in the TOUGHREACT code. In addition to the quartz and feldspar initially present in the domain, kaolinite and calcite were allowed to form, although the small time interval of these simulations negated appreciable accumulation of these secondary minerals. All thermodynamic data other than those specified for the albite-anorthite solid solution were taken from Aradóttir *et al.* [26] and Wolery [27]. Initialized fluid concentrations, mineral volume fractions and kinetic rate parameters for the current simulations are provided in Table 2.

Table 2. Initial fluid and solid compositions and rate constants [*26,27].

primary species	concentration (M)
pH	7.0
HCO_3^-	4.55e-02
AlO_2^-	7.03e-08
Cl^-	1.0
Br^-	1.0e-13
Na^+	9.92e-01
Ca^{2+}	6.49e-05
$SiO_{2(aq)}$	8.45e-04

mineral	volume fraction (m^3 min/m^3 solid)	rate constant ($mol/m^2/s$)*
quartz	0.8	1.0E-14
albite-anorthite s.s.	0.2	2.7E-13
		6.9E-11 (H^+)
		2.5E-16 (OH^-)
kaolinite	0.0	6.9E-14
		4.9E-12 (H^+)
		8.9E-18 (OH^-)
calcite	0.0	equilibrium
sealant	0.0	7.3E-08

CO_2 leak and sealant

The primary injection well for CO_2 into the lower reservoir is located outside of the current high-resolution model domain. As a result, the presence of a CO_2 source is simulated by fixing the grid cell in the lower left boundary of the domain to a constant CO_2 saturation of 75% and an elevated pressure. Two fixed pressure values of 18 MPa and 20 MPa (approximately 1 MPa and 3 MPa higher than the ambient pressure in the upper aquifer, respectively) were tested. For each combination of caprock defect permeability and initial CO_2 reservoir pressure (Table 3), a 1 year simulation was run prior to implementing any sealant or hydrodynamic mitigation in order to establish the presence of a of a CO_2 leak in the upper aquifer.

Table 3. Simulated parameter scenarios.

Scenario	Caprock defect permeability (m^2)	Fixed CO_2 reservoir pressure (MPa)
A	$1.0x10^{-15}$	20
B	$1.0x10^{-17}$	20
C	$1.0x10^{-15}$	18
D	$1.0x10^{-17}$	18

A hypothetical sealant is included in the simulations based on the thermodynamic and kinetic properties and pH dependence of amorphous silica. For lack of published data on Si-polymers, the formation of amorphous silica from $SiO_{2(aq)}$ is modeled such that the molar volume of $SiO_{2(am)}$ is increased to 500 cm^3/mol, representing a hypothetical gel or polymer that undergoes a large volume increase during gelation. This hypothetical polymer is hereafter referred to as the 'sealant' and is introduced as a dissolved solute which subsequently precipitates upon contact with acidic fluids. Aqueous sealant is introduced to the system through a pumping well along the left boundary of the upper aquifer at a uniform flow rate of 0.001 kg-H_2O/m^2/s for a total fluid injection across the boundary of 0.047 kg-H_2O/s. This injection raises the pressure near the left boundary of the upper aquifer to approximately 18.2 MPa. The injectate solution is comprised of the same initial concentrations as the aquifer formation water (Table 2) equilibrated with sealant (represented as 'waterglass' Na_2SiO_3) such that pH is increased to 10, sealant (represented as $SiO_{2(aq)}$) is increased to 0.1M and Br (an inert tracer) is increased to 1.0 mM.

The accumulation of the hypothetical sealant at the interface between the alkaline flood and acidic CO_2 plume is intended to reduce the porosity (ϕ) and thus the permeability (k) of the aquifer in the region of reactivity. This porosity-permeability relationship is difficult to constrain and is often dependent upon site-specific geometry and reactivity of a given porous media. Because this relationship is rarely known for a given system at the field scale, two options are tested in the current model. The first is a Carman-Kozeny relation [28]:

$$k = k_i \frac{(1-\phi_i)^2}{(1-\phi)^2} \left(\frac{\phi}{\phi_i}\right)^3 \qquad \text{equation 5}$$

where ϕ_i is the initial porosity prior to reactions and similarly k_i is the initial permeability. This relationship has been used extensively and is considered generally applicable to a wide range of porous media. The second relationship tested in this study is a modified Hagen-Poiseuille [29]:

$$k = C_k \frac{NP \times \pi \times d^4}{128} \qquad \text{equation 6}$$

where C_k is the number of pore throats connecting to an individual pore, NP is the number of pores in a given area of porous media and d is the average diameter of a pore throat (which will decrease with secondary mineral precipitation). This relationship is suggested as a more accurate description of the porosity-permeability relationship in conglomerates and sandstones [29]. Typical values for C_k and NP of 2 throats/pore and 1000 pores/m^2 were used, respectively [30,31].

Primary injection of CO_2 into the lower reservoir and the associated pressure build-up due to this injection continue uninterrupted regardless of the development of a leak or any remediation efforts tested in the upper aquifer. As will be shown in the subsequent sections, this continued CO_2 injection pressure is required in order to establish a reactive mixing zone sufficient to precipitate substantial quantities of sealant.

RESULTS

Evaluation of a generalized pH-dependent sealant under simplified flow conditions

As an initial demonstration of the viability of a pH-dependent sealant for substantial reduction of CO_2 flux rates into an aquifer, a simplified model domain was constructed. This domain was equivalent to that described in the "model domain" section (above) and Figure 1, except that the lower caprock layer was removed such that the entire domain comprised a single confined aquifer unit. Left and right boundary conditions remained the same, and two grid cells located at the bottom of the domain (z = -50 m) centered at x = 150 m from the left boundary were fixed to an overpressure of 17.6 MPa and a CO_2 saturation of 98%. This condition resulted in a flux of CO_2 into the surrounding aquifer at a rate of 10.28 g/s. At the same time that the leak was initiated, an alkaline sealant-bearing flood was introduced along the left boundary of the domain following the conditions described in the previous section on "leak and sealant." This scenario is less realistic than the full simulations described below because the fixed overpressure of CO_2-rich cells at the bottom of the domain can only flow into the aquifer and not in any other direction. As a result, pumping an alkaline flood into the aquifer at a higher pressure than the CO_2 leak traps the CO_2-rich fluid in these two grid cells rather than allowing this fluid to recede into the underlying reservoir. The flow of alkaline, sealant-rich water over these grid cells thus results in the development of a carapace surrounding the CO_2-rich zone. While this method is a clear simplification of the conditions encountered in an aquifer connected to an underlying CO_2 storage reservoir through a caprock defect, it serves as a means of demonstrating the ability to precipitate sealant in the vicinity of the CO_2 leak to an extent sufficient to reduce flow.

The two porosity-permeability relationships (equation 5 and 6) were tested for this scenario. Figure 2 shows the results in terms of pH change using equation 5 (Carman-Kozeny).

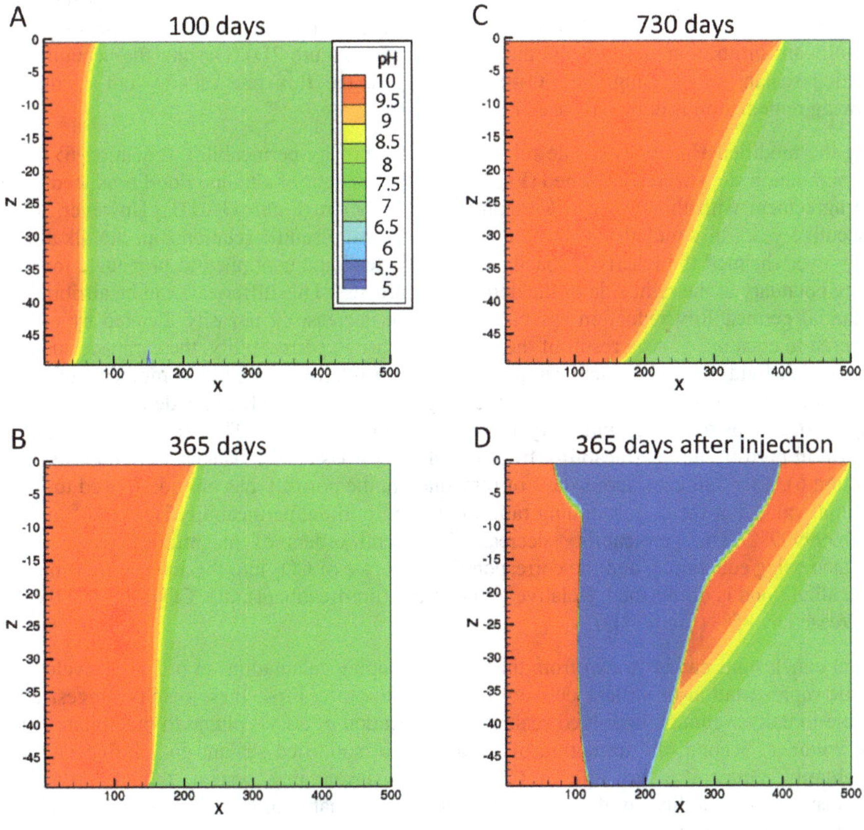

Figure 2. Time series of pH variation across domain for two years of continuous injection of alkaline sealant-bearing flood from left boundary and an additional year after the injection is shut off based on the porosity-permeability relationship in equation 5. (A) 100 days after the start of alkaline flood; (B) 365 days of continuous alkaline flood; (C) last day (730 days) of alkaline flood and (D) 365 days after the alkaline flood injection well is shut off. All distances reported in metres.

Simultaneous initiation of the CO_2 leak and the alkaline flood pressurizes the aquifer and leads to hydrodynamic management of the CO_2 leak (Figure 2A) such that CO_2 is unable to propagate into the surrounding aquifer during injection of the alkaline flood. A small low-pH zone is apparent at the base of the domain at x = 150 m indicating the presence of the CO_2. After 365 days of continuous flooding, the alkaline, sealant-bearing fluid has contacted the location of the CO_2 leak (Figure 2B) and rapid precipitation of solidified sealant occurs as a result of the acidification of the alkaline fluid in contact with CO_2. After two continuous years of flooding the top of the alkaline fluid has moved roughly 400 m across the domain, whereas the bottom of the flood is unable to progress past the CO_2 leak due to continued interaction with acidic fluid and consumption of aqueous sealant (Figure 2C). At this point in the simulation both the decrease in porosity and permeability in the four grid cells adjacent to the high pressure CO_2-rich cells were quantified. Porosity in the region immediately surrounding the CO_2 'leak' decreased by 89.5% from the initial value, and permeability correspondingly decreased by 99.9% from the initial value following equation 5. After 730 consecutive days of injection, the pumping well installed across the left boundary of the domain is shut off, returning pressure in the upper aquifer to background conditions. Now, the highest pressure in the system corresponds to the two CO_2-rich grid cells at the bottom of the domain. After an additional 365 days of simulation, this pressure gradient has resulted in significant infiltration of CO_2 into the upper aquifer (Figure 2D). Despite the accumulation of solidified sealant surrounding the CO_2-rich grid cells, the flux rate of CO_2 out of these cells following remediation is only decreased by 24%.

Using the modified Hagen-Poiseuille relationship for porosity-permeability (equation 6; [29]), the same simulation steps were performed (Figure 3). The first year of alkaline flood produced results in good agreement with those for the Carman-Kozeny relation (Figure 3A&B). However, after two consecutive years of simulation using a modified Hagen-Poiseuille relationship, the alkaline flood clearly overwhelmed the reactive capacity of the CO_2 leak and continued to propagate towards the infinite boundary at the right side of the domain (Figure 3C). This difference can be attributed to the substantial permeability reduction associated with the decrease in porosity dictated by equation 6 compared to equation 5. As a result of the larger decrease in permeability, the carapace surrounding the CO_2-rich fluid becomes quickly impermeable for a relatively small accumulation of solidified sealant such that further reaction between the CO_2 and alkaline flood is impeded. A similar result is likely to occur for a sealant with very rapid precipitation kinetics. This effect is apparent in the decrease in porosity and permeability from initial values. Using the Hagen-Poiseuille relationship (equation 6), after 730 continuous days of remediation, the porosity has only decreased to 78.2% of the initial value, but the corresponding ratio of current to initial permeability is 1.0×10^{-8}, a decrease of 99.999999%. The permeability decrease is several orders of magnitude greater than that generated using equation 5, and the corresponding flux rate of CO_2 into the aquifer after injection of the alkaline flood is diminished. Relative to the initial (pre-treatment) rate, CO_2 flux into the aquifer is decreased by 85% (Figure 3D).

Several conclusions can be drawn from these initial scoping calculations in order to develop a more realistic representation of sealant delivery and CO_2 flux rate. First, these results suggest that it is possible to deliver enough dissolved sealant to the location of a CO_2 plume to precipitate a solid in large volumes. Second, the formation of a carapace of solidified sealant does influence and may substantially reduce the flux rate of CO_2 into the surrounding system, though this is highly contingent on the ability to decrease permeability by many orders of magnitude. Third, the properties of the sealant are not the only unknowns that require optimization. In particular the porosity-permeability relationship is clearly influential to the formation of a seal, and may be further amplified by the additional coupling between porosity, permeability and capillary pressure.

Collectively these observations suggest simulation results will have a high sensitivity to the porosity-permeability relationship, which is a function of the porous media as well as the sealant.

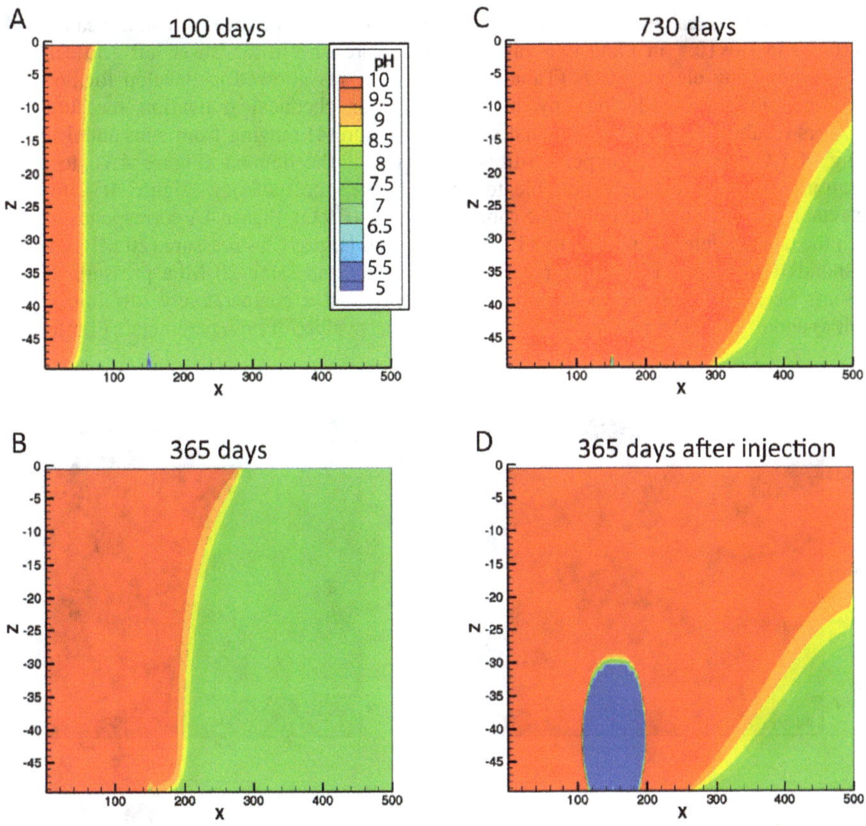

Figure 3. Time series of pH values across domain for two years of continuous injection of alkaline sealant-bearing flood from left boundary and an additional year after injection is shut off based on the porosity-permeability relationship in equation 6. (A) 100 days after start of alkaline flood; (B) 365 days of continuous alkaline flood; (C) last day (730 days) of alkaline flood and (D) 365 days after the alkaline flood injection well is shut off. All distances reported in metres.

In the following sections, a more realistic CO_2 leak will be constructed by implementing a caprock with an imposed defect (Figure 1). For all scenarios, sealant delivery will be modeled using the modified Hagen-Poiseuille relationship (equation 6), as this was shown to produce the largest decrease in permeability over a relatively short time period of sealant delivery. While long-term flooding of a sealant subject to the Carmen-Kozeny relationship (equation 5) could potentially result in development of a thick seal, the time periods necessary to establish this effect are likely prohibitive and are not considered in the current simulations.

Establishment of a CO_2 leak between reservoirs

To expand on sealant delivery from the highly simplified leakage scenario presented above, a 2 m thick, horizontal caprock at an average depth of z = -48 m was added to the domain as shown in Figure 1. The caprock separates the upper confined aquifer from the lower CO_2 injection reservoir.

Four individual CO_2 leaks were generated using two values for z-permeability in the caprock defect ($1.0x10^{-15}$ and $1.0x10^{-17}$ m^2) and two values for fixed pressure in the lower left grid cell (18 and 20 MPa) containing elevated CO_2 (Table 3). Each leak was allowed to develop for 365 days of simulation prior to introducing any hydrodynamic or chemical mitigation measures. These parameters resulted in a variety of initial scenarios (Figure 4) ranging from substantial influx and pooling of CO_2 against the upper no-flow boundary of the domain (Figure 4A), to moderate infiltration that may be barely possible to detect by seismic methods (Figure 4C), to virtually undetectable presence of CO_2 within the caprock (Figure 4B&D). Figure 4A corresponds to scenario A with the highest defect permeability ($1.0x10^{-15}$ m^2) and largest CO_2 pressure (20 MPa). Figure 4B corresponds to Scenario B the lower defect permeability and same 20 MPa pressure. Figure 4C represents Scenario C, with a high defect permeability but a comparatively low initial pressure (18 MPa) and Figure 4D represents Scenario D the lowest values of both parameters (Table 3).

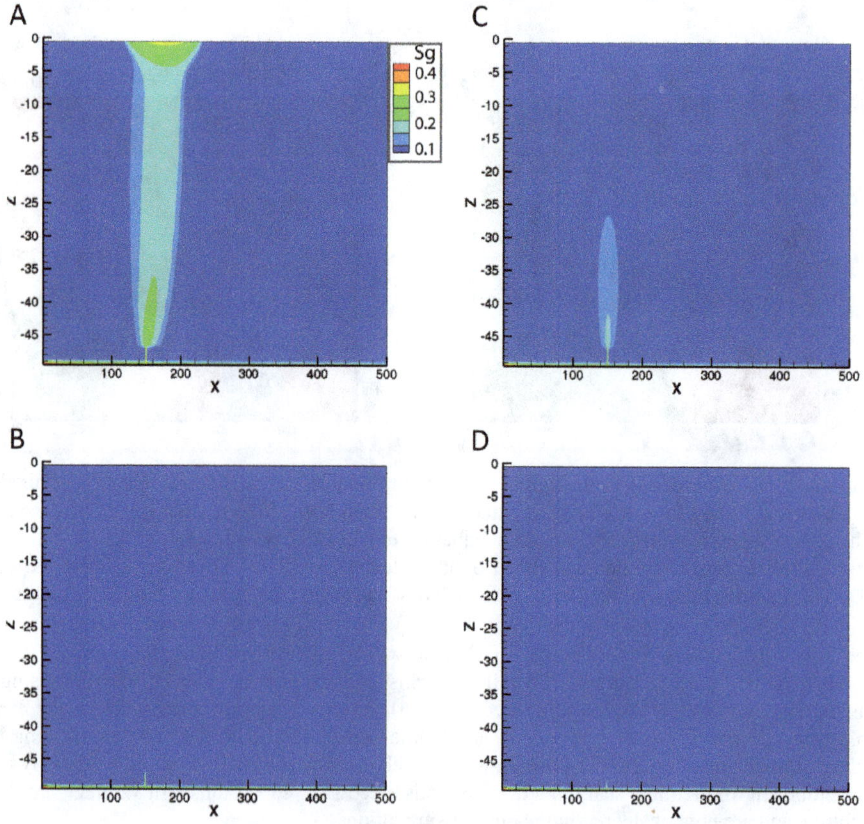

Figure 4. Four scenarios of supercritical CO_2 infiltration through a caprock defect in a confined aquifer 365 days after initial conditions were prescribed. (A) defect permeability $1.0x10^{-15}$ m^2; CO_2 pressure 20 MPa; (B) defect permeability $1.0x10^{-17}$ m^2; CO_2 pressure 20 MPa; (C) defect permeability $1.0x10^{-15}$ m^2; CO_2 pressure 18 MPa; (D) defect permeability $1.0x10^{-17}$ m^2; CO_2 pressure 18 MPa.

The flux rate of CO_2 out of the top of the caprock defect at 365 days of uninterrupted flow in Figure 4A is 3.60 g/s. This is the largest flux rate considered in these scenarios. Figure 4B corresponds to a CO_2 flux rate of 0.05 g/s, and is virtually undetectable after 365 days of CO_2 injection. The flux rate for Figure 4C is 0.56 g/s and that for Figure 4D is 0.02 g/s. Each of these scenarios will be used as the initial conditions for introduction of an alkaline sealant-bearing flood from the left boundary.

Scenario A: Defect permeability 1.0x10^{-15} m^2; CO$_2$ pressure 20MPa; CO$_2$ flux 3.60 g/s

Introduction of the alkaline sealant-bearing flood begins 365 days after the CO_2 leak is initiated (Figure 5) and continues for three consecutive years. Interaction of the alkaline flood with the acidic CO_2 plume results in some enhanced mixing at the pH boundary (Figure 5B) [e.g. 32], though this reactive front stabilizes after approximately two years of continuous flooding (Figure 5C) and forms a fairly stable reactive front thereafter (Figure 5D).

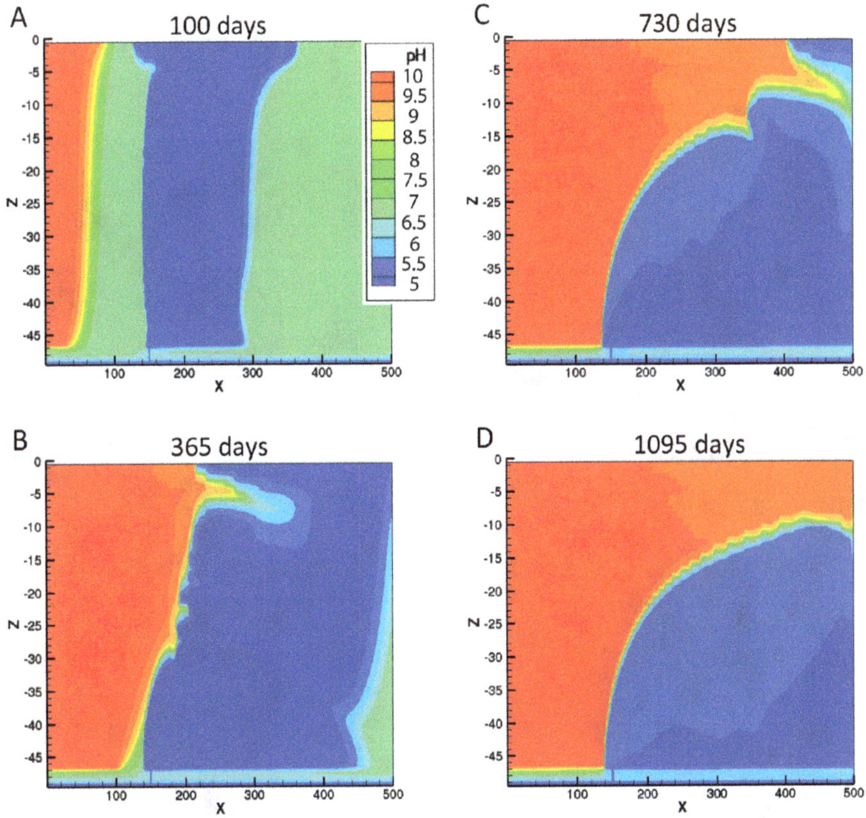

Figure 5. Scenario A time series of pH variation across domain for three years of continuous injection of alkaline sealant-bearing flood from left boundary. (A) 100 days after start of alkaline flood; (B) 365 days of continuous alkaline flood; (C) 730 days of alkaline flood and (D) last day (1095 days) before flood is shut off. All distances reported in metres.

At the specified alkaline injection rate of 0.047 kg/s, the introduction of this flood pressurizes the overlying aquifer such that CO_2 flux through the top of the caprock defect is reduced to 3.03 g/s. However, the pressure from the underlying CO_2 injection reservoir is still great enough that the shape of this stabilized profile is influenced by the balance between injection of alkaline flood at the left boundary, infiltration of CO_2 from the bottom of the domain at x = 150 m and a quasi-infinite right boundary. In contrast to the distribution of pH, the concentration of bromide introduced in the alkaline flood shows that after 1095 days of continuous delivery the inert tracer has moved well into the interior of the acidic CO_2 plume (Figure 6A). The establishment of the pH distribution shown in Figure 5D and echoed in the dissolved sealant concentration distribution 1095 days after the start of the alkaline flood (Figure 6B) in contrast to the inert bromide distribution (Figure 6A) suggest that this profile results from development of a reactive front where sealant is precipitating out of solution as a result of decreased pH. Along this reactive front, substantial solidified sealant accumulation is observed (Figure 6C), and correspondingly, permeability values decrease several orders of magnitude from the starting value of 1.0×10^{-13} m^2 after three consecutive years of remediation, in some locations as low as 5.0×10^{-17} m^2 (Figure 6D). Despite three continuous years of alkaline flood delivery, and utilizing the modified Hagen-Poiseuille porosity-permeability relationship (equation 6), the decrease in permeability associated with this simulation is only roughly two orders of magnitude.

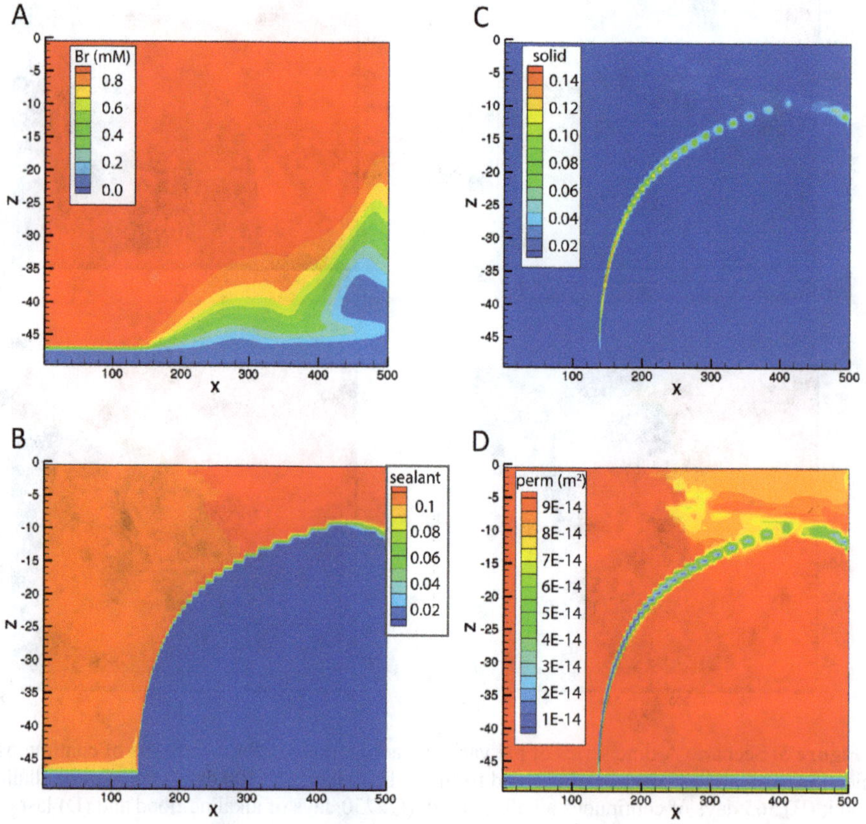

Figure 6. 1095 days after start of alkaline flood. (A) bromide concentration (mM); (B) dissolved sealant (M); (C) solidified sealant reported as change in volume fraction; (D) z-permeability (m^2).

Based on the results reflected in Figure 3, this will not be sufficient to reduce CO_2 flux into the aquifer. Furthermore, the alkaline flood was unable to form a complete carapace surrounding the CO_2 plume, such that, at best, the CO_2 leak could be diverted down-gradient, but will not be reduced or mitigated as a result of this intervention. Figure 7 shows CO_2 saturation 240 days after the alkaline flood is discontinued. The simulation shows preservation of the reactive front where maximum solidified sealant accumulation occurred, but the buoyant CO_2 phase has also permeated this barrier. No appreciable decrease in the net flux of CO_2 out of the top of the caprock has been achieved, however, relative to Figure 4A (prior to any remediation) the highest CO_2 saturation is now found just beneath the newly accumulated solidified sealant, indicating the influence of this permeability reduction on CO_2 infiltration.

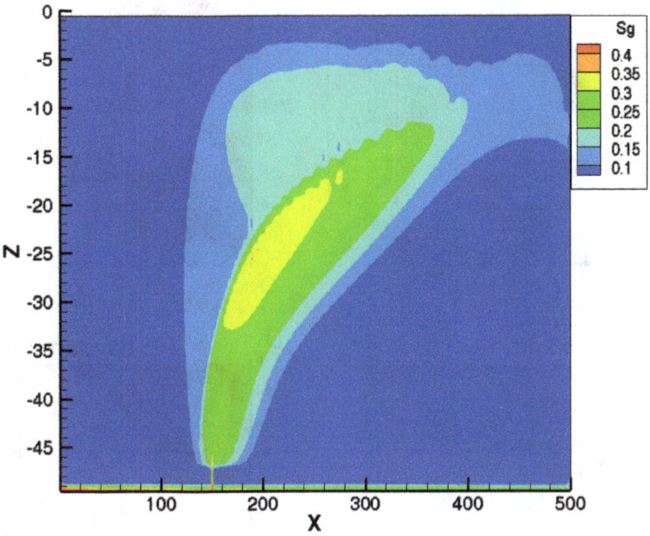

Figure 7. CO_2 saturation 240 days after the alkaline flood is discontinued.

Scenario B: Defect permeability 1.0x10^{-17} m^2; CO_2 pressure 20MPa; CO_2 flux 0.05 g/s

In contrast to the previous case, in this scenario the permeability of the caprock defect is small enough that even at a pressure differential of 3 MPa CO_2 barely begins to leak into the upper aquifer after a year of CO_2 injection (Figure 4B). This scenario represents an interesting challenge in effectively engineering a CCS project. The leak cannot be detected by any currently available technology. Even the pressure gradient in the upper aquifer as a result of this defect is too small to detect unless a well happened to be installed and instrumented directly above the leak. However, over time, this leak will develop into a pathway for CO_2 loss that will eventually become detectable and require intervention. Thus, slow-leak detection is a primary objective in long-term monitoring efforts.

Introduction of an alkaline flood to this scenario is demonstrated in Figure 8. Even 100 days into the flood (Figure 8A) it is apparent that the large pressure differential between the 18.2 MPa flood and the 20 MPa CO_2 reservoir leads to continued infiltration of CO_2 into the upper aquifer despite the remediation effort. This continued flow of CO_2 is restricted to the base of the aquifer as a result of the alkaline flood pumping wells (Figure 8B&C). The alkaline flood was continued for three years, resulting in an accumulation of solidified sealant up to 2% by volume of the total solid and a corresponding reduction in permeability surrounding the caprock defect. In several grid cells containing the mixing zone between CO_2 and the alkaline flood, permeability dropped to $1.0x10^{-29}$ m^2, resulting in a clear sealing of the porous media to further flow. However, the continued flux of CO_2 into the aquifer despite the delivery of the sealant again precluded the formation of a coherent carapace surrounding the leak and thus after the alkaline flood was ceased (following 1095 days of continuous delivery) CO_2 flux rate into the aquifer was unaltered (Figure 8D).

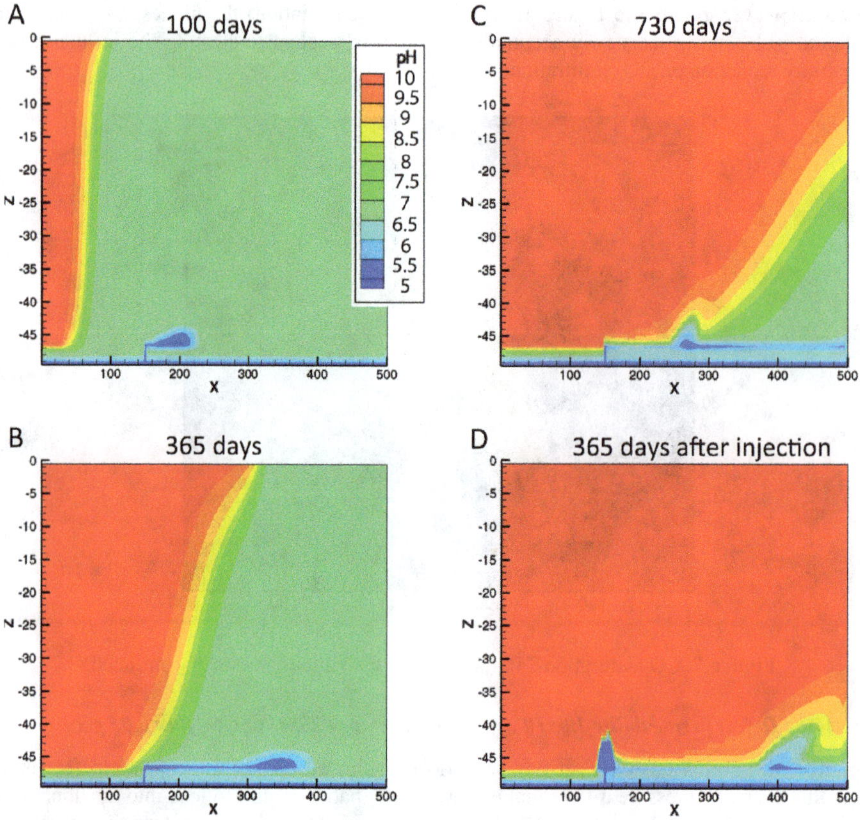

Figure 8. Scenario B time series of pH variation across domain for three years of continuous injection of alkaline sealant-bearing flood from left boundary followed by 365 days without remediation. (A) 100 days after start of alkaline flood; (B) 365 days of continuous alkaline flood; (C) 730 days of alkaline flood and (D) 365 days after alkaline flood shut off.

The combined results of scenario A and scenario B lead to several conclusions. First, the presence of a CO_2 plume in the underlying reservoir that is several MPa higher than the alkaline flood introduced in the upper aquifer leads to a balance between the flux of CO_2 and flux of dissolved sealant that precludes the formation of a coherent carapace in a reasonable period of time (several years of continuous flooding). It may be possible to eventually develop a coherent seal after many more years of constant flooding, but this is likely an unrealistic strategy. Second, the high pressure

CO_2 reservoir results in no observable decrease in CO_2 flux rate into the upper aquifer despite substantial accumulation of solidified sealant in the area surrounding the leak. Thus the accumulation of solidified sealant may influence the location and flow vectors of CO_2 in the aquifer (e.g. Figure 7), but will not decrease the loss of CO_2 from the injection reservoir while primary CO_2 delivery is still active.

Scenario C: Defect permeability 1.0x10^{-15} m^2; CO_2 pressure 18MPa; CO_2 flux 0.56 g/s

In contrast to the cases presented in Scenarios A and B, in this scenario the pressure in the underlying CO_2 reservoir is only 1MPa higher than the ambient (pre-flood) pressure of the overlying aquifer (Figure 4C). This pressure differential still leads to the development of a leak that may be detectable by surface and down-hole seismic methods. Upon initiation of the alkaline flood remediation (Figure 9A), the behavior of the system is initially quite similar to that for scenario A (Figure 5A). However, the pressurization of the upper aquifer due to the alkaline flood in this case is large enough relative to the pressure of the underlying CO_2 reservoir to completely cease further infiltration of CO_2 into the aquifer. As a result, after 365 days of flooding the head of the CO_2 plume has become disconnected from the lower reservoir (Figure 9B) and is swept down gradient and dissolved into the alkaline fluid.

After two years of continuous flooding the high pressure in the upper aquifer forces alkaline, dissolved sealant-bearing fluid into the lower CO_2 reservoir (Figure 9C). This reversal results in rapid consumption of all available reactive sealant forced through the fracture by both the contact between the alkaline fluid and the acidic CO_2 and consumption of residual CO_2 remaining in the pore space (equation 2 and 3). In contrast to the previous cases, this rapid consumption of sealant is not easily replenished by incoming alkaline flood, as the flow rate of the alkaline fluid in the lower CO_2 reservoir is very low. Since dissolved sealant is not easily supplied to this reactive front, the total accumulation of solidified sealant is quite small compared to the previous cases. From a starting value of 1.0×10^{-13} m^2 the minimum permeability in the lower reservoir after 730 days of flooding is still 85% of the original value. As a result, after the alkaline flood injection well is shut off, the flux of CO_2 into the upper aquifer immediately resumes (Figure 9D). However, because the pressure differential between the CO_2 reservoir and the upper aquifer is lower than in the previous cases, this small reduction in permeability does influence the flux rate of CO_2 into the upper aquifer, reducing it to 79% of the original 0.56 g/s.

Scenario D: Defect permeability 1.0x10^{-17} m^2; CO_2 pressure 18MPa; CO_2 flux 0.02 g/s

Similar to scenario B, in this scenario the permeability of the defect is low enough that after 365 days of initialization (Figure 4D) the leak is virtually undetectable in the upper aquifer. Continued injection of CO_2 in the lower reservoir would eventually drive CO_2 through this conduit into the upper aquifer. However, unlike scenario B, the lower pressure of the CO_2 injection reservoir arrests progression of the leak as soon as the upper aquifer is pressurized by the alkaline flood (Figure 10A). In the previous example (scenario C), this low pressure leads to infiltration of the alkaline flood into the upper section of the CO_2 reservoir. Here, the lower defect permeability prohibits infiltration of the flood, such that the contact between dissolved sealant and CO_2 occurs in a very small space at the top of the caprock defect (Figure 10B and C). In contrast to scenario C, plenty of alkaline flood is supplied to this reactive front, but the reaction is restricted by the delivery of acidic fluid through the caprock defect. As a result, the amount of solidified sealant accumulated is low, as in scenario C, but this occurs because of limited CO_2 delivery rather than limited dissolved sealant delivery.

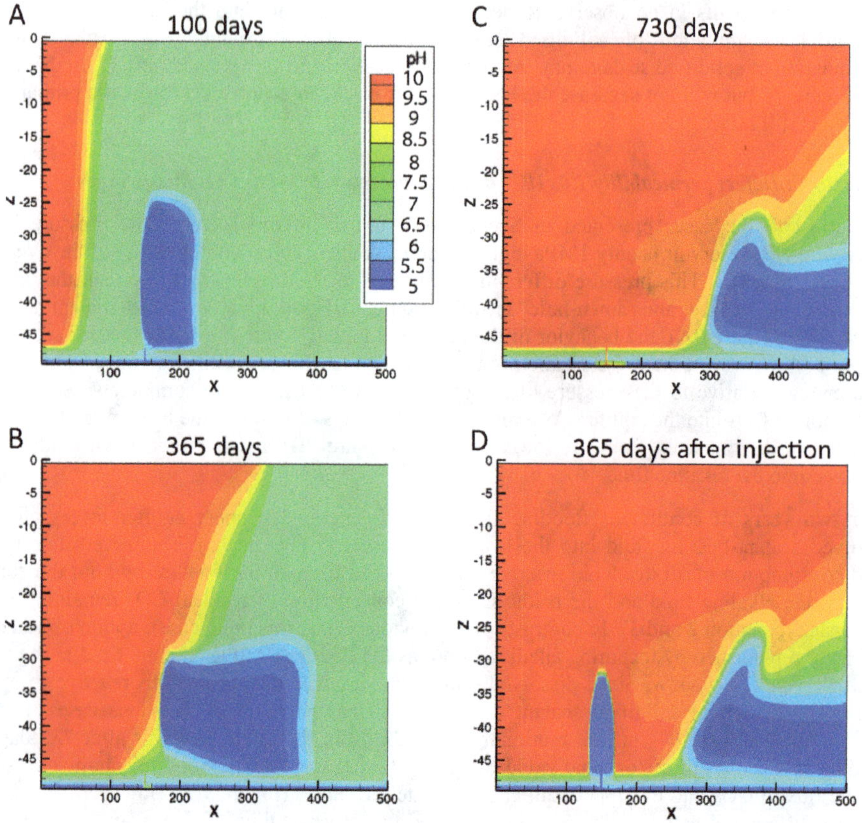

Figure 9. Scenario C time series of pH variation across domain for two years of continuous injection of alkaline sealant-bearing flood from left boundary followed by 365 days without remediation. (A) 100 days after start of alkaline flood; (B) 365 days of continuous alkaline flood; (C) 730 days of alkaline flood and (D) 365 days after alkaline flood shut off.

After two years of continuous flooding, the permeability in the grid cells just above the caprock defect is unchanged. Within the caprock defect itself the permeability has decreased to 85% of the original 1.0×10^{-17} m^2. After the alkaline flood injection well is shut off, slow infiltration of CO_2 through the caprock defect resumes (Figure 10D). However, due to the slow flow rate of CO_2 as a result of the low pressure differential and lowered defect permeability, as CO_2 enters the upper aquifer and contacts residual alkaline fluid left from the injection, permeability in the grid cells surrounding the leak begin to decline. After 365 days without alkaline injection, the permeability in the grid cells surrounding the leak has decreased to 70% of its original value. This permeability reduction as a result of the continued reaction between the slowly infiltrating CO_2 and residual alkaline sealant-bearing fluid results in a decrease in the net flux of CO_2 through the defect by 90% of the original value. This outcome represents a substantial reduction in the leakage rate of CO_2, provided that the pressure differential between the lower CO_2 reservoir and upper aquifer remains small.

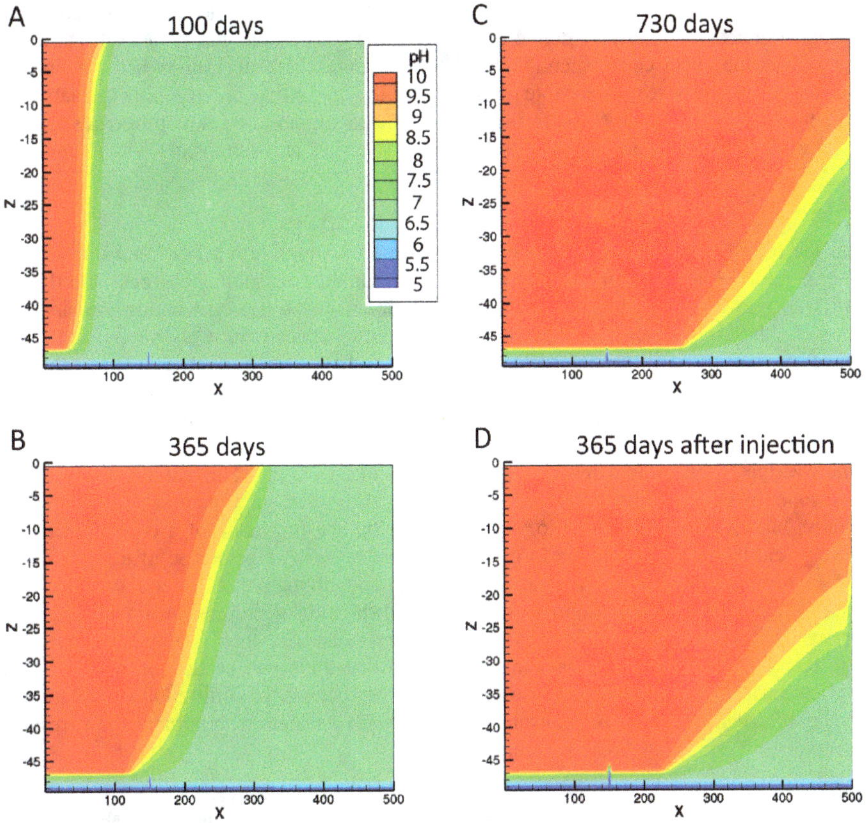

Figure 10. Scenario D time series of pH variation across domain for two years of continuous injection of alkaline sealant-bearing flood from left boundary followed by 365 days without remediation. (A) 100 days after start of alkaline flood; (B) 365 days of continuous alkaline flood; (C) 730 days of alkaline flood and (D) 365 days after alkaline flood shut off.

Further considerations

The results presented in the previous sections are intended as a generalized framework to test the relative importance of parameters and thus identify key knowledge gaps in the application of reactive sealants to arrest CO_2 leaking through a caprock defect. Several assumptions inherent in these simulations require further consideration. First, the use of a 2D domain obscures the difficulty in delivering sealant from an injection well to an implicitly high local pressure zone. The results of these simulations indicate that the pressurization of the upper aquifer via sealant injection can supersede the influence of the leak, but this requires extension to a full 3D domain to validate. Second, the importance of the porosity-permeability relationship was clearly demonstrated in these results, but the influence of reduced permeability on the capillary pressure function was not tested. Use of a Leverett scaling relation [33] to account for the influence of accumulated secondary mineral phases on capillary pressure may result in enhanced reduction of CO_2 flux rates, though this relationship may also adversely influence the extent of mixing between the alkaline flood and acidic CO_2 plume. Finally, these simulations only consider the use of a dissolved solute that reacts upon delivery to the acidic CO_2 plume. As was noted above, this method can lead to a situation in which the reactant cannot be effectively delivered inside of the CO_2 plume, but is instead restricted to a mixing zone at the interface between the flood and the CO_2. This effect might be overcome by the use of a higher viscosity sealant capable of infiltrating the CO_2 plume and reaching the damage zone even when leak rates are high. Thus further simulations using modified flow properties, potentially based off of established models of DNAPL behaviour [e.g. 34], are warranted.

SUMMARY AND CONCLUSIONS

In total the results of scenarios A – D may be summarized as follows. For a substantial caprock defect (e.g. permeability 5 orders of magnitude higher than the surrounding caprock), the flux rate of CO_2 leaking into the upper aquifer is likely high enough to be detected within the first year of injection. In these scenarios (A and C), the interaction between the CO_2 plume and a sealant introduced by pressurized injection in the overlying aquifer is highly contingent upon the vertical pressure gradient between the top of the CO_2 injection reservoir and the base of the overlying aquifer across the location of the defect. When this pressure differential increases towards the surface (i.e. CO_2 reservoir pressure is greater than sealant pressure) then CO_2 will continue to leak into the overlying aquifer and react with the sealant (scenario A). This reaction will result in a mixing zone in which substantial precipitate can accumulate and permeability can decrease drastically, but the continued influx of CO_2 can prohibit the formation of a coherent carapace to effectively seal the leak. This scenario may potentially benefit from the use of a higher viscosity sealant in order to effectively infiltrate the CO_2 plume, though this material will need to be optimized for delivery. In contrast, when the pressure differential decreases towards the surface, (i.e. CO_2 reservoir pressure is lower than sealant delivery pressure), then CO_2 will be forced back through the damage zone and sealant may even infiltrate the upper section of the CO_2 injection reservoir (scenario C). This outcome is also undesirable because the infiltration of sealant into the underlying reservoir leads to rapid consumption of dissolved sealant with minimal flow, resulting in an essentially diffusion-based reactive front.

The presence of a relatively minor caprock defect (e.g. permeability three orders of magnitude higher than the surrounding caprock) effectively dampens the results observed in the high-permeability examples. For a system in which the CO_2 reservoir pressure is higher than the overlying sealant pressure (scenario B), the continued influx of CO_2 again counteracts the formation of a coherent seal. However, the enhanced mixing zone between sealant and CO_2 can result in extreme permeability reductions. When the pressure in the overlying aquifer is greater than in the CO_2 reservoir (scenario D), this low-permeability defect inhibits infiltration of sealant into the lower reservoir and supports minor permeability reductions inside the caprock defect. While these permeability decreases are small, the combination with a small pressure differential leads to reduction in CO_2 flux rate such that continued precipitation occurs after the sealant delivery is discontinued. In total this scenario leads to the largest reduction in CO_2 flux rate after remediation.

The primary difference between the simplified conceptual model and the more realistic caprock defect models is the alternate pathway for CO_2 flow along the base of the caprock formation. In the simplified model, the lack of an alternate pathway meant that pressurization of the aquifer due to introduction of the alkaline flood essentially trapped the CO_2 in place. Thus the mixing zone between CO_2 and alkaline sealant-bearing fluid was enhanced in this model relative to the caprock defect models. Leveraging this enhanced mixing, the simplified conceptual model leads to substantial permeability reductions (up to 8 orders of magnitude using the modified Hagen-Poiseuille equation) and a substantial decrease in CO_2 flux following the alkaline injection period. Based on this contrast, a primary difficulty faced in the caprock defect models was reproduction of a mixing zone sufficient to achieve a similar permeability reduction when CO_2 is no longer trapped in a specific location.

In total these results demonstrate that in order to deliver a pH-dependent sealant, or any other sealant delivered as a dissolved solute, to a leaking damage zone during active CO_2 injection, the pressurization of the aquifer must be comparable to the pressure of the underlying CO_2 reservoir. This may be achieved by reducing the injection rate of CO_2 or alternating injection pulses of sealant in the upper aquifer and CO_2 in the lower reservoir. However, sealants that are able to penetrate the CO_2 plume, such as a higher viscosity material, may not be subject to this constraint. Total cessation of CO_2 injection is not recommended for a pH-dependent sealant, since a primary requirement for generating an effective seal is the establishment of a mixing zone between alkaline flood and acidic CO_2-rich water such that both fluids are constantly supplied to the reactive front.

All of these findings are predicated on the assumption of a porosity-permeability relationship that achieves large permeability reduction for fairly small changes in mineral volume fraction. If a relationship such as the Carmen-Kozeny or cubic law provides a better description of the behavior observed in a given reservoir, then generation of an effective seal through precipitation of a solid phase at the CO_2-flood boundary will be much harder to achieve. Therefore a fundamental result of this study is the recognition that sealant properties must be developed in conjunction with an accurate understanding of the porosity-permeability relationship of the target system. In other words, emplacement of a reactive barrier is contingent upon both the sealant and aquifer properties, and both must be well characterized in order to achieve effective results.

ACKNOWLEDGEMENTS

This study is part of a larger project, "Assessment of Leakage Detection and Intervention Scenarios for CO_2 Sequestration," supported by the CCP3; a joint industry project sponsored by BP, Chevron, Eni, Petrobras, Shell, and Suncor.

REFERENCES

1. IPCC (2005) Intergovernmental panel on climate change (IPCC) special report on carbon dioxide capture and storage. Prepared by the working group III of the intergovernmental panel on climate change (B. Metz; O. Davidson; H.D. de Coninck; M. Loos and L.A. Meyers (eds.).
2. Kharaka, Y.K. and Cole, D.R. (2011) Geochemistry of geologic sequestration of carbon dioxide. In Harmon, R. (ed.), Frontiers in Geochemistry: Contribution of Geochemistry to the Study of the Earth, chapter 8, Wiley-Blackwell, pp. 135-169.
3. Kharaka, Y.K.; Cole, D.R.; Hovorka, S.D.; Gunter, W.D.; Knauss, K.G.; Freifeld, B.M. (2006a) Gas-water-rock interactions in the Frio Formation following CO_2 injection: Implications to the storage of greenhouse gases in sedimentary basins. Geology, 34, 577-580.
4. Zhang, M. and Bacchu, S. (2011) Review of integrity of existing wells in relation to CO_2 geological storage: what do we know? International Journal of Greenhouse Gas Control, 5(4), 826-840.
5. Vasquez, J. and Eoff, L. (2010) Laboratory development and successful field application of a conformance polymer system for low-, medium-, and high- temperature applications, Society of Petroleum Engineers SPE-139308.
6. Oglesby, K.D. (2008) Novel single stage water mitigation treatment, Technical Report 3180-IT-USDOE-2098.
7. Tongwa, P.; Nygaard, R.; Blue, A.; Bai, B. (2013) Evaluation of potential fracture-sealing materials for remediating CO_2 leakage pathways during CO_2 sequestration. International Journal of Greenhouse Gas Control, 18, 128-138.

8. Xu, T.; Sonnenthal, E.; Spycher, N.; Pruess, K. (2006) TOUGHREACT – a simulation program for non-isothermal multiphase reactive geochemical transport in variably saturated geologic media: applications to geothermal injectivity and CO_2 geologic sequestration. Computational Geosciences, 32, 146-165.
9. Xu, T.; Spycher, N.; Sonnenthal, E.; Zhang, G.; Zheng, L.; Pruess, K. (2011) TOUGHREACT Version 2.0: a simulator for subsurface reactive transport under non-isothermal multiphase flow conditions. Computational Geosciences 37, 763- 774.
10. Pruess, K. and Spycher, N. (2007) ECO2N – a fluid property module for the TOUGH2 code for studies of CO_2 storage in saline aquifers. Energy Conversion and Management, 48, 1761-1767.
11. Yeh, G.T. and Tripathi, V.S. (1991) A model for simulating transport of reactive multi-species components: model development and demonstration. Water Resources Research, 27, 3075-3094.
12. Pruess, K. (2004) The TOUGH codes: a family of simulation tools for multiphase flow and transport processes in permeable media. Vadose Zone Journal, 3, 738-746.
13. Xu, T.; Apps. J.A.; Pruess, K. (2003) Reactive geochemical transport simulation to study mineral trapping for CO_2 disposal in deep Arenaceous Formations. Journal of Geophysical Research, 108, B2.
14. Xu, T.; Apps, J.A.; Pruess, K. (2004) Numerical simulation to study mineral trapping in CO_2 disposal in deep aquifers. Journal of Applied Geochemistry, 19, 917-936.
15. Xu, T.; Apps, J.A.; Pruess, K. (2005) Mineral sequestration of carbon dioxide in a sandstone-shale system. Chemical Geology, 217, 295-318.
16. Andre, L; Audigane, P.; Azaroual, M.; Menjoz, A. (2006) Numerical modeling of fluid-rock chemical interactions at the supercritical CO_2-liquid interface during CO_2 injection into a carbonate reservoir, the Dogger aquifer (Paris Basin, France). Energy Conservation and Management, 48, 1782-1797.
17. Humez, P.; Audigane, P.; Lions, J.; Chiaberge, C.; Bellenfant, G. (2011) Modeling of CO_2 leakage up through an abandoned well from deep saline aquifer to shallow fresh groundwater. Transport in Porous Media, 90, 153-181.
18. Pauline, H.; Pascal, A.; Julie, L.; Philippe, N.; Vincent, L. (2011) Tracking and CO_2 leakage from deep saline to fresh groundwaters: development of sensitive monitoring techniques. Energy Procedia, 4, 3443-3449.
19. Fahrner, S.; Schåfer, D.; Dethlefsen, F.; Dahmke, A. (2012) Reactive modeling of CO_2 intrusion into freshwater aquifers: current requirements, approaches and limitations to account for temperature and pressure effects. Environmental Earth Sciences, 67, 2269-2283.
20. Zheng, L.; Apps, J.A.; Zhang, Y.; Xu, T.; Birkholzer, J.T. (2009) On mobilization of lead and arsenic in groundwater in response to CO_2 leakage from deep geological storage. Chemical Geology, 268, 281-297.
21. Vong, C.Q.; Jacquemet, N.; Picot-Colbeaux, G.; Lions, J.; Rohmer, J.; Bouc, O. (2011) Reactive transport modeling for impact assessment of a CO_2 intrusion on trace elements mobility within fresh groundwater and its natural attenuation for potential remediation. Energy Procedia, 4, 3171-3178.
22. Jacquemet, N.; Picot-Colbeaux, G.; Vong, C.Q.; Lions, J.; Bouc, O.; Jérémy, R. (2011) Intrusion of CO_2 and impurities in a freshwater aquifer – impact evaluation by reactive transport modeling. Energy Procedia, 4, 3202-3209.
23. van Genuchten, M.T. (1980) A closed-form equation for predicting the hydraulic conductivity of unsaturated soils. Journal of American Soil Science Society, 44, 892- 898.
24. Corey, A.T. (1954) The interrelation between gas and oil relative permeabilities. Producers Monthly, 38-41.

25. Arnórsson, S. and Stefásson, A. (1999) Assessment of feldspar solubility constants in water in the range 0° to 350°C at vapor saturation pressure. Americal Journal of Science, 299, 173-209.
26. Aradóttir, E.S.P.; Sonnenthal, E.L.; Jónsson, H. (2012) Development and evaluation of a thermodynamic dataset for phases of interest in CO_2 mineral sequestration in basaltic rocks. Chemical Geology, 304-305, 26-38.
27. Wolery, T. (1992) EQ3/6: Software package for geochemical modeling of aqueous systems: package overview and installation guide (version 7.0). Report UCRL-MA-210662. Lawrence Livermore National Laboratory, Livermore, CA.
28. Bear, J. (1972) Dynamics of Fluids in Porous Media, Dover Publications, Inc., New York.
29. Ehrlich, R.; Etris, E.L.; Brumfield, D.; Yuan, L.P.; Crabtree, S.J. (1991) Petrography and reservoir physics III: physical models for permeability and formation factor. AAPG Bulletin, 75(10), 1579-1592.
30. Lucia, F.J. (2007) Carbonate Reservoir Characterization: An Integrated Approach. Springer, 336 pp.
31. Emmanuel, S.; Ague, J.J.; Walderhaug, O. (2010) Interfacial energy effects and the evolution of pore size distributions during quartz precipitation in sandstone. Geochimica et Cosmochimica Acta, 74, 3539-3552.
32. Ennis-King, J.; Paterson, L. (2007) Coupling of geochemical reactions and convective mixing in the long-term geological storage of carbon dioxide. International Journal of Greenhouse Gas Control, 1, 86-93.
33. Slider, H.C. (1976) Practical petroleum reservoir engineering methods, An Energy Conservation Science. Tulsa, Oklahoma, Petroleum Publishing Company.
34. Erning, K.; Schafer, D.; Dahmke, A.; Luciano, A.; Viotti, P.; Papini, M.P. (2011) Simulation of DNAPL distribution depending on groundwater flow velocities using TMVOC *in* GQ10: Groundwater Quality Management in a Rapidly Changing World, M. Schirmer; E. Hoehn; T. Vogt (eds.) IAHS Publications, vol. 342, pp. 128-131.

Carbon Dioxide Capture for Storage in Deep Geological Formations, Volume 4
Karl F. Gerdes (Editor)

Chapter 46

FRACTURE SEALANTS PROGRAMME: MODELLING AND FEASIBILITY

Peter Ledingham
GeoScience Limited, Falmouth Business Park,
Falmouth, Cornwall TR11 4SZ, UK

ABSTRACT: In order to assess the concept of *in situ* sealing of a leak from a CO_2 storage formation, a field test of three candidate fracture sealants is planned, to be undertaken in the Mont Terri Underground Rock Laboratory in Switzerland. The objective of the experiment is to create artificial fractures between boreholes in a highly controlled environment, establish circulation and then test the effectiveness of three candidate sealants in plugging them. As part of the concept and design development for the experiment, pre-modelling of the fracturing, circulation and sealing phases has been carried out. This modelling has provided estimates of fracturing and reopening pressures, injection characteristics, resulting apertures and circulation performance. A feasibility study into potential geophysical surveillance techniques has demonstrated that practical systems can be installed which offer the potential to monitor the fracture geometry.

KEYWORDS: fracture sealing; fracture modelling; hydrofracturing; seismic monitoring; tomography; artificial fractures

INTRODUCTION

A field test of three candidate fracture sealants is planned, to be undertaken in the Mont Terri Underground Rock Laboratory (MTURL) in Switzerland. The candidate sealants are an experimental bio-sealant, an experimental smart gel and a proprietary sealant, H2Zero, supplied by Haliburton. More details about the sealants are found in Chapter 46 of this volume. The concept and design of this field experiment (designated as CS-B) were developed over a period of 18 months between December 2012 and June 2014.

As part of the design process, pre-modelling and feasibility studies were carried out in order to develop detailed designs, specify equipment and develop procedures.

Fracture modelling

Lawrence Berkeley National Laboratory used analytical modelling and numerical simulations of coupled hydro-mechanical processes to study stress field and related fracturing processes as a means to examine and design the creation of a number of bedding-parallel hydrofractures from a single borehole. The work was divided into three tasks:

- Review of previous fracturing experience at MTURL.
- Back analysis of previous experiments (GP / GS).
- Design calculations for CS-B fracturing.

Circulation and sealing modelling

Intera Inc undertook sealant emplacement and pre/post-sealant characterisation modelling in order to refine the design for post-fracturing circulation and subsequent sealant placement. The work was divided into four tasks:

- Model development and construction.
- Definition of simulation scenarios for preliminary analyses.
- Development of test sequence for characterisation and sealant placement.
- Implementation and model executions.

Geophysical surveillance feasibility study

Semore Seismics Limited carried out a feasibility study into the potential use of two geophysical surveillance techniques for imaging the fractures created during the experiment; tomography and microseismic monitoring. Tomography, including reflection imaging, may provide an image of the effect of the fractures on the velocity field whereas microseismic monitoring should provide locations of individual events and more precise imaging of the fracture dimensions.

EXPERIMENT LOCATION

The experiment will be carried out in boreholes drilled from Gallery 04 in the MTURL, as shown in Figure 1 (plan view). The location of the experiment is represented in the figure by the two brown lines extending from the gallery towards the northwest (left in the figure). The lines are not intended to be borehole locations; they merely represent the approximate space planned for the experiment.

This location offers ample working space, is in good condition, is not too close to other current or planned experiments, does not interfere with planned developments in the laboratory and offers good access to the sandy facies of the Opalinus Clay at an appropriate drilling distance.

It was decided at an early stage to undertake the experiment in the sandy facies rather than the shaly facies, which are more common (Figure 1) because they are stronger, more stable, do not exhibit Excavation Damage Zone (EDZ) fracturing to the same extent and the created fractures would be expected to be more uniform.

In the location chosen the sandy facies will be encountered a minimum of 10m drilling depth from the gallery which means that they will be outside the EDZ and any stress disturbance. They are expected to be approximately 16m thick which is sufficient borehole length to allow the creation and testing of 7 fractures.

Several types of borehole are required:

- The main experiment boreholes.
- Extensometer boreholes.
- Sealant assessment boreholes.
- Crosshole tomography boreholes.
- Microseismic array boreholes.

These are discussed further in Chapter 46.

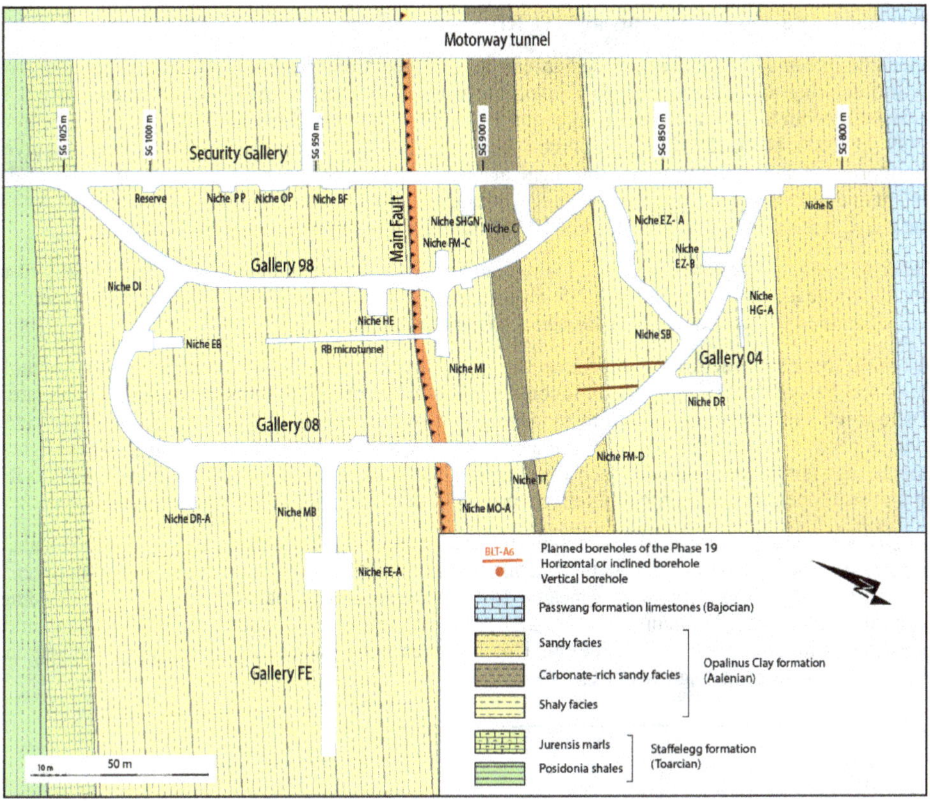

Figure 1. Plan view of the experiment location.

FRACTURE MODELLING

Each of the main experiment boreholes will be completed with a 7-zone multi-packer system. Fractures will be initiated from each of the zones in the active borehole and be 'captured' in the equivalent zones in each of the four passive boreholes. There is high confidence that the fractures will be captured because the fractures will tend to align with regular and predictable bedding planes in the Opalinus clay, there is only a short distance between the active and passive boreholes, and the passive borehole zone lengths are large compared to the active borehole zone lengths. After initially considering fracturing using gas, this option was ruled out and all fracturing will be carried out with Pearson Water.

Modelling was carried out in order to understand the significant processes affecting clay fracturing and to guide the fracture design for the experiment. In particular, it addressed the following issues:

- The expected range of fracturing pressures and volumes, hydrofracture properties and geometry.
- Options to control fracture geometry including extent of generated fractures.
- Significant uncertainties associated with fracturing activities.
- Identification and consequences of possible interactions between multiple fractures.

Previous fracturing experiments at MTURL

Valuable information for fracture design can be obtained from previous MTURL experiments on hydrofracturing and hydraulic testing involving water. The most relevant are those carried out in association with the Long-term Gas Injection Test (GP-A) and Gas Frac Self-sealing test (GS), from 1999 to 2000 [1,2]. They are important because they are the only ones that have involved monitoring of both hydraulic and mechanical responses, which makes it possible to estimate the extent and volume of the created fracture. Additional testing at this site has been performed within the HG-C (long term gas migration) and HG-D (reactive gas transport in OPA) experiments.

Other important experiments include hydro-fracturing stress measurements (HFSMs) conducted at the IS Niche in 1998 [3] and the DS Experiment conducted in 2010, which includes multiple hydrofractures performed in the sandy facies of the Opalinus Clay [4].

A typical hydraulic fracturing sequence is shown in Figure 2, from an HFSM performed in borehole BDS-4 as part of the DS experiment. It consists of an initial breakdown cycle to initiate and create a fracture followed by several fracture reopening and shut-in cycles. In this case the data indicated that a penny-shaped fracture was initiated and propagated parallel to bedding. The slightly decreasing shut-in pressure with each cycle from about 4.6MPa in the breakdown cycle to 4.2MPa in the last reopening cycle probably reflects propagation of the fracture during each injection cycle.

This is one example out of eight successful HFSMs in the BDS-4 borehole. Based on all the data it appears that the volume injected to the fractures during individual injection cycles varied between 2 and 5 litres and all the fractures appeared to be initiated and propagated parallel with bedding. Moreover, the shut-in pressures for 2nd and 3rd cycles were quite stable, around the magnitude consistent with stress normal to bedding. This indicates that fractures extended along bedding and did not deviate or rotate significantly.

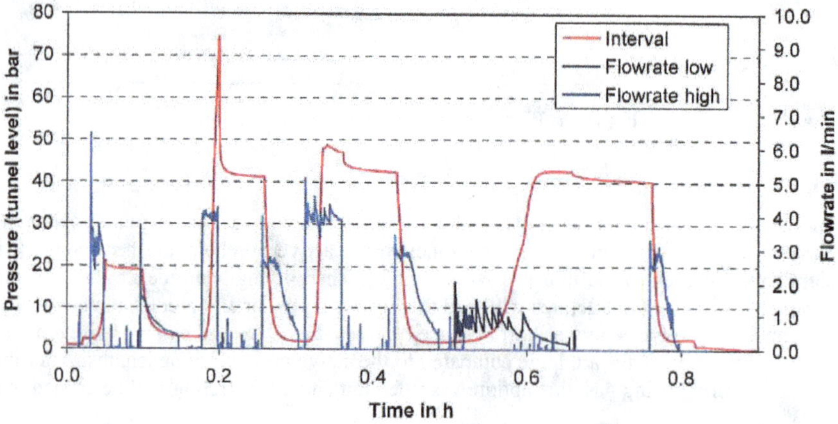

Figure 2. Typical pressure (red line) and flow rate record from an HFSM in the MTURL DS experiments, carried out in the sandy facies of the Opalinus Clay.

Stress field and expected breakdown and reopening pressures

A good understanding of the local stress field is critical in the development of fracturing design and specification of equipment. The stress field at MTURL has been measured and evaluated using various methods, but not without some difficulty [5,6,7,8]. Figure 3 presents an overview of the current best estimates of the MTURL stress field, and its orientation relative to the CS-B experiment location.

The stress field will affect the creation of the fractures and the required fluid pressure range during various hydraulic tests. Most important is the stress normal to bedding planes, which is a stress component that is quite well constrained based on all the hydraulic fracturing experiments conducted at MTURL. At the level of the tunnel system the stress normal to bedding is about 4.2MPa, which is a good indicator of the injection pressure required to keep the fractures open for water flow.

Based on experience at previous MTURL hydrofracturing experiments, the breakdown pressure is expected to be between 5.5MPa and 9MPa. It is known from common hydrofracturing practice that the breakdown pressure could depend on the rate of interval pressurisation. Among the previous MTURL experiments, the hydrofracturing at the IS Niche was conducted using a stiff system resulting in a high pressurisation rate. The breakdown pressure at one the IS Niche experiments was as high as 14.9MPa, which could indicate axial fracturing of the borehole. Fast pressurisation will increase the possibility of creating axial fractures rather radial fractures along bedding but, despite a rapid pressurisation used at all tests at the IS Niche, it appears that most fractures were still created along bedding.

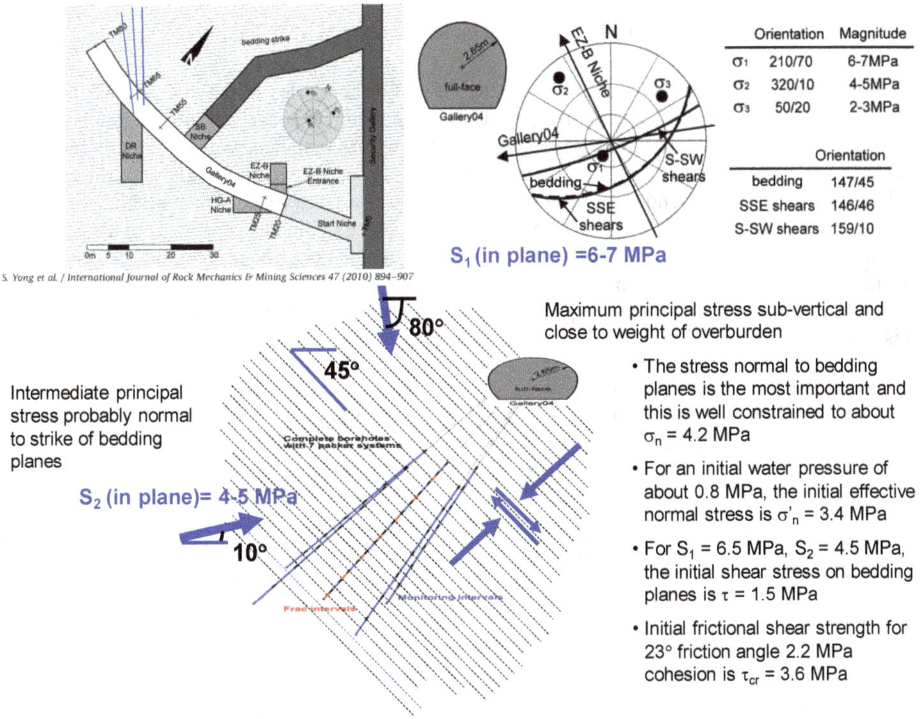

Figure 3. Current best estimate of the stress field at the proposed sealant experimental site in MTURL (pictures taken from Yong *et al*, 2010).

Figure 4 was developed to indicate the expected shut-in pressures, based on results from the DS hydrofracturing experiments, which indicated shut-in pressures increasing with depth from about 3MPa to 6MPa. Similar results from the IS Niche and GP-A/GS experiments have been added and the figure shows that for the depth range of the CS-B experiment (10 to 35m below the gallery), shut-in pressures are expected to be between 4MPa and 5MPa. This is also the pressure that will be required to keep fractures open for water flow during the sealant placement.

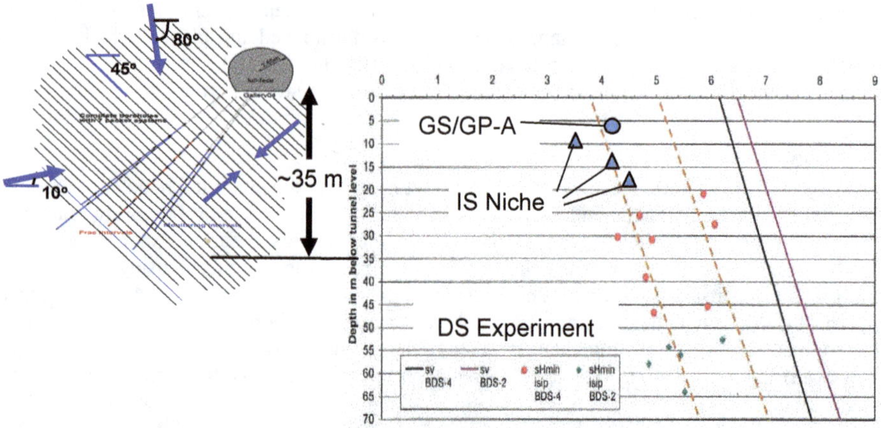

Figure 4. Summary of shut-in pressures from previous MTURL experiments.

In Figure 5, the observed pressures in two of the GP-A/GS boreholes at the end of each injection cycle are matched with the logarithmic pressure decline model according to Geertsma and de Klerk [9]. The results indicate that the average fracture aperture ranges between 16 and 24 microns, whereas fracture apertures near the injection well could be between 24 and 45 microns. In Enachescu *et al.* [1], independent estimates of transmissivity were made using pressure recovery data after each of the breakdown and reopening cycles. These estimates ranged from 1e-8 to 2e-8 m^2/s which, according to the cubic aperture-flow relation, would correspond to hydraulic conducting apertures ranging from 23 to 29 microns, which is in good agreement with Figure 5. Hence, for the CS-B experiment, hydraulic fracture apertures of 20 to 30 microns may be expected for a fracture that is kept open with a pressure above the shut-in pressure.

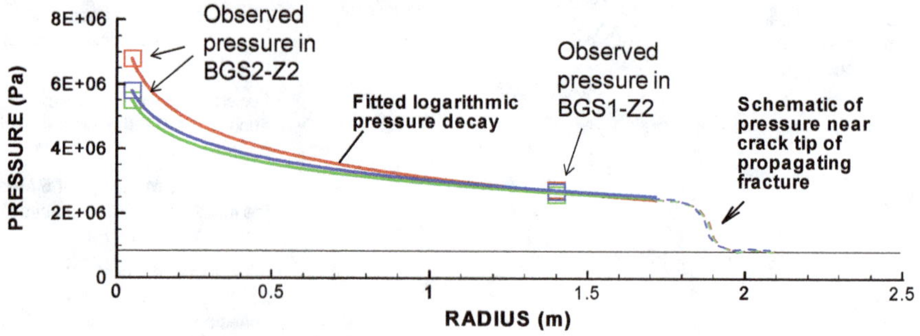

Back-calculated (approximate) hydraulic conducting aperture to produce this pressure drop:

Breakdown (red): 30 μm (average), 45 μm at well, well pressure 6.8 MPa

Reopening 1 (green): 16 μm (average), 24 μm at well, well pressure 5.5 MPa

Reopening 2 (blue): 23 μm (average), 35 μm at well, well pressure 5.8 MPa

Figure 5. Pressure decay model fitted to observed data in two of the GP-A/GS boreholes; BGS2-Z2 (injection) and BGS1-Z2 (monitoring).

It is known from the GP-A/GS experiment that it took 5 litres of injection to create a fracture of about 2m radius. To create a fracture of 3m radius, an injection volume of about 15 litres would be required and 4m radius would require more than 30 litres. The relationship was developed based on data up to 2m and the extrapolations to larger volumes are therefore uncertain. Nevertheless, it shows that substantial injection volumes might be required to create fractures of larger dimensions.

Fracture uncertainties

In terms of fracture geometry and its impact on the experiment, there are two potentially significant risks - the initiation of axial, rather than radial, fractures in the borehole; and the possibility of offset fracture propagation across bedding planes.

The pressure required to create an axial fracture depends on the tangential stress around the hole and the material properties of the rock. At the MTURL it is evident that very much higher pressures are required, and the creation of an axial fracture in the CS-B experiment is unlikely under normal circumstances. As a precaution, low pressurisation rates are recommended for the CS-B experiment.

Fracture propagation offset from the main fracture path was indicated by deformation measurements in the GP-A/GS experiments. Daneshy *et al.* [2] attributed the offset fracture growth pattern to branches consisting of tensile and shear fractures and entrapment of fluids inside the fracture. There might be no way to control or prevent it with injection parameters but it should be detectable by detailed monitoring of fluid pressure and displacement responses in the passive boreholes.

Inputs to the experiment design

The hydraulic fracturing design study has used data and information from previous relevant experiments at MTURL, together with new interpretation and applied modelling to predict and define important fracturing design parameters. These are summarised below:

- Fractures can be created and extended by Pearson Water injection using standard HFSM procedures involving several reopening and shut-in cycles.
- Diagnostics of the pressure and flow responses will help to estimate the fracture extension and to indicate whether the fracture extends along the bedding as intended.
- Fracture radius and volume can be controlled by injection volume and pressure.
- The created fracture volume and radius can be evaluated using deformation (extensometer) measurements and pressure monitoring in the passive boreholes.
- It is highly likely that fractures extending 2 to 3m radius can be successfully designed and created by injecting up to approximately 10 litres during repeated reopening and shut-in cycles.
- The pressurisation rate in the breakdown cycle should be approximately 0.1MPa/s
- The predicted breakdown pressure is between 5.5MPa and 9MPa. A much higher pressure could indicate axial fracturing.
- The predicted shut-in pressure is between 4 Mpa and 5.5 Mpa. A much lower pressure could indicate axial fracturing.
- It is expected that the fracture will propagate to about 1m radius during breakdown, and can then be propagated further during subsequent reopening cycles. The injection volume required to propagate the fracture to the desired radius can be estimated but with considerable uncertainty as it is based on only one previous MTURL experiment.
- The reopening and shut-in cycles can be repeated a number of times, perhaps with increasing injection volume, to ensure that the fracture propagates beyond the passive boreholes. For repeated reopening cycles, water pressure will be trapped inside the

fracture after each shut-in and venting, and therefore, a smaller injection volume will be required to propagate the fracture to a certain distance.

- The evolution of shut-in pressure for the repeated reopening and shut-in cycles should be monitored. Ideally, the shut-in pressure should be stable at a value around 4 to 5.5MPa.
- The fractures will need to be held open hydraulically during circulation. That is, the pressure has to be higher than the reopening pressure.
- The injection of sealant into fractures will only be possible at an injection pressure above the reopening pressure and could perhaps be executed like a regular reopening/shut-in cycle, or using a slow pumping injection rate.
- Based on observations at the GP-A/GS experiments, after a fracture has been created and repeatedly reopened, the hydraulic aperture is pressure dependent and will be about 20 to 30 microns for a fluid pressure above the fracture shut-in pressure, at which pressure the estimated transmissivity will be approximately 2e-8 m^2/s.

CIRCULATION AND SEALING MODELLING

Circulation and sealing modeling was carried out in conjunction with the fracturing study, in order to refine the design for post-fracture circulation and subsequent sealant placement, and to develop a suitable methodology for evaluating sealant performance by pre/post-sealant characterisation using hydraulic testing.

The design modeling of sealant emplacement and pre/post-sealant characterisation incorporated the relevant information from the fracture modelling in terms fracture geometry (extent, volume, properties), and potential range in fracture opening pressure and fracture closure pressure. Together with the equipment design and the test location characteristics, a numerical model was developed to simulate the effect of fracture development in terms of the fracture characteristics (aperture, extent, hydraulic property) and subsequent circulation and sealing modeling, and post-sealing characterisation.

The TOUGH2 code [10] was used, together with the special equation of state module EOS11 [11] that was originally developed to simulate the injection of a liquid with increasing viscosity that undergoes a gelling process and ultimately solidifies. Specific information on the candidate sealants, in terms of the viscosity and density, was included. The actual solidification process was reported as occurring relatively quickly, and was implemented in the model to occur instantaneously after a certain time. The resulting decrease in porosity and permeability of the fracture was then used for the post-sealing testing.

Model geometry and inputs

A three-dimensional numerical model was constructed for a single test interval. The model consisted of a multi-layer rectangular mesh which covered the injection interval length and surrounding areas with the center layer representing a potential fracture or fracture zone, implemented by the equivalent fracture porosity and permeability (Figure 6). The fracture plane was represented by an octagonal region within the central plane of about 2.6m radius.

The layers are inclined parallel to bedding to account for potential gravity effects, which are represented in the model by the corresponding angle of the flow connection parallel and perpendicular to bedding relative to the gravity vector. Figure 6 also shows the location of the boreholes with the active borehole in the center and the passive boreholes at 2m distance.

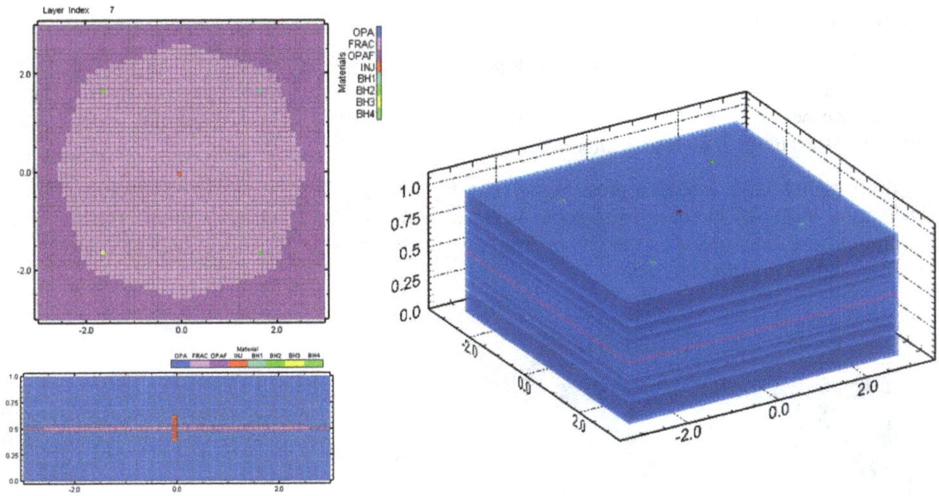

Figure 6. Geometry of 3D model in plan view through the fracture layer (upper left) and along a vertical section through the injection interval (lower left) and the 3D dimensions (right).

The physical and hydraulic characteristics of the fractures have been described above. Within the envisioned depth interval of between 10m and 35m, the fracture closure and reopening pressure is expected to be between 4MPa and 5.5MPa.

The fracture apertures of 20 to 30 microns, derived from the FIM measurements, correspond to a fracture transmissivity of between 1.E-8 and 2.E-8m2/s, which compares well with the estimated transmissivity derived from previous post-fracturing hydraulic tests. In the model, the fracture is represented by a fracture layer with a thickness of b_e = 0.02 m.

Simulations

The modeling involved several simulation steps using the results from the preceding simulation as initial conditions for the subsequent simulation. The initial conditions correspond to the approximate hydrostatic pressures at the MTURL.

1. Fracture initiation: This is implemented by a pulse injection (PI) at a fracture pressure of 9MPa, simulating the pressure decline in the created fracture, based on the measured fracture transmissivity and porosity.

2. Pre-sealant circulation: Constant-rate (RI) water injection, with prescribed pressures above the fracture-closure pressure of 4.2MPa in the observation intervals.

3. Sealant injection: Injection of sealants at the same rate as the preceding water injection, with solidification occurring after 1 day (i.e., after sealant outflow was observed in the observation intervals.

4. Post-sealant testing: Hydraulic testing phases to characterise the reduced hydraulic properties, which may include Constant-head injection or withdrawal tests, constant rate injection or multi-rate tests

Fracture initiation

The fracture response was implemented as a pulse injection (PI) at a fracture initiation pressure of 9MPa and the simulated results are shown in Figure 7, with the 3 micron case above and the 30 micron case below. The blue line shows pressure in the fracture interval while the other colours show responses in the monitoring intervals.

For the 3 micron case, injection pressure declines to about 5.5MPa after 8 hrs and pressures in the observation locations increase to a stable value of about 4.2MPa, which is above the estimated fracture-closure pressure. For the 30 micron case, there is little pressure gradient between the injection and observation intervals.

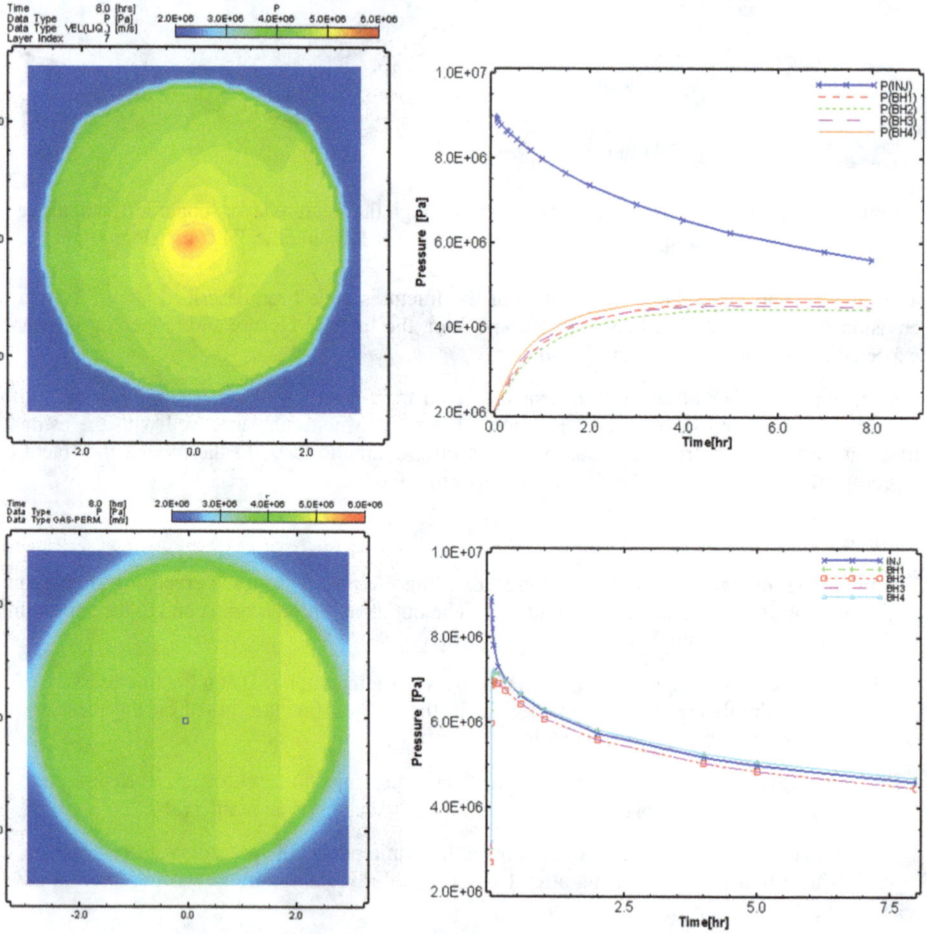

Figure 7. Simulated pressure distribution after 8 hours in the fracture plane (left) and pressure response over time in the test and observation intervals (right).

Pre-sealant circulation

Water was injected into the test interval at a constant rate, whereas pressures in the monitoring boreholes were kept above the fracture closure pressure of about 4.2MPa. To maintain the injection pressure, the water injection rate was prescribed to 2ml/min and 20ml/min for the low (3 micron) and high (30 micron) aperture cases, respectively. The results (Figure 8) show a slight pressure decline in the injection interval which stabilised after about 10 hours for the low aperture case, whereas the high aperture case shows little gradient. The flow vectors show a steady-state flow field from the injection borehole in the center to the observation intervals.

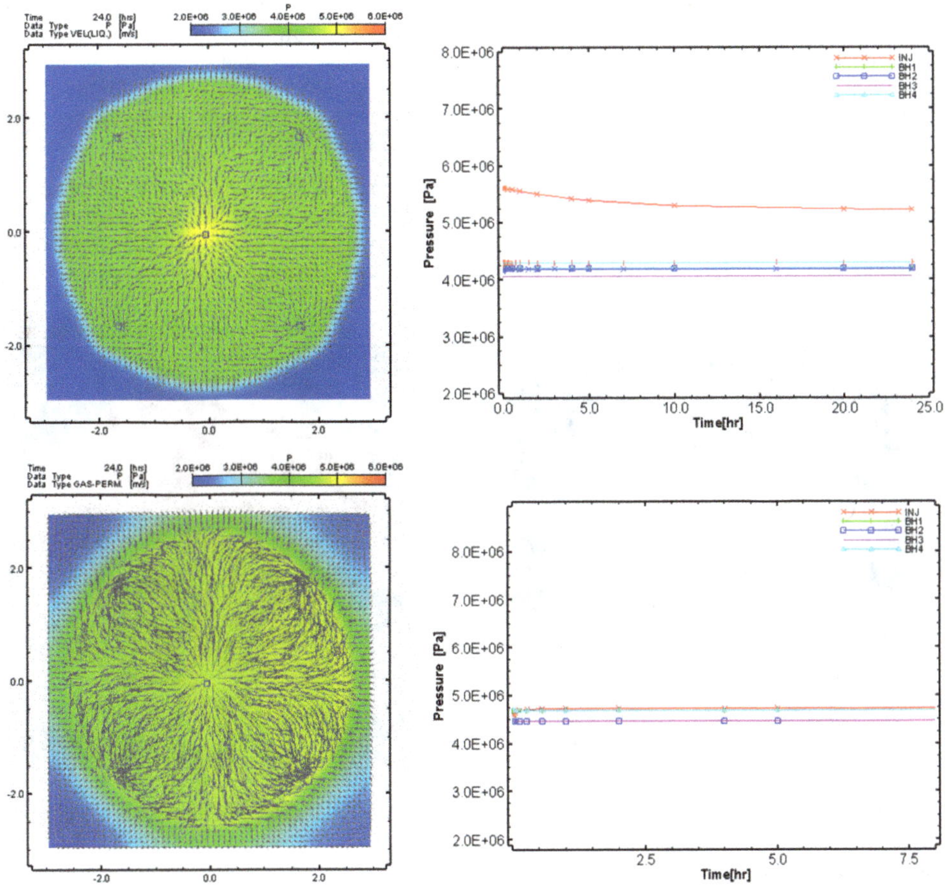

Figure 8. Simulated pressure distribution and flow velocities in the fracture plane after 24 hrs (left) and pressure response through time in the test and observation intervals (right).

Sealant placement

A separate fluid with the properties of the sealant (density and viscosity) was injected at approximately the same mass flow rate as the preceding water injection. Because of the slight difference in fluid density of the sealants, the corresponding volumetric injection rates differ somewhat. For the H2Zero sealant a density of 1006kg/m3 (8.4 lb/gal) was given and a viscosity of 25cP. The H2Zero sealant is mixed with water which reduces the viscosity according to the Einstein equation; for the simulation a relative sealant concentration of 0.3 was assumed resulting in a viscosity of the mixture of about 3cP.

For the different bio-sealant fluids, fluid densities of 1013kg/m3 and 1021kg/m3 were reported; the corresponding viscosities ranged between 1.01 and 1.05cP. For the simulations, the greater values were assumed for the final fluid injection.

In simulations of the low aperture case the H2Zero sealant was injected at a rate of 2ml/min and it was assumed that the sealant concentration reached 99% in the injection interval prior to the start of injection. The concentration of the sealant at different times is shown in Figure 9, indicating that after about 4 hrs, the fracture was fully filled with sealant. The simulated outflow at the passive boreholes indicates relatively early breakthrough of the sealant even when accounting for the effects of potential heterogeneity in the fracture. For the high aperture case the sealant was injected at a rate of 20 ml/min and the sealant filled the entire fracture after 1 hour.

Simulations of the bio-sealant injection gave very similar results apart from a small difference in injection flowrate, where there was a small peak at early time before reaching a constant rate that corresponds to the prescribed injection rate into the injection interval. The porosity and permeability distributions after solidification did not change noticeably from that of the H2Zero sealant.

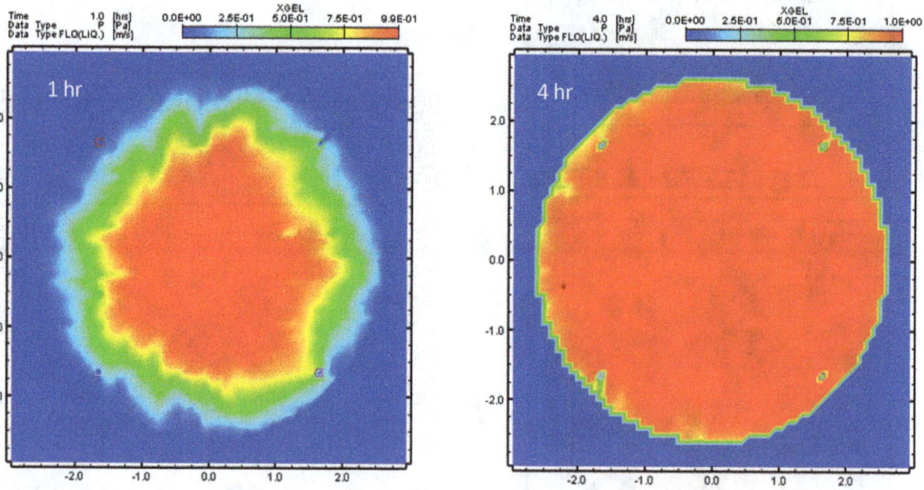

Figure 9. Simulated sealant concentration in the fracture after 1hr (left) and after 4 hrs (right).

Inputs to the experiment design

The circulation and sealing design study has used data and information from previous relevant experiments at MTURL, together with new interpretation and applied modelling to predict and define important design parameters. These are summarised below:

- For the potential range in mean fracture apertures between 3 and 30 microns, and assumed heterogeneity within the fracture, the resulting permeabilities and corresponding flow velocities differ by about two orders of magnitude (i.e. peak velocities of 3.E-7 m/s to 3.E-5 m/s). Within the fracture, the flow velocities vary by slightly more than two orders of magnitude based on the assumed heterogeneity.

- For the low aperture case (3 micron), an injection rate of 2ml/min was used, which produced a pressure difference of about 1MPa between the injection interval and the observation intervals, which were kept at a pressure of about 4.2MPa (above the inferred fracture closure pressure).
- For the high aperture case (30 micron), an injection rate of 20ml/min was used, which produced very little gradient between the injection interval and observation intervals, but maintained flow from the injection interval to the observation intervals.
- Post-sealant test simulations indicate potential interference due to transient conditions associated with the pressures during the sealant injection exceeding the fracture closure pressure, which is higher than the hydrostatic pressures at the MTURL.
- For the low-aperture case, the computed permeabilities are at or below the matrix permeability of 3.5E-20 m^2 and the hydraulic test response is dominated by the intact Opalinus clay.
- For the high-aperture case, the resulting permeabilities in the fracture after solidification of the sealant are somewhat higher than the matrix permeability and the hydraulic test response may differ from that for the intact clay.

GEOPHYSICAL SURVEILLANCE

A study was carried out into the feasibility and potential value of using active and passive geophysical monitoring techniques to map the fractures created in the experiment.

The objective of crosshole tomography would be to carry out active imaging between holes on either side of the main experiment holes so that full waveform imaging of the fracture zones could be performed.

The objective of the microseismic monitoring would be to detect and locate fracture-induced events so that the fractures can be 'mapped'.

The study included a review of previous relevant work at MTURL, the logistics of instrumentation locations, predictive modelling of results, borehole requirements and specification of tools.

Previous crosshole tomography work at MTURL

Reports of previous crosshole observations indicate that significant anisotropy can be expected between P wave velocities perpendicular and parallel to bedding.

A lack of trace repeatability between different hydrophones and even for the same hydrophone after being moved and replaced at the same location has also been a common problem, probably due to a combination of the asymmetry in the construction of the hydrophone strings and variations in coupling. A solution has been to use clamped accelerometer tools and the effects may be minimised by using axially symmetric strings and baffles between the elements to centralise the string, which has the additional benefit of attenuating noise in the fluid column.

Two sources have been used successfully, a borehole sparker and clamped piezo source. The sparker would be quicker and is likely to be more powerful but must be operated in fluid.

Vertically polarised shear wave sources are available for shallow holes but no S wave measurements were identified in previous work at MTURL. In addition, S wave tomography would only be practical if a string of clamped receivers was available.

The Geometrics Geode trace data acquisition system has been used successfully for a number of the surveys at MTURL and it is suitable for both crosshole and microseismic monitoring at the expected frequencies. The system is readily available and very well suited to this application so there is no need to consider any alternatives.

Modelling studies indicated that the resolution of Full Waveform P wave crosshole tomography could be of the order of the wavelength, but this is highly dependent on being able to acquire trace data that are repeatable with a correlation coefficient of >0.95. The usual resolution relation used for ray trace tomography is $\sqrt{(\lambda l)}$ where λ is wavelength and l is the survey dimension. Taking the P wave velocity parallel to the bedding, 3.11 km/s, the centre frequency of the sparker spectrum of ~2 kHz and setting l equal to a borehole spacing of 8m, this indicator of resolution is 3.5m. This is not a measure of the resolution, which may be closer to the wavelength of 1.5m.

Crosshole survey geometries and modelling

A system comprising two crosshole tomography image planes has been considered (Figure 10). Both planes are parallel to the main experiment boreholes, perpendicular to the bedding and pass through the active borehole. One plane is vertical and the other follows the dip of the boreholes. Two pairs of source and receiver boreholes will be positioned 4m either side of the active borehole, and 4m above and below it, so that the expected fracture length of a few metres will be largely contained within the imaging region.

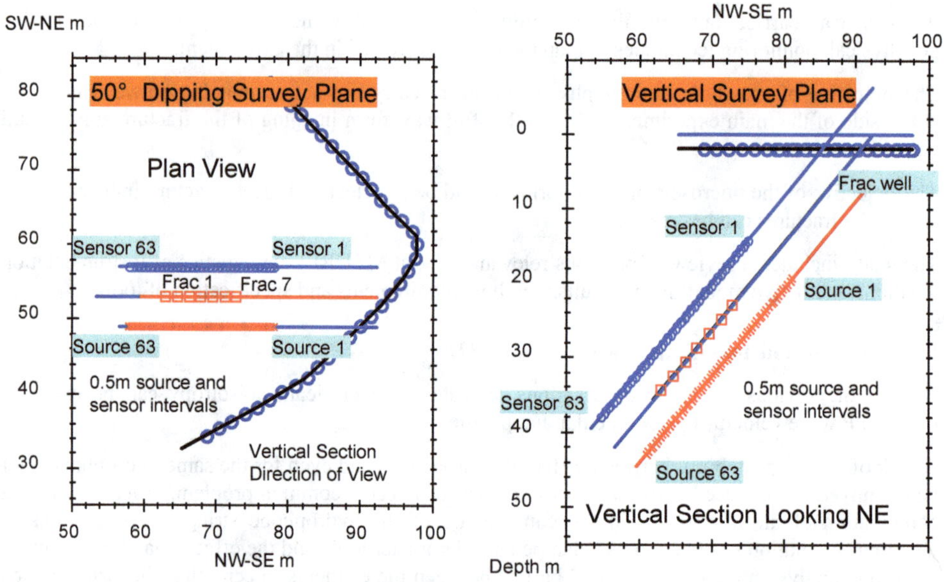

Figure 10. Perpendicular crosshole geometries shown in plan view on the left in the vertical plane on the right.

Pre-fracture crosshole surveys between both pairs of boreholes will be performed over the maximum depth extent of the holes in order to measure the vertical transverse isotropy (VTI) anisotropy and check for any azimuthal velocity variations. The source and receiver boreholes will extend 10m beyond the deepest fracture so that edge effects in the tomogram are minimised in the region of interest. The surveys will also extend 8m above the shallowest fracture.

In the idealised arrangement depicted in Figure 10 there will be 63 source and sensor positions at 0.5m intervals. Ideally, repeat crosshole surveys will be carried out following each fracturing operation to monitor changes in velocity and reflectivity with respect to the pre-fracture survey. For efficiency the depth range of these surveys may be shortened to 24m so that only two movements of the hydrophone string and two runs of the source for each survey are required to obtain a source and receiver interval of 0.5m. The strings will be left in place between successive fractures to minimise coupling variations between the last acquisition run of one survey and the first acquisition run of the next survey.

The width of the fractures is expected to be very small relative to the seismic wavelength and so even if there are a number of fractures created they are unlikely to be detectable with travel time tomography. However, the modelling work of Maurer [12] suggested that reflections may be detected from the fractures at MTURL and therefore crosshole reflection imaging offers the best prospect for imaging the fractures.

To demonstrate this technique, model P wave reflection data has been imaged using a diffraction stack method (Figure 11) which shows that it is suitable for imaging reflectors between the boreholes. The model ray paths illustrate that reflections from outside the crosshole region are not detectable by the receiver array for low dip reflectors. Only reflectors with high dip between the boreholes will be detectable outside the crosshole region.

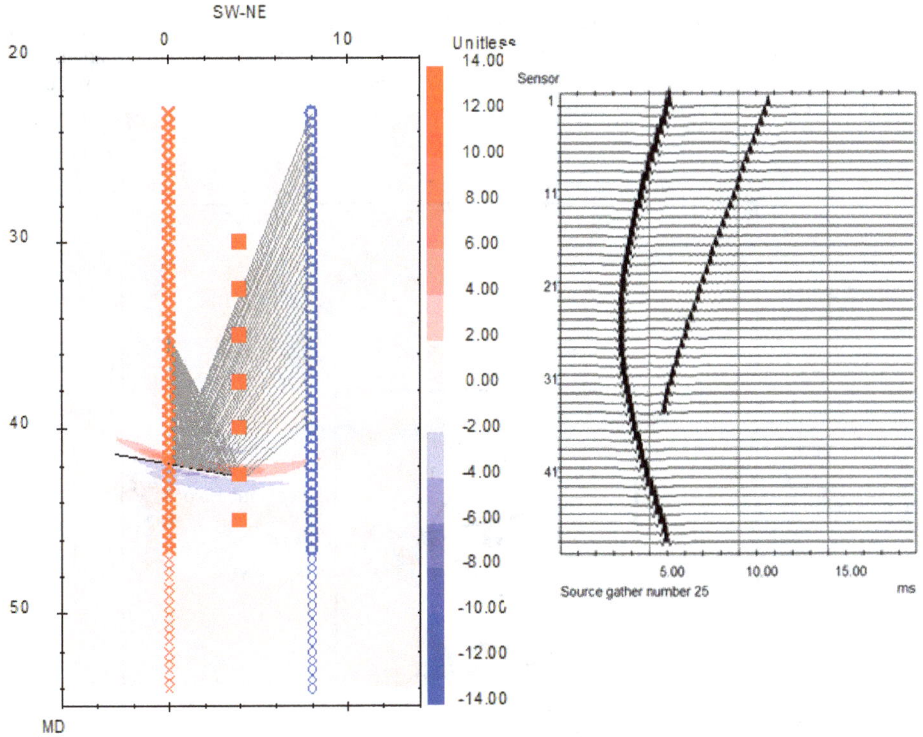

Figure 11. Fracture model and crosshole migration image created using no dip control and no aperture control. Model trace data for a band pass of 100-3500Hz are shown on the right for the reflection ray paths overlain on the migration image.

Previous microseismic monitoring work at MTURL

In the most recent report from MTURL, more than 2,500 microseismic events were reported. The dominant frequency of the first arrivals was around 3.7kHz and the following phase had a dominant frequency of around 2kHz. A further possible microseismic event in a previous experiment had a dominant frequency of 4.5kHz. In addition, a large number of acoustic emission events have been detected that have dominant frequencies in the range 1-20kHz. These high frequency events were detected on sensors within a few metres of the excavation. By analogy with the penetration and bandwidth of the sparker source it is likely that microseismic events at the lower end of the frequency range will be detectable over a few tens of metres.

There are two candidate locations for deployment of microseismic monitoring arrays; the existing tunnels and dedicated boreholes. The sensor boreholes and hydrophone string proposed for the crosshole imaging could also be used to provide high frequency microseimic monitoring. These sensors would be within a few metres of the fracture events and have bandwidths of up to 10kHz.

It has been reported that sensors coupled to the tunnel walls perform poorly, probably due to a combination of the concrete tunnel lining and the excavation damage zone where the seismic velocity is reduced and coupling may also be affected. The depth of the excavation damage zone has been investigated and reported in a number of documents and is generally considered to be in the range 1 to 2.2m.

To detect events of up to a few kHz, an array of sensors along Gallery 04 would be used. The sensors would be deployed at 2.8m in shallow holes beneath the excavation damage zone. Three further geophone stations could be considered at locations elsewhere in MTURL to provide 360° azimuthal coverage of the fracture zones (Figure 12).

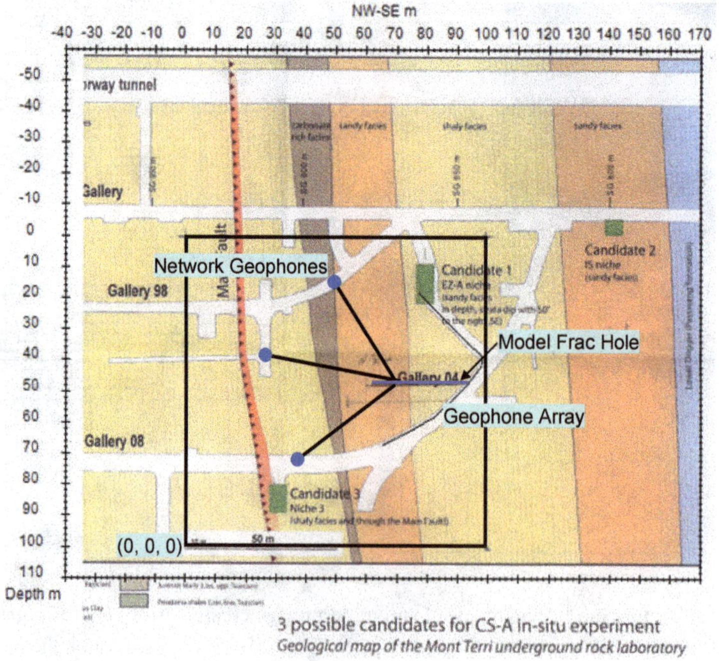

Figure 12. Possible microseismic array locations.

Microseismic resolution

The potential resolution of the two proposed arrays, along the gallery and downhole, has been estimated using typical P and S wave picking accuracies. From this analysis it was also found that the events may be located using P and S travel time picks alone. Hodogram directions of the events could be derived from 3 component sensors but in this case hodograms are not necessary to locate the events with reasonable accuracy.

Estimates of the P and S picking accuracies have been used to determine the corresponding relative hypocentre location accuracy. Constant P and S velocities of Vp = 2.34 km/s and Vs = 1.11 km/s [13] have been used in the modelling and the analysis has been performed on a 1m cubic grid. Sections through the resolution grid have been calculated at 28m depth, a plan view, and at a Y offset of 54m, a NW-SE vertical section.

The modelled tunnel array consisted of 36 sensors deployed in 2m deep boreholes (Figure 13). The analysis was performed for P and S times, P times only and S times only. For the P and S times, picking accuracies of ±0.5 and ±1.0ms for P and S were assumed. To achieve similar resolution using P times only it was found that a picking accuracy of ±0.1ms was required and for S times only a picking accuracy of ±0.2ms was required. The accuracies used in the P and S only analyses are thought to be around the limit of what might be possible. The tunnel array provides slowly varying resolution in the vertical and horizontal directions of 2-4m over a radius of 10m or more about the active borehole.

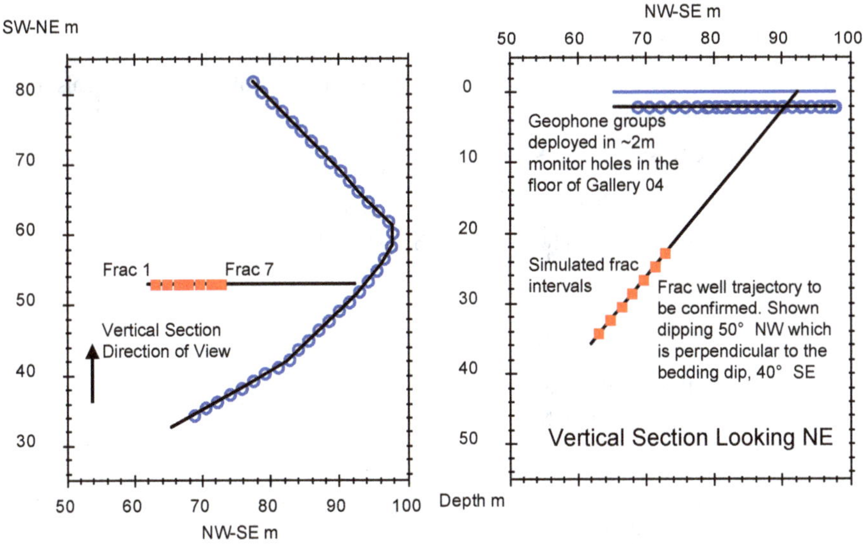

Figure 13. Geometry of the tunnel array and possible fracture depths shown on the two sections used for resolution analysis.

The bandwidth of the borehole arrays would be up to 10 kHz which would allow very small high frequency events to be detected providing the borehole noise is low enough. The analysis was performed using P and S picking accuracies of ±0.25 and ±0.5 ms, which is reasonable for high frequency events of a few kHz. As for the gallery array model, analyses were performed to determine the picking accuracies required to obtain similar resolution using P or S picks only. In this case P accuracies of ±0.1 ms and S accuracies of ±0.2ms would be required.

The resolution of the borehole arrays is more variable than for the tunnel array. It is 2-3m in the vertical and horizontal planes close to the active borehole but in some regions at distances of 5m or more it declines to 5m, and it declines further at greater distances, particularly in the horizontal plane.

Hydrophones are omni-directional so there is no possibility of using hodogram data using the borehole arrays. The event locations in the plane perpendicular to the boreholes are derived from the relative arrival times at each borehole.

Deployment of the gallery and downhole arrays together provides the opportunity for both crosshole imaging and microseismic monitoring. For both arrays some downhole shooting is required from the region of the active hole in order to calibrate the P wave velocity for microseismic processing. In principle, the downhole arrays should be able to detect very small, high frequency events as they are closest to the fracture zones, but they may suffer from noise due to the injection. The gallery array has a lower frequency response but the holes should be relatively quiet. It has potential for better resolution of S waves and for further processing such as FPS and reflection imaging.

Pre-deployment field trials

In principle the feasibility of the geophysical techniques has been established. However, field trials are required to test some aspects relating to the application of the equipment in the specific environment at MTURL. The trials are described in the following chapter.

ACKNOWLEDGEMENTS

The material presented here was carried out and overseen by the following individuals and their work is gratefully acknowledged:

Ben Dyer, Semore Seismic Ltd, UK

Bill Lanyon, Fracture Systems Ltd, UK

Jonny Rutqvist, Lawrence Berkely National Laboratory, USA

Rainer Senger, Intera Inc, Switzerland

REFERENCES

1. Enachescu C., Blümling P., Castelao A. and Steffen P. (2002) Mont Terri GP-A and GS Experiments Synthesis Report. NAGRA NIB 02-51.
2. Daneshy A., Blümling P., Marschall P. and Zuidema P. (2004) Interpretation of field experiments and observation of fracturing process. Paper presented at the SPE International Symposium and Exhibition on Formation Damage Control, Lafayette, Louisiana, U.S.A., 18–20 February 2004. Society of Petroleum Engineers, SPE 86486.
3. Evans K., Piedevache M. and Portmann F. (1999) IS Experiment: Hydrofracture Stress Tests in Boreholes BIS-C1 and BIS-C2. Mont Terri Project, Technical Note, TN99-55.
4. Enachescu C., TECHNICAL NOTE 2010-53 (April 2011) DS (Determination of stress) Experiment: Hydraulic Fracturing Tests in BDS-2 and BDS-4 at the Mt. Terri Underground Research Facility. FMT DS Experiment - Phase 15.
5. Martin C.D. and Lanyon G.W. (2003) Measurement of in-situ stress in weak rocks at Mont Terri Rock Laboratory, Switzerland. *International Journal of Rock Mechanics & Mining Sciences*, 40, 1077–1088Ref 1

6. Bossart P. and Wermeille S. (2003). The stress field in the Mont Terri region data compilation. In: Heitzmann P., Tripet J-P.(editors), Reports of the Federal Office for Water and Geology, *Geology Series*, 2003. p.65–92.
7. Corkum A.G. and Martin C.D. (2007) Modelling a mine-by test at the Mont Terri rock laboratory, Switzerland; *International Journal of Rock Mechanics & Mining Sciences*, 44, 846–859.
8. Yong S., Kaiser P.K. and Loew S., (2010) Influence of tectonic shears on tunnel-induced fracturing. International Journal of Rock Mechanics & Mining Sciences, 47, 894–907.
9. Geertsma J. and de Klerk F. (1969). A rapid method of predicting width and extent of hydraulically induced fractures. *Journal of Petroleum Technology*, (December 1969), 1571-1581.
10. Pruess, K., Oldenburg, C., Moridis, G., 1999, TOUGH2 User's Guide, Version 2.0. Lawrence Berkeley Laboratory Report LBL-43134, Berkeley, CA, 1991a.
11. Finsterle, S., G. J. Moridis, K. Pruess, A TOUGH2 Equation-of-State Module for the Simulation of Two-Phase Flow of Air, Water, and an Immiscible Gelling Liquid, Lawrence Berkeley Laboratory Report, LBL36086, Berkeley, CA, 1994.
12. Maurer H., 2013a. HG-D (Reactive gas transfer) experiment: Active and passive seismic monitoring of gas injection experiments at the HG-D test site (Mont Terri) Phase 17. TN 2012-75
13. Marelli S., Manukyan E., Maurer H., Greenhalgh S.A. and Green A.G., 2010. Appraisal of waveform repeatability for crosshole and hole-to-tunnel seismic monitoring of radioactive waste repositories. *Geophysics,* Vol 75, No. 5 pp Q21-Q34.

Carbon Dioxide Capture for Storage in Deep Geological Formations, Volume 4
Karl F. Gerdes (Editor)

Chapter 47

FRACTURE SEALANTS PROGRAMME: DESIGN

Peter Ledingham
GeoScience Limited, Falmouth Business Park, Falmouth, Cornwall TR11 4SZ, UK

ABSTRACT: In order to assess the concept of *in situ* sealing of a leak from a CO_2 storage formation, a field test of three candidate fracture sealants is planned, to be undertaken in the Mont Terri Underground Rock Laboratory in Switzerland. The concept and design has been developed over a period of 18 months between December 2012 and June 2014. Designs for the experiment location, boreholes, packer systems, surface systems, data acquisition, sealant injection, and geophysical surveillance have been developed and methodologies have been prepared.

KEYWORDS: fractures; fracture modelling; hydrofracturing; seismic monitoring; tomography; packer systems; multi-packers; fracture sealants

INTRODUCTION

A field test of three candidate fracture sealants is planned, to be undertaken in the Mont Terri Underground Rock Laboratory (MTURL) in Switzerland. The candidate sealants are an experimental bio-sealant, an experimental smart gel and a proprietary sealant, H2Zero, supplied by Haliburton. Feasibility studies for the proposed field test are reported in Chapter 45 of this volume. The concept and design of the experiment (designated as CS-B) was developed over a period of 18 months between December 2012 and June 2014 by a number of design teams, each working on one aspect of the experiment:

- SolExperts AG were responsible for designs of all the surface (i.e in the gallery) and downhole test equipment
- Swisstopo manage the MTURL and are responsible for the experiment location, site facilities and all borehole drilling
- Semore Seismics Limited were responsible for the design of two potential geophysical surveillance techniques for imaging fractures
- Montana State University undertook laboratory testing of the proposed bio-sealants
- Los Alamos National Laboratory undertook laboratory testing of the proposed smart gels
- GeoScience Limited was the design coordinator, responsible for integration of the design elements and production of the deliverables.

EXPERIMENT SUMMARY

The objective of the experiment is to create artificial fractures between boreholes in a highly controlled environment, establish circulation and then test the effectiveness of three candidate sealants in plugging them. Two fractures will be available for each candidate sealant.

The experiment will be carried out in a number of phases over a period of approximately one year:

- Equipment trials
- Drilling and equipment installation
- Stabilisation
- Fracturing
- Circulation
- Sealant placement
- Post-sealant characterisation

The experiment will be carried out in boreholes drilled from Gallery 04 in the MTURL, as described in the previous chapter and as shown in Figure 1.

Figure 1. Experiment location, looking in the approximate direction of the boreholes.

EXPERIMENT BOREHOLES

Several types of borehole are required:

- The main experiment boreholes
- Extensometer boreholes
- Sealant assessment boreholes
- Crosshole tomography boreholes
- Microseismic array boreholes

Main experiment boreholes

The experiment will be undertaken in five main boreholes; a central active hole and four passive holes at distances of between 1 and 3m away. All the holes will be inclined downwards at approximately 45°, and their azimuth will be 320° to 330° so that they are normal to the bedding direction and dip. Fractures will be created which are intended to be parallel to the bedding.

The bedding is at an oblique angle to the gallery so the borehole lengths will vary between 30m and 40m according to their exact location; generally they will be longer the further east they are. They will be 96mm diameter. Each of them will be completed with 7-zone multi-packer systems which will be installed soon after drilling to minimise the risk of borehole instability interfering with the installation.

Fracturing, circulation and sealant placement will be carried out from the active hole. The passive holes will be used for pressure monitoring, detection of fractures, the return side of circulation and detection of sealant.

Extensometer boreholes

Four boreholes will be drilled to accommodate extensometers which will be used to measure deformation due to the fracturing and the position of the induced fractures. They will be drilled in an L-shape so that two are along the dip direction and two along the strike direction, and at different distances from the central active hole, with two inside the ring of passive holes and two outside. They will be the same length as the main experiment holes, but drilled at 101mm diameter. The extensometers will be grouted into the boreholes.

Sealant assessment boreholes

Hydraulic characterisation will be used to assess the effectiveness of the sealants in terms of reduction in transmissivity. In addition, a direct assessment of the physical characteristics of the sealant will be made by coring through the sealed fractures and recovering samples.

Four 101mm diameter holes will be drilled parallel to the experiment holes at various distances from the active hole in order to investigate sealant at a range of fracture apertures. The locations of the holes will be chosen after the hydraulic results have been interpreted.

If the multi-packer systems have been removed from the experiment holes it should also be possible to overcore one or more of them to investigate sealant characteristics at the injection site, although overcoring to a depth of 30 - 40m has not been attempted at MTURL before.

Crosshole tomography boreholes

An optional addition to the experiment is an attempt to monitor and map the artificial fractures using active and passive geophysical techniques - crosshole seismic tomography and microseismic monitoring.

Four additional holes would be required; two pairs of source and receiver holes in order to allow two orthogonal planes to be imaged; one horizontal and one vertical. The holes will be approximately 40m long and lined with plastic casing, cemented in place, with a minimum ID of 59mm. They will be drilled with approximately the same azimuth and inclination as the main experiment holes.

Microseismic array boreholes

In order to detect and locate microseismic events created by the fracturing process an array of sensors is required along the gallery. However, previous experience has demonstrated that sensors coupled to the tunnel walls perform poorly, due to a combination of the concrete lining and the Excavation Damage Zone (EDZ). Therefore the 36 sensors will be deployed in individual 2.8m deep boreholes drilled from the gallery floor. The holes will be 50mm diameter or larger and be sand-filled to allow coupling between the tool and the rock.

Constraints on borehole configurations

The main constraint on borehole spacing is the headroom available at the test location. The drilling equipment will have to reach high up the wall and then be able to drill the necessary downward trajectory. The presence of the DR niche opposite the test location allows considerable flexibility and it will also be possible to drill into the floor of the gallery, so it will be feasible to achieve an equivalent available height of 3m. There is no lateral constraint, since there is ample room along the gallery wall.

Therefore it will be possible to drill the experiment holes in parallel with each other, rather than requiring a diverging configuration, which might have been necessary in a more confined space.

The precise layout of the holes is not critical to the result of the experiment as long as the passive holes and extensometer holes are at varying distances from the active hole and the extensometer hole arrays are in an orthogonal layout. Three possible configurations are shown in Figure 2.

The crosshole tomography boreholes will be outside the ring of main experiment holes but do not need to be in any particular spatial relationship in terms of layout in the gallery wall. The final configuration will be decided on site, to suit the drilling equipment used, and any other spatial constraints that are apparent.

CANDIDATE SEALANTS

Three candidate sealants will be evaluated in the experiment; a bio-sealant, a smart gel and a proprietary commercial product. In addition, Pearson Water will be tested as a control. This is an artificial formation fluid that has been shown to minimize clay swelling in previous experiments and is used in experiments throughout MTURL. All the candidate sealants are aqueous and therefore the effect of the water on its own also needs to be evaluated.

Bio-sealant

The bio-sealant proposed has been developed by Montana State University and utilises a process called ureolytic biomineralisation (e.g. [1,2]). It relies on the activity of microbes which, because they contain the enzyme urease, are capable of hydrolyzing urea. This results in an increase in the pH, ultimately exceeding the saturation level for the precipitation of calcium carbonate.

The sealing mechanism therefore involves first inoculating the formation with urealytically capable microbes (*Sporosarcina Pasteurii*), adding growth media and urea to enhance cell growth and hydrolyze the urea, and then adding calcium to stimulate precipitation of calcium carbonate. It is the precipitate that results in a permeability reduction.

The densities and viscosities of the components are very similar to those of water. Manipulation of the time required for precipitation is possible by varying the concentrations of the components.

Smart gels

The smart gel proposed has been developed by Los Alamos National Laboratory, which has expertise in synthesizing and characterizing polymer materials, including polymers that respond to environmental and induced triggers. "Smart" gels are a class of gels that swell or collapse upon sensing a specific trigger, which might be pH, temperature, concentration of metal ions or acoustic/electric/magnetic waves.

The smart gel proposed for this experiment is designed to have density and viscosity similar to water in its untriggered state but, under pH control, convert to a solid material in a matter of minutes.

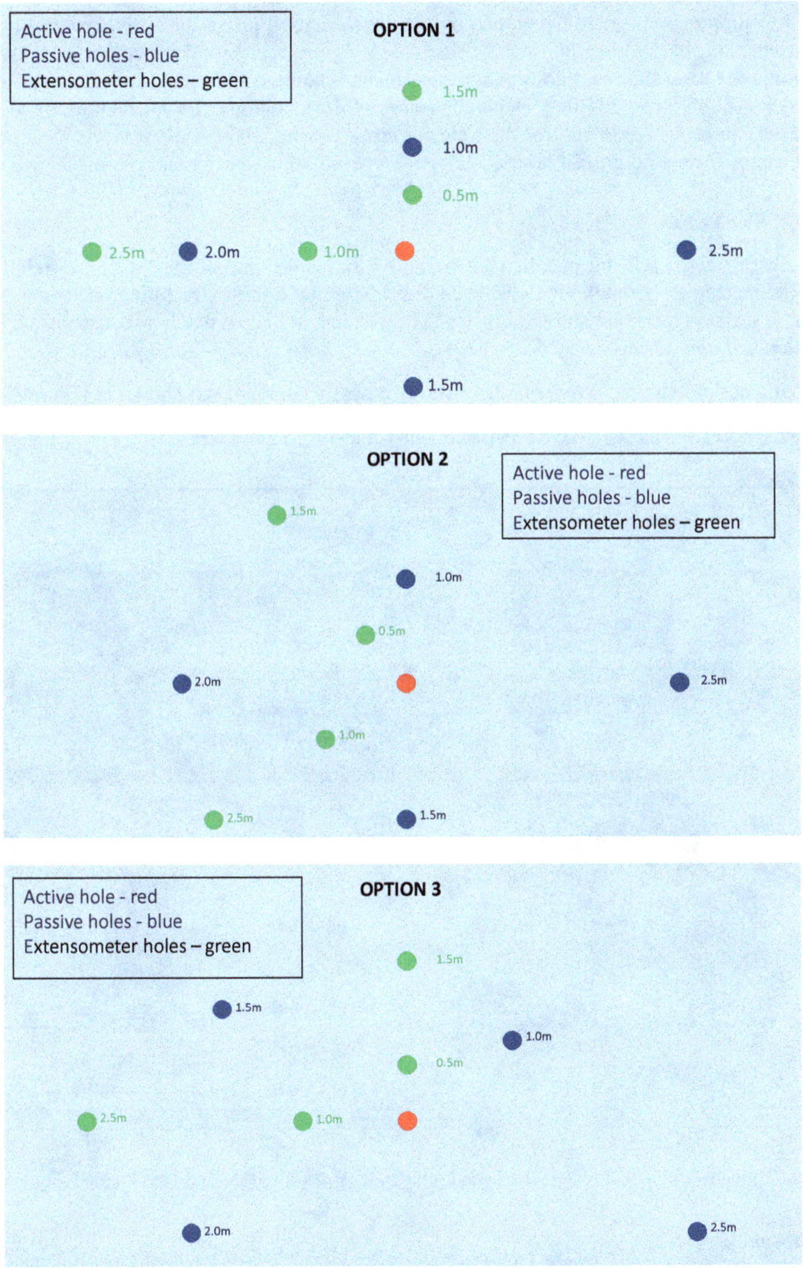

Figure 2. Possible configuration of main experiment boreholes and extensometer boreholes. View looking towards the gallery wall.

Proprietary sealant

H2Zero is a proprietary sealant developed by Halliburton, used primarily for controlling unwanted water production in hydrocarbon wells. It is a polymer-based sealant comprising two fluid components; the base fluid and an organic crosslinker. The base fluid has similar density to water and viscosity in the range 20-30cp. Depending on in-hole conditions, the concentrations of the two components can be varied to control the time for cross linking, and therefore viscosity increase, to occur. It can be from 1 hour to 24 hours.

PACKER SYSTEMS

Multi-packer systems will be installed in the active borehole and all the passive boreholes. The system in the active borehole will be used for hydraulic testing, fracturing, constant pressure circulation, sealant placement and monitoring. The systems in the passive boreholes will be used for observation and monitoring.

A schematic design with the active borehole and two passive boreholes is shown in Figure 3.

Each multi-packer system will have 7 isolated zones and 14 pressure sensors.

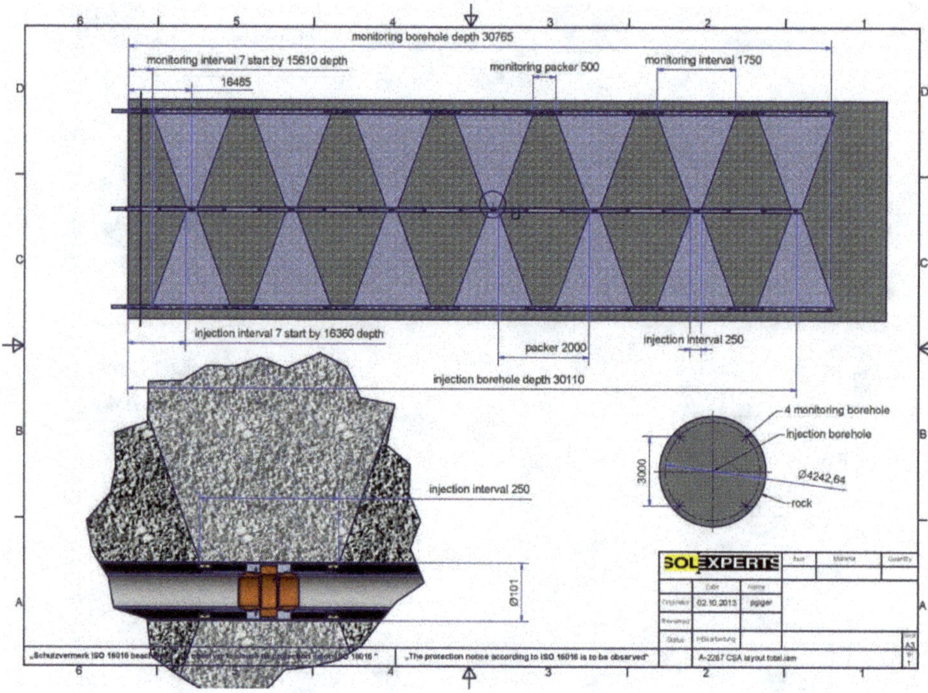

Figure 3. Schematic view of the multi-packer systems for one active and two passive boreholes.

Active borehole system

The active multi-packer system is composed of seven measuring sections, each of 0.25m length, which are isolated by one packer of 2m length (Figure 4).

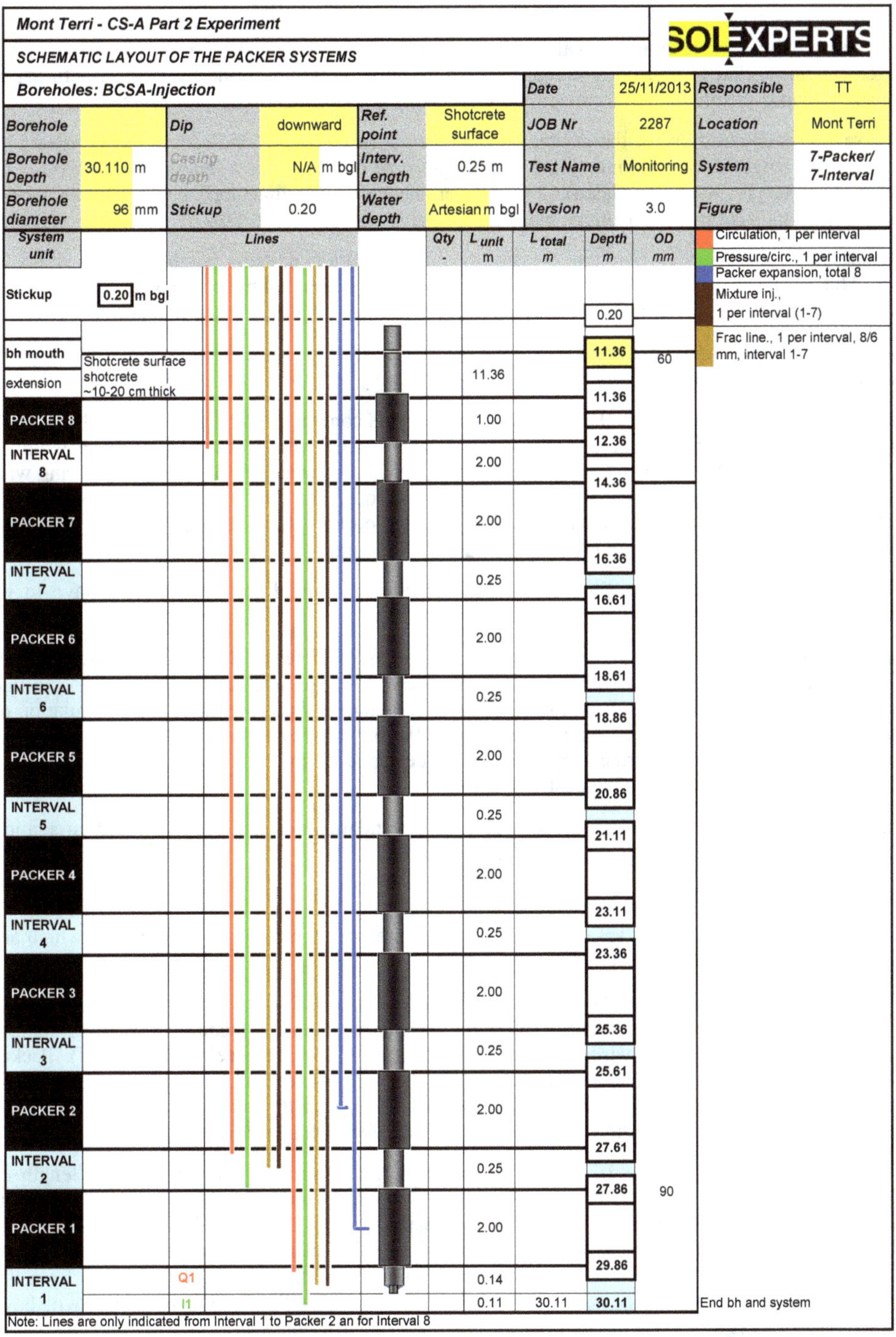

Figure 4. Schematic layout of the active borehole system.

Above packer 7, a further interval of about 2 m length is included, which is isolated by another packer of about 1 m length. Interval 8 will be pressurized to about the formation pressure and, together with packer 8, is needed to reduce the differential pressure during hydraulic fracturing in interval 7 and to ensure the proper functioning of packer 7.

The seven sealing sections of the packer systems have a length of 2m and consist of one hydraulic packer of a length of approximately 2m. The uppermost packer 8 has a length of about 1m. The outer diameter is 90mm (deflated). The packer sleeve is manufactured using natural rubber reinforced with steel wires. The metal parts are of stainless steel (1.4301 DIN).

Packer working pressure (differential pressure) is up to 12MPa in a 96mm borehole with a safety of 1.5. The packers are hydraulically compressed and the system tightness checked. All lines are conducted through the inner mandrel which protects them from the fluid in the borehole.

Eight stainless steel lines are necessary for the packer inflation, one line for each packer (OD 4mm, ID 2.4mm). The packers should be carefully inflated step by step to prevent failure.

The seven testing intervals are 0.25m long. The intervals contain a central stainless steel tube with an outer diameter of 90mm. A sintered stainless steel screen with a coarse filter mesh width is placed along the entire interval length. The filter mesh width is needed for hydraulic fracturing with high flow rates. Furthermore, the sealants should flow through the filter without precipitation in the filter and the risk of clogging. To avoid clogging of line ports by Opalinus clay, water will always be circulated through the interval at flow rates up to 200ml/min.

The appropriate width of the filter mesh and the construction of the interval still have to be evaluated during the pre-experiment bench test.

Each interval is equipped with four stainless steel lines of different diameters for the different phases in the experiment. The pressure line is connected to a pressure sensor. The circulation line serves to saturate the interval, to perform hydraulic tests and to flush the interval by circulation of Pearson Water. One line with an inner diameter of 4.9mm is needed for hydraulic fracturing at high flow rates of 3-5l/min. A further separate line is designed to inject the sealant mixtures in the fractures. Both line ports are in the centre of the interval.

Interval 8 has a length of 2m and is equipped with a standard filter and a pressure and circulation line to pressurize the interval to about formation pressure.

Passive borehole systems

The passive borehole systems are composed of seven measuring sections of 1.75m each which are isolated by one packer of 0.5m length (Figure 5).

The sealing sections of the packer systems have a length of 0.5m and consist of one hydraulic packer. The outer diameter is 90mm (deflated). The packer sleeve is manufactured using natural rubber reinforced with steel wires. The metal parts are of stainless steel (1.4301 DIN). Packer working pressure is up to 100 bar in a 96mm borehole. The packers are hydraulically compressed and the system tightness checked. All lines are conducted through the inner mandrel which protects them from the fluid in the borehole. One line is necessary for each packer for its inflation.

The monitoring intervals are 1.75m long except for the lowermost interval 1 with a length of 1.655m. The intervals contain a central stainless steel tube with an outer diameter of 90mm. A sintered stainless steel screen composed of three parts is placed along the entire interval length to ensure borehole stability and to avoid clogging of line ports. A coarse filter screen will be emplaced in the central part of the interval. This filter screen ensures the flow through of sealants from the fractures without clogging. The side parts of the filter screen with a length of 25mm will have smaller pore spaces to prevent clogging of the line ports by Opalinus clay.

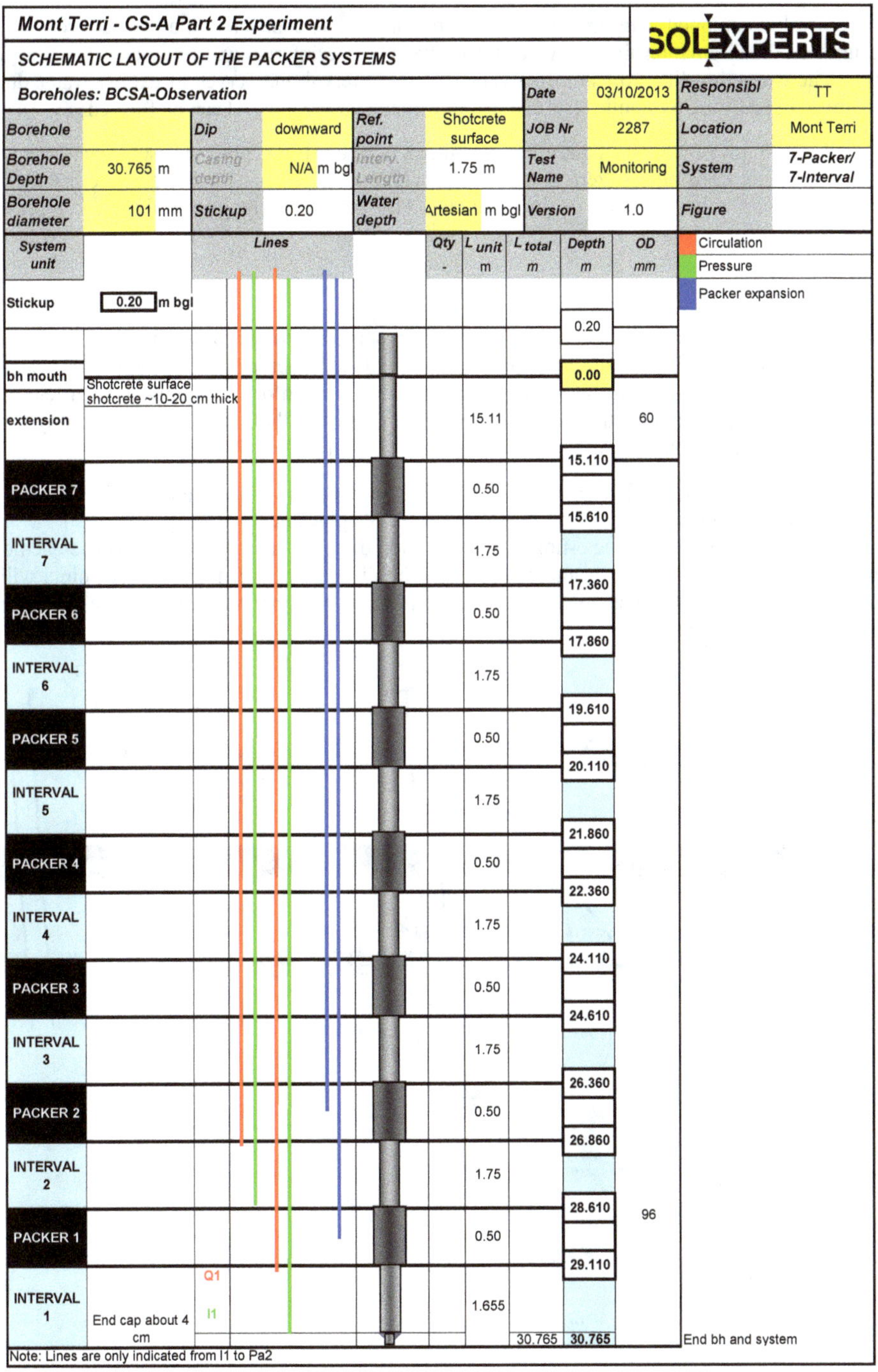

Mont Terri - CS-A Part 2 Experiment
SCHEMATIC LAYOUT OF THE PACKER SYSTEMS
SOLEXPERTS
Boreholes: BCSA-Observation
Date 03/10/2013
Responsible TT
Borehole
Dip downward
Ref. point Shotcrete surface
JOB Nr 2287
Location Mont Terri
Borehole Depth 30.765 m
Casing depth N/A m bgl
Interv. Length 1.75 m
Test Name Monitoring
System 7-Packer/ 7-Interval
Borehole diameter 101 mm
Stickup 0.20
Water depth Artesian m bgl
Version 1.0
Figure
System unit
Lines
Qty
L unit m
L total m
Depth m
OD mm
Circulation
Pressure
Packer expansion
Stickup 0.20 m bgl
0.20
bh mouth
0.00
Shotcrete surface shotcrete ~10-20 cm thick
extension
15.11
60
15.110
PACKER 7
0.50
15.610
INTERVAL 7
1.75
17.360
PACKER 6
0.50
17.860
INTERVAL 6
1.75
19.610
PACKER 5
0.50
20.110
INTERVAL 5
1.75
21.860
PACKER 4
0.50
22.360
INTERVAL 4
1.75
24.110
PACKER 3
0.50
24.610
INTERVAL 3
1.75
26.360
PACKER 2
0.50
26.860
INTERVAL 2
1.75
28.610
96
PACKER 1
0.50
29.110
INTERVAL 1
Q1
I1
End cap about 4 cm
1.655
30.765
30.765
End bh and system
Note: Lines are only indicated from I1 to Pa2

Figure 5. Schematic layout of the passive borehole systems.

Each interval is equipped with three lines. The pressure line is connected to a pressure sensor. The pressure and the circulation lines serve to inject water in the interval and to de-clog the fine filter screen. The extraction line port is in the centre of the interval and serves to extract water during circulation. In addition, hydraulic tests can be performed and the interval can be flushed.

Keller PAA-23 piezoresistive pressure transmitters are used for pore pressure measurements. Absolute pressure transmitters (zero at 0 bar abs.) with a range of 0 – 100 bar abs. are installed for the pore pressure measurements in the intervals. The measuring range for the packer pressure sensors is 0 – 100 bar.

EXTENSOMETERS

Four additional boreholes will be equipped with extensometer strings to measure the deformation due to the hydraulic fracturing and the position of the induced fractures at 14 points with respect to the anchor point / measurement point. It is planned using a chain of extensometers in a telescopic housing. A schematic setup of the extensometers with respect to the central multi-packer system is shown in Figure 6. The maximum distance from the active borehole will be about 2.5m

The displacement along the borehole is calculated by adding the displacements measured by the single chain elements. It is envisaged to use displacement transducers with a high resolution/accuracy to capture the estimated deformation of a few microns, which has been achieved before in the MTURL using the same equipment. The distance between the anchor points will be adjusted depending on the position of the extensometer. It is important to estimate the optimal position of the anchor points.

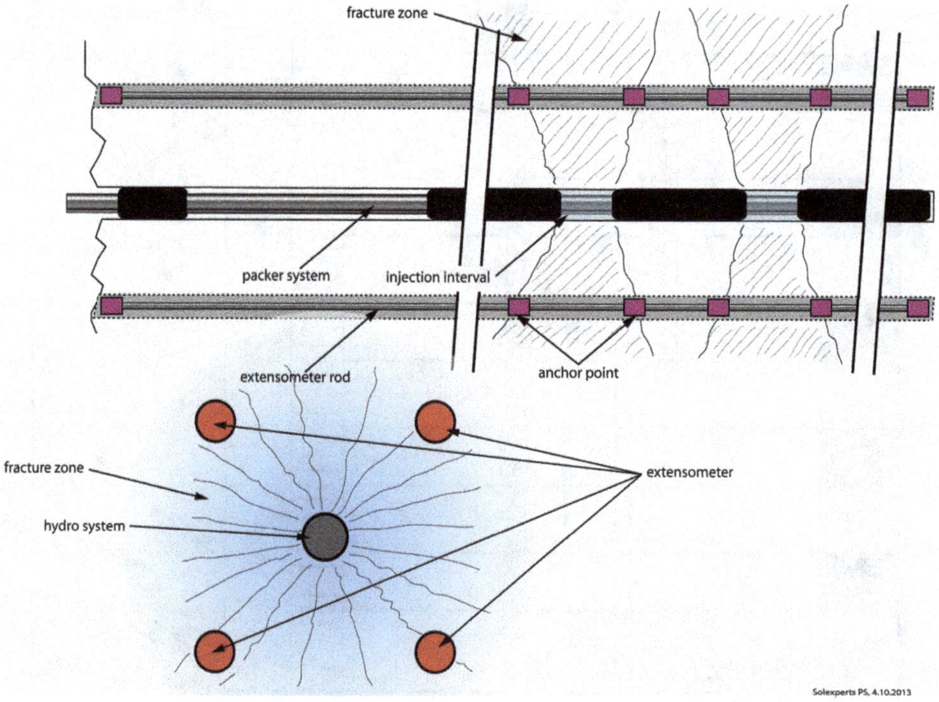

Figure 6. Schematic view of the extensometers around the central multi-packer system.

SURFACE EQUIPMENT

The surface equipment (Figure 7) has been designed to:

- Monitor interval pressures.
- Monitor packer pressures.
- Inject water in all intervals for hydraulic testing.
- Inject Pearson Water in all intervals with controlled pump pressure for hydraulic fracturing.
- Circulate Pearson Water for re-pressurization and circulation.
- Inject tracers to characterize the fractures (optional).
- Carry out controlled injection of sealants doped with tracers in the fractures while circulation with Pearson Water is taking place.

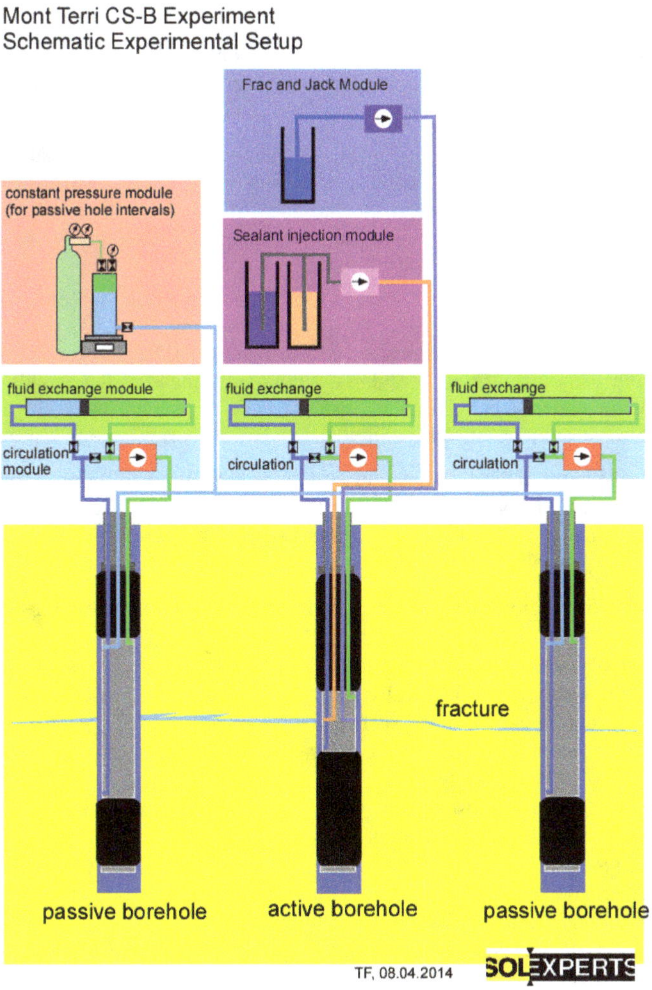

Figure 7. Schematic view of the surface system.

Injection and circulation systems

Fluids to be injected will be prepared in containers and pumped into the corresponding reservoirs. A total of four reservoirs will be installed; one for Pearson Water mixed with dye and one for each sealant component. The sealants can be mixed at the surface in a mixing tank. After the sealant injection, the mixing tank has to be flushed by circulation of fresh water to prevent clogging of the lines and ports by sealant precipitation.

The injection of the fluids will be performed with a double syringe pump with a flow range of 0.001 to 70ml/min and a maximum injection pressure of 100bar. The syringes of the pump are automatically refilled with fluid from the corresponding vessel and can be remotely controlled.

For the hydraulic fracturing, a double piston pump has to be used which produces the required flow rates of 3-5l/min.

The injection, circulation and flushing circuit is connected to a pressure vessel for pressure control under fracture opening pressure. The different fluids are mixed with different dyes. The completeness of the injection of the different injection fluids can be controlled via a sampling port by visual control or by a fluorometer (optional). If the corresponding dye is detected at the port or at the fluorometer, a complete exchange of the interval fluid can be assumed.

The flow lines of the passive borehole intervals will be connected to a circulation pump and a pressure sensor. Each flow line will be connected to a port for visual control or optionally for measurements with a fluorometer to detect the presence of the required dye in the observation interval proving the fracture opening or the arrival of the sealant component. For flushing, Pearson Water will be circulated through the interval until no dye is detected.

Data acquisition

Sensors will be connected to a permanent data acquisition system (DAS) to monitor all parameters in the system during the entire testing period. To additionally measure the signals on other data acquisition systems for a precise correlation, e.g. with seismic measurements, analogue splitters can be installed at each sensor of interest which split the signal of the corresponding sensors.

The Solexperts GeoMonitor II data acquisition system will be used to monitor all sensors. The signals are controlled and recorded with the GeoMonitor software. The proper functioning is controlled by an intelligent watchdog which performs automatic and/or semi-automatic system resets in case of failures. Details of the system features are listed below:

- On-site acquisition, conversion, visualisation and storage (in real time) of data provided by the installed instruments.
- Visualization of a maximum of 20 sensors in real-time in one graph.
- Integrated SQL database (automatically configured, no additional software packages required); data output in ASCII or Excel format possible.
- Automatic data output in SAGD format and transfer to the Mont Terri DAS via an FTP Client.
- Possibility of GeoMonitor customized alarm features, which may generate alarm notices which are distributed by e-mail, SMS or Fax.
- The recorded measurements include the date and time as well as an event code that provides information about measurement errors, measurement values exceeding their valid range and measurements exceeding an alarm limit.
- In addition, the user can create entries into a logbook. The logbook report can be filtered to show only selected events and may be exported to a Microsoft® Excel spreadsheet or

an RTF document which can be read by most word processing software (Microsoft® WordPad, Word, etc).

- Virtual real-time calculations of sensors including custom expressions, statistical values, delta values, etc.
- Remotely operated from the office using an internet access.
- Designed for true field site conditions (field proven system).
- The system is password protected, thus only users with a valid password and the proper user rights can modify the program setup.

A remote control system enables the test engineer to control the data acquisition software from distance during unattended test phases. The principal external communication line is the broadband internet connection. The internet signal is distributed through a router which is operated and administrated by Swisstopo. The PC may be reset over the internet line.

The system also incorporates backup systems and redundancy. An intelligent watchdog, uninterruptible power supplies (UPS), automatic data backup software and software alarms to assure continuous data acquisition.

The system automatically monitors the different communication lines between the interfaces and the PC (internal). The internal lines will be reset automatically by the intelligent watchdog in case of internal communication problems. System resets can be performed remotely through the internet.

In case of power failure, the integrated UPS buffer is sufficient to power the DAS during a minimum time period of about 30 minutes. One DAS channel monitors the power line. If the power fails this channel triggers an alarm in GeoMonitor. Subsequently, an alarm message is sent to the responsible site engineer (e-mail and SMS) and the PC will shut down automatically before the UPS runs out of power. (Note: for the alarm messages the internet connection needs to be active at all times). In addition, the UPS provides voltage surge protection and brown-out (low voltage) protection.

GEOPHYSICAL SURVEILLANCE

The feasibility study into the use of geophysical techniques to monitor the fracturing process was described in the in previous chapter.

The crosshole monitoring requires a high degree of repeatability between surveys, particularly for the full waveform inversion imaging technique. To achieve the most efficient data acquisition, the plan is to use a sparker source and hydrophone receiver array for the crosshole surveys, as these tools do not require clamping. To enhance the potential repeatability of the system, a wheeled weight sub on the source and hydrophones will be used to provide accurate positioning of the tools and baffles/centralisers between the hydrophones are intended to provide consistent coupling of the hydrophones.

For the microseismic monitoring, an array of 3 component geophones is proposed. These would be deployed in vertical, 2.8m deep holes drilled into the gallery floor and three separate groups. The microseismic geophone sensors will be deployed onto the bottom of the drill holes and sanded in over the length of the geophone tool. The tools will be oriented using a removable tube between the top of the geophone and the surface. The hydrophone strings would also be monitored during fracturing.

It is possible that seismic sensors, geophones, could also be built into the four extensometer tools. This arrangement would place geophones at a similar offset to the fractures as the hydrophone strings but would have a number of advantages. Geophones are expected to be more sensitive than hydrophones and may also be quieter as they would be grouted into the boreholes with the extensometers, whereas there may be noise travelling within the water column containing the hydrophones. Also, the geophones would enhance the spatial coverage of the whole network for fault plane solution analysis. A resolution analysis using two groups of 3 component geophones in each extensometer tool indicated that groups in the extensometers would provide similar resolution to the two hydrophone strings tested previously.

Although designs have been prepared for equipment and acquisition systems, it is necessary to undertake field trials ahead of the experiment to confirm their effectiveness and establish the practical constraints of their operation (see below). The working design is:

Crosshole Source	Geotomographie sparker.
Crosshole Sensors	Geometrics DHA7 24 level hydrophone array. OD 28mm
Microseismic Sensors	Sensor 3D downhole 3 component geophone package

Digital Acquisition System Specification:

- Geometrics Geode, 24 channel
- Sample interval >=0.02ms
- Continuous acquisition from 0.5ms sample interval or more. Alternatively, at smaller sample intervals there is a downtime of 1-2 seconds for 24 channels, higher channels counts would need testing

Acquisition Settings, Crosshole and Microseismic:

- Sample rate 20kHz
- Trace length 60ms (1200 samples)
- Pre-trigger 10ms (Sample 1 at -10ms, sample 201 at 0.0ms the trigger point)

In order to minimise the noise on the hydrophones it is proposed to design and build bespoke noise reduction baffles to be attached between each hydrophone sensor. These baffles will also centralise the hydrophone elements in the borehole to minimise coupling variations between deployments.

The holes are planned to dip at around 45°. To ensure accurate positioning of the elements the hydrophone string will need to be tensioned by a wheeled weight sub.

The gallery receivers will be 3 component geophone groups deployed onto the bottom of dedicated holes in the gallery floor. The geophones will be coupled by pouring sand into the hole to fill the annulus around the geophone. In order to prevent the tool being trapped by the sand a rigid pipe of the same diameter as the tool will be attached to the top of the tool. This pipe prevents the sand bridging across the top of the tool which could make it very difficult to recover the tool. It is also proposed to use a further removable rigid tube to rotate the geophone such that the horizontal components are aligned North and East.

The geophone groups in the extensometers would be permanently built into the extensometer strings and not recoverable. It will be important to ensure the groups are rigidly fixed within the arrays to ensure optimum coupling in combination with the grouting of the strings.

Both the hydrophone string and geophone tools in the galleries will be deployed by hand.

EXPERIMENTAL SEQUENCE

In advance of the main experiment, two preparatory pieces of work will be carried out: bench testing of the packer systems, and field trials of the geophysical surveillance systems.

The aim of the packer bench testing is to test each experimental step to prove its proper performance in terms of the complex functionality and test sequence, and to assess the impact of the sealants on the filters and flow lines required in the packer intervals, which is potentially a significant problem.

The specific objectives of the testing are as follows:

- Functionality check of the various surface modules in combination with the 3 sealant types under realistic conditions (pressures, flow rates, temperatures etc).
- Feasibility test of the injection (active) and monitoring (passive) interval design with two different types of screens with respect to the flushing out after contact with the various sealant types.
- Function check of the trigger mechanism for the various sealant types under "realistic" conditions, and time period for curing of the sealants.
- Optional: suitability check of the sealants for various fracture apertures simulated by filters with various mash sizes.
- Optional: suitability check of the sealants using cores instead of filters in the pressure chamber

Field trials of the geophysics surveillance systems are required to test several aspects relating to the application of the equipment in the specific environment at MTURL. The trials required are as follows:

- Gallery array, 3-component sensor testing, including deployment and recovery.
- Borehole Array testing, including connection, deployment and use of the weight sub.
- Downhole Source; test deployment and triggering to the data acquisition system.
- Crosshole Trial, including deployment, movement between tests, trigger levels and potential requirement for stacking of signals.
- Uphole trial, testing acquisition in the gallery and penetration of shots.
- Downhole (VSP) trial, testing penetration and number of shots required for stacking.

Once the experiment proper starts, drilling activity will take place in four phases:

1. Two crosshole tomography holes for the field trials
2. The active hole, passive holes, extensometer holes and remaining two crosshole tomography holes
3. Shallow microseismic array holes
4. Sealant assessment holes

Phase 1 drilling will take place sometime in advance of the main experiment. The main drilling activity will take place in Phase 2 when all the permanent equipment is installed. During this phase the drilling and packer installations will be carried out on a hole by hole basis in sequence, rather than completing all the drilling and then carrying out all the packer installations. This is to minimise installation problems caused by borehole instability.

Similarly, each extensometer string will be installed in its respective hole immediately after drilling. Therefore the sequence for drilling and equipment installation in each hole will be:

- Setting up.
- Coring to TD.
- Survey and BHTV image.
- Install equipment.

Once the packers and extensometers have been installed, there will be a period of stabilisation before the active phase of the experiment begins. In the active phase, the testing will be carried out from the lowermost interval upwards to the uppermost interval. This 'bottom-up' sequence is the safest approach to minimise the impact of any problems with test intervals and equipment.

The test sequence for each interval will comprise:

- Pre-fracture hydraulic testing.
- Fracturing.
- Circulation.
- Sealant placement.
- Post-sealant hydraulic testing.

For each test, crosshole tomography surveys will be run before and after the fracturing operation.

The experiment will be carried out in three test periods. The first period will comprise the first test (carried out in the bottom interval) with Pearson Water and the second and third tests (in intervals 2 and 3) carried out using the Haliburton H2Zero sealant, which is readily available and has established procedures for mixing and deployment, which should make it the most straightforward to test.

After the first three tests the procedures will be re-evaluated and modified as necessary in preparation for the remaining tests on the experimental sealants which are potentially more complex.

In the second period the first experimental sealant will be tested in intervals 4 and 5 and in the third period the second experimental sealant will be tested in intervals 6 and 7. In addition, a repeat test on interval No.1 will be carried out using Pearson Water.

If possible from a logistics point of view, the third test period will follow directly on from the second.

Therefore, the test sequence will be:

1 Pearson Water.
2 First H2Zero test.
3 Second H2Zero test.
4 First test of experimental sealant#1.
5 Second test of experimental sealant #1.
6 First test of experimental sealant #2.
7 Second test of experimental sealant #2.
8 Repeat Pearson Water test.

At the conclusion of this sequence, the four sealant assessment holes will be drilled at locations to be chosen based on observations from the passive holes and extensometers. If the packer systems are removed from the active and passive holes, an alternative option would be to overcore those holes to assess the sealants, although it should be noted that no overcoring of this length has been carried out at MTURL to date.

The total sequence will take approximately 1 year.

EXPERIMENT PHASES

Pre-fracture characterisation

After the installation of the packer systems, the intervals will be saturated with Pearson Water (and optionally also with dye), followed by a pressure build-up with an optional pulse injection test.

Before the start of the hydraulic testing campaigns, near-static formation pressures should be reached.

Pearson Water will be injected by the syringe pump to perform constant head injection tests. The highest pressures for the first hydraulic testing campaign should be lower than the fracturing pressure to avoid any changes in the formation properties. Therefore, a constant head injection test is considered most appropriate to control the maximum pressures. The syringe pump can manage flow rates as low as 0.001 ml/min which should be sufficient to estimate transmissivities of E-13 m^2/s order of magnitude. However, very low transmissivities require long testing times to provide reliable results for the formation properties. Therefore, the test type will be designed based on the expected transmissivities and boundary conditions and the required results.

Fracturing

A standard hydro-fracturing sequence including several re-opening and shut-in cycles will be performed. The fracture radius and volume will be controlled by controlling the injection volume and pressure and the created fracture volume and radius will be evaluated using the deformation measured by the extensometers and the pressure changes in the adjacent intervals of the passive boreholes. The presence of dye in the corresponding passive borehole intervals will confirm the proper creation of the fractures.

The planned hydraulic fracturing procedure is as follows:

1. Connect the interval to a double piston pump.
2. Circulate water in the observation intervals.
3. Monitor the extensometers and the pressures in the observation boreholes.
4. Conduct a standard hydro-fracturing sequence including several re-opening and shut-in cycles which are:
    - Breakdown cycle by the injection of Pearson Water (with dye) at a rate of about 0.1MPa/s with flowrates of 3 to 5l/min.
    - Monitor the flowrates and pressures with a very high scan rate of 1Hz. Observe the breakdown and shut-in pressures which are expected to be in the range between 5.5 to 9.0MPa and 4.0 to 5.5MPa, respectively. Higher breakdown and much lower shut-in pressures may indicate the creation of an axial (vertical) fracture. The expected radius of the fractures is about 1m at this stage.
    - Reopening cycle with the same injection rate until pressure response is observed in the passive borehole intervals.
    - Repetition of the reopening cycles possibly with increasing injection volume to ensure that the fractures propagate beyond the passive boreholes. The shut-in pressure ideally should be repeatable at about 4 to 5.5MPa. Monitor the deformation. The fracture volume should be big enough for the next test phase of sealant injection.
    - Check the dye content of the water in the observation intervals either visually or optionally with a fluorometer. The occurrence of dye proves the opening of fractures between both boreholes.

Circulation and sealant placement

This phase of the experiment is the most crucial and the most complex.

Hydraulic tests will be performed after the successful fracturing. The pressure sensitivity of the formation will be measured by a multistep head injection test. A first pressure step will be at a level below the reopening pressure to compare the transmissivity with the one determined from the first hydraulic test campaign before fracturing. Further steps will be at higher pressures until the reopening pressure is exceeded and the transmissivity increases.

Once above the reopening pressure the injection will be continued to establish circulation between the active borehole and the passive boreholes. The pressure in the passive boreholes is increased above the reopening pressure to keep the fractures open.

Tracer tests may be conducted by circulation of Pearson Water doped with tracer in the active borehole and recovery from the passive boreholes to determine the fracture volume. The measurement of the tracer concentrations at the required pressure of about 6MPa still has to be designed.

The fluids for sealant injection can be mixed in the mixing tank with the components from the reservoirs. A total of four reservoirs are planned to be installed; one for Pearson Water (optionally with dye) and one for each sealant component which has to be prepared according to the supplier's instructions. The mixing tank has to be flushed after each step to avoid the gelling of the sealant mixture.

The following procedure of sealant injection is proposed:

1. Establish circulation.
2. Circulate with component #1.
3. Pause.
4. Re-start circulation with component #2.
5. Either continuous or pulse circulations.
6. Repeat as necessary for sealant to 'trigger.'
7. Flush active and passive test intervals to clear test equipment.

Before the sealant injection, the interval volume will be exchanged with the corresponding component at flow rates up to 200ml/min. Approximately double the entire volume will be exchanged, such that complete exchange of the interval fluid can be assumed. Optionally, the complete exchange might be confirmed by pH, electrical conductivity or fluorometer measurements. However, the required high pressures represent harsh conditions for the corresponding instruments and pre-test evaluation of their feasibility is strongly recommended.

Circulation will be started and the sealant component will be injected with a flow rate of 2ml/min by the syringe pump at about 6MPa pressure. The sealant component is doped with dye. The sealant component is injected until the dye is detected at the passive borehole interval and the injection of the sealant component into the fracture is complete. However, this assumption is only valid if the sealant and dye have similar transport parameters.

Afterwards, the passive borehole intervals have to be flushed to remove the sealant component. Therefore the by-pass to the fluid exchange module is opened and the interval fluid contaminated with sealant is replaced with Pearson Water. Then the passive borehole intervals will be closed to prevent fluid circulation through the fractures. The injection interval with all lines and vessels will be connected to the fluid exchange module and flushed with Pearson Water until no dye is detected in the circuit. Afterwards, the circulation in the active borehole interval will be stopped and the bypass to the exchange module closed.

The procedure will be repeated for the next sealant components until the sealant is "triggered". Then circulation will be stopped to wait for the sealant components to react in accordance with the sealant specifications.

These steps are repeated if necessary.

Post-sealing characterisation

Hydraulic tests will be performed after the emplacement of the sealants in the active borehole intervals. Several test options exist and it is not critical which one is chosen:

- Constant head injection tests (HI) based on the injection pressure during sealant injection: Comparison with pre- and post-fracturing HI tests. Dissipation of pressure in the matrix with increasing flowrates and with pressure decrease in the observation boreholes
- Constant head withdrawal tests (HW) based on the hydrostatic pressure of 2MPa: Pressure decrease in observation boreholes related to HW pressure decrease in active borehole and dissipation of remaining overpressures through rock matrix.
- Constant rate injection test (RI): The injection would have to be stopped early to prevent reopening of fractures and creation of new fractures and is not recommended.
- Multistep constant head injection test (HI): The results of first step (below fracture re-opening pressure) can be compared with pre- and post-fracturing HI tests. Same features as HI test described above. Reopening of fractures and creation of new fractures might occur at higher pressures.

Additionally, constant head injection tests (HI) might be performed in selected intervals of the passive borehole to control the proper sealing of the fractures or to detect partially or not solidified regions.

In addition to the hydraulic characterisation a more direct assessment of the physical characteristics of the sealant will be made by coring through the sealed fractures and recovering samples.

Four 101mm diameter holes will be drilled parallel to the experiment holes at various distances from the active hole in order to investigate sealant at a range of fracture apertures. The locations of the holes will be chosen after the hydraulic results have been interpreted.

If the multi-packer systems have been removed from the experiment holes it should also be possible to overcore one or more of them to investigate sealant characteristics at the injection site, although overcoring to a depth of 30 - 40m has not been attempted at MTURL before.

Long term installation and monitoring

After the completion of the main experiment phases, the long-term performance of the sealants might be assessed by long-term monitoring and another series of hydraulic tests. For long-term monitoring the frac and jack module, the sealant injection module and the fluid exchange module could be removed. The pressures should be monitored by the DAS. For hydraulic testing, test equipment including the flow meters for the required flow rate range could be temporarily installed. The test types can be chosen based on the result from the previously performed tests

ACKNOWLEDGEMENTS

The design work presented here was carried out by the following individuals and their work is gratefully acknowledged:

- Andreas Busch, Shell, Netherlands
- Ben Dyer, Semore Seismic Ltd, UK
- Robin Gerlach, Montana State University, USA
- Bill Lanyon, Fracture Systems Ltd, UK
- Christophe Nussbaum, Swisstopo, Switzerland
- Thomas Trick, Thomas Fierz and Ursula Rosli, SolExperts AG, Switzerland

REFERENCES

1. A.C. Mitchell, F.G. Ferris, The influence of Bacillus pasteurii on the nucleation and growth of Calcium Carbonate, *Geomicrobiology Journal*, 23 (2006), pp. 213–226.
2. A.C. Mitchell, F.G. Ferris, Effect of strontium contaminants upon the size and solubility of calcite crystals precipitated by the bacterial hydrolysis of urea. *Enviro Sci & Tech* (2006) 40, 1008-14.

Carbon Dioxide Capture for Storage in Deep Geological Formations, Volume 4
Karl F. Gerdes (Editor)

Chapter 48

ASSESSING POTENTIAL FOR BIOMINERALIZATION SEALING IN FRACTURED SHALE AT THE MONT TERRI UNDERGROUND RESEARCH FACILITY, SWITZERLAND

A.B. Cunningham[1*], R. Gerlach[1,2], A. Phillips[1], E. Lauchnor[1], A. Rothman[1], R. Hiebert[1], A. Busch[4], B.P. Lomans[4] and L. Spangler[1,3]
[1]Center for Biofilm Engineering, Montana State University
[2]Chemical and Biological Engineering, Montana State University
[3]Energy Research Institute, Montana State University
[4] Shell Global Solutions
[*]Corresponding author

ABSTRACT: This report summarizes the methods and results of laboratory testing to evaluate the potential for biomineralization sealing in fractured shale. The tests were conducted using samples of Opalinus shale obtained from the Mont Terri underground testing facility in Switzerland [1]. These results will provide the basis for development of a work plan to test biomineralization sealing in a planned future Mont Terri field demonstration, if appropriate.

Important findings include:

- Porosity and pore size distribution were measured for two blocks of Mont Terri shale (OP1 and OP2) using mercury porosimetry. Mont Terri porosimetry measurements were compared with shale samples from other formations (i.e. Green River (Utah) and Bakken (North Dakota)). The majority of the porosity for all shale types tested is contained in pores less than 0.1 μm and those greater than 10 μm. All four shale samples contained very little porosity between 0.1 and 10 μm. The average porosity for the Mont Terri samples was 11.0%.
- Two cores designated as OP2-3 and OP2-4 were used to conduct biomineralization sealing tests under ambient pressure. Both cores had a single fully-penetrating fracture along the axis of flow. Sand proppant was added to the fracture in core OP2-4, resulting in an initial permeability of 2420 millidarcies. Core OP2-3 had an initial permeability of 249 millidarcies with no sand added. Both were biomineralized to approximately the same final permeability (approximately 0.2 millidarcies) after 4 days, corresponding to a four and three order of magnitude reduction in permeability, respectively.
- Another fractured shale core (OP1-4) was loaded into the high pressure testing system and permeability was measured in response to a range of overburden pressures. Applying overburden pressures in the range of 13.6 to 61.2 bar (200 - 900 psi) resulted in fractured core permeabilities ranging from 14.2μd down to 1.6μd. The highest overburden pressure (61.2 bar or 6.12 MPa) corresponds to the estimated overburden pressure at the Mont Terri site of between 6 -7 MPa.
- Following the overburden permeability tests the OP1-4 shale core was biomineralized in the high pressure system with an overburden pressure of 61.2 bar (900 psi, 6.12 MPa) to mimic Mont Terri pressure conditions. The initial permeability of 2.9 μd was reduced to approximately 0.1 μd due to biomineralization sealing (an order of magnitude reduction). This test indicates

that the biomineralization process was successful in the pressures range tested. X-ray CT scans showed very little evidence of fractures remaining in core OP1-4 after biomineralization sealing.

- In this testing, pre-biomineralization fracture aperture widths ranged from a high of 84.3 μm (OP2-4) down to 0.72 μm (OP1-4). The intermediate values of 39.6 μm (OP2.3) and 43.4 μm (OP1-2) are close to the reported hydraulic aperture range of 20-30 μm estimated for the Mont Terri shale after hydraulic fracturing—indicating that tests reported here have bracketed the range of fracture apertures expected at the Mont Terri site. Biomineralization sealing was successful for all initial fractures tested.

KEYWORDS: fracture sealant; biomineralization; Opalinus shale; Mont Terri Underground Research Laboratory; CCS leakage mitigation

INTRODUCTION & BACKGROUND

DOE is currently funding the Center for Biofilm Engineering (CBE) at Montana State University (MSU) on two CO_2 sequestration-related projects (DE-FE0004478 and DE-FE0009599) aimed at developing technology for sealing unwanted CO_2 leakage pathways. This technology is based on bacterial ureolysis-induced biomineralization, a process that promotes the deposition of carbonate minerals in porous media or fractures to seal preferential leakage pathways. Specific leakage pathways of interest include fractures, cracks and delaminations between well casing, well cement, and surrounding aquifer or caprock, and potentially the pore space of the rock. The on-going DOE projects focus entirely on proving the concept of biomineralization sealing in sandstone rock environments. This research and testing methodology has been previously published by the CBE study team [2-11, 13, 15-16].

The work for phase three of the CO_2 Capture Project (CCP3) described here was focused on extending the sandstone-based biomineralization research to evaluate the potential for biomineralization sealing using 2.54 cm (one-inch) diameter cores of fractured Opalinus shale from the Mont Terri test site.

SCOPE OF WORK

This research has demonstrated the feasibility of biomineralization technology to seal fractured shale in laboratory scale tests. This was accomplished by establishing biomineralized seals in cores at both ambient and field relevant pressure and temperature conditions. Seals were evaluated by measuring the reduction in permeability of the fractured core as biomineralization seal development progressed. X-ray computed tomography (X-ray CT) measurements before and after the tests were done to compare the original pore and fracture structure to the structure after biomineralization.

Pore size distribution and porosity were also measured on shale samples using mercury porosimetry. Porosimetry results for Opalinus shale were compared with results for shale samples from other geologic formations.

The principal tasks for this work plan were:

Task 1. Prepare shale core samples.

Task 2. Characterize shale porosity and pore size distribution.

Task 3. Perform core sealing tests at ambient pressure and temperature.

Task 4. Perform core sealing tests at field relevant pressure and temperature.

Task 5. Relate findings to field conditions at Mont Terri field site.

TASK 1. PREPARE SHALE CORE SAMPLES

Two blocks of Opalinus shale from the Mont Terri test site were used for this study. From these shale blocks multiple 2.54 cm (one-inch) diameter core samples were drilled which were subsequently used in biomineralization sealing tests. Six cores were drilled from the first shale block as shown in Figure 1 (note that the "OP1" designation refers to the cores from the first shale block while "OP2" refers to cores drilled from the second shale block). All cores were drilled parallel to the bedding planes. The attempt at wet drilling (core OP1-6) did not yield satisfactory results. All other OP1 cores were drilled dry, yielding cores which were largely intact but with a few well developed fractures. Cores OP1-1 and OP1-2 were sacrificed to develop the apparatus and experimental protocol for performing biomineralization of shale cores and simultaneous measurement of core permeability. No data from cores OP-1 and OP-2 are reported herein with the exception of the 330 md permeability for OP-2 as reported in Table 5. Biomineralization sealing test results for OP1-4 are reported under Task 4 below. The description of the OP2 shale cores is provided below under Task 3.

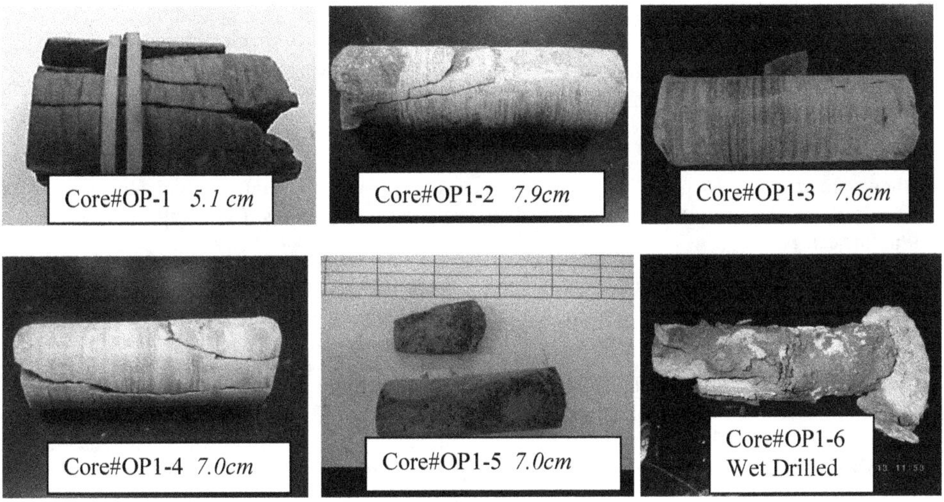

Figure 1. Cores drilled from the OP1 shale block.

TASK 2. CHARACTERIZE SHALE POROSITY AND PORE SIZE DISTRIBUTION

Characterization of Opalinus shale samples consisted primarily of using mercury porosimetry to measure porosity and pore size distribution. X-Ray CT was also used to visually characterize core fractures before and after biomineralization. CT results are reported under Task 4.

Pore Size Distribution and porosity using mercury porosimetry

Mercury porosimetry was carried out in the laboratories of Montana Emergent Technologies (MET) on the campus of Montana Tech of the University of Montana, located in Butte MT. To evaluate the pore characteristics of the Mont Terri shale, a Micromeritics Autopore IV Model 9500 Mercury Porosimeter (Figure 2) was used. This porosimeter has a 33,000 psi pressure capability which allows it to accurately measure pore sizes over a wide range, from approximately 900 μm to approximately 0.005 μm.

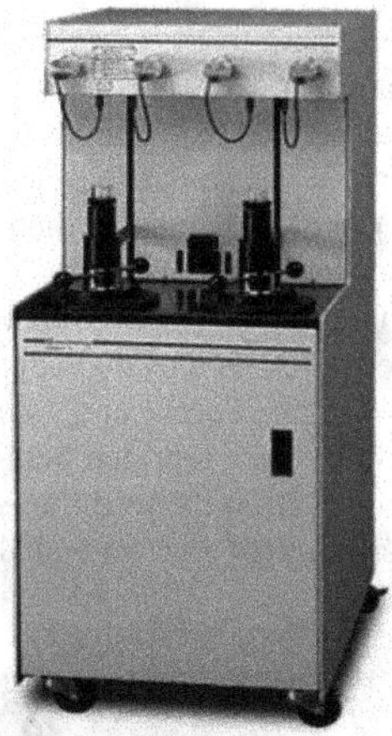

Figure 2. Micromeritics Mercury Porosimeter.

Mercury porosimetry or mercury intrusion is based on the principle that the amount of pressure required for a non-wetting fluid, such as mercury, to intrude into a porous substance is inversely proportional to the size of the pores. Material properties that can be determined from this porosimetry test include pore size distribution, total pore volume, median pore diameter, bulk density, and material density.

Samples of both OP1 and OP2 shales were tested. Shale chips with an area roughly 1 cm^2 and a few mm in thickness were placed in a bulb penetrometer which was evacuated to remove air and volatile substances. The bulb was then filled with mercury and pressurized in discrete steps. The incremental volume of mercury that intruded into the sample was recorded for each pressure point up to 33,000 psi.

The majority of the porosity for all shale types tested was contained either in pores less than 0.1 μm or those greater than 10 μm. The portion of porosity greater than 100 μm typically consists of surface roughness and indentations, rather than easily accessible pores. Cracks in the shale would fall in this category.

As shown in Figure 3, the pore size distributions for both OP1 and OP2 were very similar. However, OP2 had significant porosity around the 0.1 μm range while OP1 had pores that were an order of magnitude smaller (around 0.01 μm).

In Figure 4, the pore size distribution for both samples of Mont Terri shale were compared with shale from the Bakken formation (North Dakota) and the Green River formation (Utah). All shales contained very little porosity between 0.1 and 10 μm. However, since the Bakken Shale had almost no porosity less than about 50 μm, both OP1 and OP2 were more similar to the Green River shale.

Table 1 shows the porosity for each shale sample along with the summary pore size categories. The average porosity for the two Mont Terri shales was 11.0 %.

Table 1. Mont Terri shale pore size summary.

Core	Porosity		
	Total	**Less than 10 μm**	**Between 1-10 μm**
Mont Terri #OP1	12.31%	8.59%	0.65%
Mont Terri #OP2	9.71%	8.13%	0.27%
Green River	8.25%	4.53%	0.03%
Bakken	4.50%	0.51%	0.39%

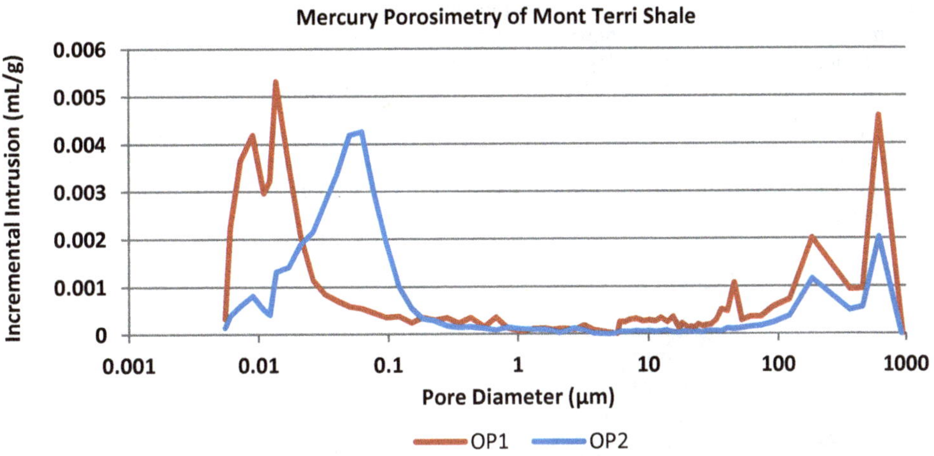

Figure 3. Pore size distribution for OP1 and OP2 shales.

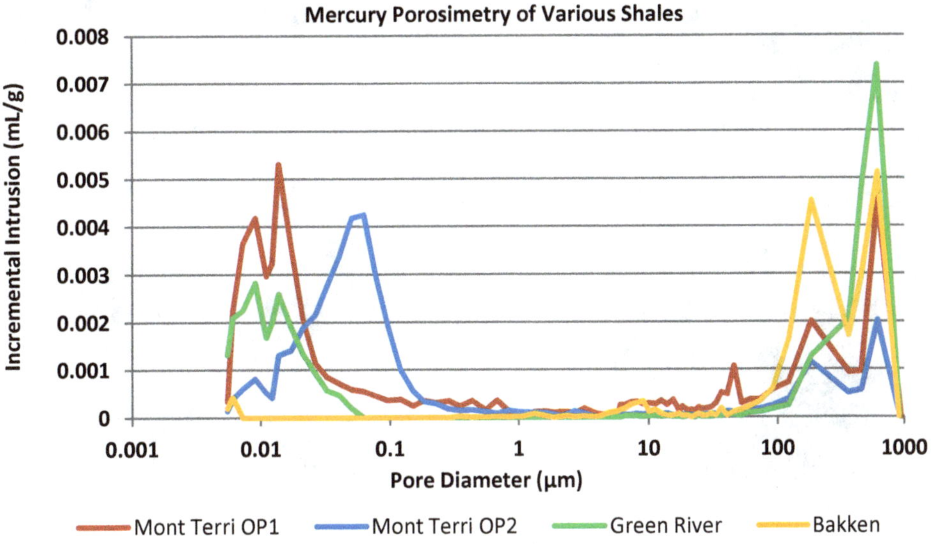

Figure 4. Comparison of Mont Terri shale pore sizes with other shales.

TASK 3. PERFORM CORE SEALING TESTS AT AMBIENT PRESSURE

Core Preparation

Four 2.54 cm (one inch) diameter cores were drilled from the second Opalinus shale block (designated OP2). Two of these cores (OP2-3 and OP2-4) were used to conduct biomineralization sealing tests under ambient pressure.

Using a hammer and a sharp object, both cores were split longitudinally. Core OP2-3 fractured into three pieces as shown in Figure 5. Core OP2-4 split cleanly into two pieces (Figure 6). Both cores were soaked in Mont Terri Brine overnight (about 16 hours). The liquid caused OP2-3 to separate into many additional pieces as shown in Figure 5. These pieces were pieced together as well as possible and wrapped in several layers of Polytetrafluoroethylene (PTFE) tape. The wrapped core was then inserted into a reinforced PVC tube for permeability testing.

The overnight soak in Mont Terri Brine caused less damage to OP2-4; the two pieces stayed intact. As shown in Figure 7, approximately one layer of 30/50 proppant sand was placed on one piece and the core was reassembled and wrapped in PTFE tape. The wrapped core was then inserted into a reinforced PVC tube for permeability testing.

Figure 5. Split OP2-3 core and core wrapped in PTFE tape. Core length was 5.77 cm (2.27 inches).

Figure 6. Split OP2-4 Core. Core length was 4.01 cm (1.58 inches).

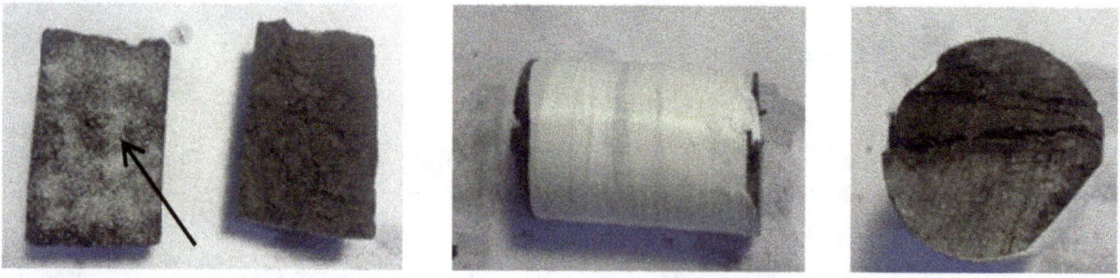

Figure 7. OP2-4 Core showing proppant placement (arrow) and wrapping pieces with PTFE tape.

System for developing biomineralization seal and measuring permeability

The flow-through system used to develop the biomineralization seal while simultaneously measuring permeability through 2.54 cm (one-inch) diameter shale cores is shown in Figure 8 and Figure 9.

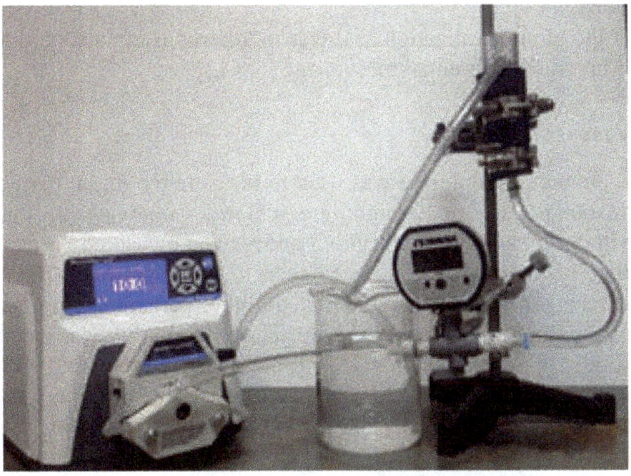

Figure 8. Apparatus for performing biomineralization sealing of shale cores and simultaneously measuring core permeability.

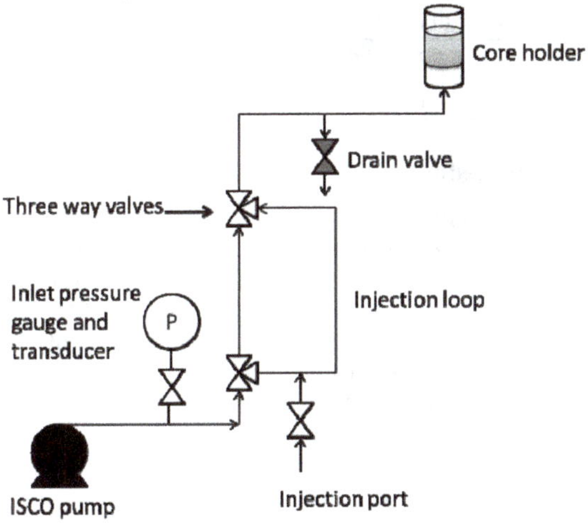

Figure 9. Process flow diagram (schematic) of apparatus for performing biomineralization sealing of shale cores and simultaneously measuring core permeability.

This system consisted of a pump for fluid injection, a pressure transducer (Omega) for determination of differential pressure across the core, and a holder around the core with sufficient overburden pressure to prevent bypass of flow around the core. A Cole Palmer peristaltic pump was used to inject fluid for initial permeability measurements. A KD Scientific syringe pump was used for injection of biomineralization fluids. Pump flow rate and influent pressure were recorded periodically and logged. The core holder consisted of PVC reinforced tubing with an inner diameter of one inch and the overburden pressure was provided by a heavy duty hose clamp placed around the core, which was tightened with a torque wrench to 20-40 in-lbs. The core holder was connected to the system via ¼-inch plastic tubing and oriented vertically with upward fluid flow. The effluent fluid that collected at the top of the core overflowed to a waste vessel.

The establishment of the biomineralization seal was monitored using standard pH measurements as well as flow rate and pressure differential monitoring [6,8,13].

Initial permeability measurements

A solution of 10 g/L ammonium chloride was used to measure the initial permeability of the cores. This solution corresponds to the ammonium chloride content of the media used for biomineralization, without urea and nutrient broth added. The measured permeability was based on the cross sectional area of the core face perpendicular to flow and the average of the shortest and longest measured core length. Fluid was recirculated with a peristaltic pump at a rate sufficient to measure a reasonable pressure drop across the core. The permeability was calculated using Darcy's Law and results for both OP2 cores are shown in Table 2. The permeability OP2-4 was an order of magnitude greater (249 millidarcies vs 2420 millidarcies) than that of OP2-3, presumably due to the presence of proppant in the fracture. This is as expected since proppant is used to for instance enhance the permeability of fractured shale for oil recovery.

Table 2. Initial Permeability values prior to biomineralization.

Core	Area (cm^2)	Average Length (cm)	Flow Rate (ml/min)	Pressure drop (meters H_2O)	Calculated Permeability (md)
OP2-3	5.43	5.77	10.0	7.3	249
OP2-4	5.43	4.01	40.0	2.2	2,420

Biomineralization sealing procedure

The two OP2 cores (OP2-3 and OP2-4) cores were biomineralized using procedures developed by Montana State University. The biomineralization procedure consisted of first inoculating the core with the ureolytic organism *Sporosarcina pasteurii*, then pumping a nutrient solution containing microbial growth medium and urea without calcium (designated CMM-) into the core, followed by the same growth medium with calcium (CMM+). Alternating pulses of CMM- and CMM+ were applied to build up the calcium carbonate sealing deposits. As the mineralization deposits built up, the flow rate was periodically reduced to keep the pressure drop at or below 25 psig as higher pressures might have caused leaks in the system.

Biomineralized core permeability measurements.

Core permeabilities following biomineralization are shown in Table 3. Although OP2-4 had an initial permeability 10 times greater than OP2-3, both were biomineralized to approximately the same final permeability (approximately 0.2 millidarcies) after 4 days. A plot of permeability vs time is shown in Figure 10. Significant calcium carbonate precipitate was observed in the fractures and on the outside of the biomineralized cores, OP2-3 and OP2-4 (Figure 11). The mineralization process effectively blocked the flow of fluid through the cracks in the shale.

Table 3. Permeability values following biomineralization

Core	Flow Rate (ml/min)	Pressure drop (meters H_2O)	Final Permeability (md)	Permeability Reduction
OP2-3	0.015	15.6	0.18	99.93%
OP2-4	0.03	19.5	0.20	99.99%

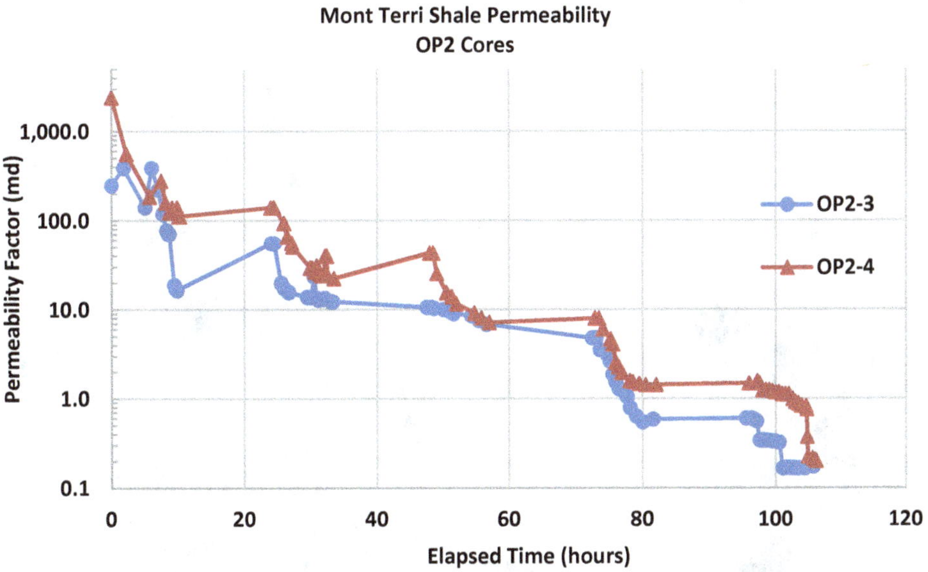

Figure 10. Permeability reduction with time during biomineralization of cores OP2-3 (blue dots) and OP2-4 (red triangles).

Figure 11. Significant calcium carbonate (arrows) was observed in the fractures and on the outside of cores OP2-3 and OP2-4 following biomineralization. All OP2 cores were drilled wet.

TASK 4. PERFORM CORE SEALING TESTS AT FIELD-RELEVANT PRESSURE

High pressure core test system

MSU/CBE has developed a high pressure, elevated temperature rock core testing system [2,3,4,17]. A Hassler-type core holder housed in an incubator to control temperature (Figure 12) has been utilized to test biofilm- and biomineral-based sealing of 2.54 cm (1 inch) diameter Berea sandstone cores. Both biofilm-only and biofilm-induced biomineralization sealing has been shown to reduce the permeability approximately 3 to 5 orders of magnitude in Berea sandstone cores under high and low pressure [9,10,13,17]. In addition, it was demonstrated that the biofilms and the deposited minerals were resistant to supercritical CO_2 exposure [9,10,12].

Figure 12. The high pressure core test system. The Hassler type core holder is in the center of the picture. Flow was from left to right. Overburden pressure was controlled via a high pressure flexible stainless steel hose connected to an Isco pump operated in constant pressure mode which was connected to the core holder (bottom of the picture).

Permeability vs. overburden pressure

A fractured shale core (OP1-4) was loaded into the high pressure testing system and permeability was measured in response to a range of overburden pressures (a picture of core OP1-4 is shown in Figure 1 and below in Figure 15).

The core was wrapped in PTFE tape to ensure a tight seal against the Viton sleeve inside the core holder. The core was loaded into the core holder and the overburden reservoir was filled. Synthetic Mont Terri brine was pumped through the core over a 5 day period. After this time the core was assumed to be saturated and permeability testing was performed. Each permeability test had a constant differential pressure across the core and a constant overburden pressure. The inlet pressure and overburden pressure were controlled by high pressure syringe pumps (Teledyne Isco 500D). The outlet of the core was open to atmospheric pressure. Both the inlet and overburden pressures were raised slowly in 50-100 psi intervals, while keeping the overburden pressure above the inlet pressure to prevent bypass of the brine solution. Once the desired inlet pressure was reached data logging for both pumps was started. Volume dispensed from the pump was monitored over time to determine an average flow rate. Permeability was calculated using Darcy's law when the flow rate through the core was stable for approximately 2 hours. The experiments were carried out in sequence from low pressures to higher pressure.

Table 4 and Figure 13 show that applying overburden pressures ranging from 13.6 to 61.2 bar (200 to 900 psi) substantially reduces the permeability of the (saturated) fractured shale core into the microdarcy range. The upper range of overburden pressure (61.2 bar), which corresponds to the estimated overburden pressure at the Mont Terri site, results in a permeability of about 1.6 microdarcies. The variability in permeability values observed for 20.4 and 61.2 bar overburden pressures indicates the possibility that the effective fracture aperture is exhibiting a dynamic response to changes in overburden pressure and flow rate. This response may be the result of random opening and closing of micro flow channels along the original fracture.

Table 4. Permeability of the fractured OP1-4 shale core in response to increasing overburden pressure.

Flow Rate (mL/hr)	Differential Pressure (bar)	Overburden Pressure (bar)	Permeability (μd)	Standard deviation (μd)
0.347	10.2	13.6	14.2	0.810
0.176	17.0	20.4	4.34	0.239
0.104	10.2	20.4	4.22	0.526
0.125	54.4	61.2	0.99	0.053
0.144	57.8	61.2	1.05	0.078
0.210	54.4	61.2	1.63	0.056

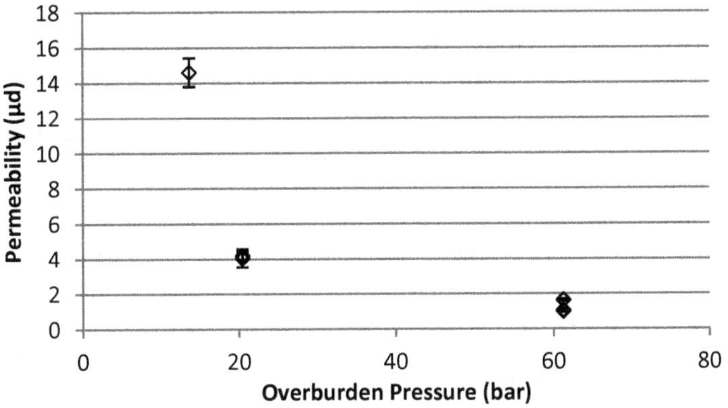

Figure 13. Observed permeability vs. overburden pressure, error bars represent the standard deviations of five measurements.

Biomineralization sealing

Following the overburden permeability tests the OP1-4 shale core was biomineralized. For this test the overburden pressure was set at 61.2 bar (900 psi) and the injection pressure was set to 54.4 bar (800 psi). Biomineralization sealing was promoted using the same methods and protocols as described in Task 3. Results shown in Figure 14 indicate that the initial permeability of 2.9 microdarcies was reduced to approximately 0.1 microdarcies during the first 50 hours. This permeability reduction was entirely due to bacterial plugging. After 50 hours calcium was added and biomineralization began, however no further permeability reduction was observed. This result is not surprising since the estimated original aperture width of approximately 0.74 μm (shown in Table 5 below) is substantially smaller than the size of the *S. pasteurii* cells (2-4 microns). This test does however indicate that the biomineralization process was not affected by pressures in the range tested as significant calcium carbonate deposits were observed in the system.

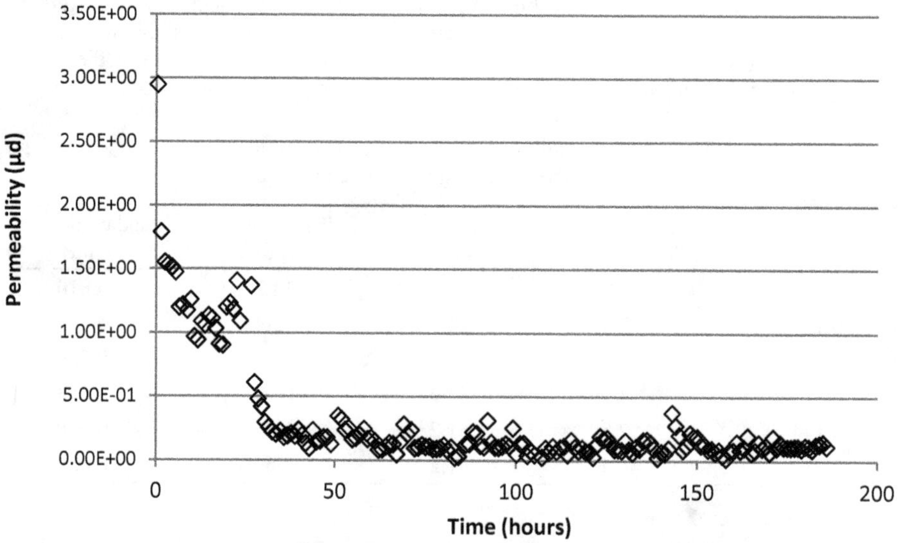

Figure 14. Permeability of core OP1-4 over time. At 24 hours the system was depressurized and switched from injecting the organisms to the growth medium. At 50 hours the system was switched from the growth medium to the calcium-containing medium. Permeability calculations accounted for the differences in fluid viscosity, since the calcium-free medium has a viscosity of 1.006 10^{-3} kg m^{-1} s^{-1} and the calcium-containing medium has a viscosity of 1.051 10^{-3} kg m^{-1} s^{-1}.

Following biomineralization sealing, core OP1-4 was scanned using X-Ray CT. Figure 15 shows the OP1-4 core before and after high pressure biomineralization along with the locations where the CT scans were performed.

Figure 16 shows the results of the X-Ray CT scans for two locations along the axis of core OP1-4 before and after biomineralization. It should be noted that the "before biomineralization" scans were taken while the core was wrapped in PTFE tape. This condition corresponds to a very small, but unknown overburden condition. As a result it is not possible to directly compare the size of the fractures before and after biomineralization as the biomineralization sealing was performed at very high (61.2 bar) overburden pressure. However the X-Ray CT scans do show very little evidence of fractures remaining after biomineralization sealing was completed.

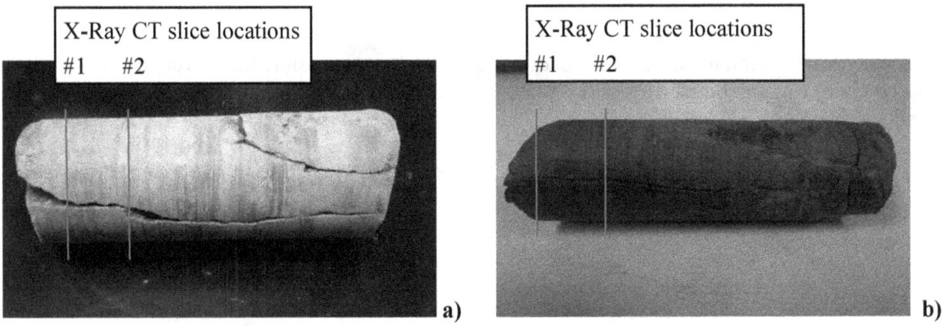

Figure 15. Core OP1-4 **a)** before biomineralization and **b)** after biomineralization.

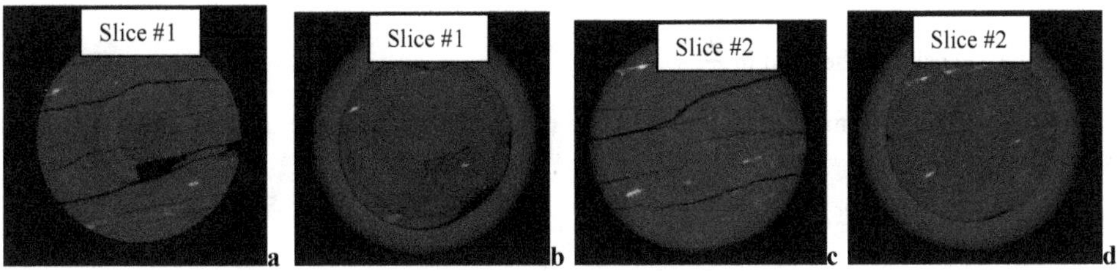

Figure 16. X-Ray CT scans of core OP1-4 before (16a and 16c,) and after (16b and 16d,) biomineralization sealing. After applying overburden pressure of 61.2 bar and completing biomineralization sealing figures 16b and 16d reveal very little evidence of residual fracturing in the shale.

TASK 5. RELATE FINDINGS TO FIELD CONDITIONS AT MONT TERRI FIELD SITE

The confining stresses at the Mont Terri site are in the range of 6-7 MPa and fracture apertures can be expected in the range of 20 to 30 μm when fractures are kept open by an overpressure (above reopening pressure). These data are reported in "Fracture Sealants Programme: Modelling and Feasibility" by Ledingham (Chapter 45 in this volume). Ambient temperature at the site is assumed to be 20 degrees C.

The high pressure permeability and biomineralization experiment described was carried out with an overburden pressure of 61.2 bar (900 psi or 6.12 MPa). This 6.12 MPa pressure is therefore in the range of possible confining stresses assumed for Mont Terri.

Core fracture aperture width calculation

Published equations were used to relate average permeability fractured cores (as measured for Task 3 and Task 4) to the corresponding (calculated) fracture aperture widths. The following derivation shows the mathematical development of the equation used to calculate fracture width from measured average fractured core permeabilities.

Derivation

The analytical solution to the Navier Stokes equation for incompressible flow through two equally spaced plates was combined with Darcy's law to yield an expression for effective fracture hydraulic conductivity, $K_{fracture}$ $[=]\, ^{m}/_{s}$.

$$K_{fracture} = \frac{w^2}{12} \cdot \frac{\rho g}{\mu}$$

Where:

w = the aperture of the fracture in metres

ρ is the density of the fluid in $\frac{kg}{m^3}$,

g is the gravitational acceleration constant in $\frac{m}{s^2}$ and

μ is the viscosity of the fluid in $\frac{kg}{m\,s}$.

A second relation, from [14], was used to relate matrix, average and fracture hydraulic conductivities:

$$K_{fracture} = \frac{K_{average}A - K_{matrix}(A - wl)}{wl} \qquad \text{Equation 1}$$

Where:

$K_{average}$ = hydraulic conductivity of the rock and core system (m/s). This was measured experimentally at different overburden and differential pressures.

A = cross sectional area of the core (m^2)

K_{matrix} = hydraulic conductivity of unfractured shale, $1.0 * 10^{-14} \frac{m}{s}$ was used. This was the observed hydraulic conductivity at the end of the first high pressure experiment.

l = width of the fracture (m)

Combining these equations leads to:

$$\frac{w^3 l}{12} \frac{\rho g}{\mu} - k_{av}A + k_m(A - wl) = 0 \qquad \text{Equation 2}$$

Which was solved for w, the aperture of the fracture.

Calculated aperture values

Calculated fracture apertures for all (pre-biomineralization) tests are shown below in Table 5.

Results shown in Table 5 indicate that the biomineralized cores exhibit pre-biomineralization fracture widths ranging from 84.3 μm (OP2-4) down to 0.74 μm (OP1-4). The intermediate values of 39.6 μm (OP2-3) and 43.4 μm (OP1-2) are close to the range of 20-30 μm estimated above for the Mont Terri shale after hydraulic fracturing. These findings indicate that the biomineralization tests described here have bracketed the range of fracture apertures expected at the Mont Terri site. Biomineralization sealing was successful for all initial fractures apertures tested.

Table 5. Results from varying overburden pressure and differential pressure in the high pressure system. Overburden pressure tests were run on core OP1-4. The permeability and fracture aperture for cores OP1-2, OP2-3, and OP2-4 measured under ambient pressure are also included in this Table.

Differential Pressure (psi)	Overburden Pressure (psi)	Permeability (μd)	Fracture Aperture (μm)
150	200	14.2	1.5
250	300	4.34	1.0
150	300	4.22	1.0
850	900	1.05	0.64
800	900	1.63	0.74

Ambient experiments	Permeability (md)	Fracture Aperture (μm)
Proppant Core (OP2-4)	2420	84.3
No Proppant (OP2-3)	250	39.6
No Proppant (OP1-2)	330	43.4

SUMMARY

Porosity and pore size distribution were measured for two blocks of Mont Terri shale (OP1 and OP2) using mercury porosimetry. Mont Terri porosimetry measurements were also compared with shale samples from other formations (i.e. Green River (Utah) and Bakken (North Dakota)). The majority of the porosity for all shale types tested is contained in pores less than 0.1 μm and those greater than 10 μm. The overall pore size distributions for both OP1 and OP2 were very similar. All four shale samples contained very little porosity between 0.1 and 10 μm. The average porosity for the two Mont Terri shale blocks tested was 11.0%

Cores OP2-3 and OP2-4 were used to conduct biomineralization sealing tests under ambient pressure. Both cores had a single fully-penetrating fracture along the axis of flow. Sand proppant was added to the fracture in core OP2-4 resulting in an initial permeability of 2490 millidarcies. Core OP2-3 had an initial permeability of 249 millidarcies. Both were biomineralized to approximately the same final permeability (approximately 0.2 millidarcies) after 4 days. This corresponds to a four and three order of magnitude reduction in permeability, respectively.

Fractured shale core (OP1-4) was inserted into the high pressure testing system and permeability was measured in response to a range of overburden pressures. Applying overburden pressures in the range of 13.6 to 61.2 bar (200 - 900 psi) resulted in fractured core permeabilities ranging from 14.2μd down to 1.6μd. The highest overburden pressure (61.2 bar or 6.12 MPa) corresponds to the estimated overburden pressure at the Mont Terri site of between 6 -7 MPa.

Following the overburden-permeability tests the OP1-4 shale core was biomineralized at an overburden pressure of 61.2 bar (900 psi). The initial permeability of 2.9 microdarcies was reduced to approximately 0.1 microdarcies due to biomineralization sealing (two order of magnitude reduction). This test indicates that the biomineralization process was not affected by pressures in the range tested. X-Ray CT scans showed very little evidence of fractures remaining in core OP1-4 after biomineralization sealing was completed.

In Phase I testing pre-biomineralization fracture widths ranged from a high of 84.3 μm (OP2-4) down to 0.74 μm (OP1-4). The intermediate values of 39.6 μm (OP2.3) and 43.4 μm (OP1-2) are close to the range of 20-30 μm estimated above for the Mont Terri shale after hydraulic fracturing, indicating that Phase I tests have bracketed the range of fracture apertures expected at the Mont Terri site. Biomineralization sealing was successful for all initial fracture apertures tested.

CONCLUSIONS

1. The two blocks of Opalinus shale tests were found to have similar pore size distributions (<0.1 μm and > 10 μm) and porosities (11% average porosity).
2. Applying overburden pressures in the range of 13.6 to 61.2 bar (200 - 900 psi) reduced the saturated, fractured shale core permeability into the microdarcy range (14.2μd to 1.6μd).
3. All shale samples tested were successfully biomineralized. Initial average fractured core permeabilities were reduced by up to four orders of magnitude. Shale fracture apertures tested bracketed the aperture range (20-30 microns) predicted at Mont Terri.
4. Results from this testing indicate that the biomineralization sealing alternative has a high probability of success in the planned Mont Terri field test.

ACKNOWLEDGEMENTS

This research was sponsored by CCP3 with supplemental funding from DOE Project #DE-FE0000397. Additional Technical review was provided by Joseph J.T. Westrich of Shell Global Solutions.

REFERENCES

1. Thury, M.; The characteristics of the Opalinus Clay investigated in the Mont Terri underground rock laboratory in Switzerland. Comptes Rendus Physique, Volume 3, Issue 7, Pages 923-933, 2002.
2. Cunningham, A.B.; Gerlach, R.; Spangler, L.; Mitchell, A.C. (2009): Microbially enhanced geologic containment of sequestered supercritical CO_2. *Energy Procedia.* 1(1):3245-3252. doi: 10.1016/j.egypro.2009.02.109
3. Cunningham, A.B., Gerlach, R.; Spangler, L.; Mitchell, A.C.; Parks, S.; Phillips, A. (2011): Reducing the risk of well bore leakage using engineered biomineralization barriers. *Energy Procedia.* 4:5178-5185. doi:10.1016/j.egypro.2011.02.49.
4. Cunningham, A.B.; Lauchnor, E.; Eldring, J. Esposito, R.; Mitchell, A.C.; Gerlach, R.; Connolly, J.; Phillips, A.J.; Ebigbo, A.; Spangler, L.H. (2013): Abandoned Well CO_2 Leakage Mitigation Using Biologically Induced Mineralization: Current Progress and Future Directions. *Greenhouse Gases: Science and Technology.*
5. Ebigbo A.; Helmig, R.; Cunningham, A.B.; Class, H.; Gerlach, R. (2010): Modelling biofilm growth in the presence of carbon dioxide and water flow in the subsurface. *Advances in Water Resources.* 33:762–781. doi: 10.1016/j.advwatres.2010.04.004
6. Ebigbo A.; Phillips, A; Gerlach, R.; Helmig, R.; Cunningham, A.B.; Class, H.; Spangler, L. (2012): Darcy-scale modeling of microbially induced carbonate mineral precipitation in sand columns. *Water Resources Research.* 48, W07519, doi:10.1029/2011WR011714.
7. Fridjonsson, E.O.; Seymour, J.D.; Schultz, L.N.; Gerlach, R; Cunningham, A.B.; Codd, S.L. (2011): NMR Measurement of Hydrodynamic Dispersion in Porous Media Subject to Biofilm Mediated Precipitation Reactions. *Journal of Contaminant Hydrology.* 120-121:79-88. doi:10.1016/j.jconhyd.2010.07.009
8. Lauchnor, E.G.; Schultz, L.; Mitchell, A.C.; Cunningham, A.B.; Gerlach, R. Bacterially Induced Calcium Carbonate Precipitation and Strontium Co-Precipitation under Flow Conditions in a Porous Media System. *Environmental Science and Technology.* es-2010-02968v. Accepted. Jan 02, 2013. http://dx.doi.org/10.1021/es304240y
9. Mitchell, A.C.; Phillips, A.J.; Hamilton, M.A.; Gerlach, R.; Hollis, K.; Kaszuba, J.P.; Cunningham, A.B. (2008): Resilience of planktonic and biofilm cultures to supercritical CO_2. *The Journal of Supercritical Fluids.* 47(2):318-325. doi:10.1016/j.supflu.2008.07.005

10. Mitchell, A.C.; Phillips, A.J.; Hiebert, R.; Gerlach, R.; Spangler, L.; Cunningham, A.B. (2009): Biofilm enhanced geologic sequestration of supercritical CO_2. *The International Journal on Greenhouse Gas Control*. 3:90-99. doi:10.1016/j.ijggc.2008.05.002
11. Mitchell, A.C.; Dideriksen, K.; Spangler, L.H.; Cunningham, A.B.; Gerlach, R. (2010): Microbially enhanced carbon capture and storage by mineral-trapping and solubility-trapping. *Environmental Science and Technology*. 44(13):5270-5276. doi: 10.1021/es903270w
12. Mitchell, A.C.; Phillips, A.J.; Schultz, L.N.; Parks, S.L.; Spangler, L.H.; Cunningham, A.B.; Gerlach, R. Microbial $CaCO_3$ mineral formation and stability in an experimentally simulated high pressure saline aquifer with supercritical CO_2. *International Journal of Greenhouse Gas Control*. Accepted February 03, 2013.
13. Phillips, A.J., R. Gerlach, R., Lauchnor, E., Mitchell, A.C, Cunningham, A.B. Spangler L. Engineered applications of ureolytic biomineralization: a review (2013). *Biofouling*, Vol. 29, No. 6, 715–733, http://dx.doi.org/10.1080/08927014.2013.796550.
14. Putra, E., V. Muralidharan, and D. S. Schechter. "Overburden pressure affects fracture aperture and fracture permeability in a fracture reservoir." *Saudi Aramco Journal of Technology* (2003): 57-63.
15. Schultz, L.; Pitts, B.; Mitchell, A.C.; Cunningham, A.B.; Gerlach, R. (2011): Imaging Biologically-Induced Mineralization in Fully Hydrated Flow Systems. *Microscopy Today*. September 2011:10-13. doi:10.1017/S1551929511000848 (with cover image feature).
16. Phillips, AJ; Lauchnor, E; Eldring, J; Esposito R; Mitchell, A.C.; Gerlach, R; Cunningham, A; and Spangler, L. (2013): Potential CO_2 Leakage Reduction through Biofilm-Induced Calcium Carbonate Precipitation. *Environmental Science & Technology*. 47 (1):142-149
17. Phillips, AJ. (2013): Biofilm-Induced Calcium Carbonate Precipitation: Application in the Subsurface, Montana State University, PhD Thesis. 223 pp

Carbon Dioxide Capture for Storage in Deep Geological Formations, Volume 4
Karl F. Gerdes (Editor)

Section 2

STORAGE MONITORING & VERIFICATION (SMV)

SMV Conclusions

Carbon Dioxide Capture for Storage in Deep Geological Formations, Volume 4
Karl F Gerdes (Editor)

Chapter 49

CCP3-STORAGE MONITORING & VERIFICATION (SMV): IMPLICATIONS OF KEY FINDINGS AND THE PATH FORWARD

Scott Imbus
Chevron Energy Technology Co., 1500 Louisiana St., Houston, Texas 77002 USA

ABSTRACT: The relevance of CCP3-SMV program results to current assurance and efficiency issues in CO_2 storage, and prospects for further progress along these lines in CCP4-SMV, are outlined in the present chapter.

KEYWORDS: CO_2 storage; well integrity; subsurface processes; Monitoring & Verification (M&V); optimization; field trialing (deployment); contingencies (intervention); subsurface integrity

INTRODUCTION: As summarized in Chapter 24, the CCP3-SMV program made considerable progress in supporting the technical case for CO_2 assurance. Accomplishments across a range of CO_2 storage aspects include: 1) raising challenges to fundamental interpretations of lab experiments in depiction of subsurface processes, 2) field trialing of several new Monitoring & Verification (M&V) technologies, 3) confirming feasibility of conducting bench to field scale experiments at an underground lab, and 4) launching a comprehensive Contingencies program to guide detection, characterization and mitigation of unexpected fluid leakage. The CCP4-SMV program plans to pursue these and other projects further with the ultimate aim of achieving technical assurance and efficiency for CO_2 storage project proponents and key stakeholders. A major disappointment in the CCP3-SMV program, however, was the unavailability of any CO_2-experienced wells to conduct an integrity survey and analysis, similar to that accomplished in CCP2-SMV [1], due to operator sensitivities. The CCP4-SMV team will continue to seek such opportunities or suitable alternatives (e.g., bench-scale field experiments in an underground lab).

KEY FINDINGS AND IMPLICATIONS FOR CO_2 STORAGE ASSURANCE

The following conveys key implications of CCP3-SMV project findings (see Chapter 24) and their implications for CO_2 storage assurance and efficacy.

Subsurface Processes

Experimental laboratory work relating to subsurface processes and their impact on storage project assurance raised new questions on procedures and the utility of such data. The relative permeability (Kr) experiments produced unexpected results that complicate efforts to estimate a realistic "trapping" number. The unanticipated experimental results were attributed primarily to the low flow rate and secondarily to processes such as CO_2 dissolution. Although a trapping number consistent with published values was determined, the results raise the question of what this parameter, which is a key input into reservoir capacity determination, means when flow rate, as well as saturation and dissolution rate, varies throughout the reservoir system. The capillary entry pressure (Pc) experiments, using two methods, show wide divergence in results among rocks with near identical facies. This suggests that artifacts are introduced by non-obvious variations in sample acquisition and handling. This study highlights the need to identify the impact of artifacts or minute facies differences and the need to acquire several Pc determinations to establish a distribution of such numbers.

Geomechanical modeling indicates top seal hysteretic effects are key to understanding the likely quality of the containment system in depleted oil fields that are candidates for CO_2 storage. Modeling and experimental work on the impact of impurities contained in injected CO_2 streams illustrate the need for an expanded area of review for pressure simulation and thus monitoring. Injection of reactive gases can markedly lower pH directly (e.g., SOx) or indirectly (O_2 dissolution of pyrite) which can undermine rock mechanical stability in carbonate cemented rocks or, in the case of O_2, result in precipitation reactions that could impede fluid flow.

Monitoring & Verification

The CCP3-SMV program supported step change improvements in LBNL's modular borehole monitoring (MBM) single completion assembly capable of downhole fluid sampling and sensing. These improvements include better integration of components for ease of installation and greater resilience to subsurface conditions. The additional incorporation of fiber optic-based distributed acoustic sensing was based on the serendipitous field trialing experience at Citronelle Dome (see Field Trialing below).

Two studies related to site specific optimization of M&V technology and decision making based on avoidance of unacceptable leakage events are important contributions to ensuring cost-effectiveness of M&V programs.

Optimization

The Certification Framework (CF) team (LBNL, UT and UT-BEG) had the opportunity to conduct a time-lapse risk assessment at In Salah, providing insights into dynamic monitoring system deployment decision making, based on current operational data (Chapter 32). The CF team also addressed a number of emerging issues in CO_2 storage, including pressure dissipation in geologic systems, fault reactivation and fracture propagation potential.

Other studies conducted for CCP3 and its member companies include a comparison of risk assessment approaches, potential for methane exsolution during CO_2 injection in methane-saturated brines, and assessment of the fate of elemental mercury entrained CO_2 injected into reservoirs. The CF risk assessment approach, with associated modeling of the potential impact of issues identified, remains a key framework for CO_2 storage assurance qualification, as well as for wider oil and gas exploration subsurface integrity assessment.

Other CCP3-SMV optimization-related work included: economics of CO_2 enhanced recovery considering CO_2 storage credits and costs of operating under US EPA UIC Class II versus VI rules (by Merchant Consulting, not included in this volume); and a white paper of technical challenges to CO_2 utilization for unconventional enhanced gas recovery (as a hydraulic fracturing fluid and recovery agent). The latter study identifies specific RD&D programs that CCP4-SMV will consider for its project portfolio.

Field Trialing

The CCP3-SMV program exceeded expectations on the number, range and successful outcomes of M&V technology deployments. A table summarizing CCP3-SMV M&V technology field deployments with objectives and outcomes appears in Chapter 24. The following outlines the significance of these deployments to improved understanding of CO_2 physico-chemical interactions in the subsurface and their potential for integrated, cost-effective M&V programs in commercial projects.

Well-based monitoring

The efficacy of the MBM tool has been alluded to above. The MBM deployment at Citronelle was important in terms of demonstrating 1) ease of installation using the "flat pack" to combine packer inflation tubing, sampling tubing and sensor cable into a compact assembly that is resilient to downhole conditions, 2) onboard diagnosis of off-depth perforations (as opposed to packer inflation failure or a cracked casing collar) using distributed temperature sensing (DTS), and 3) use of fiber optic-based distributed acoustic sensing as a possible substitute for downhole geophones. The potential for cost savings and additional data acquisition capability may be considerable compared to conventional surface (e.g., seismic) and downhole monitoring if deployed commercially.

Although the comparison of a through-casing resistivity (TCR) survey (post-CO_2 migration) to baseline open hole resistivity survey (pre-CO_2 injection) was fraught with complications (notably presence of pre-existing methane and 20% methane in the CO_2 injection stream), the study authors were able to make a key conclusion related to the base of CO_2 migration. In addition, they developed a recommended workflow to improve application of resistivity logging to quantify CO_2 saturations post-migration.

The borehole gravity logging tool deployed at Cranfield shows clear promise of detecting CO_2 saturations post-migration at modest resolution but at very low cost.

Seismic

A comparison of the repeatability, cost and field operability of 3DVSP and surface seismic at Otway showed that the latter is superior in all aspects except for coverage away from wells and for the shallower overburden. The solution proposed to realize the strengths of both techniques at the Otway site was to install a permanent shallow, buried array of geophones at the site. A similar analysis is called for at other sites, whether or not they have complications similar to the Otway site (i.e., karst and seasonal surface and shallow subsurface hydrogeologic issues).

A proposed microseismic survey at an undisclosed site was not permitted by the operator although there will likely be other opportunities for such a deployment in the future.

Other

Electromagnetic (EM) monitoring of CO_2 storage projects has emerged as a low cost approach with potentially sufficient resolution (its shortcomings on resolution compared to seismic may be compensated by shorter intervals between surveys). A new concept, BSEM (borehole to surface) takes advantage of well casings for signal transmission. A baseline acquisition at Aquistore was considered challenging but modeling indicates that the small mass of CO_2 planned for injection in this deep reservoir should be detectable. CCP4-SMV will consider acquiring a post-CO_2 injection repeat survey.

A field-based, *in situ* sensor approach to soil gas analysis (obviating the need for gas chromatography by calculating CO_2 from the difference of other gases, taking into account relative humidity, pressure and temperature) was tested at a field in Texas. Comparisons to the benchmark GC analysis were unsatisfactory. However, this result is possibly the result of inaccurate sensor accuracy specifications and/or complications in the in situ field setting.

Contingencies (Intervention)

CCP3-SMV developed the first comprehensive framework to define options and approaches to manage unexpected leakage of CO_2 from storage reservoirs. This approach entails modeling, simulation and experiments to: 1) qualify the ability to detect and characterize leakage; 2) conduct parametric analysis of CO_2 flow relative to injection as a function of receiving reservoir and fault permeabilities, and of mitigation of CO_2 leakage using passive (stop injection) versus active (hydraulic) controls; 3) characterize agents and delivery systems for sealing top and fault seal breaches; and 4) design in situ bench-scale field experiments to test sealing agents.

Among the key findings of the Contingencies program:

- Detection of incipient or small CO_2 leaks may not be possible using "conventional" CO_2 storage project surveillance systems (e.g., seismic requires a significant threshold and repeat surveys may be infrequent; above zone pressure surveillance locations may not be close enough to the leakage point).
- The rate of CO_2 leakage is expected to be highest in settings where the injection reservoir permeability is low compared to the receiving reservoir (conduit permeability is of secondary importance).
- Passive control (stop injection) will reduce flow by at least an order of magnitude, whereas active controls, particularly extraction of water from the injection reservoir and injection into the receiving reservoir (in the vicinity of the conduit) essentially stops leakage but entails continuous, possible long-term operation to maintain this state.
- Sealants might include commercial sealants designed for wells or chemically-triggered sealants. For the former, setting time and an effective delivery system is crucial and for the latter, delivery under the right physico-chemical (e.g., pH) conditions afforded by the native brine and migrating or dissolved CO_2 is required.
- A detailed design for a fracture sealing experiment on the 1-10's m scale at the Mont Terri facility has been developed. This would involve creating independent fracture sets via an active (injection) well bore, intersecting these individually with passive (production) well bores, establishing water circulation, and injecting up to three different sealants (e.g., commercial well sealant, biofilm, triggerable smart gel).

CURRENT CO_2 TECHNOLOGY GAPS AND CCP4-SMV PROGRAM PROSPECTS

Leveraging the experience of prior CCP-SMV phases and considering current real or perceived gaps in CO_2 storage readiness for commercially viable, widespread deployment at scale, the CCP4-SMV team is considering a range of options for its 2015-19 RD&D portfolio. Outlined below is an assessment of gaps and possible projects suitable for resolving them (prioritization underway as of this writing).

Well integrity

Integrity of both operating and plugged and abandoned (P&A'ed) wells remains a top concern for CO_2 containment. P&A'ed wells, particularly those wells plugged prior to implementation of modern regulations, are of particular concern, owing to their inaccessibility for monitoring and mitigation and the impracticality of re-P&A'ing numerous wells in depleted oil field storage venues. Prior CCP2 geomechanical and geochemical modeling (unpublished CCP report by LANL) illustrated the confluence of changing geomechanical stress with production activities and resultant access of reactive fluids to well materials, particularly cements. A number of studies should be considered:

- Acquisition of an additional well(s) for surveying, analysis and modeling. Alternatively, an appropriate in situ well alteration and sealing experiment at an underground laboratory (a likely CCP4 project).

- Leakage risk estimation through geomechanical modeling of well and top seal boundary damage during past development phases.
- Development of new approaches to detecting, accessing and sealing P&A well leakage.
- Development of approaches for injecting agents that would reduce permeability of the reservoir section beneath the seal and in the lower sections of degraded wells.

Subsurface Processes

CCP3-SMV studies revealed that some characterization approaches may be impacted by sample acquisition, facies variability and experimental conditions. Furthermore, output from such experiments may be inadequate to represent the reservoir (e.g., spatial differences in flow conditions) or sealing system (facies changes and natural or induced fractures). The ability to predict the location (based on structural-geomechanical analysis), aperture distribution and geometry of natural and induced fractures and faults would enable pressure control at vulnerable sites as well as informing M&V technology type and placement needed. Among the future work considered:

- Improved alternatives to experiments and models used for predicting the trapping (element of capacity) and geomechanical resilience to CO_2 columns.
- Field-based (outcrop and well penetrations) experiments to validate likely fault / fracture geometry and permeability combined with reservoir simulation / geomechanical modeling to predict nature and flow rates of natural and induced conduits.
- Estimates of worst case CO_2 volume loss due to P&A well failure or operating well blow out.

Monitoring & Verification

CCP and other organizations have made major strides in assessing the efficacy of various monitoring techniques. Further work is warranted in developing decision-making criteria for technology deployment, as well as integrating data output to provide a more comprehensive view of CO_2 plume / displaced brine migration in the reservoir and its impact on containment by the overburden and wells. Although further developments are warranted, some of the recent techniques developed to date are being commercialized. MBM warrants further development, particularly to improve its sensor integration, robustness and self-reporting capability. Development of new sensors capable of detecting cement degradation and behind casing flow would be useful for long term well maintenance and informed P&A procedures. Marine surveillance techniques should be developed and field tested (e.g., in offshore settings with naturally leaky CO_2 or hydrocarbon systems).

Optimization

The CCP2/3-SMV Certification Framework has demonstrated its use as a vehicle for risk assessment and rapid identification of key and emerging issues for further investigation. The CF should continue to be used as a tool for CCP4-SMV. The focus of the Optimization theme in CCP4-SMV, however, will likely shift towards enhanced hydrocarbon recovery, particularly from unconventional reservoirs.

Field Trialing

Whereas the CCP3-SMV program focused on trialing new M&V sensors, the CCP4-SMV Field Trialing Program will seek opportunities to understand natural and induced subsurface processes at scale, particularly as they relate to developing intervention strategies (see Contingencies, below).

Contingencies (Intervention)

As of this writing, the "fracture sealing experiment" at Mont Terri is under consideration for execution in CCP4-SMV although its expense would require additional partners. Modeling and simulation work for specific processes will likely continue, particularly if additional well and field-based data on natural and induced conduits becomes available. Further work on permeability modifiers for sealing top and fault seal breaches will likely be included in the portfolio.

SUMMARY & CONCLUSIONS

Notwithstanding the considerable progress made in CO_2 storage assurance and efficiency by CCP and other organizations, work remains to develop more realistic static and dynamic models, experimental approaches, and interpretation of monitoring and verification data. Well integrity remains a key issue and methods should be developed for P&A well assessments and mitigation. Contingency planning and intervention response technology for unexpected CO_2 migration comprise a key stakeholder assurance issue that will be the ultimate key to promoting widespread acceptance of CO_2 storage at commercial scale. Remaining RD&D relies on access to consortium-funded field projects and increasingly expensive project work. Thus CCP4 is open to collaborative relationships with project operators, governments and researchers.

REFERENCES

1. Crow W., Williams DB, Cary JW, Celia, M and Gasda, S, (2009) Well integrity evaluation of a natural CO_2 producer. In: LI Eide, ed.) *Carbon Dioxide Capture for Storage in Deep Geological Formations, Vol. 3.* CPL Press UK, pp. 317-29.

Carbon Dioxide Capture for Storage in Deep Geological Formations, Volume 4
Karl F. Gerdes (Editor)

Section 3

POLICY & INCENTIVES (P&I) AND COMMUNICATION

Carbon Dioxide Capture for Storage in Deep Geological Formations, Volume 4
Karl F. Gerdes (Editor)

Chapter 50

ISSUES, BARRIERS AND OPPORTUNITIES IN POLICIES, INCENTIVES AND STAKEHOLDER ACCEPTANCE

Arthur Lee[1], Mark Crombie[2], Dan Burt[3], Eric Beynon[3], Christhiaan van Greco[4], Sarah Bonham[5], Shahila Perumalpillai[5] and Ioannis Chrysostomidis[5]
[1]Chevron Services Company, 6001 Bollinger Canyon, San Ramon, California 94583, USA
[2]BP International Limited, ICBT Chertsey Road, Sunbury on Thames, TW16 7BP, UK
[3]Suncor Energy Services Inc, Box 2844, 150 6th Avenue Southwest Calgary, Alberta, T2P 3E3, Canada
[4]Petrobras, Av. Paulista, 500 – 9th Floor, São Paulo, Brazil
[5]ERM, 2nd Floor, Exchequer Court, 33 St Mary Axe, London, EC3A 8AA, UK

ABSTRACT: The CO_2 Capture Project (CCP) was formed in 2000 to advance technologies and improve operational approaches to help make CCS a viable option for CO_2 mitigation. Today, this partnership of major energy companies is focused on delivering results from its demonstrations, field trials and studies. The work of CCP is managed by four teams: Capture, Storage, Policy & Incentives (P&I) and Communications. The P&I Team is committed to providing technical, economic and social insights to inform the development of legal and policy frameworks and to helping public understanding. From 2009 to 2013, the team carried out three key studies that helped to inform the CCP and the broader CCS community of the challenges for real-world CCS projects in the face of regulatory schemes still under development, while going through permitting processes with government authorities, analysing stakeholders who can impact CCS project development and developing mechanisms for sharing benefits from CCS projects. This chapter will describe what the team has learned from these efforts.

KEYWORDS: CCS; regulatory development; regulatory case studies; stakeholders; benefit sharing

INTRODUCTION

Carbon capture and storage (CCS) will need to play an important role in climate change mitigation strategies. As stated by the IPCC Fifth Assessment Report (AR5), no single technology can provide all of the mitigation potential required for the stabilisation of atmospheric greenhouse gas concentrations, but CCS (along with other mitigation technologies and policies) will be key in mitigating greenhouse gas emissions over the coming decades [1]. Robust regulatory frameworks for CCS are important in helping with the successful planning, development and implementation of CCS projects.

In this chapter, the CO_2 Capture Project commissioned analyses that:

- Examined the range of regulatory issues and policy barriers encountered by real-world projects going through or were going through permitting processes
- Explored the range of issues deemed critical or important by different categories of stakeholders – from national to local governments, and from global or national environmental activist organizations to national and local media to local communities near the fence line of future CCS projects to be sited.

- Analyzed examples of benefits-sharing mechanisms which have been applied to a range of oil and gas and infrastructure projects to gain insights about how such benefits-sharing mechanisms could help CCS projects become accepted by local communities

REGULATORY CHALLENGES AND KEY LESSONS LEARNED FROM REAL WORLD DEVELOPMENT OF CCS PROJECTS

The first section of the chapter explores the real world regulatory challenges faced by projects that have undergone or were undergoing regulatory and permitting developments.

In 2012, the CO_2 Capture Project undertook an update on regulatory issues for CO_2 capture and geological storage. This study provides an update to the 2010 selected regulatory issues report by undertaking a practical and focused review of regulatory developments and issues, looking in particular at CCS projects that have undergone or progressed significantly through the regulatory process. Eight CCS case studies across Australia, Canada, Europe and the US were investigated at as part of the study. Because a number of the projects are still progressing through the regulatory approval process, individual case study responses have been kept confidential and broader findings and conclusions from the case studies have been integrated into the report.

The study found that pathways for the regulatory approval of a CCS project do exist and that, although various gaps and barriers in the regulatory frameworks in place were identified, these were not insurmountable in the cases studied, and projects have been able to progress with the relevant permits in place or expected to be granted.

Findings from the case study interviews provided the basis for:

- An update on key regulatory issues in the different jurisdictions studied;
- The analysis of broader cross-cutting findings; and
- A summary of case study lessons learnt that could be generally applicable to the development of regulatory frameworks for CCS in other regions, and the regulatory approval process for other CCS projects.

Table 1 and the text below provide an overview of key regulatory developments and issues in the jurisdictions studied. This is followed by a summary of broader crosscutting findings and case study lessons learnt.

Australia

The federal Offshore Petroleum and Greenhouse Gas Storage (OPGGS) Act was finalised in 2011 with the development of two final sets of regulations under the Act. State onshore regulations exist in Victoria, Queensland and South Australia and are under development in New South Wales and Western Australia. In addition, Victoria's state offshore regulations came into play at the start of 2012, making it the first Australian state to finalise its CCS regulatory framework for both onshore and state offshore CO_2 storage.

The Victoria onshore and offshore regulations do not provide for the transfer of tort liability for stored CO_2 to the state, and this poses a regulatory risk to project developers in the region. The case study highlighted the important role that demonstration projects can play in helping to inform the development of a CCS regulatory framework, and the importance of being able to tailor regulations to projects of different sizes: some onerous aspects of CCS regulations geared towards large-scale CCS projects could pose difficulties to smaller-scale demonstration projects.

Table 1. Key Regulatory Developments and Issues.

	Australia (Victoria)	Canada (Alberta)	Europe	US
Context	• Regulatory framework for CCS in onshore and offshore Victoria	• CCS plays a key role in Alberta's climate strategy and significant progress has been made in building a robust regulatory framework	• Transposition of the EU CCS Directive on geological storage of CO_2 (completed by 10 of 27 EU Member States)	• Regulation of various issues at the federal level under Clean Air Act (reporting) and Safe Drinking Water Act (injection & storage of CO_2) • Other issues (long-term liability; pore space) regulated at State level
Key regulatory developments	• Australian carbon tax ($23 per tonne) introduced July 2012. With the change of government in 2013, the Abbot government repealed the carbon tax policy in 2014.	• On-going Regulatory Framework Assessment (RFA) evaluating current regulation of CCS and identifying / resolving gaps	• Evolving regulatory frameworks for CO_2 storage following the transposition of the CCS Directive by Member States	• Finalisation of the Underground Injection Control (UIC) Class VI Rule for Geological Sequestration projects
Key regulatory issues	• Long-term liability for CO_2 cannot be transferred to the state (Victoria) • This differs to Common-wealth Offshore regulations – potential issues with cross-boundary storage sites	• Gaps around transfer of liability to the Crown and structure of the Post-closure Stewardship Fund, but these are being addressed through the RFA • Questions about the regulation of CO_2-EOR and transitioning from CO_2-EOR to CCS	• Insufficient detail in some areas of the regulations in some states (e.g. MMV; health & safety criteria for CO_2 transport) • Common concerns about liability for stored CO_2 – particularly climate liability and uncertain future of EUA prices	• Onerous requirements made of CCS projects under the UIC Class VI Rule • Preference to develop CO_2-EOR projects under UIC Class II Rule and questions about regulation of CO_2-EOR • Lack of carbon pricing mechanism causing funding difficulties / uncertainty amongst investors

Canada

In Canada, CCS regulatory frameworks are being developed in a number of provinces including Alberta, Saskatchewan, British Columbia and Nova Scotia. In Alberta, CCS plays a key role in Alberta's climate change strategy, and CAD $1.5bn has been allocated for the funding of three CCS projects in the province. A robust regulatory framework for CCS is evolving through the development of the Carbon Capture and Storage Statutes Amendment Act (2010), the Carbon Sequestration Tenure Regulation (2011) and the CCS Regulatory Framework Assessment (2011).

Alberta's currently on-going CCS Regulatory Framework Assessment (RFA) is helping to address any remaining gaps or issues in the regulations – for example, in relation to details of the criteria that must be met before responsibility and liability for stored CO_2 can be transferred to the Crown, and the structure of the Post-closure Stewardship Fund that can be used by the Crown in the post-closure phase for monitoring and potential remediation actions. The RFA draws together scientific, academic, regulatory, administration and industry experts, including CCS projects currently under development in the region.

Enhanced oil recovery using anthropogenic CO_2 (CO_2-EOR) is regulated according to long-standing petroleum laws and is not explicitly addressed in the existing regulatory framework for CCS in Alberta, but CO_2-EOR projects are able to gain credit for sequestered CO_2. Questions remain about the regulation of EOR-CCS projects (including provisions for a transfer from CO_2-EOR to CCS) and these are not being addressed in the RFA.

Europe

In Europe, the transposition of the 2010 Directive on the geological storage of CO_2 has triggered a number of regulatory developments at the EU Member State level. By spring 2014, only one Member State has not yet completed the transposition of the Directive. Whilst the CCS Directive helps to set a robust regulatory framework for the storage of CO_2, some of the more practical and technical details must be developed at the Member State level. In some cases this detail is currently lacking, posing some difficulties for projects first to progress through the permit application and regulatory approval process in these countries as these issues are resolved.

Other regulatory concerns relate to uncertainty about potential liability for global environmental impacts resulting in the event of a leak of CO_2. In the event of a CO_2 leak, operators would be required to surrender an amount of EU Allowances (EUAs) equal to the volume of CO_2 leaked, but there is uncertainty in the future value of EUAs, and uncertainty therefore in potential liabilities. Some concerns also exist in relation to the practicalities of being able to hand over liability for local environmental damages under the EU Environmental Liability Directive to the state, since strict liability is placed on operators and a strong 'polluter pays' principle applies.

Finally, some concerns were raised by project developers around the third-party access provisions of the Directive, with fears that the provisions for transparent and non-discriminatory third-party access to CO_2 transportation and storage infrastructure could place significant risks on CO_2 storage site operators in the event where they are required to accommodate new and additional sources of CO_2. It is likely that as the first CCS projects progress through the regulatory approval process, and as more guidance is issued across Member States, a number of these issues will be resolved and a more robust regulatory approval process will evolve for future projects.

US

In the US, the finalisation of the EPA's Underground Injection Control (UIC) Class VI rule (under the authority of the Safe Drinking Water Act) in December 2010 has important implications for the regulation of CCS. Projects injecting and storing CO_2 not for the purposes of EOR are required to gain a Class VI permit and will need to meet stringent regulatory requirements designed to minimise risk to underground sources of drinking water (USDWs). As of 2014,five Class VI permits have been issued, These include permits for the FutureGen project and the project of Archer Daniels Midland. Project developers perceive the regulatory requirements made of Class VI wells (particularly in relation to meeting requirements during the default 50-year post-injection site care period) to be particularly demanding, and some indicated a preference to develop CO_2-EOR projects using anthropogenic CO_2 that can be permitted under UIC Class II requirements. The US EPA also finalised a rule under authority of the Clean Air Act that requires facilities that conduct geologic sequestration (GS) of CO_2 and all other facilities that inject CO_2 underground to report greenhouse gas data to the EPA annually. This rule amends the regulatory framework for the US Greenhouse Gas Reporting Program.

CO_2-EOR (using anthropogenic CO_2) is likely to play an increasingly important role in providing the necessary incentives for the deployment of CCS in the US. CO_2-EOR projects are regulated under the UIC Class II regulations. Issues relating to when a project might transition from a Class II EOR well to a Class VI CCS well have been addressed to some extent. Criteria determining when such a transition might be made have been developed and further guidance on this transition is due to be published in the near future. With the lack of a federal carbon pricing mechanism, specific CCS funding initiatives such as the American Recovery and Reinvestment Act (allocating

US \$3.4bn to CCS development) play a key role in helping the development of CCS in the US, and the 45Q Tax Credit that would provide a \$10-20 / tonne CO_2 storage tax exemption has the potential to support early projects, providing uncertainties in the credit system are addressed.

Cross-cutting findings

Climate change policy context

The global climate change policy negotiations continue to play an important role that drives national policies and national behaviors under climate change, requiring reduction of emissions to some level that each nation, if subject to a global treaty, could commit to do. In the national policy context, a nation could then impose regulations that enable the adoption of CCS. The study illustrated the important role that a jurisdiction's broader policy context plays in shaping the regulatory landscape for CCS and in influencing other factors important in determining a project's success. CCS development can be incentivised by carbon pricing mechanisms placing a value on carbon and enhancing the business case for CCS, as seen, for example, with the EU Emissions Trading Scheme, the Alberta Specified Gas Emitters Regulation, and the Australia 2012 Carbon Tax, though the latter has been repealed as of July, 2014. However, market uncertainties (in terms of fluctuating carbon prices) present a risk to project developers and investors, particularly in Europe. Funding initiatives are therefore also important in incentivising CCS deployment: the American Recovery and Reinvestment Act (allocating US \$3.4bn to the development of CCS), and the Alberta

CCS Funding Act (allocating CAD \$1.5bn to the development of three CCS projects) have provided vital sources of funding to projects. Policies and regulations for industry and power generation that set energy efficiency and/or emissions targets such as Canada's recently introduced greenhouse gas performance standards for coal-fired electricity generation can also play an important role in driving the development of CCS.

As well as providing economic incentives for CCS project development, robust and ambitious climate change strategies can help to foster strong relationships between CCS project developers and regulators, as regulators have an inherent interest in the successful implementation of CCS projects and work closely with project developers to help them through the regulatory approval process. For example, in Alberta, widespread deployment of CCS (and the robust regulatory framework this requires) will be critical if targets set by the province to reduce emissions by 200 Mt by 2050 (representing a 50% reduction in business as usual emissions) are to be achieved.

Long-term liability for stored CO_2

Long-term liability for stored CO_2 is an important regulatory consideration for CCS projects, and a common concern across jurisdictions. Without a clarification and a resolution of this issue, a project developer faces critically uncertain future treatment of its liability of a geologic storage site which could completely overwhelm the future of a company as a going concern and hence stop any company from considering injecting any significant amount of CO_2. Transferring that liability ultimately to a competent authority or at least sharing and reducing the liability to a specific quantifiable level is the only way in which a commercial entity could even contemplate CCS as an option.

In Victoria, Australia, common law liability cannot be transferred to the state following the closure period, posing a risk to project developers in the region. This differs to provisions of the Australian Commonwealth offshore CCS regulations, where common law liability can be transferred after a minimum 15 year 'closure assurance period' has passed following the cessation of injection operations, raising questions about how liability would be handled with potential future cross-boundary storage projects.

In Alberta, tort liability can be assumed by the Crown following the issuance of a closure certificate, but climate change liabilities cannot currently be transferred.

In Europe, there is uncertainty about the potential extent of liability held under the EU Emissions Trading Scheme Directive with uncertain future prices of EU Allowances in the system.

In the US, long-term liability for stored CO_2 was one of the regulatory barriers to CCS deployment analysed by the Interagency Task Force on Carbon Capture and Storage in 2010. A number of options for the regulation of long-term liability were presented by the task force, including regulation at the Federal level, but to date, legislation addressing long-term liability has been developed only at the state level. States have addressed the topic in different ways and in some states such as Texas there is a lack of clarity about the extent to which a CCS trust fund enables the transfer of liability to the state.

Regulatory frameworks for CO_2-EOR

In North America, the issue of the regulation of enhanced oil recovery projects using anthropogenic CO_2 is gaining increasing attention. There is a history of CO_2-EOR operations in Alberta and in the US, and robust and long-standing regulatory frameworks for these activities are in place. However, case study interviews with both CO_2-EOR and (non-EOR) CCS projects highlighted significant differences that exist in the regulatory frameworks that apply to the different types of projects. For example, there are differences in requirements for monitoring, reporting and verification (MRV), financial responsibility, and closure period monitoring. Requirements made of CCS projects in these areas are perceived to be more stringent than those made of business-as-usual CO_2-EOR projects. Differences in regulatory treatment may mean that there are both financial and regulatory incentives to develop a CO_2-EOR project over a CCS project.

Following the cessation of a CO_2-EOR operation, a project may have the opportunity to transition to a full CCS project. In the US, the EPA UIC Program has outlined risk-based criteria that will determine when a transition from Class II to Class VI should be made, and is developing guidance on the practicalities of doing so. Class II EOR projects can opt in to the EPA's Greenhouse Gas

Reporting Program's Subpart RR and report the quantity of CO_2 sequestered at any time during the project. In Alberta, questions about the procedures that may exist for a CO_2-EOR to CCS transition have been raised but have yet to be addressed. The Alberta regulators are likely to look into this following the completion of the CCS RFA in 2013. In the meantime, questions about the regulation of CO_2-EOR in the context of a CCS project remain.

Case study lessons learnt

Interviews with project developers highlighted a number of lessons that were learnt from going through the regulatory approval process. Broadly, these can be framed in the context of lessons learnt that may be applicable to the development of regulatory frameworks in jurisdictions, and lessons learnt that may be applicable to CCS project developers going through the regulatory approval process across jurisdictions.

Development of regulatory frameworks for CCS:

- Projects already in existence (including small-scale demonstration projects) can help with the development of regulations by providing insights based on technical knowledge and experience.
- Projects first to test a newly developed regulatory framework can play an important role in working closely with regulators to help shape the development of regulations.
- Issues may arise when considering how and to what extent newly implemented regulations will be retroactively applied to projects already existing and permitted by other means.
- It is important to be able to tailor regulatory requirements to projects of different sizes and contexts; a small-scale demonstration project may not be able to meet requirements geared towards a large-scale CCS project.

Progression through the regulatory approval process:

- Projects first to progress through a newly developed regulatory framework may face a lengthy and rigorous permitting procedure if there is a lack of technical and practical detail in the regulations, and where there is a desire for a robust precedent to be set for future projects.
- Where regulatory requirements are not sufficiently detailed (for example following the introduction of a new regulatory framework), projects can take a conservative approach and go beyond perceived minimum requirements to help gain regulatory and stakeholder approval.
- In North America, CCS project developers may choose to follow a more well-defined regulatory pathway by developing CO_2–EOR projects (using anthropogenic CO_2).
- Regulatory barriers do not exist in isolation. Even when regulatory approval has been or is likely to be given, other issues are critical in influencing the successful implementation of a CCS project, including gaining stakeholder approval, being able to secure financing, and achieving commercial reality where the consumers are willing to pay for the price of the products (i.e. energy, gas, oil, electricity).
- Close relationships between CCS project developers and regulators have been crucial in helping with the development of regulations and helping with the progression of projects. Even where gaps and questions remain in CCS regulatory frameworks, project developers have been able to work closely with regulators to gain regulatory approval and progress with project development and implementation.

CCS STAKEHOLDER ISSUES REVIEW AND ANALYSIS

The second section of this chapter explores the issues that different types of stakeholders have about CCS. In 2007, the CO_2 Capture Project undertook the first stakeholders' priorities and issues review and analysis. In 2011, the CO_2 Capture Project commissioned consultancy Environmental Resources Management (ERM) to update and extend the previous work. More specifically, ERM was asked to:

- Compile and review the full range of issues raised by stakeholders of CCS projects.
- Indicate which stakeholder views appear to be most strongly held and identify what drives their sensitivity.
- Review existing surveys and studies associated with CCS issues and develop a number of selected case studies in different regions of the world.
- Provide perspectives from stakeholders on barriers and gaps associated with addressing these CCS issues from their perspectives.

The aim of this work was to identify and analyse the main stakeholder concerns and hot spots and provide an overview of options available to project developers and industry for responding to them. It should be noted that the conclusions drawn here have not been tested directly with the stakeholder groups studied in this report.

The study identified these key categories of stakeholder's priorities related to:

1. Environmental, Health and Safety Impacts;
2. Awareness and acceptance of CCS;
3. Technical aspects associated with CCS;
4. Commercial and local development benefits;
5. Policy and legal issues;
6. Diversion of resources away from renewable energy;
7. CCS as contributing to positive impacts on climate change;
8. CCS as contributing a negative impact on climate change;
9. Groups with variable positions on CCS and issues of concern.

Table 2 illustrates the main areas of concern to different stakeholder groups, highlighting those that are the focus of their attention, but also noting the full range of issues that were raised in this study.

The distribution of issues shows that the concerns of NGOs and Thought Leaders, the General Public, and Politicians and Policy makers is focused on climate change, the diversion of resources away from renewable energy projects and associated policy discussions. Local communities and regulators are particularly focused on project related environmental, social and health impacts and benefits. Industry and investors have concerns about project impacts and stakeholder opposition at the project level, and also an interest in the policy debate which may impact the commercial viability of CCS.

The focus of the interests of different stakeholders suggests that there is a continuum of stakeholder interests which are broadly directed at two different outcomes:

- Project / local level discussions associated with management of social, environmental, health and safety impacts, and delivery of local benefits;
- Global level discussions on climate change and the role of CCS.

Table 2. Areas of concern of different stakeholder groups.

	EHS Impacts	Awareness & acceptance of CCS	Technical aspects	Commercial & local development benefits	Policy & legal issues	Diversion from renewable energy	Positive impact on climate change	Variable positions on CCS
NGOs & Thought Leaders	✓		✓		✔	✔	✔	✔
General Public	✓	✓	✓	✓	✔	✔	✔	
Politicians & Policy makers	✓	✓	✓	✓	✔	✔	✔	✔
Industry	✓		✔	✔	✔	✓	✓	
Local Community	✔	✓	✓	✔	✓			✓
Regulators	✔	✔	✔		✓			
Investors	✓	✓		✔	✔			
Media	✓	✓	✓	✓	✓	✓	✓	✓
✔ Focus of interest			✓ Issue noted					

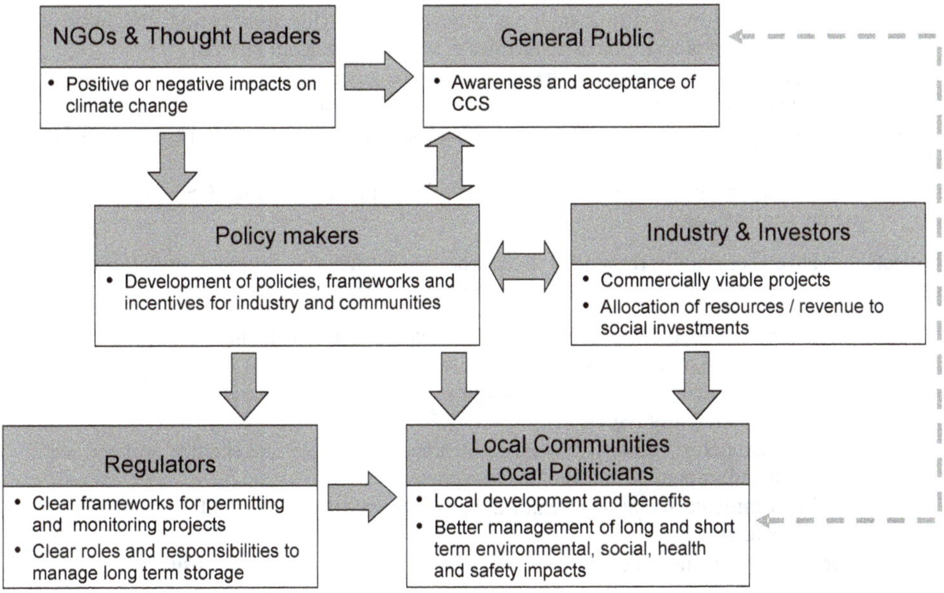

Figure 1. Global and Local Interactions.

Policy makers are at the centre of this continuum as their interest and commitment to CCS in resolving CCS concerns may influence the commitment and support provided to a project in putting in place a regulatory framework and communicating to stakeholder the value of the project.

The most important stakeholders for project development are consistently:

- Policy Makers - National Government;
- Local Community; and
- Regulators and Local Permitting Authorities.

NGOs, Thought Leaders and the Public, often did not feature as having significant influence for each project. However, it is clear that their interest and level of influence is within the wider climate change debate and the role of CCS in its resolution.

This is not to suggest that project development and the direction of policy discussions on climate change are not linked. Indeed, emissions targets, carbon taxes and other incentives may make CCS projects commercially viable and facilitate the delivery of local benefits, as companies will have more money to invest in projects. Given these different priorities, management of stakeholder issues in project development and management of stakeholder issues in the broader climate change debate may require different emphases.

At the project level there are two key areas which fundamentally aid in addressing stakeholder concerns: Communication and Engagement and addressing stakeholder issues through the Project Development Process. Key lessons learned at project level are:

- Start early to raise awareness with politicians, regulators and community.
- Educate local government and other community opinion leaders so they can answer questions about the project.
- Aim to build trust by using multiple channels to provide information and involve 'objective' stakeholders such as academics or other independent experts.
- Have good project people on the ground in the community and / or find a good representative from the community who will support the project.
- Understand community specific concerns and answer questions – don't assume what information will be needed.
- Good engagement will not necessarily result in acceptance of CCS – it is not a guarantee of success.

Projects that have successfully responded to stakeholders issues have invested more resources than usual at early stages in project development in order to:

- Demonstrate understanding of the geology, containment and monitoring feasibility to company decision review boards and regulators;
- Assess local capacity to regulate the development of the project and manage long term monitoring and liability issues;
- Identify stakeholder sensitivity, raise awareness of key stakeholder groups and understand and respond to stakeholder concerns;
- Avoid and mitigate social and health impacts or perceptions of health impacts during site selection; and
- Develop mechanisms to deliver community level benefits (a value proposition).

Key lessons learned about communication and engagement with global stakeholders are:

- The role of CCS can be discussed more meaningfully only once people (i.e. the public) have a more balanced and complete understanding of the process itself and what it can offer in the wider context of mitigating climate change; investment in broadening this understanding may be of value.

- It is important to consider the perceived trustworthiness of sources when communicating on the topic, and to take care to build and maintain the public's trust in CCS and its proponents.
- Public opinion could be strongly shaped by the media, which has yet to take a great interest in CCS.
- Working with NGOs to undertake research, or set the scope of research will help ensure studies answers the questions and concerns raised by these groups as well as CCS specialists. It can also help to demonstrate how industry is building its experience and technical capacity in CCS.
- Open and regular engagement with a range of NGOs and thought leading organisations and individuals is advisable in order to maintain an understanding of the variety of views of these stakeholders and changes in their views.

The remainder of this executive summary presents a concise overview of the main stakeholders associated with CCS projects and the issues and concerns they have about CCS based on the findings of the study.

The General Public

Public perception can have a significant influence on the success or failure of major planned projects involving new technologies and structures. If the general public is not supportive of – or is even actively opposed to – a new technology, it can become politically and/or socially unacceptable. Project developers should therefore be mindful of the potential power of the general public (and the media, as discussed below), to 'make or break' a new technology (regardless of the scientific basis for doing so).

There are two contextual conditions that serve to support acceptance of CCS. First, climate change should be recognised as a problem; and secondly, a significant reduction in CO_2 should be recognised as the only solution to the problem. An understanding of climate change and the associated need for concerted action can constitute a prerequisite for acceptance and support for CCS and other climate mitigation options by stakeholders.

The lack of knowledge about CCS in the general public could be due to the fact that there is relatively little information on CCS that is designed for the public, and CCS as a concept requires careful explanation. There can also be confusion about the difference between CCS and the broader category of carbon sequestration.

Part of the reason for the lack of general information about CCS, and the consequent lack of understanding about it, is that to date, little interest in the issue has been exhibited by the mass media in most countries.

A lack of general understanding of CCS and acceptance of its application remains a concern for those developing projects. Current perceptions can include that CCS is expensive, risky and perpetuates fossil fuel dependence.

Public understanding of technical aspects of CCS is not as important as trust in those providing information. The public will often trust universities and research institutions more than government or industry.

Local Communities

Local communities can have significant influence on the success or failure of projects. Policy makers, regulators, investors and civil society increasingly advocate for the consultation of local communities and assessment of impacts to communities in the development of major projects. Local communities can also create significant delays to projects, not only by influencing permitting processes, but also by physically restricting activities with demonstrations or blockades if there are significant levels of concern about a project.

Locals can also have direct access to media, giving them the ability to communicate their concerns to a wide audience. The media often cover high-profile aspects of CCS where a project has failed to obtain planning permission due to highly vocal local opposition.

Key insights on local community issues include:

- Concerns vary from place to place but typically involve safety and financial impact;
- It is possible to identify some 'first principles' for engagement which will help to allay some of these concerns at the outset, such as integrating public outreach into project management, conducting and applying social characterisation, developing key messages and outreach material tailored to its audience
- Engagement will not necessarily result in acceptance of CCS;
- Local opposition is an issue with CCS as with other major infrastructure projects;
- Perception of risk may not equate to actual technical risk, but it is still valid;
- Trust is a key determinant of the success of a CCS project;
- The history of a project location is a key determinant of the project's success;
- Demographic characteristics are important factors in acceptance of CCS.

Responses to CCS are very much determined by context. People tend to object less to CCS where they have already got experience of the energy industry or other large-scale industrial processes. By contrast, in cases where opposition occurs, the fossil fuel industry is generally new, and/or does not have a good long-term relationship with local stakeholders. Thus, the history of a given location can predispose people either for or against a project.

Having a value proposition for the local community from the outset of the project is vital. The value proposition needs to be developed to respond to the local context; what works in one area may not be acceptable in another.

CCS can deliver benefits to communities, e.g. if projects pay for CO_2 stored or some enhanced oil recovery revenues are re-invested locally.

Non-governmental Organisations (NGOs)

Many NGOs perceive CCS as a bridging technology, and are neutral or provide support on the condition that it is a step in moving towards a low carbon economy. Conditional support can mean NGOs vary their position from project to project, e.g. supporting CCS with regard to gas-fired power stations, but not with regard to growing reliance on coal–fired power stations. Other NGOs are still developing their positions on CCS.

Four main positions on CCS have been identified amongst NGOs:

- Positive about CCS and its contribution to addressing climate change
- CCS is a bridge to a renewable future
- CCS may help to provide a bridge but it is an unproven technology

- Against CCS as a technology to support addressing climate change

Key concerns identified amongst NGOs can include:

- Diversion of effort from renewable energy;
- Impact on ecosystems;
- Cost of deployment;
- Threat of leaks;
- Long term economic impacts;
- Continued fossil fuel use; and
- The scale of deployment.

The general public and local communities often identify with or are influenced by NGOs' viewpoints on debates like those surrounding CCS. This makes NGOs a potentially powerful lobby that can be a difficult adversary or a useful ally in project approval.

Policymakers and Politicians

Politicians at all levels are influential stakeholders in the CCS debate. Their support for the technology at large and for specific projects at regional or local level is critical to success, whilst opposition can prove to be very problematic. As policy makers, politicians set the terms under which CCS must operate and can facilitate or hinder its progress accordingly.

At local level, politicians can distance themselves from a proposed CCS project if they sense public opposition, even if their party is officially supportive of CCS at national level. It is important to develop good relationships with local politicians to try and understand their comfort with or concerns about CCS, and if possible help to avoid politicising a project.

Projects take many years to develop, so proponents should therefore engage politicians and policy makers early to help manage risks associated with government approval.

Regulators

Regulations of relevance to CCS are often not clear cut. Some jurisdictions have enacted or are working on legislation to clarify the ownership and stewardship aspects of underground pore space for CO_2 storage sites and for transfer/management of long-term liability. There are other regulatory issues beyond pore space and liability which must be dealt with as well.

Where there is a lack of regulations, the expectations of a regulator can be unclear and unpredictable. This creates uncertainty which may result in delays or complications. The fact that legislation and regulation governing CCS is still not clear cut in many contexts and countries is a problem for the regulator seeking to manage projects in this area. Governments need to develop comprehensive regulatory frameworks for CCS, and they need to support the regulator to build capacity to regulate CCS.

Investors

Fundamentally, CCS projects should not present a greater or lesser risk to investors than other infrastructure projects. Typical financial community issues will include:

- The commercial viability of CCS as an investment and potential provision of incentives for industrial deployment of CCS;

- Reputational risks when CCS is associated with coal fired power stations or because of CCS elements; and
- The extent to which employment of CCS will support a bank's climate change and energy policies.

The acceptability of a project including CCS elements to local and other stakeholders is important to investors who want to avoid financing a technology that proves to be socially or politically unacceptable. Investors will expect that these reputation risks are effectively managed by the developer. Therefore, banks will seek to understand how a project is managing stakeholder related issues as part of their investment decision-making process.

Different investors have different drivers for investment:

- Government is an important investor, as CCS is often not commercially viable; government willingness will be linked to the position of policy makers;
- Multi-lateral banks and export credit agencies may provide investment, or facilitate funding in line with international and regional policy objectives;
- Commercial banks will invest in CCS where it supports the banks' internal policies;
- Industry will invest for research and development and where CCS will be commercially viable, for example where CCS can be tied to enhanced oil recovery.

Where banks do have energy and climate change policies they may see value in:

- Project finance for activities like CCS that significantly reduce emissions;
- Corporate finance to companies that demonstrate a willingness and commitment to implementing CCS in order to reduce CO_2 emissions.

Regional Differences

Analysis at a country/region level does not appear to provide an accurate indicator of overall stakeholder views on CCS and associated sensitivity. Sensitivity will be context and location specific, reflecting a number of factors including, amongst others, location, population density, and historic issues / circumstances. In this sense, the study reflects common conclusions on broad stakeholder views. However, this is not to suggest that all stakeholder groups across the world will hold the same positions described in the following sections.

Finally, although much of the literature reflects upon negative experiences, it should be noted that there are examples of more positive stakeholder responses to CCS, particularly in Alberta, Canada.

LOCAL COMMUNITY BENEFIT SHARING MECHANISMS FOR CCS PROJECTS

The third section of the chapter explores the experiences and options available to CCS project developers for local community benefit sharing.

Project developers across the energy, mining and waste sectors are increasingly focused on enhancing local benefits, by maximising direct and indirect positive local impacts associated with a development, and also through specific community investment programs. In the context of a CCS project, benefit sharing needs to be considered particularly at the storage stage, which frequently harbours the greatest public concerns around perceived health, safety and environmental risk, yet typically receives few direct or indirect local benefits (such as employment or local procurement) that stem from project activities.

Developers need to think creatively about how to fill the 'benefits gap' at the storage stage and how, in turn, to create a value proposition for the community hosting the CO_2 storage site. A variety of approaches were identified with the potential to increase the attractiveness of a proposed CCS project to storage communities, including:

- Direct revenue sharing (for commercially driven projects);
- Distribution of direct benefits typically experienced in the capture stage (including employment, procurement of local goods and services, and infrastructure construction) across the CCS chain; and
- Development and implementation of specific community investment programs.

Community investment should be approached in a strategic way in order to ensure projects financed by the establishment of a community development fund have a positive impact on the community, and are sustainable in the long-term. Principles for a successful and sustainable community investment program include: involving multiple local stakeholders in the planning and management of the fund; having a set of principles and objectives governing the fund; allowing the community to guide how funding should be spent; and using indicators to measure the success of the fund and projects over time. Such principles for a 'strategic approach' should apply to any community investment and benefit sharing program.

Introduction and approach

In 2011, the CO_2 Capture Project conducted a review and analysis of carbon capture and storage (CCS) stakeholder issues, which is summarized in the second section of this chapter. That study identified and analyzed a number of areas of concern in relation to the development of CCS to different stakeholder groups, including local communities, which can have significant influence on the success or failure of a CCS project. One of the key findings from the paper was that local communities are more likely to become actively involved and oppose project developments when there are no apparent benefits to the local community itself.

The third section of the chapter aims to explore this finding further and investigates experience and options for local community benefit sharing. Initially, a desk-based review of local community benefit sharing experience across the energy, mining, and waste sectors was conducted. Following this, four projects in the energy sector (including one CCS project) were explored in greater detail and interviews were conducted in order to gain 'on-the-ground' insights into the benefit sharing process, and specific mechanisms employed. Findings from the review were subsequently analyzed in order to explore how community benefit sharing might apply in the context of a CCS or CCUS (carbon capture, utilization and storage) development.

Local community benefit sharing: International experience and approaches

Local community benefit sharing

The review shows that project developers are increasingly focused on enhancing local benefits associated with a project, by maximizing direct and indirect positive local impacts associated with a development (such as employment and the procurement of local goods and services), and also through specific community investment programs. It is widely recognized that the sharing of benefits with local communities in this way can help to address the potential imbalance of local costs and national or global benefits that can arise with many projects in the energy, mining, and waste sectors. Whilst projects in these sectors can have national or even (in the case of a CCS) global benefits, the negative impacts or 'costs' of the development (such as noise, visual impacts, pollution, and perceived or actual health and safety risks) are often concentrated at the local level. Benefit sharing allows benefits to be transferred to local communities and can help projects gain acceptance at the local level.

Benefit sharing within the broader social risk and impact management process

Benefit sharing should not be approached in isolation, and must be considered in the context of the broader social risk and impact management process. Within an overall framework of public consultation and stakeholder engagement, there are three important components of a social risk and impact management process. Project developers should seek to:

- Address and minimize potential direct and indirect negative impacts to the community;
- Compensate for unavoidable negative impacts to the community; and
- Enhance positive impacts to the community, by maximizing direct and indirect positive project impacts and enhancing local benefits through community investment.

Figure 2 below illustrates a framework for social risk and impact management, incorporating the above components.

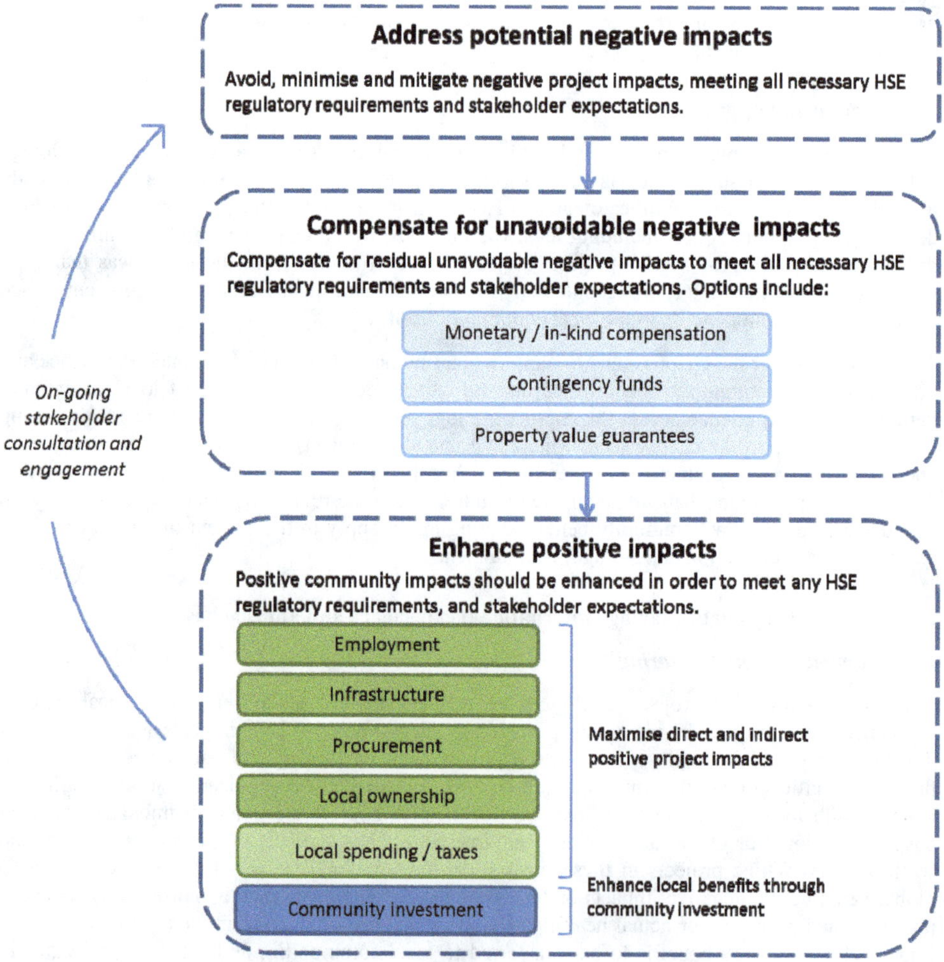

Figure 2. Framework for social risk and impact management.

Before options for benefit sharing are discussed in detail with local communities, developers should first seek to identify and manage any negative impacts from a development as far as practical. If detailed engagement on benefit sharing comes too early on in the stakeholder consultation process, communities may feel that their concerns are not being fully addressed and offers of benefits may indeed be counterproductive and increase local resistance. Community benefits must also remain distinct from agreements to meet all normal legislative and regulatory requirements: although a well-planned and implemented community benefits program has the potential to foster a sense of 'goodwill' with local legislators, community benefits should not be seen as a 'short cut' to obtain approvals.

The first priority for a project developer seeking to manage the social impacts associated with a development should be to avoid, minimize, and mitigate potential direct and indirect negative impacts. For any residual impacts, compensation measures can be applied. For example, payments or in-kind goods could be provided to communities to compensate for specific negative impacts; property value guarantee schemes could be established to compensate residents in the event that the development causes a decline in property values in the area; or contingency funds could be put in place to compensate residents from damage should there be an unexpected event or emergency linked to the development. It might be necessary to develop any or all of the above compensation measures in the context of a CCS development, depending on the project, and on specific local concerns.

Benefit sharing options and experience

After potential negative project impacts and community concerns have been addressed and managed, project developers can initiate detailed engagement on community benefits. International experience with benefit sharing approaches highlights a number of different forms of benefits associated with developments, and a number of different channels through which these benefits can be distributed to local communities, as illustrated in Figure 3.

Figure 3 shows that communities may benefit from the direct and indirect positive impacts of a project, including employment opportunities, the construction of mutually beneficial infrastructure, the procurement of local goods and services, local ownership and direct revenue sharing opportunities, and indirect positive economic impacts (including increased local spending by the workforce, and/or increased local tax revenues and expenditure on public services). The local community context, as well as the project context, will influence the extent to which a local community might benefit from these direct and indirect positive impacts. For example, local employment opportunities will only exist if workers with the required skillsets live in the nearby communities. Similarly, procurement is only likely to offer significant local benefits if the specific goods and services required by the project can actually be provided by local suppliers. Some developers, particularly with projects in developing countries, have invested in building local business capacity and/or training up the local workforce to ensure that the benefits associated with the project can be shared locally. On the project side, the scale of the development, and the timeframe for construction and operation, will influence how significant and long lasting the direct and indirect local benefits are likely to be.

Joint ownership and revenue sharing is increasingly being used as a mechanism to deliver benefits to local communities in the renewables sector. A number of wind farm projects have sold shares to communities or established cooperatives to allow local communities to have ownership in projects. Other wind developments have established revenue sharing schemes whereby local communities gain a proportion of the revenues from the sale of electricity by the project. Both approaches can help to increase local awareness and buy-in to proposed development opportunities.

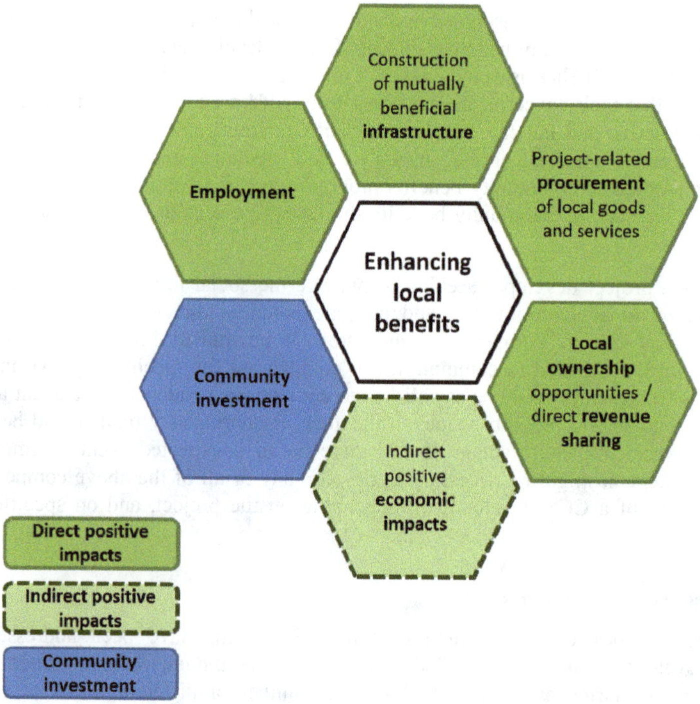

Figure 3. Community benefit sharing measures.

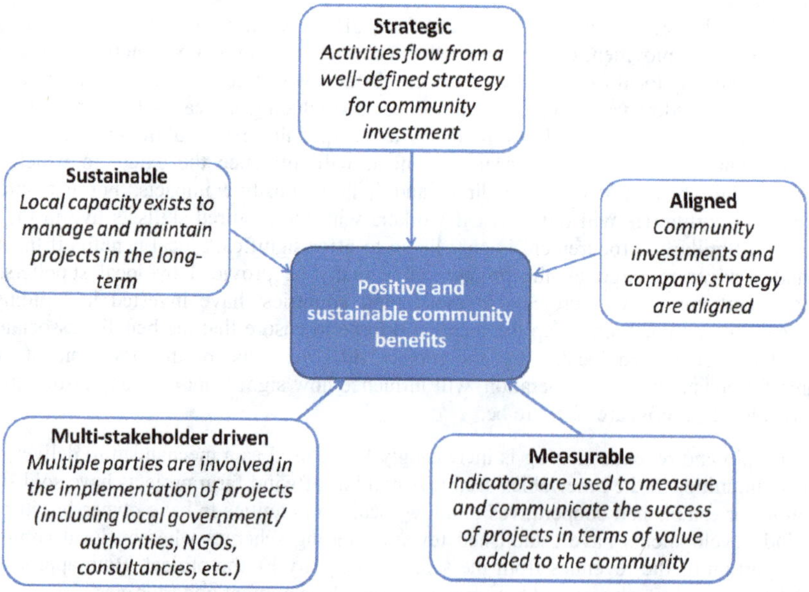

Figure 4. A strategic approach to community benefits. [2]

Another way to enhance local community benefits is through community investment. Globally, a vast number of community funds have been established by project developers, ranging in size from US $1,000-10,000 to multi-million dollar funds. Experience suggests that it is important to approach community investment in a strategic way to ensure that the projects financed by the fund have a positive impact on the community, and are sustainable in the long-term. Principles for a successful and sustainable community investment program include involving multiple local stakeholders in the planning and management of the fund; having a set of principles and objectives governing the fund; allowing the community to guide how funding should be spent; and using indicators to measure the success of the fund and projects over time. Such principles for a 'strategic approach' should apply to any community investment and benefit sharing program, as illustrated in Figure 4. [2]

Benefit sharing in the CCS context

An important factor determining the extent to which a local community might benefit from a project's direct and indirect impacts is the geographic spread of a project. This is particularly relevant in the CCS context, where the direct and indirect benefits associated with a new CCS development are likely to be concentrated at the site where a new industrial plant (with CO_2 capture) or a new CO_2 pipeline is to be installed. At the capture stage, and (to a lesser extent) at the transport stage, there may be local benefits associated with job creation, procurement needs, infrastructure upgrades and improvements, and more broadly, indirect economic benefits. However, at the storage stage, where the greatest perceived negative impacts of a CCS project often reside, there is likely to be an absence of such local benefits, although enhanced oil recovery (EOR) activities in a CCUS project can create jobs, procurement opportunities, and indirect economic benefits at the storage site.

For CCS projects where no new EOR activities are planned, developers may therefore need to think creatively about how to fill this 'benefits gap' at the storage stage and how, in turn, to create a value proposition for the community hosting the storage site. An analysis of the drivers for CCS may help developers to identify a preferred approach to benefit sharing. Commercially-driven projects may be able to consider revenue sharing as a benefit sharing option. If projects are not commercially driven and are instead government funded (e.g. with a view to progressing CCS to help meet national greenhouse gas reduction targets), opportunities to partner with local government or local authorities when engaging and consulting with local communities and other stakeholders could be explored, along with options for putting in place benefit sharing mechanisms and/or community investment programs to ensure the equal sharing of benefits across the CCS chain.

It is important to emphasize the importance of a robust stakeholder engagement process throughout the impact management and benefit sharing process. Developers should plan for and undertake targeted stakeholder engagement and consultation on the project at the earliest possible stage; take a balanced view; and should ensure that options agreed upon are carried out and stakeholder contact is maintained during the project's operation.

CONCLUSIONS

This paper has reviewed local community benefit sharing experience and approaches and has considered how such approaches might apply in the context of a CCS development, where risks and negative project impacts to local communities at the CO_2 transport and storage stages could outweigh any potential positive impacts and benefits associated with the development. Following a review of four projects and discussions with project developers and social impact management experts, the following conclusions can be drawn.

Local community benefit sharing is increasingly recognized as 'best practice' with major developments, and it can form an important part of a project's social impact management plan.

Community benefits must remain distinct from agreements to meet all normal legislative and regulatory requirements; they are not a 'short-cut' to obtain approvals.

There are some barriers to be overcome. These include:

- The need to address the imbalance of positive and negative impacts across the CCS chain. There may be a number of ways to ensure benefits are distributed across each stage of the CCS chain but they require careful appraisal.
- Potential for consultation 'burn-out'. Communities without experience of major developments may need guidance, specialist support and expertise in order to participate effectively in the community benefit sharing process. Project operators need to ensure that effective communication with the local community is maintained throughout project life.
- Determining what constitutes the 'local' community to receive the benefits to ensure that an overall sense of fairness applies throughout the process.

International experience across sectors (including the oil and gas sector) highlights a number of local community benefit sharing options associated with: 1) maximizing the direct and indirect positive impacts of a project (such as employment, local procurement, and wider economic benefits); and 2) enhancing local benefits through strategic community investment programs.

In the context of a CCS project (whilst noting that CCS project contexts may differ significantly), benefit sharing needs to be considered particularly at the storage site where frequently there is no real 'value proposition' for the local community (i.e. the real and perceived risks outweigh potential benefits).

A number of benefit sharing approaches could be used to increase the attractiveness of a proposed CCS project to communities hosting the storage site. These include revenue sharing (if the project is commercially driven), ensuring that benefits are shared across the CCS chain (i.e. distributing positive project impacts across capture, transport and storage stages as far as possible), and community investment. For non-commercial, government driven CCS projects, community investment is likely to be most applicable.

Some broad principles for community investment are summarized below.

- A strategic approach to community investment is required if community benefits are to be positive and sustainable.
- Involving multiple stakeholders in the planning and implementation of projects will help to ensure projects are effectively implemented and can be maintained in the long-term.
- A fund Committee or Board comprising the project developer and key community stakeholders can be established to manage the fund. The governance or procedural mechanisms and accountability of the process is essential to ensure that everything is transparent and beyond criticism.
- A set of Principles and Objectives governing the fund should be established to help focus how the fund is managed and spent, and how the success of the funding can be measured over time.
- The choice of funding mechanism (e.g. direct or revenue linked payments) may be influenced by the type of project being implemented, and the community context.
- Decisions on what funds should be spent on should be led by the community.

The drivers for a CCS project may influence what type of benefit sharing approach is preferable in different CCS contexts. Commercially-driven projects may be able to consider revenue sharing as a benefit sharing option. If projects are not commercially driven but instead are government funded (e.g. with a view to progressing CCS to help meet national GHG reduction targets), there may be options to partner with local government or local authorities when engaging and consulting with local communities and other stakeholders, putting in place benefit sharing mechanisms and/or community investment programs to ensure the equal sharing of benefits across the CCS chain.

It is important not to lose sight of the fact that benefit sharing is not a 'silver bullet' when it comes to local acceptance of developments, and instead must be incorporated into a robust social impact management process that incorporates targeted stakeholder engagement and consultation, and the management of project impacts.

Modern trends are towards joint ownership, profit sharing and cooperative agreements in an attempt to achieve heightened levels of community awareness and buy-in to proposed development opportunities.

SUMMARY

In this chapter, we:

- Examined the range of regulatory issues and policy barriers encountered by real-world projects going through or were going through permitting processes.
- Explored the range of issues deemed critical or important by different categories of stakeholders – from national to local governments, and from global or national environmental activist organizations to national and local media to local communities near the fence line of future CCS projects to be sited.
- Analyzed examples of benefits-sharing mechanisms which have been applied to a range of oil and gas and infrastructure projects to gain insights about how such benefits-sharing mechanisms could help CCS projects become accepted by local communities.

We confirmed again the critical nature of all three areas. In this chapter, the CO_2 Capture Project commissioned analyses that:

- Examined the range of regulatory issues and policy barriers encountered by real-world projects going through or were going through permitting processes.
- Explored the range of issues deemed critical or important by different categories of stakeholders – from national to local governments, and from global or national environmental activist organizations to national and local media to local communities near the fence line of future CCS projects to be sited.
- Analyzed examples of benefits-sharing mechanisms which have been applied to a range of oil and gas and infrastructure projects to gain insights about how such benefits-sharing mechanisms could help CCS projects become accepted by local communities. Permitting real-world projects require the practical evaluation of and the eventual removal of specific policy barriers; understand and resolve issues of concern to different levels of governments and permitting authorities, range of environmental NGOs , and target benefits in an overall business strategy that includes sharing such benefits with local stakeholders.

ACKNOWLEDGEMENTS

The authors would like to thank Professor Ray Kemp for his contributions and expert advice on risk, public perceptions of risk, and approaches to community benefit sharing. In addition, the authors are grateful for the input from the project managers and stakeholder engagement experts who provided valuable input to the case studies developed as part of the paper.

The CO_2 Capture Project (CCP) is an award-winning partnership of major energy companies, working to advance the technologies that will underpin the deployment of industrial-scale CO_2 capture and storage. Phase Three (CCP3) members are: BP, Chevron, Eni, Petrobras, Shell and Suncor.

REFERENCES

1. IPCC, 2014: Climate Change 2014: Mitigation of Climate Change. Contribution of Working Group III to the Fifth Assessment Report of the Intergovernmental Panel on Climate Change [Edenhofer, O., R. Pichs-Madruga, Y. Sokona, E. Farahani, S. Kadner, K. Seyboth, A. Adler, I. Baum, S. Brunner, P. Eickemeier, B. Kriemann, J. Savolainen, S. Schlömer, C. von Stechow, T. Zwickel and J.C. Minx (eds.)]. Cambridge University Press, Cambridge, United Kingdom and New York, NY, USA.
2. IFC (2010) Strategic Community Investment: A Good Practice Handbook for Companies Doing Business in Emerging Markets, IFC, Washington, DC. Online, available at: http://www.ifc.org/wps/wcm/connect/topics_ext_content/ifc_external_corporate_site/ifc+sustainability/publications/publications_handbook_communityinvestment__wci__1319576907570

Carbon Dioxide Capture for Storage in Deep Geological Formations, Volume 4
Karl F. Gerdes (Editor)

Chapter 51

OVERCOMING REGULATORY HURDLES TO CCUS DEPLOYMENT: A U.S. MULTI-STAKEHOLDER APPROACH

Robert F. Van Voorhees
EcoReg Matters Ltd., 1155 F Street, N.W., Washington, DC 20004

ABSTRACT: This chapter describes implementation of the EcoReg Matters Ltd. project for identifying, framing and, in some cases, resolving concerns about potential regulatory hurdles to the deployment of demonstration projects for carbon capture utilization and storage (CCUS). These hurdles can arise within the framework designed to regulate the implementation and reporting of CCUS activities in the United States. Focusing on the CCUS regulatory framework for underground injection control (UIC) and greenhouse gas (GHG) emissions reporting that are being created by the United States Environmental Protection Agency (EPA), a number of the concerns were identified through a series of multi-stakeholder processes in which representatives of industries, environmental nongovernmental organizations (ENGOs) and state environmental or oil and gas agencies participated along with representatives from EPA, the Department of Energy (DOE) National Energy Technology Laboratory (NETL), and the NETL-sponsored Regional Carbon Sequestration Partnerships (RCSPs). This Chapter summarizes issues that have been raised, explains the variety of ways in which these issues have been addressed, and identifies solutions that have been developed for many of the issues. It identifies other issues that remain unresolved and explains the approaches, alternatives considered, and current status. It also offers observations that should be helpful when addressing similar issues related to policies and regulatory frameworks for CCUS.

KEYWORDS: area of review; carbon dioxide stream characterization; experimental technology well; geologic sequestration (GS); GS testing and monitoring; monitoring reporting and verification (MRV) plan; multi-stakeholder discussion process; post-injection site care (PISC); RCRA; state primacy; storage site characterization; UIC Class V; UIC Class VI

EXECUTIVE SUMMARY

The funding for this project was provided by the CO_2 Capture Project (CCP), now in its third phase. CCP3 supported a process for identifying, framing and, in some cases, resolving concerns about potential regulatory hurdles to the demonstration and deployment of technologies for the carbon capture utilization and storage (CCUS) through individual projects. The focus is on the CCUS regulatory framework being created by the United States Environmental Protection Agency (EPA) under its existing statutory authorities. A number of the concerns have been identified through a multi-stakeholder discussion (MSD) process in which representatives of industries, environmental nongovernmental organizations (ENGOs) and state environmental or oil and gas agencies participated, beginning in November 2008 and continuing to date. Other concerns were identified in a workshop convened by EcoReg Matters Ltd. on December 15, 2011, to bring together representatives from EPA, the United States Department of Energy (DOE) National Energy Technology Laboratory (NETL), the NETL-sponsored Regional Carbon Sequestration Partnerships (RCSPs), ENGOs, state environmental and oil and gas agencies and geologic surveys, and a number of industrial project partners and trade associations.

This report identifies the issues that have been raised, explains the variety of ways in which these issues have been addressed, and identifies solutions that have been developed for many of the issues. It also identifies other issues that remain unresolved and explains the approaches, alternatives considered, and current status as the participants in this process seek additional solutions. Consistent with the agreed frameworks for many of these discussions, there will be a minimum of attribution of particular statements and expressions of positions. Instead, the focus will be on identifying the concerns, explaining the processes that have been used, identifying alternative approaches that have been discussed, and describing solutions where those have been finalized. For unresolved issues still under discussion, the report explains the current status of those discussions and offers observations that should be helpful when addressing similar issues relating to policies and regulatory frameworks for CCUS.

Carbon Dioxide Stream Characterization

The identification and characterization of the CO_2 streams to be managed through CCUS is an important step that can affect how those streams are managed under either new or existing regulatory regimes. The potential characterizations or classifications most recognized for captured CO_2 streams are purity, hazardous, waste, pollutant, and commodity. Typically, any immediate discussion of "purity" is directed at restricting the streams that can be considered for CCUS without focusing on the context in which it will be conducted and without taking a risk-based approach to regulation. For example, a blanket requirement that CO_2 streams consist of at least ninety-eight percent CO_2 would not be grounded on a practical necessity. To its credit EPA chose to avoid imposing any abstract purity limits on the streams that could be considered for CCUS. Instead, EPA has imposed requirements on content to avoid interference with geologic containment or the use of CCUS as a means for disposing of unassociated wastes. Specifically, EPA allows CO_2 streams to include "incidental associated substances derived from the source materials and the capture process, and any substances added to the stream to enable or improve the injection process". This should also be understood to include substances, such as tracers, added to improve monitoring and verification activities, as well as substances derived from naturally sourced CO_2 commingled for transportation through common pipelines.

Under the underground injection control (UIC) program of the Safe Drinking Water Act (SDWA), it is not necessary for EPA to classify CO_2 streams as "waste" for purposes of management or regulation. The UIC program covers the underground injection of fluids and applies to both the long established process of injecting CO_2 for the purpose of enhanced oil recovery (EOR) and the proposed injection of CO_2 streams for geologic sequestration (GS). This provides important flexibility to avoid potential constraints on the development of CCUS technologies. In addition to avoiding the trap of imposing abstract purity requirements on CO_2 streams, EPA defined CO_2 streams both by source ("captured from an emission source") and by the necessary circumstances of the capture and management processes. The latter was accomplished by including "incidental associated substances derived from the source" and "any substances added to the stream to enable or improve the injection process."

EPA has raised concerns, however, by taking an additional step under the waste management provisions of the Resource Conservation and Recovery Act (RCRA) of first declaring captured CO_2 streams to be "solid waste" subject to RCRA and then promulgating a "conditional exclusion" from characterization as a "hazardous waste" for CO_2 streams that can be "certified" as captured without the addition of hazardous waste, transported by methods meeting regulatory prescriptions, and then injected into Class VI wells by an operator who certifies that hazardous wastes have not been added to the injected stream. These steps have raised concerns about the potential for captured CO_2 streams to adversely affect the expectations for use in EOR operations by subjecting those operations to waste management requirements of various types, including the potential for imposing long-term liabilities under the Comprehensive Environmental Response and Liability Act (CERCLA). Absent assurances to the contrary, EOR operators may likely be inclined to avoid accepting captured anthropogenic CO_2 streams derived from power plant and other industrial flue gas in order to avoid those undesirable exposures. This issue has been identified and framed both by adverse comments and by opposing litigation, at least until the economics warrant acceptance of such regulatory and liability risks. The objections are to EPA's decision to declare CO_2 streams to be "solid waste", not to any decision to remove such streams from the "hazardous waste" category, a step that will be welcomed if the classification as "solid waste" cannot be avoided. EPA already regulates CO_2 as an air pollutant, a classification that has been confirmed by the United States Supreme Court, but the necessity of also regulating CO_2 streams as waste is being questioned. The resolution of that question awaits action by the courts or further steps by EPA.

Site Characterization

EPA's UIC regulations recognize the pivotal role played by finding and adequately characterizing a site to be used for GS or EOR. The regulatory regime for EOR has been in place since the creation of the UIC program under the SDWA, and that program was built on the foundation of programs already in place and operated by states with oil and gas exploration and production. Those programs have successfully avoided adverse health and environmental impacts from the use of CO_2 for EOR. Those EOR operations have been conducted in fields where prior experience with oil and gas production have helped significantly in the characterization and understanding of the containment properties of the geologic formations and reservoirs.

EPA's new requirements for GS have focused primarily on the saline formations that may be used for GS in areas with far less existing information about the characteristics and properties of the geologic formations to be used for storage and for containment of the injected CO_2 streams. To provide the necessary flexibility to obtain fully characterize and select appropriate sites, EPA took a number of important steps in drafting and interpreting its regulations. The siting requirements of the Class I rule are performance based and focus on storage capability and containment. EPA has accepted the desirability of gathering information about the geology of the site with stratigraphic test wells, which need not be permitted under the UIC program. After prescribing the need to conduct extensive data collection, analysis, and modeling, EPA established a process that will allow sequential development of the necessary information. By using available data and information, conducting stratigraphic tests, and other means, a project developer can obtain the permit to construct a well and then to conduct additional logging and testing, including formation testing with CO_2, to obtain the necessary additional information before seeking the authorization to inject the captured CO_2 streams. The details of this process have been fleshed out through the additional communications focused on issues of concern over uncertainties and perceived hurdles.

Project Area Definition

EPA's approach to defining the project area is tailored to the project's purpose and allows for adjustment as each project and its site's characteristics are increasingly understood. The initial delineation is based on computational modeling to determine the size and extent of the projected carbon dioxide plume and the associated pressure front. It also involves analysis of the geological structures to identify any potential pathways for migration that might allow movement of fluids into underground sources of drinking water (USDWs) that are protected from endangerment under the UIC program. Initially, the proposed Class VI regulations would have restricted GS to reservoirs below the lowermost USDW, but the final rule provides for the opportunity to demonstrate that GS can be accomplished without endangering USDWs above or below the reservoir used for storage. This purportedly provided additional flexibility, but there are concerns that this flexibility may prove illusory because the locations where that flexibility might be needed most - areas just east of the Rocky Mountains and parts of California - might also need to be able to designate exempted aquifers to use as storage reservoirs. But the final rule prohibits the use of such designations for GS even though it allows the expansion of pre-existing exempted aquifer designations to account for transitioning an oil field from production to GS. EPA acknowledges this limitation and has asked for more specific information about sites that might be considered for GS absent the prohibition on designation of exempted aquifers. EPA has committed to revisiting the Class VI regulations within six years from the final promulgation on December 10, 2010.

For the greenhouse gas emissions reporting and GS quantification rule, EPA adopted a different approach that focuses exclusively of the extent of the CO_2 plume and potential leakage pathways that might allow injected CO_2 to migrate to a point of release into the atmosphere. Although EPA's rule calls for the use of a buffer area (extending around the projected plume) that will also be covered by the monitoring, verification and reporting (MRV) plan, EPA acknowledges that in most cases the maximum monitoring area for GHG reporting purpose will be smaller than the UIC area of review (AOR). Moreover, there is flexibility to tailor the specifics of the MRV plan to reflect the assessment of risks for leakage, thus allowing for modified monitoring approaches within the buffer area itself.

Well Construction

The Class VI well construction requirements are designed to provide protection to USDWs and to prevent undue deterioration of the materials of construction by requiring compatibility with the fluids with which the materials may come into contact. Where specific requirements have been prescribed, EPA has provided flexibility through rule provisions and interpretations to allow the UIC Program Class VI Director to approve modifications and alternatives that will also provide containment and the prevention of fluid movement into or between USDWs or into any unauthorized zones. The regulations also provide an accommodation for the transitioning of wells from alternative uses to GS if the wells were engineered and constructed to fully protect USDWs and to prevent the movement of fluids into or between USDWs or into any unauthorized zones while allowing required testing, maintenance and monitoring. Publication of the final EPA Class VI Well Construction Guidance caused concerns because it appeared to undermine some of the flexibility provided for transitioning wells able to meet the regulatory requirements, but EPA revised that approach in the draft guidance for wells transitioning from Class II EOR operations to be re-permitted as Class VI wells, and the Agency has committed to revising the final Well Construction Guidance to conform to the final Transition Guidance. The ability to adapt these construction requirements to the necessary circumstances of specific site characteristics and the ability to avoid completely replacing wells that are capable of providing the necessary protections is very important to accommodating CCUS deployment.

Well Operation

The well operation requirements are performance based for the most part, although concerns have been expressed over the final injection pressure limitation, which is set at ninety percent of injection zone fracture pressure rather than being focused on avoiding any potential for movement of injected CO_2 or formation fluids through the confining zone. EPA also provided important flexibility by allowing wells to avoid having to maintain annulus pressure that exceeds injection pressure throughout the length of the well. This flexibility is necessary because operating with annulus pressure higher than injection pressure has proved in many cases to be infeasible without risking severe damage to the well when injecting the CO_2 stream under supercritical pressure through very deep wells.

Testing and Monitoring

EPA has provided flexibility for testing and monitoring by allowing the project operator to spell out the details in a testing and monitoring plan that describes how the project operator will meet the requirements of this section, including accessing sites for all necessary monitoring and testing during the life of the project. Moreover, EPA has confirmed the ultimate flexibility of the testing and monitoring plan provisions by providing authority for the Director to allow or require alternative methods. Among other things, this will allow for the development and implementation of advances in technologies and the reliance on experience regarding the conditions under which the use of specific methods is optimal. This should allow a fit for purpose approach to the selection of particular testing and monitoring methods. Some concerns remain over the fact that the rule calls for a separate testing and monitoring plan for each well even where a GS project will involve the operation of a number of wells. But EPA has taken steps and provided guidance for operators to achieve the efficiencies of project wide planning for testing and monitoring, and EPA will reconsider the desirability of using comprehensive plans in place of well-by-well planning when it conducts its six-year review.

Post-Injection Site Care and Site Closure

Great concern was expressed over EPA's initial specification of a fifty-year post-injection site care (PISC) period as a default under the proposed rule. The final rule improved this to some extent by providing that project operators can make an alternative PISC timeframe demonstration to substitute an alternative to the fifty-year default. There were initial concerns that this ability to obtain approval of an alternative PISC timeframe would only be available at the time of initial permitting, before operators had an opportunity to gain the extensive additional knowledge that would come through the collection of operational and monitoring data. That limitation would have undermined the iterative approach that EPA said it favored and that many others had advocated, including the MSD participants. After discussion of these concerns and additional consideration, EPA has confirmed and provided guidance on the use of this process to make modifications at any point in the lifetime of a GS project. This approach should serve to provide incentives for operators to gain the best understanding of the site and project and to conduct history matching of its modeling. The modeling, with any necessary adjustments, can be used to revise the project parameters as well as the PISC timeframe.

Financial Responsibility

Although EPA is very insistent on having financial responsibility demonstrations meet strict criteria for financial soundness and maintenance of coverage, the Class VI rule includes important flexibility to allow tailoring combinations of instruments to the specific circumstances of a project, changes in the types of coverage used as the project advances through its sequential phases, and the introduction of new instruments that become available in the market place. Additional flexibility is provided by spelling out the details of how instruments should be worded in a guidance document rather than in the rule itself. This will allow modifications in instruments without going through rulemaking to make changes. Moreover, the ultimate flexibility comes with the authorization to use alternative instruments satisfactory to the Director.

Experimental Permitting

When EPA recognized that the pilot and demonstration projects being conducted by the RCSPs and other project developers would require UIC permits, the Agency developed and published a guidance document providing information for applicants and UIC Program Directors to proceed with the planning and permitting of these projects under the UIC Class V permitting program provisions for experimental technology wells. Consequently, these projects were planned and budgeted for compliance with the Class V permitting requirements consistent with the GS guidance. After EPA promulgated the final Class VI rule, however, the Agency began to insist that all of these pilot and demonstration projects change their approach and obtain Class VI permits under the regulations designed for full-scale commercial operations. This insistence led initially to consternation on the part of project developers and ultimately to significant changes in the RCSP project plans, resulting in the loss of projects planned as saline injection projects and the modification of projects to divert funds originally budgeted for scientific inquiry to be spent instead on regulatory compliance. Projects have been required to conduct scientific and technical demonstrations to justify exceptions necessary to accommodate the smaller scale and timeframes of pilot and demonstration projects.

Concerns remain that this is not the best way to address the permitting of pilot and demonstration projects particularly as developers and supporters of those projects seek to move to a new generation of CCUS technologies development. Other concerns have been raised about whether EPA actually has legal authority to issue Class VI permits for pilot and demonstration projects using naturally sourced or trucked-in carbon dioxide that is not captured from emission sources. Consequently, it has been recommended that EPA should consider resuming the issuance of Class V experimental technology permits for GS projects being conducted to explore the feasibility of emerging technologies. EPA itself has noted one of the important reasons for continuing experimentation when it stated in the final Class VI rule preamble that EPA will "continue to evaluate ongoing research and demonstration projects and gather other relevant information as needed to make refinements to the rulemaking process." Unless it considers a different approach to permitting for noncommercial projects, EPA may not be in a position to benefit fully from research and demonstration projects because the regulatory requirements may stifle the level of scientific and technical knowledge that can be achieved through these projects. EPA appears committed to insisting on Class VI permits even for pilot and demonstration projects, working within that commercial project framework to accommodate smaller size and duration projects rather than resuming the use of Class V permits unless there is a uniquely experimental aspect to the project. To date, however, EPA has offered no insight as to what would be considered experimental. If EPA lacks authority to issue Class VI permits for projects using carbon dioxide that is not from emission sources, it will provide at least one criterion for defining what is likely to be experimental.

State Implementation

As EPA proceeds with the implementation of the Class VI rules, the Agency will need to address concerns that have been expressed over a tendency to require wholesale adoption of the federal rule language rather than focusing on the provision of equally stringent levels of protection for USDWs while adapting the detailed language of state regulations to the existing requirements and approaches of other state programs and to the geology and other specific conditions present in the areas of the state being used for GS. EPA will also need to address the concerns and opposition being expressed in response to EPA's proposal in the draft Transition Guidance to assign to Class VI Program Directors the role of deciding when a Class II EOR operation must transition and be repermitted as a Class VI operation. Numerous states and other stakeholders argue that this authority already resides with the Class II Program Directors - typically state oil and gas agencies - rather than with the Class VI Director, which is currently EPA itself in all states.

GHG Reporting and GS Quantification

When EPA proposed its new source performance standards for fossil fueled electric utility generating units (EGUs), EPA proposed an approach under which operators of EOR projects injecting CO_2 underground that are permitted under UIC Class II and that receive CO_2 captured from EGUs to meet the proposed performance standard will also be required to submit and receive approval of a subpart RR MRV plan and report under subpart RR. Although EPA chose to create a regulatory framework that is based on CCUS and is intended to encourage the use of CCUS, there is a perception that the Agency is creating hurdles and disincentives to the use of captured CO_2 for EOR. There are several different ways the Agency can respond to these concerns, but the primary need is to address the uncertainties currently surrounding the requirements of subpart RR reporting especially for EOR operators. Some of that can probably be done through policy statements, but there may also need to be changes to either proposed or final regulatory provisions to accommodate the needs of EOR operators. As EPA has noted, EOR operators hold a key position on the road to near term CCUS deployment.

Further Regulatory Responses

Looking forward, additional steps will be necessary to remove or overcome regulatory issues that could hinder the development and deployment of CCUS projects, whether pilot, demonstration or full-scale commercial. The ongoing process of obtaining and considering multi-stakeholder input that has been used by EPA both formally and informally has proved effective in identifying concerns, modifying approaches and implementing solutions. The challenge remains one of achieving the regulatory objectives by providing necessary environmental and human health protections while optimizing the opportunities for scientific and technical learning and advancements. Unnecessary administrative steps can detract resources from the scientific progress without benefit. UIC permits are being issued for additional projects, and improvements can be achieved through the issuance of well-designed permit conditions, additional guidance and policy determinations. Nevertheless, some of the hurdles may require rule revisions. EPA announced its intention to consider such revisions within six years of the initial promulgation of the UIC Class VI rule, which occurred in December 2010. Two-thirds of that period has passed already, and some issues that may require rule revisions have already been identified. More are likely to emerge as newly authorized projects proceed.

ACKNOWLEDGEMENTS

EcoReg Matters Ltd. acknowledges with great appreciation the support of the CO_2 Capture Project.

EcoReg Matters Ltd. also acknowledges the participation in this process of representatives of the U.S. Environmental Protection Agency, the U.S. Department of Energy, the NETL Regional Carbon Sequestration Partnerships, environmental nongovernmental organizations (ENGOs), industry oil and gas operators and electric utilities, trade associations interested in CCS development, academics, and others who will remain nameless in light of the agreements under which this project was conducted.

REFERENCES

1. Carbon Sequestration Council, Summary of the EPA - Multi-Stakeholder Discussions (MSD) Group Meeting on August 4, 2011 Regarding May 20, 2011 Comment Letter.
2. Global CCS Institute, Bridging the Gap: An Analysis and Comparison of Legal and Regulatory Frameworks for CO_2–EOR and CO_2–CCS (October 2013).
3. International Energy Agency, Carbon Capture and Storage: Model Regulatory Framework 25 (November 2010).
4. USEPA, Class V Experimental Technology Well Guidance for Pilot Geologic Sequestration Projects (UIC Program Guidance #83) (March 1, 2007).

5. USEPA, "Federal Requirements Under the Underground Injection Control (UIC) Program for Carbon Dioxide (CO_2) Geologic Sequestration (GS) Wells" to invite public comment on July 25, 2008 (73 Fed. Reg. 43492).
6. USEPA, "Federal Requirements Under the Underground Injection Control (UIC) Program for Carbon Dioxide (CO_2) Geologic Sequestration (GS) Wells; Notice of Data Availability and Request for Comment. 74 Fed. Reg. 44802 (August 31, 2009).
7. USEPA, Mandatory Reporting of Greenhouse Gases: Injection and Geologic Sequestration of Carbon Dioxide; Proposed Rule, 75 Fed. Reg. 18576 (April 12, 2010).
8. USEPA, Mandatory Reporting of Greenhouse Gases: Injection and Geologic Sequestration of Carbon Dioxide; Final Rule, 75 Fed. Reg. 75060 (December 1, 2010).
9. USEPA, "Federal Requirements Under the Underground Injection Control (UIC) Program for Carbon Dioxide (CO_2) Geologic Sequestration (GS) Wells; Final Rule," 75 Fed. Reg. 77230 (December 10, 2010).
10. USEPA, Draft Underground Injection Control (UIC) Class VI Program Financial Responsibility Guidance (EPA 816-D-10-010) (released for comment on December 14, 2010).
11. USEPA, Draft Underground Injection Control (UIC) Program Class VI Well Construction Guidance for Owners and Operators (March 2011).
12. USEPA, Underground Injection Control (UIC) Program Class VI Well Construction Guidance at 37 (July 30, 2012).
13. USEPA, Draft Underground Injection Control (UIC) Program Guidance on Transitioning Class II Wells to Class VI Wells 31-38 (December 2013) (EPA 816-P-13-004).
14. World Resources Institute (WRI). CCS Guidelines: Guidelines for Carbon Dioxide Capture, Transport, and Storage (2008).

Carbon Dioxide Capture for Storage in Deep Geological Formations, Volume 4
Karl F. Gerdes (Editor)

Chapter 52

COMMUNICATIONS SUMMARY

Kate Adlington[1], Mona Ishaq[1], Simon Taylor[1] and Mark Crombie[2]
[1]Pulse Brands, Clareville House, 26/27 Oxendon Street, London, SW1Y 4EL
[2] BP International Ltd, Sunbury-on-Thames, UK

ABSTRACT: The CCP undertook its most ambitious communications program in Phase 3. This involved continuing to share results and insights from the CCP with members and the CO_2 capture and storage (CCS) industry. However, it also involved creating digital tools to help make the science of CCS accessible to a wider, non-technical, audience. The creation of the CCS Browser (www.ccsbrowser.com) – the industry's first, multi-platform, digital tool dedicated to explaining CCS was at the centre of this.

KEYWORDS: Annual Report; CCS Browser; conferences; factsheets; Global CCS Institute; In Depth Brochure; public understanding; technical literature

CONTEXT

There has been progress in the development of CO_2 capture and storage (CCS) projects around the world during CCP3, but at a far slower pace than needed if CCS is to make a significant contribution to climate change mitigation. The impact of the financial crisis has brought about greater uncertainty and a reduced appetite for perceived risk across all major infrastructure projects and the impact on CCS, in particular, has been significant.

Given this background there has been an even greater need for the industry to communicate the potential of CCS technology as a cost-effective method for helping mitigate climate change. Keeping the oil & gas industry and wider CCS world informed of the latest research and technological work into cost effective CCS solutions therefore remained a priority in CCP3. In addition, with persistent concerns about the technical viability, the industry needs to address misunderstandings and inaccuracies with regards to the basic science of CCS. In particular, the geological principles forming the foundations of the secure storage of CO_2 or even the basic properties of CO_2 are still not widely understood. A 2011 study of the Dutch general public showed that 25% of respondents thought that people do not exhale CO_2 and over 20% thought that CO_2 causes cancer. Similarly, studies in the UK and elsewhere have shown a lack of understanding of the basics of CO_2 and CO_2 capture and storage [1].

The necessity of explaining the science of CCS to a wider audience, including politicians and the general public, has never been greater. As a 2013 report by the Global CCS Institute explains, the lack of progress has been due to "insufficient policy support exacerbated by poor public understanding" rather than because of technical impediments [2].

Understanding the central role of communications in deciding the future of CCS, the CCP undertook a far more ambitious and wide reaching communications program in Phase 3. It has continued to contribute to the technical understanding of CCS, within the oil & gas industry, as well as sharing research results and insights from the CCP Capture, Storage and Policy Teams' work with the CCS industry and academia. More importantly, the Communications Team has worked hard and with creativity to translate the knowledge of the group so it is understandable to a wider, non-technical, audience. Central to this effort has been the creation of the CCS Browser (www.ccsbrowser.com) – the first, multi-platform, digital tool to explain CCS to a non-technical audience.

SHARING TECHNICAL KNOWLEDGE: RESULTS, INSIGHTS, CHALLENGES

In the third phase, the CCP generated large amounts of data and knowledge, notably from large-scale projects such as the successful Fluid Catalytic Cracking capture demonstration (Brazil), the continuing once-through steam generator capture demonstration (Canada) and also from a number of important storage field trials and research studies.

While the detailed technical content remained confidential to CCP members, key insights from the main projects were made available by the CCP to the wider CCS industry, academia, governments, NGOs and media. This information was disseminated via a range of tools and channels, including summary reports, technical factsheets, conference presentations and through the CCP website. Some of the highlights include:

Website

The CCP website (www.co2captureproject.org) was refreshed to improve usability for industry and technical audiences. It has continued to be an important tool for knowledge sharing; ensuring CCP reports, technical papers and other publications are accessible. By the end of CCP3 almost 4,500 people were registered on the website; registrants are updated with new materials and results as they become available. The site continues to attract around 1,400 new visitors every month as well as a regular number of repeat visitors.

Technical Literature

The CCP produced a range of other communications materials to share the findings of significant projects it undertook during the period. These included technical factsheets on storage and capture demonstrations – available in print and via the CCP website (Figure 1). A number of studies undertaken by the Policy & Incentives Team were also made available on the CCP website (Figure 2).

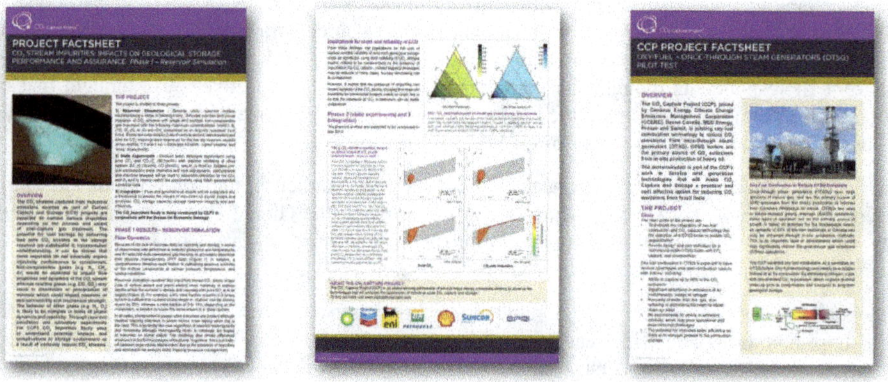

Figure 1. Project Factsheet – CO_2 impurities study and OTSG pilot test, available online.

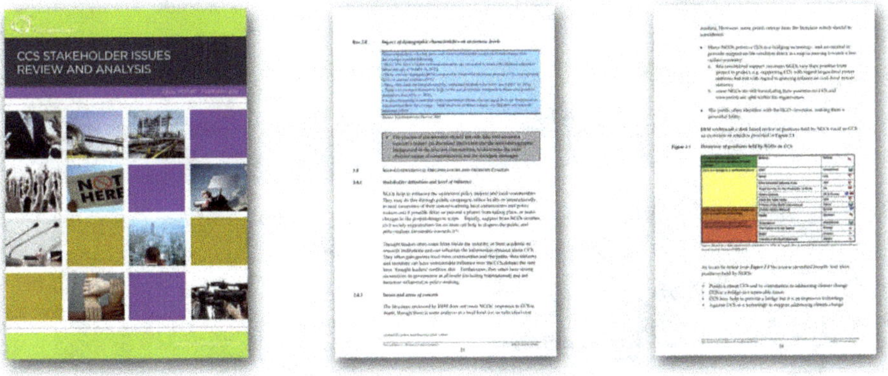

Figure 2. Policy & Incentives report on CCS Stakeholder Issues, published February 2012. A full list of P&I reports are available online.

CCP Annual Report

In 2011, the first of a series of Annual Reports was created to provide a regular update on the work of the CCP. Available in print and PDF format, this annual publication includes reports from each of the Technical Teams as well as reflections on the state of the CCS industry as a whole. It is distributed to a range of influential audiences through direct mail, the website and industry events (Figure 3).

Technical Conferences

The CCP contributed to key industry conferences including:

- Annual CCUS Conference, Pittsburgh: sponsored, exhibited and contributed papers each year throughout CCP3 (Figure 4);
- Greenhouse Gas Control Technologies (GHGT) Conference: sponsored, exhibited, contributed papers and presented posters at 2010 (Amsterdam), 2012 (Japan) and 2014 (Austin);
- Carbon Sequestration Leadership Forum Ministerial Conference: 2013 technical showcase exhibitor.

Media and NGO outreach

The CCP continued to share results with oil & gas and CCS industry publications to raise awareness and share knowledge. Publications, with an outreach of between 5k-10k readers such as Petroleum Review, Carbon Capture Journal and World Coal Magazine, carried in-depth features on the work of the CCP.

In terms of NGO engagement, CCP changed its approach during CCP3. In the second phase, the CCP had held workshops in the US and Europe to share information and discuss concerns around what was then an emerging technology – inviting attendees such as the World Resources Institute, Natural Resources Defence Council and Bellona. In the third phase, there was less direct engagement but NGOs were kept informed on progress through direct mail and e-mail updates and CCP technical and policy experts were made available for questions if required.

Figure 3. Annual Reports 2010-2013.

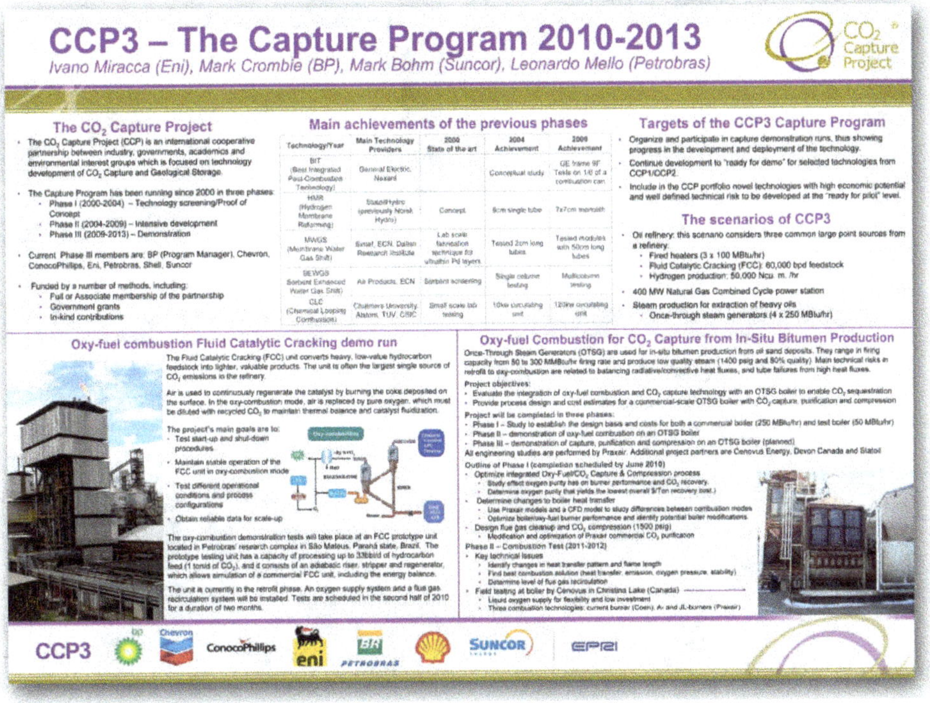

Figure 4. Project Poster from CCUS Conference, Pittsburgh 2010, providing an overview of the Capture Program.

BUILDING BROADER UNDERSTANDING: TRANSLATING THE SCIENCE

Gaining public trust is a key component in the successful delivery of CCS projects. During Phase 3, the CCP undertook one of its most significant pieces of communication work to date – the creation of the CCS Browser, a multi-platform digital resource aimed at building public understanding. The CCP also refreshed its popular In Depth brochure – offering a spatial perspective on storage – and made available an interactive digital version available online.

The Making of the CCS Browser

As the number of CCS projects around the world increases, so will the demand from the general public for accessible information on CCS. The CCP took on the challenge of creating a multi-format digital resource – the first of its kind – dedicated to helping the public learn about CCS (Figure 5). Formatted for access via tablet, PC and mobile, the CCS Browser (www.ccsbrowser.com) has been designed to allow people to explore the topic in the way best suited to them – by watching animations, listening to audio clips or by interacting with maps and diagrams (Figure 6). The site also acts as a portal to other CCP and CCS resources to allow people to explore topics in even greater depth, if needed. Given the continuing need for public reassurance around CO_2 storage, this is the main focus of the CCS Browser, with detailed animations used to explain topics such as geological trapping, capacity and containment, as well as storage site operation and CO_2 monitoring. Language has been kept deliberately non-technical, while a glossary is provided for some key terminology.

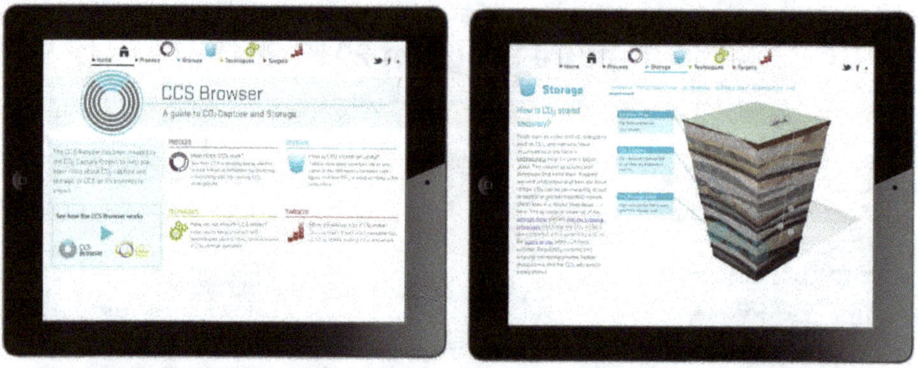

Figure 5. How CO_2 stays securely underground is one of the greatest public concerns. Animations have been used in the CCS Browser to bring technical information to life and provide insight into the scientific processes at work.

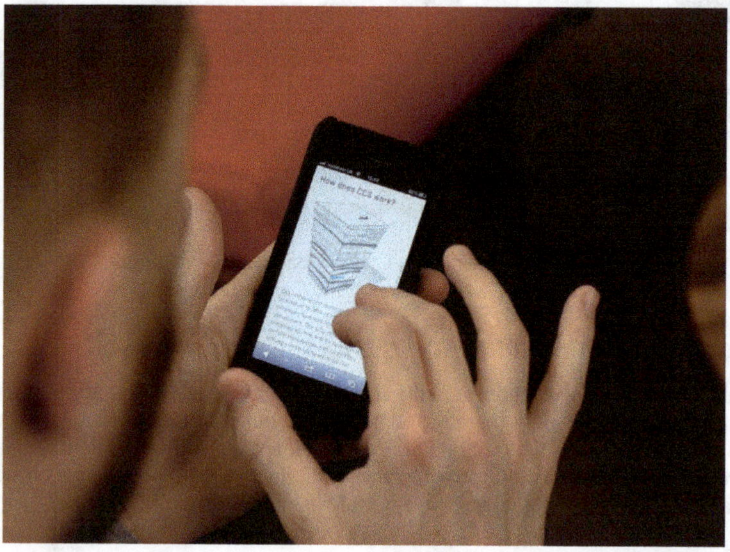

Figure 6. The CCS Browser is formatted for access via tablet, PC and mobile.

Digital In Depth – providing spatial perspective on CO_2 storage

Since its publication in the mid 2000's, the popular In Depth storage brochure has been distributed to thousands of companies and CCS educators around the world – with requests from countries including the US, Canada, Taiwan and Australia. The metre-long brochure created during CCP2 was refreshed during CCP3, using more realistic graphics, improving the layout and updating the text (Figure 7). The brochure is still one of the few resources available providing a realistic spatial perspective on CO_2 storage at depth. In response to an increasing number of requests for an online version, CCP developed a digital version of the leaflet in June 2012 – allowing visitors to scroll down on a journey almost 2000 metres underground to find out more about CO_2 capture and storage (Figure 8). The digital version is available on the project website (www.co2captureproject.org).

Figure 7. The In Depth brochure has proved popular because it provides a real insight into how CO_2 is stored and the depths involved. Spatial perspective is crucial for understanding CO_2 storage; concerns can arise when depths are inaccurately represented.

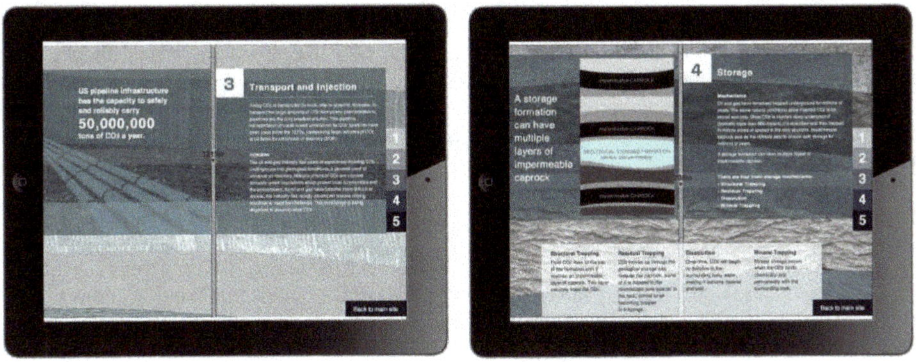

Figure 8. The In Depth brochure is available online, in a digital format – allowing visitors to take a journey underground.

Infographics – ensuring information is engaging and understandable

Infographics have been used across communications materials to make complex technical information accessible to a wider audience. The infographic in Figure 9 provides an overview of the CCP's first demonstration held in 2012 – an oxy-firing test at a pilot-scale Fluid Catalytic Cracking (FCC) unit.

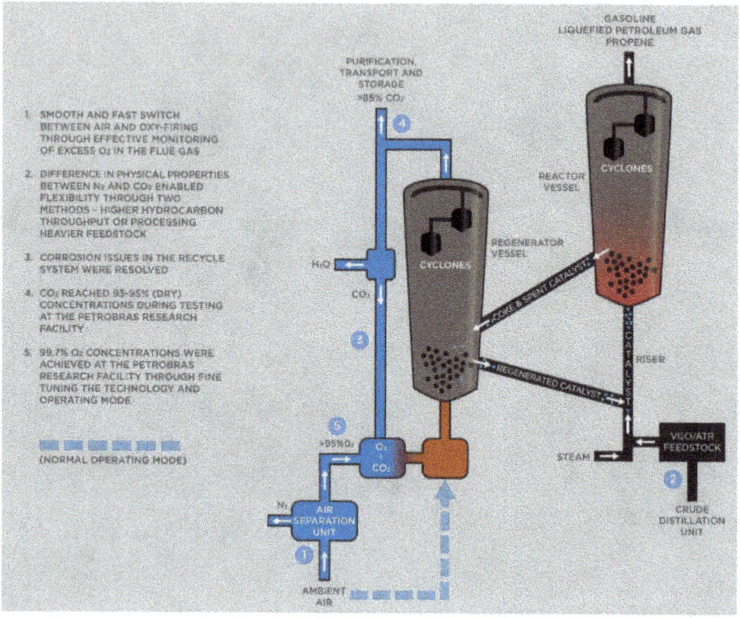

Figure 9. This infographic is a simplified representation of oxy-firing in an FCC unit.

ACKNOWLEDGEMENTS

The authors would like to thank all the communications professionals from the member companies that provided valuable input during Phase 3. We would also like to give a special thanks to the technical experts from the SMV and Capture teams that worked closely with the Communications Team to make the CCS Browser come to life. Our thanks also goes to Sarah Wade from (Wade LLC) for her ongoing support for the work of the CCP and providing invaluable feedback on core communications tools.

REFERENCES

1. K. Itaoka, A.Saito, M. Paukovic, M. de Best-Waldhober, A. Dowd, T. Jeanneret, P. Ashworth and M. James, Understanding how individuals perceive Carbon Dioxide: implications for acceptance of carbon dioxide capture and storage, 2012, p.4. Available from www.csiro.au
2. Author Anon, Global CCS Institute, The Global Status of CCS Report 2013, p.10. Available from www.globalccsinstitute.com

Carbon Dioxide Capture for Storage in Deep Geological Formations, Volume 4
Karl F Gerdes (Editor)

CONTRIBUTORS TO CCP3

Karl F. Gerdes
Editor, Karl F. Gerdes Consulting, Davis, California, US

Many people contributed to the successful completion of Phase 3 of the CO_2 Capture Project. In this section the contributions of CCP Member companies, technology providers, and chapter reviewers are acknowledged.

CCP STRUCTURE AND STAFFING

The individuals making up the management structure, advisory body and the technical teams of CCP are listed in Table 1.

TECHNOLOGY PROVIDERS & CHAPTER AUTHORS

A variety of technology providers were commissioned by CCP3 to execute the projects making up the technical program. These providers included academic institutions, research companies, CCP3 participants, national laboratories and expert consulting firms, who worked together with the CCP Technical Teams to deliver the program results. Authors of each chapter are listed in the Author Index.

CO_2 Capture

The technology providers for CCP3 Capture-related projects are:

- Amec Foster Wheeler (UK)
- Chalmers University (Sweden)
- Flemish Institute for Technological Research (VITO, Belgium)
- Instituto de Carboquímica (ICB-CSIC, Spain)
- John Zink Co. (US)
- Johnson Matthey (UK)
- Josef Bertsch GmbH (Austria)
- Pall Corporation (US)
- Praxair Inc. (US)
- Vienna University of Technology (Austria)
- University of Natural Resources – Vienna (Austria)

Storage, Monitoring and Verification

Organizations which carried out SMV projects for CCP3 are:

- Aachen University (Germany)
- Bureau of Economic Geology (BEG)-University of Texas, Austin (US)
- Lawrence Berkeley National Laboratory (US)
- Colorado School of Mines (US)
- Curtin University/CO2CRC Ltd. (Australia)
- CSIRO/CO2CRC Ltd. (Australia)

- Geoscience Australia
- GeoScience Ltd. (UK)
- GroundMetrics Inc. (US)
- Montana State University (US)
- Multiphase Technologies LLC (US)
- Schlumberger (US)
- Silixa Ltd. (UK)
- Stanford University (US)
- Taurus Reservoir Solutions (Canada)
- TRE Canada Inc.(Canada)
- University of Calgary (Canada)
- University of Texas, Austin (US)

Policy & Incentives and Communications

Work for CCP3 related to Policy and Incentives and to Communications was carried out by:

- Environmental Resources Management (UK)
- EcoReg Matters Ltd. (UK)
- Pulse Brands (UK)

REVIEWERS

On behalf of CCP3, the editor expresses profound gratitude to reviewers of the technical chapters in this volume. Their advice and perspective has measurably improved the usefulness of this volume to the CCS community. Special thanks to several reviewers who helped us with more than one chapter.

CO_2 Capture

The reviewers for CCP3 Capture-related chapters are:

- Aldo Bischi, Politecnico di Milano, Milano, Italy
- Olav Bolland, NTNU, Trondheim, Norway
- Nick Brancaccio, Chevron, US
- Nick Burke, CSIRO, Clayton, VIC, Australia
- Stefano Consonni, Politecnico di Milano, Milano, Italy
- Tom Gilmartin, BP International Ltd., UK
- Chris Higman, Consultant, Schwalbach, Germany
- Prakash Karpe, Phillips 66, US
- Jerry Lin, Arizona State University, Tempe AZ, US
- Philippe Mathieu, Consultant, Brussels, Belgium
- Torgeir Melien, Statoil ASA, Norway
- Michal Moore, University of Calgary, Calgary, AB, Canada
- Stanley Santos, IEAGHG, UK
- Dale Simbeck, SFA Pacific, US

Storage, Monitoring and Verification

Reviewers for the SMV chapters are:

- Thomas A. Buscheck, LLNL, Livermore, CA, US
- Rick Chalaturnyk, University of Alberta, Edmonton, AB, Canada
- Anozie Ebigbo, RWTH-Aachen Univ, Aachen, Germany
- David Etheridge, CSIRO, Aspendale, VIC, Australia
- Marc Fleury, IFPEN, Rueil-Malmaison Cedex, France
- Michael Godec, Advanced Resources International Inc., US
- Seyyed A. Hosseini, BEG-UT Austin, Austin, TX, US
- G. Michael Hoversten, Chevron, CA, US
- Vello Kuuskraa, Advanced Resources International, Inc., Arlington, VA, US
- Don Lawton, University of Calgary, Calgary, AB, Canada
- Paul Marschall, NAGRA, Wettingen, Switzerland
- John McBride, BYU, Provo, UT, US
- John Midgley, Energy Geoscience International Ltd., UK
- Runar Nygaard, Missouri University of Science and Technology, Rolla, MO, US
- Lincoln Paterson, CSIRO, Clayton, VIC, Australia
- Scott Ryan, Chevron, Australia
- Sandeep Sharma, Carbon Projects Pty Ltd., Bunbury, WA, Australia
- Craig Smalley, BP International Ltd., UK
- Tim Tambach, Shell Global Solutions International B.V., The Netherlands
- Michel Verliac, TOTAL S.A., France
- Max Watson, CO2CRC Ltd., Australia
- Mike Wilt, GroundMetrics, Inc., San Diego, CA, US

Policy & Incentives and Communications

Chapter reviewers for P&I and Communications:

- Howard Herzog, MIT, Cambridge, MA, US
- Juho Liponen, IEA, Paris, France

Table 1. Organizational structure of CCP3.

CCP3 EXECUTIVE BOARD	PROGRAM MANAGER	CCP3 ADVISORY BOARD		
Nigel Jenvey (Chair) *BP*	**Mark Crombie** *BP*	**Vello Kuuskraa (Chair)** *President, Advanced Resources Int'l Inc., US*		
Rodolfo Dino *Petrobras*		**Olav Bolland** Head of Dept/ Professor, *NTNU, Norway*	**Michael Celia** Prof. and Chair, *Civil & Enviro. Engr., Princeton Univ, US*	**Christopher Higman** Indep. Consultant, *Schwalbach, Germany*
Stephen Kaufman *Suncor*				
Vincent Kwong (Vice Chair) *Chevron*		**Pierpaolo Garibaldi** Indep. Consultant, *Peschiera Borromeo, Italy*	**Larry Meyer** Indep. Consultant, *Benecia, CA, US*	**Dale Simbeck** Vice President, Technology, *SFA Pacific, Inc., Palo Alto, CA, US*
John MacArthur *Shell*				
Mario Vito Marchionna *Eni*				

TECHNICAL TEAMS				
CAPTURE	**STORAGE, MONITORING & VERIFICATON**	**ECONOMIC MODELLING**	**POLICY & INCENTIVES**	**COMMUNICATIONS**
Dan Burt *Suncor*	**Stephen Bourne** *Shell*	**David Butler** *David Butler & Associates. Calgary, AB, CA*	**Eric Beynon** *Suncor*	**Rachel Barbour** *BP*
Jonathon Forsyth *BP*	**Marco Brignoli** *eni*		**Dan Burt** *Suncor*	**Mark Crombia** *BP*
Iftikhar Huq *Suncor*	**Andreas Busch** *Shell*		**Mark Crombie** *BP*	**Renato De Filippo** *eni*
Mahesh Iyer *Shell*	**Mark Chan** *Suncor*		**Renato De Filippo** *eni*	**Kurt Glaubitz** *Chevron*
Raja Jadhav *Chevron*	**Walter Crow** *BP*		**Christhiaan Greco** *Petrobras*	**Christhiaan Greco** *Petrobras*
Jamal Jamaluddin *Shell*	**Rodolfo Dino** *Petrobras*		**Arthur Lee** *Chevron*	**Tanis Shortt** *Suncor*
Leonardo De Mello *Petrobras*	**Kevin Dodds** *Formerly BP, now ANLEC, Australia*		**Charles Samuda** *Shell*	**Peter Snowden** *Shell*
Ivano Miracca *Eni*	**Bryan Dotson** *BP*			
Gustavo Moure *Petrobras*	**Grant Duncan** *Suncor*			
Betty Pun *Chevron*	**Craig Gardner** *Chevron*			
Frank Wubbolts *Shell*	**Scott Imbus** *Chevron*			
	Dan Kieke *Chevron*			
	Josephina Schembre *Chevron*			

AUTHOR INDEX

KEYWORD INDEX

www.ingramcontent.com/pod-product-compliance
Lightning Source LLC
Chambersburg PA
CBHW050647190326
41458CB00008B/2446